Series in PURE and APPLIED PHYSICS

Concepts in Quantum Mechanics

CRC SERIES *in* PURE *and* APPLIED PHYSICS

Dipak Basu
Editor-in-Chief

PUBLISHED TITLES

Handbook of Particle Physics
M. K. Sundaresan

High-Field Electrodynamics
Frederic V. Hartemann

Fundamentals and Applications of Ultrasonic Waves
J. David N. Cheeke

Introduction to Molecular Biophysics
Jack A. Tuszynski
Michal Kurzynski

Practical Quantum Electrodynamics
Douglas M. Gingrich

Molecular and Cellular Biophysics
Jack A. Tuszynski

Concepts in Quantum Mechanics
Vishu Swarup Mathur
Surendra Singh

Series in PURE and APPLIED PHYSICS

Concepts in Quantum Mechanics

Vishnu Swarup Mathur

Surendra Singh

CRC Press
Taylor & Francis Group
Boca Raton London New York

CRC Press is an imprint of the
Taylor & Francis Group, an **informa** business
A CHAPMAN & HALL BOOK

Chapman & Hall/CRC
Taylor & Francis Group
6000 Broken Sound Parkway NW, Suite 300
Boca Raton, FL 33487-2742

Printed in the United States of America on acid-free paper
10 9 8 7 6 5 4 3 2 1

International Standard Book Number-13: 978-1-4200-7872-5 (Hardcover)

Library of Congress Cataloging-in-Publication Data

Mathur, Vishnu S. (Vishnu Swarup), 1935-
Concepts in quantum mechanics / Vishnu S. Mathur, Surendra Singh.
p. cm. -- (CRC series in pure and applied physics)
Includes bibliographical references and index.
ISBN 978-1-4200-7872-5 (alk. paper)
1. Quantum theory. I. Singh, Surendra, 1953- II. Title. III. Series.

QC174.12.M3687 2008
530.12--dc22 2008044066

Visit the Taylor & Francis Web site at
http://www.taylorandfrancis.com

and the CRC Press Web site at
http://www.crcpress.com

Dedicated to the memory of
Professor P. A. M. Dirac

Contents

Preface

This book has grown out of our combined experience of teaching Quantum Mechanics at the graduate level for more than forty years. The emphasis in this book is on logical and consistent development of the subject following Dirac's classic work *Principles of Quantum Mechanics.* In this book no mention is made of *postulates of quantum mechanics* and every concept is developed logically. The alternative ways of representing the state of a physical system are discussed and the mathematical connection between the representatives of the same state in different representations is outlined. The equations of motion in Schrödinger and Heisenberg pictures are developed logically. The sequence of other topics in this book, namely, motion in the presence of potential steps and wells, bound state problems, symmetries and their consequences, role of angular momentum in quantum mechanics, approximation methods, time-dependent perturbation methods, etc. is such that there is continuity and consistency. Special concepts and mathematical techniques needed to understand the topics discussed in a chapter are presented in appendices at the end of the chapter as appropriate.

A novel inclusion in this book is a chapter on the *Three-body Problem*, a subject that has reached some level of maturity. In the chapter on *Relativistic Quantum Mechanics* an appendix has been added in which the basic concepts of special relativity and the ideas behind the covariant formulation of equations of physics are discussed. The chapter on *Quantization of Radiation Field* also covers application to topics like Rayleigh and Thomson scattering, Bethe's treatment atomic energy level shift due to the self-interaction of the electron (Lamb-Retherford shift) and Compton effect. In the chapter on *Second Quantization* the concept of *fields*, derivation of field equations from Lagrangian density, quantization of the scalar (real and complex) fields as well as quantization of Dirac field are discussed. In the section on *Interacting Fields and Covariant Perturbation Theory* the emphasis is on second order processes,such as Compton effect, pair production or annihilation, Möller and Bhabha scattering. In this context, Feynman diagrams, which delineate different electromagnetic processes, are also discussed and Feynman rules for writing out the transition matrix elements from Feynman graphs are outlined.

A number of problems, all based on the coverage in the text, are appended at the end of each chapter. Throughout this book the SI system of electromagnetic units is used. In the covariant formulation, the metric tensor $g_{\mu\nu}$ with $g_{11} = g_{22} = g_{33} = g_{44} = 1$ is used. Details of trace calculations for the cross section of Compton scattering are presented. It is hoped that this book will prove to be useful for advanced undergraduates as well as beginning graduate students and take them to the threshold of Quantum Field theory.

V. S. Mathur and Surendra Singh

Acknowledgments

The authors acknowledge the immense benefit they have derived from the lectures of their teachers and discussions with colleagues and students. One of us (VSM) was greatly influenced by lectures Prof. D. S. Kothari delivered to M. Sc. (Final) students in 1955-56 at the University of Delhi. These lectures based on Dirac's classic text, enabled him to appreciate the logical basis of quantum mechanics. VSM later taught the subject at Banaras Hindu University (BHU) for more than three decades. His approach to the subject in turn influenced the second author (SS), who attended these lectures as a student in the mid-seventies. We felt that the mathematical connection between Dirac brackets and their integral forms in the coordinate and momentum representations needs to be outlined more clearly. This has been attempted in the present text. Several other features of this text have been outlined in the preface. VSM would also like to express his gratitude to Prof. A. R. Verma, who as the Head of the Physics Department at BHU and later as the Director, National Physical Laboratory, New Delhi, always encouraged him in his endeavor. The second author (SS) would like to express his indebtedness to his many fine teachers, including the first author VSM and his graduate mentor late Prof. L. Mandel at the University of Rochester, who helped shape his attitude toward the subject. Finally we would like to acknowledge considerable assistance from Prof. Reeta Vyas with many figures and typesetting of the manuscript.

V. S. Mathur and Surendra Singh

1

NEED FOR QUANTUM MECHANICS AND ITS PHYSICAL BASIS

1.1 Inadequacy of Classical Description for Small Systems

Classical mechanics, which gives a fairly accurate description of large systems (e.g., solar system) as also of mechanical systems in our every day life, however, breaks down when applied to small (microscopic) systems such as molecules, atoms and nuclei. For example, (1) classical mechanics cannot even explain why the atoms are stable at all. A classical atom with electrons moving in circular or elliptic orbits around the nucleus would continuously radiate energy in the form of electromagnetic radiation because an accelerated charge does radiate energy. As a result the radius of the orbit would become smaller and smaller, resulting in instability of the atom. On the other hand, the atoms are found to be remarkably stable in practice. (2) Another fact of observation that classical mechanics fails to explain is *wave particle duality* in radiation as well as in material particles. It is well known that light exhibits the phenomena of *interference*, *diffraction* and *polarization* which can be easily understood on the basis of wave aspect of radiation. But light also exhibits the phenomena of *photo-electric effect*, *Compton effect* and *Raman effect* which can only be understood in terms of *corpuscular* or *quantum* aspect of radiation. The *dual* behavior of light, or radiation cannot be consistently understood on the basis of classical concepts alone or explained away by saying that light behaves as wave or particle depending on the kind of experiment we do with it (complementarity). Moreover, a beam of material particles, like electrons and neutrons, demonstrates wave-like properties (e.g., diffraction). A brief outline of phenomena that require quantum mechanics for their understanding follows.

1.1.1 Planck's Formula for Energy Distribution in Black-body Radiation

The quantum nature of radiation, that radiation is emitted or absorbed only in bundles of energy, called *quanta* (plural of *quantum*) or *photons* was introduced by Planck (1900). According to Planck, each quantum of radiation of frequency ν has energy E given by

$$E_\nu = h\nu \,, \tag{1.1.1}$$

where $h = 6.626068 \times 10^{-34}$ J-s ($\equiv 4.1357 \times 10^{-15}$ eV-s) is a universal constant known as Planck's constant. On the basis of this hypothesis, he could explain the energy distribution in the spectrum of black-body radiation. Planck derived the following formula for the energy distribution in black-body radiation:

$$u(\nu, T)d\nu = \frac{8\pi h\nu^3 d\nu}{c^3} \frac{1}{\exp(h\nu/k_B T) - 1} \tag{1.1.2}$$

where $u(\nu, T)d\nu$ is the energy density for radiation with frequencies ranging between ν and $\nu + d\nu$ and k_B is the Boltzmann constant. Equation(1.1.2) is also known as Planck's law, and has been verified in numerous experiments on the black-body radiation for all frequency ranges.

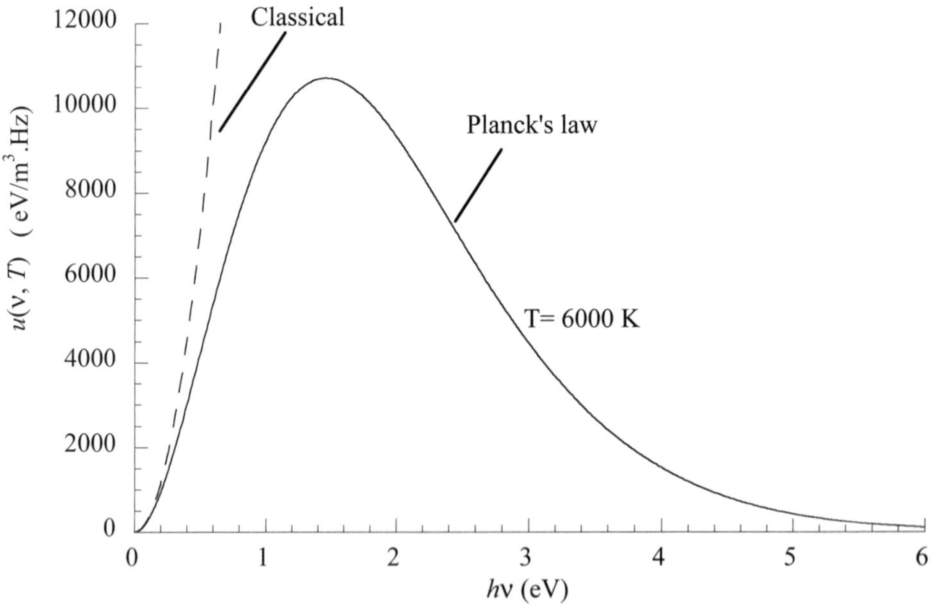

FIGURE 1.1
Energy distribution (eV/m^3·Hz) in black-body radiation. The solid curve corresponds to Planck's law and the dashed curve corresponds to the classical Rayleigh-Jean's formula [see Problem 1].

1.1.2 de Broglie Relation and Wave Nature of Material Particles

de Broglie's derivation of his famous relation

$$\lambda = \frac{h}{p} \tag{1.1.3}$$

was based on the conjecture that, if a material particle of momentum p is to be associated with a wave packet of finite extent, then the particle velocity $v = p/m$ should be identified with the group velocity v_g of the wave packet.

It may be recalled that a wave packet results from a superposition of plane waves with wavelength (or equivalently, frequency) spread over a certain range. As a result of this superposition, the amplitude of the resultant wave pattern (wave-packet) is not fixed but is subject to a wave-like variation and the velocity with which the wave packet advances in space, known as the group velocity, is given by

$$v_g \equiv \frac{d\omega}{dk} = u + k\frac{du}{dk}, \tag{1.1.4}$$

where $\omega = 2\pi\nu$ is the angular frequency, $k = 2\pi/\lambda$ is the wave number and $u(\omega) = \omega/k$ is the phase velocity. Wavelength λ, frequency ν, and phase velocity u of the wave, of course, satisfy the fundamental wave relation $u = \nu\lambda$. Thus while the wave propagates with phase velocity u, the modulation (wave packet) propagates with velocity v_g given by Eq.(1.1.4).

If we now invoke Planck's quantum condition $E = h\nu = \hbar\omega$ $(\hbar = h/2\pi)$, where E is the total energy (including rest energy) of the particle, then Eq.(1.1.4) can be written as,

$$v_g = \frac{d(E/\hbar)}{dk} = \frac{1}{\hbar}\frac{dE}{dp}\frac{dp}{dk}\,. \tag{1.1.5}$$

It is easily seen that the relation $dE/dp = v$ holds both for a relativistic $(v/c \approx 1)$ and a non-relativistic $(v/c \ll 1)$ particle[1]. Identification of v with v_g in Eq. (1.1.5) immediately leads us to de Broglie relation

$$\frac{dp}{dk} = \hbar \quad \Rightarrow \quad p = \hbar k = \frac{h}{\lambda}\,. \tag{1.1.6}$$

It is easy to see that de Broglie relation is relevant not only for a material particle but also for a quantum of radiation, i.e., a photon. Recalling that the energy of a photon of frequency ν is $E_\nu = h\nu$ and the rest mass assigned to it is zero, we have, according to the relativistic energy momentum relation [Appendix 12A1, Eq.(12A1.24)],

$$E = \sqrt{p^2c^2 + m^2c^4} = pc\,,$$

$$\text{or} \quad p = \frac{E}{c} = \frac{h\nu}{c}\,. \tag{1.1.7}$$

Using this in de Broglie relation, we find $\lambda = h/p = c/\nu$, which is the logical relation between wavelength λ and frequency ν. For material particles, such as electrons, de Broglie relation was put to test by Davisson and Germer (1926). They showed that electrons of very high energy are associated with de Broglie wavelengths of the order of X-ray wave lengths. When the beam of electrons was reflected from the surface of a Nickel crystal, they found selective maxima only for specific angles of incidence θ such that $2d\sin\theta = n\lambda$, where d is the spacing of atomic planes in the lattice [see Fig.(1.2)], and n is an integer (order of diffraction) just as in the case of X-rays. The wavelength of the electron beam found from this observation agreed with that computed from de Broglie's relation. In Thomson's experiment (1927) a collimated electron beam was incident normally on a thin gold foil. Diffraction from differently oriented crystals gives rings on a photographic plate just as obtained in the case of X-rays. In this case also computation of wave length from experimental observations agreed with calculation according to de Broglie relation.

1.1.3 The Photo-electric Effect

The quantum idea of Planck was subsequently used by Einstein (1905) to explain photo-electron emission from metals. His famous, yet simple, equation

$$h\nu = e\Phi + \frac{1}{2}mv^2\,, \tag{1.1.8}$$

where $h\nu$ is the energy of the incident photon, $e\Phi$ is the energy needed by the electron to overcome the surface barrier (Φ is called the work function) and v is the velocity acquired by the ejected electron, has been extensively verified by experiments.

---

[1]From the energy-momentum relation $E = \sqrt{p^2c^2 + m_0^2c^4}$ for a relativistic particle we have $dE/dp = pc^2/E = c\sqrt{E^2 - m_0^2c^4}/E = v$, because $E = m_0c^2/\sqrt{1 - v^2/c^2}$. For a non-relativistic particle $(v/c \ll 1, E \approx m_0c^2 + p^2/2m_0)$, this relation is obvious.

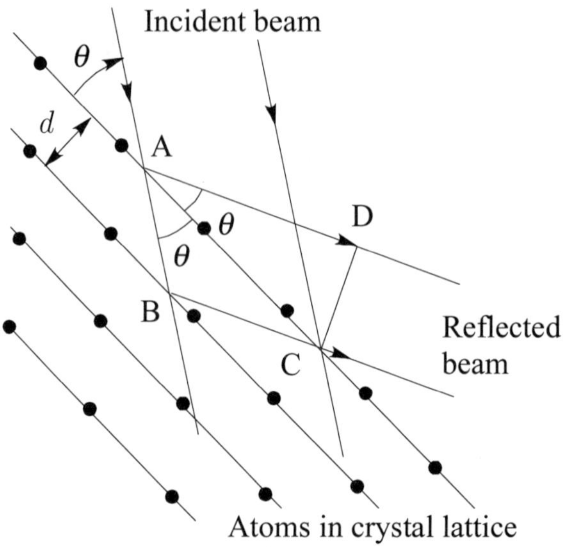

FIGURE 1.2
Bragg reflection from a particular family of atomic planes separated by a distance d. Incident and refected rays from two adjacent planes are shown. The path difference is $\overline{ABC} - \overline{AD} = 2d\sin\theta$.

According to Einstein's photo-electric equation (1.1.8) (i) the photo-electrons can be emitted only when the frequency of the incident radiation is above a certain critical value called the *threshold frequency*, (ii) The maximum kinetic energy of the electron does not depend on the intensity of light but only on the frequency of the incident radiation, and (iii) A greater intensity of the incident radiation leads to the emission of a larger number of photo-electrons or a larger photo-electric current. All these predictions have been verified.

1.1.4 The Compton Effect

The frequency of radiation scattered by an atomic electron differs from the frequency of the incident radiation and this difference depends on the direction in which the radiation is scattered. This effect, called Compton effect can again be easily understood on the basis of quantum aspect of radiation.

Consider a photon of frequency ν and energy $h\nu$ incident on an atomic electron at O and let it be scattered at an angle θ with energy $E' = h\nu'$ while the atomic electron, initially assumed to be at rest, recoils with velocity v in the direction ϕ [Fig.(1.3)]. According to the relativistic energy-momentum relation for a zero rest mass particle, the incident photon has a momentum $p = h\nu/c$ and the scattered photon has momentum $p' = h\nu'/c$. The electron with rest mass m is treated as a relativistic particle. The momentum and energy of the target electron (at rest) are given, respectively, by $p_{e0} = 0$ and $E_{e0} = mc^2$. For the recoil electron the total energy, including rest energy, is

$$E_e = \frac{mc^2}{\sqrt{1-\beta^2}}\,, \tag{1.1.9}$$

where $\beta = v/c$. The momentum of the recoil electron is in the direction ϕ and its magnitude

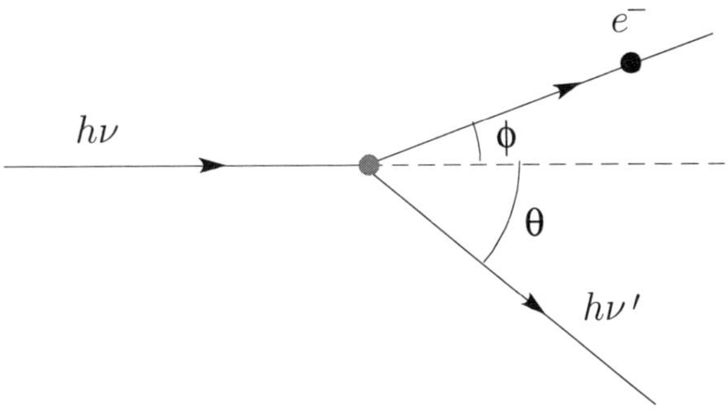

FIGURE 1.3
Compton scattering of a photon by an electron.

is

$$p_e = \frac{m\beta c}{\sqrt{1-\beta^2}}\,. \tag{1.1.10}$$

Application of the principles of conservation of energy and momentum in this process enables us to calculate the change in the wavelength, or frequency, of the scattered photon.

Conservation of energy requires

$$h\nu + mc^2 = h\nu' + \frac{mc^2}{\sqrt{1-\beta^2}}\,. \tag{1.1.11}$$

Conservation of momentum leads to

$$\frac{h\nu}{c} = \frac{m\beta c}{\sqrt{1-\beta^2}}\cos\phi + \frac{h\nu'}{c}\cos\theta\,, \tag{1.1.12a}$$

$$0 = \frac{m\beta c}{\sqrt{1-\beta^2}}\sin\phi - \frac{h\nu'}{c}\sin\theta\,. \tag{1.1.12b}$$

Eliminating ϕ from Eqs.(1.1.12) and using $\nu = c/\lambda$, we find

$$h^2\left(\frac{1}{\lambda^2} + \frac{1}{\lambda'^2} - \frac{2}{\lambda\lambda'}\cos\theta\right) = \frac{m^2c^2\beta^2}{1-\beta^2}\,. \tag{1.1.13}$$

From Eq.(1.1.11) we have:

$$h^2\left(\frac{1}{\lambda^2} + \frac{1}{\lambda'^2} - \frac{2}{\lambda\lambda'}\right) + 2mch\left(\frac{1}{\lambda} - \frac{1}{\lambda'}\right) = \frac{m^2c^2\beta^2}{1-\beta^2}\,. \tag{1.1.14}$$

Subtracting Eq. (1.1.14) from Eq.(1.1.13), we get:

$$\frac{2h^2}{\lambda\lambda'}(1-\cos\theta) = 2mch\left(\frac{1}{\lambda} - \frac{1}{\lambda'}\right)$$

$$\text{or}\qquad (\lambda' - \lambda) = \frac{h}{mc}(1-\cos\theta)\,. \tag{1.1.15}$$

The quantity h/mc, which has dimensions of length, is called the *Compton wavelength* of the electron and has the value 2.4262×10^{-3} nm. The result (1.1.15) has been verified experimentally. Thus simple particle kinematics enables us to account for both the *photoelectric effect* and the *Compton effect* provided we regard radiation to be consisting of bundles of energy called *quanta*.

In the case of Raman effect, part of the energy of the incident quantum of light may be given to the scattering molecule as energy of vibration (or of rotation). Conversely it may happen that some of the energy of vibration (or rotation) of the molecule may be transferred to the incident quantum of light. The equation of energy in this case is

$$h\nu = h\nu' \pm nh\nu_0 \,, \tag{1.1.16}$$

where ν_0 is one of the characterstic frequencies of the molecule. Hence Raman effect may also be understood on the basis of quantum aspect of radiation.

However the *corpuscular* and *wave* aspects of radiation, as well as of material particles, cannot be understood within the framework of classical mechanics. Quantum mechanics does enable us to understand the dual aspect (wave and corpuscular) of radiation and material particles, consistently [see Sec. 1.2 Interference of photons].

1.1.5 Ritz Combination Principle

Another important observation which defies classical description is *Ritz combination principle* in spectroscopy. Classically, if an atomic electron has its equilibrium disturbed in some way it would be set into oscillations and these oscillations would be impressed on the radiated electromagnetic fields whose frequencies may be measured with a spectroscope. According to classical concepts the atomic electron would emit a fundamental frequency and its harmonics. But this is not what is observed; it is found that the frequencies of all radiation emitted by an atomic electron can be expressed as difference between certain terms,

$$\nu = \nu_{mn} = T_m - T_n \,, \tag{1.1.17}$$

the number of terms T_n being much smaller than the number of spectral lines. This observation is termed as Ritz combination principle. The inevitable consequence that follows from the Ritz combination principle is that the energy content of an atom is also quantized, i.e., an atom can assume a series of definite energies only and never an energy in-between. Consequently an atom can gain or lose energy in definite amounts. When an atom loses energy, the difference between its initial and final energy is emitted in the form of a radiation quantum (photon) and if an atom absorbs a quantum of energy (i.e. a photon of appropriate freqency), its energy rises from one discrete value to another.

The results of the experiments of Frank and Hertz in which electrons in collision with atoms suffer discrete energy losses also support the view that atoms can possess only discrete sets of energies. This is unlike the classical picture of an atom as a miniature solar system (with the difference that the force law in this case is Coulomb law, instead of gravitational law). A planet in the solar system need not have discrete energies.

Bohr's *Old* Quantum Theory

Neils Bohr (1913) had suggested that the energy of an electron in an atom (say, the Hydrogen atom) may be required to take only discrete values if one is prepared to assume that (i) the electron can move only in certain discrete orbits around the nucleus and (ii) the electron *does not radiate energy when moving in these discrete orbit.* It is only when it jumps from one discrete orbit to another that it radiates (or absorbs) energy. This implies that the angular momentum of the electron about the nucleus should be quantized, i.e., allowed to take only discrete values. Bohr's model with the electron moving in an specified circular orbit around the atomic nucleus is shown in Fig. (1.4).

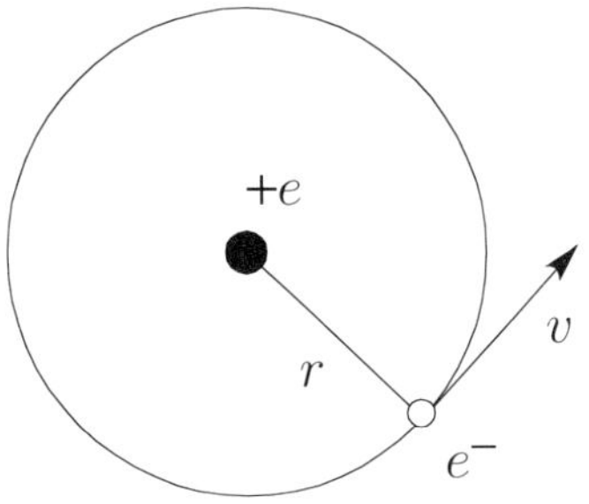

FIGURE 1.4
Bohr model of the Hydrogen atom with the electron moving in a specified circular orbit around the nucleus.

According to Bohr, the angular momentum of the electron is quantized in units of $\hbar$:

$$L = mvr = n\hbar\,, \qquad n = 1, 2, 3, \cdots \text{(an integer)} \tag{1.1.18}$$

and its total energy is given by (SI units):

$$E = \frac{1}{2}mv^2 - \frac{e^2}{4\pi\epsilon_0 r}\,. \tag{1.1.19}$$

Since the electrostatic attraction of the electron by the nucleus provides the centripetal force, we have

$$\frac{e^2}{4\pi\epsilon_0 r^2} = \frac{mv^2}{r}\,. \tag{1.1.20}$$

Eliminating v from Eqs.(1.1.18) and (1.1.19) we have

$$r_n = \frac{4\pi\epsilon_0\hbar^2 n^2}{me^2} \equiv a_0\, n^2\,, \tag{1.1.21}$$

where $a_0 = 4\pi\epsilon_0\hbar^2/me^2$ is called the radius of the first Bohr orbit, or just the *Bohr radius*. From Eqs. (1.1.19)- (1.1.21) we can express the total energy as

$$E_n = \frac{1}{2}\left(\frac{e^2}{4\pi\epsilon_0 r_n}\right) - \frac{e^2}{4\pi\epsilon_0 r_n} = -\frac{e^2}{8\pi\epsilon_0 a_0}\frac{1}{n^2}\,. \tag{1.1.22}$$

The quantity $e^2/8\pi\epsilon_0 a_0 = 13.6$ eV is called a Rydberg. The energy levels pertaining to various electron orbits, according to Bohr, are shown in Fig.(1.5)

Jumping of the electron from higher orbits to the orbits corresponding to $n = 1, n = 2, n = 3$, respectively, gives rise to the Lyman series, Balmer series, Paschen series of spectral lines in Hydrogen spectrum. Bohr's theory thus *explained* Ritz combination principle and the observed spectra of Hydrogen. However, the problem of accounting for the remarkable stability of atoms persisted. If, according to Bohr, the electron moves in a specified orbit around the nucleus and it has acceleration directed towards the centre, it would radiate energy and the orbit would get shorter and shorter, resulting in the instability of the atom. On the other hand, atoms are found to be remarkably stable!

Classical ideas also fail to explain the chemical properties of atoms of different species. For example, why are the properties of the Neon (Ne) atom, with ten electrons surrounding the nucleus, drastically different from those of Sodium (Na) atom which has just one more (eleven) electrons? The explanation can only be given in terms of a quantum mechanical

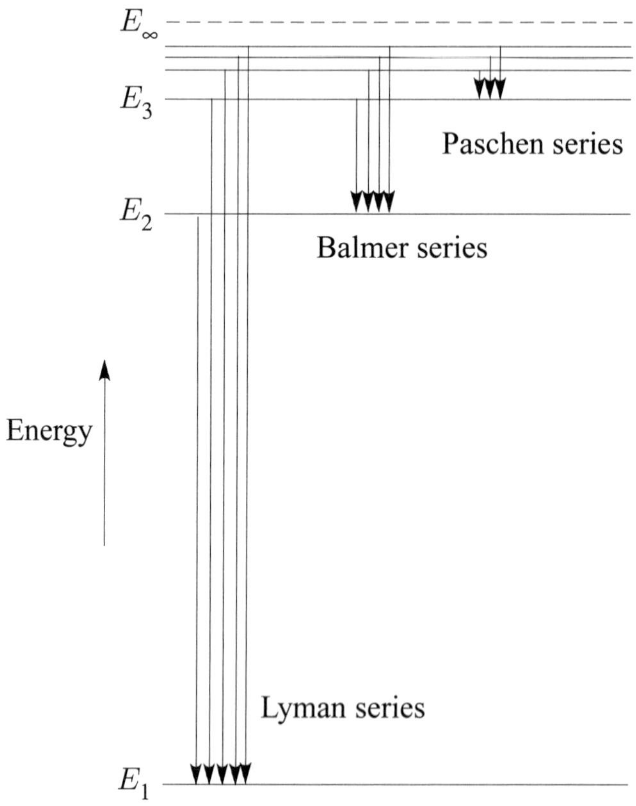

FIGURE 1.5
Energy levels of the Hydrogen atom and the spectral series.

principle, viz., Pauli exclusion principle,according to which each quantum state is either unoccupied or occupied by just one electron. In the case of Neon the first three electronic shells are completely filled, thus making the atom chemically inactive. In the case of Sodium the eleventh electron goes to the next unfilled shell. This electron called *valence electron* gives the Sodium atom valency equal to one and makes it chemically very active. Thus from the number of electrons in the atom we can generally estimate the electronic configuration in the atom to determine its valency and infer its chemical behavior. This also explains why all atoms of the same element (same Z) have identical chemical properties. This is also the principle behind the periodic classification of elements.

Classical ideas also cannot explain why an alpha particle inside a nucleus, with energy far less than the height of the Coulomb barrier at the nuclear boundary, is able to leak through the barrier.

In addition to this there are several other properties of materials which cannot be understood reasonably in terms of classical ideas. For example, solid materials have an enormous range of electrical conductivity (conductivity of silver is 10^{24} times as large as that of fused quartz). In terms of classical ideas one cannot comprehend why relative motion of negative particles (electrons) with respect to positive ions occurs more readily in silver than in quartz. Further, on the basis of classical ideas, we cannot understand why magnetic susceptibility (or permeability) of iron is much larger than that for other materials. Explanation of these phenomena, and a host of others at the atomic or molecular level, demands a new mechanics with radically new concepts. Before going into these new concepts it is important to ponder over the following question: Large and small are relative terms. Larger objects are made up of smaller objects, smaller objects are made up of still smaller objects and so on. Then why does it happen that classical concepts break down at a certain point so that they are no longer valid for still smaller objects, say atoms? The answer is that every observation is accompanied by a disturbance. This disturbance can be minimized by sophisticated instruments and devices, but this cannot be done beyond a certain point. There is always an uncertain intrinsic disturbance accompanying an observation which cannot be done away with by improved technique or increased skill of the observer. A system for which the *intrinsic disturbance* accompanying an observation is negligible is termed as large in the absolute sense. A system for which the intrinsic disturbance accompanying an observation is not negligible is termed as small in the absolute sense. It is for such systems that classical concepts break down. It is for such small objects, in the absolute sense, that a new mechanics, called quantum mechanics, based on radically new concepts, is needed.

1.2 Basis of Quantum Mechanics

1.2.1 Principle of Superposition of States

What could be the basis of the new mechanics? We shall see in the following illustrations that a new principle called the principle of *superposition of states* of a physical system could form the basis of this mechanics. In what follows we shall first elaborate on the term *state* of a physical system in quantum mechanics, and then on *superposition of states.*

The *state* of a physical system is characterized by the result of a certain observation on the system in that state being definite. We will explore this concept in subsequent sections and elaborate on how to represent the states of a physical system mathematically. Presently, to specify a physical state we shall just put some label within the symbol $|\ \rangle$ and call it a ket.

According to the principle of superposition of states a physical system in a superposition state can be looked upon as being partly in each of the two or more other states. In other words, a superposition state, say $|X\rangle$, may be looked upon as a linear combination of two (or more) other states, say $|A\rangle$ and $|B\rangle$:

$$|X\rangle = C_1\,|A\rangle + C_2\,|B\rangle\ ,$$

where C_1 and C_2 are some constants. The implication of this superposition is as follows: suppose, in the state $|A\rangle$ of the system, an observable, say α, gives a result a (i.e., the probability of getting result a is 1 and that of getting the result b is 0 and in the state $|B\rangle$ the same observable gives a result b (i.e., the probability of getting result a is zero and that of getting the result b is 1). Now what would be the result of the same measurement in the superposed state? The answer is that the result would not be something intermediate between a and b. The result will be either a or b but which one, we cannot foretell. We can only express the probability of getting the result a or b. In other words, though we expect the properties of the state $|X\rangle$ to be intermediate between the component states, the intermediate character lies not in the results of observation for α on state $|X\rangle$ being intermediate between those for the component states $|A\rangle$ and $|B\rangle$, but in the probability of getting a certain result on state $|X\rangle$ being intermediate between the corresponding probabilities for states $|A\rangle$ and $|B\rangle$. For the superposed state $|X\rangle$, the probability for getting the result a lies between 0 and 1 and the probability of getting the result b also lies between 0 and 1 and the sum of the two probabilities equals one.

We can generalize this principle to the superposition of more than two states by writing $|X\rangle = C_1\,|a_1\rangle + C_2\,|a_2\rangle + \cdots + C_n\,|a_n\rangle$, where $|a_1\rangle$, $|a_2\rangle$, $|a_3\rangle \cdots$, $|a_n\rangle$ are the states of a system in which an observation α gives results $a_1, a_2, \ldots, a_n$, respectively. If the same observation is made on the superposed state $|X\rangle$, the result would be *indeterminate*. It could be either a_1 or a_2, $\cdots$, or a_n. We *can* state the probabilities[2] of getting the results $a_1, a_2, \ldots, a_n$, viz., $|C_1|^2$, $|C_2|^2, |C_3|^2, \ldots$, $|C_n|^2$ and the sum of these probabilities must equal one. The following examples illustrate the principle of superposition of states.

1.2.1.1 Passage of a polarized photon through a polarizer

It has been found that when polarized light is used to eject photoelectrons, there is a preferential direction of emission. Since photoelectric effect needs the photon concept for its explanation, fact implies that polarization may be attributed to individual photons. Thus if the incident light is polarized in a certain sense, the associated photons may be taken to be polarized in the same sense.

To illustrate the principle of superposition of state we consider what happens to a single photon, polarized at an angle θ to the transmission axis of the polarizer when it meets the polarizer. We know that if the photon is polarized parallel to the transmission axis, it crosses the crystal and appears on the other side as a photon of the same energy (or frequency) polarized parallel to the polarizer axis (because there is complete transmission of the incident light when polarized parallel to the transmission axis). If, on the other hand, the photon is polarized perpendicular to the transmission axis of the polarizer, it is stopped and absorbed. Furthermore, classical electrodynamics tells us that if there is an incident beam polarized at an angle θ to the transmission axis then a fraction $\cos^2\theta$ of it will be transmitted and appear on the other side as light polarized paralllel to the transmission axis while a fraction $\sin^2\theta$ will be stopped and absorbed. But as regards a single photon polarized at an angle θ to the transmission axis, one cannot say that a fraction $\cos^2\theta$ of it would be transmitted and a fraction $\sin^2\theta$ of it would be stopped and absorbed because the photon, if it appears on the other side has the same energy. To maintain the indivisibility of the photon therefore, one has to sacrifice the concept of determinacy of classical mechanics and bring in indeterminacy. Thus the answer to the question as to what is the fate of a single photon (polarized at an angle θ to the transmission axis) when it crosses the crystal is that one does not know. One can only state the probability of its transmission ($\cos^2\theta$) and that of absorption ($\sin^2\theta$) and when it is transmitted it appears on the other side as photon polarized parallel to the trasnmission axis and having the original energy.

In the context of the principle of superposition of states, the state of the polarized photon (polarized at an angle θ to the polarizer axis) may be looked upon as a superposition of two states: the state of polarization parallel to the transmission axis and a state of polarization perpendicular to the transmission axis. Thus the state of a photon polarized at an angle θ to the polarizer axis can be written as

$$|\theta\rangle = C_1\,|\perp\rangle + C_2\,|\|\rangle\ .$$

When such a photon crosses the polarizer it is subjected to observation and the state is disturbed. Then it may jump to any one of the component states. Which state it will jump into is not certain; only the corresponding probability can be calculated. If it jumps to $|\|\rangle$ state, for which the probability is $\cos^2\theta$, it crosses the polarizer and appears on the other

[2]The interpretation of $|C_n|^2$ as the probability of getting a result a_n when the observation α is made on the superposed state $|X\rangle$ is justified in Sec. 1.10.

side as a photon polarized parallel to the polarizer axis with its energy (or frequencdy) unchanged. If it jumps to $|\perp\rangle$ state, for which the probability is $\sin^2\theta$, it is stopped and absorbed.

1.2.1.2 Interference of photons

Another illustration of the principle of superposition of states is provided by *interference of photons*, i.e., an attempt to understand interference of light on the basis of photon concept.

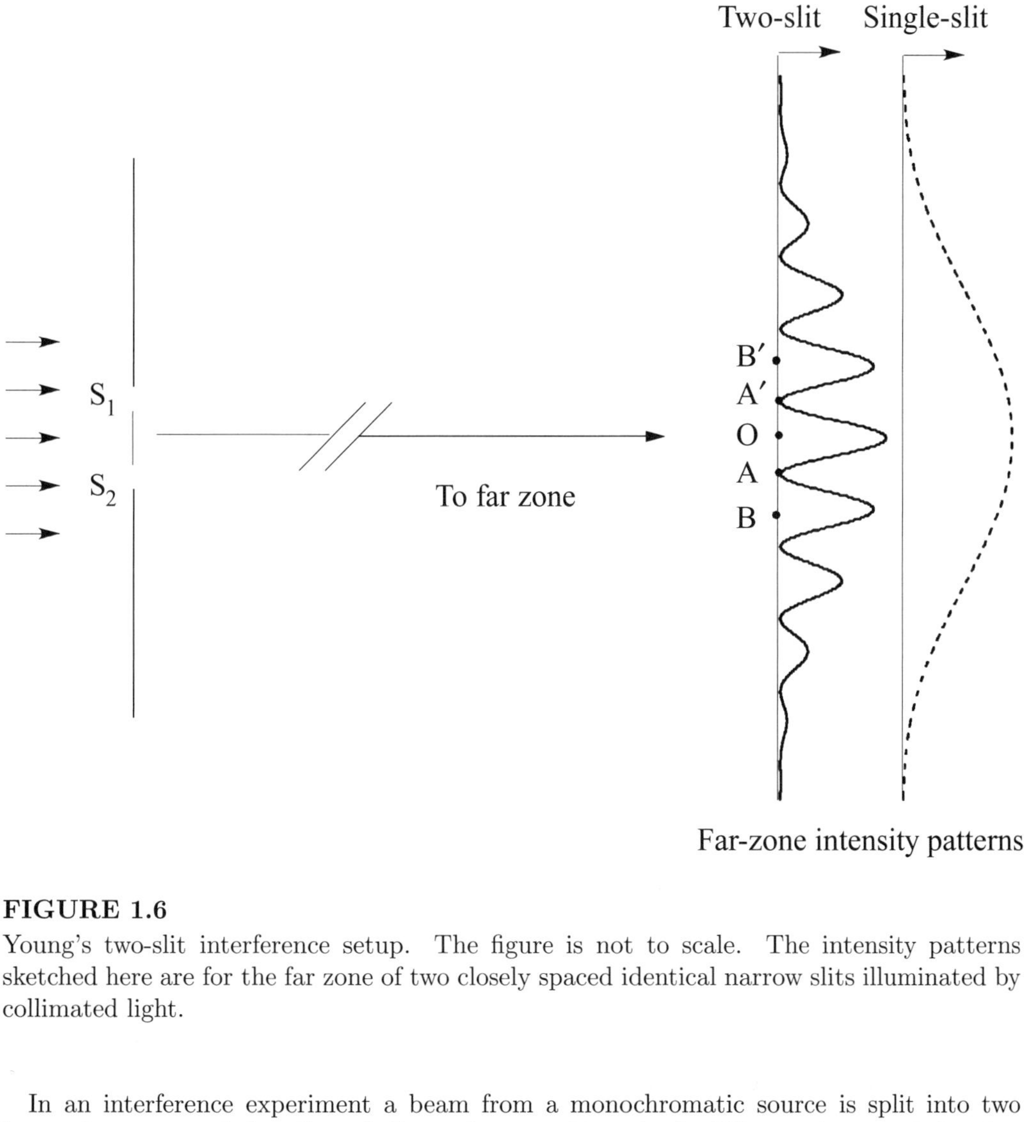

FIGURE 1.6
Young's two-slit interference setup. The figure is not to scale. The intensity patterns sketched here are for the far zone of two closely spaced identical narrow slits illuminated by collimated light.

In an interference experiment a beam from a monochromatic source is split into two beams by means of slits S_1 and S_2 as shown in Fig. (1.6). When only one of the slits (either S_1 or S_2) is opened, the intensity distribution is somewhat like the dashed curve in Fig. (1.6). Thus some photons do reach the point A i.e. A is not a totally dark point. But when both S_1 and S_2 are opened we have an interference pattern shown by the continuous curve and A is totally dark while the points O and B are bright and so on. How does one understand this? Is it that photons reaching A from S_1 and S_2 annihilate each other?

This cannot happen for it would violate the conservation of energy. To explain this we take recourse to the principle of superpostion of states. After passage through the slits the state of a photon in this experiment may be looked upon as a supersposition of two translation states of being associated with a beam through S_1 and of being associated with a beam through S_2. One cannot tell which of the two beams the photon is actually associated with unless one deliberately observes, and this observation disturbs the state of the photon. As a result of this observation, the photon will jump from the superposed state to one of the translational states, and so one will find it associated with either the beam through S_1 or through S_2. However, in this process the interference pattern will be lost.

On this basis one can easily understand the interference pattern. An incoming photon is associated with both translational states that interfere. At points O and B, where the interference is constructive, the probability of the photon reaching there is large. Where there is destructive interference the probability of photon reaching there is zero. Thus we have a probability distribution for a single photon. When instead of a single photon we have an incoming beam of photons, the probability distribution becomes the intensity distribution.

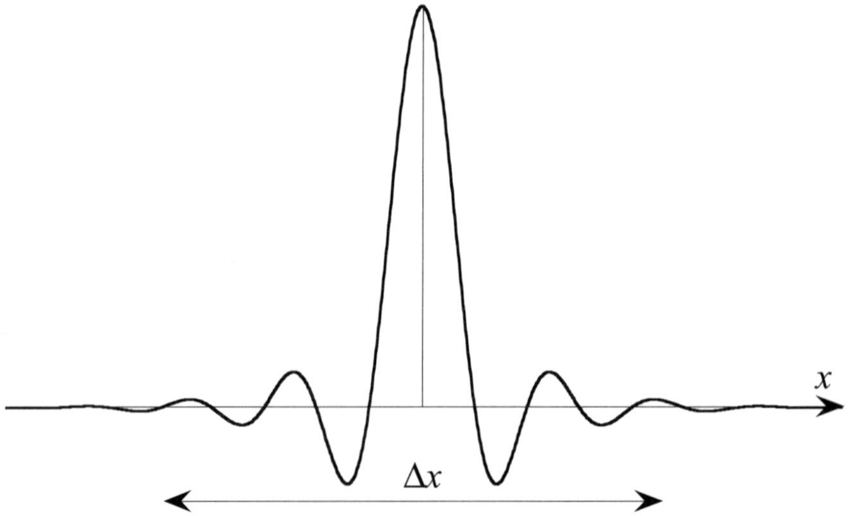

FIGURE 1.7
A wave packet associated with a moving particle. Δx, the extent of the wave packet, may be looked upon as the positional uncertainty of the particle.

1.2.2 Heisenberg Uncertainty Relations

According to de Broglie we can generally associate a particle moving with volocity v, with a wave packet so that $v = v_g$, where v_g is the group velocity of the wave packet [Eq. (1.1.3)]. If a particle happens to be associated with a continuous plane wave, characterized with a definite wavelength, then it is naturally assigned a definite momentum (since $p = h/\lambda$), i.e., uncertainty in momentum $\Delta p = 0$. Such a particle can exist anywhere within the continuous wave extending from $-\infty < x < +\infty$. Thus $\Delta x \to \infty$ for this particle. An example of a particle with precise momentum is the photon which cannot be positionally located at any time.

Now consider a particle associated with a wave packet shown in Fig.(1.7). Classically a wave packet of extent $\varDelta x$ may be looked upon as the result of superpostion of continuous waves of wavelength ranging approximately from λ to $\lambda + \varDelta\lambda$ where

$$\frac{1}{\varDelta x} \approx -2\pi \frac{\varDelta\lambda}{\lambda^2} .$$

This relation follows from $\varDelta x \varDelta k \approx 1$ and $\varDelta k = -2\pi\lambda^2 \varDelta\lambda$. Now, in the light of the principle of superpositon of states, we can look upon this state of a particle (associated with the wave packet) as one resulting from the superposition of several momentum states, with momenta ranging from p to $p + \varDelta p$, or wavelengths ranging from λ to $\lambda + \varDelta\lambda$ (each state of definite momentum can be represented by continuous plane wave of a specific wavelength by virtue of de Broglie relation). In such a superposed state a measurement of the momentum can yield any value within p and $p = p + \varDelta p$. From $p = h/\lambda$, we find $\varDelta p = -h\varDelta\lambda/\lambda^2 = h/2\pi\varDelta x$, which implies

$$\varDelta x \varDelta p \approx \hbar . \tag{1.2.23}$$

Thus if a particle is associated with the wave packet of spatial extent Δx, its position uncertainty is Δx and its momentum uncertainty is $\varDelta p$ such that $\varDelta p \varDelta x \approx \hbar$.

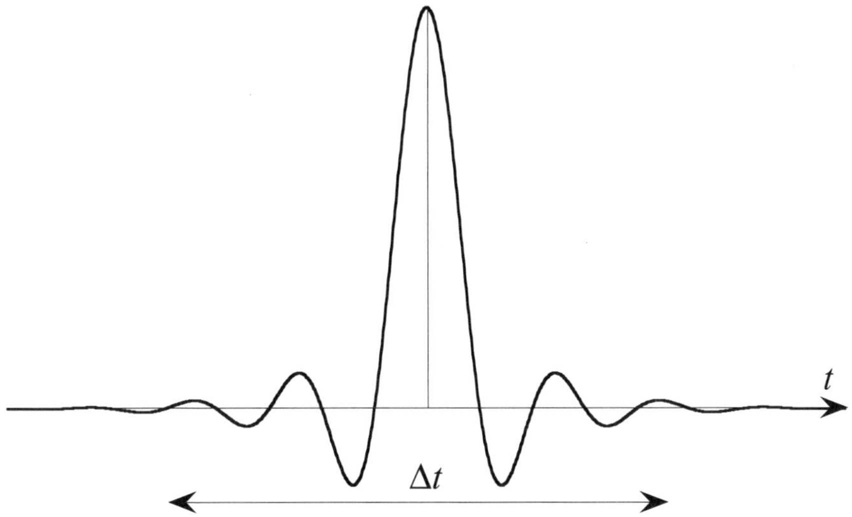

FIGURE 1.8
A wave packet of duration $\varDelta t$. Its duration $\varDelta t$ may be looked upon as the time spent on the observation of the energy of the particle.

We can alternatively consider the particle to be associated with a *time packet*, i.e., a wave packet of duration Δt in time. We can then look upon the extent Δt of the time packet to be the time spent on the observation (or assessment) of the energy of the particle. Clasically such a wave packet results from the superposition of plane waves with frequencies between ν and $\nu + \varDelta\nu$ such that $\varDelta\nu \approx 1/2\pi\varDelta t$. Using Planck's condition $E = h\nu$, we find

$\Delta E = h\Delta\nu = h/2\pi\Delta t$, which leads to

$$\Delta E \Delta t \approx \hbar\,. \tag{1.2.24}$$

Since the state of the particle associated with the time packet can be looked upon as a continuous superposition of states with frequencies between ν and $\nu + \Delta\nu$ (or of states with energies between E and $E + \Delta E$), a measurement of energy in this state can give any value ranging from E and $E + \Delta E$. Thus ΔE is the uncertainty in the energy of a particle associated with the time packet of duration Δt.

1.3 Representation of States

The states of a physical system should be represented in such a way that the underlying principle of superposition of states is incorporated in the mathematical formulation. Dirac postulated that states of a physical system might be represented by vectors in an infinite dimensional space called Hilbert space. A typical vector may be denoted by $|A\rangle$, called ket A, A being a suffix to label the state. Since a vector space has the property that two or more vectors belonging to it can be added to give a new vector in the same space, that is,

$$|X\rangle = C_1\,|A\rangle + C_2\,|B\rangle\,, \tag{1.3.1}$$

the principle of superposition of states finds the following mathematical expression:

The correspondence between a ket vector and the state of dynamical system at a particular time is such that if a state labeled by X results from the superposition of certain states labeled by A and B, the corresponding ket vector $|X\rangle$ is expressible linearly in terms of the corresponding ket vectors $|A\rangle$ and $|B\rangle$ representing the component states [Eq.(1.3.1)].

The preceding statement leads to certain properties of the superposition of states:

1. When two or more states are superposed, the order in which they occur in the superposition is unimportant, i.e., the superposition is symmetric between the states that are superposed. This means $C_1\,|A\rangle + C_2\,|B\rangle$ and $C_2\,|B\rangle + C_1\,|A\rangle$ represent the same state. This is also true in vector addition (commutativity holds).

2. If a state represented by $|A\rangle$ is superposed onto itself it gives no new state. Mathematically,

$$C_1\,|A\rangle + C_2\,|A\rangle = (C_1 + C_2)\,|A\rangle$$

represents the same state as $|A\rangle$ does. It therefore follows that if a ket vector corresponding to a state is multiplied by any nonzero complex number, the resulting ket vector corresponds to the same state. In other words, it is only the *direction* and not the *magnitude* of a ket vector that specifies the state.

3. If $|X\rangle = C_1\,|A_1\rangle + C_2\,|A_2\rangle + \cdots + C_n\,|A_n\rangle$, then the state represented by $|X\rangle$ is said to be dependent on the component states $|A_1\rangle\,, |A_2\rangle\,, \cdots, |A_n\rangle$.

An important question that may be asked is why vectors belonging to an infinite dimensional space are chosen to represent physical states of any system. This is done to include the possibility of having an infinite number of mutually orthogonal ket vectors[3] which correspond to an infinite number of mutually orthogonal states.

[3] In three dimensional space we can think of only three mutually orthogonal vectors. In an N-dimensional space we can think of N mutually orthogonal vectors.

1.4 Dual Vectors: Bra and Ket Vectors

Corresponding to a set of vectors in a vector space we can have another set of vectors in a conjugate space. Mathematicians call these two sets of vectors as *dual* vectors. Following Dirac we call one set of vectors as *ket* vectors and the second set of vectors as *bra* vectors. We have one to one correspondence in the two sets of vectors, for example the vectors $|A\rangle$ and $\langle A|$, belonging to the two different vector spaces, correspond to each other and therefore represent the same physical state. The two vectors are called *conjugate imaginaries* of each other. Similarly $|B\rangle$ and $\langle B|$ are conjugate imaginaries of each other, then the set of vectors

$$\begin{aligned} |A\rangle + |B\rangle \quad & \text{and} \quad \langle A| + \langle B| \\ C\,|A\rangle \quad & \text{and} \quad C^* \langle A| \\ C_1 |A\rangle + C_2 |B\rangle \quad & \text{and} \quad C_1^* \langle A| + C_2^* \langle B| \end{aligned}$$

are also conjugate imaginaries of each other. The whole mathematical theory which Dirac has developed is symmetric between bras and kets.

Though the two sets of vectors, kets and bras, belong to different spaces and a ket vector cannot be added to a bra vector, we can define the product of a bra and a ket vector (taken in this order so that it becomes a *bracket*) the result being a complex number[4]. It turns out that the brackets $\langle A| B\rangle$ and $\langle B| A\rangle$, both complex numbers, are in fact complex conjugates of each other, i.e.,

$$\overline{\langle A| B\rangle} = \langle B| A\rangle \,. \tag{1.4.1}$$

In Dirac notation a bar over a bracket or a complex number indicates its complex conjugate. At some places we may also use a star (*) to denote the complex conjugate of a number. From Eq. (1.4.1) we have $\overline{\langle A| A\rangle} = \langle A| A\rangle$ i.e. $\langle A| A\rangle$ is real and positive except when $|A\rangle$ is zero in which case $\langle A| A\rangle = 0$. The positive square root of $\langle A| A\rangle$ defines the length of the ket vector $|A\rangle$ (or of its conjugate imaginary $\langle A|$). The ket vector or bra vector is said to be normalized when

$$\langle A| A\rangle = 1 \,.$$

The bra vectors $\langle A|$ and $\langle B|$ or the ket vectors $|A\rangle$ and $|B\rangle$ are said to be orthogonal if the scalar product

$$\langle A| B\rangle = 0 \quad \text{or} \quad \langle B| A\rangle = 0 \,.$$

1.5 Linear Operators

A linear operator[5] $\hat{\alpha}$ has the property that it can operate on a ket vector from left to right to give another ket vector

$$\hat{\alpha}\,|A\rangle = |F\rangle \,. \tag{1.5.1}$$

[4]This is somewhat like the inner producdt of a covariant vector A_i and a contravariant vector B^i. The two vectors cannot be added but an inner product $\sum_i A_i B^i$ (or simply $A_i B^i$) exists.

[5]In our notation we use a caret over the symbols representing linear operators to distinguish them from numbers.

A linear operator can also operate on a bra vector from right to left to give another bra vector

$$\langle B|\,\hat{\alpha} = \langle C|\ . \tag{1.5.2}$$

We shall first state the properties of a linear operator, develop an algebra for ket and bra vectors and linear operators, and subsequently take up their physical interpretation.

1.5.1 Properties of a Linear Operator

A linear operator $\hat{\alpha}$ is considered to be known if the result of its operation on every ket vector or every bra vector is known. If a linear operator, operating on every ket vector gives 0 (null vector), then the linear operator is a null operator. Further, two linear operators are said to be equal if both produce the same result when applied to every ket vector. Some other properties of linear operators are

$$\hat{\alpha}\,\{|A\rangle + |A'\rangle\} = \hat{\alpha}\,|A\rangle + \hat{\alpha}\,|A'\rangle\ , \tag{1.5.3}$$

$$\hat{\alpha}\,\{C\,|A\rangle\} = C\hat{\alpha}\,|A\rangle\ , \tag{1.5.4}$$

$$(\hat{\alpha} + \hat{\beta})\,|A\rangle = \hat{\alpha}\,|A\rangle + \hat{\beta}\,|A\rangle\ , \tag{1.5.5}$$

$$\hat{\alpha}\hat{\beta}\,|A\rangle = \hat{\alpha}\left\{\hat{\beta}\,|A\rangle\right\} \neq \hat{\beta}\,\{\hat{\alpha}\,|A\rangle\}\ . \tag{1.5.6}$$

In general, the commutative axiom of multiplication does not hold for the product of two operators $\hat{\alpha}$ and $\hat{\beta}$ ($\hat{\alpha}\hat{\beta} \neq \hat{\beta}\hat{\alpha}$). In particular, if two operators, say $\hat{\alpha}$ and $\hat{\beta}$ commute, then this is stated explicitly as

$$(\hat{\alpha}\hat{\beta} - \hat{\beta}\hat{\alpha}) = \hat{0} \quad \text{or} \quad [\hat{\alpha},\,\hat{\beta}] = \hat{0}\,.$$

1.6 Adjoint of a Linear Operator

If $\hat{\alpha}$ is a linear operator, its adjoint $\hat{\alpha}^\dagger$ is defined such that the ket $\hat{\alpha}\,|P\rangle$ and the bra $\langle P|\,\hat{\alpha}^\dagger$ are conjugate imaginaries of each other. That is, if

$$\hat{\alpha}\,|P\rangle = |A\rangle \tag{1.6.1}$$

then the adjoint of $\hat{\alpha}$ is defined by

$$\langle P|\,\hat{\alpha}^\dagger = \langle A|\ . \tag{1.6.2}$$

Since $\langle A|\,B\rangle = \overline{\langle B|\,A\rangle}$, this implies

$$\langle P|\,\hat{\alpha}^\dagger\,|B\rangle = \overline{\langle B|\,\hat{\alpha}\,|P\rangle} \tag{1.6.3}$$

This is a general result, which holds for any linear operator $\hat{\alpha}$ and for any set of vectors $|A\rangle$ and $|B\rangle$. It expresses a frequently used property of the adjoint of an operator. Several important corrolaries follow from the definition of the adjoint of a linear operator.

1. *The adjoint of the adjoint of a linear operator is the original operator:* $\left(\hat{\alpha}^\dagger\right)^\dagger = \hat{\alpha}$.

   By replacing the linear operator $\hat{\alpha}$ by its adjoint $\hat{\alpha}^\dagger$ in Eq.(1.6.3) we have

$$\langle P|\,(\hat{\alpha}^\dagger)^\dagger\,|B\rangle = \overline{\langle B|\,\hat{\alpha}^\dagger\,|P\rangle}\,, \tag{1.6.4}$$

and by interchanging the labels of bra and ket vectors in Eq.(1.6.3) we find

$$\langle B|\,\hat{\alpha}^\dagger\,|P\rangle = \overline{\langle P|\,\hat{\alpha}\,|B\rangle}\,. \tag{1.6.5}$$

On taking the complex conjugate of this equation we obtain

$$\overline{\langle B|\,\hat{\alpha}^\dagger\,|P\rangle} = \langle P|\,\hat{\alpha}\,|B\rangle\,. \tag{1.6.6}$$

A comparison of Eqs. (1.6.4) and (1.6.6) shows $\langle P|\,(\hat{\alpha}^\dagger)^\dagger\,|B\rangle = \langle P|\,\hat{\alpha}\,|B\rangle$ which implies $(\hat{\alpha}^\dagger)^\dagger = \hat{\alpha}$.

2. *The adjoint of the product of two operators is equal to the products of the adjoints in the reversed order:* $(\hat{\alpha}\hat{\beta})^\dagger = \hat{\beta}^\dagger\hat{\alpha}^\dagger$.

   Let $\langle P|\,\hat{\alpha} = \langle A|$ and $\langle Q|\,\hat{\beta}^\dagger = \langle B|$ so that $\hat{\alpha}^\dagger\,|P\rangle = |A\rangle$ and $\hat{\beta}\,|Q\rangle = |B\rangle$. Then we have the inner products

$$\langle B|\,A\rangle = \langle Q|\,\hat{\beta}^\dagger\hat{\alpha}^\dagger\,|P\rangle\,, \tag{1.6.7}$$

$$\langle A|\,B\rangle = \langle P|\,\hat{\alpha}\hat{\beta}\,|Q\rangle\,. \tag{1.6.8}$$

   But from the definition of inner product, we have

$$\begin{aligned}\langle B|\,A\rangle &= \overline{\langle A|\,B\rangle}\\ &= \overline{\langle P|\,\hat{\alpha}\hat{\beta}\,|Q\rangle}\\ &= \langle Q|\,(\hat{\alpha}\hat{\beta})^\dagger\,|P\rangle\,.\end{aligned} \tag{1.6.9}$$

   Comparing Eq.(1.6.7) and (1.6.9) we have

$$(\hat{\alpha}\hat{\beta})^\dagger = \hat{\beta}^\dagger\hat{\alpha}^\dagger\,. \tag{1.6.10}$$

   This can be generalized to the product of three or more operators as

$$(\hat{\alpha}\hat{\beta}\hat{\gamma})^\dagger = \hat{\gamma}^\dagger(\hat{\alpha}\hat{\beta})^\dagger = \hat{\gamma}^\dagger\hat{\beta}^\dagger\hat{\alpha}^\dagger \tag{1.6.11}$$

$$(\hat{\alpha}\hat{\beta}\cdots\hat{\delta})^\dagger = \hat{\delta}^\dagger\cdots\hat{\beta}^\dagger\hat{\alpha}^\dagger\,. \tag{1.6.12}$$

3. $|A\rangle\langle B|$ *behaves like a linear operator and that* $(|A\rangle\langle B|)^\dagger = |B\rangle\langle A|$

   It is easy to see that $|A\rangle\langle B|$ behaves like a linear operator because operating on a ket from left, it gives another ket and operating on a bra from right, it gives another bra,

$$|A\rangle\langle B|\,P\rangle = \langle B|\,P\rangle\,|A\rangle = \text{a complex number} \times |A\rangle\,,$$

   and $\langle Q|\,A\rangle\langle B| = \text{a complex number} \times \langle B|\,.$

   Now $\langle Q|\,A\rangle\langle B|$ is a bra vector whose conjugate imaginary is the ket $\overline{\langle Q|\,A\rangle}\,|B\rangle = \langle A|\,Q\rangle\,|B\rangle = |B\rangle\langle A|\,Q\rangle$. Since conjugate imaginary of $\langle Q|\,\hat{\alpha}$ is $\hat{\alpha}^\dagger\,|Q\rangle$ it follows that if we identify $|A\rangle\langle B|$ as an operator $\hat{\alpha}$ then $|B\rangle\langle A|$ should be identified with $\hat{\alpha}^\dagger$. Hence the operators $|A\rangle\langle B|$ and $|B\rangle\langle A|$ are adjoint of each other,

$$(|A\rangle\langle B|)^\dagger = |B\rangle\langle A|\,. \tag{1.6.13}$$

A linear operator that is equal to its own adjoint

$$\hat{\alpha} = \hat{\alpha}^\dagger\,, \tag{1.6.14}$$

is called a *self-adjoint* or *Hermitian operator*. The term real operator is also used sometimes. If, on the other hand, a linear operator satisfies $\hat{\alpha} = -\hat{\alpha}^\dagger$, then the operator $\hat{\alpha}$ is called an *anti-Hermitian* or *pure imaginary* operator.

1.7 Eigenvalues and Eigenvectors of a Linear Operator

In general, a linear operator $\hat{\alpha}$ operating on a ket $|B\rangle$ gives another ket, say, $|F\rangle$

$$\hat{\alpha}\,|B\rangle = |F\rangle\ .$$

The *direction* of the new ket vector $|F\rangle$ is, in general, different from that of $|B\rangle$. In a particular situation (for a specific ket vector $|A\rangle$) it may happen that

$$\hat{\alpha}\,|A\rangle = \alpha\,|A\rangle\ , \tag{1.7.1}$$

where α may be a complex number. When this happens, we say that $|A\rangle$ is an eigen ket vector of $\hat{\alpha}$, belonging to the eivenvalue α. It can happen that a linear operator α admits a set of eigen kets belonging to different eigenvalues:

$$\begin{aligned}\hat{\alpha}\,|\alpha'\rangle &= \alpha'\,|\alpha'\rangle\\ \hat{\alpha}\,|\alpha''\rangle &= \alpha''\,|\alpha''\rangle\\ &\vdots\\ \hat{\alpha}\,\left|\alpha^{(s)}\right\rangle &= \alpha^{(s)}\,\left|\alpha^{(s)}\right\rangle\end{aligned}$$

where we have labeled the eigen kets by writing the eigenvalues inside them. Eigenvalues of operators are denoted by removing the carets and putting primes or suffixes on the symbols representing the corresponding operators. Several corollaries follow from the definition of an eigenvalue.

1. *The eigenvalues of a self-adjoint (Hermitian) operator ($\hat{\alpha}^\dagger = \hat{\alpha}$) are real.*

   Let $|\alpha\rangle$ be an eigen ket of a Hermitian operator $\hat{\alpha}$ belonging to eigenvalue α,

$$\hat{\alpha}\,|\alpha\rangle = \alpha\,|\alpha\rangle\ . \tag{1.7.2}$$

   Multiplying on the left by $\langle\alpha|$ we get

$$\langle\alpha|\,\hat{\alpha}\,|\alpha\rangle = \alpha\,\langle\alpha|\,\alpha\rangle\ . \tag{1.7.3}$$

   Taking the complex conjugate of both sides we have

$$\overline{\langle\alpha|\,\hat{\alpha}\,|\alpha\rangle} = \alpha^*\,\overline{\langle\alpha|\,\alpha\rangle} = \alpha^*\,\langle\alpha|\,\alpha\rangle\,. \tag{1.7.4}$$

   But from the definition (1.6.3) of adjoint operator the left hand side of this equation $\overline{\langle\alpha|\,\hat{\alpha}\,|\alpha\rangle} = \langle\alpha|\,\hat{\alpha}^\dagger\,|\alpha\rangle$. Using this in Eq.(1.7.4) we find that $\langle\alpha|\,\hat{\alpha}^\dagger\,|\alpha\rangle = \alpha^*\langle\alpha|\,\alpha\rangle$ or $\langle\alpha|\,\hat{\alpha}\,|\alpha\rangle = \alpha^*\,\langle\alpha|\,\alpha\rangle$, since $\hat{\alpha}^\dagger = \hat{\alpha}$. On comparing this result with Eq. (1.7.3) we conclude

$$\alpha^* = \alpha\,, \tag{1.7.5}$$

   which implies that α is real. This can be generalized to all eigenvalues of a self-adjoint operator $\hat{\alpha}$ and we conclude that the eigenvalues of a self-adjoint operator are real.

2. *The eigenvalues associated with the eigen kets of a self-adjoint (Hermitian) operator are the same as the eigenvalues associated with eigen bras of the same operator.*

Let $|\alpha'\rangle$ be an eigen ket belonging to eigenvalue α' of a self-adjoint (Hermitian) operator:

$$\hat{\alpha}\,|\alpha'\rangle = \alpha'\,|\alpha'\rangle\,, \tag{1.7.6}$$

where α' is real, since $\alpha^\dagger$ is Hermitian. Taking the conjugate imaginary of both sides leads to

$$\langle\alpha'|\,\hat{\alpha}^\dagger = \alpha'^*\,\langle\alpha|\;. \tag{1.7.7}$$

But, since $\hat{\alpha}^\dagger = \hat{\alpha}$ and $\alpha'^* = \alpha'$ (real), Eq. (1.7.4) leads to

$$\langle\alpha'|\,\hat{\alpha} = \alpha'\,\langle\alpha'|\;. \tag{1.7.8}$$

Thus the conjugate imaginary of an eigen ket of a Hermitian operator $\hat{\alpha}$ is an eigen bra of the same operator belonging to the same eigenvalue. The converse also holds.

Proceeding in this manner we can show that this result holds for other eigen kets and eigenvalues as well.

3. *Eigenvectors belonging to different eigenvalues of a Hermitian aperator are orthogonal.*

   If $|\alpha'\rangle$ and $|\alpha''\rangle$ are the eigen kets belonging, respectively, to the eigenvalues α' and α'' of a Hermitian operator $\hat{\alpha}$, then

$$\hat{\alpha}\,|\alpha'\rangle = \alpha'\,|\alpha'\rangle \tag{1.7.9}$$
$$\hat{\alpha}\,|\alpha''\rangle = \alpha''\,|\alpha''\rangle\;. \tag{1.7.10}$$

   Taking conjugate imaginary of each side of Eq. (1.7.9) we obtain

$$\langle\alpha'|\,\hat{\alpha}^\dagger = \alpha'^*\,\langle\alpha'|\;,$$

   or

$$\langle\alpha'|\,\hat{\alpha} = \alpha'\,\langle\alpha'|\;, \tag{1.7.11}$$

   where we have used $\hat{\alpha}^\dagger = \hat{\alpha}$ (Hermitian operator) and $\alpha'^* = \alpha'$ (eigenvalue of a Hermitian operator). Multiplying Eq. (1.7.14) on the right by $|\alpha''\rangle$ we get

$$\langle\alpha'|\,\hat{\alpha}\,|\alpha''\rangle = \alpha'\,\langle\alpha'|\,\alpha''\rangle\;, \tag{1.7.12}$$

   and multiplying Eq (1.7.15) on the left by $\langle\alpha'|$ we get

$$\langle\alpha'|\,\hat{\alpha}\,|\alpha''\rangle = \alpha''\,\langle\alpha'|\,\alpha''\rangle\;, \tag{1.7.13}$$

   where α' and α'', being a numbers, have been taken out of the brackets. Subtracting Eq. (1.7.18) from (1.7.17) we obtain

$$(\alpha' - \alpha'')\,\langle\alpha'|\,\alpha''\rangle = 0\,. \tag{1.7.14}$$

   This equation implies that, if $\alpha' \neq \alpha''$, then $\langle\alpha'|\,\alpha''\rangle = 0$. In other words, eigenvectors (eigen kets or eigen bras) of a self-adjoint operator $\hat{\alpha}$, belonging to different eigenvalues are orthogonal in the sense that their inner product is zero, i.e., $\langle\alpha'|\,\alpha''\rangle = 0$.

   If $\alpha'' = \alpha'$ then $\langle\alpha'|\,\alpha''\rangle = \langle\alpha'|\,\alpha'\rangle \neq 0$ and is taken to be unity if $|\alpha'\rangle$ is normalized. In general, if the eigenvectors are normalized

$$\left\langle \alpha^{(r)} \middle| \alpha^{(s} \right\rangle = \delta_{rs}\,. \tag{1.7.15}$$

1.8 Physical Interpretation

So far we have defined a linear operator and its adjoint and have developed an algebra involving ket and bra vectors and linear operators. We have also mathematically defined eigenvectors (eigen kets and eigen bras) and eigenvalues of a linear operator and have established some corollaries. We shall now explore if we can give a physical meaning to Hermitian (self-adjoint) linear operators, their eigenvalues and their eigen kets or eigen bras.

Self-adjoint operators may correspond to real dynamical variables of classical mechanics. Such operators may be expressed in terms of the position and momentum operators and may represent some physical quantity, e.g., energy or angular momentum. Because of their correspondence with some observable physical quantities, these operators may also be termed as observables[6]. We can now give a physical meaning to eigenvectors (kets or bras) and eigenvalues and also the orthogonality of eigenvectors.

1.8.1 Physical Interpretation of Eigenstates and Eigenvalues

The equation $\hat{\alpha}\,|\alpha'\rangle = \alpha'\,|\alpha'\rangle$ can be given the following physical interpretation. When the system is in the state specified by $|\alpha'\rangle$ then an observation pertaining to the operator $\hat{\alpha}$ made on the system is certain to give the result α'. Likewise, the eigenvalue equation

$$\hat{\alpha}\left|\alpha^{(r)}\right\rangle = \alpha^{(r)}\left|\alpha^{(r)}\right\rangle \tag{1.8.1}$$

implies that when the physical system is in the state represented by $\left|\alpha^{(r)}\right\rangle$, the result of measurement of a physical quantity pertaining to $\hat{\alpha}$ is completely predictable and it is $\alpha^{(r)}$. Various eigenvalues of the self-adjoint operator represent the results of measurement pertaining to the Hermitian operator $\hat{\alpha}$ on the respective eigenstates. If the physical system is not in any one of the eigenstates of $\hat{\alpha}$ but in an arbitrary state $|X\rangle$ which is expressible as a superposition of these eigenstates:

$$|X\rangle = C_1\,|\alpha'\rangle + C_2\,|\alpha''\rangle + \cdots + C_N\left|\alpha^{(N)}\right\rangle \equiv \sum_{r=1}^{N} C_r\left|\alpha^{(r)}\right\rangle, \tag{1.8.2}$$

then the result of a measurement pertaining to $\hat{\alpha}$ will be indeterminate. It will, of course, be one of the eigenvalues of $\hat{\alpha}$ but which one we cannot foretell. We can thus interpret the eigenvalues of $\hat{\alpha}$ as the possible results of measurements of $\hat{\alpha}$ on the system in any state. The results of measurement of $\hat{\alpha}$ will never be different from the eigenvalues α', α'' $\cdots$ $\alpha^{(r)} \cdots \alpha^{(N)}$. If the state of the system happens to be an eigenstate of $\hat{\alpha}$, the results of measurement of $\hat{\alpha}$ is predictable and is the corrsesponding eigenvalue. If the state is not an eigenstate of $\hat{\alpha}$ then the result is indeterminate, but it is one of the eigenvalues. It is now also evident why we have required an observable to be a self-adjoint operator. Only a self-adjoint operator can admit real eigenvalues and we do require the eigenvalues to be real because the results of any physical observation must be real.

[6] In addition to being self-adjoint linear operators, observables must conform to the requirement of *completeness* [see Sec. 1.10].

1.8.2 Physical Meaning of the Orthogonality of States

The eigenstates of a Hermitian (self-adjoint) operator $\hat{\alpha}$ belonging to different eigenvalues, say $\alpha^{(r)}$ and $\alpha^{(s)}$ are orthogonal in the sense that

$$\left\langle \alpha^{(r)} \middle| \alpha^{(s)} \right\rangle = \delta_{rs}, \tag{1.8.3}$$

where we have assumed that all eigen kets or bras are normalized. This is how we express orthogonality condition mathematically. It means that the *overlap* between two orthogonal states is zero or the projection of one orthogonal state on the other is zero.

Physically a set of states $|\alpha'\rangle$, $|\alpha''\rangle \cdots \left|\alpha^{(r)}\right\rangle$ are said to be orthogonal to each other if there exists an observation (in this case pertaining to $\hat{\alpha}$) which, when made on the system in each one of these states, is destined to give different results.

1.9 Observables and Completeness Criterion

We have seen that for an operator $\hat{\alpha}$ to be an *observable*, that is, for it to correspond to a measurable quantity, it must be a Hermitian (self-adjoint) operator because its eigenvalues, which represents the possible results of measurement, must be real. Consequently, the eigenvectors of $\hat{\alpha}$ must satisfy the orthogonality condition

$$\left\langle \alpha^{(r)} \middle| \alpha^{(s)} \right\rangle = \delta_{rs}. \tag{1.9.1}$$

However, to be classed as an observable, a Hermitian operator $\hat{\alpha}$ must, in addition, satisfy the completeness criterion, i.e., the eigenvectors of $\hat{\alpha}$ must form a complete set of ket vectors so that an arbitrary state $|X\rangle$ can be expressed in terms of them:

$$|X\rangle = \sum_{r=1}^{N} C_r \left|\alpha^{(r)}\right\rangle. \tag{1.9.2}$$

It is only when an arbitrary state $|X\rangle$ is expressible in terms of the eigenstates of $\hat{\alpha}$ that $\hat{\alpha}$ can be measured in the state $|X\rangle$ and the result is one of the eigenvalues of $\hat{\alpha}$. If the eigenstates of $\hat{\alpha}$ do not form a complete set and an arbitrary state $|X\rangle$ of the system is not expressible in terms of them, then $\hat{\alpha}$ cannot be called an observable. We shall see an example of a Hermitian operator that is not an observable in Chapter 3.

The completeness condition can be expressed mathematically as follows. Let an operator $\sum_{r=1}^{N} \left|\alpha^{(s)}\right\rangle \left\langle \alpha^{(s)}\right|$ operate on an arbitrary state $|X\rangle$, which can be expressed in terms of the eigen kets of $\hat{\alpha}$ by means of Eq.(1.9.2), since the set of eigen kets $\left|\alpha^{(s)}\right\rangle$ form a complete set. Then

$$\sum_{s=1}^{N}\left|\alpha^{(s)}\right\rangle\left\langle\alpha^{(s)}\right|X\big\rangle=\sum_{s=1}^{N}\left|\alpha^{(s)}\right\rangle\left\langle\alpha^{(s)}\right|\sum_{r=1}^{N}C_r\left|\alpha^{(r)}\right\rangle$$
$$=\sum_{r=1}^{N}C_r\sum_{s=1}^{N}\left|\alpha^{(s)}\right\rangle\left\langle\alpha^{(s)}\middle|\alpha^{(r)}\right\rangle$$
$$=\sum_{r=1}^{N}C_r\sum_{s=1}^{N}\left|\alpha^{(s)}\right\rangle\delta_{rs}$$
$$=\sum_{r=1}^{N}C_r\left|\alpha^{(r)}\right\rangle=|X\rangle\,.$$

Since $|X\rangle$ is an arbitrary state, it follows that

$$\sum_{s=1}^{N}\left|\alpha^{(s)}\right\rangle\left\langle\alpha^{(s)}\right|=\hat{1}\,, \tag{1.9.3}$$

is a unit operator. Equation (1.9.3) can be looked upon as the mathematical expression of the completeness condition.

Thus, if an operator $\hat{\alpha}$ is to be an observable, its eigenstate must satisfy the twin conditions of

(i) orthogonality: $\left\langle\alpha^{(s)}\middle|\alpha^{(r)}\right\rangle=\delta_{rs}$ (1.9.1)

and (ii) completeness: $\sum_{r=1}^{N}\left|\alpha^{(r)}\right\rangle\left\langle\alpha^{(r)}\right|=\hat{1}\,.$ (1.9.3)

If the eigenvalues of $\hat{\alpha}$ are not discrete but continuous so that α varies continuously in a certain range, then the completeness relation can be written as

$$\int|\alpha\rangle\,d\alpha\,\langle\alpha|=\hat{1}\,. \tag{1.9.3a}$$

It could also be that eigenvalues of $\hat{\alpha}$ are discrete in a certain range and continuous in another. In such a case the completeness condition can be written as

$$\sum_{r=1}^{N}\left|\alpha^{(r)}\right\rangle\left\langle\alpha^{(r)}\right|+\int|\alpha\rangle\,d\alpha\,\langle\alpha|=\hat{1}\,. \tag{1.9.3b}$$

The operator $\left|\alpha^{(r)}\right\rangle\left\langle\alpha^{(r)}\right|$ may be regarded as the projection operator for the state $\left|\alpha^{(r)}\right\rangle$ for it is easily seen that, operating on any arbitrary state $|X\rangle$, it projects out the state $\left|\alpha^{(r)}\right\rangle$ apart from a multiplying constant. Physically the completeness condition (1.9.3) implies that the sum of the projection operators for all eigenstates of an observable $\hat{\alpha}$ is a unit operator.

It may be noted that *observables*, that is, the operators corresponding to some physical measurements, are not the only operators used in quantum mechanics. Apart from observables, we use several other linear operators which have different functions to perform but they all have the common property that their operation on a state, in general, yields a new state.

1.10 Commutativity and Compatibility of Observables

If two observables $\hat{\alpha}$ and $\hat{\xi}$ do not commute, $\hat{\alpha}\hat{\xi} \neq \hat{\xi}\hat{\alpha}$, then the eigenstates of $\hat{\alpha}$ are not the eigenstates of $\hat{\xi}$ and vice versa. This means that if $\left|\xi^{(s)}\right\rangle$ is an eigenstate of $\hat{\xi}$ belonging to the eigenvalue $\xi^{(s)}$, then a measurement pertaining to the observable $\hat{\alpha}$ on the system in the state $\left|\xi^{(s)}\right\rangle$, will disturb the state. To see this we express the state $\left|\xi^{(s)}\right\rangle$ as a superposition of the eigenstates of $\hat{\alpha}$ as

$$\left|\xi^{(s)}\right\rangle = \sum_r \left|\alpha^{(r)}\right\rangle \left\langle \alpha^{(r)} \middle| \xi^{(s)} \right\rangle \equiv \sum_r C_r \left|\alpha^{(r)}\right\rangle , \tag{1.10.1}$$

where

$$C_r = \left\langle \alpha^{(r)} \middle| \xi^{(s)} \right\rangle . \tag{1.10.2}$$

We arrived at this expansion by inserting a unit operator $\hat{1} = \sum_r \left|\alpha^{(r)}\right\rangle \left\langle \alpha^{(r)}\right|$ before $\left|\xi^{(s)}\right\rangle$. A measurement of $\hat{\alpha}$ on the system in the state $\left|\xi^{(s)}\right\rangle$ will throw it into one of the eigenstates of $\hat{\alpha}$; which one, we cannot foretell. Thus the result of measurement of $\hat{\alpha}$ will be indeterminate although it will be one of the eigenvalues of $\hat{\alpha}$. We can interpret $|C_r|^2$ as the probability of getting the result $\alpha^{(r)}$ when an observation for $\hat{\alpha}$ is made on the system in the state $\left|\xi^{(s)}\right\rangle$. This interpretation is justified because

$$\sum_r |C_r|^2 = \sum_r C_r^* C_r = \sum_r \left\langle \xi^{(s)} \middle| \alpha^{(r)} \right\rangle \left\langle \alpha^{(r)} \middle| \xi^{(s)} \right\rangle = \left\langle \xi^{(s)} \middle| \xi^{(s)} \right\rangle = 1 ,$$

which conforms to the fact that the probability of getting any one of the eigenvalues of $\hat{\alpha}$ as the result, when an observation for $\hat{\alpha}$ is made on the system in the state $\left|\xi^{(s)}\right\rangle$, is one. In view of the interpretation of $|C_r|^2$ as a probability, the coefficient $C_r \equiv \left\langle \alpha^{(r)} \middle| \xi^{(s)} \right\rangle$ is referred to as a *probability amplitude*.

Conversely, an eigenstate of $\hat{\alpha}$, say $\left|\alpha^{(r)}\right\rangle$, may be expanded in terms of the eigenstates of the observable $\hat{\xi}$ by introducing the unit operator $\hat{1} = \sum_s \left|\xi^{(s)}\right\rangle \left\langle \xi^{(s)}\right|$ before $\left|\alpha^{(r)}\right\rangle$:

$$\left|\alpha^{(r)}\right\rangle = \sum_s \left|\xi^{(s)}\right\rangle \left\langle \xi^{(s)} \middle| \alpha^{(r)} \right\rangle \equiv \sum_s C'_s \left|\xi^{(s)}\right\rangle \tag{1.10.3}$$

where

$$C'_s = \left\langle \xi^{(s)} \middle| \alpha^{(r)} \right\rangle . \tag{1.10.4}$$

A measurement of $\hat{\xi}$ in the state $\left|\alpha^{(r)}\right\rangle$ will throw the system into any one of the eigenstates of $\hat{\xi}$; which one, again, we cannot foretell. Thus the result of measurement of $\hat{\xi}$ on the state $\left|\alpha^{(r)}\right\rangle$ is indeterminate although it will be one of the eigenvalues of $\hat{\xi}$. We can interpret $|C'_s|^2$ as the probability of getting the result $\xi^{(s)}$ when an observation for $\hat{\xi}$ is made on the system in the state $\left|\alpha^{(r)}\right\rangle$. Since C'_s [Eq.(1.10.4)] and C_r [Eq. (1.10.2)] are complex conjugates of each other $|C'_s|^2 = |C_r|^2$, it follows that the probability of getting the result $\xi^{(s)}$ when an observation for $\hat{\xi}$ is made on the system in a state in which a measurement of $\hat{\alpha}$ is destined to give a result $\alpha^{(r)}$ equals the probability of getting the result $\alpha^{(r)}$ when an observation for $\hat{\alpha}$ is made on the system in a state in which $\hat{\xi}$ is destined to give the result $\xi^{(s)}$.

In short, if the observables $\hat{\alpha}$ and $\hat{\xi}$ do not commute, their measurements are not compatible in the sense that we cannot find a single state in which both observables can be simultaneously measured without disturbing the state.

When, however two observables say $\hat{\xi}$ and $\hat{\eta}$ commute: $[\hat{\xi}, \hat{\eta}] = \hat{\xi}\hat{\eta} - \hat{\eta}\hat{\xi} = 0$, then they admit a set of simultaneous eigenstates $\left|\xi', \eta'\right\rangle, \left|\xi'', \eta''\right\rangle, \cdots, \left|\xi^{(s)}, \eta^{(s)}\right\rangle$. In any one of these states both observables can be measured without disturbing the state.

As we shall see in the next section, the position and momentum observables do not commute: $[\hat{x}, \hat{p}_x] \neq 0$. Consequently, if there is a state in which a measurement of $\hat{p}_x$ gives a precise result, the measurement of $\hat{x}$ is uncertain and vice versa. Thus the measurement of $\hat{x}$ and $\hat{p}_x$ are not compatible. On the other hand, in the hydrogen atom problem, as we shall see later, the set of observable $\hat{H}$ (Hamiltonian), $\hat{L}^2$ (angular momentum squared), and $\hat{L}_z$ (z-component of angular momentum) commute:

$$[\hat{H}, \hat{L}^2] = 0 = [\hat{L}^2, \hat{L}_z] = [\hat{H}, \hat{L}_z]\,.$$

This set of commuting observable admits the set of simultaneous eigenstates $|n\,,\ell\,,m\rangle$. In any one of these states the measurements corresponding to the observables $\hat{H}$, $\hat{L}^2$, $\hat{L}_z$ may be simultaneously performed, without disturbing the state, giving the results E_n, $\hbar^2\ell(\ell+1)$ and $m\hbar$, respectively.

1.11 Position and Momentum Commutation Relations

In classical mechanics, if we consider a system of particles with f degrees of freedom then we need to specify f generalized position coordinates, $q_1\,, q_2\,, \cdots\,, q_f$ and f generalized momenta $p_1\,, p_2\,, \cdots\,, p_f$ where the quantities q_i and p_i are said to be canonically conjugate to each other. After expressing the Hamiltonian in terms of these coordinates one can write the classical equations of motion in the canonical form and solve them, subject to given boundary conditions. Subsequently one can accurately determine any quantity relevant to the system at any time [see Appendix 1A1], provided all the initial conditions of the system are known.

In quantum mechanics we treat the generalized position coordinates and generalized momenta as observables or self-adjoint or Hermitian operators conforming to the *completeness criterion*. Thus we have the coordinate observables $\hat{q}_1\,, \hat{q}_2\,, \cdots\,, \hat{q}_f$ and momentum observables $\hat{p}_1\,, \hat{p}_2\,, \cdots\,, \hat{p}_f$. All position observables commute with each other:

$$[\hat{q}_i, \hat{q}_j] = 0\,, \qquad i, j = 1, 2, \cdots f\,. \tag{1.11.1}$$

Their measurements are compatible with another and there exist simultaneous eigenstates in each of which all positon coordinates can be measured without disturbing the state. Similarly, all momentum observables commute with each other:

$$[\hat{p}_i, \hat{p}_j] = 0\,, \qquad i, j = 1, 2, \cdots f\,. \tag{1.11.2}$$

Measurements of all momentum observables are compatible with one another; therefore, there exist simultaneous eigenstates where all momenta can be measured without disturbing the state.

Now, the principle of uncertainty tells us that the measurements of a position observable $\hat{x}$ and its canonically conjugate momentum $\hat{p}_x$ are not compatible. If there is a state in which $\hat{p}_x$ is measured with complete determinacy, a measurement of x will give indeterminate result and vice versa. Thus $\hat{x}$ and $\hat{p}_x$ do not commute

$$[\hat{x}, \hat{p}_x] \neq 0 \tag{1.11.3}$$

and, in general,

$$[\hat{q}_i, \hat{p}_i] \neq 0\,. \tag{1.11.4}$$

To work out this commutator we note that $[\hat{q}_i, \hat{p}_i]$ is a pure imaginary (anti-Hermitian) operator[7] and can be put equal to $i\hat{X}$ where $\hat{X}$ is a self-adjoint (Hermitian) operator. For the operator $\hat{X}$ we can make the choice $\hat{X} = \hbar\hat{1}$, based on dimensional considerations (the product of momentum and coordinate has dimensions of action or $\hbar$). This choice yields results in agreement with observations. Thus the commutator of position and momentum observables is

$$[\hat{q}_i, \hat{p}_j] = i\hbar\,\hat{1}\,. \tag{1.11.5}$$

This commutator, together with the relations,

$$[\hat{q}_i, \hat{q}_j] = [\hat{p}_i, \hat{p}_j] = \hat{0} \tag{1.11.6}$$

gives the set of commutation relations for the set of observables $\hat{q}_i$ and $\hat{p}_i$. On the basis of these commutation relations (choosing the operator $\hat{X} = \hbar\,\hat{1}$) we are able to derive results which are consistent with observations. This is one reason for accepting the commutation relations Eq. (1.11.5) and (1.11.6) for the position and momentum observables.

Another reason for accepting these commutation relations is the close analogy between the classical Poisson bracket (CPB) of two classical dynamical variables u and v defined by[8]

$$\{u,\, v\} = \sum_i \left\{ \frac{\partial u}{\partial q_i}\frac{\partial v}{\partial p_i} - \frac{\partial u}{\partial p_i}\frac{\partial v}{\partial q_i} \right\}, \tag{1.11.7}$$

and $(ih)^{-1}$ times the quantum commutator brackets of the correspomding observables $\hat{u}$ and $\hat{v}$, viz. $(i\hbar)^{-1}[\hat{u},\, \hat{v}]$. Both of these brackets, which otherwise have completely different definitions, satisfy some common identities:

Classical Poisson Bracket	Quantum Commutator Bracket
$\{u,\, v\} = -\{v,\, u\}$	$[\hat{\alpha},\, \hat{\beta}] = -[\hat{\beta},\, \hat{\alpha}]$
$\{u,\, C\} = 0$	$[\hat{u},\, C] = \hat{0}$
$\{u_1 + u_2,\, v\} = \{u_1,\, u\} + \{u_2,\, v\}$	$[\hat{u}_1 + \hat{u}_2,\, \hat{v}] = [\hat{u}_1,\, \hat{v}] + [\hat{u}_2,\, \hat{v}]$
$\{u,\, v_1 + v_2\} = \{u,\, v_1\} + \{u,\, v_2\}$	$[\hat{u},\, \hat{v}_1 + \hat{v}_2] = [\hat{u},\, \hat{v}_1] + [\hat{u},\, \hat{v}_2]$
$\{u_1 u_2,\, v\} = \{u_1,\, v\}\, u_2 + u_1\, \{u_2,\, v\}$	$[\hat{u}_1\hat{u}_2,\, \hat{v}] = [\hat{u}_1,\, \hat{v}]\, \hat{u}_2 + \hat{u}_1\, [\hat{u}_2,\, \hat{v}]$
$\{u,\, v_1 v_2\} = \{u,\, v_1\}\, v_2 + v_1\, \{u,\, v_2\}$	$[\hat{u},\, \hat{v}_1\hat{v}_2] = [\hat{u},\, \hat{v}_1]\, \hat{v}_2 + \hat{v}_1\, [\hat{u},\, \hat{v}_2]$
Jacobi identity: $\{u, \{v,\, w\}\} + \{v, \{w,\, u\}\} + \{w, \{u,\, v\}\} = 0$	Jacobi identity: $[\hat{u}, [\hat{v},\, \hat{w}]] + [\hat{v}, [\hat{w},\, \hat{u}]] + [\hat{w}, [\hat{u},\, \hat{v}]] = \hat{0}$

Now just as in classical mechanics the CPB of two real dynamical variables is real, so also in quantum mechanics we expect the analogous brackets for two observables (self-adjoint operators) to be self-adjoint. Now $[\hat{u}, \hat{v}]$ is pure imaginary. To make it self-adjoint (Hermitian), we may divide it by $i\hbar$. Thus the quantum analogue of CPB $\{u, v\}$ is $(i\hbar)^{-1}[\hat{u}, \hat{v}]$.

[7]We can easily see that the commutator of two self-adjoint operators $\hat{\alpha}$ and $\hat{\beta}$ is anti-Hermitian: $[\hat{\alpha},\, \hat{\beta}]^\dagger = -[\hat{\alpha},\, \hat{\beta}]$.
Proof: $[\hat{\alpha},\, \hat{\beta}]^\dagger = (\hat{\alpha}\hat{\beta} - \hat{\beta}\hat{\alpha})^\dagger = (\hat{\alpha}\hat{\beta})^\dagger - (\hat{\beta}\hat{\alpha})^\dagger = \hat{\beta}^\dagger\hat{\alpha}^\dagger - \hat{\alpha}^\dagger\hat{\beta}^\dagger = \hat{\beta}\hat{\alpha} - \hat{\alpha}\hat{\beta} = -[\hat{\alpha},\, \hat{\beta}]$.

[8]For the properties and significance of classical Poisson brackets, see Appendix (1A1).

Incidentally, there also exists a close analogy between the classical equations of motion for a real dynamical variable α

$$\frac{d\alpha}{dt} = \frac{\partial \alpha}{\partial t} + \{\alpha, H\} \tag{1.11.8}$$

and the Heisenberg equation of motion[9] for a Heisenberg observable in quantum mechanics

$$\frac{d\hat{\alpha}^H}{dt} = \frac{\partial \hat{\alpha}^H}{\partial t} + \frac{1}{i\hbar}\left[\hat{\alpha}^H, \hat{H}\right]. \tag{1.11.9}$$

We observe that in the classical equation, the CPB stands in analogy with the quantum commutator bracket divided by $i\hbar$ in the Heisenberg equation of motion in quantum mechanics.

From the definition of classical Poisson brackets, we can easily work out the fundamental brackets and write

$$\{q_i, p_i\} = 1, \tag{1.11.10}$$

$$\{q_i, q_j\} = \{p_i, p_j\} = 0. \tag{1.11.11}$$

Writing similar conditions for quantum commutator brackets, in analogy with the classical Poisson brackets, we get

$$\frac{1}{i\hbar}[\hat{q}_i, \hat{p}_i] = \hat{1},$$

$$\frac{1}{i\hbar}[\hat{q}_i, \hat{q}_j] = \frac{1}{i\hbar}[\hat{p}_i, \hat{p}_j] = \hat{0},$$

which are precisely the quantum commutator relations (1.11.5) and (1.11.6). These equations are also referred to as the fundamental quantum conditions for the position and momentum observables. These conditions are based on logical reasoning, apart from an arbitrariness in the choice of the self-adjoint operator $\hat{X}$ as $\hbar\,\hat{1}$. This choice is made so that the results of calculations agree with observations.

1.12 Commutation Relation and the Uncertainty Product

On the basis of the commutation relation between two observables $\hat{A}$ and $\hat{B}$

$$[\hat{A}, \hat{B}] = i\hat{C}, \tag{1.12.1}$$

it is possible to work out the product of minimum uncertainty in the measurements of $\hat{A}$ and $\hat{B}$ for any (arbitrary) state, say $|\psi\rangle$. For this state the *quantum mechanical expectation values* of $\hat{A}$ and $\hat{B}$ are given as

$$\bar{A} = \langle\hat{A}\rangle = \langle\psi|\,\hat{A}\,|\psi\rangle, \tag{1.12.2}$$

and

$$\bar{B} = \langle\hat{B}\rangle = \langle\psi|\,\hat{B}\,|\psi\rangle. \tag{1.12.3}$$

(Quantum mechanical *expectation value* of an observable $\hat{A}$ for a state $|\psi\rangle$, which is not an eigenstate of $\hat{A}$, is the average value of the result when $\hat{A}$ is measured on a large number

[9]In Chapter 3, Sec. 3.8 on *Equations of Motion* we shall discuss the Schrödinger and Heisenberg pictures and derive Heisenberg equations of motion.

of identical systems, each one in the state $|\psi\rangle$). We can treat this quantity as the average value of $\hat{A}$ for the state $|\psi\rangle$. The square of the uncertainty ΔA in the measurement of $\hat{A}$ is given by

$$(\Delta A)^2 \equiv \langle\psi|\,(\hat{A}-\bar{A})^2\,|\psi\rangle = \langle\hat{A}^2\rangle - \langle\hat{A}\rangle^2\,. \tag{1.12.4}$$

Similarly, the square of the uncertainty ΔB in the measurement of $\hat{B}$ is given by

$$(\Delta B)^2 \equiv \langle\psi|\,(\hat{B}-\bar{B})^2\,|\psi\rangle = \langle\hat{B}^2\rangle - \langle\hat{B}\rangle^2\,. \tag{1.12.5}$$

These quantities are like *mean square deviations* and can be looked upon as the expectation values of and $\hat{A}'^2$ and $\hat{B}'^2$, respectively, where $\hat{A}' = (\hat{A}-\bar{A})$ and $\hat{B}' = (\hat{B}-\bar{B})$. Now let us define an operator $\hat{O}$ as

$$\hat{O} = \hat{A} + (\alpha + i\beta)\hat{B} \tag{1.12.6}$$

where α and β are real numbers and $\hat{A}$ and $\hat{B}$ are observables and, therefore, self-adjoint operators. Now the expectation value of $\hat{O}\hat{O}^\dagger$ is nonnegative

$$\langle\psi|\,\hat{O}^\dagger\hat{O}\,|\psi\rangle \geq 0\,, \tag{1.12.7}$$

because the square of the length of the ket vector $\hat{O}\,|\psi\rangle$ cannot be negative. Equation (1.12.6) then implies that

$$\langle\psi|\,\hat{A}^2 + (\alpha^2+\beta^2)\hat{B}^2 + \alpha(\hat{A}\hat{B}+\hat{B}\hat{A}) + i\beta(\hat{A}\hat{B}-\hat{B}\hat{A}\,|\psi\rangle \geq 0\,.$$

Using the substitution $\hat{A}\hat{B}+\hat{B}\hat{A} \equiv \hat{S}$ and the commutator (1.12.1) to write $\hat{A}\hat{B}-\hat{B}\hat{A} = i\hat{C}$, the inequality (1.12.8) preceding inequality can be written as

$$\langle\hat{A}^2\rangle + \langle\hat{B}^2\rangle\left[\alpha + \frac{1}{2}\frac{\langle\hat{S}\rangle}{\langle\hat{B}^2\rangle}\right]^2 + \langle\hat{B}^2\rangle\left[\beta - \frac{1}{2}\frac{\langle\hat{C}\rangle}{\langle\hat{B}^2\rangle}\right]^2 - \frac{1}{4}\frac{\langle\hat{S}\rangle^2}{\langle\hat{B}\rangle^2} - \frac{1}{4}\frac{\langle\hat{C}\rangle^2}{\langle\hat{B}\rangle^2} \geq 0 \tag{1.12.8}$$

This inequality holds for all real values of α and β. In particular, we can choose these numbers in such a way that the quantities inside the parentheses are both zero:

$$\alpha = -\frac{1}{2}\frac{\langle\hat{S}\rangle}{\langle\hat{B}^2\rangle}\,, \quad \beta = \frac{1}{2}\frac{\langle\hat{C}\rangle}{\langle\hat{B}^2\rangle}\,.$$

With these choices we have

$$\langle\hat{A}^2\rangle - \frac{1}{4}\frac{\langle\hat{S}\rangle^2}{\langle\hat{B}^2\rangle} - \frac{1}{4}\frac{\langle\hat{C}\rangle^2}{\langle\hat{B}^2\rangle} \geq 0$$

or

$$\langle\hat{A}^2\rangle\langle\hat{B}^2\rangle \geq \frac{1}{4}\left(\langle\hat{S}\rangle^2 + \langle\hat{C}\rangle^2\right) \geq \frac{1}{4}\langle\hat{C}\rangle^2\,. \tag{1.12.9}$$

If $\hat{A}' \equiv \hat{A} - \langle\hat{A}\rangle$ and $\hat{B}' \equiv \hat{B} - \langle\hat{B}\rangle$, then obviously $[\hat{A}',\, \hat{B}'] = [\hat{A},\, \hat{B}] = i\hat{C}$. Therefore, following Eq. (1.12.9), we have the inequality

$$\langle\hat{A}'^2\rangle\langle\hat{B}'^2\rangle \geq \frac{1}{4}\langle\hat{C}\rangle^2\,. \tag{1.12.10}$$

Using Eqs. (1.12.4) and (1.12.5) in this equation we obtain

$$\Delta A \Delta B \geq \frac{1}{2}\langle\hat{C}\rangle\,. \tag{1.12.11}$$

The uncertainty relation for position and momentum observables follows from this equation. Recalling that the commutation relation for the observables $\hat{x}$ and $\hat{p}$ is $[\hat{x}, \hat{p}] = i\hbar\,\hat{1}$, we have the important result

$$\Delta x \Delta p \geq \frac{\hbar}{2}\,. \tag{1.12.12}$$

This is the product of uncertainties in the measurement of position and momentum coordinates of a particle in any state. The minimum value of this product is $\hbar/2$.

Problems

In computing various physical quantities the following table of constants and their combinations may be useful. A good reference for most recent values of constants is NIST website: http://physics.nist.gov/cgi-bin/cuu/

Constant Name	Usual symbol and current value
Speed of light	$c=2.997\,924\,58\times10^{8}$ m/s
Elementary charge	$e = 1.602\,177 \times 10^{-19}$ C
Electron mass	$m_e = 9.109 \times 10^{-31}$ kg
Proton mass	$m_p = 1.673 \times 10^{-27}$ kg
Proton-to-electron mass ratio	$m_p/m_e = 1836$
Neutron mass	$m_n = 1.675 \times 10^{-27}$ kg
Bohr magneton	$\mu_B = 9.274 \times 10^{-24}$ J/T $= 5.788 \times 10^{-5}$ eV/T
Nuclear magneton	$\mu_N = 5.051 \times 10^{-27}$ J/T $= 3.152 \times 10^{-8}$ eV/T
Planck's constant	$h = 6.626\,069 \times 10^{-34}$ J-s $\hbar \equiv h/2\pi = 1.054\,572 \times 10^{-34}$ J-s
Gravitational constant	$G = 6.67428(67)\,6 \times 10^{-11}$ N-m^2/kg^2
Fine structure constant	$\alpha = e^2/4\pi\epsilon_0\hbar c = 1/137.036 \approx 1/137$
Stefan-Boltzmann constant	$\sigma = 5.670\,4 \times 10^{-8}$ W/m^2-K^4
Boltzmann constant	$k_B = 1.380\,65 \times 10^{-23}$ J/K $\equiv 8.6173 \times 10^{-5}$ eV/K
Avogadro's number	$N_A = 6.022\,141\,79(30) \times 10^{23}$ /mol

In practice, it is easier and more useful to remember certain combinations of fundamental constants rather than the constants themselves.

Combination of Constants

Fine structure constant $\alpha = e^2/4\pi\epsilon_0\hbar c$	$1/137.036 \approx 1/137$
$\hbar c$	197.3271 eV- nm (MeV-fm)
$k_B T$	1/40 eV at 293 K; 1/39 eV at 300K
Electron rest mass energy $m_e c^2$	0.5110 MeV
Proton rest mass energy $m_p c^2$	938.28 MeV
Neutron rest energy $m_n c^2$	939.57 MeV
Proton-electron mass ratio m_p/m_e	1836.15
Bohr radius $a_0 = \hbar/m_e c\alpha$	0.5292×10^{-10} m
Planck time $\sqrt{\hbar G/c^5}$	5.4×10^{-44} s
Compton wavelength of electron $\hbar/m_e c$	3.8616×10^{-13} m
1 degree	1.745×10^{-2} rad
1 eV	1.602×10^{-19} J

Energy of a photon can be calculated by using the formula

$$E = h\nu = \frac{2\pi\hbar c}{\lambda} = \frac{1238}{\lambda(\text{in nm})}\,\text{eV}\,.$$

1. Given that there are $8\pi\nu^2/c^3 d\nu$ modes (Planck considered each mode to be an oscillator, see Chapter 13) per unit volume in the frequency range ν and $\nu + d\nu$, calculate the average radiation energy density for a cavity in equilibrium at temperature T by assuming that the energy of an oscillator E is (i) continuous (ii) quantized ($E = nh\nu$, $n = 0, 1, 2, \cdots$) in units of $h\nu$. Note that the probability that an oscillator has energy E is $\propto \exp(-E/k_B T)$. Does this help you appreciate Planck's insight?

2. The energy of a proton beam is 1000 eV. Calculate the associated de Broglie wave length. Given: 1 eV=1.602×10^{-19} J. For other constants see the table above.

3. Calculate the de Broglie wavelength of electrons in a beam of energy 20 keV. How does this compare to the wavelength of X rays of 20 keV?

4. The work function of Barium is 2.11 eV. Calculate the *threshold frequency* and *threshold wavelength* for the light which can emit photo-electrons from Barium.

5. The threshold wavelength of radiation for photo-electric emission from tungsten is 230 nm. Find the kinetic energy of photo-electrons emitted from its surface by ultra-violet light of wave length 180 nm.

6. A photon of wave length 331 nm falls on a photo-cathode and an electron of energy 3×10^{-19} J is emitted. In another case a photon of wave length 100 nm is incident on the same photo-cathode and an electron of energy 0.972×10^{-19} J is emitted. Calculate from this data,

   (a) the work function and the threshold wave length of the photo-cathode.

   (b) the value of Planck constant h.

7. A photon of ultra-violet radiation of wave length 300 nm is incident on a photo-cathode. The work function of the material of the photo cathode is 2.26 eV. Calculate the velocity of the photo electron emitted.

8. The work function of the surface of a photo-cathode is 3.30 eV. Calculate the threshold frequency of radiation, and the corresponding wavelength, which can emit photo electrons when incident on the surface.

9. In an X-ray scattering experiment the Compton shift was found to be 2.4×10^{-3} nm for scattering at 90°. What would be the Compton shift for scattering at 45°?

10. Calculate the radius of the first Bohr orbit a_0 for the Hydrogen atom.

11. According to classical electrodynamics an accelerated electron will lose energy at the rate $dE/dt = -e^2a^2/6\pi\epsilon_0 c^3$. Calculate the acceleration of an electron in a circular Bohr orbit of radius $a_0 = 4\pi\epsilon_0\hbar^2/m_e e^2$. How long does it take the electron to reduce its energy from -13.6 eV (ground state energy) to -2×13.6 eV assuming that it continues to radiate at this rate?

12. According to Bohr, the Balmer series of spectral lines of Hydrogen arise due to the jump of the electron from higher orbits, $n = 3, 4, 5$, etc to the orbit $n = 2$. Calculate the wave lengths of the first three Balmer lines in nm. [The energy of the electron in the nth orbit is given by $E_n = -\frac{1}{2}\frac{e^2}{4\pi\epsilon_0 a_0}\frac{1}{n^2}$.]

13. What is the frequency of the light emitted by an electron in a transition from the first excited state to the ground state in hydrogen? Compare this with the frequency of the orbital motion of the electron in the first excited state. [Given: The velocity of an electron for the nth energy level is $v_n = \alpha c/n$ and the radius its orbit is given by $a_o n^2$, where $a_o = 0.0529$ nm is the Bohr radius and α is the fine structure constant.]

Section 1.2 onwards

1. If $\hat{F}$ is a linear operator show that $\hat{F}+\hat{F}^\dagger$ is a self-adjoint (Hermitian) operator and $\hat{F}-\hat{F}^\dagger$ is a pure imaginary (anti-Hermitian) operator.

2. Prove that
$$\left[\hat{A}\,,\frac{1}{\hat{B}}\right] = -\frac{1}{\hat{B}}\left[\hat{A}\,,\hat{B}\right]\frac{1}{\hat{B}}\,.$$

3. Prove that
$$\text{(i) } \exp(\hat{L})\hat{\alpha}\exp(-\hat{L}) = \hat{\alpha} + \left[\hat{L}\,,\hat{\alpha}\right] + \tfrac{1}{2!}\left[\hat{L},\left[\hat{L}\,,\hat{\alpha}\right]\right] + \cdots$$
$$\text{(ii) } e^{\lambda\hat{A}}\hat{B}e^{-\lambda\hat{A}} = \hat{B} + \lambda\left[\hat{A}\,,\hat{B}\right] + \tfrac{\lambda^2}{2!}\left[\hat{A}\,,\left[\hat{A}\,,\hat{B}\right]\right] + \cdots$$

4. Prove that a linear operator that commutes with an observable ξ also commutes with a function of $\hat{\xi}$.

5. Show that $\left(\hat{\alpha}\hat{\beta}\hat{\gamma}\hat{\delta}\right)^\dagger = \delta^\dagger\hat{\gamma}^\dagger\hat{\beta}^\dagger\hat{\alpha}^\dagger$.

6. If $\hat{A}$ is a self-adjoint (Hermitian) operator show that $\langle\psi|\,\hat{A}^2\,|\psi\rangle \geq 0$ where $|\psi\rangle$ is an arbitrary state.

7. Prove that if $\hat{A}$ and $\hat{B}$ are self-adjoint (Hermitian) operators, the product $\hat{A}\hat{B}$ is also self-adjoint (Hermitian) provided $\hat{A}$ and $\hat{B}$ commute.

8. Two observables $\hat{\xi}$ and $\hat{\alpha}$ do not commute. Show that in a state in which the observable $\hat{\alpha}$ is certain to have the value $\alpha^{(n)}$, the probability of $\hat{\xi}$ having the value $\xi^{(r)}$ is the same as the probability of $\hat{\alpha}$ having the value $\alpha^{(n)}$ in a state in which $\hat{\xi}$ is certain to have the values $\xi^{(r)}$.

9. If $\left|\alpha^{(r)}\right\rangle$ represents a complete set of eigenstates of observable $\hat{\alpha}$, belonging to the eigenvalues $\alpha^{(r)}$, $r = 1, 2, 3\cdots$, so that an arbitrary state may be expanded in terms of them, i.e. $|\psi\rangle = \sum_r c_r\left|\alpha^{(r)}\right\rangle$, then show that $\langle\psi|\,\hat{\alpha}\,|\psi\rangle = \sum_r |c_r|^2\alpha^{(r)}$. Hence give a physical interpretation to the quantum mechanical expectation value $\langle\hat{A}\rangle = \langle\psi|\,\hat{A}\,|\psi\rangle$ of an observable $\hat{A}$ for the state $|\psi\rangle$.

10. The eigenstates $\left|\xi^{(r)}\right\rangle$ of an observable $\hat{\xi}$ satisfy, apart from the orthogonality condition, the completeness criterion as well,
$$\sum_r \hat{P}_r = \hat{1},, \qquad \text{where} \quad \hat{P}_r = \left|\xi^{(r)}\right\rangle\left\langle\xi^{(r)}\right|\,.$$
Show that $\hat{P}_r$ may be regarded as the projection operator for the state $\left|\xi^{(r)}\right\rangle$ in the sense that operating on an arbitrary state it projects out this particular state. Hence interpret physically the completeness criterion in terms of the projection operators $\hat{P}_r$.

11. Show that the commutator $[\hat{x}\,,\hat{p}^n] = i\hbar\, n\hat{p}^{n-1}$.

12. Show that the commutator $[\hat{x}^n\,,\hat{p}] = i\hbar\, n\hat{x}^{n-1}$.

13. Show that if $\hat{A}$ and $\hat{B}$ are self-adjoint operators, then $[\hat{A},\hat{B}]$ is a pure imaginary operator.

Appendix 1A1: Basic Concepts in Classical Mechanics

In this appendix some of the basic concepts in classical mechanics, which are required for Chapter 1 are outlined. In order to specify the configurations of N particles $3N$ coordinates are needed. If there are, say n, constraints on the system, then we have to specify $3N$-coordinates and n constraints on them. If the coordinates are so chosen that n restrictions amount to holding n of the coordinates as constants, then we have to specify only $3N - n$ coordinates. Then the number of coordinates $f = 3N - n$ is referred to as the number of *degrees of freedom of the system*; the relevant coordinates are called the *generalized coordinates*. Generalized coordinates do not pertain necessarily to individual particles and they may not necessarily have dimension of length. For example, for a two-particle system the generalized coordinates can be the coordinates of the center of mass (R, Θ, Φ) and the relative coordinates (r, θ, φ). If the only restriction happens to be that the distance between the two particles is fixed then this amounts to holding r constant. The number of degrees of freedom is $f = 5$ and the relevant generalized coordinates then are

$$q_1 = R,\ q_2 = \Theta,\ q_3 = \Phi,\ q_4 = \theta,\ q_5 = \phi\,.$$

1A1.1 Lagrange Equations of Motion

Consider a system with f degrees of freedom. The configuration can be specified by f generalized coordinates, say q_1, q_2, $\cdots q_f$. The time rates of change of these generalized coordinates $\dot{q}_1$, $\dot{q}_2$, $\cdots$, $\dot{q}_f$ are called generalized velocities. The generalized velocities also need not have dimensions of velocity. It is possible to express the total kinetic energy of the system as a function of all q_i and $\dot{q}_i$. Also the total potential energy can be expressed as a function of all q_i's. The Lagrangian of the system defined by

$$L = T - V \tag{1A1.1.13}$$

is also a function of generalized coordinates and generalized velocities. In the Lagrangian formulation of mechanics the equations of motion can be written as

$$\frac{d}{dt}\left(\frac{\partial L(q_i,\, \dot{q}_i)}{\partial \dot{q}_i}\right) - \frac{\partial L(q_i,\, \dot{q}_i)}{\partial q_i} = 0\,. \tag{1A1.1.14}$$

There is an alternative formulation called the Hamiltonian formulation. This formulation is in terms of the generalized coordinates q_i and the generalized momentum coordinates p_i, defined by

$$p_i = \frac{\partial L(q_i,\, \dot{q}_i)}{\partial \dot{q}_i} = \frac{\partial T}{\partial \dot{q}_i}\,. \tag{1A1.1.15}$$

The Hamiltonian H of the system is defined by

$$H(q_i,\, p_i) = T + V\,. \tag{1A1.1.16}$$

The transformation from the Lagrangian formulation (in which the description is in terms of q_i and $\dot{q}_i$'s) to the Hamiltonian formulation(in which the description is in terms of q_i and p_i) can be brought about by what is known as the słLegendre transformations. These equations of motion in Hamiltonian formulation, also called Hamilton's equations of motion (or equations of motion in the canonical form), are

$$\dot{q}_i = \frac{\partial H}{\partial p_i}, \tag{1A1.1.17}$$

$$\dot{p}_i = -\frac{\partial H}{\partial q_i}. \tag{1A1.1.18}$$

It may be seen that in the Hamiltonian formulation the generalized position and momentum coordinates are treated on a reciprocal basis. The position and momentum coordinates, q_i and $\dot{p}_i$ are said to be *canonically conjugate* to each other. One can see that the equations of motion in the canonical form are symmetric between q_i and p_i (except for a difference of sign).

1A1.2 Classical Dynamical Variables

A function of generalized position and momentum coordinate and also of time (explicit dependence), which may also correspond to a physical quantity is called a classical dynamical variable. Examples are: Hamiltonian, total angular momentum, total linear momentum etc. Even the generalized position and momentum coordinates q_i and p_i may also be regarded as classical dynamical variables.

The equation of motion of a classical dynamical variables $\alpha(q_i, p_i, t)$ can be derived as follows:

$$\begin{aligned}\frac{d\alpha}{dt} &= \frac{\partial\alpha}{\partial t} + \sum_i \left(\frac{\partial\alpha}{\partial q_i}\dot{q}_i + \frac{\partial\alpha}{\partial p_i}\dot{p}_i\right)\\ &= \frac{\partial\alpha}{\partial t} + \sum_i \left(\frac{\partial\alpha}{\partial q_i}\frac{\partial H}{\partial p_i} - \frac{\partial\alpha}{\partial p_i}\frac{\partial H}{\partial q_i}\right),\end{aligned}$$

where we have used the equations of motion in the canonical form (1A2.5) and (1A2.6). Thus

$$\frac{d\alpha}{dt} = \frac{\partial\alpha}{\partial t} + \{\alpha, H\}. \tag{1A1.1.19}$$

Here the curly bracket $\{\alpha, H\}$ is called the classical Poisson bracket (CPB)of the dynamical variables α and H. In general, for any two dynamical variables α and β we define CPB by

$$\{\alpha, \beta\} = \sum_i \left(\frac{\partial\alpha}{\partial q_i}\frac{\partial\beta}{\partial p_i} - \frac{\partial\alpha}{\partial p_i}\frac{\partial\beta}{\partial q_i}\right). \tag{1A1.1.20}$$

Since q_i and p_i are also to be treated as classical dynamical variables we can also define the Poisson brackets for them (known as fundamental Poisson brackets). We can easily verify that

$$\{q_i, q_j\} = 0 = \{p_i, p_j\}, \tag{1A1.1.21}$$

$$\{q_i, p_j\} = \delta_{ij}. \tag{1A1.1.22}$$

It may be observed that the basic feature of classical mechanics is *determinacy* in the sense that if all details about a system of particles (interactions etc.) are known and the initial configuration of a system is known, then one can, in principle, write the classical equations of motion [Eq. (1A2.5) and (1A2.6)] and solve them for $q_i(t)$ and $p_i(t)$ subject to initial conditions. Subsequently one can calculate precisely any quantity relevant to the system at any time.

References

[1] P. A. M. Dirac, *Principles of Quantum Mechanics*, Third Edition (Clarendon Press, 1971) Chapters I and II.

2

REPRESENTATION THEORY

2.1 Meaning of Representation

The *bra* and *ket* vectors, as well as linear operators, are somewhat abstract mathematical quantities. Although the logic behind representing physical states and observables, respectively, by ket (or bra) vectors and self adjoint linear operators is perfect and, having developed the algebra for these quantities, we have been able to deduce therefrom some general results, identities and corollaries, yet these abstract quantities are not the ones in terms of which we can carry out numerical calculations and arrive at numbers to be compared with the experimental results. So we ask if it is possible to find more convenient mathematical quantities to represent ket vectors and linear operators that can be handled more conveniently. The answer is in the affirmative.

Each of the various ways in which an abstract mathematical quantities, like a ket vector or a linear operator, can be represented by a set of numbers is called a *representation*. The set of numbers that represent the abstract quantity are called the *representatives* of the abstract quantity in that *representation*.

2.2 How to Set up a Representation

To set up a representation, we choose a *complete set of commuting observables* (CSCO). A set of commuting observables is said to be complete if specifying the eigenvalues of all the observables determines a unique (to within a multiplicative phase factor) common eigenvector. From a given CSCO, we can obtain another CSCO by adding an observable that commutes with all the observables in the original set. For this reason, one usually confines to the *minimal* set of observables needed to construct a unique orthonormal basis of common eigenvectors. It is clear that for a given system several sets of CSCO may exist. The choice of the complete set of commuting observables characterizes the representation.

Consider the CSCO consisting of observables $\hat{\xi}, \hat{\eta}, \hat{\zeta}, \cdots$. These observables naturally admit a set of simultaneous eigenstates [eigen *bras* or eigen *kets*, also called the basis states] of the representation:

$$\begin{aligned}
\langle \xi' , \eta' , \zeta' \cdots | &\equiv \langle \xi' | \\
\langle \xi'' , \eta'' , \zeta'' \cdots | &\equiv \langle \xi'' | \\
&\vdots \\
\left\langle \xi^{(r)} , \eta^{(r)} , \zeta^{(r)} \cdots \right| &\equiv \left\langle \xi^{(r)} \right| \\
&\vdots \\
\left\langle \xi^{(N)} , \eta^{(N)} , \zeta^{(N)} \cdots \right| &\equiv \left\langle \xi^{(N)} \right| .
\end{aligned}$$

Here N is the total number of basis states, which form a complete set. In what follows we will use an abbreviated notation for denoting the simultaneous eigenstates by suppressing eigenvalues of all operators except one. The ket to be represented is multiplied on the left by each one of the basis bras in succession to get a set of numbers which may be arranged in the form of a column vector (matrix):

$$|A\rangle \to \begin{pmatrix} \langle \xi' | A\rangle \\ \langle \xi'' | A\rangle \\ \vdots \\ \left\langle \xi^{(r)} \middle| A\right\rangle \\ \vdots \\ \left\langle \xi^{(N)} \middle| A\right\rangle \end{pmatrix} = \begin{pmatrix} C_1 \\ C_2 \\ \vdots \\ C_r \\ \vdots \\ C_N \end{pmatrix} \equiv C . \tag{2.2.1}$$

The column vector C thus represents $|A\rangle$ in the $\hat{\xi}, \hat{\eta}, \hat{\zeta}, \cdots$ representation. In other words, the set of numbers $C_1 , C_2 , \cdots C_r , \cdots , C_N$, where $C_r = \left\langle \xi^{(r)} \middle| A\right\rangle$, can be looked upon as the representative of $|A\rangle$ in this representation. Likewise, the state vector $\langle A|$ is represented by a row matrix:

$$\begin{aligned}
\langle A| &\to \Big(\langle A| \xi' \rangle \quad \langle A| \xi'' \rangle \quad \cdots \quad \langle A| \xi^{r} \rangle \cdots \langle A|\xi^{(N)}\rangle \Big) \\
&\equiv (C_1^* \quad C_2^* \cdots C_r^* \cdots C_N) \equiv C^\dagger .
\end{aligned} \tag{2.2.2}$$

Since $|A\rangle$ and $\langle A|$ represent the same physical state, the column vector C and the row vector $C^\dagger$ can be regarded as representatives of the same state[1].

We can give a physical interpretation to the representatives of $|A\rangle$ (or $\langle A|$)

1. Since the basis states $\left|\chi^{(r)}\right\rangle$ satisfy the completeness criterion $\sum_{r=1}^{N} \left|\xi^{(r)}\right\rangle \left\langle \xi^{(r)}\right| = \hat{1}$, we can expand the state $|A\rangle$ in terms of basis states as

$$|A\rangle = \sum_{r=1}^{N} \left|\xi^{(r)}\right\rangle \left\langle \xi^{(r)}\right| |A\rangle = \sum_{r=1}^{N} C_r \left|\xi^{(r)}\right\rangle . \tag{2.2.3}$$

From this equation we see that we can look upon the representatives C_r of $|A\rangle$ as the coefficients of expansion when $|A\rangle$ is expanded in terms of the basis kets of the representation.

[1]A dagger on a matrix, in our notation, means Hermitian conjugate of the matrix.

2. From the normalization condition

$$\langle A|A\rangle = 1\,, \tag{2.2.4}$$

we have $$\langle A|\left(\sum_{r=1}^{N}\left|\xi^{(r)}\right\rangle\left\langle\xi^{(r)}\right|\right)|A\rangle = \sum_{r=1}^{N} C_r^* C_r = 1\,. \tag{2.2.5}$$

We can look upon $C_r^* C_r = |C_r|^2$ as the probability of getting the result $\xi^{(r)}$ when an observation $\hat{\xi}$ is made on the system in state $|A\rangle$. The sum of these probabilities, i.e., the probability of getting any one of the results $\xi\prime, \xi\prime\prime \cdots \xi^{(r)} \cdots \xi^{(N)}$, is naturally equal to 1. This is represented by Eq. (2.2.5).

2.3 Representatives of a Linear Operator

A linear operator $\hat{\alpha}$ is known, if the result of its operation on an arbitrary ket, say $|X\rangle$, is known. Now, since $|X\rangle$ may be expanded in terms of the basis kets ($|X\rangle = \sum_{r=1}^{N} C_r \left|\xi^{(r)}\right\rangle$), we can say that $\hat{\alpha}$ is known if the result of its operation on each one of the basis kets is known, i.e., if the set of ket vectors $\hat{\alpha}|\xi'\rangle, \hat{\alpha}\,|\xi''\rangle, \cdots, \hat{\alpha}\left|\xi^{(r)}\right\rangle, \cdots, \hat{\alpha}\left|\xi^{(N)}\right\rangle$ is known. But each one of this set of N ket vectors is known, respectively, by its representatives given in the following columns:

$$\begin{array}{ccccccc}
\langle\xi'|\,\hat{\alpha}\,|\xi'\rangle & \langle\xi'|\,\hat{\alpha}\,|\xi''\rangle & \cdots & \langle\xi'|\,\hat{\alpha}\left|\xi^{(r)}\right\rangle & \cdots & \langle\xi'|\,\hat{\alpha}\left|\xi^{(N)}\right\rangle \\
\langle\xi''|\,\hat{\alpha}\,|\xi'\rangle & \langle\xi''|\,\hat{\alpha}\,|\xi''\rangle & \cdots & \langle\xi''|\,\hat{\alpha}\left|\xi^{(r)}\right\rangle & \cdots & \langle\xi''|\,\hat{\alpha}\left|\xi^{(N)}\right\rangle \\
\vdots & & & & & \\
\left\langle\xi^{(s)}\right|\hat{\alpha}\,|\xi'\rangle & \left\langle\xi^{(s)}\right|\hat{\alpha}\,|\xi''\rangle & \cdots & \left\langle\xi^{(s)}\right|\hat{\alpha}\left|\xi^{(s)}\right\rangle & \cdots & \left\langle\xi^{(r)}\right|\hat{\alpha}\left|\xi^{(N)}\right\rangle \\
\vdots & & & & & \\
\left\langle\xi^{(N)}\right|\hat{\alpha}\,|\xi'\rangle & \left\langle\xi^{(N)}\right|\hat{\alpha}\,|\xi''\rangle & \cdots & \left\langle\xi^{(N)}\right|\hat{\alpha}\left|\xi^{(r)}\right\rangle & \cdots & \left\langle\xi^{(N)}\right|\hat{\alpha}\left|\xi^{(N)}\right\rangle\,.
\end{array}$$

Thus, a linear operator $\hat{\alpha}$ may be represented by a set of numbers forming a square $(N \times N)$ matrix $M(\hat{\alpha}) \equiv M$ given by

$$M = \begin{pmatrix}
M_{11} & M_{12} & \cdots & M_{1r} & \cdots & M_{1N} \\
M_{21} & M_{22} & \cdots & M_{2r} & \cdots & M_{2N} \\
\vdots & & & & & \\
M_{s1} & M_{12} & \cdots & M_{sr} & \cdots & M_{sN} \\
\cdots & & & & & \\
M_{N1} & M_{N2} & \cdots & M_{Nr} & \cdots & M_{NN}
\end{pmatrix}, \tag{2.3.1a}$$

where a typical matrix element M_{sr} is given by

$$M_{sr} = \left\langle\xi^{(s)}\right|\hat{\alpha}\left|\xi^{(r)}\right\rangle\,. \tag{2.3.1b}$$

If the number of basis states is infinite ($N \to \infty$), then the representative matrix for $\hat{\alpha}$ is infinite dimensional.

Some corollaries result from matrix representation of operators:

1. *The matrix representing the product of two operators $\hat{\alpha}$ and $\hat{\beta}$ equals the product of the matrices (taken in the same order) representing these operators:*

$$M(\hat{\alpha}\hat{\beta}) = M(\hat{\alpha})M(\hat{\beta}) \tag{2.3.2}$$

with
$$M_{sr}(\hat{\alpha}\hat{\beta}) = \sum_{k=1}^{N} M_{sk}(\hat{\alpha})M_{kr}(\hat{\beta})\,. \tag{2.3.3}$$

From the definition of a typical element of the matrix representing the product operator $\hat{\alpha}\hat{\beta}$ we have

$$\begin{aligned} M_{sr}(\hat{\alpha}\hat{\beta}) &\equiv \left\langle \xi^{(s)}\right| \hat{\alpha}\hat{\beta} \left|\xi^{(r)}\right\rangle \\ &= \left\langle \xi^{(s)}\right| \hat{\alpha} \sum_{k=1}^{N} \left|\xi^{(k)}\right\rangle \left\langle \xi^{(k)}\right| \hat{\beta} \left|\xi^{(r)}\right\rangle \\ &= \sum_{k=1}^{N} M_{sk}(\hat{\alpha})M_{kr}(\hat{\beta}) \end{aligned}$$

where we have used the *completeness condition* for the basis states $|\xi^{(k)}\rangle$. Thus the matrix $M(\hat{\alpha}\hat{\beta})$ is the product $M(\hat{\alpha}) \times M(\hat{\beta})$. Note that, in general, $M(\hat{\alpha})M(\hat{\beta}) \neq M(\hat{\beta})M(\hat{\alpha})$.

2. *The matrix $M(\hat{\alpha})$, representing a self adjoint operator $\hat{\alpha}$ ($\hat{\alpha}^\dagger = \hat{\alpha}$) is a Hermitian matrix:*

$$M^\dagger = M \quad \text{or} \quad \left[M^\dagger\right]_{sr} \equiv M^*_{rs}(\hat{\alpha}) = M_{sr}(\hat{\alpha})\,. \tag{2.3.4}$$

From the definition of a typical matrix element

$$\begin{aligned} \left[M^\dagger\right]_{sr} \equiv M^*_{rs}(\hat{\alpha}) &= \overline{\left\langle \xi^{(r)}\right| \hat{\alpha} \left|\xi^{(s)}\right\rangle} \\ &= \left\langle \xi^{(s)}\right| \hat{\alpha}^\dagger \left|\xi^{(r)}\right\rangle \\ &= \left\langle \xi^{(s)}\right| \hat{\alpha} \left|\xi^{(r)}\right\rangle = M_{sr}(\hat{\alpha})\,. \end{aligned}$$

Thus the matrix elements satisfy the condition for the matrix equality $M^\dagger = M$ to hold.

3. *The matrix representing the observable $\hat{\xi}$ itself or $\hat{\eta}$ or $\hat{\zeta}, \cdots$ in the $\hat{\zeta}, \hat{\xi}, \hat{\eta}, \cdots$ representation is diagonal, the diagonal elements being the respective eigenvalues.*

Again, by using the definition of matrix elements, we have

$$M_{sr}(\hat{\xi}) = \left\langle \xi^{(s)}\right| \hat{\xi} \left|\xi^{(r)}\right\rangle = \xi^{(r)} \left\langle \xi^{(s)}\middle| \xi^{(r)}\right\rangle = \xi^{(r)}\delta_{sr}\,. \tag{2.3.5}$$

In other words,

$$M = \begin{pmatrix} \xi' & 0 & \cdots & 0 \\ 0 & \xi'' & 0\cdots & 0 \\ \vdots & & & \\ 0 & 0 & \cdots & \xi^{(N)} \end{pmatrix}. \tag{2.3.6}$$

Similarly, for any other observable, say $\hat{\eta}$, characterizing the representation, we have

$$\begin{aligned} M_{sr}(\hat{\eta}) &= \left\langle \xi^{(s)} \right| \hat{\eta} \left| \xi^{(r)} \right\rangle \\ &= \left\langle \xi^{(s)}, \eta^{(s)}, \zeta^{(s)} \cdots \right| \hat{\eta} \left| \xi^{(r)}, \eta^{(r)}, \zeta^{(r)} \cdots \right\rangle \\ &= \eta^{(r)} \delta_{sr} \,. \end{aligned} \tag{2.3.7}$$

The $\hat{\xi}, \hat{\eta}, \hat{\zeta}, \cdots$ representation is, therefore, also referred to as one in which the observables $\hat{\zeta}, \hat{\xi}, \hat{\eta} \cdots$ are diagonal (i.e., represented by diagonal matrices).

4. *If an operator $\hat{\alpha}$ admits the set of eigenvalues $\alpha^{(1)}, \alpha^{(2)}, \cdots, \alpha^{(N)}$ then the matrix, representing $\hat{\alpha}$ in any representation, also admits the same set of eigenvalues.*[2]

Let the observable $\hat{\alpha}$ admit a set of eigenstates $\left|\alpha^{(1)}\right\rangle, \left|\alpha^{(2)}\right\rangle, \cdots, \left|\alpha^{(N)}\right\rangle$ belonging, respectively, to the eigenvalues $\alpha^{(1)}, \alpha^{(2)}, \cdots, \alpha^{(N)}$:

$$\hat{\alpha} \left| \alpha^{(n)} \right\rangle = \alpha^{(n)} \left| \alpha^{(n)} \right\rangle . \tag{2.3.8}$$

Consider a representation in which a set of observables $\hat{\xi}, \hat{\eta}, \hat{\zeta}, \cdots$ are diagonal and the observable $\hat{\alpha}$ is represented by the matrix

$$M = M(\hat{\alpha}) = \begin{pmatrix} M_{11} & M_{12} & \cdots & M_{1r} & \cdots & M_{1N} \\ M_{21} & M_{22} & \cdots & M_{2r} & \cdots & M_{2N} \\ \vdots & & & & & \\ M_{s1} & M_{12} & \cdots & M_{sr} & \cdots & M_{sN} \\ \vdots & & & & & \\ M_{N1} & M_{N2} & \cdots & M_{Nr} & \cdots & M_{NN} \end{pmatrix}, \tag{2.3.9}$$

where $M_{sr} = \left\langle \xi^{(s)} \right| \hat{\alpha} \left| \xi^{(r)} \right\rangle$. Also, in the same representation, the state $\left|\alpha^{(n)}\right\rangle$ is represented by the column vector A_n given by

$$A_n = \begin{pmatrix} A_{1n} \\ A_{2n} \\ \vdots \\ A_{rn} \\ \vdots \\ A_{Nn} \end{pmatrix} \quad \text{where} \quad A_{rn} = \left\langle \xi^{(r)} \middle| A_n \right\rangle .$$

[2] In general, a matrix M multiplying a column vector X gives a new column vector $Y = MX$. But it is possible to find a set of column vectors $X_1, X_2, \cdots, X_n, \cdots, X_N$, such that $MX_n = \lambda_n X_n$, $n = 1, 2, \cdots, N$. When this is so, X_n is called an eigenvector of M, belonging to the eigenvalue λ_n and the set of numbers $\lambda_1, \lambda_2, \cdots, \lambda_N$ are called the eigenvalues of the matrix M. The number of eigenvalues a $N \times N$ square matrix can admit, equals its order N [see Appendix 3A1].

Let us see the result of operating matrix M on the column vector A_n,

$$MA_n = \begin{pmatrix} M_{11} & M_{12} & \cdots & M_{1r} & \cdots & M_{1N} \\ M_{21} & M_{22} & \cdots & M_{2r} & \cdots & M_{2N} \\ \vdots & & & & & \\ M_{s1} & M_{12} & \cdots & M_{sr} & \cdots & M_{sN} \\ \cdots & & & & & \\ M_{N1} & M_{N2} & \cdots & M_{Nr} & \cdots & M_{NN} \end{pmatrix} \begin{pmatrix} A_{1n} \\ A_{2n} \\ \vdots \\ A_{rn} \\ \vdots \\ A_{Nn} \end{pmatrix}$$

$$= \begin{pmatrix} \sum_{r=1}^{N} M_{1r}A_{rn} \\ \sum_{r=1}^{N} M_{2r}A_{rn} \\ \vdots \\ \sum_{r=1}^{N} M_{sr}A_{rn} \\ \vdots \\ \sum_{r=1}^{N} M_{Nr}A_{rn} \end{pmatrix} = \alpha^{(n)} \begin{pmatrix} A_{1n} \\ A_{2n} \\ \vdots \\ A_{sn} \\ \vdots \\ A_{Nn} \,. \end{pmatrix} .$$

Thus, in concise form,

$$MA_n = \alpha^{(n)} A_n \,. \tag{2.3.10}$$

This result follows because

$$\begin{aligned} \sum_{r=1}^{N} M_{sr}A_{rn} &= \sum_{r=1}^{N} \left\langle \xi^{(s)} \middle| \hat{\alpha} \middle| \xi^{(r)} \right\rangle \left\langle \xi^{(r)} \middle| \alpha^{(n)} \right\rangle \\ &= \left\langle \xi^{(s)} \middle| \hat{\alpha} \sum_{r=1}^{N} \middle| \xi^{(r)} \right\rangle \left\langle \xi^{(r)} \middle| \alpha^{(n)} \right\rangle = \left\langle \xi^{(s)} \middle| \hat{\alpha}\hat{1} \middle| \alpha^{(n)} \right\rangle \\ &= \alpha^{(n)} A_{sn} \,. \end{aligned} \tag{2.3.11}$$

Thus A_n is an eigenvector of the matrix M belonging to the eigenvalue $\alpha^{(n)}$. We can prove this for any index n. In other words, the eigenvalues of the matrix M, which represents the observable $\hat{\alpha}$ in some representation are the same as the eigenvalues admitted by the observable itself. Further, the eigenvectors of the matrix M are the column vectors which represent the eigenstates $\left|\alpha^{(1)}\right\rangle, \left|\alpha^{(2)}\right\rangle, \cdots, \left|\alpha^{(N)}\right\rangle$ in the $\hat{\xi}, \hat{\eta}, \hat{\zeta}, \cdots$ representation. To summarize, if

$$\hat{\alpha} \left|\alpha^{(n)}\right\rangle = \alpha^{(n)} \left|\alpha^{(n)}\right\rangle , \tag{2.3.12}$$

then

$$MA_n = \alpha^{(n)} A_n \qquad n = 1, 2, \cdots, N \,. \tag{2.3.13}$$

The operator equation (2.3.8), as well as the matrix equation (2.3.10), represents the physical fact that if, on a system in a quantum state represented by either $|\alpha^{(n)}\rangle$ or the column matrix A_n, an observation pertaining to the observable $\hat{\alpha}$ (or M) is made, the result is destined to be the eigenvalue $\alpha^{(n)}$.

2.4 Change of Representation

It is possible to set up alternative representations for physical states (ket vectors) and observables (self-adjoint operators conforming to completeness criterion). We shall see that column vectors representing the same physical state and square matrices representing the same observable in two different representations are connected through a unitary transformation. In other words, one can change from one representation to another through a unitary transformation[3].

Consider a representation in which the set of observables $\hat{\xi}, \hat{\eta}, \hat{\zeta}, \cdots$ are diagonal and the basis states are $\left|\xi^{(1)}\right\rangle, \left|\xi^{(2)}\right\rangle, \cdots, \left|\xi^{(N)}\right\rangle$, where the abbreviated notation $\left\langle \xi^{(r)}\right| \equiv \left\langle \xi^{(r)}, \eta^{(r)}, \zeta^{(r)}, \cdots\right|$ is used. Let ket $|A\rangle$ be represented in this representation by a column vector A

$$|A\rangle \to A = \begin{pmatrix} A_1 \\ A_2 \\ \vdots \\ A_r \\ \vdots \\ A_N \end{pmatrix} \tag{2.4.1}$$

where

$$A_r = \left\langle \xi^{(r)} \middle| A \right\rangle . \tag{2.4.2}$$

Also, let an observable $\hat{\alpha}$ be represented by an $N \times N$ square matrix K

$$\hat{\alpha} \to K(\hat{\alpha}) = \begin{pmatrix} K_{11} & K_{12} & \cdots & K_{1r} & \cdots & K_{1N} \\ K_{21} & K_{22} & \cdots & K_{2r} & \cdots & K_{2N} \\ \vdots & & & & & \\ K_{s1} & K_{s2} & \cdots & K_{sr} & \cdots & K_{sN} \\ \cdots & & & & & \\ K_{N1} & K_{N2} & \cdots & K_{Nr} & \cdots & K_{NN} \end{pmatrix} \tag{2.4.3}$$

$$K_{rs} = \left\langle \xi^{(r)} \middle| \hat{\alpha} \middle| \xi^{(s)} \right\rangle . \tag{2.4.4}$$

Now, consider an alternative representation in which a set of observable $\hat{P}, \hat{Q}, \hat{R}, \cdots$ are diagonal and basis states are $\left|P^{(1)}\right\rangle, \left|P^{(2)}\right\rangle, \cdots, \left|P^{(N)}\right\rangle$. We assume for simplicity that the number of basis states is the same in both representations. In the second representation

$$|A\rangle \to C = \begin{pmatrix} C_1 \\ C_2 \\ \vdots \\ C_r \\ \vdots \\ C_N \end{pmatrix} \tag{2.4.5}$$

---

[3] A unitary transformation S transforms a square $(N \times N)$ matrix K to another square matrix R, and a column matrix A to another column matrix C, such that

$$K \xrightarrow{S} SKS^{\dagger} = R$$

and

$$A \xrightarrow{S} \quad SA = C \, .$$

Here the transforming matrix S is unitary: $S\,S^{\dagger} = I$ and $S^{\dagger}\,S = I$, where $S^{\dagger}$ is the Hermitian adjoint of S. It can be shown that (i) eigenvalues of a square matrix are not changed if the matrix is subjected to a unitary transformation, i.e., if $K\,A = a\,A$, then $RC = a\,C$. Thus, matrix K and R have the same set of eigenvalues. (ii) Any relationship between the matrices is preserved under a unitary transformation (see Appendix 3A1).

where

$$C_r = \left\langle P^{(r)} \middle| A \right\rangle . \tag{2.4.6}$$

while

$$\hat{\alpha} \to L(\hat{\alpha}) = \begin{pmatrix} L_{11} & L_{12} & \cdots & L_{1n} & \cdots & L_{1N} \\ L_{21} & L_{22} & \cdots & L_{2n} & \cdots & L_{2N} \\ \vdots & & & & & \\ L_{m1} & L_{m2} & \cdots & L_{mn} & \cdots & L_{mN} \\ \cdots & & & & & \\ L_{N1} & L_{N2} & \cdots & L_{Nn} & \cdots & L_{NN} \end{pmatrix} \tag{2.4.7}$$

$$L_{mn} = \left\langle P^{(m)} \middle| \hat{\alpha} \middle| P^{(n)} \right\rangle . \tag{2.4.8}$$

The mathematical relationship between the square matrices K and L representing the same observable $\hat{\alpha}$ in the two different representations and that between the column vectors A and C representing the physical state $|A\rangle$ in the two representations is easily established. From the definition of matrix L, we have reperesenting

$$\begin{aligned} L_{mn} &\equiv \left\langle P^{(m)} \middle| \hat{\alpha} \middle| P^{(n)} \right\rangle \\ &= \sum_r \sum_s \left\langle P^{(m)} \middle| \xi^{(r)} \right\rangle \left\langle \xi^{(r)} \middle| \hat{\alpha} \middle| \xi^{(s)} \right\rangle \left\langle \xi^{(s)} \middle| P^{(n)} \right\rangle \\ &= \sum_r \sum_s S_{mr} K_{rs} S^*_{ns} = \sum_r \sum_s S_{mr} K_{rs} (S^\dagger)_{sn} = (SKS^\dagger)_{mn} . \end{aligned} \tag{2.4.9}$$

Here $S_{mr} = \left\langle P^{(m)} \middle| \xi^{(r)} \right\rangle$ and $S_{ns} = \left\langle P^{(n)} \middle| \xi^{(s)} \right\rangle$, which implies that

$$(S^\dagger)_{sn} = S^*_{ns} = \overline{\left\langle P^{(n)} \middle| \xi^{(s)} \right\rangle} = \left\langle \xi^{(s)} \middle| P^{(n)} \right\rangle . \tag{2.4.10}$$

Hence matrix L and K are related by

$$L = SKS^\dagger . \tag{2.4.11}$$

Also, from the definition of column matrix C, we have

$$\begin{aligned} C_m &\equiv \left\langle P^{(m)} \middle| A \right\rangle = \sum_s \left\langle P^{(m)} \middle| \xi^{(s)} \right\rangle \left\langle \xi^{(s)} \middle| A \right\rangle \\ &= \sum_s S_{ms} A_s = (SA)_m , \end{aligned} \tag{2.4.12}$$

which means that C and A are related by

$$C = SA . \tag{2.4.13}$$

Thus we can pass on from a representation in which the set of commuting observables $\hat{\xi}, \hat{\eta}, \hat{\zeta}, \cdots$ are diagonal, to the one in which $\hat{P}, \hat{Q}, \hat{R}, \cdots$ are diagonal, through the transformation

$$K \xrightarrow{S} SKS^\dagger = L$$

and $\quad A \xrightarrow{S} SA = C .$

We can also show that the square matrix S, which brings about the transformation, is unitary $SS^\dagger = S^\dagger S = I$. By considering a typical matrix element of $SS^\dagger$, we have

$$\begin{aligned}(SS^\dagger)_{mn} &= \sum_r (S)_{mr}(S^\dagger)_{rn} = S_{mr}S^*_{nr} \\ &= \sum_r \left\langle P^{(m)}\middle|\xi^{(r)}\right\rangle \overline{\left\langle P^{(n)}\middle|\xi^{(r)}\right\rangle} = \sum_r \left\langle P^{(m)}\middle|\xi^{(r)}\right\rangle\left\langle \xi^{(r)}\middle|P^{(n)}\right\rangle \\ &= \left\langle P^{(m)}\middle|P^{(n)}\right\rangle = \delta_{mn}\,. \end{aligned} \tag{2.4.14}$$

This implies that

$$SS^\dagger = I\,. \tag{2.4.15}$$

Similarly, by considering a typical element of $S^\dagger S$, we have

$$\begin{aligned}(S^\dagger S)_{rs} &= \sum_m (S^\dagger)_{rm}(S)_{ms} = S^*_{mr}S_{ms} \\ &= \sum_m \overline{\left\langle P^{(m)}\middle|\xi^{(r)}\right\rangle}\left\langle P^{(m)}\middle|\xi^{(s)}\right\rangle = \sum_m \left\langle \xi^{(r)}\middle|P^{(m)}\right\rangle\left\langle P^{(m)}\middle|\xi^{(s)}\right\rangle \\ &= \left\langle \xi^{(r)}\middle|\xi^{(s)}\right\rangle = \delta_{rs} \end{aligned} \tag{2.4.16}$$

and this implies that

$$S^\dagger S = I\,. \tag{2.4.17}$$

It could be that the total number of basis states in the $\hat{\xi}\,,\hat{\eta}\,,\hat{\zeta}$, representation is N, while that in the $\hat{P}\,,\hat{Q}\,,\hat{R}$ representation is $M(\neq N)$. Consequently, the square matrix K (representing $\hat{\alpha}$ in the former representation) has dimension $(N \times N)$, while the square matrix L (representing $\hat{\alpha}$ in the latter representation) has dimension $(M \times M)$. Also, the column vector A, has N elements while the column vector C, has M elements. The transforming matrix S $(S_{mr} = \langle P^{(m)}|\xi^{(r)}\rangle)$ is then a rectangular matrix of dimension $(M \times N)$, while $S^\dagger$ is a rectangular matrix of dimension $(N \times M)$. Obviously, $SS^\dagger = I$ (unit matrix of dimension $M \times M$), while $S^\dagger S = I$ (unit matrix of dimension $N \times N$). The transformation relations

$$K \to SKS^\dagger = L$$

and

$$A \to SA = C$$

are still valid. However, S being a rectangular matrix, does not admit an inverse.

2.5 Coordinate Representation

As all the position observables $\hat{q}_1\,,\hat{q}_2\,,\cdots\hat{q}_f$, for a system with f degrees of freedom, commute with each other, we can set up a representation in which the position observables are diagonal. This representation is called the *coordinate representation*. In this representation, the basis states $\langle q_1\,,q_2\cdots q_f|$ are the simultaneous eigenstates of $\hat{q}_1\,,\hat{q}_2\cdots\hat{q}_f$ belonging to the eigenvalues $q_1\,,q_2\cdots q_f$, each eigenvalue varying continuously over a certain range. Thus the basis bras, in this case, do not form a denumerable set of states (as, for example $\left|\xi^{(1)}\right\rangle$, $\left|\xi^{(2)}\right\rangle$, $\cdots$ etc.) but a continuous set of states. Consequently, the representatives $\langle q_1\,,q_2\cdots q_f|\,A\rangle$ of a state $|A\rangle$, in this representation do not form a discrete set of numbers which could be written in the form of a column vector. The representative numbers, in this case, vary continuously with the continuous variation of the eigenvalues of the position observables. So we can regard $\langle q_1, q_2, \cdots, q_f | A\rangle$, the coordinate representative of state $|A\rangle$, as a function of the eigenvalues of the position observables. This function

$$\langle q_1\,, q_2 \cdots q_f | A\rangle \equiv \Psi_A(q_1\,, q_2 \cdots q_f)$$

is called the *wave function* and is the most common means of representing the physical state $|A\rangle$ of a system.

2.5.1 Physical Interpretation of the Wave Function

We have seen that, in the $\hat{\xi}\,, \hat{\eta}\,, \hat{\zeta}\,, \cdots$ representation, the quantity $|C_r|^2 = |\langle \xi^{(r)}|A\rangle|^2$ is interpreted as the probability of getting the result $\xi^{(r)}$, when a measurement of $\hat{\xi}$ is made on the system in state $|A\rangle$. Likewise, we can interpret

$$|\langle q_1, q_2, \cdots, q_f | A\rangle|^2 dq_1 dq_2 \cdots dq_f \equiv |\Psi_A(q_1, q_2, \cdots, q_f)|^2 dq_1 dq_2 \cdots dq_f$$

to be the probability that the results of measurements of $\hat{q}_1\,, \hat{q}_2\,, \cdots \hat{q}_f$ on the system in state $|A\rangle$ will lie between q_1 and $q_1 + dq_1\,, q_2$ and $q_2 + dq_2, \cdots, q_f$ and $q_f + dq_f$, respectively. Consequently, $\int \cdots \int |\Psi_A|^2 dq_1 dq_2 \cdots dq_f$ gives the probability that the results of measurements of coordinate observables on the system in state $|A\rangle$ will lie within the respective ranges of the eigenvalues. This probability is obviously equal to 1:

$$\int \cdots \int |\Psi_A(q_1, q_2, \cdots, q_f)|^2 dq_1 dq_2 \cdots dq_f = 1\,. \tag{2.5.1}$$

This condition is called the wave function normalization condition. This condition also follows from the normalization of the ket vector $|A\rangle$:

$$\langle A|A\rangle = 1\,. \tag{2.5.2}$$

By using the completeness criterion

$$\int \cdots \int |q_1, q_2, \cdots, q_f\rangle\, dq_1 dq_2 \cdots dq_f\, \langle q_1, q_2, \cdots, q_f| = \hat{1} \tag{2.5.3}$$

for basis vectors in the coordinate representation, we obtain from Eq. (2.5.2),

$$\begin{aligned}\int \cdots \int \langle A| q_1, q_2, \cdots, q_f\rangle\, dq_1 dq_2 \cdots dq_f\, \langle q_1, q_2, \cdots, q_f| A\rangle \\ = \int \cdots \int \Psi^*(q_1, q_2, \cdots, q_f) dq_1 dq_2 \cdots dq_f \Psi(q_1, q_2, \cdots, q_f) = 1\,,\end{aligned}$$

which is the wave function normalization condition given by Eq. (2.5.1).

Normalization condition on the wave function with one degree of freedom.

The number of position observables to be used to set up a coordinate representation is obviously equal to the number of degrees of freedom of the physical system. For a system with one degree of freedom, for example, a particle constrained to move along a straight line (say, the x-axis), the position observable is $\hat{q}_1 = \hat{x}$. In a representation in which $\hat{x}$ is diagonal, the state $|A\rangle$ is represented by the wave function $\langle x|A\rangle = \Psi_A(x)$, where x is the continuously varying eigenvalue of the position observable, and $|\Psi_A(x)|^2 dx = |\langle x| A\rangle|^2 dx$

gives the probability that a measurement of $\hat{x}$ in the state $|A\rangle$ will yield a result between x and $x + dx$. This interpretation follows from the fact that the state $|A\rangle$ can be expanded in terms of the eigenstates $|x\rangle$ of the observable $\hat{x}$ as:

$$|A\rangle = \int |x\rangle\, dx\, \langle x|\, A\rangle = \int \Psi_A(x) dx\, |x\rangle\, . \tag{2.5.4}$$

Since the kets $|x\rangle$ form a continuum of states, this expansion is not discrete, but continuous. In the case of expansion of an arbitrary state $|A\rangle$ in terms of a complete set of discrete eigenstates of an observable [Eq. (2.2.11)], we interpreted $|C_r|^2$ as the probability of getting the result $\xi^{(r)}$ when an observation for $\hat{\xi}$ is made on the system in the state $|A\rangle$. In the present context we can interpret $|\Psi_A(x)|^2 dx$ as the probability that a measurement for $\hat{x}$ gives a result between x and $x+dx$ on the system in state $|A\rangle$. The normalization condition for $\psi_A(x)$ is given by

$$\int \Psi^*(x)\Psi(x)dx = 1\, . \tag{2.5.5}$$

where the integration is over the whole range of eigenvalues of $\hat{x}$. This interpretation also implies that $\Psi_A(x)$ should be a continuous function of x.

2.5.1.1 Normalization condition on the wavefunction with three degrees of freedom

For a system with three degrees of freedom, for example, a particle moving in three-dimensional space, we naturally choose the three commuting position observables to be

$$\hat{q}_1 = \hat{x}\, , \qquad \hat{q}_2 = \hat{y}\, , \qquad \hat{q}_3 = \hat{z}\, .$$

In this representation, the set of commuting observables $\hat{x}, \hat{y}, \hat{z}$ are diagonal and the basis states $|x, y, z\rangle = |\boldsymbol{r}\rangle$, where x, y, z denote the continuously varying eigenvalues of $\hat{x}, \hat{y}, \hat{z}$ form a continuum of states. The basis states correspond to various locations of the particle in three-dimensional space. In this representation, a state $|A\rangle$ of the system is represented by the wave function $\Psi_A(x, y, z) = \langle x, y, z|A\rangle$ and $|\langle x, y, z|A\rangle|^2 dxdydz$ represents the probability that a measurement of position observables $\hat{x}, \hat{y}, \hat{z}$ on the system in this state, will yield results between x and $x + dx$, y and $y + dy$, and z and $z + dz$, respectively. The normalization condition[4]

$$\iiint |\Psi_A(x, y, z)|^2 dxdydz \equiv \iiint |\Psi_A(\vec{r})|^2 d^3r = 1 \tag{2.5.6}$$

simply expresses the fact that the probability that the particle is found somewhere within the available space is one. This interpretation of $\Psi_A(\vec{r})$ naturally implies that $\Psi_A(\vec{r})$ should be continuous at all points of available space.

2.6 Replacement of Momentum Observable $\hat{p}$ by $-i\hbar\frac{d}{d\hat{q}}$

Before justifying this replacement, it is necessary to specify the meaning of the operator $\frac{d}{d\hat{q}}$. To start with, consider a system with only one degree of freedom. The operator $\frac{d}{d\hat{q}}$ may be

[4]It may be noted that a change in the coordinates from x, y, z to r, θ, φ does not mean a change in the representation because the two sets of coordinates are related.

regarded a linear operator which, like any linear operator, can operate on a ket vector to the right to give another ket

$$\frac{d}{d\hat{q}}\left|P\right\rangle = \left|R\right\rangle ,$$

and it can also operate on a bra vector to the left to give another bra vector,

$$\left\langle Q\right|\frac{d}{d\hat{q}} = \left\langle S\right| ,$$

such that

$$\left\{\left\langle Q\right|\frac{d}{d\hat{q}}\right\}\left|P\right\rangle = \left\langle Q\right|\left\{\frac{d}{d\hat{q}}\left|P\right\rangle\right\} . \tag{2.6.1}$$

Now let the coordinate representative of $|P\rangle$ be $\Psi(q)$ so that

$$\langle q|P\rangle = \Psi(q) \tag{2.6.2}$$

which implies that

$$\langle P|q\rangle = \Psi^*(q) \tag{2.6.3}$$

and the coordinate representative of $|Q\rangle$ be $\Phi(q)$ which implies that

$$\langle Q|q\rangle = \Phi^*(q) . \tag{2.6.4}$$

Now, if we accept the definition that the coordinate representative of $\frac{d}{d\hat{q}}\left|P\right\rangle$ equals the partial derivative with respect to q, of the coordinate representative of the state $|P\rangle$:

$$\left\langle q\right|\frac{d}{d\hat{q}}\left|P\right\rangle = \frac{\partial}{\partial q}\left\langle q\right|P\rangle = \frac{\partial}{\partial q}\Psi(q) , \tag{2.6.5}$$

then we can show, using the condition (2.6.1), that the linear operator $\frac{d}{d\hat{q}}$ is pure imaginary operator. To show this, we introduce in Eq. (2.6.1) the unit operator

$$\hat{1} = \int \left|q\right\rangle dq \left\langle q\right| ,$$

after the curly bracket on the left hand side and before the curly bracket on the right-hand side, to get

$$\begin{aligned}
\int \left\langle Q\right|\frac{d}{d\hat{q}}\left|q\right\rangle dq \left\langle q\right|P\rangle &= \int \left\langle Q\right|q\rangle dq \left\langle q\right|\frac{d}{d\hat{q}}\left|P\right\rangle \\
&= \int \Phi^*(q)dq\frac{\partial}{\partial q}\Psi(q) \\
&= \Phi^*(q)\Psi(q)|_{q_{min}}^{q_{max}} - \int \frac{\partial\Phi^*(q)}{\partial q}\Psi(q)dq .
\end{aligned}$$

The first term on the right is zero, since the function $\Phi(q)$ and $\Psi(q)$ satisfy the boundary conditions. This leads to

$$\int \left\langle Q\right|\frac{d}{d\hat{q}}\left|q\right\rangle dq\Psi(q) = -\int \frac{\partial\Phi^*(q)}{\partial q}\Psi(q)dq . \tag{2.6.6}$$

Since $\Psi(q)$ is arbitrary, by comparing the integrands on the two sides, we have

$$-\left\langle Q\right|\frac{d}{d\hat{q}}\left|q\right\rangle = \frac{\partial\Phi^*(q)}{\partial q} . \tag{2.6.7}$$

Also, from Eq. (2.6.5), we have

$$\langle q|\,\frac{d}{d\hat{q}}\,|Q\rangle = \frac{\partial\Phi(q)}{\partial q}\,. \tag{2.6.8}$$

Since the right hand sides of these equations are complex conjugates of each other, we have

$$\langle q|\,\frac{d}{d\hat{q}}\,|Q\rangle^* = -\,\langle Q|\,\frac{d}{d\hat{q}}\,|q\rangle$$

or
$$\langle Q|\left(\frac{d}{d\hat{q}}\right)^\dagger|q\rangle = -\,\langle Q|\,\frac{d}{d\hat{q}}\,|q\rangle\,. \tag{2.6.9}$$

Since this holds for arbitrary $|q\rangle$ and $\langle Q|$ we have the operator relation [5]

$$\left(\frac{d}{d\hat{q}}\right)^\dagger = -\frac{d}{d\hat{q}} \tag{2.6.10}$$

showing that $\frac{d}{d\hat{q}}$ is a pure imaginary (anti-Hermitian) operator.

We can also easily work out the commutation relations between the operators $\frac{d}{d\hat{q}}$ and $\hat{q}$. By letting $\frac{d}{d\hat{q}}\hat{q}$ operate on $|P\rangle$ and taking the coordinate representative of the resulting ket, we have

$$\begin{aligned}\langle q|\,\frac{d}{d\hat{q}}\hat{q}\,|P\rangle &= \frac{\partial}{\partial q}\,\{q\Psi(q)\}\\ &= \Psi(q) + q\frac{\partial\Psi(q)}{\partial q}\\ &= \langle q|\,P\rangle + \langle q|\,\hat{q}\frac{d}{d\hat{q}}\,|P\rangle\,.\end{aligned}$$

Since this holds for an arbitrary ket $|P\rangle$, we have the operator relation

$$\frac{d}{d\hat{q}}\hat{q} = \hat{1} + \hat{q}\frac{d}{d\hat{q}}$$

or
$$\left[\frac{d}{d\hat{q}}\,,\hat{q}\right] = \hat{1}$$

or
$$\left[-i\hbar\frac{d}{d\hat{q}}\,,\hat{q}\right] = -i\hbar\,\hat{1}\,. \tag{2.6.11}$$

Since $\frac{d}{d\hat{q}}$ is anti-Hermitian, the operator $-i\hbar\frac{d}{d\hat{q}}$ is a Hermitian (self-adjoint) operator. According Eq. (2.6.11), the self-adjoint operator $-i\hbar\frac{d}{d\hat{q}}$ satisfies the same commutation relation with $\hat{q}$ as $\hat{p}$ does. This is one argument in support of the replacement of the observable $\hat{p}$ by the self-adjoint operator $-i\hbar\frac{d}{d\hat{q}}$.

Another argument in support of the identification of the momentum observable $\hat{p}$ with the operator $-i\hbar\frac{d}{d\hat{q}}$ is that, operating on an eigenstate $|p\rangle$ of momentum, it reproduces the eigenvalue p multiplied by the same state $|p\rangle$:

$$-\,i\hbar\frac{d}{d\hat{q}}\,|p\rangle = p\,|p\rangle\,,$$

or
$$\langle q|\,-\,i\hbar\frac{d}{d\hat{q}}\,|p\rangle = p\,\langle q|\,p\rangle\,. \tag{2.6.12}$$

[5] To show this, Dirac introduced the concepts of *standard ket* and *standard bra*. While his arguments are quite logical, this can also be shown without having to introduce these concepts.

To show the validity of Eq. (2.6.12), consider the left hand side of this equation, which gives

$$\langle q| - i\hbar\frac{d}{d\hat{q}}\, |p\rangle = -i\hbar\frac{\partial}{\partial q}\, \langle q|\, p\rangle\ . \tag{2.6.13}$$

Now the momentum eigenstate in the coordinate representation is just the plane wave,

$$\langle q|\, p\rangle = Ce^{ipq/\hbar}\, . \tag{2.6.14}$$

Using Eq. (2.6.14) in Eq. (2.6.13), we immediately get Eq. (2.6.12). Note that consideration of the time dependence of states will not make any difference in this result. Thus we have the important result:

If we use $\hat{x}$ to denote the position observable and $\hat{p}_x$ to denote the corresponding momentum observable, the preceding arguments justify the replacement of the observable $\hat{p}_x$ by $-i\hbar\frac{d}{d\hat{x}}$ in the coordinate represenation.

This result is very useful when we wish to express the Schrödinger equation in the coordinate representation.[6]

We can easily generalize the results of this section to a system with f degrees of freedom. Let the coordinate representative of state $|A\rangle$ of such a system be denoted by $\Psi_A(q_1, q_2, \cdots, q_f) \equiv \langle q_1, q_2, \cdots, q_f|\, A\rangle$. This definition implies $\Psi_A^*(q_1, q_2, \cdots, q_f) = \langle A|\, q_1, q_2, \cdots,$ We can now introduce linear operators $\frac{d}{d\hat{q}_1}, \frac{d}{d\hat{q}_2}, \cdots, \frac{d}{d\hat{q}_f}$. Each of these operators operating to the right on a ket yields another ket vector and operating to the left on a bra vector yields another bra vector. Thus $\frac{d}{d\hat{q}_r}$ operating on $|A\rangle$ and $\langle B|$ vectors yields

$$\frac{d}{d\hat{q}_r}\, |A\rangle \equiv |C\rangle \quad \text{and} \quad \langle B|\, \frac{d}{d\hat{q}_r} \equiv \langle D|$$

such that

$$\left\{\langle B|\, \frac{d}{d\hat{q}_r}\right\} |A\rangle = \langle B| \left\{\frac{d}{d\hat{q}_r}\, |A\rangle\right\}\, . \tag{2.6.15}$$

If we again accept the definition that the coordinate representative of $\frac{d}{d\hat{q}_r}\, |A\rangle$ equals the partial derivative with respect to q_r of the coordinate representative of $|A\rangle$:

$$\langle q_1, q_2, \cdots, q_f|\, \frac{d}{d\hat{q}_r}\, |A\rangle = \frac{\partial}{\partial q_r}\, \langle q_1, q_2, \cdots, q_f|\, A\rangle \equiv \frac{\partial \Psi_A(q_1, q_2, \cdots, q_f)}{\partial q_r}\, , \tag{2.6.16}$$

then, using the condition (2.6.15), we can show that $\frac{d}{d\hat{q}_r}$ is a pure imaginary operator.

To show this, we again introduce the unit operator

$$\hat{1} = \int \cdots \int |q_1\, , q_2\, , \cdots q_f\rangle\, dq_1 dq_2 \cdots dq_f\, \langle q_1\, , q_2\, , \cdots q_f|\ ,$$

[6] The corresponding result for the momentum representation that the position observable $\hat{x}$ can be replaced by $i\hbar\frac{d}{d\hat{p}}$, although correct in principle as this also conforms to the basic commutation relation, is of use only in a limited number of cases. In Chapter 3 we will learn that it is more convenient to use Fourier transforms of the interaction potential instead.

after the curly bracket on the left hand side of Eq. (2.6.15) and before the curly bracket on the right hand side of Eq. (2.6.15), to get

$$\int\cdots\int \langle B|\,\frac{d}{d\hat{q}_r}\,|q_1,q_2,\cdots,q_f\rangle\,dq_1dq_2\cdots dq_f\,\langle q_1,q_2,\cdots,q_f|\,A\rangle$$
$$=\int\cdots\int \langle B|\,q_1,q_2,\cdots,q_f\rangle\,dq_1dq_2\cdots dq_f\,\langle q_1,q_2,\cdots,q_f|\,\frac{d}{d\hat{q}_r}\,|A\rangle$$
$$=\int\cdots\int \Psi_B^*(q_1,q_2,\cdots,q_f)\frac{\partial\Psi_A(q_1,q_2,\cdots,q_f)}{\partial q_r}dq_1dq_2\cdots dq_f\,.$$

Integrating the right hand side with respect to q_r by parts, and letting the first term equal to zero on account of boundary conditions, we get

$$\int\cdots\int \langle B|\,\frac{d}{d\hat{q}_r}\,|q_1,q_2,\cdots,q_f\rangle\,\Psi_A(q_1,q_2,\cdots,q_f)dq_1dq_2\cdots dq_f$$
$$=-\int\cdots\int \frac{\partial\Psi_B^*(q_1,q_2,\cdots,q_f)}{\partial q_r}\Psi_A(q_1,q_2,\cdots,q_f)dq_1dq_2\cdots dq_f\,.$$

Since $\Psi_A(q_1,q_2,\cdots,q_f)$ represents an arbitrary state we have, on comparing the integrands on both sides,

$$-\langle B|\,\frac{d}{d\hat{q}_r}\,|q_1\,,q_2\,,\cdots,q_f\rangle=\frac{\partial\Psi_B^*(q_1\,,q_2\,,\cdots,q_f)}{\partial q_r}\,. \tag{2.6.17}$$

Also, according to Eq. (2.6.16), we have

$$\langle q_1\,,q_2\,,\cdots,q_f|\,\frac{d}{d\hat{q}_r}\,|B\rangle=\frac{\partial\Psi_B(q_1\,,q_2\,,\cdots,q_f)}{\partial q_r}\,. \tag{2.6.18}$$

Since the right hand sides of Eqs.(2.6.17) and (2.6.18) are complex conjugates of each other, $-\langle B|\,\frac{d}{d\hat{q}_r}\,|q_1,q_2,\cdots,q_f\rangle$ and $\langle q_1,q_2,\cdots,q_f|\,\frac{d}{d\hat{q}_r}\,|B\rangle$ are also complex conjugates of each other. This leads to the result

$$\left(\frac{d}{d\hat{q}_r}\right)^\dagger=-\frac{d}{d\hat{q}_r}\,,$$

which shows that the operator $\frac{d}{d\hat{q}_r}$ is a pure imaginary operator and, therefore, the operator $-i\hbar\frac{d}{d\hat{q}_r}$ is a self-adjoint (Hermitian) operator.

Now, if we let $\frac{d}{d\hat{q}_r}\hat{q}_s$ operate on the ket $|A\rangle$, and take the coordinate representative of the resulting ket, we obtain

$$\begin{aligned}\langle q_1\,,q_2\,,\cdots,q_f|\,\frac{d}{d\hat{q}_r}\hat{q}_s\,|A\rangle &= \frac{\partial}{\partial q_r}\,\langle q_1\,,q_2\,,\cdots,q_f|\,\hat{q}_s\,|A\rangle\\ &= \frac{\partial}{\partial q_r}\,[q_s\Psi_A(q_1\,,q_2\,,\cdots,q_f)]\\ &= \delta_{rs}\Psi_A(q_1\,,q_2\,,\cdots,q_f)+q_s\frac{\partial\Psi_A(q_1\,,q_2\,,\cdots,q_f)}{\partial q_r}\\ &= \delta_{rs}\Psi_A(q_1\,,q_2\,,\cdots,q_f)+\langle q_1\,,q_2\,,\cdots,q_f|\,\hat{q}_s\frac{d}{d\hat{q}_r}\,|A\rangle\,.\end{aligned}$$

Comparing the operators sandwiched between the states $\langle q_1\,,q_2\,,\cdots,q_f|$ and $|A\rangle$ on the two sides, we get

$$\frac{d}{d\hat{q}_r}\hat{q}_s=\hat{1}\delta_{rs}+\hat{q}_s\frac{d}{d\hat{q}_r}$$

or

$$\frac{d}{d\hat{q}_r}\hat{q}_s-\hat{q}_s\frac{d}{d\hat{q}_r}=\hat{1}\delta_{rs}\,. \tag{2.6.19}$$

Similarly, in accordance with Eq. (2.6.16), we have

$$\begin{aligned}\langle q_1, q_2, \cdots, q_f| \frac{d}{d\hat{q}_r}\frac{d}{d\hat{q}_s}|A\rangle &= \frac{\partial}{\partial q_r}\frac{\partial}{\partial q_s}\Psi_A(q_1, q_2, \cdots, q_f) \\ &= \frac{\partial}{\partial q_s}\frac{\partial}{\partial q_r}\Psi_A(q_1, q_2, \cdots, q_f) \\ &= \langle q_1, q_2, \cdots, q_f| \frac{d}{d\hat{q}_s}\frac{d}{d\hat{q}_r}|A\rangle ,\end{aligned} \tag{2.6.20}$$

where we have interchanged the order of differrentiation in the second step. Equation (2.6.20) hold for an arbitrary ket $|A\rangle$. It follows therefore that

$$\frac{d}{d\hat{q}_r}\frac{d}{d\hat{q}_s} = \frac{d}{d\hat{q}_s}\frac{d}{d\hat{q}_r} . \tag{2.6.21}$$

Multiplying both sides of Eqs.(2.6.19) and (2.6.21) by $-i\hbar$, we can write

$$\left(-i\hbar\frac{d}{d\hat{q}_r}\right)\hat{q}_s - \hat{q}_s\left(-i\hbar\frac{d}{d\hat{q}_r}\right) = \delta_{rs}(-i\hbar)\hat{1}$$

$$\left(-i\hbar\frac{d}{d\hat{q}_r}\right)\left(-i\hbar\frac{d}{d\hat{q}_s}\right) - \left(-i\hbar\frac{d}{d\hat{q}_s}\right)\left(-i\hbar\frac{d}{d\hat{q}_r}\right) = 0$$

which can be compared to the commutation relations satisfied by $\hat{p}_r$ and $\hat{q}_s$:

$$\hat{p}_r\hat{q}_s - \hat{q}_s\hat{p}_r = -i\hbar\,\delta_{rs}\hat{1} \tag{2.6.22}$$

$$\hat{p}_r\hat{p}_s - \hat{p}_s\hat{p}_r = 0 . \tag{2.6.23}$$

This gives justification for the replacement of the operator $\hat{p}_r$ by $-i\hbar\frac{d}{\hat{q}_r}$.

2.7 Integral Representation of Dirac Bracket $\langle A_2|\,\hat{F}\,|A_1\rangle$

For simplicity, we confine our attention to a system with three degrees of freedom. Let

$$\hat{F} = \hat{F}(\hat{x}, \hat{y}, \hat{z}, \hat{p}_x, \hat{p}_y\,\hat{p}_z) = \hat{F}(\hat{x}\,\hat{y}, \hat{z}, -i\hbar\frac{d}{d\hat{x}}, -i\hbar\frac{d}{d\hat{y}}, -i\hbar\frac{d}{d\hat{z}})$$

be an operator in Dirac space so that it can operate to the right on a ket as also to the left on a bra vector so that

$$\left\{\langle A_2|\,\hat{F}\right\}|A_1\rangle = \langle A_2|\left\{\hat{F}\,|A_1\rangle\right\} = \langle A_2|\,\hat{F}\,|A_1\rangle . \tag{2.7.1}$$

In the coordinate representation, in which the coordinate observables $\hat{x}, \hat{y}, \hat{z}$ are diagonal and the basis states are $|x, y, z\rangle \equiv |\boldsymbol{r}\rangle$, the states $|A_1\rangle$ and $|A_2\rangle$ are represented by the wave functions $\Psi_1(\boldsymbol{r})$ and $\Psi_2(\boldsymbol{r})$, respectively, where

$$\Psi_1(\boldsymbol{r}) = \langle \boldsymbol{r}|A_1\rangle \tag{2.7.2a}$$

$$\Psi_2(\boldsymbol{r}) = \langle \boldsymbol{r}|A_2\rangle . \tag{2.7.2b}$$

Now, the Dirac bracket $\langle A_2|\hat{F}|A_1\rangle$ can be written as

$$\begin{aligned}\langle A_2|\,\hat{F}\,|A_1\rangle &= \int \langle A_2|\,\boldsymbol{r}\rangle\, d^3r\, \langle \boldsymbol{r}|\,\hat{F}(\hat{x},\hat{y},\hat{z},-i\hbar\frac{d}{d\hat{x}},-i\hbar\frac{d}{d\hat{y}},-i\hbar\frac{d}{d\hat{z}})\,|A_1\rangle \\ &= \int \Psi_2^*(\boldsymbol{r})F(x,y,z,-i\hbar\frac{\partial}{\partial x},-i\hbar\frac{\partial}{\partial y},-i\hbar\frac{\partial}{\partial z})\Psi_1(\boldsymbol{r})d^3r \\ &= \int \Psi_2^*(\boldsymbol{r})\mathcal{F}\Psi_1(\boldsymbol{r})\, d^3r\,,\end{aligned} \tag{2.7.3}$$

where we have used Eq. (2.6.5) or Eq. (2.6.17), and put $\mathcal{F} = F(x,y,z,-i\hbar\frac{\partial}{\partial x},-i\hbar\frac{\partial}{\partial y},-i\hbar\frac{\partial}{\partial z}) = F(\boldsymbol{r},-i\hbar\boldsymbol{\nabla})$. Thus the differential operator $\mathcal{F}$, representing Dirac linear operator $\hat{F}$ in the coordinate representation, is specified by the equation:[7]

$$\langle\boldsymbol{r}|\hat{F}|A\rangle = \mathcal{F}\langle\boldsymbol{r}|A\rangle = \mathcal{F}\Psi_A(\boldsymbol{r}). \tag{2.7.4}$$

Hermitian Adjoint of a Differential Operator in the Coordinate Space

Let the coordinate representative of a Dirac linear operator $\hat{F}$ be $\mathcal{F}$, and the coordinate representative of its adjoint $\hat{F}^\dagger$ be $\mathcal{F}^\dagger$. The latter may also be called the Hermitian adjoint (also referred to as Hermitian conjugate) of the differential operator $\mathcal{F}$. Now, from the definition of the adjoint of a linear operator, we have

$$\langle A_1|\hat{F}^\dagger|A_2\rangle = \Big[\langle A_2|\hat{F}|A_1\rangle\Big]^*\,, \tag{2.7.5}$$

for arbitrary $\langle A_1|$ and $|A_2\rangle$. In the coordinate representation the two sides of this take the form

$$\begin{aligned}\int \Psi_1^*(\boldsymbol{r})\mathcal{F}^\dagger\Psi_2(\boldsymbol{r})d^3r &= \Big[\int \Psi_2^*(\boldsymbol{r})\,(\mathcal{F}\Psi_1(\boldsymbol{r}))\,d^3r\Big]^* \\ &= \int (\mathcal{F}\Psi_1(\boldsymbol{r}))^*\,\Psi_2(\boldsymbol{r})d^3r\end{aligned} \tag{2.7.6}$$

where $\Psi_1(\boldsymbol{r}) = \langle\boldsymbol{r}|A_1\rangle$ and $\Psi_2(\boldsymbol{r}) = \langle\boldsymbol{r}|A_2\rangle$. Equation (2.7.6) may be looked upon as the definition of the Hermitian conjugate $\mathcal{F}^\dagger$ of a differential operator $\mathcal{F}$ in the coordinate space.

[7]The linear operator $\frac{d^n}{d\hat{x}^n}$, operating on a ket means $\frac{d}{d\hat{x}}$ operating n times in succession on the ket to the right. Thus

$$\begin{aligned}\langle\boldsymbol{r}|\,\frac{d^2}{d\hat{x}^2}\,|A\rangle &= \langle\boldsymbol{r}|\,\frac{d}{d\hat{x}}\frac{d}{d\hat{x}}\,|A\rangle = \langle\boldsymbol{r}|\,\frac{d}{d\hat{x}}\,|B\rangle\,, \qquad \text{where} \quad |B\rangle = \frac{d}{d\hat{x}}\,|A\rangle \\ &= \frac{\partial}{\partial x}\,\langle\boldsymbol{r}|\,B\rangle = \frac{\partial}{\partial x}\,\langle\boldsymbol{r}|\,\frac{d}{d\hat{x}}\,|A\rangle = \frac{\partial^2}{\partial x^2}\,\langle\boldsymbol{r}|\,A\rangle = \frac{\partial^2}{\partial x^2}\Psi_A(\boldsymbol{r})\,.\end{aligned}$$

Similarly,

$$\langle\boldsymbol{r}|\,\frac{d^2}{d\hat{x}^2}+\frac{d^2}{d\hat{y}^2}+\frac{d^2}{d\hat{z}^2}\,|A\rangle = \left(\frac{\partial^2}{\partial x^2}+\frac{\partial^2}{\partial y^2}+\frac{\partial^2}{\partial z^2}\right)\langle\boldsymbol{r}|\,A\rangle = \nabla^2\Psi_A(\boldsymbol{r})$$

and $$\langle\boldsymbol{r}|\,-\frac{\hbar^2}{2\mu}\left(\frac{d^2}{d\hat{x}^2}+\frac{d^2}{d\hat{y}^2}+\frac{d^2}{d\hat{z}^2}\right)+\hat{V}(\hat{x},\hat{y},\hat{z})\,|A\rangle = \left(-\frac{\hbar^2}{2\mu}\nabla^2+V(\boldsymbol{r})\right)\Psi_A(\boldsymbol{r})\,.$$

In general,

$$\begin{aligned}&\langle\boldsymbol{r}|\,\hat{F}(\hat{x},\hat{y},\hat{z},-i\hbar\frac{d}{d\hat{x}},-i\hbar\frac{d}{d\hat{y}},-i\hbar\frac{d}{d\hat{z}})\,|A\rangle \\ &= F(x,y,z,-i\hbar\frac{\partial}{\partial x},-i\hbar\frac{\partial}{\partial y},-i\hbar\frac{\partial}{\partial z})\,\langle\boldsymbol{r}|\,A\rangle = \mathcal{F}\Psi_A(\boldsymbol{r})\,,\end{aligned}$$

which is Eq. (2.7.4).

If the linear operator $\hat{F}$ is Hermitian (self-adjoint) ($\hat{F}^\dagger = \hat{F}$) or, equivalently, the differential operator $\mathcal{F}$ is Hermitian[8] ($\mathcal{F}^\dagger = \mathcal{F}$), then the condition for $\mathcal{F}$ to be called Hermitian becomes

$$\int \Psi_1^*(\boldsymbol{r})\mathcal{F}\Psi_2(\boldsymbol{r})d^3r = \int [\mathcal{F}\Psi_1(\boldsymbol{r})]^*\Psi_2(\boldsymbol{r})d^3r\,, \tag{2.7.7}$$

where the functions $\Psi_1(\boldsymbol{r})$ and $\Psi_2(\boldsymbol{r})$ are the coordinate representative of arbitrary states $|A_1\rangle$ and $|A_2\rangle$.

2.8 The Momentum Representation

The momentum observables $\hat{p}_1\,,\hat{p}_2\,,\cdots\hat{p}_f$, like the position observables $\hat{q}_1\,,\hat{q}_2\,,\cdots\hat{q}_f$, also form a set of commuting observables. So, we can as well set up an alternative representation in which the basis states are the simultaneous eigenstates of the momentum observables, $\langle p_1, p_2, \cdots, p_f|$, where $p_1, p_2, \cdots, p_f$ represent the continuously varying eigenvalues of the momentum observables. This representation is called the *momentum representation.* In this representation, the representatives of a state $|A\rangle$ are the set of numbers $\langle p_1\,,p_2\,,\cdots p_f|A\rangle$, which vary continuously with the eigenvalues $p_1\,,p_2\,,\cdots p_f$. In other words, an arbitrary state $|A\rangle$ can be represented by the function $\Phi_A(p_1\,,p_2\,,\cdots p_f)$ of the eigenvalues of the momentum observables:

$$\Phi(p_1\,,p_2\,,\cdots p_f) = \langle p_1\,,p_2\,,\cdots p_f|A\rangle\,. \tag{2.8.1}$$

The function $\Phi_A(p_1\,,p_2\,,\cdots p_f)$ is called the momentum representative of the state $|A\rangle$, or the wave function of the system in momentum space.

2.8.1 Physical Interpretation of $\Phi(p_1\,,p_2\,,\cdots p_f)$

The normalization condition $\langle A|A\rangle = 1$ for state $|A\rangle$ implies

$$\begin{aligned}\int\cdots\int \langle A|\,p_1\,,p_2\,,\cdots p_f\rangle\, dp_1dp_2\cdots dp_f\,\langle p_1\,,p_2\,,\cdots p_f|\,A\rangle \\ = \int\cdots\int |\Phi_A(p_1\,,p_2\,,\cdots p_f)|^2\, dp_1dp_2\cdots dp_f = 1\end{aligned} \tag{2.8.2}$$

which means that $|\Phi_A(p_1\,,p_2\,,\cdots p_f)|^2\, dp_1dp_2\cdots dp_f$ could be interpreted as the probability that the results of measurements of momentum observables $\hat{p}_1\,,\hat{p}_2\,,\cdots,\hat{p}_f$ on state $|A\rangle$ will lie between p_1 and p_1+dp_1, p_2 and p_2+dp_2, and so on. The normalization condition (2.8.2) expresses the fact that the probability that measurements of $\hat{p}_1\,,\hat{p}_2\,,\cdots,\hat{p}_f$ on the state $|A\rangle$ will yield results that lie within the respective ranges of their eigenvalues is one.

[8] In our notation (†) put on a Dirac linear operator means the adjoint of the linear operator; when '$\dagger$' is put on a differential operator, it means its Hermitian conjugate, and when put on a matrix, it means the Hermitian adjoint of the matrix.

2.9 Dirac Delta Function

The need for delta function arises when we wish to write the orthogonality condition for the eigenstates of an observable which admits continuous eigenvalues. We know that if an observable α admits a set of discrete eigenvalues, say $\alpha^{(r)}$, then its eigenvectors belonging to different eigenvalues are orthogonal:

$$\left\langle \alpha^{(r)} \middle| \alpha^{(s)} \right\rangle = \delta_{rs} \tag{2.9.1}$$

and

$$\sum_s \left\langle \alpha^{(r)} \middle| \alpha^{(s)} \right\rangle = 1\,. \tag{2.9.2}$$

If there is an observable, say $\hat{q}$, which admits continuous eigenvalues q, then the orthogonality of eigenstates $|q\rangle$ of the observable $\hat{q}$ demands that

$$\langle q|\, q'\rangle = 0 \quad \text{if} \quad q \neq q' \tag{2.9.3}$$

$$\langle q|\, q'\rangle \to \infty \quad \text{as} \quad q \to q' \tag{2.9.4}$$

with

$$\int \langle q|\, q'\rangle\, dq' = 1\,, \tag{2.9.5}$$

where the integration is over the entire range of the eigenvalue q. Dirac invented a function $\delta(x-a)$ to describe such a behavior. Formal definition and properties of Dirac delta function $\delta(x-a)$ are as follows:

1. Dirac delta function is defined by

$$\delta(x-a) = \begin{cases} 0 & \text{if} \quad x \neq a \\ \infty & \text{if} \quad x = a \end{cases} \tag{2.9.6}$$

such that

$$\int \delta(x-a)dx = 1\,. \tag{2.9.7}$$

2. Dirac delta function is a symmetric function of its argument:

$$\delta(x-a) = \delta(a-x).$$

3. $(x-a)\delta(x-a) = 0.$

4. $\delta[c(x-a)] = \frac{1}{c}\delta(x-a).$

5. Assuming that $f(x)$ has a single zero x_0,

$$\delta(f(x)) = \frac{\delta(x-x_0)}{\left|\dfrac{df}{dx}\right|_{x=x_0}}.$$

6. $f(x)\delta(x-a) = f(a)\delta(x-a).$

7. $\int f(x)\delta(x-a)dx = f(a).$

Graphically, the delta function $\delta(x-a)$ may be regarded as a symmetric function with a peak at $r = a$ with the height of the peak tending to infinity and its width tending to zero, such that the area under the curve is finite $(= 1)$ [see Fig. 2.1]. Mathematically, the delta function can be represented in a number of ways:

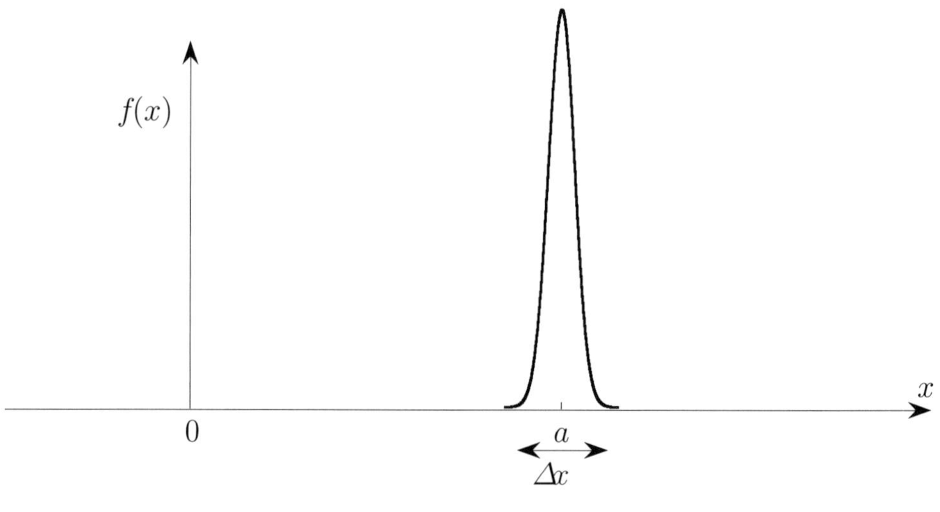

FIGURE 2.1
A symmetric function $f(x)$ centered at $x = a$ leads to a delta function $\delta(x-a)$ when its width $\Delta x \to 0$ and peak height $\to \infty$ such that the area under the curve is equal to 1.

(i)

$$\delta(x-a) = \lim_{g\to\infty} \frac{\sin[g(x-a)]}{\pi(x-a)} \quad \text{or} \quad \delta(x) = \lim_{g\to\infty} \frac{\sin(gx)}{\pi x}. \tag{2.9.8}$$

To prove this identity, let us examine the function

$$\frac{\sin(gx)}{\pi x} = \frac{g}{\pi}\frac{\sin(gx)}{gx}.$$

We note that this function has a principal peak of height g/π at $x = 0$. The width of its principal maximum is $2\pi/g$ and the function oscillates with a period π/g [see Fig. 2.2]. As $g \to \infty$, the height of the peak tends to infinity and its width tends to zero. Also, the area under the curve

$$\int_{-\infty}^{\infty} \frac{\sin(gx)}{\pi x} dx = \frac{1}{\pi}\int_{-\infty}^{\infty} \frac{\sin(t)}{t}\, dt = 1$$

independent of the value of g. So the function $\lim_{g\to\infty} \frac{\sin(gx)}{\pi x}$ has the properties of the δ function and can furnish a representation for it.

(ii) Delta function also has a Fourier representation. To establish the Fourier representation, we recall that the Fourier transform $\phi(k)$ of a function $f(x)$ is given by

$$\phi(k) = \frac{1}{\sqrt{2\pi}}\int_{-\infty}^{\infty} f(x)e^{-ikx}dx\,,$$

whereas $f(x)$, the inverse Fourier transform of $\phi(k)$ is expressed as

$$f(x) = \frac{1}{\sqrt{2\pi}}\int_{-\infty}^{\infty} \phi(k)e^{ikx}dk\,.$$

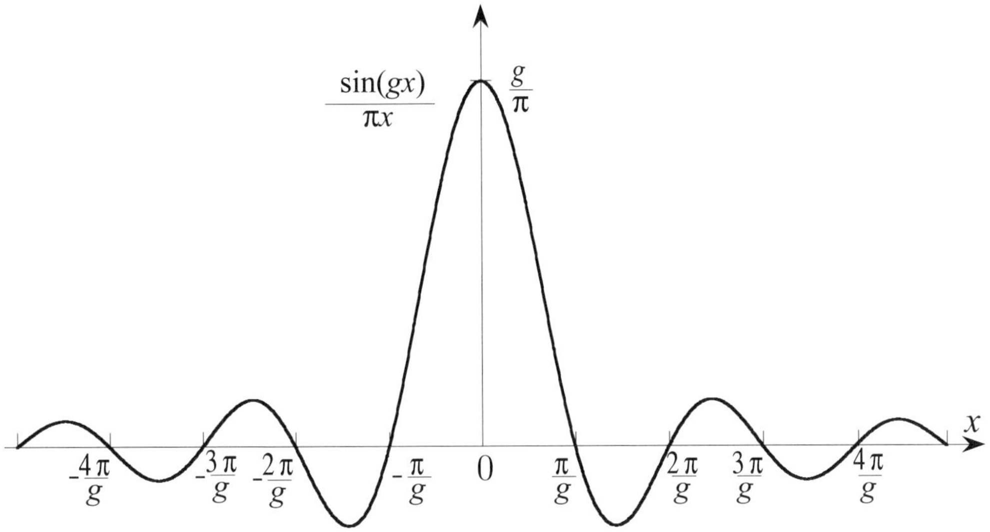

FIGURE 2.2
Graphical representation of the function $f(x) = \sin(gx)/\pi x$.

Combining these two equations, we have the Fourier integral theorem:

$$f(x) = \frac{1}{2\pi}\int_{-\infty}^{\infty} f(x')dx' \int_{-\infty}^{\infty} e^{-ik(x'-x)}dx\,. \tag{2.9.9}$$

Comparing this result with the property (7) of delta function

$$f(x) = \int dx'\, f(x')\delta(x'-x)\,, \tag{2.9.10}$$

we have the representation

$$\delta(x'-x) = \frac{1}{2\pi}\int dk\, e^{-ik(x'-x)}$$

$$\text{or}\qquad \delta(x-a) = \frac{1}{2\pi}\int dk\, e^{-ik(x-a)}\,. \tag{2.9.11}$$

2.9.1 Three-dimensional Delta Function

The product of three delta functions

$$\delta^3(\boldsymbol{r}-\boldsymbol{a}) = \delta(x-a_x)\delta(y-a_y)\delta(z-a_z)$$

is called the three-dimensional delta function. With the help of Eq. (2.9.11) it follows that

$$\begin{aligned}\delta^3(\boldsymbol{r}-\boldsymbol{a}) &= \frac{1}{(2\pi)^3}\int dk_x\, e^{-ik_x(x-a_x)}\int dk_y\, e^{-ik_y(y-a_y)}\int dk_z\, e^{-ik_z(z-a_z)} \\ &\equiv \frac{1}{(2\pi)^3}\int d^3k\, e^{-i\boldsymbol{k}\cdot(\boldsymbol{r}-\boldsymbol{a})}\,. \end{aligned} \tag{2.9.12}$$

Apart from enabling one to write the orthogonality conditions for the eigenstates of an observable, which admits continuous eigenvalues, the delta function has other uses also.

For example, it can be used to express the charge density $\rho(\boldsymbol{r})$ on account of a point charge Q at $\boldsymbol{r} = \boldsymbol{a}$ as

$$\rho(\boldsymbol{r}) = Q\delta^3(\boldsymbol{r} - \boldsymbol{a})\,. \tag{2.9.13}$$

We can easily see that the integrated charge is Q

$$\int \rho(\boldsymbol{r})d^3r = Q\,. \tag{2.9.14}$$

With this brief digression on Dirac delta function, we can resume our discussion of the momentum representation.

2.9.2 Normalization of a Plane Wave

The eigenstate $|p\rangle$ [or $|k\rangle$, where $k = p/\hbar$] of momentum observable $\hat{p}$ can be expressed in the coordinate representation as

$$|p\rangle \quad \rightarrow \quad \langle x|\, p\rangle = C\, e^{ipx/\hbar} \tag{2.9.15}$$

or

$$|k\rangle \quad \rightarrow \quad \langle x|\, k\rangle = C'\, e^{ikx}\,. \tag{2.9.16}$$

Here we consider the particle to have only one degree of freedom. We shall see that the normalization constants C and C' are different. This means the normalization constant for the plane wave depends on whether we use p or k to specify the continuously varying eigenvalue of the momentum observable.

Now the normalization condition for the momentum state $|p\rangle$ is

$$\langle p|\, p'\rangle = \delta(p - p')\,. \tag{2.9.17}$$

On the other hand

$$\begin{aligned}\langle p|\, p'\rangle &= \int \langle p|\, x\rangle\, dx\, \langle x|\, p'\rangle \\ &= \int C^* e^{-ipx/\hbar} C e^{ip'x/\hbar} dx = |C|^2\, 2\pi\, \delta[(p - p')/\hbar] \\ &= |C|^2\, 2\pi\, \hbar\, \delta(p - p')\,.\end{aligned} \tag{2.9.18}$$

Comparing this to the right hand side of Eq. (2.9.17), we have $2\pi\, \hbar\, |C|^2 = 1$, which gives

$$C = \frac{1}{\sqrt{h}}\,, \tag{2.9.19}$$

where we have chosen the normalization constant to be real and positive. Similarly, the normalization condition $\langle k|\, k'\rangle = \delta(k - k')$ gives

$$C' = \frac{1}{\sqrt{2\pi}}\,. \tag{2.9.20}$$

2.10 Relation between the Coordinate and Momentum Representations

For simplicity, consider a system with one degree of freedom. Then the coordinate representative of a state $|A\rangle$ is given by the wave function $\Psi(x) = \langle x|A\rangle$ and the momentum representative of the same state is given by $\Phi(k) = \langle k|A\rangle$, where we use $k = \frac{p}{\hbar}$ to represent the continuously varying eigenvalue of the momentum observable. Now the momentum representative of the state can be written as

$$\Phi(k) = \langle k|\, A\rangle = \int \langle k|\, x\rangle\, dx\, \langle x|\, A\rangle = \frac{1}{\sqrt{2\pi}} \int \psi(x)\, e^{-ikx} dx\,. \tag{2.10.1}$$

Thus the momentum representative is the Fourier transform of the coordinate representative.

Similarly, the coordinate representative can be written as

$$\Psi(x) = \langle x|\, A\rangle = \int \langle x|\, k\rangle\, dk\, \langle k|\, A\rangle = \frac{1}{\sqrt{2\pi}} \int \Phi(k)\, e^{ikx} dx\,. \tag{2.10.2}$$

Thus the coordinate representative is the inverse Fourier transform of the momentum representative. If we use p to denote the continuously varying momentum eigenvalue, then we can rewrite these very relations as

$$\Phi(p) = \frac{1}{\sqrt{h}} \int \Psi(x)\, e^{-ipx/\hbar} dx\,, \tag{2.10.3}$$

and

$$\Psi(x) = \frac{1}{\sqrt{h}} \int \Phi(p)\, e^{ipx/\hbar} dp\,. \tag{2.10.4}$$

For a system with three degrees of freedom, we can take the three commuting coordinate observables to be $\hat{x}\,, \hat{y}\,, \hat{z}$. Then the coordinate representative of the state $|A\rangle$ is the wave function $\Psi_A(\boldsymbol{r}) \equiv \Psi_A(x, y, z) = \langle x, y, z|A\rangle$.

For the momentum representation, we take the three commuting momentum observables to be $\hat{p}_x\,, \hat{p}_y\,, \hat{p}_z$ (or $\hat{k}_x, \hat{k}_y, \hat{k}_z$). Then the state $|A\rangle$ is represented by the function $\Phi_A(\boldsymbol{k}) \equiv \Phi_A(k_x, k_y, k_z) = \langle k_x, k_y, k_z|\, A\rangle$. Using the resolution of the unit operator in the coordinate representation, we can express the momentum representative of state $|A\rangle$ as

$$\begin{aligned}
\Phi_A(\boldsymbol{k}) &= \iiint \langle k_x, k_y, k_z|\, x, y, z\rangle\, dxdydz\, \langle x, y, z|\, A\rangle \\
&= \iiint \langle k_x|\, x\rangle\, \langle k_y|\, y\rangle\, \langle k_z|\, z\rangle\, dxdydz \Psi_A(x, y, z) \\
&= \frac{1}{(2\pi)^{3/2}} \iiint e^{-i\boldsymbol{k}\cdot\boldsymbol{r}} \Psi_A(\boldsymbol{r}) d^3 r\,.
\end{aligned} \tag{2.10.5}$$

This equation shows that the momentum representatives of state $|A\rangle$ is the three-dimensional Fourier transform of the coordinate representative of the same state. Likewise, the coordinate representative of state $|A\rangle$

$$\begin{aligned}
\Psi_A(\boldsymbol{r}) &= \iiint \langle x, y, z|\, k_x, k_y, k_z\rangle\, dk_x dk_y dk_z\, \langle k_x, k_y, k_z|\, A\rangle \\
&= \frac{1}{(2\pi)^{3/2}} \iiint e^{i\boldsymbol{k}\cdot\boldsymbol{r}} \Phi_A(\boldsymbol{k}) d^3 k\,,
\end{aligned} \tag{2.10.6}$$

is found to be the three-dimensional inverse Fourier transform of its momentum representative.

If we choose to express the continuously varying eigenvalues of momentum observables by $(p_x, p_y, p_z) = \boldsymbol{p}$ instead of $(\boldsymbol{k}_x, \boldsymbol{k}_y, \boldsymbol{k}_z) = \boldsymbol{k}$, then these relations can be written as

$$\Phi(\boldsymbol{p}) = \frac{1}{h^{3/2}} \iiint e^{-i\boldsymbol{p}\cdot\boldsymbol{r}/\hbar} \Phi(\boldsymbol{r}) d^3 r\,, \tag{2.10.7}$$

$$\Psi(\boldsymbol{r}) = \frac{1}{h^{3/2}} \iiint e^{i\boldsymbol{p}\cdot\boldsymbol{r}/\hbar} \Phi(\boldsymbol{p}) d^3 p\,. \tag{2.10.8}$$

These relations can be generalized to a system with f degrees of freedom in a straightforward manner:

$$\begin{aligned}\Phi_A(p_1, p_2, \cdots, p_f) &\equiv \langle p_1, p_2, \cdots, p_f | A\rangle \\ &= \frac{1}{\hbar^{f/2}} \int \cdots \int e^{-i(p_1 q_1 + p_2 q_2 + \cdots + p_f q_f)/\hbar} \Psi(q_1, q_2, \cdots, q_f) dq_1 dq_2 \cdots dq_f \,, \end{aligned} \tag{2.10.9}$$

$$\begin{aligned}\Psi(q_1, q_2, \cdots, q_f) &\equiv \langle q_1, q_2, \cdots, q_f | A\rangle \\ &= \frac{1}{\hbar^{f/2}} \int \cdots \int e^{(p_1 q_1 + p_2 q_2 + \cdots + p_f q_f)/\hbar} \Phi(p_1, p_2, \cdots, p_f) dp_1 dp_2 \cdots dp_f \,. \end{aligned} \tag{2.10.10}$$

Comparison of Coordinate and Momentum Representations

Although the coordinate and momentum representations are different, there are some similarities. The table below compares the two representations.

Coordinate Representation	**Momentum Representation**
1. In the coordinate representation, the position observables $\hat{q}_1, \hat{q}_2, \cdots, \hat{q}_f$ are taken to be diagonal and the basis states are taken to be the simultaneous eigenstates $\vert q_1, q_2, \cdots, q_f\rangle$ of these observables, where $q_1, q_2, \cdots, q_f$ represent the continuously varying eigenvalues of these observables. So the basis states are not denumerable, but form a continuous set of states.	1. In the momentum representation, the mo mentum observables $\hat{p}_1, \hat{p}_2, \cdots, \hat{p}_f$ are taken t be diagonal and the basis states are taken t be the simultaneous eigenstates $\vert p_1, p_2, \cdots p_f$ of these observables, where $p_1, p_2, \cdots p_f$ rep resent the continuously varying eigenvalue of the momentum observables. So the basi states are not enumerable, but form a contin uous set of states.
2. An arbitary state $\vert X\rangle$ is represented by the set of numbers $\langle q_1, q_2, \cdots, q_f \vert X\rangle \equiv \Psi(q_1, q_2, \cdots, q_f)$ which vary continuously with the eigenvalues of the position observables. So $\Psi(q_1, q_2, \cdots, q_f)$ may be looked upon as a function of the eigenvalues. It is generally referred to as the wave function of the system in the coordinate space.	2. An arbitary state $\vert X\rangle$ is represente by the set of numbers $\langle p_1, p_2, \cdots, p_f \vert X\rangle \equiv$ $\Phi(p_1, p_2, \cdots, p_f)$ which vary continuously wit the eigenvalues of the momentum observables So $\Phi(p_1, p_2, \cdots, p_f)$ may be looked upon as function of the eigenvalues. It is generally re ferred to as the wave function of the syste in the momentum space.
3. The expression $\vert\Psi(q_1, \cdots q_f)\vert^2 dq_1 \cdots dq_f$ may be interpreted as the probability that a measurement of the observables $\hat{q}_1, \hat{q}_2, \cdots \hat{q}_f$ on the system in state $\vert X\rangle$ (or Ψ) will lead to results between q_1 and $q_1 + dq_1$, q_2 and $q_2 + dq_2 \cdots$, and $q_f + dq_f$. The normalization of the state $\langle X\vert X\rangle = 1$ implies $\int \cdots \int \vert\Psi(q_1, \cdots, q_f)\vert^2 dq_1 \cdots dq_f = 1$.	3. The expression $\vert\Phi(p_1, \cdots p_f)\vert^2 dp_1 \cdots dp$ may be interpreted as the probability that measurement of the observables $\hat{p}_1, \hat{p}_2, \cdots \hat{p}$ on the system in state $\vert X\rangle$ (or Φ) will lea to results between p_1 and $p_1 + dp_1$, p_2 an $p_2 + dp_2 \cdots$, and $p_f + dp_f$. The normaliza tion of the state $\langle X\vert X\rangle = 1$ implies tha $\int \cdots \int \vert\Phi(p_1, \cdots, p_f\vert^2 dp_1 \cdots dp_f = 1$

Problems

1. A two-dimensional vector space is spanned by two orthogonal vectors $|1\rangle$ and $|2\rangle$. A linear operator $\hat{A}$ has the following effect on these vectors:

$$\hat{A}\,|1\rangle = 2\,|1\rangle + i\sqrt{2}\,|2\rangle\,,$$
$$\hat{A}\,|2\rangle = -i\sqrt{2}\,|1\rangle + 3\,|2\rangle\,.$$

   (a) Obtain the matrix representation of the operator $\hat{A}$ in this basis. Is this matrix Hermitian? If so, find the eigenvalues and eigenvectors of this matrix.

   (b) Express the eigenstates $|x_1\rangle$ and $|x_2\rangle$ of $\hat{A}$ in terms of the basis vectors $|1\rangle$ and $|2\rangle$. For this first express the column matrices representing the states $|x_1\rangle$ and $|x_2\rangle$ in terms of those representing the basis states.

   (c) Write out the projection operators that will project the eigenstates $|x_1\rangle$ and $|x_2\rangle$.

   (d) Obtain the matrices representing these projection operators in the basis spanned by the basis states. Show that these matrices satisfy the closure condition.

2. Prove that $[\hat{x},\,\hat{f}(\hat{p})] = i\hbar\dfrac{d\hat{f}(\hat{p})}{d\hat{p}}$, where $\dfrac{d\hat{f}(\hat{p})}{d\hat{p}}$ means differentiating the function $f(p)$ with respect to p and replacing the variable p by the linear operator $\hat{p}$.

   [Hint: Following Dirac, $\hat{x} = i\hbar\dfrac{d}{d\hat{p}}$ where $\dfrac{d}{d\hat{p}}$ can be treated as a linear operator with the property $\langle p|\,\dfrac{d}{d\hat{p}}\,|A\rangle = \dfrac{\partial}{\partial p}\,\langle p|\,A\rangle\,.$]

3. Prove that $[\hat{F}(\hat{x}),\,\hat{p}] = i\hbar\dfrac{d\hat{F}(\hat{x})}{d\hat{x}}$, where $\dfrac{d\hat{F}(\hat{x})}{d\hat{x}}$ means differentiating the function $F(x)$ with respect to x and replacing the variable x by the linear operator $\hat{x}$.

   [Hint: Following Dirac, $\hat{p} = -i\hbar\dfrac{d}{d\hat{x}}$ where $\dfrac{d}{d\hat{x}}$ can be treated as a linear operator with the property $\langle x|\,\dfrac{d}{d\hat{x}}\,|A\rangle = \dfrac{\partial}{\partial x}\,\langle x|\,A\rangle\,.$]

4. Show that in a representation in which a set of commuting observables $\hat{\xi}$, $\hat{\eta}$, $\hat{\zeta}$ are diagonal, the operators $\hat{\xi}^2$, $\hat{\eta}^2$, $\hat{\zeta}^2$ and $\hat{\zeta}\,\hat{\eta}$ are also represented by diagonal matrices.

5. If the Hamiltonian of a system

$$\hat{H} = \frac{\hat{p}^2}{2\mu} + \hat{V}(\hat{x},\hat{y},\hat{z})\,,$$

   admits a set of eigenstates $|n\rangle$ with energies $E_n\,(n = 1,2,\cdots,N)$ show that

$$\sum_n (E_n - E_m)|x_{nm}|^2 = \frac{\hbar^2}{2\mu}\,,$$

   where the summation is over all the eigenstates of $\hat{H}$ and $\hat{x}$ is a Cartesian component of $\hat{\mathbf{r}}$ with $x_{nm} = \langle n|\,\hat{x}\,|m\rangle$ is an element of the matrix representing $\hat{x}$ in the representation in which $\hat{H}$ is diagonal.

[Hint: Use the identity $\hat{x}^2\hat{H} - 2\hat{x}\hat{H}\hat{x} + \hat{H}\hat{x}^2 = [\hat{x},\, [\hat{x},\, \hat{H}]]$ and evaluate the commutator bracket on the right hand side using the basic commutation relations. Then take expectation value of both sides in state $|m\rangle$.]

6. An element of the matrix, representing an observable $\hat{\alpha} = \alpha(\hat{q}, -i\hbar\partial/\partial\hat{q})$ in the $\hat{\xi}, \hat{\eta}, \hat{\zeta}\cdots$ representation, is given by $\alpha_{ps} = \langle \xi^{(p)} | \, \hat{\alpha} \, | \xi^{(s)} \rangle$, where the basis states $|\xi^{(s)}\rangle \equiv |\xi^{(s)}, \eta^{(s)}, \zeta^{(s)}, \ldots\rangle$. If the basis states $|\xi^{(s)}\rangle$ are represented by the wave functions $\Psi_s(\boldsymbol{r}) = \langle \boldsymbol{r} | \xi^{(s)} \rangle$ in the coordinate representation, re-express this matrix element in the integral form in terms of the wave functions.

   [Hint: Use Eq. (2.6.5)].

7. Show that the eigenvalues of a matrix are not changed if it is subjected to a unitary transformation.

8. In problem (5) show also that

$$\sum_m (E_n - E_m)\, |\boldsymbol{p}_{mn}|^2 = \frac{1}{2} \langle n | \, \hat{p}^2\hat{V} + \hat{V}\hat{p}^2 - 2\hat{\boldsymbol{p}}\hat{V}\cdot\hat{\boldsymbol{p}} \, | n \rangle$$

$$= -\frac{\hbar^2}{2} \int \psi_n^*(\mathbf{r}) \left(\nabla^2 V(\mathbf{r})\right) \psi_n(\boldsymbol{r})\, d\tau$$

   where $|\boldsymbol{p}_{mn}|^2 = |\langle m | \, \hat{\boldsymbol{p}} \, | n \rangle|^2$ and $\hat{\boldsymbol{p}}$ is the momentum operator $-i\hbar\boldsymbol{\nabla}$.

   [Hint: Since $\left[\hat{\boldsymbol{p}}, \hat{H}\right] = \left[\hat{\boldsymbol{p}}, \hat{V}\right]$, we have $(E_n - E_m)\,\langle m | \, \hat{\boldsymbol{p}} \, | n \rangle = \langle m | \, [\hat{\boldsymbol{p}},\, \hat{V}] \, | n \rangle$. Pre-multiply both sides scalarly by $\langle n | \, \hat{\boldsymbol{p}} \, | m \rangle$ and sum over m to get

$$\sum_m (E_n - E_m)\, |\boldsymbol{p}_{mn}|^2 = \langle n | \, \hat{p}^2\hat{V} - \hat{\boldsymbol{p}}\hat{V}\cdot\hat{\boldsymbol{p}} \, | n \rangle = \langle n | \, \hat{V}\hat{p}^2 - \hat{\boldsymbol{p}}\hat{V}\cdot\hat{\boldsymbol{p}} \, | n \rangle$$

$$= \frac{1}{2} \langle n | \, \hat{p}^2\hat{V} + \hat{V}\hat{p}^2 - 2\hat{\boldsymbol{p}}\hat{V}\cdot\hat{\boldsymbol{p}} \, | n \rangle$$

   Use Eq. (2.6.5) to write the right hand side in the integral form and simplify.]

9. Show that the matrix representing a unitary operator $\hat{U}$ $[\hat{U}^{-1} \equiv \hat{U}^\dagger]$ in any representation is a unitary matrix. What is the significance of unitary transformations in quantum mechanics?

10. A certain observable is represented in a certain representation, by the following matrix

$$M = \frac{1}{\sqrt{2}} \begin{pmatrix} 0 & 1 & 0 \\ 1 & 0 & 1 \\ 0 & 1 & 0 \end{pmatrix}.$$

    Find the eigenvalues and normalized eigenvectors of this matrix. Does any physical quantity correspond to this operator?

11. The eigen kets $|\xi^{(n)}\rangle$ of an observable, belonging to different eigenvalues $\xi^{(n)}$ $(n = 1, 2, \cdots N)$ are orthogonal in the sense that

$$\left\langle \xi^{(m)} \middle| \xi^{(n)} \right\rangle = \delta_{mn}\,.$$

    If these states be represented by a wave function in the coordinate representation (consider the system to have three degrees of freedom), then express this orthogonality condition in terms of the wave functions.

12. The complete set of eigenstates $\left|\xi^{(n)}\right\rangle$ $(n = 1, 2, \cdots N)$ of an observable satisfy the completeness criterion

$$\sum_{n=1}^{N} \left|\xi^{(n)}\right\rangle \left\langle \xi^{(n)}\right| = \hat{I}\,.$$

If this set of states is represented by the set of wave functions $\psi_n(\boldsymbol{r}) = \left\langle \boldsymbol{r}\right| \xi^{(n)}\rangle$ in the coordinate representation (assume the system to have three degrees of freedom), then re-express the completeness criterion in terms of wave functions.

13. In problems (11) and (12), write the orthogonality condition and the completeness criterion in the coordinate representation when the number of degrees of freedom is f and the set of commuting coordinate observables are $\hat{q}_1,\ \hat{q}_2,\ \hat{q}_3,\ \cdots, \hat{q}_f$.

14. Find the momentum representative of the plane wave state which is expressed by the wave function

$$\Psi_{k_0}(\boldsymbol{r}) = \frac{1}{\sqrt{2\pi}}\, e^{i\boldsymbol{k}_0\cdot\boldsymbol{r}}\,,$$

in the coordinate representation.

15. The ground state wave function of the Hydrogen atom is represented by

$$\langle \boldsymbol{r}|\, n = 1, \ell = 1, m = 0\rangle \equiv \Psi_{100}(\boldsymbol{r}) = \frac{1}{\sqrt{\pi a_o^3}}\, e^{-r/a_o}\,.$$

Find the corresponding wave function in the momentum space.
[Ans: $\phi(p) = (2\pi a_o)^{3/2}\pi^{-5/2}\hbar^{5/2}/(a_o^2p^2 + \hbar^2)^2$]

16. A linear harmonic oscillator in its ground state has the wave function

$$\psi(x) = \left(\frac{1}{a\sqrt{\pi}}\right)^{1/2} e^{-x^2/2a^2}\,,$$

in the coordinate space. Show that its momentum representative is

$$\phi(p) = \left(\frac{a}{\hbar\sqrt{\pi}}\right)^{1/2} e^{-a^2p^2/2\hbar^2}\,.$$

17. A particle is in a state described by the wave function

$$\psi(x) = \frac{1}{(2\pi\sigma^2)^{1/4}}\, e^{-x^2/4\sigma^2}\, \exp(ik_o x)$$

at time $t = 0$ in the coordinate space.

(a) Find the corresponding wave function $\phi(p)$ in the momentum space.

(b) Determine the expectation values of $\hat{x}$ and $\hat{x}^2$ for this state,

$$\langle x\rangle = \int_{-\infty}^{\infty} dx\psi^*(x)x\psi(x) \quad \text{and} \quad \langle x^2\rangle = \int_{-\infty}^{\infty} dx\psi^*(x)x^2\psi(x)$$

and hence the quantity $\Delta x = \sqrt{\langle x^2\rangle - \langle x\rangle^2}$ using $\psi(x)$.

(c) Determine the expectation values of $\hat{p}$ and $\hat{p}^2$ for this state

$$\langle p \rangle = \int_{-\infty}^{\infty} dp \psi^*(p) p \psi(p) \quad \text{and} \quad \langle p^2 \rangle = \int_{-\infty}^{\infty} dp \psi^*(p) p^2 \psi(p)$$

and hence the quantity $\Delta p = \sqrt{\langle p^2 \rangle - \langle p \rangle^2}$.

(d) Find the product $(\Delta x)(\Delta p)$ for this state.

18. Show that the eigenvalues of the matrix representing an observable $\hat{\alpha}$ in any representation are the same as the eigenvalues of $\hat{\alpha}$ itself.

19. The $2s$ state of the Hydrogen atom has the wave function

$$\Psi_{200}(\boldsymbol{r}) = \frac{1}{\sqrt{32\pi a_o^3}}(2-\rho)e^{-\rho/2} .$$

where ρ is equal to r/a_o and a_o = Bohr radius. Find the momentum representative of this state.

20. The ground state wave function of a linear harmonic oscillator is given by

$$\psi(x) = N_o \, e^{-x^2\sqrt{mk}/2\hbar} .$$

Show that its ground state wave function in the momentum space is

$$\phi(p) = M_o \, e^{-p^2/2\hbar\sqrt{mk}} .$$

Determine the normalization constants N_o and M_o.

21. Three (2×2) matrices $\sigma_x, \sigma_y, \sigma_z$ satisfy the following equations:

$$\begin{aligned} \sigma_x^2 = \sigma_2^2 = \sigma_3^2 &= I , \\ \sigma_x\sigma_y - \sigma_y\sigma_x &= 2i\sigma_z , \\ \sigma_y\sigma_z - \sigma_z\sigma_y &= 2i\sigma_x , \\ \sigma_z\sigma_x - \sigma_x\sigma_z &= 2i\sigma_y . \end{aligned}$$

Find the three matrices if σ_z is to be diagonal.

22. The state of a particle with one degree of freedom is described by the wave function

$$\psi(x) = \phi(x)\exp(-ip_o x/\hbar) ,$$

where the function $\phi(x)$ is real. What is the physical meaning of the quantity p_o?

23. Normalize the following function in momentum space

$$\phi(\boldsymbol{p}) = N\exp(-\alpha p/\hbar) , \quad p = |\boldsymbol{p}|$$

and show that its coordinate representative is

$$\psi(\boldsymbol{r}) = \frac{(2\alpha)^{3/2}}{\pi}\frac{\alpha}{(\alpha^2 + r^2)} .$$

24. A wave packet may be looked upon as a result of superposition of plane continuous waves with wavelength λ (or wave numbers $k = 2\pi/\lambda$) varying in a certain range. Let a wave packet at $t = 0$ be given by

$$\Psi(x,0) = \frac{1}{\sqrt{2\pi}} \int_{-\infty}^{\infty} dk\phi(k)e^{ikx} ,$$

where $\phi(k)$ has a Gaussian shape

$$\phi(k) = a_o e^{-(k-k_0)^2\sigma^2} ,$$

$\psi(x)$ and $\phi(k)$ being the coordinate and momentum representatives of the same state. Evaluate $|\psi(x,0)|^2$ and $|\phi(k)|^2$ and estimate the product of maximum uncertainties in the measurements of x and p in the state of a particle represented by the wave packet at $t = 0$. Given: $p = h/\lambda = \hbar k$.

25. Show that
(i) the eigenvalues of a Hermitian matrix are real
(ii) the eigenvalues of a unitary matrix are uni-modular
(iii) eigenvalues of an orthogonal matrix are ± 1

26. Show that the eigenvalues of a matrix are invariant under a unitary transformation.

27. Show that the trace of a matrix is unchanged under a unitary transformation.

28. Show that the eigenvector X_k and X_ℓ of a matrix M, belonging to different eigenvalues λ_k and λ_ℓ are orthogonal in the sense that $X_k^\dagger X_\ell = 0$ if $\lambda_k \neq \lambda_\ell$, where $X_k^\dagger$ represents the complex conjugate of the transpose of matrix X_k.

29. At a given time the normalized wave-function of a particle moving along x direction is given by

$$\Psi(x) = \frac{1}{(\sigma^2\pi)^{1/4}} e^{-x^2/2\sigma^2} e^{ipx/\hbar} .$$

For this state find

(a) the most probable location of the particle
(b) the expectation value $\langle x \rangle$ for x
(c) the expectation value of its momentum
(d) $\langle p^2 \rangle$ and $\langle x^2 \rangle$ for this state and hence $(\Delta x)^2 \equiv \langle x^2 \rangle - \langle x \rangle^2$ and $(\Delta p)^2 \equiv \langle p^2 \rangle - \langle p \rangle^2$

30. The state of an oscillator of angular frequency ω is represented by the wave function:

$$\Psi(x) = e^{-m\omega x^2/\hbar} .$$

Find the momentum representative of this state. What are the quantum mechanical expectation values of its momentum and position?

Find the probability that the magnitude of momentum is larger than $(m\hbar\omega)^{1/2}$.

31. Define Dirac delta function $\delta(x)$. Show that

$$\int_{-x_1}^{x_2} dx f(x)\delta(x) = f(0) ,$$

where x_1 and x_2 are arbitrary positive numbers.

32. Show that

$$\delta(x^2 - a^2) = \frac{1}{2|a|}\left[\delta(x-a) + \delta(x+a)\right] .$$

33. Show that

$$\delta(f(x)) = \frac{\delta(x - x_0)}{\left|\frac{\partial f}{\partial x}\right|_{x=x_0}} .$$

where x_0 is the zero of $f(x)$.

34. Show that

$$\delta'(x) = -\frac{\delta(x)}{x} .$$

35. In a three-dimensional vector space spanned by orthonormal basis vectors $|1\rangle$, $|2\rangle$, $|3\rangle$ a linear operation $\hat{R}$ has the following effect on these vectors:

$$\hat{R}\,|1\rangle = |1\rangle , \quad \hat{R}\,|2\rangle = |3\rangle , \quad \text{and} \quad \hat{R}\,|3\rangle = -\,|2\rangle .$$

(a) Write the matrix representation of $\hat{R}$ in this basis.

(b) Calculate the eigenvalues and eigenvectors of $\hat{R}$.

36. The ground state of the Hydrogen atom is represented by the wave function

$$\Psi_{100}(\boldsymbol{r}) = \frac{1}{\sqrt{\pi a_o^3}} e^{-r/a_o} ,$$

where a_o = Bohr radius. Find:
(a) the quantum mechanical expectation value of r: $\langle r\rangle_{100}$
(b) the most probable value of r
(c) the root-mean square value of r: $\left[\langle r^2\rangle_{100}\right]^{1/2}$

37. The first excited state of the Hydrogen atom is represented by

$$\Psi_{200}(\boldsymbol{r}) = \frac{1}{\sqrt{32\pi a_o^3}}(r/a_o - 2)e^{-r/2a_o} .$$

For this state find
(a) the quantum mechanical expectation value of r
(b) the most probable value of r
(c) the root mean square value of r

38. If $\hat{\alpha}^\dagger$ is the adjoint of a linear operator $\hat{\alpha}$, then show that

$$(\langle P|\,\hat{\alpha}\,|Q\rangle)^* = \langle Q|\,\hat{\alpha}^\dagger\,|P\rangle .$$

Also show that, in the coordinate representation this criterion is expressed as

$$\int (\alpha_q \phi_1(q))^* \; \phi_2(q)\, d\tau \;=\; \int \varphi_1^*(q)\, \alpha_q^\dagger\, \phi_2(q)\, d\tau$$

where $\phi_1(q) = \langle q|\,Q\rangle$ and $\phi_2(q) = \langle q|\,P\rangle$ are the coordinate representatives of the states $|Q\rangle$ and $|P\rangle$ and α_q and $\alpha_q^\dagger$ are, respectively, the coordinate representatives of the linear operators $\hat{\alpha}$ and $\hat{\alpha}^\dagger$ given by

$$\langle q|\,\hat{\alpha}\,|Q\rangle = \alpha_q\,\langle q|\,Q\rangle = \alpha_q\phi_1(q)$$

and

$$\langle q|\,\hat{\alpha}^\dagger\;|P\rangle = \alpha_q^\dagger\;\langle q|\,P\rangle = \alpha_q^\dagger\phi_2(q) .$$

Express the condition for $\hat{\alpha}$ to be a self-adjoint operator (or α_q to be a Hermitian operator).

39. Show that for any state of the Hydrogen atom

$$\langle T\rangle = -\frac{1}{2}\langle V\rangle$$

where T and V represent, respectively, the kinetic energy and the Coulomb potential energy in the Hamiltonian. Check this for the ground state and the first excited state of the Hydrogen atom. The relevant wave functions are given in problems (36) and (37).

40. In dealing with the coordinate representation of the Schrödinger equation in three dimensions it is convenient to introduce a radial momentum operator. Show that the radial momentum operator

$$\hat{p}_r = \frac{1}{2}\left[\hat{r}^{-1}\hat{\boldsymbol{r}}\cdot\hat{\boldsymbol{p}} + \hat{\boldsymbol{p}}\cdot\hat{\boldsymbol{r}}\hat{r}^{-1}\right],$$

satisfies the following commutation relation

$$[\hat{x},\hat{p}_r] = i\hbar\frac{\hat{x}}{\hat{r}},\ [\hat{y},\hat{p}_r] = i\hbar\frac{\hat{y}}{\hat{r}},\ [\hat{z},\hat{p}_r] = i\hbar\frac{\hat{z}}{\hat{r}},\ [\hat{r},\hat{p}_r] = i\hbar$$

where the operator $\hat{r}$ is defined by $\hat{r}^2 = \hat{\boldsymbol{r}}\cdot\hat{\boldsymbol{r}}$ and $\hat{r}\hat{r}^{-1} = 1 = \hat{r}^{-1}\hat{r}$.

[Hint: Express $\hat{p}_r$ in terms of Cartesian observables $\hat{x}$, $\hat{p}_x$, etc. and use their commutation relations. Finally, to establish the last relation use $[f(\hat{r}),\hat{\boldsymbol{p}}] = i\hbar\boldsymbol{\nabla}f(\hat{r})$ (see problems 2 and 3)].

Using the coordinate representative of radial momentum operator

$$p_r = -i\hbar\frac{1}{r}\frac{\partial}{\partial r}r = -i\hbar\left(\frac{\partial}{\partial r}+\frac{1}{r}\right),$$

show that it is Hermitian provided that the wave functions satisfy the following constraint

$$\lim_{r\to 0} r\psi(\boldsymbol{r}) = 0\,.$$

Why is $\hat{p}_r$ not an observable?

[Hint: Solve the eigenvalue problem for p_r in the coordinate representation with the constraint just stated.]

References

[1] P. A. M. Dirac, *Principles of Quantum Mechanics*, Third Edition (Clarendon Press, Oxford, 1971), Chapters III and IV.

[2] C. Cohen-Tannoudji, B. Diu, and F. Laloë, *Quantum Mechanics*, Vol. I (John Wiley & Sons, New York, 1977).

[3] T. F. Jordan, *Quantum Mechanics in Simple Matrix Form* (Dover Publications, New York, 2007).

3

EQUATIONS OF MOTION

3.1 Schrödinger Equation of Motion

We shall now consider the connection between the state of a system at one time and that at a subsequent time, provided that, in between, the state is not disturbed by an observation. It may be recalled that making an observation on a dynamical system will, in general, disturb the state of the system unless the state happens to be an eigenstate of the observable pertaining to the observation. If the state is not disturbed in between, the time evolution of the state of the system is governed by an equation of motion which would enable one to determine the state at a later time if the state at an earlier time is known. Such an equation of motion would enable one to get a complete dynamical picture of the system.

To determine such an equation of motion, we are guided by two principles:

1. The principle of superposition of states holds good at all times. This means that if the state of the system initially can be regarded as a superposition of other states, then the superposition relationship holds throughout the time evolution of the state unless it is disturbed by an observation.

2. The length or the *norm* of the ket vector representing the state of the system is preserved at all times.

According to the first requirement, if the initial state is expresssed as $|R(t_0)\rangle = C_1\,|A(t_0)\rangle + C_2\,|B(t_0)\rangle$, then the subsequent state can be expressed as $|R(t)\rangle = C_1\,|A(t)\rangle + C_2\,|B(t)\rangle$. Now this is possible only if the state at time t results from the operation of a linear (time-dependent) operator $\hat{T}$ on the state at t_0:

$$|R(t)\rangle = \hat{T}\,|R(t_o)\rangle\ , \tag{3.1.1}$$

where the operator $\hat{T}$ cannot depend on the state in question (to preserve linearity). It can, however, depend on the initial and final times t_o and t and the nature of the system. We will make this explicit by writing $\hat{T} \equiv \hat{T}(t, t_o)$. It is obvious that $\hat{T}(t_o, t_0) = \hat{1}$ is the identity operator.

The second requirement, $\langle R(t)|R(t)\rangle = \langle R(t_0)|R(t_0)\rangle$, implies that the operator $\hat{T}$ must be unitary:

$$\hat{T}^\dagger\hat{T} = \hat{T}\hat{T}^\dagger = \hat{1}\,. \tag{3.1.2}$$

Now consider an infinitesimal time translation operation when $t{\to}t_o$ or $\Delta t \equiv\!\to 0$. For physical continuity, we assume that the limit

$$\lim_{t\to t_o}\frac{|R(t)\rangle - |R(t_o)\rangle}{t - t_o} = \lim_{\Delta t\to 0}\frac{|R(t_o+\Delta t)\rangle - |R(t_o)\rangle}{\Delta t}$$

exists and can be represented by $\left[\frac{d}{dt}\,|R(t)\rangle\right]_{t=t_o}$. By virtue of Eq. (3.1.1) this derivative is given by

$$\left[\frac{d}{dt}\,|R(t)\rangle\right]_{t=t_o} = \lim_{t\to t_o}\frac{\hat{T}(t,t_o) - \hat{1}}{t - t_o}\,|R(t_o)\rangle \equiv \hat{\alpha}\,|R(t_o)\rangle\ , \tag{3.1.3}$$

where the liner operator $\hat{\alpha}$ is given by

$$\hat{\alpha} \equiv \lim_{t \to t_o} \frac{\hat{T(t, t_o)} - \hat{1}}{t - t_o} . \tag{3.1.4}$$

It can be seen that $\hat{\alpha}$ is a pure imaginary operator and, therefore, $i\hbar\,\hat{\alpha}$ is a Hermitian (or self-adjoint) operator.[1]

Now there are reasons for identifying the operator $i\hbar\,\hat{\alpha}$ with the Hamiltonian $\hat{H}$ of the system and we shall discuss them shortly. If we make this identification, Eq. (3.1.3) can be written as

$$i\hbar \frac{d}{dt} \left| R(t) \right\rangle = \hat{H} \left| R(t) \right\rangle , \tag{3.1.5}$$

where we have dropped the label t_o since t_o is an arbitrary initial time. Equation (3.1.5), called the *Schrödinger equation of motion*, gives a causal connection between the state of the system at some initial time and at a subsequent time. The justifications for identifying the operator $i\hbar\,\hat{\alpha}$, with $\hat{H}$ are as follows.

1. Theory of special relativity puts (E/c^2) in the same relation to time t as momenta p_x, p_y, p_z are to position coordinates x, y, z. In other words, $p_x, p_y, p_z, E/c^2$ transform exactly as x, y, z, t do, or $p_x, p_y, p_z, iE/c$ form a four-vector just as x, y, z and ict do. Now in quantum mechanics p_x, p_y, and p_z are replaced by $-i\hbar\frac{\partial}{\partial x}$, $-i\hbar\frac{\partial}{\partial y}$, $-i\hbar\frac{\partial}{\partial z}$. So iE/c must be replaced by $-i\hbar\frac{\partial}{\partial(ict)}$ or E by $i\hbar\frac{\partial}{\partial t}$. So the operator $i\hbar(\partial/\partial t)$ or $i\hbar\hat{\alpha}$ is essentially the energy operator – the Hamiltonian.

2. The Schrödinger equation of motion (3.1.5) refers to the so called *Schrödinger picture*, in which the states change with time and observables are independent of time. From this picture, we can go over to another picture called the *Heisenberg picutre*, in which the states are independent of time and the time dependence is passed on to the observables, called the Heisenberg observables [see Sec. 3.8]. In this picture, the equation of motion for an observable $\hat{F}^H$ is given by

$$\frac{d\hat{F}^H}{dt} = \frac{\partial \hat{F}^H}{\partial t} + \frac{1}{i\hbar}[\hat{F}^H, \hat{H}^H] . \tag{3.1.6}$$

This equation is analogous to the classical equation of motion for a real dynamical variables F_c:

$$\frac{dF_c}{dt} = \frac{\partial F_c}{\partial t} + \{F_c, H_c\} ,$$

where the curly bracket stands for classical Poisson Bracket and H_c is the classical Hamiltonian. We have seen that $1/i\hbar$ times the commutator bracket is the quantum analog of classical Poisson bracket. Now, since H_c is the classical Hamiltonian, the operator $\hat{H}^H$ could be identified as the Hamiltonian in the Heisenberg picture. Therefore, the operator $\hat{H}$ in Eq.(3.1.5) could be identified as the Hamiltonian in the Schrödinger picture[2]. When the system has a classical analog then we may also assume that the Hamiltonian operator occuring in the Schrödinger equation of motion (3.1.5) is the same function of the coordinate and momentum observables as its classical counterpart is of the canonical coordinates and momenta. However, as Dirac

[1]For small time translation, $t = t_o + \Delta t$, $(\Delta t \to 0)$, we can write $\hat{T} = \hat{\alpha}\Delta t + \hat{1}$ and $\hat{T}^\dagger = \alpha^\dagger \Delta t + 1$. Then the unitarity of $\hat{T}$ implies $(\hat{\alpha}\Delta t + \hat{1})(\hat{\alpha}^\dagger \Delta t + \hat{1}) = \hat{1}$. Expanding and neglecting terms of of order $(\Delta t)^2$, we get $(\hat{\alpha} + \hat{\alpha}^\dagger)\Delta t + \hat{1} = \hat{1}$, which leads to $\hat{\alpha}^\dagger = -\hat{\alpha}$.

[2]In fact, $\hat{H}^H$ is identical with $\hat{H}$.

has observed,[3] this assumption holds good only when the Hamiltonian is expressed in terms of Cartesian coordinates and momenta and not in terms of more general curvilinear coordinates and canonically conjugate momenta.

3.2 Schrödinger Equation in the Coordinate Representation

The Schrödinger equation of motion

$$i\hbar\frac{d}{dt}|R(t)\rangle = \hat{H}|R(t)\rangle \tag{3.2.1}$$

may easily be written in the coordinate representation. Consider the system to have three degrees of freedom and take the basis states to be the simultaneous eigenstates $|x, y, z\rangle$ of three commuting observables $\hat{x}, \hat{y}, \hat{z}$, where x, y, z represent the continuously varying eigenvalues of the coordinate observables. The state $|R(t)\rangle$ can be represented by the wave function $\Psi(x, y, z, t) = \langle x, y, z|R(t)\rangle$. Premultiplying Eq. (3.2.1) on both sides by $\langle x, y, z|$ and expressing the Hamiltonian explicitly, we obtain

$$i\hbar\frac{d}{dt}\langle x, y, z|\,R(t)\rangle = \langle x, y, z|\left[\frac{\hat{p}_x^2+\hat{p}_y^2+\hat{p}_z^2}{2m} + \hat{V}(\hat{x}, \hat{y}, \hat{z})\right]|R(t)\rangle$$

or

$$i\hbar\frac{d}{dt}\Psi(x, y, z, t) = \langle x, y, z|\left[-\frac{\hbar^2}{2m}\left(\frac{d^2}{d\hat{x}^2}+\frac{d^2}{d\hat{y}^2}+\frac{d^2}{d\hat{z}^2}\right) + \hat{V}(\hat{x}, \hat{y}, \hat{z})\right]|R(t)\rangle$$

$$= \left[-\frac{\hbar^2}{2m}\left(\frac{\partial^2}{\partial x^2}+\frac{\partial^2}{\partial y^2}+\frac{\partial^2}{\partial z^2}\right) + V(x, y, z)\right]\Psi(x, y, z, t)$$

where we have made use of Eq. (2.6.5). Writing $\boldsymbol{r}$ for x, y, z we obtain

$$i\hbar\frac{d}{dt}\Psi(\boldsymbol{r}, t) = \left[-\frac{\hbar^2}{2m}\nabla^2 + V(\boldsymbol{r})\right]\Psi(\boldsymbol{r}, t)\,. \tag{3.2.2}$$

This is the Schrödinger equation of motion in the coordinate representation for a system with three degrees of freedom.

We can as well write Schrödinger equation of motion in the coordinate representation for a system with many degrees of freedom, for example, for a system in which more than one particles are involved. This equation is

$$i\hbar\frac{d}{dt}\Psi(q_1, q_2, \cdots, q_f, t) = \mathcal{H}\Psi(q_1, q_2, \cdots, q_f, t) \tag{3.2.3}$$

where the differential operator $\mathcal{H}$, which is the coordinate representative of the Hamiltonian operator, can be obtained just by replacing the coordinate and momentum observables, occurring in the expression for the Hamiltonian operator $\hat{H}$ by $q_1, q_2, \cdots, q_f$ and $-i\hbar\frac{\partial}{\partial q_1}, -i\hbar\frac{\partial}{\partial q_2}, \cdots, -i\hbar\frac{\partial}{\partial q_f}$, respectively, where $q_1, q_2, \cdots, q_f$ are the continuously varying eigenvalues of the Cartesian coordinate observables of the system.

[3] See *Principles of Quantum Mechanics*, P. A. M. Dirac, Third Edition, footnote, page 144.

3.3 Equation of Continuity

We have given the coordinate representative $\Psi(\boldsymbol{r}\,,t) \equiv \langle r\,|R(t)\rangle$ of state $|R(t)\rangle$ the interpretation that

$$\rho(\boldsymbol{r},t)d^3r = \Psi^*(\boldsymbol{r},t)\Psi(\boldsymbol{r},t)d^3r \tag{3.3.1}$$

represents the probability that at time t, the particle will be found in the region of space $d^3r = r^2dr\,\sin\theta d\theta\, d\phi$ around the point $\boldsymbol{r}$ (Sec.2.5). So $\rho(\boldsymbol{r},t)$ may be interpreted as the probability density at time t and the normalization condition

$$\int \Psi^*(\boldsymbol{r},t)\Psi(\boldsymbol{r},t)d^3r = \int \rho(\boldsymbol{r},t)d^3r = 1\,, \tag{3.3.2}$$

expresses the notion that the particle will be found somewhere in the space available to the system. We now show that the time-dependent Schrödinger equation (3.2.2) leads to the *equation of continuity*, which expresses the conservation of probability everywhere in space accessible to the system. The time-dependent Schrödinger equation and its complex conjugate are given by

$$\begin{aligned} i\hbar\frac{\partial}{\partial t}\Psi(\boldsymbol{r},t) &= \left[-\frac{\hbar^2}{2m}\nabla^2 + V(\boldsymbol{r})\right]\Psi(\boldsymbol{r},t) \\ -i\hbar\frac{\partial}{\partial t}\Psi^*(\boldsymbol{r},t) &= \left[-\frac{\hbar^2}{2m}\nabla^2 + V(\boldsymbol{r})\right]\Psi^*(\boldsymbol{r},t)\,. \end{aligned} \tag{3.2.2*}$$

Premultiplying the first equation by $\Psi^*(\boldsymbol{r},t)$ and post-multiplying the second equation by $\Psi(\boldsymbol{r},t)$, and subtracting the results, we get

$$\frac{\partial}{\partial t}\Psi^*\Psi = -\frac{\hbar}{2im}\boldsymbol{\nabla}\cdot\left[\Psi^*(\boldsymbol{\nabla}\Psi) - (\boldsymbol{\nabla}\Psi)^*\Psi\right]\,. \tag{3.3.3}$$

If we interpret

$$\boldsymbol{S}(\boldsymbol{r},t) = \frac{\hbar}{2im}\left[\Psi^*(\boldsymbol{\nabla}\Psi) - (\boldsymbol{\nabla}\Psi)^*\Psi\right] = \mathcal{R}e\left[\frac{\hbar}{im}\Psi^*(\boldsymbol{\nabla}\Psi)\right] \tag{3.3.4}$$

as the *probability current density*, then Eq. (3.3.3) can be written as

$$\frac{\partial\rho(\boldsymbol{r},t)}{\partial t} + \boldsymbol{\nabla}\cdot\boldsymbol{S}(\boldsymbol{r},t) = 0\,. \tag{3.3.5}$$

This equation is of the same form as the equation continuity in electrodynamics

$$\frac{\partial\rho_{\rm ch}(\boldsymbol{r},t)}{\partial t} + \boldsymbol{\nabla}\cdot\boldsymbol{j}(\boldsymbol{r},t) = 0\,, \tag{3.3.6}$$

where $\rho_{ch}(\boldsymbol{r},t)$ is the electric charge density and $\boldsymbol{j}(\boldsymbol{r},t)$ is the electric current density. The equation of continuity in electrodynamics represents *conservation of charge*. Equation (3.3.5) is the quantum analog of the equation of continuity representing the conservation of probability (or the norm) $N{\equiv}\int\rho(\boldsymbol{r}\,,t)d^3r = 1$. Recalling that the divergence represents net probability current out of a small volume element, the equation of continuity implies that if the probability density $\rho(\boldsymbol{r}\,,t)$ in a certain region of space increases with time, then there must be a net in-flow of probability into this region ($\boldsymbol{\nabla}\cdot\boldsymbol{S} < 0$) and if the probability density decreases with time, then there must be a net out-flow of probability ($\boldsymbol{\nabla}\cdot\boldsymbol{S} > 0$).

3.4 Stationary States

We can relate the state of a system $|R(t)\rangle$ at time t to its state $|R(0)\rangle$ at time $t = 0$ through a linear (time evolution) operator $\hat{T}(t)$:

$$|R(t)\rangle = \hat{T}(t)|R(0)\rangle\,. \tag{3.4.1}$$

Substituting Eq. (3.4.1) into Schrödinger equation (3.1.5), we get

$$i\hbar\frac{d\hat{T}(t)}{dt} = \hat{H}\hat{T}(t)\,. \tag{3.4.2}$$

If the Hamiltonian $\hat{H}$ is independent of time (or $\hat{H}$ is invariant under the time translation operation), then Eq. (3.4.2) may be integrated to give

$$\hat{T}(t) = e^{-i\hat{H}t/h}\,. \tag{3.4.3}$$

It is easy to chaeck that the time evolution operator T commutes with the Hamiltonian:

$$[\hat{H}\,, \hat{T}(t)] = 0\,. \tag{3.4.4}$$

It follows from this equation that, if the system is initially ($t = 0$) in an eigenstate of the Hamiltonian, then at a later time also, it continues to be in the eigenstate of the Hamiltonian. To see this consider an eigenstate $|R_n(0)\rangle$ at $t = 0$ of the Hamiltonian belonging to the eigenvalue E_n

$$\hat{H}|R_n(0)\rangle = E_n|R_n(0)\rangle\,, \tag{3.4.5}$$

where the index n labels eigenstates and eigenvalues of the Hamiltonian. According to Eq. (3.4.1), this state evolves into state $|R_n(t)\rangle = \hat{T}\;|R_(0)\rangle$ such that

$$\begin{aligned}\hat{H}|R_n(t)\rangle &= \hat{H}\hat{T}|R_n(0)\rangle = \hat{T}\hat{H}\;|R_n(0)\rangle \\ &= E_n\hat{T}\;|R_n(0)\rangle = E_n\;|R_n(t)\rangle\;.\end{aligned}$$

Thus $|R_n(t)\rangle$ is also an eigenstate of $\hat{H}$ belonging to the same eigenvalue E_n. Moreover, the time evolution of an eigenstate is especially simple:

$$|R_n(t)\rangle = e^{-i\hat{H}t/\hbar}\;|R_n(0)\rangle = e^{-iE_nt/\hbar}\;|R_n(0)\rangle\;. \tag{3.4.6}$$

Such states as $|R_n(t)\rangle$, which systems with time-independent Hamiltonians admit and whose time dependence is given by Eq. (3.4.6), are called *stationary states*. The stationary states $|R_n(t)\rangle$ given by Eq. (3.4.6) are so called because (1) the probability density $\rho(\boldsymbol{r}, t)$ is independent of time and (2) the expectation value $\langle\hat{\alpha}\rangle = \langle R_n(t)|\hat{O}|R_n(t)\rangle$ of an observable $\hat{O}$ for such states is independent of time.

By substituting Eq. (3.4.6) in the time-dependent Schrödinger equation (3.1.5), we find that the time-independent part $|R_n(0)\rangle$ of the stationary state satisfies the time-independent Schrödinger equation:

$$\hat{H}|R_n(0)\rangle = E_n|R_n(0)\rangle\,. \tag{3.4.7}$$

Thus, if the Hamiltonian of a system is time-independent, then we can find the allowed energies (energy eigenvalues) and the corresponding states (energy eigenstates) by solving the time-independent Schrödinger equation (3.4.7), usually in some representation such as the coordinate representation. Since the time-independent Schrödinger equation also has the form of an eigenvalue equation, we will also refer to it as the Schrödinger eigenvalue equation as distinct from the time dependent Schrödinger equation. Once the possible energies E_n of the stationary states have been determined, the time dependence of these states is simply expressed via Eq. (3.4.6).

3.5 Time-independent Schrödinger Equation in the Coordinate Representation

The time-independent Schrödinger equation (3.4.7), since it is an eigenvalue equation for the Hamiltonian, may be rewritten as

$$\hat{H}\,|E_n\rangle = E_n\,|E_n\rangle\,, \tag{3.5.1}$$

where we have designated the eigenstates by their corresponding energy eigenvalues ($|E_n\rangle \equiv |R_n(0)\rangle$).

The form of equation (3.5.1) in the coordinate representation will depend on the number of degrees of freedom of the system. Consider a system with one degree of freedom. Such a system is characterized by one position observable $\hat{x}$ and one momentum observable $\hat{p}$ ($\equiv -i\hbar\frac{d}{d\hat{x}}$). The Hamiltonian for such a system is expressed as[4]

$$\hat{H} = \frac{\hat{p}^2}{2m} + \hat{V}(\hat{x}) = -\frac{\hbar^2}{2m}\frac{d^2}{d\hat{x}^2} + \hat{V}(\hat{x})\,. \tag{3.5.2}$$

To express this equation in the coordinate representation, we choose the basis states to be the eigenstates $\langle x|$ of the position observable $\hat{x}$, where x denotes the continuously varying eigenvalues of $\hat{x}$. Then the state $|E_n\rangle$ is represented by the wave function

$$\langle x|E_n\rangle = \Psi_n(x)\,. \tag{3.5.3}$$

Premultiply both sides of Eq. (3.5.1) by $\langle x|$ and using the expression for $\hat{H}$ given by Eq. (3.5.2), we get

$$\langle x|\; -\frac{\hbar^2}{2m}\frac{d^2}{d\hat{x}^2} + \hat{V}(\hat{x})\,|E_n\rangle = E_n\,\langle x|\,E_n\rangle\,,$$

or

$$-\frac{\hbar^2}{2m}\frac{d^2}{dx^2}\Psi_n(x) + V(x)\Psi_n(x) = E_n\Psi_n(x)\,, \tag{3.5.4}$$

where we have used Eq. (2.6.5) for the first term and let $\hat{V}(\hat{x})$ operate to the left on $\langle x|$ in the second term. Thus the time-independent Schrödinger equation written in the coordinate representation (or in the coordinate space) is second order differential equation. The differential operator

$$\hat{\mathcal{H}} = \left(-\frac{\hbar^2}{2m}\frac{d^2}{dx^2} + V(x)\right) \tag{3.5.5}$$

[4]The linear operator $\frac{d^2}{d\hat{x}^2}$ stands for $\frac{d}{d\hat{x}}\frac{d}{d\hat{x}}$ or the operator $\frac{d}{d\hat{x}}$ operating two times on the state on the right, in succession.

is the Hamiltonian operator in the coordinate space and the wave function $\Psi_n(x) = \langle x|E_n\rangle$ is the coordinate representative of the state $|E_n\rangle$. To minimize the proliferation of symbols we will often use the symbol H to denote the Hamiltonian operator in the coordinate representation.

We can write, similarly, the time-independent Schrödinger equation in the coordinate space for a system with three degrees of freedom. In this case, the basis states are the simultaneous eigenstates $|x, y, z\rangle$ of the three commuting coordinate observables $\hat{x}\,, \hat{y}\,, \hat{z}$, where x, y, z are the continuously varying eigenvalues of the three observables. The momentum observables, which are not diagonal in this representation, are given, respectively,

$$\hat{p}_x = -i\hbar\frac{d}{d\hat{x}}\,, \quad \hat{p}_y = -i\hbar\frac{d}{d\hat{y}}\,, \quad \hat{p}_z = -i\hbar\frac{d}{d\hat{z}}\,.$$

For such a system the Hamiltonian operator can be written explicitly as

$$\hat{H} = -\frac{\hbar^2}{2m}\left(\frac{d^2}{d\hat{x}^2} + \frac{d^2}{d\hat{y}^2} + \frac{d^2}{d\hat{z}^2}\right) + \hat{V}(\hat{x}, \hat{y}, \hat{z})\,. \tag{3.5.6}$$

Using this in the time-independent Schrödinger equation (3.5.1) and pre-multiplying both sides of the resulting equation by $\langle x, y, z|$, we get

$$\langle x, y, z| -\frac{\hbar^2}{2m}\left(\frac{d^2}{d\hat{x}^2} + \frac{d^2}{d\hat{y}^2} + \frac{d^2}{d\hat{z}^2}\right) + \hat{V}(\hat{x}, \hat{y}, \hat{z})\,|E_n\rangle = E_n\,\langle x, y, z|\,E_n\rangle$$

or

$$\left[-\frac{\hbar^2}{2m}\left(\frac{\partial^2}{\partial x^2} + \frac{\partial^2}{\partial y^2} + \frac{\partial^2}{\partial z^2}\right) + V(x, y, z)\right]\Psi_n(x, y, z) = E_n\Psi_n(x, y, z) \tag{3.5.7}$$

Here we have used Eq. (2.6.5) for the first term and let $\hat{V}(x)$ operate to the left on $\langle x, y, z|$ in the second term. $\Psi_n(x, y, z) = \langle x, y, z|E_n\rangle$ is the wave function representing the state $|E_n\rangle$ in the coordinate representation. This may also be called the coordinate representative of the state $|E_n\rangle$. The differential operator

$$-\frac{\hbar^2}{2m}\left(\frac{\partial^2}{\partial x^2} + \frac{\partial^2}{\partial y^2} + \frac{\partial^2}{\partial z^2}\right) + V(x, y, z) \equiv -\frac{\hbar^2}{2m}\nabla^2 + V(\boldsymbol{r}) \tag{3.5.8}$$

represents the Hamiltonian operator in the coordinate space. Thus the time-independent Schrödinger equation in the coordinate space, for a system with three degrees of freedom, is a second order partial differential equation for the wave function.

Working in the coordinate representation in three dimensions, we can as well use polar coordinates $(r\,, \theta\,, \psi)$ instead of the Cartesian coordinates (x, y, z). This leads to

$$\Psi_n(x, y, z) \equiv \Psi_n(\boldsymbol{r}) = \psi_n(r, \theta, \varphi) \tag{3.5.9}$$

$$V(x, y, z) \equiv V(\boldsymbol{r}) = V(r, \theta, \varphi) \tag{3.5.10}$$

$$\frac{\partial^2}{\partial x^2} + \frac{\partial^2}{\partial y^2} + \frac{\partial^2}{\partial z^2} \equiv \nabla^2 = \frac{1}{r^2}\frac{\partial}{\partial r}r^2\frac{\partial}{\partial r} + \frac{1}{r^2\sin\theta}\frac{\partial}{\partial\theta}\sin\theta\frac{\partial}{\partial\theta} + \frac{1}{r^2\sin^2\theta}\frac{\partial^2}{\partial\varphi^2}\,. \tag{3.5.11}$$

Similar, expressions in cylindrical and other coordinate systems can be written down. It is important to note that *transformation from one set of space coordinates to another, does not imply a change of representation* because the two sets of coordinates are mutually related.

Finally, the time-independent Schrödinger equation in the coordinate representation for a physical system with f degrees of freedom can be written by choosing the basis states to be the simultaneous eigenstates $\langle q_1, q_2, \cdots q_f|$ of f commuting Cartesian coordinate observables $\hat{q}_1\,, \hat{q}_2\,, \cdots \hat{q}_f$, where $q_1\,, q_2\,, \cdots, q_f$ represent the continuously varying eigenvalues of the coordinate observables. Now the Hamiltonian operator $\hat{H}$ of such a system may be expressed in terms of the position and momentum observables as

$$\hat{H} = \hat{H}\left(\hat{q}_1\,,\hat{q}_2\,,\cdots\hat{q}_f\,,-i\hbar\frac{d}{d\hat{q}_1}\,,-i\hbar\frac{d}{d\hat{q}_2}\,,\cdots\,,-i\hbar\frac{d}{d\hat{q}_f}\right)\,. \tag{3.5.12}$$

Using this in the time-independent Schrödinger equation (3.5.1) and multiplying both sides of the resulting equation, from left, by $\langle q_1, q_2, \cdots|$, we get

$$\langle \hat{q}_1\,,\hat{q}_2\,,\cdots,\hat{q}_f|\,\hat{H}\,|E_n\rangle = E_n\,\langle \hat{q}_1\,,\hat{q}_2\,,\cdots\hat{q}_f|\,E_n\rangle$$

or $$\mathcal{H}(q_1,\cdots q_f,-i\hbar\frac{d}{dq_1},\cdots\,,-i\hbar\frac{d}{dq_f})\Psi_n(q_1,\cdots,q_f) = E_n\Psi_n(q_1,\cdots,q_f)\,. \tag{3.5.13}$$

Here $\Psi_n(q_1\cdots q_f) = \langle q_1\,,q_2\,,\cdots,q_f|E_n\rangle$ is the coordinate representative of the state $|E_n\rangle$ and Eq. (2.7.4) has been used.

3.6 Time-independent Schrödinger Equation in the Momentum Representation

Let us first consider a system with only one degree of freedom. This system is characterized with one position observable $\hat{x}$ and one momentum observable $\hat{p}$. The Hamiltonian can be explicitly written as

$$\hat{H} = \frac{\hat{p}^2}{2m} + \hat{V}(\hat{x})\,. \tag{3.6.1}$$

In the momentum representation, the observable $\hat{p}$ is diagonal and the basis states are $\langle p|$, where p stands for the continuously varying eigenvalue of the observable $\hat{p}$. In this representation, the state $|E_n\rangle$ is represented by the function

$$\Phi_n(p) = \langle p|\,E_n\rangle\;, \tag{3.6.2}$$

which is referred to as the wave function in the momentum space. With the Hamiltonian given by Eq. (3.6.1), the time-independent Schrödinger equation (3.5.1) becomes

$$\left[\frac{\hat{p}^2}{2m} + \hat{V}(\hat{x})\right]\,|E_n\rangle = E_n\,|E_n\rangle\,.$$

Pre-multiplying both sides of this equation by $\langle p|$, we get

$$\left[\frac{p^2}{2m}\,\langle p|\,E_n\rangle + \langle p|\,\hat{V}(\hat{x})\,|E_n\rangle\right] = E_n\,\langle p|\,E_n\rangle$$

or $$\left(\frac{p^2}{2m} - E_n\right)\Phi_n(p) = -\,\langle p|\,\hat{V}(\hat{x})\,|E_n\rangle = -\int \langle p|\,\hat{V}(\hat{x})\,|p'\rangle\,dp'\,\langle p'|\,E_n\rangle$$

or $$\left(\frac{p^2}{2m} - E_n\right)\Phi_n(p) = -\int v(p,p')\Phi_n(p')dp'\,. \tag{3.6.3}$$

Thus, in the momentum representation, the time-independent Schrödinger equation $\hat{H}|E_n\rangle = E_n|E_n\rangle$ reduces to an integral equation. The kernel $v(p,p')$ can be expressed explicitly as

$$\begin{aligned} v(p,p') &= \langle p|\, \hat{V}(\hat{x})\, |p'\rangle \\ &= \iint \langle p|\, x\rangle\, dx\, \langle x|\, \hat{V}(\hat{x})\, |x'\rangle\, dx'\, \langle x'|\, p'\rangle \\ &= \frac{1}{2\pi\hbar} \iint e^{-ipx/\hbar} dx V(x) \delta(x-x') dx'\, e^{ip'x'/\hbar} \\ &= \frac{1}{2\pi\hbar} \int e^{-i(p-p')x/\hbar} V(x) dx\,. \end{aligned} \tag{3.6.4}$$

Thus $v(p,p')$ is the Fourier transform of the potential function $V(x)$.

Consider now a system with three degrees of freedom. In this case, the system is characterized by three coordinate observables $\hat{x}\,, \hat{y}\,, \hat{z}$ and three momentum observables $\hat{p}_x\,, \hat{p}_y\,, \hat{p}_z$. The time-independent Schrödinger equation in three dimension can be written explicitly as

$$\left(\frac{\hat{p}_x^2 + \hat{p}_y^2 + \hat{p}_z^2}{2m} + \hat{V}(\hat{x},\hat{y},\hat{z}) \right) |E_n\rangle = E_n\, |E_n\rangle\;. \tag{3.6.5}$$

To express this equation in the momentum representation, we choose the basis states to be the simultaneous eigenstates $|p_x, p_y, p_z\rangle \equiv |\boldsymbol{p}\rangle$ of three mutually commuting momentum observables $\hat{p_x}\,, \hat{p_y}\,, \hat{p_z}$. Here $p_x\,, p_y\,, p_z$ represent the continuously varying eigenvalues of the three momentum observables. In this representation, the state $|E_n\rangle$ is represented by the function $\Phi_n(\boldsymbol{p}) = \langle p_x, p_y, p_z|E_n\rangle = \langle \boldsymbol{p}|E_n\rangle$. Pre-multiplying both sides of Eq. (3.6.5) by $\langle \boldsymbol{p}|$, we get

$$\langle \boldsymbol{p}|\, \frac{\hat{p}_x^2 + \hat{p}_y^2 + \hat{p}_z^2}{2m} + \hat{V}(\hat{x},\hat{y},\hat{z})\, |E_n\rangle = E_n\, \langle \boldsymbol{p}|\, E_n\rangle\,,$$

or

$$\left(\frac{p^2}{2m} - E_n\right) \Phi_n(p) = -\iiint \langle \boldsymbol{p}|\, \hat{V}\, |\boldsymbol{p}'\rangle\, \Phi_n(\boldsymbol{p}') d^3p'\,. \tag{3.6.6}$$

The quantity $\langle \boldsymbol{p}|\hat{V}|\boldsymbol{p}'\rangle$ is easily be seen to be the three-dimensional Fourier transform of the potential function $V(\boldsymbol{r})$:

$$v(\boldsymbol{q}) \equiv \langle \boldsymbol{p}|\, \hat{V}\, |\boldsymbol{p}'\rangle = \frac{1}{(2\pi\hbar)^3} \iiint e^{-i(\boldsymbol{p}-\boldsymbol{p}')\cdot \boldsymbol{r}/\hbar} V(\boldsymbol{r}) d^3r\,, \tag{3.6.7}$$

where $\boldsymbol{q} = (\boldsymbol{p} - \boldsymbol{p}')/\hbar$. This can be worked out if $V(\boldsymbol{r})$ is known.

For a central potential, $V(\boldsymbol{r}) = V(r)$, we can carry out integration over the angles of $\boldsymbol{r}$ by using spherical polar coordinates to express the integral in Eq. (3.6.4) as

$$v(\boldsymbol{q}) == \frac{1}{(2\pi\hbar)^3} \iiint e^{-i\boldsymbol{q}V(r)\cdot \boldsymbol{r}}\, r^2 dr \sin\theta\, d\theta\, d\varphi\,. \tag{3.6.8}$$

To perform angular integration, we choose the polar axis to be along the direction of $\boldsymbol{q}$. Then using the substitution $\xi = \cos\theta$ we obtain

$$\begin{aligned} v(q) &= \frac{2\pi}{(2\pi\hbar)^3} \int_0^\infty V(r) r^2 dr \int_{-1}^1 e^{-iqr\xi} d\xi \\ &= \frac{1}{2\pi^2\hbar^3 q} \int_0^\infty V(r) r \sin(qr)\, dr\,, \end{aligned} \tag{3.6.9}$$

where we have written $v(\boldsymbol{q}) = v(q)$, since the right-hand side of Eq. (3.6.9) depends only on the magnitude of $\boldsymbol{q}$. It can be easily seen that, for Yukawa, exponential and Gaussian forms of (spherically symmetric) potentials

$$V_{\rm Y}(r) = -V_0 \frac{e^{-\mu r}}{r}\,, \tag{3.6.10}$$

$$V_{\rm exp}(r) = -V_0\, e^{-\mu r}\,, \tag{3.6.11}$$

$$V_{\rm G}(r) = -V_0\, e^{-\mu^2 r^2}\,, \tag{3.6.12}$$

$v(q)$ has the following expressions

$$v_{\rm Y}(q) = -\frac{V_0}{2\pi^2\hbar^3\mu(\mu^2+q^2)}\,, \tag{3.6.13}$$

$$v_{\rm exp}(q) = -\frac{V_0\mu}{\pi^2\hbar^3(\mu^2+q^2)^2}\,, \tag{3.6.14}$$

$$v_{\rm G}(q) = -\frac{V_0}{(2\hbar\sqrt{\pi}\mu)^3}\, e^{-q^2/4\mu^2}\,, \tag{3.6.15}$$

where $q \equiv |\boldsymbol{p} - \boldsymbol{p}\,'|/\hbar$.

3.6.1 Two-body Bound State Problem (in Momentum Representation) for Non-local Separable Potential

A simple application of the time-independent Schrödinger equation in the momentum representation is the two-body bound state problem for nonlocal separable potential. The two-body Schrödinger equation in the center of mass frame [see Chapter 5] in momentum representation is

$$\left(\frac{p^2}{2\mu} - E_n\right)\Phi_n(\boldsymbol{p}) = -\int \langle \boldsymbol{p}|\,\hat{V}\,|\boldsymbol{p}\,'\rangle\,\Phi_n(\boldsymbol{p}\,')d^3p' \tag{3.6.16}$$

where μ is the reduced mass of the particles, $\Phi_n(\boldsymbol{p}) = \langle \boldsymbol{p}|E_n\rangle$ is the momentum representative of the two-body state $|E_n\rangle$ with energy E_n. If the potential is local, the potential at a point $\boldsymbol{r}$ depends only on its cordinate $\boldsymbol{r}$:

$$\langle \boldsymbol{r}|\hat{V}|\boldsymbol{r}\,'\rangle = V(\boldsymbol{r})\,\delta^3(\boldsymbol{r} - \boldsymbol{r}\,')\,. \tag{3.6.17}$$

For a such a potential, the matrix element

$$\langle \boldsymbol{p}|\,\hat{V}\,|\boldsymbol{p}\,'\rangle = \frac{1}{(2\pi\hbar)^3}\int e^{-i(\boldsymbol{p}-\boldsymbol{p}\,')\cdot\boldsymbol{r}/\hbar}V(\boldsymbol{r})d^3r \equiv V(\boldsymbol{p} - \boldsymbol{p}\,') \tag{3.6.18}$$

is a function of the difference $\boldsymbol{p} - \boldsymbol{p}'$ only. We can then substitute this into the Schrödinger equation (3.6.16) in the momentum representation and solve it for E_n and $\phi_n(\boldsymbol{p})$.

Yamaguchi[5] introduced a non-local potential for the two-body interaction, with a separable form

$$\langle \boldsymbol{p}|\,\hat{V}\,|\boldsymbol{p}\,'\rangle = -\frac{\lambda}{2\mu}g(\boldsymbol{p})g(\boldsymbol{p}\,')\,, \tag{3.6.19}$$

[5] Y. Yamaguchi, Phys. Rev. **95**, 1628 (1954).

where μ is the reduced mass of the two-body system and λ is a parameter. For such a potential Eq. (3.6.16) assumes the form

$$\frac{1}{2\mu}\left(p^2+\alpha^2\right)\Phi(\boldsymbol{p})=\frac{\lambda}{2\mu}\int g(\boldsymbol{p})g(\boldsymbol{p}')\Phi(\boldsymbol{p}')d^3p'\,, \tag{3.6.20}$$

where $E_n=-\frac{\alpha^2}{2\mu}$ is a negative quantity, being the energy of a bound state. For spherically symmetric interaction, we can assume that $g(\boldsymbol{p})$ depends only on the magnitude $p=|\boldsymbol{p}|$ of $\boldsymbol{p}$:

$$g(\boldsymbol{p})=g(p)\,. \tag{3.6.21}$$

The Schrödinger equation (3.6.16) then has a simple solution

$$\Phi(\boldsymbol{p})=N\frac{g(p)}{\alpha^2+p^2}\,, \tag{3.6.22}$$

where N is a normalization constant. Substituting into Eq. (3.6.20), we get

$$\frac{1}{\lambda}=\frac{1}{\lambda(\alpha)}=\int\frac{g^2(p')d^3p'}{\alpha^2+p'^2}\,. \tag{3.6.23}$$

The potential parameter λ, thus, depends on the two-body binding energy $\frac{\alpha^2}{M}$. In other words, the energy with which the two-body potential is to bind the two-body system dictates the parameter λ. It can also be easily seen that the normalization constant N of the two-body wave function $\Phi(\boldsymbol{p})$ in the momentum space is given by

$$\frac{1}{N^2}=\int\frac{g^2(p)\,d^3p}{(\alpha^2+p^2)^2}\,. \tag{3.6.24}$$

3.7 Time-independent Schrödinger Equation in Matrix Form

We can alternatively choose a representation in which a set of observables, say $\hat{\xi}\,,\hat{\eta}\,,\hat{\zeta}\cdots$ are diagonal. The choice of these observables is suggested by the problem. The basis states of this representation are (using the abbreviated notation): $|\xi'\,,\eta'\,,\zeta'\cdots\rangle\equiv|\xi'\rangle$, $|\xi''\,,\eta''\,,\zeta''\cdots\rangle\equiv|\xi''\rangle$, $\cdots$, $\left|\xi^{(N)}\,,\eta^{(N)}\,,\zeta^{(N)}\cdots\right\rangle\equiv\left|\xi^{(N)}\right\rangle$. We assume that the number of basis states is finite. If the number of basis states is infinite, we can truncate the basis to retain only the first N states.

In this representation, the Hamiltonian $\hat{H}$ of the system is represented by an $N\times N$ matrix

$$H=\begin{pmatrix} H_{11} & H_{12} & \cdots & H_{1j} & \cdots & H_{1N}\\ H_{21} & H_{22} & \cdots & H_{2j} & \cdots & H_{2N}\\ \vdots & & & & & \\ H_{i1} & H_{i2} & \cdots & H_{ij} & \cdots & H_{mN}\\ \vdots & & & & & \\ H_{N1} & H_{N2} & \cdots & H_{Nj} & \cdots & H_{NN}\end{pmatrix}, \tag{3.7.1}$$

where a typical matrix element H_{ij} is given by

$$H_{ij}=\left\langle \xi^{(i)}\right|\hat{H}\left|\xi^{(j)}\right\rangle\,. \tag{3.7.2}$$

If we choose to represent the basic states $\left|\xi^{(j)}\right\rangle$ in the coordinate representation

$$\left|\xi^{(j)}\right\rangle \to \left\langle \boldsymbol{r} \middle| \xi^{(j)}\right\rangle = \chi_j(\boldsymbol{r}), \tag{3.7.3}$$

then the elements of Hamiltonian matrix H can be expressed in integral form [see Eq. (2.7.3)]

$$H_{ij} = \int \chi_i^*(\boldsymbol{r}) \left(-\frac{\hbar^2}{2m}\nabla^2 + V(\boldsymbol{r})\right) \chi_j(\boldsymbol{r}) d^3r \tag{3.7.4}$$

and can be evaluated numerically. Further, in this $(\hat{\xi}, \hat{\eta}, \hat{\zeta}, \cdots)$ representation, the eigenstate $|E_n\rangle$ of $\hat{H}$, belonging to the energy value E_n can be represented by a column matrix A_n

$$A_n = \begin{pmatrix} A_{1n} \\ A_{2n} \\ \vdots \\ A_{jn} \\ \vdots \\ A_{Nn} \end{pmatrix} \tag{3.7.5}$$

where

$$A_{jn} = \langle \xi^{(j)} | E_n \rangle . \tag{3.7.6}$$

Thus the state $|E_n\rangle$ can be expressed as

$$|E_n\rangle = \sum_j A_{jn} \left|\xi^{(j)}\right\rangle \tag{3.7.7}$$

or

$$\Psi_n(\boldsymbol{r}) \equiv \langle \boldsymbol{r} | E_n \rangle = \sum_j A_{jn}\chi_j(\boldsymbol{r}), \tag{3.7.8}$$

where $\Psi_n(\boldsymbol{r})$ is the coordinate representative of the state $|E_n\rangle$.

Now we can see that the time-independent Schrödinger equation (3.5.1) can be written as a matrix eigenvalue equation:

$$HA_n = E_n A_n , \tag{3.7.9}$$

or, explicitly,

$$\begin{pmatrix} H_{11} & H_{12} & \cdots & H_{it} & \cdots & H_{1N} \\ H_{21} & H_{22} & \cdots & H_{2t} & \cdots & H_{2N} \\ & & & \vdots & & \\ H_{s1} & H_{s2} & \cdots & H_{st} & \cdots & H_{sN} \\ & & & \vdots & & \\ H_{N1} & H_{N2} & \cdots & H_{Nt} & \cdots & H_{NN} \end{pmatrix} \begin{pmatrix} A_{1n} \\ A_{2n} \\ \vdots \\ A_{tn} \\ \vdots \\ A_{Nn} \end{pmatrix} = E_n \begin{pmatrix} A_{1n} \\ A_{2n} \\ \vdots \\ A_{tn} \\ \vdots \\ A_{Nn} \end{pmatrix} . \tag{3.7.10}$$

In the standard eigenvalue problem, the elements of the Hamiltonian matrix H are known, while its eigenvalues E_n and eigenvectors A_n are to be determined. This is a routine mathematical problem and can be solved manually by setting up the secular equation and solving it, if the matrix dimension is small. When the matrix dimension is large, it can be done on a computer by feeding all the elements of the Hamiltonian matrix into the computer and running the program for finding the eigenvalues and eigenvectors of the matrix. This procedure is referred to as the program for diagonalizing the matrix H because the problem of finding the eigenvalues and eigenvectors of a matrix is equivalent to that of finding a unitary transformation, represented by a unitary matrix S, ($S^\dagger S = SS^\dagger = I$) which would reduce H to the diagonal form

$$H \xrightarrow{S} S\,H\,S^\dagger = \begin{pmatrix} E_1 & 0 & \cdots & 0 \\ 0 & E_2 & \cdots & 0 \\ \vdots & & \cdots & \\ 0 & 0 & \cdots & E_N \end{pmatrix} . \tag{3.7.11}$$

The diagonal elements of the diagonalized matrix are the eigenvalues of the Hamiltonian matrix (or the eigenvalues of the Hamiltonian $\hat{H}$) while various columns of $S^\dagger$ yield the eigenvectors A_n of H, belonging to various eigenvalues E_n [see Appendix 3A1]. Thus the eigenvalues $E_1, E_2, \cdots E_n$ of $\hat{H}$, can be determined by solving the matrix eigenvalue problem. The corresponding wave functions Ψ_n are obtained via Eq. (3.7.8) since the eigenvectors

$$A_n = \begin{pmatrix} A_{1n} \\ A_{2n} \\ \vdots \\ A_{tn} \\ \vdots \\ A_{Nn} \end{pmatrix}$$

are also known.

3.8 The Heisenberg Picture

There are two ways of looking at the observables and states of a physical system. They are referred to as

1. Schrödinger picture
2. Heisenberg picture

In the Schrödinger picture, we view the state of a physical system to be evolving with time, while observables are taken to be time-independent. In this picture, the equation of motion governs the time evolution of the state and is referred as the Schrödinger equation of motion:

$$i\hbar \frac{d}{dt} \left| R(t) \right\rangle = \hat{H} \left| R(t) \right\rangle . \tag{3.8.1}$$

In the Heisenberg picture, we view the state of a dynamical system (the Heisenberg state) to be independent of time, while the time dependence is passed on to the observables. Consequently, the equation of motion in the Heisenberg picture pertains to the observables, the states in this picture being independent of time.

The change from Schrödinger to Heisenberg picture may be brought about by a unitary transformation.[6] To identify this unitary transformation, we note that the expectation value $\langle R(t) | \, \hat{\alpha} \, | R(t) \rangle$ of an observable $\hat{\alpha}$ (in the Schrödinger picture) can be rewritten, by

[6] A unitary transformation changes a state $|X\rangle \to |X'\rangle \equiv \hat{U}\,|X\rangle$ and an observable $\hat{\alpha} \to \hat{\alpha}' \equiv \hat{U}\hat{\alpha}\hat{U}^\dagger$, the transformation operator $\hat{U}$ being unitary: $\hat{U}\hat{U}^\dagger = \hat{U}^\dagger\hat{U} = \hat{1}$. Such a transformation preserves the norm of

using the time evolution operator $T(t)$, as $\langle R(0)|\,\hat{T}^\dagger(t)\hat{\alpha}\hat{T}(t)\,|R(0)\rangle$. This equation can be interpreted as the expectation value of a time-dependent observable $\hat{T}^\dagger(t)\hat{\alpha}\hat{T}(t)$ in the time-independent state $|R(0)\rangle$. This suggests that the unitary transformation needed to pass from the Schrödinger picture to Heisenberg picture is the adjoint (or inverse) $\hat{T}^\dagger(t)$ of the time-evolution operator $\hat{T}(t)$ defined by Eqs.(3.4.1) and (3.4.3). Consequently, the passage from Schrödinger picture state $|R(t)\rangle$ and observables $\hat{\alpha}$ to Heisenberg picture state $|R^H\rangle$ and obsevables $\alpha^H(t)$ is defined by

$$|R^H\rangle = \hat{T}^\dagger(t)\,|R(t)\rangle = \hat{T}^\dagger(t)\hat{T}(t)\,|R(0)\rangle = |R(0)\rangle \tag{3.8.2}$$

and

$$\hat{\alpha}^H = \hat{T}^\dagger(t)\,\hat{\alpha}\,\hat{T}(t)\,. \tag{3.8.3}$$

So we see that the state of the system in the Heisenberg picture is time-independent; the time dependence is passed on to the observable $\hat{\alpha}^H$. Here the superscript H distinguishes the Heisenberg picture states and observables from those in the Schrödinger picture.

To obtain the equation of motion for an observable in the Heisenbeg picture, we multiply Eq. (3.8.3) by $\hat{T}(t)$, from the left and use its unitarity ($\hat{T}(t)\hat{T}^\dagger(t) = 1 = \hat{T}^\dagger(t)\hat{T}(t)$ to get

$$\hat{T}\hat{\alpha}^H = \hat{\alpha}\hat{T}\,. \tag{3.8.4}$$

Differentiating this equation with respect to t and multiplying both sides by $i\hbar$, we get

$$i\hbar\left(\frac{d\hat{T}}{dt}\hat{\alpha}^H + \hat{T}\frac{d\hat{\alpha}^H}{dt}\right) = i\hbar\left(\frac{d\hat{\alpha}}{dt}\hat{T} + \hat{\alpha}\frac{d\hat{T}}{dt}\right)$$

or

$$\hat{H}\hat{T}\hat{\alpha}^H + i\hbar\hat{T}\frac{d\hat{\alpha}^H}{dt} = i\hbar\frac{d\hat{\alpha}}{dt}\hat{T} + \hat{\alpha}\hat{H}\hat{T} \tag{3.8.5}$$

where we have used Eq. (3.4.2) to replace $i\hbar\frac{d\hat{T}}{dt}$ by $\hat{H}\hat{T}$. Now, since $\hat{\alpha}$ is a Schrödinger observable, it has either no time dependence at all ($\frac{d\hat{\alpha}}{dt} = 0$) or at most, an explicit time dependence ($\frac{d\hat{\alpha}}{dt} = \frac{\partial\hat{\alpha}}{\partial t}$). Premultiplying both sides of Eq. (3.8.5) by $T^\dagger$ and inserting $\hat{T}^\dagger(t)\hat{T}(t) = 1$ between $\hat{\alpha}$ and $\hat{H}$ on the right hand side, we get

$$\hat{T}^\dagger\hat{H}\hat{T}\hat{\alpha}^H + i\hbar\frac{d\hat{\alpha}^H}{dt} = i\hbar\hat{T}^\dagger\frac{\partial\hat{\alpha}}{\partial t}\hat{T} + \hat{T}^\dagger\hat{\alpha}\hat{T}\hat{T}^\dagger\hat{H}\hat{T}$$

or

$$\frac{d\hat{\alpha}^H}{dt} = \frac{\partial\hat{\alpha}^H}{\partial t} + \frac{1}{i\hbar}\left[\hat{\alpha}^H, \hat{H}^H\right]\,, \tag{3.8.6}$$

where $\left[\hat{\alpha}^H, \hat{H}^H\right]$ is the commutator of Heisenberg operators[7] $\hat{\alpha}^H$ and $\hat{H}^H \equiv \hat{H}$. This equation is called the Heisenberg equation of motion.

Working in the Schrödinger picture, we could not see any semblance between the classical equations of motion and the equation of motion in quantum mechanics. But in Heisenberg picture we immediately see the close connection between quantum mechanical equation of motion (3.8.4) and its classical counterpart:

$$\frac{d\alpha_{cl}}{dt} = \frac{\partial\alpha_{cl}}{\partial t} + \{\alpha_{cl}, H_{cl}\}\,, \tag{3.8.7}$$

---

a state $|X\rangle$

$$\langle X|X\rangle = \langle X'|X'\rangle\,,$$

and also, the expectation value of an observable for any state

$$\langle X|\hat{\alpha}|X\rangle = \langle X'|\hat{\alpha}'|X'\rangle\,.$$

[7]It is easily seen that $\hat{H}^H \equiv \hat{T}^\dagger\hat{H}\hat{T}$ is the same as $\hat{H}$.

where α_{cl} is a real classical dynamical variable, and H_{cl} is the classical Hamiltonian. We find that the classical Poisson bracket, occurring in classical equation of motion, corresponds to $\frac{1}{i\hbar}$ times the quantum commutator bracket, in the Heisenberg equation of motion.

For a Schrödinger observable $\hat{\alpha}$ that has no explicit time dependence ($\frac{d\hat{\alpha}}{dt} = \frac{\partial\hat{\alpha}}{\partial t} = 0$), that is, it has no time dependence at all (not even explicit time dependence) $\frac{\partial\hat{\alpha}^H}{dt} \equiv \hat{T}^\dagger \frac{\partial\hat{\alpha}}{\partial t}\hat{T} = 0$. Then the Heisenberg equation of motion reduces to

$$\frac{d\hat{\alpha}^H}{dt} = \frac{1}{i\hbar}\left[\hat{\alpha}^H, \hat{H}^H\right]. \tag{3.8.8}$$

If an observable $\hat{\alpha}$ commutes with the Hamiltonian: $[\hat{\alpha}, \hat{H}] = [\hat{\alpha}^H, \hat{H}^H] = 0$, the Hesienberg equation of motion leads to $d\hat{\alpha}^H/dt = 0$, which implies that $\hat{\alpha}^H$ is a constant of motion and its expectation value in any state is constant in time:

$$\langle\hat{\alpha}\rangle = \langle R(t)|\,\hat{\alpha}\,|R(t)\rangle = \langle R(0)|\,\hat{T}^\dagger\hat{\alpha}\hat{T}\,|R(0)\rangle = \left\langle R^H\right|\hat{\alpha}^H\left|R^H\right\rangle.$$

From this equation we imediately see

$$\frac{d\langle\hat{\alpha}\rangle}{dt} = \left\langle R^H\right|\frac{d\hat{\alpha}^H}{dt}\left|R^H\right\rangle = 0.$$

Thus, commutativity of $\hat{\alpha}$ and $\hat{H}$ implies not only that their measurements are compatible and there exist simultaneous eigenstates of $\hat{\alpha}$ and $\hat{H}$, but also that $\hat{\alpha}$ is a conserved quantity and its expectation value for any state is constant in time.

3.9 The Interaction Picture

Another picture, called the *interaction picture* is useful in such problems in which the Hamiltonian $\hat{H}$ has a dominant time-independent part $\hat{H}_0$ and an interaction part $\hat{H}_I$ which depends on time. In the Schrödinger picture, the equation of motion for such a system may be written as

$$i\hbar\frac{d}{dt}\,|\Psi(t)\rangle = (\hat{H}_0 + \hat{H}_I)\,|\Psi(t)\rangle\,. \tag{3.9.9}$$

To go to the interaction picture we invoke a unitary transformation through the operator $\hat{T}^\dagger(t) \equiv e^{i\hat{H}_0 t/\hbar}$ so that

$$|\Psi(t)\rangle \to e^{i\hat{H}_0 t/\hbar}\,|\Psi(t)\rangle \equiv \left|\Psi^{\rm int}(t)\right\rangle, \tag{3.9.10}$$

while any observable $\hat{\alpha}$ in the Schrödinger picture is transformed to $\hat{\alpha}^{\rm int}(t)$

$$\hat{\alpha} \to e^{i\hat{H}_0 t/\hbar}\hat{\alpha}\,e^{-i\hat{H}_0 t/\hbar} \equiv \hat{\alpha}^{\rm int}(t)\,. \tag{3.9.11}$$

This implies that[8]

$$\hat{H}_I \to e^{i\,H_0\,t/\hbar}\hat{H}_I e^{-i\,\hat{H}_0 t/\hbar} \equiv H_I^{\rm int}\,. \tag{3.9.12}$$

[8]In our notation the suffix 'I' on $\hat{H}$ is used to denote the interaction part of the Hamiltonian while the superscript '*int*' indicates that the operator pertains to the interaction picture.

We can easily see that the state $\left|\Psi^{\mathrm{int}}(t)\right\rangle$ in the interaction picture satisfies a Schrödinger -like equation of motion

$$i\hbar\frac{d}{dt}\left|\Psi^{\mathrm{int}}(t)\right\rangle = \hat{H}_I^{\mathrm{int}}(t)\left|\Psi^{\mathrm{int}}(t)\right\rangle . \tag{3.9.13}$$

Thus while in the Schrödinger picture the state is time-dependent and in the Heisenberg picture the state is time-independent and time dependence is passed on to the observables, in the interaction picture the time dependence caused by the full Hamiltonian is split into two parts. A part of the time dependence is assigned the state vector $\left|\Psi^{\mathrm{int}}(t)\right\rangle$ while the *residual* time dependence is retained by $\hat{H}_I^{\mathrm{int}}(t)$. This picture is not relevant when we deal with stationary states but can be useful when the interaction part of the Hamiltonian depends on time. This picture is very useful in the theory of *interacting quantum fields* and is discussed in more detail in Chapter 14 (Section 14.9).

Problems

1. By expressing the state $|R(t)\rangle$ as a column vector and the Hamiltonian operator by a square matrix in a certain representation, say ξ, η, $\zeta, \cdots$ representation, write the Schrödinger equation of motion

$$i\hbar\frac{d}{dt}\,|R(t)\rangle = |R(t)\rangle$$

as a matrix equation.

2. Take $\psi(x) = A\exp[i(kx-\omega t)]$, where $\omega = E/\hbar$, and show that the probability current density $\boldsymbol{S}$ is given by $\boldsymbol{S} = \frac{\hbar\boldsymbol{k}}{m}\rho(x)$, where $\rho(x) = \psi^*(x,t)\psi(x,t)$ is the probability density. Note the similarity of this equation with the corresponding equation in fluid mechanics $\boldsymbol{j} = \rho(x)\boldsymbol{v}$.

3. Calculate the probability current corresponding to the wave function

$$\psi(r,\theta) = f(\theta)\frac{\exp(ikr)}{r} .$$

Examine $\boldsymbol{S}$ for large values of r and interpret your result.

4. Use Heisenberg equation of motion to derive the time rate of change of the expectation value of an observable $\hat{\alpha}$ given by

$$\int \psi^*(\boldsymbol{r},t)\alpha_{\boldsymbol{r}}\psi(\boldsymbol{r},t)d^3r$$

where $\alpha_{\boldsymbol{r}}$ is the coordinate representative of the observable $\hat{\alpha}$, defined by

$$\langle \boldsymbol{r}|\,\hat{\alpha}\,|R(t)\rangle = \alpha_{\boldsymbol{r}}\,\langle \boldsymbol{r}|\,R(t)\rangle = \alpha_{\boldsymbol{r}}\psi(\boldsymbol{r},t) .$$

5. Write the time-independent Schrödinger equation for a system with Hamiltonian $\hat{H} = \hat{H}_o + \hat{H}'$ in a matrix representation in which the unperturbed Hamiltonian $\hat{H}_o$ is diagonal and the basis states $|E_n\rangle$ are the eigenstates of the unperturbed Hamiltonian $\hat{H}_o$, belonging to the eigenvalues E_n.

6. A particle (with one degree of freedom) is moving in a potential $V(x) = \frac{1}{2}kx^2$. Write down the time-independent Schrödinger equation in the momentum representation.

7. A particle with three degrees of freedom is moving in a spherically symmetric potential $V(r)$. Write down the time-independent Schrödinger equation for the particle when $V(r)$ has the following alternative forms:

   (a) Yukawa form: $V(r) = -V_o \frac{e^{-\mu r}}{r}$

   (b) Exponential form: $V(r) = -V_o e^{-\mu r}$

   (c) Gaussian form: $V(r) = -V_o e^{-\alpha r^2}$

   where V_o and μ and α are real constants. The Fourier transforms of these potential forms are given by Eqs. (3.6.13) through (3.6.15).

8. Show that the problem of finding the eigenvalues and eigenvectors of a square matrix is equivalent to that of diagonalizing the matrix.

9. If a unitary matrix U diagonalizes a square matrix M so that $U\,M\,U^\dagger = D$, then show that the elements of the diagonal matrix D are just the eigenvalues of M and the various columns of $U^\dagger$ (Hermitian adjoint of U) are the eigenvectors of M belonging to these eigenvalues.

10. Show that the necessary and sufficient condition that two square matrices be diagonalized by the same unitary transformation is that they commute. What is the significance of this result in quantum mechanics?

11. Using the Schrödinger equation in the matrix form, work out the unitary transformation matrix $\hat{T}(t)$ which would bring about a transformation from the Schrödinger picture to the Heisenberg picture. Finally derive Heisenberg equations of motion for the Heisenberg operators in the matrix form.

12. Show that the commutator bracket of position and momentum observables has the same value $i\hbar\,\hat{1}$ in both the Schrödinger and Heisenberg pictures.

13. Show that if two operators commute in the Schrödinger picture, they also do so in the Heisenberg picture.

14. Show that the basic commutator $[\hat{x}^H, \hat{p}^H]$ is a constant of motion.

15. Show that the expectation value of an observable for any state is the same in both the Schrödinger and Heisenberg pictures.

16. Write down the Heisenberg equation of motion in the energy representation (a matrix representation in which the Hamiltonian is diagonal).

17. Show that if an observable $\hat{\alpha}$ (in the Schrödinger picture) does not explicitly depend on time then

$$\frac{d}{dt}\langle\alpha\rangle = \frac{1}{i\hbar}\,\langle\psi|\,[\hat{\alpha}, \hat{H}]\,|\psi\rangle\,,$$

where $\langle\alpha\rangle = \langle\psi|\,\hat{\alpha}\,|\psi\rangle$ denotes the quantum mechanical expectation value of $\hat{\alpha}$ for the state $|\psi\rangle$.

18. Given the Hamiltonian of a particle:

$$\hat{H} = \frac{1}{2m}\hat{\boldsymbol{p}}^2 + \hat{V}(\hat{r}),$$

show, by using the Heisenberg equations of motion and the basic commutation relations for $\hat{x}, \hat{y}, \hat{z}$ and $\hat{p}_x, \hat{p}_y, \hat{p}_z$, that the expectation values of $\hat{x}$ and $\hat{p}_x$ (as also of $\hat{y}$, $\hat{p}_y$ and $\hat{z}$ and $\hat{p}_z$) satisfy the classical equations of motion:

$$m\frac{d\langle x\rangle}{dt} = \langle p_x\rangle$$

and

$$\frac{d\langle p_x\rangle}{dt} = -\left\langle \frac{\partial V}{\partial x}\right\rangle.$$

19. Find the equations of motion for the Heisenberg operators $\hat{x}^H, \hat{p}^H, (\hat{x}^H)^2, (\hat{p}^H)^2$ if the Hamiltonian of the system is given by:

$$\hat{H} = \frac{\hat{p}^2}{2m} + \hat{V}(\hat{x}, \hat{y}, \hat{z}).$$

20. A particle is subject to a central potential such that its Hamiltonian is

$$\hat{H} = \frac{\hat{p}^2}{2m} + \hat{V}(\hat{r}),$$

where $\hat{p}^2 = \hat{p}_x^2 + \hat{p}_y^2 + \hat{p}_z^2$ and $\hat{r}^2 = \hat{\boldsymbol{r}} \cdot \hat{\boldsymbol{r}}$. Which of the following observables are constants of motion: $\hat{p}_x, \hat{L}_z, \hat{L}^2$? Prove your answer.

21. Prove that in the Schrödinger picture, the evolution of a state is given by

$$|R(t)\rangle = \sum_{k=0}^{\infty} \frac{1}{k!}(-i\hat{H}\,t/\hbar)^k \; |R(0)\rangle\,.$$

22. The wave function representing the state of a free particle at time $t = 0$ is given by:

$$\psi(x, t = 0) = A\exp(ip_0 x/\hbar)\exp(-(x - x_0)^2/4\sigma^2)$$

where p_o and x_o are real parameters.

(a) Determine the normalization constant A and, using the Schrödinger equation for a free particle, show that $\psi(x,t)$ for $t > 0$ is given by:

$$\psi(x,t) = \frac{(2\pi)^{-1/4}}{\sqrt{\sigma + \frac{i\hbar t}{2m\sigma}}}\exp(ik_0x_0)\exp(-\sigma^2 k_o^2)$$
$$\times \exp\left[\frac{-(x-x_0)^2 + 4\sigma^4 k_0^2 + +4i(x-x_0)\sigma^2 k_o}{4\sigma^2 + (2i\hbar t/m)}\right],$$

where $k_o = p_o/\hbar$.

(b) Calculate $\langle x\rangle$ and $\langle p\rangle$ as functions of time.

(c) Calculate $\langle x^2\rangle$ and $\langle p^2\rangle$ hence Δx and Δp given by $(\Delta x)^2 = \langle x^2\rangle - \langle x\rangle^2$ and $(\Delta p)^2 = \langle p^2\rangle - \langle p\rangle^2$.

[Hint: The general solution of Schrodinger equation of motion for a free particle is:

$$\psi(x,t) = \int a(k) \frac{1}{\sqrt{2\pi}} \exp(ikx) \exp(-i\hbar k^2 t/2m) dk \,.$$

Putting $t = 0$ on both sides would not change $a(k)$ which can be determined because the form of $\psi(x, 0)$ is given.]

23. For a free particle in one dimension, show that

    (a) $\langle p \rangle$ is a constant in time while $\langle x \rangle$ increases linearly with time.

    (b) Derive the equation of motion satisfied by $\langle x^2 \rangle$.

    (c) Solving this equation show that

    $$[\Delta x(t)]^2 = \frac{1}{m^2} [\Delta p]^2 t^2 + [\Delta x(0)]^2 \,.$$

    where $\Delta x(t)$ and $\Delta p(t)$ have been defined in Problem 22.

24. Show that, in the interaction picture, the state $\left|\Psi^{\rm int}(t)\right\rangle$ satisfies the Schrödinger -like equation

$$i\hbar \frac{d}{dt} \left|\Psi^{\rm int}(t)\right\rangle = \hat{H}_I^{\rm int}(t) \left|\Psi^{\rm int}(t)\right\rangle \,,$$

where $\left|\Psi^{\rm int}(t)\right\rangle \equiv e^{i\hat{H}_0 t/\hbar} \left|\Psi(t)\right\rangle$ and $\hat{H}_I^{\rm int}(t) \equiv e^{i\,\hat{H}_0 t/\hbar} \hat{H}_I e^{-i\,\hat{H}_0\,t/\hbar}$.

Appendix 3A1: Matrices

Determination of Eigenvalues and Diagonalization of a Matrix

3A1.1 Characteristic Equation of a Matrix

Let M be a $N \times N$ square matrix and X a $N \times 1$ column matrix (also a vector)

$$X = \begin{pmatrix} x_1 \\ x_2 \\ \vdots \\ x_N \end{pmatrix} .$$

In general, the product $MX = Y$ is a column vector Y that is different from from X. However, if for a certain choice of X, it so happens that

$$MX = \lambda X , \tag{3A1.1.1}$$

then the column vector X is called an eigenvector of M belonging to the eigenvalue λ. A square matrix of order $N \times N$ can have a set of N eigenvalues and N eigenvectors.

To determine the possible eigenvalues and eigenvectors of a matrix M, we rewrite Eq. (3A1.1.1) as as a set of N linear equations:

$$\begin{aligned} (M_{11} - \lambda)x_1 + M_{12}x_2 + \cdots + M_{1N}x_N &= 0 , \\ M_{21}x_1 + (M_{22} - \lambda)x_2 + \cdots + M_{1N}x_N &= 0 , \\ &\vdots \\ M_{N1}x_1 + M_{N2}x_2 + \cdots + (M_{NN} - \lambda)x_N &= 0 . \end{aligned} \tag{3A1.1.2}$$

This set of linear equations admits a non-trivial solution only if the determinant

$$\det(M - \lambda I) \equiv \begin{vmatrix} (M_{11} - \lambda) & M_{12} & M_{13} & \cdots & M_{1N} \\ M_{21} & (M_{22} - \lambda) & M_{23} & \cdots & M_{2N} \\ M_{31} & M_{32} & (M_{33} - \lambda) & \cdots & M_{2N} \\ \vdots & & & & \\ M_{N1} & M_{N2} & M_{N3} & \cdots & (M_{NN} - \lambda) \end{vmatrix} = 0 . \tag{3A1.1.3}$$

This equation, called the characteristic equation of matrix M, reduces to an Nth degree equation in λ:

$$\lambda^N + b_1\lambda^{N-1} + b_2\lambda^{N-2} + \cdots + b_{N-1}\lambda + b_N = 0 , \tag{3A1.1.4}$$

and, therefore, admits N roots: $\lambda_1 , \lambda_2, \cdots, \lambda_N$, which are the N eigenvalues of the matrix M. To determine the eigenvector

$$X_r = \begin{pmatrix} x_{1r} \\ x_{2r} \\ \vdots \\ x_{Nr} \end{pmatrix}$$

corresponding to the eigenvalue λ_r, we replace λ in Eq. (3A1.1.2) by the eigenvalue λ_r, and solve the resulting equations for $x_{1r}, x_{2r}, \cdots, x_{Nr}$. By repeating it for all $1 \le r \le N$, we can determine all N eigenvectors of M belonging, respectively, to the eigenvalues $\lambda_1, \lambda_2, \cdots \lambda_N$:

$$X_1 = \begin{pmatrix} x_{11} \\ x_{21} \\ \vdots \\ x_{N1} \end{pmatrix}, X_2 = \begin{pmatrix} x_{12} \\ x_{22} \\ \vdots \\ x_{N2} \end{pmatrix}, \cdots, X_N = \begin{pmatrix} x_{1N} \\ x_{2N} \\ \vdots \\ x_{NN} \end{pmatrix}.$$

3A1.2 Similarity (and Unitary) Transformation of Matrices

A similarity transformation is brought about by a square non-singular matrix S. It transforms a given matrix A into A' such that

$$A \xrightarrow{S} A' = SAS^{-1}. \tag{3A1.1.5}$$

The same matrix transforms a column vector X to X' such that

$$X \xrightarrow{S} X' = SX. \tag{3A1.1.6}$$

This transformation preserves the relationship between matrices and column vectors and leaves the eigenvalues of the matrix invariant (unchanged).

When the transforming matrix S is unitary ($SS^\dagger = S^\dagger S = I$), the inverse transformation is $S^{-1} = S^\dagger$, and the transformation S is termed as unitary similarity transformation:

$$A' = S\,A\,S^\dagger \quad \text{and} \quad X' = SX. \tag{3A1.1.7}$$

If S is a real orthogonal matrix ($S^T S = SS^T = I$), the inverse transformation $S^{-1} = S^T$, where S^T is the transpose of matrix S, the similarity transformation reduces to

$$A' = SAS^T \quad \text{and} \quad X' = SX. \tag{3A1.1.8}$$

This is a special case of unitary transformation.

3A1.3 Diagonalization of a Matrix

Let a matrix M admit the set of eigenvalues $\lambda_1, \lambda_2, \cdots \lambda_j, \cdots \lambda_N$ and eigenvectors $X_1, X_2, \cdots, X_N$ such that that $MX_r = \lambda_r X_r$. Let us form a square matrix S from all these eigenvectors according to

$$S \equiv (X_1\ X_2\ \cdots\ X_N) = \begin{pmatrix} X_{11} & X_{12} & \cdots & X_{1N} \\ X_{21} & X_{22} & \cdots & X_{2N} \\ & \vdots & \vdots & \\ X_{N1} & X_{N2} & \cdots & X_{NN} \end{pmatrix}, \tag{3A1.1.9}$$

and let D be the diagonal matrix whose diagonal elements are the eigenvalues of M

$$D = \begin{pmatrix} \lambda_1 & 0 & 0 \cdots & 0 \\ 0 & \lambda_2 & 0 \cdots & 0 \\ & & \vdots & \\ 0 & 0 & 0 \cdots & \lambda_N \end{pmatrix}. \tag{3A1.1.10}$$

Then we can easily verify that

$$SD = MS \quad \text{or} \quad D = S^{-1}MS. \tag{3A1.1.11}$$

If S is unitary (which will happen if the eigenvectors X_r of the matrix M satisfy the twin conditions of orthogonality $X_r{}^{\dagger} X_k = \delta_{rk}$ and completeness $\sum_r X_r X_r^{\dagger} = I$), then

$$D = S^{\dagger} \, M \, S \,. \tag{3A1.1.12}$$

The unitary matrix $S^{\dagger}$ thus transforms the matrix M to the diagonal form

$$D = S^{\dagger} \, M \, S \,. \tag{3A1.1.13}$$

The diagonal elements of D are simply the eigenvalues of the matrix M and various columns of the matrix S (the adjoint of the transforming matrix $S^{\dagger}$) are just the eigenvectors X_r of M belonging to the eigenvalues λ_r $(r = 1, 2, \cdots, N)$.

For a matrix M of large dimensionality, the determination of its eigenvalues λ_r and eigenvectors X_r $(r = 1, 2, \cdots, N)$ (or the determination of its diagonalizing matrix $S^{\dagger}$ and S, and the diagonalized matrix D) can be done on a computer.

References

[1] P. A. M. Dirac, *Principles of Quantum Mechanics*, Third Edition (Clarendon Press, Oxford, 1971) Chapter V.

[2] T. F. Jordan, *Quantum Mechanics in Simple Matrix Form* (Dover Publications, New York, 2005).

4

PROBLEMS OF ONE-DIMENSIONAL POTENTIAL BARRIERS

Among the simplest applications of Schrödinger 's time-independent equation for stationary states are the problems pertaining to the motion of a particle in the presence of potential steps. In the present chapter we shall consider the following problems:

(a) Motion of a particle across a potential step of finite height and infinite extent.

(b) Motion of a particle through a potential barrier of finite height and finite extent.

(c) Leakage of a particle within a potential well, through a potential barrier of finite height and finite extent.

(d) Motion of a particle in a periodic potential.

Before proceeding further we note that physically acceptable wave functions must satisfy the time-independent Schrödinger equation and preserve the probabilistic interpretation of the wave function. These requirements lead to certain conditions on the wave function:

(i) The wave function $\psi(x)$ must be a single-valued function of x so that a unqiue probability $|\psi(x)|^2 dx$ of locating the particle between x and $x + dx$ can be assigned to each point x. For a bound state (localized) wave function $\psi(x)$, the probablity of finding the particle somewhere in space must equal one: $\int_{-\infty}^{\infty} |\psi(x)|^2 dx = 1$. This is a statement of the normalization of bound state wave function. Once normalized, the wave function remains normalized for all times since the Schrödinger equation preserves normalization.

For the normalization integral to be finite, the wave function must be finite everywhere and satisfy the boundary condition $\psi(x) \to 0$ as $|x| \to \infty$ for a bound state.[1] As a consequence of boundary conditions, the bound state spectrum in one dimension is discrete and nondegenerate. This means that each bound state energy eigenvalue belongs to one (and only one) wave function.

(ii) The wave function as well as its first derivative must be continuous at each point in space even if the potential has (finite) discontinuity. This is because the wave function must be twice differentiable as it satisfies a second order differential equation. For the wave function's second derivative to exist, its first derivative must be continuous, which in turn requires the wave function itself to be continuous.

When the potential has infinite discontinuity (as in an infinite square well or Dirac delta function potential), the first derivative may be discontinuous and the wave function may have a kink.

[1]For scattering solutions, the wave function approaches a constant as $|x| \to \infty$ corresponding to incident or scattered particle flux $\propto |\psi(x)|^2$.

4.1 Motion of a Particle across a Potential Step

Consider the motion of a particle of energy E along a single axis [number of degrees of freedom of the system $f = 1$)] in the presence of a potential step of height V_o. In this case, the potential $V(x)$ has the form [Fig. 4.1]

$$V(x) = \begin{cases} 0\,, & x < 0 : \text{Region I} \\ V_o\,, & x > 0 : \text{Region II}\,. \end{cases} \tag{4.1.1}$$

The particle approaches the potential step from the left. We consider two cases separately:

(i) Particle energy greater than the potential step: $E > V_o$

Let the quantum state of the particle in regions I and II be represented by $\psi_1(x)$ and $\psi_2(x)$ in the coordinate representation. Then for the potential of Eq. (4.1.1), time-independent Schrödinger equation [Eq. (3.5.4)] takes the following forms for the two regions

$$\frac{d^2\psi_1(x)}{dx^2} + k^2\psi_1(x) = 0 \qquad x < 0 : \text{Region I} \tag{4.1.2}$$

$$\frac{d^2\psi_2(x)}{dx^2} + k'^2\psi_2(x) = 0 \qquad x > 0 : \text{Region II} \tag{4.1.3}$$

where

$$k = \sqrt{\frac{2mE}{\hbar^2}} \quad \text{and} \quad k' = \sqrt{\frac{2m(E - V_o)}{\hbar^2}}\,. \tag{4.1.4}$$

The respective solutions in the two regions are

$$\psi_1(x) = Ae^{ikx} + Be^{-ikx}\,, \tag{4.1.5}$$

$$\psi_2(x) = Ce^{ik'x} + De^{-ik'x}\,. \tag{4.1.6}$$

If we assume a particle incident from the left, there is no possibility of a wave traveling from right to left in region II. This implies that $D = 0$. For the continuity of the wave function, it is necessary to match the wave function and its first derivative at the point of discontinuity of the potential at $x = 0$. This requires

$$\psi_1(x = 0) = \psi_2(x = 0) \tag{4.1.7}$$

and

$$\left.\frac{d\psi_1}{dx}\right|_{x=0} = \left.\frac{d\psi_2}{dx}\right|_{x=0}\,. \tag{4.1.8}$$

These conditions, with the help of Eqs. (4.1.5) and (4.1.6) yield

$$B = \frac{k - k'}{k + k'}A \tag{4.1.9a}$$

and

$$C = \frac{2k}{k + k'}A\,. \tag{4.1.9b}$$

We have seen in Chapter 3, Sec. 3.4 that the probability current density

$$S(x,t) = \mathcal{R}e\left[\psi^*\frac{\hbar}{im}\frac{d\psi}{dx}\right] \tag{4.1.10}$$

can be regarded as the probability per unit area that the particle will pass across a given point per unit time. We can easily calculate the probability current density for the incident wave $\psi_i(x) = Ae^{ikx}$, the reflected wave $\psi_r(x) = Be^{-ikx}$ or the transmitted wave $\psi_t(x) = Ce^{ik'x}$ in accordance with Eq. (4.1.10) to obtain

$$S_i = \frac{\hbar k}{m}|A|^2 , \tag{4.1.11}$$

$$S_r = \frac{\hbar k}{m}|B|^2 , \tag{4.1.12}$$

$$S_t = \frac{\hbar k'}{m}|C|^2 . \tag{4.1.13}$$

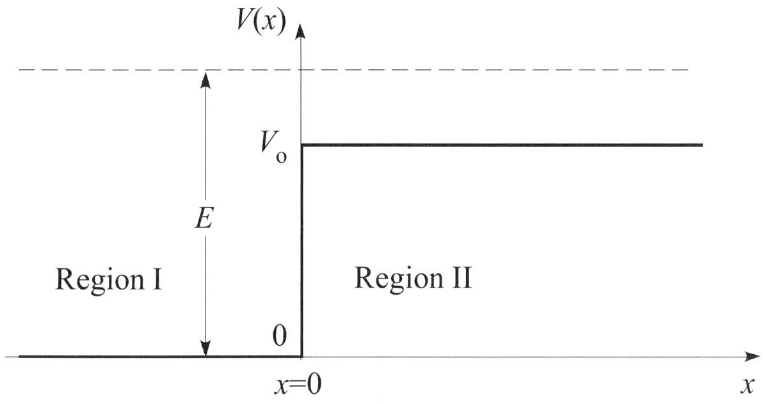

FIGURE 4.1
Potential step with height V_o less than the energy E of the particle.

Now, the transmittivity T equals the ratio of the transmitted to incident current

$$T = \frac{S_t}{S_i} = \frac{4kk'}{(k+k')^2} = \frac{4\sqrt{E(E-V_o)}}{(\sqrt{E}+\sqrt{E-V_o})^2} . \tag{4.1.14}$$

Similarly, the reflectivity R equals the ratio of reflected current to the incident current.

$$R = \frac{S_r}{S_i} = \left(\frac{k-k'}{k+k'}\right)^2 = \left[\frac{\sqrt{E}-\sqrt{E-V_o}}{\sqrt{E}+\sqrt{E-V_o}}\right]^2 . \tag{4.1.15}$$

It is easily verified that $T+R=1$. This means the particle is either reflected or transmitted at the step. The variation of transmittivity and reflectivity as functions of E/V_o is shown in Fig. 4.2.

We note in this case ($E > V_o$) that although the kinetic energy of the particle is large enough to overcome the potential barrier and classically the particle should be transmitted over the potential step, according to quantum mechanics [Fig. 4.2], there is a finite probability that the particle will turn back (reflected). Thus even in this relatively simple problem, quantum mechanics predicts a nonclassical behavior for a material particle. This behavior is a direct consequence of the wave-like character that quantum mechanics ascribes to material particles. The probability of reflection, of course, decreases with increasing E/V_o.

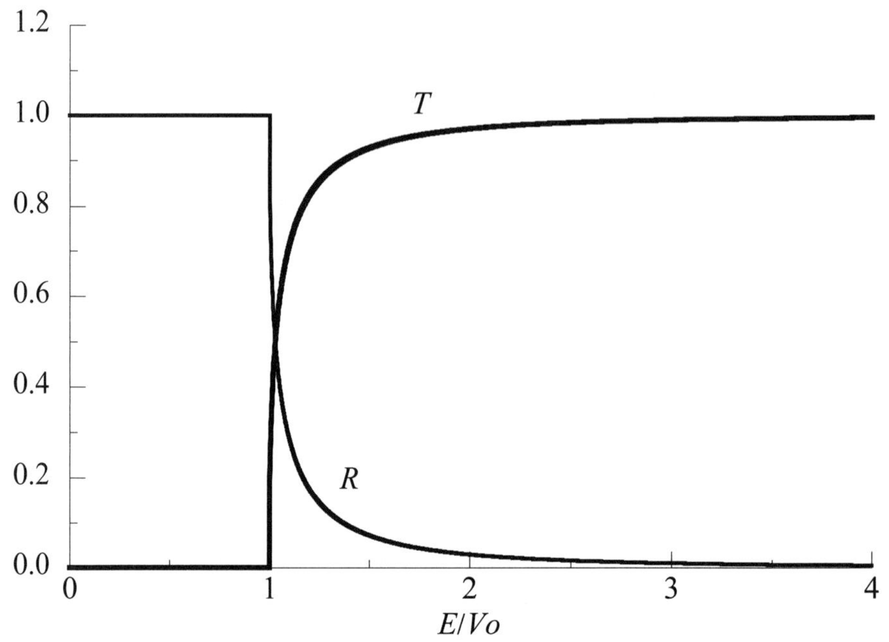

FIGURE 4.2
Transmittivity T and reflectivity R as functions of E/V_o.

(ii) Particle energy less than the potential step: $E < V_o$

The potential step and particle energy relative to it are shown in Fig. 4.3 for this case. The time-independent Schrödinger equation for regions I and II, in the coordinate representation, now takes the form

$$\frac{d^2\psi_1(x)}{dx^2} + k^2\psi_1(x) = 0\,, \quad x < 0 : \text{Region I} \tag{4.1.16}$$

$$\frac{d^2\psi_2(x)}{dx^2} - \kappa^2\psi_2(x) = 0\,, \quad x > 0 : \text{Region II} \tag{4.1.17}$$

where

$$k = \sqrt{\frac{2mE}{\hbar^2}} \quad \text{and} \quad \kappa = \sqrt{\frac{2m(V_o - E)}{\hbar^2}}\,. \tag{4.1.18}$$

These equations are similar to those in the previous case ($E > V_o$), except that in this case k'^2 is replaced by $-\kappa^2 = -2m(V_o - E)/\hbar^2$. Since $V_o > E$ in this case, κ^2 is positive and κ is real. The solutions in regions I and II may easily be written as

$$\psi_1(x) = Ae^{ikx} + Be^{-ikx} \tag{4.1.19}$$

$$\psi_2(x) = Ce^{\kappa x} + De^{-\kappa x}\,. \tag{4.1.20}$$

We discard the positive exponential term ($C = 0$) in $\psi_2(x)$ because we expect $\psi_2(x) \to 0$ as $x \to \infty$. Again, we match the solutions in the two regions at the boundary $x = 0$ (point of discontinuity of the potential). This leads to the following expressions for the amplitudes

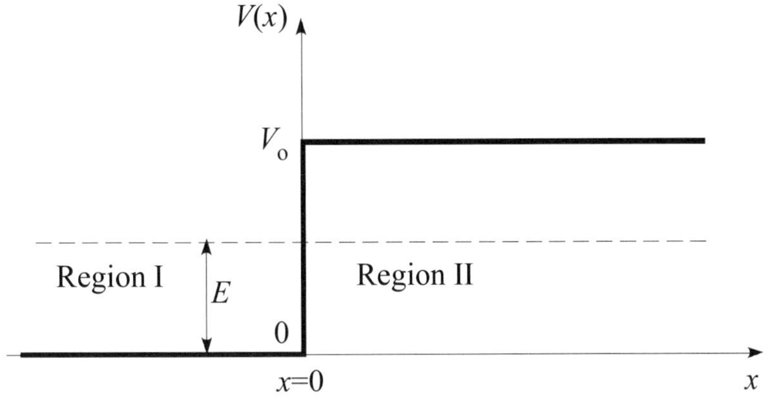

FIGURE 4.3
Potential step with height V_o greater than the energy E of incident particle.

B adb D:

$$B = \frac{k - i\kappa}{k + i\kappa}\, A\,, \tag{4.1.21}$$

and

$$D = \frac{2k}{k + i\kappa}\, A\,. \tag{4.1.22}$$

The probability current densities for the incident and reflected waves $\psi_i(x) = Ae^{ikx}$ and $\psi_r(x) = Be^{-ikx}$ are, respectively,

$$S_i = \frac{\hbar k}{m}|A|^2\,, \tag{4.1.23}$$

$$S_r = \frac{\hbar k}{m}|B|^2\,. \tag{4.1.24}$$

The reflectivity $R = S_r/S_i$ is then given by

$$R = \frac{|B|^2}{|A|^2} = \frac{|1 - i\kappa/k|^2}{|1 + i\kappa/k|^2} = 1\,. \tag{4.1.25}$$

Thus the particle is totally reflected at the step. This result is to be expected, even classically, since there can be no transmission of particles through a potential step of infinite extent if $E < V_o$. Yet there is a new feature which is not present in the classical theory. The wave function in the region II ($x > 0$) is finite and exponentially decaying. This implies that the particle can be found in Region II, whereas, classically, it is forbidden to enter this region.

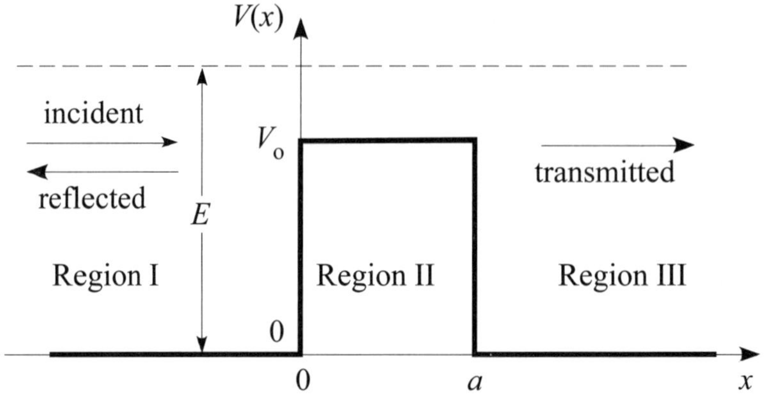

FIGURE 4.4
Potential barrier of finite extent a with height of the barrier V_o less than the energy E of the incident particle.

4.2 Passage of a Particle through a Potential Barrier of Finite Extent

Consider a particle of energy E incident from the left on a potential barrier of height V_o and width a. The potential corresponding to such a barrier has the form

$$V(x) = \begin{cases} 0\,, & -\infty < x < 0 \ \ : \text{Region I} \\ V_o\,, & 0 < x < a \ \ : \text{Region II} \\ 0\,, & a < x < \infty \ \ : \text{Region III}\,. \end{cases} \tag{4.2.1}$$

This potential is shown graphically in Fig. 4.4. As before, we consider two cases separately: (i) particle energy greater than the height of the potential barrier ($E > V_o$) and (ii) particle energy less than the height of the potential barrier ($E < V_o$).

(i) Particle energy greater than the barrier height: $E > V_o$

In this case, the time-independent Schrödinger equation [Eq. (3.5.4)] in the coordinate representation for regions I, II and III can be written as

$$\frac{d^2\psi_1(x)}{dx^2} + k^2\psi_1(x) = 0 \qquad \text{Region I} \tag{4.2.2}$$

$$\frac{d^2\psi_2(x)}{dx^2} + k'^2\psi_2(x) = 0 \qquad \text{Region II} \tag{4.2.3}$$

$$\frac{d^2\psi_3(x)}{dx^2} + k^2\psi_3(x) = 0 \qquad \text{Region III} \tag{4.2.4}$$

where

$$k = \sqrt{\frac{2mE}{\hbar^2}} \quad \text{and} \quad k' = \sqrt{\frac{2m(E - V_o)}{\hbar^2}}\,. \tag{4.2.5}$$

The solutions in the three regions are, respectively,

$$\psi_1(x) = Ae^{ikx} + Be^{-ikx} \qquad \text{Region I} \tag{4.2.6}$$

$$\psi_2(x) = Ce^{ik'x} + De^{-ik'x} \qquad \text{Region II} \tag{4.2.7}$$

$$\psi_3(x) = Ge^{ikx} + Fe^{-ikx} \qquad \text{Region III}. \tag{4.2.8}$$

Since, for a particle incident from left, there cannot be any possibility of the particle moving from right to left in region III, we choose $F = 0$. In this case, two sets of boundary conditions have to be obeyed at the potential steps at $x = 0$ and $x = a$. These are given by

$$\psi_1(x=0) = \psi_2(x=0)\,, \tag{4.2.9a}$$

$$\left.\frac{d\psi_1(x)}{dx}\right|_{x=0} = \left.\frac{d\psi_2(x)}{dx}\right|_{x=0}, \tag{4.2.9b}$$

and

$$\psi_2(x=a) = \psi_3(x=a)\,, \tag{4.2.10a}$$

$$\left.\frac{d\psi_2(x)}{dx}\right|_{x=a} = \left.\frac{d\psi_3(x)}{dx}\right|_{x=a}. \tag{4.2.10b}$$

These conditions lead, respectively, to

$$A + B = C + D\,, \tag{4.2.11a}$$

$$k(A - B) = k'(C - D)\,, \tag{4.2.11b}$$

and

$$Ce^{ik'a} + De^{-ik'a} = Ge^{ika}\,, \tag{4.2.12a}$$

$$k'(Ce^{ik'a} - De^{-ik'a}) = kGe^{ika}\,. \tag{4.2.12b}$$

If we determine the expressions for C/G and D/G from the last two conditions and substitute them into the expressions for A/G and B/G, determined from the first two conditions, we get

$$\frac{A}{G} = \frac{1}{4}\left[\left(1+\frac{k}{k'}\right)\left(1+\frac{k'}{k}\right)e^{i(k-k')a} + \left(1-\frac{k}{k'}\right)\left(1-\frac{k'}{k}\right)e^{i(k+k')a}\right]. \tag{4.2.13}$$

Now, the probability current density S_i for the incident wave $\psi_i(x) = Ae^{ikx}$ and S_t for the transmitted wave $\psi_t(x) = Ge^{ikx}$ can be obtained from Eq. (4.1.10) to be

$$S_i = \frac{\hbar k}{m}|A|^2\,,$$

$$S_t = \frac{\hbar k}{m}|G|^2\,.$$

Taking the ratio of the transmitted and incident probability current densities, using Eq. (4.2.13) for the ratio G/A and simplifying the result, we find the transmission coefficient (transmittivity) to be

$$T \equiv \frac{S_t}{S_i} = \frac{|G|^2}{|A|^2} = \frac{1}{1 + \frac{1}{4}\left(\frac{k}{k'} - \frac{k'}{k}\right)^2 \sin^2(ak')}\,. \tag{4.2.14}$$

From this expression we see that, in general, $T < 1$, so that there is always some possibility of reflection. However, when $\sin(ak') = 0$ or $ak' = N\pi$ (N = integer) then $T = 1$, which

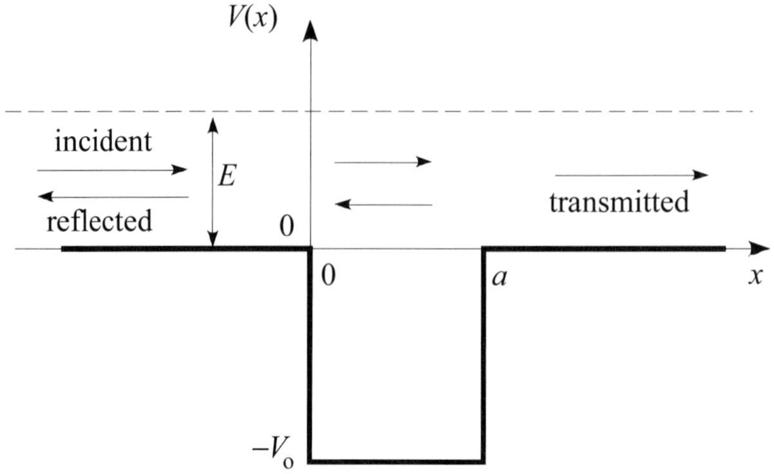

FIGURE 4.5
Transmission of a particle through a potential well of depth V_o and of finite extent a.

implies complete transmission. This is phenomenon is called *transmission resonance*. In terms of de Broglie wavelength inside the barrier $\lambda' = 2\pi/k'$, the condition for transmission resonance can be written as $a = N\pi/k' = N\lambda'/2$. Hence the transmission resonances occur whenever an integer number of half (de Broglie) wavelengths can fit inside the barrier. We also note that if $k' = k$ ($V_o = 0$), there is complete transmission, which is to be expected, even classically.

The same formula for the transmission coefficient is applicable if there is a potential well instead of a potential barrier, i.e., the height of the potential barrier $+V_o$ is replaced by $-V_o$ (Fig. 4.5) so that the potential is given by

$$V(x) = \begin{cases} 0, & -\infty < x < 0 \\ -V_o, & 0 < x < a \\ 0, & a < x < \infty\,. \end{cases}$$

In this case the transmission coefficient is given by

$$T = \frac{S_t}{S_i} = \frac{1}{1 + \frac{1}{4}(\frac{k}{k''} - \frac{k''}{k})^2 \sin^2(ak'')}\,, \tag{4.2.15}$$

where

$$k'' = \sqrt{\frac{2m(E + V_o)}{\hbar^2}}\,. \tag{4.2.16}$$

Here also we have $T < 1$, in general. So there is a possibility of reflection at the boundary $x = 0$. This is totally unexpected according to classical mechanics. On the other hand, when $\sin(ak'') = 0$ or $ak'' = N\pi$ (N is an integer), then there is complete transmission. This is called *transmission resonance*. Such transmission resonances for specific energies occur in the scattering of electrons from noble gasses like Argon and Neon. This effect is also known as the *Ramsauer-Townsend effect*.

(ii) Particle energy less than the barrier height: $E < V_o$

Consider case of a particle incident from the left with energy less than the barrier height as shown in Fig. 4.6. According to classical mechanics there cannot be any transmission

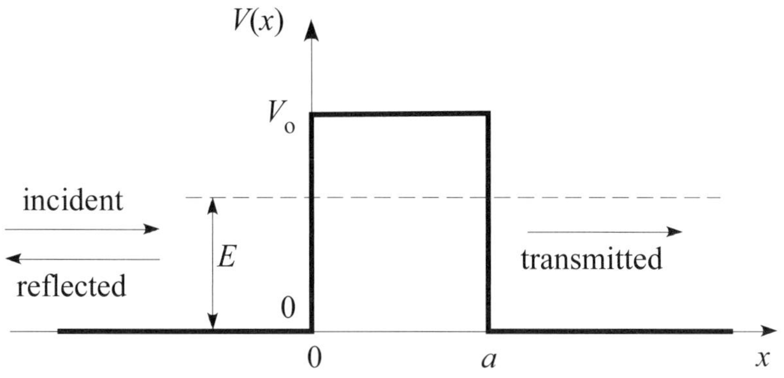

FIGURE 4.6
Transmission of a particle through a potential barrier of finite extent and height V_o greater than the energy E of the incident particle.

of particles in this case since $E < V_o$. However, as we shall see, according to quantum mechanics, there exists a finite (non-zero) probability that the particle with $E < V_o$ would cross the barrier if it is of finite extent.

As before, the time-independent Schrödinger equation in the coordinate representation in the three regions can be written as

$$\frac{d^2\psi_1(x)}{dx^2} + k^2\psi_1(x) = 0\,, \qquad -\infty < x < 0 \;:\; \text{Region I} \tag{4.2.17}$$

$$\frac{d^2\psi_2(x)}{dx^2} - \kappa^2\psi_2(x) = 0\,, \qquad 0 < x < a \;:\; \text{Region II} \tag{4.2.18}$$

$$\frac{d^2\psi_3(x)}{dx^2} + k^2\psi_3(x) = 0\,, \qquad a < x < \infty \;:\; \text{Region III} \tag{4.2.19}$$

where

$$k = \sqrt{\frac{2mE}{\hbar^2}} \quad \text{and} \quad \kappa = \sqrt{\frac{2m(V_o - E)}{\hbar^2}}\,. \tag{4.2.20}$$

The solutions in the three regions are of the form

$$\psi_1(x) = Ae^{ikx} + Be^{-ikx} \qquad \text{Region I} \tag{4.2.21}$$

$$\psi_2(x) = Ce^{\kappa x} + De^{-\kappa x} \qquad \text{Region II} \tag{4.2.22}$$

$$\psi_3(x) = Ge^{ikx} + Fe^{-ikx} \qquad \text{Region III}\,. \tag{4.2.23}$$

We must choose $F = 0$ because for particles incident on the barrier from the left, we cannot have a wave traveling from right to left in Region III. We have two sets of boundary conditions to be satisfied at $x = 0$ and $x = a$:

$$\psi_1(0) = \psi_2(0)$$

$$\left.\frac{d\psi_1}{dx}\right|_{x=0} = \left.\frac{d\psi_2}{dx}\right|_{x=0}$$

$$\psi_2(a) = \psi_3(a)$$

$$\left.\frac{d\psi_2}{dx}\right|_{x=a} = \left.\frac{d\psi_3}{dx}\right|_{x=a}\,.$$

These boundary conditions give us, after some simplification,

$$\frac{A}{G} = e^{iak}\left[\cosh(\kappa a) - \frac{i}{2}\left(\frac{k}{\kappa} - \frac{\kappa}{k}\right)\sinh(\kappa a)\right]. \quad (4.2.24)$$

The probability current densities for the incident wave $\psi_i(x) = Ae^{ikx}$ and the transmitted wave $\psi_t(x) = Ge^{ikx}$ are, respectively, according to Eq. (4.1.10),

$$S_i = \frac{\hbar k}{m}|A|^2 \quad \text{and} \quad S_t = \frac{\hbar k}{m}|G|^2 .$$

Using these probability current densities we find that the transmission coefficient $T = S_t/S_i$ is finite and given by

$$T = \frac{S_t}{S_i} = \frac{|G|^2}{|A|^2} = \frac{1}{1 + \frac{1}{4}\left(\frac{k}{\kappa} + \frac{\kappa}{k}\right)^2 \sinh^2(\kappa a)} . \quad (4.2.25)$$

Even in the case when the height of the barrier $V_o \gg E$ or $\kappa/k \gg k/\kappa$ and $\sinh \kappa a \approx e^{\kappa a}/2$, we find finite transmission given by

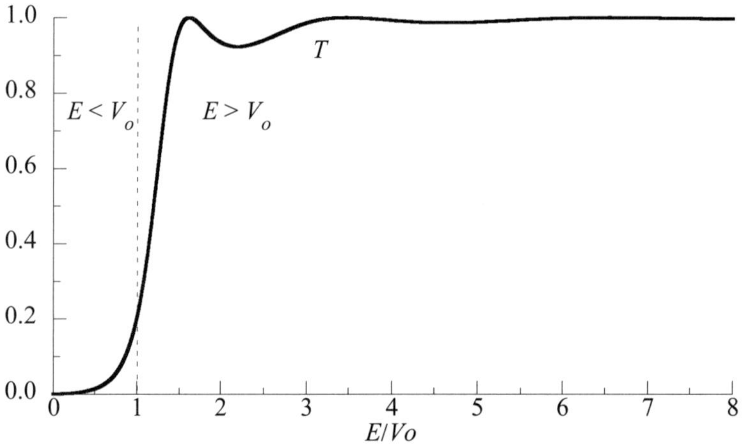

FIGURE 4.7
Transmission coefficient as function of E/V_o.

$$T \approx 16(E/V_o) \times e^{-2\kappa a} . \quad (4.2.26)$$

Hence there is a finite probability of tunneling of a particle through a potential barrier of height V_o provided the width a of the barrier is finite. This probability, of course, decreases rapidly as the barrier gets thicker (a increases) or height V_o increases.

The transmission coefficient of a particle, for a barrier of finite width a and height V_o is plotted in Fig. 4.7 as a function of particle energy E. We see that even for particle energies insufficient to surmount the potential barrier ($E < V_o$), the transmission coefficient is non-zero. For particle energies exceeding the barrier height ($E > V_o$) perfect transmission ($T = 1$) occurs at resonance energies given by the relation

$$k'a = \sqrt{\frac{2m(E - V_o)}{\hbar^2}}\, a = N\pi , \quad N = 0, 1, 2, \cdots \quad (4.2.27)$$

The conclusion that a particle could tunnel through a potential barrier of finite width even if its energy is less than the height of the barrier led Gamow to provide an explanation for the α-decay of a nucleus, in which an α-particle with energy less than the height of the Coulomb barrier could leak through the barrier and come out of the nucleus.

4.3 Tunneling of a Particle through a Potential Barrier

It is well known that α-particles are held inside the nucleus by strong attractive forces of short range of the order of nuclear radius R. Outside the range the only force is the Coulomb repulsive force which decreases with increasing distance. The potential energy curve for this problem is shown in Fig. (4.8). Mathematically, we can represent this potential as

$$V(r) = \begin{cases} -U\,, & r < R \\ \dfrac{2Ze^2}{4\pi\epsilon_0 r}\,, & r > R \end{cases} \tag{4.3.1}$$

where R is the nuclear radius. The attractive nuclear force dominates inside the nucleus ($r < R$). Outside the nucleus Coulomb interaction between the α particle (charge $+2e$) and the daughter nucleus (charge Ze) dominates.

In alpha-decay, the height of the Coulomb barrier at $r = R$

$$V_c(R) = \frac{2Ze^2}{4\pi\epsilon_0 R} \equiv V_o \tag{4.3.2}$$

is found to be much larger than the energy E of the α-particle escaping through the potential barrier. Classical mechanics cannot explain how an α-particle with energy $E < V_o$ is able to overcome the barrier. Gamow explained this on the basis of quantum mechanics and this effect came to be known as *quantum mechanical tunneling effect.*

For the physically realistic potential (potential well of depth U for $r < R$ and Coulomb barrier for $r > R$) [Fig. 4.8], this problem can actually be treated by using an approximation called WKBJ approximation [Chapter 8]. However, this problem was first qualitatively treated by Gamow for an idealized potential that has the shape shown in Fig. 4.9(a). It consists of a potential well of depth U surrounded by a potential barrier of finite height V_o and finite width $a = b - R$. Mathematically this potential can be expressed as

$$V(r) = \begin{cases} -U\,, & 0 \le r < R & : \text{Region I} \\ V_o\,, & R < r < b & : \text{Region II} \\ 0\,, & b < r < \infty & : \text{Region III}\,. \end{cases} \tag{4.3.3}$$

This is a three-dimensional problem with a spherically symmetric potential. It may be noted here (and it will be discussed in detail in the next chapter) that in a problem with spherically symmetric potential, the Schrödinger equation in the coordinate representation [Eq. (3.5.7), Chapter 3], may be separated into a radial and an angular equation. The substitution $u(r) = rR(r)$ reduces the radial equation further to the form of a one-dimensional Schrödinger equation with an additional term (the centrifugal potential) in the potential. This additional term arises due to the angular momentum of the particle. By considering only the particle motion with zero angular momentum, the radial can be written as

$$\left[-\frac{\hbar^2}{2m}\frac{d^2}{dr^2} + V(r)\right] u(r) = Eu(r)\,. \tag{4.3.4}$$

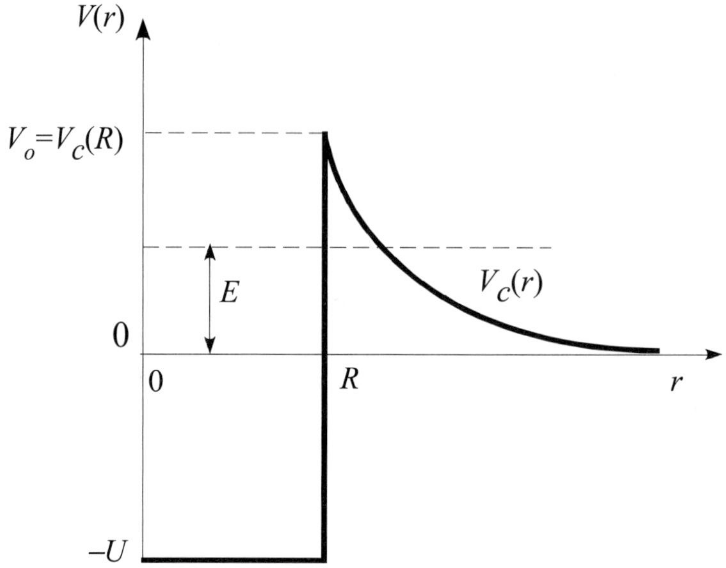

FIGURE 4.8
The nuclear potential well and the Coulomb barrier for an α-particle. R is the nuclear radius.

This is exactly of the form of one-dimensional Schrödinger equation [Eq. (3.5.4), Chapter 3] with x replaced by r and $\psi(x)$ replaced by $u(r)$.

Let us denote the radial function in the three regions by $u_1(r)$, $u_2(r)$ and $u_3(r)$. Then the Schrödinger equation for the three regions may be written as

$$\frac{d^2u_1(r)}{dr^2} + \delta^2 u_1(r) = 0\,, \qquad 0 < r < R \ : \ \text{Region I} \tag{4.3.5}$$

$$\frac{d^2u_2(r)}{dr^2} - \kappa^2 u_2(r) = 0\,, \qquad R < r < b \ : \ \text{Region II} \tag{4.3.6}$$

$$\frac{d^2u_3(r)}{dr^2} + k^2 u_3(r) = 0\,, \qquad b < r < \infty \ : \ \text{Region III} \tag{4.3.7}$$

where

$$\delta = \sqrt{\frac{2m(E+U)}{\hbar^2}}\,, \quad \kappa = \sqrt{\frac{2m(V_o - E)}{\hbar^2}}\,, \quad \text{and} \quad k = \sqrt{\frac{2mE}{\hbar^2}}\,. \tag{4.3.8}$$

The solutions in the three regions can be written as

$$u_1(r) = A\sin\delta r + A_1\cos\delta r\,, \tag{4.3.9}$$

$$u_2(r) = B_+ e^{\kappa r} + B_- e^{-\kappa r}\,, \tag{4.3.10}$$

$$u_3(r) = Ce^{ik(r-b)} + De^{-ik(r-b)}\,, \tag{4.3.11}$$

where we have written the constants in u_3 as Ce^{-ikb} and De^{ikb} in anticipation of the boundary condition at $r = b$. Since $u_1(r)$ must vanish at $r = 0$ so that the radial wave function is finite at the origin, we must choose $A_1 = 0$. In region III, we cannot have an incoming wave (corresponding to particle flux incident from the right). Therfore, we must put $D = 0$.

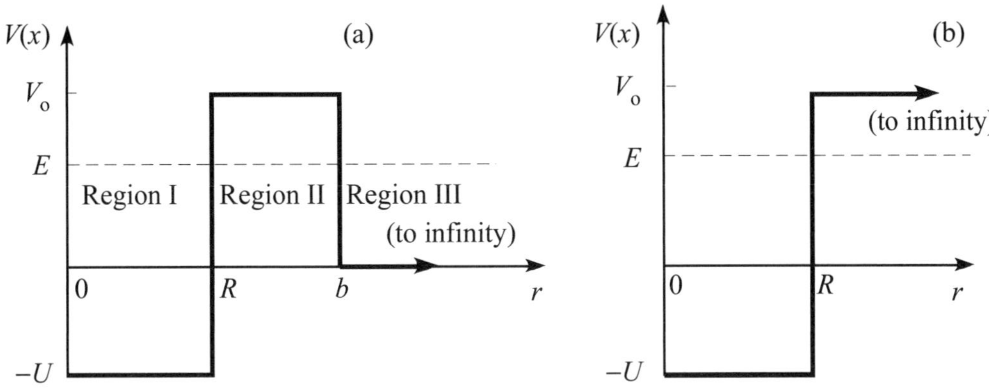

FIGURE 4.9
(a) A potential well of depth U and adjacent potential barrier of height V_o and width $a = b - R$; (b) In the limit $b \to \infty$ the barrier extends to infinity.

The solutions in the three regions, as well as their first derivatives, must satisfy the following boundary conditions at $r = R$ and $r = b$

$$u_1(R) = u_2(R)\,, \tag{4.3.12a}$$

$$\left.\frac{du_1}{dr}\right|_{r=R} = \left.\frac{du_2}{dr}\right|_{r=R}\,, \tag{4.3.12b}$$

$$u_2(b) = u_3(b)\,, \tag{4.3.13a}$$

$$\left.\frac{du_2}{dr}\right|_{r=b} = \left.\frac{du_3}{dr}\right|_{r=b}\,. \tag{4.3.13b}$$

These boundary conditions lead to the following equations

$$A\sin\delta R - B_+e^{\kappa R} - B_-e^{-\kappa R} = 0\,, \tag{4.3.14a}$$

$$A\frac{\delta}{\kappa}\cos\delta R - B_+e^{\kappa R} + B_-e^{-\kappa R} = 0\,, \tag{4.3.14b}$$

$$B_+e^{\kappa b} + B_-e^{-\kappa b} - C = 0\,, \tag{4.3.15a}$$

$$B_+e^{\kappa b} - B_-e^{-\kappa b} - \frac{ik}{\kappa}C = 0\,. \tag{4.3.15b}$$

The condition that these equations give a nontrivial solution for $A\,, B_+\,, B_-\,$, and C is that the determinant

$$\begin{vmatrix} \sin\delta R & -e^{\kappa R} & -e^{-\kappa R} & 0 \\ \frac{\delta}{\kappa}\cos\delta R & -e^{\kappa R} & e^{-\kappa R} & 0 \\ 0 & e^{\kappa b} & e^{-\kappa b} & -1 \\ 0 & e^{\kappa b} & -e^{-\kappa b} & -\frac{ik}{\kappa} \end{vmatrix} = 0\,. \tag{4.3.16}$$

On simplification, this gives

$$e^{\kappa a}(k + i\kappa)\,(\kappa\sin\delta R + \delta\cos\delta R) + e^{-\kappa a}(k - i\kappa)\,(\kappa\sin\delta R - \delta\cos\delta R) = 0\,, \tag{4.3.17}$$

where $a = b - R$ is the width of the potential barrier. This transcendental equation determines particle energy E permitted by the potential well.

To gain some insight into the problem, let us consider particle energy E not too close to the top of the barrier or a barrier that is not too thin. Then, as a first approximation, we can neglect $e^{-\kappa a}$ compared to $e^{\kappa a}$ in Eq. (4.3.17). This gives us $\kappa \sin \delta R + \delta \cos \delta R = \kappa \sin \delta R\,[1 + (\delta/\kappa) \cot \delta R] = 0$. Using the definition of κ and δ in terms of particle energy [Eq. (4.3.8)], this leads us to

$$1 + \frac{\delta}{\kappa} \cot \delta R = 1 + \sqrt{\frac{E+U}{V_o - E}} \cot \left(\sqrt{\frac{2m(E+U)}{\hbar^2}} R \right) = 0\,. \tag{4.3.18}$$

The energies E_n, determined by this equation, would be the exact energy levels of the particle if $b \to \infty$ or $e^{-\kappa a} \to 0$. In other words these are the exact energy levels of a particle in a potential well of the kind given by Fig. 4.9(b) when the barrier, of height V_o, is of infinite extent. However, when b is finite and $e^{-\kappa(b-R)}$ is small but not equal to zero, the energy of the particle in the potential well, as given by the above expression, is only an approximation. To find better estimates for the allowed energies than those given by Eq. (4.3.18), let us denote a particular solution of Eq. (4.3.18) by parameters κ_0, δ_0, k_0 which correspond to the approximate energy level E_0. To obtain an improved approximation we put

$$E = E_0 + \Delta E\ ,\ \kappa = \kappa_0 + \Delta\kappa\ ,\ k = k_0 + \Delta k\ ,\ \delta = \delta_0 + \Delta\delta\,,$$

where we have $k_0 \Delta k = -\kappa_0 \Delta\kappa = \delta_0 \Delta\delta = m\Delta E/\hbar^2$ in view of Eq. (4.3.8).

Substituting for κ, k, and δ in the exact Eq. (4.3.17) and simplifying with the help of Eq. (4.3.18), we obtain a complex value for E with

$$\Delta E = E' - iE''\,, \tag{4.3.19}$$

where

$$E' = \frac{2\hbar^2}{m} e^{-2\kappa_0 a} \left(\frac{k_0^2 - \kappa_0^2}{k_0^2 + \kappa_0^2} \right) \left(\frac{1}{R\kappa_0 + 1} \right) \left(\frac{\delta_0^2 \kappa_0^2}{\delta_0^2 + \kappa_0^2} \right), \tag{4.3.20}$$

$$E'' = \frac{2\hbar^2}{m} e^{-2\kappa_0 a} \left(\frac{1}{R\kappa_0 + 1} \right) \left(\frac{\delta_0^2 \kappa_0^2}{\delta_0^2 + \kappa_0^2} \right) \left(\frac{2\kappa_0 k_0}{k_0^2 + \kappa_0^2} \right). \tag{4.3.21}$$

We can interpret the complex energy as follows. In a stationary state, energy E is real and the probability $P = \int_0^R dr |u_1(r) e^{-iEt/\hbar}|^2 = \int_0^R dr |u_1(r)|^2$ of finding the particle within the region of the potential well is constant in time. But, if for a certain state energy $E = E_0 + \Delta E = E_0 + E' - iE''$ is complex, then this probability varies with time according to

$$P(t) = \int_0^R dr |u_1(r) e^{-iEt/\hbar}|^2 = \int_0^R dr |u_1(r)|^2 e^{-2E''t/\hbar}\,. \tag{4.3.22}$$

Thus the probability of finding the particle in the potential well decays exponentially with time as it should do for radioactive decay. We can write the probability

$$P(t) = P(0) e^{-t/\tau} = P(0) e^{-\lambda t}\,, \tag{4.3.23}$$

where τ is the mean lifetime and $\lambda = 1/\tau$ is the decay constant. The mean lifetime $\tau = 1/\lambda$ is given

$$\tau = \frac{1}{\lambda} = \frac{m}{4\hbar} e^{2\kappa_0 a} (\kappa_0 R + 1) \left(\frac{1}{\delta_0^2} + \frac{1}{\kappa_0^2} \right) \left(\frac{k_0^2 + \kappa_0^2}{2k_0 \kappa_0} \right), \tag{4.3.24}$$

where

$$k_0 = \sqrt{\frac{2mE_0}{\hbar^2}}\,,\ \kappa_0 = \sqrt{\frac{2m}{\hbar^2}(V_o - E_0)}\,,\ \text{and}\ \delta_0 = \sqrt{\frac{2m}{\hbar^2}(E_0 + U)}\,. \tag{4.3.25}$$

Thus the probability for the particle to be in the nucleus decays exponentially in time in accordance with the observed law of radio-active decay. Thus Gamow was able to explain why an α-particle is able to leak out of the potential well even if its energy E is less than the height V_o of the barrier of finite thickness a. If, instead of the barrier shown in Fig. 4.9(a), we have the Coulomb barrier of Fig. 4.8, then qualitatively the same explanation will hold. A more realistic treatment of α-decay, considering the Coulomb barrier, is given using the WKBJ approximation in Chapter 8.

4.4 Bound States in a One-dimensional Square Potential Well

Complex energy encountered in the previous example means that states with energy $E > V_\infty$, where V_∞ is the potential far from the origin, that are localized in the well do not exist. For localized solutions, particle motion and therefore the integral $\int_0^\infty |\psi(r)|^2 dr$ must be bounded. This is clearly not the case with the wave function given by Eqs. (4.3.11) for a finite barrier. On the other hand, if the barrier extends all the way to infinity, the solution $u_2(r)$ with $B_+ = 0$ is valid throughout the region $R \le r < \infty$; it vanishes as $r \to \infty$ and represents bounded motion. This is a necessary but not sufficient condition for stationary solutions which are localized in the well (bound state solutions). To explore bound state solutions let us consider a one dimensional potential well [Fig 4.10]

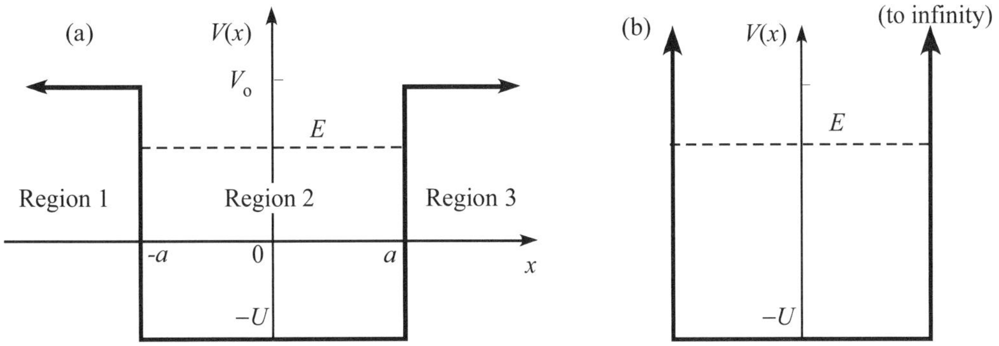

FIGURE 4.10
(a) One-dimensional potential well of finite depth and (b) one-dimensional potential with infinitely repulsive walls (one-dimensional box).

$$V(x) = \begin{cases} -U\,, & -a \le x \le a \\ V_0\,, & |x| > a\,. \end{cases} \tag{4.4.1}$$

For bound states we must look for solutions that vanish as $|x| \to \infty$. This is possible only for $E < V_0$. We are therefore looking for stationary solutions with $-U \leq E \leq V_0$. Once again we have to solve the Schrödinger equation with the potential (4.4.1) in three different regions. Let us denote the wave function in the three regions by $\psi_1(x)$, $\psi_2(x)$ and $\psi_3(x)$. Then the Schrödinger for the three regions may be written as

$$\frac{d^2\psi_1(x)}{dx^2} - \kappa^2\psi_1(x) = 0\,, \qquad -\infty < x < -a \ : \text{Region I} \tag{4.4.2}$$

$$\frac{d^2\psi_2(x)}{dx^2} + k^2\psi_2(x) = 0\,, \qquad -a < x < a \ : \text{Region II} \tag{4.4.3}$$

$$\frac{d^2\psi_3(x)}{dx^2} - \kappa^2\psi_3(x) = 0\,, \qquad a < x < \infty \ : \text{Region III} \tag{4.4.4}$$

where

$$\kappa = \sqrt{\frac{2m(V_o - E)}{\hbar^2}} \ \text{and} \ \ k = \sqrt{\frac{2m(E+U)}{\hbar^2}}\,. \tag{4.4.5}$$

The solutions in the three regions can be written as

$$\psi_1(x) = A_1 e^{\kappa x} + B_1 e^{-\kappa x}\,, \tag{4.4.6a}$$

$$\psi_2(x) = A_2 \sin kx + B_2 \cos kx\,, \tag{4.4.6b}$$

$$\psi_3(x) = A_3 e^{\kappa x} + B_3 e^{-\kappa x}\,. \tag{4.4.6c}$$

The boundedness of the solution as $x \to -\infty$ requires that $B_1 = 0$. Similarly, the boundedness as $x \to \infty$ requires $A_3 = 0$. Thus the bounded solution has the form

$$\psi_1(x) = A_1 e^{\kappa x} \qquad -\infty < x < -a\,, \tag{4.4.7a}$$

$$\psi_2(x) = A_2 \sin kx + B_2 \cos kx, \qquad -a < x < a\,, \tag{4.4.7b}$$

$$\psi_3(x) = B_3 e^{-\kappa x} \qquad a < x < \infty\,. \tag{4.4.7c}$$

As seen before, the solutions of the Schrödinger equation and their derivatives must be continuous across the boundaries. Matching the solutions and their derivatives at $x = \pm a$ leads us to

$$A_1 e^{-\kappa a} = -A_2 \sin ka + B_2 \cos ka\,, \tag{4.4.8a}$$

$$\kappa A_1 e^{-\kappa a} = kA_2 \cos ka + kB_2 \sin ka\,, \tag{4.4.8b}$$

$$B_3 e^{-\kappa a} = A_2 \sin ka + B_2 \cos ka\,, \tag{4.4.8c}$$

$$-\kappa B_3 e^{-\kappa a} = kA_2 \cos ka - kB_2 \sin ka\,. \tag{4.4.8d}$$

Thus we have a set of four coupled linear equations in the unknown coefficients A_1, A_2, B_2 and B_3. The conditions for a nontrivial solution of these equations is that the determinant

$$\begin{vmatrix} -\sin ka & \cos ka & -e^{-\kappa a} & 0 \\ k\cos ka & k\sin ka & -\kappa e^{-\kappa a} & 0 \\ \sin ka & \cos ka & 0 & -e^{-\kappa a} \\ k\cos ka & -k\sin ka & 0 & \kappa e^{-\kappa a} \end{vmatrix}$$

$$= e^{-2\kappa a}\left(2\kappa \sin ka + 2k\cos ka\right)\left(-2\kappa\cos ka + 2k\sin ka\right) = 0\,. \tag{4.4.9}$$

This leads to the condition

$$k \tan ka = \kappa\,, \tag{4.4.10}$$

or

$$k \cot ka = -\kappa\,. \tag{4.4.11}$$

It is not possible to satisfy these conditions simultaneously. Hence we have two classes of solutions governed, respectively, by (4.4.10) and (4.4.11). However, in either of these cases, energy E is the only unknown quantity since both κ and k depend on it via Eq. (4.4.5). Since Eqs. (4.4.10) and (4.4.11) can be satisfied only for certain values of E, it is clear that quantum mechanically, localized solutions are possible only for certain special values of energy. Thus in contrast to unbounded motion discussed in Secs. 4.1 through 4.3, bounded motion is allowed only for certain discrete values of energy. These values correspond to bound states, where the particle is localized in the well.

An inspection of Eqs. (4.4.8) shows that the first constraint (4.4.10) corresponds to $A_2 = 0$ and the second constraint (4.4.11) corresponds to $B_2 = 0$, leading to the two classes of solutions of the form

$$\text{Class I}: \quad k\tan ka = \kappa \qquad \psi_I(x) = \begin{cases} Ae^{-\kappa|x|}\,, & |x| > a \\ Ae^{-\kappa a}\dfrac{1}{\cos ka}\cos kx\,, & -a < x < a \end{cases} \tag{4.4.12}$$

$$\text{Class II}: \quad k\cot ka = -\kappa \qquad \psi_{II}(x) = \begin{cases} Be^{-\kappa|x|}\,, & |x| > a \\ Be^{-\kappa a}\dfrac{1}{\sin ka}\sin kx\,, & -a < x < a\,. \end{cases} \tag{4.4.13}$$

It can be seen that class I solutions are even functions $\psi_I(-x) = \psi_I(x)$ of x and are said to have even parity, whereas class II solutions are odd functions $\psi_{II}(-x) = -\psi_{II}(x)$ of x and are said to have odd parity. Thus to specify a bound state uniquely, we need not only the energy but also the parity. The origin of this fact is that the potential, and therefore the Hamiltonian under consideration is symmetric (invariant) under the reflection of the coordinate (the operation of changing $x \to -x$). Whenever the Hamiltonian possesses such a symmetry, the eigenstates of the Hamiltonian are labeled by the appropriate *quantum number* reflecting this symmetry. This will be discussed more fully in Chapter 6 on symmetries in quantum mechanical systems.

It is worth noting that, if we let $b \to \infty$ in the three-dimensional tunneling problem considered in Sec. 4.4, we obtain Eq. (4.3.18), which coincides with Eq. (4.4.11) for odd parity solutions. Solutions of this equation will be considered in Chapter 5 in the context of bound states in an attractive square well potential in three dimensions. Here we concentrate on the even parity solutions.

The allowed energy values can be obtained by solving the transcendental Eqs. (4.4.10) graphically. In the limit $V_o \to \infty$, we have a particle confined to a box with infinitely repulsive walls [Fig. 4.10(b)]. Equations (4.4.10) and (4.4.12) in this case lead to

$$\cos ka = 0 \;\Rightarrow\; ka = \left(n + \frac{1}{2}\right)\pi\,, \quad n = 0, 1, 2, 3\cdots \tag{4.4.14}$$

with eigenfunctions and energy eigenvalues given by

$$\psi_I(x) = A\cos(2n+1)\pi x/2a \qquad -a \le x \le a \tag{4.4.15}$$

and
$$E_n = -U + \frac{\hbar^2}{2m}\frac{(2n+1)^2\pi^2}{4a^2}\,. \tag{4.4.16}$$

The eigenfunction vanishes outside the box. Odd parity solutions (Type II) in this limit, $V_0 \to \infty$, are obtained from Eqs. (4.4.11) and (4.4.13) to be

$$\psi_{II}(x) = A\sin n\pi x/a \qquad -a \le x \le a \tag{4.4.17}$$

with
$$E_n = -U + \frac{\hbar^2}{2m}\frac{n^2\pi^2}{4a^2}\,. \tag{4.4.18}$$

To find the energy eigenvalues in the general case, we write Eqs.(4.4.10) as

$$\tan ka = \frac{\kappa a}{ka} = \frac{\sqrt{k_o^2 a^2 - k^2 a^2}}{ka}, \tag{4.4.19}$$

where
$$k = \sqrt{\frac{2m(E+U)}{\hbar^2}}, \quad k_o = \sqrt{\frac{2m(V_o+U)}{\hbar^2}}. \tag{4.4.20}$$

Then the energy eigenvalues are obtained by plotting the functions

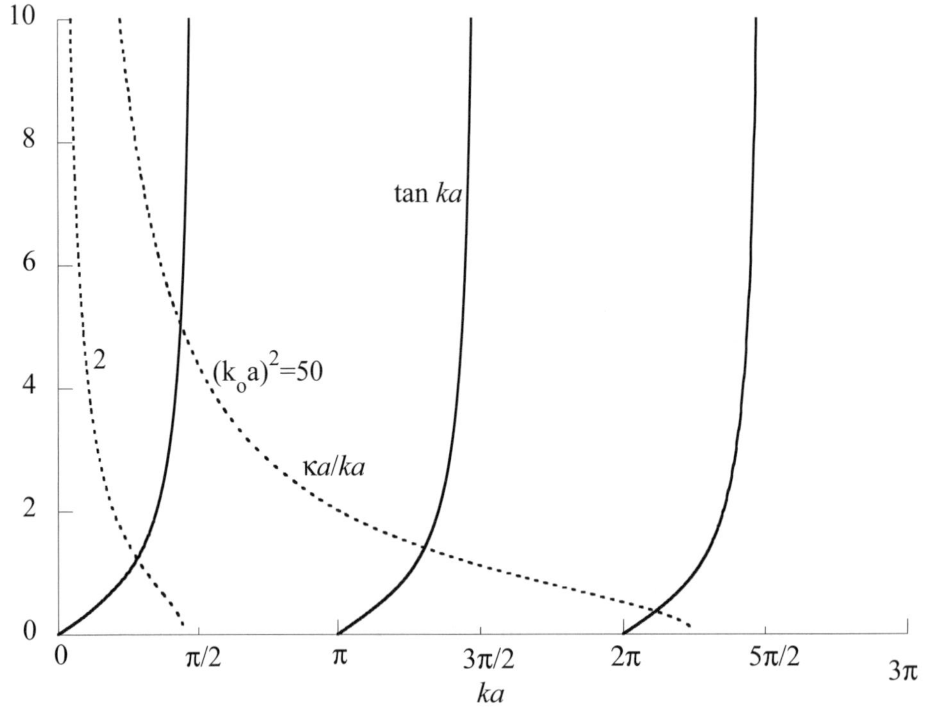

FIGURE 4.11
Graphical solution of the eigenvalue equation for the one-dimensional square potential well for even parity states. The solid curves are $y = \tan ka$ and the dashed curves are $y = \sqrt{k_o^2 a^2 - k^2 x^2}/ka$.

$$y = \tan ka \quad \text{and} \quad y = \frac{\sqrt{k_o^2 a^2 - k^2 a^2}}{ka}, \tag{4.4.21}$$

and finding their points of intersection. Figure 4.11 shows a plot of these functions. The number of intersections and, therefore, the number of bound states increases as the parameter $\frac{2m(V_o+U)a^2}{\hbar^2}$ increases. We see that there is always at least one intersection no matter how shallow or narrow the well is. Therefore at least one even-parity bound state will always exist in a one-dimensional potential well. A similar analysis for the odd-parity states shows that for at least one odd-parity bound state to exist, the well depth and range product $2m(V_0 + U)a^2/\hbar^2$ must exceed a certain value. In Chapter 5, we shall see that the condition for a bound state to exist in a three-demensional potential well coincides with Eq. (4.4.11).

It is clear that the wave function for bound states has different characteristics; it is localized inside the well and is normalizable.

4.5 Motion of a Particle in a Periodic Potential

We now consider the quantum mechanical motion of a particle in a periodic potential. An example of this would be the motion of an electron in a solid, where atomic nuclei occupy relatively fixed positions on a regular lattice. An electron moving in any direction will see a potential which is a periodic function of position. Although the actual form of this potential would be quite complex, novel features that arise in this problem can be extracted by considering a simpler model for the potential due to Kronig and Penney. In the Kronig-Penney model, we consider one-dimensional motion of a particle of mass m in a periodic potential with rectangular sections of length a (barriers) and b (valleys), such that the potential has period $L = a + b$ [Fig. (4.12)]:

$$V(x) = \begin{cases} V_o\,, & -a/2 < x < a/2 \\ 0\,, & a/2 < x < a/2 + b \end{cases} \tag{4.5.1}$$

and $$V(x + nL) = V(x)\,, \tag{4.5.2}$$

where n is an integer.

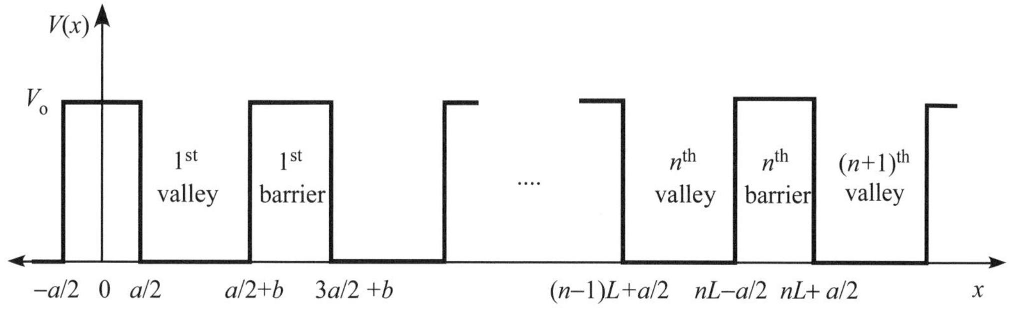

FIGURE 4.12
Periodic potential with period L, where each barrier is of height V_o and width a and each valley is of width b with $a + b = L$.

For this potential, the time-independent Schrödinger equation in, say, the n-th valley, is

$$\frac{d^2\psi(x)}{dx^2} + k^2\psi(x) = 0\,, \quad (n-1)L + a/2 < x < nL - a/2 \tag{4.5.3}$$

where $$k^2 = 2mE/\hbar^2\,. \tag{4.5.4}$$

The solution in the n-th valley will be a linear combination of $e^{\pm ikx}$, which we write as

$$\psi_n(x) = A_n e^{ik(x-nL)} + B_n e^{-ik(x-nL)}\,. \tag{4.5.5}$$

The solution in the $(n+1)$-th valley can be written by replacing n in the preceding equation by $n+1$ as

$$\psi_{n+1}(x) = A_{n+1}e^{ik[x-(n+1)L]} + B_{n+1}e^{-ik[x-(n+1)L]} \,. \tag{4.5.6}$$

The time-independent Schrödinger equation in, say, the n-th barrier, will be

$$\frac{d^2\phi(x)}{dx^2} - \kappa^2\phi(x) = 0 \,, \quad nL - a/2 < x < nL + a/2 \tag{4.5.7}$$

where $$\kappa^2 = 2m(V_o - E)/\hbar^2 = (2mV_o/\hbar^2) - k^2 \tag{4.5.8}$$

and we have assumed $E < V_o$. The solution of Eq (4.4.7) within the n-th barrier will be a linear combination of exponential functions $e^{\pm\kappa x}$, which we write as

$$\phi_n(x) = C_n e^{\kappa(x-nL)} + D_n e^{-\kappa(x-nL)} \,. \tag{4.5.9}$$

Matching the solutions and their derivatives at the boundaries $x = nL - a/2$ and $x = nL + a/2$ of the n-th barrier, we obtain the following equations

$$\psi_n(nL - a/2) = \phi_n(nL - a/2) \,, \tag{4.5.10a}$$

$$\left.\frac{d\psi_n}{dx}\right|_{x=nL-a/2} = \left.\frac{d\phi_n}{dx}\right|_{x=nL-a/2} \,, \tag{4.5.10b}$$

$$\psi_{n+1}(nL + a/2) = \phi_n(nL + a/2) \,, \tag{4.5.11a}$$

$$\left.\frac{d\psi_{n+1}}{dx}\right|_{x=nL+a/2} = \left.\frac{d\phi_n}{dx}\right|_{x=nL+a/2} \,. \tag{4.5.11b}$$

Using Eqs. (4.4.5), (4.4.6) and (4.4.9), the first two conditions give

$$\begin{pmatrix} C_n \\ D_n \end{pmatrix} = \frac{1}{2}\begin{pmatrix} e^{-ika/2}e^{-\kappa a/2}\left(1 - i\frac{k}{\kappa}\right) & e^{ika}e^{-\kappa a/2}\left(1 + i\frac{k}{\kappa}\right) \\ e^{-ika/2}e^{\kappa a/2}\left(1 + i\frac{k}{\kappa}\right) & e^{ika}e^{\kappa a/2}\left(1 - i\frac{k}{\kappa}\right) \end{pmatrix}\begin{pmatrix} A_n \\ B_n \end{pmatrix} \,, \tag{4.5.12}$$

while the last two conditions give

$$\begin{pmatrix} A_{n+1} \\ B_{n+1} \end{pmatrix} = \frac{1}{2}\begin{pmatrix} e^{-\kappa a/2}e^{ik(b+a/2)}\left(1 - \frac{\kappa}{ik}\right) & e^{\kappa a/2}e^{ik(b+a/2)}\left(1 + \frac{\kappa}{ik}\right) \\ e^{-\kappa a/2}e^{-ik(b+a/2)}\left(1 + \frac{\kappa}{ik}\right) & e^{\kappa a/2}e^{-ik(b+a/2)}\left(1 - \frac{\kappa}{ik}\right) \end{pmatrix}\begin{pmatrix} C_n \\ D_n \end{pmatrix} \,. \tag{4.5.13}$$

Eliminating C_n and D_n from Eqs. (4.4.12) and (4.4.13), we get a relation connecting the solutions in the n-th and $(n+1)$-th valley

$$\begin{pmatrix} A_{n+1} \\ B_{n+1} \end{pmatrix} = \begin{pmatrix} R_{11} & R_{12} \\ R_{21} & R_{22} \end{pmatrix}\begin{pmatrix} A_n \\ B_n \end{pmatrix} \,, \tag{4.5.14}$$

where

$$R_{12} = R_{21}^* = \frac{1}{2}e^{ik(a+b)}(-i\tau)\sinh\kappa a \,, \tag{4.5.15}$$

$$R_{11} = R_{22}^* = e^{ikb}\left(\cosh\kappa a - i\frac{\sigma}{2}\sinh\kappa a\right) \,, \tag{4.5.16}$$

and the constants τ and σ are given by

$$\sigma = \frac{\kappa}{k} - \frac{k}{\kappa} \,, \quad \tau = \frac{\kappa}{k} + \frac{k}{\kappa} \,. \tag{4.5.17}$$

Note that the matrix R, known as the transfer matrix, is independent of the integer n. It can be seen that the determinant of the matrix is unity

$$\det(R) = |R_{11}|^2 - |R_{12}|^2 = 1\,, \tag{4.5.18}$$

and its trace is given by

$$\eta \equiv \text{trace}(R) = R_{11} + R_{22} = 2\left(\cos kb \cosh \kappa a + \frac{\sigma}{2} \sin kb \sinh \kappa a\right). \tag{4.5.19}$$

Eigenvalues of the matrix R are obtained by solving the characteristic equation

$$\begin{vmatrix} R_{11} - \lambda & R_{12} \\ R_{12} & R_{11} - \lambda \end{vmatrix} = \lambda^2 - \eta\lambda + 1 = 0\,, \tag{4.5.20}$$

where we have used Eqs.(4.4.18) and (4.4.19). The roots of Eq. (4.4.20) are given by

$$\lambda_\pm = \frac{1}{2}\left(\eta \pm \sqrt{\eta^2 - 4}\right), \tag{4.5.21}$$

where $\eta = \text{trace}(R)$ is given by Eq. (4.4.19). It is easily checked that

$$\lambda_+ + \lambda_- = \text{trace}(R) = \eta\,, \tag{4.5.22a}$$
$$\lambda_+\lambda_- = \det(R) = 1\,. \tag{4.5.22b}$$

Since η is real, it follows from these conditions that the eigenvalues of R are either complex conjugates of one another and have unit magnitude (unimodular) or they are real and one of them exceeds unity. Let the eigenvectors of R corresponding to $\lambda_\pm$ be $\begin{pmatrix} a_+ \\ b_+ \end{pmatrix}$ and $\begin{pmatrix} a_- \\ b_- \end{pmatrix}$ so that

$$R\begin{pmatrix} a_+ \\ b_+ \end{pmatrix} = \lambda_+ \begin{pmatrix} a_+ \\ b_+ \end{pmatrix} \tag{4.5.23a}$$

and

$$R\begin{pmatrix} a_- \\ b_- \end{pmatrix} = \lambda_- \begin{pmatrix} a_- \\ b_- \end{pmatrix}. \tag{4.5.23b}$$

Then the column vector $\begin{pmatrix} A_0 \\ B_0 \end{pmatrix}$, which determines the solution of Schrödinger equation in the zeroth valley (around $x = 0$) can be expressed as a linear combination of the eigenvectors of R as

$$\begin{pmatrix} A_0 \\ B_0 \end{pmatrix} = C_1 \begin{pmatrix} a_+ \\ b_+ \end{pmatrix} + C_2 \begin{pmatrix} a_- \\ b_- \end{pmatrix}. \tag{4.5.24}$$

Using Eq. (4.4.14) recursively, we can express the column vector $\begin{pmatrix} A_n \\ B_n \end{pmatrix}$, which determines the solution of the Schrödinger equation in the n-th valley, in terms of the solution in the zeroth valley:

$$\begin{pmatrix} A_n \\ B_n \end{pmatrix} = R\begin{pmatrix} A_{n-1} \\ B_{n-1} \end{pmatrix} = R^2\begin{pmatrix} A_{n-2} \\ B_{n-2} \end{pmatrix} = \cdots = R^n\begin{pmatrix} A_0 \\ B_0 \end{pmatrix} = R^n\left[C_1\begin{pmatrix} a_+ \\ b_+ \end{pmatrix} + C_2\begin{pmatrix} a_- \\ b_- \end{pmatrix}\right],$$

or
$$\begin{pmatrix} A_n \\ B_n \end{pmatrix} = C_1(\lambda_+)^n \begin{pmatrix} a_+ \\ b_+ \end{pmatrix} + C_2(\lambda_-)^n \begin{pmatrix} a_- \\ b_- \end{pmatrix}. \tag{4.5.25}$$

This equation expresses the wave function in the n-th valley in terms of the wave function in the zeroth valley.[1] In order for the resulting wave function to be physically acceptable,

[1]Solution in one valley can be related to that in any other using the transfer matrix R.

it must remain finite in all regions of space including $n \to \infty$. For this to happen, the eigenvalues $\lambda_\pm$ must be less than unity. From Eqs. (4.4.21) and (4.4.22) and the comments following them, it follows that this condition will be met provided that η is restricted to the range $|\eta| < 2$ or $-2 < \eta < 2$. If $|\eta| > 2$, then $\lambda_\pm$ are real and $\lambda_+ > 1$. This implies that $\lambda_+^n \to \infty$ as $n \to \infty$ so that the wave function grows without limit as $n \to \infty$. Thus physically acceptable solutions are possible if $-2 < \eta < 2$ which, with the help of Eq. (4.4.19), leads to

$$-1 < \cosh \kappa a \cos kb - \frac{k^2 - \kappa^2}{2k\kappa} \sinh \kappa a \sin kb < 1 \,. \tag{4.5.26}$$

For values of κ and k satisfying this inequality, since $-1 < \eta/2 < 1$, we may write $\eta/2 = \cos KL$, where K is some wave number. Using this in Eq. (4.4.21), we see that the eigenvalues of the transfer matrix for allowed values of k can be written as

$$\lambda_+ = e^{iKL} = \lambda_-^* \,, \tag{4.5.27}$$

showing that they are unimodular and complex conjugates of one another.

So far we have considered the case $E < V_o$. The results for $E > V_o$ can be obtained by replacing $\kappa \to i\mathcal{K}$ and $\frac{\hbar^2 \mathcal{K}^2}{2m} = (E - V_o)$. This gives us

$$-1 \leq \cos \mathcal{K}a \cos kb - \frac{\mathcal{K}^2 + k^2}{2\mathcal{K}k} \sin \mathcal{K}a \sin kb \leq 1 \,. \tag{4.5.28}$$

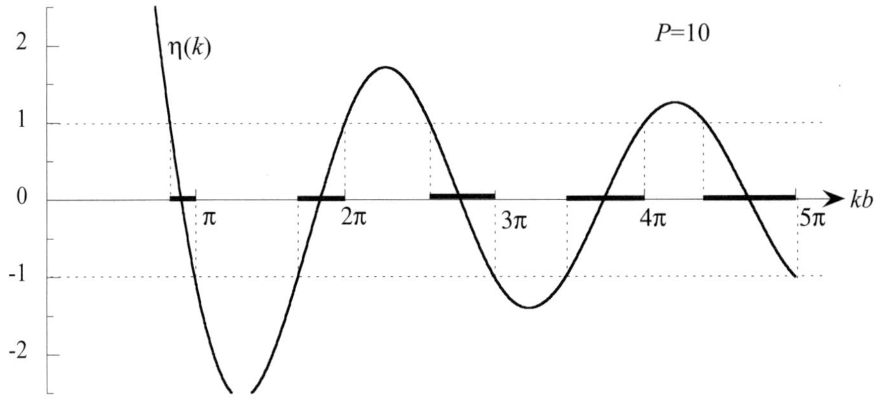

FIGURE 4.13
Plot of $\eta(k)$ as a function of kb. Allowed values of k are shown by the bold portions of the kb axis.

The new feature in this problem is that the condition for allowed energies is an inequality (4.4.26) or (4.4.28), which may be satisfied for a range, possibly continuous, of energies. By the same argument, certain other range-of-energy values, for which the inequality is violated, may be forbidden. For example, since $\cosh \kappa a \geq 1$, energy values for which $kb = N\pi$, where N is an integer, violate condition (4.4.26) and are, therefore, forbidden or lie at the edge of allowed bands. Furthermore, since sines and cosines are bounded oscillatory functions, several ranges of allowed and forbidden energy may exist. Thus the energy spectrum of a particle in a periodic potential will consist of allowed bands of energy separated by forbidden bands of energy (gaps).

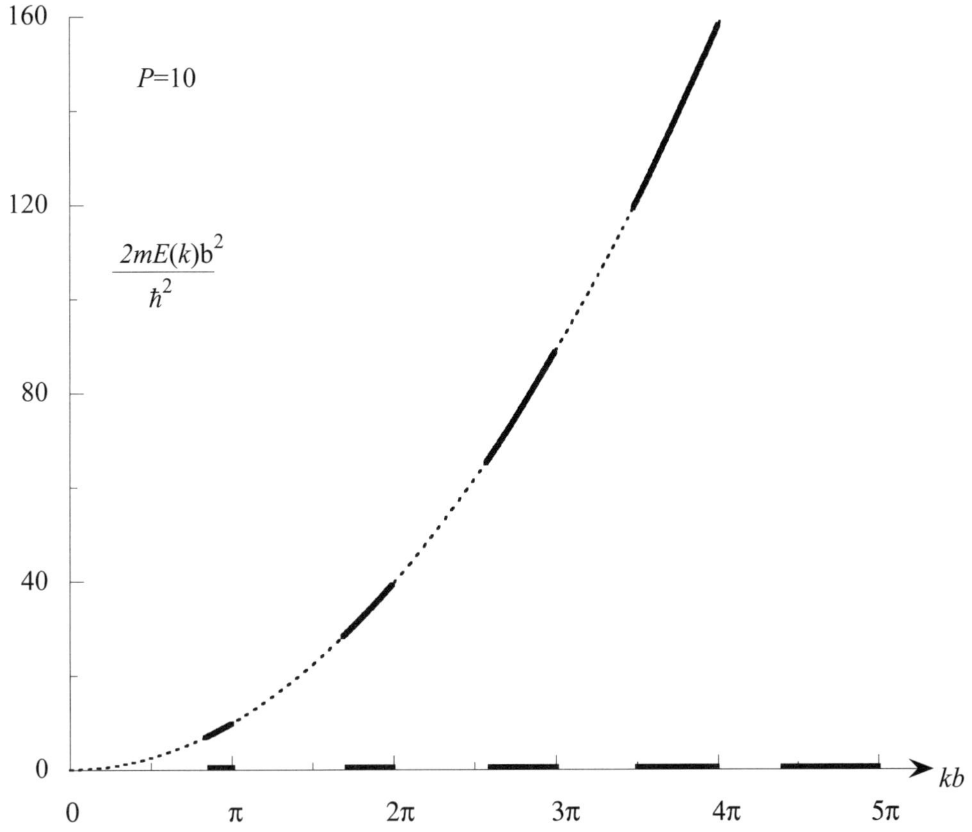

FIGURE 4.14
Plot of $E(k)$ as a function of allowed values of k. Forbidden energies are shown by the dotted portion of the curve.

We can get a clearer picture by considering the limiting case $V_o \to \infty$, $a \to 0$ such that $mbV_oa/\hbar^2 \to P$ (a finite number). In this limit $\kappa^2 \approx 2mV_o/\hbar^2$, $\sigma \approx \frac{\kappa}{k} \approx \frac{\sqrt{P}}{k}\sqrt{\frac{2}{ab}}$, and $\kappa a = \sqrt{\frac{2Pa}{b}} \ll 1$ so that Eq. (4.4.26) leads to $|\eta(k)|/2 \equiv |\cosh(\kappa a)\cos kb + (\sigma/2)\sinh(\kappa a)\sin kb| \approx |\cos kb + P\,\frac{\sin kb}{kb}| < 1$ or

$$-1 \leq \cos kb + P\,\frac{\sin kb}{kb} \leq 1 \tag{4.5.29}$$

as the condition for allowed k values (and therefore the allowed energy values $E(k) =$

$\hbar^2k^2/2m$). Here we have used $\sinh \kappa a \approx \kappa a \approx \sqrt{\frac{2Pa}{b}}$. If we plot $\cos kb + P(\sin kb/kb)$ as a function of kb for a fixed value of P, say $P = 10$, we find that for $kb = N\pi$, ($N = 0, 1, 3, 5, \cdots$) its magnitude exceeds 1. These values of k therefore correspond to forbidden energies. Figure 4.13 shows such a plot. The bold regions along the kb axis correspond to allowed values of k ($|\eta(k)|/2 < 1$) and, therefore, to allowed energies $E(k) = \hbar^2k^2/2m$. Energy levels corresponding to k values outside these regions are not allowed.

We may look at this situation in another way. For a free electron of momentum $\hbar k$, a plot of $E(k) = \hbar^2k^2/2m$ vs k is a parabola. However, for an electron moving in a periodic potential, certain values of energies are disallowed. In such a case a plot of E vs k is not a smooth parabola. The parabolic relation is interrupted for values for k close to an integral multiple of π/b [Fig. 4.14].

We have thus seen how the band structure of energy levels of an electron, in a periodic potential arises. Although the Kronig-Penney model makes a drastic assumption in treating the potential due to atoms in a lattice as a square well, the fundamental result that a periodic potential leads to allowed energy bands and forbiden energy gaps is independent of this idealization.

Problems

1. Consider the solution of time-independent Schrödinger equation in the presence of a potential step [Sec. 4.1] when particle energy is less than the step height ($E < V_o$). Examine the continuity of the wave function and its derivative across the potential step in the limit $V_o \to \infty$. Is the wave function continuous? What about the derivative of the wave function?

2. A particle of mass $M = 939$ MeV with energy E is incident from the left on a potential barrier of height V_o and of finite extent a represented by

$$V(x) = \begin{cases} 0 & x < 0 \\ V_o & 0 < x < a \\ 0 & x > a\,. \end{cases}$$

   If $V_o = 2.0$ MeV and the extent $a = 3$ fm, calculate the first two *resonant* energies for perfect transmission. (Given: $\hbar^2/M = 41.6$ Mev·fm^2.)

3. In problem 2, if the potential of height 2.0 MeV is replaced by a potential well of depth 2.0 MeV of the same extent (3.0 fm) then what would be the first two resonant energies for perfect transmission.

4. Calculate the probability of transmission of a particle of mass $M = 939$ MeV/c^2 and kinetic energy 1.0 MeV through a potential barrier of height (a) 3.0 MeV (b) 30.0 MeV and of extent 3.0 fm in both cases.

5. A particle of mass m is incident from the left on a potential barrier of finite extent represented by

$$V(x) = \begin{cases} 0 & x < 0 \\ V_o & 0 < x < a \\ 0 & x > a. \end{cases}$$

Calculate the first three resonant energies for complete transmission given $V_o a^2 = \frac{9\pi^2\hbar^2}{8m}$. Plot the transmission coefficient T as a function of E/V_o for $0 < E/V_o < 3$.

6. Consider a periodic potential

$$V(x) = \begin{cases} V_o & -a/2 < x < a/2 \\ 0 & a/2 < x < b \end{cases}$$

and $V(x + nL) = V(x)$ where $L = (a + b)$ and n is an integer. The solutions in the n-th and $(n+1)$-th valleys on the two sides of the n-th barrier are given by

$$\psi_n(x) = A_n e^{ik(x-nL)} + B_n e^{-ik(x-nL)}, \qquad (n-1)L + a/2 < x < (n-1)L + a/2 + b$$

$$\psi_{n+1}(x) = A_{n+1} e^{ik[x-(n+1)L]} + B_{n+1} e^{-ik[x-(n+1)L]}, \qquad nL + a/2 < x < nL + a/2 + b\,.$$

The solution within the barrier itself is given by

$$\phi_n(x) = C_n e^{\kappa(x-nL)} + D_n e^{-\kappa(x-nL)}, \quad (n-1)L + a/2 + b < x < nL + a/2\,.$$

Here $k^2 = 2mE/\hbar^2$ and $\kappa^2 = 2m(V_o - E)/\hbar^2$. By matching the solutions at the boundaries of the barrier, show that

$$\begin{pmatrix} A_{n+1} \\ B_{n+1} \end{pmatrix} = \begin{pmatrix} R_{11} & R_{12} \\ R_{21} & R_{22} \end{pmatrix} \begin{pmatrix} A_n \\ B_n \end{pmatrix}$$

where the matrix R is independent of n, being given by

$$R_{12} = R_{21}^* = (1/2)e^{ik(a+b)}(-i\tau)\sinh(\kappa a)$$

and

$$R_{11} = R_{22}^* = e^{ikb}\cosh(\kappa a) - i(\sigma/2)\sinh(\kappa a)$$

with

$$\tau = (\kappa/k + k/\kappa) \text{ and } \sigma = (\kappa/k - k/\kappa)\,.$$

7. In problem 6 show that the eigenvalues λ_+ and λ_- of matrix R are given by

$$\lambda_\pm = \frac{1}{2}\left(\eta \pm \sqrt{\eta^2 - 4}\right)$$

where $\eta = 2Re(R_{11}) = 2[\cosh(\kappa a)\cos kb + \frac{\sigma}{2}\sinh(\kappa a)\,\sin kb]$.

Relate the solution in the n-th valley, i.e., the column vector $\begin{pmatrix} A_n \\ B_n \end{pmatrix}$ to the solution in the zeroth valley $\begin{pmatrix} A_0 \\ B_0 \end{pmatrix}$. Hence show that if the solution in the n-th valley, for any large value of n, is to be physical then the condition $|\eta| \leq 2$ or

$$\frac{1}{2}|\eta| = |\cosh(\kappa a)\cos kb + \frac{\sigma}{2}\sinh(\kappa a)\sin kb| \leq 1 \qquad (4.5.30)$$

must hold. Hence show that energies $E = \hbar^2 k^2/2m$ of the particle, corresponding to kb in the vicinity of $n\pi$, are forbidden.

8. Consider the barrier in problem 5 and 6 to become infinite ($V_o \to \infty$) with the width a tending to zero so that $V_o a$ tends to a finite quantity, say, $V_o a \to \hbar^2 P/m$. Choose P equal to π and plot $|\eta|/2$ as a function of kb and show the forbidden regions of kb.

9. What boundary conditions must be satisfied by the bound state wave function in one dimension? Show that the bound states are nondegenerate. [Hint: Assuming that two different eigenfunctions belong to the same energy leads to a contradiction.] What can you say about the degeneracy of unbound (scattering) solutions?

10. Consider the bound states of a one-dimensional square well potential

$$V(x) = \begin{cases} -V_o & |x| < a \\ 0 & |x| > a\,. \end{cases}$$

Determine the allowed bound state energies. [Hint: You should get two conditions - one for the states that are even functions of x and the other for the states that are odd functions of x.] What is the condition for at least one (a) even-parity bound state, (b) odd-parity bound state to exist?

References

[1] L. R. B. Elton, *Introductory Nuclear Theory* (Sir Isaac Pitman and Sons Ltd., London, 1965).

[2] E. Merzbacher, *Quantum Mechanics* (John Wiley and Sons, New York, 1970).

[3] L. I. Schiff, *Quantum Mechanics*, Third Edition (McGraw Hill Book Company, Inc., New York, 1968).

5

BOUND STATES OF SIMPLE SYSTEMS

5.1 Introduction

We shall now apply the time-independent Schrödinger equation to study the bound states of simple systems.

1. A free particle in a box with sharp boundaries.
2. Particle moving in a one-dimensional harmonic potential well (a simple harmonic oscillator).
3. Two-body system with mutual central interaction between its constituents. Under this heading we shall consider (i) the problem of the Hydrogen (or Hydrogen-like) atoms and (ii) the bound state of the neutron-proton system (the deuteron).
4. A particle in three-dimensional (a) square well potential (b) harmonic oscillator potential.

As discussed in Chapter 4, a physically acceptable wave function in the coordinate space satisfies the following conditions:

(i) **Continuity**: The wave function must be single-valued and both the wave function and its first derivative must be continuous even if the potential has a (finite) discontinuity at some point. When the potential has infinite discontinuity the first derivative may be discontinuous and the wave function may have a kink.

(ii) **Boundary Conditions:** For a bound state of the system, the wave function must be finite everywhere and decrease to zero as $r \to \infty$. This condition follows from the requirement that bound state wave function must be normalized to unity in order to maintain the probabilistic interpretation of $\psi(\boldsymbol{r})$.

5.2 Motion of a Particle in a Box

Consider a particle which moves freely inside a cubical box of dimension L and volume $V = L^3$. If we choose the origin of the coordinates at one corner of the cube as in Fig. 5.1, then the particle confinement to the cube means that the walls are infinite potential steps so that the wave function must vanish at the boundary surfaces $x = 0\,, x = L\,;\ y = 0\,, x = L\,;$ $z = 0$, and $x = L$ shown in Fig. 5.1. For this problem the time-independent Schrödinger equation in the coordinate representation [see Eq. (3.5.7)] is

$$-\frac{\hbar^2}{2m}\left[\frac{\partial^2}{\partial x^2} + \frac{\partial^2}{\partial y^2} + \frac{\partial^2}{\partial z^2}\right]\psi(x, y, z) = E\psi(x, y, z)\,. \qquad (5.2.1)$$

Since the boundary surfaces coincide with the coordinate surfaces in the Cartesian coordinate system, we look for solutions separable in these coordinates by assuming the wave function $\psi(x, y, z)$ to have the form

$$\psi(x, y, z) = u_1(x)u_2(y)u_3(z)\,. \tag{5.2.2}$$

Substituting this in Eq. (5.2.1) and separating the variables we obtain

$$\frac{d^2u_1(x)}{dx^2} + k_1^2 u_1(x) = 0\,, \tag{5.2.3}$$

$$\frac{d^2u_2(y)}{dy^2} + k_2^2 u_2(y) = 0\,, \tag{5.2.4}$$

$$\frac{d^2u_3(z)}{dz^2} + k_3^2 u_3(z) = 0\,, \tag{5.2.5}$$

where

$$k_1^2 + k_2^2 + k_3^2 \equiv k^2 = \frac{2mE}{\hbar^2}\,. \tag{5.2.6}$$

The solutions of these equations can be written easily

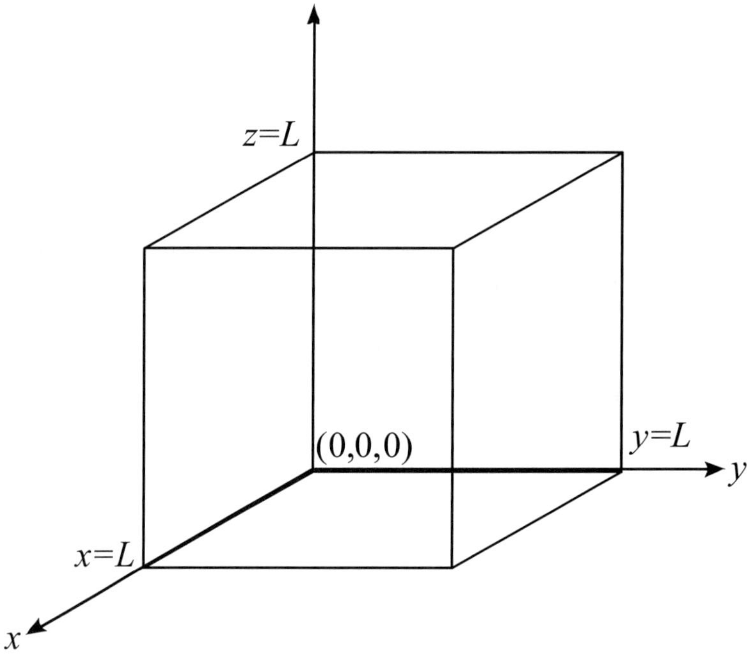

FIGURE 5.1
A three-dimensional box with impenetrable walls corresponds to a potential function which vanishes everywhere inside the box but has an infinite positive step at the walls.

$$u_1(x) = C_1 \sin k_1 x + D_1 \cos k_1 x\,, \tag{5.2.7}$$

$$u_2(y) = C_2 \sin k_2 y + D_2 \cos k_2 y\,, \tag{5.2.8}$$

$$u_3(z) = C_3 \sin k_3 x + D_3 \cos k_3 z\,. \tag{5.2.9}$$

Application of the boundary condition that the wave function vanish at the bounding surfaces of the box requires $u_1(x)$, $u_2(y)$, and $u_3(z)$ to satisfy $u_1(0) = 0 = u_1(L)$, $u_2(0) = 0 = u_2(L)$, and $u_3(0) = 0 = u_3(L)$. These requirements on Eqs. (5.2.7) – (5.2.9) lead to

$$D_1 = D_2 = D_3 = 0\,, \tag{5.2.10}$$

$$k_1 L = n_1\pi\,,\ k_2 L = n_2\pi\,,\ k_3 L = n_3\pi\,, \tag{5.2.11}$$

where n_1, n_2 and n_3 can take positive integer values 1,2,3, $\cdots$. The normalized wave functions are given by

$$\psi_{n_1,n_2,n_3}(x,y,z) = \sqrt{\frac{8}{L^3}}\ \sin\left(\frac{n_1\pi x}{L}\right)\sin\left(\frac{n_2\pi y}{L}\right)\sin\left(\frac{n_3\pi z}{L}\right). \tag{5.2.12}$$

From Eqs. (5.2.6) and (5.2.11), the energy levels are given by

$$E_n \equiv E_{n_1,n_2,n_3} = \frac{\hbar^2 k^2}{2m} = \frac{\hbar^2\pi^2}{2mL^2}(n_1^2+n_2^2+n_3^2) \equiv \frac{\hbar^2\pi^2}{2mL^2}n^2\,. \tag{5.2.13}$$

The wave function (5.2.12) may be considered the coordinate respresentative of the stationary state $|n_1, n_2, n_3\rangle$ characterized by three integer quantum numbers. Each state may be represented by a point in the positive octant of three-dimensional (n_1, n_2, n_3) space (Fig. 5.2).

Note that several states may correspond to the same energy. For example, the states $(2,1,1)$, $(1,2,1)$ and $(1,1,2)$ belong to the same value of energy $6\hbar^2\pi^2/2mL^2$. These states are said to be degenerate in energy. The number of states that have the same or very nearly the same energy increases as the energy increases. Furthermore, the spacing between the successive energy levels $\Delta E = |E_{n_1,n_2,n_3} - E_{n_1+1,n_2,n_3}| = \hbar^2\pi^2(2n_1+1)/2mL^2$ can be made as small as desired by making L sufficiently large. We shall see in later chapters that when a system makes a transition from one energy state to another, the rate of transition depends on the density of states in the vicinity of the final energy state.

5.2.1 Density of States

We may find the number of energy levels $dN(E)$ within energy interval E and $E + dE$ (or with n between n and $n + dn$) by counting the number of points that lie in a spherical shell or radius E and thickness dE. It may be noted here that although the energy E (and therefore n) is quantized, the spacing between the successive energy levels becomes infinitesimally small as L becomes large. We can then treat E and therefore n as continuous variables. Then from Eq. (5.2.13) an energy interval dE corresponds to an interval dn given by

$$dE = \frac{\hbar^2\pi^2}{2mL^2} \times 2n\,dn\,. \tag{5.2.14}$$

Now the number of points in the positive octant of a spherical shell of radius n and thickness dn is $dN = \frac{1}{8} \times 4\pi n^2 dn$. With the help of Eq. (5.2.13) and (5.2.14) we find

$$dN(E) = \frac{1}{4\pi^2}\left[\frac{2m}{\hbar^2}\right]^{3/2} L^3\sqrt{E}dE = \frac{1}{4\pi^2}\left[\frac{2m}{\hbar^2}\right]^{3/2} V\sqrt{E}dE\,, \tag{5.2.15}$$

where V is the volume of the box (occupied by the system). This is an important quantity that will be used in later chapters.

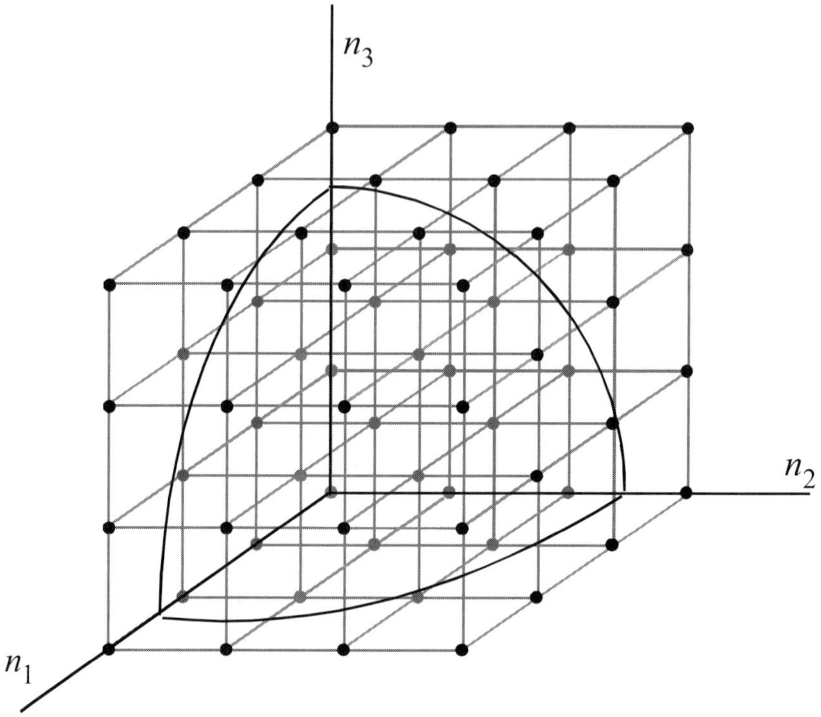

FIGURE 5.2
When a particle is enclosed in cubical box of dimension L the energy levels are discrete. Each energy state corresponds to a point in three-dimensional (n_1, n_2, n_3) space in the positive octant. The spacing between the successive energy levels can be made as small as desired by choosing L sufficiently large.

5.3 Simple Harmonic Oscillator

Among the bound state problems, one of the simplest and most important is that of a linear harmonic oscillator, that is, aparticle of mass m moving in a one-dimensional quadratic potential

$$V(x) = \frac{1}{2}kx^2 . \tag{5.3.1}$$

Classically, such a particle executes simple harmonic motion with angular frequency

$$\omega = \sqrt{\frac{k}{m}} . \tag{5.3.2}$$

To treat this problem quantum mechanically we first write the Hamiltonian operator

$$\hat{H} = \frac{\hat{p}^2}{2m} + \frac{1}{2}m\omega^2\hat{x}^2 , \tag{5.3.3}$$

where $\hat{x}$ and $\hat{p}$ are the position and momentum operators (observables) satisfying the commutation relations

$$[\hat{x}, \hat{p}] = i\hbar\hat{1} . \tag{5.3.4}$$

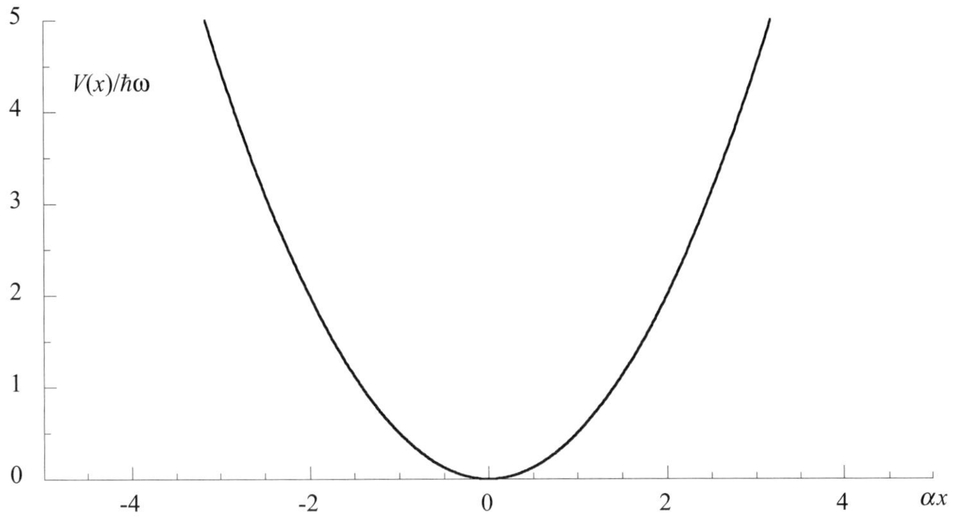

FIGURE 5.3
Harmonic oscillator potential in one dimension.

The time-independent Schrödinger equation [Eq. (3.5.1)] is

$$\hat{H}\,|E\rangle = E\,|E\rangle\,. \tag{5.3.5}$$

This equation physically means that a measurement of energy in state $|E\rangle$ will definitely give the result E. We shall see that the system can exist in stationary states belonging to certain specific energies $E_0, E_1, E_2, \cdots$. The problem is to find the allowed values of energy (energy eigenvalues) and the corresponding states (eigenstates). To do this, we write the time-independent Schrödinger equation (5.3.5) in some representation, say, the coordinate representation. Then Eq. (5.3.5) with the Hamiltonian given by Eq. (5.3.3) assumes the form

$$\left[-\frac{\hbar^2}{2m}\frac{d^2}{dx^2} + \frac{1}{2}m\omega^2 x^2\right]\psi(x) = E\psi(x)\,, \tag{5.3.6}$$

where $\psi(x) = \langle x|\,E\rangle$ is the coordinate representative of the state $|E\rangle$ and x stands for the continuously varying eigenvalue of the observable $\hat{x}$. Equation (5.3.6) may be rewritten as

$$\frac{d^2\psi(x)}{dx^2} + \left(\frac{2mE}{\hbar^2} - \frac{m^2\omega^2}{\hbar^2}x^2\right)\psi(x) = 0\,. \tag{5.3.7}$$

Introducing the variable $\xi = \sqrt{m\omega/\hbar}\; x$ and substituting $\psi(x) = u(\xi)\left(\frac{m\omega}{\hbar}\right)^{1/4}$ in Eq. (5.3.7), we find[1]

$$\frac{d^2u(\xi)}{d\xi^2} + \left(\frac{2E}{\hbar\omega} - \xi^2\right)u(\xi) = 0\,. \tag{5.3.8}$$

For large ξ (large x) this equation assumes the form

$$\frac{d^2u(\xi)}{d\xi^2} + (\mp 1 - \xi^2)u(\xi) = 0\,. \tag{5.3.9}$$

[1]Since $|\psi(x)|^2$ is interpreted as a probability density, it must conform to the transformation law for probability densities, i.e., under a change of variable $\xi = \xi(x)$, the wave function $\psi(x)$ is transformed to function $u(\xi)$ such that $|u(\xi)|^2 d\xi = |\psi(x)|^2 dx$.

Note that we have retained a term $\mp u(\xi)$ [from the $(2E/\hbar\omega)u(\xi)$ term], which is negligible anyway compared to $\xi^2 u(\xi)$ in the limit considered here, because this allows us to integrate Eq. (5.3.9) to give

$$u(\xi) \sim e^{\pm\xi^2/2} . \tag{5.3.10}$$

The boundary condition

$$\lim_{x\to\infty} \psi(x) = 0 \tag{5.3.11}$$

demands that $u \to 0$ as $\xi \to \infty$. This means we have to discard the positive exponential and write $u \sim e^{-\xi^2/2}$ as $|\xi| \to \infty$. The complete solution of Eq. (5.3.8) is therefore of the form

$$u(\xi) = e^{-\xi^2/2} v(\xi) . \tag{5.3.12}$$

Substituting this into Eq. (5.3.8), we find that $v(\xi)$ satisfies the equation

$$\frac{d^2 v(\xi)}{d\xi^2} - 2\xi \frac{dv(\xi)}{d\xi} + \left(\frac{2E}{\hbar\omega} - 1\right) v(\xi) = 0 . \tag{5.3.13}$$

In order to satisfy the boundary condition at infinity, Eq. (5.3.13) must admit a polynomial solution. This is possible only if the coefficient of $v(\xi)$ is a nonnegative even integer [see Appendix 5A1 Sec. 4 for the solution of this equation]

$$\frac{2E}{\hbar\omega} - 1 = 2n , \quad \text{where } n = 0, 1, 2, 3 \cdots \tag{5.3.14}$$

With this constraint Eq. (5.3.13) reduces to Hermite equation

$$\frac{d^2 v(\xi)}{d\xi^2} - 2\xi \frac{dv(\xi)}{d\xi} + 2nv(\xi) = 0 , \tag{5.3.15}$$

while the allowed values of energy, labeled by index n, are given by

$$E_n = \left(n + \frac{1}{2}\right) \hbar\omega . \tag{5.3.16}$$

The Hermite equation admits normalizable solutions

$$v_n(\xi) = H_n(\xi) , \tag{5.3.17}$$

called Hermite polynomials. Explicitly,

$$H_n(\xi) = (-1)^n e^{\xi^2} \frac{d^n}{d\xi^n} e^{-\xi^2} \tag{5.3.18}$$

Thus the eigenfunctions of the linear harmonic oscillator are

$$\psi_n(x) = N_n e^{-\alpha^2 x^2/2} H_n(\alpha x) , \tag{5.3.19}$$

where $\alpha = \sqrt{m\omega/\hbar}$ and the normalization constant N_n is given by

$$N_n = \left[\frac{\alpha}{2^n n! \sqrt{\pi}}\right]^{1/2} . \tag{5.3.20}$$

It may be noted that eigenfunctions of the harmonic oscillator $\psi_n(x)$ are normalized orthogonal functions. This is expressed by writing

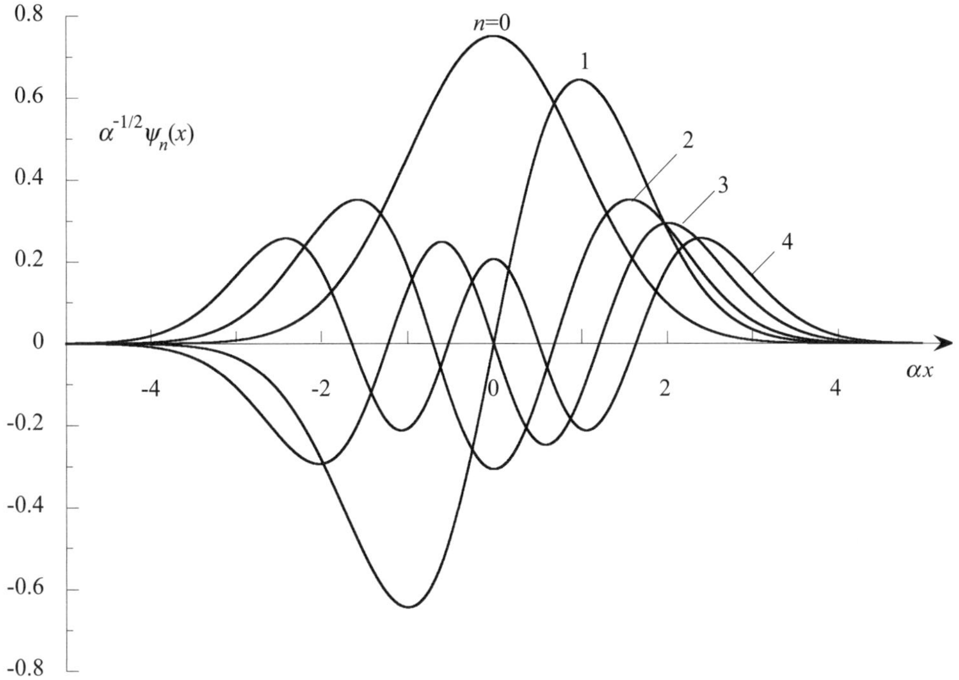

FIGURE 5.4
Harmonic oscillator wave functions for $n = 0, 1, 2, 3, 4$. Here $\alpha = \sqrt{m\omega/\hbar}$.

$$\int_{-\infty}^{\infty} dx \psi_n(x)\psi_m(x) = \delta_{mn} \,. \tag{5.3.21}$$

Figure 5.4 shows harmonic oscillator wave functions for $n = 0 - 4$. The wave function for the ground state $n = 0$ is a Gaussian centered at $x = 0$. All other wave functions oscillate as functions of x and have n zeros. Figures 5.5 and 5.6 show the probability density $|\psi_n(x)|^2$ as a function of x for $n = 0, 1$ and 10. The horizontal lines show the allowed range of a classical oscillator with the same total energy as the quantum mechanical oscillator. The classical probability distribution $P_{cl}(x)$ for finding the particle between x and $x + dx$ is easily computed as the fraction of each period $2\pi/\omega$ that the particle spends in the interval x and $x + dx$. This is given by

$$P_{cl}(x)dx = \frac{\omega}{2\pi}\frac{2dx}{v(x)} \,, \tag{5.3.22}$$

where $v(x)$ is the speed of the particle and we have used the fact that in each period the particle is found in any interval twice. If we write the position of the classical oscillator as $x(t) = A \sin \omega t$, where the amplitude A is related to the energy E by $A = \sqrt{2E/m\omega^2}$, we can express the speed of the oscillator as a function of position as

$$v(x) = \omega\sqrt{A^2 - x^2} \,. \tag{5.3.23}$$

Using this, we find the classical probability distribution for the position of the oscillator is given by

$$P_{cl}(x) = \frac{1}{\pi\sqrt{A^2 - x^2}} \,. \tag{5.3.24}$$

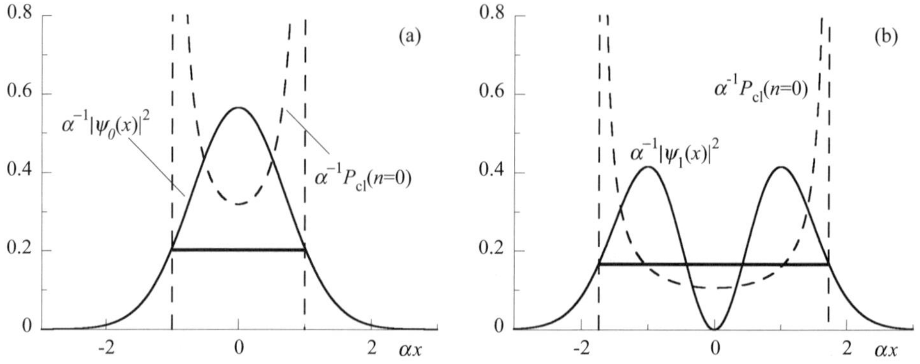

FIGURE 5.5
Harmonic oscillator probability density (a) $|\psi_0(x)|^2$ and (b) $|\psi_1(x)|^2$ compared with the classical probability density for the same average energy $P_{cl}(x) = 1/\pi\sqrt{A^2 - x^2}$.

As expected this distribution is nonzero only for $-A \leq x \leq A$ since the oscillator is confined to the region between classical turning points $x = \pm A$. In contrast to this, quantum mechanically, there is significant probability for the particle to be found in the classically forbidden region. It may be noted that the classical and quantum mechanical distributions differ signficantly for small n (See Fig. 5.5(a) and 5(b) for $n = 0, 1$) but as n increases, the average probability distribution according to quantum mechanics approaches the classical probability distribution (shown by the dashed curve for $n = 10$).

5.4 Operator Formulation of the Simple Harmonic Oscillator Problem

In the preceding section we solved the eigenvalue problem for the simple harmonic oscillator by using the coordinate representation of the time-independent Schrödinger equation. This can be done with equal ease in the momentum representation as well because of the symmetric role played by the position and momentum observables in the Hamiltonian for a simple harmonic oscillator. In fact it is possible to solve the eigenvalue problem for the harmonic oscillator without using any specific representation. To see this, we begin by writing the Hamiltonian of a linear harmonic oscillator as

$$\hat{H} = \hbar\omega\left(\frac{\hat{p}^2}{2m\hbar\omega} + \frac{m\omega}{2\hbar}\hat{x}^2\right), \tag{5.4.1}$$

where ω is the classical frequency $\omega = \sqrt{k/m}$ and the observables $\hat{x}$ and $\hat{p}$ satisfy the commutation relations (5.3.4). By factoring out $\hbar\omega$ from the Hamiltonian, we have made its coefficient in Eq. (5.4.1) dimensionless. It is clear that $\sqrt{2m\hbar\omega}$ has dimensions of momentum and $\sqrt{2\hbar/m\omega}$ has dimensions of length.

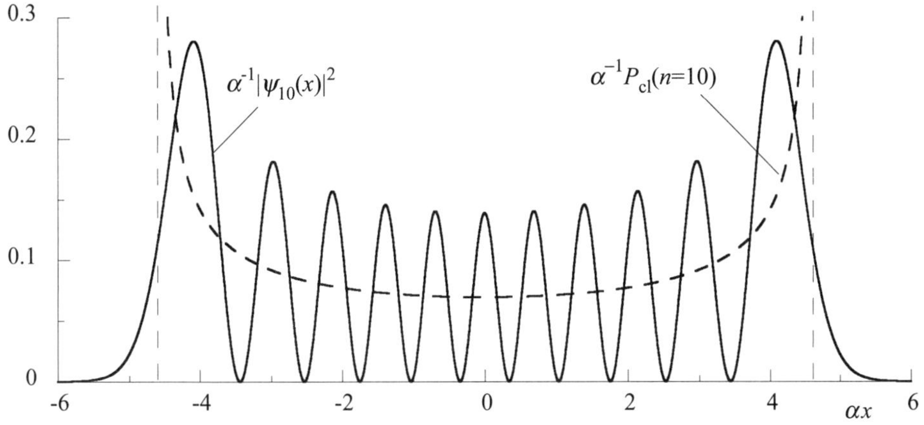

FIGURE 5.6
Probability density $|\psi_{10}(x)|^2$ as a function of x. The dotted curve shows the classical distribution. We can see that the average probability distribution according to quantum mechanics approaches the classical probability distribution, shown by dotted curve, for large n.

Let us introduce dimensionless operator $\hat{a}$ and its Hermitian adjoint $\hat{a}^\dagger$ by

$$\hat{a} = \frac{1}{\sqrt{2}}\left[\sqrt{\frac{m\omega}{\hbar}}\,\hat{x} + \frac{i}{\sqrt{m\hbar\omega}}\,\hat{p}\right], \tag{5.4.2}$$

$$\hat{a}^\dagger = \frac{1}{\sqrt{2}}\left[\sqrt{\frac{m\omega}{\hbar}}\,\hat{x} - \frac{i}{\sqrt{m\hbar\omega}}\,\hat{p}\right]. \tag{5.4.3}$$

With the help of the commutation relations of $\hat{x}$ and $\hat{p}$ [Eq. (5.3.4)], we find that $\hat{a}$ and $\hat{a}^\dagger$ satisfy the commutation relation

$$[\hat{a}^\dagger, \hat{a}] \equiv \hat{a}\hat{a}^\dagger - \hat{a}^\dagger\hat{a} = 1\,. \tag{5.4.4}$$

From the definition of $\hat{a}$ and $\hat{a}^\dagger$ we also find that

$$\hat{a}\hat{a}^\dagger = \hat{a}^\dagger\hat{a} + 1 = \frac{1}{\hbar\omega}\left[\frac{1}{2}m\omega^2\,\hat{x}^2 + \frac{\hat{p}^2}{2m} + \frac{1}{2}\hbar\omega\right]. \tag{5.4.5}$$

A comparison of Eqs. (5.4.5) and (5.4.1) allows us to write the Hamiltonian in terms of $\hat{a}$ and $\hat{a}^\dagger$ as

$$\hat{H} = \hbar\omega\left(\hat{a}^\dagger\hat{a} + \frac{1}{2}\right) = \frac{1}{2}\hbar\omega\left(\hat{a}^\dagger\hat{a} + \hat{a}\hat{a}^\dagger\right). \tag{5.4.6}$$

The commutation relations of $\hat{a}$ and $\hat{a}^\dagger$ with the Hamiltonian (5.4.6) are found, with the help of Eq. (5.4.4), to be

$$[\hat{a}, \hat{H}] = \hbar\omega[\hat{a}, \hat{a}^\dagger\hat{a} + \frac{1}{2}] = \hbar\omega(\hat{a}\hat{a}^\dagger - \hat{a}^\dagger\hat{a})\hat{a} = \hbar\omega\,\hat{a}\,, \tag{5.4.7}$$

$$[\hat{a}^\dagger, \hat{H}] = \hbar\omega[\hat{a}, \hat{a}^\dagger\hat{a} + \frac{1}{2}] = \hbar\omega\hat{a}^\dagger(\hat{a}^\dagger\hat{a} - \hat{a}\hat{a}^\dagger) = -\hbar\omega\,\hat{a}^\dagger\,. \tag{5.4.8}$$

5.4.1 Physical Meaning of the Operators $\hat{a}$ and $\hat{a}^\dagger$

Let $|E_n\rangle$ be an eigenstate of $\hat{H}$ belonging to the eigenvalue $E_n = (n + 1/2)\hbar\omega$ so that

$$\hat{H}\,|E_n\rangle = E_n\,|E_n\rangle\,. \tag{5.4.9}$$

To find out what the states $\hat{a}\,|E_n\rangle$ and $\hat{a}^\dagger\,|E_n\rangle$ represent, we let $\hat{H}$ operate on them. Operating on $\hat{a}\,|E_n\rangle$ by $\hat{H}$ and using the commutation relation (5.4.7), we find

$$\hat{H}\,(\hat{a}\,|E_n\rangle) = (\hat{a}\hat{H} - \hbar\omega\,\hat{a})\,|E_n\rangle = (E_n - \hbar\omega)\,(\hat{a}\,|E_n\rangle)\,. \tag{5.4.10}$$

From this equation we see that $\hat{a}\,|E_n\rangle$ is an eigenstate of $\hat{H}$ belonging to eigenvalue $E_n - \hbar\omega \equiv E_{n-1}$. Hence $\hat{a}$ may be called a *lowering operator* since operating on the eigenstate $|E_n\rangle$ by $\hat{a}$ results in lowering of energy by one quantum of energy ($\hbar\omega$). The term *annihilation operator* is also used since lowering of energy by $\hbar\omega$ may also be interpreted as annihilation of one quantum of energy. Similarly,

$$\hat{H}\,\big(\hat{a}^\dagger\,|E_n\rangle\big) = (\hat{a}^\dagger\hat{H} + \hbar\omega\,\hat{a}^\dagger)\,|E_n\rangle = (E_n + \hbar\omega)\,\big(\hat{a}^\dagger\,|E_n\rangle\big) \tag{5.4.11}$$

shows that $\hat{a}^\dagger\,|E_n\rangle$ is an eigenstate of $\hat{H}$, belonging to energy $E_n + \hbar\omega \equiv E_{n+1}$. The operator $\hat{a}^\dagger$ is, therefore, called a *raising operator* (or *creation operator*) since $\hat{a}^\dagger$ acting on an energy eigenstate raises the energy by one quantum (or creates one quantum of energy).

It is clear that each time we operate on an energy eigenstate $|E_n\rangle$ by $\hat{a}$ we get a new energy eigenstate with one quantum of energy $\hbar\omega$ less than the previous one and every time we operate by $\hat{a}^\dagger$ we get a new energy eigenstate with one quantum of energy more than the previous one. It also follows from Eqs. (5.4.10) and (5.4.11) that successive energy levels of a harmonic oscillator are separated by one quantum of energy $\hbar\omega$.

What is the lowest energy state of the oscillator? If we assume that $|E_0\rangle$ is the lowest energy state *ground state* with energy E_0, then it must satisfy

$$\hat{H}\,|E_0\rangle = E_0\,|E_0\rangle\,, \tag{5.4.12}$$

$$\hat{a}\,|E_0\rangle = 0\,, \tag{5.4.13}$$

where the last equation follows from the fact that there are no states with energy lower than E_0. Pre-multiplying Eq. (5.4.13) with $\hat{a}^\dagger$, we obtain

$$\hat{a}^\dagger\hat{a}\,|E_0\rangle = 0\,. \tag{5.4.14}$$

Using $\hat{H}$ given by Eq. (5.4.6) in Eq. (5.4.12), we find

$$\left(\hat{a}^\dagger\hat{a} + \frac{1}{2}\right)\hbar\omega\,|E_0\rangle = E_0\,|E_0\rangle\,. \tag{5.4.15}$$

The first term on the left-hand side of this equation is zero from Eq. (5.4.14). This implies that the ground state energy E_0 is given by

$$E_0 = \frac{1}{2}\hbar\omega\,. \tag{5.4.16}$$

We can now ascertain what the energy eigenvalues are. In view of Eq. (5.4.11), $\hat{a}^\dagger\,|E_0\rangle$ is a state with energy eigenvalue $E_1 = E_0 + \hbar\omega$. Repeated applications of $\hat{a}^\dagger$ on the ground state then show that $\hat{a}^{\dagger n}\,|E_0\rangle$ is a state with energy $E_n = n\hbar\omega + E_0$. Hence the energy levels of a simple harmonic oscillator are

$$E_n = n\hbar\omega + E_0 = (n + 1/2)\hbar\omega\,, \qquad n = 0, 1, 2, \cdots \tag{5.4.17}$$

As already noted, successive energy levels differ in energy by $\hbar\omega$. The unit of energy $\hbar\omega$ or spacing between successive energy levels is referred to as a quantum of energy and n is referred to the number of quanta (photons or phonons). The ground state is known as the zero-quantum state and its energy E_o is referred to as zero-point energy. First excited state is one-quantum state, and so on.

With the help of Eq. (5.4.17), the eigenvalue equation $\hat{H}\,|E_n\rangle = E_n\,|E_n\rangle$ [Eq. (5.4.9)] can be rewritten as

$$\left(\hat{a}^\dagger\hat{a} + \frac{1}{2}\right)|E_n\rangle = \left(n + \frac{1}{2}\right)|E_n\rangle$$

or

$$\hat{a}^\dagger\hat{a}\,|E_n\rangle = n\,|E_n\rangle\,. \tag{5.4.18}$$

The operator $\hat{a}^\dagger\hat{a} = \hat{n}$ may be called the number operator since its eigenvalue n is the number of quanta in the state. The state $|E_n\rangle = |n+1/2\rangle$ may as well be represented by the number n written within the ket so that

$$\hat{n}\,|n\rangle = n\,|n\rangle \quad \text{and} \quad \hat{H}\,|n\rangle = (n+1/2)\hbar\omega\,|n\rangle\,. \tag{5.4.19}$$

Furthermore, with the help of Eqs. (5.4.10) and (5.4.11) we see that $\hat{a}^\dagger\,|n\rangle$ and $\hat{a}\,|n\rangle$ are essentially the states $|n+1\rangle$ and $|n-1\rangle$ so that we can write

$$\begin{aligned}\hat{a}^\dagger\,|n\rangle &= C_n\,|n+1\rangle\,,\\ \hat{a}\,|n\rangle &= D_n\,|n-1\rangle\,.\end{aligned}$$

To determine the constants C_n and D_n, we multiply each equation by its Hermitian conjugate from the left. This gives us

$$\begin{aligned}\langle n|\,\hat{a}\hat{a}^\dagger\,|n\rangle &= |C_n|^2\,\langle n+1|\,n+1\rangle \quad \Rightarrow |C_n|^2 = (n+1)\,,\\ \langle n|\,\hat{a}^\dagger\hat{a}\,|n\rangle &= |D_n|^2\,\langle n-1|\,n-1\rangle \quad \Rightarrow |D_n|^2 = n\,.\end{aligned}$$

Thus apart from a phase factor, $C_n = \sqrt{n+1}$ and $D_n = \sqrt{n}$, so that the effect of $\hat{a}$ and $\hat{a}^\dagger$ on state $|n\rangle$ is specified by

$$\hat{a}^\dagger\,|n\rangle = \sqrt{n+1}\,|n+1\rangle\,, \tag{5.4.20}$$

$$\hat{a}\,|n\rangle = \sqrt{n}\,|n-1\rangle\,. \tag{5.4.21}$$

5.4.2 Occupation Number Representation (ONR)

The representation in which the observables $\hat{n}$ and $\hat{H}$ are diagonal is called the occupation number representation. The basis states of this representation are $|0\rangle, |1\rangle, |2\rangle, \cdots, |n\rangle \cdots$. Since both operators $\hat{n}$ and $\hat{H}$ are diagonal in this basis,

$$\langle s|\,\hat{n}\,|r\rangle = r\delta_{sr}\,, \tag{5.4.22}$$

$$\langle s|\,\hat{H}\,|r\rangle = (r+1/2)\delta_{sr}\,, \tag{5.4.23}$$

their matrix representatives are diagonal matrices

$$\hat{n} \equiv \begin{pmatrix} 0 & 0 & 0 & 0 & 0\cdots \\ 0 & 1 & 0 & 0 & \cdots \\ 0 & 0 & 2 & 0 & \cdots \\ \cdots & \cdots & \cdots & \cdots & \end{pmatrix}, \tag{5.4.24}$$

$$\hat{H} \equiv \begin{pmatrix} \frac{1}{2} & 0 & 0 & 0 & 0\cdots \\ 0 & \frac{3}{2} & 0 & 0 & \cdots \\ 0 & 0 & \frac{5}{2} & 0 & \cdots \\ \cdots & \cdots & \cdots & \cdots & \end{pmatrix}. \tag{5.4.25}$$

A typical element of the matrix representing the raising operator $\hat{a}^\dagger$ in the ONR is given by

$$\langle m| \hat{a}^\dagger |n\rangle = \sqrt{n+1}\,\delta_{m,n+1}\,, \tag{5.4.26}$$

while that representing the lowering operator $\hat{a}$ is given by

$$\langle m| \hat{a} |n\rangle = \sqrt{n}\,\delta_{m,n-1}\,. \tag{5.4.27}$$

These matrices are not diagonal; they are one step off diagonal

$$\hat{a}^\dagger \equiv \begin{pmatrix} 0 & 0 & 0 & 0 \cdots \\ \sqrt{1} & 0 & 0 & 0 \cdots \\ 0 & \sqrt{2} & 0 & 0 \cdots \\ 0 & 0 & \sqrt{3} & 0 \cdots \\ \vdots & \vdots & \vdots & \vdots \ddots \end{pmatrix}, \tag{5.4.28}$$

$$\hat{a} \equiv \begin{pmatrix} 0 & \sqrt{1} & 0 & 0 & \cdots \\ 0 & 0 & \sqrt{2} & 0 & \cdots \\ 0 & 0 & 0 & \sqrt{3} & \cdots \\ \vdots & \vdots & \vdots & \vdots & \ddots \end{pmatrix}. \tag{5.4.29}$$

With the help of Eqs. (5.4.2) and (5.4.3) together with Eqs. (5.4.26) and (5.4.27), we can easily see that the typical elements of the matrices representing the position and momentum operators

$$\hat{x} = \sqrt{\frac{\hbar}{2m\omega}}\left(\hat{a} + \hat{a}^\dagger\right), \tag{5.4.30}$$

and

$$\hat{p} = -i\sqrt{\frac{\hbar m\omega}{2}}\left(\hat{a} - \hat{a}^\dagger\right), \tag{5.4.31}$$

will be given, respectively, by

$$x_{n'n} \equiv \langle n'| \hat{x} |n\rangle = \sqrt{\frac{\hbar}{2m\omega}}\left(\sqrt{n}\,\delta_{n',n-1} + \sqrt{n+1}\,\delta_{n',n+1}\right), \tag{5.4.32}$$

$$p_{n'n} \equiv \langle n'| \hat{p} |n\rangle = -i\sqrt{\frac{\hbar m\omega}{2}}\left(\sqrt{n}\,\delta_{n',n-1} - \sqrt{n+1}\,\delta_{n',n+1}\right). \tag{5.4.33}$$

5.5 Bound State of a Two-particle System with Central Interaction

The interaction between two particles is referred to as central interaction if the interaction potential depends only on the magnitude of the vector $\boldsymbol{r}_{12} = \boldsymbol{r}_2 - \boldsymbol{r}_1$ from one particle to the other and not on its direction.

An example of a two-particle system is the Hydrogen atom where the two particles are the proton (charge $+e$) and the electron (charge $-e$) interacting via the Coulomb potential

$$V(r) = -\frac{e^2}{4\pi\epsilon_0 r}, \tag{5.5.1}$$

where r is the distance between them. Other examples include singly ionized Helium, doubly ionized Lithium, and, in general, a multiply ionized atom with a single electron where the two interacting particles are the nucleus and the electron. The mutual interaction between the nucleus with charge $+Ze$ (Z being the atomic number or the number of protons inside the nucleus) and electron is the Coulomb interaction (5.5.2) with e^2 replaced by Ze^2. Under this heading we may also consider the deuteron, which is a bound system of the neutron and proton) interacting via a central nuclear potential $V_{np}(r)$.

For such two-body systems the Hamiltonian can be written as

$$\hat{H} = \frac{\widehat{\boldsymbol{p}}_1^2}{2m_1} + \frac{\widehat{\boldsymbol{p}}_2^2}{2m_2} + \hat{V}(\hat{\boldsymbol{r}}_2 - \hat{\boldsymbol{r}}_1) \tag{5.5.2}$$

where m_1 and m_2 are the masses of the particles and $\hat{V}(\hat{\boldsymbol{r}}_2 - \hat{\boldsymbol{r}}_1)$ is the interaction between them. Since $\hat{H}$ is independent of time, the time dependence of the state $|R(t)\rangle$ can be factored out as

$$|R(t)\rangle = e^{-i\mathcal{E}t/\hbar}\,|R(0)\rangle\,, \tag{5.5.3}$$

where $|R(0)\rangle$ is the state at $t = 0$. Substituting it in the Schrödinger equation of motion,

$$i\hbar\frac{d}{dt}\,|R(t)\rangle = \hat{H}\,|R(t)\rangle\,, \tag{5.5.4}$$

we find that $|R(0)\rangle$ satisfies the time-independent Schrödinger equation (eigenvalue equation)

$$\hat{H}\,|R(0)\rangle = \mathcal{E}\,|R(0)\rangle\,. \tag{5.5.5}$$

It is clear that $\mathcal{E}$ is the eigenvalue of the two-body Hamiltonian $\hat{H}$. To determine the possible states and the possible energies (eigenvalues) that the system can have, we write the eigenvalue equation in, say, the coordinate representation. In this representation the position observables $\hat{\boldsymbol{r}}_1$ and $\hat{\boldsymbol{r}}_2$, which commute with each other, are taken to be diagonal. The basis states are $|\boldsymbol{r}_1, \boldsymbol{r}_2\rangle$ where $\boldsymbol{r}_1$ and $\boldsymbol{r}_2$ are the continuously varying eigenvalues of the respective coordinate observables. In this representation, the momentum observables $\hat{\mathbf{p}}_1$ and $\hat{\mathbf{p}}_2$ may be replaced by the operators $-i\hbar\widehat{\boldsymbol{\nabla}}_1$ and $-i\hbar\widehat{\boldsymbol{\nabla}}_2$, respectively. To write Eq. (5.5.5) in the coordinate representation, we write the Hamiltonian operator explicitly, as in Eq. (5.5.2) and premultiply both sides of Eq. (5.5.5) by the basis bra vector $\langle \boldsymbol{r}_1, \boldsymbol{r}_2|$. This gives [see Chapter 3, Eq. (3.5.7)]

$$\left[-\frac{\hbar^2}{2m_1}\nabla_1^2 - \frac{\hbar^2}{2m_1}\nabla_2^2 + V(\boldsymbol{r}_2 - \boldsymbol{r}_1)\right]\Psi(\boldsymbol{r}_1, \boldsymbol{r}_2) = \mathcal{E}\Psi(\boldsymbol{r}_1, \boldsymbol{r}_2) \tag{5.5.6}$$

where

$$\nabla_1^2 = \frac{\partial^2}{\partial x_1^2} + \frac{\partial^2}{\partial y_1^2} + \frac{\partial^2}{\partial z_1^2}\,, \tag{5.5.7}$$

$$\nabla_2^2 = \frac{\partial^2}{\partial x_2^2} + \frac{\partial^2}{\partial y_2^2} + \frac{\partial^2}{\partial z_2^2}\,, \tag{5.5.8}$$

and

$$\Psi(\boldsymbol{r}_1, \boldsymbol{r}_2) = \langle \boldsymbol{r}_1, \boldsymbol{r}_2|\, R(0)\rangle\,, \tag{5.5.9}$$

is the coordinate representative of the state $|R(0)\rangle$.

To proceed further we note that the potential depends only on the relative coordinates $\boldsymbol{r}_2 - \boldsymbol{r}_1 \equiv (x_2 - x_1, y_2 - y_1, z_2 - z_1)$ of the particles, which suggests that the problem might be simpler in terms of relative coordinates instead of the coordinates of the two particles. Accordingly, we introduce the relative and center-of-mass coordinates, $\boldsymbol{r} \equiv (x, y, z)$ and $\boldsymbol{R} \equiv (X, Y, Z)$ via

$$\boldsymbol{r} = \boldsymbol{r}_2 - \boldsymbol{r}_1\,, \tag{5.5.10}$$

$$\boldsymbol{R} = \frac{m_1\boldsymbol{r}_1 + m_2\boldsymbol{r}_2}{m_1 + m_2}\,. \tag{5.5.11}$$

Note that these equations represent a coordinate transformation, not a change of representation. We are still working in the coordinate representation in terms of a new set of coordinates.

By expressing the differential operators $\frac{\partial^2}{\partial x_1^2}, \frac{\partial^2}{\partial x_2^2}$, etc. in terms of new coordinates $\boldsymbol{R} = (X, Y, Z)$ and $\boldsymbol{r} = (x, y, z)$, we can write Eq. (5.5.6) as

$$\left[-\frac{\hbar^2}{2\mu}\left(\frac{\partial^2}{\partial x^2} + \frac{\partial^2}{\partial y^2} + \frac{\partial^2}{\partial z^2}\right) - \frac{\hbar^2}{2M}\left(\frac{\partial^2}{\partial X^2} + \frac{\partial^2}{\partial Y^2} + \frac{\partial^2}{\partial Z^2}\right) + V(\boldsymbol{r})\right]\Psi = \mathcal{E}\Psi \qquad (5.5.12)$$

where $M = m_1 + m_2$ is the total mass and $\mu = m_1 m_2/(m_1 + m_2)$ is the reduced mass. From Eq. (5.5.12) we see that the kinetic energy term has separated into the sum of kinetic energy of center of mass motion and kinetic energy of relative motion. Since the potential V depends only on the relative coordinate $\boldsymbol{r}$, the overall Hamiltonian itself separates into two independent parts, one representing the center of mass motion as a free particle of mass M and the other representing the relative motion as a particle of mass μ moving in a potential $V(\boldsymbol{r})$. Hence we look for the wave function $\Psi(\boldsymbol{r}_1, \boldsymbol{r}_2)$ as the product

$$\Psi(\boldsymbol{r}_1, \boldsymbol{r}_2) = \Phi_{cm}(\boldsymbol{R})\psi(\boldsymbol{r}), \qquad (5.5.13)$$

where $\Phi_{cm}(\boldsymbol{R})$ describes the motion of the center of mass and $\psi(\boldsymbol{r})$ describes the relative motion of the particles. Substituting this form of the wave function in Eq. (5.5.12) and dividing both sides by $\Phi_{cm}(\boldsymbol{R})\psi(\boldsymbol{r})$ we obtain two independent equations

$$\left[-\frac{\hbar^2}{2M}\left(\frac{\partial^2}{\partial X^2} + \frac{\partial^2}{\partial Y^2} + \frac{\partial^2}{\partial Z^2}\right)\right]\Phi_{cm}(\boldsymbol{R}) = E_0\Phi_{cm}(\boldsymbol{R}), \qquad (5.5.14)$$

$$\left[-\frac{\hbar^2}{2\mu}\left(\frac{\partial^2}{\partial x^2} + \frac{\partial^2}{\partial y^2} + \frac{\partial^2}{\partial z^2}\right) + V(\boldsymbol{r})\right]\psi(\boldsymbol{r}) = E\psi(\boldsymbol{r}), \qquad (5.5.15)$$

where the total energy has been written as the sum of center of mass energy E_0 and energy of relative motion E as $\mathcal{E} = E_0 + E$. The first equation represents the eigenvalue equation for the center of mass motion of the two-body system. This motion is usually not of much interest but plays an important role in cooling and trapping of atoms and ions in optical or magneto-optical traps. The motion of the center of mass (free or confined) is easily discussed in terms of problems already considered once the form of the confining potential is known. For example, the center of mass motion of ions confined to a magneto-optical trap can be described as the motion of a particle in a harmonic oscillator potential.

The second equation [Eq. (5.5.15)], represents the relative motion of the two particles. It is the equation of motion for a particle of effective mass $\mu = m_1 m_2/(m_1+m_2)$ with energy E moving in a potential $V(\boldsymbol{r})$. In what follows we consider relative motion in the presence of a central potential $V(\boldsymbol{r}) = V(r)$, which depends only on r and not on the relative orientation of the particles. Such a potential is said to be spherically symmetric.

Introducing spherical polar coordinates r, θ, φ (because of the spherical symmetry of the potential) and expressing the differential operators in terms of these variables [see Appendix 5A2], we can write Eq. (5.5.15) as

$$\left[-\frac{\hbar^2}{2\mu}\nabla^2 + V(r)\right]\psi(r, \theta, \varphi) = E\psi(r, \theta, \varphi), \qquad (5.5.16)$$

where $$\nabla^2 \equiv \frac{1}{r^2}\frac{\partial}{\partial r}\left(r^2\frac{\partial}{\partial r}\right) + \frac{1}{r^2 \sin\theta}\frac{\partial}{\partial \theta}\left(\sin\theta\frac{\partial}{\partial \theta}\right) + \frac{1}{r^2 \sin^2\theta}\frac{\partial^2}{\partial \varphi^2}. \qquad (5.5.17)$$

We can separate the radial and angular dependence of the wave function by writing

$$\psi(r, \theta, \varphi) = R(r)Y(\theta, \varphi). \qquad (5.5.18)$$

Substituting this in Eq. (5.5.16) and separating the terms that depend on radial and angular variables, we arrive at

$$\frac{1}{R(r)}\left[\frac{d}{dr}r^2\frac{dR}{dr}+\frac{2\mu r^2}{\hbar^2}(E-V(r))R\right]=-\frac{1}{Y}\left[\frac{1}{\sin\theta}\frac{\partial}{\partial\theta}\sin\theta\frac{\partial Y}{\partial\theta}+\frac{1}{\sin^2\theta}\frac{\partial^2 Y}{\partial\varphi^2}\right]. \quad (5.5.19)$$

Since radial coordinate r and angular coordinates (θ,φ) are independent variables, a function of r (on the left side) and a function of angles (on the right side) cannot be equal to one another for all values of r and (θ,φ) unless both are separately equal to a constant, say λ. Thus Eq. (5.5.19) separates into two equations

$$\frac{d}{dr}\left(r^2\frac{dR}{dr}\right)+\frac{2\mu\, r^2}{\hbar^2}(E-V(r))R(r)=\lambda\, R(r)\,, \quad (5.5.20)$$

and

$$-\left\{\frac{1}{\sin\theta}\frac{\partial}{\partial\theta}\left(\sin\theta\frac{\partial}{\partial\theta}\right)+\frac{1}{\sin^2\theta}\frac{\partial^2}{\partial\varphi^2}\right\}Y(\theta,\varphi)=\lambda\, Y(\theta,\varphi)\,. \quad (5.5.21)$$

The angular equation may again be separated into two independent equations by the substitution

$$Y(\theta,\varphi)\equiv\Theta(\theta)\Phi(\varphi) \quad (5.5.22)$$

in Eq. (5.5.21). Dividing the resulting equation throughout by $\Theta(\theta)\Phi(\varphi)$ and separating the variables we get

$$\frac{d^2\Phi}{d\varphi^2}=-m^2\Phi\,, \quad (5.5.23)$$

and

$$\sin\theta\frac{d}{d\theta}\sin\theta\frac{d\Theta}{d\theta}+\lambda\sin^2\theta\,\Theta=m^2\Theta\,, \quad (5.5.24)$$

where we have written the separation constant as m^2. The physically acceptable normalized solutions of Eq. (5.5.23) are

$$\Phi_m(\varphi)=\frac{1}{\sqrt{2\pi}}\exp(im\varphi)\,, \quad (5.5.25)$$

where m, called *magnetic quantum number*, must be a positive or negative integer including zero for $\Phi_m(\varphi)$ to be a single-valued function of φ. The normalization condition for $\Phi_m(\varphi)$ is

$$\int_0^{2\pi}d\varphi\;\Phi^*_{m'}(\varphi)\Phi_m(\varphi)=\int_0^{2\pi}d\varphi\left(\frac{e^{im'\varphi}}{\sqrt{2\pi}}\right)^*\frac{e^{im\varphi}}{\sqrt{2\pi}}=\delta_{mm'}\,. \quad (5.5.26)$$

Equation (5.5.24) for Θ can be transformed to a more recognizable form by the substitution $w=\cos\theta$

$$\frac{d}{dw}(1-w^2)\frac{dP}{dw}+\left(\lambda-\frac{m^2}{1-w^2}\right)P(w)=0 \quad (5.5.27)$$

where $P(w)=\Theta(\theta)$. This equation resembles the associated Legendre equation. For physically acceptable solutions of this equation, which remain finite for all w, including $w=\pm 1$ (corresponding to the polar axis $\theta=\pm\pi/2$), we require $\lambda=\ell(\ell+1)$, where ℓ is a positive integer, including zero. With this restriction on λ, this equation admits polynomial solutions, called *associated Legendre polynomials*, denoted by$P_\ell^{|m|}(w)$ and given by

$$P_\ell^{|m|}(w)=(1-w^2)^{|m|/2}\frac{d^{|m|}}{dw^{|m|}}P_\ell(w)\,, \quad (5.5.28)$$

where $P_\ell(w)$ is Legendre polynomial of order ℓ. Since $P_\ell(w)$ is polynomial of order ℓ in w, it is clear from the definition (5.5.28) that $|m| \leq \ell$ or $-\ell \leq m \leq \ell$. The normalization integral for the associated Legendre polynomial is given by

$$\int_{-1}^{1} dw P_\ell^{|m|}(w) P_{\ell'}^{|m|}(w) = \int_0^\pi \sin\theta d\theta P_\ell^{|m|}(\cos\theta) P_{\ell'}^{|m|}(\cos\theta)$$
$$= \sqrt{\frac{(\ell+|m|)!}{(\ell-|m|)!}\frac{2}{(2\ell+1)}}\,\delta_{\ell\ell'}\,. \tag{5.5.29}$$

With the help of Eqs. (5.5.25), (5.5.26), (5.5.28), and (5.5.29), the normalized solution of the angular equation (5.5.21) is given by

$$Y_{\ell m}(\theta,\varphi) = \sqrt{\frac{(\ell-|m|)!(2\ell+1)}{4\pi(\ell+|m|)}}\,P_\ell^{|m|}(\cos\theta)\exp(im\varphi)\,. \tag{5.5.30}$$

These functions, called spherical harmonics, are normalized and satisfy the orthogonality condition

$$\int_0^{2\pi} d\varphi \int_0^\pi \sin\theta d\theta\; Y^*_{\ell m}(\theta,\phi) Y_{\ell m'}(\vartheta,\varphi) = \delta_{\ell\ell'}\delta_{mm'}\,. \tag{5.5.31}$$

Spherical harmonics are eigenfunctions of the square of the angular momentum operator $\hat{L}^2$ and also of $\hat{L}_z$, the z-component of the angular momentum operator. To see this we express the components of the orbital angular momentum operator (in coordinate representation) in spherical polar coordinates as

$$L_x = -i\hbar\left(y\frac{\partial}{\partial z} - z\frac{\partial}{\partial y}\right) = i\hbar\left(\sin\theta\frac{\partial}{\partial\theta} + \cot\theta\cos\varphi\frac{\partial}{\partial\varphi}\right), \tag{5.5.32}$$

$$L_y = -i\hbar\left(z\frac{\partial}{\partial x} - x\frac{\partial}{\partial z}\right) = i\hbar\left(-\cos\varphi\frac{\partial}{\partial\theta} + \cot\theta\sin\varphi\frac{\partial}{\partial\varphi}\right), \tag{5.5.33}$$

$$L_z = -i\hbar\left(x\frac{\partial}{\partial y} - y\frac{\partial}{\partial x}\right) = -i\hbar\frac{\partial}{\partial\varphi}, \tag{5.5.34}$$

$$L^2 = L_x^2 + L_y^2 + L_z^2 = -\hbar^2\left(\frac{1}{\sin\theta}\frac{\partial}{\partial\theta}\sin\theta\frac{\partial}{\partial\theta} + \frac{1}{\sin^2\theta}\frac{\partial^2}{\partial\varphi^2}\right). \tag{5.5.35}$$

Comparing Eq. (5.5.35) with the left-hand side of the angular equation (5.5.21) we immediately see that

$$L^2\, Y_{\ell m}(\theta,\varphi) = \ell(\ell+1)\hbar^2\, Y_{\ell m}(\theta,\varphi)\,. \tag{5.5.36}$$

Finally, using Eq. (5.5.34) for L_z and the definition of $Y_{\ell m}(\theta,\varphi)$ we see

$$L_z Y_{\ell m}(\theta,\varphi) = m\hbar\, Y_{\ell m}(\theta,\varphi)\,. \tag{5.5.37}$$

Hence spherical harmonics $Y_{\ell m}(\theta,\varphi)$ are simultaneous eigenstates of L^2 and L_z belonging to the eigenvalues $\ell(\ell+1)\hbar^2$ and $m\hbar$, respectively.

We can summarize the results of this section by saying that the relative motion of two-body systems involving spherically symmetric potential is described by a wave function of the form

$$\psi(r,\theta,\varphi) = R_\ell(r) Y_{\ell m}(\theta,\varphi)\,, \tag{5.5.38}$$

where the angular part of the wave function is the same for all spherically symmetric central potentials, while the radial function $R_\ell(r)$ satisfies Eq. (5.7.21) with $\lambda = \ell(\ell+1)$

$$\frac{1}{r^2}\frac{d}{dr}r^2\frac{dR_\ell}{dr} + \left[\frac{2\mu}{\hbar^2}(E - V(r)) - \frac{\ell(\ell+1)}{r^2}\right] R_\ell(r) = 0 \,. \tag{5.5.39}$$

The radial function is different depending on the form of the potentials $V(r)$.

For bound states of a given potential $V(r)$, if any, the normalization condition, in terms of R_ℓ reads

$$\int_0^\infty dr\,|rR_\ell(r)|^2 = 1 \,. \tag{5.5.40}$$

For this integral to be finite the radial function must satisfy the boundary conditions: (i) for large distances from the origin, it must vanish sufficiently rapidly that $rR_\ell(r) \to 0$ as $r \to \infty$ and (ii) near the origin $r = 0$, its behavior must be such that $rR_\ell(r) \to 0$ as $r \to 0$. The last requirement can be justified as follows. By noting $\frac{1}{r^2}\frac{d}{dr}r^2\frac{dR_\ell}{dr} = \frac{1}{r}\frac{d^2}{dr^2}rR_\ell(r)$, we see that the radial equation (5.5.39) can be cast into an equation for $u_\ell(r) = rR_\ell$:

$$\frac{d^2}{dr^2}u_\ell(r) + \frac{2\mu}{\hbar^2}\left[E - V(r) - \frac{\hbar^2\ell(\ell+1)}{2\mu r^2}\right] u_\ell(r) = 0 \,. \tag{5.5.41}$$

Thus the radial equation reduces to a one-dimensional Schrödinger equation for $u_\ell(r) = rR_\ell(r)$ for a particle of mass μ subject to an effective potential $V(r) + \frac{\hbar^2\ell(\ell+1)}{2\mu r^2}$ in the region $r \geq 0$ and an infinitely repulsive potential in the region $r < 0$. It follows that the wave function $u(r) = rR_\ell(r)$ must vanish for $r < 0$. The condition $rR_\ell(r) \to 0$ as $r \to 0$ then ensures the continuity of the wave function. Thus the boundary conditions satisfied by the radial function of a bound state wave function are

$$u_\ell(r) \equiv rR_\ell(r) \to 0 \quad \text{as} \quad r \to 0 \,, \tag{5.5.42}$$

$$u_\ell(r) \equiv rR_\ell(r) \to 0 \quad \text{as} \quad r \to \infty \,. \tag{5.5.43}$$

Extending the analogy of the radial equation to one dimensional Schrödinger equation further, we conclude that the bound state energy spectrum of a central potential is discrete and each bound state energy E_n corresponds to one (and only one) radial function. In other words E_n and ℓ uniquely determine the radial function. We may then label the radial wave function by $R_{n\ell}$. Note that these comments refer only to the radial part of the wave function.

5.6 Bound States of Hydrogen (or Hydrogen-like) Atoms

For this problem the two-body interaction is of the form

$$V(r) = -\frac{Ze^2}{4\pi\epsilon_0 r} \,. \tag{5.6.1}$$

For $Z = 1$, this interaction refers to the Hydrogen atom, for $Z = 2$ it refers to the singly ionized He atom and so on. For the Hydrogen atom problem, the reduced mass is given by

$$\mu = m_e m_p/(m_e + m_p) = m_e[1 - m_e/(m_e + m_p)] \tag{5.6.2}$$

where m_e and m_p are, respectively, the mass of the electron and the proton. Using $m_e = 0.511$ Mev/c^2 and $m_p = 938$ MeV/c^2, we find that $\mu = 0.999 m_e$. For this reason we can refer(inaccurately) to the relative motion as motion of the electron around the proton or the nucleus. Such a picture would be highly inaccurate for positronium (bound state of an electron and positron), which is also described by the Coulomb potential (5.6.1) with $Z = 1$, and reduced mass $\mu = m_e/2$.

As seen in the previous section, the solution of the Schrödinger equation for the relative motion has the form

$$\psi(r,\theta,\varphi) = R_\ell(r) Y_{\ell m}(\theta,\varphi)\,, \tag{5.5.38*}$$

where $R_\ell(r)$ satisfies the radial equation

$$\frac{1}{r^2}\frac{d}{dr}r^2\frac{dR_\ell}{dr} + \left[\frac{2\mu}{\hbar^2}\left(E - \frac{Ze^2}{4\pi\epsilon_0 r}\right) - \frac{\ell(\ell+1)}{r^2}\right] R_\ell(r) = 0\,. \tag{5.6.3}$$

This equation depends on ℓ but not m. Therefore we expect energy levels to be $2\ell + 1$-fold degenerate with respect to the direction of orbital angular momentum in space. Note that for bound states in Coulomb potetial, $E = -|E| < 0$ is a negative quantity. We can cast radial equation (5.6.3) in a more convenient form by introducing parameters κ and γ by

$$\kappa = \sqrt{\frac{8\mu|E|}{\hbar^2}} \equiv \sqrt{-\frac{8\mu E}{\hbar^2}}\,, \tag{5.6.4}$$

$$\gamma = Z\frac{2\mu e^2}{4\pi\epsilon_0\hbar^2\kappa} \equiv Z\frac{e^2}{4\pi\epsilon_0\hbar c}\sqrt{\frac{\mu c^2}{2|E|}}\,. \tag{5.6.5}$$

It can be seen that the parameter κ has dimensions of inverse length, whereas γ is dimensionless. Using the scaled dimensionless radial variable

$$\rho = \kappa\, r\,, \tag{5.6.6}$$

we can write the radial equation as

$$\frac{1}{\rho^2}\frac{d}{d\rho}\rho^2\frac{dF_\ell}{d\rho} + \left(-\frac{1}{4} + \frac{\gamma}{\rho} - \frac{\ell(\ell+1)}{\rho^2}\right) F_\ell(\rho) = 0\,, \tag{5.6.7}$$

where $F_\ell(\rho)$ is the radial function in terms of the scaled radial variable. To ensure the normalization condition (5.5.40) let us examine the behavior of the radial function for $\rho \to 0$ and $\rho \to \infty$. Multiplying Eq. (5.6.7) by ρ^2 and taking the limit $\rho \to 0$ we obtain the equation satisfied by $F_\ell(\rho)$ for small ρ

$$\frac{d}{d\rho}\rho^2\frac{dF_\ell}{d\rho} - \ell(\ell+1)F_\ell(\rho) = 0\,. \tag{5.6.8}$$

We seek solutions of the form $F_\ell(\rho) =$ constant $\times\, \rho^c$, which may be thought of as the first term of the series solution for small ρ. Substituting this in Eq. (5.6.7) we find

$$c(c+1) = \ell(\ell+1)\,. \tag{5.6.9}$$

This gives $c = -(\ell+1)$ or ℓ. The solution with $c = -(\ell+1)$ does not satisfy the boundary condition $\rho F_\ell(\rho) \to 0$ [Eq. (5.5.42)] as $\rho \to 0$. Thus for a given ℓ, the acceptable solution near $\rho = 0$ has the the form

$$F_\ell(\rho) \approx \text{constant} \times \rho^\ell\,. \tag{5.6.10}$$

To calculate the asymptotic behavior of the radial function, we take the limit $\rho \to \infty$ of Eq. (5.6.6). By dropping terms with $1/\rho$ and $1/\rho^2$ dependence we obtain

$$\frac{d^2 F_\ell}{d\rho^2} - \frac{1}{4} F = 0\,, \tag{5.6.11}$$

which implies that $F_\ell \approx e^{\pm\rho/2}$, independent of ℓ. Once again, the normalizable solution, which vanishes as $\rho \to \infty$ [Eq. (5.5.43)] is the negative exponential

$$F_\ell(\rho) \sim e^{-\rho/2}\,. \tag{5.6.12}$$

Incorporating the behavior of $F_\ell(\rho)$ for small and large ρ as specified in Eqs. (5.6.10) and (5.6.12) we seek solutions of the form

$$F_\ell(\rho) = e^{-\rho/2}\rho^\ell u(\rho)\,. \tag{5.6.13}$$

Using this in Eq. (5.6.7), we find that $u(\rho)$ satisfies the equation

$$\frac{d^2 u}{d\rho^2} + (2\ell + 2 - \rho)\frac{du}{d\rho} + (\gamma - \ell - 1)u(\rho) = 0\,. \tag{5.6.14}$$

This equation, called the associated Laguerre equation, admits a polynomial solution only if $\gamma - \ell - 1$ is a positive integer, including zero [see Appendix 5A1 Sec. 3]. If this condition is not met, normalizable solutions satisfying the condition (5.5.43) are not possible. Hence we conclude that γ must be a positive integer n which for a given ℓ must satisfy the condition

$$\gamma \equiv n \geq \ell + 1\,. \tag{5.6.15}$$

Combining this with the definition of γ [Eq. (5.6.5)] we find the allowed bound state energies (eigenvalues) are given by

$$E_n = -\frac{Z^2 e^4 \mu}{2(4\pi\epsilon_0 \hbar)^2 n^2} \equiv -\frac{Z^2 e^2}{8\pi\epsilon_0 a_0 n^2}\,, \tag{5.6.16}$$

where

$$a_0 = \frac{4\pi\epsilon_0\hbar^2}{\mu e^2} = \frac{4\pi\epsilon_0 \hbar c}{e^2}\frac{\hbar}{\mu c}\,. \tag{5.6.17}$$

The parameter a_0, with dimensions of length, is known as the Bohr radius.[2] It emerges as a natural unit for atomic lengths. The integer n is called the *principal quantum number.* This solves the bound state energy eigenvalue problem for the Coulomb potential. As concluded, on general grounds, in the preceding section, the bound state spectrum is indeed discrete. Moreover, since the potential $V(r) \to 0$ as $r \to \infty$, the number of bound states ($E_n < 0$) is infinite.

The solutions of the radial equation (5.6.14) with $\gamma = n$ [Eq. (5.6.15)], are the associated Laguerre polynomials[3] defined by

$$L_{n+\ell}^{2\ell+1}(\rho) = \sum_{s=0}^{n-\ell-1} \frac{(-1)^s[(n+\ell)!]^2 \rho^s}{(n-\ell-1-s)!(2\ell+1+s)!\, s!}\,. \tag{5.6.18}$$

---

[2]We neglect the difference between a_0, defined here, and Bohr radius $\left(\frac{4\pi\epsilon_0\hbar^2}{m_e e^2}\right)$ since the reduced mass $\mu = 0.999 m_e$ [Eq. (5.6.2)] differs negligibly even for the lightest of the nuclei in the hydrogenic atoms.

[3]Different notations for the associated Laguerre polynomial may be found in other texts. Our L_α^β in the notation used in Merzbacher, Messiah and Liboff would be denoted by $L_{\alpha-\beta}^\beta$.

Note that $L_{n+\ell}^{2\ell+1}(\rho)$ is a polynomial of degree $n-\ell-1$. The radial function is thus labeled by two indices n and ℓ. It is important to remember that the radial function depends on index n via the scale factor κ as well, which from Eqs. (5.6.5) and (5.6.15) is given by

$$\kappa_n = \frac{2Z}{na_0}\,. \tag{5.6.19}$$

Reverting back to the unscaled radial variable via the relation $\rho = \kappa_n r = 2Zr/na_0$ [Eq. (5.6.6)], the normalized solution of the radial equation (5.6.3) is

$$R_{n\,\ell}(r) = \left(\frac{2Z}{na_0}\right)^{3/2} \sqrt{\frac{(n+\ell)!}{2n[(n+\ell)!]^3}}\; e^{-Zr/na_0}\; \rho^\ell L_{n+\ell}^{2\ell+1}(2Zr/na_0)\,. \tag{5.6.20}$$

Collecting all the information of this section we can, finally, write down the normalized Hydrogen atom wave function as

$$\psi_{n\ell m}(r,\theta,\varphi) = \left(\frac{2Z}{na_0}\right)^{3/2} \sqrt{\frac{(n-\ell-1)!}{2n[(n+\ell)!]^3}}\; e^{-Zr/na_0} \left(\frac{2Zr}{na_0}\right)^{\ell} L_{n+\ell}^{2\ell+1}(2Zr/na_0) Y_{\ell m}(\theta,\varphi)\,. \tag{5.6.21}$$

We see that the bound state wave function is completely determined by the principal quantum number n, orbital angular momentum quantum number ℓ, and magnetic quantum number m. In other words, the energy ($\hat{H}$), orbital angular momentum squared ($\hat{L}^2$), and the z-component of the orbital angular momentum ($\hat{L}_z$) form a complete set of observables for the Hydrogen atom. The wave function (5.6.21) can then be looked upon as the coordinate representative $\psi_{\ell mn}(\boldsymbol{r}) = \langle \boldsymbol{r} | n,\ell,m\rangle$ of the bound state $|n,\ell,m\rangle$ of the Hydrogen atom characterized by the principal quantum number n, orbital angular momentum quantum number ℓ and magnetic quantum number m. The state $|n,\ell,m\rangle$ is a simultaneous eigenstate of $\hat{H}$, $\hat{L}^2$, and $\hat{L}_z$

$$\hat{H}\,|n,\ell,m\rangle = E_n\,|n,\ell,m\rangle\,, \tag{5.6.22}$$

$$\hat{L}^2\,|n,\ell,m\rangle = \ell(\ell+1)\hbar^2\,|n,\ell,m\rangle\,, \tag{5.6.23}$$

$$\hat{L}_z\,|n,\ell,m\rangle = m\hbar\,|n,\ell,m\rangle\,. \tag{5.6.24}$$

The energy corresponding to state $|n,\ell,m\rangle$, written in terms of Bohr radius a_0 is

$$E_n = -\frac{Z^2e^2}{8\pi\epsilon_0 a_0 n^2} = -\frac{Z^2}{n^2}\,\frac{e^2}{8\pi\epsilon_0 a_0}\,. \tag{5.6.25}$$

Using the last expression, the bound state energy can be written as $E_n = -Z^2/n^2$ in units of energy called Rydberg (1 Rydberg$=e^2/8\pi\epsilon_0 a_0 \approx 13.6$ eV). From the dependence of bound state energy on n, we see that there are an infinite number of bound states. The distance between successive energy levels decreases as n increases and the levels become crowded as we approach the limit $E_\infty = 0$.

The bound state energy depends only on the principal quantum number n and not on the quantum numbers ℓ and m. This means there are several linearly independent states that correspond to the same value of energy. The number of such states associated with an energy level is referred to as its degree of degeneracy. To calculate the degree of degeneracy of an energy level we note that, according to Eq. (5.6.14), for a given value of the principal quantum number n, the angular momentum quantum number ℓ can take the values

$$\ell = 0, 1, 2, \cdots n-1. \tag{5.6.26}$$

These states with different ℓ but the same n have the same energy. But each value of ℓ corresponds to $2\ell+1$ values of m. Hence the degree of degeneracy of the n-th energy level is given by

$$\sum_{\ell=0}^{n-1}\sum_{m=-\ell}^{\ell} m = \sum_{\ell=0}^{n-1}(2\ell+1) = n^2 . \tag{5.6.27}$$

The bound state energy levels of the Hydrogen atom are thus n^2-fold degenerate. The occurrence of degeneracy is associated with symmetry properties of the system. The degeneracy of energy levels with respect to m reflects the invariance of the Coulomb system under rotations about the origin. This is a property common to all centrally symmetric potentials. The degeneracy with respect to ℓ corresponds to another symmtery property that is peculiar to the Coulomb potential [see Chapter 6]. In the spectroscopic notation, the ℓ values are denoted by lower case letters s, p, d, f, $\cdots$ corresponding, respectively, to $\ell = 0$, 1, 2, 3, $\cdots$ so that the levels of the Hydrogen atom are denoted by n followed by the symbol for ℓ. Thus the first level ($n=1$) consists of one 1s state. The second level ($n=2$) consists of one 2s and three 2p states. The third level ($n=3$) consists of one 3s, three 3p, and five 3d states and so on. If we include electron spin (not considered so far), which can take two values $\pm\frac{1}{2}$, then the degree of degeneracy of each energy level doubles to $2n^2$. For example, the first level consists of two 1s states. The second level consists of two 2s and six 2p states.

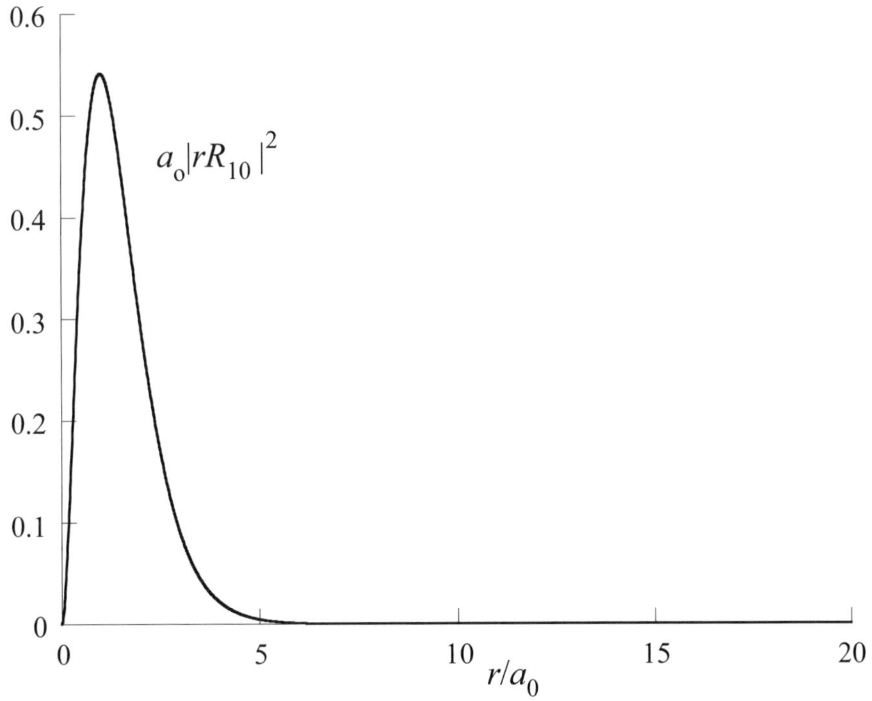

FIGURE 5.7
Radial probability distribution $r^2R_{10}^2(r)$ for the Hydrogen atom.

The radial functions $R_{n\ell}(r)$ for the first few states are given below:

$$\begin{aligned}
R_{10} &= \left(\frac{Z}{a_0}\right)^{3/2} 2e^{-Zr/a_0}\,, \\
R_{20} &= \left(\frac{Z}{2a_0}\right)^{3/2} 2\left(1-\frac{r}{2a_0}\right) e^{-Zr/2a_0}\,, \\
R_{21} &= \left(\frac{Z}{2a_0}\right)^{3/2} \frac{1}{\sqrt{3}}\left(\frac{r}{a_0}\right) e^{-Zr/2a_0}\,, \\
R_{30} &= \left(\frac{Z}{3a_0}\right)^{3/2} 2\left[1-\frac{2}{3}\frac{r}{a_0}+\frac{2}{27}\left(\frac{r}{a_0}\right)^2\right] e^{-Zr/3a_0}\,, \\
R_{31} &= \left(\frac{Z}{3a_0}\right)^{3/2} \frac{8}{9\sqrt{2}}\left(\frac{r}{a_0}\right)\left(1-\frac{1}{6}\frac{r}{a_0}\right) e^{-Zr/3a_0}\,, \\
R_{32} &= \left(\frac{Z}{3a_0}\right)^{3/2} \frac{4}{27\sqrt{10}}\left(\frac{r}{a_0}\right)^2 e^{-Zr/3a_0}\,.
\end{aligned} \tag{5.6.28}$$

These expressions confirm the role of Bohr radius a_0 as a natural unit for atomic lengths. Thus small and large atomic distances are defined relative to a_0. For small and large distances, the form of the radial function is given by

$$R_{n\ell}(r) \approx \left(\frac{Z}{a_0}\right)^{3/2} \begin{cases} \dfrac{2}{n^2(2\ell+1)!}\sqrt{\dfrac{(n+\ell)!}{(n-\ell-1)}}\left(\dfrac{2Zr}{na_0}\right)^{\ell} & r \ll a_0 \\ \dfrac{2(-1)^{n-\ell-1}}{n^2\sqrt{(n+\ell)!(n-\ell-1)!}}\left(\dfrac{2Zr}{na_0}\right)^{n-1} e^{-Zr/na_0} & r \gg a_0\,. \end{cases} \tag{5.6.29}$$

The probability of finding the electron in a spherical shell of radius r and thickness dr, centered at the nucleus, when the system is in the state specified by the quantum numbers n, ℓ, m is given by $r^2R_{n\ell}^2(r)dr$. Figures (5.7) through (5.9) show $r^2R_{n\ell}^2(r)$ as a function of r/a_0 for the 1s, 2s and 2p states of the Hydrogen atom ($Z = 1$).

From the expression for the radial function and the integrals involving associated Laguerre polynomials, the mean values

$$\langle r^k\rangle_{n\ell m} = \int_0^\infty dr r^{k+2} R_{n\ell}^2(r)\,, \tag{5.6.30}$$

for $k = -3, -2, -1, 1$, and 2 are given by

$$\langle r^{-3}\rangle_{n\ell m} = \frac{Z^3}{a_0^3 n^3 \ell(\ell+1/2)(\ell+1)}\,, \tag{5.6.31}$$

$$\langle r^{-2}\rangle_{n\ell m} = \frac{Z^2}{a_0^2 n^3(\ell+1/2)}\,, \tag{5.6.32}$$

$$\langle r^{-1}\rangle_{n\ell m} = \frac{Z}{a_0 n^2}\,, \tag{5.6.33}$$

$$\langle r\rangle_{n\ell m} = \frac{a_0}{2Z}\left[3n^2-\ell(\ell+1)\right]\,, \tag{5.6.34}$$

$$\langle r^2\rangle_{n\ell m} = \frac{a_0^2 n^2}{2Z^2}\left[5n^2+1-3\ell(\ell+1)\right]\,. \tag{5.6.35}$$

In closing the discussion of bound states of hydrogenic atoms we mention that the exact treatment given here is possible because we have considered only the Coulomb interaction

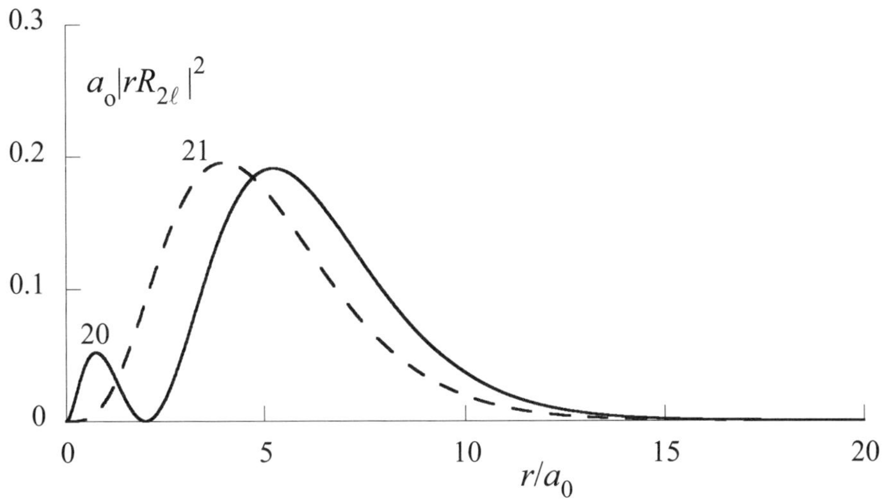

FIGURE 5.8
Radial probability distribution $r^2R_{2\ell}^2(r)$ for the Hydrogen atom.

between the electron and the nucleus. This is not the whole story. The electron carries an intrinsic spin and magnetic moment. The magnetic moment of the moving electron interacts with the Coulomb field of the nucleus, giving rise to the so-called spin-orbit interaction term in the Hamiltonian [see Chapter 8]. The proton in the Hydrogen atom (and the nucleus in many hydrogenic atoms) also has an intrinsic magnetic moment, which intreacts with the magnetic moment of the electron. These and other interactions lead to additional terms in the Hamiltonian. Fortunately, these additional terms are small compared to the Coulomb interaction and, as discussed in Chapter 8, their effect on energy levels can be calculated by treating them as perturbations to the dominant Coulomb interaction. Their most important effect is to remove (at least partially) the degeneracy of energy levels.

5.7 The Deuteron Problem

Another important two-body problem in physics is the deuteron which is the bound state of a neutron (n) and a proton (p). This is the most fundamental problem in nuclear physics and is an important source of information on the nature of the $n-p$ interaction. Unlike the H-atom problem where the electron-nucleus interaction was precisely known and our aim was to determine the energy levels of the atom, in the deuteron problem the basic input (the $n-p$ interaction V_{np}) is not precisely known and so our aim is to infer the nature of the $n-p$ interaction by using the empirical data about the deuteron:

(a) Experimentally it is known that the deuteron exists in a weakly bound state of binding energy $B = 2.23$ MeV. Besides the ground state no stable excited states of the deuteron have been found to exist.

(b) The total angular momentum of the deuteron is $J = 1$ (in units of $\hbar$).

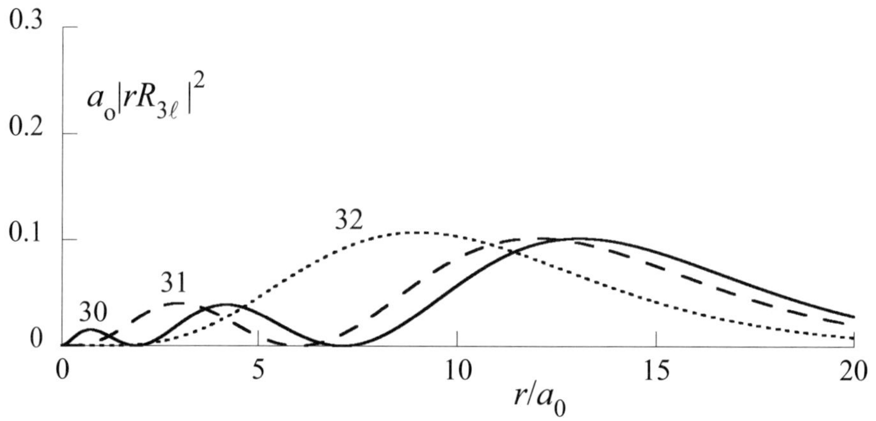

FIGURE 5.9
Radial probability distribution $r^2R_{3\ell}^2(r)$ for the Hydrogen atom.

(c) The rms charge radius of the deuteron is 2.14 fm (1 fermi $\equiv$ 1fm=10^{-15} m). It has zero electric dipole moment and a very small electric quadrupole moment $Q = 0.286$ e-fm^2.

(e) The deuteron magnetic moment $\mu_d = 0.8574\ \mu_N$ is very close to the sum of neutron and proton magnetic moments $\mu_p + \mu_n = 2.7928\ \mu_N - 1.9132\ \mu_N = 0.8796\ \mu_N$, where nuclear magneton $\mu_N = 5.0507866 \times 10^{-27}$ joules/tesla is the natural unit for nuclear magnetic moments.

From its vanishing electric dipole moment and a very small value of electric quadrupole moment[4] it follows that the deuteron state is primarily a spherically symmetric s-state($\ell = 0$). This is supported also by the fact that the magnetic moment of the deuteron is the sum of the proton and the neutron magnetic moments, indicating that their spins are parallel ($S = 1$) and there is no orbital motion of the proton relative to the neutron. This is consistent with the total angular momentum of the ground state being $J = 1$.

The $n - p$ interaction must be attractive for a bound state to exist and like all nuclear interaction it must be short ranged. The spatial extent of a wave function is determined by its binding energy and not by the range of the potential. The small binding energy of the deuteron indicates that the deuteron is an extended object and is not sensitive to the details of the potential. We assume that the $n - p$ interaction is central and derivable from a potential function. To start with, we take a very simple potential in form of a square well as shown Fig. 5.10

$$V(r) = \begin{cases} -V_o & r \leq r_o \\ 0 & r > r_o \end{cases} \tag{5.7.1}$$

where r_o is the range of the interaction. We ask what the energy level structure is and what values of V_o and r_o are consistent with a bound state ($\ell = 0$) at energy $B = 2.226$ Mev.

[4]The non-zero quadrupole moment tells us that the deuteron is, strictly speaking, not a pure s-state. An s-state has spherical symmetry and therefore cannot have an electric quadrupole moment. A non-zero quadrupole moment requires a deformation proportional to Y_{20}, which corresponds to a d-state ($\ell = 2$). However, since the quadrupole moment is small, to a first approximation, we can ignore non-zero ℓ states.

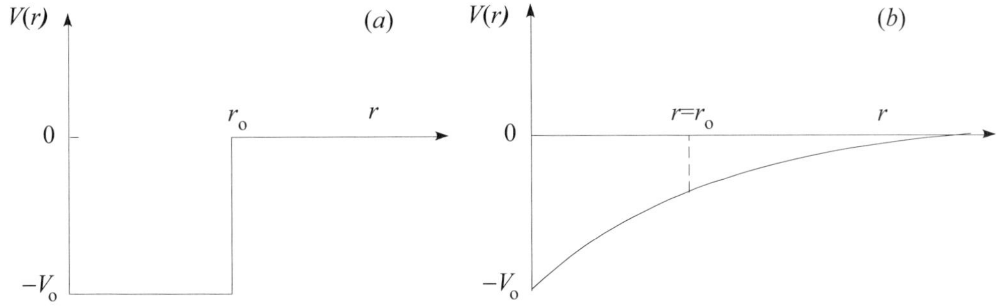

FIGURE 5.10
Examples of two-body central potentials in three dimensions: (a) spherically symmetric square well potential, (b) exponential potential well.

As in the case of the Hydrogen atom, the Schrödinger equation (5.5.16) representing the relative motion of the two particles (n and p) can be separated into the radial and angular parts since the potential is central. The radial equation (5.5.39) in this case ($\ell = 0$) can be written as

$$\frac{1}{r^2}\frac{d}{dr}r^2\frac{dR}{dr} + \frac{2\mu}{\hbar^2}[E - V(r)]R(r) = 0 \tag{5.7.2}$$

where $\mu = m_p m_n/(m_n + m_n)$ is the reduced mass for the $n-p$ system. Using the substitution $u(r) = rR(r)$ and $E = -B$ (B being the binding energy) we can write the radial equation in the two regions as

$$\frac{d^2u_1}{dr^2} + k^2 u_1(r) = 0 \qquad r \le r_o \text{ (Region I)} \tag{5.7.3}$$

$$\frac{d^2u_2}{dr^2} - \kappa^2 u_2(r) = 0 \qquad r > r_o \text{ (Region II)} \tag{5.7.4}$$

where $$k^2 = \frac{2\mu}{\hbar^2}(V_o - B) \quad \text{and} \quad \kappa^2 = \frac{2\mu B}{\hbar^2} \equiv \frac{2\mu}{\hbar^2}V_o - k^2\,. \tag{5.7.5}$$

The solutions in the two regions are of the form

$$u_1(r) = A_1\sin(kr) + D_1\cos(kr)\,, \qquad r \le r_o\,, \tag{5.7.6}$$

$$u_2(r) = A_2\exp(-\kappa r) + D_2\exp(\kappa r), \qquad r > r_o\,, \tag{5.7.7}$$

where A_1, D_1, A_2, and D_2 are constants to be determined. Since $u(r) = rR(r)$ is required to satisfy the boundary conditions for a bound state wave function, it must vanish at the origin and fall off to zero as $r \to \infty$ [Eqs. (5.5.42) and (5.5.43)]. To satisfy these conditions we must choose $D_1 = 0 = D_2$. The continuity of the wave function and its derivative requires that the sinusoidal solution in region 1 and the exponential solution in region 2 and their derivatives must match at $r = r_o$:

$$u_1(r_o) = u_2(r_o) \qquad \Rightarrow \quad A_1 \sin kr_o = A_2 e^{-\kappa r_o}\,, \tag{5.7.8}$$

$$\left.\frac{du_1}{dr}\right|_{r_o} = \left.\frac{du_2}{dr}\right|_{r_o} \qquad \Rightarrow \quad A_1 k \cos kr_o = -A_2 \kappa e^{-\kappa r_o}\,. \tag{5.7.9}$$

These conditions lead to the transcendental equation

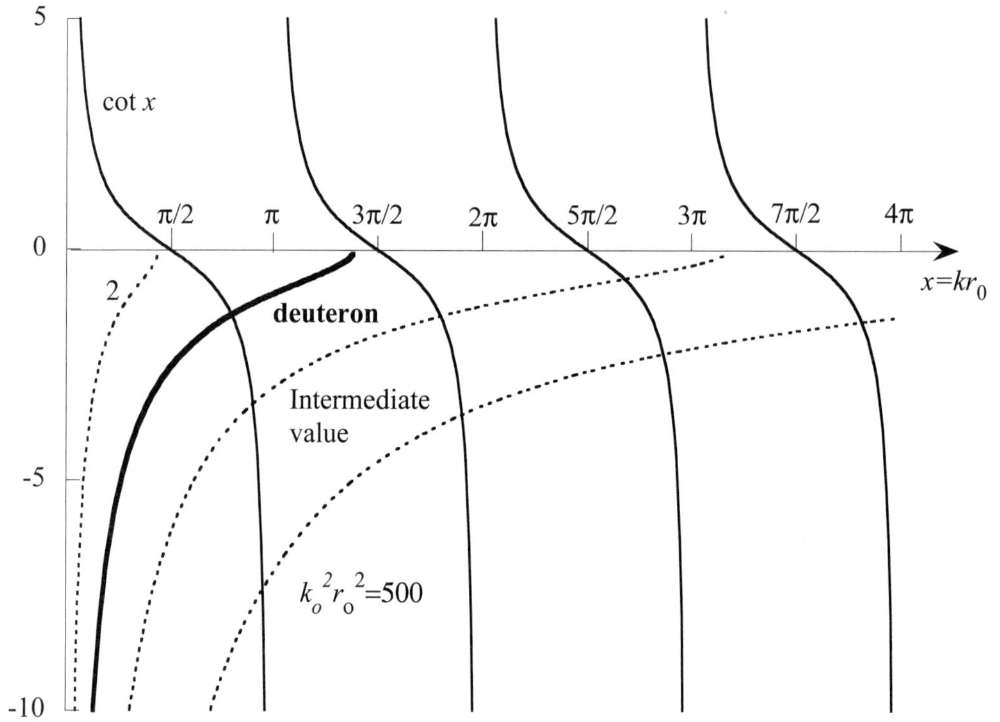

FIGURE 5.11
Curves showing the functions $y = \cot x$ (full curve) and $y = -\sqrt{(k_o^2 r_o^2 - x^2)}/x$ (dotted curves), where $x = kr_o$ and $k_o^2 r_o^2 = 2\mu V_o r_o^2/\hbar^2$ for small $k_o^2 r_o^2 = 2$, intermediate and large $k_o^2 r_o^2 = 500$. The thick curve is for the deuteron (see the text).

$$(kr_o)\cot(kr_o) = -\kappa\, r_o\,, \tag{5.7.10}$$

which must be satisfied by the allowed values of binding energy. To find the allowed binding energies we must solve Eq. (5.7.10) or, equivalently, the set of two simultaneous equations

$$y = \cot x, \tag{5.7.11}$$

$$y = -\frac{\kappa r_o}{x} = -\frac{\sqrt{k_o^2 r_o^2 - x^2}}{x} \tag{5.7.12}$$

for $kr_o = x$, where $k_o^2 r_o^2 = 2\mu V_o r_o^2/\hbar^2$. A plot of these curves is shown in Fig. 5.11. The bound state energies are determined by the values of $x = kr_o$, where the two curves intersect. It can be seen that for a given depth V_o and range r_o of the potential well, the two curves intersect at a finite number of points. This means the bound states are discrete and their number is finite. Denoting the number of bound states by N and labeling them according to their binding energies in descending order $V_o > B_1 > B_2 > B_3 \cdots > B_N > 0$, we see that the first bound state lies deepest in the well and the last one just below the top of the well. From Fig. 5.11, we can see that the n-th intersection occurs in the range $n\pi > x \equiv kr_o > (2n-1)\pi/2$. Using the definition (5.7.5) of k, we find the binding energy B_n of the n-th state satisfies

$$V_o - \frac{\hbar^2 n^2 \pi^2}{2\mu r_o^2} < B_n < V_o - \frac{\hbar^2 (2n-1)^2 \pi^2}{8\mu r_o^2}\,. \tag{5.7.13}$$

Since the minimum value of the binding energy is zero, it follows from Eq. (5.7.13) that to support N bound states the potential V_o must satisfy the condition

$$V_o r_o^2 \geq \frac{\hbar^2}{2\mu} \frac{(2N-1)^2 \pi^2}{4} . \tag{5.7.14}$$

This is a relationship between the range and depth of the potential well to support N bound states. The larger the product $V_o r_o^2$, the larger the number of bound states. This means there are more bound states if the well is deep or broad or both. In particular, no bound state can exist unless

$$V_o r_o^2 \geq \frac{\hbar^2 \pi^2}{8\mu} . \tag{5.7.15}$$

Let us apply these results to the deuteron. Using the binding energy $B = 2.226$ MeV (from the data presented at the beginning of this section) and reduced mass $\mu = m_n m_p/(m_n + m_p) \approx M/2$, where M is the nucleon mass (assuming neutron and proton masses to be equal: $m_n \approx m_p \equiv M = 938$ MeV), the parameter $\kappa = \sqrt{\frac{MB}{\hbar^2}} \approx 0.232 \;\; \text{fm}^{-1}$. For the range of interaction we may take $r_o = 3$ fm, a value comparable to the charge radius of the deuteron. This gives us $\kappa r_o \approx 0.70$.

To find the parameters of the potential to fit the deuteron binding energy, we need the first (taking the deuteron to be the ground state of the $n-p$ system) solution of the transcendental equation (5.7.10) with $\kappa r_o \approx 0.70$. Assuming that the well depth is large compared to the binding energy, the required solution for x is slightly larger than $\pi/2$. Using the substitution $x = \pi/2 + \epsilon$, and assuming ϵ to be small compared to $\pi/2$, in Eq. (5.7.10) we estimate $\epsilon \approx 2\kappa r_o/\pi = 0.44$ (a more careful estimate gives $\epsilon = 0.350$). Using this estimate $x \equiv k r_o = \frac{\pi}{2} + 0.44 \approx 2.01$. Comparing the magnitudes of $k r_o$ and κr_o we find $k^2/\kappa^2 \equiv V_o/B - 1 \approx 8.25$ or $V_o = 9.25B = 20.6$ MeV. Thus the depth of the potential V_o is indeed large compared to the binding energy of the deuteron consistent with our assumption. That is why the deuteron is called a weakly bound system.

The normalized ground state radial wave function for the deuteron is given by

$$u(r) = \begin{cases} \sqrt{\dfrac{2\kappa}{\kappa r_o + 1}} \sin(kr) & \text{for } r \leq r_o \\[2ex] \sqrt{\dfrac{2\kappa}{\kappa r_o + 1}} \sin(k r_o) \exp[-\kappa(r - r_o)], & \text{for } r > r_o . \end{cases} \tag{5.7.16}$$

Figure 5.12 shows $u(r)$ as a function of r. The spatial extent of the wave function, approximately $1/\kappa \approx 4.3$ fm, is larger than the range of the potential r_o. This means the particles spend significant time outside the range of interaction as expected of a weakly bound system. The fraction of the time for which the separation is greater than r_o is easily calculated as the integral

$$\int_{r_o}^{\infty} dr |u(r)|^2 = \frac{1}{[(\kappa r_o + 1)(1 + (\kappa r_o/k r_o)^2)]} , \tag{5.7.17}$$

which yields a value of 52%. So the particles spend more than half their time outside the range of the potential.

Note that the knowledge of the binding energy allows us to determine only the depth-range combination $V_o r_o^2$, not the depth V_o and range r_o separately. If r_o can be measured independently, for example, in scattering experiments, then the well depth V_o can be determined.

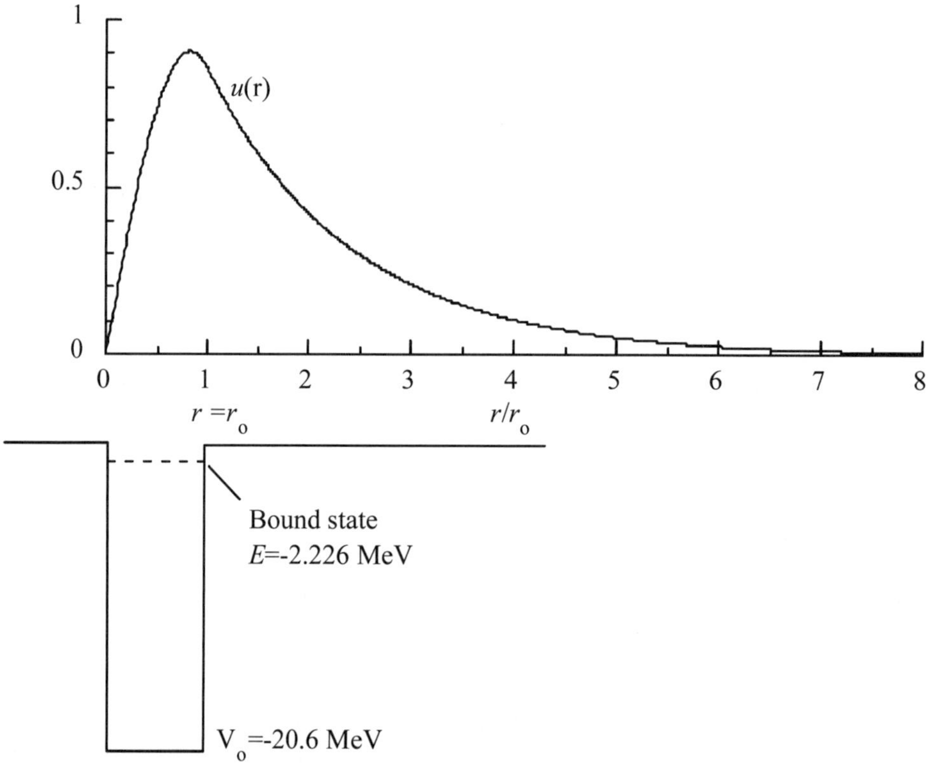

FIGURE 5.12
Deuteron wave function $u(r)$ as a function of r. Here r_o is the range of the interaction.

On the question of excited states of the deuteron

We may now ask whether excited s-states ($\ell = 0$) are possible for the deuteron. We find that the range and depth combination for the deuteron $V_o r_o^2 = 185$ MeV$\cdot$ fm^2 is less than $\frac{\hbar^2 9\pi^2}{8\mu} = 460$ MeV$\cdot$ fm^2, the minimum needed for at least one excited s-state to exist [Eq. (5.7.14)]. Hence we conclude that the excited s-states of the deuteron are not possible with the depth and range combination that fits the observed data.

Let us also examine whether the deuteron can exist as a bound system in a higher angular momentum state. We will show that for the depth-range combination that we have estimated, the deuteron cannot exist as a bound system in the p-state. Still higher angular momentum states would then be ruled out automatically, since the effective binding potential becomes less and less attractive as ℓ increases. Let us assume that in the p-state the system is just bound so that $E = -B = 0$. Then the radial equation (5.5.39) for the potential $V(r)$ given by Eq. (5.7.1), and $\ell = 1$, assumes the form:

$$\frac{d^2u}{dr^2} + \left(k_0^2 - \frac{2}{r^2}\right) u(r) = 0\,, \qquad r \leq r_o \tag{5.7.18}$$

$$\frac{d^2u}{dr^2} - \frac{2}{r^2} u(r) = 0\,, \qquad r > r_o \tag{5.7.19}$$

where $k_o = \sqrt{2\mu V_o/\hbar^2}$. With the change of variables

$$\xi = k_0 r \quad \text{and} \quad u(r) = \sqrt{k_o}\,(v(\xi)/\xi)\,, \tag{5.7.20}$$

in (5.7.18), we find that $v(\xi)$ satisfies the equation

$$\frac{1}{\xi}\frac{d^2v(\xi)}{d\xi^2} - \frac{2}{\xi^2}\frac{dv(\xi)}{d\xi} + \frac{v(\xi)}{\xi} = 0 . \tag{5.7.21}$$

Multiplying throughout by ξ and differentiating the resulting equation with respect to ξ we can cast this equation in the form

$$\left[\frac{d^2}{d\xi^2} + 1\right]\frac{1}{\xi}\frac{dv}{d\xi} = 0 . \tag{5.7.22}$$

From this the solution for $\frac{1}{\xi}\frac{dv}{d\xi}$ can be immediately recognized as a combination of $\sin\xi$ and $\cos\xi$. The solution for $v(\xi)$ which vanishes at $\xi = 0$ [consistent with Eq. (5.5.42)] is then

$$v(\xi) = A_1(\sin\xi - \xi\cos\xi) \quad \text{for } r < r_o . \tag{5.7.23}$$

The solution of Eq. (5.7.19) is $u(r) = C/r$ for $r > r_o$, which via Eq. (5.7.20) leads to

$$v(\xi) = A_2 \quad \text{for } r > r_o . \tag{5.7.24}$$

Here A_1 andA_2 are constants. Matching the solutions as well as their first derivatives at $r = r_o$ or at $\xi = k_o r_o$, we get

$$k_o r_o \sin(k_o r_o) = 0 , \tag{5.7.25}$$

which is possible only if $k_o r_o = \pi$ or $V_o r_o^2 = \frac{\hbar^2\pi^2}{2\mu}$. Thus the minimum value of the range and depth combination needed for at least one p-state exceeds that for the s-state [Eq. (5.7.15)]. This is to be expected as a deeper potential well is needed to overcome the cetrifugal contribution to the effective potential. For the deuteron $\mu \approx= M/2$, $M = 938$ MeV being the nucleon mass, this leads to $V_o r_o^2 = 408$ MeV, which exceeds the range-depth combination (185 MeV·fm^2) for the deuteron. Thus this range-depth combination for the deuteron rules out a bound p-state.

The question of nonzero angular momentum bound states of the deuteron is relevant since its nonzero quadrupole moment would require an $\ell = 2$ state. Indeed there is some evidence that the deuteron state is an admixture containing a large $\ell = 0$ component and a small $\ell = 2$ component. Such an admixture would be ruled out if central forces were the only forces acting between the neutron and proton. This suggests that noncentral forces also play a role in the $n - p$ interaction.

Deuteron problem with an exponential potential

We have seen that the deuteron is a weakly bound system. It is therefore not expected to be sensitive to the details of the potential. To explore this further, let us consider another attractive potential of the form

$$V(r) = -V_o \exp(-r/r_o) , \tag{5.7.26}$$

where r_o defines the range of the potential. Substituting this into the radial equation (5.7.2) with $u(r) = rR(r)$ for the s-state ($\ell = 0$), we get

$$\frac{d^2u}{dr^2} + \frac{2\mu}{\hbar^2}\left(E + V_o e^{-r/r_o}\right)u(r) = 0 . \tag{5.7.27}$$

Introducing the independent variable z by

$$z = \sqrt{\frac{8\mu V_o}{\hbar^2}}\, r_o \exp(-r/2r_o) , \tag{5.7.28}$$

we can transform the radial equation to the Bessel form[5]

$$\frac{d^2\chi}{dz^2} + \frac{1}{z}\frac{d\chi}{dz} + \left(1 - \frac{\nu^2}{z^2}\right)\chi(z) = 0\,, \tag{5.7.29}$$

where $$\nu^2 \equiv \frac{8\mu B}{\hbar^2} r_o^2\,, \tag{5.7.30}$$

and $\chi(z)$ denotes the transformed radial function. This equation has two independent solutions: $J_\nu(z)$ and $J_{-\nu}(z)$ [see Appendix 5A1 Sec. 5]. In the limit $z \to 0$ (or $r \to \infty$),

$$J_\nu(z) \approx \frac{z^\nu}{2^n\Gamma(\nu+1)} \quad \text{and} \quad J_{-\nu}(z) \approx \frac{z^{-\nu}}{2^{-\nu}\Gamma(-\nu+1)}\,. \tag{5.7.31}$$

We rule out $J_{-\nu}$ because it diverges for $z \to 0$ ($r \to \infty$). So the acceptable solution Eq. (5.7.29) is

$$\chi(z) = AJ_\nu(z)\,, \tag{5.7.32}$$

or $$u(r) = A\, J_\nu\left(\sqrt{\frac{8\mu V_o}{\hbar^2}}\, r_o \exp(-\frac{r}{2r_o})\right)\,. \tag{5.7.33}$$

Further, $u(r)$ must satisfy the boundary condition $u(0) = 0$ [Eq. (5.5.42)] at the origin. This means $J_\nu\left(\sqrt{\frac{8\mu V_o}{\hbar^2}}\, r_o\right) = 0$, which implies that $\left(\sqrt{\frac{8\mu V_o}{\hbar^2}}\, r_o\right)$ must be a zero of J_ν. For the deuteron problem, the value of $\nu = \sqrt{\frac{8\mu B}{\hbar^2}}\, r_o \approx 1.4$ if the range r_o is taken be of the order of 3 fm. If z_1 is the first zero of $J_{1.4}(z)$ then the above condition demands that $\sqrt{\frac{8\mu V_o}{\hbar}}\, r_o = z_1$ or

$$r_o^2 V_o = \frac{\hbar^2}{8\mu} z_1^2. \tag{5.7.34}$$

This is the relationship between the depth and range, if the potential has an exponential shape. Once again we see that a knowledge of the binding energy allows us to determine only the range-depth combination for the potential as was the case for the square well potential. The behavior of the wave function is also qualitatively similar to that discussed for the s-state in a square well.

5.8 Energy Levels in a Three-dimensional Square Well: General Case

In Sec. 5.7, we considered the problem of bound state energy levels in a square well for $\ell = 0$. Here we consider the bound states in a spherical square well potential for arbitrary values of ℓ. Obviously, the results of this section will reduce to those of Sec. 5.7 for $\ell = 0$.

Consider a particle of mass m moving in a spherical potential well of the form:

$$V(r) = \begin{cases} -V_o\,, & r \le a \\ 0\,, & r > a. \end{cases} \tag{5.7.1*}$$

[5]See Appendix 5A1 Sec. 5 for Bessel equation.

Such a potential could describe, for example, a nucleon moving in a nucleus, independently of others, in a common potential $V(r)$. The potential $V(r)$ takes into account the effect of other nucleons in an average sense. The radial equation (5.5.39) for the spherical square well potential of Eq. (5.7.1*) may be written as

$$\frac{d^2R_\ell}{dr^2} + \frac{2}{r}\frac{dR_\ell}{dr} - \frac{2m}{\hbar^2}(V_o - B)R_\ell(r) - \frac{\ell(\ell+1)}{r^2}R_\ell(r) = 0, \qquad \text{for } r < a\,, \tag{5.8.1}$$

and

$$\frac{d^2R_\ell}{dr^2} + \frac{2}{r}\frac{dR_\ell}{dr} - \frac{2mB}{\hbar^2}R_\ell(r) - \frac{\ell(\ell+1)}{r^2}R_\ell(r) = 0, \qquad \text{for } r > a\,, \tag{5.8.2}$$

where we have put $E = -B$, since we are interested in the spectrum of bound states, which will have negative energies.

With the substitution

$$\rho = kr\,, \tag{5.8.3}$$

and

$$R_\ell(r) = k^{3/2}\frac{y_\ell(\rho)}{\rho^{1/2}}\,, \tag{5.8.4}$$

where

$$k^2 = \frac{2m}{\hbar^2}(V_o - B)\,, \tag{5.8.5}$$

Eq. (5.8.1) reduces to the Bessel form

$$\frac{d^2y_\ell(\rho)}{d\rho^2} + \frac{1}{\rho}\frac{dy_\ell(\rho)}{d\rho} + \left(1 - \frac{(\ell+1/2)^2}{\rho^2}\right)y_\ell(\rho) = 0\,. \tag{5.8.6}$$

The general solution of this equation can be written as a linear combination of $J_{\ell+1/2}(\rho)$ and $J_{-(\ell+1/2)}(\rho)$ as

$$y_\ell(\rho) = AJ_{\ell+1/2}(\rho) + A'J_{-(\ell+1/2)}(\rho)\,. \tag{5.8.7}$$

Since $R_\ell(r)$ must be finite for all values of r, we must rule out $J_{-(\ell+1/2)}$, which is singular at the origin. Then choosing $A' = 0$ in Eq. (5.8.7), we can write the radial wave function (5.8.5) inside the well as

$$R_\ell(r) = Ak^{3/2}\frac{1}{(kr)^{1/2}}J_{\ell+1/2}(kr) \equiv Cj_\ell(kr) \quad \text{for} \quad r < a \tag{5.8.8}$$

where we have replaced $Ak^{3/2}$ by another constant C by writing $Ak^{3/2} = C\sqrt{\pi/2}$. This allows us to write the solution inside the well in terms of the spherical Bessel function $j_\ell(kr) = \sqrt{\frac{\pi}{2kr}}\,J_{\ell+1/2}(kr)$ [see Appendix 5A1 Sec. 5].

The radial equation (5.8.2) for the exterior region ($r > a$) may similarly be reduced to the Bessel form by using the substitution

$$\rho = i\kappa r\,, \quad \kappa = (2mB/\hbar^2)^{1/2} \tag{5.8.9}$$

and $y_\ell(\rho)$ as defined in Eq. (5.8.4). In this case, since the domain of ρ does not include the origin $\rho = 0$, both solutions of Eq. (5.8.6) are acceptable. The general solution outside the well can then be written as

$$\begin{aligned} R_\ell(r) &= D\sqrt{\frac{\pi}{2i\kappa r}}J_{\ell+1/2}(i\kappa r) + D'(-1)^{\ell+1}\sqrt{\frac{\pi}{2i\kappa r}}J_{-(\ell+1/2)}(i\kappa r) \\ &\equiv Dj_\ell(i\kappa r) + D'\eta_\ell(i\kappa r) \end{aligned} \tag{5.8.10}$$

where $j_\ell(\rho) \equiv \sqrt{\pi/2\rho}J_{\ell+1/2}(\rho)$ and $\eta_\ell(\rho) \equiv (-1)^{\ell+1}\sqrt{\pi/2\rho}J_{-(\ell+1/2)}(\rho)$ are, respectively, the *spherical Bessel* and *spherical Neumann* functions [see Appendix 5A1]. Instead of the spherical Bessel functions, another set of functions, known as *spherical Hankel functions* of the first and the second kind, are more convenient for boundary conditions at infinity. They are defined as linear combinations[6]

$$h_\ell^{(1)}(\rho) \equiv j_\ell(\rho) + i\eta_\ell(\rho) \to \frac{1}{\rho}\exp[i(\rho - (\ell+1)\pi/2)]\,, \tag{5.8.11}$$

$$h_\ell^{(2)}(\rho) = j_\ell(\rho) - i\eta_\ell(\rho) \to \frac{1}{\rho}\exp[-i(\rho - (\ell+1)\pi/2)]\,, \tag{5.8.12}$$

where $\rho = i\kappa r$ and the arrow specifies the asymptotic ($\rho \to \infty$) behavior. In the exterior region $(r > a)$ we must choose the solution to be $h_\ell^{(1)}$ because asymptotically it will fall off as $e^{-\kappa r}$, which is the desired behavior.

Having identified the acceptable solutions inside and outside the well we can write the interior and exterior radial wave function in terms of these solutions. The energy levels of the system may be obtained from the continuity condition[7]

$$\left.\frac{1}{R_{\text{int}}}\frac{dR_{\text{int}}}{dr}\right|_{r=a} = \left.\frac{1}{R_{\text{ext}}}\frac{dR_{\text{ext}}}{dr}\right|_{r=a}. \tag{5.8.13}$$

For the s-state ($\ell = 0$) we obtain

$$R_{\text{int}}(r) = j_0(kr) = \frac{\sin(kr)}{kr}\,, \qquad r < a \tag{5.8.14}$$

and

$$R_{ext}(r) = h_0^{(1)}(i\kappa r) = -\frac{e^{-\kappa r}}{\kappa r}\,, \qquad r > a\,. \tag{5.8.15}$$

Using the matching condition (5.8.13) we have,

$$ka\,\cot(ka) = -\kappa a\,. \tag{5.8.16}$$

As expected this is the same as Eq. (5.7.10) with r_o replaced by a. This is the equation we discussed in Sec. 5.7.

Similarly, for the p-state ($\ell = 1$), we have

$$R_{\text{int}}(r) = Aj_1(kr) = \frac{\sin(kr)}{k^2r^2} - \frac{\cos(kr)}{kr}\,, \qquad (r < a) \tag{5.8.17}$$

and

$$R_{\text{ext}}(r) = Ch_1^{(1)}(i\kappa r) = Ci\left(\frac{1}{\kappa r} + \frac{1}{\kappa^2 r^2}\right)e^{-\kappa r}\,, \qquad (r > a)\,. \tag{5.8.18}$$

Matching condition (5.8.13) gives us, for this case,

$$\frac{\cot(ka)}{ka} - \frac{1}{k^2a^2} = \frac{1}{\kappa a} + \frac{1}{\kappa^2 a^2}\,. \tag{5.8.19}$$

We can show in this case that there is no bound state for $V_o a^2 < \frac{\pi^2\hbar^2}{2\mu}$, one bound state for $\frac{\pi^2\hbar^2}{2\mu} \leq V_o a^2 < \frac{(2\pi)^2\hbar^2}{2\mu}$ and so on.

This can be continued for higher values of ℓ. In each case we obtain a transcendental equation for binding energy. These equations have to be solved numerically to find the allowed energy values. Solutions of Eq. (5.8.16) give us energies for $1s$, $2s$, $3s$,$\cdots$ states, while the solutions of Eq. (5.8.19) give us the energies for $1p$, $2p$, $3p$, $\cdots$ states. For each value of ℓ only a finite number of excited states are possible, depending on the values of V_o and a.

---

[6]From the explicit expressions for j_ℓ and η_ℓ given in Appendix 5A1, the forms of the corresponding spherical Hankel functions of the first kind may be derived.

[7]The continuity of the radial wave function and its first derivative expressed by $R_{\text{int}}(r)|_{r=a} = R_{\text{ext}}(r)|_{r=a}$ and $\left.\frac{dR_{\text{int}}}{dr}\right|_{r=a} = \left.\frac{dR_{\text{ext}}}{dr}\right|_{r=a}$ may be combined into a single equation given by Eq. (5.8.13).

Energy levels in a spherical square well potential of infinite depth

The eigenvalue problem for the bound state of a square can be solved for the case $V_o \to \infty$. In this case the potential function is $V(r) \to -\infty$ for $r < a$ and 0 for $r > a$. If we measure the energies from the bottom of the well, we may as well take the potential to be

$$V(r) = \begin{cases} 0\,, & \text{for} \quad 0 < r < a \\ \infty\,, & \text{for} \quad a < r < \infty\,. \end{cases} \tag{5.8.20}$$

In this case the radial function vanishes for $r > a$. Inside the well ($r < a$) the radial equation (5.5.41) for $u_\ell(r) = rR_\ell(r)$ can be written as

$$\frac{d^2u_\ell}{dr^2} - \frac{\ell(\ell+1)}{r^2}u_\ell(r) + k^2u_\ell(r) = 0\,, \quad \text{for} \quad r < a, \tag{5.8.21}$$

where $$k^2 = \frac{2mE}{\hbar^2}\,. \tag{5.8.22}$$

Note that since the bottom of the well is chosen to be the zero of energy, the bound state energy E is positive ($E > 0$). With the substitution $\rho = kr$ and $u_\ell(r) = k^{1/2}\rho^{1/2}y_\ell(\rho)$, Eq. (5.8.21) reduces to the Bessel form

$$\frac{d^2y_\ell(\rho)}{d\rho^2} + \frac{1}{\rho}\frac{dy_\ell(\rho)}{d\rho} + \left(1 - \frac{(\ell+1/2)^2}{\rho^2}\right)y_\ell(\rho) = 0\,. \tag{5.8.23}$$

The regular solution of this equation, which goes to zero as $\rho \to 0$ ($r \to 0$), is

$$y_\ell(\rho) = J_{\ell+1/2}(\rho)\,. \tag{5.8.24}$$

Then the radial function, with the help of Eq. (5.8.4), can be written as

$$R_\ell(r) \equiv \frac{u_\ell(r)}{r} = A\frac{J_{\ell+1/2}(kr)}{\sqrt{kr}} = Cj_\ell(kr)\,. \tag{5.8.25}$$

The boundary condition at $r = a$ requires $R_\ell(a) = 0$ or

$$j_\ell(ka) = 0\,. \tag{5.8.26}$$

If $\omega_{n\ell}$ denotes the n-th zero of $j_\ell(ka)$, then the allowed values of k and E are given by

$$k_{n\ell} = \frac{\omega_{n\ell}}{a}\,, \tag{5.8.27}$$

$$E_{n\ell} = \frac{\hbar^2}{2m}\frac{\omega_{n\ell}^2}{a^2}\,. \tag{5.8.28}$$

The radial part of the corresponding wave function is given by

$$R_{n\ell}(r) = C_{n\ell}\, j_\ell(k_{n\ell}r) = C_{n\ell}\, j_\ell(\omega_{n\ell}r/a)\,, \tag{5.8.29}$$

where the normalization constant $C_{n\ell}$ is determined by the condition (5.5.40).

5.9 Energy Levels in an Isotropic Harmonic Potential Well

Consider a particle of mass m moving a three-dimensional potential well of the shape

$$V(r) = \frac{1}{2}m\omega^2r^2 \tag{5.9.1}$$

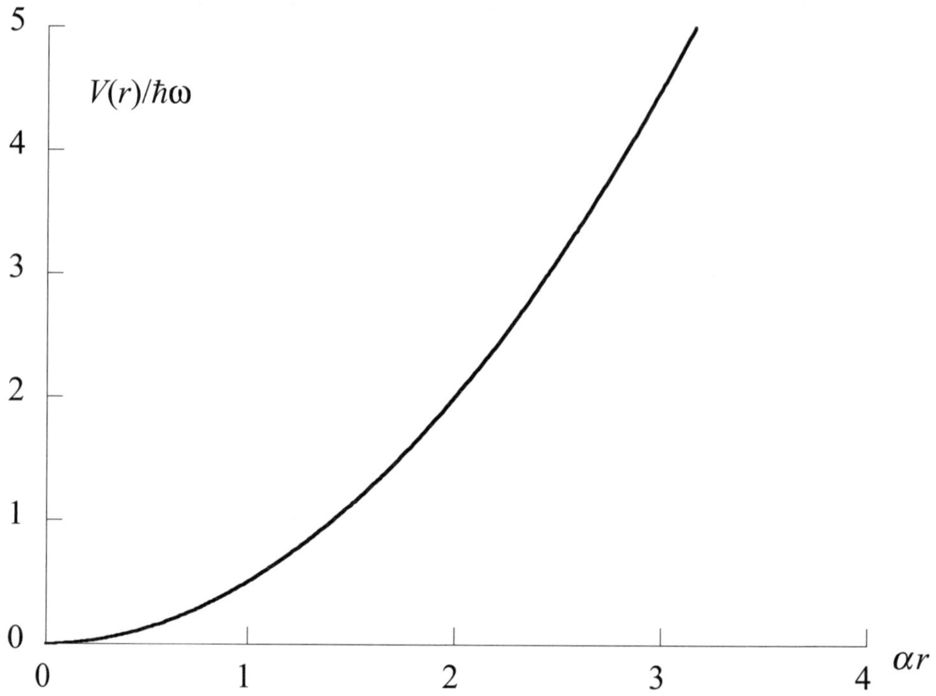

FIGURE 5.13
Isotropic harmonic potential well in three dimensions.

where ω is the natural frequency of a classical oscillator. A constant potential term can be added to this potential, which amounts to a change in the reference level from where the energy is measured. We will measure the energy levels from the bottom of the well so that $V(0) = 0$.

By writing the potential as $V(r) = \frac{1}{2}m\omega^2(x^2+y^2+z^2)$ we can see that particle motion in this potential may be considered as a superposition of independent simple harmonic motion in x, y and z directions. Since the natural frequency is the same in all directions, the choice of x, y, and z axes is arbitrary and the oscillator is called an isotropic oscillator. The shape of the potential is shown in Fig. 5.13.

With the potential (5.9.1), the radial equation (5.5.41) for $u_\ell(r) = rR_\ell(r)$ takes the form:

$$\frac{d^2u_\ell(r)}{dr^2} - \left[\frac{\ell(\ell+1)}{r^2} + \frac{m^2\omega^2r^2}{\hbar^2} - \frac{2mE}{\hbar^2}\right]u_\ell(r) = 0\,. \tag{5.9.2}$$

By expressing the energy in units of $\hbar\omega$ by writing $E = \epsilon\hbar\omega$ and distance in units of $\sqrt{\hbar/m\omega}$ by writing $r = \sqrt{\frac{\hbar}{m\omega}}\,\rho$, we find the radial equation can be written as

$$\frac{d^2v_\ell(\rho)}{d\rho^2} - \left[\frac{\ell(\ell+1)}{\rho^2} + \rho^2 - 2\epsilon\right]v_\ell(\rho) = 0 \tag{5.9.3}$$

where $v_\ell(\rho)$ is related to the radial function by $u_\ell(r) = (m\omega/\hbar)^{1/4}v_\ell(\rho)$. To solve this equation we note that we are looking for solutions that have the correct behavior, $v_\ell \to 0$ as $\rho \to \infty$ (or $r \to \infty$) [Eq. (5.5.43)]. In the limit $\rho \to \infty$ this equation assumes the approximate form

$$\frac{d^2 v_\ell}{d\rho^2} - (\rho^2 \mp 1) v_\ell = 0 \,. \tag{5.9.4}$$

Here we have retained a term $\mp v_\ell(\rho)$ from $2\epsilon v_\ell$, which is negligible compared to $\rho^2 u_\ell$ but allows this equation to be integrated. Equation (5.9.4) has solutions $e^{\pm\rho^2/2}$. Out of these, only the negative exponential

$$v_\ell \sim \exp(-\rho^2/2) \tag{5.9.5}$$

has the correct asymptotic behavior. The radial function $u_\ell(r)$ must also have the correct behavior $u_\ell(0) = 0$ [Eq. (5.5.42)] near the origin $r = 0$. In this limit, Eq. (5.9.3) assumes the form

$$\frac{d^2 v_\ell(\rho)}{d\rho^2} - \frac{\ell(\ell+1)}{\rho^2} v_\ell(\rho) = 0 \,. \tag{5.9.6}$$

Assuming a series solution, the leading term of which is of the form $u_\ell \sim \rho^\nu$, we find $\nu = \ell+1$ or $\nu = -\ell$. Of the two solutions corresponding to these values of ν, only

$$v_\ell(\rho) \sim \rho^{\ell+1} \tag{5.9.7}$$

has the requisite behavior for $\rho \to 0$. Incorporating the small and large ρ behavior, we look for the solutions of Eq. (5.9.3) in the form

$$v_\ell(r) = \rho^{\ell+1} \exp(-\rho^2/2) f_\ell(\rho) \,. \tag{5.9.8}$$

Substituting this into the radial equation (5.9.3), we find that f_ℓ must satisfy the equation

$$\frac{d^2 f_\ell}{d\rho^2} + \left(\frac{2\ell+2}{\rho} - 2\rho\right)\frac{df_\ell}{d\rho} - (3 + 2\ell - 2\epsilon) f_\ell(\rho) = 0 \,. \tag{5.9.9}$$

Since $f_\ell(\rho)$ must be a regular function, we try a series solution of the form

$$f_\ell(\rho) = \sum_{s=0}^{\infty} a_s \rho^s \,. \tag{5.9.10}$$

Susbtituting this in Eq. (5.9.9), we obtain

$$\sum_{s=0}^{\infty} \left[(3 + 2\ell - 2\epsilon + 2s)a_s - ((s+2)(s+1) + (2\ell+2)(s+2))a_{s+2}\right] \rho^s = 0 \,. \tag{5.9.11}$$

This leads to a two-step recursion relation for the coefficients a_n

$$a_{s+2} = -\frac{(2\epsilon - 2\ell - 3 - 2s)}{(s+2)(s+2\ell+3)} a_s \,. \tag{5.9.12}$$

This recursion relation connects coefficients of even powers of ρ to coefficients of even powers and coefficients of odd powers to coefficients of odd powers. With the help of this recursion relation all even power coefficients can be expressed in terms of a_0 and odd power coefficents can be expressed in terms of a_1. Hence we get two possible solutions. The choice $a_1 \neq 0$ gives the odd power solution which, by factoring out ρ, can be written as $\rho \times F(\rho^2)$, where $F(\rho^2)$ is a function of ρ^2. This results in changing the power of ρ in Eq. (5.9.8) from $\ell+1$ to $\ell+2$ and thus modifying the small ρ behavior of the wave function. Therefore we choose $a_1 = 0$, leaving only the even power solution based on $a_0 \neq 0$. With the substitution $s = 2p$, the recursion relation for the even power solution can be written as

$$a_{p+1} = -\frac{\left[\frac{1}{2}(\epsilon - \ell - 3/2) - p\right]}{(p+1)(p+\ell+3/2)} a_p \,. \tag{5.9.13}$$

For normalizable solutions, this series must terminate. An inspection of Eq. (5.9.13) shows that for this to be the case $\epsilon-\ell-\frac{3}{2}$ must equal a positive even integer $2n_r$ ($n_r=0,1,2,3,\cdots$). Since ℓ takes values $0,1,2,\cdots$, this means $\epsilon-\frac{3}{2}$ itself must equal a positive integer $n\geq\ell$ such that $n-\ell=2n_r$. These conditions can be written as

$$\left(\epsilon-\frac{3}{2}\right)=n\,,\quad n=0,1,2,3,\cdots \tag{5.9.14}$$

$$n-\ell=2n_r=0,2,4,\cdots \tag{5.9.15}$$

Recalling that $\epsilon\equiv E/\hbar\omega$, these conditions imply that the allowed energy values are given by

$$E_n=\left(n+\frac{3}{2}\right)\hbar\omega\,,\quad n=0,1,2,3\cdots \tag{5.9.16}$$

and since $n-\ell$ is a non-negative even integer, it follows that for a given n, the allowed values of ℓ are given by

$$\ell=n,n-2,n-4,\cdots,1\text{ or }0\,. \tag{5.9.17}$$

Here the smallest value of ℓ is 1 for odd values of n and 0 for even values. This determines the allowed energy levels in a three-dimensional isotropic harmonic potential well.

For these values of energy, the series in Eq. (5.9.13) terminates at $p=n_r=(n-\ell)/2$ and the solution is a polynomial in ρ^2 of order $(n-\ell)/2$. Equation (5.9.13) can be recognized as the recursion relation for the associated Laguerre polynomial [Appendix 5A1, Eq. (5A1.3.5)]

$$L^{\ell+1/2}_{(n+\ell+1)/2}(\rho^2)=\sum_{p=0}^{(n-\ell)/2}(-1)^p\frac{[\{(n+\ell+1)/2\}!]^2}{\{(n-\ell)/2-p\}!p!(p+\ell+1/2)!}\rho^{2p}\,. \tag{5.9.18}$$

Collecting all the information, the radial function $R_{nl}(r)$, labeled by the energy index n and angular momentum index ℓ is given by

$$R_{n\ell}(r)=A_{n\ell}\left(\sqrt{\frac{m\omega}{\hbar}}\,r\right)^{\ell}e^{-m\omega r^2/2\hbar}L^{\ell+1/2}_{(n+\ell+1)/2}\left(\frac{m\omega r^2}{\hbar}\right)\,, \tag{5.9.19}$$

$$A_{n\ell}=\left(\frac{m\omega}{\hbar}\right)^{3/2}\sqrt{\frac{2\times((n-\ell)/2)!}{[((n+\ell+1)/2)!]^3}}\,. \tag{5.9.20}$$

The energy levels for the isotropic harmonic oscillator are highly degenerate as the energy depends only on n and not on ℓ and m. Rotational invariance of the Hamiltonian can explain the degeneracy relative to the index m. The degeneracy relative to ℓ is indicative of a larger symmtery of the Hamiltonian; rotational invariance of the Hamiltonian alone is not sufficient to explain this degeneracy. To compute the degree of degeneracy we note that for a given n, the ℓ values are given by

$$\ell=\begin{cases}0,2,4,\cdots,n: & \text{total }(n+2)/2\text{ terms for even }n\\ 1,3,5,\cdots,n: & \text{total }(n+1)/2\text{ terms for odd }n\,.\end{cases} \tag{5.9.21}$$

For each value of ℓ there are $2\ell+1$ values that m can take. The degree of degeneracy is then

$$\sum_{\text{even } \ell}^{n}(2\ell+1) = 1+5+9+\cdots+(2n+1) = \frac{(n+2)}{2}\cdot\frac{1}{2}(2n+1+1))$$
$$= \frac{1}{2}(n+1)(n+2) \qquad (5.9.22)$$
$$\sum_{\text{odd } \ell}^{n}(2\ell+1) = 3+7+11+\cdots+(2n+1) = \frac{(n+1)}{2}\cdot\frac{1}{2}(2n+1+3)$$
$$= \frac{1}{2}(n+1)(n+2)\,. \qquad (5.9.23)$$

Thus the degeneracy of level E_n is given by $(n+1)(n+2)/2$ for all n. Each state $|n,\ell,m\rangle$ is specified by three quantum numbers corresponding to the eigenvalues $(n+3/2)\hbar\omega$, $\hbar^2\ell(\ell+1)$ and m of the operators $\hat{H}$, $\hat{L}^2$, and $\hat{L}_z$, which form a complete set of observables for this problem. The wave function $\psi_{n\ell m}(\boldsymbol{r})$ is then a representative of the state $|n,\ell,m\rangle$ in the coordinate representation:

$$< \boldsymbol{r}|n,\ell,m> \equiv \psi_{n\ell m}(\boldsymbol{r}) = R_{n\ell}(r)Y_{\ell m}(\theta,\varphi)$$
$$= A_{n\ell}\left(\sqrt{\frac{m\omega}{\hbar}}\,r\right)^{\ell} e^{-m\omega r^2/2\hbar} L^{\ell+1/2}_{(n+\ell+1)/2}\left(\frac{m\omega r^2}{\hbar}\right) Y_{\ell m}(\theta,\varphi)\,, \qquad (5.9.24)$$

where $A_{n\ell}$ is given by Eq. (5.9.20).

Problems

1. The quantum states of a linear mechanical harmonic oscillator are given by

   $$< x|\psi_n >= \psi_n(x) = N_n H_n(\alpha x)\exp(-\alpha^2x^2/2)$$

   where $\alpha = (mk/\hbar^2)^{1/4} = \sqrt{m\omega/\hbar}$ and $\omega = \sqrt{k/m}$ is the classical frequency of the oscillator.

   Use this wave function to evaluate the matrix elements x_{nm} and p_{nm} and check your results with Eqs. (5.4.32) and (5.4.33).

2. Show that $H_n'(\xi) = 2nH_{n-1}$ and $H_{n+1} = 2\xi H_n - 2nH_{n-1}$ where $H_n(\xi)$ is a Hermite polynomial of order n and $H_n'(\chi) = \frac{d}{d\chi}H_n(\chi)$.

3. In a one-dimensional harmonic oscillator problem, the Hamiltonian may also be expressed as $\hat{H} = \hbar\omega(\hat{a}^\dagger\hat{a} + \frac{1}{2})$ where $\hat{a}^\dagger$ and $\hat{a}$ are the creation and annihilation operators so that $\hat{a}\,|n\rangle = \sqrt{n}\;|n-1\rangle$ and $\hat{a}^\dagger\,|n\rangle = \sqrt{n+1}\;|n+1\rangle$. Determine the following matrix elements: $(x^2)_{mn}$, $(p^2)_{mn}$, $(x^3)_{mn}$, $(p^3)_{mn}$. The operators $\hat{x}$ and $\hat{p}$ are defined in terms of $\hat{a}$ and $\hat{a}^\dagger$ by Eqs. (5.4.30) and (5.4.31).

4. Write the time-independent Schrödinger equation for the linear harmonic oscillator in the momentum representation for the wave function $\phi(p)$. Justify any boundary conditions that must be imposed. Find the normalized wave function for the ground state and the first excited state and show that they are the Fourier transforms of the corresponding wave functions in the coordinate representation.

5. Prove the addition theorem for spherical harmonics

$$P_\ell(\cos\alpha) = \frac{4\pi}{2\ell+1}\sum_{m=-\ell}^{\ell} Y_{\ell m}(\theta,\varphi)Y_{\ell m}(\theta',\varphi'),$$

where θ, φ and θ', φ' are the angles of $\boldsymbol{r}$ and $\boldsymbol{r}'$ and α is the angle between $\boldsymbol{r}$ and $\boldsymbol{r}'$.

6. Using the Harmonic oscillator wave function given in Problem 1 to evaluate

$$< x^2 >_{nn} = \int \psi_n^*(x)x^2\psi_n(x)dx$$

using the generating function of Hermite polynomials,

$$S(\xi,s) = \exp\left(-s^2+2s\xi\right) = \sum_{n=0}^{\infty}\frac{H_n(\xi)}{n!}s^n.$$

Check your result with that in Problem 3.

7. Work out Problem 6 also using the matrix elements $x_{mn} = \langle m|\,\hat{x}\,|n\rangle$ derived in Problem 1 [also in Eq. (5.4.32)].

8. Consider the problem of a particle of mass m moving in a two-dimensional harmonic oscillator potential well

$$V(x,y) = \frac{1}{2}m\omega_1^2x^2 + \frac{1}{2}m\omega_2^2y^2.$$

(a) Write the energy eigenvalues and the corresponding coordinate wave functions. If $\omega_1 = \omega = \omega_2$, write down the states that have the energy $10\hbar\omega$.

(b) Find the energy eigenvalues and eigenfunctions for a two-dimensional isotropic harmonic oscillator ($\omega_1 = \omega_2 = \omega$) in cylindrical coordinates. Discuss the degeneracy of energy levels. What is the significance of quantum numbers needed to uniquely specify the states of the oscillator? Hence identify the complete set of observables for this problem.

9. A charged particle (charge e, mass m) moves in a constant magnetic field $\boldsymbol{B} = \boldsymbol{e}_z\,B$ corresponding to the vector potential $\boldsymbol{A} = \boldsymbol{e}_x(-yB)$. The Hamiltonian for this problem can be writen as

$$\hat{H} = \frac{1}{2m}\left(\hat{p}_x + eyB\right)^2 + \frac{\hat{p}_y^2}{2m} + \frac{\hat{p}_z^2}{2m}.$$

(a) Show that $\hat{p}_x$ and $\hat{p}_z$ commute with the Hamiltonian. Hence $\hat{H}$, $\hat{p}_x$, and $\hat{p}_z$ have simultaneous eigenstates $|k_x, k_z, E\rangle$ with eigenvalues $\hbar k_x$, $\hbar k_y$, E, respectively, corresponding to the coordinate space wave function

$$\langle x, y, z|\, k_x, k_z, E\rangle = \frac{e^{i(k_x x + k_z z)}}{2\pi}\psi(y).$$

(b) Show that $\psi(y)$ satisfies the equation for a linear harmonic oscillator

$$\left[\frac{1}{2m}\frac{\partial^2}{\partial y^2} + \frac{1}{2}m\Omega_c^2(y+y_o)^2\right]\psi(y) = \left(E - \frac{\hbar^2k_z^2}{2m}\right)\psi(y),$$

where $y_o = \hbar k_x/eB$ and $\Omega_c = eB/m$ is the cyclotron frequency. Find the eigen energies of this equation.

10. The ground state of Hydrogen atom is represented by the coordinate space wave function

$$\psi_{100}(r) = \frac{1}{(\pi a_0^3)^{1/2}} e^{-r/a_0} ,$$

where $a_0 = 4\pi\epsilon_0\hbar/m_e e^2$ is the first Bohr radius. Find the quantum mechanical expectation value of r for this state. Compare it with the most probable value and the root mean square value $[\langle r^2\rangle_{100}]^{1/2}$ of r for this state.

11. The first excited state of Hydrogen atom is represented by

$$\psi_{200} = \frac{1}{(32\pi a_0^3)^{1/2}} (r/a_0 - 2)\, e^{-r/2a_0} .$$

Find for this state (a) quantum mechanical expectation value of r, (b) the most probable value of r, and (c) root mean square value of r.

12. Calculate the probability current density using the stationary state Hydrogen atom wave function $\psi_{n\ell m}(r,\theta,\varphi,t)$ given that

$$\boldsymbol{\nabla} = \boldsymbol{e}_r\frac{\partial}{\partial r} + \boldsymbol{e}_\theta\frac{1}{r}\frac{\partial}{\partial\theta} + \boldsymbol{e}_\varphi\frac{1}{r\sin\theta}\frac{\partial}{\partial\varphi} .$$

Show that only the azimuthal component of the current density is nonzero and is given by

$$\boldsymbol{j} = \boldsymbol{e}_\varphi\frac{|\psi_{n\ell m}|^2 m\hbar}{\mu r\sin\theta} .$$

By treating this current density as a classical current density compute the z-component) of the magnetic moment arising from this current density.

13. Show that for any state of the Hydrogen atom, $\langle\hat{T}\rangle = \frac{1}{2}\langle\hat{V}\rangle$, where $\hat{T}$ and $\hat{V}$ are respectively, the kinetic and potential energy operators. Check this for the $1s$ and $2s$ states of the Hydrogen atom. The relevant wave functions are given in Problems 7 and 8.

14. The radial wave functions for the $|n=3,\ell,m\rangle$ states of the Hydrogen atom ($\ell \le n-1$ and $-\ell \le m \le +\ell$) are given by

$$R_{30}(r) = 2(1/3a_0)^{3/2}\left(1 - \frac{2r}{3a_0} + \frac{2r^2}{27a_0^2}\right) e^{-r/3a_0} ,$$
$$R_{31}(r) = \frac{8}{9\sqrt{2}}(1/3a_0)^{3/2}\left(r/a_0 - \frac{r^2}{6a_0^2}\right) e^{-r/3a_0} ,$$
$$R_{32}(r) = \frac{4}{27\sqrt{10}}(1/3a_0)^{3/2}(r/a_0)^2 e^{-r/3a_0} .$$

Calculate the most probable value of r in the $3s$ ($|n=3,l=0,m=0\rangle$) state. Also determine the quantum mechanical expectation value of r for all the three states.

15. In the deuteron problem take the n-p interaction to be spherically symmetric and of the square well shape

$$V(r) = \begin{cases} -V_o , & 0 < r < r_o \\ 0 , & r > r_o . \end{cases}$$

Matching the solutions for $\ell = 0$ in the two regions at $r = r_o$, establish the condition:

$$kr_o \cot(kr_o) = -\kappa r_o \,,$$

where $k = \sqrt{2\mu(V_o - B)/\hbar^2}$, $\kappa = \sqrt{2\mu B/\hbar^2}$ and $E = -B$. Taking $V_o = 40$ MeV ($\gg B$) and $r_o = 1.85$ fm estimate the binding energy B of the deuteron ground state.

Show that the experimental information about B enables one only to find the depth range relationship for the potential $V_o r_o^2 = \frac{\pi^2}{8\mu}\hbar^2$.

16. Assuming the $n - p$ interaction to have an exponential form

$$V(r) = -V_o \exp(-r/r_o)\,,$$

reduce the radial equation for the s-state ($\ell = 0$)

$$\frac{d^2}{dr^2}u(r) + \frac{2\mu}{\hbar^2}\left[E + V_o \exp(-r/r_o)\right]u(r) = 0$$

to Bessel form by the substitution: $z = 2(\sqrt{2\mu V_o}/\hbar) r_o \exp(-r/2r_o)$.

Show that in this case the boundary condition enables one to establish the following depth-range relation

$$V_o r_o^2 = \hbar^2 y_1^2/8\mu$$

where y_1 is the first zero of $J_\nu(z)$ where $\nu = \sqrt{(8\mu B/\hbar^2)}$.

17. Show that

$$\int_0^\infty j_l(kr) j_l(kr') k^2 dk = (\pi/2)\frac{\delta(r - r')}{r^2}\,.$$

18. Solve the three-dimensional harmonic oscillator problem by writing the Hamiltonian as the sum of the Hamiltonian for three linear harmonic oscillators:

$$\hat{H} = \frac{\hat{p_x}^2}{2m} + \frac{1}{2}m\omega^2\hat{x}^2 + \frac{\hat{p_y}^2}{2m} + \frac{1}{2}m\omega^2\hat{y}^2 + \frac{\hat{p_z}^2}{2m} + \frac{1}{2}m\omega^2\hat{z}^2 \equiv \hat{H}_1 + \hat{H}_2 + \hat{H}_3\,.$$

Find the eigenfunctions in the form $\psi(x, y, z) = X(x)Y(y)Z(z)$ and the corresponding energy eigenvalues. How are these levels related to those discussed in the text? Discuss energy level degeneracy in terms of these levels.

19. Show that the three-dimensional delta function can be expressed as

$$\delta^3(r - r') = \frac{\delta(r - r')}{r^2}\frac{\delta(\theta - \theta')}{\sin\theta}\delta(\varphi - \varphi')\,.$$

20. Establish the recursion relations

(a) $2J_n'(x) = J_{n-1}(x) - J_{n+1}(x)$

(b) $\frac{2n}{x}J_n(x) = J_{n-1}(x) + J_{n+1}(x)$

where $J_n'(x)$ denotes the derivative of $J_n(z)$ with respect to its argument x.

21. Prove the following relations for the Legendre polynomials

(a) $nP_{n-1}(x) = nxP_n(x) + (1-x^2)\frac{dP_n}{dx}$

(b) $(n+1)P_{n+1}(x) = (n+1)xP_n(x) - (1-x^2)\frac{dP_n}{dx}$

(c) $(n+1)P_{n+1}(x) - (2n+1)xP_n(x) + nP_{n-1}(x) = 0$

22. Prove the following relations for the spherical Bessel functions

(a) $j_{l-1}(x) + j_{l+1}(x) = \frac{2l+1}{x} j_l(x)$

(b) $\frac{d}{dx} j_l(x) = \frac{1}{(2l+1)}[l\, j_{l-1}(x) - (l+1) j_{l+1}(x)]$

23. If $F(k)$ is the Fourier transform of $f(x)$ and $G(k)$ is the Fourier transform of $g(x)$ then show that the Fourier transform of $f(x)g(x)$ is

$$\int F(k')G(k-k')dk' .$$

References

[1] H. Eyring, J. Walter and G. E. Kimball, *Quantum Chemistry* (John Wiley and Sons, Inc., New York).

[2] L. D. Landau and E. M. Lishitz, *Quantum Mechanics*, (Pergamon Press, New York, 1976).

[3] L. Pauling and E. B. Wilson, *Introduction to Quantum Mechanics* (McGraw Hill Book Company, New York).

[4] L. I. Schiff, *Quantum Mechanics*, Third Edition (McGraw Hill Book Company, Inc., New York, 1968).

[5] G. Arfken, *Mathematical Methods for Physicists* (Academic Press, Inc., New York, 1985).

[6] I. S. Gradshteyn and I. M. Ryzhik, *Table of Integrals, Series and Products* (Academic Press, Inc., New York, 1965).

Appendix 5A1: Special Functions

In this appendix we consider some special functions and the pertinent differential equations which are used in Chapter 5.

5A1.1 Legendre and Associated Legendre Equations

The equation

$$(1-x^2)\frac{d^2y}{dx^2} - 2x\frac{dy}{dx} + \ell(\ell+1)y = 0\,, \tag{5A1.1.1}$$

is called the **Legendre equation**. The solution to this equation may be found by series method by assuming that

$$y = x^c \sum_{s=0}^{\infty} a_s x^s\,. \tag{5A1.1.2}$$

Substituting this into Eq. (5A1.1.1) and equating the coeffiecient of each power of x, say, $x^{(c+s-2)}$ to zero we can express a_s in term of a_{s-2}. For $s=0$, $s=1$ and arbitrary s we get, respectively,

$$\left.\begin{aligned} a_0 c(c-1) &= 0 \\ a_1(c+1)c &= 0 \end{aligned}\right\} \tag{5A1.1.3}$$

$$a_{s+2} = \frac{(c+s)(c+s+1) - \ell(\ell+1)}{(c+s+1)(c+s+2)} a_s\,, \quad s \geq 0\,. \tag{5A1.1.4}$$

If $a_0 \neq 0$ then $c(c-1) = 0$ and if $a_1 \neq 0$ then $c(c+1) = 0$. By considering various possibilities, we find that it is sufficient to consider either $a_0 \neq 0$ or $a_1 \neq 0$ but not both. We choose $a_0 \neq 0$ so that $c = 0$ or $c = 1$. Then the solution y can be written as the sum of two power series, one involving even powers ($c = 0$) and the other involving odd powers ($c = 1$) of x:

$$\begin{aligned} y = a_{0even}&\left[1 - \frac{\ell(\ell+1)}{2!}x^2 + \frac{\ell(\ell+1)(\ell-2)(\ell+3)}{4!}x^4 + \cdots\right] \\ + a_{0odd}&\left[x - \frac{(\ell-1)(\ell+2)}{3!}x^3 + \frac{(\ell-1)(\ell+2)(\ell-3)(\ell+4)}{5!}x^5 + \cdots\right]. \end{aligned} \tag{5A1.1.5}$$

Both even and odd power series are solutions of the Legendre equation and converge for $0 < x^2 < 1$ regardless of the value of ℓ. The series diverge at $x^2 = 1$ unless they terminate. An inspection of the recurrence relation (5A1.1.4) shows that it will terminate only if ℓ is zero or a positive integer. When ℓ is an even integer the $c = 0$ series becomes a polynomial and when it is an odd integer the $c = 1$ series becomes a polynomial. These polynomials are called the **Legendre polynomials** and are denoted by $P_\ell(x)$. The constants a_{0even} and a_{0odd} are chosen such that $P_\ell(1) = 1$. The first few Legendre polynomials are:

$$\begin{aligned} P_0(x) &= 1\,, & P_1(x) &= x\,, \\ P_2(x) &= \frac{1}{2}\left(3x^2 - 1\right), & P_3(x) &= \frac{1}{2}\left(5x^3 - 3x\right), \\ P_4(x) &= \frac{1}{8}\left(35x^4 - 30x^2 + 3\right), & P_5(x) &= \frac{1}{8}\left(63x^5 - 70x^3 + 15x\right). \end{aligned}$$

A convenient expression for the Legendre polynomials is the so-called Rodrigue's formula

$$P_\ell(x) = \frac{1}{2^\ell \ell!}\left(\frac{d}{dx}\right)^\ell (x^2-1)^\ell\,, \tag{5A1.1.6}$$

which is useful in establishing many properties of Legendre polynomials.

The orthogonality of Legendre polynomials may easily be established. Since $P_\ell(x)$ satisfies the Legendre equation (5A1.1.1), we have

$$\frac{d}{dx}\left[(1-x^2)\frac{d}{dx}P_\ell(x)\right] + \ell(\ell+1)P_\ell(x) = 0\,. \tag{5A1.1.7}$$

Multiplying this equation on the left by $P_{\ell'}(x)$ and integrating over x from -1 to +1, we get

$$P_{\ell'}(x)(1-x^2)\frac{dP_\ell(x)}{dx}\Bigg|_{-1}^{1} - \int_{-1}^{1} dx\left[\frac{dP_{\ell'}(x)}{dx}(1-x^2)\frac{dP_\ell(x)}{dx}\right] + \ell(\ell+1)\int_{-1}^{1} dx P_{\ell'}(x)P_\ell(x) = 0$$

or
$$\int_{-1}^{1} dx\left[\frac{dP_{\ell'}}{dx}(1-x^2)\frac{dP_\ell(x)}{dx}\right] = \ell(\ell+1)\int_{-1}^{1} dx P_{\ell'}(x)P_\ell(x)\,.$$

Interchanging ℓ and ℓ' and subtracting the resulting equation from the equation above, we get

$$(\ell-\ell')(\ell+\ell'+1)\int_{-1}^{1} dx P_{\ell'}(x)P_\ell(x) = 0\,. \tag{5A1.1.8}$$

This implies that

$$\int_{-1}^{1} dx P_{\ell'}(x)P_\ell(x) = 0 \qquad \text{if} \quad \ell \neq \ell'\,. \tag{5A1.1.9}$$

The value of this integral for $\ell = \ell'$, may be obtained by using the fact that the function

$$(1-2xz+z^2)^{-1/2} = \sum_{\ell=0}^{\infty} P_\ell(x) z^\ell \tag{5A1.1.10}$$

serves as the generating function for Legendre polynomials. Squaring both sides of this equation and integrating them over x from -1 to +1, we get

$$\int_{-1}^{1} dx(1-2xz+z^2)^{-1} = \sum_\ell \sum_{\ell'} z^{\ell+\ell'}\int_{-1}^{1} dx P_\ell(x)P_{\ell'}(x)\,,$$

or
$$\frac{1}{z}\left[\ln(1+z) - \ln(1-z)\right] = \sum_\ell z^{2\ell}\int_{-1}^{1} dx\,[P_\ell(x)]^2\,,$$

where we have used the orthogonality condition for Legendre polynomials on the right-hand side. Writing the series expansion for the left-hand side and equating the coefficients of $z^{2\ell}$ on both sides we get

$$\int_{-1}^{1} dx\,[P_\ell(x)]^2 = \frac{2}{2\ell+1}\,. \tag{5A1.1.11}$$

Equations (5A1.1.9) and (5A1.1.11) may be combined as

$$\int_{-1}^{1} dx P_\ell(x)P_{\ell'}(x) = \frac{2}{2\ell+1}\delta_{\ell\ell'}\,. \tag{5A1.1.12}$$

Another set of polynomials closely associated with the Legendre polynomials are the so-called **associated Legendre polynomials**. If we differentiate the Legendre equation (5A1.1.1) m times and put

$$(1-x^2)^{m/2}\frac{d^m y}{dx^m} = w(x)\,, \tag{5A1.1.13}$$

we get, after some simplification, the *associated Legendre equation*

$$(1-x^2)\frac{d^2w(x)}{dx^2} - 2x\frac{dw(x)}{dx} + \left[\ell(\ell+1) - \frac{m^2}{1-x^2}\right]w(x) = 0\,. \tag{5A1.1.14}$$

The polynomial solutions of this equation for integer ℓ, called associated Legendre polynomials $P_\ell^m(x)$, are given by

$$P_\ell^m(x) = (1-x^2)^{m/2}\frac{d^m P_\ell(x)}{dx^m}\,. \tag{5A1.1.15}$$

Obviously $P_\ell^m(x) = 0$ for $m > \ell$ since P_ℓ is a polynomial in x of order ℓ.

To establish the orthogonality condition for the associated Legendre polynomials, we start with Eq. (5A1.1.14) for $P_\ell^m(x)$ and $P_{\ell'}^m(x)$:

$$\frac{d}{dx}\left[(1-x^2)\frac{dP_\ell^m(x)}{dx}\right] + \left[\ell(\ell+1) - \frac{m^2}{1-x^2}\right]P_\ell^m(x) = 0\,,$$

and
$$\frac{d}{dx}\left[(1-x^2)\frac{dP_{\ell'}^m(x)}{dx}\right] + \left[\ell'(\ell'+1) - \frac{m^2}{1-x^2}\right]P_{\ell'}^m(x) = 0\,.$$

Multiplying the first equation by $P_{\ell'}^m(x)$, the second by $P_\ell^m(x)$ and subtracting the results, we get

$$\frac{d}{dx}\left[(1-x^2)\left(P_{\ell'}^m\frac{dP_\ell^m(x)}{dx} - P_\ell^m(x)\frac{dP_{\ell'}^m(x)}{dx}\right)\right] + (\ell-\ell')(\ell+\ell'+1)P_\ell^m(x)P_{\ell'}^m(x) = 0\,.$$

On integrating both sides of this equation over x from -1 to +1, the first term contributes nothing, leaving us

$$(\ell-\ell')(\ell+\ell'+1)\int_{-1}^{1}dx P_\ell^m(x)P_{\ell'}^m(x) = 0\,.$$

This equation implies

$$\int_{-1}^{1}dx P_\ell^m(x)P_{\ell'}^m(x) = 0 \qquad \text{for} \quad \ell \neq \ell'\,. \tag{5A1.1.16}$$

For $\ell = \ell'$ the integral is not zero and gives the normalization integral for the associated Legendre polynomials

$$N_{\ell m} \equiv \int_{-1}^{1}dx\,[P_\ell^m(x)]^2\,. \tag{5A1.1.17}$$

To evaluate this integral, we use Eq. (5A1.1.15) to write it as

$$N_{\ell m} = \int_{-1}^{1}dx(1-x^2)^m\frac{d^m P_\ell(x)}{dx^m}\frac{d^m P_\ell(x)}{dx^m}\,. \tag{5A1.1.18}$$

Integration by parts once gives us

$$N_{\ell m} = -\int_{-1}^{1}dx\frac{d}{dx}\left[(1-x^2)^m\frac{d^m P_\ell(x)}{dx^m}\right]\frac{d^{m-1}P_\ell(x)}{dx^{m-1}}\,. \tag{5A1.1.19}$$

Now take Eq. (5A1.1.1) with $y(x) = P_\ell(x)$ and differentiate it $(m-1)$ times using the Leibnitz differentiation theorem. This gives, on multiplying both sides with $(1-x^2)^{m-1}$ and simplifying,

$$\frac{d}{dx}\left[(1-x^2)^m \frac{d^m P_\ell(x)}{dx^m}\right] + (\ell+m)(\ell-m+1)(1-x^2)^{m-1}\frac{d^{m-1}P_\ell(x)}{dx^{m-1}} = 0\,.$$

Using this result in Eq. (5A1.1.19) we get

$$N_{\ell m} = (\ell+m)(\ell-m+1)\int_{-1}^{1}(1-x^2)^{m-1}\frac{d^{m-1}P_\ell(x)}{dx^{m-1}}\frac{d^{m-1}P_\ell(x)}{dx^{m-1}}\,.$$

Treating this as a recurrence relation, we get after m iterations,

$$\begin{aligned} N_{\ell m} &= (\ell+m)(\ell-m+1)(\ell+m-1)(\ell-m+2)\cdots(\ell+1)\ell\int_{-1}^{1}dx\,[P_\ell(x)]^2 \\ &= \frac{(\ell+m)!}{\ell!}\frac{\ell!}{(\ell-m)!}\frac{2}{2\ell+1} \end{aligned}$$

$$\text{or}\quad N_{\ell m} \equiv \int_{-1}^{1}dx\,[P_\ell^m(x)]^2 = \frac{(\ell+m)!}{(\ell-m)!}\frac{2}{2\ell+1}\,. \tag{5A1.1.20}$$

By combining Eqs. (5A1.16) and (5A1.20), the orthogonality of associated Legendre polynomial can be written as

$$\int_{-1}^{1}dx P_\ell^m(x)P_{\ell'}^m(x) = \frac{(\ell+m)!}{(\ell-m)!}\frac{2}{2\ell+1}\delta_{\ell\ell'}\,. \tag{5A1.1.21}$$

From our discussion of the associated Legendre polynomials it might appear that m is restricted to nonnegative integers $m \le \ell$. However, if we express $P_\ell(x)$ by Rodrigue's formula [Eq. (5A1.1.6)], both positive and negative values of m $(-\ell \le m \le \ell)$ may be permitted. Using this definition and the Leibnitz differentiation formula we have

$$P_\ell^m(x) = \frac{1}{2^\ell \ell!}(1-x^2)^{m/2}\frac{d^{\ell+m}}{dx^{\ell+m}}(x^2-1)^\ell\,, \tag{5A1.1.22}$$

$$P_\ell^{-m}(x) = (-1)^m\frac{(n-m)!}{(n+m)!}P_\ell^m(x)\,. \tag{5A1.1.23}$$

Table 5.1 below lists the first few associated Legendre polynomials and the corresponding spherical harmonics.

5A1.2 Spherical Harmonics

In solving the time-independent Schrödinger equation for a spherically symmetric potential $V(\boldsymbol{r}) = V(r)$

$$-\frac{\hbar^2}{2\mu}\nabla^2\Psi(\boldsymbol{r}) + V(r)\Psi(\boldsymbol{r}) = E\Psi(\boldsymbol{r})\,, \tag{5A1.2.1}$$

or even the Laplace equation $\nabla^2\Psi(\boldsymbol{r}) = 0$, we can express the function $\Psi(\boldsymbol{r})$ as a product of radial and angular functions as

$$\Psi(\boldsymbol{r}) = R(r)Y(\theta,\varphi)\,. \tag{5A1.2.2}$$

TABLE 5.1
The first few associated Legendre polynomials and spherical harmonics

$P_0 = 1$	$Y_0^0 = \left(\frac{1}{4\pi}\right)^{1/2}$
$P_1^1 = -\sin\theta$	$Y_1^1 = -\frac{1}{2}\left(\frac{3}{2\pi}\right)^{1/2} \sin\theta e^{i\varphi}$
$P_1^0 = \cos\theta$	$Y_1^0 = \frac{1}{2}\left(\frac{3}{\pi}\right)^{1/2} \cos\theta$
$P_1^{-1} = \frac{1}{2}\sin\theta$	$Y_1^{-1} = \frac{1}{2}\left(\frac{3}{2\pi}\right)^{1/2} \sin\theta e^{-i\varphi}$
$P_2^2 = 3\sin^2\theta$	$Y_2^2 = \frac{1}{4}\left(\frac{15}{2\pi}\right)^{1/2} \sin^2\theta e^{2i\varphi}$
$P_2^1 = -3\sin\theta\cos\theta$	$Y_2^1 = -\frac{1}{2}\left(\frac{15}{2\pi}\right)^{1/2} \sin\theta\cos\theta e^{i\varphi}$
$P_2^0 = \frac{1}{2}(3\cos^2\theta - 1)$	$Y_2^0 = \frac{1}{4}\left(\frac{5}{\pi}\right)^{1/2} (3\cos^2\theta - 1)$
$P_2^{-1} = \frac{1}{2}\sin\theta\cos\theta$	$Y_2^{-1} = \frac{1}{2}\left(\frac{15}{2\pi}\right)^{1/2} \sin\theta\cos\theta e^{-i\varphi}$
$P_2^{-2} = \frac{1}{8}\sin^2\theta$	$Y_2^2 = \frac{1}{4}\left(\frac{15}{2\pi}\right)^{1/2} \sin^2\theta e^{-2i\varphi}$
$P_3^3 = -15\sin^3\theta$	$Y_3^3 = -\frac{1}{8}\left(\frac{35}{\pi}\right)^{1/2} \sin^3\theta e^{i3\varphi}$
$P_3^2 = 15\sin^2\theta\cos\theta$	$Y_3^2 = \frac{1}{4}\left(\frac{105}{2\pi}\right)^{1/2} \sin^2\theta\cos\theta e^{2i\varphi}$
$P_3^1 = -\frac{3}{2}\sin\theta(5\cos^2\theta - 1)$	$Y_3^1 = -\frac{1}{8}\left(\frac{21}{\pi}\right)^{1/2} \sin\theta(5\cos^2\theta - 1)e^{i\varphi}$
$P_3^0 = \frac{1}{2}(5\cos^3\theta - 3\cos\theta)$	$Y_3^0 = \frac{1}{4}\left(\frac{7}{\pi}\right)^{1/2} (5\cos^3\theta - 3\cos\theta)$
$P_3^{-1} = \frac{1}{8}\sin\theta(5\cos^2\theta - 1)$	$Y_3^{-1} = \frac{1}{8}\left(\frac{21}{\pi}\right)^{1/2} \sin\theta(5\cos^2\theta - 1)e^{-i\varphi}$
$P_3^{-2} = \frac{1}{8}\sin^2\theta\cos\theta$	$Y_3^{-2} = \frac{1}{4}\left(\frac{105}{2\pi}\right)^{1/2} \sin^2\theta\cos\theta e^{-2i\varphi}$
$P_3^{-3} = \frac{1}{48}\sin^3\theta$	$Y_3^{-3} = -\frac{1}{8}\left(\frac{35}{\pi}\right)^{1/2} \sin^3\theta e^{-i3\varphi}$

Since the Laplacian operator ∇^2 in spherical polar coordinates may be expressed as [see Appendix 5A2, Eqs. (5A2.4.13) and (5A2.4.15)]

$$\nabla^2 = \frac{1}{r^2}\frac{\partial}{\partial r}r^2\frac{\partial}{\partial r} + \frac{1}{r^2\sin\theta}\frac{\partial}{\partial\theta}\sin\theta\frac{\partial}{\partial\theta} + \frac{1}{r^2\sin^2\theta}\frac{\partial^2}{\partial\varphi^2}, \qquad (5A1.2.3)$$

the original equation (5A1.16) may be separated into radial and angular parts by the method of separation of variables. The form of the radial equation will, of course, depend on the form of the potential $V(r)$. However, irrespective of the form of the spherically symmetric potential $V(r)$, the angular equation has the same form:

$$-\left[\frac{1}{\sin\theta}\frac{\partial}{\partial\theta}\sin\theta\frac{\partial}{\partial\theta}+\frac{1}{\sin^2\theta}\frac{\partial^2}{\partial\varphi^2}\right]Y(\theta,\varphi)=\lambda Y(\theta,\varphi)\,, \tag{5A1.2.4}$$

where λ is a constant. The angular equation (5A1.2.4) admits physically acceptable solution only if $\lambda = \ell(\ell+1)$ where ℓ is a non-negative integer $[\ell = 0, 1, 2, 3, \cdots]$. The solutions of Eq. (5A1.2.4) are referred to as **spherical harmonics**.

To find the form of spherical harmonics we write $Y(\theta,\varphi)$ as a product of a function of θ and a function of φ as

$$Y(\theta,\varphi)=\Theta(\theta)\Phi(\varphi)\,. \tag{5A1.2.5}$$

With this substitution we can rewrite the angular equation (5A1.2.4), as

$$-\frac{1}{\Phi}\frac{d^2\Phi}{d\varphi^2}=\frac{\sin\theta}{\Theta(\theta)}\frac{d}{d\theta}\sin\theta\frac{d\Theta}{d\theta}+\ell(\ell+1)\sin^2\theta\,. \tag{5A1.2.6}$$

Since φ and θ are independent variables, a function of φ cannot be equal to a function of θ for all values of θ and φ unless both functions are equal to a constant. Equating both sides of Eq. (5A2.27) to a constant m^2, we can separate the angular equation into an equation involving θ and another involving φ:

$$\frac{d^2\Phi}{d\varphi^2}+m^2\Phi=0\,, \tag{5A1.2.7}$$

$$\left[\sin\theta\frac{d}{d\theta}\sin\theta\frac{d}{d\theta}+\ell(\ell+1)\sin^2\theta\right]\Theta(\theta)=m^2\Theta(\theta)\,. \tag{5A1.2.8}$$

The normalized solutions of the Φ-equation are

$$\Phi_m(\varphi)=\frac{1}{\sqrt{2\pi}}e^{im\varphi}\,. \tag{5A1.2.9}$$

For single-valued solutions $[\Phi(\varphi+2\pi)=\Phi(\varphi)]$ valid over the entire angular range $0<\varphi<2\pi$, the parameter m must be a positive or negative integer including 0 $[m=0,\pm1,\pm2,\pm3,\cdots]$. With this restriction, the Φ-solutions satisfy the orthonormality condition

$$\int_0^{2\pi}d\varphi\Phi_m(\varphi)\Phi^*_{m'}(\varphi)=\delta_{mm'}\,. \tag{5A1.2.10}$$

The Θ-equation admits a physically acceptable (polynomial) solution only if $\lambda=\ell(\ell+1)$, where ℓ is a non-negative integer. With the change of variables

$$\cos\theta=x \quad \text{and} \quad \Theta(\theta)\to w(x)$$

the Θ-equation can be put in the form

$$(1-x^2)\frac{d^2w(x)}{dx^2}-2x\frac{dw(x)}{dx}+\left[\ell(\ell+1)-\frac{m^2}{1-x^2}\right]w(x)=0\,. \tag{5A1.2.11}$$

This is the associated Legendre equation [Eq. (5A1.1.14)] whose normalized solutions are the normalized associated Legendre polynomials $\mathcal{P}_\ell^m(x)$

$$\mathcal{P}_\ell^m(x)=\sqrt{\frac{(\ell-m)!(2\ell+1)}{(\ell+m)!2}}\,P_\ell^m(x)\,, \tag{5A1.2.12}$$

$$\int_{-1}^{1}dx\,\mathcal{P}_\ell^m(x)\mathcal{P}_{\ell'}^m(x)=\delta_{\ell\ell'}\,. \tag{5A1.2.13}$$

The normalized spherical harmonics can then be written as[8]

$$Y_{\ell m}(\theta,\varphi)=\sqrt{\frac{(\ell-m)!(2\ell+1)}{(\ell+m)!4\pi}}\,P_\ell^m(\cos\theta)e^{im\varphi}\,. \tag{5A1.2.14}$$

In view of the orthonormality of angular harmonics [Eqs. (5A1.2.10) and (5A1.2.13)], the orthogonality of spherical harmonics

$$\int_0^{2\pi}d\varphi\int_0^{\pi}\sin\theta d\theta Y^*_{\ell m}(\theta,\varphi)Y_{\ell m}(\theta,\varphi)=\delta_{\ell\ell'}\delta_{mm'} \tag{5A1.2.15}$$

is established.

5A1.3 Laguerre and Associated Laguerre Equations

The equation

$$x\frac{d^2y(x)}{dx^2}+(1-x)\frac{dy(x)}{dx}+\alpha y(x)=0 \tag{5A1.3.1}$$

is called Laguerre equation, and

$$x\frac{d^2u(x)}{dx^2}+(\beta+1-x)\frac{du(x)}{dx}+(\alpha-\beta)u(x)=0\,, \tag{5A1.3.2}$$

is called the associated Laguerre equation. The latter can be obtained by differentiation both sides of Laguerre equation β times and putting $d^\beta y(x)/dx^\beta=u(x)$. Equation (5A1.3.2) is of significance in the Hydrogen atom problem. It can be solved by the usual series method by assuming

$$y(x)=x^c\sum_{s=0}^{\infty}a_sx^s\,. \tag{5A1.3.3}$$

Substituting into Eq. (5A1.3.2) and equating the coefficient of each power of x, say x^{c+s-1}, to zero we get

$$a_s(c+s)(c+s+\beta)-a_{s-1}(c+s-1-\alpha+\beta)=0\,. \tag{5A1.3.4}$$

Putting $s=0$ and using the fact that $a_{-1}=0$, we get

$$a_0c(c+\beta)=0\,.$$

For nontrivial soultions ($a_0\neq0$), we must have $c=0$ or $c=-\beta$. The choice $c=-\beta$ will make the solution singular at the origin. So $c=0$ is the acceptable solution. On using Eq. (5A1.3.4) as a recurrence relation, we can express all coefficients in terms of a_0 as

$$a_s=-\frac{(\alpha-\beta-s+1)}{s(s+\beta)}a_{s-1}=(-1)^s\frac{\beta!(\alpha-\beta)!}{s!(s+\beta)!(\alpha-\beta-s)!}a_0\,. \tag{5A1.3.5}$$

If we choose

$$a_0=\frac{(\alpha!)^2}{\beta!(\alpha-\beta)!} \tag{5A1.3.6}$$

[8]Sometimes a factor $(-1)^m$, called the Condon-Shortley phase, is included in the definition of spherical harmonics.

then we can write

$$a_s = \frac{(-1)^s(\alpha!)^2}{s!(\beta+s)!(\alpha-\beta-s)!} \tag{5A1.3.7}$$

and

$$y(x) = \sum_{s=0}^{\infty} \frac{(-)^s(\alpha!)^2 x^s}{s!(\beta+s)!(\alpha-\beta-s)!}\,. \tag{5A1.3.8}$$

If $\alpha - \beta$ is a positive integer then Eq. (5A1.3.2) admits a polynomial solution, called associated Laguerre polynomial $L_\alpha^\beta(x)$ given by

$$y(x) = L_\alpha^\beta(x) = \sum_{s=0}^{\alpha-\beta} \frac{(-)^s(\alpha!)^2 x^s}{s!(\beta+s)!(\alpha-\beta-s)!}\,. \tag{5A1.3.9}$$

Analytic Forms of Laguerre and Associated Laguerre Polynomials

Consider the simple equation

$$x\frac{dw}{dx} + (x-\alpha)w = 0\,, \tag{5A1.3.10}$$

whose solution is $w(x) = ce^{-x}x^a$. If we differentiate both sides of this equation $\alpha + 1$ times and put $d^\alpha w/dx^\alpha = z(x)$, we get

$$x\frac{d^2z}{dx^2} + (x+1)\frac{dz}{dx} + (\alpha+1)z = 0\,. \tag{5A1.3.11}$$

The solution of this equation is

$$z(x) = \frac{d^\alpha}{dx^\alpha}\left(ce^{-x}x^\alpha\right)\,. \tag{5A1.3.12}$$

If we further write

$$z(x) = e^{-x}y(x)\,, \tag{5A1.3.13}$$

then $y(x)$ satisfies the equation

$$x\frac{d^2y}{dx^2} + (1-x)\frac{dy}{dx} + \alpha y = 0\,, \tag{5A1.3.14}$$

which is the Laguerre equation. Its solution can, therefore, be written as

$$y(x) = e^x\, z(x) = e^x\frac{d^\alpha}{dx^\alpha}ce^{-x}x^\alpha\,. \tag{5A1.3.15}$$

Since α is an integer the function $y(x)$ is a polynomial in x. The solutions of Eq. (5A1.3.14) are called the Laguerre polynomials and denoted by $L_\alpha(x)$:

$$L_\alpha(x) = e^x\frac{d^\alpha}{dx^\alpha}ce^{-x}x^\alpha\,. \tag{5A1.3.16}$$

Since the associated Laguerre equation results when both sides of Laguerre equation are differentiated β time and $d^\beta y/dx^\beta$ is put equal to u we can express the associated Laguerre polynomial as

$$L_\alpha^\beta(x) = \frac{d^\beta}{dx^\beta}\left[e^x\frac{d^\alpha}{dx^\alpha}ce^{-x}x^\alpha\right] \equiv \frac{d^\beta}{dx^\beta}L_\alpha(x)\,. \tag{5A1.3.17}$$

It can be checked that when the constant c in Eq. (5A1.3.17) is chosen to be $(-1)^\beta$, the resulting polynomials $L_\alpha^\beta(x)$ are identical with those given by Eq. (5A1.3.9).

Orthonormality of Associated Laguerre Functions

We now show that the associated Laguerre functions

$$\mathcal{L}_\alpha^\beta(x) = e^{-x/2}x^{(\beta-1)/2}L_\alpha^\beta(x) \tag{5A1.3.18}$$

where α and β are integers, form a set of orthogonal functions with the normalization integral given by

$$I_N = \int_0^\infty e^{-x}x^{\beta-1}\left[L_\alpha^\beta(x)\right]^2 x^2 dx = \frac{(\alpha!)^3(2\alpha-\beta+1)}{(\alpha-\beta)!}\,. \tag{5A1.3.19}$$

To prove this result in several steps. First consider the integral

$$I_1 \equiv \int_0^\infty e^{-x}x^\gamma L_\alpha(x)dx\,. \tag{5A1.3.20}$$

Using the expression for $L_\alpha(x)$ given by Eq. (5A1.3.16) and integrating by parts γ times we find

$$I_1 = c(-1)^\gamma\gamma! \int_0^\infty \frac{d^{\alpha-\gamma}}{dx^{\alpha-\gamma}}\left(e^{-x}x^\alpha\right)dx\,. \tag{5A1.3.21}$$

It follows from this equation that the integral

$$I_1 \equiv \int_0^\infty e^{-x}x^\gamma L_\alpha(x)dx = \begin{cases} 0 & \gamma<\alpha \\ c(-1)^\alpha(\alpha!)^2 & \gamma=\alpha\,. \end{cases} \tag{5A1.3.22}$$

Next consider the integral

$$I_2 \equiv \int_0^\infty e^{-x}x^\gamma L_\alpha^\beta(x)dx = \int_0^\infty e^{-x}x^\gamma\frac{d^\beta}{dx^\beta}L_\alpha(x)dx\,, \tag{5A1.3.23}$$

where in the last step, we have used definition of the associated Laguerre polynomial. Integrating by parts β times we find

$$\begin{aligned} I_2 &= (-1)^\beta\int_0^\infty\left[\frac{d^\beta}{dx^\beta}\left(e^{-x}x^\gamma\right)\right]L_\alpha(x)dx \\ &= (-1)^\beta\int_0^\infty\left[(-1)^\beta e^{-x}x^\gamma+\beta(-1)^{\beta-1}e^{-x}\gamma x^{\gamma-1}+\cdots\right]L_\alpha(x)dx\,. \end{aligned} \tag{5A1.3.24}$$

Using the result (5A1.3.22) we find that this integral vanishes for $\gamma<\alpha$ while for $\gamma=\alpha$ we get

$$I_2 = (-1)^{2\beta}\int_0^\infty L_\alpha(x)e^{-x}x^\alpha dx = (-1)^\alpha(\alpha!)^2\,,\quad \gamma=\alpha\,. \tag{5A1.3.25}$$

Thus we have the result

$$I_2 \equiv \int_0^\infty dx\, e^{-x}x^\gamma L_\alpha^\beta(x) = \begin{cases} 0\,, & \gamma<\alpha \\ c(-1)^\alpha(\alpha!)^2\,, & \gamma=\alpha\,. \end{cases} \tag{5A1.3.26}$$

Now consider the integral

$$I_3 = \int_0^\infty e^{-x}x^\beta L_\alpha^\beta(x)L_\gamma^\beta(x)dx\,. \tag{5A1.3.27}$$

Since $L_\alpha^\beta(x)$ is a polynomial in x of degree $\alpha-\beta$, it follows that $x^\beta L_\gamma^\beta(x) = x^\beta \frac{d^\beta}{dx^\beta} L_\gamma(x)$ is a polynomial in x of degree γ. Using this result in combination with (5A1.3.22), we conclude $I_3 = 0$ if $\gamma < \alpha$. Similarly, since $x^\beta L_\alpha^\beta(x) = x^\beta \frac{d^\beta}{dx^\beta} L_\alpha(x)$ is a polynomial in x of degree α, it follows that $I_3 = 0$ if $\alpha < \gamma$. Thus we have

$$I_3 \equiv \int_0^\infty e^{-x}\, x^\beta\, L_\alpha^\beta\, L_\gamma^\beta(x)\, dx = 0 \quad \text{if} \;\; \gamma \neq \alpha\,. \tag{5A1.3.28}$$

By a similar argument we can show that

$$I_4 \equiv \int_0^\infty e^{-x} x^{\beta+1} L_\alpha^\beta(x) L_\gamma^\beta(x) dx = 0\;, \text{ if } \gamma \neq \alpha\,. \tag{5A1.3.29}$$

For $\gamma = \alpha$, the integral I_3 [Eq. (5A3.26)] becomes,

$$I_{3(\gamma=\alpha)} = \int_0^\infty e^{-x}\, x^\beta\, L_\alpha^\beta(x)\;\, L_\alpha^\beta(x) dx\,. \tag{5A1.3.30}$$

To evaluate this integral we express $x^\beta\, L_\alpha^\beta(x)$ in the integrand as a polynomial using the definition (5A.3.17)

$$\begin{aligned}
x^\beta\, L_\alpha^\beta(x) &= x^\beta\, \frac{\partial^\beta}{\partial x^\beta} \left[c e^x \,\frac{d^\alpha}{dx^\alpha}\, \left(x^\alpha\, e^{-x}\right)\right] \\
&= c\, x^\beta\, \frac{\partial^\beta}{\partial x^\beta}\, \left[e^x \left\{ (-1)^\alpha e^{-x}\, x^\alpha + \alpha(-1)^{\alpha-1}\, e^{-x}\, \alpha x^{\alpha-1} + \,...\, \right\} \right] \\
&= c\, x^\beta\, (-1)^\alpha\, \alpha(\alpha-1) \ldots (\alpha-\beta+1)\, x^{\alpha-\beta} + \text{smaller powers of } \; x \\
&= c\, (-1)^\alpha \frac{\alpha!}{(\alpha-\beta)!} x^\alpha + \text{smaller powers of } x\,.
\end{aligned}$$

Using this result in Eq. (5A1.3.30) for $\gamma = \alpha$, we find by virtue of the result (5A1.3.26), that only the x^α term gives nonzero contribution

$$\begin{aligned}
I_{3(\gamma=\alpha)} &= c\,(-1)^\alpha \frac{\alpha!}{(\alpha-\beta)!} \int_0^\infty e^{-x}\, x^\alpha\, L_\alpha^\beta(x)\, dx \\
&= c\,(-1)^\alpha\, \frac{\alpha!}{(\alpha-\beta)!}\; I_2(\gamma=\alpha) \\
&= c^2\,(-1)^{2\alpha}\, \frac{(\alpha!)^3}{(\alpha-\beta)!} \quad \text{using Eq. (5A1.3.26)}\,.
\end{aligned}$$

Since $c^2 = (-1)^{2\beta}$ and both 2α and 2β are even integers, we have

$$\left(I_3\right)_{\gamma=\alpha} = \frac{(\alpha!)^3}{(\alpha-\beta)!}\,. \tag{5A1.3.31}$$

Finally, take up the normalization integral

$$I_N \equiv \int_0^\infty e^{-x} x^{\beta+1}\, \left[L_\alpha^\beta(x)\right]^2\, dx\,.$$

Replacing one of the factors $L_\alpha^\beta(x)$ in the integrand by the series (5A1.3.9) starting from the highest power of x, we get

$$I_N = \int_0^\infty dx\, e^{-x} L_\alpha^\beta(x) x^{\beta+1} \left[\frac{(-1)^{\alpha-\beta}(\alpha!)^2\, x^{\alpha-\beta}}{0!\alpha!(\alpha-\beta)!} + \frac{(-1)^{\alpha-\beta-1}(\alpha!)^2 x^{\alpha-\beta-1}}{1!(\alpha-1)!(\alpha-\beta-1)!} + O(x^{\alpha-\beta-2}) \right]$$

$$= \frac{(-1)^{\alpha-\beta}\alpha!}{(\alpha-\beta)!} \int_0^\infty dx\, e^{-x}\, x^{\alpha+1} L_\alpha^\beta(x) + \frac{(-1)^{\alpha-\beta-1}\, \alpha^2((\alpha-1)!)}{(\alpha-\beta-1)!}\, c\,(-1)^\alpha(\alpha!)^2 \tag{5A1.3.32}$$

where we have used the fact that, by virtue of the result (5A1.3.26), the contribution from $x^{\alpha-\beta-2}$ and smaller powers vanishes. To evaluate the remaining integral

$$I_5 \equiv \int_0^\infty dx\, e^{-x}\, x^{\alpha+1}\, L_\alpha^\beta(x) \;=\; \int_0^\infty dx\, e^{-x}\, x^{\alpha+1} \frac{d^\beta}{dx^\beta} L_\alpha(x) \tag{5A1.3.33}$$

we integrate by parts β times in succession, and then expand $\frac{d^\beta}{dx^\beta}\left(x^{-x}\, x^{\alpha+1}\right)$ by the Leibnitz differentiation theorem and use the result (5A1.3.26) to get

$$I_5 = (-1)^{2\beta}\, c \int_0^\infty dx\, x^{\alpha+1} \frac{d^\alpha}{dx^\alpha}\left(e^{-x}\, x^\alpha\right) + c\,\beta(-1)^{\alpha+2\beta-1}\,(\alpha+1)(\alpha!)^2 . \tag{5A1.3.34}$$

The value of the integral on the right-hand side of this equation may be obtained by integrating by parts α times in succession to get the result $c\,(-1)^{\alpha+2\beta}[(\alpha+1)!]^2$, so that

$$I_5 = c\,(-1)^{\alpha+2\beta}\alpha!(\alpha+1)!(\alpha-\beta+1)! . \tag{5A1.3.35}$$

Susbtituting this in Eq. (5A1.3.32) above, and combining the two terms in the resulting equation, we get

$$I_N = \frac{(\alpha!)^3}{(\alpha-\beta)!}\,(2\alpha-\beta+1) \tag{5A1.3.36}$$

where we have taken c to be equal to $(-1)^\beta$ and α and β to be non-negative integers.

5A1.4 Hermite Equation

The second order equation

$$\frac{d^y}{dx^2} - 2x\frac{dy}{dx} + 2ny = 0 \tag{5A1.4.1}$$

is called Hermite equation. To solve this equation we follow the usual method of series expansion by writing $y(x)$

$$y(x) = x^c \sum_{s=0}^\infty a_s x^s .$$

Substituting this in Eq. (5A1.4.1) and equating the coefficient of each power of x, say of x^{c+s-2}, to zero we get

$$a_s(c+s)(c+s-1) = 2a_{s-2}(c+s-2-n) . \tag{5A1.4.2}$$

Putting $s=0$ and 1 and using the fact that $a_{-1} = 0 = a_{-2}$, we get, respectively,

$$\begin{aligned} a_0 c(c-1) &= 0 , \\ a_1(c+1)c &= 0 . \end{aligned} \tag{5A1.4.3}$$

If $a_0 \neq 0$ then $c(c-1) = 0$ and if $a_1 \neq 0$ then $c(c+1) = 0$. By considering various possibilities, as in the case of Legendre polynomials, it is sufficient to consider $c = 0$ and a_0 and a_1 to be arbitrary. Then Eq. (5A.4.38) leads to the recursion relation

$$a_s = -\frac{2(-s+2+n)}{s(s-1)} a_{s-2}\,, \qquad s \geq 2\,. \tag{5A1.4.4}$$

With the help of this equation, even coefficients $a_0, a_2, a_4, \cdots$ can be expressed in terms of a_0 and the odd coefficients $a_1, a_3, a_5, \cdots$ can be expressed in terms of a_1. Thus the general solution of the Hermite equation can be written as a sum of two series, one involving even power of x, and the other involving odd powers of x as

$$\begin{aligned} y = a_0 &\left(1 - \frac{2n}{2!} x^2 + 2^2 \frac{n(n-2)}{4!} x^4 \;\ldots.\right) \\ &+ a_1 \left(x - \frac{2(n-1)}{3!} x^3 + 2^2 \frac{(n-1)(n-3)}{5!} x^5 \;\ldots.\right). \end{aligned}$$

Since a_0 and a_1 are arbitrary, each of the two series is a solution of the Hermite equation. It can be seen from these series that when n is a non-negative even integer, the first series becomes a polynomial and when n is a non-negative odd integer, the second series becomes a polynomial.

For integer n it is convenient to rewrite the polynomial solution by expressing all the coefficients in terms of the coefficient of the highest power of x. To do this we rewrite Eq. (5A1.4.4) as

$$a_{s-2} = -\frac{s(s-1)}{2(n-s+2)}\, a_s\,.$$

This gives

$$\begin{aligned} a_{n-2} &= -\frac{n(n-1)}{2\cdot 2}\, a_n \;=\; -\frac{n(n-1)}{2^2\,1!}\, a_n \\ a_{n-4} &= \frac{n(n-1)(n-2)(n-3)}{4^2\cdot 2!}\, a_n \end{aligned}$$

and so on. The polynomial solution is thus

$$\begin{aligned} y = \frac{a_n}{2^n} &\left[(2x)^n - \frac{n(n-1)}{1!}(2x)^{n-2} + \frac{n(n-1)(n-2)(n-3)}{2!}(2x)^{n-4}\right. \\ &\left. - \frac{n(n-1)(n-2)(n-3)(n-4)(n-5)}{3!}(2x)^{n-6} + \cdots\right]. \end{aligned} \tag{5A1.4.5}$$

With the choice $a_n = 2^n$, the series is called the Hermite polynomial of order n. The first few Hermite polynomials are:

$$\begin{aligned} H_0(x) &= 1 \\ H_1(x) &= 2x \\ H_2(x) &= 4x^2 - 2 \\ H_3(x) &= 8x^3 - 12x \\ H_4(x) &= 16x^4 - 48x^2 + 12 \\ H_5(x) &= 32x^5 - 160x^3 + 120x\,. \end{aligned} \tag{5A1.4.6}$$

Analytic Expression for Hermite Polynomial

Consider a simple differential equation

$$\frac{du(x)}{dx} + 2x\,u(x) = 0\,,$$

whose solution is $u(x) = Ce^{-x^2}$. If we differentiate this equation n times with respect to x and put $d^n u(x)/dx^n = z(x)$, then this equation transforms to

$$\frac{d^2 z}{dx^2} + 2x\,\frac{dz}{dx} + 2(n+1)z = 0\,.$$

It is obvious that the solution of this equation is given by

$$z(x) = \frac{d^n u(x)}{dx^n} = \frac{d^n}{dx^n}\left(Ce^{-x^2}\right).$$

By carrying out the differentiation, we find the $z(x)$ can be written as the product of e^{-x^2} and a polynomial of order n in x

$$z(x) = e^{-x^2}\,y(x).$$

Substituting this in the differential equation for $z(x)$, we find the polynomial $y(x)$ satisfies the equation

$$\frac{d^2 y}{dx^2} - 2x\,\frac{dy}{dx} + 2ny = 0 \tag{5A1.4.7}$$

which is the Hermite equation. Obviously, the polynomial $y(x)$ is given by

$$y(x) = e^{x^2}\,\frac{d^n}{dx^n}(C\,e^{-x^2})\,, \tag{5A1.4.8}$$

where C is a constant. This polynomial can be identified with the Hermite polynomial by the choice $C = (-1)^n$ so that

$$H_n(x) = (-1)^n\,e^{x^2}\,\frac{d^n}{dx^n}(\,e^{-x^2})\,. \tag{5A1.4.9}$$

Orthogonality of Hermite Polynomials

Hermite polynomials $H_n(x)$ form a set of orthogonal functions in the sense that

$$I_{mn} \equiv \int_{-\infty}^{\infty} e^{-x^2}\,H_m(x)\,H_n(x)dx = 0 \quad \text{for } m \neq n\,. \tag{5A1.4.10}$$

To show this, we first observe that

$$\frac{d}{dx}H_n(x) \;=\; 2n\,H_{n-1}(x)\,. \tag{5A1.4.11}$$

This can be seen by simply differentiating the series (5A1.4.5). By expressing $H_n(x)$ in the integrand by means of Eq. (5A1.4.9) and integrating once, we find

$$\begin{aligned} I_{mn} &= \int_{-\infty}^{\infty} H_m(x)\,H_n(x)\,e^{-x^2}\,dx = (-1)^n \int_{-\infty}^{\infty} H_m(x)\,\frac{d^n}{dx^n}\,e^{-x^2}\,dx \\ &= (-1)^n\,H_m(x)\,\frac{d^{n-1}}{dx^{n-1}}\,e^{-x^2}\Big|_{-\infty}^{\infty} - (-1)^n \int_{-\infty}^{\infty} \frac{dH_m(x)}{dx}\,\frac{d^{n-1}}{dx^{n-1}}\,(\,e^{-x^2})\,dx\,. \end{aligned}$$

The integrated term vanishes since

$$\frac{d^{n-1}}{dx^{n-1}}\,(e^{-x^2}) = e^{-x^2} \times \text{polynomial of order } (n-1) \text{ in } x\,.$$

Using Eq. (5A1.4.11) to express the first derivative of $H_m(x)$ in terms of $H_{m-1}(x)$ we find the integral I_{mn} is given by

$$I_{mn} = (-1)^{n+1}\,2m \int_{-\infty}^{\infty} H_{m-1}(x)\,\frac{d^{n-1}}{dx^{n-1}}\,(e^{-x^2})\,dx\,. \tag{5A1.4.12}$$

Using this equation for I_{mn} as a recurrence relation m times, we get

$$I_{mn} = (-1)^{n+m}\,2^m\,m! \int_{-\infty}^{\infty} H_0(x)\,\frac{d^{n-m}}{dx^{n-m}}\,(e^{-x^2})\,dx \tag{5A1.4.13}$$

$$= (-1)^{n+m}\,2^m\,m!\,\left[\frac{d^{n-m-1}}{dx^{n-m-1}}\,(e^{-x^2})\right]_{-\infty}^{+\infty}\,. \tag{5A1.4.14}$$

This vanishes if $n > m$. Similarly, if $m > n$, we can interchange the role of $H_m(x)$ and $H_n(x)$ and show that $I_{mn} = 0$ if $m > n$. Thus

$$I_{mn} = 0 \quad \text{for } m \neq n\,. \tag{5A1.4.15}$$

The normalization integral I_{nn}, from Eq. (5A1.4.13), is given by

$$I_{nn} = (-1)^{2n}\,2^n\,n! \int_{-\infty}^{\infty} e^{-x^2}\,dx = 2^n n!\,\sqrt{\pi}\,. \tag{5A1.4.16}$$

Using this result we can define normalized Hermite polynomials by

$$\mathcal{H}_n(x) = \frac{1}{\sqrt{2^n\,n!\,\sqrt{\pi}}}\,e^{-x^2/2}\,H_n(x)\,. \tag{5A1.4.17}$$

These functions satisfy the orthonormality condition

$$\int_{-\infty}^{\infty} \mathcal{H}_m(x)\mathcal{H}_m(x)dx = \frac{1}{2^n n!\,\sqrt{\pi}} \int_{-\infty}^{\infty} e^{-x^2} H_m(x)H_m(x)dx = \delta_{mn}\,. \tag{5A1.4.18}$$

5A1.5 Bessel Equation

The equation

$$x^2\frac{d^2y}{dx^2} + x\frac{dy}{dx} + (x^2 - n^2)y = 0 \tag{5A1.5.1}$$

is called Bessel equation of order n. For its solution we again use the series method and assume that

$$y(x) = x^c \sum_{s=0}^{\infty} a_s x^s\,. \tag{5A1.5.2}$$

Substituting this series into Eq. (5A1.5.1) and equating the coefficient of each power of x, say, x^{c+s}, to zero we get

$$a_s\left[(c+s)^2 - n^2\right] = -a_{s-2}\,. \tag{5A1.5.3}$$

This enables all even coefficients to be expressed in terms of a_0 and all odd coefficients to be expressed in terms of a_1. Putting $s = 0$ into Eq. (5A1.5.3) we get the indicial equation

$$a_0\left(c^2 - n^2\right) = 0$$

which implies $c = \pm n$ if $a_0 \neq 0$. Putting $s = 1$ into Eq. (5A1.5.3) we get

$$a_1\left[(c+1)^2 - n^2\right] = 0\,, \tag{5A1.5.4}$$

which implies $c = (n-1)$ or $-(n+1)$ if $a_1 \neq 0$. A careful consideration of possible scenarios shows that the two conditions are equivalent and it is sufficient to choose either a_0 or a_1 nonzero, but not both. Making the first choice ($a_0 \neq 0$), we have $c = \pm n$. Expressing all the coefficients in terms of a_0 we can write the series solution for Bessel equation as

$$\begin{aligned} y = a_0\, x^c \Bigg[1 &- \frac{x^2}{(c+2)^2 - n^2} + \frac{x^4}{((c+4)^2 - n^2)((c+2)^2 - n^2)} \\ &+ \frac{x^6}{((c+6)^2 - n^2)((c+4)^2 - n^2)((c+2)^2 - n^2} + \ldots\Bigg] \end{aligned} \tag{5A1.5.5}$$

where $c = \pm n$. Both $[y]_{c=n}$ and $[y]_{c=-n}$ are solutions of Eq. (5A1.5.1). For $c = n$, the s-th term of the series can be written[9] as

$$\begin{aligned} t_s &= \frac{(-1)^s\, x^{2s+n}}{(2n+2s)(2s)(2n+2s-2)(2s-2)\cdots(2n+2)\,2}\, a_0 \\ &= \frac{(-1)^s\, x^{2s+n}\Gamma(n+1)}{2^{2s}\,\Gamma(n+s+1)\,\Gamma(s+1)}\, a_0 \end{aligned}$$

If a_0 is chosen to be $1/2^n\Gamma(n+1)$, the series is denoted by $J_n(x)$ and called Bessel function of order n

$$[y]_{c=n} = \sum_{s=0}^{\infty} \frac{(-1)^s}{\Gamma(n+s+1)\,\Gamma(s+1)} \left(\frac{x}{2}\right)^{n+2s} \equiv J_n(x)\,. \tag{5A1.5.6}$$

Similarly, the choice $c = -n$ leads to $J_{-n}(x)$ (Bessel function of order $-n$)

$$[y]_{c=-n} = \sum_{s=0}^{\infty} \frac{(-1)^s}{\Gamma(s-n+1)\,\Gamma(s+1)} \left(\frac{x}{2}\right)^{-n+2s} \equiv J_{-n}(x)\,. \tag{5A1.5.7}$$

Bessel functions $J_n(x)$ and $J_{-n}(x)$ constitute two independent solutions of the Bessel equation for noninteger values of n. When n is an integer, $J_n(x)$ and $J_{-n}(x)$ are related by

$$J_{-n}(x) = (-1)^n\, J_n(x)\,. \tag{5A1.5.8}$$

For integer n, the independent solutions of Bessel equation are denoted by $J_n(x)$ and $Y_n(x)$, where $Y_n(x)$, called Bessel function of order n of the second kind, is defined by

$$Y_n(x) = \lim_{r\to n} \frac{1}{\sin r\pi}\left[\cos r\pi\, J_r(x) - J_{-r}(x)\right]. \tag{5A1.5.9}$$

[9] For integer argument, the Γ-function is defined by $\Gamma(n+1) = n\Gamma(n) = n!$, $\Gamma(1) = 1$, $\Gamma(0) = \infty$. For negative integer values of its argument also, it is infinity. For positive half-integer argument $\Gamma(n+\frac{1}{2}) = (n-\frac{1}{2})\Gamma(n-\frac{1}{2})$ with $\Gamma(1/2) = \sqrt{\pi}$.

Bessel functions for integer and half integer order are built-in functions in many popular mathematical programs such as Mathcad and Mathematica. When plotted as functions of x, they look like damped sine or cosine functions. Of the two independent solutions, $J_n(x)$ is regular at the origin, whereas $Y_n(x)$ (or $J_{-n}(x)$ for non-integer n) is singular at the origin.

It can be easily seen that the series for Bessel functions of order 1/2 and −1/2 assume the following forms:

$$J_{\frac{1}{2}}(x) = \sqrt{\frac{2x}{\pi}}\,\frac{\sin x}{x}\,, \qquad (5A1.5.10)$$

$$J_{-\frac{1}{2}}(x) = \sqrt{\frac{2x}{\pi}}\,\frac{\cos x}{x}\,. \qquad (5A1.5.11)$$

Using the recurrence formula

$$(2n/x)J_n(x) = J_{n+1}(x) + J_{n-1}(x)\,, \qquad (5A1.5.12)$$

we can also derive the expressions for Bessel functions of order ±3/2 and ±5/2:

$$J_{\frac{3}{2}}(x) = \sqrt{\frac{2}{\pi x}}\left(\frac{\sin x}{x} - \cos x\right)\,, \qquad (5A1.5.13)$$

$$J_{-\frac{3}{2}}(x) = -\sqrt{\frac{2}{\pi x}}\left(\frac{\cos x}{x} + \sin x\right)\,, \qquad (5A1.5.14)$$

$$J_{\frac{5}{2}}(x) = \sqrt{\frac{2}{\pi x}}\left(\frac{3-x^2}{x^2}\sin x - \frac{3}{x}\cos x\right)\,, \qquad (5A1.5.15)$$

$$J_{-\frac{5}{2}}(x) = \sqrt{\frac{2}{\pi x}}\left(\frac{3}{x}\sin x + \frac{3-x^2}{x^2}\cos x\right)\,. \qquad (5A1.5.16)$$

Bessel functions of half-integer order are useful in the discussion of many quantum mechanical problems involving central potentials.

Spherical Bessel Functions

The half-integer Bessel functions $J_{\ell+\frac{1}{2}}(x)$, when multiplied by $\sqrt{\pi/2x}$ give what are known as spherical Bessel functions and are denoted by $j_\ell(x)$:

$$j_\ell(x) = \sqrt{\frac{\pi}{2x}}\,J_{\ell+\frac{1}{2}}(x)\,. \qquad (5A1.5.17)$$

The negative half-integer Bessel functions $J_{-\ell-\frac{1}{2}}(x)$ when multiplied by $\sqrt{\pi/2x}$ are called spherical Neumann functions and denoted by $\eta_\ell(x)$:

$$\eta_\ell(x) = \sqrt{\frac{\pi}{2x}}\,(-1)^{\ell+1}\,J_{-\ell-\frac{1}{2}}(x)\,. \qquad (5A1.5.18)$$

The first few spherical Bessel and spherical Neumann functions are, explicitly,

$$\begin{aligned} j_0(x) &= \frac{\sin x}{x}\,, & \eta_0(x) &= -\frac{\cos x}{x}\,, \\ j_1(x) &= \frac{\sin x}{x^2} - \frac{\cos x}{x}\,, & \eta_1(x) &= -\frac{\cos x}{x^2} - \frac{\sin x}{x}\,, \\ j_2(x) &= \left(\frac{3}{x^3} - \frac{1}{x}\right)\sin x - \frac{3}{x^2}\cos x\,, & \eta_2(x) &= -\left(\frac{3}{x^2} - \frac{1}{x}\right)\cos x - \frac{3}{x^2}\sin x\,. \end{aligned} \qquad (5A1.5.19)$$

For small values of their arguments, spherical Bessel functions yield

$$j_\ell(x) \xrightarrow{x\to 0} \frac{2^\ell \ell!}{(2\ell+1)!} x^\ell = \frac{x^\ell}{(2\ell+1)!!}, \tag{5A1.5.20a}$$

$$\eta_\ell(x) \xrightarrow{x\to 0} -\frac{(2\ell)!}{2^\ell \ell!} x^{-\ell-1} = -(2\ell-1)!! x^{-\ell-1}. \tag{5A1.5.20b}$$

The asymptotic ($x \to \infty$) forms of $j_\ell(x)$ and $\eta_\ell(x)$, useful in scattering and radiation problems, are given by

$$j_\ell(x) \xrightarrow{x\to\infty} \frac{1}{x}\cos\left[x-(\ell+1)\pi/2\right] = \frac{1}{x}\sin(x-\ell\pi/2), \tag{5A1.5.21a}$$

$$\eta_\ell(x) \xrightarrow{x\to\infty} \frac{1}{x}\sin\left[x-(\ell+1)\pi/2\right] = -\frac{1}{x}\cos(x-\ell\pi/2). \tag{5A1.5.21b}$$

Another set of functions, the so-called spherical Hankel functions, are useful in scattering problems. They are introduced by the relations

$$h_\ell^{(1)}(x) = j_\ell(x) + i\eta_\ell(x), \tag{5A1.5.22a}$$

$$h_\ell^{(2)}(x) = j_\ell(x) - i\eta_\ell(x) \tag{5A1.5.22b}$$

Their asymptotic forms, with the help of Eqs. (5A1.5.21) and (5A1.5.22), are given by

$$h_\ell^{(1)}(x) \xrightarrow{x\to\infty} \frac{1}{x} e^{i[x-(\ell+1)\pi/2]} = -\frac{i}{x} e^{i(x-\ell\pi/2)}, \tag{5A1.5.23a}$$

$$h_\ell^{(2)}(x) \xrightarrow{x\to\infty} \frac{1}{x} e^{-i[x-(\ell+1)\pi/2]} = \frac{i}{x} e^{-i(x-\ell\pi/2)}. \tag{5A1.5.23b}$$

From the asymptotic forms (5A1.5.21) and (5A1.5.23) we see that whereas spherical Bessel and Neumann functions $j_\ell(x)$ and $\eta_\ell(x)$ correspond to standing spherical waves, spherical Hankel functions correspond to traveling spherical waves.

Modified Bessel Equation

The differential equation

$$x^2\frac{d^2y}{dx^2} + x\frac{dy}{dx} - \left(x^2+n^2\right)y = 0 \tag{5A1.5.24}$$

is called the modified Bessel equation. This equation reduces to Bessel form with the substitution $\xi = ix$. The function $I_n(x) = i^{-n}J_n(ix)$ is thus a solution of the modified Bessel equation. Using the series expression for Bessel function [Eq. (5A.5.6)] we obtain

$$I_n(x) = \sum_{r=0}^{\infty} \frac{\left(\frac{x}{2}\right)^{n+2r}}{\Gamma(r+1)\,\Gamma(n+r+1)}. \tag{5A1.5.25}$$

If n is a fraction, $I_n(x)$ and $I_{-n}(x)$ are independent solutions of the modified Bessel equation of order n. For small and large values of its argument, $I_n(x)$ has the following limiting forms:

$$I_n(x) \xrightarrow{x\to 0} \frac{1}{\Gamma(n+1)}\left(\frac{x}{2}\right)^n, \tag{5A1.5.26a}$$

$$I_n(x) \xrightarrow{x\to\infty} \frac{1}{\sqrt{2\pi x}}\left[e^x + e^{-x}\,e^{-i(n+1/2)\pi}\right]. \tag{5A1.5.26b}$$

A second independent solution of the modified Bessel equation (5A1.5.24) is defined as

$$K_n(x) = \frac{\pi}{2}\frac{I_{-n}(x) - I_n(x)}{\sin n\pi}\,, \tag{5A1.5.27}$$

which for integer n must be treated as a limit. This functions takes the following limiting forms for small and large values of its argument

$$K_0(x) \xrightarrow{x\to 0} -\ln(x/2) - 0.5772\cdots\,, \tag{5A1.5.28a}$$

$$K_n(x) \xrightarrow{x\to 0} \frac{\Gamma(n)}{2}\left(\frac{2}{x}\right)^n\,,\ n > 0 \tag{5A1.5.28b}$$

$$K_n(x) \xrightarrow{x\to \infty} \frac{1}{\sqrt{2\pi x}}\,e^{-x}\,. \tag{5A1.5.28c}$$

In contrast to Bessel function, modified Bessel functions are not oscillatory. Instead, their behavior (for large values their argument) is similar to the exponential functions. For this reason I_n and K_n are sometimes referred to as hyperbolic Bessel functions.

We can also introduce the modified spherical Bessel functions (for integer n) by

$$i_n(x) = \sqrt{\frac{\pi}{2x}}\,I_{n+\frac{1}{2}}(x) \equiv i^{-n}j_n(ix)\,, \tag{5A1.5.29a}$$

$$k_n(x) = \sqrt{\frac{\pi}{2x}}\,K_{n+\frac{1}{2}}(x) \equiv -i^n h_n^{(1)}(ix)\,. \tag{5A1.5.29b}$$

For small argument, these have the following limiting forms

$$i_n(x) \xrightarrow{x\to 0} \frac{x^n}{(2n+1)!!}\,,\ k_n(x) \xrightarrow{x\to 0} \frac{(2n-1)!!}{x^{n+1}}\,, \tag{5A1.5.30a}$$

and for large argument

$$i_n(x) \xrightarrow{x\to \infty} \frac{e^x}{2x}\,,\ k_n(x) \xrightarrow{x\to \infty} \frac{e^{-x}}{x}\,. \tag{5A1.5.31a}$$

Orthogonality of Bessel Functions

Let α_{nk} $(k = 1, 2, 3, \cdots)$ be the zeros of Bessel function of order $n > -1$, i.e., α_{nk} are the positive roots of the equation

$$J_n(\alpha) = 0\,. \tag{5A1.5.32}$$

Then Bessel functions $J_n(\alpha_{nk}x)$ ($k = 1, 2, 3, \cdots$) of order n form a set of orthogonal functions satisfying

$$\int_0^1 xJ_n(\alpha_{nk}x)J_n(\alpha_{nk'}x)dx = \frac{1}{2}J_{n+1}^2(\alpha_{nk})\,\delta_{kk'}\,. \tag{5A1.5.33}$$

Appendix 5A2: Orthogonal Curvilinear Coordinate Systems

The position of a particle in space can be specified, not only by the *Cartesian system of coordinates* (x, y, z), with range $-\infty < x, y, z < \infty$, but also by other sets of coordinates: (a) spherical polar (r, θ, φ),(b) cylindrical (ρ, φ, z), (c) parabolic (ξ, η, φ) and so on. These different sets of coordinates of a point are mathematically related to each other. They are called *orthogonal curvilinear coordinate systems* because in any such system the curves, along which one (and only one) of the coordinates varies, are mutually orthogonal at a particular point. We now consider specific coordinate systems.

5A2.1 Spherical Polar Coordinates

Let OX, OY, OZ be the set of Cartesian axes and P be a point whose Cartesian coordinates are (x, y, z) as shown in Fig. (5A2.1). The spherical polar coordinates of P are:

(i) $r = \overline{OP}$ is the distance of point P from the origin, which varies in the range $0 \leq r < \infty$.

(ii) $\theta = \angle ZOP$ is the angle that $\overline{OP}$ makes with the Z-axis. It varies in the range $0 \leq \theta \leq \pi$, where $\theta = 0$ on the $+Z$ polar axis and $\theta = -\pi$ on the $-Z$ polar axis.

(iii) $\varphi = \angle P'OX$ is the angle that the plane containing $\overline{OP}$ and $\overline{OZ}$ makes with the XZ-plane. It varies in the range $0 \leq \varphi \leq 2\pi$, where $\varphi = 0$ if P lies in the XZ-plane and $\varphi = \pi/2$ if it lies in the YZ-plane.

Mathematically, they are related to x, y, z via the transformation

$$\begin{aligned} x &= r \sin\theta \cos\varphi\,, \\ y &= r \sin\theta \sin\varphi\,, \\ z &= r \cos\theta\,. \end{aligned} \tag{5A2.1.1}$$

Conversely, we have

$$\begin{aligned} r &= \sqrt{x^2 + y^2 + z^2}\,, \\ \theta &= \arccos(z/r)\,, \\ \varphi &= \tan^{-1}(y/x)\,. \end{aligned} \tag{5A2.1.2}$$

Coordinate Curves

1. Along the line OP, only r increases. This is called the r-curve and $\boldsymbol{e}_r$ is a unit vector along this curve.

2. Along the circle $PQS'Q'P'SP$ only θ changes. This is called the θ-curve. A unit vector along the tangent to this curve at P is denoted by $\boldsymbol{e}_\theta$. If θ changes by $d\theta$, the distance moved along this curve is $rd\theta$.

3. Along the curve $PGP'G'P$ only φ changes. This is called the φ-curve. A unit vector tangential to this curve at P is denoted by $\boldsymbol{e}_\varphi$. If φ changes by $d\varphi$, the distance moved along this curve is $r \sin\theta\, d\varphi$.

   It may be noted that, unlike the unit vectors $\boldsymbol{e}_x, \boldsymbol{e}_y, \boldsymbol{e}_z$ in the Cartesian system, the unit vectors $\boldsymbol{e}_r, \boldsymbol{e}_\theta, \boldsymbol{e}_\varphi$ change their directions from point to point. However, at any point these three unit vectors are mutually orthogonal.

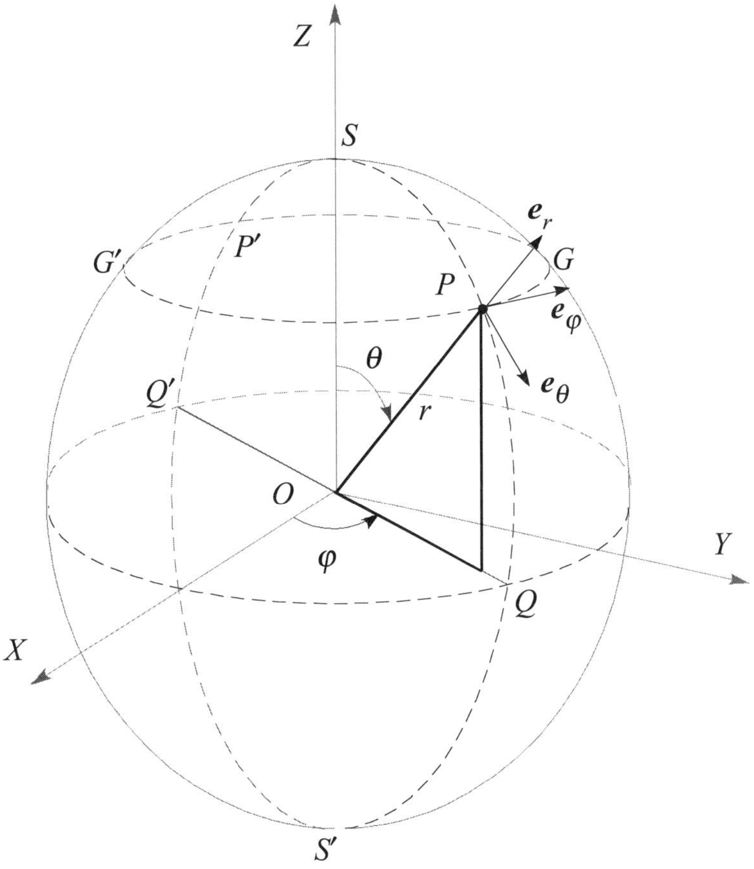

FIGURE 5A2.1
Spherical polar coordinates of point P. $\hat{e}_r\,, \hat{e}_\theta\,, \hat{e}_\varphi$ denote, respectively, the unit vectors along the r-curve, θ-curve and φ-curve at the point P.

Coordinate Surfaces

1. The surface of a sphere of radius r, with its center at the origin, is the surface of constant r.
2. The curved surface of a cone with its axis as the z-axis and semi-vertical angle θ is the surface of constant θ.
3. The plane containing the lines PO and OZ is a surface of constant φ.

5A2.2 Cylindrical Coordinates

The cylindrical coordinates of a point P in terms of its Cartesian coordinates $(x\,, y\,, z)$ are given by [see Fig. 5A2.2],

$$
\begin{aligned}
\rho &= \overline{OQ} = \overline{SP} = \sqrt{x^2 + y^2}\,,\\
\varphi &= \angle XOQ = \angle LSP = \tan^{-1}\frac{y}{x}\,,\\
z &= \overline{OS} = \overline{QP}\,.
\end{aligned}
\tag{5A2.1.3}
$$

Conversely,

$$x = \rho\cos\varphi\,,$$
$$y = \rho\sin\varphi\,, \qquad (5A2.1.4)$$
$$z = z\,.$$

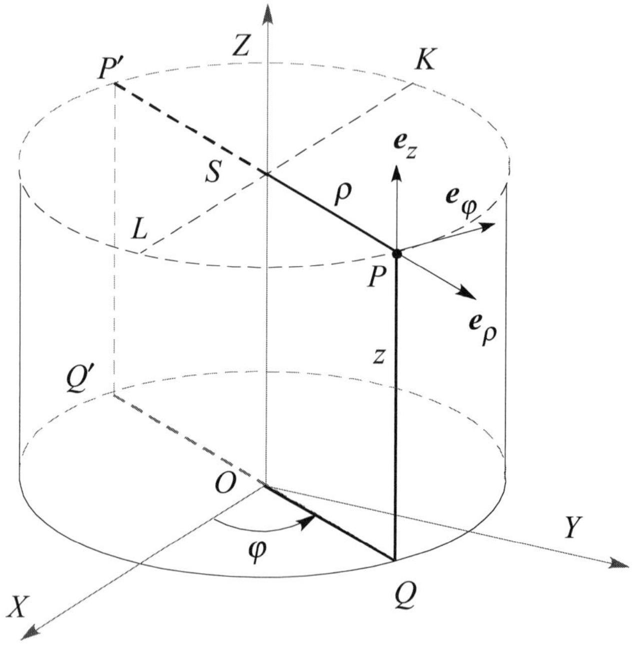

FIGURE 5A2.2
Cylindrical coordinates of a point P: $\rho = \overline{SP} = \overline{OQ}$; $z = \overline{OS} = \overline{QP}$; $\varphi = \angle XOQ = \angle LSP$.

Coordinate Curves

(i) Along the line SP (or OQ), only ρ increases. This is the ρ-curve. A unit vector at P, along this curve is denoted by $\hat{e}_\rho$.

(ii) Along the line QP only z increases. This is called the z-curve. A unit vector at P, tangential to this curve, is denoted by $\hat{e}_z$.

(iii) Along the circle $PKP'LP$ only φ changes. This is the φ-curve. A unit vector at P, along this curve is denoted by $\hat{e}_\varphi$. If φ changes by $d\varphi$, the distance moved along this curve is $\rho\, d\varphi$.

It may be seen that, although the directions of these unit vectors change from point to point, at any point P the unit vectors $\hat{e}_\rho\,, \hat{e}_\varphi\,, \hat{e}_z$ are orthogonal.

Coordinate Surfaces

(i) The surface of a cylinder with axis OZ and radius ρ, is the surface of constant ρ .

(ii) The plane passing through the point P, and parallel to XY-plane is the surface of constant z.

(iii) The plane containing the lines OQ or SP and OZ is the surface of constant φ.

5A2.3 Parabolic Coordinates

The parabolic coordinates (ξ, η, φ) of a point P are mathematically defined in terms of the spherical polar coordinates as

$$\begin{aligned} \xi &= r(1-\cos\theta) = r - z\,, \\ \eta &= r(1+\cos\theta) = r + z\,, \\ \varphi &= \varphi \text{ azimuthal angle}\,. \end{aligned} \tag{5A2.1.5}$$

To see this geometrically, let OX, OY, OZ be the Cartesian axes [Fig. 5A2.3] and through the point P, let us draw two parabolas PRP' and $PR'S'$ with the focus at the origin, and with z-axis as the common axis. (Only two parabolas with a common axis and focus can be drawn through a point in a plane.) The plane in which these parabolas are drawn may be characterized by the azimuth angle φ, i.e., the plane containing the vectors OP and OZ. The angle φ is defined in the same way as the angle between the plane containing the lines OP and OZ and the XZ-plane. Thus $\varphi = 0$ if P lies in the $X-Z$ plane.

The equations to the parabolas in polar coordinates may be written as

$$\frac{\xi}{r} = 1 - \cos\theta \tag{5A2.1.6}$$

and

$$\frac{\eta}{r} = 1 + \cos\theta\,. \tag{5A2.1.7}$$

The latus rectums of the two parabolas, ξ and η , and the azimuth φ angle constitute the parabolic coordinates of the point P.

Coordinate Curves

(i) Along the curve PRP', the coordinates ξ and φ are fixed and only η varies. This is the η-curve. A unit vector along this curve at any point may be denoted by $\hat{e}_\eta$. If η changes by $d\eta$, the distance moved along this curve is $\frac{\sqrt{\xi+\eta}}{2\sqrt{\eta}}d\eta$.

(ii) Along the curve $PR'P'$, η and φ are fixed and only ξ changes. This is called the ξ-curve. A unit vector along this curve at any point may be denoted by $\hat{e}_\xi$. If ξ changes by $d\xi$, the distance moved along this curve is $\frac{\sqrt{\xi+\eta}}{2\sqrt{\xi}}d\xi$.

(iii) Along the circle $PGP'G$ only φ changes. This is called the φ-curve. A unit vector at P, tangential to this curve is denoted by $\hat{e}_\varphi$. If φ changes by $d\varphi$ then the distance moved along this curve is $\sqrt{\xi\eta}\, d\varphi$.

The unit vectors $\hat{e}_\xi, \hat{e}_\eta, \hat{e}_\varphi$ have different directions at different points, but at any point P they are mutually orthogonal.

Coordinate Surfaces

(i) The surface generated by the revolution of the parabola PRP' about its axis (z-axis) is a surface of constant ξ.

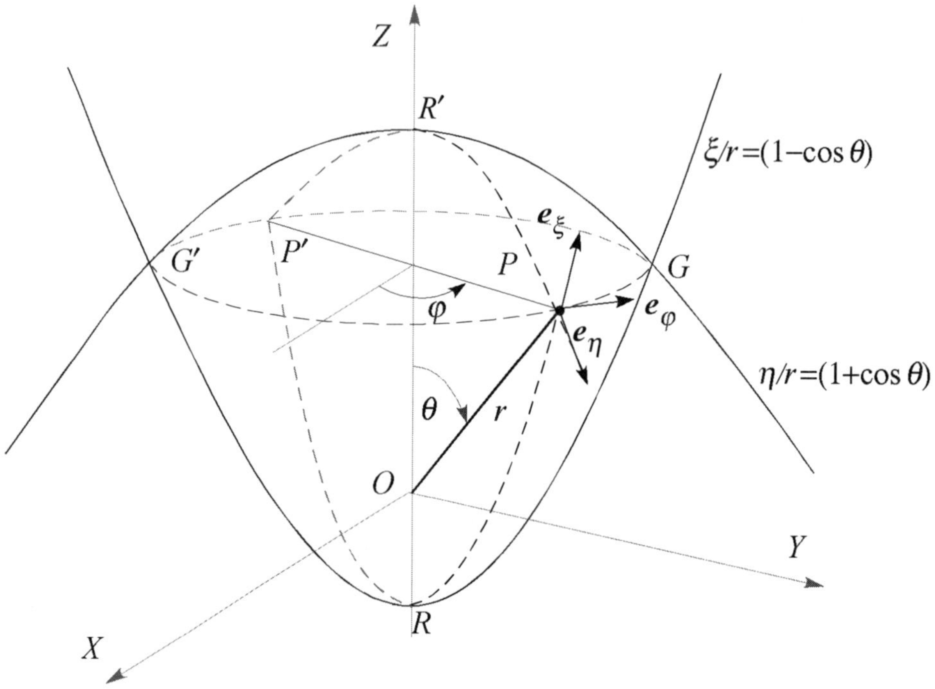

FIGURE 5A2.3
This figure shows the parabolic coordinates of a point P. ξ is the latus rectum of the parabola PRP' and equals $2 \times \overline{OR}$. η is the latus rectum of the parabola $PR'P'$ and equals $2 \times \overline{OR'}$. φ is the angle which the plane containing OP and OZ makes with the XZ plane.

(ii) The surface generated by the revolution of the parabola $PR'P'$ about its axis is a surface of constant η.

(iii) The plane containing the line OP and OZ is a surface of constant φ.

5A2.4 General Features of Orthogonal Curvilinear System of Coordinates

Let P be a point whose curvilinear coordinates are u_1, u_2, u_3 and P' be its infinitesimally close neighbor whose curvilinear coordinates are $u_1 + du_1, u_2 + du_2, u_3 + du_3$ [Fig. (5A2.4)]. Then the vector $\vec{PP'}$ is given by

$$\overline{PP'} \equiv d\boldsymbol{r} = \boldsymbol{e}_1 h_1 du_1 + \boldsymbol{e}_2 h_2 du_2 + \boldsymbol{e}_3 h_3 du_3 , \qquad (5A2.1.8)$$

where h_1, h_2, h_3 are the *scale factors*. This means that to find the length traversed along the u_1 -curve when u_1 changes by du_1, du_1 is to be multiplied by the scaling factor h_1, and so on. From the definition of infinitesimal displacement $d\boldsymbol{r}$ in Eq. (5A2.1.8),

$$dr^2 = d\boldsymbol{r}.d\boldsymbol{r} = h_1^2 du_1^2 + h_2^2 du_2^2 + h_3^2 du_3^2 . \qquad (5A2.1.9)$$

The volume enclosed by a rectangular parallelepiped, whose adjacent sides are $\boldsymbol{e}_1 h_1 du_1, \boldsymbol{e}_2 h_2 du_2$, and $\boldsymbol{e}_3 h_3 du_3$ is given by;

$$dV = (\boldsymbol{e}_1 h_1 du_1 \times \boldsymbol{e}_2 h_2 du_2) \cdot \boldsymbol{e}_3 h_3 du_3 = h_1 h_2 h_3 du_1 du_2 du_3 . \qquad (5A2.1.10)$$

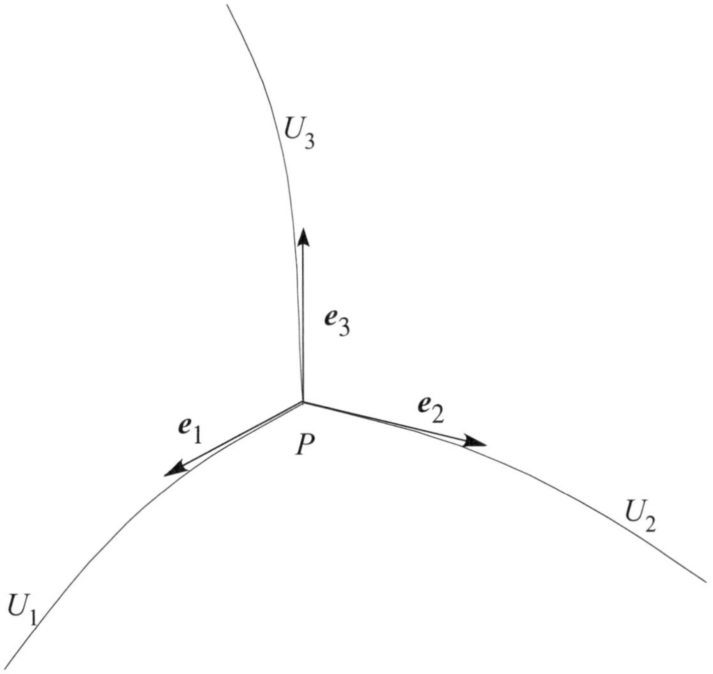

FIGURE 5A2.4
A general orthogonal curvilinear system of coordinates. The coordinates of a point P are u_1, u_2, u_3. PU_1, PU_2, PU_3 are the coordinate curves at P. For an infinitesimal change du_1 in u_1 the infinitesimal distance moved along this coordinate curve is $h_1 du_1$.

In this system of coordinates, the gradient of a scalar function $\psi(x, y, z)$ is expressed as

$$\nabla\psi = \frac{1}{h_1}\frac{\partial\psi}{\partial u_1}\boldsymbol{e}_1 + \frac{1}{h_2}\frac{\partial\psi}{\partial u_2}\boldsymbol{e}_2 + \frac{1}{h_3}\frac{\partial\psi}{\partial u_3}\boldsymbol{e}_3 \,. \tag{5A2.1.11}$$

Also, if $\boldsymbol{A} = A_1\boldsymbol{e}_1 + A_2\boldsymbol{e}_2 + A_3\boldsymbol{e}_3$ is a vector, then its divergence is given by

$$\nabla\cdot\boldsymbol{A} = \frac{1}{h_1h_2h_3}\left[\frac{\partial(A_1h_2h_3)}{\partial u_1} + \frac{\partial(A_2h_1h_3)}{\partial u_2} + \frac{\partial(A_3h_1h_2)}{\partial u_3}\right] . \tag{5A2.1.12}$$

Hence if we replace $\boldsymbol{A}$ by $\nabla\psi$, and identify the components of $\boldsymbol{A}$ by the components of $\nabla\psi$, then

$$\nabla.\nabla\psi \equiv \nabla^2\psi = \frac{1}{h_1h_2h_3}\left[\frac{\partial(A_1h_2h_3)}{\partial u_1} + \frac{\partial(A_2h_1h_3)}{\partial u_2} + \frac{\partial(A_3h_1h_2)}{\partial u_3}\right] . \tag{5A2.1.13}$$

In particular, for Cartesian coordinates we have

$$\begin{aligned} d\boldsymbol{r} &= \boldsymbol{e}_x dx + \boldsymbol{e}_y dy + \boldsymbol{e}_z dz \,, \\ \Rightarrow \quad & h_1 = h_2 = h_3 = 1 \,. \end{aligned} \tag{5A2.1.14}$$

For spherical polar coordinates, we have

$$\begin{aligned} d\boldsymbol{r} &= \boldsymbol{e}_r dr + \boldsymbol{e}_\theta r d\theta + \boldsymbol{e}_\varphi r\sin\theta d\varphi \,, \\ \Rightarrow \quad & h_1 = 1 \,, h_2 = r \,, h_3 = r\sin\theta \,. \end{aligned} \tag{5A2.1.15}$$

For cylindrical coordinates we have

$$\begin{aligned} d\boldsymbol{r} &= \boldsymbol{e}_\rho d\rho + \boldsymbol{e}_\varphi \rho d\varphi + \boldsymbol{e}_z dz\,, \\ \Rightarrow \quad h_1 &= 1\,, h_2 = \rho\,, h_3 = 1\,. \end{aligned} \tag{5A2.1.16}$$

For parabolic coordinates, we have

$$\begin{aligned} d\boldsymbol{r} &= \boldsymbol{e}_\xi \frac{\sqrt{\xi+\eta}}{2\sqrt{\xi}} d\xi + \boldsymbol{e}_\eta \frac{\sqrt{\xi+\eta}}{2\sqrt{\eta}} d\eta + \boldsymbol{e}_\phi \sqrt{\xi\eta} d\varphi\,, \\ \Rightarrow \quad h_\xi &= \frac{\sqrt{\xi+\eta}}{2\sqrt{\xi}}; \quad h_\eta = \frac{\sqrt{\xi+\eta}}{2\sqrt{\eta}}; \quad h_\varphi = \sqrt{\xi\eta}\,. \end{aligned} \tag{5A2.1.17}$$

Once the scaling factors $h_1\,, h_2\,, h_3$ for an orthogonal curvilinear system of coordinates are known, the operator ∇^2 can be expressed in terms of the corresponding coordinates.

6

SYMMETRIES AND CONSERVATION LAWS

6.1 Symmetries and Their Group Properties

A physical system is set to possess a symmetry, if it remains invariant (unchanged) under each one of the symmetry operations of the system. As a simple example, we may consider a system consisting of three identical particles placed at the vertices A, B, C of an equilateral triangle [Fig. 6.1]. The symmetry operations which leave the system unchanged are as follows.

Rotations[1] of the triangle about an axis passing through its centroid O and perpendicular to its plane, through (i) 0, (ii) $2\pi/3$, and (iii) $4\pi/3$ and reflections about the perpendiculars (iv) AP (v) BQ and and (vi) CR. This set of six symmetry operations, which we may call, respectively, E, D, F, A, B, C forms a group under *multiplication* in the sense that

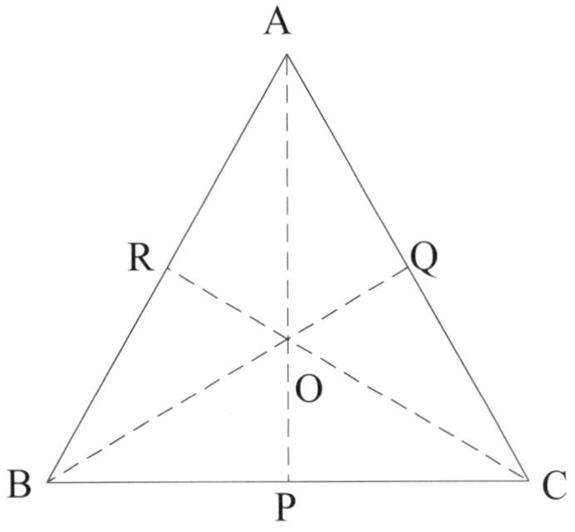

FIGURE 6.1
Symmetry operations of an equilateral triangle.

1. Closure holds. Two operations done in succession are equivalent to a single operation which is also contained in the group. (Example: $D \times F \equiv E$).

2. Identity exists. The identity operation E has the property that $E \times X = X \times E \equiv X$, where X is any element of the group [example: $E \times D = D \times E = D$].

[1]We limit our discussion to counterclockwise rotations.

3. Existence of the inverse. Every element X of the group has a unique inverse, $X^{-1} = Y$, which is contained in the group, such that $XY = YX = E$ [example $D^{-1} = F$ so that $D \times F = F \times D = E$].

4. Associativity holds. The associative law of multiplication holds among the elements, so that $A \times (B \times C) = (A \times B) \times C$.

The symmetry group of the equilateral triangle, also called the group D_3, is characterized by the group multiplication table [Table 6.1]

TABLE 6.1
Multiplication Table for Group D_3

×	E	A	B	C	D	F
E	E	A	B	C	D	F
A	A	E	D	F	B	C
B	B	F	E	D	C	A
C	C	D	F	E	A	B
D	D	C	A	B	F	E
F	F	B	C	A	E	D

Likewise four identical particles placed at the vertices of a square constitute a system with eight symmetry operations. This is called the D_4 group.

6.2 Symmetries in a Quantum Mechanical System

A physical system is characterized by its Hamiltonian $\hat{H}$. If the latter is independent of time its eigenstates and corresponding energies are given by the time-independent Schrödinger equation

$$\hat{H}\,|E_n\rangle = E_n\,|E_n\rangle\ . \tag{6.2.1}$$

If this system, i.e., its Hamiltonian, remains invariant under a set of unitary symmetry operations $\hat{E}$, $\hat{R}$, $\hat{S}$, $\hat{U}$, $\cdots$, then these operations form a group, called the symmetry group of the Hamiltonian, in the sense that, among the symmetry operations (i) closure holds, (ii) identity exists,(iii)inverse exists and (iv) associative law of multiplication holds. These symmetries lead to the following consequences:

1. *The Hamiltonian commutes with each one of the unitary symmetry operators*

   Under a symmetry operation $\hat{R}$

$$\hat{H} \to \hat{R}\hat{H}\hat{R}^\dagger \equiv \hat{H}'\ . \tag{6.2.2}$$

   Since this symmetry operation leaves the Hamiltonian invariant, we have, $\hat{H}' = \hat{H}$ or $\hat{R}\hat{H}\hat{R}^\dagger = \hat{H}$. Post-multiplying both sides of this equation with $\hat{R}$, we get, since $\hat{R}^\dagger\hat{R} = \hat{R}\hat{R}^\dagger = \hat{1}$,

$$\hat{R}\hat{H} = \hat{H}\hat{R} \quad \text{or} \quad [\hat{R}, \hat{H}] = 0\ . \tag{6.2.3}$$

2. *The energy levels of the system are degenerate, the degeneracy being a consequence of the invariance of $\hat{H}$ under these symmetry operations.*

   Consider the eigenvalue equation (6.2.1) and let $\hat{R}$ be one of the symmetry operators, which leaves the Hamiltonian invariant so that $\hat{R}\hat{H}\,|E_n\rangle = E_n\hat{R}\,|E_n\rangle$. In view of Eq. (6.2.3) we can as well write

$$\hat{H}\hat{R}\,|E_n\rangle = E_n\hat{R}\,|E_n\rangle\ . \tag{6.2.4}$$

   This equation shows that $\hat{R}\,|E_n\rangle$ is also an eigenstate of $\hat{H}$, belonging to the same energy E_n.

   Similarly, $\hat{S}\,|E_n\rangle\,, \hat{U}\,|E_n\rangle\,, \cdots$ are all eigenstates of $\hat{H}$ belonging to the same eigenvalue. This degeneracy is thus a consequence of the symmetry of the Hamiltonian.

3. *Unitary symmetry operations give rise to conservation laws.*

   Let the symmetry operation $\hat{U}(a)$ be unitary so that we can write it as $\hat{U}(a) = \exp(-ia\hat{A})$, where a is a real parameter and $\hat{A}$ is a Hermitian operator. If the Hamiltonian remains invariant under the symmetry operation $\hat{U}(a)$, then according to Eq. (6.2.3),

$$\begin{aligned}[\hat{U},\hat{H}] = 0 \quad &\Rightarrow \quad [\exp(-ia\hat{A}),\hat{H}] = 0 \\ &\Rightarrow \quad [\hat{A},\hat{H}] = 0\,.\end{aligned} \tag{6.2.5}$$

   This result combined with Heisenberg equation of motion, implies that the observable $\hat{A}$ is a constant of motion and corresponds to a conserved quantity.

6.3 Basic Symmetry Groups of the Hamiltonian and Conservation Laws

Every physical system has three basic symmetries. Apart from these symmetries, physical systems may have specific symmetries as well. These basic symmetries are:

1. Space translation symmetry
2. Time translation symmetry
3. Rotational symmetry

We begin by clarifying the meaning of *transformation of a physical system* under symmetry operations. We describe the effect of symmetry operation $\mathcal{R}$ on a system in quantum state $|\psi\rangle$ by a linear operator $\hat{R}$ such that the new state of the system is described by $|\psi'\rangle$

$$|\psi'\rangle = \hat{R}\,|\psi\rangle\ . \tag{6.3.1}$$

It is important to remember that while $\mathcal{R}$ operates in ordinary space (spanned by coordinates $\boldsymbol{r}$, momenta $\boldsymbol{p}$, and time t), operator $\hat{R}$ acts in state space. Equation (6.3.1) means that the transformed wave function $\psi'(\boldsymbol{r}) = \langle\boldsymbol{r}|\,\psi'\rangle$ at $\boldsymbol{r}_o$ is obtained from the value of the original wave function $\psi(\boldsymbol{r}) = \langle\boldsymbol{r}|\,\psi\rangle$ at the point $\boldsymbol{r}'_o$ which under $\mathcal{R}$ transforms to $\boldsymbol{r}_o = \mathcal{R}\boldsymbol{r}'_0$:

$$\psi'(\boldsymbol{r}_o) = \psi(\boldsymbol{r}'_o) = \psi(\mathcal{R}^{-1}\boldsymbol{r}_o)\,, \quad \boldsymbol{r}_o = \mathcal{R}\boldsymbol{r}'_o\ (\text{or } \boldsymbol{r}'_o = \mathcal{R}^{-1}\boldsymbol{r}_o)\,. \tag{6.3.2}$$

Since this holds for any arbitrary point $\boldsymbol{r}_0$, we can write

$$\langle \boldsymbol{r} | \psi' \rangle = \langle \mathcal{R}^{-1}\boldsymbol{r} | \psi \rangle . \tag{6.3.3}$$

Combining this with Eq. (6.3.1) we have the relation

$$\langle \boldsymbol{r} | \hat{R} | \psi \rangle = \langle \boldsymbol{r} | \psi' \rangle \equiv \langle \mathcal{R}^{-1}\boldsymbol{r} | \psi \rangle , \tag{6.3.4}$$

defining the linear operator $\hat{R}$ in the coordinate representation. Operator $\hat{R}$ is unitary, since the norms of $|\psi\rangle$ and $|\psi'\rangle$ are equal.

Having defined the transformation of state vectors, we can deduce the transformation law for observables. Let $\hat{X}$ be an observable corresponding to a measurement of some property of the system by a device fixed to the system. Under the action of $\mathcal{R}$, the $\hat{X}$ is transformed into an operator $\hat{X}'$ corresponding to a measurement by the transformed device, while the state $|\psi\rangle$ of the system is transformed into $|\psi'\rangle$. It is clear that measurement of $\hat{X}$ in state $|\psi\rangle$ must yield the same value as the measurement of $\hat{X}'$ on state $|\psi'\rangle$, i.e.,

$$\langle \psi | \hat{X} | \psi \rangle = \langle \psi' | \hat{X}' | \psi' \rangle = \langle \psi | \hat{R}^\dagger \hat{X}' \hat{R} | \psi \rangle . \tag{6.3.5}$$

Since this must hold for every $|\psi\rangle$, we have $\hat{R}^\dagger \hat{X}' \hat{R} = \hat{X}$ or

$$\hat{X}' = \hat{R}\hat{X}\hat{R}^\dagger . \tag{6.3.6}$$

These ideas will become clearer as we consider specific examples of these transformations.

6.3.1 Space Translation Symmetry

According to this symmetry, the Hamiltonian of any physical system is invariant under a spatial translation through an arbitrary displacement vector $\boldsymbol{a}$ and, in fact, under each one of the spatial translation operations which form a 'continuous group'. These transformations are defined by $\mathcal{U}\boldsymbol{r} = \boldsymbol{r} + \boldsymbol{a}$ and, conversely, $\mathcal{U}^{-1}\boldsymbol{r} = \boldsymbol{r} - \boldsymbol{a}$.

Let $\hat{U}(\boldsymbol{a})$ denote the corresponding translation operator in the space of state vectors $|\psi\rangle$. Then a state transform according to

$$|\psi'\rangle = \hat{U}(\boldsymbol{a}) |\psi\rangle . \tag{6.3.7}$$

In analogy with Eqs. (6.3.3) and (6.3.4), the coordinate representatives of this state yields

$$\psi'(\boldsymbol{r}) \equiv \langle \boldsymbol{r} | \hat{U}(\boldsymbol{a}) | \psi \rangle = \psi(\mathcal{U}^{-1}\boldsymbol{r}) = \psi(\boldsymbol{r} - \boldsymbol{a}) . \tag{6.3.8}$$

Making a Taylor exapansion around $\boldsymbol{r}$, we obtain

$$\begin{aligned} \psi'(\boldsymbol{r}) \equiv \psi(\boldsymbol{r} - \boldsymbol{a}) &= \psi(\boldsymbol{r}) - \left(a_x \frac{\partial}{\partial x} + a_y \frac{\partial}{\partial y} + a_z \frac{\partial}{\partial z} \right) \psi(\boldsymbol{r}) \\ &\quad + \frac{1}{2!} \left(a_x \frac{\partial}{\partial x} + a_y \frac{\partial}{\partial y} + a_z \frac{\partial}{\partial z} \right)^2 \psi(\boldsymbol{r}) + \cdots , \\ &= \exp\left[-\boldsymbol{a} \cdot \boldsymbol{\nabla} \right] \psi(\boldsymbol{r}) . \end{aligned} \tag{6.3.9a}$$

Using Eq. (6.3.8), this can be written as

$$\begin{aligned} \langle \boldsymbol{r} | \hat{U}(\boldsymbol{a}) | \psi \rangle &= \exp\left[-\boldsymbol{a} \cdot \boldsymbol{\nabla} \right] \langle \boldsymbol{r} | \psi \rangle \\ &= \langle \boldsymbol{r} | \exp\left[-\frac{i}{\hbar} \boldsymbol{a} \cdot (-i\hbar \boldsymbol{\nabla}) \right] | \psi \rangle \\ &= \langle \boldsymbol{r} | \exp(-\frac{i}{\hbar} \hat{\boldsymbol{p}} \cdot \boldsymbol{a}) | \psi \rangle . \end{aligned} \tag{6.3.9b}$$

In the last step we have used Eq. (2.6.5) of Chapter 2: $\frac{\partial}{\partial x}\langle \boldsymbol{r}|\psi\rangle = \langle \boldsymbol{r}|\frac{d}{d\hat{x}}|\psi\rangle$ and introduced the mometum operator $\hat{\boldsymbol{p}} = -i\hbar\boldsymbol{\nabla}$. Since the state $|\psi\rangle$ is arbitrary, we have the transformation law for state vectors

$$|\psi'\rangle = \exp\left[-\frac{i}{\hbar}\hat{\boldsymbol{p}}\cdot\hat{\boldsymbol{a}}\right]|\psi\rangle\,. \tag{6.3.10}$$

It follows, since $\hat{p}$ is Hermitian, that the operator $\hat{U}(\boldsymbol{a})$ is unitary, as it should be.

Now, according to Eq. (6.3.6), the Hamiltonian will transform to $\hat{H}' = \hat{U}(\boldsymbol{a})\hat{H}\hat{U}^{\dagger}(\boldsymbol{a})$ under space translation. If the Hamiltonian of the system is to remain invariant under the space translation operation, we must have $\hat{H}' \equiv \hat{U}(\boldsymbol{a})\hat{H}\hat{U}^{\dagger}(\boldsymbol{a}) = \hat{H}$. By virtue of the unitarity of $\hat{U}(\boldsymbol{a})$, this leads to

$$\hat{H}\hat{U}(\boldsymbol{a}) - \hat{U}(\boldsymbol{a})\hat{H} \equiv \left[\hat{H}\,,\hat{U}(\boldsymbol{a})\right] = 0\,,$$

or

$$\left[\hat{H}\,,\exp(-i\hat{\boldsymbol{p}}\cdot\boldsymbol{a}/\hbar)\right] = 0\,. \tag{6.3.11}$$

The last equation implies that

$$\left[\hat{H}\,,\hat{p}_x\right] = 0 = \left[\hat{H}\,,\hat{p}_y\right] = \left[\hat{H}\,,\hat{p}_z\right]\,, \tag{6.3.12}$$

or that the momentum of an isolated system is conserved or a constant of the motion.

It may be noted that the set of spatial translation operations $\hat{U}(\boldsymbol{a})$, where the displacement vector $\boldsymbol{a}$ may vary continuously, form a continuous group. Hence the invariance of the Hamiltonian of a system, under a set of spatial translation operations which form a continuous group, leads to the conservation of linear momentum.

6.3.2 Time Translation Symmetry

The invariance of the Hamiltonian $\hat{H}$ of an isolated system, under time translation operation, obviously implies that its energy does not change with the passage of time. The implication of the invariance of $\hat{H}$ under time translation operation $\hat{T}(t)$ as conservation of energy, or vice versa, may formally be seen as follows:

Conservation of energy implies, according to the Heisenberg equation of motion, that $[\hat{H}\,,\hat{H}] = 0$, which is obvious. This implies that

$$\left[\exp(-i\hat{H}t/\hbar)\,,\hat{H}\right] = 0\,,$$

or

$$\left[\hat{T}(t)\,,\hat{H}\right] = 0\,, \tag{6.3.13}$$

where $\hat{T}(t) = \exp(-i\hat{H}t/\hbar)$ is the time translation operator [Eq. (3.4.3), Chapter 3]. Eq. (6.3.13) implies invariance of the Hamiltonian under the time translation operation.

It may be noted that the set of operations $\hat{T}(t)$, where t is a continuous parameter, form a continuous group, called the time translation group.

Thus the invariance of the Hamiltonian of a system, under the group of time-translation operations, leads to the conservation of energy.

6.3.3 Spatial Rotation Symmetry

Isotropy of space requires that a physical system should behave in the same way if it is rotated in space in an arbitrary manner. The rotation about an arbitrary axis, say about the direction $\boldsymbol{n}$, through an arbitrary angle θ may be specified by the operator $\hat{R}_{\boldsymbol{n}}(\theta)$. This means that the rotated state $|\psi'\rangle$ may be obtained by the operation of the rotation operator on the original state $|\psi\rangle$:

$$|\psi\rangle \to |\psi'\rangle = \hat{R}_{\boldsymbol{n}}(\theta)\,|\psi\rangle\,. \tag{6.3.14}$$

Since the parameters θ and $\boldsymbol{n}$, specifying the rotation operation, can vary continuously, these set of rotation operations constitute a continuous set. Moreover, they constitute a continuous group, called *rotation group* or $R(3)$ group, as these operators also satisfy the conditions of closure, existence of identity and inverse, and associative law of multiplication. It may be pointed out here that the rotation group belongs to a class of groups called *Lie groups*.

Let us first identify the rotation operator $\hat{R}_z(\delta\theta)$, corresponding to anticlockwise rotation $\mathcal{R}$ of the coordinate system about z-axis through an infinitesimal angle $\delta\theta$ [Fig. 6.2]. Then we can see from Fig. 6.2(b) that under $\mathcal{R}^{-1}$, a point $P(x,y,z)$ is transformed to $P'(x',y',z')$ with

$$\left.\begin{aligned} x' &= x + y\delta\theta \\ y' &= y - x\delta\theta \\ z' &= z \end{aligned}\right\} \equiv \mathcal{R}^{-1}\boldsymbol{r}\,. \tag{6.3.15}$$

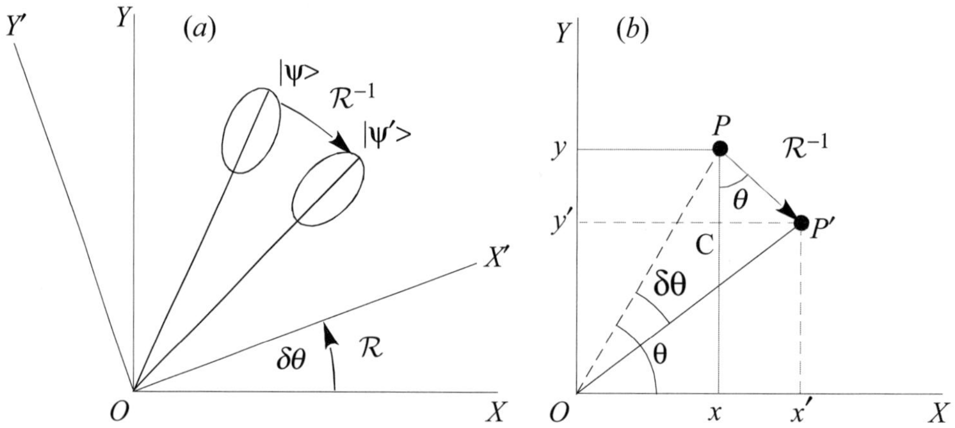

FIGURE 6.2
Schematic representation of the passive and active transformation of the state of a system [Fig. 6.2(a)]. A positive rotation $\mathcal{R}$ of coordinate axes through $\delta\theta$ about the z-axis with the system being fixed (passive rotation) is equivalent to keeping the coordinate axes unchanged and rotating the system (active rotation) through $-\delta\theta$ ($\mathcal{R}^{-1}$). Under $\mathcal{R}^{-1}$ a point P is transformed to P'. The state $|\psi'\rangle$ which arises from $|\psi\rangle$ through rotation $\mathcal{R}^{-1}$, appears in the original coordinate system exactly as $|\psi\rangle$ does in the rotated coordinate system. From the figure [Fig 6.2(b)] it can be seen that $\overline{PP'} = \overline{OP}\cdot\delta\theta$ and $x'-x = \overline{CP} = \overline{PP'}\sin\theta = \overline{OP}\delta\theta\cdot\sin\theta = y\delta\theta$. Similarly, $y'-y = -\overline{CP} = -\overline{OP}\delta\theta\cdot\cos\theta = -x\delta\theta$.

As a result of this rotation let the state of the system change from $|\psi\rangle$ to $|\psi'\rangle$, where

$$|\psi'\rangle = \hat{R}_z(\delta\theta)\,|\psi\rangle\,. \tag{6.3.16}$$

Then, according to Eq. (6.3.4), we have

$$< \boldsymbol{r}|\hat{R}_z(\delta\theta)|\psi> =< \boldsymbol{r}|\psi'>= \psi(\mathcal{R}^{-1}\boldsymbol{r}) = \psi(x+y\delta\theta,\, y-x\delta\theta,\, z)\,,$$
$$= \psi(x,y,z) + y\delta\theta\frac{\partial\psi}{\partial x} - x\delta\theta\frac{\partial\psi}{\partial y} + \cdots\,,$$
$$= \left[1+\delta\theta\left(y\frac{\partial}{\partial x} - x\frac{\partial}{\partial y}\right)\right]\psi(r)\,,$$

or

$$< \boldsymbol{r}|\hat{R}_z(\delta\theta)|\psi> =< \boldsymbol{r}|\left[\hat{1}+\delta\theta\left(\hat{y}\frac{d}{d\hat{x}} - \hat{x}\frac{d}{d\hat{y}}\right)\right]|\psi>\,, \tag{6.3.17}$$

where, in the last step we have used Eq.(2.7.4), Chapter 2 to replace differential operators by observables. Hence the operator for corresponding to infinitesimal rotation about z-axis is given by

$$\hat{R}_z(\delta\theta) = \left(\hat{1} - \frac{i}{\hbar}\delta\theta\,\hat{L}_z\right)\,, \tag{6.3.18}$$

where

$$\hat{L}_z = \hat{x}\hat{p}_y - \hat{y}\hat{p}_x = -i\hbar\left(\hat{x}\frac{d}{d\hat{y}} - \hat{y}\frac{d}{d\hat{x}}\right)\,. \tag{6.3.19}$$

A finite rotation by an angle θ about the z-axis is equivalent to N rotations, each of magnitude $\delta\theta = \theta/N$ about the z-axis, in succession. When N tends to infinity, $\delta\theta$ becomes infinitesimal and we have

$$\hat{R}_z(\theta) = \lim_{N\to\infty}\left[1 - i\frac{\delta\theta}{\hbar}\hat{L}_z\right]^N = \exp\left(-\frac{i\theta\,\hat{L}_z}{\hbar}\right)\,. \tag{6.3.20}$$

If, instead of the z-axis, the rotation is about an arbitrary direction $\boldsymbol{n}$, the rotation operator is

$$\hat{R}_{\boldsymbol{n}}(\theta) = e^{-i\theta\hat{\boldsymbol{L}}.\boldsymbol{n}/\hbar}\,. \tag{6.3.21}$$

From this expression we can see that the rotation operator $\hat{R}_{\boldsymbol{n}}(\theta)$ is unitary, since $\hat{L}$ is Hermitian.

The invariance of the Hamiltonian under rotation implies $\hat{R}_{\boldsymbol{n}}(\theta)\hat{H}\hat{R}^\dagger_{\boldsymbol{n}}(\theta) = \hat{H}$ or, since the rotation operator $\hat{R}_{\boldsymbol{n}}(\theta)$ is unitary,

$$[\hat{R}_{\boldsymbol{n}}(\theta), \hat{H}] = [\exp(-i\theta\hat{\boldsymbol{L}}\cdot\boldsymbol{n}/\hbar), \hat{H}] = 0\,,$$

or

$$[\hat{L}_x, \hat{H}] = [\hat{L}_y, \hat{H}] = [\hat{L}_z, \hat{H}] = 0\,. \tag{6.3.22}$$

The conservation of orbital angular momentum (being a consequence of invariance of Hamiltonian under arbitrary rotation) holds good if the particle is not endowed with intrinsic angular momentum (or spin) and its states may be specified by a one-component (scalar) wave function. The electron, however, has an intrinsic angular momentum.

According to Uhlenbeck and Goudsmit, the indirect evidence for intrinsic angular momentum, (or spin) in the case of electron came from several observations like anomalous Zeeman effect, doublet-structure of the spectra of alkali atoms. The famous Stern-Gerlach experiment provided direct and decisive evidence for the existence of electron spin, $\frac{1}{2}$ (in units of $\hbar$), and for the fact that an electron can exist in two alternative spin states, viz., the *spin up* state in which the spin is along the z-direction and the *spin down* state in which the spin is directed opposite to the z-direction. The z-direction is arbitrary and is used here only for definiteness.

When a particle is endowed with spin,the total state $|\Psi\rangle$ of the particle may be regarded as the product of two states

$$|\Psi\rangle = |\psi\rangle\,|\chi\rangle\,, \tag{6.3.23}$$

where the state $|\psi\rangle$ characterizes the orbital angular momentum of the particle and, for this state, a coordinate representation is possible. The state $|\chi\rangle$ denotes the spin state of the particle, and as we shall see later, the spin state cannot admit a coordinate-representation. It can only admit a matrix representation, i.e., it can only be represented by a column matrix. Hence the total wave function in this case is a multi-component wave function.

Now, under a rotation θ about z-axis, $|\psi\rangle$ transforms to $|\psi'\rangle = \exp(-i\hat{L}_z\theta/\hbar)\,|\psi\rangle$. Likewise we can assert that the spin state $|\chi\rangle$ transforms to $|\chi'\rangle = \exp(-i\hat{S}_z\theta/\hbar)\,|\chi\rangle$. Since the transformation operator has to be unitary, $\hat{S}_z$ is hermitian. By analogy $\hat{S}_z$ may be called the spin angular momentum of the particle. Thus under spatial rotation of θ about the z-axis,

$$|\Psi\rangle \to |\Psi'\rangle = \exp[-i(\hat{L}_z + \hat{S}_z)\theta/\hbar]\,|\psi\rangle\,|\chi\rangle \equiv \exp(-i\hat{J}_z\theta/\hbar)\,|\Psi\rangle\,, \tag{6.3.24}$$

where $\hat{J}_z = \hat{L}_z + \hat{S}_z$ is called the z-component of the total angular momentum of the system. For rotation about an arbitrary direction $\boldsymbol{n}$, the rotation operator is given by:

$$\hat{R}_n(\theta) = \exp(-i\hat{\boldsymbol{J}}\cdot\boldsymbol{n}\,\theta/\hbar) \tag{6.3.25}$$

so that

$$|\Psi\rangle \;\to\; |\Psi'\rangle = \hat{R}_{\boldsymbol{n}}(\theta)\,|\Psi\rangle\,.$$

The set of rotation operators $\hat{R}_{\boldsymbol{n}}(\theta)$ forms a group. The components $\hat{J}_x$, $\hat{J}_y$, $\hat{J}_z$ of the vector operator $\hat{J}$ (the total angular momentum operator) are called the *generators* of the rotation group because any element of the group may be constructed in terms of the generators and the continuous parameters θ and $\boldsymbol{n}$.

In accordance with the postulate of isotropy of space we assume that the Hamiltonian of the system is invariant under arbitrary rotation in space or under each one of the set of rotation operations which form the rotation group. This implies that

$$[\exp(-i\,\hat{\boldsymbol{J}}\cdot\boldsymbol{n}\theta/\hbar), \hat{H}] = 0\,,$$

or

$$[\hat{J}_x, \hat{H}] = [\hat{J}_y, \hat{H}] = [\hat{J}_z, \hat{H}] = 0\,. \tag{6.3.26}$$

Thus if a particle is endowed with spin, isotropy of space implies conservation of total angular momentum $\hat{\boldsymbol{J}}$.

6.4 Lie Groups and Their Generators

A *Lie group* is a group of continuous transformations, each element of which leaves the Hamiltonian of a system, invariant. In what follows, we shall study the general properties of Lie groups and their *generators*. In particular, we shall consider (1) the three-dimensional rotation group R(3) [also called the *special orthogonal group* SO(3)] and (ii) the *special unitary group* SU(2).

Continuous Groups and Lie Groups

A set of coordinate transformation operations $g(a)$, under which

$$x \rightarrow x' = g(a)x\,, \tag{6.4.1}$$

where x stands for a set of coordinates $(x_1\,, x_2\,, x_3 \cdots, x_n)$ of the system and a stands for a set of parameters $(a_1\,, a_2\,, a_3\,, \cdots, a_r)$, is said to form a *continuous group*, if

(1) Closure holds,

$$g(a) \otimes g(b) = g(c)\,, \tag{6.4.2}$$

where $g(c)$ is also an element of the group.

(2) A unique identity element $g(a_0)$ exists, such that

$$g(a_0) \otimes g(a) = g(a) = g(a) \otimes g(a_0)\,. \tag{6.4.3}$$

(3) For every element $g(a)$, a unique inverse $g(\bar{a})$ exists, which also belongs to the group such that

$$g(\bar{a}) \otimes g(a) = g(a_0) = g(a) \otimes g(\bar{a})\,. \tag{6.4.4}$$

(4) Associative law of multiplication holds:

$$\{g(a) \otimes g(b)\} \otimes g(d) = g(a) \otimes \{g(b) \otimes g(d)\}\,. \tag{6.4.5}$$

The multiplication sign $\otimes$ between the two operators means that the operations are done in succession.

If from the closure condition (6.4.2) one implies that c is an analytic function of a and b such that $c = c(a, b)$ or, explicitly,

$$c_k = \phi_k(a_1\,, a_2\,, \cdots, a_r; b_1\,, b_2\,, \cdots, b_r)\,, \quad k = 1, 2, 3, \cdots r \tag{6.4.6}$$

then the continuous group is called a *Lie group*.

Generators of a Lie Group

Consider a Lie group of transformation matrices of order of order $(n \times n)$,

$$g(a) = \begin{pmatrix} g_{11}(a) & g_{12}(a) & \cdots & g_{1n}(a) \\ g_{21}(a) & g_{22}(a) & \cdots & g_{2n}(a) \\ \vdots & \vdots & \vdots & \vdots \\ g_{n1}(a) & g_{n2}(a) & \cdots & g_{nn}(a) \end{pmatrix}.$$

Each of these matrices transforms an n-dimensional vector

$$x \equiv \begin{pmatrix} x_1 \\ x_2 \\ \vdots \\ x_n \end{pmatrix} \rightarrow x' \equiv \begin{pmatrix} x'_1 \\ x'_2 \\ \vdots \\ x'_n \end{pmatrix},$$

according to

$$x \rightarrow x' = g(a)\, x\,, \tag{6.4.7}$$

or, explicitly,

$$x'_k = \sum_{j=1}^{n} g_{kj}(a) x_j \equiv f_k(x_1, x_2, \cdots, x_n;\, a_1, a_2, \cdots, a_r)\,. \tag{6.4.8}$$

Recall that in Eqs. (6.4.7) a stands for $\{a_1, a_2, \cdots a_r\}$. We may adopt the convention that the identity transformation operator $g(a_0)$ corresponds to $a \equiv a_0 = 0$ or $a_1 = a_2 = \cdots a_r = 0$ so that

$$x_k = f_k(x_1, x_2, \cdots, x_n;\ a_1 = 0, a_2 = 0, \cdots, a_r = 0) \equiv f_k(x, 0)\,. \tag{6.4.9}$$

We can make a small change in x by varying a by a small amount da around the identity a_0. Then $x = x + dx = f(x, da)$ or, explicitly,

$$x_k + dx_k = f_k(x_1, \cdots, x_n; da_1, da_2, \cdots da_\nu \cdots da_r)\,. \tag{6.4.10}$$

This implies that

$$dx_k = f_k(x_1, x_2, \cdots, x_n;\ da_1, da_2, \cdots, da_r) - f_k(x_1, x_2, \cdots, x_n;\ 0, 0, \cdots, 0)$$

or

$$dx_k = \sum_{\nu=1}^{r} \left(\frac{\partial f_k}{\partial a_\nu}\right) da_\nu = \sum_{\nu=1}^{r} u_{k\nu} da_\nu \tag{6.4.11}$$

where

$$u_{k\nu} \equiv \left(\frac{\partial f_k}{\partial a_\nu}\right)_{a=0}\,. \tag{6.4.12}$$

Let $\psi(x)$ be the wave function representing the state of a physical system. Under the infinitesimal transformation $x \to x' = g(da)x$, let the wave function $\psi(x) \to \psi'(x) = \psi(x') = \psi(x + dx) \equiv \psi(x) + d\psi(x)$. Then the change $d\psi(x)$ is given by

$$\begin{aligned} d\psi &= \sum_{k=1}^{n} \frac{\partial \psi}{\partial x_k} dx_k = \sum_{k=1}^{n} \frac{\partial \psi}{\partial x_k} \sum_{\nu=1}^{r} u_{k\nu}(x) da_\nu\,, \\ &= \sum_{\nu=1}^{r} da_\nu \left\{ \sum_{k=1}^{n} u_{k\nu} \frac{\partial}{\partial x_k} \right\} \psi\,, \end{aligned}$$

or

$$d\psi = -i \sum_{\nu=1}^{r} da_\nu X_\nu \psi\,, \tag{6.4.13}$$

where

$$X_\nu = \sum_{k=1}^{n} i \frac{\partial f_k}{\partial a_\nu} \frac{\partial}{\partial x_k}\,. \tag{6.4.14}$$

The set of operators $\hat{X}_1, \hat{X}_2, \cdots, \hat{X}_r$ are called the generators of the Lie group and are as many in number as the number of parameters a_ν. Thus under the infinitesimal transformation $x \to x' = g(da)x$, the wave function transforms according to $\psi(x) \to \psi'(x)$, where

$$\psi'(x) = \psi + d\psi = \left(1 - i \sum_{\nu=1}^{r} da_\nu X_\nu\right) \psi(x)\,. \tag{6.4.15}$$

With the choice of all the parameters a_ν to be real, and all the generators X_ν made Hermitian [Eq. (6.4.14)], the transformation equation (6.4.15) for an infinitesimal transformation may easily be generalized for a finite transformation [as in Eq. (6.3.18)] by constructing a set of unitary operators

$$\hat{U}_\nu = \exp(-i a_\nu X_\nu)\,, \tag{6.4.16}$$

for $\nu = 1, 2, \cdots, r$, where the parameters a_ν can vary continuously from the identity ($a_\nu = 0$) to any finite values within the respective ranges. An arbitrary element of the Lie group may now be expressed as the product of unitary operators, so that

$$U = \exp\left(-i \sum_{\nu=1}^{r} a_\nu X_\nu\right)\,. \tag{6.4.17}$$

To summarize, the coordinate transformation $x \to x' = g(a)x$ leads to the following transformation for the state function:

$$\psi(x) \to \psi'(x) = U\psi(x) = \exp\left(-i\sum_{\nu=1}^{r} a_\nu X_\nu\right)\psi(x)\,. \tag{6.4.18}$$

It is obvious that if this set of transformation operators, which form the Lie group, leave the Hamiltonian of a system invariant, then the Hamiltonian must commute with each one of the generators: $X_1\,, X_2\cdots\,, X_r$, so every generator corresponds to a conserved quantity for the system.

6.5 Examples of Lie Group

6.5.1 Proper Rotation Group $R(3)$ (or Special Orthogonal Group $SO(3)$)

This is a group of transformations wherein a typical 3×3 transformation matrix A, which is real, orthogonal and unimodular, transforms a three-dimensional vector $x \to x'$:

$$x \to x' = Ax\,. \tag{6.5.1}$$

The three-dimensional vector in this case represents the Cartesian coordinates of a particle and the state of the particle may be specified by the wave function $\psi(x_1, x_2, x_3)$. The conditions satisfied by the real matrix A are:

(i) $AA^T = I$, where A^T is the transpose of A.

(ii) $\det(A) = |A| = 1$.

For the identity operator we have

$$A_0 \equiv I = \begin{pmatrix} 1&0&0\\ 0&1&0\\ 0&0&1 \end{pmatrix}\,. \tag{6.5.2}$$

Let $A = I + dg$ be an infinitesimal transformation which carries the column vector $x \to x'$, where

$$x' \equiv x + dx = (I+dg)x \quad \text{or} \quad dx = dg\,x\,. \tag{6.5.3}$$

The orthogonality condition on $A = I + dg$ then implies that dg is an anti-symmetric matrix: $dg^T = -dg$. In view of this observation, the diagonal elements of dg are zero and the off-diagonal elements can be expressed in terms of three independent parameters as

$$dg = \begin{pmatrix} 0 & da_3 & -da_2\\ -da_3 & 0 & da_1\\ da_2 & -da_1 & 0 \end{pmatrix}, \tag{6.5.4}$$

where da_1, da_2, da_3 are real and infinitesimal parameters. Using this in Eq. (6.5.3) we can write the infinitesimal transformation as

$$\begin{pmatrix} dx_1\\ dx_2\\ dx_3 \end{pmatrix} = \begin{pmatrix} 0 & da_3 & -da_2\\ -da_3 & 0 & da_1\\ da_2 & -da_1 & 0 \end{pmatrix}\begin{pmatrix} x_1\\ x_2\\ x_3 \end{pmatrix}$$

or

$$\left.\begin{aligned} dx_1 &= da_3\,x_2 - da_2\,x_3\\ dx_2 &= -da_3\,x_1 + da_1\,x_3\\ dx_3 &= da_2\,x_1 - da_1\,x_2 \end{aligned}\right\}. \tag{6.5.5}$$

According to Eq. (6.4.11), this implies that

$$f_1(x,a) \equiv x_1' = x_1 + a_3x_2 - a_2x_3\,, \tag{6.5.6a}$$
$$f_2(x,a) \equiv x_2' = x_2 - a_3x_1 + a_1x_3\,, \tag{6.5.6b}$$
$$f_3(x,a) \equiv x_3' = x_3 + a_2x_1 - a_1x_2\,. \tag{6.5.6c}$$

There will be three generators in this case as there are three parameters. They can be worked out, according to Eq. (6.4.14), as follows

$$\begin{aligned} X_1 &= i\left\{\frac{\partial f_1}{\partial a_1}\frac{\partial}{\partial x_1} + \frac{\partial f_2}{\partial a_1}\frac{\partial}{\partial x_2} + \frac{\partial f_3}{\partial a_1}\frac{\partial}{\partial x_3}\right\}, \\ &= i\left(x_3\frac{\partial}{\partial x_2} - x_2\frac{\partial}{\partial x_3}\right) = i\left(z\frac{\partial}{\partial y} - y\frac{\partial}{\partial z}\right) = L_x/\hbar\,. \end{aligned} \tag{6.5.7}$$

$$\begin{aligned} X_2 &= i\left\{\frac{\partial f_1}{\partial a_2}\frac{\partial}{\partial x_1} + \frac{\partial f_2}{\partial a_2}\frac{\partial}{\partial x_2} + \frac{\partial f_3}{\partial a_2}\frac{\partial}{\partial x_3}\right\}, \\ &= i\left(-x_3\frac{\partial}{\partial x_1} + x_1\frac{\partial}{\partial x_3}\right) = i\left(x\frac{\partial}{\partial z} - z\frac{\partial}{\partial x}\right) = L_y/\hbar\,. \end{aligned} \tag{6.5.8}$$

$$\begin{aligned} X_3 &= i\left\{\frac{\partial f_1}{\partial a_3}\frac{\partial}{\partial x_1} + \frac{\partial f_2}{\partial a_3}\frac{\partial}{\partial x_2} + \frac{\partial f_3}{\partial a_3}\frac{\partial}{\partial x_3}\right\}, \\ &= i\left(x_2\frac{\partial}{\partial x_1} - x_1\frac{\partial}{\partial x_2}\right) = i\left(y\frac{\partial}{\partial x} - x\frac{\partial}{\partial y}\right) = L_z/\hbar\,. \end{aligned} \tag{6.5.9}$$

The generators of $R(3)$ group satisfy the same commutation relations as the components of orbital angular momentum:

$$[X_k, X_\ell] = i\epsilon_{k\ell m}X_m\,, \tag{6.5.10}$$

whereas $[L_k\,, L_\ell] = i\hbar\epsilon_{k\ell m}L_m$ where $\epsilon_{k\ell m}$ is the alternating symbol[2]. In terms of these generators and continuous parameters, a typical $R(3)$ operator for a finite transformation may be constructed from Eq. (6.4.17) as

$$U = \exp\left(-i\sum_{\nu=1}^{3} a_\nu X_\nu\right) \tag{6.5.11}$$

which transforms $\psi(x) \to \psi'(x) = U\psi(x) = \exp\left[-i\sum_{\nu=1}^{3} a_\nu X_\nu\right]\psi(x)$.

[2]The alternating symbol is defined by

$$\epsilon_{ijk} = \begin{cases} 1 & \text{if } ijk \text{ is a cyclic permutation of } (123) \\ -1 & \text{if } ijk \text{ is a cyclic permutation of } (213) \\ 0 & \text{in all other cases} \end{cases}.$$

Each index takes on values 1, 2 and 3. It is also easy to establish the following identity

$$\epsilon_{ijk}\epsilon_{imn} = \delta_{jm}\delta_{kn} - \delta_{jn}\delta_{km}\,.$$

6.5.2 The SU(2) Group

A group of transformations, each of which transforms a two-dimensional column vector $x \to x' = Ax$, so that in component form

$$\begin{pmatrix} x_1' \\ x_2' \end{pmatrix} = \begin{pmatrix} a_{11} & a_{12} \\ a_{21} & a_{22} \end{pmatrix} \begin{pmatrix} x_1 \\ x_2 \end{pmatrix} , \tag{6.5.12}$$

where A is a unitary matrix ($AA^\dagger = A^\dagger A = I$) of determinant unity, is called an $SU(2)$ group. Let us take an infinitesimal transformation operator to be

$$A = I + dg\,. \tag{6.5.13}$$

Then the unitary nature of A implies that da is an anti-Hermitian matrix

$$(dg)^\dagger = -dg\,. \tag{6.5.14}$$

Then if we choose dg to be a traceless matrix of the form

$$\begin{pmatrix} ida_1 & da_2 + ida_3 \\ -da_2 + ida_3 & -ida_1 \end{pmatrix} \tag{6.5.15}$$

where $da_1\,, da_2\,, da_3$ are infinitesimal parameters, the requirements of unitarity of A, as well as its unimodular character ($\det A = 1$), are automatically met.

For an infinitesimal transformation $x \to x' \equiv x + dx = (I + dd)x$, we have $dx = dg\,x$ or

$$\begin{pmatrix} dx_1 \\ dx_2 \end{pmatrix} = \begin{pmatrix} ida_1 & da_2 + ida_3 \\ -da_2 + ida_3 & -ida_1 \end{pmatrix} \begin{pmatrix} x_1 \\ x_2 \end{pmatrix} . \tag{6.5.16}$$

This together with (6.4.11) gives

$$f_1(x, a) \equiv x_1' = x_1 + ia_1x_1 + (a_2 + ia_3)x_2\,, \tag{6.5.17a}$$

and

$$f_2(x, a) \equiv x_2' = x_2 + (-a_2 + ia_3)x_1 - ia_1x_2\,. \tag{6.5.17b}$$

Then according to Eq. (6.4.14), the generators of this group are:

$$X_1 = i\left\{\frac{\partial f_1}{\partial a_1}\frac{\partial}{\partial x_1} + \frac{\partial f_2}{\partial a_1}\frac{\partial}{\partial x_2}\right\} = -\left(x_1\frac{\partial}{\partial x_1} - x_2\frac{\partial}{\partial x_2}\right) , \tag{6.5.18}$$

$$X_2 = i\left\{\frac{\partial f_1}{\partial a_2}\frac{\partial}{\partial x_1} + \frac{\partial f_2}{\partial a_2}\frac{\partial}{\partial x_2}\right\} = i\left(x_2\frac{\partial}{\partial x_1} - x_1\frac{\partial}{\partial x_2}\right) , \tag{6.5.19}$$

$$X_3 = i\left\{\frac{\partial f_1}{\partial a_3}\frac{\partial}{\partial x_1} + \frac{\partial f_2}{\partial a_3}\frac{\partial}{\partial x_2}\right\} = -\left(x_2\frac{\partial}{\partial x_1} + x_1\frac{\partial}{\partial x_2}\right) . \tag{6.5.20}$$

It can be seen that the generators X_1, X_2, X_3 satisfy the following commutation relations

$$[X_k\,, X_\ell] = -2i\epsilon_{k\ell m}X_m\,, \tag{6.5.21}$$

where (k, ℓ, m) are a cyclic permutation of (1,2,3). Redefining the generators to be

$$\tau_k \equiv -\frac{1}{2}X_k\,, \tag{6.5.22}$$

we can rewrite the commutation relations as

$$[\tau_k\,, \tau_\ell] = i\epsilon_{k\ell m}\tau_m\,. \tag{6.5.23}$$

Thus we see that the three generators of the $SU(2)$ group satisfy commutation relations similar to those satisfied by the generators of the rotation group $R(3)$ [Eq. (6.5.10)].

Though the two groups are not *isomorphic*, in that there is no one-to-one correspondence between the elements of $SU(2)$ and $R(3)$ groups, there is a *homomorphic* mapping of $SU(2)$ onto $R(3)$. This means that every $SU(2)$ transformation corresponds to a unique rotation while every 3-dimensional rotation corresponds to a pair of $SU(2)$ transformations. A two-dimensional matrix representation of the generators τ_k can be in terms of the Pauli matrices

$$\tau_1 \equiv -\frac{X_1}{2} \rightarrow \frac{1}{2}\begin{pmatrix} 0 & 1 \\ 1 & 0 \end{pmatrix} \equiv \frac{1}{2}\sigma_1 \qquad (6.5.24a)$$

$$\tau_2 \equiv -\frac{X_2}{2} \rightarrow \frac{1}{2}\begin{pmatrix} 0 & -i \\ i & 0 \end{pmatrix} \equiv \frac{1}{2}\sigma_2 \qquad (6.5.24b)$$

$$\tau_3 \equiv -\frac{X_3}{2} \rightarrow \frac{1}{2}\begin{pmatrix} 1 & 0 \\ 0 & -1 \end{pmatrix} \equiv \frac{1}{2}\sigma_3 \qquad (6.5.24c)$$

Higher dimensional matrix representations are also possible.

In terms of the generators X_k or τ_k and the continuous parameters a_k, a typical SU(2) transformation operator (for a finite transformation) is given according to Eq. (6.4.16) by

$$U = \exp\left(-i\sum_{k=1}^{3} a_k X_k\right) . \qquad (6.5.25)$$

This transforms a state $\psi(x) \rightarrow \psi'(x) = U\psi(x)$.

SU(2) symmetry, i.e., invariance of the Hamiltonian under the group of SU(2) transformations, implies that each one of the generators of SU(2) corresponds to a conserved quantity.

6.5.3 Isospin and SU(2) Symmetry

The *isospin formalism* was developed by Heisenberg to incorporate the following observations into a mathematical formalism:

(a) If Coulomb interaction could be ignored, then the strong interaction between two nucleons or between a nucleon and a pion or between two pions is independent of the charge states of the interacting particles.

(b) The members of a charge multiplet, e.g., the neutron and proton or the pions (π^+, π^0, π^-) have very nearly the same mass.

To develop this formalism Heisenberg postulated the existence of an *isospin vector* in the fictitious isospin space, in analogy with the angular momentum vector $\boldsymbol{J}$ in the real space. In the quantum mechanical treatment, both these vectors correspond to vector operators. Heisenberg further postulated that the components of the isospin vector operator, $\hat{T}_1, \hat{T}_2, \hat{T}_3$, satisfy commutation relations similar to those satisfied by the components of angular momentum operator

$$[\hat{T}^2, \hat{T}_k] = 0 , \qquad (6.5.26a)$$

$$[\hat{T}_k, \hat{T}_\ell] = i\epsilon_{k\ell m}\hat{T}_m , \qquad (6.5.26b)$$

where k, ℓ, m are again a cyclic permutations of (1,2,3). (It may noted here that while spin or angular momentum has dimension of $\hbar$, isospin is dimensionless.)

Now, in physical space, rotational symmetry implies that the rotation leaves the Hamiltonian invariant, and the Hamiltonian commutes with the components of angular momentum. Since $\hat{H}$, $\hat{J}^2$, and $\hat{J}_z$ form a set of commuting observables, their simultaneous eigenstates $|n, j, m\rangle$, with $-j \leq m \leq j$, are given by

$$\hat{H} \; |n, j, m\rangle = E_{n,j} \; |n, j, m\rangle \; , \tag{6.5.27a}$$

$$\hat{J}^2 \; |n, j, m\rangle = \hbar^2 j(j+1) \; |n, j, m\rangle \; , \tag{6.5.27b}$$

$$\hat{J}_z \; |n, j, m\rangle = \hbar m \; |n, j, m\rangle \; . \tag{6.5.27c}$$

All these states have energies, which depend on the quantum numbers n and j but not on m. This degeneracy of the quantum states is a consequence of rotational symmetry. As is well known, this symmetry is violated when an external magnetic field, say, in z-direction, is applied. In that case the energies depend on m as well.

Carrying the analogy between isospin and angular momentum further, Heisenberg postulated that rotation of axes in the fictitious isospin space leaves the strong interaction Hamiltonian $\hat{H}_s$ (without Coulomb interaction) invariant and that this symmetry (called isospin symmetry) is violated when Coulomb or electromagnetic interaction is *switched on*. The invariance of the strong interaction Hamiltonian $\hat{H}_s$ under rotation in isospin space implies that $\hat{H}_s$ commutes with the rotation operators in the isospin space and hence with the components of the isospin vector $\hat{T}_1$, $\hat{T}_2$, $\hat{T}_3$. We can assume that $\hat{H}_s$, $\hat{T}^2$, $\hat{T}_3$ form a set of commuting observables and that they admit a set of simultaneous eigenstates $|E_t, t, t_3\rangle$ with $-t \leq t_3 \leq t$ which satisfy

$$\hat{H}_s \; |E_t, t, t_3\rangle = E_t \; |E_t, t, t_3\rangle \; , \tag{6.5.28a}$$

$$\hat{T}^2 \; |E_t, t, t_3\rangle = t(t+1) \; |E_t, t, t_3\rangle \; , \tag{6.5.28b}$$

$$\hat{T}_3 \; |E_t, t, t_3\rangle = t_3 \; |E_t, t, t_3\rangle \; . \tag{6.5.28c}$$

The set of particle states $|E_t, t, t_3\rangle$ forms an isospin (or charge) multiplet. These particle states have very nearly the same energy (mass), but different charges. This is so because, according to isospin symmetry, energy (mass) depends on t but not on t_3. This symmetry is, however, violated when the electromagnetic interaction is *switched on*, in which case the energy (mass) would depend on t_3 as well.

It may be pointed out here that the charge Q associated with a member of an isospin (or charge) multiplet is related to the third component of isospin (t_3) assigned to it by the relation

$$Q = t_3 + \frac{Y}{2} \, , \tag{6.5.29}$$

where Y is the *hypercharge* associated with the multiplet. The isospin t and the hypercharge Y have the same values for the whole multiplet, while different members of the isospin (or charge) multiplet are distinguished by different values of t_3. The assignment of isospin t to a charge multiplet can be made on the basis of its multiplicity $(2t+1)$. The nucleon doublet is assigned $t = 1/2$ while the pion triplet is assigned $t = 1$. There are several examples of charge (or isospin) multiplets in strongly interacting particles.

Coming back to the analogy between spin (angular momentum) and isospin, we note that, whereas in the former case the external magnetic field may be switched on or off, the electromagnetic interactions in the latter case are always there. But since the electromagnetic interaction is much weaker than the strong interaction, the actual energies (masses) of particle states belonging to an isospin multiplet would differ only slightly, compared to the ideal (hypothetical) case when the electromagnetic interaction is assumed to be *switched off* and there is no mass difference between the members of the isospin multiplet. That is why the masses of the particles associated with an isospin multiplet are nearly, but not exactly, equal.

Now, what has $SU(2)$ symmetry to do with isospin symmetry? We note that the generators of $SU(2)$ symmetry group and the generators of the 3-dimensional rotation group in the isospin space, viz. the components $\hat{T}_1$, $\hat{T}_2$, $\hat{T}_3$ of isospin, satisfy identical commutation relations [Eqs. (6.5.10) and (6.5.26)]

Further, according to isospin symmetry, the strong interaction does not distinguish between the members of an isospin multiplet, for example between the states of the nucleon doublet consisting of the proton (p) and the neutron (n)

$$|p\rangle \equiv |t = 1/2,\ t_3 = 1/2\rangle \text{ represented by } \begin{pmatrix} 1 \\ 0 \end{pmatrix} , \tag{6.5.30a}$$

$$|n\rangle \equiv |t = 1/2,\ t_3 = -1/2\rangle \text{ represented by } \begin{pmatrix} 0 \\ 1 \end{pmatrix} , \tag{6.5.30b}$$

or a linear combination of them, say

$$|N\rangle \equiv c_1\ |p\rangle + c_2\ |n\rangle \text{ represented by } \begin{pmatrix} c_1 \\ c_2 \end{pmatrix} . \tag{6.5.30c}$$

We can as well say that strong interaction is invariant under an operation which brings about the following transformation of states

$$\begin{pmatrix} 1 \\ 0 \end{pmatrix} \to \begin{pmatrix} 0 \\ 1 \end{pmatrix} \to \begin{pmatrix} c_1 \\ c_2 \end{pmatrix} . \tag{6.5.31}$$

But such a transformation may be brought about by a general SU(2) transformation

$$U = \exp\left(\sum_{k=1}^{3} a_k \tau_k\right) , \tag{6.5.32}$$

where the generators τ_1 , τ_2 , τ_3 have representations in terms of Pauli matrices [Eq. (6.5.24)].

Similarly, for the isospin triplet of pions π^+, π^0, and π^-,

$$\left|\pi^+\right\rangle \equiv |t = 1, t_3 = 1\rangle \qquad \text{represented by } \begin{pmatrix} 1 \\ 0 \\ 0 \end{pmatrix} , \tag{6.5.33a}$$

$$\left|\pi^0\right\rangle \equiv |t = 1, t_3 = 0\rangle \qquad \text{represented by } \begin{pmatrix} 0 \\ 1 \\ 0 \end{pmatrix} , \tag{6.5.33b}$$

$$\left|\pi^-\right\rangle \equiv |t = 1, t_3 = -1\rangle \qquad \text{represented by } \begin{pmatrix} 0 \\ 0 \\ 1 \end{pmatrix} , \tag{6.5.33c}$$

or a linear combination of these, say,

$$|\pi\rangle \equiv c_1|\pi^+ > +c_2|\pi^0 > +c_3|\pi^- > \text{ represented by } \begin{pmatrix} c_1 \\ c_2 \\ c_3 \end{pmatrix} , \tag{6.5.33d}$$

isospin symmetry implies that a unitary operation can bring about the transformation of states

$$\begin{pmatrix} 1 \\ 0 \\ 0 \end{pmatrix} \to \begin{pmatrix} 0 \\ 1 \\ 0 \end{pmatrix} \to \begin{pmatrix} 0 \\ 0 \\ 1 \end{pmatrix} \to \begin{pmatrix} c_1 \\ c_2 \\ c_3 \end{pmatrix} .$$

would leave the strong interaction Hamiltonian invariant. But this transformation again may be brought about by an $SU(2)$ transformation operator U [Eq. (6.5.32)] where, for the generators τ_k, we may now have a three dimensional matrix representation:

$$\tau_1 = \frac{1}{\sqrt{2}}\begin{pmatrix} 0 & 1 & 0 \\ 1 & 0 & 1 \\ 0 & 1 & 0 \end{pmatrix}, \quad \tau_2 = \frac{i}{\sqrt{2}}\begin{pmatrix} 0 & -1 & 0 \\ 1 & 0 & -1 \\ 0 & 1 & 0 \end{pmatrix}, \quad \tau_3 = \begin{pmatrix} 1 & 0 & 0 \\ 0 & 0 & 0 \\ 0 & 0 & -1 \end{pmatrix}. \tag{6.5.34}$$

The invariance of the strong interaction Hamiltonian under rotation of axes in the fictitious isospin space (isospin symmetry) or under an SU(2) transformation (SU(2)-symmetry) are equivalent statements and imply the commutation of the Hamiltonian with each one of the components of isospin or with each one of the generators of SU(2) group. This is also referred to as *conservation of isospin* in strong interactions.

Problems

1. Show that the following set of eight matrices

$$E = \begin{pmatrix} 1 & 0 \\ 0 & 1 \end{pmatrix}, \quad A = \begin{pmatrix} 0 & 1 \\ -1 & 0 \end{pmatrix}, \quad B = \begin{pmatrix} -1 & 0 \\ 0 & -1 \end{pmatrix}, \quad C = \begin{pmatrix} 0 & -1 \\ 1 & 0 \end{pmatrix},$$

$$D = \begin{pmatrix} 1 & 0 \\ 0 & -1 \end{pmatrix}, \quad F = \begin{pmatrix} -1 & 0 \\ 0 & 1 \end{pmatrix}, \quad G = \begin{pmatrix} 0 & -1 \\ -1 & 0 \end{pmatrix}, \quad H = \begin{pmatrix} 0 & 1 \\ 1 & 0 \end{pmatrix},$$

form a group under matrix multiplication. Construct the group multiplication table.

2. Define *homomorphism* in groups. Show that the symmetry group D of equilateral triangle [or the matrix group E', A', B', C', D', F' (Appendix 6A1)] is homomorphic onto the group consisting of $\{1, -1\}$ under multiplication.

3. Show that the invariance of the Hamiltonian of a system, under the group of rotation operations $\hat{R}(\alpha, \beta, \gamma)$, where α, β, γ are the Euler angles, results in degeneracy so that the simultaneous eigenstates $|n, j, m\rangle$ of $\hat{H}$, $\hat{J}^2$, $\hat{J}_z$, with $-j \leq m \leq j$ all correspond to the same energy.

4. The $SO(2)$ group is a group of all transformations wherein a transformation matrix A, which is real orthogonal and unimodular, transforms a vector

$$X = \begin{pmatrix} x_1 \\ x_2 \end{pmatrix} \text{ to } X' = \begin{pmatrix} x'_1 \\ x'_2 \end{pmatrix}.$$

Show that a typical transformation matrix

$$A = \begin{pmatrix} \cos\varphi & \sin\varphi \\ -\sin\varphi & \cos\varphi \end{pmatrix},$$

where φ is a parameter, can be expressed as $A = \exp(i\sigma_2\varphi)$, where $\sigma_2 = \begin{pmatrix} 0 & -i \\ i & 0 \end{pmatrix}$ is a Pauli matrix.

5. As a result of an $SO(2)$ transformation, a function $\psi(x,y)$ transforms to $\psi'(x,y) = \psi(x',y')$, so that $\psi'(x,y) = R(\varphi)\psi(x,y)$. Show that the generator of this transformation is given by $X = (x_2\frac{\partial}{\partial x_1} - x_1\frac{\partial}{\partial x_2})$. To make the generator Hermitian we can write it as $i\hbar X = -i\hbar(x_1\frac{\partial}{\partial x_2} - x_2\frac{\partial}{\partial x_1}) = L_3$. Then show that for a finite transformation, $R(\varphi)$ is given by $R(\varphi) = \exp(\varphi X) = \exp(-iL_3\varphi/\hbar)$.

6. Find the isospin t and the hypercharge Y associated with the charge quadruplet $(\Delta^{++}, \Delta^{+}, \Delta^{0}, \Delta^{-})$. Find the third component of isospin t_3 associated with each one of the charge states so as to justify the charge on them in accordance with the formula $Q = t_3 + \frac{Y}{2}$.

References

[1] J. P. Elliott and P. G. Dawber, *Symmetry in Physics* (Oxford University Press, New York, 1979), Chapter 3.

[2] M. Leon, *Particle Physics: An Introduction* (Academic Press, New York, 1973).

[3] D. P. Lichtenberg, *Unitary Symmetries and Elementary Particles* (Academic Press, New York, 1978).

[4] E. Merzbacher, *Quantum Mechanics* (John Wiley and Sons, New York, 1970), Chapter 12.

[5] J. McL. Emmerson, *Symmetry Principles in Particle Physics* (Clarendon Press, Oxford, 1972).

[6] A. Messiah, *Quantum Mechanics* (John Wiley and Sons, New York, 1977), Vol II, Chapters 13 and 15.

[7] L. I. Schiff, *Quantum Mechanics* (McGraw-Hill Book Company, New York, 1968), Chapter 7.

Appendix 6A1: Groups and Representations

A set of elements, $E, A, B, C, D, \cdots, X, Y$, which may all be numbers, matrices, vectors, symmetry operations, or some such entities, are said to form a group if, among them, a kind of operation, called *group multiplication* $\otimes$, is defined such that

(1) On *multiplying* an element A of the set with another element B, one gets an element C belonging to the same set: $A \otimes B = C$. This property is called *closure*.

(2) One of the elements of the set, say E, is the *identity* such that $E \otimes X = X = X \otimes E$ where X can be any element of the set. This property is called the existence of the identity.

(3) Every element X of the set admits an inverse $Y \equiv X^{-1}$, which is contained in the set, such that $X \otimes X^{-1} = E = X^{-1} \otimes X$. This property is called the existence of the *inverse*.

(4) The associative law of multiplication holds, i.e., $A \otimes (B \otimes C) = (A \otimes B) \otimes C$. This property is called *associativity* (of group multiplication).

Examples of groups include:

(a) The set of all positive and negative integers including zero forms a group under the operation of simple addition.

(b) The set of all $(n \times n)$ non-singular matrices including the unit matrix forms a group under matrix multiplication.

(c) The set of all three-dimensional vectors forms a group under vector addition.

(d) The symmetry operations of an equilateral triangle (E, A, B, C, D, F) form a group.

Multiplication of two operations means performing the two operations in succession.

Isomorphism and Homomorphism in Groups

Two groups are said to be *isomorphic* when a unique one-to-one correspondence exists between their elements so that the product of any two elements in the first group corresponds to the product of the corresponding elements in the second group. The elements of the two groups may consist of different entities with different rules of combination (multiplication), but they are said to be isomorphic if the above mentioned one-to-one correspondence exists between them. Obviously, two isomorphic groups have to be of the same order, i.e., both have to have the same number of elements. It follows that two isomorphic groups have the same multiplication table.

For example, the group D_3 of the symmetry elements of an equilateral triangle consisting of the elements E, A, B, C, D, F and the group of 2×2 matrices $M : \{E', A', B', C', D', F'\}$ under matrix multiplication, where,

$$E' = \begin{pmatrix} 1 & 0 \\ 0 & 1 \end{pmatrix}, \quad A' = \begin{pmatrix} 1 & 0 \\ 0 & -1 \end{pmatrix}, \quad B' = \begin{pmatrix} -\frac{1}{2} & \frac{\sqrt{3}}{2} \\ \frac{\sqrt{3}}{2} & \frac{1}{2} \end{pmatrix},$$
$$C' = \begin{pmatrix} -\frac{1}{2} & -\frac{\sqrt{3}}{2} \\ -\frac{\sqrt{3}}{2} & \frac{1}{2} \end{pmatrix}, \quad D' = \begin{pmatrix} -\frac{1}{2} & \frac{\sqrt{3}}{2} \\ -\frac{\sqrt{3}}{2} & -\frac{1}{2} \end{pmatrix}, \quad F' = \begin{pmatrix} -\frac{1}{2} & -\frac{\sqrt{3}}{2} \\ \frac{\sqrt{3}}{2} & -\frac{1}{2} \end{pmatrix}, \tag{6A1.1.35}$$

are isomorphic, and it may be verified that they have the same multiplication table [Table 6.1].

Homomorphism is a less sharp correspondence between two groups. A group G : $\{E, A, A', B, C, C', \cdots .\}$ is said to be homormorphic onto another group H : $\{\bar{E}, \bar{A}, \bar{C}\}$, if to every element of G there corresponds one and only one element of H, while to every element of H there corresponds at least one element of G but more than one element of G may also correspond to one element of H. The correspondence between the two groups has to be such that products correspond to products. Unlike isomorphism, homomorphism is not a reciprocal relationship. The number of elements in G must be greater than the number of elements in H if G is homomorphic onto H. One may also say that isomorphism is a special case of homomorphism.

Matrix Representation of a Group

The matrix representation of a group G is a matrix group M onto which the group G to be represented is homomorphic. In particular when the group G to be represented, and the matrix group, are isomorphic then the matrix representation is said to be faithful. For example, the matrix group M : $\{E'A'B'C'D'F'\}$ is a faithful representation of the symmetry group of an equilateral triangle. Each matrix group is clearly its own faithful representation. On the other hand, one may regard all elements of the group to correspond to a one-dimensional matrix group consisting of only the identity element $\{I\}$. This is because all the elements of the first group can be mapped on to the group consisting of the identity $\{I\}$. The latter may be called an unfaithful representation of the whole group D_3. We can also regard the elements E, F, D of the group D_3 to be mapped to one-dimensional matrix $\{I\}$ while the elements A, B, C are mapped on the one-dimensional matrix $\{-I\}$. This is because the operators (group elements) E, D, F stand for the rotations of the equilateral triangle about an axis through the center and perpendicular to the plane of the triangle through $0, 2\pi/3$ and $4\pi/3$, while A, B, C stand for reflections about the medians. The matrix group consisting of $+I$ and $-I$ is also an unfaithful representation of the whole group D_3.

The dimension of a matrix representation of a group is equal to the number of rows (or columns) in the matrices representing the group.

Reducible and Irreducible Representations of the Group

Starting from a matrix representation M : $\{D(A_1), D(A_2), \cdots, D(A_h)\}$ of a group G : $\{A_1, A_2, A_3, \cdots, A_h\}$ we can get another representation by subjecting all the matrices of this representation to the same similarity transformation. The new representation is equivalent to the former representation.

Now, from two different representations of the group, of different dimensions m and n, viz.,

$$M : \{D(A_1), D(A_2), \cdots, D(A_h)\} \qquad (m - \text{dimensional}), \tag{6A1.1.36}$$

and

$$M' : \{D'(A_1), D'(A_2), \cdots, D'(A_h)\} \qquad (n - \text{dimensional}), \tag{6A1.1.37}$$

it is possible to form an $(m+n)$-dimensional representation

$$\begin{pmatrix} D(A_1) & 0 \\ 0 & D'(A_1) \end{pmatrix}, \quad \begin{pmatrix} D(A_2) & 0 \\ 0 & D'(A_2) \end{pmatrix}, \cdots, \begin{pmatrix} D(A_h) & 0 \\ 0 & D'(A_h) \end{pmatrix}.$$

Furthermore, the matrices of this $(m+n)$-dimensional representation may all be subjected to a similarity transformation to give an equivalent representation. The resulting $(m+n)$-dimensional representation said to be *reducible* for the simple reason that the matrices of this representation may all be put in a block diagonal form by a similarity transformation and from this one can get two sets of representation matrices of (reduced) orders m and n.

A reducible representation of a group is one of which all the representative matrices can be reduced to a block diagonal form through a similarity (or unitary) transformation. When all the representative matrices of a group cannot be reduced to a block diagonal form by a similarity transformation then the matrix representation is said to be *irreducible*.

7

ANGULAR MOMENTUM IN QUANTUM MECHANICS

7.1 Introduction

The invariance of the Hamiltonian of a physical system under spatial rotation, together with the connection between the rotation operator $\hat{R}_{\boldsymbol{n}}(\theta)$ and the angular momentum operator $\hat{J} \equiv (\hat{J}_x, \hat{J}_y, \hat{J}_z)$, leads to the principle of conservation of angular momentum [Chapter 6, Sec. 6.3] and to the status of total angular momentum as a constant of motion. This makes angular momentum a very important quantity in quantum mechanics.

Commutation Rules for the Components of Angular Momentum

It is easy to verify that components of orbital angular momentum

$$\hat{L}_x = \hat{y}\hat{p}_z - \hat{z}\hat{p}_y \,, \tag{7.1.1a}$$

$$\hat{L}_y = \hat{z}\hat{p}_x - \hat{x}\hat{p}_z \,, \tag{7.1.1b}$$

$$\hat{L}_z = \hat{x}\hat{p}_y - \hat{y}\hat{p}_x \,, \tag{7.1.1c}$$

do not commute with one another. In fact, using the basic commutation relations between the observables $\hat{x}, \hat{y}, \hat{z}$ and $\hat{p}_x, \hat{p}_y, \hat{p}_z$, we find

$$[\hat{L}_x, \hat{L}_y] = i\hbar\hat{L}_z \,, \tag{7.1.2a}$$

$$[\hat{L}_y, \hat{L}_z] = i\hbar\hat{L}_x \,, \tag{7.1.2b}$$

$$[\hat{L}_z, \hat{L}_x] = i\hbar L_y \,. \tag{7.1.2c}$$

In vector notation, these relations may be written as

$$\hat{\mathbf{L}} \times \hat{\mathbf{L}} = i\hbar\hat{\mathbf{L}} \,. \tag{7.1.3}$$

These commutation relations are characteristic, not only of the components of orbital angular momentum, but also of the components of total angular momentum. We can in fact show, quite independently, that the generators $\hat{J}_x, \hat{J}_y, \hat{J}_z$ of the rotation group [a typical operator being $\hat{R}_{\mathbf{n}}(\theta) \equiv \exp(-i\mathbf{n}\cdot\hat{\mathbf{J}}\,\theta/\hbar)$], which are the components of total angular momentum, obey the same commutation relations as the components of orbital angular momentum do. To see this, let us consider the system to be consisting of a particle P, whose distance from the origin is unity. Let us apply two infinitesimal clockwise rotations about the x and y-axes, respectively, and investigate the difference which arises by applying them in different order. We adopt the active view of rotation. This means positive rotations of the system are taken to be clockwise. Let P be a point situated on x-axis a unit distance from the origin so that its Cartesian coordinates are $(1, 0, 0)$. First consider the case when the rotation about the x-axis precedes that about the y-axis [Fig. 7.1]. The first rotation

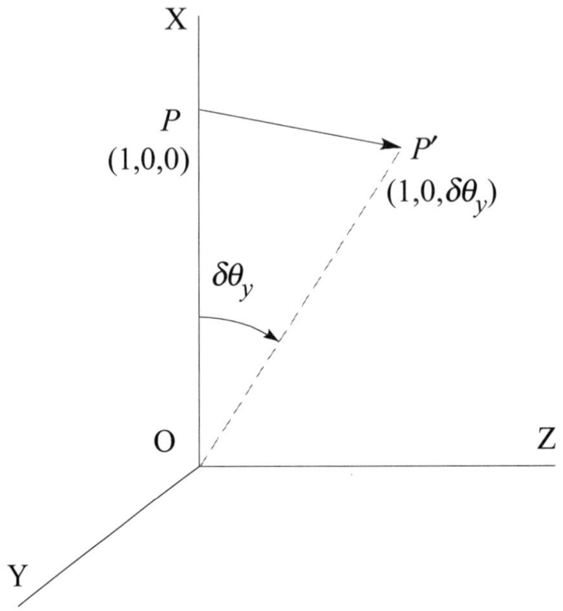

FIGURE 7.1
Displacement of the particle from $P \rightarrow P'$ when rotation about the x-axis through angle $\delta\theta_x$ precedes that about the y-axis through angle $\delta\theta_y$. Note that we are using the active view of rotations in which the axes stay fixed but the system is rotated. This means a counter-clockwise rotation of coordinate axes (with the system fixed) is equivalent to a clockwise rotation of the system by the same amount (with the axes fixed).

about the x-axis leaves P unchanged while the second about the y-axis brings it P' in the xz plane. Next consider the case when rotation about the y-axis precedes that about the x-axis [Fig. 7.1], so that $P \rightarrow P' \rightarrow P''$. It is clear from the figures that the difference between the two sets of operations, with the order of rotations interchanged, is a net displacement of the particle of magnitude $\delta\theta_x\delta\theta_y$ in the xy plane. This displacement is equivalent to a counter-clockwise rotation about the z-axis of magnitude $\delta\theta_x\delta\theta_y$.

Mathematically, the difference between these infinitesimal rotations can be expressed as

$$\left[\hat{R}_y(\delta\theta_y)\hat{R}_x(\delta\theta_x) - \hat{1}\right] - \left[\hat{R}_x(\delta\theta_x)\hat{R}_y(\delta\theta_y) - \hat{1}\right] = \left[\hat{R}_z(-\delta\theta_x\delta\theta_y) - \hat{1}\right] . \tag{7.1.4}$$

Using $\hat{R}_x(\delta\theta_x) = \left(1 - i\delta\theta_x\hat{J}_x/\hbar\right)$ for an infinitesimal rotation, and similar expressions for $\hat{R}_y$ and $\hat{R}_z$ in Eq.(7.1.4), and expanding the resuting expressions on both sides, being careful about the order of operator multiplication, we obtain

$$\frac{1}{\hbar^2}\delta\theta_x\delta\theta_y(\hat{J}_x\hat{J}_y - \hat{J}_y\hat{J}_x) = \frac{i}{\hbar}\delta\theta_x\delta\theta_y\hat{J}_z ,$$

or

$$\hat{J}_x\hat{J}_y - \hat{J}_y\hat{J}_x = i\hbar\hat{J}_z . \tag{7.1.5a}$$

Similarly, by considering the difference between two sets of infinitesimal rotations about y and z-axes in different order (and also about z- and x-axes) we can get the commutation

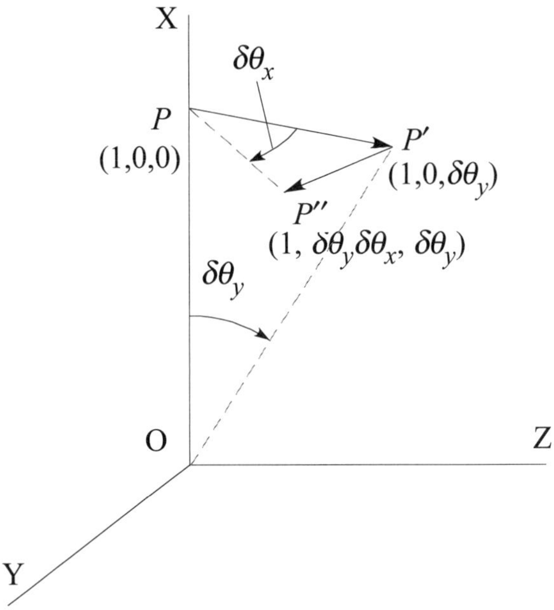

FIGURE 7.2
Displacement of the particle from $P \to P' \to P''$ when rotation about the y-axis through angle $\delta\theta_y$ precedes that about the x-axis through angle $\delta\theta_x$.

relations

$$[\hat{J}_y\hat{J}_z - \hat{J}_z\hat{J}_y] = i\hbar\hat{J}_x\,, \tag{7.1.5b}$$

and

$$[\hat{J}_y\hat{J}_z - \hat{J}_z\hat{J}_y] = i\hbar\hat{J}_x\,. \tag{7.1.5c}$$

Although the components of angular momentum do not commute among themselves, it is easily verified, using the commutation relations (7.1.5), that each Cartesian component of angular momentum operator commutes with square of the angular momentum operator $\hat{\mathbf{J}}^2 = \hat{J}_x^2 + \hat{J}_y^2 + \hat{J}_z^2$:

$$[\hat{J}^2, \hat{J}_x] = [\hat{J}^2, \hat{J}_y] = [\hat{J}^2, \hat{J}_z] = 0\,. \tag{7.1.6}$$

In view of the commutation relations (7.1.5) and (7.1.6), we can choose one of the components of angular momentum, say $\hat{J}_z$, to construct simultaneous eigenstates of the set of commuting observables $\hat{J}^2$ and $\hat{J}_z$. If we label the simultaneous eigenstates of $\hat{J}^2$ and $\hat{J}_z$ as $|\eta, m\rangle$, then

$$\hat{J}^2\,|\eta, m\rangle = \eta\hbar^2\,|\eta, m\rangle\;, \tag{7.1.7}$$

$$J_z\,|\eta, m\rangle = m\hbar\,|\eta, m\rangle\;. \tag{7.1.8}$$

What are the sets of values that η and m can take? We shall discuss this in the next section.

In what follows, we will use $\hbar$ as unit of angular momentum so that it will not appear explicitly in equations involving angular momentum operators. We can restore it at the end of the calculations using dimensional arguments.

7.2 Raising and Lowering Operators

Let us introduce two operators $\hat{J}_+$ and $\hat{J}_-$ defined by

$$\hat{J}_+ = \hat{J}_x + i\hat{J}_y \,, \tag{7.2.1}$$

$$\hat{J}_- = \hat{J}_x - i\hat{J}_y \,. \tag{7.2.2}$$

It is easy to check that $\hat{J}_+$ and $\hat{J}_-$ are not self-adjoint operators. On the other hand $\hat{J}_-$ is the adjoint of $\hat{J}_+$ and vice versa. Using the definition of $\hat{J}_\pm$ and Eq. (7.1.5), the following commutation relations can be established

$$[\hat{J}^2, \hat{J}_+] = [\hat{J}^2, \hat{J}_-] = 0 \,, \tag{7.2.3}$$

$$[\hat{J}_z, \hat{J}_+] = +\hat{J}_+ \,, \tag{7.2.4}$$

$$[\hat{J}_z, \hat{J}_-] = -\hat{J}_- \,, \tag{7.2.5}$$

$$[\hat{J}_+, \hat{J}_-] = 2\hat{J}_z \,. \tag{7.2.6}$$

According to the properties of linear operators, $\hat{J}_+$ and $\hat{J}_-$ operating on the state $|\eta, m\rangle$ would yield states which are different from the original state. Let us examine whether the new states $\hat{J}_\pm |\eta, m\rangle$ are still the eigenstates of $\hat{J}^2$ and $\hat{J}_z$.

From Eqs. (7.2.3), we have

$$\hat{J}^2\hat{J}_+ |\eta, m\rangle = \hat{J}_+\hat{J}^2 |\eta, m\rangle = \eta J_+ |\eta, m\rangle \,, \tag{7.2.7}$$

and

$$\hat{J}^2\hat{J}_- |\eta, m\rangle = \hat{J}_-\hat{J}^2 |\eta, m\rangle = \eta J_- |\eta, m\rangle \,. \tag{7.2.8}$$

Thus $J_+ |\eta, m\rangle$ and $J_- |\eta, m\rangle$ remain eigenstates of $\hat{J}^2$ with the same eigenvalue η. However, from Eqs. (7.2.4) and (7.2.5) we have

$$\hat{J}_z\hat{J}_+ |\eta, m\rangle = (\hat{J}_+\hat{J}_z + \hat{J}_+) |\eta, m\rangle = (m+1)\hat{J}_+ |\eta, m\rangle \,, \tag{7.2.9}$$

and

$$\hat{J}_z\hat{J}_- |\eta, m\rangle = (\hat{J}_-\hat{J}_z - \hat{J}_-) |\eta, m\rangle = (m-1)\hat{J}_- |\eta, m\rangle \,. \tag{7.2.10}$$

This means that $\hat{J}_+ |\eta, m\rangle$ is an eigenstate of $\hat{J}_z$ belonging to the eigenvalue $(m+1)$ while $\hat{J}_- |\eta, m\rangle$ is an eigenstate of $\hat{J}_z$ belonging to the eigenvalue $(m-1)$. Thus apart from a multiplying constant, we can interpret the states $\hat{J}_+ |\eta, m\rangle$ and $\hat{J}_- |\eta, m\rangle$ as $|\eta, m+1\rangle$ and $|\eta, m-1\rangle$, respectively. The operators $\hat{J}_+$ and $\hat{J}_-$ are termed as the 'raising' and 'lowering' operators, respectively. By applying the raising operator to a state $|\eta, m\rangle$ in succession we can construct states with m increasing in steps of unity. Similarly, by applying the lowering operator to the state $|\eta, m\rangle$ in succession, we can construct successive states with m decreasing in steps of unity. Hence, in the set of states $|\eta, m\rangle$, the m-values (eigenvalues of $\hat{J}_z$) are spaced by unity. It is natural to ask if the sequence of m values is bounded. Let us investigate this further.

Using the definition $\hat{J}^2 = \hat{J}_x^2 + \hat{J}_y^2 + \hat{J}_z^2$ in Eq. (7.1.6) we get

$$(\hat{J}_x^2 + \hat{J}_y^2) |\eta, m\rangle \equiv (\hat{J}^2 - \hat{J}_z^2) |\eta, m\rangle = (\eta - m^2) |\eta, m\rangle \,. \tag{7.2.11}$$

Since $\hat{J}_x^2 + \hat{J}_y^2$ is the sum of the squares of Hermitian operators, its eigenvalue $(\eta - m^2)$ must be non-negative, which means

$$m^2 \leq \eta \quad \text{or} \quad -\sqrt{\eta} \leq m \leq \sqrt{\eta} \,. \tag{7.2.12}$$

Thus the spectrum of m is bounded both from above and below. Let us denote the upper and lower limits of m by m_2 and m_1, respectively. Then, according to the concept of raising and lowering operators, the lower and upper limits of m must be such that

$$\hat{J}_+ |\eta, m_2\rangle = 0 \quad \text{or} \quad \hat{J}_-\hat{J}_+ |\eta, m_2\rangle = 0\,, \tag{7.2.13a}$$

$$\hat{J}_- |\eta, m_1\rangle = 0 \quad \text{or} \quad \hat{J}_+\hat{J}_- |\eta, m_-\rangle = 0\,. \tag{7.2.13b}$$

Using the commutation relations for the components of angular momentum, we can establish that

$$\hat{J}_-\hat{J}_+ = \hat{J}_x^2 + \hat{J}_y^2 + i(\hat{J}_x\hat{J}_y - \hat{J}_y\hat{J}_x) = \hat{J}^2 - \hat{J}_z^2 - \hat{J}_z\,, \tag{7.2.14}$$

$$\hat{J}_+\hat{J}_- = \hat{J}_x^2 + \hat{J}_y^2 - i(\hat{J}_x\hat{J}_y - \hat{J}_y\hat{J}_x) = \hat{J}^2 - \hat{J}_z^2 + \hat{J}_z\,. \tag{7.2.15}$$

Hence Eq. (7.2.13) imply that

$$\eta - m_2^2 - m_2 = 0 = \eta - m_1^2 + m_1\,. \tag{7.2.16}$$

This yields $m_1 = -m_2$ or $m_1 = m_2 + 1$. The latter value for m_1 is obviously absurd since the lower limit cannot exceed the upper limit. Hence the lower and upper bounds of m must satisfy $m_1 = -m_2$. If we rename the upper limit m_2 as j, then $m_1 = -j$. Using this in Eq. (7.2.16), the eigenvalue of $\hat{J}^2$ can be written as

$$\eta = j(j+1)\,, \tag{7.2.17}$$

and j can be called the total angular momentum quantum number for the state while m can be called the magnetic quantum number.The state itself can be labeled by the quantum numbers j and m as $|j,m\rangle$ with

$$\hat{J}^2\, |j,m\rangle = j(j+1)\, |j,m\rangle\,, \tag{7.2.18a}$$

$$\hat{J}_z\, |j,m\rangle = m\, |j,m\rangle\,. \tag{7.2.18b}$$

Since we can go from $m = j$ to $m = -j$ in a total of $2j+1$ steps, it follows that $2j+1$ must be a positive integer and therefore, j must be an integer or a half integer. The magnetic quantum number m can then take any value from $+j$ to $-j$ in integer steps.

We can now express the states $\hat{J}_+\, |j,m\rangle$ and $\hat{J}_-\, |j,m\rangle$ from Eqs. (7.2.9) and (7.2.10) as $|j, m+1>$ and $|j, m-1>$, respectively, apart from some multiplying constants

$$\hat{J}_+\, |j,m\rangle = \Gamma_+^{jm}\, |j,m+1\rangle\,, \tag{7.2.19a}$$

$$\hat{J}_-\, |j,m\rangle = \Gamma_-^{jm}\, |j,m-1\rangle\,. \tag{7.2.19b}$$

To determine the constant Γ_+^{jm}, we multiply each side of Eq. (7.2.19a) by its conjugate imaginary from the left and use Eq. (7.2.14) to get

$$\begin{aligned} |\Gamma_+^{jm}|^2 &= \langle jm|\, \hat{J}_-\hat{J}_+\, |jm\rangle = \langle jm|\, \hat{J}^2 - \hat{J}_z^2 - \hat{J}_z\, |jm\rangle\,, \\ &= j(j+1) - m^2 - m = (j-m)(j+m+1)\,, \end{aligned}$$

or

$$|\Gamma_+^{jm}| = \Gamma_+^{jm} = \sqrt{(j-m)(j+m+1)}\,, \tag{7.2.20}$$

where we have taken the constant Γ_+^{jm} to be real assuming the phase factor to be 1. Proceeding in a similar manner with Eq. (7.2.19b), the constant Γ_-^{jm} is found to be

$$\Gamma_-^{jm} = \sqrt{(j+m)(j-m+1)}\,, \tag{7.2.21}$$

where we have chosen Γ_-^{jm} to be real, again assuming the phase factor to be 1.

7.3 Matrix Representation of Angular Momentum Operators

We can set up a matrix representation for angular momentum operators. To set up this representation we choose the set of commuting observables $\hat{J}^2$ and $\hat{J}_z$ to be diagonal and the basis states to be the set of their simultaneous eigenstates $|j,m\rangle$, where $-j \le m \le j$. In this representation any observable $\hat{X}$, which may stand for any of the operators $\hat{J}_x, \hat{J}_y, \hat{J}_z, \hat{J}_+, \hat{J}_-, \hat{J}^2$ or any combination of them, may be represented by a $(2j+1)\times(2j+1)$ matrix as

$$\hat{X} \to [X_{mm'}] = \begin{pmatrix} X_{j,j} & X_{j,j-1} & X_{j,j-2} & \cdots & X_{j,-j} \\ X_{j-1,j} & X_{j-1,j-1} & X_{j-1,j-2} & \cdots & X_{j-1,-j} \\ \vdots & \vdots & \vdots & \vdots & \vdots \\ X_{-j,j} & X_{-j,j-1} & X_{-j,j-2} & \cdots & X_{-j,-j} \end{pmatrix}, \tag{7.3.1}$$

where a typical matrix element is $X_{mm'} = \langle jm|\, \hat{X}\, |jm'\rangle$. Obviously, the operators $\hat{J}^2$ and $\hat{J}_z$ are represented by the diagonal matrices,

$$\hat{J}^2 \to [J^2_{mm'}] = \begin{pmatrix} j(j+1) & 0 & \cdots & 0 \\ 0 & j(j+1) & \cdots & 0 \\ \vdots & \vdots & \vdots & \vdots \\ 0 & \cdots & j(j+1) & 0 \\ 0 & \cdots & 0 & j(j+1) \end{pmatrix} \equiv j(j+1)\, I\,, \tag{7.3.2a}$$

with

$$J^2_{mm'} \equiv \langle jm|\, \hat{J}^2\, |jm'\rangle = j(j+1)\, \delta_{mm'}\,, \tag{7.3.2b}$$

and

$$\hat{J}_z \to [(J_z)_{mm'}] = \begin{pmatrix} j & 0 & 0 & \cdots & 0 \\ 0 & j-1 & 0 & \cdots & 0 \\ \vdots & \vdots & \vdots & \vdots & \vdots \\ 0 & 0 & \cdots & -(j-1) & 0 \\ 0 & 0 & \cdots & 0 & -j \end{pmatrix}, \tag{7.3.3a}$$

with

$$(J_z)_{mm'} \equiv \langle jm|\, \hat{J}_z\, |jm'\rangle = m'\, \delta_{mm'} \tag{7.3.3b}$$

where I is the identity matrix.

To construct the matrices representing $\hat{J}_x$ and $\hat{J}_y$ in this representation we first construct the matrices representing $\hat{J}_+$ and $\hat{J}_-$. These follow immediately from Eqs. (7.2.19), (7.2.20) and (7.2.21) to be

$$(J_+)_{mm'} \equiv \langle jm|\, \hat{J}_+\, |jm'\rangle = \sqrt{(j-m')(j+m'+1)}\;\delta_{m,m'+1}\,, \tag{7.3.4}$$

$$(J_-)_{mm'} \equiv \langle jm|\, \hat{J}_-\, |jm'\rangle = \sqrt{(j+m')(j-m'+1)}\;\delta_{m,m'-1}\,. \tag{7.3.5}$$

In terms of these matrices, those representing $\hat{J}_x$ and $\hat{J}_y$ may easily be determined by using the relations

$$\hat{J}_x = \frac{1}{2}\left(\hat{J}_+ + \hat{J}_-\right), \tag{7.3.6a}$$

$$\hat{J}_y = \frac{1}{2i}\left(\hat{J}_+ - \hat{J}_-\right). \tag{7.3.6b}$$

Pauli Spin Matrices

As a simple example let us determine the matrices representing the angular momentum operators for the total angular momentum quantum number $j = 1/2$. In this case $2j+1 = 2$ so that the matrices representing the angular momentum operators are 2×2 matrices [with the notation $X_{mm'} = \langle \frac{1}{2}, m | \hat{X} | \frac{1}{2}, m' \rangle$]

$$\hat{J}^2 \to \begin{pmatrix} j(j+1) & 0 \\ 0 & j(j+1) \end{pmatrix} = \begin{pmatrix} 3/4 & 0 \\ 0 & 3/4 \end{pmatrix} = \frac{3}{4}\begin{pmatrix} 1 & 0 \\ 0 & 1 \end{pmatrix} = \frac{3}{4} I \,, \tag{7.3.7}$$

$$\hat{J}_z \to \begin{pmatrix} j & 0 \\ 0 & -j \end{pmatrix} = \begin{pmatrix} 1/2 & 0 \\ 0 & -1/2 \end{pmatrix} = \frac{1}{2}\begin{pmatrix} 1 & 0 \\ 0 & -1 \end{pmatrix} \equiv \frac{1}{2}\sigma_z \,, \tag{7.3.8}$$

$$\hat{J}_+ \to \begin{pmatrix} \langle \frac{1}{2}, \frac{1}{2} | \hat{J}_+ | \frac{1}{2}, \frac{1}{2} \rangle & \langle \frac{1}{2}, \frac{1}{2} | \hat{J}_+ | \frac{1}{2}, -\frac{1}{2} \rangle \\ \langle \frac{1}{2}, -\frac{1}{2} | \hat{J}_+ | \frac{1}{2}, \frac{1}{2} \rangle & \langle \frac{1}{2}, -\frac{1}{2} | \hat{J}_+ | \frac{1}{2}, \frac{1}{2} \rangle \end{pmatrix} = \begin{pmatrix} 0 & 1 \\ 0 & 0 \end{pmatrix} , \tag{7.3.9}$$

$$\hat{J}_- \to \begin{pmatrix} \langle \frac{1}{2}, \frac{1}{2} | \hat{J}_- | \frac{1}{2}, \frac{1}{2} \rangle & \langle \frac{1}{2}, \frac{1}{2} | \hat{J}_- | \frac{1}{2}, -\frac{1}{2} \rangle \\ \langle \frac{1}{2}, -\frac{1}{2} | \hat{J}_- | \frac{1}{2}, \frac{1}{2} \rangle & \langle \frac{1}{2}, -\frac{1}{2} | \hat{J}_- | \frac{1}{2}, -\frac{1}{2} \rangle \end{pmatrix} = \begin{pmatrix} 0 & 0 \\ 1 & 0 \end{pmatrix} . \tag{7.3.10}$$

From Eqs. (7.3.6), (7.3.9) and (7.3.10), we find that the matrix representations of $\hat{J}_x$ and $\hat{J}_y$ for $j = 1/2$ are

$$J_x = \frac{1}{2}\left(\hat{J}_+ + \hat{J}_-\right) \to \frac{1}{2}\begin{pmatrix} 0 & 1 \\ 1 & 0 \end{pmatrix} \equiv \frac{1}{2}\sigma_x \,, \tag{7.3.11}$$

and

$$J_y = \frac{1}{2i}\left(\hat{J}_+ - \hat{J}_-\right) \to \frac{1}{2}\begin{pmatrix} 0 & -i \\ i & 0 \end{pmatrix} \equiv \frac{1}{2}\sigma_y \,. \tag{7.3.12}$$

The matrices $\sigma_x, \sigma_y, \sigma_z$ are called Pauli spin matrices. For a spin-half particle, the spin operator $\hat{\mathbf{S}} \equiv (\hat{S}_x, \hat{S}_y, \hat{S}_z)$ can be represented (in a representation in which $\hat{S}^2$ and $\hat{S}_z$ are diagonal) by

$$\hat{\mathbf{S}} = \frac{\hbar}{2}\boldsymbol{\sigma} \,, \tag{7.3.13}$$

where $\boldsymbol{\sigma}$ is called the Pauli spin vector, whose components are the Pauli spin matrices. Hence $S_x = \hbar\sigma_x/2, S_y = \hbar\sigma_y/2, S_z = \hbar\sigma_z/2$. It may be noted here that for spin operators and spin states, coordinate representation is not possible; the only possible representation is the matrix representation.

7.4 Matrix Representation of Eigenstates of Angular Momentum

A simultaneous eigenstate $|j, m_o\rangle$ of $\hat{J}^2$ and $\hat{J}_z$ may be represented by a column vector in a representation in which $\hat{J}^2$ and $\hat{J}_z$ are diagonal and the basis states are $|j, m\rangle$ with $-j \leq m \leq j$ as

$$|j, m_o\rangle \to \begin{pmatrix} \langle j, j | j, m_o \rangle \\ \langle j, j-1 | j, m_o \rangle \\ \langle j, j-2 | j, m_o \rangle \\ \vdots \\ \langle j, -j | j, m_o \rangle \end{pmatrix} \equiv \begin{pmatrix} \delta_{j, m_o} \\ \delta_{j-1, m_o} \\ \delta_{j-2, m_o} \\ \vdots \\ \delta_{-j, m_o} \end{pmatrix} . \tag{7.4.1}$$

It is obvious that only one of the elements of the representative column matrix is non-zero and its value is unity; all others are zero. For example, the states $|j, j\rangle$ and $|j, -j\rangle$ are represented as

$$|j, m = j\rangle \rightarrow \begin{pmatrix} 1 \\ 0 \\ 0 \\ \vdots \\ 0 \end{pmatrix} \quad \text{and} \quad |j, m = -j\rangle \rightarrow \begin{pmatrix} 0 \\ 0 \\ 0 \\ \vdots \\ 1 \end{pmatrix} .$$

As a special case, consider the representation of angular momentum states for $j = 1/2$. These states, referred to as the spin states, have the following representation

$$\left| j = \frac{1}{2}, m = \frac{1}{2} \right\rangle \rightarrow \begin{pmatrix} 1 \\ 0 \end{pmatrix} \equiv \alpha \,, \tag{7.4.2}$$

$$\left| j = \frac{1}{2}, m = -\frac{1}{2} \right\rangle \rightarrow \begin{pmatrix} 0 \\ 1 \end{pmatrix} \equiv \beta \,. \tag{7.4.3}$$

The spin states $\alpha \equiv \begin{pmatrix} 1 \\ 0 \end{pmatrix}$ and $\beta \equiv \begin{pmatrix} 0 \\ 1 \end{pmatrix}$ are also referred to as *spin-up* and *spin-down* states, respectively. Since $\hat{J}^2$ and $\hat{J}_z$ for $j = 1/2$ have the matrix representation $\hat{J}^2 \rightarrow \frac{3}{4} I$ and $\hat{J}_z \rightarrow \frac{1}{2}\sigma_z$, corresponding to the eigenvalue equations in Dirac notation,

$$\hat{J}^2 \left| \frac{1}{2}, \pm\frac{1}{2} \right\rangle = \frac{3}{4} \left| \frac{1}{2}, \pm\frac{1}{2} \right\rangle , \tag{7.4.4}$$

$$\hat{J}_z \left| \frac{1}{2}, \pm\frac{1}{2} \right\rangle = \pm\frac{1}{2} \left| \frac{1}{2}, \pm\frac{1}{2} \right\rangle , \tag{7.4.5}$$

we have the matrix equations

$$\hat{J}^2 \alpha \equiv \frac{3}{4} I \alpha = \frac{3}{4} \alpha \,, \tag{7.4.6a}$$

$$\hat{J}^2 \beta \equiv \frac{3}{4} I \beta = \frac{3}{4} \beta \,, \tag{7.4.6b}$$

$$\hat{J}_z \alpha \equiv \frac{1}{2} \sigma_z \alpha = \frac{1}{2} \alpha \,, \tag{7.4.7a}$$

$$\hat{J}_z \beta \equiv \frac{1}{2} \sigma_z \beta = -\frac{1}{2} \beta \,. \tag{7.4.7b}$$

The physical content of the two sets of equations, (7.4.4) and (7.4.5) on the one hand and Eqs. (7.4.6) and (7.4.7) on the other, is the same. Thus the state $|1/2, 1/2\rangle$ represented by matrix α is the simultaneous eigenstate of $\hat{J}^2$ and $\hat{J}_z$, belonging to the eigenvalues $j(j+1) = 3/4$ and $m = 1/2$, respectively. Likewise the state $|1/2, -1/2\rangle$ represented by matrix β is the simultaneous eigenstate of $\hat{J}^2$ and $\hat{J}_z$ belonging to the eigenvalues $j(j+1) = 3/4$ and $m = -1/2$, respectively, the angular momentum being expressed in units of $\hbar$.

Thus for an electron or any spin-half particle, there are two alternative spin states: the spin up state $|1/2, 1/2\rangle$, represented by $\begin{pmatrix} 1 \\ 0 \end{pmatrix} \equiv \alpha$ and spin down state $|1/2, -1/2\rangle$ represented by matrix $\begin{pmatrix} 0 \\ 1 \end{pmatrix} \equiv \beta$. The direct evidence for the spin of the electron and for the fact that it can exist in two alternative spin states came from the Stern-Gerlach experiment.

Stern-Gerlach Experiment

As is well known, the magnetic moment of an electron due to its *orbital motion* is

$$\boldsymbol{\mu}_L = -\frac{e}{2m_e} g_L \boldsymbol{L} \tag{7.4.8}$$

where $\boldsymbol{L}$ is the orbital angular momentum of the electron and the gyromagnetic ratio g_L [1] pertaining to orbital motion is 1. The negative sign in $\boldsymbol{\mu}_L$ reflects the fact that for electrons $\boldsymbol{\mu}$ is directed opposite to $\boldsymbol{L}$. On account of its intrinsic (spin) angular momentum, the electron also has an intrinsic magnetic moment, which can be similarly written down as

$$\boldsymbol{\mu}_S = -\frac{e}{2m_e} g_S \boldsymbol{S}, \tag{7.4.9}$$

where g_S is the gyromagnetic ratio pertaining to spin and is assigned the value 2 in accordance with experimental data and Dirac theory [Chapter 12]. Hence the total magnetic moment of the electron is given by

$$\boldsymbol{\mu} = \boldsymbol{\mu}_L + \boldsymbol{\mu}_S = -\frac{e}{2m_e}(2\boldsymbol{S} + \boldsymbol{L}). \tag{7.4.10}$$

For a silver atom in its ground state, the angular momentum as well as the magnetic moment, is due to the spin of the valence electron. Putting $\boldsymbol{L} = 0$ and $\boldsymbol{S} = \hbar\boldsymbol{\sigma}/2$, we have

$$\boldsymbol{\mu} = -\mu_B \boldsymbol{\sigma}, \tag{7.4.11}$$

where $\mu_B = e\hbar/mc$ is Bohr magneton (atomic unit of magnetic moment). In the Stern-

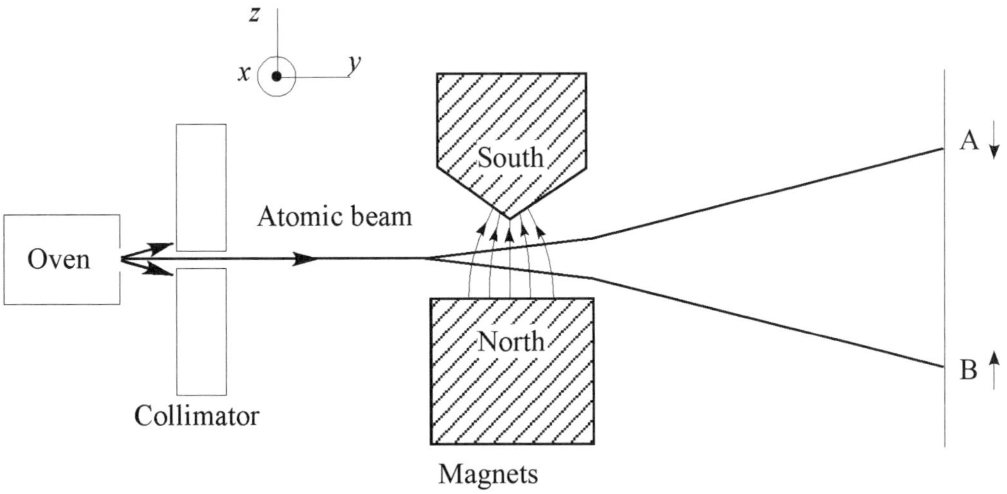

FIGURE 7.3
The set-up for the Stern-Gerlach experiment.

[1]The gyromagnetic ratio g is defined to be the ratio of the magnetic moment in units of Bohr magneton ($\mu_B = e\hbar/2m_e$) to the corresponding angular momentum (in units of $\hbar$). Thus corresponding to orbital and spin angular momenta we have $g_L \equiv \frac{\mu_L/\mu_B}{L/\hbar}$ and $g_S = \frac{\mu_S/\mu_B}{S/\hbar}$.

Gerlach experiment an inhomogeneous magnetic field is produced predominantly along the z-direction, perpendicular to the direction of a collimated beam of silver atoms. The gradient of the magnetic field created by the special design of magnetic poles is along the z-direction as shown in Fig. 7.3. Then the inhomogeneous field between the pole pieces in the vicinity of some suitably chosen origin can be written as

$$\boldsymbol{B}(z) = \boldsymbol{B}_0 + z\frac{\partial B}{\partial z}\boldsymbol{e}_z \,, \tag{7.4.12}$$

where $\boldsymbol{B}_0 = \boldsymbol{B}(0)$ is the value of the field at the origin and the field gradient $\frac{\partial B}{\partial z} \equiv \left(\frac{\partial B}{\partial z}\right)_{z=0}$ is assumed to be constant between the pole pieces. If a particle with magnetic moment $\boldsymbol{\mu}$ [Eq. (7.4.11)] enters the region of the inhomogeneous magnetic field, it will experience a force given by

$$\mathbf{F} = -\nabla(-\boldsymbol{\mu}\cdot\boldsymbol{B}) = \mu_z\left(\frac{\partial B}{\partial z}\right) = -\mu_B\sigma_z\left(\frac{\partial B}{\partial z}\right) . \tag{7.4.13}$$

The force on the particle, which lasts for a short time while the particle passes through the inhomogeneous magnetic field, will direct particles in different directions depending on the value of σ_z (the projection of spin along the z-axis). An atom with its magnetic moment pointing along the $+z$ axis will be deflected upward, whereas an atom whose magnetic moment was pointed in the $-z$ would be deflected downward. Thus the atoms exiting the magnetic field would be spread out according to the z-component their magnetic moment leaving a record on the photographic plate. Classically, since all orientations of the magnetic moment are possible, we would expect a smear of silver atoms on the photographic plate. However, it was found that the silver atoms accumulated only at two distinct positions A and B on the photographic plate indicating that there are only two possible values of σ_z : $+1$ and -1 [Fig. 7.3]. Silver atoms with $\sigma_z = +1$ (spin up) were deflected toward B and those with $\sigma_z = -1$ were deflected toward A (since for the electrons the magnetic moment is directed opposite to their spin).

When the Stern-Gerlach experiment was repeated with the field and field gradient along x-direction, which was also perpendicular to the beam of silver atoms, the silver atoms accumulated at two distinct spots along the x-direction. What this experiment demonstrated was the quantization of $\hat{J}_z$; given the direction of magnetic field gradient, the magnetic force on a particle of magnetic moment $\boldsymbol{\mu}$ may take on only a discrete set of values depending on the projection of its angular momentum in the direction of the field gradient, thus implying the quantization of angular momentum. If the orbital angular momentum of the particle is zero, we can determine the magnitude of its spin from the number of spots $(2S+1)$ on the photographic plate.

7.5 Coordinate Representation of the Orbital Angular Momentum Operators and States

For the orbital angular momentum states and operators we can have a matrix representation as well as a coordinate representation, in addition to the option of representing these entities in the Dirac notation. On the other hand, as already mentioned, for the spin states and operators, the only possible representation, other than that in the Dirac notation, is the matrix representation.

In the coordinate representation, the set of coordinate observables $\hat{x}\,,\hat{y}\,,\hat{z}$ are taken to be diagonal (we assume the system to have three degrees of freedom), and the basis states are the simultaneous eigenstates $|x, y, z\rangle$ of these observables. The state $|\ell, m\rangle$, which is the simultaneous eigenstate of $\hat{L}^2$ and $\hat{L}_z$ and is characterized by the orbital angular momentum quantum number ℓ and magnetic quantum number m, may then be represented by the function

$$\langle x, y, z|\, \ell, m\rangle = \widetilde{Y}_{\ell m}(x, y, z) \equiv \widetilde{Y}_{\ell m}(\mathbf{r})\,. \tag{7.5.1}$$

The eigenvalue equations

$$\hat{L}^2|\ell, m> = \hbar^2\, \ell(\ell+1)|\ell, m> \tag{7.5.2a}$$

$$\hat{L}_z|\ell, m> = \hbar m\,|\ell, m> \tag{7.5.2b}$$

may be written in the coordinate representation by pre-multiplying both sides of these two equations on the left by $\langle x, y, z|$. Using Eqs. (7.1.1), (2.7.4) and (7.5.1), these equations reduce to the set of differential equations:

$$-\hbar^2\left[\left(y\frac{\partial}{\partial z} - z\frac{\partial}{\partial y}\right)^2 + \left(z\frac{\partial}{\partial x} - x\frac{\partial}{\partial z}\right)^2 + \left(x\frac{\partial}{\partial y} - y\frac{\partial}{\partial x}\right)^2\right]\widetilde{Y}_{\ell\, m}(x, y, z)$$

$$\equiv L^2\widetilde{Y}_{\ell\, m}(x, y, z) = \hbar^2\ell(\ell+1)\widetilde{Y}_{\ell\, m}(x, y, z)\,, \tag{7.5.3}$$

and $$-i\hbar\left(x\frac{\partial}{\partial y} - y\frac{\partial}{\partial x}\right)\widetilde{Y}_{\ell\, m}(x, y, z) \equiv L_z\widetilde{Y}_{\ell\, m}(x, y, z) = m\hbar\widetilde{Y}_{\ell\, m}(x, y, z)\,. \tag{7.5.4}$$

Here L^2 and L_z are the differential operators, representing Dirac linear operators $\hat{L}^2$ and $\hat{L}_z$, in the coordinate space and $\widetilde{Y}_{\ell m}(x, y, z) \equiv \langle x, y, z|\, \ell, m\rangle$.

We can as well use the spherical polar coordinates r, θ, φ, instead of the Cartesian coordinates to express the states and observables; this does not imply a change in representation but merely a coordinate transformation. With this change in in coordinates, the angular momentum observables may be re-expressed as

$$L^2 = -\hbar^2\left[\frac{1}{\sin\theta}\frac{\partial}{\partial\theta}\sin\theta\frac{\partial}{\partial\theta} + \frac{1}{\sin^2\theta}\frac{\partial^2}{\partial\varphi^2}\right]\,, \tag{7.5.5}$$

and $$L_z = -i\hbar\frac{\partial}{\partial\varphi}\,, \tag{7.5.6}$$

while the function $\widetilde{Y}_{\ell m}(x, y, z) \equiv \widetilde{Y}_{\ell m}(\mathbf{r})$ may be rewritten as $\widetilde{Y}_{\ell m}(r, \theta, \varphi)$. Since the differential operators L^2 and L_z involve only the angles θ and φ, their simultaneous eigenfunctions will also depend only on these variables. These functions, called spherical harmonics, are denoted by $Y_{\ell m}(\theta, \varphi)$ and satisfy the eigenvalue equations [see Sec. 5.5, Eqs. (5.5.36) and (5.5.37) and also Appendix 5A1],

$$L^2 Y_{\ell m}(\theta, \varphi) = \hbar^2\ell(\ell+1)Y_{\ell m}(\theta, \varphi)\,, \tag{7.5.7a}$$

$$L_z Y_{\ell m}(\theta, \varphi) = \hbar m Y_{\ell m}(\theta, \varphi)\,. \tag{7.5.7b}$$

The functions $\widetilde{Y}_{\ell m}(\mathbf{r})$ occurring in Eqs. (7.5.3) and (7.5.4) are called *harmonic polynomials* and are expressed in terms of spherical harmonics as

$$\widetilde{Y}(\boldsymbol{r}) = r^\ell Y_{\ell m}(\theta, \varphi)\,. \tag{7.5.8}$$

According to Eqs. (7.5.3) and (7.5.4) the harmonic polynomials are also simultaneous eigenfunctions of L^2 and L_z, belonging to the eigenvalues $\ell(\ell+1)\hbar^2$ and $m\hbar$.

To summarize, we use the coordinate representation for the orbital angular momentum states, whereas for the spin states the column matrix representation is adopted. The product of an orbital angular momentum state and a spin state can thus be represented by the product of a wave function and a column matrix or by a multi-component wave function.

7.6 General Rotation Group and Rotation Matrices

In considering spatial rotations in three-dimensional space, we may consider either (a) the rotation of the coordinate system $(O,\ X,\ Y,\ Z)$ (characterized by its origin O and axes OX, OY and OZ) with the physical system fixed in space (passive viewpoint) or (b) the rotation of the physical system in the opposite sense with the coordinate system fixed in space (active viewpoint). We will adopt the active viewpoint. Thus a positive rotation of the coordinate system (the physical system being fixed) about a certain axis through an angle θ will be described as rotation of the physical system (the coordinate system being fixed) about the same axis through an angle $-\theta$.

As we have seen in Sec. 6.3, we can specify rotation in terms of a unit vector $\boldsymbol{n}$ along the axis of rotation (which requires two independent parameters for its specification) and the angle of rotation θ. Alternatively, a general rotation can also be specified by three Euler angles which take the original axes $(O,\ X,\ Y,\ Z)$ into the new set of axes $(O,\ X''',\ Y''',Z''')$ as follows. We first define a general rotation in terms of the rotations of coordinate axes by adopting the passive viewpoint and then write down the corresponding rotation operator for the active viewpoint.

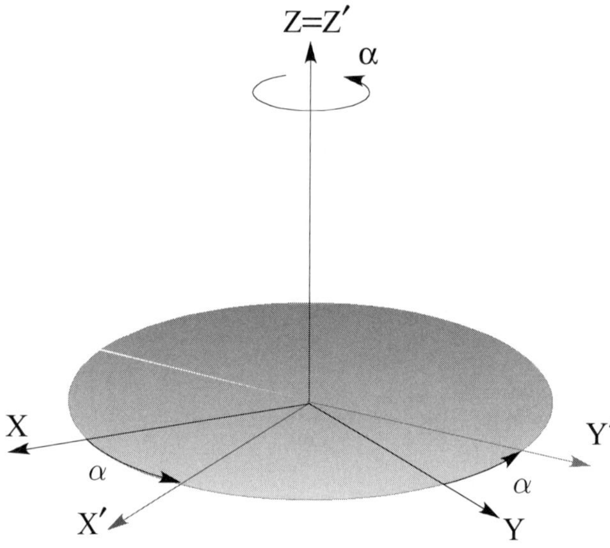

FIGURE 7.4
The first step of Euler rotation involves rotation of axes about OZ through angle α (positive when measured counterclockwise).

(i) A rotation of the coordinate axes about the original z-axis through angle α in the counter clockwise sense, taking the axes $(O,\ X,\ Y,\ Z)$ to $(O,\ X',\ Y',\ Z')$ [Fig. 7.4].

(ii) Next, a rotation is brought about the new y-axis (OY') through angle β so that the new coordinate axes are $(O,\ X'',\ Y'',\ ,Z'')$ [Fig. 7.5].

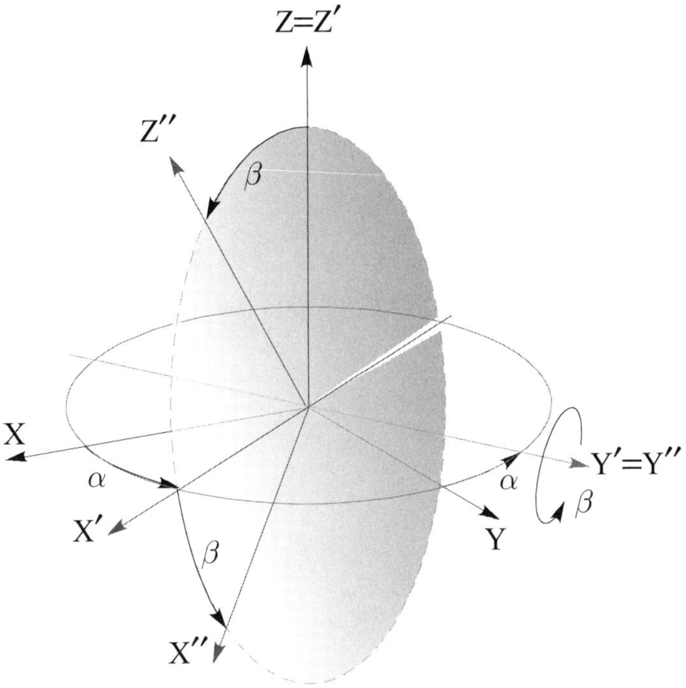

FIGURE 7.5
The second step of Euler rotation involves rotation of axes about OY' through angle β (positive when measured counterclockwise).

(iii) Finally a rotation is brought about OZ'' axis through angle γ, the new coordinate axes being $(O,\, X''',\, Y''',\, Z''')$ [Fig. 7.6].

The rotation operator describing the rotation of the physical system (active viewpoint) corresponding to the Euler rotation outlined above is given by

$$\begin{aligned}\hat{R}(\alpha,\beta,\gamma) &= \hat{R}_{z''}(\gamma)\,\hat{R}_{y'}(\beta)\,\hat{R}_z(\alpha)\,,\\ &= e^{-i\gamma\hat{J}_{z''}/\hbar}e^{-i\beta\hat{J}_{y'}/\hbar}e^{-i\alpha\hat{J}_z/\hbar}\,.\end{aligned} \tag{7.6.1}$$

Note the angular momentum operators and the order in which they appear in this equation. We can put this equation in a form where only the components of angular momentum along the coordinate axes OX, OY, and OZ (which stay fixed in the active viewpoint) appear by using the transformation properties of operators under rotation. Under the rotation $\hat{R}_z(\alpha)$, the operator $\hat{J}_y$ is transformed into $\hat{J}_{y'} = \hat{R}_z(\alpha)\hat{J}_y\,\hat{R}_z^\dagger(\alpha) = e^{-i\alpha\hat{J}_z/\hbar}\hat{J}_y e^{i\alpha\hat{J}_z/\hbar}$ so that

$$\hat{R}_{y'}(\beta) \equiv e^{-i\beta\hat{J}_{y'}/\hbar} = e^{-i\alpha\hat{J}_z/\hbar}\,e^{-i\beta\hat{J}_y/\hbar}\,e^{i\alpha\hat{J}_z/\hbar}\,. \tag{7.6.2}$$

Similarly, noting that $\hat{J}_{z''}$ is obtained from $\hat{J}_z$ by successive application of the rotations $\hat{R}_z(\alpha)$ and $R_{y'}(\beta)$ so that $\hat{J}_{z''} = e^{-i\beta\hat{J}_{y'}/\hbar}\,e^{-i\alpha\hat{J}_z/\hbar}\hat{J}_z e^{i\alpha\hat{J}_z/\hbar}e^{i\beta\hat{J}_{y'}/\hbar}$, which allows us to write

$$\begin{aligned}\hat{R}_{z''}(\gamma) &= \hat{R}_{y'}(\beta)\,\hat{R}_z(\alpha)\,\hat{R}_z(\gamma)\,\hat{R}_z^\dagger(\alpha)\,\hat{R}_{y'}^\dagger(\beta)\,,\\ &= e^{-i\beta\hat{J}_{y'}/\hbar}\,e^{-i\alpha\hat{J}_z/\hbar}\,e^{-i\gamma\hat{J}_z/\hbar}\,e^{i\alpha\hat{J}_z/\hbar}\,e^{i\beta\hat{J}_{y'}/\hbar}\,.\end{aligned} \tag{7.6.3}$$

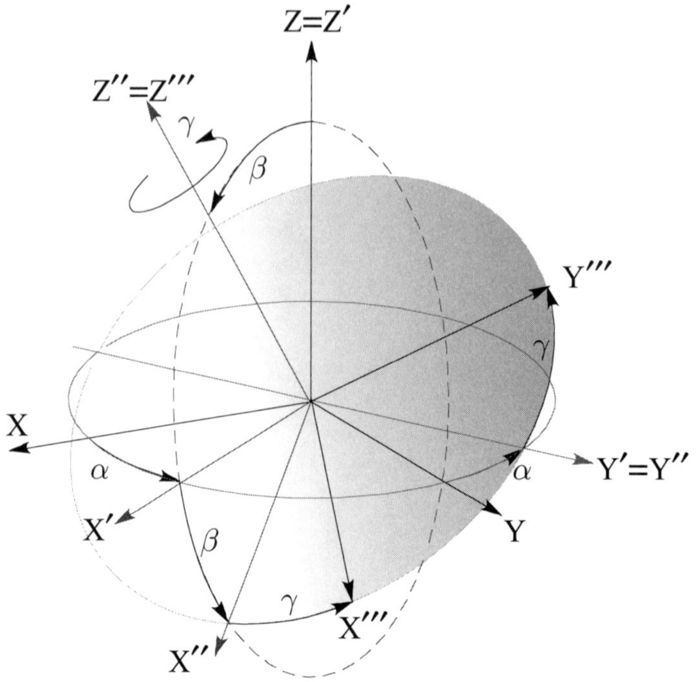

FIGURE 7.6
The third step of Euler rotation is the rotation of axes about OZ'' through angle γ (positive when measured counterclockwise).

Substituting Eqs. (7.6.2) and (7.6.3) into Eq. (7.6.1), we can eliminate $\hat{J}_{y'}$ and $\hat{J}_{z''}$ to arrive at

$$\hat{R}(\alpha,\beta,\gamma) = \exp\left[-\frac{i}{\hbar}\alpha\,\hat{J}_z\right]\exp\left[-\frac{i}{\hbar}\beta\,\hat{J}_y\right]\exp\left[-\frac{i}{\hbar}\gamma\,\hat{J}_z\right] . \tag{7.6.4}$$

Thus in the active view of rotation, wherein the coordinate axes are fixed, the rotation of the system (particle) corresponding to Euler angles (α, β, γ) may be achieved by a rotation γ about z-axis first, followed by a rotation β about y-axis and finally a rotation α about z-axis, all in the clockwise sense.

Group Property of Rotation Operators

The infinite set of rotation operators, corresponding to continuously varying angles α ($0 \le \alpha \le 2\pi$), β ($0 \le \beta \le \pi$) and γ ($0 \le \gamma \le 2\pi$) constitutes a continuous (Lie) group. Every rotation operator $\hat{R}(\alpha,\beta,\gamma)$, for given values of α, β, and γ is a member of this group and the group elements obviously satisfy (i) closure, (ii) existence of identity, (iii) existence of inverse, and (iv) associativity under group multiplication.

7.6.1 Rotation Matrices

A matrix representation for the rotation operators $\hat{R}(\alpha,\beta,\gamma)$ may be set up by choosing a representation in which the commuting observables $\hat{J}^2$ and $\hat{J}_z$ are diagonal and the basis states are $2j+1$ simultaneous eigenstates $|j,m\rangle$ with $-j \le m \le j$. In this representation, the rotation operator $\hat{R}(\alpha,\beta,\gamma)$ is represented by a $(2j+1)(2j+1)$ matrix D^j whose typical matrix element is given by

$$D^j_{mm'}(\alpha,\beta,\gamma) = \langle j,m|\,\hat{R}(\alpha,\beta,\gamma)\,|j,m'\rangle\ . \tag{7.6.5}$$

This matrix is called the *rotation matrix* (or simply the D-matrix) and each matrix element is a function of Euler angles.

Under the rotation specified by the Euler angles (α,β,γ), the eigenstate $|j,m\rangle$ of $\hat{J}^2$ and $\hat{J}_z$ transforms to $\hat{R}(\alpha,\beta,\gamma)|j,m>$, which may be called the *rotated state.* It is clear from the structure of $\hat{R}(\alpha,\beta,\gamma)$ that the rotated state will have the same j but different m, in general because $\hat{J}_x$, $\hat{J}_y$, and $\hat{J}_z$ commute with $\hat{J}^2$ but not among themseleves. In fact, the rotated state may be expanded by inserting a unit operator $\sum_{m'=-j}^{j} |j,m'\rangle\langle j,m'|$ as

$$\begin{aligned}\hat{R}(\alpha,\beta,\gamma)\,|j,m\rangle &= \sum_{m'=-j}^{j} |j,m'\rangle\langle j,m'|\,\hat{R}(\alpha,\beta,\gamma)|j,m>,\\ &= \sum_{m'=-j}^{j} D^j_{m'm}(\alpha,\beta,\gamma)\,|j,m'\rangle\ ,\end{aligned} \tag{7.6.6}$$

which shows that the rotated state is an eigenstate of $\hat{J}^2$ belonging to the same eigenvalue $j(j+1)$ as the original state $|j,m\rangle$. The value for $\hat{J}_z$, however, is indeterminate for this state and can have any value in the range $-j \le m \le +j$.

Properties of Rotation Matrices

(i) Since the rotation operator is unitary

$$\hat{R}(\alpha,\beta,\gamma)\hat{R}^\dagger(\alpha,\beta,\gamma) = \hat{1} = \hat{R}^\dagger(\alpha,\beta,\gamma)\hat{R}(\alpha,\beta,\gamma)\,, \tag{7.6.7}$$

the representative matrix is also unitary

$$D^j(D^j)^\dagger = I = (D^j)^\dagger D^j\,, \tag{7.6.8}$$

or $$\sum_{m'=-j}^{j} D^j_{mm'}(\alpha,\beta,\gamma)D^{j*}_{km'}(\alpha,\beta,\gamma) = \delta_{mk} = \sum_{m'=-j}^{j} D^{j*}_{m'm}(\alpha,\beta,\gamma)D^j_{m'k}(\alpha,\beta,\gamma)\,. \tag{7.6.9}$$

(ii) The set of functions $D^j_{mm'}(\alpha,\beta,\gamma)$ may also be treated as a complete set of functions in the space of Euler angles and they satisfy the orthonormality condition,

$$\int_0^{2\pi} d\alpha \int_0^{\pi} \sin\beta d\beta \int_0^{2\pi} d\gamma D^{j*}_{mk}(\alpha,\beta,\gamma)D^{j'}_{m'k'}(\alpha,\beta,\gamma) = \delta_{jj'}\delta_{mm'}\delta_{kk'}\frac{8\pi^2}{2j+1}\,. \tag{7.6.10}$$

Consequently these functions may be used as a basis of expansion so that any function of Euler angles may be expanded in terms of these functions.

(iii) The functions $D^j_{mm'}(\alpha, \beta, \gamma)$ for integral values of j ($j = L$) are also the eigenfunctions $\psi_{LKM}(\alpha, \beta, \gamma)$ of a *symmetric top*:

$$\psi_{LKM}(\alpha, \beta, \gamma) = \left(\frac{2L+1}{8\pi^2}\right)^{1/2} D^L_{-K,-M}(\alpha, \beta, \gamma)\,. \tag{7.6.11}$$

In this state, the square of the orbital angular momentum is $L(L+1)$ (in units of $\hbar^2$), its z-component is M (in units of $\hbar$) and its component along the body-fixed axis is K (in units of $\hbar$).

Transformation of Spherical Harmonics

We can now easily see how spherical harmonics transform under a rotation specified by Euler angles. Let θ, φ be the angular coordinates of a point $\boldsymbol{r}$ and θ', φ' be the angular coordinates of the point $\mathcal{R}^{-1}\boldsymbol{r}$, then under the rotation $\mathcal{R}$ a state $|\ell, m\rangle$ is transformed according to

$$\hat{R}(\alpha\beta, \gamma)\,|\ell, m\rangle = \sum_{m'=-\ell}^{\ell} |\ell, m'\rangle\,\langle \ell, m'|\,\hat{R}(\alpha, \beta, \gamma)\,|\ell, m\rangle\,. \tag{7.6.12}$$

Taking the coordinate representative of both sides, by premultiplying both sides by $\langle \boldsymbol{r}|$, and using $\langle \boldsymbol{r}|\,\hat{R}(\alpha, \beta, \gamma)\,|\ell, m\rangle \equiv \langle \mathcal{R}^{-1}\boldsymbol{r}|\,\ell, m\rangle = r^\ell Y_{\ell m}(\theta', \varphi')$ and $\langle \boldsymbol{r}|\,|\ell, m\rangle = r^\ell Y_{\ell m}(\theta, \varphi)$, we obtain the transformation

$$Y_{\ell m}(\theta', \varphi') = \sum_{m'=-\ell}^{\ell} D^\ell_{m'm}(\alpha, \beta, \gamma) Y_{\ell m'}(\theta, \varphi)\,, \tag{7.6.13}$$

where the coordinates θ', φ' may be regarded as the coordinates of $\boldsymbol{r}$ relative to the new set of axes obtained by the rotation $\mathcal{R}$ on the original set of axes. Equation (7.6.13) shows how spherical harmonics of the rotated cordinates θ', φ' are expressd in terms of spherical harmonics of the original coordinates θ, φ.

7.7 Coupling of Two Angular Momenta

Suppose two angular momenta $\hat{\boldsymbol{J}}_1$ and $\hat{\boldsymbol{J}}_2$ are to be added (or coupled) to give the total angular momentum $\hat{\boldsymbol{J}} = \hat{\boldsymbol{J}}_1 + \hat{\boldsymbol{J}}_2$. The component angular momenta could, for example, represent the orbital angular momentum and spin of a single particle, or the spins of two particles or the total angular momenta of two particles. Now the Cartesian components of the angular momenta $\hat{\boldsymbol{J}}_1$, $\hat{\boldsymbol{J}}_2$ and $\hat{\boldsymbol{J}}$ satisfy the usual commutation relations. With the help of these commutation relations, it may be easily verified that out of all these angular momentum operators, we can form two different sets of mutually commuting observables: (i) $\hat{J}_1^2, \hat{J}_{1z}, \hat{J}_2^2, \hat{J}_{2z}$ and (ii) $\hat{J}^2, \hat{J}_1^2, \hat{J}_2^2$ and $\hat{J}_z$.

The simultaneous eigenstates of the first set of commutating variables are $|j_1 m_1 j_2 m_2\rangle \equiv |j_1 m_1\rangle\,|j_2 m_2\rangle$, where $-j_1 \le m_1 \le j_1$ and $-j_2 \le m_2 \le j_2$. These states are referred to as the *uncoupled states*. The total number of uncoupled states is obviously $(2j_1+1)(2j_2+1)$. The simultaneous eigenstates of the second set of commuting observables are denoted by $|(j_1, j_2)jm >$ with $|j_1 - j_2| \le j \le j_1 + j_2$ and are called the *coupled states*. The total number of the coupled states, $\sum_{j=|j_1-j_2|}^{j_1+j_2} \sum_{m=-j}^{j} m = \sum_{j=|j_1-j_2|}^{j_1+j_2} (2j+1) = (2j_1+1)(2j_2+1)$, is the same as the number of uncoupled states as it should be.

In quantum mechanics, coupling of two angular momenta $\hat{\boldsymbol{J}}_1$ and $\hat{\boldsymbol{J}}_2$ implies construction of the the coupled states $|(j_1, j_2)jm>$, which are simultaneous eigenstates of $\hat{\boldsymbol{J}}^2, \hat{J}_z$ (and also of $\hat{\boldsymbol{J}}_1^2$, $\hat{\boldsymbol{J}}_2^2$) in terms of the uncoupled states $|j_1 m_1 j_2 m_2\rangle$, which are the products of the simultaneous eigenstates of $\hat{J}_1^2, \hat{J}_{1z}$ and $\hat{J}_2^2, \hat{J}_{2z}$. Since the uncoupled states form a complete set of states, any coupled state can be expressed in terms of them. Thus for the ket $|j,m\rangle \equiv |(j_1 j_2)jm\rangle$, we have

$$|j,m\rangle \equiv |(j_1 j_2)jm >= \sum_{m_1}\sum_{m_2} |j_1 m_1 j_2 m_2 > \langle j_1 m_1 j_2 m_2 |\, jm\rangle \,. \tag{7.7.1}$$

The coefficients $\langle j_1 m_1 j_2 m_2|\, jm\rangle$ are called the vector coupling coefficients or Clebsch-Gordan coefficients and are taken to be real

$$\langle j_1 m_1 j_2 m_2|\,(j_1 j_2)jm\rangle = \langle\,(j_1 j_2)jm|\, j_1 m_1 j_2 m_2\rangle \,. \tag{7.7.2}$$

We shall denote these coefficients by $(j_1 m_1 j_2 m_2|jm)$ so that Eq. (7.7.1) may be rewritten as

$$|j,m\rangle = \sum_{m_1=-j_1}^{j_1}\sum_{m_2=-j_2}^{j_2} (j_1 m_1 j_2 m_2|jm)\, |j_1 m_1\rangle\, |j_2 m_2\rangle \,. \tag{7.7.3}$$

Since the coupled states also form a complete set of states, being the simultaneous eigenstates of a set of commuting observables $\hat{J}^2, \hat{J}_1^2, \hat{J}_2^2$ and $\hat{J}_z$, it is also possible to express an uncoupled $|j_1 m_1 j_2 m_2\rangle$ state in terms of the coupled states as

$$|j_1 m_1 j_2 m_2\rangle = \sum_{j=|j_1-j_2|}^{j_1+j_2}\sum_{m=-j}^{j} |(j_1 j_2)jm\rangle\, \langle\,(j_1 j_2)jm|\, j_1 m_1 j_2 m_2\rangle \,, \tag{7.7.4}$$

where, as will be seen in the next section, the limits of j are given by $|j_1 - j_2| \leq j \leq |j_1 + j_2|$. The summation over m is redundant because the uncoupled state $|j_1 m_1 j_2 m_2\rangle$ is an eigenstate of $\hat{J}_z = \hat{J}_{1z} + \hat{J}_{2z}$ belonging the eigenvalue $m_1 + m_2$ so only the coupled states with $m = m_1 + m_2$ will contribute. In other words, the Clebsch-Gordan coefficient $(j_1 m_1 j_2 m_2|jm)$ vanishes unless $m = m_1 + m_2$. Making use of this explicitly and the reality of Clebsch-Gordan coefficients [Eq. (7.7.2)] we can write Eq. (7.7.4) as

$$|j_1 m_1\, j_2 m_2\rangle = \sum_{j=|j_1-j_2|}^{j_1+j_2} (j_1 m_1 j_2 m_2|j, m_1 + m_2)\; |(j_1 j_2)j, m_1 + m_2\rangle \,. \tag{7.7.5}$$

From Eq. (7.7.3) we note that in the coupled state $\hat{J}_{1z}$ and $\hat{J}_{2z}$ are not well defined (although their sum $\hat{J}_z = \hat{J}_{1z} = \hat{J}_{2z}$ is) while from Eq. (7.7.5) we note that in the uncoupled state $\hat{J}^2$ is not well defined.

7.8 Properties of Clebsch-Gordan Coefficients

(a) *The Clebsch-Gordan coefficient* $(j_1 m_1 j_2 m_2|jm)$ *vanishes unless* $m = m_1 + m_2$.

Operating on both sides of Eq. (7.7.3) with $\hat{J}_z = \hat{J}_{1z} + \hat{J}_{2z}$, and using $\hat{J}_z\,|jm\rangle = m\,|jm\rangle$ and $(\hat{J}_{1z} + \hat{J}_{2z})\,|j_1 m_1\rangle\,|j_2 m_2\rangle = (m_1 + m_2)\,|j_1 m_1\rangle\,|j_2 m_2\rangle$, we get

$$m|jm >= \sum_{m_1}\sum_{m_2} (j_1 m_1 j_2 m_2|jm)(m_1 + m_2)\, |j_1 m_1\rangle\, |j_2 m_2\rangle \,. \tag{7.8.1a}$$

Transposing the right-hand side of this equation to the left-hand side and combining the two terms by expressing $|jm\rangle$ in terms of uncoupled states, we get

$$\sum_{m_1}\sum_{m_2}(m-m_1-m_2)(j_1m_1j_2m_2|jm)|j_1m_1>|j_2m_2>=0\,. \tag{7.8.1b}$$

Since the uncoupled states are orthogonal to each other and no linear relationship can exist between them, each one of the terms in the sum must vanish. Hence $(m-m_1-m_2)(j_1m_1j_2m_2|jm)=0$, which implies the Clebsch-Gordan coefficient

$$(j_1m_1j_2m_2|jm)=0 \quad \text{if} \quad m\neq m_1+m_2\,. \tag{7.8.2}$$

As a result of this property, the double summation in Eq. (7.7.3) (expansion of a coupled state in terms of uncoupled states) may be replaced by a single summation:

$$|(j_1j_2)jm\rangle=\sum_{m_2=-j_2}^{j_2}(j_1,(m-m_2),j_2m_2|jm)\,|j_1,m-m_2\rangle\,|j_2,m_2\rangle\,. \tag{7.8.3}$$

(b) *Orthogonality of Clebsch-Gordan Coefficients*

From the orthogonality of the coupled states, $\langle(j_1j_2)jm|\,(j_1j_2)j'm'\rangle=\delta_{jj'}\delta_{mm'}$, and the expansions

$$|(j_1j_2)j'm'\rangle=\sum_{m_1'}\sum_{m_2'}(j_1m_1'j_2m_2'|j'm')\,|j_1m_1'\rangle\,|j_2m_2'\rangle\,,$$

and

$$\langle(j_1j_2)jm|=\sum_{m_1}\sum_{m_2}(j_1m_1j_2m_2|jm)\,\langle j_1m_1|\,\langle j_2m_2|\,,$$

we have

$$\sum_{m_1}\sum_{m_2}\sum_{m_1'}\sum_{m_2'}(j_1m_1j_2m_2|jm)(j_1m_1'j_2m_2'|j'm')\delta_{m_1m_1'}\delta_{m_2m_2'}=\delta_{jj'}\delta_{mm'}\,,$$

or

$$\sum_{m_2=-j_2}^{j_2}(j_1,m-m_2,\,j_2m_2|jm)(j_1,m'-m_2,j_2m_2|j'm')=\delta_{j\,j'}\delta_{m\,m'}\,. \tag{7.8.4}$$

This expresses the orthogonality condition for Clebsch-Gordan coefficients.

(c) *The Δ-Condition*

The Clebsch-Gordan coefficient is zero unless j_1, j_2, j satisfy the Δ-condition that the sum of any two of the j's is greater than the third and the absolute difference of any two of the j's is less than the third. Mathematically, for the Clebsch-Gordan coefficient to be non-zero, the following conditions must be satisfied

$$\begin{aligned}|j_1-j_2|\le j\le j_1+j_2\,,\\ |j-j_1|\le j_2\le j+j_1\,,\\ |j-j_2|\le j_1\le j+j_2\,.\end{aligned} \tag{7.8.5}$$

The coupled state $|(j_1j_2)jm>$ does not exist if the Δ-condition is not satisfied. Since the coupled state can be expressed in terms of the uncoupled states, multiplied by Clebsch-Gordan coefficients, it follows that the Clebsch-Gordan coefficient vanishes if the Δ-condition is not satisfied.

(d) *Symmetry Relations*

In the Clebsch Gordan coefficient $(j_1m_1j_2m_2|jm)$, the positions of j_1m_1, j_2m_2 and jm may be interchanged and the signs of magnetic quantum numbers be so adjusted that the equality $m = m_1 + m_2$ is satisfied. The relationship between these coefficients can be given by the symmetry relations

$$(j_1m_1j_2m_2|jm) = \begin{cases} (-)^{j_1+j_2-j}(j_2m_2j_1m_1|jm) \\ (-)^{j_1+j_2-j}(j_1,-m_1,j_2,-m_2|j,-m) \\ (-)^{j_1-m_1}\left(\frac{2j+1}{2j_2+1}\right)^{1/2}(j_1,m_1,j,-m|j_2,-m_2) \\ (-)^{j_2+m_2}\left(\frac{2j+1}{2j_1+1}\right)^{1/2}(j,-m,j_2,m_2|j_1,-m_1) \end{cases} \tag{7.8.6}$$

Appendix (7A1) gives, without derivation, the analytic expression for the Clebsch-Gordan coefficients $(j_{1,}(m-m_2),j_2,m_2|jm)$. Simplified analytic expressions for the Clebsch-Gordan coefficients are also given for $j_2 = 1/2$ and 1 in Tables 7.1 and 7.2.

7.8.1 The Vector Model of the Atom

The vector model is a means to visualize the coupled and uncoupled atomic angular momentum states in a semi-classical way. To visualize the coupled state $|(j_1j_2)jm>$ we may assume that the (classical) angular momentum vectors $\boldsymbol{J}_1$ (of length $j_1^* = \sqrt{j_1(j_1+1)}$) and $\boldsymbol{J}_2$ (of length $j_2^* = \sqrt{j_2(j_2+1)}$) both precess about the total angular momentum $\mathbf{J}$ (with the same angular frequency) and the vector $\boldsymbol{J}$ itself precesses about the z-direction. The vectors $\boldsymbol{J}_1$ and $\boldsymbol{J}_2$ can, for example, represent the orbital angular momentum and the spin of a valence electron in an atom. According to this model the projections of the vectors along the z-direction (direction of the applied magnetic field), m_1 and m_2, are uncertain while the projection of $\boldsymbol{J}$ along the z-axis is well-defined [Fig. 7.7]. The angle θ between the vectors $\boldsymbol{J}_1$ and $\boldsymbol{J}_2$ is fixed

$$\cos\theta = \frac{j^{*2} - j_1^{*2} - j_2^{*2}}{2j_1^*\,j_2^*} = \frac{j(j+1) - j_1(j_1+1) - j_2(j_2+1)}{2\sqrt{j_1(j_1+1)j_2(j_2+1)}} \tag{7.8.7}$$

as are the lengths of the vectors $\mathbf{J}_1$, $\mathbf{J}_2$ and $\mathbf{J}$.

The uncoupled state $|j_1m_1j_2m_2>$, which occurs when the applied magnetic field along the z-direction is so high that the coupling between the orbital angular momentum and spin is broken, can be visualized by assuming that the vectors $\mathbf{J}_1$ and $\mathbf{J}_2$ precess independently about the z-direction so that their projections m_1, m_2 along the z-axis are well-defined [Fig. **??**]. In this case there is uncertainty, both in magnitude and direction about the total angular momentum $\mathbf{J}$.

7.8.2 Projection Theorem for Vector Operators

A typical matrix element for an observable in the angular momentum operator basis is of the form $\langle n',j',m'|\,\hat{O}\,|n,j,m\rangle$, where n and n' represent quantum numbers needed to complete the basis states. The observable $\hat{O}$ can be a scalar, vector or a tensor operator. Operators can be classified according to their transformation under rotation. Thus an observable $\hat{O}$ is said to be a scalar if it commutes with the angular momentum operator. This follows from the law of transformation of operators (6.3.6) under a rotation $\hat{R}$. For an infinitesimal rotation through an angle $\delta\theta$ about an axis $\boldsymbol{n}$, we have $\hat{R}_{\boldsymbol{n}}(\delta\theta) = 1 - i\hbar^{-1}\delta\theta\ \boldsymbol{n}\cdot\hat{\boldsymbol{J}}$. The

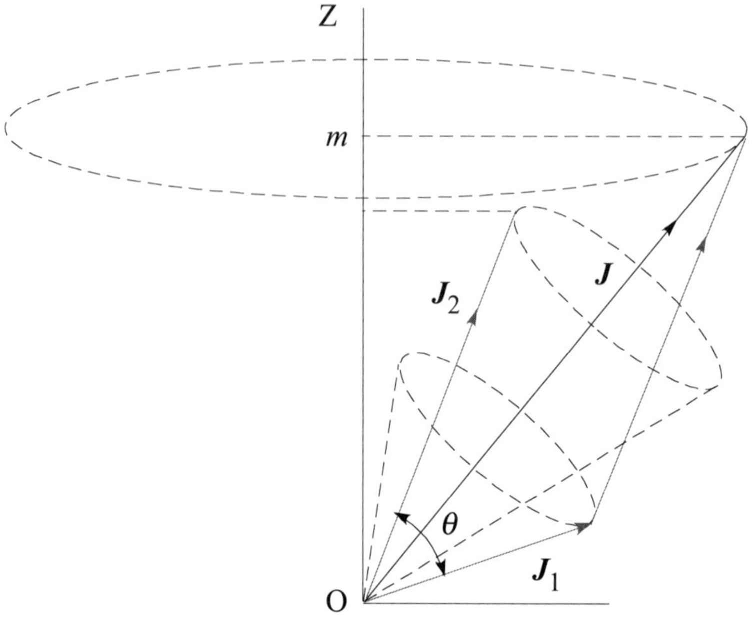

FIGURE 7.7
Vector model of the atom showing the coupled state in which the vectors $\boldsymbol{J}_1$ and $\boldsymbol{J}_2$ precess about $\boldsymbol{J}$ while $\boldsymbol{J}$ precesses about the z-direction.

operator $\hat{O}$ is transformed according to

$$\begin{aligned}\hat{O}' \equiv R(\delta\theta)\,\hat{O}\,\hat{R}^\dagger(\delta\theta) &= \left(1 - \frac{i}{\hbar}\delta\theta\,\boldsymbol{n}\cdot\hat{\boldsymbol{J}}\right)\,\hat{O}\,\left(1 + \frac{i}{\hbar}\delta\theta\,\boldsymbol{n}\cdot\hat{\boldsymbol{J}}\right) \\ &= \hat{O} - \frac{i}{\hbar}\delta\theta\,[\boldsymbol{n}\cdot\hat{\boldsymbol{J}}, \hat{O}]\,,\end{aligned} \tag{7.8.8}$$

where we have kept only the first order terms. But for a scalar, we must have $\hat{O}' = \hat{O}$ for all rotations $\hat{R}$. Comparing this with Eq. (7.8.8), we find that a scalar operator satisfies the commutation relation

$$[\hat{\boldsymbol{J}}, \hat{O}] = 0\,. \tag{7.8.9}$$

Examples of scalar operators include $\hat{J}^2$, $\hat{p}^2$, $\hat{\boldsymbol{r}}\cdot\hat{\boldsymbol{p}}$ and the Hamiltonian $\hat{H}$ of an isolated system. For such operators the matrix element

$$\langle n', j', m'|\,\hat{O}\,|n, j, m\rangle = \alpha_j(n', n)\delta_{jj'}\delta_{mm'} \tag{7.8.10a}$$

where $\alpha_j(n', n)$ is a constant which depends on j, n, n' but not m. In particular, for fixed n and j we have

$$\langle n, j, m'|\,\hat{O}\,|n, j, m\rangle = \alpha_j(n)\delta_{mm'}\,, \tag{7.8.10b}$$

that is, the matrix representation of $\hat{O}$ is a diagonal matrix and all the diagonal elements are equal. Thus in the subspace with fixed (n, j) spanned by $2j+1$ basis states $|n, j, m\rangle$ $(-j \le m \le j)$, a scalar operator is proportional to the identity operator. This relation is an example of the *Wigner-Eckart theorem*, which can be applied to a whole class of operators known as the *irreducible tensor operators*. Scalar operators are a special, and the simplest, case of these operators. In addition to the scalar operators already considered, we consider

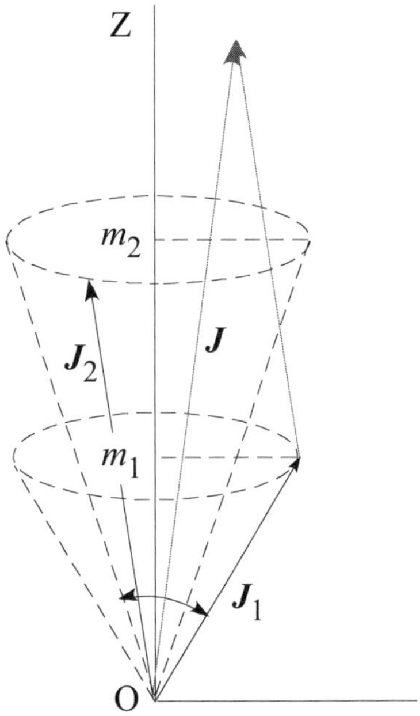

FIGURE 7.8
In the uncoupled state both $\boldsymbol{J}_1$ and $\boldsymbol{J}_2$ precess independently about the z-direction. $\boldsymbol{J}$ is undefined.

vector operators, which are another class of irreducible tensor operators and derive results analogous to Eq. (7.8.10).

A vector observable $\hat{\boldsymbol{V}}$ is a set of three operators $\hat{V}_x$, $\hat{V}_y$, $\hat{V}_z$, which satisfy the following commutation relations

$$\left[\hat{J}_x, \hat{V}_x\right] = 0\,, \qquad \left[\hat{J}_y, \hat{V}_y\right] = 0\,, \qquad \left[\hat{J}_z, \hat{V}_z\right] = 0\,, \tag{7.8.11a}$$

$$\left[\hat{J}_x, \hat{V}_y\right] = i\hbar\,\hat{V}_z\,, \qquad \left[\hat{J}_y, \hat{V}_z\right] = i\hbar\,\hat{V}_x\,, \qquad \left[\hat{J}_z, \hat{V}_x\right] = i\hbar\,\hat{V}_y\,, \tag{7.8.11b}$$

$$\left[\hat{J}_x, \hat{V}_z\right] = -i\hbar\,\hat{V}_y\,, \qquad \left[\hat{J}_y, \hat{V}_x\right] = -i\hbar\,\hat{V}_z\,, \qquad \left[\hat{J}_z, \hat{V}_y\right] = -i\hbar\,\hat{V}_x\,. \tag{7.8.11c}$$

Note that the relations in the second and third columns are obtained by cyclic permutations of the indices x, y, and z. These relations follow from the transformation law for the components of a vector under rotation. Consider, for example, an infinitesimal rotation through an angle $\delta\varphi$ about the z-axis. Then, according to Eq. (6.3.6), a vector operator $\hat{\boldsymbol{V}}$ is transformed to

$$\begin{aligned} \hat{\boldsymbol{V}}' &\equiv R_{\boldsymbol{n}}(\delta\varphi)\hat{\boldsymbol{V}}R_{\boldsymbol{n}}^{\dagger}(\delta\varphi) = \left(1 - i\hbar^{-1}\delta\varphi\,\hat{J}_z\right)\,\hat{\boldsymbol{V}}\,\left(1 - i\hbar^{-1}\delta\varphi\,\hat{J}_z\right)\,, \\ &= \hat{\boldsymbol{V}} - i\hbar^{-1}\delta\varphi\,[\hat{J}_z, \hat{\boldsymbol{V}}]\,. \end{aligned} \tag{7.8.12}$$

The vector operator $\hat{\boldsymbol{V}}$ is also a vector, whose components transform, according to Eq. (6.3.15), as

$$\hat{V}_x' = V_x + \delta\varphi\,\hat{V}_y\,, \tag{7.8.13a}$$

$$\hat{V}_y' = -\delta\varphi\,\hat{V}_x + V_y\,, \tag{7.8.13b}$$

$$\hat{V}_z' = V_z\,. \tag{7.8.13c}$$

A component-by-component comparison of Eqs. (7.8.12) and (7.8.13) leads to the commutator relations in the third column of Eq. (7.8.11). Similarly, by considering rotations about x and y axes, other commutation relations can be established.

For a single particle with orbital and spin angular momentum observables $\hat{\boldsymbol{L}}$ and $\hat{\boldsymbol{S}}$, the angular momentum operator is $\hat{\boldsymbol{L}} + \hat{\boldsymbol{S}}$. It is then easy to show that, beside $\hat{\boldsymbol{J}}$, $\hat{\boldsymbol{L}}$ and $\hat{\boldsymbol{S}}$, position and momentum observables $\hat{\boldsymbol{r}}$ and $\hat{\boldsymbol{p}}$ are vector observables satisfying the commutation relations (7.8.11). For a system consisting of two electrons, the angular momentum operator is $\hat{\boldsymbol{J}} = \hat{\boldsymbol{J}}_1 + \hat{\boldsymbol{J}}_2$ and $\hat{\boldsymbol{L}}_1$, $\hat{\boldsymbol{S}}_1$, $\hat{\boldsymbol{r}}_1$, $\hat{\boldsymbol{L}}_2$, etc. are all vector operators.

If we introduce the operators $\hat{V}_\pm$ and $\hat{J}_\pm$ by

$$\hat{V}_\pm = \hat{V}_x \pm i\,\hat{V}_y, \quad \text{and} \quad \hat{J}_\pm = \hat{J}_x \pm i\,\hat{J}_y\,, \tag{7.8.14}$$

then, with the help of Eqs. (7.8.11), we can show that

$$\left[\hat{J}_z, \hat{V}_\pm\right] = \pm\hbar\hat{V}_\pm\,, \tag{7.8.15}$$

$$\left[\hat{J}_+, \hat{V}_+\right] = 0 = \left[\hat{J}_-, \hat{V}_-\right]\,, \tag{7.8.16}$$

$$\left[\hat{J}_+, \hat{V}_-\right] = 2\hbar\hat{V}_3\,, \tag{7.8.17}$$

$$\left[\hat{J}_-, \hat{V}_+\right] = -2\hbar\hat{V}_3\,. \tag{7.8.18}$$

Consider now the basis states $|n, j, m\rangle$ where n denotes quantum numbers, in addition to the angular momentum quantum numbers j and m, that are needed to specify the states of the system completely. Operating with both sides of Eq. (7.8.15) on $|n, j, m\rangle$, and rearranging, we obtain

$$\hat{J}_z\left(\hat{V}_\pm\,|n, j, m\rangle\right) = \hat{V}_\pm\hat{J}_z\,|n, j, m\rangle \pm \hbar\hat{V}_\pm\,|n, j, m\rangle = (m \pm 1)\hbar\hat{V}_\pm\,|n, j, m\rangle\,. \tag{7.8.19}$$

From this we see that $\hat{V}_\pm\,|n, j, m\rangle$ is an eigenstate of $\hat{J}_z$ with eigenvalue $(m \pm 1)\hbar$. Since $\hat{J}_z$ is a Hermitian operator, states belonging to its different eigenvalues are orthogonal. From this we have the selection rules

$$\langle n', j', m'|\,\hat{V}_z\,|n, j, m\rangle = 0\,, \quad \text{if } m' \neq m\,, \tag{7.8.20}$$

$$\langle n', j', m'|\,\hat{V}_+\,|n, j, m\rangle = 0\,, \quad \text{if } m' \neq m + 1\,, \tag{7.8.21}$$

$$\langle n', j', m'|\,\hat{V}_-\,|n, j, m\rangle = 0\,, \quad \text{if } m' \neq m - 1\,. \tag{7.8.22}$$

Restricting now to the $2j + 1$ dimensional subspace spanned by $|n, j, m\rangle$ $(-j \leq m \leq j)$, with fixed values of n and j, we obtain from Eq. (7.8.16)

$$\langle n, j, m+2|\,\hat{J}_+\hat{V}_+\,|n, j, m\rangle = \langle n, j, m+2|\,\hat{V}_+\hat{J}_+\,|n, j, m\rangle\,. \tag{7.8.23}$$

Inserting the unit operator $\sum_{m=-j}^{j} |n, j, m\rangle\langle n, j, m| = \hat{1}$ between $\hat{J}_+$ and $\hat{V}_+$ and using Eqs. (7.8.20) through (7.8.22), we obtain

$$\begin{aligned}&\langle n, j, m+2|\,\hat{J}_+\,|n, j, m+1\rangle\langle n, j, m+1|\,\hat{V}_+\,|n, j, m\rangle \\ &\quad = \langle n, j, m+2|\,\hat{V}_+\,|n, j, m+1\rangle\langle n, j, m+1|\,\hat{J}_+\,|n, j, m\rangle\,,\end{aligned}$$

or

$$\frac{\langle n, j, m+1|\,\hat{V}_+\,|n, j, m\rangle}{\langle n, j, m+1|\,\hat{J}_+\,|n, j, m\rangle} = \frac{\langle n, j, m+2|\,\hat{V}_+\,|n, j, m+1\rangle}{\langle n, j, m+2|\,\hat{J}_+\,|n, j, m+1\rangle}\,. \tag{7.8.24}$$

Using this relation recursively for $-j \leq m \leq j-2$, we have

$$\begin{aligned}
\frac{\langle n,j,-j+1|\,\hat{V}_+\,|n,j,-j\rangle}{\langle n,j,-j+1|\,\hat{J}_+\,|n,j,-j\rangle} &= \frac{\langle n,j,-j+2|\,\hat{V}_+\,|n,j,-j+1\rangle}{\langle n,j,-j+2|\,\hat{J}_+\,|n,j,-j+1\rangle} \\
&= \frac{\langle n,j,-j+3|\,\hat{V}_+\,|n,j,-j+2\rangle}{\langle n,j,-j+3|\,\hat{J}_+\,|n,j,-j+2\rangle} \\
&\cdots \\
&= \frac{\langle n,j,j|\,\hat{V}_+\,|n,j,j-1\rangle}{\langle n,j,j|\,\hat{J}_+\,|n,j,j-1\rangle}\,.
\end{aligned} \tag{7.8.25}$$

Denoting the common ratio in Eq. (7.8.25), which can depend only on n and j, by $\lambda_+(n,j)$, we can express the matrix element of $\hat{V}_+$ in terms of the matrix element of $\hat{J}_+$ as

$$\langle n,j,m+1|\,\hat{V}_+\,|n,j,m\rangle = \lambda_+(n,j)\,\langle n,j,m+1|\,\hat{J}_+\,|n,j,m\rangle\,. \tag{7.8.26}$$

Using the selection rule (7.8.21), which implies $\langle n,j,m'|\,\hat{V}_+\,|n,j,m\rangle$ and $\langle n,j,m'|\,\hat{J}_+\,|n,j,m\rangle$ are nonzero only if $m'-m=1$, we can write Eq. (7.8.26) as

$$\langle n,j,m'|\,\hat{V}_+\,|n,j,m\rangle = \lambda_+(n,j)\,\langle n,j,m'|\,\hat{J}_+\,|n,j,m\rangle\,, \tag{7.8.27}$$

for arbitrary values of m and m'. Similarly, by taking the matrix element of $[\hat{J}_-,\hat{V}_-]=0$ between $\langle n,j,m-2|$ and $|n,j,m\rangle$, and using the selection rule (7.8.22), we find the matrix element of $\hat{V}_-$ is proportional to that of $\hat{J}_-$,

$$\langle n,j,m'|\,\hat{V}_-\,|n,j,m\rangle = \lambda_-(n,j)\,\langle n,j,m'|\,\hat{J}_-\,|n,j,m\rangle\,. \tag{7.8.28}$$

Finally, by taking the matrix element of commutation relation (7.8.18) between $\langle n,j,m|$ and $|n,j,m\rangle$ and using the selection rule (7.8.20) we find

$$\begin{aligned}
-2\hbar\,\langle n,j,m|\,\hat{V}_z\,|n,j,m\rangle &= \langle n,j,m|\,\hat{J}_-\hat{V}_+ - \hat{V}_+\hat{J}_-\,|n,j,m\rangle \\
&= \hbar\sqrt{j(j+1)-m(m+1)}\,\langle n,j,m+1|\,\hat{V}_+\,|n,j,m\rangle \\
&\quad - \hbar\sqrt{j(j+1)-m(m-1)}\,\langle n,j,m|\,\hat{V}_+\,|n,j,m-1\rangle\,.
\end{aligned}$$

The matrix elements on the right can be replaced by the matrix elements of $\hat{J}_+$ with the help of Eq. (7.8.27). Evaluating the resulting matrix elements of $\hat{J}_+$ and simplifying we find

$$\langle n,j,m|\,\hat{V}_z\,|n,j,m\rangle = m\hbar\,\lambda_+(n,j)\,. \tag{7.8.29}$$

Similar argument, starting with (7.8.17), leads to

$$\langle n,j,m|\,\hat{V}_z\,|n,j,m\rangle = m\hbar\,\lambda_-(n,j)\,. \tag{7.8.30}$$

A comparison of Eqs. (7.8.29) and (7.8.30), shows that we must have

$$\lambda_-(n,j) = \lambda_+(n,j) \equiv \lambda(n,j)\,. \tag{7.8.31}$$

Since any component of $\hat{\boldsymbol{V}}$ can be expressed as a combination of $V_\pm$, and V_z, it follows from Eqs. (7.8.27) through (7.8.31) that

$$\langle n,j,m'|\,\hat{\boldsymbol{V}}\,|n,j,m\rangle = \lambda(n,j)\,\langle n,j,m'|\,\hat{\boldsymbol{J}}\,|n,j,m\rangle\,. \tag{7.8.32}$$

Thus inside the subspace spanned by the states $|n,j,m\rangle$ $(-j \le m \le j)$, the matrix elements of $\hat{\boldsymbol{V}}$ are proportional to those of $\hat{\boldsymbol{J}}$. This result constitutes the Wigner-Eckart theorem for vector operators. The quantity $\lambda(n,j)$ is known as the reduced matrix element.

It follows from Eq. (7.8.34) that the matrix element of $\hat{\boldsymbol{J}} \cdot \hat{\boldsymbol{V}}$ is

$$\begin{aligned}\langle n,j,m'|\, \hat{\boldsymbol{J}} \cdot \hat{\boldsymbol{V}}\, |n,j,m\rangle &= \sum_{m''} \langle n,j,m'|\, \hat{J}_i\, |n,j,m''\rangle \langle n,j,m''|\, \hat{V}_i\, |n,j,m\rangle \\ &= \lambda(n,j) \sum_{m''} \langle n,j,m'|\, \hat{J}_i\, |n,j,m''\rangle \langle n,j,m''|\, \hat{J}_i\, |n,j,m\rangle \\ &= \lambda(n,j)\, \langle n,j,m'|\, \hat{J}^2\, |n,j,m\rangle = \lambda(n,j)\, \hbar^2\, j(j+1)\delta_{mm'}\,. \end{aligned} \tag{7.8.33}$$

Note that this result is similar to Eq. (7.8.10), which is to be expected since $\hat{\boldsymbol{J}} \cdot \hat{\boldsymbol{V}}$ is a scalar operator. Using this in Eq. (7.8.32) we can write

$$\langle n,j,m'|\, \hat{\boldsymbol{V}}\, |n,j,m\rangle = \frac{\langle n,j,m'|\, \hat{\boldsymbol{J}} \cdot \hat{\boldsymbol{V}}\, |n,j,m\rangle}{\hbar^2\, j(j+1)} \langle n,j,m'|\, \hat{\boldsymbol{J}}\, |n,j,m\rangle\,. \tag{7.8.34}$$

Thus within the $(2j+1)$-dimensional space spanned by $|n,j,m\rangle$, all vector operators are proportional to $\hat{\boldsymbol{J}}$. This result is known as the *projection theorem*. Physically, this result can be understood in terms of a (classical) vector model of the system. For an isolated physical system, the total angular momentum $\boldsymbol{J}$ is a constant of motion. So all physical quantities associated with the system rotate about the vector $\boldsymbol{J}$. In particular, for a vector quantity $\boldsymbol{V}$, only its component parallel to vector $\boldsymbol{J}$ has nonzero time average value [Fig. 7.9].

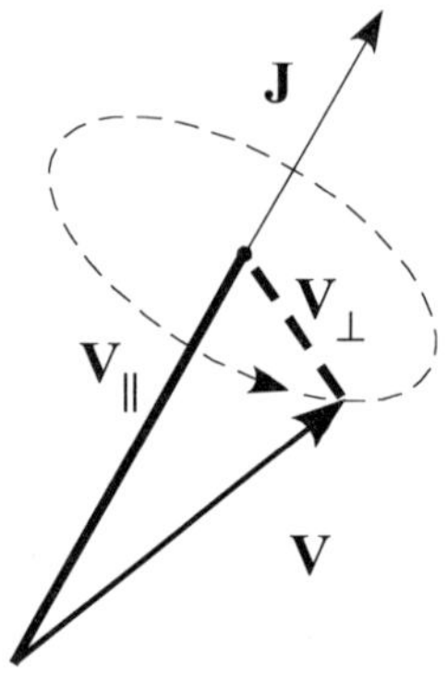

FIGURE 7.9
Classical vector model of the projection theorem. Since the vector $\boldsymbol{V}$ rotates rapidly about the total angular momentum vector $\boldsymbol{J}$, only its projection $\boldsymbol{V}_{\|}$ parallel to $\boldsymbol{J}$ has nonzero (expectation) value.

$$\boldsymbol{V}_{\|} = \frac{\boldsymbol{J} \cdot \boldsymbol{V}}{J^2}\, \boldsymbol{J}\,, \tag{7.8.35}$$

which is the classical analog of Eq. (7.8.34). It must be kept in mind that the proportionality of vector $\hat{\boldsymbol{V}}$ and $\hat{\boldsymbol{J}}$ holds as long as we restrict to the states belonging to the same subspace with fixed n and j. A vector operator $\hat{\boldsymbol{V}}$ may possess nonzero matrix element between states belonging to different subspaces, whereas the corresponding matrix elements of $\hat{\boldsymbol{J}}$ always vanish. We shall see examples of this in Sec. 8.4.

7.9 Coupling of Three Angular Momenta

The formalism of preceding section for the coupling of two angular momenta can be extended to the coupling of three or more angular momenta. We shall illustrate this for the coupling of three and four angular momenta, which are useful in discussing two-body problems.

Consider three angular momentum operators $\hat{\boldsymbol{J}}_1$, $\hat{\boldsymbol{J}}_2$, $\hat{\boldsymbol{J}}_3$, which combine to give the total angular momentum operator $\hat{\boldsymbol{J}} = \hat{\boldsymbol{J}}_1 + \hat{\boldsymbol{J}}_2 + \hat{\boldsymbol{J}}_3$. Let us construct the coupled states, which are the simultaneous eigenstates of $\hat{J}^2$ and $\hat{J}_z$ from the uncoupled states, which are the simultaneous eigenstates of $\hat{J}_1^2$, $\hat{J}_{1z}$, $\hat{J}_2^2$, $\hat{J}_{2z}$, $\hat{J}_3^2$, $\hat{J}_{3z}$. There are three alternative schemes to do this.

One scheme is to add $\hat{\boldsymbol{J}}_1$ and $\hat{\boldsymbol{J}}_2$ first and then add the resultant to the $\hat{\boldsymbol{J}}_3$:

$$\hat{\boldsymbol{J}}_1 + \hat{\boldsymbol{J}}_2 = \hat{\boldsymbol{J}}_{12} \,, \tag{7.9.1a}$$

and

$$\hat{\boldsymbol{J}}_{12} + \hat{\boldsymbol{J}}_3 = \hat{\boldsymbol{J}} \,. \tag{7.9.1b}$$

According to this scheme, simultaneous eigenstates of $\hat{J}_1^2$, $\hat{J}_2^2$, $\hat{J}_{12}^2$, $\hat{\boldsymbol{J}}_3^2$, $\hat{J}^2$, and $\hat{J}_z$ can be constructed as

$$\begin{aligned} |(j_1 j_2) j_{12} j_3; jm\rangle = &\sum_{m_1}\sum_{m_2}\sum_{m_{12}}\sum_{m_3} (j_1 m_1 j_2 m_2 | j_{12} m_{12})(j_{12} m_{12} j_3 m_3 | jm) \\ &\times |j_1 m_1\rangle \, |j_2 m_2\rangle \, ||j_3 m_3\rangle \,. \end{aligned} \tag{7.9.2}$$

Two additional schemes are

$$\left.\begin{aligned} \hat{\boldsymbol{J}}_2 + \boldsymbol{J}_3 &= \hat{\boldsymbol{J}}_{23} \\ \hat{\boldsymbol{J}}_1 + \hat{\boldsymbol{J}}_{23} &= \hat{\boldsymbol{J}} \end{aligned}\right\}, \tag{7.9.3}$$

or

$$\left.\begin{aligned} \hat{\boldsymbol{J}}_1 + \hat{\boldsymbol{J}}_3 &= \hat{\boldsymbol{J}}_{13} \\ \hat{\boldsymbol{J}}_{13} + \hat{\boldsymbol{J}}_2 &= \hat{\boldsymbol{J}} \end{aligned}\right\}. \tag{7.9.4}$$

The first of these lead to the simultaneous eigenstates of $\hat{J}_1^2$, $\hat{J}_2^2$, $\hat{J}_3^2$, $\hat{J}_{23}^2$, $\hat{J}^2$, $\hat{J}_z$ as

$$|j_1(j_2 j_3) j_{23} jm\rangle = \sum_{m_1}\sum_{m_{23}}\sum_{m_2}\sum_{m_3} (j_2 m_2 j_3 m_3 | j_{23} m_{23})(j_1 m_1 j_{23} m_{23} | jm) \, |j_1 m_1\rangle \, |j_2 m_2\rangle \, |j_3 m_3\rangle \,. \tag{7.9.5}$$

Simultaneous eigenstates of $\hat{J}_1^2$, $\hat{J}_3^2$, $\hat{J}_{13}^2$, $\hat{J}_2^2$, $\hat{J}^2$, $\hat{J}_z$ may be similarly expressed.

All these sets of orthonormal states, say, the first two $|(j_1 j_2) j_{12} j_3 jm\rangle$ and $|j_1(j_2 j_3) j_{23} jm\rangle$, form a complete set of states and any one state from the former set may be expressed in terms of the latter set of states and vice versa. For example, the state $|(j_1 j_2) j_{12} j_3 \; jm\rangle$ can be expressed as

$$|(j_1 j_2) j_{12} j_3 \; jm\rangle = \sum_{j_{23}} U(j_1 j_2 j \; j_3 j_{12} j_{23}) \; |j_1(j_2 j_3) j_{23} jm\rangle \,. \tag{7.9.6}$$

The coefficients $U(j_1j_2jj_3j_{12}j_{23})$ are called the normalized Racah coefficients which can be expressed in terms of Racah coefficients W or the so-called $6-j$ symbols as follows

$$\begin{aligned} U(j_1j_2j\,j_3j_{12}j_{23}) &= \sqrt{(2j_{12}+1)(2j_{23}+1)}\,W(j_1j_2j\,j_3j_{12}j_{23})) \\ &= \sqrt{(2j_{12}+1)(2j_{23}+1)}\,(-1)^{-j_1-j_2-j_3-j}\begin{Bmatrix} j_1 & j_2 & j_{12} \\ j_3 & j & j_{23} \end{Bmatrix}. \end{aligned} \tag{7.9.7}$$

Analytic expression for Racah coefficients are given at the end of this chapter.

7.10 Coupling of Four Angular Momenta ($L-S$ and $j-j$ Coupling)

In dealing with two particles with spin, we have to add four angular momenta corresponding to their orbital motion and spin. Let us denote the four angular momenta by $\hat{\boldsymbol{L}}_1, \hat{\boldsymbol{L}}_2$ and $\hat{\mathbf{S}}_1, \hat{\mathbf{S}}_2$, where $\hat{\boldsymbol{L}}_1, \hat{\boldsymbol{L}}_2$ may denote the orbital angular momenta of two electrons and $\hat{\mathbf{S}}_1, \hat{\mathbf{S}}_2$ their spins, so that $\hat{\boldsymbol{L}}_1 + \hat{\boldsymbol{L}}_2 + \hat{\boldsymbol{S}}_1 + \hat{\boldsymbol{S}}_2 = \hat{\boldsymbol{J}}$. There are many ways of doing this. Two common alternatives are the so-called $L-S$ coupling and $j-j$ coupling schemes.

(i) The L-S coupling scheme, where the orbital and spin angular momenta are added and the resultants are added to obtain the total angular momentum operator,

$$\left.\begin{aligned} \hat{\boldsymbol{L}}_1 + \hat{\boldsymbol{L}}_2 &= \hat{\boldsymbol{L}} \\ \hat{\boldsymbol{S}}_1 + \hat{\boldsymbol{S}}_2 &= \hat{\boldsymbol{S}} \\ \hat{\boldsymbol{L}} + \hat{\boldsymbol{S}} &= \hat{\boldsymbol{J}} \end{aligned}\right\}. \tag{7.10.1}$$

(ii) The $j-j$ coupling scheme, where the orbital and spin angular momenta of each particle are added and the results are added to obtain the total angular momentum operator,

$$\left.\begin{aligned} \hat{\boldsymbol{L}}_1 + \hat{\boldsymbol{S}}_1 &= \hat{\boldsymbol{J}}_1 \\ \hat{\boldsymbol{L}}_2 + \hat{\boldsymbol{S}}_2 &= \hat{\boldsymbol{J}}_2 \\ \hat{\boldsymbol{J}}_1 + \hat{\boldsymbol{J}}_2 &= \hat{\boldsymbol{J}} \end{aligned}\right\}. \tag{7.10.2}$$

In the $L-S$ scheme we construct the coupled state $|(\ell_1\ell_2)L(s_1s_2)S;jm\rangle$, which is a simultaneous eigenstate of the set of commuting observables $\hat{L}_1^2$, $\hat{L}_2^2$, $\hat{L}^2$, $\hat{S}_1^2$, $\hat{S}_2^2$, $\hat{S}^2$, $\hat{J}^2$, $\hat{J}_z$ as

$$|(\ell_1\ell_2)L(s_1s_2)S;jm\rangle = \sum_{m_L}\sum_{m_S}(Lm_LSm_S|j,m)\;|(\ell_1\ell_2)Lm_L\rangle\;|(s_1s_2)Sm_S\rangle \tag{7.10.3}$$

where
$$|(\ell_1\ell_2)Lm_L\rangle = \sum_{m_{\ell_1}}\sum_{m_{\ell_2}}(\ell_1m_{\ell_1}\ell_2m_{\ell_2}\,|Lm_L)|\ell_1m_{\ell_1}\rangle\;|\ell_2m_{\ell_2}\rangle \tag{7.10.4}$$

and
$$|(s_1s_2)Sm_s\rangle = \sum_{m_{s_1}}\sum_{m_{s_2}}(s_1m_{s_1}s_2m_{s_2}|Sm_s)\;|s_1m_{s_1}\rangle\;|s_2m_{s_2}\rangle\,. \tag{7.10.5}$$

In the $j-j$ coupling scheme we construct the $j-j$ coupled state $|(\ell_1s_1)j_1(\ell_2s_2)j_2;jm>$, the simultaneous eigenstate of the set of commuting observables $\hat{L}_1^2, \hat{S}_1^2, \hat{J}_1^2, \hat{L}_2^2, \hat{S}_2^2, \hat{J}_2^2, \hat{J}^2, \hat{J}_z$ as

$$|(\ell_1s_1)j_1(\ell_2s_2)j_2;jm\rangle = \sum_{m_1}\sum_{m_2}(j_1m_1j_2m_2|jm)\;|(\ell_1s_1)j_1m_1\rangle\;|(\ell_2s_2)j_2m_2\rangle \tag{7.10.6}$$

where
$$|(\ell_1 s_1)j_1 m_1\rangle = \sum_{m_{\ell_1}}\sum_{m_{s_1}} (\ell_1 m_{\ell_1} s_1 m_{s_1}|j_1 m_1)\, |\ell_1 m_{\ell_1}\rangle\, |s_1 m_{s_1}\rangle \tag{7.10.7}$$
and
$$|(\ell_2 s_2)j_2 m_2\rangle = \sum_{m_{\ell_2}}\sum_{m_{s_2}} (\ell_2 m_{\ell_2} s_2 m_{s_2}|j_2 m_2)\, |\ell_2 m_{\ell_2}\rangle\, |s_2 m_{s_2}\rangle\,. \tag{7.10.8}$$

The set of $L-S$ coupled states form a complete set of states and so do the set of $j-j$ coupled states. The overlap $\langle(\ell_1\ell_2)L(s_1s_2)S;\, jm|\,(\ell_1 s_1)j_1(\ell_2 s_2)j_2; jm\rangle$ between an $L-S$ and $j-j$ coupled state can be expressed in terms of the the so called $LS-jj$ coupling coefficient

$$\begin{aligned}\langle(\ell_1\ell_2)L(s_1s_2)S;\, jm|\,(\ell_1 s_1)j_1(\ell_2 s_2)j_2; jm\rangle &= \begin{bmatrix}\ell_1 & \ell_2 & L\\ s_1 & s_2 & S\\ j_1 & j_2 & j\end{bmatrix}\\ &= \sqrt{(2L+1)(2S+1)(2j_1+1)(2j_2+1)}\begin{Bmatrix}\ell_1 & \ell_2 & L\\ s_1 & s_2 & S\\ j_1 & j_2 & j\end{Bmatrix}\,,\end{aligned} \tag{7.10.9}$$

where the square bracket is called the $LS-jj$ coupling coefficient while the curly bracket is called the $9-j$ symbol. Equation (7.10.9) implies that an $L-S$ coupled state may be expressed in terms of the set of $j-j$ coupled states and a $j-j$ coupled state may be expressed in terms of $L-S$ coupled states. The $9-j$ symbol has some symmetry properties:

(i) The effect of interchanging two rows or two columns of the $9-j$ symbol is to multiply by a phase factor $(-1)^{\Sigma}$, where $\Sigma = \ell_1+\ell_2+L+s_1+s_2+S+j_1+j_2+j =$ sum of all the nine j's.

(ii) The 9-j symbol is unchanged under reflection about the principal diagonal,

$$\begin{Bmatrix}\ell_1 & \ell_2 & L\\ s_1 & s_2 & S\\ j_1 & j_2 & j\end{Bmatrix} = \begin{Bmatrix}\ell_1 & s_1 & j_1\\ \ell_2 & s_2 & j_2\\ L & S & j\end{Bmatrix}. \tag{7.10.10}$$

Analytic Expression for the Clebsch-Gordan Coefficient

Racah (1942) gave an analytic expression for the Clebsch-Gordan coefficient $(j_1m_1j_2m_2|jm)$ if the $\Delta(j_1j_2j)$ condition is fulfilled:

$$\begin{aligned}(j_1m_1j_2m_2|jm) =& \delta_{m,\,m_1+m_2}\left[\frac{(2j+1)(j_1+j_2-j)!(j+j_1-j_2)!(j+j_2-j_1)!}{(j_1+j_2+j+1)!}\right]^{1/2}\\ &\times[(j_1+m_1)!(j_1-m_1)!(j_2+m_2)!(j_2-m_2)!(j+m)!(j-m)!]^{1/2}\\ &\times\sum_k \frac{(-)^k}{k!}[(j_1+j_2-j-k)!(j_1-m_1-k)!(j_2+m_2-k)!\\ &\times(j_1-j_2+m_1+k)!(j-j_1-m_2+k)!]^{-1}.\end{aligned} \tag{7.10.11}$$

Here k takes all integral values, consistent with the factorial notation.

Analytic Expression for the Racah Coefficient

$$W(a,b,c,d,e,f) = (-)^{-a-b-c-d}\begin{Bmatrix} a & b & e \\ d & c & f \end{Bmatrix} = \Delta(a\,b\,c)\Delta(c\,d\,e)\Delta(a\,c\,f)\Delta(b\,d\,f)$$
$$\times \sum_k \left[\frac{(-)^{k+a+b+c+d}(k+1)!}{(k-a-b-c)!(k-c-d-e)!(k-a-c-f)!(k-b-d-f)!(a+b+c+d-k)!} \right.$$
$$\left. \times \frac{1}{(a+d+e+f-k)!(b+c+e+f-k)!} \right] \tag{7.10.12}$$

where

$$\Delta(a\,b\,c) = \begin{cases} \dfrac{(a+b-c)!(a-b+c)!(-a+b+c)!}{(a+b+c+1)!}, & \text{if a,b,c satisfy } \Delta \text{ condition.} \\ 0, & \text{if a,b,c do not satisfy } \Delta \text{ condition.} \end{cases} \tag{7.10.13}$$

Here k takes all integral values consistent with the factorial notation.

TABLE 7.1
Clebsch-Gordan coefficients for $j_2 = 1/2$.

$m_2 \rightarrow$	$m_2 = \frac{1}{2}$	$m_2 = -\frac{1}{2}$
$j = j_1 + \frac{1}{2}$	$\left(\frac{j_1+m+1/2}{2j_1+1}\right)^{1/2}$	$\left(\frac{j_1-m+1/2}{2j_1+1}\right)^{1/2}$
$j = j_1 - \frac{1}{2}$	$-\left(\frac{j_1-m+1/2}{2j_1+1}\right)^{1/2}$	$\left(\frac{j_1+m+1/2}{2j_1+1}\right)^{1/2}$

TABLE 7.2
Clebsch-Gordan coefficients for $j_2 = 1$.

$m_2 \rightarrow$	$m_2 = 1$	$m_2 = 0$	$m_2 = -1$
$j = j_1 + 1$	$\left(\frac{(j_1+m)(j_1+m+1)}{(2j_1+1)(2j_1+2)}\right)^{1/2}$	$\left(\frac{(j_1-m+1)(j_1+m+1)}{(2j_1+1)(j_1+1)}\right)^{1/2}$	$\left(\frac{(j_1-m)(j_1-m+1)}{(2j_1+1)(2j_1+2)}\right)^{1/2}$
$j = j_1$	$-\left(\frac{(j_1+m)(j_1-m+1)}{2j_1(j_1+1)}\right)^{1/2}$	$\left(\frac{m}{\{j_1(j_1+1)\}^{1/2}}\right)$	$\left(\frac{(j_1+m+1)(j_1-m)}{2j_1(j_1+1)}\right)^{1/2}$
$j = j_1 - 1$	$\left(\frac{(j_1-m)(j_1-m+1)}{2j_1(2j_1+1)}\right)^{1/2}$	$-\left(\frac{(j_1-m)(j_1+m)}{j_1(2j_1+1)}\right)^{1/2}$	$\left(\frac{(j_1+m+1)(j_1+m)}{2j_1(2j_1+1)}\right)^{1/2}$

Problems

1. Prove the commutation relations:

   (a) $[\hat{L}_j, \hat{x}_k] = i\,\hbar \epsilon_{jk\ell} x_\ell$.

   (b) $[L_j, p_k] = i\,\hbar \epsilon_{jk\ell} p_\ell$, where $\hat{L}_i$ is the ith component of the orbital angular momentum and $\epsilon_{ik\ell}$ is the alternating symbol. Also show that $[\hat{L}_x, \hat{p}^2] = 0$ and $[\hat{L}_x, \hat{r}^2] = 0$, where $\hat{p}^2 = \hat{p}_x^2 + \hat{p}_y^2 + \hat{p}_z^2$ and $\hat{r}^2 = \hat{x}^2 + \hat{y}^2 + \hat{z}^2$.

2. Show that, for a state $|j, m\rangle$, corresponding to a definite value of $\hat{J}_z$, the quantum mechanical expectation values of $\hat{J}_x$ and $\hat{J}_y$ are zero.

3. The spin component of the electron along z-axis is $+1/2$ (in units of $\hbar$) when it is in the state $|s = 1/2, m = 1/2\rangle$. If there is a z'-axis, making angle θ with the z-axis, calculate the probabilities that the component of spin along z'-axis will have values $+\hbar/2$ and $-\hbar/2$. Also calculate the quantum mechanical expectation value of the projection of spin along z'-axis, for this state.

4. In a representation in which $\hat{J}^2$ and $\hat{J}_z$ are diagonal, deduce matrices representing the operators $\hat{J}_+, \hat{J}_-, \hat{J}_x, \hat{J}_y$ for the cases (a) $j = 1$ and (b) $j = 3/2$.

5. In the state $|j, m = j\rangle$ the square of the total angular momentum of the particle is $\hbar^2\ j(j+1)$ and its projection along the z-axis is maximum $m = j$. Along a direction making an angle θ with the z-axis the angular momentum can have different projections with various probabilities. Calculate the probabilities of the projections $m = +j$ and $m = -j$.

6. Show that the operator $\boldsymbol{\sigma}_1 \cdot \boldsymbol{\sigma}_2$ has eigenvalue $+1$ for spin triplet state χ_1^m and -3 for the spin singlet state χ_0^0 of two spin-half particles. Construct the projection operators for the triplet and singlet states.

7. Couple the spins of two spin-half particles and express the spin triplet state ($S = 1$) and spin singlet state ($S = 0$) in terms of the uncoupled states. For the Clebsch-Gordan coefficients you can use Table 7.1.

8. Couple the orbital angular momentum $l = 1$ of an electron with its spin ($s = 1/2$) and write the wave functions of the coupled states ($P_{3/2}$ and $P_{1/2}$) in terms of the uncoupled states. For the Clebsch-Gordan coefficients you may use Table 7.2. For m take the maximum value (3/2 and 1/2) in the two cases.

9. Couple the orbital angular momentum $\ell = 2$ of an electron to its spin ($S = 1/2$)to construct the coupled states $|j = 5/2, m = 5/2\rangle$ and $|j = 3/2, m = 3/2\rangle$. Take the Clebsch-Gordan coefficients from Table 7.2.

10. Couple the spin of a two-Fermion system ($S = 1$) to their relative orbital angular momenta (i) $L = 0$ and (ii) $L = 2$ to construct ${}^3S_1$ and ${}^3D_1$ states, respectively. Use the Clebsch-Gordan coefficients from Table 7.2 and for m take the maximum value.

11. If $\hat{\boldsymbol{V}}$ is a vector operator then $\langle jm|\, \hat{V}_z\, |jm\rangle = \frac{m}{j(j+1)} \langle jm|\, \hat{\boldsymbol{J}} \cdot \hat{\boldsymbol{V}}\, |jm\rangle$. This is called the projection theorem for vector operators. Use this theorem to find the quantum mechanical expectation value for the operator μ_z for the coupled state $|(j_1 j_2)j, m = j\rangle$. Given: $\boldsymbol{\mu} = g_1 \hat{\boldsymbol{J}}_1 + g_2 \hat{\boldsymbol{J}}_2$.

12. In the coordinate representation, the orbital angular momentum operators $\hat{L}_x$, $\hat{L}_y$, $\hat{L}_z$, $\hat{L}^2$ are represented, respectively, by the differential operators

$$\begin{aligned} L_x &= -i\,\hbar \left(y\frac{\partial}{\partial z} - z\frac{\partial}{\partial y} \right) , \\ L_y &= -i\,\hbar \left(z\frac{\partial}{\partial x} - x\frac{\partial}{\partial z} \right) , \\ L_z &= -i\,\hbar \left(x\frac{\partial}{\partial y} - y\frac{\partial}{\partial x} \right) \\ L^2 &= L_x^2 + L_y^2 + L_z^2 . \end{aligned}$$

Re-express these differential operators in spherical polar coordinates.

13. Taking $\hat{L}^2$ to be represented in the coordinate representation by the differential operator

$$\hat{L}^2 \equiv -\hbar^2 \left[\frac{1}{\sin\theta}\frac{\partial}{\partial\theta}\left(\sin\theta\frac{\partial}{\partial\theta}\right) + \frac{1}{\sin^2\theta}\frac{\partial^2}{\partial\varphi^2} \right] ,$$

solve the eigenvalue equation

$$\hat{L}^2 Y(\theta,\varphi) = \lambda Y(\theta,\varphi)$$

to find the set of eigenvalues and eigenfunctions of $\hat{L}^2$. Show that these eigenfunctions are also the eigenfunctions of $\hat{L}_z = -i\,\hbar\frac{\partial}{\partial\varphi}$.

14. Show that the rotation operator pertaining to a rotation of axes, first by angle α about z-axis and then by angle β about the new y-axis and finally by angle γ about the new z-axis is given by: $\hat{R}(\alpha,\beta,\gamma) = \exp(-i\alpha\hat{J}_z)\exp(-i\beta\hat{J}_y)\exp(-i\gamma\hat{J}_x)$ where $\hat{J}_x$, $\hat{J}_y$, and $\hat{J}_z$ are components of the vector operator $\hat{\boldsymbol{J}}$ (in units of $\hbar$).

15. If A and B are two vectors and $\boldsymbol{\sigma} = (\sigma_x, \sigma_y, \sigma_z)$ is the Pauli spin vector then show that

$$(\boldsymbol{\sigma}\cdot\boldsymbol{A})(\boldsymbol{\sigma}\cdot\boldsymbol{B}) = \boldsymbol{A}\cdot\boldsymbol{B} + i\boldsymbol{\sigma}\cdot(\boldsymbol{A}\times\boldsymbol{B}) .$$

16. In the spherical polar coordinate system, the z-component $\hat{L}_z$ of the orbital angular momentum is given by

$$L_z = -i\,\hbar\frac{\partial}{\partial\varphi} .$$

Prove the commutation relation $[\hat{L}_z, \varphi] = i\hbar 1$, and show that the product of uncertainties in the simultaneous measurements of $\hat{L}_z$ and $\hat{\varphi}$ is given by

$$\Delta L_z \Delta\varphi \geq \hbar/2 .$$

17. Show that the total angular momentum quantum number j, defined by

$$\hat{J}^2\,|j,m\rangle = \hbar^2\,j(j+1)\,|j,m\rangle$$

can take only positive integer or half-integer values.

References

[1] M. E. Rose, *Elementary Theory of Angular Momentum* (John Wiley and Sons, New York, 1957).

[2] D. M. Brink and G. R. Satchler, *Angular Momentum* (Clarendon Press, Oxford, 1962).

[3] A. R. Edmonds, *Angular Momentum in Quantum Mechanics* (Princeton University Press, 1957).

8

APPROXIMATION METHODS

8.1 Introduction

We have seen that to determine the possible energies E_n (energy eigenvalues) which a physical system can have and the corresponding eigenfunctions (wave functions) ψ_n, we have to set up and solve the time-independent Schrödinger equation

$$H\psi_n = E_n\psi_n \,. \tag{8.1.1}$$

In this equation, the wave function ψ_n, representing the n-th eigenstate of the system in the coordinate representation, is a function of the continuously varying eigenvalues of the position observables of the system, the number of position observables being equal to number of degrees of freedom of the system. The Hamiltonian operator H, in the coordinate representation, is a differential operator involving the continuously varying eigenvalues of the position observables and the derivatives with respect to them. Since we will be dealing with the operators mostly in the coordinate representation, we shall denote them without the caret.

Now, the exact solutions of the time-independent Schrödinger equation are possible only in a limited number of cases such as the Hydrogen or Hydrogen-like atoms, linear harmonic oscillator, and other problems discussed in Chapter V. Even only slightly more complicated problems lead to equations that cannot be solved exactly. In such cases we have to take recourse to approximation methods. Different approximation methods are useful in different situations. Some of these methods discussed in this chapter are:

1. Perturbation methods
   In many problems, quantities of different order of magnitude appear in the Hamiltonian so that it can be written as $H_0+\lambda H'$, where the second term $\lambda H'$ is small (to be defined more precisely later) compared to H_0 such that when it is neglected, the problem can be solved excatly. H_0 is referred to as the unperturbed Hamiltonian and the small term $\lambda H'$ is referred to as a perturbation (correction) to the Hamiltonian H_0. In such cases, the first step is to solve the simplified problem exactly and then calculate approximately the corrections due to the small term that were neglected in the simplified problem. The methods for calculating these corrections are referred to as perturbation methods. These methods may be further classified into two groups.

   (a) Time-independent perturbation theory deals with perturbations that do not depend on time. Two cases, when the unperturbed Hamiltonian admits (i) non-degenerate states and (ii) degenerate states, will be considered separately.

   (b) Time-dependent perturbation theory deals with perturbations that depend on time and will be discussed in Chapter 10.

2. Variational method
3. Wentzel-Kramers-Brillouin-Jeffreys (WKBJ) approximation
4. Born approximation (used in scattering theory) discussed in Chapter 9
5. Adiabatic and Sudden approximations (Chapter 10)

   In addition, we will also discuss approximation methods pertaining to many Fermion (or many electron) systems. These include Hartree and Hartree-Fock equations for N-electron atoms, statistical model of the atom and *occupation number representation* for dealing with many Fermion systems.

8.2 Non-degenerate Time-independent Perturbation Theory

Consider the Hamiltonian H to be time-independent in which case the time-independent Schrödinger equation is:

$$H\psi_n = E_n\psi_n\,, \tag{8.1.1*}$$

where E_n is the energy for the eigenstate ψ_n. Suppose that the total Hamiltonian H can be split up into two parts:

$$H = H_0 + \lambda H'\,, \tag{8.2.1}$$

where H_0 is the unperturbed part and $\lambda H'$ is the perturbation. Here λ is a parameter, introduced purely as a device to keep track of different orders of corrections. By varying it between 0 and 1 we can also use it to *tune* the strength of the perturbation. At the end of the calculations we simply put $\lambda = 1$. The unperturbed part is such that the time-independent Schrödinger equation

$$H_0\psi_n^{(0)} = E_n^{(0)}\psi_n^{(0)} \tag{8.2.2}$$

is exactly solvable. We shall further assume in this section that the energy levels for H_0 are non-degenerate. This means that there exists only one state $\psi_n^{(0)}$ corresponding to each energy $E_n^{(0)}$. The unperturbed states form an orthonormal complete set satisfying

$$\int d^3r\psi_m^{(0)*}(\boldsymbol{r})\psi_n(0)(\boldsymbol{r}) = \delta_{mn} \tag{8.2.3}$$

and

$$\sum_n \psi_n^{(0)*}(\boldsymbol{r})\psi_n^{(0)}(\boldsymbol{r}') = \delta^3(\boldsymbol{r}-\boldsymbol{r}')\,. \tag{8.2.4}$$

The eigenstate ψ_n of the full Hamiltonian can now be expanded in terms of the eigenstates of H_0 as

$$\psi_n(\boldsymbol{r}) = \sum_m a_{mn}\psi_n^{(0)}(\boldsymbol{r})\,. \tag{8.2.5}$$

Now, since the perturbation is small, we expect that the changes in the wave functions as well as energies will be small. Accordingly, we assume that the coefficients a_{mn} have a *perturbation expansion* in powers of λ (or the perturbation)

$$a_{mn} = a_{mn}^{(0)} + \lambda a_{mn}^{(1)} + \lambda^2 a_{mn}^{(2)} + \cdots\,. \tag{8.2.6a}$$

The coefficients $a_{mn}^{(s)}$ are chosen so that

$$a_{nn}^{(0)} = 1 \quad \text{and} \quad a_{nn}^{(s)} = 0 \quad \text{for } s \geq 1\,. \tag{8.2.6b}$$

Without this last requirement, the coefficients are not uniquely defined. Similarly, we assume that the energy E_n also has a perturbation expansion

$$E_n = E_n^{(0)} + \lambda E_n^{(1)} + \lambda^2 E_n^{(2)} + \cdots . \tag{8.2.7}$$

Substituting these expansions into the perturbed Schrödinger equation (8.1.1*), we get

$$(H_0 + \lambda H') \sum_m (a_{mn}^{(0)} + \lambda a_{mn}^{(1)} + \lambda^2 a_{mn}^{(2)} + \cdots) \psi_m^{(0)}$$
$$= (E_n^{(0)} + \lambda E_n^{(1)} + \lambda^2 E_n^{(2)} + \cdots) \sum_m (a_{mn}{}^{(0)} + \lambda a_{mn}^{(1)} + \lambda^2 a_{mn}^{(2)} + \ldots) \psi_m^{(0)} . \tag{8.2.8}$$

Collecting and equating the zero, first, and second order terms in λ from both sides, we get

$$\sum_m a_{mn}^{(0)} H_0 \psi_m^{(0)} = E_n^{(0)} \sum_m a_{mn}^{(0)} \psi_m^{(0)} , \tag{8.2.9}$$

$$\sum_m H_0 a_{mn}^{(1)} \psi_m^{(0)} + \sum_m H' a_{mn}^{(0)} \psi_m^{(0)} = \sum_m E_n^{(0)} a_{mn}^{(1)} \psi_m^{(0)} + \sum_m E_n^{(1)} a_{mn}^{(0)} \psi_m^{(0)} , \tag{8.2.10}$$

$$\sum_m \left(H_0 a_{mn}^{(2)} + H' a_{mn}^{(1)} \right) \psi_m^{(0)} = \sum_m \left(E_n^{(0)} a_{mn}^{(2)} + E_n^{(1)} a_{mn}^{(1)} + E_n^{(2)} a_{mn}^{(0)} \right) \psi_m^{(0)} . \tag{8.2.11}$$

Similar relations for terms involving higher powers of λ can be obtained. The first equation (8.2.9) gives us

$$\sum_m a_{mn}^{(0)} \left(E_m^{(0)} - E_n^{(0)} \right) \psi_m^{(0)} = 0 . \tag{8.2.12a}$$

Multiplying both sides by $\psi_k^{(0)*}$, integrating over the coordinate space and using the orthogonality of unperturbed wave functions, we get

$$a_{kn}^{(0)} \left(E_k^{(0)} - E_n^{(0)} \right) = 0 . \tag{8.2.12b}$$

Now, since the states are non-degenerate, the difference $(E_k^{(0)} - E_n^{(0)})$ is nonzero only for $k = n$. It follows that for Eq. (8.2.12b) to hold all $a_{kn}^{(0)}$ with $k \neq n$ must vanish and for $k = n$ we must have $a_{nn}^{(0)} = 1$. This can be summarized as

$$a_{mn}^{(0)} = \delta_{mn} . \tag{8.2.13}$$

Using this result in Eq. (8.2.10), we obtain

$$\sum_m \left(E_m^{(0)} - E_n^{(0)} \right) a_{mn}^{(1)} \psi_m^{(0)} = \left(E_n^{(1)} - H' \right) \psi_n^{(0)} . \tag{8.2.14a}$$

Multiplying both sides of this equation by $\psi_k^{(0)*}$, integrating over the coordinate space, and using the orthogonality and linear independence of unperturbed wave functions, we get

$$(E_k^{(0)} - E_n^{(0)}) a_{kn}^{(1)} = E_n^{(1)} \delta_{kn} - H'_{kn} \tag{8.2.14b}$$

where

$$H'_{kn} \equiv \int d^3 r \psi_k^{(0)*}(\boldsymbol{r}) H' \psi_n^{(0)}(\boldsymbol{r}) \tag{8.2.14c}$$

is the kn-th element of the matrix representing H' in the basis in which H_0 is diagonal. If we put $k = n$ on both sides of Eq. (8.2.14b), we obtain

$$E_n^{(1)} = H'_{nn} = \int d^3 r \psi_n^{(0)*}(\boldsymbol{r}) \, H' \, \psi_n^{(0)}(\boldsymbol{r}) . \tag{8.2.15}$$

Thus we have the important result that the first order correction to the energy E_n is simply the expectation value of the perturbation H' in the unperturbed state $\psi_n^{(0)}$.

For $k \neq n$, Eq. (8.2.14b) leads to the first order correction to the state amplitude

$$a_{mn}^{(1)} = \frac{H'_{mn}}{E_n^{(0)} - E_m^{(0)}}\,, \qquad m \neq n\,, \tag{8.2.16}$$

where we have replaced k by m, since it is an arbitrary index. Thus, correct to first order in the perturbation H', the energy and the wave function for the n-th state are given by

$$E_n = E_n^{(0)} + H'_{nn}\,, \tag{8.2.17}$$

$$\psi_n = \sum_m \left(a_{mn}^{(0)} + a_{mn}^{(1)}\right)\psi_m^{(0)} = \psi_n^{(0)} + \sum_{m\neq n} \frac{H'_{mn}}{E_n^{(0)} - E_m^{(0)}}\,\psi_m^{(0)}\,. \tag{8.2.18}$$

To find the second order correction to the energy, we rearrange Eq. (8.2.11) to get

$$\sum_m a_{mn}^{(2)}(E_m^{(0)} - E_n^{(0)})\psi_m^{(0)} + \sum_m a_{mn}^{(1)}(H' - E_n^{(1)})\psi_m^{(0)} = E_n^{(2)}\psi_n^{(0)}\,. \tag{8.2.19a}$$

Again multiplying both sides by $\psi_m^{(0)*}$ and integrating over the coordiates space, we obtain

$$a_{kn}^{(2)}(E_k^{(0)} - E_n^{(0)}) + \sum_m a_{mn}^{(1)} H'_{km} - E_n^{(1)} a_{kn}^{(1)} = E_n^{(2)}\delta_{kn}\,. \tag{8.2.19b}$$

Putting $k = n$ on both sides of this equation, we get

$$\sum_{m\neq n} a_{mn}^{(1)} H'_{nm} = E_n^{(2)}\,. \tag{8.2.19c}$$

Substituting the expression for $a_{mn}^{(1)}$ given in Eq. (8.2.16), the second order correction to the energy E_n can be written as

$$E_n^{(2)} = \sum_{m\neq n} \frac{H'_{mn}\,H'_{nm}}{E_n^{(0)} - E_m^{(0)}}\,. \tag{8.2.20}$$

For $k \neq n$, Eq. (8.2.18b) give the second order correction to the wave function which, with the help of Eq. (8.2.16), can be written as

$$a_{kn}^{(2)} = \sum_{m\neq n} \frac{H'_{km}\,H'_{mn}}{\left(E_n^{(0)} - E_m^{(0)}\right)\left(E_n^{(0)} - E_k^{(0)}\right)} - \frac{H'_{nn}H'_{kn}}{\left(E_n^{(0)} - E_k^{(0)}\right)^2}\,, \qquad k \neq n\,. \tag{8.2.21}$$

Combining Eqs. (8.2.17) and (8.2.20), and (8.2.18) and (8.2.21), we obtain the energy and wave function of state ψ_n correct to second order in the perturbation

$$E_n = E_n^{(0)} + H'_{nn} + \sum_{m\neq n} \frac{H'_{mn}\,H'_{nm}}{E_n^{(0)} - E_m^{(0)}}\,, \tag{8.2.22}$$

$$\psi_n = \psi_n^{(0)} + \sum_{m\neq n} \left[\frac{H'_{mn}}{E_n^{(0)} - E_m^{(0)}}\left(1 - \frac{H'_{nn}}{E_n^{(0)} - E_m^{(0)}}\right) + \sum_{k\neq n} \frac{H'_{mk}\,H'_{kn}}{\left(E_n^{(0)} - E_m^{(0)}\right)\left(E_n^{(0)} - E_k^{(0)}\right)}\right]\psi_m^{(0)}\,. \tag{8.2.23}$$

This procedure can be continued to calculate higher order corrections. It will be seen that the calculation of energy correction $E_n^{(s)}$ requires knowlege of ψ_n only to order $s - 1$ in the perturbation. Let us consider some applications of the results so far.

Perturbation by a Linear Potential

A harmonic oscillator of charge e described by the Hamiltonian operator

$$\hat{H}_o = \frac{\hat{p}_x^2}{2m} + \frac{1}{2}m\omega^2\hat{x}^2 \,, \tag{8.2.24}$$

is subject to an external electric field in $+x$ direction. This causes a perturbation $\hat{H}' = -e\mathcal{E}\hat{x} = -\beta\hat{x}$. We are interested in calculating the effect of this perturbation on the eigenstates and eigenvalues. First of all note that this problem can be solved exactly by writing the total Hamiltonian as

$$\hat{H} \equiv \hat{H}_0 + \hat{H}' = \frac{\hat{p}_x^2}{2m} + \frac{1}{2}m\omega^2\hat{x}^2 - \beta\hat{x} = \frac{\hat{p}_x^2}{2m} + \frac{1}{2}m\omega^2(\hat{x} - \beta/m\omega^2)^2 - \frac{\beta^2}{2m\omega^2} \,. \tag{8.2.25}$$

By introducing new observables $\hat{X} = \hat{x} - \beta/m\omega^2$ and $\hat{p}_X = \hat{p}_x$ so that $[\hat{X}, \hat{p}_X] = i\hbar$, the total Hamiltonian can be written as

$$\hat{H} = \frac{\hat{p}_X^2}{2m} + \frac{1}{2}m\omega^2\hat{X}^2 - \frac{\beta^2}{2m\omega^2} \,. \tag{8.2.26a}$$

This Hamiltonian has exact energy eigenvalues

$$E_n = (n + 1/2)\hbar\omega - \frac{\beta^2}{2m\omega^2} \tag{8.2.26b}$$

and eigenfunctions $\Psi_n(X) = \psi_n(x - \beta/m\omega^2)$, where $\psi_n(z)$ are the linear harmonic oscillator eigenfunctions discussed in Sec. 5.3.

We now use perturbation theory to calculate the correction to energy levels due to $\hat{H}'$. The unperturbed Hamiltonian $\hat{H}_0$ has eigenvalues $E_n^{(0)} = (n + 1/2)\hbar\omega$ and eigenfunctions $\Psi_n^{(0)}(x) = \psi_n(x)$. Since the levels of a harmonic linear oscillator are non-degenerate, we can use non-degenerate perturbation theory to calculate the change in E_n. The first order correction to the energy of the n-th level is given by

$$E_n^{(1)} = -\beta \langle n | \, \hat{x} \, | n \rangle = -\beta \int_{-\infty}^{\infty} \psi_n(x) \, x \, \psi_n(x) dx, \tag{8.2.27}$$

which vanishes, since the wave functions of linear oscillator have definite parity $\psi_n(-x) = (-1)^n\psi(x)$. We shall see that symmetry arguments play an important role in simplifying the calculations.

So the first order correction to energy vanishes. In search of leading nonzero corrections we calculate the second order correction given by Eq. (8.8.20). To evaluate this, we work in the occupation number represenation [Sec. 5.4] and express the perturbation in terms of annihilation and creation operators as

$$\hat{H}' = -\beta\sqrt{\frac{\hbar}{2m\omega}}\,(\hat{a} + \hat{a}^\dagger) \,. \tag{8.2.28}$$

Its nonzero matrix elements are [see Chapter 5, Eq. (5.4.32)]

$$\langle n+1 | \, \hat{H}' \, | n \rangle = H'_{n+1,n} = -\beta\sqrt{\frac{\hbar}{2m\omega}}\,\sqrt{n+1} \,, \tag{8.2.29a}$$

$$\langle n-1 | \, \hat{H}' \, | n \rangle = H_{n-1,n} = -\beta\sqrt{\frac{\hbar}{2m\omega}}\sqrt{n} \,. \tag{8.2.29b}$$

Using these and the unperturbed energy $E_n^{(0)} = (n + 1/2)\hbar\omega$, we find the second order energy correction [Eq. (8.2.20)] is

$$
\begin{aligned}
E_n^{(2)} &= \sum_{m \neq n} \frac{H'_{mn}\, H'_{nm}}{E_n^{(0)} - E_m^{(0)}} = \frac{|H'_{n+1,n}|^2}{E_n^{(0)} - E_{n+1}^{(0)}} + \frac{|H'_{n-1,n}|^2}{E_n^{(0)} - E_{n-1}^{(0)}} \\
&= -\frac{\beta^2}{2m\omega^2}\,.
\end{aligned} \tag{8.2.30}
$$

Hence to second order in the perturbation the energies are

$$
E_n = E_n^{(0)} - \frac{\beta^2}{2m\omega^2}\,. \tag{8.2.31}
$$

A comparison of this to the exact energy (8.2.26b) shows that this is exact. Indeed it can be shown that corrections of order higher than 2 in perturbation are zero.

Ground State Energy of Helium Atom

As another application of the results of this section, we calculate the ground state energy of a Helium atom. In this calculation we ignore electron spin. The Hamiltonian for a Helium atom [Fig. 8.1] (in the coordinate representation) may be written as

$$
\begin{aligned}
H &= -\frac{\hbar^2}{2m_e}\nabla_1^2 - \frac{\hbar^2}{2m_e}\nabla_2^2 - \frac{2e^2}{4\pi\epsilon_0 r_1} - \frac{2e^2}{4\pi\epsilon_0 r_2} + \frac{e^2}{4\pi\epsilon_0 r_{12}} \\
&\equiv H_0 + \frac{e^2}{4\pi\epsilon_0 r_{12}} \equiv H_0 + H'\,,
\end{aligned} \tag{8.2.32}
$$

where m_e is the mass of the electron. The suffixes 1 and 2 refer to the two electrons and we regard the electron-electron interaction as the perturbation. The unperturbed Hamiltonian H_0 is the sum of the Hamiltonians H_{01} and H_{02} for the two electrons, each interacting with a nucleus of charge $+2e$:

$$
H_0 = \left(-\frac{\hbar^2}{2m_e}\nabla_1^2 - \frac{2e^2}{4\pi\epsilon_0 r_1}\right) + \left(-\frac{\hbar^2}{2m_e}\nabla_2^2 - \frac{2e^2}{4\pi\epsilon_0 r_2}\right) \equiv H_{01} + H_{02}\,.
$$

The unperturbed ground state wave function and energy for a Helium atom are determined by

$$
(H_{01} + H_{02})\psi_{gs}^0(\boldsymbol{r}_1, \boldsymbol{r}_2) = E_{gs}^0 \psi_{gs}^0(\boldsymbol{r}_1, \boldsymbol{r}_2)\,, \tag{8.2.33}
$$

where the unperturbed ground state wave function can be written as the product of single-electron wave functions

$$
\psi_{gs}^{(0)}(\boldsymbol{r}_1, \boldsymbol{r}_2) = \psi_{1gs}(\boldsymbol{r}_1)\psi_{2gs}(\boldsymbol{r}_2) \tag{8.2.34}
$$

where

$$
H_{01}\psi_{1gs}(\boldsymbol{r}_1) = E_{1gs}\psi_{1gs}(\boldsymbol{r}_1)\,, \tag{8.2.35}
$$

and

$$
H_{02}\psi_{2gs}(\boldsymbol{r}_2) = E_{2gs}\psi_{2gs}(\boldsymbol{r}_2)\,. \tag{8.2.36}
$$

Now the single-electron ground state energy and wave function are given by

$$
E_{1gs} = -\frac{Z^2 e^2}{8\pi\epsilon_0 a_o} = E_{2gs}\,, \qquad \text{with } Z = 2\,, \tag{8.2.37}
$$

and

$$
\psi_{1gs}(\boldsymbol{r}_1) = \left(\frac{Z^3}{\pi a_o^3}\right)^{1/2} \exp(-\rho_1/2), \qquad \text{where } \rho_1 = 2Zr_1/a_o \tag{8.2.38}
$$

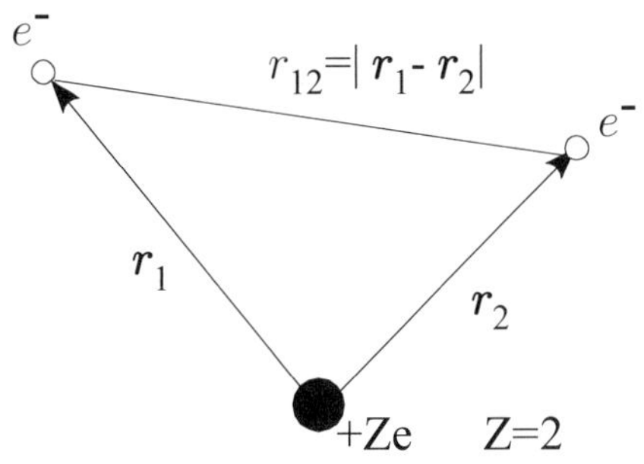

FIGURE 8.1
Helium atom has two-electrons interacting with a nucleus of charge $+2e$. The nucleus can be considered infinitely heavy compared to the electrons. We can then consider the electrons to be moving in the electrostatic potential of the nucleus.

and $a_o = 4\pi\epsilon_0\hbar^2/m_e e^2$ is the Bohr radius. The expression for $\psi_{2gs}(\boldsymbol{r}_2)$ can be obtained from $\psi_{1gs}(\boldsymbol{r}_1)$ by replacing the index 1 by 2. With the help of these equations, we find that the unperturbed ground state energy of the He atom is

$$E_{gs}^{(0)} = -\frac{2Z^2e^2}{8\pi\epsilon_0 a_o} \equiv -2Z^2 \text{ Rydberg}, \tag{8.2.39}$$

and the unperturbed ground state wave function is

$$\psi_{gs}^{(0)}(\boldsymbol{r}_1, \boldsymbol{r}_2) = \left(\frac{Z^3}{\pi a_o^3}\right) \exp\left[-(\rho_1 + \rho_2)/2\right] . \tag{8.2.40}$$

From Eq. (8.2.15), the first order perturbation correction $E_{gs}^{(1)}$ to the ground state energy is simply the expectation value of the perturbation in the unperturbed state:

$$\begin{aligned} E_{gs}^{(1)} &= \int d^3r_1 \int d^3r_2 \psi_{gs}^0 * (\boldsymbol{r}_1, \boldsymbol{r}_2) \frac{e^2}{4\pi\epsilon_0 r_{12}} \psi_{gs}^0(\boldsymbol{r}_1, \boldsymbol{r}_2) \\ &= \frac{Ze^2}{128\pi^3\epsilon_0 a_o} \iint \exp[-(\rho_1 + \rho_2)] \frac{1}{|\boldsymbol{\rho}_1 - \boldsymbol{\rho}_2|} d^3\rho_1 d^3\rho_2 . \end{aligned} \tag{8.2.41}$$

Taking the value of the integral[1] to be $20\pi^2$, we obtain

$$E_{gs}^{(1)} = \frac{5Z}{4} \frac{e^2}{8\pi\epsilon_0 a_o} . \tag{8.2.42}$$

[1] An integral of the kind

$$I = \frac{1}{4\pi\epsilon_0} \iint \frac{F(r_1)F(r_2)}{|\boldsymbol{r}_1 - \boldsymbol{r}_2|} d^3r_1 d^3r_2$$

may be looked upon as twice the potential energy of a spherically symmetric charge distribution $F(r) = (q_o/a^3)f(r/a)$, where a has dimension of length and q_o is the total charge in Coulomb. To calculate the work done in building up the charge distribution, we divide the charge distribution into thin concentric spherical shells of radius r and thickness dr and imagine the charge to be built up by adding these shells of charge starting with $r = 0$ to all the way up to $r \to \infty$. At a certain stage of this process, when a charge distribution of radius r already exists, the electrostatic potential at the surface of this charge distribution is given by

$$V(r) = \frac{1}{4\pi\epsilon_0 r} \int_0^r 4\pi r'^2 dr'\, F(r') = \frac{q_o}{\epsilon_0 a\rho} \int_0^\rho \rho'^2 d\rho' f(\rho') ,$$

where $\rho = r/a$. To extend this charge distribution of radius r to one of radius $r + dr$, with the same charge

Total ground state energy of the normal Helium atom, correct to first order is, therefore, with $Z = 2$

$$E_{gs} = \left(-2Z^2 + \frac{5}{4}Z\right) \frac{e^2}{8\pi\epsilon_0 a_o} = -5.5 \text{ Rydberg} = -74.8\,\text{eV}\,. \tag{8.2.43}$$

For comparison, the observed value of the ground state energy of the Helium atom is -78.9 eV.

8.3 Time-independent Degenerate Perturbation Theory

We shall now consider the perturbation method for stationary states when the unperturbed Hamiltonian H_0 admits degenerate states. Let the energy levels of H_0 be denoted by $E_1^{(0)}, E_2^{(0)}, \cdots, E_k^{(0)}, \cdots$. and let the level $E_k^{(0)}$ be α-fold degenerate such that it corresponds to a set of states $\psi_{k1}^0, \psi_{k2}^0, \cdots, \psi_{k\alpha}^0$. In such a case, we can construct a suitable linear combination of the degenerate states

$$\chi_{k\ell}^0 = \sum_{\ell'=1}^{\alpha} K_{\ell\ell'} \psi_{k\ell'}^0 \tag{8.3.1}$$

so that the ℓ-th perturbed wave function tends to the above combination when the perturbation vanishes. Thus, to first order in perturbation, the perturbed wave function may be expressed as

$$\psi_{k\ell} = \chi_{k\ell}^0 + \lambda \psi'_{k\ell}\,. \tag{8.3.2}$$

Expanding the correction $\psi'_{k\ell}$ in terms of the orthogonal set of functions $\psi_{k\ell}^{(0)}$ as

$$\psi'_{k\ell} = \sum_{\ell' k'} a_{k\ell,k'\ell'} \psi_{k'\ell'}^0\,, \tag{8.3.3}$$

and substituting this in the perturbed Schrödinger equation we obtain

$$(H_0 + \lambda H')(\chi_{k\ell}^0 + \lambda \psi'_{k\ell}) = (E_k^0 + \Lambda E'_{k\ell})(\chi_{k\ell}^0 + \lambda \psi'_{k\ell}) \tag{8.3.4}$$

---

distribution, we must bring in an amount of charge $dq = F(r)4\pi r^2 dr = q_o f(\rho) 4\pi \rho^2 d\rho$ from infinity to a distance $r = a\rho$ from the center. The amount of work done in this process is given by

$$dW = V(r)dq = \frac{4\pi q_o^2}{\epsilon_0 a} \rho d\rho\, f(\rho) \int_0^{\rho} \rho'^2 d\rho'\, f(\rho')\,.$$

Hence the total work in building up, from the origin outward, a spherical charge distribution of density $f(\rho)$ is given by

$$W = \frac{1}{2} I = \frac{4\pi q_o^2}{\epsilon_0 a} \int_0^{\infty} \rho\, d\rho\, f(\rho) \int_0^{\rho} \rho'^2\, d\rho'\, f(\rho')\,.$$

For $f(\rho) = e^{-\rho}$ we have $W = \frac{1}{2}\, I = \frac{4\pi q_o^2}{\epsilon_0 a} \frac{5}{8}$, which gives

$$I = \frac{1}{4\pi\epsilon_0} \frac{q_o^2}{a} \iint \frac{f(\rho_1) f(\rho_2)}{|\boldsymbol{\rho}_1 - \boldsymbol{\rho}_2|} d^3\rho_1 d^3\rho_2 = \frac{5\pi q_o^2}{\epsilon_o a} \quad \text{or} \quad \iint \frac{e^{-(\rho_1+\rho_2)}}{|\boldsymbol{\rho}_1 - \boldsymbol{\rho}_2|} d^3\rho_1 d^3\rho_2 = 20\pi^2\,.$$

where we have considered terms only up to the first order in the perturbation. Collecting zero order terms from both sides, we get

$$(H_0 - E_k^0)\chi_{k\ell}^0 = 0\,, \tag{8.3.5}$$

which is the unperturbed equation.

Collecting the first order terms we get

$$H_0\psi'_{k\ell} + H'\chi_{k\ell}^0 - E_k^0\psi'_{k\ell} - E'_{k\ell}\chi_{k\ell}^0 = 0\,. \tag{8.3.6}$$

Using the expansions (8.3.1) and (8.3.3), we can rewrite this equation as

$$\sum_{k'\ell'}(H_0 - E_k^0)a_{k\ell,k'\ell'}\psi_{k'\ell'}^0 + \sum_{\ell'}(H' - E'_{k\ell})K_{\ell\ell'}\psi_{k\ell'}^0 = 0\,.$$

Pre-multiplying both sides on the left by $\psi_{kj}^{(0)*}$, and integrating over the coordinate space, we get

$$\sum_{k'\ell'}(E_{k'}^0 - E_k^0)a_{k\ell,k'\ell'}\delta_{kk'}\Delta_{j\ell'} + \sum_{\ell'}\int \psi_{kj}^{0*}(H' - E'_{k\ell})K_{\ell\ell'}\psi_{k\ell'}^0 d\tau = 0\,.$$

The first term vanishes because of the degeneracy of energy levels. Then we get

$$\sum_{\ell'}(H'_{j\ell'} - E'_{k\ell}\Delta_{j\ell'})K_{\ell\ell'} = 0\,, \tag{8.3.7}$$

where
$$H'_{j\ell'} \equiv \int \psi_{kj}^{0*}\, H'\, \psi_{k\ell'}^0 d\tau\,, \tag{8.3.8}$$

and
$$\Delta_{j\ell'} \equiv \int \psi_{kj}^{0*}\psi_{k\ell'}^0 d\tau\,. \tag{8.3.9}$$

Note that if the degenerate eigenfunctions ψ_{kj}^0 for $j = 1, 2, \cdots, \alpha$, belonging to the same energy $E_k{}^{(0)}$ are orthogonal to each other, then $\Delta_{j\ell'}$ may be replaced by $\delta_{j\ell'}$.

Thus we get a set of linear equations in $K_{\ell 1}, K_{\ell 2}, \cdots, K_{\ell\alpha}$:

$$\sum_{\ell'=1}^{\alpha} A_{j\ell'}K_{\ell\ell'} = 0 \quad j = 1, 2, \cdots, \tag{8.3.10}$$

where
$$A_{j\ell'} \equiv H'_{j\ell'} - E'_{k\ell}\Delta_{j\ell'}\,. \tag{8.3.11}$$

The condition that the set of equations (8.3.10) gives a non-trivial solution for the set of quantities $K_{\ell\ell'}$ is that the determinant of the matrix of the coefficients $A_{j\ell'}$ must vanish:

$$\begin{vmatrix} H'_{11} - E'_{k\ell}\Delta_{11} & H'_{12} - E'_{k\ell}\Delta_{12} & \cdots & H'_{1\alpha} - E'_{k\ell}\Delta_{1\alpha} \\ H'_{21} - E'_{k\ell}\Delta_{21} & H'_{22} - E'_{k\ell}\Delta_{22} & \cdots & H'_{2\alpha} - E'_{k\ell}\Delta_{2\alpha} \\ \vdots & \vdots & \vdots & \vdots \\ H'_{\alpha 1} - E'_{k\ell}\Delta_{\alpha 1} & H'_{\alpha 2} - E'_{k\ell}\Delta_{\alpha 2} & \cdots & H'_{\alpha\alpha} - E'_{k\ell}\Delta_{\alpha\alpha} \end{vmatrix} = 0\,. \tag{8.3.12}$$

This secular equation is an α-degree polynomial in $E'_{k\ell}$ and by solving it we may find α roots and label these roots by ℓ and call them $E_{k\ell}^{(1)}$ where $\ell = 1, 2, 3, \cdots, \alpha$. Thus, as a result of the first order correction to the energy, the degeneracy is removed, partly or completely, and we have

$$E_k^{(0)} \to E_k^0 + E'_{k\ell}\,, \quad \ell = 1, 2, \cdots, \alpha\,.$$

For any one of the roots $E'_{k\ell}$ of the secular equation (8.3.12), we may solve the set of linear equations Eq. (8.3.10) to determine the set of coefficients $K_{\ell\ell'}$ to get the corresponding zero order wave function $\chi_{k\ell}^{(0)}$ to which the perturbed wave function $\psi_{k\ell}$ tends when the perturbation vanishes. Thus

to zero order
$$\psi_{k\ell} = \chi_{k\ell}^{0}, \tag{8.3.13}$$

and, to first order
$$E_{k\ell} = E_k^0 + E_{k\ell}^{(1)}. \tag{8.3.14}$$

In a very special case when the degenerate wave functions $\psi_{k\ell}^{(0)}$ can be identified with the correct combination $\chi_{k\ell}^{(0)}$, we have $K_{\ell\ell'} = \delta_{\ell\ell'}$. Also $\Delta_{j\ell} = \delta_{j\ell}$ if the set of degenerate functions are orthogonal to each other. In such a case the set of linear equations (8.3.10) implies that $A_{j\ell} \equiv H'_{j\ell} - E'_{k\ell}\delta_{j\ell} = 0$ or

$$E_{k\ell}^{(1)} = H'_{\ell\ell} = \int \psi_{k\ell}^{0*} H' \psi_{k\ell}^{0} d\tau\,. \tag{8.3.15}$$

Some of the most important application of the degenerate perturbation theory are in atomic physics. A few of these will be considered next.

Stark Effect in Hydrogen

We have seen that with the exception of the ground state, all energy levels of the Hydrogen atom with pure Coulomb interaction are highly degenerate. Some of this degeneracy is removed when an electric field is applied. This is called the *Stark effect.*

We can see this as follows. In the presence of an external electric field $\boldsymbol{\mathcal{E}} = \mathcal{E}\boldsymbol{e}_z$, where the z-axis is chosen along the direction of the field, the perturbed Hamiltonian of the Hydrogen atom may be written as

$$H = -\frac{\hbar^2}{2m}\nabla^2 - \frac{e^2}{4\pi\epsilon_0 r} - \boldsymbol{\mathcal{E}}\cdot\boldsymbol{d} \equiv H_0 + H'\,. \tag{8.3.16}$$

where $\boldsymbol{d} = -e\boldsymbol{r}$ is electric dipole moment of the atom and the unperturbed Hamiltonian H_0 and the perturbation H' can be written as

$$H_0 = -\frac{\hbar^2}{2m}\nabla^2 - \frac{e^2}{4\pi\epsilon_0 r}, \tag{8.3.17a}$$

$$H' = -\boldsymbol{\mathcal{E}}\cdot(-e\boldsymbol{r}) = e\mathcal{E}z = e\mathcal{E}r\cos\theta\,. \tag{8.3.17b}$$

The unperturbed problem for the Hydrogen atom was solved in Chapter 5 and the unperturbed wave functions are simply the functions $\psi_{n\ell m}^{(0)}(r,\theta,\varphi)$ given there. We can now discuss the effect of the perturbation on various states. We will consider only the $n = 1$ and $n = 2$ states here.

For $n = 1$ we have the ground state wave function

$$\psi_{100}(r,\theta,\varphi) = \frac{1}{\sqrt{\pi a_o^3}}\, e^{-r/a_o}\,. \tag{8.3.18}$$

The ground state is nondegenerate. The first order correction $\langle \psi_{100}|\, \hat{H}'\, |\psi_{100}\rangle$ to the energy of the ground state vanishes because the ground state wave function (8.3.18) is an even function of z whereas the perturbation (in the form $e\mathcal{E}z$) is an odd function of z.

For $n = 2$ we have four states which are degenerate in energy. The corresponding wave functions are

$$\psi_{200} = \frac{1}{\sqrt{32\pi a_o^3}}\left(\frac{r}{a_o} - 2\right) e^{-r/2a_o}\,,$$
$$\psi_{211} = \frac{1}{\sqrt{64\pi a_o^3}}\frac{r}{a_o} e^{-r/2a_o}\, e^{i\varphi}\sin\theta\,,$$
$$\psi_{21\bar{1}} = \frac{1}{\sqrt{64\pi a_o^3}}\frac{r}{a_o} e^{-r/2a_o}\, e^{-i\varphi}\sin\theta\,,$$
$$\psi_{210} = \frac{1}{\sqrt{32\pi a_o^3}}\frac{r}{a_o} e^{-r/2a_o}\,\cos\theta\,.$$

To determine the first order perturbation correction to the energy in this case we set up the secular determinant and equate it to zero. Since the degenerate functions $\psi_{200}, \psi_{211}, \psi_{21\bar{1}}$ and ψ_{210} are orthogonal $\Delta_{j\ell'} = \delta_{j\ell'}$, where j or ℓ' stand for the set of quantum numbers 200, 211 etc.

The secular equation is ($\bar{1} \equiv -1$):

$$\begin{vmatrix} (H'_{200,200} - E') & H'_{200,211} & H'_{200,210} & H'_{200,21\bar{1}} \\ H'_{211,200} & (H'_{211,211} - E') & H'_{211,210} & H'_{211,21\bar{1}} \\ H'_{210,200} & H'_{210,211} & (H'_{210,210} - E') & H'_{210,21\bar{1}} \\ H'_{21\bar{1},200} & H'_{21\bar{1},211} & H'_{21\bar{1},210} & H'_{21\bar{1},21\bar{1}} - E' \end{vmatrix} = 0\,, \tag{8.3.19}$$

where E' denotes the first order perturbation correction to the energy of $n = 2$ state. We can easily check that the only two nonzero matrix elements of H' in Eq. (8.3.19) are

$$H'_{200,210} = H'_{210,200} = e\mathcal{E}\int d^3r\psi^*_{210} r\cos\theta\psi_{200} = -3e\mathcal{E}a_o \tag{8.3.20}$$

Using this result in Eq. (8.3.19) we find, on simplification, the following equation

$$E'^2\left(E'^2 - (3e\mathcal{E}a_o)^2\right) = 0\,, \tag{8.3.21a}$$

which gives four roots

$$E' = 0,\ 0,\ 3e\mathcal{E}a_o,\ -3e\mathcal{E}a_o\,. \tag{8.3.21b}$$

Thus the four-fold degeneracy of the $n = 2$ state is partly lifted in first order of the perturbation. The first two values (both zero) of E' lead to energy $E_2 = E_2^{(0)} = -e^2/32\pi\epsilon_0 a_o$ and correspond to any two linearly independent combinations of ψ_{211} and $\psi_{21\bar{1}}$. This level is two-fold degenerate. The remaining two values correspond, respectively, to antisymmetric and symmetric combinations of ψ_{200} and ψ_{210}:

$$E_2^{(0)} \quad\rightarrow\quad \begin{cases} E_2^{(0)} + 3eEa_o = -\dfrac{e^2}{32\pi\epsilon_0 a_o} + 3eEa_o \leftrightarrow \chi_2^- = \frac{1}{\sqrt{2}}(\psi_{200} - \psi_{210})\,, \\ E_2^{(0)} - 3eEa_o = -\dfrac{e^2}{32\pi\epsilon_0 a_o} - 3eEa_o \leftrightarrow \chi_2^+ = \frac{1}{\sqrt{2}}(\psi_{200} + \psi_{210})\,. \end{cases} \tag{8.3.22}$$

The functions on the right represent the combinations of degenerate states to which the perturbed states corresponding to these energies will tend as the perturbation is withdrawn.

We note that the perturbation mixes only states with the same m value. This is a consequence of the fact that the perturbation commutes with $\hat{L}_z$. Hence the perturbed states can still be required to be eigenstates of $\hat{L}_z$. This means that it was necessary to consider only the linear combination of ψ_{210} and ψ_{200} to construct orthonormal basis of degenerate states, which would have resulted in a quadratic secular equation. The states ψ_{211} and $\psi_{21\bar{1}}$ stay unmixed. Recognition of symmetries of $\hat{H}_0$ and $\hat{H}'$ can simplify the calculations considerably. We shall see many examples of this in the next few applications of degenerate perturbation theory.

Fine Structure of Hydrogen

In Chapter 5 we solved the Hydrogen-like atom problem with Hamiltonian $\hat{H}_0 = \frac{\hat{p}^2}{2m_e} - \frac{Ze^2}{4\pi\epsilon_0 r}$. This Hamiltonian ignored the spin of the electron (as also the nucleus) and assumed electron motion to be non-relativistic. Hence there are several corrections to this Hamiltonian, which include the relativistic correction to the kinetic energy due to the fast motion of the electron and the spin-orbit interaction due to the spin of the electron. These corrections are of the same order of magnitude. They follow naturally in the non-relativistic limit of Dirac theory discussed in Chapter 13.

From the relativistic expression $E = \sqrt{p^2c^2 + m_e^2c^2}$ for the energy of a particle we obtain the kinetic energy K after subtracting the rest energy m_ec^2. For non-relativistic speeds $v/c = pc/E \ll 1$, the kinetic energy is given by

$$K = \sqrt{p^2c^2 + m_e^2c^4} - m_ec^2 = m_ec^2\left[\sqrt{1 + \frac{p^2c^2}{m_e^2c^4}} - 1\right] = \frac{p^2}{2m_e} - \frac{p^4}{8m_e^3c^2} + \cdots . \tag{8.3.23}$$

The first term is the non-relativistic kinetic energy already included in the Hamiltonian $\hat{H}_0$. The second term gives relativistic correction to the kinetic energy

$$\hat{H}_{kin} = -\frac{\hat{p}^4}{8m_e^3c^2} . \tag{8.3.24}$$

The spin-orbit term arises from the interaction of electron spin with the electric field produced by the nucleus. Electron spin arises naturally in the Dirac theory of the electron. It suffices to say at this stage that there is wealth of spectroscopic evidence for the existence of an intrinsic angular momentum (spin) and associated magnetic moment of the electron. This can form the basis for introducing a spin-dependent term in the Hamiltonian.

The spin of the electron participates in the conservation of angular momentum on equal footing with the orbital angular momentum and like any angular momentum associated with a charge particle, gives rise to an intrinsic magnetic moment

$$\boldsymbol{\mu} = -\frac{e}{2m_e} g_S \boldsymbol{S} , \tag{8.3.25}$$

where $\boldsymbol{S}$ is the spin angular momentum of the electron and g_S is the corresponding gyromagnetic ratio (the ratio of the magnetic moment to the spin angular momentum). For the electron spin $g_S = 2$. This magnetic moment moving in the electric field of the nucleus sees a magnetic field $\boldsymbol{B}_n = -\boldsymbol{v} \times \boldsymbol{E}/c^2$, where $\boldsymbol{v} = \boldsymbol{p}/m_e$ is the velocity of the electron and $\boldsymbol{E}$ is the electric field of the nucleus given by

$$\boldsymbol{E} = -\boldsymbol{\nabla}\left[-\frac{Ze}{4\pi\epsilon_0 r}\right] = -\frac{Ze}{4\pi\epsilon_0 r^3}\boldsymbol{r} . \tag{8.3.26a}$$

With the help of this equation, the magnetic field of the nucleus as seen by the electron can be written as

$$\boldsymbol{B}_n = -\boldsymbol{v} \times \boldsymbol{E}/c^2 = \frac{\boldsymbol{p}}{m_e} \times \left[\frac{Ze}{4\pi\epsilon_0 c^2 r^3}\boldsymbol{r}\right] = -\frac{Ze}{4\pi\epsilon_0 m_e c^2 r^3}\boldsymbol{L} , \tag{8.3.26b}$$

where $\boldsymbol{L} = \boldsymbol{r} \times \boldsymbol{p}$ is the orbital angular momentum of the electron. The potential energy associated with the interaction of the intrinsic magnetic moment with this magnetic field is

$$H_{LS} = -\boldsymbol{\mu} \cdot \boldsymbol{B}_n = -\frac{e}{2m_e} g_s \boldsymbol{S} \cdot \left[-\frac{Ze}{4\pi\epsilon_0 m_e c^2 r^3} \boldsymbol{L} \right] = \frac{Ze^2}{4\pi\epsilon_0 m_e^2 c^2 r^3} \boldsymbol{S} \cdot \boldsymbol{L} \,. \tag{8.3.26c}$$

Thus the interaction of the moving intrinsic magnetic moment of the electron with the electric field of the nucleus is really an interaction between the spin and orbital angular momentum of the electron. For this reason, this interaction is also referred to as spin-orbit interaction.

We have used the nonrelativistic kinematics to arrive at the spin-orbit interaction, which has the correct form but it is twice as large as a relativistically correct treatment gives. This factor is explained by the *Thomas precession.* Incorporating this we get the final form for the spin-orbit interaction

$$\hat{H}_{LS} = \frac{Ze^2}{8\pi\epsilon_0 m_e^2 c^2 r^3} \hat{\boldsymbol{S}} \cdot \hat{\boldsymbol{L}} \,. \tag{8.3.27}$$

Both $\hat{H}_{kin}$ and $\hat{H}_{LS}$ are small terms compared to H_0. Therefore a first order perturbative treatment is adequate to calculate their effect on the energy levels. The total Hamiltonian can be written as

$$\hat{H} = \hat{H}_0 + \hat{H}' \,, \tag{8.3.28a}$$

where

$$\hat{H}_0 = \frac{\hat{p}^2}{2m_e} - \frac{Ze^2}{4\pi\epsilon_0 r} \,, \tag{8.3.28b}$$

and

$$\hat{H}' = \hat{H}_{kin} + \hat{H}_{LS} \,. \tag{8.3.28c}$$

Because of the introduction of spin, the basis states must now include the information about the spin and the orbital angular momentum of the electron. The unperturbed states may thus be labelled as $\left|E_n^0\right\rangle = \left|n\ell m_\ell \, s m_s\right\rangle$. This implies that every state of the H atom without spin, is now doubled due to the two possible states of the spin $m_s = \pm\frac{1}{2}$ and for each n we now have $2n^2$ degenerate states. We are thus dealing with degenerate perturbation theory and even for moderate values of n we have to diagonalize faily large energy matrices. We can avoid a lot of this work if we notice that $\hat{H}'$ commutes with $\hat{J}^2$, $\hat{L}^2$, $\hat{S}^2$, and $\hat{J}_z$, where $\hat{\boldsymbol{J}} = \hat{\boldsymbol{L}} + \hat{\boldsymbol{S}}$ is the total angular momentum observable. Hence a suitable basis is $|n(\ell s)jm\rangle$. These states are easily constructed from uncoupled (angular momentum) states $|n\ell m_\ell \, s m_s\rangle$ with the help of Clebsch-Gordan coefficients. In this scheme atomic states are written in rather arcane (but standard) spectroscopic notation as $\left|n\,{}^{2S+1}L_J\right\rangle$, where n denotes the principal quantum number, the total orbital angular momentum L is denoted by capital letters $S, P, D, F, \cdots$, which correspond, respectively, to $L = 0, 1, 2, 3, 4\cdots$ and S and J indicate numerical values of total spin and total angular momenta. (We use upper case letters to denote total angular momenta and lower case letters to denote angular momenta of individual electrons.) For single electron atoms the lower case and upper case letters can be used interchangeably. This also holds for alkali atoms: Lithium, Sodium, Potassium, Cesium, Rubidium etc. as long as we restrict to the excitation of the single electron outside the closed shell. In this notation, the lowest lying states of hydrogenic atoms are labelled as follows.

For $n = 1$, the orbital angular momentum is $L = \ell = 0$ and the electron spin $S = s = \frac{1}{2}$ leading to total angular momentum $J = \frac{1}{2}$ and $m = \pm\frac{1}{2}$. These states are labeled as $1^2S_{1/2}$ (m values are suppressed).

For $n=2$, the orbital angular momentum is $L=\ell=0,1$ and spin $S=s=\frac{1}{2}$ giving us $J=j=\frac{1}{2}(\ell=0); \frac{1}{2}(\ell=1), \frac{3}{2}(\ell=1)$ for a total of eight ($2n^2$) states. These states are denoted by ${}^2S_{1/2}$, ${}^2P_{1/2}$, and ${}^2P_{3/2}$, respectively. The m values have been suppressed.

For $n=3$ we have $L=\ell=0,1,2$ and $S=s=\frac{1}{2}$ giving us $J=j=1/2(\ell=0); 1/2, 3/2(\ell=1)$; and $3/2, 5/2(\ell=2)$ for a total of eighteen states. These are denoted by ${}^2S_{1/2}$; ${}^2P_{1/2}$, ${}^2P_{3/2}$; ${}^2D_{3/2}$, ${}^2D_{5/2}$.

By expressing $\hat{\boldsymbol{L}} \cdot \hat{\boldsymbol{S}}$ as

$$\hat{\boldsymbol{L}} \cdot \hat{\boldsymbol{S}} = \frac{1}{2}\left[(\hat{\boldsymbol{L}}+\hat{\boldsymbol{S}})^2 - \hat{\boldsymbol{L}}^2 - \hat{\boldsymbol{S}}^2\right] = \frac{1}{2}\left[\hat{\boldsymbol{J}}^2 - \hat{\boldsymbol{L}}^2 - \hat{\boldsymbol{S}}^2\right], \tag{8.3.29}$$

we see that with basis vectors $|n(\ell s)jm\rangle$, both $\hat{H}_0$ and $\hat{H}'$ are diagonal.

The first order correction $E^{(1)}_{LS} \equiv \langle n(\ell s)jm|\, \hat{H}_{LS}\, |n(\ell s)jm\rangle$ to the energy of state $|n(\ell s)jm\rangle$ due to spin-orbit interaction is then

$$E^{(1)}_{LS} = \frac{Ze^2}{16\pi\epsilon_0 m_e^2 c^2}\left\langle \frac{1}{r^3}\right\rangle_{n\ell} \hbar^2\,[j(j+1)-\ell(\ell+1)-s(s+1)]\,. \tag{8.3.30}$$

Substituting the value of $\left\langle \frac{1}{r^3}\right\rangle_{n\ell}$ from Eq. (5.6.31) in this equation and rewriting the resulting expression in terms of unperturbed energy

$$E^{(0)}_n = -\frac{Z^2e^2}{2a_o n^2}, \quad a_o = \frac{4\pi\epsilon_0\hbar^2}{m_e e^2}, \tag{8.3.31}$$

we find the first order correction due to spin-orbit interaction is

$$E^{(1)}_{LS} = m_e c^2\, \frac{(Z\alpha)^2}{4n^3}\, \frac{[j(j+1)-\ell(\ell+1)-s(s+1)]}{\ell(\ell+1/2)(\ell+1)}, \tag{8.3.32}$$

where

$$\alpha = \frac{e^2}{4\pi\epsilon_0\hbar c} \approx \frac{1}{137}. \tag{8.3.33}$$

The dimensionless constant α, because of its origin in this calculation, is called the *fine structure constant.* From Eq. (8.3.32) we can see that spin-orbit correction is indeed a small fraction $|E^{(1)}_{LS}|/|E^{(0)}_n| \approx \mathcal{O}[(Z\alpha)^2/n]$ of the unperturbed energy. However, for each energy level with a given n, the degeneracy with respect to ℓ is lifted but in a special way; for each nonzero orbital angular momentum ℓ, the total angular momentum can take two values $j=\ell+\frac{1}{2}$ and $\ell-\frac{1}{2}$ so that each energy level with given n and $\ell \neq 0$ splits into two with the $j=\ell+\frac{1}{2}$ level having higher energy than $j=\ell-\frac{1}{2}$ level. Formula (8.3.32) holds for $\ell=0$ as well. In this case ($\ell=0$), there is no spin-orbit splitting as we only have one value of $j=\frac{1}{2}$.

To calculate the correction due to the kinetic energy $\hat{H}_{kin}$ term we note that from $\hat{H}_0 = \frac{\hat{p}^2}{2m_e} - \frac{Ze^2}{4\pi\epsilon_0 r}$, which holds for hydrogenic atoms, we can write the kinetic energy term as

$$\hat{H}_{kin} = -\frac{1}{2m_e c^2}\left(\frac{p^2}{2m_e}\right)^2 = -\frac{1}{2m_e c^2}\left[\hat{H}_0 + \frac{Ze^2}{4\pi\epsilon_0 r}\right]^2. \tag{8.3.34}$$

This term is also diagonal in $|n(\ell s)jm\rangle$ basis, giving the first order correction $E^{(1)}_{kin} \equiv \langle n(\ell s)jm| \hat{H}_{kin} |n(\ell s)jm\rangle$

$$\begin{aligned} E^{(1)}_{kin} &= -\frac{1}{2m_e c^2} \langle n(\ell s)jm| \left(\hat{H}_0 + \frac{Ze^2}{4\pi\epsilon_0 r}\right)\left(\hat{H}_0 + \frac{Ze^2}{4\pi\epsilon_0 r}\right) |n(\ell s)jm\rangle \\ &= -\frac{1}{2m_e c^2}\left[E^{(0)2}_n + \frac{2Ze^2 E^{(0)}_n}{4\pi\epsilon_0}\left\langle\frac{1}{r}\right\rangle_{n\ell} + \left(\frac{Ze^2}{4\pi\epsilon_0}\right)^2\left\langle\frac{1}{r^2}\right\rangle_{n\ell}\right] . \end{aligned} \quad (8.3.35)$$

Substituting the expressions for the unperturbed energy [Eq. (8.3.31)] and the radial integrals from Eqs. (5.6.32) and (5.6.33), we obtain the first order relativistic correction

$$E^{(1)}_{kin} = -m_e c^2 \frac{(Z\alpha)^4}{2n^4}\left[\frac{2n}{(2\ell+1)} - \frac{3}{4}\right] . \quad (8.3.36)$$

Note that this correction is always negative since $\ell \leq n-1$, so that the relativistic correction always lowers the energy levels. It is of the same order $mc^2 \times \alpha^4$ as the spin-orbit correction. However, since it depends only on n and ℓ but not j, it changes the overall dependence of energy on n and ℓ but does not contribute to the splitting of the energy levels.

Combining the spin-orbit and relativistic corrections, both being of the same order ($|E^{(0)}_n|\alpha^4$), for a level with quantum number n, we obtain the overall change in energy

$$E^{(1)}_{FS} \equiv E^{(1)}_{LS} + E^{(1)}_{kin} = -\frac{m_e c^2 (Z\alpha)^4}{2n^4}\left[\frac{n}{j+\frac{1}{2}} - \frac{3}{4}\right] = E^{(0)}_n \frac{(Z\alpha)^2}{n^2}\left[\frac{n}{j+\frac{1}{2}} - \frac{3}{4}\right] , \quad (8.3.37)$$

for both $\ell = j \pm \frac{1}{2}$. The overall energy change in energy thus depends only on n and j. Its net effect is to split each unperturbed $n, \ell \neq 0$ level into two with $n, j = \ell + \frac{1}{2}$ and $n, j = \ell - \frac{1}{2}$, the level with lower j having the lower energy. The splitting of a level with given ℓ is due entirely to the spin-orbit interaction, while the kinetic energy correction decreases the energy of both states relative to the unperturbed energy level. Figure 8.2 shows the fine structure of the $n = 2$ level in hydrogen when spin-orbit and kinetic energy corrections are included. Note that ${}^2S_{1/2}$ and ${}^2P_{1/2}$ levels are still degenerate. This degeneracy is removed when further (small) corrections due to the vacuum quantum fluctuations of the electromagnetic field are taken into account (not shown). The split between ${}^2S_{1/2}$ and ${}^2P_{1/2}$ levels is called the *Lamb shift*. It can only be calculated using quantum field theory, which explains the Lamb shift to 12 decimal places in agreement with experiment!

In closing this section we note that by a judicious choice of unperturbed basis states based on a recognition of appropriate symmetries, we were able to turn the calculation of degenerate perturbation theory into one of nondegenerate perturbation theory.

8.4 The Zeeman Effect

The change in the energy levels of an atom caused by the application of a uniform external magnetic field is called the *Zeeman effect*. Again we illustrate this for hydrogenic atoms. If such an atom is placed in a uniform external magnetic field $\boldsymbol{B}$, represented by the vector potential

$$\boldsymbol{A} = \frac{1}{2}\boldsymbol{B} \times \boldsymbol{r} , \text{ with } \boldsymbol{\nabla} \cdot \boldsymbol{A} = 0, \text{ and } \boldsymbol{B} = \boldsymbol{\nabla} \times \boldsymbol{A} , \quad (8.4.1)$$

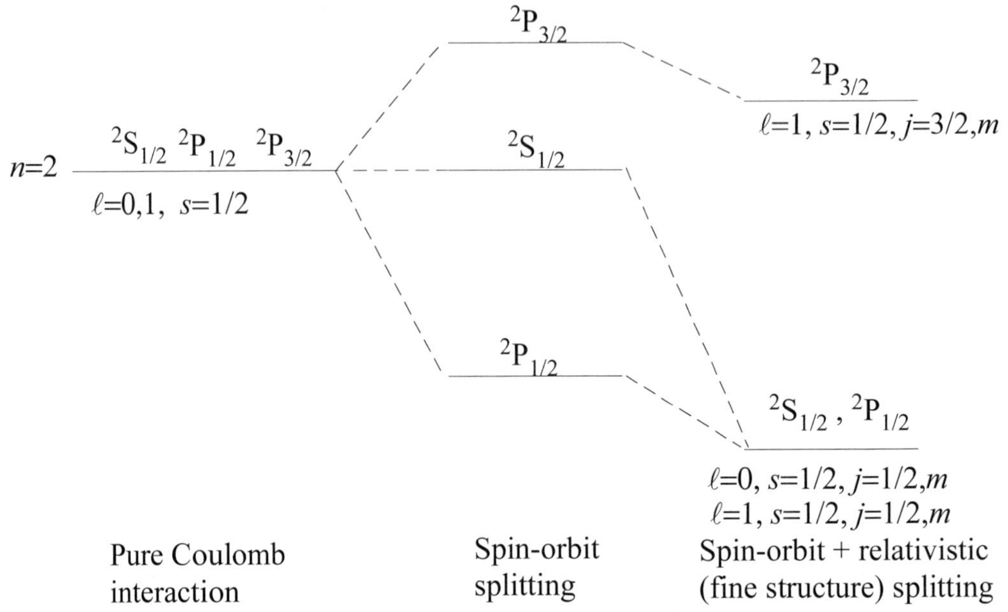

FIGURE 8.2
Fine structure splitting $n = 2$ level in hydrogen atom. The figure is not to scale. The middle column shows the effect of spin-orbit correction alone, while the third column shows the effect of both spin-orbit and relativistic kinetic energy correction. Lamb shift splitting of $^2S_{1/2}$ and $^2P_{1/2}$ levels is not shown.

its kinetic energy must be modified by replacing the momentum $\hat{p} \to (\hat{p} + e\boldsymbol{A})$ (electronic charge being $-e$). In addition, the interaction of the intrinsic magnetic moment of the electron [Eq. (8.3.25)] with the external magnetic field adds a term $-\hat{\boldsymbol{\mu}} \cdot \boldsymbol{B}$ to the Hamiltonian. Thus the overall Hamiltonian becomes

$$\hat{H} = \frac{(\hat{\boldsymbol{p}} + e\boldsymbol{A})^2}{2m_e} - \frac{Ze^2}{4\pi\epsilon_0 r} + \hat{H}_{FS}\,, \tag{8.4.2}$$

where the fine structure Hamiltonian $\hat{H}_{FS} = \hat{H}_{LS} + \hat{H}_{kin}$ represents the sum of the spin-orbit interaction and the relativistic correction to the kinetic energy. In Eq. (8.4.3), written in the coordinate representation, $\boldsymbol{A}$ is an ordinary function and $\hat{\boldsymbol{p}}$ ($\equiv -i\hbar\boldsymbol{\nabla}$) is a differential operator acting to the right on a wave function. Keeping this in mind, the first term of the Hamiltonian (8.4.2) can be written as

$$\begin{aligned}\frac{1}{2m_e}(\hat{\boldsymbol{p}} + e\boldsymbol{A})^2\psi(\boldsymbol{r}) &= \frac{1}{2m_e}(-i\hbar\boldsymbol{\nabla} + e\boldsymbol{A}) \cdot (-i\hbar\boldsymbol{\nabla} + e\boldsymbol{A})\psi(\boldsymbol{r})\,,\\ &= \left(-\frac{\hbar^2}{2m_e}\nabla^2 - \frac{ie\hbar}{m_e}\boldsymbol{A}\cdot\boldsymbol{\nabla}\right)\psi(\boldsymbol{r}) + \left(\frac{ie\hbar}{2m_e}\boldsymbol{\nabla}\cdot\boldsymbol{A} + \frac{e^2A^2}{2m_e}\right)\psi(\boldsymbol{r})\,.\end{aligned} \tag{8.4.3a}$$

Using $\boldsymbol{\nabla} \cdot \boldsymbol{A} = 0$ and

$$-\frac{ie\hbar}{m_e}\boldsymbol{A}\cdot\boldsymbol{\nabla}\psi(\boldsymbol{r}) = \frac{e}{2m_e}(\boldsymbol{B}\times\boldsymbol{r})\cdot\hat{\boldsymbol{p}}\psi(\boldsymbol{r}) = \frac{e}{2m_e}\boldsymbol{B}\cdot(\boldsymbol{r}\times\hat{\boldsymbol{p}})\psi(\boldsymbol{r}) = \frac{e}{2m_e}\boldsymbol{B}\cdot\hat{\boldsymbol{L}}\psi(\boldsymbol{r})\,, \tag{8.4.3b}$$

where $\hat{\boldsymbol{L}} = \boldsymbol{r} \times \hat{\boldsymbol{p}}$, in Eq. (8.4.3a) we have

$$\frac{1}{2m_e}(\hat{\boldsymbol{p}} + e\boldsymbol{A})^2 = \frac{\hat{\boldsymbol{p}}^2}{2m_e} + \frac{e}{2m_e}\boldsymbol{B}\cdot\hat{\boldsymbol{L}} + \frac{e^2(\boldsymbol{r}\times\boldsymbol{B})^2}{8m_e}. \tag{8.4.4}$$

Using this in Eq. (8.4.2) together with the expression for the intrinsic magnetic moment of the electron (8.3.25), we find the overall Hamiltonian for a hydrogenic atom in an external uniform field is

$$\hat{H} = \frac{\hat{\boldsymbol{p}}^2}{2m_e} - \frac{Ze^2}{4\pi\epsilon_0 r} + \hat{H}_{FS} + \frac{e}{2m_e}(\hat{\boldsymbol{L}} + 2\hat{\boldsymbol{S}})\cdot\boldsymbol{B} + \frac{e^2(\boldsymbol{r}\times\boldsymbol{B})^2}{8m_e}. \tag{8.4.5}$$

Recall that the contributions from $\hat{H}_{LS}$ are of order $m_e c^2\alpha^2 \approx 10^{-4}$ eV. The order of magnitude of the fourth term is $\frac{e\hbar}{2m_e}B = 5.8\times 10^{-9}\ \frac{\text{eV}}{\text{gauss}} \times\ B(\text{in gauss})$. Thus unless the external field is as large as 10^4 gauss, an extremely large field (the earth's magnetic field is about 1 gauss), the fourth term is small compared with the $\hat{H}_{FS}$ term. The fifth term, quadratic in the field, is completely negligible, being of the order of 10^{-15} to 10^{-16} eV/gauss^2. Ignoring the fifth term we can then write the Hamiltonian for a hydrogenic atom in an external magnetic field as

$$\hat{H} = \frac{\hat{\boldsymbol{p}}^2}{2m_e} - \frac{Ze^2}{4\pi\epsilon_0 r} + \hat{H}_{FS} + \hat{H}_{mag}\,, \tag{8.4.6a}$$

where

$$\hat{H}_{FS} = \hat{H}_{LS} + \hat{H}_{kin}\,, \tag{8.4.6b}$$

$$\hat{H}_{mag} = \frac{eB}{2m_e}(\hat{L}_z + 2\hat{S}_z) = \frac{eB}{2m_e}(\hat{J}_z + \hat{S}_z)\,, \tag{8.4.6c}$$

$\hat{L}_z$, $\hat{S}_z$, and $\hat{J}_z(=\hat{L}_z + \hat{S}_z)$ are, respectively, the z-components of orbital, spin and total angular momentum operators and we have taken the direction of $\boldsymbol{B}$ field to be the z-axis.

The next step depends on the what choice we make for the unperturbed Hamiltonian and that depends on the relative size of $\hat{H}_{FS}$ and $\hat{H}_{mag}$. From the discussion in the preceding paragraph, it follows that we have to consider weak ($B < 10^4$ gauss) and strong magnetic fields separately.

Weak Field Zeeman Effect

In this case the unperturbed Hamiltonian is $\hat{H}_0 = \frac{\hat{\boldsymbol{p}}^2}{2m_e} - \frac{Ze^2}{4\pi\epsilon_0 r} + \hat{H}_{FS}$. We have seen that fine structure interaction forces the atom into eigenstates of $\hat{J}^2$ and $\hat{J}_z$. The unperturbed Hamiltonian $\hat{H}_0$ and $\hat{L}^2, \hat{S}^2, \hat{J}^2$, and $\hat{J}_z$ form are mutually commuting observables. Hence the basis states are $|\ell s j m\rangle$ (suppressing n). Of these observables only $\hat{J}^2$ fails to commute with the perturbation $\hat{H}_{mag}$. This means $\hat{H}_{mag}$ is diagonal in ℓ, s, m. Besides the diagonal elements only nonzero elements of $\hat{H}_{mag}$ are

$$\frac{eB}{2m_e}\langle n\ell s j' m|\,\hat{J}_z + \hat{S}_z\,|n\ell s j m\rangle = \frac{eB}{2m_e}\Big[m\hbar\delta_{j'j} + \langle \ell s j' m|\,\hat{S}_z\,|\ell s j m\rangle\Big]\,. \tag{8.4.7}$$

The remaining matrix element can be evaluated by expressing the coupled angular momentum states $|\ell s j m\rangle$ in terms of uncoupled states $|m_\ell m_s\rangle$ using the Clebsch-Gordan coefficients [Chapter 7] as

$$\begin{aligned}\langle(\ell s)j'm|\,\hat{S}_z\,|(\ell s)jm\rangle &= \sum_{m_s}\langle j'm|\,\hat{S}_z\,|\ell, m_\ell = m - m_s, s, m_s\rangle\langle \ell, m_\ell = m - m_s, s, m_s|\,jm\rangle \\ &= \hbar\sum_{m_s} m_s(\ell, m_\ell = m - m_s, s, m_s|j'm)(\ell, m_\ell = m - m_s, s, m_s|jm)\,,\end{aligned} \tag{8.4.8}$$

where we have suppressed ℓ and s values in writing the coupled states for the ease of writing and treated the CG coefficients to be real. For our case of hydrogenic atoms $s = \frac{1}{2}$ we have using Table 7.1

$$\langle (\ell, 1/2)\ell \pm 1/2, m| \, \hat{S}_z \, |(\ell, 1/2)\ell \pm 1/2, m\rangle = \pm \frac{m\hbar}{2\ell + 1}, \tag{8.4.9}$$

$$\langle (\ell, 1/2)\ell \pm 1/2, m| \, \hat{S}_z \, |(\ell, 1/2)\ell \mp 1/2, m\rangle = -\frac{\hbar}{2\ell + 1}\sqrt{\left(\ell + \frac{1}{2} + m\right)\left(\ell + \frac{1}{2} - m\right)}. \tag{8.4.10}$$

Thus $\hat{H}_{mag}$ mixes certain states with $j = \ell \pm \frac{1}{2}$. The mixed states are still the eigenstates of $\hat{L}^2, \hat{S}^2, \hat{J}_z$. For weak fields ($B \ll 10^4$ gauss), the mixing is small and $|n\ell sjm\rangle$ are still good basis states. Thus to first order in B, the correction to the energy is simply the expectation value of $\hat{H}_{mag}$ in the state $|n\ell sjm\rangle$. Combining (8.4.7) and (8.4.9), we find

$$\begin{aligned} E^{(1)}_{mag} &\equiv \frac{eB}{2m_e} \langle (\ell, 1/2)\ell \pm 1/2, m| \, \hat{J}_z + \hat{S}_z \, |(\ell, 1/2)\ell \pm 1/2, m\rangle \,, \\ &= \frac{eB}{2m_e}\left[m\hbar \pm \frac{m\hbar}{2\ell + 1}\right] = \frac{e\hbar B}{2m_e}\left(\frac{2j+1}{2\ell + 1}\right) m \,, \end{aligned} \tag{8.4.11}$$

for both values of j. The first order shift (weak field) of energy levels can thus be written as

$$E^{(1)}_{mag} = \frac{e\hbar B}{2m_e} g_j m \,, \tag{8.4.12}$$

where, in analogy with the gyromagnetic ratio g_s for spin and and g_ℓ for orbital angular momentum, we have introduced the gyromagnetic ratio $g_j = \frac{2j+1}{2\ell+1}$, called the *Landé g-factor.* Unlike g_s and g_ℓ, however, Landé g-factor is not constant but depends on j and ℓ. The weak field energy of a hydrogenic level is then

$$E_{n\ell sjm} = E^{(0)}_n \left[1 + \frac{(Z\alpha)^2}{n^2}\left(\frac{n}{j + \frac{1}{2}} - \frac{3}{4}\right)\right] + \frac{e\hbar B}{2m_e} g_j m \,. \tag{8.4.13}$$

The weak field Zeeman shift is proportional to m and lifts the remaining degeneracy of energy levels with respect to m. Figure 8.3 shows the splitting of ^2P ($n = 2, \ell = 1$) level due to spin-orbit and weak field magnetic interaction. The magnetic splitting (Zeeman splitting) of levels is uniform within each multiplet ($-j \le m \le j$ sublevels) belonging to a given value of j but differs from one multiplet to another due to different Landé g-factors. For historical reasons, this is referred to as *anomalous Zeeman effect.*

The weak field Zeeman splitting for a multiplet of levels belonging to a given J can be calculated quite generally (even for multi-electron atoms) using the projection theorem [see Chapter 7], according to which the expectation value of any vector operator $\hat{\boldsymbol{V}}$ in the subspace spanned by $2j + 1$ magnetic levels of total angular momentum J (which is obtained by summing two angular momenta L and S) is proportional to the expection value of $\hat{\boldsymbol{J}}$. In our case, the vector operator is $\hat{\boldsymbol{L}} + 2\hat{\boldsymbol{S}} = \hat{\boldsymbol{J}} + \hat{\boldsymbol{S}}$. Then the weak-field Zeeman shift is given by

$$\begin{aligned} E^{(1)}_{mag} &= \left\langle (\ell s)jm \middle| \left(\hat{1} + \frac{\hat{\boldsymbol{J}} \cdot \hat{\boldsymbol{S}}}{\hat{J}^2}\right) \hat{J}_z \middle| (\ell s)jm \right\rangle \frac{eB}{2m_e}, \\ &= \frac{eB}{2m_e}\left\langle n(\ell s)jm \middle| \left(\hat{1} + \frac{\hat{J}^2 + \hat{S}^2 - \hat{L}^2}{2\hat{J}^2}\right) \hat{J}_z \middle| n(\ell s)jm \right\rangle \\ &= \frac{e\hbar B}{2m_e} \, g_j \, m, \end{aligned} \tag{8.4.14}$$

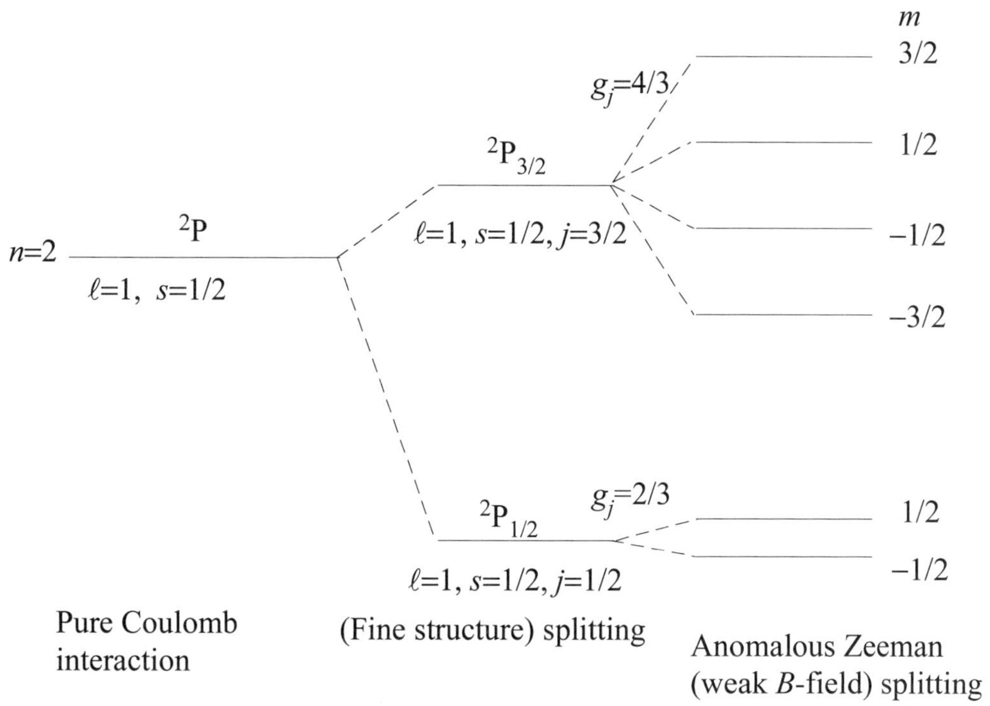

FIGURE 8.3
Weak B-field (anomalous) Zeeman splitting for ^2P $(n = 2, \ell = 1)$ level in hydrogen.

where

$$g_j = 1 + \frac{j(j+1) + s(s+1) - \ell(\ell+1)}{2j(j+1)},$$

is called Lande's splitting factor (g-factor). It is easy to check that for $^2P_{1/2}$and $^2P_{3/2}$ this gives Landé g-factors of 2/3 and 4/3 in agreement with the previous calculation. It should be kept in mind that projection theorem gives level splitting in the weak field limit and does not rule out magnetic interaction mixing states with different j-values.

Strong Field Magnetic Splitting: Paschen-Back Effect

For very large external magnetic fields (large compared to the internal magnetic field), the magnetic interaction $\hat{H}_{mag}$ dominates $\hat{H}_{FS}$. In that case we can ignore $\hat{H}_{FS}$. The appropriate commuting obsevables in this case are $\hat{L}^2$, $\hat{L}_z$, $\hat{S}^2$, $\hat{S}_z$ and $\hat{J}_z = \hat{L}_z + \hat{S}_z$, which commute with both $\hat{H}_0 = \frac{\hat{\mathbf{p}}^2}{2m_e} - \frac{Zc^2}{4\pi\epsilon_0 r}$ and $\hat{H}_{mag}$. The unperturbed states $|n\ell s m_\ell m_s\rangle$ are degenerate but $\hat{H}_{mag}$ is diagonal in them! Hence the first order correction to energy levels is

$$E^{(1)}_{mag} = \frac{e\hbar B}{2m_e}(m_\ell + 2m_s)\,. \tag{8.4.15}$$

This regime of the Zeeman splitting is referred to as the *strong-field Zeeman effect* or the *Paschen-Back effect*. The strong field drives the atom into states of definite ℓ, m_ℓ, s, m_ℓ. The level structure in this case is different from the weak field case. This is shown in Fig. 8.4 for the ^2P state of the hydrogen atom. Calculation of the energy shift due to $\hat{H}_{FS}$ is left to the problem at the end of this chapter.

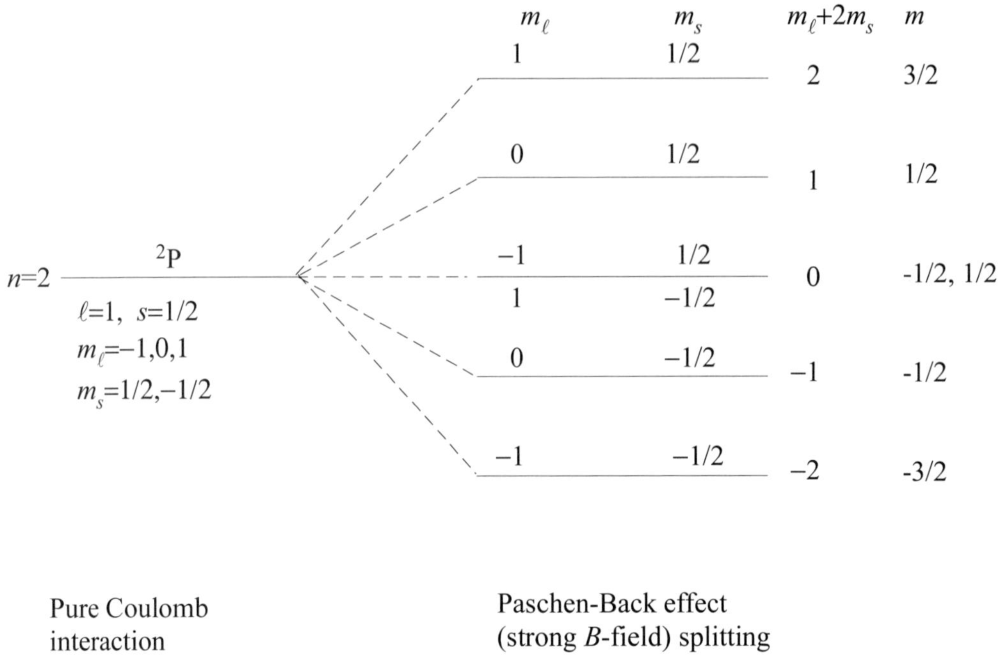

FIGURE 8.4
Strong B-field Zeeman splitting for ^2P $(n = 2, \ell = 1)$ level in hydrogen.

The transition from the weak field to strong field is smooth. We have focused on the calculation of energies. If we calculate the perturbed states we find that in weak fields $\hat{H}_{mag} \ll \hat{H}_{FS}$ the basis states have strong $|\ell sjm\rangle$ flavor. For strong magnetic fields $\hat{H}_{mag} \gg \hat{H}_{FS}$ the basis states are $|\ell m_\ell s m_s\rangle$. For intermediate fields, mixing of states with different j occurs and the states have a character intermediate between the two extremes. As the field strength increases starting from zero, the character of the perturbed states changes smoothly from $|\ell sjm\rangle$ to $|\ell m_\ell s m_s\rangle$. The operator $\hat{J}_z$ commutes with $\hat{H}_{mag}$ for all field strengths so that $m = m_\ell + m_s$ is always a good quantum number.

8.5 WKBJ Approximation

This method of approximation, known after Wentzel, Kramers, Brillouin and Jeffreys, for solving one-dimensional Schrödinger equation is applicable in situations where the potential $V(x)$ is a slowly varying function of x. This method is expected to work best for states with large quantum numbers (semiclassical limit). However, in certain cases the WKBJ method works better than naively expected and may even yield the exact result (such as the harmonic oscillator energy levels). In any case, one may trust the WKBJ approximation for highly excited states and then extend its limits toward the ground state with less confidence. In this sense it is complementary to the variational method that works well near the ground state. These two techniques together give quick but not very precise information for low lying states and highly excited states.

In a region where where the potential varies slowly the local de Broglie wavelength

$$\lambda(x) = \frac{h}{p(x)} = \frac{2\pi\hbar}{[2m(E - V(x))]^{1/2}} \tag{8.5.1}$$

also varies slowly with x so that the wave function $\psi(x)$ does not deviate significantly from the form it would take if $V(x)$ or $p(x)$ were independent of x, i.e., the wave function will be of the plane wave form $\psi(x) \sim e^{ikx}$ $(k = p/\hbar)$. In such cases, without loss of generality, we may assume $\psi(x)$ to have the form

$$\psi(x) = \exp[iS(x)/\hbar]\,, \tag{8.5.2}$$

where the normalization constant has been absorbed in the definition of $S(x)$. To see how the WKBJ method works, consider the time-independent Schrödinger equation in one dimension in the presence of a potential $V(x)$. We express the function $S(x)$ as a power series in $\hbar$ as

$$S(x) = S_0(x) + \hbar S_1(x) + \hbar^2 S_2(x) + \cdots\,. \tag{8.5.3}$$

Substituting this in the time-independent Schrödinger equation we find

$$E = \frac{1}{2m}\left(\frac{dS}{dx}\right)^2 - \frac{i\hbar}{2m}\frac{dS}{dx} + V(x)\,. \tag{8.5.4}$$

In the semi-classical limit $\hbar \to 0$, we have

$$E = \frac{1}{2m}\left(\frac{dS_0}{dx}\right)^2 + V(x)\,. \tag{8.5.5}$$

This equation is the same as $E = \frac{p^2}{2m} + V(x)$ provided that

$$\frac{dS_0}{dx} = \pm\sqrt{E - V(x)} = \pm p\,. \tag{8.5.6}$$

It is clear that we can interpret $p(x)$ as the momentum of the particle if $E > V(x)$. The point where $V(x) = E$ is called a *classical turning point*. Thus in the semiclassical limit $\hbar \to 0$, the wave function $\psi_0(x) = e^{iS_o(x)/\hbar}$ contains all the information needed for the classical motion. To go beyond the classical limit we need to consider higher order terms in $\hbar$ in the expansion of $S(x)$. It is also clear that the behavior of the solution will be different in different regions depending on whether $E < V(x)$ or $E > V(x)$.

So as a first step, we establish the form of the solution for a general potential $V(x)$, in the two regions called Region I $[E > V(x)]$ and Region II $[E < V(x)]$. Figure 8.5 shows E and $V(x)$ and the two regions I and II. Classically, a particle moving from right (Region I) to left will turn back as soon as it reaches the turning point x_0, where $V(x_0) = E$, because it does not have enough energy to cross the potential barrier and move into the region II. According to quantum mechanics, the particle has a finite probability to appear in the region II as well. The Schrödinger equation, and its solution, of course, have different forms on the two sides of the turning point.

The time-independent Schrödinger equation for a one-dimensional problem [Eq. (3.5.9)]

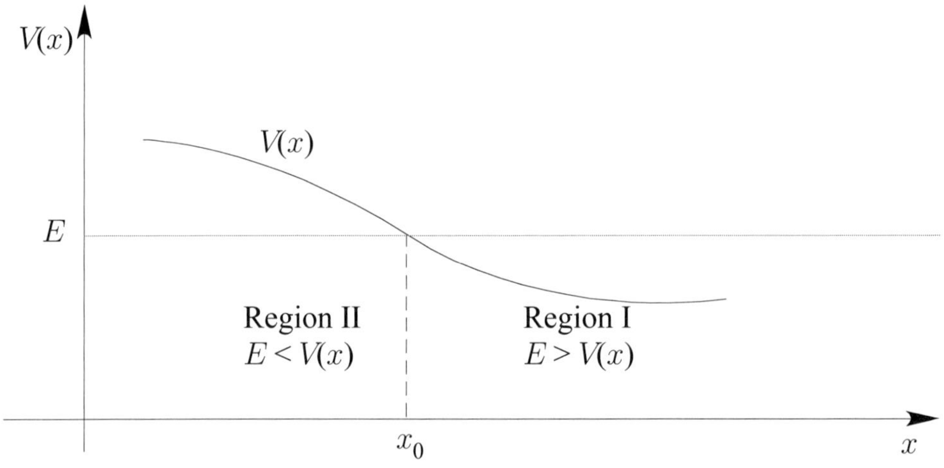

FIGURE 8.5
The point $x = x_0$ at which $V(x) = E$ is the classical turning point for a particle moving from right to left.

may then be written as[2]

$$\frac{d^2\psi}{dx^2} + k^2(x)\psi(x) = 0, \qquad k^2 = \frac{2m}{\hbar^2}[E - V(x)] \qquad \text{Region I: } V(x) < E\,, \tag{8.5.7}$$

$$\frac{d^2\psi}{dx^2} - \kappa^2(x)\psi(x) = 0\,, \qquad \kappa^2 = \frac{2m}{\hbar^2}[V(x) - E] \qquad \text{Region II : } V(x) > E\,. \tag{8.5.8}$$

Once we get the solution of Eq. (8.5.7) for the region I $[V(x) < E]$, we can easily deduce the solution of Eq. (8.5.8) for region II $[V(x) > E]$ by replacing $ik(x)$ by $\kappa(x)$.

Assuming a solution of the form (8.5.2) with an expansion of $S(x)$ in powers of $\hbar$ [Eq. (8.5.3)] and substituting it in Eq. (8.5.7), we obtain

$$i\hbar(S_0'' + \hbar S_1'' + \cdots) - (S_0' + \hbar S_1' + \cdots)^2 + p^2(x) = 0\,, \tag{8.5.9}$$

where the primes denote derivatives with respect to x such that $S' = dS/dx$, $S'' = d^2S/dx^2$, $\cdots$. Equating the terms of the same order in $\hbar$ from both sides, we have

$$-(S_0')^2 + p^2(x) = 0\,, \tag{8.5.10a}$$

$$iS_0'' - 2S_0'S_1' = 0\,, \tag{8.5.10b}$$

$$S_0'S_2' + \frac{1}{2}(S_1')^2 - \frac{i}{2}S_1'' = 0\,, \tag{8.5.10c}$$

etc.

On integration, the first equation (8.5.10a) leads to

$$S_0(x) = \pm\int^x p(x')dx' = \pm\hbar\int^x k(x')dx'\,, \tag{8.5.11}$$

[2]It may be noted that even if the problem is three-dimensional, as long as the potential $V(r)$ is spherically symmetric, the radial equations may still be put in the forms of Eqs. (8.5.7) and (8.5.8) by writing the radial function as $R_\ell(r) = u_\ell(r)/r$. The dependent variable in these equations would be $u_\ell(r)$ instead of $\psi(x)$, the independent variable would be r instead of x, and the effective potential would be $V(r) + \frac{\hbar^2}{2m}\frac{\ell(\ell+1)}{r^2}$ instead of $V(x)$.

where the constant of integration is taken into account by leaving the lower limit of integration unspecified. The second equation (8.5.10b) gives

$$S_1(x) = \frac{i}{2}\ln S_0'(x) + \text{constant} = \frac{i}{2}\ln[k(x)] + C_1\,, \tag{8.5.12}$$

where C_1 is the constant of integration. The third equation (8.5.10c) gives

$$S_2(x) = \mp\frac{1}{4}\frac{p'(x)}{p^2(x)} \mp \frac{1}{8}\int^x \frac{(p')^2}{p^3}dx\,. \tag{8.5.13}$$

Restricting only to zero and first order terms in $\hbar$, we find that the wave function in region I $[V(x) < E]$ must be of the form

$$\begin{aligned}\psi(x) &= \exp[i(S_0 + \hbar S_1)/\hbar] = \exp\left[\pm i\int^x k(x')dx' - \frac{1}{2}\ln k(x) + C_1\right]\\ &= \frac{A}{\sqrt{k(x)}}\exp\left[\pm i\int_{x_0}^{x} k(x')dx'\right],\end{aligned} \tag{8.5.14}$$

where the constants have all been absorbed into the definition of A.

To find the form of the solution for region II $[V(x) > E]$, we have to consider Eq. (8.5.8) which may be obtained by replacing $ik(x)$ by $\kappa(x)$ in Eq. (8.5.7). So the WKBJ solution in the region II $[V(x) > E]$ has the form

$$\psi(x) = \frac{B}{\sqrt{\kappa(x)}}\exp\left[\pm\int_{x_0}^{x}\kappa(x')dx'\right] = \frac{B}{\sqrt{\kappa(x)}}\exp\left[\mp\int_{x}^{x_0}\kappa(x')dx'\right]. \tag{8.5.15}$$

Thus the WKBJ solutions of the Schrödinger equation on the two sides of the turning point are, respectively,

$$\psi(x) \approx \frac{A}{\sqrt{k(x)}}e^{\pm i\xi_1}, \qquad \xi_1 \equiv \int_{x_0}^{x} k(x')dx', \text{ Region I: } E > V(x)\,, \tag{8.5.16}$$

$$\psi(x) \approx \frac{B}{\sqrt{\kappa(x)}}e^{\mp\xi_2}, \qquad \xi_2 \equiv \int_{x}^{x_0} \kappa(x')dx', \text{ Region II: } E < V(x)\,. \tag{8.5.17}$$

Physical boundary conditions must be considered in order to decide which combination of solutions is valid in the different regions of type I and II. Such considerations also determine the constants A and B in the appropriate regions.It may noted that ξ_1 and ξ_2 so defined are both positive. (Since $k(x)$ and $\kappa(x)$ are both positive in the respective regions, the upper limit should be greater than the lower limit in the integrals for ξ_1 and ξ_2 if they are to be positive.) It is clear that these two solutions are, respectively, approximations to the actual wave functions in the region I (on the right of the turning point where $V(x) < E$) and in the region II (on the left of the turning point where $V(x) > E$. At or near the turning point $x = x_0$, however, these solutions are not valid because at this point $\lambda = h/h \to \infty$. This can be seen more clearly by investigating the conditions for the validity of WKBJ solutions.

The Validity of WKBJ Approximation

The validity of the WKBJ approximation may be judged by comparing the magnitudes of the successive terms, S_0 and $\hbar S_1$ in the expansion of $S(x)$. For this approximation to be valid,

$$\left|\frac{\hbar S_1}{S_0}\right| \ll 1\,. \tag{8.5.18}$$

Since $S_0 = \hbar \int_{x_0}^{x} k(x')dx$ is a monotonically increasing function of x, as long as $k(x')$ does not vanish, the ratio $|\hbar S_1/S_0|$ is small if

$$\left|\frac{\hbar S_1'}{S_0'}\right| \ll 1\,. \tag{8.5.19}$$

Using $S_0'(x) = \pm\hbar k(x)$ from Eq. (8.5.10a) and $S_1' = iS_0''/2S_0' = ik'(x)/2k(x)$ from Eq. (8.5.10b), condition (8.5.19) becomes

$$\left|\frac{\hbar}{2}\frac{k'(x)}{k(x)}\frac{1}{\hbar k(x)}\right| = \left|\frac{\lambda}{4\pi}\frac{1}{k(x)}\frac{dk}{dx}\right| \ll 1\,. \tag{8.5.20}$$

This means that as long as the fractional change in $k(x)$ (momentum) over a distance $\lambda(x)/4\pi$ is small, the approximation is valid in that domain of x. This condition is, obviously, satisfied for a constant potential and also for a slowly varying potential, in regions far away from the turning point. But it breaks down at or near the turning point where $\lambda \to \infty$. For this reason the WKBJ solutions are also referred to as *asymptotic solutions* (valid far from the turning point). Since there may be several regions of type I and II, we need to connect the solutions in different regions to one another. This is done by means of the conection formulas.

The Connection Formulas

The WKBJ solutions on the two sides of the classical turning point (CTP) have different forms. In the region $V(x) < E$ the solutions are oscillatory and can be written as some linear combinations of solutions (8.5.16). In the region $V(x) > E$, they are combinations of increasing and decreasing exponential forms (8.5.17). To connect these WKBJ solutions in the two regions across the CTP (where the WKBJ solutions are not valid), we solve the Schrödinger equation exactly for a linear slowly varying potential,

$$V(x) = E - \frac{\hbar^2}{2m}Cx\,, \tag{8.5.21}$$

where E is the energy of the particle and we have written the slope of the potential near the turning point as $\frac{\hbar^2 C}{2m}$ in terms of a positive constant ($C > 0$) for later convenience. We may look upon the assumed form of the potential as a Taylor series expansion of the potential near the classical turning point. The exact solution of the Schrödinger equation with this potential can then be worked out on both sides of the turning point and can be matched to the WKBJ solutions on the two sides. This procedure gives rise to the so called *connection formulas* and the corresponding WKBJ solutions are said to be connected.

For the potential given by Eq. (8.5.21), the Schrödinger equation near the turning point $x = 0$ can be written as [3]

$$\left.\begin{aligned} &\frac{d^2\psi_1}{dx^2} + k^2(x)\psi_1(x) = 0 \\ &k^2(x) = \frac{2m}{\hbar^2}[E - V(x)] = Cx \end{aligned}\right\} \qquad \text{Region I: } x > 0\,, \tag{8.5.22}$$

$$\left.\begin{aligned} &\frac{d^2\psi_2}{dx^2} - \kappa^2(x)\psi_2(x) = 0 \\ &\kappa^2(x) = \frac{2m}{\hbar^2}[V(x) - E] = -Cx = C|x| \end{aligned}\right\} \qquad \text{Region II: } x < 0\,. \tag{8.5.23}$$

[3] In the vicinity of a turning point at $x = x_0$, the potential has the form $V(x) = E - \frac{\hbar^2}{2m}C(x - x_0)$. By a change of independent variable $X = x - x_0$, the potential can be brought to the form $V(X) = E - \frac{\hbar^2}{2m}C\,X$ with a turning point at $X = 0$.

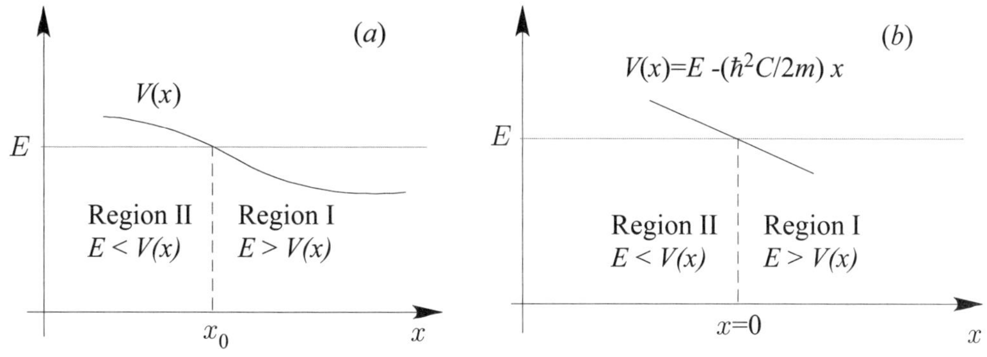

FIGURE 8.6
For a linear potential of the form $V(x) = E - \frac{\hbar^2 C}{2m}\, x$, the classical turning point is at $x = 0$.

Introducing the independent variable ξ_1 and the dependent variable $v(\xi_1)$ in Eq. (8.5.22) by

$$\xi_1 = \int_0^x k(x')dx' = \sqrt{C}\frac{2}{3}x^{3/2}\,, \tag{8.5.24}$$

and

$$\psi_1(x) = \sqrt{\frac{\xi_1(x)}{k(x)}}v(\xi_1) = \left(\frac{2}{3C}\right)^{1/6} \xi_1^{1/3}\, v(\xi_1)\,, \tag{8.5.25}$$

we find that the Schrödinger equation in region I reduces to the Bessel equation of order 1/3:

$$\frac{d^2 v(\xi_1)}{d\xi_1^2} + \frac{1}{\xi_1}\frac{dv(\xi_1)}{d\xi_1} + \left(1 - \frac{1}{3^2}\frac{1}{\xi_1^2}\right) v(\xi_1) = 0\,. \tag{8.5.26}$$

Hence $v(\xi_1) = J_{\pm\frac{1}{3}}(\xi_1)$ and the solution in region I is given by

$$\psi_1^{\pm}(x) = A_{\pm}\sqrt{\frac{\xi_1}{k}}J_{\pm\frac{1}{3}}(\xi_1)\,. \tag{8.5.27}$$

Similarly, in region II $[E < V(x)\,; x < 0]$, we introduce the independent variable ξ_2 and the dependent variable $w(\xi_2)$ through the equations

$$\xi_2 = \int_x^0 \kappa(x')dx' = (2\sqrt{C}/3)|x|^{3/2}\,, \tag{8.5.28}$$

$$\psi_2(x) = \sqrt{\frac{\xi_2}{\kappa(x)}}\; w(\xi_2) = \left(\frac{2}{3C}\right)^{1/6} \xi_2^{1/3}\; w(\xi_2)\,. \tag{8.5.29}$$

Note that ξ_2 increases as we move away from the turning point into region II. These substitutions reduce the Schrödinger equation in region II to the form of modified Bessel equation of order 1/3:

$$\frac{d^2 w(\xi_2)}{d\xi_2^2} + \frac{1}{\xi_2}\frac{dw(\xi_2)}{d\xi_2} + \left(-1 - \frac{1}{3^2}\frac{1}{\xi_2^2}\right) w(\xi_2) = 0\,. \tag{8.5.30}$$

The solution of this equation is the modified Bessel functions $I_{\pm 1/3}(\xi_2)$ so that the WKBJ solution in region II $[V(x) > E]$ is

$$\psi_2^{\pm}(x) = B_{\pm}\sqrt{\frac{\xi_2}{\kappa}}\; I_{\pm\frac{1}{3}}(\xi_2)\,. \tag{8.5.31}$$

The modified Bessel function is also called Bessel function of *imaginary argument* because it is related to the Bessel function by $I_\nu(\xi_2) = e^{-i\pi\nu/2} J_\nu(i\xi_2)$ [see Chapter 5, Appendix 5A1 Sec. 5]. Thus the two forms of the solution are analytic continuations of each other as x changes sign and with the choice $B_\pm = \mp A_\pm$, the two functions ψ_1 (in region I) and ψ_2 (in region II) constitute a continuous solution of the Schrödinger equation near the turning point.

Now Eqs.(8.5.27) and (8.5.31) are valid for all values of x only for the potential $E - \frac{\hbar^2 C}{2m}\, x$ [Eq. (8.5.21)]. What then is gained by writing the solution [Eqs. (8.5.27) and (8.5.31)] in terms of $k(x)$ and $\kappa(x)$ for a general potential $V(x)$ that is strictly not equal to $E - \frac{\hbar^2 C}{2m}\, x$ but which merely behaves like it near the turning point? The reason is the remarkable observation that the asymptotic forms of $J_{\pm 1/3}\nu(\xi_1)$ and $I_{\pm 1/3}(\xi_2)$ for large arguments $\xi_1, \xi_2 \to \infty$ [See Appendix 5A1 Sec. 5]:

$$
\begin{aligned}
J_{\pm 1/3}(\xi_1) &\xrightarrow{\xi_1 \to \infty} \sqrt{\frac{2}{\pi \xi_1}}\, \cos(\xi_1 \mp \frac{\pi}{6} - \frac{\pi}{4})\,, \\
I_{\pm 1/3}(\xi_2) &\xrightarrow{\xi_2 \to \infty} \sqrt{\frac{1}{2\pi \xi_2}}\, \left[e^{\xi_2} + e^{-\xi_2}\, \exp^{-i\pi(1/2 \pm 1/3)}\right]\,, \\
I_{-1/3} - I_{1/3} &\xrightarrow{\xi_2 \to \infty} \sin(\pi/3)\sqrt{\frac{2}{\pi \xi_2}} e^{-\xi_2}\,,
\end{aligned}
\tag{8.5.32}
$$

are such that away from the turning point $x = 0$, i.e., for large $\xi_1 = \int_0^x k(x')dx'$ and $\xi_2 = \int_x^0 \kappa(x')dx'$, appropriate combinations of Eqs. (8.5.27) and (8.5.31) agree with the WKBJ forms (8.5.16) and (8.5.17). This allows us to use Eqs. (8.5.27) and (8.5.31) with $B_\pm = \mp A_\pm$ to interpolate between WKBJ solutions in regions of type I and II for any potential. Using the asymptotic form of the solution near the turning point we find that

$$
\psi_1^+ \;\to\; A_+\sqrt{\frac{2}{\pi k}}\, \cos(\xi_1 - 5\pi/12)\,, \tag{8.5.33}
$$

and

$$
\psi_2^+ \;\to\; -A_+\sqrt{\frac{1}{2\pi \kappa}}\, \left[e^{\xi_2} + e^{-\xi_2} e^{-i5\pi/6}\right]\,, \tag{8.5.34}
$$

are connected asymptotic solutions. Similarly, the asymptotic forms of ψ_1^- and ψ_2^-

$$
\psi_1^- \;\to\; A_-\sqrt{\frac{2}{\pi k}}\, \cos(\xi_1 - \pi/12)\,, \tag{8.5.35}
$$

and

$$
\psi_2^- \;\to\; A_-\sqrt{\frac{1}{2\pi \kappa}}\, \left[e^{\xi_2} + e^{-\xi_2}\, e^{-i\pi/6}\right]\,, \tag{8.5.36}
$$

are also connected asymptotic solutions. It is obvious that a linear combination of the asymptotic forms of ψ_2^+ and ψ_2^- on the far left of the turning point will connect with the corresponding linear combination of ψ_1^+ and ψ_1^- on the far right of the turning point. We will illustrate it by two examples.

Consider the combination $\psi^+ + \psi^-$ with $A_+ = A_- = 1$. Using the asymptotic forms (8.5.33) through (8.5.36), we obtain

$$
\psi^+ + \psi^- \;\to\; \begin{cases} \left(\frac{2}{\pi\, k}\right)^{1/2} \sqrt{3} \cos\left(\xi_1 - \pi/4\right) & \text{Region I} \\[2ex] \left(\frac{1}{2\pi\, k}\right)^{1/2} \sqrt{3}\, e^{-\xi_2} & \text{Region II}\,. \end{cases} \tag{8.5.37}
$$

Hence we have the connection formula

$$
\frac{e^{-\xi_2}}{2\sqrt{\kappa}} \xrightarrow{II \to I} \frac{\cos(\xi_1 - \frac{\pi}{4})}{\sqrt{k}}\,. \tag{8.5.38}
$$

Let us form another linear combination and work out its asymptotic forms in the two regions:

$$\frac{1}{\sqrt{3}}\left[(\psi^+ + \psi^-)\cos\eta - (\psi^+ - \psi^-)\sin\eta\right]$$
$$\longrightarrow \begin{cases} \left(\frac{2}{\pi k(x)}\right)^{1/2} \cos(\xi_1 - \pi/4 + \eta) & \text{Region I} \\ 2\sin\eta \left(\frac{1}{2\pi\kappa(x)}\right)^{1/2} e^{\xi_2} & \text{Region II}\,. \end{cases} \tag{8.5.39}$$

This gives the connection formula

$$\frac{e^{\xi_2}}{\sqrt{\kappa(x)}}\sin\eta \quad \xleftarrow{II \leftarrow I} \quad \frac{\cos(\xi_1 - \pi/4 + \eta)}{\sqrt{k(x)}}\,. \tag{8.5.40}$$

Other connection formulas may be derived by choosing different combinations. What combinations to choose depends on the boundary conditions to be satisfied or some other information we have about the system. We will see an example of this in the discussion of α decay.

The connection formulas must be used with caution; the arrow indicates the direction in which the formulas can be used reliably to connect the asymptotic solutions. For example, the arrow in Eq. (8.5.38) means that the WKBJ solution $e^{-\xi_2}/2\sqrt{\kappa(x)}$ in region II $[E < V(x)]$ connects to the WKBJ solution $\cos(\xi_1 - \pi/4)/\sqrt{k(x)}$ in the region I $[E > V(x)]$. This is because we know the boundary condition on the wave function in the inaccessible region, viz., that the wave function has to vanish far from the turning point. The reverse connection can lead to errors. For example, if we use the arrow in the reverse direction, a small error in the phase of the cosine term would introduce a sine term, whose contribution in region I might be considered negligible, but according to (8.5.40) it would connect to a positive exponential function in region II. In general, we always start with the wave function in the region of space where the boundary conditions on the wave function are known and then match it onto the adjacent region through the connection formula. We will see an example of this in nuclear α-decay.

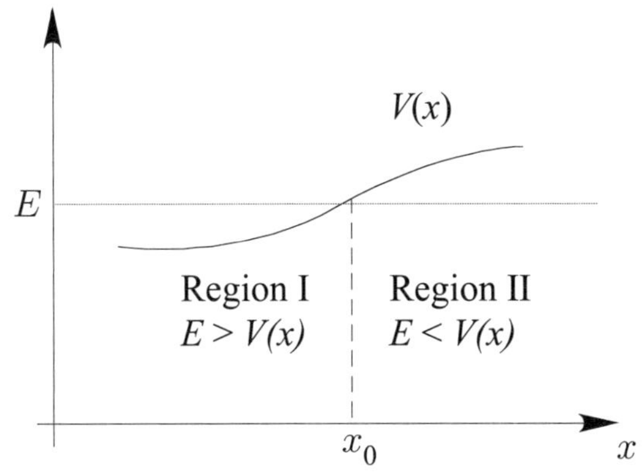

FIGURE 8.7
Potential increasing through the classical turning point.

If the potential $V(x)$ is increasing through the turning point so that the region I $[E >$

$V(x)]$ is on the left and region II $[E < V(x)]$ is on the right [Fig. 8.7], the connection formulas are

$$\frac{1}{\sqrt{k(x)}}\cos\left[\int_x^0 k(x')dx' - \frac{\pi}{4}\right] \xleftarrow{I\leftarrow II} \frac{1}{\sqrt{\kappa(x)}}\exp[-\int_0^x dx'\kappa(x')]\,, \tag{8.5.41}$$

$$\frac{1}{\sqrt{k(x)}}\cos\left[\int_x^0 k(x')dx' - \frac{\pi}{4} + \eta\right] \xrightarrow{I\rightarrow II} \frac{\sin\eta}{\sqrt{k(x)}}\exp[\int_0^x dx'\kappa(x')]\,. \tag{8.5.42}$$

Finally, the case of the potential where two (or more) classical turning points converge or are not sufficiently separated requires special treatment because the region of validity of one solution may overlap with that of the next. Other connection formulas by assuming the potential to have the form $V(x) = E - \frac{\hbar^2 C}{2m}(x - x_o)^n$ near the turning point can be devloped in terms of Bessel functions of order $\pm\frac{1}{n+2}$. For some potentials, such as the infinite square well, the wave function must vanish in region I at the infinite potential barrier. In such cases the connection formulas are simply

$$0 \leftrightarrow \frac{\sin\left[\int_0^x k(x')dx'\right]}{\sqrt{k(x)}}\,. \tag{8.5.43}$$

8.6 Particle in a Potential Well

We can now apply the WKBJ approximation to find the energy levels of a particle in a potential well $V(x)$ where $V(x)$ is a slowly varying function of x. Consider a particle moving in a one-dimensional potential well as shown in Fig. 8.8. For any energy level E there are two 'classical turning points' x_1 and x_2, given by

$$V(x_1) = V(x_2) = E\,. \tag{8.6.1}$$

The regions $x < x_1$ and $x > x_2$, referred to as regions II and II′, respectively, are the classically forbidden regions where $E < V(x)$, while the region $x_1 \le x \le x_2$, referred to as region I, is characterized by $E > V(x)$. Since the state we are referring to is a bound state, the WKBJ wave function in regions II and II′ should contain a decreasing exponential term.

Let us introduce variables ξ_1 and ξ_2 on the two sides of the CTP x_1 by

$$\xi_2 = \int_x^{x_1} \kappa(x')dx', \qquad \kappa^2(x') = \frac{2m}{\hbar^2}[V(x') - E]\ ;\ \text{Region II}, \tag{8.6.2}$$

and

$$\xi_1 = \int_{x_1}^{x} k(x')dx', \qquad k^2(x') = \frac{2m}{\hbar^2}[E - V(x')]\ ;\ \text{Region I}. \tag{8.6.3}$$

The limits are chosen to keep both ξ_1 and ξ_2 positive. Using the first connection formula (8.5.38), we can connect the WKBJ wave function with the negative exponential term $\exp(-\xi_2)/2\sqrt{\kappa(x)}$ in region II_L with the WKBJ wave function $\cos(\xi_1 - \pi/4)/\sqrt{k(x)}$ in region I.

Similarly, on the two sides of the CTP x_2, we define the variables ξ_1' and ξ_2' by

$$\xi_2' = \int_{x_2}^{x} \kappa(x')dx', \qquad \kappa^2(x') = \frac{2m}{\hbar^2}[V(x') - E]\ ;\ \text{Region II}', \tag{8.6.4}$$

and

$$\xi_1' = \int_x^{x_2} k(x')dx', \qquad k^2(x') = \frac{2m}{\hbar^2}[E - V(x')]\ ;\ \text{Region I}, \tag{8.6.5}$$

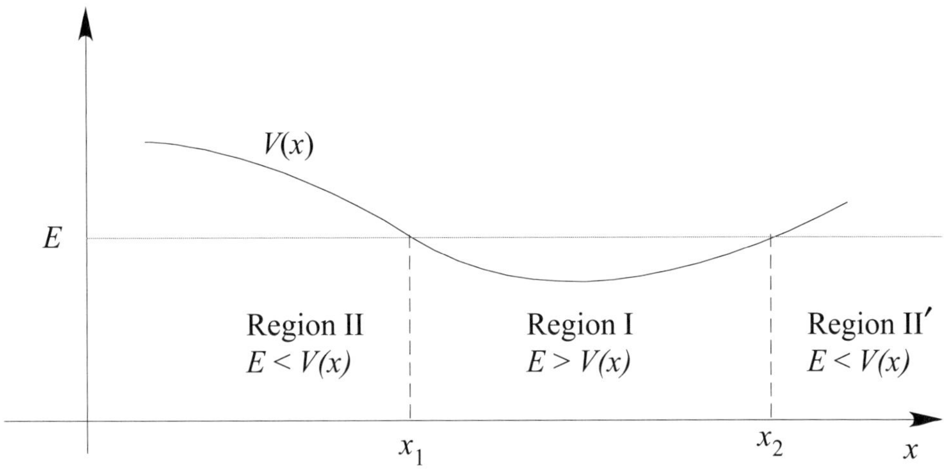

FIGURE 8.8
A particle with energy E in a potential well $V(x)$, where $V(x)$ is a slowly varying function of x. The points x_1 and x_2 are the classical turning points at which $E = V(x_1) = V(x_2)$.

where the limits are again such that both ξ_1' and ξ_2' are positive away from the turning point. Using the same connection formula, we can connect the WKBJ wave function with the negative exponential term $\exp(-\xi_2')/2\sqrt{\kappa(x)}$ in region II′ with the WKBJ wave function $\cos(\xi_1' - \pi/4)/\sqrt{k(x)}$ in region I. Thus we have

$$\psi_I(x) = \begin{cases} \psi_A(x) = \frac{A}{\sqrt{k(x)}}\cos\left(\int_{x_1}^{x} k(x')dx' - \pi/4\right) & I \leftarrow II\,, \\ \\ \psi_B(x) = \frac{B}{\sqrt{k(x)}}\cos\left(\int_{x}^{x_2} k(x')dx' - \pi/4\right) & I \leftarrow II'\,. \end{cases} \tag{8.6.6}$$

At any point x in region I, in between and away from the turning points x_1 and x_2, these WKBJ solutions must be the same function. This requirement can be written in terms of its logarithmic derivative as

$$\frac{1}{\psi_A(x)}\frac{d\psi_A}{dx} = \frac{1}{\psi_B(x)}\frac{d\psi_B}{dx}\,. \tag{8.6.7}$$

This condition, with the help of the relation $\int_x^{x_2} k(x')dx' = \int_{x_1}^{x_2} k(x')dx' - \int_{x_1}^{x} k(x')dx'$, leads to

$$\tan\left(\int_{x_1}^{x} k(x')dx' - \pi/4\right) = \tan\left(\int_{x_1}^{x} k(x')dx' - \Phi + \pi/4\right)\,, \tag{8.6.8}$$

where
$$\Phi \equiv \int_{x_1}^{x_2} k(x')dx'\,. \tag{8.6.9}$$

Equation (8.6.8) is solved by

$$\int_{x_1}^{x} k(x')dx' - \pi/4 = n\pi + \int_{x_1}^{x} k(x')dx' - \Phi + \pi/4\,,$$

or

$$\Phi \equiv \int_{x_1}^{x_2} k(x)dx = \left(n + \frac{1}{2}\right)\pi\,. \tag{8.6.10}$$

Using (8.6.3) or (8.6.5), this can be written as

$$\int_{x_1}^{x_2} \sqrt{2m[E - V(x)]}\; dx = \left(n + \frac{1}{2}\right)\hbar\pi\,, \tag{8.6.11}$$

which tells us that there is a bound state with energy E provided that the condition (8.6.11) is satisfied. Note that x_1 and x_2 also depend on energy. By solving this equation we can determine the energy levels in a potential well. It should be kept in mind that because of the assumptions made in arriving at the WKBJ solutions we expect these energy levels to be more accurate for large n than for small n. In fact, using $p = \hbar k = (h/2\pi)k$, the condition (8.6.10) can also be written as the Bohr-Sommerfeld's quantization condition

$$2\int_{x_1}^{x_2} p(x)dx \equiv \oint p(x)dx = \left(n + \frac{1}{2}\right)h\,, \tag{8.6.12}$$

which is expected to hold in the limit of large quantum numbers.

As a simple application of these results consider the linear harmonic oscillator with $V(x) = \frac{1}{2}m\omega^2x^2$. Then for an energy E there are two turning points $x_{1,2}(E) = \pm\sqrt{2E/m\omega^2}$. The integral (8.6.11) in this case is easily evaluated to give

$$\int_{-\sqrt{2E/m\omega^2}}^{\sqrt{2E/m\omega^2}} \sqrt{2m(E^2 - m^2\omega^2x^2)}\; dx = E\frac{\pi}{\omega} = \hbar\pi\left(n + \frac{1}{2}\right)\,, \tag{8.6.13}$$

which gives $E_n = \hbar\omega(n + 1/2)$, a result we already know so well. The fact that we get energy levels correct all the way down to the ground state is accidental.

8.7 Application of WKBJ Approximation to α-decay

An important application of the WKBJ method is to nuclear α-decay, in which a parent nucleus of charge $(Z + 2)$ decays into an α-particle of charge $2e$ and a daughter nucleus of charge Ze. To describe this we imagine that the α-particle moves in the field of the daughter nucleus.

Inside the nucleus the α-particle is attracted to the center by the nuclear force which dominates the electrostatic repulsion due to the daughter nucleus. Once it is outside of the range of the nuclear force, only the electrostatic repulsion and the angular momentum barrier remain. We will assume the nuclear potential to be a short range square well of V_0.

Then the potential has the shape shown in Fig. 8.9 and can be represented mathematically by

$$V(r) = \begin{cases} -V_0 & \text{for } r < R\,, \\ \dfrac{2Ze^2}{4\pi\epsilon_0 r} & \text{for } r > R\,. \end{cases} \tag{8.7.1}$$

It seems that the depth V_0 of the potential well is not very large because even the lowest state

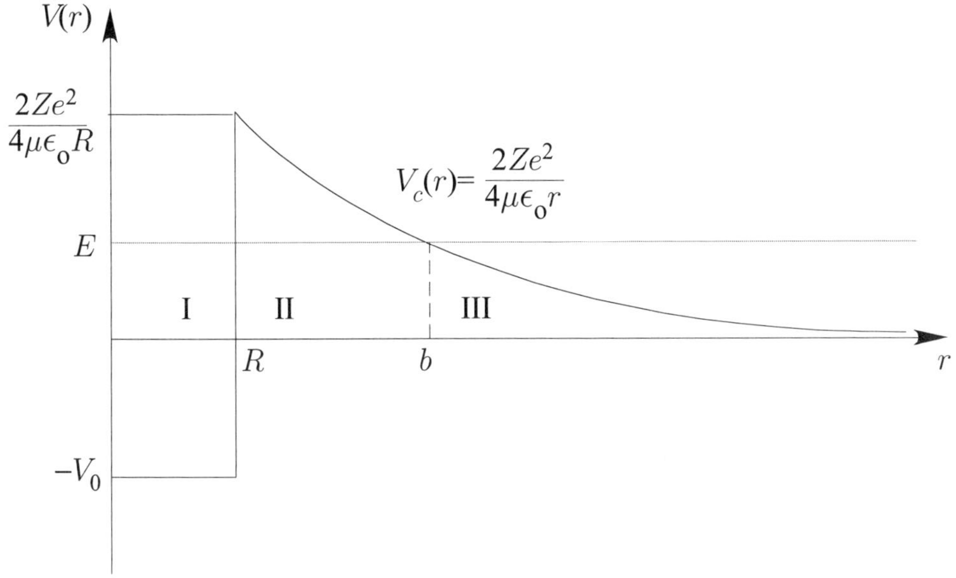

FIGURE 8.9
Nuclear potential as seen by an alpha particle. The depth of the potential well of radius R is V_0. The height of the adjacent Coulomb barrier $V_c = 2Ze^2/4\pi\epsilon_0 R$ at $r = R$ is much larger than the energy E of the particle. At the CTP ($r = b$), the Coulomb potential $V_c(b) = 2Ze^2/4\pi\epsilon_0 b = E$. The regions I, II, III are as indicated.

of the alpha particle is unbound. We are required to explain how an alpha particle within the nucleus, with energy E much less than the height of the Coulomb barrier $2Ze^2/4\pi\epsilon_0 R$, is able to leak through the barrier and come out with the same energy E. A qualitative explanation for this was provided by Gamow [Chapter 4]. To give a realistic explanation we will use WKBJ approximation. We divide the space into three regions [see Fig. 8.9]:

(a) Region I, where $r < R$ and $V(r) = -V_0$. In this region we define

$$\alpha^2 = \frac{2\mu}{\hbar^2}(E + V_0)\,. \tag{8.7.2}$$

(b) Region II, where $R \le r \le b$ and $V(r) = 2Ze^2/4\pi\epsilon_0 r > E$. In this region we define

$$\kappa^2(r) \equiv \frac{2\mu}{\hbar^2}\left(\frac{2Ze^2}{4\pi\epsilon_0 r} - E\right) \tag{8.7.3}$$

where b is the CTP so that $E = 2Ze^2/4\pi\epsilon_0 b$.

(c) Region III, where $r > b$ and $V(r) = 2Ze^2/4\pi\epsilon_0 r < E$. In this region we define

$$k^2(r) \equiv \frac{2\mu}{\hbar^2}\left(E - \frac{2Ze^2}{4\pi\epsilon_0 r}\right). \qquad (8.7.4)$$

Then the radial Schrödinger equation for $u(r) = rR(r)$ for $\ell = 0$ [Eq. (5.5.41)] in the three regions has the forms:indexWKBJ approximation!applied to radial equation

$$\frac{d^2u_1}{dr^2} + \alpha^2 u_1(r) = 0\,, \qquad \alpha^2 = \frac{2\mu}{\hbar^2}(E + V_0) \qquad \text{Region I: } r < R\,, \qquad (8.7.5)$$

$$\frac{d^2u_2}{dr^2} - \kappa^2(r)u_2(r) = 0, \qquad \kappa^2(r) = \frac{2\mu}{\hbar^2}\left(\frac{2Ze^2}{4\pi\epsilon_0 r} - E\right) \qquad \text{Region II: } R \le r \le b\,, \qquad (8.7.6)$$

$$\frac{d^2u_3}{dr^2} + k^2(r)u_3(r) = 0, \qquad k^2(r) = \frac{2\mu}{\hbar^2}\left(E - \frac{2Ze^2}{4\pi\epsilon_0 r}\right) \qquad \text{Region III: } r > b\,. \qquad (8.7.7)$$

The solution in region I is

$$u_1(r) = A_1 \sin(\alpha\, r)\,, \qquad (8.7.8)$$

where the constant A_1 is to be determined by applying the continuity condition to the wave function and its derivative at $r = b$. This function already satisfies the boundary condition at $r = 0$. In region III we need a solution whose asymptotic form represents a wave traveling away from the barrier

$$u_3(r) = \frac{A_3}{\sqrt{k}} \exp\left[i\left(\int_b^r k(r')dr' - \frac{\pi}{4}\right)\right], \qquad (8.7.9)$$

where the phase $\pi/4$ has been included to simplify the calculation. To use the connection formulas to extend this into region II, we write this equation as

$$u_3(r) = \frac{A}{\sqrt{k}}\left[\cos\left(\int_b^r k(r')dr' - \frac{\pi}{4}\right) + i\sin\left(\int_b^r k(r')dr' - \frac{\pi}{4}\right)\right]. \qquad (8.7.10)$$

According to the connection formulas (8.5.41) and (8.5.42), this solution extends into region II in the form

$$\begin{aligned} u_2(r) &= \frac{A_3}{\sqrt{\kappa(r)}}\left[\frac{1}{2}\exp\left(-\int_r^b \kappa(r')dr'\right) - i\exp\left(\int_r^b \kappa(r')dr'\right)\right], \quad R < r \ll b\,, \\ &= \frac{A_3}{\sqrt{\kappa(r)}}\left[\frac{1}{2}e^{-s}\exp\left(\int_R^r \kappa(r')dr'\right) - ie^{s}\exp\left(-\int_R^r \kappa(r')dr'\right)\right]. \end{aligned} \qquad (8.7.11)$$

Here we have introduced a parameter s by

$$\begin{aligned} s &= \int_R^b \kappa(r')dr' = \int_R^b \sqrt{\frac{2\mu}{\hbar^2}\left(\frac{2Ze^2}{4\pi\epsilon_0 r'} - E\right)}\, dr'\,, \\ &= \sqrt{\frac{4\mu Ze^2 b}{4\pi\epsilon_0\hbar^2}}\left[\cos^{-1}\sqrt{\frac{R}{b}} - \sqrt{\frac{R}{b} - \frac{R^2}{b^2}}\right], \end{aligned} \qquad (8.7.12)$$

which will be a large number as $E = V_c(b) = 2Ze^2/4\pi\epsilon_0 b$ is assumed to be small compared to the height of the barrier $V_c(R) = 2Ze^2/4\pi\epsilon_0 R$. (That is why we can apply the WKBJ method.) This means we can neglect the first term on the right hand side in Eq. (8.7.11). Then, applying the boundary conditions at $r = R$ we get

$$A_1 \sin(\alpha R) = -i\frac{e^s A_3}{\sqrt{\kappa(R)}}\,, \tag{8.7.13a}$$

and

$$\alpha \cot\,(\alpha R) \approx -\kappa(R)\,. \tag{8.7.13b}$$

Now the flux of the transmitted wave is $\frac{\hbar}{\mu}|A_3|^2$ and that of the wave incident wave at $r = R$ is $\frac{\hbar\alpha}{4\mu}|A_1|^2$. Using the definition of transmission coefficient [Sec. 4.2], we find

$$T = \frac{(\hbar/\mu)|A_3|^2}{(\hbar\alpha/4\mu)|A_1|^2} = \frac{4\kappa(R)}{\alpha[1+\cot^2(\alpha R)]}\,e^{-2s} = \frac{4\alpha\kappa(R)}{\alpha^2+\kappa^2(R)}\,e^{-2s} \tag{8.7.14}$$

where s given by Eq. (8.7.12). This is the probability that the α-particle will escape every time it hits the barrier. The frequency with which the α-particle hits the barrier is $\frac{v}{2R} = \frac{\hbar\alpha}{2R}$. Hence the probability of escape per second (the rate of nuclear decay) is given by

$$\lambda = T\,\frac{\hbar\alpha}{\mu 2R} = \frac{4\alpha^2\hbar\kappa(R)}{2\mu R[\alpha^2+\kappa^2(R)]}\,e^{-2s} = \frac{4(E+V_0)}{(V_0+V_c(R))}\sqrt{\frac{V_c(R)-E}{2\mu R^2}}\,e^{-2s}\,. \tag{8.7.15}$$

Then the lifetime τ of the nucleus is given by

$$\tau \equiv \lambda^{-1} = \frac{[V_0+V_c(R)]}{4(E+V_0)}\sqrt{\frac{2\mu R^2}{V_c(R)-E}}\,e^{2s} \equiv \tau_0 e^{2s}\,. \tag{8.7.16}$$

For α-particle energies small compared to the barrier height $V_c(R)$, we have $b \gg R$, so that

$$2s \approx 2\sqrt{\frac{4\mu Ze^2 b}{4\pi\epsilon_0\hbar^2}}\,\frac{\pi}{2} = \frac{4\pi Ze^2}{4\pi\epsilon_0\hbar c}\sqrt{\frac{\mu c^2}{2E}} \approx \frac{4Z}{\sqrt{E(\text{in MeV})}}\,, \tag{8.7.17a}$$

$$\ln\frac{1}{\tau} = \ln\frac{1}{\tau_0} - \frac{4Z}{\sqrt{E(\text{in MeV})}} \tag{8.7.17b}$$

where we have used reduced $\mu = \frac{m_\alpha m_Z}{m_\alpha + m_Z} \approx m_\alpha \approx 3.8 \times 10^3$ MeV, where m_Z is the daughter nucleus mass. The numerical constants are hard to compute reliably because of large uncertainties in the shape of the nuclear potential. Nevertheless, the last equation predicts the Z and E dependence accurately, which fits the experimental data over a large range of lifetimes.

8.8 The Variational Method

The variational method is another widely used approximation method in quantum mechanics. This method is frequently used in atomic, molecular and nuclear problems to estimate the ground state energy and ground state wave function of a system. The basis of this method is the *variational principle* explained below.

Consider a system whose Hamiltonian H is known and the system is understood to admit a set of discrete energy levels E_i, and normalized and orthogonal eigenstates $|\psi_i\rangle$, such that

$$\hat{H}\psi_i = E_i\,|\psi_i\rangle\,, \quad i = 1, 2, 3, \cdots \tag{8.8.1}$$

The energy levels E_i and the eigenfunctions $|\psi_i\rangle$, which in the coordinate representation are functions of the continuously varying position coordinates of the system, are not known and need to be determined. Now consider any state $|\phi\rangle$ be of the system, which is normalized

$$\langle\phi|\,\phi\rangle = 1\,, \tag{8.8.2}$$

and may depend on a number of adjustable parameters. Then the *variational principle* states that the expectation value

$$E(\phi) \equiv \langle\phi|\,\hat{H}\,|\phi\rangle$$

is always greater than, or at most equal to, the ground state energy E_0 of the system, i.e., $E(\phi) \geq E_0$; the equality being obtained if $|\phi\rangle$ is, somehow, chosen to be $|\psi_0\rangle$.

According to this principle, we can obtain an estimate for the ground state energy by making an educated choice for the ground state. We choose a trial ground state $|\phi(\lambda_i)\rangle$ which includes some parameters λ_i that relate to the physical properties of the system and work out the expectation value $E(\phi) = E(\lambda_i)$ in terms of these parameters, called the variational parameters. Note that as a function of these parameters, we are really considering a family of trial states. By choosing these parameters to minimize $E(\phi) = E(\lambda_i)$, we can get the *best* $E(\lambda_i^o)$ for the ground state energy E_0 as also the best ground state $|\phi(\lambda_i^o)\rangle$ (for the optimal values λ_i^o of the variational parameters), within the family of trial states.

To prove the variational principle, we expand the trial state in terms of the complete set of states $|\psi_i\rangle$. No generality is lost in assuming that the states $|\psi_i\rangle$ are orthogonal since, if any of these functions are degenerate, orthogonal linear combinations of them can always be constructed. Thus we have

$$|\phi\rangle = \sum_i a_i\,|\psi_i\rangle\,. \tag{8.8.3}$$

The normalization condition $\langle\phi|\,\phi\rangle = 1$ on ϕ implies

$$\sum_i a_i^* a_i = 1\,. \tag{8.8.4}$$

Using Eq. (8.8.3), the energy expectation value can be evaluated as

$$E(\phi) = \langle\phi|\,H\,|\phi\rangle = \sum_i\sum_j a_j^* a_i E_i \delta_{ij} = \sum_i |a_i|^2 E_i\,. \tag{8.8.5}$$

Since E_0 is the lowest energy of the system, being the ground state energy ($E_0 \leq E_1, E_2, E_3, \cdots$), if we replace all the energies under the sum in Eq. (8.8.5) by E_0 and use the normalization condition $\sum_i |a_i|^2 = 1$ we immediately obtain

$$E(\phi) \equiv \sum_i |a_i|^2 E_i \geq \left(\sum_i |a_i|^2\right) E_0 = E_0\,. \tag{8.8.6}$$

This is the variational principle, which says that the energy expectation value in any trial state is an upper bound to the ground state energy. In case all a_is, except a_o, are zero so that $\phi = \psi_o$ then only the equality holds and $E(\phi) = E_0$. Thus unless we are able to guess the trial function $\phi(\boldsymbol{r}) = \langle\boldsymbol{r}|\,\phi\rangle$, which is identical with the ground state wave function $\psi_0(\boldsymbol{r}) = \langle\boldsymbol{r}|\,\psi_0\rangle$, the expectation value $E(\phi) > E_0$. By adjusting the variational parameters to minimize $E(\phi) \equiv E(\lambda_i)$ we can get the best estimate to the ground state energy and the best approximation to the ground state wave function. We may even try alternative forms of the trial wave function and find the one that with the adjustment of the variation parameters gives the minimum E.

Good physical intuition about the system is the key to the accuracy of the variational method in estimating the ground state energy. It is even possible in many cases to obtain estimates of the next few excited states above the ground state. The variational method is complementary to the WKBJ method in that while the former is used for estimating the energy of the ground state or at most the first few low lying states, the WKBJ is reliable for large quantum number states.

As an application of the variational method let us estimate the ground state energy of the normal Helium atom.

Ground State Energy of the Helium Atom

The complete Hamiltonian of the normal Helium atom, without spin degree of freedom, is given by

$$\begin{aligned} H &= \left(-\frac{\hbar^2}{2m}\nabla_1^2 - \frac{2e^2}{4\pi\epsilon_0 r_1}\right) + \left(-\frac{\hbar^2}{2m}\nabla_2^2 - \frac{2e^2}{4\pi\epsilon_0 r_2}\right) + \frac{e^2}{4\pi\epsilon_0|\boldsymbol{r}_1 - \boldsymbol{r}_2|} \\ &\equiv H_1 + H_2 + H_{12}\,. \end{aligned} \tag{8.8.7}$$

In the ground state both electrons must occupy the lowest hydrogenic state $|\psi_{100}\rangle$ [see Sec. 5.6]. Therefore, for the trial wave function we use the product of the ground state hydrogenic wave functions with nuclear charge Ze for the two electrons and treat Z as a variational parameter. Treating Z as a variational parameter allows us to take into account the fact that the full nuclear charge Ze is not seen by either of the electrons because of the screening effect of the other. (In the Hamiltonian, of course, we put $Z = 2$.) Thus, for the trial wave function, we have

$$\phi(\boldsymbol{r}_1, \boldsymbol{r}_2; Z) = \psi_{100}(\boldsymbol{r}_1; Z)\psi_{100}(\boldsymbol{r}_2; Z) = \left(\frac{Z^3}{\pi a_o^3}\right) e^{-Z(r_1+r_2)/a_o}\,, \tag{8.8.8}$$

where Z will be treated as a variational parameter. Then the energy integral is

$$\begin{aligned} E(Z) &= < \phi|H|\phi > \\ &= \langle \psi_{100}(1;Z)|\, H_1\, |\psi_{100}(1;Z)\rangle \times \langle \psi_{100}(2;Z)|\, \psi_{100}(2;Z)\rangle \\ &\quad + \langle \psi_{100}(1;Z)|\, \psi_{100}(1;Z)\rangle \times \langle \psi_{100}(2;Z)|\, H_2\, |\psi_{100}(2;Z)\rangle \\ &\quad + \langle \phi(1,2;Z)|\, H_{12}\, |\phi(1,2;Z)\rangle \\ &= 2 \times \langle \psi_{100}(1;Z)|\, H_1\, |\psi_{100}(1;Z)\rangle + \langle \phi(1,2;Z)|\, H_{12}\, |\phi(1,2;Z)\rangle\,, \end{aligned} \tag{8.8.9}$$

where we have used the fact that hydrogenic wave functions are normalized, $\langle \psi_{100}|\, \psi_{100}\rangle = 1$ and that the expectation value of H_1 in the state $|\psi_{100}(1;Z)\rangle$ is equal to the expectation value of H_2 in state $|\psi_{100}(2;Z)\rangle$. To evaluate the first term we write

$$H_1 = -\frac{\hbar^2}{2m}\nabla_1^2 - \frac{Ze^2}{4\pi\epsilon_0 r_1} + \frac{(Z-2)e^2}{4\pi\epsilon_0 r_1}\,. \tag{8.8.10a}$$

Then with the help of Eq. (5.6.33) [$\langle r^{-1}\rangle_{n\ell m} = Z/a_o n^2$] we have

$$\left\langle \frac{(Z-2)e^2}{4\pi\epsilon_0 r} \right\rangle_{100} = \frac{(Z-2)e^2}{4\pi\epsilon_0}\frac{Z}{a_o}\,, \tag{8.8.10b}$$

$$\begin{aligned} \langle \psi_{100}(1;Z)|\, H_1\, |\psi_{100}(1;Z)\rangle &= -\frac{Z^2e^2}{8\pi\epsilon_0 a_o} + \frac{(Z-2)e^2}{4\pi\epsilon_0}\frac{Z}{a_o}\,, \\ &= \langle \psi_{100}(2;Z)|\, H_2\, |\psi_{100}(2;Z)\rangle\,. \end{aligned} \tag{8.8.10c}$$

There remains the last term in Eq. (8.8.9), which in the coordinate representation can be written as

$$\langle \phi(1,2;Z)|\, H_{12}\, |\phi(1,2;Z)\rangle = \left(\frac{Z^3}{\pi a_o^3}\right)^2 \frac{e^2}{4\pi\epsilon_0} \iint \frac{\exp[-2Z(r_1+r_2)/a_o]}{|r_1 - r_2|} d^3r_1 d^3r_2 \,. \quad (8.8.11)$$

With the change of variables $\boldsymbol{\rho}_i = \frac{2Z\boldsymbol{r}_i}{a_o}$ and $\rho_{12} = |\boldsymbol{\rho}_1 - \boldsymbol{\rho}_2|$, this integral takes the form

$$\left(\frac{Z^3}{\pi a_o^3}\right)^2 \frac{e^2}{4\pi\epsilon_0} \frac{a_o^5}{32Z^5} \iint \frac{\exp[-(\rho_1+\rho_2)]}{\rho_{12}} d^3\rho_1 d^3\rho_2 \,. \quad (8.8.12)$$

The ρ-integral was evaluated in Sec. 8.2 [Footnote 1] and has the value $20\pi^2$. Thus we have finally

$$E(Z) = -2\left[\frac{Z^2e^2}{8\pi\epsilon_0 a_o}\right] + 2\left[\frac{(Z-2)Ze^2}{4\pi\epsilon_0 a_o}\right] + \frac{5}{8}\frac{Ze^2}{4\pi\epsilon_0 a_o} = \frac{e^2}{4\pi\epsilon_0 a_o}\left(Z^2 - \frac{27Z}{8}\right) . \quad (8.8.13)$$

To minimize the energy with respect to Z, we put $\frac{dE}{dZ} = 0$, which yields $Z_o = 27/16$. Putting this optimum value of the variational parameter in Eq. (8.8.13) we get

$$E = -5.7\frac{e^2}{8\pi\epsilon_0 a_o} = -5.7 \text{ Rydberg} \quad (8.8.14)$$

This compares favorably with the experimentally measured value $E_{\text{exp}} = -5.808$ Rydberg.

We note that if we simply put $Z = 2$, we get -5.5 Rydberg, showing clearly that treating Z as a variational parameter results in a better estimate for the ground state energy. Also note that the variational method gives a better value for the ground state energy of the normal Helium atom than the first order perturbation calculation.

8.9 The Problem of the Hydrogen Molecule

The basic question as to what binds two neutral Hydrogen atoms into a molecule cannot be answered classically. This type of binding of identical atoms, called *homopolar binding*, was first explained by Heitler and London on the basis of quantum mechanics. They showed that the *homopolar bond* is due to a typical quantum effect called the *exchange effect.*

The Hydrogen molecule consists of two nuclei (protons) A and B a definite distance R apart and two electrons 1 and 2 [Fig. 8.10]. In the coordinate representation the Hamiltonian operator of the system can be written as

$$H = -\frac{\hbar^2}{2m}(\nabla_1^2+\nabla_2^2) + \frac{e^2}{4\pi\epsilon_0 R} + \frac{e^2}{4\pi\epsilon_0 r_{12}} - \frac{e^2}{4\pi\epsilon_0 r_{1A}} - \frac{e^2}{4\pi\epsilon_0 r_{2B}} - \frac{e^2}{4\pi\epsilon_0 r_{2A}} - \frac{e^2}{4\pi\epsilon_0 r_{1B}} \,. \quad (8.9.1)$$

If we write $\boldsymbol{r}_{1A} \equiv \boldsymbol{r}_1$ and $\boldsymbol{r}_{2B} = \boldsymbol{r}_2$ then $\boldsymbol{r}_{2A} = \boldsymbol{R} + \boldsymbol{r}_2$ and $\boldsymbol{r}_{1B} = -\boldsymbol{R} + \boldsymbol{r}_1$ [Fig. 8.10]. It is not possible to solve the equation exactly for this problem. The most convenient method for this problem is the variational method.

To write the trial wave function we note that if we regard this system as consisting of two independent Hydrogen atoms in the ground state with *electron 1 attached to nucleus A* and *electron 2 attached to nucleus B* (neglecting the interactions $-e^2/4\pi\epsilon_0 r_{2A}$, $-e^2/4\pi\epsilon_0 r_{1B}$,

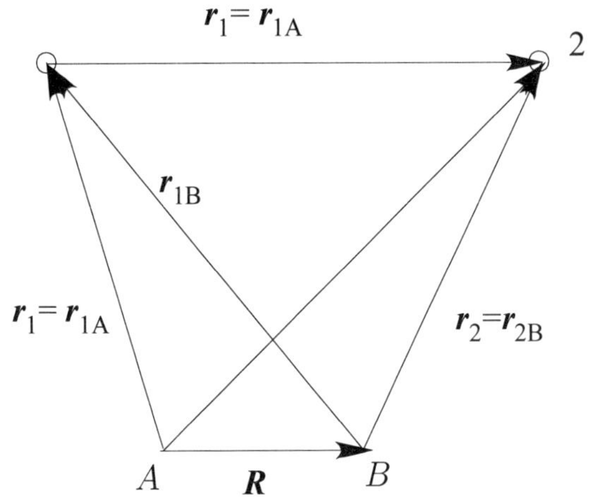

FIGURE 8.10
In the Hydrogen molecule, the labels A and B denote the protons; 1 and 2 denote the electrons.

$e^2/4\pi\epsilon_0 R$ and $e^2/4\pi\epsilon_0 r_{12}$), then the ground state wave function of this system may be written as

$$u_1(\boldsymbol{r}_1, \boldsymbol{r}_2) = u_A(\boldsymbol{r}_1) u_B(\boldsymbol{r}_2)\,, \tag{8.9.2}$$

where
$$u_A(\boldsymbol{r}_1) = \frac{1}{\sqrt{\pi a_o^3}} \exp(-r_{1A}/a_o) = \frac{1}{\sqrt{\pi a_o^3}} \exp(-r_1/a_o) \tag{8.9.3}$$

and
$$u_B(\boldsymbol{r}_2) = \frac{1}{\sqrt{\pi a_o^3}} \exp(-r_{2B}/a_o) = \frac{1}{\sqrt{\pi a_o^3}} \exp(-r_2/a_o)\,. \tag{8.9.4}$$

Now, since the two electrons are identical, we can regard this system as consisting of two independent Hydrogen atoms in the ground state with *electron 1 attached to nucleus B* and *electron 2 attached to nucleus A* (neglecting the interaction terms in the Hamiltonian). In that case the ground state wave function of the system can be written as

$$u_2(\boldsymbol{r}_1, \boldsymbol{r}_2) = u_A(\boldsymbol{r}_{2A}) u_B(\boldsymbol{r}_{1B})\,, \tag{8.9.5}$$

where
$$u_A(\boldsymbol{r}_{2A}) = \frac{1}{\sqrt{\pi a_o^3}} \exp(-\boldsymbol{r}_{2A}/a_o) \tag{8.9.6}$$

and
$$u_B(\boldsymbol{r}_{1B}) = \frac{1}{\sqrt{\pi a_o^3}} \exp(-\boldsymbol{r}_{1B}/a_o)\,. \tag{8.9.7}$$

Therefore, for the ground state trial wave function of the Hydrogen molecule, we choose a linear combination of $u_1(\boldsymbol{r}_1, \boldsymbol{r}_2)$ and $u_2(\boldsymbol{r}_1, \boldsymbol{r}_2)$

$$\psi(\boldsymbol{r}_1, \boldsymbol{r}_2) = u_1(\boldsymbol{r}_1, \boldsymbol{r}_2) + C u_2(\boldsymbol{r}_1, \boldsymbol{r}_2) \tag{8.9.8}$$

where C is regarded as a variational parameter. Note that the trial wave function $\psi(\boldsymbol{r}_1, \boldsymbol{r}_2)$ is, in general, not normalized even if $u_1(\boldsymbol{r}_1, \boldsymbol{r}_2)$ and $u_2(\boldsymbol{r}_1, \boldsymbol{r}_2)$ are. Then according to the variational principle, the best approximation to the ground state energy of the Hydrogen molecule is obtained by minimizing the integral

$$E(C) = \frac{\int \psi^* H \psi d\tau}{\int \psi^* \psi d\tau}\,, \tag{8.9.9}$$

with respect to the variation parameter C. Using Eq. (8.9.8) we find the energy integral $E(C)$ can be written as

$$E(C) = \frac{H_{11} + C\,H_{12} + C\,H_{21} + C^2\,H_{22}}{1 + C^2 + 2C\gamma}\,, \tag{8.9.10}$$

where

$$H_{ij} \equiv \int u_i(\boldsymbol{r}_1, \boldsymbol{r}_2) H\, u_j(\boldsymbol{r}_1, \boldsymbol{r}_2) d\tau_1 d\tau_2\,, \tag{8.9.11}$$

$$\gamma \equiv \int u_1(\boldsymbol{r}_1, \boldsymbol{r}_2) u_2(\boldsymbol{r}_1, \boldsymbol{r}_2) d\tau_1 d\tau_2\,, \tag{8.9.12}$$

and the functions u_1 and u_2 have been assumed to be normalized to unity. Since H is symmetric between the two electrons, we conclude $H_{11} = H_{22}$ and $H_{12} = H_{21}$. Explicitly, we have[4]

$$H_{22} = \iint u_A(\boldsymbol{r}_{2A}) u_B(\boldsymbol{r}_{1B}) H u_A(\boldsymbol{r}_{2A}) u_B(\boldsymbol{r}_{1B}) d^3 r_{1B} d^3 r_{2A}\,, \tag{8.9.13}$$

$$H_{11} = \iint u_A(\boldsymbol{r}_{1A}) u_B(\boldsymbol{r}_{2B}) H u_A(\boldsymbol{r}_{1A}) u_B(\boldsymbol{r}_{2B}) d^3 r_{1A} d^3 r_{2B}\,, \tag{8.9.14}$$

$$H_{12} = \iint u_A(\boldsymbol{r}_1) u_B(\boldsymbol{r}_2)\, H\, u_A(\boldsymbol{r}_2) u_B(\boldsymbol{r}_1) d^3 r_1 d^3 r_2\,, \tag{8.9.15}$$

and

$$H_{21} = \iint u_A(\boldsymbol{r}_2) u_B(\boldsymbol{r}_1) H u_A(\boldsymbol{r}_1) u_B(\boldsymbol{r}_2) d^3 r_1 d^3 r_2\,. \tag{8.9.16}$$

The equalities $H_{11} = H_{22}$ and $H_{12} = H_{21}$ are obvious because interchanging 1 and 2 leaves these integrals unchanged but transforms $H_{11} \leftrightarrow H_{22}$ and $H_{12} \leftrightarrow H_{21}$. Using these symmetry properties we can write the energy integral as

$$E(C) = \frac{(1 + C^2) H_{11} + 2C\,H_{12}}{1 + C^2 + 2C\,\gamma}\,. \tag{8.9.17}$$

The optimum value of the variation parameter C, which minimizes $E(C)$ can be found from the condition $\partial I/\partial C = 0$, which gives

$$(1 - C^2)(H_{12} - H_{11}) = 0 \;\text{ or }\; C = \pm 1\,. \tag{8.9.18}$$

Thus we obtain two solutions

For $C = +1$:

$$E_s(R) = (H_{11} + H_{12})/(1 + \gamma)\,, \tag{8.9.19}$$

For $C = -1$:

$$E_a(r) = (H_{11} - H_{12})/(1 - \gamma)\,. \tag{8.9.20}$$

One of these solutions corresponds to the minimum of $E(C)$ and the other to the maximum. We must investigate which of the two values of C gives the minimum value of $E(C)$.

We note that $C = +1$ corresponds to the wave function

$$\psi = \psi_s(\boldsymbol{r}_1, \boldsymbol{r}_2) = u_1(\boldsymbol{r}_1, \boldsymbol{r}_2) + u_2(\boldsymbol{r}_1, \boldsymbol{r}_2)\,, \tag{8.9.21}$$

which is a symmetric function of the spatial coordinates $\boldsymbol{r}_1$ and $\boldsymbol{r}_2$. Since electrons, being Fermions, obey the Pauli principle, the overall space-spin wave function of the two electrons must be anti-symmetric in space-spin coordinates. Hence it follows that if the space function

[4] $d\tau \equiv d\tau_1 d\tau_2 \equiv d^3 r_1 d^3 r_2 \equiv d^3 r_{1A} d^3 r_{2B} = d^3 r_{1B} d^3 r_{2A}$.

is symmetric, the spin function of the two electrons must be anti-symmetric so that the two electrons must find themselves in the spin singlet state

$$\chi_0^0 = \frac{1}{\sqrt{2}}\left[\alpha(1)\beta(2) - \beta(1)\alpha(2)\right] . \tag{8.9.22}$$

On the other hand, $C = -1$ corresponds to the wave function

$$\psi = \psi_a(\boldsymbol{r}_1, \boldsymbol{r}_2) = u_1(\boldsymbol{r}_1, \boldsymbol{r}_2) - u_2(\boldsymbol{r}_1, \boldsymbol{r}_2) , \tag{8.9.23}$$

which is an anti-symmetric (under the interchange of spatial coordinates $\boldsymbol{r}_1$ and $\boldsymbol{r}_2$) wave function. Hence, according to Pauli, the spin state in this case must be symmetric so that the two electrons must find themselves in one of the spin triplet states

$$\chi_1^1 = \alpha(1)\alpha(2), \tag{8.9.24}$$

$$\chi_1^0 = \frac{1}{\sqrt{2}}\left[\alpha(1)\beta(2) + \beta(1)\alpha(2)\right], \tag{8.9.25}$$

$$\chi_1^{-1} = \beta(1)\beta(2). \tag{8.9.26}$$

To answer the question whether the space symmetric (and spin anti-symmetric) or the space anti-symmetric (and spin symmetric) ground state wave function corresponds to the bound state of the Hydrogen molecule, we have to work out the integrals $H_{11} = H_{22}$, $H_{12} = H_{21}$ and γ explicitly. Using the complete Hamiltonian (8.9.1) in Eqs. (8.9.13) through (8.9.16), we find that the matrix elements H_{11} and H_{22} are given by

$$H_{22} = H_{11} = 2E_0 + \frac{e^2}{4\pi\epsilon_0 R} + 2J + J' , \tag{8.9.27}$$

where $E_0 = -e^2/8\pi\epsilon_0 a_o$ is the ground state energy of the Hydrogen atom and the integrals represented by J and J' are given by

$$\begin{aligned} J &\equiv \int {u_A}^2(\boldsymbol{r}_{1A})\left(-\frac{e^2}{4\pi\epsilon_0 r_{1B}}\right)d^3r_1 = \int {u_B}^2(\boldsymbol{r}_{2B})\left(-\frac{e^2}{4\pi\epsilon_0 r_{2A}}\right)d^3r_2 \\ &= \frac{e^2}{4\pi\epsilon_0 a_o}\left[-\frac{1}{D} + e^{-2D}\left(1 + \frac{1}{D}\right)\right], \end{aligned} \tag{8.9.28}$$

and

$$\begin{aligned} J' &\equiv \iint u_A^2(\boldsymbol{r}_1)u_B^2(\boldsymbol{r}_2)\left(\frac{e^2}{4\pi\epsilon_0 r_{12}}\right)d^3r_1d^3r_2 \\ &= \frac{e^2}{4\pi\epsilon_0 a_o}\left[\frac{1}{D} - e^{-2D}\left(\frac{1}{D} + \frac{11}{8} + \frac{3D}{4} + \frac{D^2}{6}\right)\right] . \end{aligned} \tag{8.9.29}$$

Here $D = R/a_o$ is the nuclear separation in units of Bohr radius a_o. Similarly, the elements $H_{12} = H_{21}$ can be written as

$$H_{21} = H_{12} = \left(2E_0 + \frac{e^2}{4\pi\epsilon_0 R}\right)\gamma + 2K\gamma^{1/2} + K' \tag{8.9.30}$$

where γ given by

$$\begin{aligned} \gamma^{1/2} &= \int u_A(\boldsymbol{r}_{1A})u_B(\boldsymbol{r}_{1B})d^3r_1 = \int u_B(\boldsymbol{r}_{2B})u_A(\boldsymbol{r}_{2A})d^3r_2 \\ &= e^{-D}\left(1 + D + \frac{D^3}{3}\right) \end{aligned} \tag{8.9.31}$$

is consistent with Eq. (8.9.12) and K and K' are given by

$$K = \int u_A(\boldsymbol{r}_{1A})\left[-\frac{e^2}{4\pi\epsilon_0 r_{1A}}\right]u_B(\boldsymbol{r}_{1B})d^3r_1 = \int u_B(\boldsymbol{r}_{2B})\left[-\frac{e^2}{4\pi\epsilon_0 r_{2B}}\right]u_A(\boldsymbol{r}_{2A})d^3r_2$$
$$= -\frac{e^2}{4\pi\epsilon_0 a_o}e^{-D}(1+D) \tag{8.9.32}$$

and $$K' = \iint u_A(\boldsymbol{r}_{1A})u_B(\boldsymbol{r}_{2B})\left[\frac{e^2}{4\pi\epsilon_0 r_{12}}\right]u_A(\boldsymbol{r}_{2A})u_B(\boldsymbol{r}_{1B})d^3r_1 d^3r_2$$
$$= \frac{e^2}{20\pi\epsilon_0 a_o}\left[-e^{-2D}\left(-\frac{25}{8}+\frac{23}{4}D+3D^2+D^3\right)\right.$$
$$\left.+\frac{6}{D}\{\gamma(0.5772+\ln D)+\Delta^2\mathrm{Ei}(-4D)-2\gamma\Delta\mathrm{Ei}(-2D)\}\right], \tag{8.9.33}$$

where $$\Delta \equiv \exp(-D)(1-D+D^3/3), \tag{8.9.34}$$
$$E_i(x) \equiv \int_{-\infty}^{x}\frac{\exp(-u)}{u}du. \tag{8.9.35}$$

Using these expressions we find

For C = +1 $$E_s(R) = \frac{H_{11}+H_{12}}{1+\gamma} = 2\varepsilon_0 + \frac{e^2}{4\pi\epsilon_0 R} + \frac{2J+J'+2K\gamma^{1/2}+K'}{1+\gamma}. \tag{8.9.36}$$

For C = -1 $$E_a(R) = \frac{H_{11}-H_{12}}{1-\gamma} = 2\varepsilon_0 + \frac{e^2}{4\pi\epsilon_0 R} + \frac{2J+J'-2K\gamma^{1/2}-K'}{1-\gamma}. \tag{8.9.37}$$

By plotting $E_s(R)$ and $E_a(R)$ as functions of R [Fig. 8.11], we find that the curve for $E_a(R)$ corresponds to repulsion at all distances, there being no equilibrium position for the nuclei. The curve for $E_s(R)$ corresponds to attraction of Hydrogen atoms resulting in the formation of a stable molecule. $E_s(R)$ shows a minimum at $R = 0.080$ nm. We can interpret this to be the equilibrium distance between the two protons in the Hydrogen molecule (the experimental value of the equilibrium distance is 0.074 nm). The value of $E(R) - 2\varepsilon_0$ at this value of R is equal to -3.14 eV. This implies that the energy of dissociation of the Hydrogen molecule into atoms is 3.14 eV. (This value is smaller than the experimental value of 4.52 eV.)

In answer to the question as to what constitutes the chemical bond between two neutral atoms (like Hydrogen atoms), we can say that the bond is due to a pair of electrons held jointly by two atoms. As to what causes the attraction between two Hydrogen atoms when the spins of the two electrons are anti-parallel, we may say that the binding comes about because of the *exchange energies* K and K'.

8.10 System of n Identical Particles: Symmetric and Anti-symmetric States

It can be seen that the Hamiltonian of n identical particles is invariant under the exchange of coordinates of any pair of particles (permutation operation) or under the exchange of

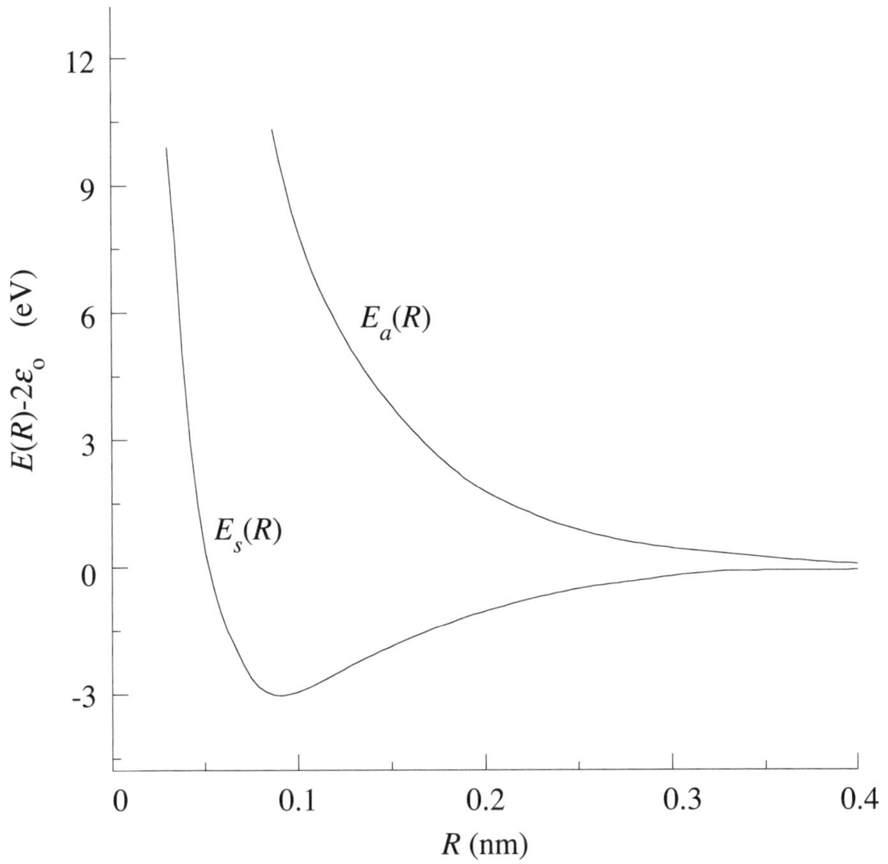

FIGURE 8.11
Plot of $(E(R) - 2\varepsilon_0)$ as a function of R. E_a corresponds to the case when the electrons are in spin triplet (symmetric) state and the spatial wave function is anti- symmetric and E_s to the case when the electrons are in spin singlet (anti-symmetric) state and the spatial part of the wave function is symmetric.

coordinates of more than one pair of particles. Mathematically,

$$\hat{H} \rightarrow \hat{P}\hat{H}\hat{P}^{-1} = \hat{H} \;\Rightarrow\; \hat{P}\hat{H} = \hat{H}\hat{P} \;\text{ or }\; [\hat{P}, \hat{H}] = 0\,. \tag{8.10.1}$$

So the Hamiltonian of n identical particles commutes with the permutation operator.

Let us investigate the effect of the permutation operation on an n-particle wave function $\Psi(X_1, X_2, \cdots, X_n)$, where X stands for x, y, z, ζ with ζ being the spin coordinate of the particle which can take values $+1$ or -1 (corresponding to spin up or down). Suppose Ψ is an eigenstate of $\hat{H}$, belonging to the eigenvalue E,

$$H\Psi(X_1, X_2, \cdots, X_n) = E\Psi(X_1, X_2 \cdots, X_n)\,. \tag{8.10.2}$$

Now $PH\Psi(X_1, X_2, ...X_n) = PE\Psi = EP\Psi(X_1, X_2, ...X_n)$ and, since P and H commute, we can have

$$HP\Psi(X_1, X_2 \cdots, X_n) = E\,P\Psi(X_1, X_2, \cdots, X_n)\,. \tag{8.10.3}$$

Thus the state $P\Psi$ corresponds to the same energy as the state Ψ.

To identify the state $P\Psi$, we note that since the particles are identical, any observable quantity which depends on the particle coordinates must remain unchanged as a result of the permutation of particle coordinates. Now the wave function $\Psi(X_1, X_2, ...X_n)$ is not an observable quantity, but $|\Psi(X_1, X_2, ...X_n)|^2$ is, as this has the interpretation of probability density. Hence we must have $|P\Psi(X_1, X_2, ...X_n)|^2 = |\Psi(X_1, X_2, ...X_n)|^2$, which is possible only if

$$P\Psi(X_1, X_2, ...X_n) = \pm\Psi(X_1, X_2, ...X_n) . \tag{8.10.4}$$

In other words, *the wave function of an n-particle system should either be symmetric or anti-symmetric with respect to a permutation operation.*

Now, whether the wave function of a system of n identical particles is symmetric or anti-symmetric, depends on the spin of the particles or the statistics obeyed by them, i.e., on whether the particles are Bosons obeying Bose-Einstein statistics or Fermions obeying Fermi-Dirac statistics. If the n identical particles are Bosons, that is, they all have an integer spin (including zero), then the total wave function is symmetric with respect to permutation of the coordinates of the particles. Hence in this case

$$\Psi(X_1,X_2, ...X_n) = \frac{1}{\sqrt{n!}} \sum_P P\{\phi_1(X_1)\phi_2(X_2)\cdots\phi_n(X_n)\} , \tag{8.10.5}$$

where $\phi_1, \phi_2, \cdots, \phi_n$ are the single particle states occupied by n particles, and the permutation operator P permutes the coordinates $X_1, X_2, X_3 \cdots, X_n$. The sum is over all the $n!$ permutations and the factor $1/\sqrt{n!}$ is to normalize the symmetric n-particle wave function. It is assumed that the single-particle wave functions are normalized.

If, on the other hand the n identical particles are Fermions, i.e., they all have a half-integer spin, then the total wave function is anti-symmetric. In this case

$$\Psi(X_1, X_2, ..., .X_n) = \frac{1}{\sqrt{n!}} \sum_P \epsilon_P P\{\phi_1(X_1)\phi_2(X_2)\cdots\phi_n(X_n)\} , \tag{8.10.6}$$

where $\epsilon_P = (-1)^s$, s being the number of interchanges in the permutation P to bring it to the original sequence, and the sum is over all the $n!$ permutations. The anti-symmetric n-particle wave function (8.10.6) may also be written in a determinant form (also called Slater determinant)

$$\Psi(X_1, X_2, ..., X_n) = \frac{1}{\sqrt{n!}} \begin{vmatrix} \phi_1(X_1) & \phi_1(X_2) & \cdots & \phi_1(X_n) \\ \phi_2(X_1) & \phi_2(X_2) & \cdots & \phi_2(X_n) \\ \vdots & \cdots & \cdots & \vdots \\ \phi_n(X_1) & \phi_n(X_2) & \cdots & \phi_n(X_n) \end{vmatrix} . \tag{8.10.7}$$

If, in the determinant form of n-particle wave function, two single-particle states are the same, say $\phi_1 = \phi_2$, i.e., both states are characterized by the same set of quantum numbers, then the determinant vanishes. So the determinant form explicitly incorporates the Pauli exclusion principle, i.e., two identical particles (Fermions) cannot occupy the same state.

Ground State of Helium Revisited

As the simplest example of an n-particle Fermion system, we consider the two-electron system in a Helium atom. The Hamiltonian of the system can be written as

$$H = \left(-\frac{\hbar^2}{2m}\nabla_1^2 - \frac{2e^2}{4\pi\epsilon_0 r_1}\right) + \left(-\frac{\hbar^2}{2m}\nabla_2^2 - \frac{2e^2}{4\pi\epsilon_0 r_2}\right) + \frac{e^2}{4\pi\epsilon_0 r_{12}} \equiv H_1^0 + H_2^0 + H_{12} . \tag{8.10.8}$$

If the interaction H_{12} is ignored, then to the first approximation the two-electron wave function may be written as

$$\Psi(X_1, X_2) = \frac{1}{\sqrt{2}} \begin{vmatrix} \phi_1(X_1) & \phi_1(X_2) \\ \phi_2(X_1) & \phi_2(X_2) \end{vmatrix},$$

where ϕ_i represents a single-particle wave function, which is the product of a spatial part and a spin part. In the ground state of the Helium atom, both the electrons occupy the lowest spatial state corresponding to $n = 1, \ell = 0, m = 0$ so that the spatial part of the wave function is $\psi_{100}(\boldsymbol{r}) \equiv \psi_0(\boldsymbol{r})$. It follows from Pauli's principle that the spin quantum numbers (m_s) of the two electrons must be different, i.e., the two electrons should be in opposite spin states (spin up and spin down). Hence the single-particle wave functions are $\phi_1(X) = \psi_0(\boldsymbol{r})\alpha(\zeta)$ and $\phi_2(X) = \psi_0(\boldsymbol{r})\beta(\zeta)$ and therefore

$$\begin{aligned} \Psi_0(X_1, X_2) &= \frac{1}{\sqrt{2}} \begin{vmatrix} \psi_0(\boldsymbol{r}_1)\alpha(\zeta_1) & \psi_0(\boldsymbol{r}_2)\alpha(\zeta_2) \\ \psi_0(\boldsymbol{r}_1)\beta(\zeta_1) & \psi_0(\boldsymbol{r}_2)\beta(\zeta_2) \end{vmatrix} \\ &= \psi_0(\boldsymbol{r}_1)\psi_0(\boldsymbol{r}_2)\frac{1}{\sqrt{2}}[\alpha(\zeta_1)\beta(\zeta_2) - \beta(\zeta_1)\alpha(\zeta_2)] \\ &\equiv \psi_0(\boldsymbol{r}_1)\psi_0(\boldsymbol{r}_2)\chi_0^0(\zeta_1, \zeta_2). \end{aligned} \tag{8.10.9}$$

Thus the spatial part of the wave function is symmetric, while the spin part is anti-symmetric. The overall wave function, which is the product of spatial and spin wave functions, is anti-symmetric under the exchange of electron (spin and spatial) coordinates. Using this wave function we find the unperturbed energy is ($Z = 2$)

$$E^{(0)} = -2\frac{Z^2e^2}{8\pi\epsilon_0 a_o} = -8 \text{ Rydberg} = -108.8 \text{ eV}. \tag{8.10.10}$$

This estimate, which is too high, can be improved by treating the interaction term as a perturbation. The first order perturbation correction $E^{(1)}$ to the ground state energy is equal to the expectation value of the perturbation part of the Hamiltonian (H_{12}) in the unperturbed state

$$\begin{aligned} E^{(1)} &= \sum_{\zeta_1}\sum_{\zeta_2} \iint \psi_0^*(\boldsymbol{r}_1)\psi_0^*(\boldsymbol{r}_2)\chi_0^{0*} H_{12}\psi_0(\boldsymbol{r}_1)\psi_0(\boldsymbol{r}_2)\chi_0^0 d^3r_1 d^3r_2 \\ &= \iint \psi_0^*(\boldsymbol{r}_1)\psi_0^*(\boldsymbol{r}_2)H'\psi_0(\boldsymbol{r}_1)\psi_0(\boldsymbol{r}_2)d^3r_1 d^3r_2 = \frac{5}{8} \times 2\,\frac{e^2}{4\pi\epsilon_0 a_o} \end{aligned}$$

where we have made use of the normalization condition for the singlet spin state χ_0^0. The evaluation of the spatial integral has already been done in Sec. 8.2 [footnote 1].

The ground state energy of the Helium atom is therefore given by

$$E_{gs} = 2\left(-\frac{2^2e^2}{8\pi\epsilon_0 a_o}\right) + \frac{5}{2}\frac{e^2}{8\pi\epsilon_0 a_o} = -\frac{11}{2}\frac{e^2}{8\pi\epsilon_0 a_o} \tag{8.10.11}$$

where $\varepsilon_0 = -\frac{2^2e^2}{8\pi\epsilon_0 a_o}$ is the ground state energy of the Hydrogen-like atom with $Z = 2$. This result is the same as that obtained in the perturbation calculation without considering the electron spin and anti-symmetric states. However, the effects of spin and total anti-symmetrization become important for the excited states of the normal Helium atom, which we consider next.

8.11 Excited States of the Helium Atom

In this case the two electrons can exist in different orbitals so that the Pauli principle does not restrict the spin quantum numbers to be different, i.e., the two electrons can exist in any spin state, α or β. We have the following possibilities for the states ϕ_1 and ϕ_2

I	$\phi_1(X) = \psi_a(\boldsymbol{r})\alpha(\zeta)$	$\phi_2(X) = \psi_b(\boldsymbol{r})\alpha(\zeta)$
II	$\phi_1(X) = \psi_a(\boldsymbol{r})\beta(\zeta)$	$\phi_2(X) = \psi_b(\boldsymbol{r})\beta(\zeta)$
III	$\phi_1(X) = \psi_a(\boldsymbol{r})\alpha(\zeta)$	$\phi_2(X) = \psi_b(\boldsymbol{r})\beta(\zeta)$
III	$\phi_1(X) = \psi_a(\boldsymbol{r})\beta(\zeta)$	$\phi_2(X) = \psi_b(\boldsymbol{r})\alpha(\zeta)\,.$

The anti-symmetric wave functions in the four cases are:

$$\Psi_I(X_1, X_2) = \frac{1}{\sqrt{2}}\begin{vmatrix} \psi_a(\boldsymbol{r}_1)\alpha(\zeta_1) & \psi_a(\boldsymbol{r}_2)\alpha(\zeta_2) \\ \psi_b(\boldsymbol{r}_1)\alpha(\zeta_1) & \psi_b(\boldsymbol{r}_2)\alpha(\zeta_2) \end{vmatrix}$$

$$= \frac{1}{\sqrt{2}}[\psi_a(\boldsymbol{r}_1)\psi_b(\boldsymbol{r}_2) - \psi_b(\boldsymbol{r}_1)\psi_a(\boldsymbol{r}_2)]\chi_1^{+1}(\zeta_1, \zeta_2)\,, \tag{8.11.1}$$

where
$$\chi_1^{+1}(\zeta_1, \zeta_2) = \alpha(\zeta_1)\alpha(\zeta_2)\,, \tag{8.11.2}$$

$$\Psi_{II}(X_1, X_2) = \frac{1}{\sqrt{2}}\begin{vmatrix} \psi_a(\boldsymbol{r}_1)\beta(\zeta_1) & \psi_a(\boldsymbol{r}_2)\beta(\zeta_2) \\ \psi_b(\boldsymbol{r}_1)\beta(\zeta_1) & \psi_b(\boldsymbol{r}_2)\beta(\zeta_2) \end{vmatrix}$$

$$= \frac{1}{\sqrt{2}}[\psi_a(\boldsymbol{r}_1)\psi_b(\boldsymbol{r}_2) - \psi_b(\boldsymbol{r}_1)\psi_a(\boldsymbol{r}_2)]\,\chi_1^{-1}(\zeta_1, \zeta_2)\,, \tag{8.11.3}$$

where
$$\chi_1^{-1}(\zeta_1, \zeta_2) = \beta(\zeta_1)\beta(\zeta_2)\,, \tag{8.11.4}$$

$$\Psi_{III}(X_1, X_2) = \frac{1}{\sqrt{2}}\begin{vmatrix} \psi_a(\boldsymbol{r}_1)\alpha(\zeta_1) & \psi_a(\boldsymbol{r}_2)\alpha(\zeta_2) \\ \psi_b(\boldsymbol{r}_1)\beta(\zeta_1) & \psi_b(\boldsymbol{r}_2)\beta(\zeta_2) \end{vmatrix}$$

$$= \frac{1}{\sqrt{2}}[\psi_a(\boldsymbol{r}_1)\psi_b(\boldsymbol{r}_2)\alpha(\zeta_1)\beta(\zeta_2) - \psi_b(\boldsymbol{r}_1)\psi_a(\boldsymbol{r}_2)\beta(\zeta_1)\alpha(\zeta_2)]\,,$$

$$\Psi_{IV}(X_1, X_2) = \frac{1}{\sqrt{2}}\begin{vmatrix} \psi_a(\boldsymbol{r}_1)\beta(\zeta_1) & \psi_a(\boldsymbol{r}_2)\beta(\zeta_2) \\ \psi_b(\boldsymbol{r}_1)\alpha(\zeta_1) & \psi_b(\boldsymbol{r}_2)\alpha(\zeta_2) \end{vmatrix}$$

$$= \frac{1}{\sqrt{2}}[\psi_a(\boldsymbol{r}_1)\psi_b(\boldsymbol{r}_2)\beta(\zeta_1)\alpha(\zeta_2) - \psi_b(\boldsymbol{r}_1)\psi_a(\boldsymbol{r}_2)\alpha(\zeta_1)\beta(\zeta_2)]\,.$$

From Ψ_{III} and Ψ_{IV} one can form alternative combinations:

$$\frac{1}{\sqrt{2}}(\Psi_{III} + \Psi_{IV}) = \frac{1}{\sqrt{2}}\{\psi_a(\boldsymbol{r}_1)\psi_b(\boldsymbol{r}_2) - \psi_b(\boldsymbol{r}_1)\psi_a(\boldsymbol{r}_2)\}\chi_1^0(\zeta_1, \zeta_2)\,, \tag{8.11.5}$$

where
$$\chi_1^0(\zeta_1, \zeta_2) = \frac{1}{\sqrt{2}}[\alpha(\zeta_1)\beta(\zeta_2) + \beta(\zeta_1)\alpha(\zeta_2)]\,, \tag{8.11.6}$$

and
$$\frac{1}{\sqrt{2}}(\Psi_{III} - \Psi_{IV}) = \frac{1}{\sqrt{2}}\{\psi_a(\boldsymbol{r}_1)\psi_b(\boldsymbol{r}_2) + \psi_b(\boldsymbol{r}_1)\psi_a(\boldsymbol{r}_2)\}\chi_0^0(\zeta_1, \zeta_2)\,, \tag{8.11.7}$$

where
$$\chi_0^0(\zeta_1, \zeta_2) = \frac{1}{\sqrt{2}}[\alpha(\zeta_1)\beta(\zeta_2) - \beta(\zeta_1)\alpha(\zeta_2)]\,. \tag{8.11.8}$$

We note that the states $\Psi_I(X_1, X_2)$, $\Psi_{II}(X_1, X_2)$, $\frac{1}{\sqrt{2}}(\Psi_{III} + \Psi_{IV})$ are all anti-symmetric in space coordinates and symmetric in spin coordinates. The spin states χ_1^m with $m =$

$+1, -1, 0$ are referred to as the spin triplet states. The last combination $\frac{1}{\sqrt{2}}(\Psi_{III} - \Psi_{IV})$ is symmetric in space coordinates and anti-symmetric in spin coordinates. The spin state χ_1^0 is called the spin singlet state.

When we calculate the energy of the Helium atom for the above states, to the first order of perturbation, we find that the energy when the space function is anti-symmetric and the spin function is symmetric is different from the energy when the space function is symmetric and the spin function is anti-symmetric. When the space function is anti-symmetric (and the spin state is triplet), the energy is given by

$$E_{asym} = \frac{1}{2}\iint [\psi *_a (\boldsymbol{r}_1)\psi_b^*(\boldsymbol{r}_2) - \psi_b^*(\boldsymbol{r}_1)\psi_a^*(\boldsymbol{r}_2)] \\ \times \left(H_1^0 + H_2^0 + \frac{e^2}{4\pi\epsilon_0 r_{12}}\right)[\psi_a(\boldsymbol{r}_1)\psi_b(\boldsymbol{r}_2) - \psi_b(\boldsymbol{r}_1)\psi_a(\boldsymbol{r}_2)]d^3r_1 d^3r_2 \,. \quad (8.11.9)$$

The spin functions χ_1^{+1}, χ_1^{-1} or χ_1^0, being normalized, contribute a factor 1 to the integral.

When the space function is symmetric (and the spin state is singlet), the energy is

$$E_{sym} = \frac{1}{2}\iint [\psi_a^*(\boldsymbol{r}_1)\psi_b^*(\boldsymbol{r}_2) + \psi_b^*(\boldsymbol{r}_1)\psi_a^*(\boldsymbol{r}_2)] \\ \times \left(H_1^0 + H_2^0 + \frac{e^2}{4\pi\epsilon_0 r_{12}}\right)[\psi_a(\boldsymbol{r}_1)\psi_b(\boldsymbol{r}_2) + \psi_b(\boldsymbol{r}_1)\psi_a(\boldsymbol{r}_2)]d^3r_1 d^3r_2 \,. \quad (8.11.10)$$

In this case also, the spin function χ_0^0, being normalized, contributes a factor 1. We can combine Eqs. (8.11.9) and (8.9.10) into one with the understanding that the upper signs pertain to spatially symmetric (spin singlet) state while the lower signs pertain to spatially anti-symmetric (spin triplet) state. Further, if ε_a and ε_b are the energies of the Hydrogen-like atom with $Z = 2$ in the states designated by a and b, then

$$E = \varepsilon_a + \varepsilon_b + \frac{1}{2}\iint [(\psi_a^*(\boldsymbol{r}_1)\psi_b^*(\boldsymbol{r}_2) \pm \psi_b^*(\boldsymbol{r}_1)\psi_a^*(\boldsymbol{r}_2)]\frac{e^2}{4\pi\epsilon_0 r_{12}} \\ \times [\psi_a(\boldsymbol{r}_1)\psi_b(\boldsymbol{r}_2) \pm \psi_b(\boldsymbol{r}_1)\psi_a(\boldsymbol{r}_2)]d^3r_1 d^3r_2 \\ = \varepsilon_a + \varepsilon_b + \frac{1}{2}\iint \frac{e^2}{4\pi\epsilon_0 r_{12}}\left[|\psi_a(\boldsymbol{r}_1)|^2|\psi_b(\boldsymbol{r}_2)|^2 + |\psi_b(\boldsymbol{r}_1)|^2|\psi_a(\boldsymbol{r}_2)|^2\right] d^3r_1 d^3r_2 \\ \pm \frac{1}{2}\iint \frac{e^2}{4\pi\epsilon_0 r_{12}}\left[\psi_a^*(\boldsymbol{r}_1)\psi_b^*(\boldsymbol{r}_2)\psi_b(\boldsymbol{r}_1)\psi_a(\boldsymbol{r}_2) + \psi_b^*(\boldsymbol{r}_1)\psi_a^*(\boldsymbol{r}_2)\psi_a(\boldsymbol{r}_1)\psi_b(\boldsymbol{r}_2)\right] d^3r_1 d^3r_2 \,.$$

Since $\boldsymbol{r}_1$ and $\boldsymbol{r}_2$ are dummy variables and $r_{12} = |\boldsymbol{r}_1 - \boldsymbol{r}_2|$ is symmetric between $\boldsymbol{r}_1$ and $\boldsymbol{r}_2$ the contribution of the first two integrals in the above expression is the same and so is the contribution of the last two integrals. Hence the energy levels of the Helium atom are given by

$$E = \varepsilon_a + \varepsilon_b + C \ \pm J \,, \quad (8.11.11)$$

where

$$C \equiv \iint \frac{e^2}{r_{12}}|\psi_a(\boldsymbol{r}_1)|^2|\psi_b(\boldsymbol{r}_2)|^2 d^3r_1 d^3r_2 \,, \quad (8.11.12)$$

and

$$J \equiv \iint \frac{e^2}{r_{12}}\psi_a^*(\boldsymbol{r}_1)\psi_b^*(\boldsymbol{r}_2)\psi_b(\boldsymbol{r}_1)\psi_a(\boldsymbol{r}_2)d^3r_1 d^3r_2 \,. \quad (8.11.13)$$

Here C is the energy of Coulomb interaction between the charge clouds of the two electrons. The last term J has no classical analog for it appears only because anti-symmetric total wave function has been used. The sign of this term is $+$ or $-$, depending on whether the space part of the total wave function is symmetric and spin state is singlet or the space part of the wave function is anti-symmetric and the spin state of the electron pair is triplet.

If the excited state of the Helium atom is such that the orbitals of the two electrons are the same ($a = b$), then the exchange term J is zero and the energy is

$$E = 2\varepsilon_a + C\,. \tag{8.11.14}$$

But when the orbitals of the two electrons are different ($a \neq b$), then the energy levels

$$E = \varepsilon_a + \varepsilon_b + C - J \tag{8.11.15}$$

corresponding to the spin triplet state (spatial function anti-symmetric) are different from the energy level

$$E = \varepsilon_a + \varepsilon_b + C + J \tag{8.11.16}$$

corresponding to the spin singlet state (spatial function symmetric). Thus a normal Helium atom has two distinct spectra, viz., triplet spectra given by Eq. (8.11.15) and singlet spectra given by Eq. (8.11.16). Obviously, the singlet energy levels, which are higher, are non-degenerate and the triplet energy levels, which are lower, are degenerate since each level corresponds to three spin states.

8.12 Statistical (Thomas-Fermi) Model of the Atom

The statistical model of the atom is an approximate method for dealing with multi-electron atoms. This model was introduced by Thomas and Fermi as a means for obtaining a self-consistent potential $V(\boldsymbol{r})$ and electron density distribution $\rho(\boldsymbol{r})$ around an atomic nucleus. This model assumes that $V(\boldsymbol{r})$ varies slowly enough over an electron wavelength so that electrons may be *localized* within a volume over which the potential changes by a small fraction of itself. In other words an electron in a statistical atom may be represented locally by a plane wave just like a *free* electron. The electrons then can be treated according to statistical mechanics, and can be regarded as obeying Fermi-Dirac statistics which requires the electron states to fill in order of increasing energy such that, according to Pauli, the occupancy for each state is zero or one. At low temperatures one may regard the electrons to be *completely degenerate* which means that all energy levels up to a certain maximum E_F (Fermi energy) are completely occupied and those above E_F are completely unoccupied.

The number of particle states dn, corresponding to momentum between $\boldsymbol{p}$ and $\boldsymbol{p} + d\boldsymbol{p}$ within a volume $d\tau$ is given by Eq. (5.2.15) with $E = p^2/2m$,

$$dn = 2 \times \frac{4\pi p^2 dp d\tau}{h^3}\,. \tag{8.12.1}$$

The factor of 2 takes account of two spin states of the electron for the same energy.

Since it is the case of *complete degeneracy*, the number of electrons in a volume $d\tau$ around a point $\boldsymbol{r}$ is given by

$$\rho(\boldsymbol{r})d\tau = \frac{8\pi}{h^2}\int_0^{P_F} p^2 dp d\tau = \frac{8\pi}{3h^2}P_F^3 d\tau \tag{8.12.2}$$

where P_F is the maximum momentum of the Fermi distribution for electrons *located* around the point $\boldsymbol{r}$. In other words around the point $\boldsymbol{r}$ the electrons can have momentum up to a certain maximum given by

$$E_F = \frac{P_F^2}{2m} - eV(r) = \frac{P_F^2}{2m} - \left[\frac{Ze^2}{4\pi\epsilon_0 r} - eB(r)\right] \tag{8.12.3}$$

where $-Ze^2/4\pi\epsilon_0 r$ is the potential energy of the electron on account of the nucleus of charge Ze and $eB(r)$ is its potential energy on account of the electron distribution — the total potential energy of the electron at the point $\boldsymbol{r}$ being $-eV(r)$. We assume spherically symmetric electron distribution as well as potential. E_F is independent of r, being the maximum total energy of the Fermi distribution. According to Poisson's equation which ensures self-consistency between the electronic charge distribution, $\rho_{ch}(r) = -e\rho(r)$, and the potential $V(r)$ (due to nuclear charge as well electronic charge distribution) [March (1957)], we have

$$\nabla^2 V(r) = -\frac{\rho_{ch}(r)}{\epsilon_0} = \frac{e\rho(r)}{\epsilon_0}$$

or

$$\nabla^2 \left(\frac{P_F^2}{2me} - \frac{E_F}{e}\right) = \frac{e}{\epsilon_0}\frac{8\pi}{3h^3}P_F^3 . \tag{8.12.4}$$

If we write

$$\frac{P_F^2}{2m} = eV(r) + E_F \equiv \frac{Ze^2}{4\pi\epsilon_0 r}\phi(\xi) , \tag{8.12.5}$$

and

$$r = b\xi , \tag{8.12.6}$$

where

$$b \equiv \left(\frac{3\pi}{4}\right)^{2/3} \frac{4\pi\epsilon_0\hbar^2}{2me^2Z^{1/3}} , \tag{8.12.7}$$

and assume the atom to be spherically symmetric so that $\nabla^2 \to \frac{1}{r^2}\frac{d}{dr}(r^2\frac{d}{dr})$, then Eq. (8.12.4) reduces to the form

$$\frac{d^2\phi}{d\xi^2} = \frac{\phi^{3/2}}{\xi^{1/2}} . \tag{8.12.8}$$

This is the Thomas-Fermi equation. Bush and Caldwell (1938) solved this equation numerically, for the first time, for the boundary conditions

(i) $V(r) \to \frac{Ze}{4\pi\epsilon_0 r}$ or $\phi \to 1$ as $r \to 0$

(ii) $\rho(r) \to 0$ or $\phi \to 0$ as $r \to \infty$

Solutions for other boundary conditions appropriate to different physical problems were obtained by several other workers [March (1957)]. Modifications in this model to include exchange energy of the electrons were made by Dirac (1930). Modifications for higher temperatures, when the electrons are no longer completely degenerate, were made by Feynman, Metropolis and Teller (1949). Relativistic generalization of Thomas-Fermi equation was done by Mathur (1957, 1960).

8.13 Hartree's Self-consistent Field Method for Multi-electron Atoms

The most general problem in atomic physics is that of n electrons in an atom. For such a system one can write the total Hamiltonian as

$$H = \sum_{k=1}^{n} h_k + \sum_{k<\ell}^{n}\sum_{\ell}^{n} v_{k\ell} , \tag{8.13.1}$$

where h_k is the single-electron Hamiltonian given by

$$h_k = -\frac{\hbar^2}{2m}\nabla_k^2 - \frac{Ze^2}{4\pi\epsilon_0 r_k} \tag{8.13.2}$$

and

$$v_{k\ell} = \frac{e^2}{4\pi\epsilon_0|\boldsymbol{r}_k - \boldsymbol{r}_\ell|}\,. \tag{8.13.3}$$

The exact n-particle Schrödinger equation for this problem

$$H\Phi(\boldsymbol{r}_1, \boldsymbol{r}_2, \cdots, \boldsymbol{r}_n) = E\Phi(\boldsymbol{r}_1, \boldsymbol{r}_2, \cdots, \boldsymbol{r}_n) \tag{8.13.4}$$

is exceedingly complicated because Φ is a function of position coordinates of all the electrons and H involves, apart from kinetic energy and interaction of individual electrons with the nucleus, all the inter-electron interactions.

A considerable simplification of the problem was introduced by Hartree (1928). According to Hartree, each electron moves independently of others and its wave function satisfies an equation which takes account of its interaction with all other electrons as well as the nucleus. Let $\psi_\ell(\boldsymbol{r}_\ell)$ be the wave function of the ℓ-th electron. Then the average interaction of the k-th electron with all other electrons is

$$v_k(\boldsymbol{r}_k) = \sum_{\ell\neq k}\int \psi_\ell^*(\boldsymbol{r}_\ell)\frac{e^2}{4\pi\epsilon_0|\boldsymbol{r}_k - \boldsymbol{r}_\ell|}\psi_\ell(\boldsymbol{r}_\ell)d^3r_\ell \tag{8.13.5}$$

and the Schrödinger equation satisfied by the k-th electron wave function is

$$\left[-\frac{\hbar^2}{2m}\nabla_k^2 - \frac{Ze^2}{4\pi\epsilon_0 r_k} + v_k(\boldsymbol{r}_k)\right]\psi_k(\boldsymbol{r}_k) = \varepsilon_k\psi_k(\boldsymbol{r}_k) \tag{8.13.6}$$

or

$$\left[-\frac{\hbar^2}{2m}\nabla_k^2 - \frac{Ze^2}{4\pi\epsilon_0 r_k} + \sum_{\ell\neq k}\int\frac{e^2|\psi_\ell(\boldsymbol{r}_\ell)|^2}{4\pi\epsilon_0|\boldsymbol{r}_k - \boldsymbol{r}_\ell|}d^3r_\ell\right]\psi_k(\boldsymbol{r}_k) = \varepsilon_k\psi_k(\boldsymbol{r}_k)\,. \tag{8.13.7}$$

Since similar single-particle equations will hold for all other electrons, we have a set of n independent single-particle equations for $k = 1, 2, \cdots, n$.

To solve these equations, we begin by choosing a set of n approximate single-particle wave functions $\psi_k(\boldsymbol{r}_k) = \psi_{n'\ell'm'}(\boldsymbol{r}_k) = R_{n'\ell'}(\boldsymbol{r}_k)Y_{\ell'm'}(\theta_k, \varphi_k)$. The guiding principle for assigning the quantum numbers n', ℓ', m' to the k-th electron can be the electronic configuration in the atom. Using the initial choice for single-particle wave functions, $v_k(\boldsymbol{r}_k)$ [Eq. (8.13.5)] are determined and Hartree equations (8.13.6) solved to get a new set of functions $\psi_k(\boldsymbol{r}_k)$. The new set of functions can then be used in Eq. (8.13.5) to obtain a more accurate estimate of the average potential $v_k(\boldsymbol{r}_k)$ experienced by the k-th electron due to the presence of the other electrons. Using this improved average potential, we can again solve the set of equations (8.13.6) to get more accurate single-particle wave functions. We can repeat the whole cycle of operations until self-consistency is achieved, i.e., until the latest set of $v_k(\boldsymbol{r}_k)$ differs negligibly from the previous set.

Total Energy of the Atom

In Dirac notation, Hartree equations (8.13.6) may be written as

$$\left[\hat{h}_k + \sum_{\ell \neq k} \langle \ell | \, \hat{v}_{k\ell} \, | \ell \rangle \right] |k\rangle = \varepsilon_k \, |k\rangle \,, \tag{8.13.8a}$$

where $$\hat{v}_{k\ell} = \frac{e^2}{4\pi\epsilon_0 \hat{r}_{k\ell}} = \frac{e^2}{4\pi\epsilon_0 |\hat{\boldsymbol{r}}_k - \hat{\boldsymbol{r}}_\ell|} \,, \tag{8.13.8b}$$

$$\hat{h}_k = \frac{\hat{\boldsymbol{p}}_k^2}{2m} - \frac{Ze^2}{4\pi\epsilon_0 \hat{r}_k} \quad \text{and} \quad |k\rangle \equiv |\psi_k\rangle \,. \tag{8.13.8c}$$

Pre-multiplying both sides of Eq. (8.13.8a) by $\langle k|$ we obtain

$$\langle k| \, \hat{h}_k \, |k\rangle + \sum_{\ell \neq k} \langle k\ell | \, \hat{v}_{k\ell} \, | \ell k\rangle = \varepsilon_k \,,$$

where the states $\langle k\ell|$ and $|\ell k\rangle$ are conjugate imaginaries of each other. By summing the last equation over all electrons, we obtain the total energy of the atom

$$\begin{aligned} E &= \sum_k \langle k| \, \hat{h}_k \, |k\rangle + \sum_{k<\ell}\sum \langle k\ell| \, \hat{v}_{k\ell} \, |\ell k\rangle \\ &= \sum_k \varepsilon_k - \sum_{k<\ell}\sum \langle k\ell| \, \hat{v}_{k\ell} \, |\ell k\rangle \,, \end{aligned} \tag{8.13.9}$$

where $\sum_{k<\ell}\sum$ means summation over all electron pairs when each pair is counted only once. The subtraction of the second term in the last step compensates for the double counting of the interactions in the first term $\sum_k \varepsilon_k$.

Variational Method of Deriving Hartree Equations

We have seen that, if H represents a one-particle Hamiltonian in coordinate representation

$$H = -\frac{\hbar^2}{2m}\nabla^2 + V(r) \,, \tag{8.13.10}$$

then, according to the variational principle, a function $\psi(\boldsymbol{r})$, which minimizes the integral,

$$E \equiv \int \psi^*(\boldsymbol{r}) H \psi(\boldsymbol{r}) d\tau \tag{8.13.11}$$

and, at the same time, preseves the normalization integral

$$N \equiv \int \psi^*(\boldsymbol{r}) \psi(\boldsymbol{r}) d\tau \,, \tag{8.13.12}$$

is given by[5]

$$H\psi(\boldsymbol{r}) = \Lambda \psi(\boldsymbol{r}) \,. \tag{8.13.13}$$

The constant Λ turns out to be the stationary value of E. In other words, Eq. (8.13.13) serves the purpose of selecting the best function ψ which makes the integral E stationary and at the same time holds the normalization integral N constant.

[5]For calculus of variation see Appendix 14A1. Work out Problem 17 in this chapter.

This principle may be generalized to many particle systems. Accordingly, the best ground state function for an n-electron atom is the one which minimizes the energy,

$$E = \int \Psi^* H \Psi d\tau$$

subject to the condition that Ψ remains normalized, $\int \Psi^* \Psi d\tau = 1$. In the n-electron case, the Hamiltonian is

$$H = \sum_k \left(-\frac{\hbar^2}{2m} \nabla_k^2 - \frac{Ze^2}{r_k} \right) + \frac{1}{2} \sum_k \sum_{\ell \neq k} \frac{e^2}{4\pi\epsilon_0 r_{k\ell}} \,. \tag{8.13.14}$$

Let us choose the trial wave function to be a product of single-electron wave functions

$$\Psi = \psi_1(\boldsymbol{r}_1)\psi_2(\boldsymbol{r}_2)\cdots\psi_k(\boldsymbol{r}_k)\cdots\psi_n(\boldsymbol{r}_n)\,. \tag{8.13.15}$$

Then the integral to be minimized is

$$\begin{aligned} E = \iint \cdots \int d^3r_1 d^3r_2 \cdots d^3r_n \psi_1^*(\boldsymbol{r}_1)\psi_2^*(\boldsymbol{r}_2)\cdots\psi_n^*(\boldsymbol{r}_n) \\ \times \left\{ \sum_k \left(-\frac{\hbar^2}{2m}\nabla_k^2 - \frac{Ze^2}{4\pi\epsilon_0 r_k} \right) + \frac{1}{2}\sum_k \sum_{\ell\neq k} \frac{e^2}{4\pi\epsilon_0 r_{k\ell}} \right\} \psi_1(\boldsymbol{r}_1)\psi_2(\boldsymbol{r}_2)\cdots\psi_n(\boldsymbol{r}_n)\,. \end{aligned} \tag{8.13.16}$$

where we seek $\delta E = 0$ while keeping the normalization integral equal to 1.

Consider the variation of $\psi_k(\boldsymbol{r}_k)$ for a specific k. This function must be chosen to minimize the above integral while preserving the normalization integral $\int \psi_k^*(\boldsymbol{r}_k)\psi_k(\boldsymbol{r}_k)d^3r_k = 1$. The variational principle now tells us that the function ψ_k which satisfies this condition is given by

$$\begin{aligned} &\delta \left\{ \int d^3r_k \psi_k^*(\boldsymbol{r}_k) \left(-\frac{\hbar^2}{2m}\nabla_k^2 - \frac{Ze^2}{4\pi\epsilon_0 r_k} \right) \psi_k(\boldsymbol{r}_k) \right. \\ &\qquad \left. + \sum_{\ell\neq k} \iint d^3r_k d^3r_\ell \psi_k^*(\boldsymbol{r}_k)\psi_\ell^*(\boldsymbol{r}_\ell) \frac{e^2}{4\pi\epsilon_0 r_{k\ell}} \psi_k(\boldsymbol{r}_k)\psi_\ell(\boldsymbol{r}_\ell) \right\} \\ &= \delta \int \psi_k^*(\boldsymbol{r}_k) \left\{ \left(-\frac{\hbar^2}{2m}\nabla_k^2 - \frac{Ze^2}{4\pi\epsilon_0 r_k} \right) + \sum_{\ell\neq k} \iint \psi_\ell^*(\boldsymbol{r}_\ell) \frac{e^2}{4\pi\epsilon_0 r_{kl}} \psi_\ell(\boldsymbol{r}_\ell) d^3r_\ell \right\} \psi_k(\boldsymbol{r}_k) d^3r_k \\ &\equiv \delta\varepsilon_k = 0 \end{aligned} \tag{8.13.17}$$

subject to the condition that

$$\int \psi_k^*(\boldsymbol{r}_k)\psi_k(\boldsymbol{r}_k)d^3r_k = 1\,. \tag{8.13.18}$$

Note that other terms in the sum over k do not contribute to this variation since variation is given only to to one ψ_k. It may also be noted that the factor 1/2 in the sum over inter-electron interactions in Eq. (8.13.16) takes care of double counting. But in Eq. (8.13.17), since variation is given only to one $\psi_k(\boldsymbol{r}_k)$ for a specific k, there is no question of double counting and so there is no factor of 1/2.

The variational principle now tells us that the function $\psi_k(\boldsymbol{r}_k)$, which satisfies conditions (8.13.17) and (8.13.18) is given by

$$\left\{-\frac{\hbar^2}{2m}\nabla_k^2 - \frac{Ze^2}{4\pi\epsilon_0 r_k} + \frac{e^2}{4\pi\epsilon_0}\sum_{\ell\neq k}\int\frac{|\psi_\ell(\boldsymbol{r}_\ell)|^2}{r_{k\ell}}d^3r_\ell\right\}\psi_k(\boldsymbol{r}_k) = \varepsilon_k\psi_k(\boldsymbol{r}_k)\,. \tag{8.13.19}$$

This gives us the set of Hartree equations for $k = 1, 2, \cdots, n$.

8.14 Hartree-Fock Equations

Hartree equations were modified by Fock, who considered the fact that because electrons are Fermions, they require the n-electron wave function to be anti-symmetric:

$$\Psi(X_1, X_2\cdots, X_n) = \frac{1}{(n!)^{1/2}}\begin{vmatrix}\phi_1(X_1) & \phi_1(X_2) & \cdots & \phi_1(X_n)\\ \phi_2(X_1) & \phi_2(X_2) & \cdots & \phi_2(X_n)\\ \vdots & & \cdots & \vdots\\ \phi_n(X_1) & \phi_n(X_2) & \cdots & \phi_n(X_n)\end{vmatrix} \tag{8.14.1}$$

where $X_i \equiv (\boldsymbol{r}_i, \zeta_i)$ specifies the space and spin coordinates of the i-th electron and $\phi_s(X_i) = \psi_s(\boldsymbol{r}_i)\alpha(\zeta_i)$ or $\psi_s(\boldsymbol{r}_i)\beta(\zeta_i)$, depending on whether in the s-th state the i-th electron is in spin up or spin down state. The determinant wave function comprises $n!$ terms and can be written more compactly as

$$\Psi(X_1, X_2, \cdots, X_n) = \frac{1}{\sqrt{n!}}\sum_P \epsilon_P\ P[\phi_1(X_1)\phi_2(X_2)\cdots\phi_n(X_n)]\,, \tag{8.14.2}$$

where summation is over all permutations of $1, 2, \cdots, n$ and $\epsilon_P = +1$ or -1 depending on whether it is an even or odd permutation. The permutation operator may act either upon the subscripts of ϕ_s (the quantum numbers of single-particle states) or upon the subscripts of X_i.

Let us calculate E given by

$$E = \iint\cdots\int\Psi^* H\Psi dX_1 dX_2\cdots dX_n \tag{8.14.3}$$

using the determinant form (8.14.1) for Ψ. Since the Hamiltonian H, in the coordinate representation [Eq. (8.13.14)], is a symmetric operator and is unchanged by any permutation of electronic coordinates, we can write

$$E = \sqrt{n!}\iint\cdots\int\Psi_0^* H\,\Psi_0 dX_1 dX_2\cdots dX_n\,, \tag{8.14.4}$$

where $\Psi_0 = \phi_1(X_1)\phi_2(X_2)\cdots\phi_n(X_n)$. Using Eq. (8.14.2) in this equation, we find

$$E = \iint\cdots\int\sum_P \epsilon_P\ P\{\phi_1^*(X_1)\phi_2^*(X_2)\cdots\phi_n^*(X_n)\}\left\{\sum_k\left(-\frac{\hbar^2}{2m}\nabla_k^2 - \frac{Ze^2}{4\pi\epsilon_0 r_k}\right)\right.$$
$$\left. + \frac{1}{2}\sum_k\sum_{\ell\neq k}\frac{e^2}{4\pi\epsilon_0 r_{k\ell}}\right\}\phi_1(X_1)\phi_2(X_2)\cdots\phi_n(X_n)dX_1 dX_2\cdots dX_n\,. \tag{8.14.5}$$

Now, when the single-particle operators are sandwiched between the states Ψ^* and Ψ, only the identity permutation in Ψ^* will give non-zero result. But when a two-particle operator e^2/r_{12} is sandwiched between Ψ^* and Ψ then the identity permutation, as well as the one in which the coordinates of the k-th and ℓ-th particles are exchanged, will give a non-zero contribution. This is because the single-particle wave functions are ortho-normal. This leads us to

$$E = \sum_{k=1}^{n} \int \phi_k^*(X_k) \left\{ -\frac{\hbar^2}{2m}\nabla_k^2 - \frac{Ze^2}{4\pi\epsilon_0 r_k} \right\} \phi_k(X_k) dX_k$$
$$+ \frac{1}{2}\sum_{k}^{n}\sum_{\ell\neq k}^{n} \iint \{\phi_k^*(X_k)\phi_\ell^*(X_\ell) - \phi_k^*(X_\ell)\phi_\ell^*(X_k)\} \frac{e^2}{4\pi\epsilon_0 r_{k\ell}} \phi_k(X_k)\phi_\ell(X_\ell) dX_k dX_\ell\,.$$

Expressing the single-particle states ϕ_k and ϕ_ℓ in terms of the space and spin parts, and considering the fact no part of the Hamiltonian depends on spins and that the spin functions are normalized and orthogonal, we can do away with spin functions. In the last term the spins of the k-th and ℓ-th electrons must be parallel, otherwise the term will vanish. This gives us

$$E = \sum_{k=1}^{n} \int d^3r_1 \psi_k^*(\boldsymbol{r}_1) \left\{ -\frac{\hbar^2}{2m}\nabla_1^2 - \frac{Ze^2}{4\pi\epsilon_0 r_1} \right\} \psi_k(\boldsymbol{r}_1)$$
$$+ \frac{1}{2}\sum_{k=1}^{n}\sum_{\ell\neq k}^{n} \iint d^3r_1 d^3r_2 \frac{e^2}{4\pi\epsilon_0 r_{12}} |\psi_k(\boldsymbol{r}_1)|^2 |\psi_\ell(\boldsymbol{r}_2)|^2$$
$$- \frac{1}{2}\sum_{k=1}^{n}\sum_{\ell\neq k\,,||\,spins}^{n} \iint \frac{e^2}{4\pi\epsilon_0 r_{12}} \psi_k^*(\boldsymbol{r}_1)\psi_\ell^*(\boldsymbol{r}_2)\psi_k(\boldsymbol{r}_2)\psi_\ell(\boldsymbol{r}_1) d^3r_1 d^3r_2\,. \tag{8.14.6}$$

In this expression we have replaced the dummy variables $\boldsymbol{r}_k$ and $\boldsymbol{r}_\ell$ by $\boldsymbol{r}_1$ and $\boldsymbol{r}_2$ respectively, in the first two integrals and by $\boldsymbol{r}_2$ and $\boldsymbol{r}_1$, respectively, in the third integral. The last term is the *exchange energy*.

Now we apply the variational principle to find the best single-particle wave functions (for $k = 1, 2, \cdots, n$) which minimize E and at the same time keep the normalization integral constant. Then, if we we vary only one single-particle wave function $\psi_k(\boldsymbol{r}_1)$, the integral to be minimized is

$$\varepsilon_k = \int \psi_k(\boldsymbol{r}_1)\, h_k \psi_k(\boldsymbol{r}_1) d^3r_1\,, \tag{8.14.7}$$

where
$$h_k \equiv \left\{ -\frac{\hbar^2}{2m}\nabla_1^2 - \frac{Ze^2}{4\pi\epsilon_0 r_1} \right\} + \frac{e^2}{4\pi\epsilon_0}\sum_{\ell\neq k}^{n} \int \frac{|\psi_\ell(\boldsymbol{r}_2)|^2}{r_{12}} d^3r_2$$
$$- \frac{e^2}{4\pi\epsilon_0}\sum_{\ell\neq k,\,||\,spins} \int \frac{\psi_\ell^*(\boldsymbol{r}_2)\psi_k(\boldsymbol{r}_2)\psi_\ell(\boldsymbol{r}_1)}{4\pi\epsilon_0 r_{12}\psi_k(\boldsymbol{r}_1)} d^3r_2 \tag{8.14.8}$$

such that the normalization integral $\int \psi_k^*(\boldsymbol{r}_1)\psi_k(\boldsymbol{r}_1) d^3r_1$ is constant. The factor of $1/2$ in the summations is removed as there is no double counting now since k is fixed.

According to the variational principle, the single-particle wave function ψ_k must satisfy

$h_k\psi_k(\boldsymbol{r}_1) = \varepsilon_k\psi_k(\boldsymbol{r}_1)$, where ε_k is the stationary value of the integral (8.14.8), or explicitly,

$$\left\{-\frac{\hbar^2}{2m}\nabla_1^2 - \frac{Ze^2}{4\pi\epsilon_0 r_1} + \frac{e^2}{4\pi\epsilon_0}\sum_{\ell\neq k}^{n}\int\frac{|\psi_\ell(\boldsymbol{r}_2)|^2}{r_{12}}d^3r_2\right\}\psi_k(\boldsymbol{r}_1)$$
$$-\frac{e^2}{4\pi\epsilon_0}\sum_{\ell\neq k}^{n}\psi_\ell(\boldsymbol{r}_1)\int\frac{\psi_\ell^*(\boldsymbol{r}_2)\psi_k(\boldsymbol{r}_2)}{r_{12}}d^3r_2 = \varepsilon_k\psi_k(\boldsymbol{r}_1)\,. \tag{8.14.9}$$

We can take equations like this to hold for all single-particle wave functions. We can as well remove the restriction $\ell \neq k$ from both summations. These equations are called Hartree-Fock equations.

Exchange Energy

The *exchange energy* has no classical analog. It arises because of the exchange of the electronic coordinates in various terms of the expanded determinant wave function and takes care of the correlations between the positions of electrons with parallel spins . The exchange energy of an n-electron system may easily be calculated in the plane wave approximation, assuming that the electrons are free. Such a calculation is useful in the theory of metals and also in the statistical model of the atom in which one treats electrons as *localized* within a volume over which the fractional change in potential is small. Considering the last term in Eq. (8.14.6), we have for the total exchange energy

$$E_{exch} = \sum_{i=1}^{n}\varepsilon_{exch,i} \tag{8.14.10}$$

where $$\varepsilon_{exch,i} = -\sum_{\ell=1,\,||\,spins}^{n}\iint\frac{e^2}{4\pi\epsilon_0 r_{12}}\psi_i^*(\boldsymbol{r}_1)\psi_\ell^*(\boldsymbol{r}_2)\psi_i(\boldsymbol{r}_2)\psi_\ell(\boldsymbol{r}_1)d^3r_1d^3r_2\,. \tag{8.14.11}$$

We may regard $\varepsilon_{exch,i}$ as the exchange energy of the i-th electron due to all other electrons. Let us calculate this term when the wave functions $\psi_i(\boldsymbol{r}_1)$ and $\psi_\ell(\boldsymbol{r}_2)$ can be approximated by plane waves

$$\psi_i(\boldsymbol{r}_1) = \frac{1}{\sqrt{V}}e^{i\boldsymbol{k}_i\cdot\boldsymbol{r}_i}, \quad \psi_\ell(\boldsymbol{r}_2) = \frac{1}{\sqrt{V}}e^{i\boldsymbol{k}_\ell\cdot\boldsymbol{r}_2} \tag{8.14.12}$$

in Eq. (8.14.11). This gives, on transforming to the relative and center-of-mass coordinates in the double integral in Eq. (8.14.11), and evaluating the integral,

$$\varepsilon_{exch,i} = \sum_{\ell=1,\,||\,spins}^{n} -\frac{e^2}{4\pi\epsilon_0 V}\frac{4\pi}{|\boldsymbol{k}_i-\boldsymbol{k}_\ell|^2}\,. \tag{8.14.13}$$

Since the momenta $\boldsymbol{p}_\ell = \hbar\boldsymbol{k}_\ell$ pertaining to the plane wave states of the ℓ-th electron are continuously varying, we can replace the summation over ℓ by integration over the $\boldsymbol{k}_\ell$-space so that

$$\sum_{\ell,\,||\,spins} \rightarrow \int\frac{Vd^3p_\ell}{h^3} \equiv \int\frac{Vd^3k_\ell}{(2\pi)^3}\,. \tag{8.14.14}$$

Multiplication by 2 in the phase space Vd^3p_ℓ is avoided because we are considering the exchange between electrons of like (parallel) spins. Thus Eq. (8.14.13) gives us the exchange energy contribution

$$\varepsilon_{exch,i} = -\frac{e^2}{4\pi\epsilon_0 V}\frac{4\pi V}{(2\pi)^3}\int_0^{k_F} 2\pi k_\ell^2 dk_\ell \int_{-1}^{+1} \frac{d(\cos\theta)}{(k_i^2 + k_\ell^2 - 2k_i k_\ell \cos\theta)}$$
$$= -\frac{e^2 k_F}{8\pi^2\epsilon_0}\left[\frac{k_F^2 - k_i^2}{k_F k_i}\ln\left(\frac{k_F + k_1}{k_F - k_i}\right) + 2\right] \qquad (8.14.15)$$

where $\hbar k_F = P_F$ is the maximum momentum of the Fermi distribution.

Thomas-Fermi-Dirac Model of the Atom

Dirac (1930) introduced the exchange correction into the Thomas-Fermi model of the atom; the modified model is known as the Thomas-Fermi-Dirac model. When we include the exchange energy, the total energy of the electron on the top of the Fermi distribution may be re-expressed as [see Eq. (8.12.3)]

$$E_F = \frac{\hbar^2 k_F^2}{2m} - eV(r) - \frac{e^2 k_F}{4\pi^2\epsilon_0}\,. \qquad (8.14.16)$$

Using Poisson's equation, $\nabla^2 V(r) = -\rho_{ch}/\epsilon_0 = e\rho(r)/\epsilon_0$, where $\rho(r)$ is the electron number density, we have

$$\nabla^2\left(\frac{P_F^2}{2m} - \frac{e^2 P_F}{4\pi^2\epsilon_0\hbar}\right) = \frac{32\pi^2 e^2}{12\pi\epsilon_0 h^3}P_F^3\,. \qquad (8.14.17)$$

This is the Thomas-Fermi-Dirac equation.

Average Exchange Energy per Electron

The total exchange energy of an n-electron system is found from Eqs. (8.14.10) and (8.14.13) to be

$$E_{exch} = -\frac{e^2}{\epsilon_0 V}\sum_i^n \sum_{\ell,\parallel\, spins}^n \frac{1}{|\boldsymbol{k}_i - \boldsymbol{k}_\ell|^2}\,. \qquad (8.14.18)$$

Changing the summations into integrations we obtain

$$E_{exch} = -\frac{e^2}{\epsilon_0 V}\left(\frac{V}{(2\pi)^3}\right)^2 \int d^3k_i \int \frac{d^3k_\ell}{|\boldsymbol{k}_i - \boldsymbol{k}_\ell|^2}\,.$$

Introducing the relative momentum $\boldsymbol{k}$ and center-of mass momentum $\boldsymbol{K}$ variables by

$$\boldsymbol{k} = \frac{1}{2}(\boldsymbol{k}_i - \boldsymbol{k}_\ell) \quad \text{and} \quad \boldsymbol{K} = \boldsymbol{k}_i + \boldsymbol{k}_\ell\,, \qquad (8.14.19)$$

we can write the exchange contribution as

$$E_{exch} = -\frac{e^2}{\epsilon_0 V}\left(\frac{V}{(2\pi)^3}\right)^2 \int \frac{d^3k}{4k^2}\int d^3K\,. \qquad (8.14.20)$$

If we consider it to be a case of complete degeneracy, both k_i and k_ℓ are restricted by

$$k_i \equiv |\boldsymbol{k} + \frac{\boldsymbol{K}}{2}| = \sqrt{k^2 + (K^2/4) + Kk\cos\alpha} \;\le k_F \qquad (8.14.21a)$$

$$k_\ell \equiv |-\boldsymbol{k} + \frac{\boldsymbol{K}}{2}| = \sqrt{k^2 + (K^2/4) - Kk\cos\alpha} \;\le k_F \qquad (8.14.21b)$$

where α is the angle between $\boldsymbol{K}$ and $\boldsymbol{k}$.

To carry out the integration we note that if (i) $k + \frac{K}{2} \leq k_F$ or $\frac{K}{2} \leq (k_F - k)$, then both conditions (8.14.21) are satisfied for all values of α. On the other hand, if (ii) $k_F - k \leq \frac{1}{2}K \leq \sqrt{k_F^2 - k^2}$, then both conditions are satisfied for restricted range of α given by

$$-\frac{k_F^2 - k^2 - K^2/4}{Kk} \leq \cos\alpha \leq \frac{k_F^2 - k^2 - K^2/4}{Kk}.$$

Finally, if (iii) $\frac{1}{2}K > \sqrt{k_F^2 - k^2}$, the above conditions [Eq. (8.14.21)] are not satisfied. Hence for integration over K-space, we divide the integral into three regions

$$(i)\quad \frac{1}{2}K < k_F - k$$

$$(ii)\quad k_F - k \leq \frac{1}{2}K \leq \sqrt{k_F^2 - k^2}$$

$$\text{and } (iii)\quad \frac{1}{2}K > \sqrt{k_F^2 - k^2}.$$

Then the integral can be carried out as

$$\int d^3K = \int_0^{2(k_f - k)} K^2 dK \int_{-1}^{1} 2\pi d(\cos\alpha) + \int_{2(k_F - k)}^{2\sqrt{k_F^2 - k^2}} K^2 dK \int_{-(k_F^2 - k^2 - K^2/4)/Kk}^{(k_f^2 - k^2 - K^2/4)/Kk} 2\pi d(\cos\alpha)$$

$$= \frac{32\pi}{3} k_F^3 \left(1 - \frac{3}{2}\frac{k}{k_F} + \frac{1}{2}\frac{k^3}{k_F^3}\right).$$

Exchange energy contribution then becomes

$$E_{exch} = -\frac{e^2}{\epsilon_0 V}\left(\frac{V}{(2\pi)^3}\right)^2 \int_0^{k_F} \frac{4\pi k^2 dk}{4k^2} \frac{32\pi}{3} k_F^3 \left(1 - \frac{3k}{2k_F} + \frac{1}{2}\frac{k^3}{k_F^3}\right) = -\frac{e^2 V}{16\pi^4 \epsilon_0} k_F^4. \quad (8.14.22)$$

Because of complete degeneracy, the total number of states must equal the total number of electrons

$$n = \frac{2V}{h^3} \int_0^{P_F} 4\pi p^2 dp = \frac{2V}{h^3}\frac{4\pi}{3} P_F^3. \quad (8.14.23)$$

Using this to define the average volume per electron V/n and equating it to a spherical volume of radius r_s via $V/n = 3h^3/8\pi P_F^3 \equiv 4\pi r_s^3/3$, the average exchange energy per electron can be written as

$$\varepsilon_{exch} = E_{exch}/n = -\frac{0.916}{r_s}\ \text{Rydberg}, \quad (8.14.24)$$

where 1 Rydberg $= e^2/8\pi\epsilon_0 a_o$ and r_s is expressed in units of the Bohr radius a_o.

8.15 Occupation Number Representation

The occupation number representation (ONR) allows one to deal with a system of n indistinguishable particles in a simple way and applies to both particles obeying Fermi-Dirac statistics (Fermions) and particles obeying Bose-Einstein statistics (Bosons). However, in the present section we shall develop this formalism essentially for Fermions.

A non-interacting n-Fermion system is represented by the Slater determinant wave function:

$$\Psi_S(X_1, X_2, \cdots, X_n) = \frac{1}{\sqrt{n!}} \begin{vmatrix} \phi_{\nu_1}(X_1) & \phi_{\nu_1}(X_2) & \cdots & \phi_{\nu_1}(X_n) \\ \phi_{\nu_2}(X_1) & \phi_{\nu_2}(X_2) & \cdots & \phi_{\nu_2}(X_n) \\ \vdots & & \cdots & \vdots \\ \phi_{\nu_n}(X_1) & \phi_{\nu_n}(X_2) & \cdots & \phi_{\nu_n}(X_n) \end{vmatrix} \tag{8.15.1}$$

where $\phi_{\nu_1}, \phi_{\nu_2}, \cdots, \phi_{\nu_n}$ are the single-particle states, obtained on the basis of the assumption that the particles are non-interacting, moving independently of each other in a common potential. The subscript S denotes the ordered set of single-particle states characterized by labels $\nu_1, \nu_2, \cdots, \nu_n$, where each label ν stands for a set of quantum numbers and X_i denotes the position and spin coordinates of the i-th particle. If it is a system in which the Fermions are interacting with each other then the actual anti-symmetric wave function of the system may be expanded linearly in terms of the determinant wave functions Ψ_S as

$$\Psi(X_1,\, X_2,\, \cdots,\, X_n) = \sum_S C_S \Psi_S(X_1,\, X_2,\, \cdots,\, X_n) \tag{8.15.2}$$

where the sum over S extends over various ordered set of levels. One can see that the set $\Psi(X_1,\, X_2,\, \cdots,\, X_n)$ for all ordered sets of levels S, is indeed a complete set of totally anti-symmetric n-particle wave functions provided the single-particle wave functions $\phi_\nu(X)$ form a complete orthonormal set.

Now, if the number of particles is large, the use of Slater determinants is cumbersome. Moreover, the information the determinant contains is redundant. Since the particles are indistinguishable, it is not meaningful to say which particle occupies which single-particle state. All that matters is whether the single-particle state ν_1 is occupied or unoccupied. Thus, for Fermions, an ordered set of occupied single-particle states,

$$|S\rangle \equiv |\nu_1, \nu_2, \cdots, \nu_n\rangle \tag{8.15.3}$$

specifies the n-particle state $\Psi(X_1, X_2, \cdots, X_n)$ and reference to the coordinates is not necessary. The representation of an n-particle state in terms of an ordered set of occupied single-particle states is called occupation number representation (ONR).

Creation and Annihilation Operators

In ONR, a single-particle state of a Fermion is to represented by $|\nu_i\rangle$. This state can be thought of as arising from a state with no particle at all (the vacuum state $|0\rangle$) by adding a particle in the state characterized by ν_i . The addition of a particle in the state (or level) is described by an operator $\hat{b}^\dagger_{\nu_i}$ operating on the vacuum state $|0\rangle$:

$$|\nu_i\rangle = \hat{b}^\dagger_{\nu_i}|0\rangle\,. \tag{8.15.4}$$

The operator $\hat{b}^\dagger_{\nu_i}$ is called the *creation operator* for a single-particle in the state ν_i.

In general, we define a creation operator $\hat{b}_\nu^\dagger$ by

$$\hat{b}_\nu^\dagger |\nu_1, \nu_2, \cdots, \nu_n\rangle = |\nu_1, \nu_2, \cdots, \nu_n\rangle \tag{8.15.5}$$

where the state ν of the $(n+1)$-th particle may not be necessarily ordered. We can restore the order by interchanging the order of single-particle states. This interchange results in a change of sign if an odd number of interchanges are required to reduce the disordered state to an ordered state and no sign change if an even number of interchanges are required. For example, an ordered state $|\nu_1, \nu_2\rangle$ represents

$$\begin{vmatrix} \phi_{\nu_1}(X_1) & \phi_{\nu_1}(X_2) \\ \phi_{\nu_2}(X_1) & \phi_{\nu_2}(X_2) \end{vmatrix} .$$

The state $|\nu_2, \nu_1\rangle$ then represents

$$\begin{vmatrix} \phi_{\nu_2}(X_1) & \phi_{\nu_2}(X_1) \\ \phi_{\nu_1}(X_2) & \phi_{\nu_1}(X_2) \end{vmatrix} = - \begin{vmatrix} \phi_{\nu_1}(X_1) & \phi_{\nu_1}(X_1) \\ \phi_{\nu_2}(X_2) & \phi_{\nu_2}(X_2) \end{vmatrix} .$$

Hence $|\nu_2, \nu_1\rangle = -|\nu_1, \nu_2\rangle$. It follows that the ket vector $|\nu_1, \nu_2, \cdots, \nu_n\rangle$ representing the n-particle state may be obtained by a succession of creation operators operating upon the vacuum state via

$$|\nu_1, \nu_2, \cdots, \nu_n\rangle = \hat{b}_{\nu_1}^\dagger \hat{b}_{\nu_2}^\dagger \cdots \hat{b}_{\nu_n}^\dagger |0\rangle . \tag{8.15.6}$$

From the anti-symmetry of the Slater determinant (or ordered state) we can see that the creation operators for Fermions satisfy anti-commutation relations, for we have

$$\hat{b}_\nu^\dagger \hat{b}_{\nu'}^\dagger |0\rangle = |\nu, \nu'\rangle$$

and $$\hat{b}_{\nu'}^\dagger \hat{b}_\nu^\dagger |0\rangle = |\nu', \nu\rangle = -|\nu, \nu'\rangle .$$

These relations imply that

$$\hat{b}_\nu^\dagger \hat{b}_{\nu'}^\dagger = -\hat{b}_{\nu'}^\dagger \hat{b}_\nu^\dagger \quad \text{or} \quad \hat{b}_\nu^\dagger \hat{b}_{\nu'}^\dagger + \hat{b}_{\nu'}^\dagger \hat{b}_\nu^\dagger \equiv [\hat{b}_\nu^\dagger, \hat{b}_{\nu'}^\dagger]_+ = 0 . \tag{8.15.7}$$

We can also see that the adjoint of $\hat{b}_\nu^\dagger$, that is $\hat{b}_\nu$ can be interpreted as the annihilation operator, for we have

$$\hat{b}_\nu^\dagger |0\rangle = |\nu\rangle$$

and writing the conjugate imaginary of both sides we have,

$$\langle 0|\hat{b}_\nu = \langle \nu| .$$

From these relations we find

$$\langle 0|\, \hat{b}_\nu \hat{b}_\nu^\dagger \,|0\rangle = \langle \nu|\, \nu\rangle = 1 .$$

But $\langle 0|\, 0\rangle$ also equals one, so we conclude

$$\hat{b}_\nu \hat{b}_\nu^\dagger \,|0\rangle = |0\rangle \quad \text{or} \quad \hat{b}_\nu \,|\nu\rangle = |0\rangle . \tag{8.15.8}$$

Hence $\hat{b}_\nu$ may be regarded as the annihilation (or destruction) operator for a particle in the state $|\nu\rangle$.

It can be seen that the annihilation operators also satisfy anti-commutation relations. To see this we take the adjoint of the operator products on both sides of Eq. (8.15.7) and obtain

$$\hat{b}_{\nu'} \hat{b}_\nu = -\hat{b}_\nu \hat{b}_{\nu'} \quad \text{or} \quad \hat{b}_\nu \hat{b}_{\nu'} + \hat{b}_{\nu'} \hat{b}_\nu \equiv [\hat{b}_\nu, \hat{b}_{\nu'}]_+ = 0 . \tag{8.15.9}$$

It can also be see that $[\hat{b}_\nu\,, \hat{b}_\nu^\dagger]_+ = \hat{1}$, for we have

$$\hat{b}_\nu \hat{b}_\nu^\dagger \,|S\rangle = \begin{cases} 0 & \text{if } \nu \text{ is contained in } S \equiv (\nu_1, \nu_2, \cdots, \nu_n)\,, \\ |S\rangle & \text{if } \nu \text{ is not contained in } S\,. \end{cases}$$

Here we have used the fact that each single-particle state is occupied at most by one Fermion. We also have

$$\hat{b}_\nu^\dagger \hat{b}_\nu \,|S\rangle = \begin{cases} |S\rangle & \text{if } \nu \text{ is contained in } S\,, \\ 0 & \text{if } \nu \text{ is not contained in } S\,. \end{cases}$$

Hence $(\hat{b}_\nu \hat{b}_\nu^\dagger + \hat{b}_\nu^\dagger \hat{b}_\nu)\,|S\rangle = |S\rangle$, irrespective of whether ν is or is not contained in S. Hence we have the result

$$\hat{b}_\nu \hat{b}_\nu^\dagger + \hat{b}_\nu^\dagger \hat{b}_\nu \equiv [\hat{b}_\nu\,, \hat{b}_\nu^\dagger]_+ = \hat{1}\,.$$

Furthermore, since $\hat{b}_\nu \hat{b}_{\nu'}^\dagger = -\hat{b}_{\nu'}^\dagger \hat{b}_\nu$, if $\nu \neq \nu'$ we have

$$[\hat{b}_\nu\,, \hat{b}_{\nu'}^\dagger]_+ = \delta_{\nu\nu'}\,. \tag{8.15.10}$$

Fock Space

We can visualize a Hilbert space spanned by the single-particle states $|\nu_i\rangle$ for all ν_i and another Hilbert space spanned by two particle states $|\nu_i, \nu_j\rangle$. Both of these have been encountered before. In general we can consider a Hilbert space spanned by n-particle states $|\nu_1, \nu_2, \cdots, \nu_n\rangle$, where $\nu_1, \nu_2, \cdots, \nu_n$ represents an ordered set of states. As we have seen these states can be obtained from the vacuum state by applying appropriate creation operators. We can as well think of a great Hilbert space, called a Fock space, spanned by the vacuum state, all single-particle states, all two particle states, all three particle states and so on.

One-body and Two-body Operators

Having established the correspondence between the Slater determinants $\Psi_S(X_1, X_2, \cdots, X_n)$ and the ONR vectors $|S\rangle \equiv |\nu_1, \nu_2, \cdots, \nu_n\rangle$, we now establish a similar correspondence between the operators $O(X_1, X_2, \cdots, X_n)$ in the configuration space (or in coordinate representation) and the operators $\hat{O}$ in the Fock space by the requirement

$$\begin{aligned} \int \cdots \int \Psi_{S'}(X_1, X_2, \cdots, X_n) O(X_1, X_2, \cdots, X_n) \Psi_S(X_1, X_2, \cdots, X_n) dX_1 dX_2 \cdots dX_n \\ = \langle S'|\hat{O}|S\rangle \end{aligned} \tag{8.15.11}$$

for any two many-particle states S' and S. Because the particles are indistinguishable, all many-body operators $O(X_1, X_2, \cdots, X_n)$ in the coordinate space must be symmetric under the permutation of coordinates. Essentially we shall be interested in one and two-body operators

$$O^{(I)}(X_1, X_2, \cdots, X_n) = \sum_{i=1}^{n} O^{(1)}(X_i) \tag{8.15.12}$$

$$\text{and} \quad O^{(II)}(X_1, X_2, \cdots, X_n) = \sum_{i<j} \sum O^{(2)}(X_i, X_j)\,, \tag{8.15.13}$$

where $O^{(1)}$ and $O^{(2)}$ depend, respectively, on the coordinates of one and two particles. For instance, $O^{(1)}(X)$ could be a single-particle Hamiltonian in some external potential while $O^{(2)}(X_1, X_2)$ could represent the potential energy of interaction between a pair of particles. In the Fock space, the operators corresponding to $O^{(I)}(X_1, X_2, \cdots, X_n)$ and $O^{(II)}(X_1, X_2, \cdots, X_n)$ are, respectively, $\hat{O}^{(I)}$ and $\hat{O}^{(II)}$. According to the requirement in Eq. (8.15.11), it follows that

$$\hat{O}^{(I)} = \sum_{\nu}\sum_{\nu'} O^{(1)}_{\nu\nu'}\, \hat{b}^{\dagger}_{\nu}\hat{b}_{\nu'}\,, \tag{8.15.14}$$

where
$$O^{(I)}_{\nu\nu'} \equiv \int \phi^*_\nu(X) O^{(1)}(X)\phi_{\nu'}(X)dX\,, \tag{8.15.15}$$

$$\hat{O}^{(II)} = \frac{1}{2}\sum_{\nu\nu'}\sum_{\mu\mu'} O^{(2)}_{\nu\mu\,\mu'\nu'}\hat{b}^{\dagger}_{\nu}\hat{b}^{\dagger}_{\mu}\hat{b}_{\nu'}\hat{b}_{\mu'}\,, \tag{8.15.16}$$

and
$$O^{(2)}_{\nu\mu\,\mu'\nu'} \equiv \int dX_1 \int dX_2 \phi^*_\nu(X_1)\phi^*_\mu(X_2) O^{(2)}(X_1, X_2)\phi_{\mu'}(X_1)\phi_{\nu'}(X_2)\,. \tag{8.15.17}$$

It is easy to check that in the simple case when the system consists of two identical Fermions, the bracket $\langle pq|\, \hat{O}^{(I)}\, |\ell k\rangle$ gives the same answer as the integral

$$\iint \frac{1}{\sqrt{2}}\begin{vmatrix} \phi^*_q(X_1) & \phi^*_p(X_1) \\ \phi^*_q(X_2) & \phi^*_p(X_2)\end{vmatrix} \left(O^{(1)}(X_1) + O^{(1)}(X_2)\right) \frac{1}{\sqrt{2}}\begin{vmatrix} \phi_\ell(X_1) & \phi_k(X_1) \\ \phi_\ell(X_2) & \phi_k(X_2)\end{vmatrix} dX_1 dX_2\,,$$

viz., $O^{(1)}_{q\ell}\delta_{pk} + O^{(1)}_{pk}\delta_{q\ell} - O^{(1)}_{p\ell}\delta_{qk} - O^{(1)}_{qk}\delta_{p\ell}$, when Eq. (8.15.14) is used. Also the bracket $\langle pq|\hat{O}^{(II)}|\ell k\rangle$ gives the same answer as the integral

$$\iint \frac{1}{\sqrt{2}}\begin{vmatrix} \phi^*_q(X_1) & \phi^*_p(X_1) \\ \phi^*_q(X_2) & \phi^*_p(X_2)\end{vmatrix} O^{(2)}(X_1, X_2) \frac{1}{\sqrt{2}}\begin{vmatrix} \phi_\ell(X_1) & \phi_k(X_1) \\ \phi_\ell(X_2) & \phi_k(X_2)\end{vmatrix} dX_1 dX_2\,,$$

viz., $O^{(2)}_{qp\ell\,k} - O^{(2)}_{pq\ell\,k}$, when Eq. (8.15.16) is used.

Here it may be noted that $|\ell k\rangle$ denotes the ordered state of two particles and the interchange of ℓ and k amounts to a change of sign. It may also be noted that the conjugate imaginary of the ket $|qp\rangle$ is bra $\langle pq|$.

Using the occupation number representation it is possible to work out the expectation values of the operators of the type $O^{(I)}(X_1, X_2, \cdots, X_n)$ and $O^{(II)}(X_1, X_2, \cdots, X_n)$ for the n-particle anti-symmetric states [Eq. (8.15.1)] with relative ease. ONR is quite useful in the microscopic theories of nuclear structure where nuclear properties are understood in terms of the motions of individual nucleons. The main virtue of ONR lies in its mathematical convenience. It provides a simple shorthand formulation for taking into account Pauli's principle in many body calculations of any degree of complexity by observing the rules of algebra of particle creation and annihilation operators.

Problems

1. There is another internal magnetic interaction term in the Hydrogen atom, which arises due to the intrinsic magnetic dipole moment of the proton (nucleus)

$$\boldsymbol{\mu}_p = \frac{eg_p}{2m_p}\boldsymbol{I}\,,$$

where $g_p \approx 5.6$ is the gyromagnetic ratio for the proton and $\boldsymbol{I}$ is its spin. The electron is perturbed by the magnetic field produced by this dipole. Write down the Hamiltonian for this interaction. This interaction is responsible for the *hyperfine* splitting of hydrogenic levels and is called hyperfine interaction. Which of these terms contributes to the ground state? Calculate the effect of this interaction on the ground state of the Hydrogen atom. This splitting is the origin of the famous "21–cm line" of the Hydrogen.

Ans: $\left[-\frac{\mu_o e}{4\pi m_e r^3}\frac{\boldsymbol{L}\cdot\boldsymbol{\mu}_p}{r^3} - \frac{\mu_o}{4\pi r^3}\left\{3(\boldsymbol{\mu}_e\cdot\boldsymbol{e}_r)(\boldsymbol{\mu}_p\cdot\boldsymbol{e}_r) - \boldsymbol{\mu}_e\cdot\boldsymbol{\mu}_p\right\} - \frac{2\mu_o}{3}\boldsymbol{\mu}_e\cdot\boldsymbol{\mu}_p\delta(\boldsymbol{r})\right]$,
where $\boldsymbol{\mu}_p$ and $\boldsymbol{\mu}_e$ are the magnetic (dipole) moments of the proton and the electron, $\boldsymbol{e}_r = \boldsymbol{r}/r$ is a unit vector from the nucleus to the electron and m_e and m_p are, respectively, the electron and proton masses.

2. Show that the first order perturbation energy equals the quantum mechanical expectation value of the perturbation for the unperturbed state.

3. A one-dimensional harmonic oscillator is perturbed by an external potential energy, so that

$$\hat{H} = \hat{H}_0 + \hat{H}' = \left(\frac{\hat{p}^2}{2m} + \frac{1}{2}k\hat{x}^2\right) + b\hat{x}^3 + c\hat{x}^4 .$$

Calculate the change in each of the first three energy levels as a result of the perturbation given

$$\psi_n(x) = N_n\, H_n(\alpha x)\, e^{-\alpha^2 x^2/2}$$

$$N_n = \left(\frac{\alpha}{2^n\, n!\, \sqrt{\pi}}\right)^{1/2}, \qquad \alpha = \sqrt{\frac{m\omega}{\hbar}}$$

$$H_0(\xi) = 1\ ;\ H_1(\xi) = 2\xi\ ;\ H_2(\xi) = 4\xi^2 - 2 .$$

4. Use the properties of Bessel function for small and large arguments to show that the solutions (8.5.27) and (8.5.31) (as well as their derivatives) are continuous across the turning point. Hence derive the connecting formulas (8.5.38) and (8.5.40).

5. A one-dimensional potential well is given by $V(x) = -V_0(1 - |x|/a)$ for $|x| < a$ and zero for $|x| > a$. How many bound states can exist for this potential?

6. Calculate the (bound) energy levels for the Hydrogen atom by applying the WKBJ method to the radial equation for the function $rR(r)$ [Eq. (8.5.41)]. Compare them with the exact energy levels.
Ans: $-\frac{m_e c^2 Z^2 \alpha^2}{2}\left[\sqrt{\ell(\ell+1} + n + \frac{1}{2}\right]$, where α is the fine structure constant.

7. Find the energy levels for $H = \frac{p^2}{2m} + Cr$ in the WKBJ approximation.

8. Use the following trial functions to estimate the ground state energy of the Hydrogen atom by the variational method.

   (i) $\psi_1(\alpha, r) = A/(r^2 + \alpha^2)$, where α is the variation parameter.

   (ii) $\psi_2(\alpha, r) = Ce^{-\alpha r^2/2}$, where α is the variation parameter.

9. A trial function ψ differs from the actual eigenfunction u_E of H ($Hu_E = Eu_E$) by a small amount, so that $\psi = u_E + \epsilon\,\psi_1$ where u_E and ψ_1 are normalized and orthogonal and the parameter $\epsilon \ll 1$. Show that the expectation value of H for the state ψ differs from E by a term of the order of ϵ^2.

10. A particle of mass m is bound by a potential $V(r) = -V_0 e^{-r/a}$ where $\frac{\hbar^2}{mV_o a^2} = \frac{3}{4}$. Using the trial wave function $\phi = Ne^{-\alpha r}$ find the upper bound to the energy of the ground state.

11. Find the ground state energy for the Hamiltonian

$$H = -\frac{\hbar^2}{2m}\frac{d^2}{dx^2} + \frac{1}{2}m\omega^2 x^2$$

using the trial wave function $\psi(x) = N\, e^{-\alpha x^2}$ and α as the variational parameter.

12. Use the variational principle to estimate the energy of the *first excited state* of a linear harmonic oscillator. How would you choose the trial function?

13. In a Hydrogen-like atom problem consider the atomic nucleus to be a uniformly charged sphere of radius R instead of being a point charge. This would modify the Coulomb potential to

$$V_c(r) = \begin{cases} -\frac{Ze^2}{4\pi\epsilon_0 r} & \text{for } r > R \\ \frac{Ze^2}{8\pi\epsilon_0 R}\left[\left(\frac{r}{R}\right)^2 - 3\right] & \text{for } r \leq R\,. \end{cases}$$

Use first order perturbation theory to calculate the energy shift caused by this modification for the $|n = 1\,, \ell = 0\,, m = 0\rangle$ state.

14. The unperturbed Hamiltonian for two identical linear oscillators is given by

$$H_0 = \left(-\frac{\hbar^2}{2m}\frac{d}{dx_1^2} + \frac{1}{2}kx_1^2\right) + \left(-\frac{\hbar^2}{2m}\frac{d}{dx_2^2} + \frac{1}{2}kx_2^2\right).$$

The oscillators interact via the Hamiltonian $H_1 = \frac{\epsilon}{2}\,kx_1\,x_2\,(x_1^2 + x_2^2)$ where ϵ is small. Show that this interaction does not change the ground state energy of the system while the energy of the unperturbed first excited state (which is degenerate) undergoes a splitting given by

$$\left(2 \pm \frac{3}{4}\frac{\epsilon\,\hbar}{\sqrt{mk}}\right)\hbar\omega\,,$$

where $\omega = \sqrt{k/m}$ is the classical frequency. (Use perturbation theory for degenerate energy levels.)

15. The Hamiltonian of the Hydrogen atom placed in a uniform magnetic field $\boldsymbol{B} = B\boldsymbol{e}_z$ is (apart from its internal Hamiltonian)

$$H = -\mu_B\, B\sigma_{ez} + g\,\boldsymbol{\sigma}_e \cdot \boldsymbol{\sigma}_p\,,$$

where $\boldsymbol{\sigma}_e$ and $\boldsymbol{\sigma}_p$ are the Pauli spin vectors for the electron and proton, respectively, μ_B is Bohr magneton and the magnetic field is applied in the z-direction. When $\boldsymbol{B}$ is zero, the energy for all the three triplet states $|S = 1\,,\ S_z = \pm 1\,, 0\rangle$ of the electron-proton system is g since the value of $\boldsymbol{\sigma}_e \cdot \boldsymbol{\sigma}_p$ for the triplet states is $+1$. Using the degenerate state perturbation theory calculate the energy shifts when $\boldsymbol{B}$ is non zero.

16. Estimate the ground state energy of a particle moving in a potential $V(r) = -V_o e^{-\mu^2 r^2}$ using the variational method and the trial wave function $\psi(r) = Ae^{-\beta^2 r^2}$, given $V_o = \frac{\pi^4}{4}\frac{\hbar^2\mu^2}{m}$, where m is the mass of the particle.

17. Given the Hamiltonian operator for a particle, $H = -K\nabla^2 + V(x, y, z)$, where $K = \hbar^2/2m$, show that the function $\psi(x, y, z)$, which makes the integral $\iiint \psi^* H\psi \, dx\, dy\, dz$ stationary (minimum), while keeping the normalization integral $\int \psi^*\psi \, dx\, dy\, dz$ a constant, is given by:

$$H\psi(x,\, y,\, z) = \lambda\psi(x,\, y,\, z).$$

(It is to be assumed that the permissible function is to vanish at the boundary of volume integration.)

[HINT: From the boundary conditions we have

$$\iiint \psi^*\nabla^2\psi \, dx\, dy\, dz = -\iiint (\psi_x^*\psi_x + \psi_y^*\psi_y + \psi_z^*\psi_z) dx\, dy\, dz\,,$$

where ψ_x denotes the derivative $\psi_x \equiv \frac{\partial\psi}{\partial x}$ of ψ. So the integral to be made stationary (minimum) is

$$\iiint \left[K\left(\psi_x^*\psi_x + \psi_y^*\psi_y + \psi_z^*\psi_z\right) + \psi^* V\psi\right] dx\, dy\, dz\,.$$

Since the normalization integral is to be kept constant, we may construct the function, whose integral is to be minimized, as

$$F = \left[K(\psi_x^*\psi_x + \psi_y^*\psi_y + \psi_z^*\psi)\right] + \psi^* V\psi + \lambda\psi^*\psi\,,$$

where λ is an undetermined multiplier. Using this function in the Euler-Lagrange equations:

$$\frac{\partial F}{\partial \psi} - \left(\frac{\partial}{\partial x}\frac{\partial F}{\partial \psi_x} + \frac{\partial}{\partial y}\frac{\partial F}{\partial \psi_y} + \frac{\partial}{\partial z}\frac{\partial F}{\partial \psi_z}\right) = 0$$

and

$$\frac{\partial F}{\partial \psi^*} - \left(\frac{\partial}{\partial x}\frac{\partial F}{\partial \psi_x^*} + \frac{\partial}{\partial y}\frac{\partial F}{\partial \psi_y^*} + \frac{\partial}{\partial z}\frac{\partial F}{\partial \psi_z^*}\right) = 0\,,$$

where ψ, ψ^* and their derivatives are to be treated as independent variables, gives the desired equations.]

18. Show that if the Hamiltonian of a system changes suddenly from H_o to $H_o + V$ in time Δt which is short compared to all relevant periods, the changed state of the system is given by

$$|\psi(t+\Delta t)\rangle = \exp\left[-\frac{i}{\hbar}\int_t^{t+\Delta t} V(t')dt'\right] |\psi(t)\rangle\,.$$

19. From the orthonormality and completeness of single-particle wave functions $\psi_\nu(X)$ (where $X \equiv \boldsymbol{r}, \zeta$) show that the set of anti-symmetric functions $\Psi_S(X_1, X_2, ...X_N)$ given by

$$\Psi_S(X_1, X_2, ...X_N) = \frac{1}{\sqrt{N!}}\begin{vmatrix} \psi_{\nu_1}(X_1) & \psi_{\nu_2}(X_1) & \cdots & \psi_{\nu_N}(X_1) \\ \psi_{\nu_1}(X_2) & \psi_{\nu_2}(X_2) & \cdots & \psi_{\nu_N}(X_2) \\ \vdots & & \vdots & \\ \psi_{\nu_1}(X_N) & \psi_{\nu_2}(X_N) & \cdots & \psi_{\nu_N}(X_N) \end{vmatrix}\,,$$

for all permuted sets of N labels

$$S \equiv \nu_1, \nu_2, \cdots, \nu_N\,,$$

$$S' \equiv \nu_2, \nu_1, \cdots, \nu_N\,,$$

and so on, form a complete orthonormal set of totally anti-symmetrized N-particle wave functions.

20. Using the anti-symmetry of N-particle wave functions, expressed as Slater determinants, show that the creation operators $\hat{b}_\nu^\dagger$ satisfy the condition

$$\hat{b}_\nu^\dagger \hat{b}_{\nu'}^\dagger = -\hat{b}_{\nu'}^\dagger \hat{b}_\nu^\dagger\,.$$

Hence show that the adjoints of the creation operators, $\hat{b}_\nu$ can be interpreted as annihilation operators and show that $\hat{b}_\nu \hat{b}_{\nu'} = -\hat{b}_{\nu'} \hat{b}_\nu$. Interpret this as conforming to Pauli's principle. Also show that

$$\hat{b}_\nu \hat{b}_\nu^\dagger \,|S\rangle \equiv \hat{b}_\nu \hat{b}_\nu^\dagger \,|\nu_1, \nu_2, \cdots, \nu_N\rangle = \begin{cases} 0 & \text{if } \nu \in S \\ |S\rangle & \text{if } \nu \notin S \end{cases},$$

and

$$\hat{b}_\nu^\dagger \hat{b}_\nu \,|S\rangle \equiv \hat{b}_\nu^\dagger \hat{b}_\nu \,|\nu_1, \nu_2, \cdots, \nu_N\rangle = \begin{cases} |S\rangle & \text{if } \nu \in S \\ 0 & \text{if } \nu \notin S \end{cases}.$$

Hence $\hat{b}_\nu \hat{b}_\nu^\dagger + \hat{b}_\nu^\dagger \hat{b}_\nu \equiv [\hat{b}_\nu\,, \hat{b}_\nu^\dagger]_+ = \hat{1}$ is an identity.

21. In the configuration space, one-body operator $O^I(X_1,\, X_2,\, \cdots,\, X_N)$ and two-body operator $O^{II}(X_1,\, X_2,\, \cdots,\, X_N)$ are defined by

$$O^I(X_1,\, X_2,\, \cdots,\, X_N) = \sum_i O^{(1)}(X_i)\,,$$

and

$$O^{II}(X_1,\, X_2,\, \cdots,\, X_N) = \sum_{i<j} O^{(2)}(X_i,\, X_j) = \frac{1}{2}\sum_{i\neq j} O^{(2)}(X_i,\, X_j)\,.$$

Show that, in the occupation number representation, one- and two-particle operators can be represented by

$$\hat{O}^{(I)} = \sum_{\nu,\nu'} O^{(1)}_{\nu\nu'}\, \hat{b}_\nu^\dagger \hat{b}_{\nu'}\,,$$

$$\hat{O}^{(II)} = \sum_{\nu\nu'\mu\mu'} O^{(2)}_{\nu\,\mu\,\mu'\,\nu'}\, \hat{b}_\nu^\dagger \hat{b}_\mu^\dagger \hat{b}_{\nu'} \hat{b}_{\mu'}\,,$$

where

$$O^{(1)}_{\nu\nu'}\,, \equiv \int \psi_\nu * (X) O^{(1)}(X) \psi_{\nu'}(X) dX$$

and

$$O^{(2)}_{\nu\mu\,\mu'\nu'} = \int dX \int dX' \psi_\nu^*(X) \psi_\mu^*(X') O^{(2)}(X, X') \psi_{\mu'}(X) \psi_{\nu'}(X')\,.$$

22. Show that, for the two-Fermion system, the integral

$$\iint \Psi^*_{pq}(X_1, X_2) O^I(X_1, X_2) \Psi_{\ell k}(X_1, X_2) dX_1 dX_2$$

has the same value as the bracket $\langle p, q|\; \hat{O}^I \;|\ell, k\rangle$ and the integral

$$\iint \Psi^*_{pq}(X_1, X_2) \hat{O}^{II}(X_1, X_2) \Psi_{\ell k}(X_1, X_2) dX_1 dX_2$$

has the same value as the bracket $\langle pq|\; \hat{O}^{II} \;|\ell k\rangle$, where $\Psi_{\ell k}$ is the determinant wave function for the ordered state $|\ell k\rangle$ and the ONR operators $\hat{O}^I$ and $\hat{O}^{II}$ are defined as in Problem 21.

23. The ordered state $|S\rangle = |\nu_1, \nu_2, \cdots, \nu_n\rangle$ of a system of n Fermions may be represented, in the coordinate representation, by a Slater determinant given in Problem 19. Show that the expectation value of one-body operator $\hat{O}^{(I)} \equiv \sum_{i=1}^{n} O^{(i)}(X_i)$ for this state, obtained by evaluating the integral

$$\iint \cdots \int \Psi_S^*(X_1, X_2, \cdots, X_n) \sum_{i=1}^{n} O^{(1)}(X_i) \Psi_S(X_1, X_2, \cdots, X_n)\; dX_1 dX_2 \cdots dX_n$$

gives the same result as obtained by evaluating the bracket

$$\langle S|\, \hat{O}^I\, |S\rangle \equiv \langle \nu_n, \nu_{n-1}, \cdots, \nu_1|\, \hat{O}^I\, |\nu_1, \nu_2, \cdots, \nu_n\rangle\ .$$

The one particle operator $\hat{O}^I$ in the occupation number representation is defined as in Problem 21.

References

[1] N. H. March, *Advances in Physics* (Taylor and Francis, Ltd., London) **6**, p. 1 (1957).

[2] P. A. M. Dirac, Proc. Camb. Phil. Soc. **26**, 376 (1930).

[3] R. P. Feynman, N. Metropolis, and E. Teller, Phys. Rev. **15**, 1561 (1949).

[4] V. S. Mathur, Proc. Nat. Inst. Sci. (India) **23A**, 430 (1957); V. S. Mathur, Prog. Theor. Phys. (Japan) **23**, 391 (1960).

[5] H. Eyring, J. Walter and G. E. Kimball, *Quantum Chemistry* (John Wiley and Sons, New York, 1944).

[6] E. C. Kemble, *Fundamental Principles of Quantum Mechanics* (McGraw Hill Book Company, New York, 1937).

[7] L. I. Schiff, *Quantum Mechanics*, Third Edition (McGraw Hill Book Company, New York, 1968).

[8] E. M Corson, *Perturbation Methods in Quantum Mechanics of N-Electron Systems* (Hafner Publishing Company, New York, 1951).

9

QUANTUM THEORY OF SCATTERING

9.1 Introduction

Most of our knowledge about the structure of matter and interaction between particles is derived from scattering experiments. From a theoretical point of view, scattering problems are concerned with the continuous (and positive) energy eigenvalues and unbound eigenfunctions of the Schrödinger equation. We have already encountered one-dimensional examples of scattering in Chapter 4, in the discussion of reflection and transmission of an incident particle with definite momentum from a potential step or barrier. In this chapter we consider scattering from a more formal point of view. We will confine our discussion to elastic scattering (scattering without loss or gain of energy by the projectile) from central potentials although many of the concepts and results are applicable to inelastic scattering.

In a typical scattering experiment, a target particle of mass m_2 is bombarded with another particle (projectile) of mass m_1 and carrying a momentum $\boldsymbol{p}_1$. After interaction with the target particle the projectile is scattered at some angle θ_0 with respect to the incident direction (z-direction). Since we will be dealing with elastic scattering from central potentials, we can infer from the symmetry of the problem that the scattering will be axially symmetric. This means the probability of the projectile being scattered in a direction θ_0, φ_0 will not depend on the azimuth angle φ_0. According to quantum mechanics, the angle θ_0 at which a projectile is scattered in a particular case cannot be predicted. However, if a beam consisting of a large number of identical particles, each carrying the same momentum $\boldsymbol{p}_1$, is incident on a target consisting of a large number of scatterers, we can use quantum mechanics to predict the angular distribution of scattered particles provided the interaction between the incident and target particles is known. It may be noted here that the scattering problem, despite the involvement of many particles in the incident beam and the target, remains essentially a two-body problem. Accordingly, we treat the problem as that of a single incident particle striking a single target particle and endeavor to calculate the probability of the incident particle being scattered in a direction (θ_0, φ_0) into a solid angle $d\Omega_0 = \sin\theta_0 d\theta_0 d\varphi_0$. This probability is related to the number of particles scattered per second in the direction (θ_0, φ_0), within the solid angle $d\Omega_0$ and is usually quantified in terms of the *differential cross-section* $\sigma(\theta_0)$ for scattering defined as

$$d\sigma(\theta_0) \equiv \sigma(\theta_0)d\Omega_0 = \frac{\begin{array}{c}\text{Number of particles scattered per second into}\\ \text{the solid angle } d\Omega_0 \text{ in direction } (\theta_0, \varphi_0)\end{array}}{\text{Incident flux density} \times \text{Number of targets}}\,.$$

The quantity $\sigma(\theta_0)$ has the dimensions of area/steradian. Hence $\sigma(\theta_0)d\Omega_0$ can be thought of as an effective cross-sectional area of the incident beam that would contain the number of particles scattered by a single scatterer into the solid angle $d\Omega_0$ in direction (θ_0, φ_0).

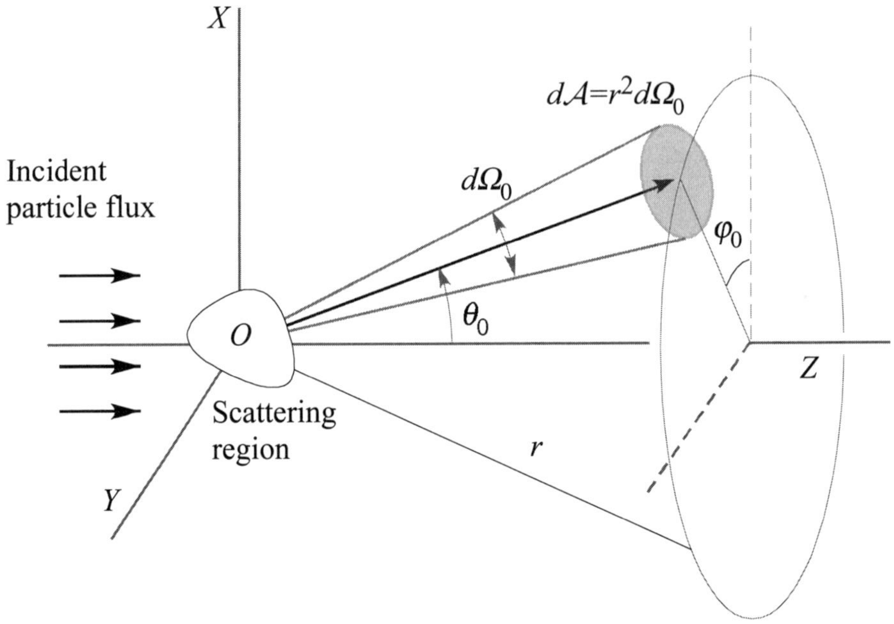

FIGURE 9.1
In a typical scattering experiment, particle flux emerging through a small area $d\mathcal{A} = r^2 d\Omega_0$ located far from the scattering region and subtending a solid angle $d\Omega_0$ at the scattering region is measured.

9.2 Laboratory and Center-of-mass (CM) Reference Frames

In describing the scattering we have to use different reference frames. A frame of reference in which the observer and the target are at rest is called the *laboratory frame of reference.* Obviously, the experimental measurements for the scattering cross-sections are carried out in the laboratory frame. On the other hand the theoretical calculations for the scattering cross-sections are conveniently made in another frame called the *center-of-mass (CM) frame of reference* in which the center of mass of the two particles, target and the projectile, is at rest. The advantage of the center-of-mass reference frame is that we consider only the relative motion of the two particles. As seen in the case of bound state problems, by working in the center-of-mass frame, we are able to reduce the two-body problem to a one-body problem. To compare the results of the theoretical calculations with those of the experiments, of course, we have to transform from the CM frame to the laboratory frame where the observations are made. To find this transformation let us consider the kinematics of scattering.

Let m_1 be the mass of the incident particle whose velocity is $\boldsymbol{v}$ [Fig. 9.2(a)] in the laboratory frame, in which the target of mass m_2 is at rest. We take the direction of the momentum of the incident particle to be the polar axis (z-axis). After the collision, let the incident particle be scattered into the direction (θ_0, φ_0) with velocity $\boldsymbol{v}_1$ and let the target recoil in the direction $(\phi_0, \pi + \varphi_0)$ with velocity $\boldsymbol{v}_2$. Since the collision is elastic, we can apply the laws of conservation of momentum and kinetic energy to the scattering process.

By equating the momentum and kinetic energy before and after the collision, we obtain the following equations

$$\begin{aligned} m_1 v &= m_1 v_1 \cos\theta_0 + m_2 v_2 \cos\phi_0 \,, \\ 0 &= m_1 v_1 \sin\theta_0 - m_2 v_2 \sin\phi_0 \,, \\ \frac{1}{2} m_1 v^2 &= \frac{1}{2} m_1 v_1^2 + \frac{1}{2} m_2 v_2^2 \,. \end{aligned} \tag{9.2.1}$$

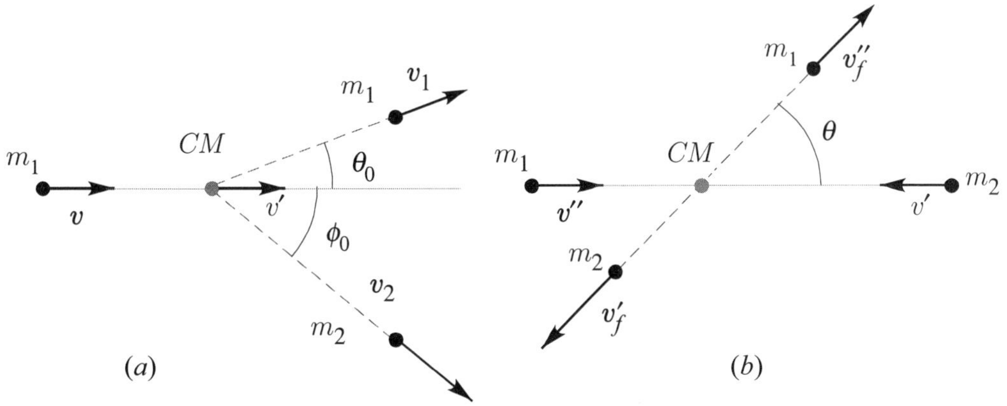

FIGURE 9.2
(a) Elastic scattering of a particle of mass m_1 from a target of mass m_2 in the laboratory frame. θ_0 is the scattering angle in the laboratory frame and ϕ_0 is the recoil angle of the target. (b) Elastic scattering as viewed in the center-of-mass frame. θ is the scattering angle in the CM frame. Initially the particles approach the center-of-mass with velocities $\boldsymbol{v}$ and $\boldsymbol{v}'$ and, after collision, recede away with velocities $\boldsymbol{v}''_f$ and $\boldsymbol{v}'_f$ such that $|\boldsymbol{v}''_f| = |\boldsymbol{v}''|$ and $|\boldsymbol{v}'_f| = |\boldsymbol{v}'|$.

Let us view this same process in the center-of-mass frame [Fig. 9.2(b)]. We note that, in the laboratory frame the center-of-mass frame moves with velocity[1]

$$\boldsymbol{v}' = \frac{m_1 \boldsymbol{v}}{m_1 + m_2} \,. \tag{9.2.2}$$

In this frame, the particles of mass m_1 and m_2 approach the center-of-mass with velocities $\boldsymbol{v}'' = \boldsymbol{v} - \boldsymbol{v}'$ and $\boldsymbol{v}'$, respectively [Fig. 9.2]. The conservation of momentum requires that, after the collision, the two particles will recede away from the center of mass in opposite directions with velocities $\boldsymbol{v}''_f$ and $\boldsymbol{v}'_f$, respectively, so that $|\boldsymbol{v}''_f| = |\boldsymbol{v}''|$ and $|\boldsymbol{v}'_f| = |\boldsymbol{v}'|$ Let this direction be θ. Thus while θ_0 is the angle of scattering in the laboratory frame, θ is the corresponding angle of scattering in the CM frame.

The relationship between θ_0 and θ is easily determined using the law of addition of velocities. According to this law the vector sum of the velocity $\boldsymbol{v}'$ of the center of mass in the laboratory frame and the velocity $\boldsymbol{v}''_f$ of mass m_1 after collision in the center-of-mass frame, should be equal to the velocity $\boldsymbol{v}_1$ of mass m_1 in the laboratory frame after collision [Fig. 9.3]. Working with the components of velocities, we obtain

[1] This relation is obtained by taking the time derivative of the center-of-mass coordinate $\boldsymbol{R} = \frac{m_1 \boldsymbol{r}_1 + m_2 \boldsymbol{r}_2}{m_1 + m_1}$.

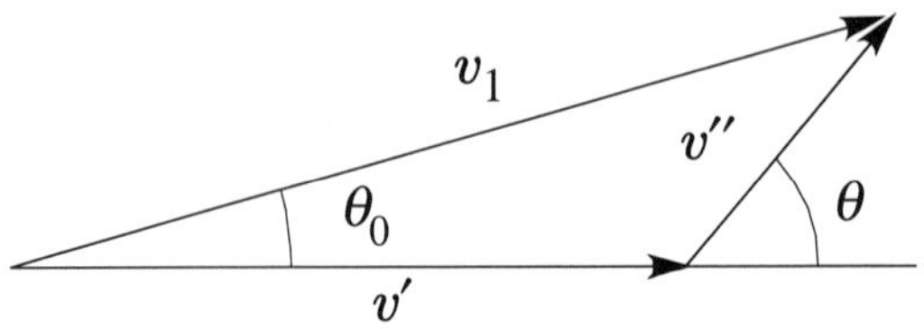

FIGURE 9.3
Relation between the scattering angles in the laboratory frame and the CM frame.

$$\begin{aligned} v_1 \cos\theta_0 &= v' + v''_f \cos\theta = v' + v'' \cos\theta\,, \\ v_1 \sin\theta_0 &= v''_f \sin\theta = v'' \sin\theta\,, \end{aligned}$$

so that

$$\tan\theta_0 = \frac{v''\sin\theta}{v' + v''\cos\theta} = \frac{\sin\theta}{\gamma + \cos\theta}\,, \tag{9.2.3}$$

where

$$\gamma \equiv \frac{v'}{v''} = \frac{v'}{v - v'} = \frac{m_1}{m_2}. \tag{9.2.4}$$

From this we see that if $m_1 \ll m_2$ (target much more massive than projectile) then $\theta \approx \theta_0$. On the other hand, if $m_1 \approx m_2$ (projectile and target equally massive) then $\gamma \approx 1$ and

$$\tan\theta_0 \approx \frac{\sin\theta}{1+\cos\theta} = \tan(\theta/2) \;\Rightarrow\; \theta_0 \approx \theta/2. \tag{9.2.5}$$

9.2.1 Cross-sections in the CM and Laboratory Frames

The number of particles scattered per second into a solid angle $d\Omega_0$ about the direction (θ_0, φ_0) in the laboratory frame is exactly the same as the number of particles scattered into a solid angle $d\Omega$ about direction (θ, φ) in the center-of-mass frame. This means

$$\mathcal{F} n \sigma_0(\theta_0) d\Omega_0 = \mathcal{F} n \sigma(\theta) d\Omega\,, \tag{9.2.6}$$

where $\mathcal{F}$ is the flux density (number of particles per unit area per second) of the incident particles and n is the number of scatterers in the target. Thus the cross-sections in the center-of-mass frame and the laboratory frame are related by

$$\frac{\sigma_0(\theta_0)}{\sigma(\theta)} = \frac{\sin\theta\, d\theta}{\sin\theta_0\, d\theta_0}\,. \tag{9.2.7}$$

Using the relation (9.2.3) between the angles θ and θ_0 we find

$$\frac{\gamma\cos\theta + 1}{(\gamma+\cos\theta)^2} d\theta = \sec^2\theta_0\, d\theta_0 = (1+\tan^2\theta_0) d\theta_0 = \frac{(1+\gamma^2+2\gamma\cos\theta)}{(\gamma+\cos\theta)^2} d\theta_0\,. \tag{9.2.8}$$

With the help of this equation we find

$$\frac{\sin\theta}{\sin\theta_0}\frac{d\theta}{d\theta_0} = \frac{\sin\theta}{\sin\theta_0}\frac{(1+\gamma^2+2\gamma\cos\theta)}{(\gamma\cos\theta+1)} = \frac{\sin\theta\sec\theta_0}{\tan\theta_0}\frac{(1+\gamma^2+2\gamma\cos\theta)}{(\gamma\cos\theta+1)}$$

or

$$\frac{\sin\theta}{\sin\theta_0}\frac{d\theta}{d\theta_0} = \frac{(1+\gamma^2+2\gamma\cos\theta)^{3/2}(\gamma+\cos\theta)^2}{(\gamma\cos\theta+1)}\,. \tag{9.2.9}$$

This leads us to the following relation between the cross-sections in the laboaratory frame and the CM frame

$$\sigma_0(\theta_0) = \left[\frac{(1+\gamma^2+2\gamma\cos\theta)^{3/2}}{\gamma\cos\theta+1}\right]\sigma(\theta)\,. \tag{9.2.10}$$

9.3 Scattering Equation and the Scattering Amplitude

As already noted, elastic scattering is a two-body problem involving a projectile of mass m_1 and a target of mass m_2. The wave function $\Psi(\boldsymbol{r}_1, \boldsymbol{r}_2)$, representing the scattering state of this system in the laboratory frame may, therefore, be obtained by solving the time-independent Schrödinger equation

$$H\Psi(\boldsymbol{r}_1, \boldsymbol{r}_2) = E_0\Psi(\boldsymbol{r}_1, \boldsymbol{r}_2)\,, \tag{9.3.1}$$

where H is the Hamiltonian operator of the two-body system. For central interaction between the particles, the Hamiltonian in the coordinate representation has the form

$$H = -\frac{\hbar^2}{2m_1}\nabla_1^2 - \frac{\hbar^2}{2m_2}\nabla_2^2 + V(|\boldsymbol{r}_1 - \boldsymbol{r}_2|)\,. \tag{9.3.2}$$

As in the case of two-body bound state problem discussed in Sec. 5.5, we can reduce the scattering problem to center-of-mass motion corresponding a particle of mass $M = m_1 + m_2$ moving freely and the motion of a fictitious particle of reduced mass $\mu = m_1 m_2/(m_1 + m_2)$ moving in a potential V. Accordingly, we introduce the center-of-mass coordinate $\boldsymbol{R}$ and the relative coordinate $\boldsymbol{r}$ by

$$\boldsymbol{R} = \frac{m_1\boldsymbol{r}_1 + m_2\boldsymbol{r}_2}{m_1 + m_2}\,, \quad \text{and} \quad \boldsymbol{r} = \boldsymbol{r}_1 - \boldsymbol{r}_2\,. \tag{9.3.3}$$

Since the potential depends only on the relative coordinate $\boldsymbol{r}$, it is possible to express the total wave function $\Psi(\boldsymbol{r}_1, \boldsymbol{r}_2)$ as a product of two functions $\Phi_{cm}(\boldsymbol{R})$ and $\psi(\boldsymbol{r})$. Substituting $\Psi(\boldsymbol{r}_1, \boldsymbol{r}_2) = \Phi_{cm}(\boldsymbol{R})$ and $\psi(\boldsymbol{r})$ in Eq. (9.3.1), with the Hamiltonian (9.3.2) expressed in terms of $\boldsymbol{R}$ and $\boldsymbol{r}$, and dividing the result by $\Phi_{cm}(\boldsymbol{R})\psi(\boldsymbol{r})$ we obtain after separating the variables two independent equations

$$-\frac{\hbar^2}{2(m_1 + m_2)}\nabla_{\boldsymbol{R}}^2\, \Phi_{cm}(\boldsymbol{R}) = E_{cm}\, \Phi_{cm}(\boldsymbol{R}) \tag{9.3.4}$$

and

$$\left[-\frac{\hbar^2}{2\mu}\nabla_{\boldsymbol{r}}^2 + V(\boldsymbol{r})\right]\psi(\boldsymbol{r}) = E\,\psi(\boldsymbol{r}) \tag{9.3.5}$$

where

$$E_{cm} = E_0\frac{m_1}{m_1 + m_2} \tag{9.3.6a}$$

is the energy associated with the motion of the center-of-mass and

$$E = E_0 - E_{cm} = E_0\frac{m_2}{m_1 + m_2} \tag{9.3.6b}$$

is the energy associated with the relative motion of the particles in the CM frame.

In scattering problems our main interest is in Eq. (9.3.5), which describes the relative motion of the two particles in the center-of-mass frame. It may be noted that while Eq. (9.3.5) for the scattering problem looks the same as Eq. (5.5.15) for the bound state problem, the energy spectrum and the asymptotic behavior of the wave function are different in the two cases. Whereas the bound state problem deals with the discrete part of the energy spectrum, which is determined by the asymptotic (large distance) behavior $\lim_{r\to\infty}\,\psi(\boldsymbol{r}) \to 0$ of the wave function, the scattering problem concerns the continuous part of the energy spectrum, where the energy E is specified in advance and the asymptotic behavior of the wave functions is sought in terms of E. In the scattering problem, the wave function $\psi(\boldsymbol{r})$ has the following asymptotic behavior [$r \to \infty$ where $V(r) \to 0$]

$$\lim_{r\to\infty} \psi(\boldsymbol{r}) \equiv \psi^{(+)}(\boldsymbol{r}) = A\left[e^{i\boldsymbol{k}\cdot\boldsymbol{r}} + f(\theta)\,\frac{e^{ikr}}{r}\right], \tag{9.3.7}$$

which is a superposition of a plane wave and an outgoing spherical wave. Here A is a normalization constant. The reason for this prescribed asymptotic behavior is simple. When the colliding particles are far apart they no longer interact so that we want the wave function to be a superposition of a plane wave representing the incident particle moving with momentum $\hbar\boldsymbol{k}$ and an outgoing spherical wave representing the scattered particle. The amplitude $f(\theta)$ (also called scattering amplitude) of the outgoing spherical wave depends on θ (the φ-dependence being ruled out because of axial symmetry) and falls off as $1/r$ since the radial flux density must fall off as $1/r^2$. The asymptotic form of the wave function satisfies the Schrödinger equation in the force-free ($V = 0$) region through terms of order $1/r$.

Scattering equation (9.3.5) admits another solution $\psi^{(-)}(\boldsymbol{r})$, which behaves asymptotically as a plane wave plus an incoming wave. This solution is not relevant to scattering problems, since the scattered particle is expected to travel outward from the region of interaction.

Using the definition of probability current density $\boldsymbol{S}(\boldsymbol{r}) = \frac{\hbar}{2i\mu}\left[\psi^*\boldsymbol{\nabla}\psi - (\boldsymbol{\nabla}\psi)^*\psi\right]$ [Eq. (3.3.4)], we find that the plane wave term

$$\psi_{inc}(\boldsymbol{r}) = Ae^{i\boldsymbol{k}\cdot\boldsymbol{r}} = Ae^{ikz} \tag{9.3.8}$$

associated with the incident particle corresponds to a incident particle flux density

$$\boldsymbol{S}_{inc}(\boldsymbol{r},t) = |A|^2\frac{\hbar\boldsymbol{k}}{\mu}. \tag{9.3.9}$$

Similarly, the scattered wave

$$\psi_{sc}(r) = A\,f(\theta)\frac{\exp(ikr)}{r} \tag{9.3.10}$$

corresponds to scattered particle flux density

$$\boldsymbol{S}_{sc}(\boldsymbol{r},t) = |A|^2\frac{\hbar\boldsymbol{k}'}{\mu\,r^2}|f(\theta)|^2 \tag{9.3.11}$$

where $\boldsymbol{k}'$ is the scattering vector and we have retained only the lowest order term in $1/r$, higher order terms being negligible compared to the leading $1/r^2$-term in the asymptotic region. Note that for elastic scattering considered here $|\boldsymbol{k}| = k = |\boldsymbol{k}'|$

From the scattered particle flux density (the number of particles crossing a unit area per second), the number of scattered particles crossing a small area element $d\boldsymbol{\mathcal{A}}$, located at a distance r in the direction (θ,φ), per second is found to be $\boldsymbol{S}_{sc}\cdot d\boldsymbol{\mathcal{A}} = S_{sc}r^2 d\Omega$, where $d\Omega = \frac{\boldsymbol{k}'}{|\boldsymbol{k}'|}\cdot d\boldsymbol{\mathcal{A}}/r^2$ is the solid angle subtended by the area element $d\boldsymbol{\mathcal{A}}$ at the scattering center [Fig. 9.4]. By definition of the scattering cross-section, this number $S_{sc}r^2 d\Omega$, the number of particles scattered into the solid angle $d\Omega$ per second, must equal the number of particles contained in a cross-sectional area $d\sigma$ of the incident beam

$$d\sigma \times \text{ Incident flux density} = \boldsymbol{S}_{sc}\cdot d\boldsymbol{\mathcal{A}} = S_{sc}r^2 d\Omega$$

or

$$d\sigma \times \frac{|A|^2\hbar k}{\mu} = |A|^2|f(\theta)|^2\frac{\hbar k'}{\mu}d\Omega$$

or

$$\frac{d\sigma}{d\Omega} \equiv \sigma(\theta) = |f(\theta)|^2. \tag{9.3.12}$$

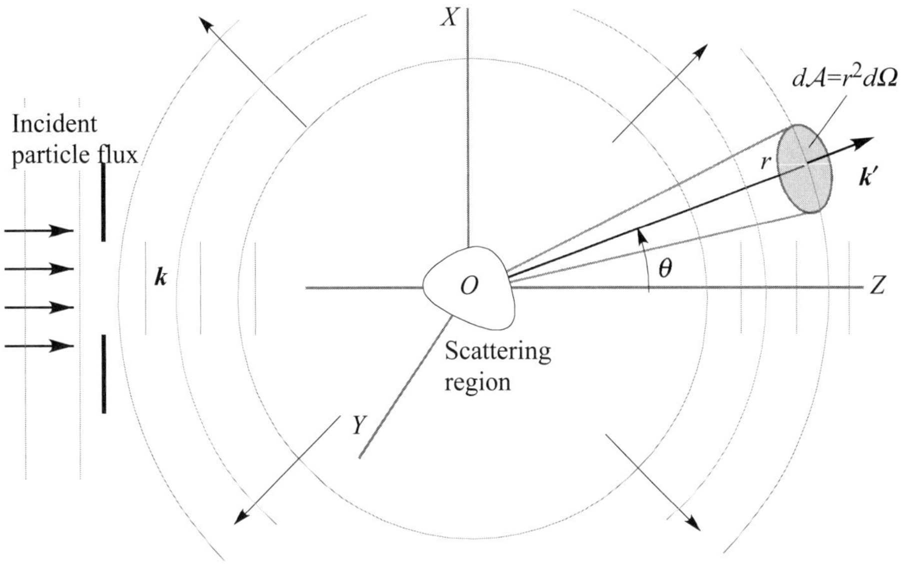

FIGURE 9.4
The incident particle with momentum $\hbar\boldsymbol{k}$ in center-of-mass frame is associated with a plane wave and the scattered particle is associated with an outgoing spherical wave ψ_{sc}. To calculate the flux of particles scattered in the direction $\boldsymbol{k}'$ (or θ, φ), we need to consider only the scattered wave because of the collimation of the incident beam.

Note that the normalization constant A has disappeared from the expression for $\sigma(\theta)$. We may normalize the wave function to unit incident flux or normalize it over a large box that has periodic boundary conditions. We will simply choose $A = 1$. Finally, it should be pointed out that theoretically, at any point P [Fig. 9.4] there exists a superposition of the incident plane wave and the scattered (outgoing spherical) wave [Eq. (9.3.7)]. However, in the experiment, the incident wave can be eliminated at all angles except in a narrow range around the forward direction ($\theta = 0$) by using a directional detector or using a collimated beam. Therefore, at nonzero scattering angles, we can consider only the scattered wave for the purpose of calculating the flux of particles emerging from the scattering region.

Determination of Scattering Amplitude $f(\theta)$

The problem of calculating the differential cross-section of elastic scattering, in the center-of-mass frame, now reduces to solving the scattering equation (9.3.5), which may be rewritten as

$$\left[\nabla^2 + k^2 - U(r)\right] \psi(\boldsymbol{r}) = 0\,, \tag{9.3.13a}$$

where

$$k^2 \equiv \frac{2\mu E}{\hbar^2}\,, \quad \text{and} \quad U(r) \equiv \frac{2\mu}{\hbar^2} V(r)\,. \tag{9.3.13b}$$

By solving this equation we determine the asymptotic form of the solution $\psi^{(+)}(\boldsymbol{r})$ and extract the scattering amplitude $f(\theta)$. This can be done either by using

(i) the method of *partial waves*

(ii) the *Born approximation*

The method of partial waves is exact and involves no approximation. However, it is convenient to use this method only when the energy E is low; with increasing energy, the method becomes increasingly complex because, as we shall see later, an increasingly large number of *partial waves* have to be considered. At high energies such that the interaction between the two particles may be treated as a perturbation compared to the energy E, we can use the Born approximation. We shall discuss the method of partial waves for spherically symmetric potential first followed by a discussion of the Born approximation.

9.4 Partial Waves and Phase Shifts

When the interaction potential is spherically symmetric, $V(\boldsymbol{r}) = V(r)$, then the scattering equation (9.3.5) may be separated into radial and angular equations by the substitution, $\psi(\boldsymbol{r}) = R_\ell(r)Y_{\ell m}(\theta,\varphi)$ and expressing the Laplacian operator ∇^2 in terms of polar angles θ,φ. We have already gone through this exercise in connection with the bound state problems [see Chapter 5, Eqs. (5.5.17) and (5.5.18)]. The angular solutions $Y_{\ell m}(\theta,\varphi)$ are the familiar functions called spherical harmonics. The radial equation (5.5.39) in the present context can be rewritten as

$$\frac{d^2R_\ell(r)}{dr^2} + \frac{2}{r}\frac{dR_\ell(r)}{dr} + \left[k^2 - U(r) - \frac{\ell(\ell+1)}{r^2}\right]R_\ell(r) = 0\,. \tag{9.4.1}$$

This equation is to be solved for the given potential $V(r)$ and energy $E > 0$ to find the asymptotic form of the radial function. As discussed in Chapter 5, Sec. 5.5, the regular solution of the radial equation (9.4.1) which vanishes $[rR_\ell(r) \to 0]$ as $r \to 0$ is uniquely determined once $V(r)$ and E are specified. The solution is labeled by the angular momentum eigenvalue ℓ and energy E via the relation $k = \sqrt{2\mu E/\hbar^2}$. The general solution of the scattering equation (9.3.13) then has the form

$$\psi(\boldsymbol{r}) = \sum_{\ell=0}^{\infty}\sum_{m=-\ell}^{\ell} A_{\ell m}R_\ell(r)Y_{\ell m}(\theta,\varphi)\,. \tag{9.4.2}$$

Since the scattering solution has no φ-dependence, only the $m = 0$ terms contribute. Recalling that $Y_{\ell 0} \to P_\ell(\cos\theta)$, the complete solution of the scattering equation (5.3.13) can then be written as

$$\psi(\boldsymbol{r}) = \sum_{\ell=0}^{\infty} B_\ell R_\ell(r)P_\ell(\cos\theta)\,. \tag{9.4.3}$$

To compute the general asymptotic form of the wave function we consider the radial equation in the limit $r \to \infty$. With the substitution $u_\ell(r) = rR_\ell(r)$ [see Sec. 5.5], the radial equation becomes

$$\frac{d^2u_\ell(r)}{dr^2} + \left[k^2 - U(r) - \frac{\ell(\ell+1)}{r^2}\right]u_\ell(r) = 0\,. \tag{9.4.4}$$

If the potential $V(r)$ falls off sufficiently fast with distance, we can neglect both the potential and the centrifugal term so that the radial equation (9.4.4) assumes the form

$$\frac{d^2u_\ell(r)}{dr^2} + k^2u_\ell(r) = 0\,, \tag{9.4.5}$$

with solutions $u_\ell(r) \sim e^{\pm ikr}$. Factoring out this phase term, we write the radial function as $u_\ell(r) = v_\ell(r)e^{\pm ikr}$. Substituting this in Eq. (9.4.4), we find that $v_\ell(r)$ obeys the equation

$$\frac{d^2 v_\ell(r)}{dr^2} \pm 2ik\frac{dv_\ell(r)}{dr} - \left[U(r) + \frac{\ell(\ell+1)}{r^2}\right] v_\ell = 0\,. \tag{9.4.6a}$$

It must still satisfy the condition $v_\ell(0) = 0$ to conform to the requirement $u_\ell(0) = 0$ on radial function. If we assume that $v_\ell(r)$ is slowly varying, we can neglect $|d^2 v_\ell/dr^2|$ compared with $|k\frac{dv_\ell}{dr}|$, Eq. (9.4.6a) yields

$$\frac{1}{v_\ell}(r)\frac{dv_\ell(r)}{dr} = \mp\frac{i}{2k}\left[U(r) + \frac{\ell(\ell+1)}{r^2}\right]. \tag{9.4.6b}$$

This equation is readily integrated to give

$$v_\ell(r) \approx \exp\left[\mp\frac{i}{2k}\int^r dr'\left(U(r') + \frac{\ell(\ell+1)}{r'^2}\right)\right] \tag{9.4.6c}$$

and

$$u_\ell(r) \approx \exp\left[\pm i\left\{kr - \frac{1}{2k}\int^r dr'\left(U(r') + \frac{\ell(\ell+1)}{r'^2}\right)\right\}\right]. \tag{9.4.6d}$$

Now the intergral in $v_\ell(r)$ converges in the asymptotic limit $r \to \infty$ if the potential $U(r) \to 0$ falls off faster than $1/r$ with increasing r [or $rU(r) \to 0$ as $r \to \infty$]. Hence letting $r \to \infty$ we can put the integral equal to some finite constant ϵ_ℓ

$$\frac{1}{2k}\int^r dr'\left[U(r') + \frac{\ell(\ell+1)}{r'^2}\right] = \epsilon_\ell\,. \tag{9.4.7}$$

So the asymptotic behavior of the radial functions $R_\ell(r) = u_\ell(r)/r$ is given by the linear combination of $e^{i(kr-\epsilon_\ell)}$ and $e^{-i(kr-\epsilon_\ell)}$ as

$$R_\ell(r) \sim \frac{\sin(kr - \ell\pi/2 + \delta_\ell)}{kr}, \tag{9.4.8}$$

where we have written $\epsilon_\ell = \ell\pi/2 - \delta_\ell$ so that $\delta_\ell = 0$ when the potential is absent[2]. Thus the constant δ_ℓ is the difference in phase of the actual wave function and the wave function of free motion in the absence of the potential. For this reason it is called the *phase shift* and is determined by the behavior of $u_\ell(r)$ in the region of finite r where the potential is nonzero.

Substituting the asymptotic form of the radial function (9.4.8) in Eq. (9.4.3), we find the general asymptotic form of the wave function for a potential that falls off faster than $1/r$ as

$$\psi(\boldsymbol{r}) \sim \sum_{\ell=0}^{\infty} B_\ell \frac{\sin(kr - \ell\pi/2 + \delta_\ell)}{kr} P_\ell(\cos\theta)\,. \tag{9.4.9}$$

This equation is to be compared with Eq. (9.3.7) to determine the scattering amplitude. To facilitate this comparison, we express the incident plane wave $\psi_{in}(\boldsymbol{r}) = e^{ikz}$ in terms of spherical harmonics. This is easily done by noting that e^{ikz} is the solution of the Schrödinger equation for a free particle

$$(\nabla^2 + k^2)\psi(\boldsymbol{r}) = 0 \tag{9.4.10}$$

[2] In the absence of any potential [$U(r) = 0$], the regular (at $r = 0$) solution of the radial equation is $R_\ell(r) \sim j_\ell(kr)$ with asymptotic behavior $\sim \frac{\sin(kr-\ell\pi/2)}{kr}$ [Sec. 5.8].

in Cartesian coordinates when the particle has momentum $\hbar k$ in $+z$ direction. On the other hand, solutions of this equation which are regular as $r \to 0$ in spherical polar coordinates are of the form [cf. Chapter 5, Eq. (5.8.11)]

$$\psi_{k\ell}(r) \sim j_\ell(kr) Y_{\ell m}(\theta, \varphi) , \tag{9.4.11}$$

where $j_\ell(kr)$ is the spherical Bessel. Using the solutions in the spherical polar coordinates as a basis and noting that the plane wave solution $e^{ikz} = e^{ikr\cos\theta}$ is independent of the azimuthal angle φ, we can write the incident plane wave as the sum

$$\psi_{inc}(r) = e^{ikz} = \sum_{\ell=0}^{\infty} C_\ell j_\ell(kr) P_\ell(\cos\theta) . \tag{9.4.12}$$

To determine the constants C_ℓ we multiply both sides of this equation with $P_{\ell'}(\cos\theta)\sin\theta d\theta$ and integrate over θ to get,

$$\int_0^\pi e^{ikr\cos\theta} P_{\ell'}(\cos\theta)\sin\theta d\theta = A_{\ell'} j_{\ell'}(kr) \frac{2}{2\ell'+1} \tag{9.4.13}$$

where the orthogonality condition for Legendre's polynomials has been used. This equation holds for all values of r. Hence replacing ℓ by ℓ' in Eq. (9.4.13), integrating by parts, and taking the limit $r \to \infty$, we get

$$2e^{i\ell\pi/2} \frac{\sin(kr - \ell\pi/2)}{kr} = C_\ell \frac{\sin(kr - \ell\pi/2)}{kr} \frac{2}{2\ell+1} , \tag{9.4.14}$$

which immediately leads to

$$C_\ell = e^{i\ell\pi/2}(2\ell+1) . \tag{9.4.15}$$

Thus a plane wave can be expanded in terms of the solutions of the free-particle Schrödinger equation in the spherical polar cordinates as

$$\begin{aligned} \psi_{inc}(\boldsymbol{r}) \equiv e^{ikz} &= \sum_{\ell=0}^{\infty} e^{i\ell\pi/2}(2\ell+1) j_\ell(kr) P_\ell(\cos\theta) \\ &\xrightarrow{r\to\infty} \sum_{\ell=0}^{\infty} e^{i\ell\pi/2}(2\ell+1) \frac{\sin(kr-\ell\pi/2)}{kr} P_\ell(\cos\theta) . \end{aligned} \tag{9.4.16}$$

This expansion amounts to expressing the state of a particle with a definite linear momentum $\hbar\boldsymbol{k}$, as a superposition of *partial waves* each of which represents a state of definite angular momentum $\hbar\sqrt{\ell(\ell+1)}$ of the particle about the scattering center. Obviously, when the momentum of the particle is well defined, its angular momentum about the scattering center is indeterminate since the corresponding observables do not commute.

Comparing the asymptotic form of the solution of the scattering equation (9.4.9) with that for a free particle (9.4.16) we find that the presence of a potential introduces a phase shift δ_ℓ in each partial wave. Phase shift $\delta_\ell(k)$ depends on (i) ℓ, (ii) momentum $\hbar k$ or energy E, and (iii) the nature of the interaction $V(r)$.

Scattering amplitude in terms of phase shifts

Having expressed the incident plane wave in terms of partial waves, we can now determine the scattering amplitude $f(\theta)$ in terms of phase shifts by equating the asymptotic form

(9.4.9) of the solution of the scattering equation with Eq. (9.3.7):

$$\sum_{\ell=0}^{\infty} B_\ell \frac{\sin(kr - \ell\pi/2 + \delta_\ell)}{kr} P_\ell(\cos\theta) = e^{ikz} + f(\theta)\frac{e^{ikr}}{r}$$

$$= \sum_{\ell=0}^{\infty} e^{i\ell\pi/2}(2\ell+1)\frac{\sin(kr - \ell\pi/2)}{kr} P_\ell(\cos\theta) + f(\theta)\frac{e^{ikr}}{r}. \tag{9.4.17}$$

Equating the coefficients of $\frac{e^{-ikr}}{r}$ and $\frac{e^{ikr}}{r}$ from both sides, we get

$$B_\ell = e^{i\delta_\ell + i\ell\pi/2}(2\ell+1) \tag{9.4.18}$$

and

$$f(\theta) = \frac{1}{k}\sum_{\ell=0}^{\infty}(2\ell+1)e^{i\delta_\ell}\sin\delta_\ell\, P_\ell(\cos\theta). \tag{9.4.19}$$

From these equations we find the differential scattering cross-section is

$$\sigma(\theta) = |f(\theta)|^2 = \frac{1}{k^2}\left|\sum_{\ell=0}^{\infty}(2\ell+1)e^{i\delta_\ell}\sin\delta_\ell P_\ell(\cos\theta)\right|^2. \tag{9.4.20}$$

By integrating the differential scattering cross-section over all angles, we obtain the total scattering cross-section as

$$\sigma_{tot} = \int_0^\pi |f(\theta)|^2 2\pi\sin\theta d\theta$$

$$= \frac{2\pi}{k^2}\sum_{\ell=0}^{\infty}\sum_{\ell'=0}^{\infty}(2\ell+1)(2\ell'+1)e^{i(\delta_\ell - \delta_{\ell'})}\sin\delta_\ell\sin\delta_{\ell'}\int_0^\pi P_\ell(\cos\theta)P_{\ell'}(\cos\theta)\sin\theta d\theta$$

or

$$\sigma_{tot} = \frac{4\pi}{k^2}\sum_{\ell=0}^{\infty}(2\ell+1)\sin^2\delta_\ell, \tag{9.4.21}$$

where we have used the orthogonality property of the Legendre polynomials. From Eq. (9.4.19) we note that the imaginary part of the scattering amplitude in the forward direction ($\theta = 0$) is given by

$$\mathcal{Im}\ f(\theta = 0) = \frac{1}{k}\sum_{\ell=0}^{\infty}(2\ell+1)\ \sin^2\delta_\ell. \tag{9.4.22}$$

Hence the total scattering cross-section σ_{tot} may also be expressed as

$$\sigma_{tot} = \frac{4\pi}{k}\mathcal{Im}\ f(\theta = 0). \tag{9.4.23}$$

This result is called the *optical theorem.*

From the expressions for $\sigma(\theta)$ and σ_{tot} [Eqs. (9.4.20) and (9.4.21)] it might appear that an infinite number of phase shifts, δ_0, δ_1, $\delta_2, \cdots \delta_\ell \cdots$ are needed for the calculation of the differential and total cross-sections. However, we can see from a semi-classical argument that if the interaction has a finite range, i.e., if the potential vanishes beyond a certain range [$V(r) = 0$ for $r > b$], then only a finite number of partial waves undergo phase shifts. To see this consider a particle with momentum $\boldsymbol{p} = \hbar\boldsymbol{k}$ approaching a target O so that its impact parameter is s [Fig. 9.5]. It is then obvious that if $s > b$, the particle will

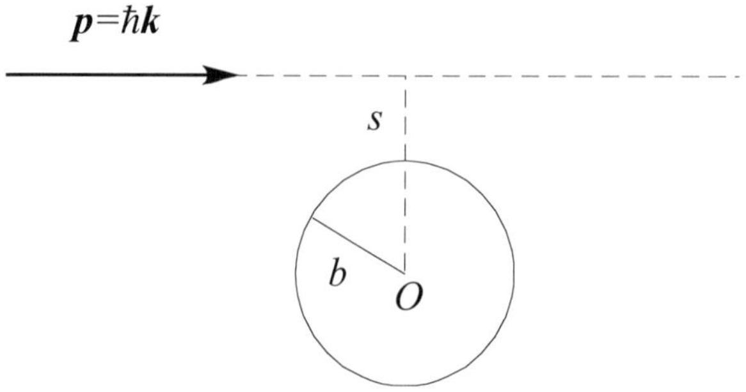

FIGURE 9.5
Impact parameter and the range of interaction. When the impact parameter s is larger than the range of interaction b, the incident particle remains unaffected by the potential.

go unscattered while if $s \leq b$, then the particle will undergo scattering. Now, classically, the angular momentum of the particle with respect to the target is $\hbar ks$. It follows that particles with angular momentum $\hbar\sqrt{\ell(\ell+1)} \approx \hbar\ell > \hbar kb$ will remain unscattered. This result means that partial waves with $\ell > kb$ will not undergo scattering or

$$\delta_\ell = 0 \quad \text{for} \ \ \ell > kb \,. \tag{9.4.24}$$

Hence if the energy of the incident particle $E = \hbar^2k^2/2\mu$ in the center-of-mass frame is known and the range b of the interaction is given, we need to calculate the phase shifts only for the partial waves with $\ell \leq kb$.

Sign of the phase shift and the nature of the potential

So far we have merely expressed the scattering cross-section in terms of partial wave phase shifts. The phase shifts must be determined by solving the Schrödinger equation for a given potential. In practice we often face the inverse problem, i.e., we seek to infer the potential from the phase shifts measured in an experiment. A simple relationship between the sign of the phase shift and the overall nature (attractive or repulsive) of the potential can be established generally. Consider two potentials $V(r)$ and $V'(r)$. Then the radial functions $u_\ell(r) = rR_\ell(r)$ and $u'_\ell(r) = rR'_\ell(r)$ in the two cases satisfy

$$\frac{d^2u_\ell(r)}{dr^2} + \left[k^2 - U(r) - \frac{\ell(\ell+1)}{r^2}\right]u_\ell(r) = 0 \tag{9.4.25}$$

and

$$\frac{d^2u'_\ell(r)}{dr^2} + \left[k^2 - U'(r) - \frac{\ell(\ell+1)}{r^2}\right]u'_\ell(r)\,, = 0. \tag{9.4.26}$$

where $U(r) = 2\mu V(r)/\hbar^2$, $U'(r) = 2\mu V'(r)/\hbar^2$ and $k^2 = 2\mu E/\hbar^2$. The asymptotic solutions of these equations are

$$u_\ell(r) \sim \sin(kr - \ell\pi/2 + \delta_\ell) \tag{9.4.27}$$

and

$$u'_\ell(r) \sim \sin(kr - \ell\pi/2 + \delta'_\ell)\,. \tag{9.4.28}$$

Multiplying Eq. (9.4.25) by $u'_\ell(r)$, Eq. (9.4.26) by $u_\ell(r)$ and taking the difference of the resulting equations, we get

$$\frac{d}{dr}\left[u_\ell(r)\frac{du'_\ell(r)}{dr} - u'_\ell(r)\frac{du_\ell(r)}{dr}\right] = -[U(r) - U'(r)]u_\ell(r)u'_\ell(r)\,. \tag{9.4.29a}$$

Integrating this equation with respect to r from $r = 0$ to some value $r = R$, and recalling that both radial functions satisfy the boundary condition $u_\ell(0) = 0 = u'_\ell(0)$, we get

$$\left[u_\ell(r)\frac{du'_\ell(r)}{dr} - u'_\ell(r)\frac{du_\ell(r)}{dr}\right]_{r=R} = -\int_0^R [U(r) - U'(r)]u_\ell(r)u'_\ell(r)dr. \tag{9.4.29b}$$

This equation holds for all values of R. By choosing R to be sufficiently large, we can use the asymptotic forms for the functions $u_\ell(r)$ and $u'_\ell(r)$ on the left-hand side of Eq. (9.4.29b). We then find

$$k\sin(\delta'_\ell - \delta_\ell) = -\frac{2\mu}{\hbar^2}\int_0^R [V'(r) - V(r)]u_\ell u'_\ell dr\,. \tag{9.4.30}$$

If the difference $V'(r) - V(r) \equiv \Delta V(r)$ between the two potentials is small, we expect the two radial functions to be similar. We can then ignore the difference between them in the integral on the right-hand side and write

$$k\ \sin(\Delta\delta_\ell) = -\frac{2\mu}{\hbar^2}\int_0^\infty \Delta V(r)\ u_\ell^2(r)dr\,. \tag{9.4.31}$$

Since $u_\ell^2(r)$ is a positive quantity, we see that the change in the phase shift is in opposite direction to the average (weighted by the radial function squared) change in the potential. Thus if the potential changes to become more positive (repulsive), the phase shift will decrease and if it changes to be more negative (attractive) the phase shift will increase. As a specific example consider $V(r) = 0$ so that $\delta_\ell = 0$. Then for a repulsive potential $V'(r) > 0$ the phase shift δ'_ℓ is negative [Eq. (9.4.31)]. This means the radial function is pushed out in comparison with the radial function $j_\ell(kr)$ for $V = 0$ [Fig. 9.6]. For an attractive potential $(V'(r) < 0)$ δ'_ℓ is positive, which means the radial function is pulled in compared to $j_\ell(kr)$ [Fig.(9.6)].

9.5 Calculation of Phase Shift

We now come to the calculation of phase shifts. For a potential that has a finite range r_o, the condition $rV(r) \to 0$ as $r \to \infty$ is clearly satisfied. Therefore the method of partial waves is applicable. The radial equation in the interior region $r \le r_o$ is [see Sec. 9.4]

$$\frac{d^2R_\ell^{\rm in}(r)}{dr^2} + \frac{2}{r}\frac{dR_\ell^{\rm int}(r)}{dr} + \left[k^2 - U(r) - \frac{\ell(\ell+1)}{r^2}\right]R_\ell^{\rm in}(r) = 0\,, \quad r < r_o\,, \tag{9.4.1*}$$

where $U(r) = 2\mu V(r)/\hbar^2$ and $k^2 = 2\mu E/\hbar^2$ and we have denoted the solution of this equation by $R_\ell^{\rm int}(r)$. The radial equation in the exterior region $(r > r_o)$ where $V(r) = 0$ has the form

$$\frac{d^2R_\ell^{\rm ex}(r)}{dr^2} + \frac{2}{r}\frac{dR_\ell^{\rm ex}(r)}{dr} + \left[k^2 - \frac{\ell(\ell+1)}{r^2}\right]R_\ell^{\rm ex}(r) = 0\,, \quad r > r_o\,, \tag{9.4.1*}$$

whose solutions are $j_\ell(kr)$ and $\eta_\ell(kr)$ [Chapter 5, Sec. 5.8]. Since $r = 0$ is excluded from the exterior region $[r > r_o]$ both are acceptable solutions. We choose the exact solution in the external region to be the combination

$$R_\ell^{\rm ex}(r) = \cos\delta_\ell\ j_\ell(kr) - \sin\delta_\ell\ \eta_\ell(kr)\,, \tag{9.5.1}$$

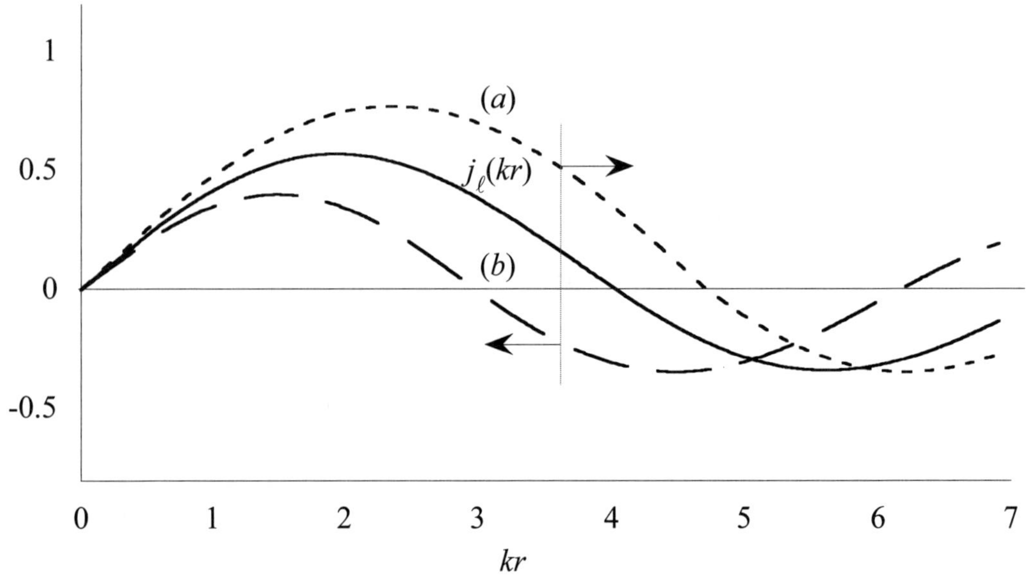

FIGURE 9.6
A comparison of the radial function $R_\ell(r)$ for a potential $V(r)$ with the radial function $j_\ell(kr)$ in the absence of interaction. When the potential is positive (repulsive) δ_ℓ is negative and the radial function is pushed out [curve (a)]. When the potential is negative (attractive) then $\delta_\ell > 0$ and the radial function is pulled in [curve (b)] as compared to $j_\ell(kr)$.

which conforms to the prescribed asymptotic form

$$R_\ell^{\rm ex}(r) \sim \frac{\sin(kr - \ell\pi/2 + \delta_\ell)}{kr}, \tag{9.4.8*}$$

since asymptotic forms of $j_\ell(kr)$ and $\eta_\ell(kr)$ are

$$j_\ell(kr) \sim \frac{\sin(kr - \ell\pi/2)}{kr}, \text{ and } \eta_\ell(kr) \sim -\frac{\cos(kr - \ell\pi/2)}{kr}. \tag{9.5.2}$$

Requiring the interior and exterior solutions and their derivatives to match at the boundary $r = r_o$, we obtain

$$R_\ell^{\rm in}(r_o) = R_\ell^{\rm ex}(r_o) \tag{9.5.3a}$$

and

$$\left(\frac{dR_\ell^{\rm in}}{dr}\right)_{r=r_o} = \left(\frac{dR_\ell^{\rm ex}}{dr}\right)_{r=r_o}. \tag{9.5.3b}$$

These can be combined to yield the condition for the continuity of the logarithmic derivatives of the solutions in the two regions

$$\frac{1}{R_\ell^{\rm in}(r_o)}\left[\frac{dR_\ell^{\rm in}}{dr}\right]_{r=r_o} = \frac{1}{R_\ell^{\rm ex}(r_o)}\left[\frac{dR_\ell^{\rm ex}}{dr}\right]_{r=r_o}. \tag{9.5.4}$$

Denoting the logarithmic derivative of the interior wave function at r_o by γ_ℓ and using the exterior solution (9.5.1) in Eq. (9.5.4), we find

$$\gamma_\ell = \frac{k\, j'_\ell(kr_o)\cos\delta_\ell - k\eta'_\ell(kr_o)\sin\delta_\ell}{j_\ell(kr_o)\cos\delta_\ell - \eta_\ell(kr_o)\sin\delta_\ell}, \tag{9.5.5a}$$

where

$$j'_\ell(\rho) \equiv \frac{dj_\ell(\rho)}{d\rho} \quad \text{and} \quad \eta'_\ell(\rho) \equiv \frac{d\eta_\ell(\rho)}{d\rho}. \tag{9.5.5b}$$

By rearranging this equation we find that the phase shift is given by

$$\tan\delta_\ell = \frac{k\, j'_\ell(kr_o) - \gamma_\ell j_\ell(kr_o)}{k\eta'_\ell(kr_o) - \gamma_\ell\eta_\ell(kr_o)} \tag{9.5.6}$$

Thus, if the logarithmic derivative of the internal wave function at the boundary $r = r_o$ is known for a given value of ℓ, then the phase shift δ_ℓ can be calculated.

Phase shifts as meeting grounds for theory and experiment

Experimentally measured differential cross-sections $\sigma(\theta_0)$ for different laboratory energies E_0 can be converted to center-of-mass cross-section for the corresponding CM energies E (or k) and this data can be analyzed in terms of s-wave ($\ell = 0$), p-wave ($\ell = 1$), d-wave ($\ell = 2$), $\cdots$ phase shifts $\delta_0(k)$, $\delta_1(k)$, $\delta_2(k), \cdots$, with the help of Eq. (9.4.20). These experimentally measured phase shifts can then be compared with the theoretically calculated phase shifts based on model potentials and the use of Eq. (9.5.6). We may thus regard the phase shifts as the meeting grounds for the theory and experiment.

9.6 Phase Shifts for Some Simple Potential Forms

We illustrate the method for calculating the phase shifts by considering two simple potentials.

(a) Hard Sphere Potential

This potential describes the scattering when the target and projectile particles cannot come closer than a certain relative distance distance r_o. Thus for $r < r_o$ the particles see an infinitely repulsive barrier. This interaction corresponds to a potential of the form

$$V(r) = \begin{cases} \infty, & \text{for} \quad r < r_0 \\ 0, & \text{for} \quad r > r_0, \end{cases} \tag{9.6.1}$$

where $\boldsymbol{r}$ is the relative coordinate. In this case the radial wave function $R^{\text{in}}_\ell(r)$ in the interior region ($r \le r_o$) vanishes. So from Eq. (9.5.3a), we find

$$\tan\delta_\ell = \frac{j_\ell(kr_o)}{\eta_\ell(kr_o)}. \tag{9.6.2}$$

The calculation of the phase shift is particularly simple for $kr_o << 1$ (low energies or short range of potential or both), where

$$j_\ell(kr_o) \approx \frac{(kr_o)^\ell}{(2\ell+1)!!} \quad \text{and} \quad \eta_\ell(kr_o) \approx -\frac{(2\ell-1)!!}{(kr_o)^{\ell+1}}. \tag{9.6.3}$$

With the help of these expressions we find

$$\tan\delta_\ell \approx \delta_\ell = -\frac{(kr_o)^{2\ell+1}}{(2\ell+1)!!\,(2\ell-1)!!}\,. \tag{9.6.4}$$

From this equation we see that (i) all phase shifts are negative as expected for a repulsive potential, (ii) δ_ℓ falls off rapidly as ℓ increases, and (ii) all phase shifts tend to zero as $kr_o \to 0$.

Substituting the phase shifts from Eq. (9.6.4) into Eq. (9.4.21), we find the total scattering cross-section is given by

$$\sigma_{tot} = 4\pi r_o^2 \sum_{\ell=0}^{\infty} \frac{(kr_o)^{4\ell}}{[2\ell+1)!!(2\ell-1)!!]^2}\,. \tag{9.6.5}$$

In the very low energy limit, $kr_o \to 0$, only the $\ell = 0$ term (s-wave) contribution survives, giving

$$\delta_0 = -kr_o\,, \tag{9.6.6}$$

$$\sigma_{tot} = \frac{4\pi}{k^2}\sin^2\delta_0^2 = 4\pi r_o^2\,. \tag{9.6.7}$$

Note that this is four times the classical scattering cross-section from a hard sphere. In the classical scattering, the incident particle sees the geometrical crosss-section πa^2 of the sphere. In the quantum mechanical scattering, the wave-like character of the projectile causes diffraction at the edges, resulting in a larger cross-section.

(b) Attractive Square-well Potential

The attractive square-well potential has the form

$$V(r) = \begin{cases} -V_o\,, & \text{for } r < r_o \\ 0\,, & \text{for } r > r_o\,. \end{cases} \tag{9.6.8}$$

Recall that this potential was used in Chapter 5 to describe the short range neutron-proton interaction. In this case the radial equation (9.4.1) for the interior region ($r < r_o$) is

$$\frac{d^2R_\ell^{\rm in}(r)}{dr^2} + \frac{2}{r}\frac{dR_\ell^{\rm in}(r)}{dr} + \left[\alpha^2 - \frac{\ell(\ell+1)}{r^2}\right]R_\ell^{\rm in}(r) = 0\,, \tag{9.6.9a}$$

where

$$\alpha^2 \equiv k^2 + k_o^2 \quad \text{and} \quad k_0^2 \equiv \frac{2\mu V_o}{\hbar^2}\,. \tag{9.6.9b}$$

As seen in Sec. 9.5, the independent solutions of this equation are $j_\ell(\alpha r)$ and $\eta_\ell(\alpha r)$. Since the solution η_ℓ is singular at $r = 0$ (which is part of the interior region), we drop it and write the regular solution [cf. Eq. (9.4.11)]

$$R_\ell^{\rm in}(r) = A_\ell j_\ell(\alpha r)\,. \tag{9.6.10}$$

The logarithmic derivative of the interior wave function is given by

$$\gamma_\ell = \left[\frac{1}{R_\ell^{\rm in}}\left(\frac{dR_\ell^{\rm in}}{dr}\right)\right]_{r=r_o} = \frac{\alpha\, j_\ell'(\alpha r_o)}{j_\ell(\alpha r_o)}\,. \tag{9.6.11}$$

Using the explicit form of spherical Bessel functions we can calculate the logarithmic derivative of the interior radial function. For example, for the s-wave ($\ell = 0$) and p-wave ($\ell = 1$) we get[3]

$$\gamma_0 = \frac{\alpha\, j_0'(\alpha\, r_o)}{j_0(\alpha\, r_o)} = \alpha \cot(\alpha\, r_o) - \frac{1}{r_o} \tag{9.6.12}$$

and

$$\gamma_1 = \frac{\alpha\, j_1'(\alpha r_o)}{j_1(\alpha r_o)} = \alpha \frac{(\alpha^2 r_o^2 - 2) + 2\alpha r_o \cot(\alpha r_o)}{\alpha r_o - \alpha^2 r_o^2 \cot(\alpha r_o)}\,. \tag{9.6.13}$$

Results like these, when used in Eq. (9.5.6), for different ℓ will yield the corresponding phase shifts. In general, the exact results must be obtained numerically. However, in certain limits, we can obtain analytically tractable results. Thus using Eqs. (9.6.12) and (9.6.13) in Eq. (9.5.6), we find that low energy ($kr_o \ll 1$) phase shifts for the s and p waves are given by

$$\tan \delta_0 \approx -\frac{\gamma_0 k r_o^2}{1 + \gamma_0 r_o} \tag{9.6.14}$$

and

$$\tan \delta_1 \approx \frac{(kr_o)^3}{3} \frac{1 - \gamma_1 r_o}{2 + \gamma_1 r_o}\,. \tag{9.6.15}$$

In the low energy limit or when short range potentials (as in neutron-proton scattering) are involved, scattering is dominated by the s-wave. For example consider the scattering of a 1 MeV neutron from a proton-rich target (water, for example). As seen in Chapter 5, the range of nuclear force is $r_o \approx 2$ fm (2×10^{-15} m). Neglecting the motion of the protons, the CM energy is $E = 0.5$ MeV. This corresponds to $k = \sqrt{\frac{2\mu E}{\hbar^2}} = \sqrt{\frac{2\mu c^2 E}{\hbar^2 c^2}} \approx 0.1\ \text{fm}^{-1}$ and therefore to $k_o r_o = 0.2$. Since only the partial waves with $\ell < kr_o$ contribute to scattering [Eq. (9.4.24)], it is clear that $\ell = 0$ partial wave will dominate scattering.

Let us therefore consider the contribution of the s-wave phase shift to the total scattering cross-section. Using Eq. (9.6.12) for γ_0 in Eq. (9.6.14), we find

$$\tan \delta_0 \approx -\frac{\gamma_0 k r_o^2}{1 + \gamma_0 r_o} = -kr_0 \left[\frac{\alpha r_o \cot \alpha r_o - 1}{\alpha r_o \cot \alpha r_o}\right] = kr_o \left[\frac{\tan \alpha r_o}{\alpha r_o} - 1\right]. \tag{9.6.16}$$

Then the total scattering cross-section in the low energy limit $kr_o \ll 1$ is given by

$$\sigma_{tot} = \frac{4\pi}{k^2} \sin^2 \delta_0 \approx 4\pi r_o^2 \left[\frac{\tan \alpha r_o}{\alpha r_o} - 1\right]^2. \tag{9.6.17}$$

Recalling that $\alpha^2 = k^2 + k_o^2 = \frac{2\mu(E+V_o)}{\hbar^2}$ [Eq. (9.6.9b)], we see that when $k \ll k_o$ ($E \ll V_o$), scattering is almost independent of the energy of the incident particle and, since the s-wave dominates, scattering is spherically symmetric.

This formula is not applicable if the range and depth of the potential are such that the condition $\alpha\, r_o \approx (2N + 1)\pi/2$, where N is an integer, is satisfied. In this case $\tan(\alpha\, r_o) \to \infty$, indicating a large increase in the cross-section. This case will be considered next.

---

[3]Explicit expressions for $j_\ell(\rho)$ and $\eta_\ell(\rho)$ for $\ell = 0, 1, 2$ are given in Appendix 5A1. For $\rho \ll 1$, we have

$$j_0(\rho) \approx 1 - \rho^2/6 \cdots \qquad \eta_0(\rho) = -1/\rho$$
$$j_1(\rho) \approx \rho/3 \cdots \qquad \eta_1(\rho) = -1/\rho^2 \cdots$$

Using explicit expressions for j_ℓ we get Eqs. (9.6.12) and (9.6.13), and using Eq. (9.5.6) for phase shifts, we can get Eqs. (9.6.14) and (9.6.15), where we let $kr_o \to 0$.

Resonance Scattering

From Eq. (9.6.17), we see that low energy scattering cross-section for a square well potential increases significantly when

$$\alpha r_o \equiv \sqrt{\frac{2\mu(E+V_o)r_o^2}{\hbar^2}} = \frac{(2N+1)\pi}{2} . \tag{9.6.18}$$

For these values we get, from Eq. (9.6.16), $\tan\delta_0 \to \infty$ or $\delta_0 \approx (2m+1)\pi/2$ so that at these values the scattering cross-section is given by

$$\sigma_{tot} = \frac{4\pi}{k^2}\sin^2\delta_0 = \frac{4\pi}{k^2} \gg 4\pi r_o^2 . \tag{9.6.19}$$

From our discussion of bound states in a square well for $\ell = 0$ [see Sec. 5.7], we know that this is the condition for the occurrence of a $\ell = 0$ bound state with $E \approx 0$. Thus whenever αr_o approaches any of the values given by (9.6.18), the cross-section peaks at the value $4\pi/k^2$.

To examine the behavior of cross-section as $\alpha r_o \to (2N+1)\pi/2$, we see from Eq. (9.6.12) that near a resonance $\gamma_0\, r_o = -1$ and $(\alpha\, r_o)\cot(\alpha\, r_o) \ll 1$. Hence the phase shift δ_0 is given by [Eq. (9.6.14)]

$$\cot\delta_0 = -\frac{1+\gamma_0 r_o}{\gamma_0 k r_o^2} = \frac{\alpha r_o \cot(\alpha r_o)}{k r_o[\alpha r_o \cot(\alpha r_o) - 1]} \approx \frac{\alpha}{k}\cot(\alpha r_o) , \tag{9.6.20}$$

whence

$$\sigma_{tot} = \frac{4\pi}{k^2}\sin^2\delta_0 = \frac{4\pi}{k^2}\frac{1}{1+\cot^2\delta_0} \approx \frac{4\pi}{k^2+\alpha^2\cot^2(\alpha\, r_o)} . \tag{9.6.21}$$

From Eqs. (9.6.17) and (9.6.21), we see that for most low energies the incident particle is scattered as if the potential were a hard sphere potential of radius r_o and scattering almost independent of energy. But for certain energies given by

$$\cot(\alpha r_o) = 0 \quad \Rightarrow \quad \sqrt{k_0^2 + k^2} r_o \approx k_0 + \frac{k^2}{2k_0})r_o = (2N+1)\pi/2,$$

where $k_o^2 \equiv \frac{2\mu V_o}{\hbar^2}$, the particle interacts strongly, enhancing the cross-section from the value $4\pi\, r_o^2$ to a very large value $4\pi/k^2$ [Fig.(9.7)] and, unlike off-resonance scattering, has strong energy dependence. From a physical viewpoint, resonance scattering means that whenever the potential well has a bound state with a binding energy close to zero, the (low energy) particle has a tendency to be bound to the well. Since a bound state with positive energy does not really exist, the particle interacts strongly, which causes enhanced scattering. This is called *resonance scattering.*

Resonances like the one just discussed for $\ell = 0$ also exist for nonzero values of ℓ, whenever the potential well supports a bound state with nearly zero energy for that value of ℓ. When this happens the incident particle tends to *stick around* the well producing a large distortion of its wave function leading to a large amount of scattering.

Ramsauer-Townsend Effect

A rare (noble) gas atom consists of closed electronic shells so its atomic size is small and the combined force of the atom on an external electron is strong and has a short range. Hence if a beam of slow electrons (k small) is incident on such atoms and $V(r)$ has a short range and is strongly attractive, then only the s-wave suffers an appreciable phase shift. If the attractive potential is strong enough, it can change the phase of the s-wave significantly

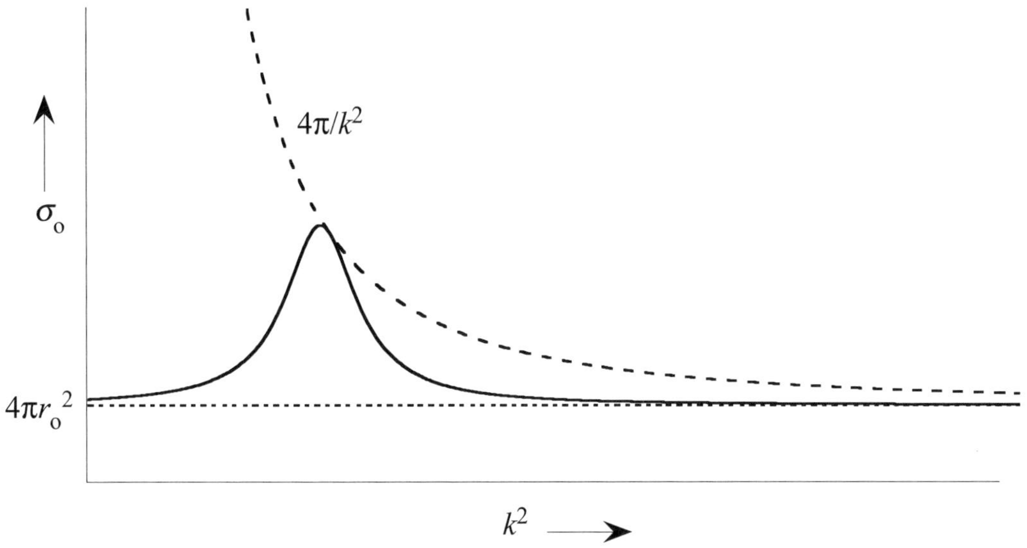

FIGURE 9.7
Resonance scattering. For low energy the total cross-section of scattering is close to $4\pi\, r_o^2$ (r_o is the range of interaction) and rather insensitive to energy. In a small neighborhood of certain specific energies, the cross-section may show strong dependence on energy increasing to a very large value $4\pi/k^2$ before returning to the background value $4\pi r_o^3$ outside the neighborhood.

while the phase shifts δ_ℓ for higher partial waves are still zero. In particular, if the depth and range of the potentials are such that the phase shift $\delta_0 \approx \pi$, then the scattering cross-section

$$\sigma_{tot} = \frac{4\pi}{k^2}\sum_{\ell=0}^{\infty}(2\ell+1)\sin^2\delta_\ell \approx 0.$$

This accounts for the extremely low minimum observed in the elastic scattering of electrons by rare gas atoms at about 0.7 eV of incident energy. This is called the Ramsauer- Townsend effect.

(c) *s*-Wave Phase Shift for a Potential of Exponential Shape

Let us consider another short range potential, which does not have a sharp cut-off. A potential that has this feature is the attractive exponential potential

$$V(r) = -V_o \exp(-\alpha r). \tag{9.6.22}$$

The radial part of the scattering equation for $\ell = 0$ assumes the form

$$\frac{d^2R_o(r)}{dr^2} + \frac{2}{r}\frac{dR_o(r)}{dr} + [k^2 - U(r)]R_o(r) = 0\,, \tag{9.6.23}$$

where

$$U(r) = \frac{2\mu}{\hbar^2}V(r) = -\frac{2\mu V_o}{\hbar^2}e^{-\alpha r} = -U_o e^{-\alpha r}\,. \tag{9.6.24}$$

With the substitution $R_o(r) = u(r)/r$, we can rewrite Eq. (9.6.23) as

$$\frac{d^2u(r)}{dr^2} + [k^2 + U_o \exp(-\alpha r)]u(r) = 0 \tag{9.6.25}$$

If we introduce the change of variable in Eq. (9.6.25) by

$$z = \frac{2\sqrt{U_o}}{\alpha} e^{-\alpha r/2}, \tag{9.6.26}$$

and denote the solution of the resulting equation by $\Phi(z)$, we obtain

$$\frac{d^2\Phi(z)}{dz^2} + \frac{1}{z}\frac{d\Phi(z)}{dz} + \left(1 + \frac{\nu^2}{z^2}\right)\Phi(z) = 0, \tag{9.6.27}$$

where
$$\nu^2 = \frac{4k^2}{\alpha^2}. \tag{9.6.28}$$

This equation may be looked upon as Bessel equation of imaginary order ($\pm i\nu$) and admits the following two solutions[4]

$$J_{+i\nu}(z) = \cos(\nu \ln z) + \sum_{m=1}^{\infty} \frac{(-1)^m z^{2m} \cos(u_m - \nu \ln z)}{2^{2m} m!(1^2+\nu^2)^{1/2}(2^2+\nu^2)^{1/2}\cdots(m^2+\nu^2)^{1/2}}, \tag{9.6.29}$$

$$J_{-i\nu}(z) = \sin(\nu \ln z) - \sum_{m=1}^{\infty} \frac{(-1)^m z^{2m} \sin(u_m - \nu \ln z)}{2^{2m} m!(1^2+\nu^2)^{1/2}(2^2+\nu^2)^{1/2}\cdots(m^2+\nu^2)^{1/2}}, \tag{9.6.30}$$

where

$$u_m = \sum_{s=1}^{m} \tan^{-1}(\nu/s). \tag{9.6.31}$$

A complete solution of Eq. (9.6.25), which conforms to the requirements for $r \to 0$ and $r \to \infty$,

$$u(0) = 0 \quad \text{and} \quad u(r) \xrightarrow{r\to\infty} \frac{\sin(kr)}{k} + Ae^{ikr}, \tag{9.6.32}$$

may be obtained from the combination $\{J_{i\nu}(z)J_{-i\nu}(z_0) - J_{-i\nu}(z)J_{i\nu}(z_0)\}$ and is of the form

$$u(r) = \frac{e^{i\delta_0(k)}}{k}\left(\sin[kr + \delta_0(k)] + \sum_{m=1}^{\infty} C_m e^{-m\alpha r} \sin[u_m + kr + \delta_0(k)]\right), \tag{9.6.33}$$

where

$$\delta_0(k) = \theta - kb, \tag{9.6.34}$$

$$\theta = \tan^{-1}\left(J_{-i\nu}(z_0)/J_{i\nu}(z_0)\right), \tag{9.6.35}$$

$$z_0 \equiv z(r=0) = 2\sqrt{U_0}/\alpha, \tag{9.6.36}$$

$$b = \frac{2}{\alpha}\ln(2\sqrt{U_0}/\alpha), \tag{9.6.37}$$

$$C_m = \frac{(-4U_0/\alpha^2)^m}{2^m m![(1^2+\nu^2)(2^2+\nu^2)\cdots(m^2+\nu^2)]^{1/2}}. \tag{9.6.38}$$

[4]For details see R. Vyas and V. S. Mathur, Phys. Rev. C **18**, 1537 (1978).

We can identify $\delta_0(k)$ with the s-wave phase shift because

$$R_o(r) = \frac{u(r)}{r} \xrightarrow{r\to\infty} \frac{e^{i\delta_0(k)}}{kr}\sin[kr + \delta_0(k)]\,. \tag{9.6.39}$$

Thus the s-wave phase shift is given by

$$\tan\delta_0(k) = \frac{\tan\theta - \tan kb}{1+\tan\theta\tan kb}, \tag{9.6.40}$$

where

$$\tan\theta = \frac{\sin(\nu\,\ln z_0) - \sum\limits_{m=1}^{\infty} C_m \sin(u_m - \nu\,\ln z_0)}{\cos(\nu\,\ln z_0) + \sum\limits_{m=1}^{\infty} C_m \cos(u_m - \nu\,\ln z_0)}\,. \tag{9.6.41}$$

If the exponential potential corresponds to the realistic binding of the $n-p$ system (the deuteron with binding energy B=2.226 MeV), then z_0 is a zero of $J_\mu(z)$ with $\mu \equiv 2\sqrt{MB}/\hbar\alpha \approx 1.4$ [see Sec. 5.7]. Let us calculate the low energy scattering cross-section in this case. Once again, since scattering is dominated by the s-wave, we have

$$\begin{aligned}\sigma_{tot} &= \frac{4\pi}{k^2}\sin^2\delta_o = \frac{4\pi}{\alpha^2}\left[\frac{4}{\nu^2}\,\frac{\tan^2\theta}{1+\tan^2\theta}\right] \\ &= \frac{4\pi}{\alpha^2}\left[\frac{4}{\nu^2}\,\frac{J^2_{-i\nu}(z_0)}{J^2_{i\nu}(z_0)+J^2_{-i\nu}(z_0)}\right] \equiv X^2\,.\end{aligned} \tag{9.6.42}$$

Now in the limit $k \to 0$ (or $\nu \to 0$) we can neglect $J_{-i\nu}(z_0)$ as compared to $J_{i\nu}(z_0)$ [see Eqs. (9.6.29) and (9.6.30)]. It is to be kept in mind that z_0 is a zero of $J_\mu(z)$ with $\mu = 2\sqrt{MB}/\hbar\alpha$, and not of $J_{i\nu}(z)$ or $J_0(z)$ as $\nu \to 0$. Using this we find

$$\lim_{\nu\to 0} X \approx 2\frac{\sqrt{4\pi}}{\alpha}\lim_{\nu\to 0}\frac{J_{-i\nu}(z_0)}{\nu J_{i\nu}(z_0)}, \tag{9.6.43}$$

which is of the form 0/0. To evaluate this limit we use L'Hospital's rule. We differentiate both numerator and denominator with respect to ν using Eqs. (9.6.25) and (9.6.26) and finally put $\nu = 0$ and $z = z_0$. On simplify the result we get

$$\frac{\alpha}{\sqrt{4\pi}}\lim_{\nu\to 0} X = 2\,\ln(z_0)\left[1 - \frac{\sum\limits_{m=1}^{\infty}(-z_0^2/4)^m\frac{1}{(m!)^2}\sum\limits_{s=1}^{m}(1/s)}{(\ln\,z_0)J_0(z_0)}\right]. \tag{9.6.44}$$

Substituting this in Eq. (9.6.42), we find

$$\sigma_{tot} \xrightarrow{k\to 0} \frac{4\pi}{\alpha^2}(2\ln\,z_0)^2\left[1 - \frac{\sum\limits_{m=1}^{\infty}(-z_0^2/4)^m\frac{1}{(m!)^2}\sum\limits_{s=1}^{m}(1/s)}{(\ln\,z_0)J_0(z_0)}\right]^2, \tag{9.6.45}$$

where $z_0 \equiv (2\sqrt{U_0}/\alpha)$ is the first zero of $J_\mu(z)$ with $\mu = 2\sqrt{MB}/\hbar\alpha \approx 1.4$. Since the range of the exponential potential $r_0 \approx 1/\alpha$, this result is comparable to the value $4\pi\,r_o^2 = 4\pi/\alpha^2$ which we obtained for the hard sphere or the square-well potentials of range r_o in the low energy scattering limit.

9.7 Scattering due to Coulomb Potential

The Coulomb potential falls off as $1/r$ (not faster than $1/r$). So the method of partial waves is not quite applicable to Coulomb scattering. The problem, however, can be solved in an alternative way. By finding the asymptotic form of the Coulomb wave function $\psi_c(\boldsymbol{r})$ for the scattering state and identifying the form of the Coulomb scattering amplitude $f_c(\theta)$, one can determine the differential cross-section of Coulomb scattering. The use of parabolic coordinates instead of polar coordinates enables one to write the scattering equation for Coulomb potential in a convenient form.

Let us consider the general case in which an incident particle of charge Ze is scattered by a target of charge $Z'e$. The Schrödinger equation for this system, describing the relative motion of the two particles in the center-of-mass frame, may be written as

$$\left[-\frac{\hbar^2}{2\mu}\nabla^2 + \frac{ZZ'e^2}{4\pi\epsilon_0 r}\right]\psi_c(\boldsymbol{r}) = \frac{\hbar^2 k^2}{2\mu}\psi_c(\boldsymbol{r})\,, \tag{9.7.1}$$

where $\mu = \frac{m_1 m_2}{m_1+m_2}$ is the reduced mass, and $E = \frac{\hbar^2 k^2}{2\mu}$ is the energy associated with the relative motion in the CM frame. We rewrite Eq. (9.7.1) using the system of parabolic coordinates (ξ, η, φ) defined by [5]

$$\begin{aligned} \xi &= r(1-\cos\theta) = r - z\,, \\ \eta &= r(1+\cos\theta) = r + z\,, \\ \varphi &= \text{azimuthal angle } \varphi\,, \end{aligned} \tag{9.7.2a}$$

instead of polar coordinates (r, θ, φ). As in the case of spherical polar coordinates, the azimuth is just the angle between planes containing the point P and the z-axis and the XZ plane. If, in the former plane, we draw two parabolas passing through the point P with z-axis as the common axis and the origin as the focus, then the latus rectums of the two parabolas are the parabolic coordinates ξ and η of the point P. This is an orthogonal curvilinear system of coordinates and at any point the tangents to the curves along which ξ, η, φ increase are mutually orthogonal to each other.

In the parabolic system of coordinates the Laplacian ∇^2 is expressed as

$$\nabla^2 \equiv \frac{4}{\xi+\eta}\left[\frac{\partial}{\partial\xi}\left(\xi\frac{\partial}{\partial\xi}\right) + \frac{\partial}{\partial\eta}\left(\eta\frac{\partial}{\partial\eta}\right)\right] + \frac{1}{\xi\eta}\frac{\partial^2}{\partial\varphi^2}, \tag{9.7.2b}$$

so the scattering equation (9.7.1) takes the form

$$\frac{1}{\xi+\eta}\left\{\frac{\partial}{\partial\xi}\left(\xi\frac{\partial\psi_c}{\partial\xi}\right) + \frac{\partial}{\partial\eta}\left(\eta\frac{\partial\psi_c}{\partial\eta}\right)\right\} + \frac{k^2}{4}\psi_c = \frac{nk}{(\xi+\eta)}\psi_c\,, \tag{9.7.3}$$

where

$$n \equiv \frac{\mu\, ZZ'e^2}{\hbar^2 4\pi\epsilon_0 k}\,, \tag{9.7.4}$$

and the term $\frac{1}{\xi\eta}\frac{\partial^2\psi_c}{\partial\varphi^2}$ has been dropped because ψ_c does not depend on φ due to axial symmetry of the problem.

[5] For details regarding the parabolic coordinate system see Appendix 5A2.

Recalling that in the scattering solution in the asymptotic limit has an incident plane wave part $e^{ikz} = e^{ik(\xi-\eta)/2}$ and a spherically outgoing wave part e^{ikr}/r, we look for Coulomb scattering solution ψ_c in the form

$$\psi_c \equiv \psi_c(\boldsymbol{r}) = e^{[ik(\eta-\xi)/2]}f(\xi) = e^{ikz}f(\xi)\,. \tag{9.7.5}$$

Substituting this form into Eq. (9.7.3) we find that $f(\xi)$ satisfies the equation

$$\xi\frac{d^2f}{d\xi^2} + (1-ik\xi)\frac{df}{d\xi} - nkf(\xi) = 0\,. \tag{9.7.6}$$

With the substitution

$$\zeta = ik\xi \qquad \text{and} \quad G(\zeta) = f(\xi)\,, \tag{9.7.7}$$

Eq. (9.7.6) assumes the form

$$\zeta\frac{d^2G}{d\zeta^2} + (1-\zeta)\frac{dG}{d\zeta} + inG(\zeta) = 0\,. \tag{9.7.8}$$

This equation resembles the confluent hypergeometric equation which has the standard form

$$Z\frac{d^2F}{dZ^2} + (b-Z)\frac{dF}{dZ} - aF(Z) = 0\,. \tag{9.7.9}$$

The independent solutions of this equation are

$$W_1(a,b;Z) = \frac{\Gamma(b)}{\Gamma(b-a)}(-Z)^{-a}g(a,a-b+1;-Z) \tag{9.7.10}$$

and

$$W_2(a,b;Z) = \frac{\Gamma(b)}{\Gamma(a)}\exp(Z)Z^{a-b}g(1-a,b-a;Z) \tag{9.7.11}$$

and the asymptotic form of the function $g(\alpha,\beta;Z)$ is

$$g(\alpha,\beta;Z) \sim 1 + \frac{\alpha\beta}{Z} + \mathcal{O}(1/Z^2)\,. \tag{9.7.12}$$

The general solution of the confluent hypergeometric equation (9.7.9) is a linear combination of the W_1 and W_2

$$F(a,b;Z) = C_1W_1(a,b;Z) + C_2W_2(a,b;Z)\,. \tag{9.7.13}$$

Comparing Eq. (9.7.8) with the standard form (9.7.9), we find $a = -in$ and $b = 1$. Then the solution, which is regular at the origin, is the linear combination with ($C_1 = C_2 = C$):

$$G(\zeta) = f(\xi) = C\left[W_1(-in,\, 1,\, ik\xi) + W_2(-in,\, 1,\, ik\xi)\right]\,. \tag{9.7.14}$$

Its asymptotic behavior is

$$\begin{aligned} f(\xi) \xrightarrow{r\to\infty} C\Bigg[&\frac{\Gamma(1)}{\Gamma(1+in)}(-ik\xi)^{in}\left(1+\frac{n^2}{ik\xi}\right) \\ &+\frac{\Gamma(1)e^{ik\xi}}{\Gamma(-in)}(ik\xi)^{-in-1}\left(1+\frac{(1+in)^2}{ik\xi}\right)\Bigg]\,. \end{aligned} \tag{9.7.15}$$

Reverting back to r and z coordinates by writing $\xi = r - z$, $(k\xi)^{in} = e^{in\ln k(r-z)}$, $\pm i = \exp(\pm i\pi/2)$ and substituting the result in Eq. (9.7.5), we find the asymptotic behavior, correct up to order $1/r$,

$$\begin{aligned} \psi_c(\boldsymbol{r}) = e^{ikz}f(\xi) \sim \frac{Ce^{ikz}}{\Gamma(1+in)}\Bigg[&e^{n\pi/2}e^{in\ln k(r-z)}\left(1+\frac{n^2}{ik(r-z)}\right) \\ &+\frac{\Gamma(1+in)}{\Gamma(1-in)}(-in)e^{ik(r-z)}e^{n\pi/2}\frac{e^{-in\ln k(r-z)}}{ik(r-z)}\Bigg]\,. \end{aligned}$$

Expressing $r - z = r(1-\cos\theta) = 2r\sin^2(\theta/2)$ in terms of spherical polar coordinates, we can write the asymptotic form of the scattering solution as

$$\psi_c(\boldsymbol{r}) \sim \frac{Ce^{n\pi/2}}{\Gamma(1+in)}\left[e^{i(kz+n\ \ln 2kr\sin^2(\theta/2)}\left(1+\frac{n^2}{i2kr\sin^2(\theta/2)}\right)\right.$$
$$\left. -\frac{n}{(2k\sin^2\theta/2)}\frac{\Gamma(1+in)}{\Gamma(1-in)}e^{-in\ \ln\sin^2(\theta/2)}\frac{e^{i(kr-n\ \ln 2kr)}}{r}\right]. \tag{9.7.16}$$

We are thus able to express the asymptotic behavior of the Coulomb wave function as a distorted plane wave, represented by the first term, plus Coulomb scattering amplitude multiplied by distorted outgoing spherical wave, represented by the second term. From this asymptotic form, the scattering amplitude $f_c(\theta)$ for the Coulomb potential is easily identified to be the coefficient of the spherically expanding wave $e^{i(kr-n\ \ln 2kr)}/r$

$$f_c(\theta) \equiv \frac{n}{2k\sin^2(\theta/2)}e^{-in\ \ln\sin^2(\theta/2)+2i\eta_0+i\pi}, \tag{9.7.17}$$

where
$$\exp(2i\eta_0) \equiv \frac{\Gamma(1+in)}{\Gamma(1-in)}. \tag{9.7.18}$$

Using this expression for the scattering amplitude, we find the Coulomb scattering cross-section $\sigma_c(\theta)$ is given by

$$\sigma_c(\theta) = |f_c(\theta)|^2 = \frac{n^2}{4k^2\sin^4(\theta/2)} = \left(\frac{ZZ'e^2}{16\pi\epsilon_0 E}\right)^2 \text{cosec}^4(\theta/2), \tag{9.7.19}$$

where $E = \hbar^2k^2/2\mu$ is the center-of-mass energy of the projectile. This is the classical Rutherford's formula. Thus the quantum mechanical result for the differential cross-section for Coulomb scattering agrees with the classical result. However, the angle-dependent phase factor $e^{-in\ln\sin^2(\theta/2)}$ can lead to nonclassical features in the scattering of identical particles as we shall see next.

Pure Coulomb Scattering of Identical Particles

In pure Coulomb scattering of identical particles, say, proton on proton $(p-p)$ scattering when the nuclear part of the interaction can be ignored, the overall wave function, which is the product of space and spin parts $\Psi_{tot} = \Psi_{space} \times \Psi_{spin}$ must be anti-symmetric. (For scattering of identical Bosons the overall wave function must be symmetric under an exchange of particle coordinates.) The spin part of the wave function for the two particles (projectile and target) can be anti-symmetric (spin singlet χ_0^0) or symmetric (spin triplet χ_1^m). This means that the space function is symmetric under an exchange of particle coordinates if the particles are in spin singlet state and anti-symmetric if the particles are in spin triplet state.

Now the space part of the wave function can be written as the product of center-of-mass wave function and the wave function for relative motion

$$\Psi_{space}(\boldsymbol{r}_1, \boldsymbol{r}_2) = \psi(\boldsymbol{r})\Phi_{cm}(\boldsymbol{R}), \tag{9.7.20}$$

where the center-of-mass coordinate $\boldsymbol{R}$ and relative coordinate $\boldsymbol{r}$ are given by

$$\boldsymbol{R} = \frac{1}{2}(\boldsymbol{r}_1 + {}_2), \qquad \boldsymbol{r} = \boldsymbol{r}_1 - \boldsymbol{r}_2. \tag{9.7.21}$$

Under an exchange of space coordinates of the particles

$$\boldsymbol{R} \to \boldsymbol{R}, \qquad \Phi_{cm}(\boldsymbol{R}) \to \Phi_{cm}(\boldsymbol{R}), \tag{9.7.22a}$$
$$\boldsymbol{r} \to -\boldsymbol{r}, \qquad \psi(\boldsymbol{r}) \to \psi(-\boldsymbol{r}). \tag{9.7.22b}$$

Thus the center-of-mass wave function Φ_{cm} is always symmetric under an exchange of particle coordinates. It follows that for the total wave function to be anti-symmetric under an exchange of particle coordinates, $\psi(\boldsymbol{r})$ must be symmetric ($\psi(-\boldsymbol{r}) = \psi(\boldsymbol{r})$) under exchange when the particles are in spin singlet state and anti-symmetric ($\psi(-\boldsymbol{r}) = -\psi(\boldsymbol{r})$) when the particles are in spin triplet state. Recalling that $\boldsymbol{r} \to -\boldsymbol{r}$ means $r, \theta, \varphi) \to (r, \pi - \theta, \pi + \varphi)$ and the asymptotic form of the unsymmetrized wave function is

$$\psi(\boldsymbol{r}) \sim e^{ikz} + f(\theta)\frac{e^{ikr}}{r}\,, \tag{9.7.23}$$

we can write the symmetric and antisymetric space wave functions as

$$\psi_{sym} \sim \left[e^{ikz} + e^{-ikz}\right] + [f(\theta) + f(\pi - \theta)]\frac{e^{ikr}}{r}\,, \tag{9.7.24a}$$

$$\psi_{anti} \sim \left[e^{ikz} - e^{-ikz}\right] + [f(\theta) - f(\pi - \theta)]\frac{e^{ikr}}{r}\,. \tag{9.7.24b}$$

Thus we can define symmetric and anti-symmetric scattering amplitudes as

$$f_{sym}(\theta) = f(\theta) + f(\pi - \theta)\,, \tag{9.7.25a}$$

$$f_{anti}(\theta) = f(\theta) - f(\pi - \theta)\,. \tag{9.7.25b}$$

Hence the Coulomb scattering cross-sections for proton-proton scattering in the spin singlet and triplet states are, respectively,

$$\sigma_{singlet} = |f_c(\theta) + f_c(\pi - \theta)|^2\,, \tag{9.7.26a}$$

$$\sigma_{triplet} = |f_c(\theta) + f_c(\pi - \theta)|^2\,. \tag{9.7.26b}$$

If the incident proton and the target proton are unpolarized, then the probability of their

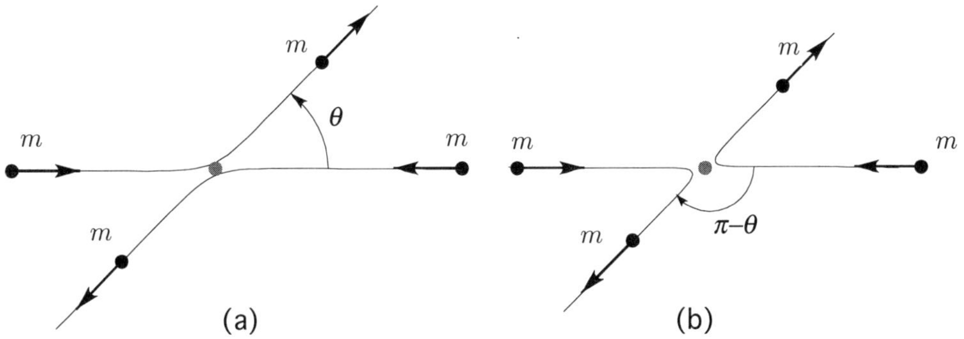

FIGURE 9.8
When the incident particle and the target are identical, it is not possible to distinguish between the two situations when the scattering angle is θ and $\pi - \theta$ in the centre-of-mass frame of reference.

being in the spin singlet ($S = 0$) state is 1/4 and that of their being in the spin triplet ($S = 1$) state is 3/4. Hence, in this case,

$$\begin{aligned}\sigma_{un}(\theta) &= \frac{3}{4}|f_c(\theta) - f_c(\pi - \theta)|^2 + \frac{1}{4}|f_c(\theta) + f_c(\pi - \theta)|^2 \\ &= |f_c(\theta)|^2 + |f_c(\pi - \theta)|^2 - \frac{1}{2}\left[f_c(\theta)f_c^*(\pi - \theta) + f_c^*(\theta)f_c(\pi - \theta)\right]\,.\end{aligned} \tag{9.7.27}$$

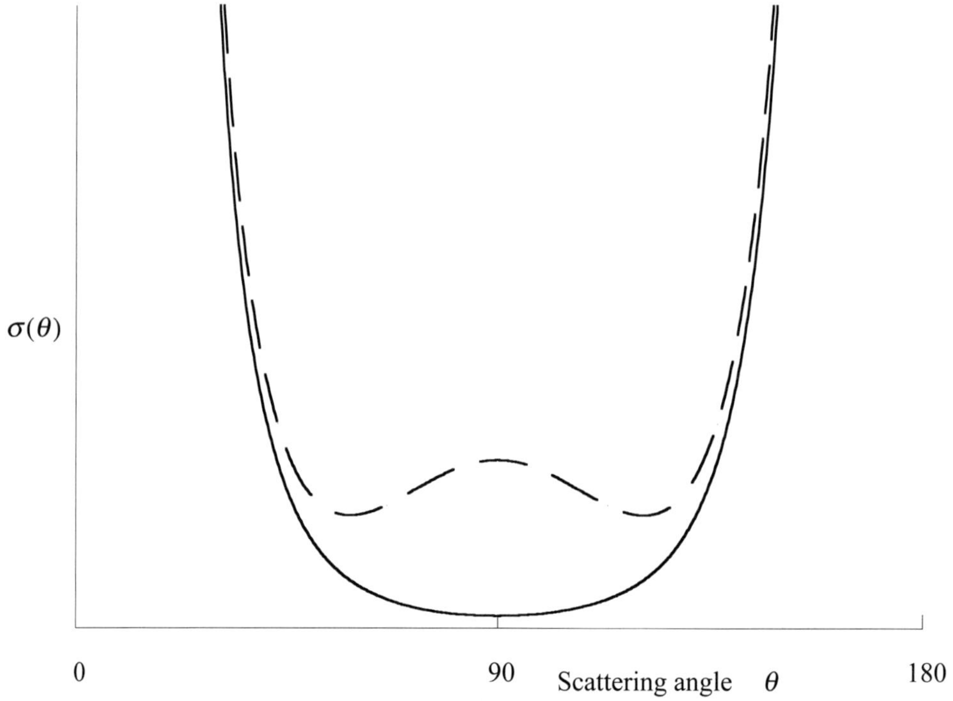

FIGURE 9.9
Pure Mott scattering cross-section (continuous curve) and measured (qualitative) $p-p$ scattering cross-section (dashed curve) in the center-of-mass frame.

Substituting the expression for $f_c(\theta)$ given by Eq. (9.7.17) we get, after simplification

$$\sigma(\theta) = \left(\frac{e^2}{16\pi\epsilon_0 E}\right)^2 \left[\frac{1}{\sin^4(\theta/2)} + \frac{1}{\cos^4(\theta/2)} - \frac{\cos[n\,\ln(\tan^2\theta/2)]}{\sin^2(\theta/2)\cos^2(\theta/2)}\right]. \tag{9.7.28}$$

This is the *Mott scattering formula.* We can see that Mott scattering is symmetric about $\theta = \pi/2$ [Fig. 9.9]. Thus the nonclassical phase shift of the scattering amplitude shows up in the interference term. The observed $p-p$ scattering cross-section deviates considerably from Mott scattering cross-section, particularly around $\theta = \pi/2$ because the actual $p-p$ interaction consists of Coulomb interaction as well as short range nuclear interaction.

9.8 The Integral Form of Scattering Equation

The Schrödinger equation for scattering $E > 0$

$$(\nabla^2 + k^2)\psi(\boldsymbol{r}) = U(\boldsymbol{r})\psi(\boldsymbol{r}), \tag{9.8.1a}$$

where

$$k^2 = \frac{2\mu E}{\hbar^2}, \quad \text{and} \quad U(\boldsymbol{r}) = \frac{2\mu V(\boldsymbol{r})}{\hbar^2} \tag{9.8.1b}$$

can be written in an integral form

$$\psi(\boldsymbol{r}) = e^{i\boldsymbol{k}\cdot\boldsymbol{r}} + \int G_k(\boldsymbol{r},\boldsymbol{r}')U(\boldsymbol{r}')\psi(\boldsymbol{r}')d^3r' . \tag{9.8.2}$$

Here the first term $e^{i\boldsymbol{k}\cdot\boldsymbol{r}}$ is a solution of the Schrödinger equation for a free particle

$$\left(\nabla^2 + k^2\right)\psi(\boldsymbol{r}) = 0 , \tag{9.8.3}$$

and the function $G_k(\boldsymbol{r},\boldsymbol{r}')$ is the Green's function, satisfying the equation

$$(\nabla^2 + k^2)G_k(\boldsymbol{r},\boldsymbol{r}') = \delta^3(\boldsymbol{r}-\boldsymbol{r}') . \tag{9.8.4}$$

Equation (9.8.2) is called an integral equation because the unknown function $\psi(\boldsymbol{r})$ also appears in the integral on the right-hand side.

It is straightforward to check that when Eq. (9.8.2) is substituted on the left-hand side of Eq. (9.8.1) and Eqs. (9.8.3) and (9.8.4) are used we recover the right-hand side of Eq. (9.8.1). Equation (9.8.2) does not yet represent a solution to the scattering problem because the energy E only specifies the magnitude of $\boldsymbol{k}$ and the Green's function could be any one of infinitely many solutions of Eq. (9.8.4). This arbitrariness is removed once we impose the appropriate boundary conditions on the wave function.

In the scattering problem, the first term $e^{i\boldsymbol{k}\cdot\boldsymbol{r}}$, which is a solution of the Schrödinger equation for a free particle, may be taken to represent a plane corresponding to an incident particle of energy $E = \hbar^2k^2/2\mu$ and momentum $\hbar\boldsymbol{k}$, while the second term, which depends on the potential $V(\boldsymbol{r})$, must represent, asymptotically, an outgoing wave corresponding to scattered particle. Thus the problem of solving the scattering equation has been reduced to the problem of finding the appropriate Green's function, which leads to the correct asymptotic behavior (outgoing spherical wave) for the wave function.

Expression for the Green's Function $G_k(\boldsymbol{r},\boldsymbol{r}')$

Introducing the Fourier transform $g(\boldsymbol{k}',\boldsymbol{r}')$ of Green's function $G_k(\boldsymbol{r},\boldsymbol{r}')$ by

$$G_k(\boldsymbol{r},\boldsymbol{r}') = \frac{1}{(2\pi)^{3/2}}\int d^3k' g(\boldsymbol{k}',\boldsymbol{r}')e^{i\boldsymbol{k}'\cdot\boldsymbol{r}} , \tag{9.8.5}$$

and the following representation of the three-dimensional delta function [Chapter 2, Eq. (2.9.12)]

$$\delta^3(\boldsymbol{r}-\boldsymbol{r}') = \frac{1}{(2\pi)^3}\int d^3k'\, e^{i\boldsymbol{k}'\cdot(\boldsymbol{r}-\boldsymbol{r}')} , \tag{9.8.6}$$

in Eq. (9.8.4), we find

$$g(\boldsymbol{k}',\boldsymbol{r}') = \frac{1}{(2\pi)^{3/2}}\frac{e^{-i\boldsymbol{k}'\cdot\boldsymbol{r}'}}{k^2 - k'^2} . \tag{9.8.7}$$

Substituting this into Eq. (9.8.5), we get

$$G_k(\boldsymbol{r},\boldsymbol{r}') = \frac{1}{(2\pi)^3}\int \frac{\exp[i\boldsymbol{k}'\cdot(\boldsymbol{r}-\boldsymbol{r}')]}{k^2-k'^2}d^3k' . \tag{9.8.8}$$

Taking the direction of unit vector $(\boldsymbol{r}-\boldsymbol{r}') = |\boldsymbol{r}-\boldsymbol{r}'|\boldsymbol{n} \equiv \rho\boldsymbol{n}$ to be the direction of polar axis and the direction of $\boldsymbol{k}'$ to be given by the angles θ', φ', we can carry out the angular integration

$$G_k(\boldsymbol{r},\boldsymbol{r}') = \frac{2\pi}{(2\pi)^3}\int_0^\infty 2\pi k'^2 dk' \int_0^\pi \sin\theta' d\theta' \frac{e^{ik'\rho\cos\theta'}}{k^2-k'^2},$$
$$= \frac{1}{4\pi^2 i\rho}\left[\int_0^\infty \frac{k'dk'\, e^{ik'\rho}}{k^2-k'^2} - \int_0^\infty \frac{k'dk'\, e^{-ik'\rho}}{k^2-k'^2}\right].$$

Changing the variable k' to $-k'$ in the second integral, we may rewrite this equation as

$$G_k(\boldsymbol{r},\boldsymbol{r}') = \frac{1}{4\pi^2 i\rho}\int_{-\infty}^{\infty} \frac{k'dk'\, e^{ik'\rho}}{k^2-k'^2} \tag{9.8.9}$$

where $\rho \equiv |\boldsymbol{r}-\boldsymbol{r}'|$. This integral has singularities at $k' = \pm k$ and is not defined as an ordinary integral. However, by treating it as the limiting value of certain integrals defined in the complex k' plane, we can find Green's functions appropriate for different boundary conditions satisfied by the wave function. Here we consider two Green's functions, $G_k^{(+)}(\boldsymbol{r},\boldsymbol{r}')$ and $G_k^{(-)}(\boldsymbol{r},\boldsymbol{r}')$, defined by

$$G_k^{(+)}(\boldsymbol{r},\boldsymbol{r}') = \lim_{\epsilon\to 0}\frac{1}{(2\pi)^3}\int \frac{d^3k'\, e^{i\boldsymbol{k}'\cdot(\boldsymbol{r}-\boldsymbol{r}')}}{k^2-k'^2+i\epsilon} = \lim_{\epsilon\to 0}\frac{1}{4\pi^2 i\rho}\int_{-\infty}^{\infty}\frac{k'dk'\, e^{ik'\rho}}{k^2-k'^2+i\epsilon} \tag{9.8.10}$$

and $$G_k^{(-)}(\boldsymbol{r},\boldsymbol{r}') = \lim_{\epsilon\to 0}\frac{1}{(2\pi)^3}\int \frac{d^3k'\, e^{i\boldsymbol{k}'\cdot(\boldsymbol{r}-\boldsymbol{r}'}}{k^2-k'^2-i\epsilon} = \lim_{\epsilon\to 0}\frac{1}{4\pi^2 i\rho}\int_{-\infty}^{\infty}\frac{k'dk'\, e^{ik'\rho}}{k^2-k'^2-i\epsilon}. \tag{9.8.11}$$

It turns out that these Green's functions lead to very different asymptotic behavior for the wave function.

The integrals over k' in $G_k^{(+)}(\boldsymbol{r},\boldsymbol{r}')$ and $G_k^{(-)}(\boldsymbol{r},\boldsymbol{r}')$ can be evaluated by the method of contour integration [see Appendix 9A1]. The integrand has simple poles at $k' = \pm\sqrt{k^2+i\epsilon} \approx \pm(k+i\epsilon/2k)$. Choosing the path of integration to run along the real axis from $-\infty$ to $+\infty$ and along a semi-circle of large radius in the upper half plane ($\rho > 0$) as shown in Fig. 9.10, we find that only the residue at the simple pole $k' = k + i\epsilon/2k$ in the upper half of the complex-k' plane contributes to the integral leading to

$$G_k^{(+)}(\boldsymbol{r},\boldsymbol{r}') = \lim_{\epsilon\to 0}\frac{1}{4\pi^2 i\rho}\int_{-\infty}^{\infty}\frac{k'dk'\, e^{ik'\rho}}{(k+\frac{i\epsilon}{2k}+k')(k+\frac{i\epsilon}{2k}-k')}$$
$$= \lim_{\epsilon\to 0}\frac{1}{4\pi^2 i\rho}\,2\pi i \times \left[-\frac{e^{ik\rho-\epsilon\rho/2k}}{2}\right] = -\frac{1}{4\pi}\frac{e^{ik\rho}}{\rho},$$

or $$G_k^{(+)}(\boldsymbol{r},\boldsymbol{r}') = -\frac{1}{4\pi}\frac{e^{ik|\boldsymbol{r}-\boldsymbol{r}'|}}{|\boldsymbol{r}-\boldsymbol{r}'|}. \tag{9.8.12}$$

Similarly, the integrand in Eq. (9.8.11) has simple poles at $k' = k - i\epsilon/2k$ and $k' = -k + i\epsilon/2k$. Choosing the path of integration as in Fig 9.11, we find only the residue from the pole at $k' = -k + i\epsilon/2k$ in the upper half of the complex-k' plane contributes, leading

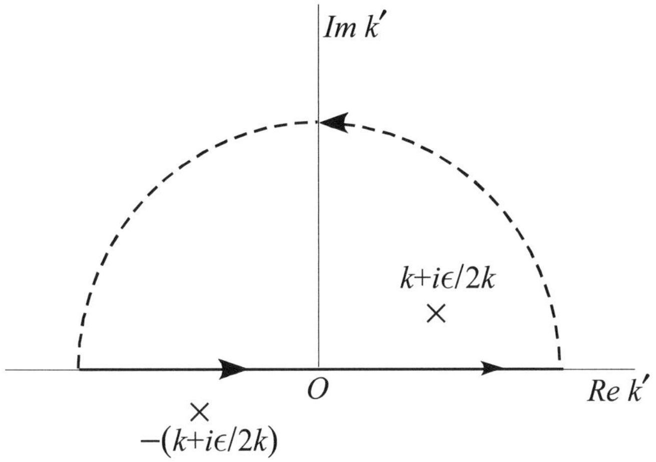

FIGURE 9.10
For the calculation of $G_k^{(+)}(\boldsymbol{r},\boldsymbol{r}')$ the pole of the integrand at $k' = k + i\frac{\epsilon}{2k}$ lies in the upper half of the complex k'-plane while that at $k' = -k - i\frac{\epsilon}{2k}$ lies in the lower half.

us to

$$G_k^{(-)}(\boldsymbol{r},\boldsymbol{r}') = \lim_{\epsilon\to 0}\frac{1}{4\pi^2 i\rho}\int_{-\infty}^{\infty}\frac{k'dk'\,e^{ik'\rho}}{(k-\frac{i\epsilon}{2k}+k')(k-\frac{i\epsilon}{2k}-k')}$$

$$= \lim_{\epsilon\to 0}\frac{1}{4\pi^2 i\rho}\,2\pi\, i\times\left[-\frac{e^{-ik\rho-\epsilon\rho/2k}}{2}\right] = -\frac{1}{4\pi}\frac{e^{-ik\rho}}{\rho},$$

or

$$G_k^{(-)}(\boldsymbol{r},\boldsymbol{r}') = -\frac{1}{4\pi}\frac{e^{-ik|\boldsymbol{r}-\boldsymbol{r}'|}}{|\boldsymbol{r}-\boldsymbol{r}'|}. \tag{9.8.13}$$

We shall see that $G_k^{(+)}$ leads to the scattering wave function $\psi^{(+)}(\boldsymbol{r})$, which behaves asymptotically as a plane wave plus an outgoing spherical wave. On the other hand, $G_k^{(-)}$ leads to the scattering function $\psi^{(-)}(\boldsymbol{r})$ which behaves asymptotically as a plane wave plus an ingoing spherical wave. It is straightforward to check that while both Green's functions satisfy Eq. (9.8.4), they solve very different physical problems. Green's functions $G_k^{(+)}$ and $G_k^{(+)}$ are also referred to, respectively, as retarded and advanced Green's functions.

9.8.1 Scattering Amplitude

With $G^{(+)}(\boldsymbol{r},\boldsymbol{r}')$ given by Eq. (9.8.12) as the Green's function, the integral equation for the scattering wave function reads

$$\psi_k^{(+)}(\boldsymbol{r}) = e^{i\,\boldsymbol{k}\cdot\boldsymbol{r}} + \int G_k{}^{(+)}(\boldsymbol{r},\boldsymbol{r}')U(\boldsymbol{r}')\psi_k^{(+)}(\boldsymbol{r}')d^3r'$$

$$= e^{i\,\boldsymbol{k}\cdot\boldsymbol{r}} - \int\frac{e^{i\,k|\boldsymbol{r}-\boldsymbol{r}'|}}{4\pi|\boldsymbol{r}-\boldsymbol{r}'|}U(\boldsymbol{r}')\psi_k^{(+)}(\boldsymbol{r}')d^3r'. \tag{9.8.14}$$

The contribution to the integral comes from the values or r' where the potential is nonzero. This means that if the potential vanishes sufficiently rapidly, then in the asymptotic limit

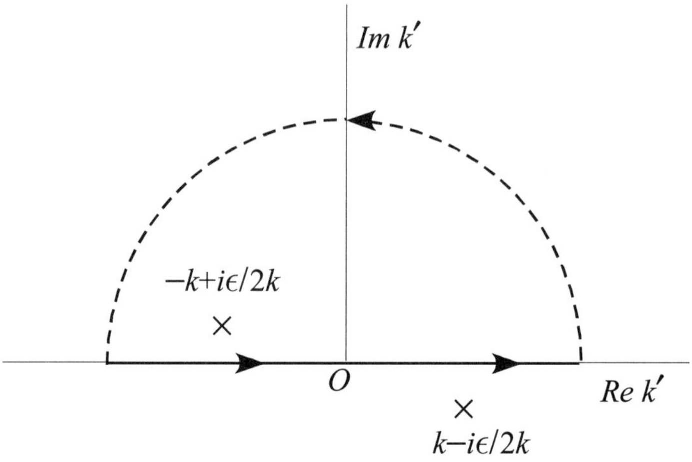

FIGURE 9.11
For the calculation of $G_k^{(-)}(\boldsymbol{r}, \boldsymbol{r}')$ the pole of of the integrand at $k' = -k + i\frac{\epsilon}{2k}$ lies in the upper half of the k'-plane while that at $k' = k - i\frac{\epsilon}{2k}$ lies in the lower half.

$r \to \infty$ we can approximate the exponent in the integrand as

$$k|\boldsymbol{r} - \boldsymbol{r}'| \approx kr - k\boldsymbol{n} \cdot \boldsymbol{r}' \equiv kr - \boldsymbol{k}' \cdot \boldsymbol{r}' , \quad \boldsymbol{n} = \boldsymbol{r}/r , \tag{9.8.15}$$

where $\boldsymbol{n}$ is a unit vector in the direction of $\boldsymbol{r}$ (the direction of scattering) or the final momentum $\boldsymbol{k}' = k\boldsymbol{n}$. In the denominator we can neglect r' compared to r. The reason for keeping the second term in Eq. (9.8.15) is that it can lead to a significant variation of the phase $e^{-i\boldsymbol{k}' \cdot \boldsymbol{r}'}$ as r' varies over the region of nonzero potential. Hence $G^{(+)}$ leads to the asymptotic expression

$$\psi_k^{(+)}(\boldsymbol{r}) \sim e^{i\boldsymbol{k} \cdot \boldsymbol{r}} - \frac{1}{4\pi} \frac{2\mu}{\hbar^2} \frac{e^{ikr}}{r} \int e^{-i\boldsymbol{k}' \cdot \boldsymbol{r}'} V(r') \psi_k^{(+)}(\boldsymbol{r}') d^3 r' . \tag{9.8.16}$$

This has precisely the form needed for describing the scattering of an incident plane wave $e^{i\boldsymbol{k} \cdot \boldsymbol{r}}$ [corresponding to a particle with center-of-mass momentum $\hbar\boldsymbol{k}$] by a potential $V(r)$ producing a spherically outgoing wave (corresponding to the scattered particle). Comparing this to the asymptotic form of the scattering solution

$$\psi_k^{(+)}(\boldsymbol{r}) \sim e^{i\boldsymbol{k} \cdot \boldsymbol{r}} + f(\theta) \frac{e^{ikr}}{r} , \tag{9.8.17}$$

we find that the scattering amplitude $f(\theta)$ is given by

$$f(\theta) = -\frac{1}{4\pi} \frac{2\mu}{\hbar^2} \int e^{i\,\boldsymbol{k}' \cdot \boldsymbol{r}'} V(r') \psi_k^{(+)}(\boldsymbol{r}') d^3 r' . \tag{9.8.18}$$

This expression for the scattering amplitude is exact but only formal since we must still solve the integral equation (9.8.14) for $\psi_k^{(+)}(\boldsymbol{r})$ appearing in the integrand. However, it can form the basis for obtaining approximate expressions for the scattering amplitude. Some of these will be discussed in Sec. 9.10.

9.9 Lippmann-Schwinger Equation and the Transition Operator

The integral scattering equation (9.8.2)

$$\psi_k^{(+)}(\boldsymbol{r}) = e^{i\boldsymbol{k}\cdot\boldsymbol{r}} + \int d^3r'\, G_k^{(+)}(\boldsymbol{r},\boldsymbol{r}')U(\boldsymbol{r}')\psi_k^{(+)}(\boldsymbol{r}'), \tag{9.8.2*}$$

where $$G_k^{(+)}(\boldsymbol{r},\boldsymbol{r}') = \frac{1}{(2\pi)^3}\int d^3k'\, \frac{e^{i\boldsymbol{k}'\cdot(\boldsymbol{r}-\boldsymbol{r}')}}{-k'^2+k^2+i\epsilon}, \tag{9.8.12*}$$

may be written, in Dirac notation, as[6]

$$|\psi_k^{(+)}\rangle = |k\rangle + \hat{g}_0(k^2+i\epsilon)\hat{v}|\psi_k^{(+)}\rangle, \tag{9.9.1}$$

where $$\hat{g}_0(z) = (z-\hat{h}_0)^{-1}, \tag{9.9.2}$$

is called the resolvent operator for the free Hamiltonian $\hat{h}_0 \equiv \hat{p}^2/2\mu$ (For convenience, we will use the units $\hbar = 2\mu = 1$ throughout this section.) To see this we first note that the Green's function $G_k^{(+)}(\boldsymbol{r},\boldsymbol{r}')$ is only the matrix element of the resolvent operator $\hat{g}_0(k^2+i\epsilon)$ between the coordinate states $\langle\boldsymbol{r}|$ and $|\boldsymbol{r}'\rangle$

$$\langle\boldsymbol{r}|\hat{g}_0(k^2+i\epsilon)|\boldsymbol{r}'\rangle = \int\int \langle\boldsymbol{r}|\boldsymbol{k}'\rangle d^3k'\langle\boldsymbol{k}'|(k^2+i\epsilon-\hat{h}_0)^{-1}|\boldsymbol{k}''\rangle d^3k''\langle\boldsymbol{k}''|\boldsymbol{r}'\rangle,$$

where we have introduced unit operators in accordance with the completeness condition for momentum eigenfunctions, before and after the resolvent operator. On substituting for the coordinate representatives of momentum states and simplifying, we obtain

$$\langle\boldsymbol{r}|\hat{g}_0(k^2+i\epsilon)|\boldsymbol{r}'\rangle = \frac{1}{(2\pi)^3}\int \frac{e^{i\boldsymbol{k}\cdot(\boldsymbol{r}-\boldsymbol{r}')}}{k^2+i\epsilon-k'^2}d^3k' = G_k^{(+)}(\boldsymbol{r},\boldsymbol{r}'). \tag{9.9.3}$$

Now, we can rewrite Eq. (9.8.2) as[7]

$$\psi_k^{(+)}(\boldsymbol{r}) = e^{i\boldsymbol{k}\cdot\boldsymbol{r}} + \iint d^3r'd^3r''\, G_k^{(+)}(\boldsymbol{r},\boldsymbol{r}')v(\boldsymbol{r}')\delta^3(\boldsymbol{r}'-\boldsymbol{r}'')\psi_k^{(+)}(\boldsymbol{r}'')$$

or

$$\begin{aligned}\left\langle \boldsymbol{r}\,|\,\psi_k^{(+)}\right\rangle &= \langle\boldsymbol{r}|\boldsymbol{k}\rangle + \iint \langle\boldsymbol{r}|\hat{g}_0(k^2+i\epsilon)|\boldsymbol{r}'\rangle d^3r'\langle\boldsymbol{r}'|\hat{v}|\boldsymbol{r}''\rangle d^3r''\langle\boldsymbol{r}''|\psi_k^{(+)}\rangle \\ &= \langle\boldsymbol{r}|\boldsymbol{k}\rangle + \langle\boldsymbol{r}|\hat{g}_0(k^2+i\epsilon)\hat{v}|\psi_k^{(+)}\rangle,\end{aligned} \tag{9.9.4}$$

where we have removed unit operators between $\hat{g}_0$ and $\hat{v}$ and between $\hat{v}$ and $|\psi_k^{(+)}\rangle$. Since this holds for any state $|\boldsymbol{r}\rangle$, this implies

$$\left|\psi_k^{(+)}\right\rangle = |\boldsymbol{k}\rangle + \hat{g}_o(k^2+i\epsilon)\hat{v}\left|\psi_k^{(+)}\right\rangle. \tag{9.9.1*}$$

[6] We use small letters with a caret - $\hat{v}$, $\hat{g}_o$, $\hat{g}$, $\hat{h}_0$, $\hat{h}$, etc. to denote two-body operators in a space spanned by two-body states. The capital letters will be used (Chapter 11) to denote operators in space spanned by three-body states.

[7] For a local two-body central potential $\hat{v}$,

$$\langle\boldsymbol{r}'|\,\hat{v}\,|\boldsymbol{r}''\rangle = v(\boldsymbol{r}')\delta^3(\boldsymbol{r}'-\boldsymbol{r}'') = v(r')\delta^3(\boldsymbol{r}'-\boldsymbol{r}'').$$

The scattering amplitude $f(\theta)$, given in the integral form by Eq. (9.8.10), may also be written in the concise Dirac notation as

$$f(\theta) = -\frac{1}{4\pi}\iint e^{-i\boldsymbol{k}'\cdot\boldsymbol{r}'}v(\boldsymbol{r}')\delta^3(\boldsymbol{r}'-\boldsymbol{r})\psi_k(\boldsymbol{r}'')d^3r'd^3r''$$

$$= -\frac{(2\pi)^3}{4\pi}\iint \langle\boldsymbol{k}'|\,\boldsymbol{r}'\rangle\, d^3r'\,\langle\boldsymbol{r}'|\,\hat{v}\,|\boldsymbol{r}''\rangle\, d^3r''\left\langle \boldsymbol{r}''|\,\psi_k^{(+)}\right\rangle$$

or

$$f(\theta) = -2\pi^2\langle\boldsymbol{k}'|\hat{v}|\psi_k^{(+)}\rangle, \tag{9.9.5}$$

where we have removed the unit operators between $\langle\boldsymbol{k}'|$ and $\hat{v}$ and between $\hat{v}$ and $|\psi_k^{(+)}\rangle$.

The generalization of the scattering equation (9.9.1) to off-shell energies (i.e., replacement of $(k^2+i\epsilon)$ to any complex energy $z \equiv s+i\epsilon$, where s is not necessarily equal to k^2), results in a very important equation in the formal theory of scattering, called the Lippmann-Schwinger equation:

$$\left|\psi_s^{(+)}\right\rangle = |\boldsymbol{k}\rangle + \hat{g}_0(z)\hat{v}\left|\psi_s^{(+)}\right\rangle. \tag{9.9.6}$$

This equation can be written in any representation (coordinate or momentum) and it can be shown that, for any potential $\hat{v}$, this equation admits a unique solution for all complex values of z, except at the bound state poles (i.e., at $z = -E_n^B$, which are the energies of the bound states of the two-body system), and at the right-hand cut (i.e., at $z=$ real positive energies).

The Transition Operator

The Lippmann-Schwinger equation can also be written in the operator form by introducing the *transition operator* $\hat{t}(z)$ by the equation

$$\hat{t}(z)|\boldsymbol{k}\rangle = \hat{v}|\psi_k^{(+)}\rangle. \tag{9.9.7}$$

To see this, we pre-multiply both sides of Eq. (9.9.6) by $\hat{v}$, then we get

$$\hat{v}|\psi_s^{(+)}\rangle = \hat{v}|\boldsymbol{k}\rangle + \hat{v}\hat{g}_0(z)\hat{v}|\psi_s^{(+)}\rangle$$

or

$$\hat{t}(z)|\boldsymbol{k}\rangle = \hat{v}|\boldsymbol{k}\rangle + \hat{v}\hat{g}_0(z)\hat{t}(z)|\boldsymbol{k}\rangle.$$

Since this holds for any state $|k\rangle$, we obtain the operator equality

$$\hat{t}(z) = \hat{v} + \hat{v}\hat{g}_0(z)\,\hat{t}(z), \tag{9.9.8}$$

which is the operator version of the Lippmann-Schwinger equation. The following operator identities may easily be established

(i) $$\hat{g}(z) = \hat{g}_0(z) + \hat{g}_0(z)\hat{v}\hat{g}(z), \tag{9.9.9}$$

(ii) $$\hat{g}(z) = \hat{g}_0(z) + \hat{g}(z)\hat{v}\hat{g}_0(z), \tag{9.9.10}$$

(iii) $$\hat{g}(z)\hat{v} = \hat{t}(z)\hat{g}_0(z), \tag{9.9.11}$$

(iv) $$\hat{v}\hat{g}(z) = \hat{g}_0(z)\hat{t}(z) \tag{9.9.12}$$

where

$$\hat{g}(z) \equiv (z-\hat{h})^{-1} \equiv (z-\hat{h}_0-\hat{v})^{-1}. \tag{9.9.13}$$

To prove the first identity, we multiply both sides of Eq. (9.9.2) from the right by the unit operator $\hat{1} = (z-\hat{h}_0-\hat{v})\hat{g}(z)$ and then rearrange we get the identity (9.9.9).

If we pre-multiply both sides of Eq. (9.9.2) by the unit operator $\hat{1} = \hat{g}(z)(z-\hat{h}_0-\hat{v})$, we get $\hat{g}_0(z) = \hat{g}(z)(z-\hat{h}_0-\hat{v})(z-\hat{h}_0)^{-1} = \hat{g}(z) - \hat{g}(z)\hat{v}\hat{g}_0(z)$ which, on rearrangement, leads to Eq. (9.9.10).

To establish the identity (9.9.11) we write the Lippmann-Schwinger equation (9.9.8) as

$$\begin{aligned}\hat{v} &= [\hat{1} - \hat{v}\hat{g}_0(z)]\hat{t}(z) = [\hat{1} - \{(z - \hat{h}_0) - (z - \hat{h}_0 - \hat{v})\}\hat{g}_0(z)]\hat{t}(z) \\ &= (z - \hat{h}_0 - \hat{v})\hat{g}_0(z)\hat{t}(z)\,.\end{aligned}$$

Multiply both sides from the left by $\hat{g}(z)$ to get the identity (9.9.11). Also, from Eqs. (9.9.10) and (9.9.11), we have $\hat{g}(z) = \hat{g}_0(z) + \hat{g}_0(z)\hat{t}(z)\hat{g}_0(z)$, which, when compared with Eq. (9.9.9), yields $\hat{t}(z)\hat{g}_0(z) = \hat{v}\hat{g}(z)$, viz., the identity (9.9.12).

Using the identities (9.9.11) and (9.9.12), the Lippmann-Schwinger equation (9.9.8) may be written alternatively as

$$\hat{t}(z) = \hat{v} + \hat{v}\hat{g}(z)\hat{v} \tag{9.9.14}$$

or

$$\hat{t}(z) = \hat{v} + \hat{t}(z)\hat{g}_o(z)\hat{v}\,. \tag{9.9.15}$$

Equations (9.9.8), (9.9.14) and (9.9.15) may all be looked upon as versions of the Lippmann-Schwinger equations in the operator form.

In the momentum representation Eq. (9.9.8) may be written in the integral form

$$\langle \boldsymbol{k}|\hat{t}(z)|\boldsymbol{k}'\rangle \;=\; \langle \boldsymbol{k}|\hat{v}|\boldsymbol{k}'\rangle \;+\; \int \frac{\langle \boldsymbol{k}|\hat{v}|\boldsymbol{q}\rangle d^3q\langle \boldsymbol{q}|\hat{t}(z)|\boldsymbol{k}'\rangle}{z - q^2}\,. \tag{9.9.16}$$

The quantity $\langle \boldsymbol{k}|\, \hat{t}(z)\, |\boldsymbol{k}'\rangle$ (or $\langle \boldsymbol{q}|\, \hat{t}(z)\, |\boldsymbol{k}'\rangle$) is the matrix element of the two-body transition operator in the momentum representation.

If $z = k^2 + i\epsilon$ and $\epsilon \to 0$ and $|\boldsymbol{k}| = |\boldsymbol{k}'|$, then the matrix element is termed an on-energy shell or simply as *on-shell* t-matrix element.

If $z = k^2 + i\epsilon$ and $\epsilon \to 0$ but $|\boldsymbol{k}| \neq |\boldsymbol{k}'|$, then the matrix element is termed a *half off-shell* t-matrix element.

If $z \neq k^2 + i\epsilon$ and $|\boldsymbol{k}| \neq |\boldsymbol{k}'|$, then the matrix element is termed a *full off-shell* t-matrix element.

The on-shell t-matrix element $\left\langle \boldsymbol{k}'\right|\, \hat{t}(k^2 + i\epsilon)\, |\boldsymbol{k}\rangle = \left\langle \boldsymbol{k}'\right|\, \hat{v}\, \left|\psi_k^{(+)}\right\rangle$ where $|\boldsymbol{k}| = |\boldsymbol{k}'|$ is related to the scattering amplitude $f(\theta)$ [Eq. (9.9.5)] and has a bearing only on two-body scattering. Conversely, from the two-body scattering, we can get information only about the on-shell elements of the two-body t-matrix. The half off-shell and full off-shell t-matrix elements remain undetermined from the two-body data. The off-shell behavior of the t-matrix can be put to test only in a three-body calculation.

It may be seen that if the two-body potential is non-local and separable[8],

$$\langle \boldsymbol{k}|\hat{v}|\boldsymbol{k}'\rangle = -\frac{\lambda}{M}g(\boldsymbol{k})g(\boldsymbol{k}'), \tag{9.9.17}$$

where $g(\boldsymbol{k})$ is a real function of $\boldsymbol{k}$ and λ and M are parameters characterizing the interaction, then the corresponding t-matrix element also has a separable form,

$$\langle \boldsymbol{k}|\, \hat{t}(z)\, |\boldsymbol{k}'\rangle = -\frac{g(\boldsymbol{k})g(\boldsymbol{k}')}{D(z)}\,. \tag{9.9.18}$$

To check this we may put both separable forms (9.9.17) and (9.9.18) into the Lippmann-Schwinger equation in the momentum representation [Eq. (9.9.16)] and find that $\boldsymbol{k}$ and $\boldsymbol{k}'$ dependence of both sides of the equation is consistent and

$$D(z) = \frac{1}{(\lambda/M)} + \int \frac{g^2(k')d^3k'}{s + i\epsilon - k'^2}\,, \tag{9.9.19}$$

[8] Y. Yamaguchi, Phys. Rev. **95**, 1628 (1954). The function $g(\boldsymbol{k})$ should not be confused with the resolvant operator $\hat{g}(z)$ defined in Eq. (9.9.13).

where $z = s + i\epsilon$.

9.10 Born Expansion

The integral expression for the scattering amplitude $f(\theta)$ [Eq. (9.8.10)] involves the total scattering wave function $\psi_k^{(+)}(\boldsymbol{r}')$, which is the solution of the integral equation

$$\psi_k^{(+)}(\boldsymbol{r}) = e^{i\boldsymbol{k}\cdot\boldsymbol{r}} + \int G_k^{(+)}(\boldsymbol{r},\boldsymbol{r}')U(\boldsymbol{r}')\psi_k^{(+)}(\boldsymbol{r}')d^3r'\,. \qquad (9.8.14^*)$$

Exact solutions of this equation are rare. However we can develop an approximation scheme by solving the integral equation iteratively. As a first approximation, the function $\psi_k^{(+)}(\boldsymbol{r}')$ in Eq. (9.8.14*) may be replaced by $e^{i\boldsymbol{k}\cdot\boldsymbol{r}'}$, and we may write

$$\psi_k^{(+)}(\boldsymbol{r}) = e^{i\boldsymbol{k}\cdot\boldsymbol{r}} + \int K_1(\boldsymbol{r},\boldsymbol{r}')e^{i\boldsymbol{k}\cdot\boldsymbol{r}'}d^3r'\,, \qquad (9.10.1)$$

where $$K_1(\boldsymbol{r},\boldsymbol{r}') \equiv G_k^{(+)}(\boldsymbol{r},\boldsymbol{r}')U(\boldsymbol{r}')\,. \qquad (9.10.2)$$

Now the function $e^{i\boldsymbol{k}\cdot\boldsymbol{r}'} + \int K_1(\boldsymbol{r}',\boldsymbol{r}'')e^{i\boldsymbol{k}\cdot\boldsymbol{r}''}d^3r''$ would be a better approximation to the function $\psi_k^{(+)}(\boldsymbol{r}')$ under the integral in Eq. (9.8.14*) than the plane wave function $e^{i\boldsymbol{k}\cdot\boldsymbol{r}'}$. So if we use this in (9.8.14*), we obtain

$$\begin{aligned}\psi_k^{(+)}(\boldsymbol{r}) &= e^{i\boldsymbol{k}\cdot\boldsymbol{r}} + \int G_k^{(+)}(\boldsymbol{r},\boldsymbol{r}')\left\{e^{i\boldsymbol{k}\cdot\boldsymbol{r}'} + \int K_1(\boldsymbol{r}',\boldsymbol{r}'')e^{i\boldsymbol{k}\cdot\boldsymbol{r}''}d^3r''\right\}U(\boldsymbol{r}')d^3r' \\ &= e^{i\boldsymbol{k}\cdot\boldsymbol{r}} + \int K_1(\boldsymbol{r},\boldsymbol{r}')e^{i\boldsymbol{k}\cdot\boldsymbol{r}'}d^3r' + \int K_2(\boldsymbol{r},\boldsymbol{r}')e^{i\boldsymbol{k}\cdot\boldsymbol{r}'}d^3r' \qquad (9.10.3)\end{aligned}$$

where $$K_2(\boldsymbol{r},\boldsymbol{r}') \equiv \int K_1(\boldsymbol{r},\boldsymbol{r}'')K_1(\boldsymbol{r}'',\boldsymbol{r}')d^3r''\,, \qquad (9.10.4)$$

where in the second integral, we have interchanged the dummy variables $\boldsymbol{r}'$ and $\boldsymbol{r}''$. Repeating this iterative process indefinitely, we get

$$\psi_k^{(+)}(\boldsymbol{r}) = e^{i\boldsymbol{k}\cdot\boldsymbol{r}} + \sum_{n=1}^{\infty}\int K_n(\boldsymbol{r},\boldsymbol{r}')e^{i\boldsymbol{k}\cdot\boldsymbol{r}'}d^3r'\,, \qquad (9.10.5)$$

where $$K_n(\boldsymbol{r},\boldsymbol{r}') \equiv \int K_1(\boldsymbol{r},\boldsymbol{r}'')K_{n-1}(\boldsymbol{r}'',r')d^3r''\,. \qquad (9.10.6)$$

This is called the *Born series* or Born expansion. If the iteration is carried out indefinitely, then this convergent series represents the exact solution of the integral equation (9.8.14). The convergence and the validity of approximations based on this series are discussed in books on scattering theory.

9.10.1 Born Approximation

When the energy of the incident particles is very high then there is justification in retaining only the first term in the Born series (9.10.5). This amounts to replacement of the function

$\psi_k^{(+)}(\boldsymbol{r}')$ by $e^{i\boldsymbol{k}\cdot\boldsymbol{r}'}$ in the integral expression [Eq. (9.8.18)] for the scattering amplitude

$$f(\theta) \approx f_{BA}(\theta) = -\frac{1}{4\pi}\frac{2\mu}{\hbar^2}\int e^{-i\boldsymbol{k}'\cdot\boldsymbol{r}'}V(r')e^{i\boldsymbol{k}\cdot\boldsymbol{r}'}d^3r'$$

or

$$f_{BA}(\theta) = -\frac{\mu}{2\pi\hbar^2}\int e^{i\boldsymbol{K}\cdot\boldsymbol{r}'}V(\boldsymbol{r}')d^3r', \tag{9.10.7}$$

where

$$\boldsymbol{K} = \boldsymbol{k} - \boldsymbol{k}', \quad K = |\boldsymbol{K}| = 2k\sin(\theta/2) \tag{9.10.8}$$

is the momentum transfer vector and θ is the angle of scattering (angle between the vectors $\boldsymbol{k}$ and $\boldsymbol{k}'$) in the center-of-mass frame. This approximation is known as the *first Born approximation* in which the scattering amplitude is proportional to the three-dimensional Fourier transform of the scattering potential.

If the potential is spherically symmetric $V(\boldsymbol{r}) = V(r)$ then the integration over angles of $\boldsymbol{r}'$ is easily carried out by choosing the polar axis to be along the momentum transfer vector $\boldsymbol{K}$. This gives us

$$f(\theta) \approx f_{BA}(\theta) = -\frac{2\mu}{\hbar^2}\int_0^\infty V(r')\frac{\sin(Kr')}{Kr'}\,r'^2dr'\,. \tag{9.10.9}$$

If the form of scattering potential $V(r')$ is known, the scattering amplitude $f(\theta)$ and the differential cross-section for elastic scattering in the center-of-mass frame may easily be worked out.

Scattering from the Screened Coulomb Potential

Let the interaction between two charged particles of charge Ze and $Z'e$, respectively, be represented by the screened Coulomb potential

$$V(r) = \frac{ZZ'e^2}{4\pi\epsilon_0 r}\exp(-r/r_o), \tag{9.10.10}$$

where r_o is the screening radius. This potential takes into account the screening of Coulomb interaction between the atomic nucleus and a charged projectile due to the presence of electrons in the atom. Note that the screened Coulomb potential, apart from the multiplicative factors, has the form of the Yukawa potential, originally introduced to describe the nuclear interaction between protons and neutrons due to pion exchange. Using this in Eq. (9.10.10) we get

$$\begin{aligned} f_{BA}(\theta) &= -\frac{\mu ZZ'e^2}{2\pi\epsilon_0\hbar^2}\int_0^\infty \frac{1}{r'}\exp(-r'/r_o)r'^2\frac{\sin Kr'}{Kr'}dr', \\ &= -\frac{\mu ZZ'e^2}{2\pi\epsilon_0\hbar^2}\frac{1}{K}\left[\frac{K}{(K^2+1/r_o^2)}\right] = -\frac{\mu ZZ'e^2}{2\pi\epsilon_0\hbar^2}\frac{r_o^2}{1+4k^2r_o^2\sin^2(\theta/2)}\,. \end{aligned} \tag{9.10.11}$$

From this expression we obtain the differential scattering cross-section due to screened Coulomb potential

$$\sigma(\theta) = \frac{(2\mu ZZ'e^2r_o^2/4\pi\epsilon_0\hbar^2)^2}{(1+4k^2r_o^2\sin^2(\theta/2))^2}\,. \tag{9.10.12}$$

At low energies $kr_o \ll 1$, the scattering cross-section has a weak dependence on the angle of scattering. At higher energies $kr_o \gg 1$ it is highly peaked in the forward direction $\theta = 0$.

For $r_o \to \infty$, the screened potential approaches the Coulomb potential and we get

$$f^c_{BA}(\theta) = -\frac{\mu Z Z' e^2}{2\pi\epsilon_o\hbar^2}\frac{1}{K^2} = -\frac{\mu Z Z' e^2}{2\pi\epsilon_o\hbar^2}\frac{1}{4k^2\sin^2(\theta/2)}$$

and

$$\sigma^c(\theta) = |f^c_{BA}(\theta)|^2 = \left(\frac{\mu Z Z' e^2}{8\pi\epsilon_o\hbar^2 k^2}\right)^2 \frac{1}{\sin^4(\theta/2)}\,. \tag{9.10.13}$$

We recognize this to be the Rutherford formula for Coulomb scattering. This agreement happens to be an accident because the Born amplitude is not the correct Coulomb scattering amplitude [see Sec. 9.7].

Scattering from the Square Well Potential

For a potential of the shape

$$V(r) = \begin{cases} -V_o\,, & \text{for } r \le r_o \\ 0\,, & \text{for } r > r_o\,, \end{cases} \tag{9.10.14}$$

the scattering amplitude in the Born approximation is calculated to be [Eq. (9.10.9)]

$$f_{BA}(\theta) = -\frac{\mu V_o}{\hbar^2}\int_0^{r_o}\frac{\sin Kr'}{Kr'}r'^2 dr' = \frac{2\mu V_o r_o}{\hbar^2 K^2}\left[\cos(Kr_o) - \frac{\sin(Kr_o)}{Kr_o}\right],$$
$$K = |\boldsymbol{k} - \boldsymbol{k}'| = 2k\sin\theta/2\,. \tag{9.10.15}$$

From this we can calculate the scattering cross-section $\sigma_{BA}(\theta) = |f_{BA}(\theta)|^2$. Note that the angular dependence is hidden in K.

In the low energy limit $kr_o \ll 1$ or $Kr = 2kr\sin\theta/2 \ll 1$, we obtain

$$f_{BA} = \frac{2\mu V_o r_o}{\hbar^2 K^2}\left[-\frac{1}{2}(Kr_o)^2\right] = -\frac{2\mu V_o r_o^3}{3\hbar^2}\,, \tag{9.10.16}$$

$$\sigma = \int d\Omega |f_{BA}|^2 = 4\pi r_o^2\left[\frac{2\mu V_o r_o^2}{3\hbar^2}\right]^2. \tag{9.10.17}$$

It is easy to check that phase shift analysis [Eq. (9.6.17)] leads to the same result in the low energy limit.

9.10.2 Validity of Born Approximation

We have seen that the total scattering solution is given by

$$\begin{aligned}\psi^{(+)}(\boldsymbol{r}) &= e^{i\boldsymbol{k}\cdot\boldsymbol{r}} + \psi_{sc}(\boldsymbol{r}) \\ &= e^{i\boldsymbol{k}\cdot\boldsymbol{r}} - \frac{1}{4\pi}\frac{2\mu}{\hbar^2}\int\frac{e^{ik|\boldsymbol{r}-\boldsymbol{r}'|}}{|\boldsymbol{r}-\boldsymbol{r}'|}V(\boldsymbol{r}')\psi^{(+)}(\boldsymbol{r}')d^3r'\,.\end{aligned} \tag{9.10.18}$$

The (first) Born approximation consists in replacing the function $\psi^{(+)}_k(\boldsymbol{r}')$ by the plane wave function $\exp(i\boldsymbol{k}\cdot\boldsymbol{r}')$ in the integrand of Eq. (9.10.18). We expect this approximation to be justified if the scattered wave

$$\psi^{(1)}_{sc}(\boldsymbol{r}) = -\frac{1}{4\pi}\frac{2\mu}{\hbar^2}\int\frac{e^{ik|\boldsymbol{r}-\boldsymbol{r}'|}}{|\boldsymbol{r}-\boldsymbol{r}'|}V(\boldsymbol{r}')e^{i\boldsymbol{k}\cdot\boldsymbol{r}'}d^3r'\,, \tag{9.10.19}$$

is small compared to the plane wave amplitude in the coordinate range where the potential is nonzero, i.e., near the origin $r = 0$. From Eq. (9.10.19), we estimate the scattered wave near $r = 0$ to be

$$\psi_{sc}(0) = -\frac{1}{4\pi}\frac{2\mu}{\hbar^2}\int \frac{e^{ikr'}}{r'} V(\boldsymbol{r}')e^{i\boldsymbol{k}\cdot\boldsymbol{r}'}\, d^3r' \,. \tag{9.10.20}$$

If we assume the potential to be spherically symmetric, $V(\boldsymbol{r}) = V(r)$, and carry out the angular integration, the condition for the Born approximation to be valid, $|\psi_{sc}(0)| \ll |e^{i\boldsymbol{k}\cdot\boldsymbol{r}}| = 1$, can be written as

$$\frac{2\mu}{k\hbar^2}\left|\int dr'\, e^{ikr'} \sin kr' V(r')\right| \ll 1\,. \tag{9.10.21}$$

Some qualitative conclusions can be drawn from this even without detailed knowledge of the potential.

At low energies $kr' \ll 1$, we can approximate $e^{ikr'} \approx 1$ and $\sin kr' \approx kr'$ under the integral in Eq. (9.10.21) giving us the condition for the valdity of Born approximation to be

$$\frac{2\mu}{\hbar^2}\left|\int r'V(r')dr'\right| \ll 1\,. \tag{9.10.22}$$

Thus if the potential has depth (or height) V_o and range r_o, the condition for Born approximation to be good is

$$\frac{\mu V_o r_o^2}{\hbar^2} \ll 1\,. \tag{9.10.23}$$

This condition means the Born approximation is valid if the potential is too weak to support a bound state [see Sec. 5.7].

At high energies $kr' \gg 1$, where r' is at most of the order of the range of the potential, the function $e^{kr'}\sin kr' = (e^{2ikr'} - 1)/2i$ in the integrand of Eq. (9.10.16) oscillates rapidly, so that the principal contribution to the integral comes from values of $kr' < 1$. If the potential has depth V_o and range r_o, the condition for the validity of the Born approximation in the high energy limit becomes

$$\frac{2\mu}{\hbar^2 k}\left|\int_o^{1/k} dr'\; V(r')\frac{1}{2i}(e^{2ikr} - 1)\right| = \frac{\mu V_o}{\hbar^2 k^2} = \frac{\mu V_o r_o^2}{\hbar^2 k^2 r_o^2} \ll 1\,. \tag{9.10.24}$$

This can be written as

$$\frac{\mu V_o r_o^2}{\hbar^2} \ll k^2 r_o^2\,. \tag{9.10.25}$$

It may be noted that this condition (in the high energy limit) may be satisfied even when the potential supports bound states. It also follows from (9.10.23) and (9.10.25) that if the Born approximation is valid at low energies it is also valid at high energies.

Born Approximation for the Square Well Potential

For the attractive square well potential [Eq. (9.10.14)], the scattering wave $\psi_{sc}(0)$ [Eq. (9.10.20)] is given by

$$\psi_{sc}(0) = \frac{\mu V_o}{2\hbar^2 k^2}\left[\cos 2kr_o - 1 - i(\sin 2kr_o - 2kr_o)\right]\,. \tag{9.10.26}$$

The condition for the validity of the Born approximation, and therefore the scattering amplitude (9.10.15) to be a good approximation to the exact result, then takes the form

$$\frac{\mu V_o}{2\hbar^2 k^2}\left|4k^2 r_o^2 - 4kr_o\sin(2kr_o) + 2 - 2\cos(2kr_o)\right|^{1/2} \ll 1\,. \tag{9.10.27}$$

At small energies ($kr_o \ll 1$), this leads to

$$\frac{\mu V_o}{2\hbar^2 k^2} \times 2k^2 r_o^2 = \frac{\mu V_o r_o^2}{\hbar^2} \ll 1\,. \tag{9.10.28}$$

Thus we can expect the Born approximation to be valid at low energies for potential too weak (depth-range combination too small) to support a bound state.

For high energies ($kr_o >> 1$) the first term inside the absolute value sign in Eq. (9.10.27) dominates. Then we can expect the Born aproximation to be valid if

$$\frac{\mu V_o r_o^2}{\hbar^2} \ll kr_o\,. \tag{9.10.29}$$

If the square well potential is strong enough to bind the two-body system and conforms to the depth-range relationship (to support at least one bound state)

$$V_o r_o^2 \approx \frac{\pi^2 \hbar^2}{8\mu}\,, \tag{9.10.30}$$

the condition (9.10.29) is still satisfied for sufficiently high energies $kr_o \gg 1$. Born approximation thus supplements the method of partial waves which would have required a large number of partial waves to calculate the cross-section in this case [see Sec. 9.4].

The above discussion in terms of the bound state should not be taken too far because if we have a barrier ($V_o < 0$) instead of a well, there are no bound states but there is still a criterion for the validity of the Born approximation, which in fact is the same as Eq. (9.10.17).

Validity of Born Approximation for Screened Coulomb Potential

Let us consider the conditions under which the scattering amplitude (9.10.11) under Born approximation is expected to be valid for the screened Coulomb potential [cf. Eq. (9.10.10)]

$$V(r) = \frac{ZZ'e^2}{4\pi\epsilon_0}\frac{e^{-\alpha r}}{r} \equiv V_o \frac{e^{-\alpha r}}{\alpha r}\,, \quad V_o = \frac{ZZ'e^2\alpha}{4\pi\epsilon_0}\,. \tag{9.10.31}$$

According to Eq. (9.10.21), we can expect Eq. (9.10.11) to be valid if

$$\left(\frac{\mu ZZ'e^2}{4\pi\epsilon_0\hbar^2 k}\right)\left|\int_0^\infty \frac{\exp(2ikr')-1}{r'} e^{-\alpha r'} dr'\right| \ll 1\,. \tag{9.10.32}$$

By evaluating the integral, we can write this condition as [9]

$$\frac{\mu ZZ'e^2}{4\pi\epsilon_0\hbar^2 k}\sqrt{\left(\ln\sqrt{(1+4k^2r_o^2)}\right)^2 + \left(\tan^{-1} 2kr_o\right)^2} \ll 1\,, \tag{9.10.33}$$

where $\quad r_o = \alpha^{-1}$ is the effective range of the potential. (9.10.34)

[9] The integral to be evaluated is $I = \int_0^\infty \frac{\exp(2ikr')-1}{r'} e^{-\alpha r'} dr'$. By differentiating I with respect to α, we find

$$\frac{\partial I}{\partial \alpha} = -\int_0^\infty (e^{2ikr'} - 1)e^{-\alpha r'} dr' = \frac{1}{\alpha - 2ik} + \frac{1}{\alpha}\,,$$

which is readily integrated to yield $I = \ln\left(1 + \frac{2ik}{\alpha - 2ik}\right) = -\ln[1 - i(2k/\alpha)]$. The constant of integration in the last step is fixed by noting that $I \to 0$ as $\alpha \to \infty$.

For low energies $kr_o \ll 1$ this criterion becomes $\frac{2\mu ZZ'e^2 r_o}{4\pi\epsilon_0 \hbar^2} \ll 1$. If we define an effective barrier (since we have assumed both the projectile and the target particles to have like charges) $V_o = \frac{ZZ'e^2}{4\pi\epsilon_0 r_o}$ and effective range r_o, this condition is simply $V_o r_0^2 \ll 1$, which is similar to that for a spherical square well. Hence Born approximation does not hold for $kr_o \ll 1$ and cannot be used for scattering of slow electrons from atoms.

For high energies $kr_o >> 1$, the criterion (9.10.33) becomes $\frac{\mu ZZ'e^2}{4\pi\epsilon_0 \hbar^2 k} \ln(2kr_o) \ll 1$ which can always be satisfied provided the energy is sufficiently large.

9.10.3 Born Approximation and the Method of Partial Waves

The method of phase shift analysis discussed in Sec. 9.4 gives exact expression for the scattering cross-section in terms of the phase shifts experienced by angular momentum partial waves. The phase shifts are expressed in terms of the logarithmic derivative of the radial function which is obtained by solving the radial Schrödinger equation for the given scattering potential. However, the relation between the phase shifts and the scattering potential remains recondite. A partial wave analysis of the integral equation provides a direct connection bewteen the phase shifts and the scattering potential.

We have seen the following expansions of the incident plane wave (solution of the scattering equation in the absence of scattering potential) and of the solution of the scattering equation with the scattering potential [Eqs. (9.4.16) and(9.4.3)]:

$$\psi_{inc}(r) = e^{i\,\boldsymbol{k}\cdot\boldsymbol{r}} = \sum_{\ell=0}^{\infty} i^\ell (2\ell+1) R_\ell(r) P_\ell(\cos\theta) \tag{9.10.35}$$

and

$$\psi_k^{(+)}(r) = \sum_{\ell=0}^{\infty} B_\ell R'_\ell(r) P_\ell(\cos\theta) \tag{9.10.36}$$

where the radial functions

$$R_\ell(r) = \frac{u_\ell(r)}{kr} \quad \text{and} \quad R'_\ell(r) = \frac{u'_\ell(r)}{kr} \tag{9.10.37}$$

are determined by the following differential equations

$$\frac{d^2 u_\ell}{dr^2} + \left(k^2 - \frac{\ell(\ell+1)}{r^2}\right) u_\ell(r) = 0 \tag{9.10.38}$$

and

$$\frac{d^2 u'_\ell}{dr^2} + \left(k^2 - U(r) - \frac{\ell(\ell+1)}{r^2}\right) u'_\ell(r) = 0\,. \tag{9.10.39}$$

The solution of the first equation and its asymptotic form are given by

$$u_\ell(r) = kr\, j_\ell(kr) \;\sim\; \sin(kr - \ell\pi/2), \tag{9.10.40}$$

and, as we have seen in Sec. 9.4, if $V(r)$ falls off faster than $1/r$ as $r \to \infty$, the radial function $u'_\ell(r)$ has the asymptotic form

$$u'_\ell(r) \;\sim \sin(kr - \ell\pi/2 + \delta_\ell)\,. \tag{9.10.41}$$

Multiplying Eq. (9.10.38) by $u'_\ell(r)$ and Eq. (9.10.39) by $u_\ell(r)$ and subtracting we get

$$\frac{d}{dr}\left[u'_\ell(r)\frac{du_\ell}{dr} - u_\ell(r)\frac{du'_\ell}{dr}\right] = -U(r) u_\ell(r) u'_\ell(r). \tag{9.10.42a}$$

Integrating both sides over r from 0 to R we get

$$\left[u'_\ell(r)\frac{du_\ell}{dr} - u_\ell(r)\frac{du'_\ell}{dr}\right]_{r=R} = -\frac{2\mu}{\hbar^2}\int_0^R u_\ell(r)V(r)u'_\ell(r)dr, \tag{9.10.42b}$$

because both radial functions must vanish at the origin $u_\ell(0) = u'_\ell(0) = 0$. If we let $R \to \infty$ we can use the asymptotic forms for $u_\ell(r)$ and $u'_\ell(r)$ on the left-hand side and find

$$k\sin\delta_\ell = -\frac{2\mu}{\hbar^2}\int_0^\infty u_\ell(r)V(r)u'_\ell(r)dr\,. \tag{9.10.43}$$

This is an explicit expression for the phase shift in terms of the scattering potential and the radial function. Now, if we invoke the Born approximation, we can use the replacement $u'_\ell(r) \approx u_\ell(r) = krj_\ell(r)$ under the integral in Eq. (9.10.43) to get

$$\sin\,\delta_\ell = -\frac{2\mu\, k}{\hbar^2}\int_0^\infty V(r)[j_\ell(kr)]^2 r^2 dr\,. \tag{9.10.44}$$

Using the expression (9.4.19) for the scattering amplitude in terms of the phase shifts, we get

$$f(\theta) = -\frac{2\mu}{\hbar^2}\sum_{\ell=0}^\infty (2\ell+1)P_\ell(\cos\theta)\int_0^\infty [j_\ell(kr)]^2 V(r) r^2 dr \tag{9.10.45}$$

where we have approximated $e^{i\delta_\ell}\sin\delta_\ell$ by $\sin\delta_\ell$. By using the mathematical identity

$$\sum_{\ell=0}^\infty (2\ell+1)[j_\ell(kr)]^2 P_\ell(\cos\theta) = \frac{\sin(Kr)}{Kr}\,, \tag{9.10.46}$$

where $K = 2k\sin(\theta/2)$, in Eq. (9.10.45), we recover the Born expression (9.10.9) for the scattering amplitude

$$f(\theta) = -\frac{2\mu}{\hbar^2}\int_0^\infty \frac{\sin(Kr)}{Kr}V(r)r^2 dr\,. \tag{9.10.47}$$

As a simple application of Eq. (9.10.44), consider a short range potential so that the contribution to the integral comes form small values of $r \sim 0$. Then using the small argument expansion for $j_\ell(kr) \approx (kr)^\ell/(2\ell+1)!!$, we obtain the expression

$$\delta_\ell \approx -\frac{2\mu k^{2\ell+1}}{\hbar^2[(2\ell+1)!!]^2}\int_0^\infty r^{2\ell+1}V(r)dr\,. \tag{9.10.48}$$

If $V(r) = V_o$ for $r < r_o$ and vanishes for $r > r_o$ we obtain for the ℓth partial wave phase shift

$$\delta_\ell = -\frac{2\mu V_o r_o^2}{\hbar^2}\frac{(kr_o)^{2\ell+1}}{(2\ell+3)[(2\ell+1)!!]^2}\,. \tag{9.10.49}$$

This expression shows that the phase shifts decrease rapidly with ℓ and beyond $\ell \approx kr_o$, the phase shifts are negligible.

Problems

1. A particle of mass m_1 collides elastically with a particle of mass m_2 which is initially at rest in the laboratory frame of reference. Show that all recoil (mass m_2) particles are scattered in the forward hemisphere in the laboratory frame. If the angular distribution is symmetric in the center-of-mass system, what is it for m_1 in the laboratory system.

2. A plane wave can be expanded in term of the partial waves as

$$e^{ikz} = \sum_{\ell=0}^{\infty} i^\ell (2\ell+1)\, j_\ell(kr)\, P_\ell(\cos\theta).$$

Interpret this expansion in terms of the principle of superposition of states.

3. Assuming the range of the $n-p$ interaction to be of the order of 3 fm (3×10^{-15} m), estimate which partial waves will undergo phase shifts if neutrons of energy
   (i) 10 Mev
   (ii) 1.0 MeV
   (iii) 0.01 MeV
   (iv) 1.0 kev
   are incident on a proton target given $\frac{\hbar^2}{2\mu} = \frac{\hbar^2}{M} = 41.47$ MeV·fm^2?

4. Derive the following relation

$$k\,\sin(\delta_\ell - \delta_{\ell'}) \;=\; \frac{-2\mu}{\hbar^2}\int [V(r) \,-\, V'(r)]\; u_\ell(r)\, u'_\ell(r)\, dr$$

where $\frac{\hbar^2 k^2}{2\mu}$ is the energy of relative motion of colliding particles in the center-of-mass frame, δ_ℓ and $\delta'_{\ell'}$ are the ℓth partial wave phase shifts for potentials $V(r)$ and $V'(r)$, respectively, and $u_\ell(r) = rR_\ell(r)$ and $u'_\ell(r) = R'_\ell(r)r$ are the radial functions for the two potentials. From this infer what the sign of the phase shift has to do with the nature (attractive or repulsive) of the scattering potential.

5. Prove the optical theorem

$$\sigma_{tot} \;=\; \frac{4\pi}{k}\;\mathcal{I}m\,[f(0)]\,,$$

where $\mathcal{I}m\,[f(0)]$ is the imaginary part of the forward scattering amplitude $f(0)$.

6. Use optical theorem to show that the Born approximation cannot be expected to give the correct scattering amplitude or the correct differential scattering cross-section in the forward direction.

7. Considering the potential to have a square well shape

$$V(r) = \begin{cases} -V_o\,, & \text{for } r \le r_o \\ 0\,, & \text{for } r > r_o\,, \end{cases}$$

show that, for low energies $E = \hbar^2 k^2/2\mu$, the condition of resonance scattering can be met if $\cot\alpha_o r_o \approx 0$, where $\alpha_o^2 \equiv U_o = 2\mu V_o/\hbar^2$, in which case

$$\sigma_{tot} = \frac{4\pi}{k^2 + \alpha_o^2\cot^2(\alpha_o r_o)}\,.$$

8. Considering the scattering potential to be of three-dimensional square well shape, find the limit for the total scattering cross-section when the center-of-mass energy, or $k = (2\mu E/\hbar^2)^{1/2}$, tends to zero. You may consider only s or p-wave phase shifts. For what values of $V_o r_o^2$ is the scattering cross-section zero (Ramsauer-Townsend effect).

9. Calculate the center-of-mass scattering amplitude in the Born approximation where the interaction between the two particles is

   (i) Exponential: $V(r) = -V_o\, e^{-r/a}$

   (ii) Yukawa: $V(r) = -\,V_o\,\frac{e^{-\mu\, r}}{r}$

   The Fourier transforms of these potential forms are given in Eqs. (3.6.13) through (3.6.15).

10. Assuming the $n-p$ interaction to be of square well shape for a short range potential,

$$V(r) = \begin{cases} -V_o\,, & \text{for} \quad r \leq r_o \\ 0\,, & \text{for} \quad r > r_o\,, \end{cases}$$

    set up the radial equations for scattering in the internal $r < r_o$ and external $r > r_o$ regions. Obtain expressions for the s- and p-wave phase shifts when the energy is small ($kr_o \ll 1$).

11. Calculate the differential cross-section of scattering $\frac{d\sigma}{d\Omega}$ for a square well potential

$$V(r) = \begin{cases} -V_o\,, & \text{for} \quad r \leq r_o \\ 0\,, & \text{for} \quad r > r_o\,, \end{cases}$$

    in the Born approximation. Show that (a) the scattering is peaked in the forward direction, and (b) at large energies the cross-section is inversely proportional to the energy.

12. Show that in the Born approximation the differential scattering cross-section for a potential $V(r)$ is exactly the same as that obtained from the potential $-V(r)$.

13. Show that the first Born amplitude is essentially the three-dimensional Fourier transform of the potential

$$f_{BA}(\theta) = -\sqrt{\frac{\pi}{2}\frac{2\mu}{\hbar^2}}\,\frac{1}{(2\pi)^{3/2}}\int V(r) e^{i\,\boldsymbol{K}\cdot\boldsymbol{r}} d^3r\,,$$

    where $\boldsymbol{K} = \boldsymbol{k} - \boldsymbol{k}'$.

14. Calculate the s-wave phase shift for a spinless particle scattered by a target with a hard sphere interaction between them. What is the extreme low energy limit for the total scattering cross-section? Explain the result.

15. For the screened Coulomb potential,

$$V(r) = \frac{ZZ'e^2}{4\pi\epsilon_0 r}\exp(-r/r_o)$$

    work out the differential cross-section of scattering in Born approximation. Consider the limit when the range r_o tends to infinity. Show that the total cross-section obtained from the differential cross-section diverges.

16. Calculate the Coulomb scattering cross-section for two identical spinless charged particles (Bosons) and compare it to that for two identical spin half particles in a state with total spin 1. Discuss the angular dependence in the two cases.

17. Assuming the neutron-neutron strong interaction potential to be the same as the neutron-proton (deuteron) potential [see Chapter 5] which supports a single bound state of $\ell = 0$, discuss the low energy limit of neutron-neutron scattering cross-section.

18. Consider the scattering from a delta-shell potential $V(r) = (V_o/r_o)\delta(r - r_o)$.

    (i) Use the first Born approximation to compute the scattering amplitude and investigate the conditions under which the approximation is valid.

    (ii) Compute the phase shift for the s-wave in the first Born approximation and compare it with the exact result

$$\tan\delta_0 = \frac{(2\mu V_o/\hbar^2 k)\ \sin^2 kr_o}{1 + (\mu V_o/\hbar^2 k)\sin 2kr_o}\,.$$

19. Express the form of Coulomb scattering wave function as a function of r and θ. What conclusions can you draw regarding its form at forward angles?

References

[1] M. L. Goldberger and K. M. Watson, *Collision Theory* (John Wiley and Sons, New York, 1964).

[2] R. G. Newton, *Scattering of Waves and Particles* (McGraw Hill Book Company, New York, 1966).

[3] L. I. Schiff, *Quantum Mechanics*, Third Edition (McGraw Hill Book Company, Inc., New York, 1968).

[4] Ta-You Wu and Takashi Ohmura, *Quantum Theory of Scattering* (Prentice Hall, Englewood Cliffs, NJ, 1962).

Appendix 9A1: Calculus of Residues

Function of a Complex Variable

A complex number, $z = x + iy$, may be represented by a point in the complex z-plane or the $X - Y$ plane where OX is called the real axis and OY the imaginary axis. This representation of complex numbers is referred to as an *Argand diagram*. When z denotes any one of the set of complex numbers, we call it a complex variable. If, for each value of z, the value of a second complex variable w is prescribed, then w is called a function of the complex variable z and is written as

$$w = f(z)\,. \tag{9A1.1.1}$$

If each value of z corresponds to only one value of w, then the function $w = f(z)$ is said to be single-valued. If a single value of z corresponds to more than one value of w, then the function $w = f(z)$ is said to be multi-valued. It is possible to resolve $w = f(z)$ into real and imaginary parts as

$$w = f(z) = u(x, y) + i\,v(x, y) \tag{9A1.1.2}$$

where u and v are real functions of the real variables x and y. The function $f(z)$ is continuous at a point $z = z_0$ if and only if all the following three conditions are satisfied:

(i) $f(z_0)$ exists

(ii) $\lim_{z\to z_0} f(z)$ exists

(iii) $\lim_{z\to z_0} f(z) = f(z_0)$

Differentiability

If the limit $\lim_{\Delta z\to 0} \frac{f(z+\Delta z)-f(z)}{\Delta z}$ exists, and is independent of the manner in which $\Delta z \equiv \Delta x + i\,\Delta y \to 0$, then the function $f(z)$ is said to be differentiable. The necessary twin conditions, called *Cauchy-Riemann conditions*, for the function $f(z)$ to be differentiable are

$$\frac{\partial u}{\partial x} = \frac{\partial v}{\partial y} \quad \text{and} \quad \frac{\partial u}{\partial y} = -\frac{\partial v}{\partial x}\,. \tag{9A1.1.3}$$

Contour Integral of a Function of a Complex Variable

The integral $\int_{z_0}^{z_n} f(z)\,dz$ of a function of a complex variable along a curve Γ may be defined as follows [Fig. 9A1.1]. We divide the curve into n segments by means of points $z_0, z_1, z_2, \cdots, z_n$ and denote the interval between points z_{r-1} and z_r by Δz_r and the average value of the function in this interval by $\bar{f}(z_r)$. Then the sum $L_n = \sum_{r=1}^{n} \bar{f}_r(z_r)\Delta z_r$ in the limit $n \to \infty$ equals the integral $\int_{z_0}^{z_n} f(z)\,dz$ along the curve (contour) Γ.

Cauchy's Theorem

It can be shown that if a function $f(z)$ of a complex variable is analytic (i.e., single-valued, continuous and differentiable) on and every point inside a closed curve (contour) C, then

$$\oint_C f(z)dz = 0. \tag{9A1.1.4}$$

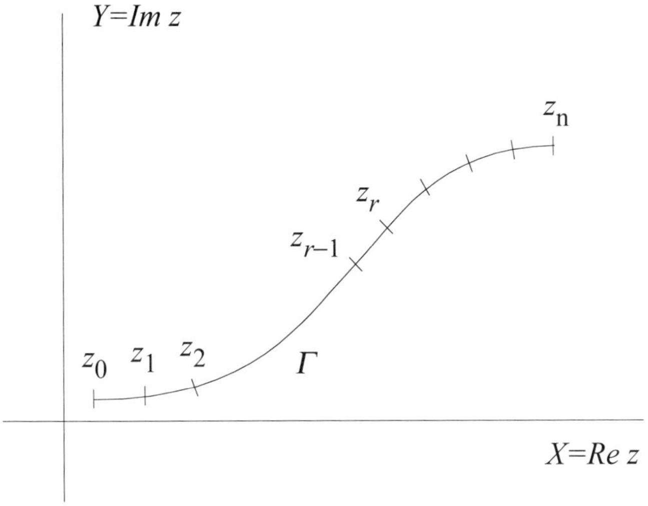

FIGURE 9A1.1
The integral of $f(z)$ along a curve from z_0 to z_n in the complex z-plane depends on the curve and on the direction the curve is traversed.

By convention, the integration along a closed contour in the counter-clockwise sense is taken to be positive.

A corollary that follows from this theorem is that if $f(z)$ is analytic at all points on and inside a closed contour C except in regions bounded by closed curves $C_1, C_2, \cdots, C_n$, then

$$\oint_C f(z)dz = \oint_{C_1} f(z)dz + \oint_{C_2} f(z)dz + \cdots + \oint_{C_n} f(z)dz. \tag{9A1.1.5}$$

Cauchy's Integral Formula

If a function $f(z)$ is analytic at all points on and inside a closed contour C, then the function defined by $\phi(z) = \frac{f(z)}{(z-a)}$ is not analytic at the point $z = a$. Let C_1 be a small closed circle with center at $z = a$ and radius r [Fig. 9A1.2]. According to Cauchy's theorem (corollary),

$$\oint_C \phi(z)dz = \oint_{C_1} \phi(z)dz\,. \tag{9A1.1.6}$$

The contour integral on the right-hand side may be easily evaluated by putting $z - a = re^{i\theta}$ and $dz = re^{i\theta}\, i\, d\theta$. In the limit $r \to 0$, its value is $2\pi i\, f(a)$.

Using Eq. (9A1.1.6) we have

$$\oint_C \frac{f(z)}{z-a} dz = 2\pi i\, f(a) \quad \text{or} \quad f(a) = \frac{1}{2\pi i} \oint_C \frac{f(z)dz}{z-a}\,. \tag{9A1.1.7}$$

It is remarkable that Cauchy's integral formula enables one to compute the value of the function $f(z)$ at any point a inside the contour C from the knowledge of the function on the boundary of the contour.

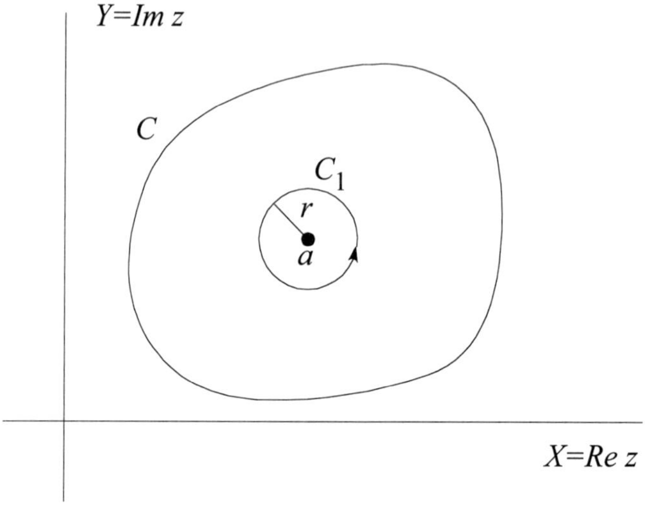

FIGURE 9A1.2
C is a closed contour and C_1 is a small circle within it with its center at a.

Taylor Series Expansion

We can rewrite Cauchy's integral formula as

$$f(z) = \frac{1}{2\pi i} \oint_C \frac{f(t)dt}{t-z}, \tag{9A1.1.8}$$

where the function $f(t)$ of the complex variable t is analytic at every point on and inside the closed contour C. At a point $a = z + h$, within the closed contour C, we have

$$f(z+h) = \frac{1}{2\pi i} \oint_C \frac{f(t)dt}{t-(z+h)}. \tag{9A1.1.9}$$

Using the identity

$$\frac{1}{t-(z+h)} = (t-z)^{-1}\left[1 + \frac{h}{t-z} + \frac{h^2}{(t-z)^2} + \cdots + \frac{h^n}{(t-z)^n}\right] + \frac{h^{n+1}}{(t-z)^{n+1}} \frac{1}{[t-(z+h)]},$$

and substituting this series in Eq. (9A1.1.9), we find

$$f(z+h) = \sum_r C_r h^r, \tag{9A1.1.10}$$

where

$$C_r = \frac{1}{2\pi i} \oint_C \frac{f(t)}{(t-z)^{r+1}} dt. \tag{9A1.1.11}$$

It can be shown that Taylor's expansion series is convergent.

Laurent Series Expansion

If a function $f(z)$ of a complex variable z is analytic at all points on two closed contours C and C_1 and within the annular space between them but not within C_1 (which surrounds a point z_0), then the function at any point a within the annular region is given by

$$f(a) = \frac{1}{2\pi i} \oint_C \frac{f(z)dz}{z-a} - \frac{1}{2\pi i} \oint_{C_1} \frac{f(z)dz}{z-a} . \tag{9A1.1.12}$$

This can be seen easily by considering a closed contour Γ, which encloses the annular region

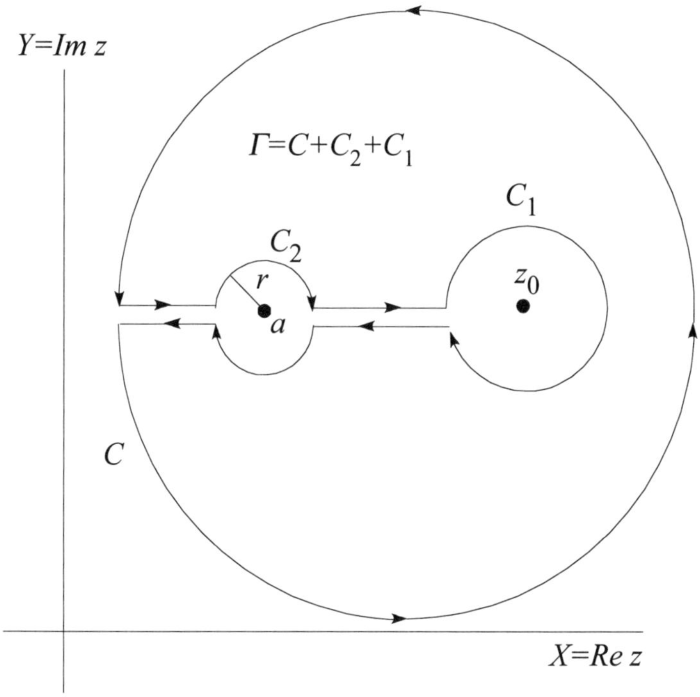

FIGURE 9A1.3
The closed contour Γ encloses the ring space between the closed contours C and C_1 but excludes a small circle C_2 surrounding the point a in the ring space.

but excludes a small circle surrounding the point a [Fig. 9A1.3]. Within the space enclosed by Γ, the function $\phi(z) = \frac{f(z)}{z-a}$ is analytic. Therefore, the contour integral $\oint_\Gamma \phi(z)dz = 0$ implies

$$\oint_C \phi(z)dz - \oint_{C_1} \phi(z)dz - 2\pi i\, f(a) = 0 ,$$

which leads to Eq. (9A1.1.12).

If the point a within the annular space is put equal to $z_0 + h$, where z_0 is a point within the inner contour C_1 at which the function $f(z)$ is not analytic, then we can rewrite

Eq. (9A1.1.12) as

$$f(z_0 + h) = \frac{1}{2\pi i} \oint_C \frac{f(z)dz}{z - (z_0 + h)} - \frac{1}{2\pi i} \oint_{C_1} \frac{f(z)dz}{z - (z_0 + h)} . \tag{9A1.1.13}$$

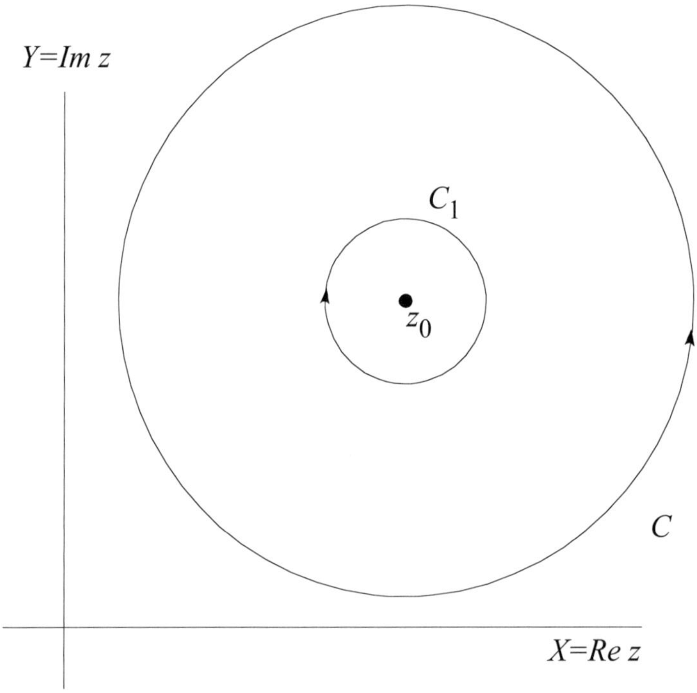

FIGURE 9A1.4
The concentric circles about $z = z_0$ represent the closed contours C and C_1.

For simplicity we may visualize the contours C and C_1 as represented by concentric circles about the point z_0 [Fig. 9A1.4]. Since the point $z_0 + h$ lies in the annular space, $|z - z_0| \geq h$ for the first integral along C in Eq. (9A1.1.13) while $|z - z_0| \leq h$ for the second integral along C_1. So in the first case we should expand $(z - (z_0 + h))^{-1}$ in powers of $\frac{h}{z - z_0}$ as

$$\frac{1}{z - (z_0 + h)} = \frac{1}{z - z_0} \left[1 + \frac{h}{z - z_0} + \frac{h^2}{(z - z_0)^2} + \cdots \frac{h^n}{(z - z_0)^n} + \cdots \right] ,$$

and in the second case $-(z - (z_0 + h))^{-1}$ should be expanded in powers of $\frac{z - z_0}{h}$ as

$$\frac{-1}{z - (z_0 + h)} = \frac{1}{h} \left[1 + \frac{z - z_0}{h} + \frac{(z - z_0)^2}{h^2} + \cdots + \frac{(z - z_0)^n}{h^n} + \cdots \right] .$$

Substituting the respective series in Eq. (9A1.1.13), we get the following series expansion for the function at a point in the annular space:

$$f(z_0 + h) = \sum_{n=0}^{\infty} a_n h^n + \sum_{m=1}^{\infty} b_m h^{-m}\,, \tag{9A1.1.14}$$

where

$$a_n = \frac{1}{2\pi\, i} \oint_C \frac{f(z)}{(z - z_0)^{n+1}} dz\,, \tag{9A1.1.15}$$

and

$$b_m = \frac{1}{2\pi\, i} \oint_{C_1} \frac{f(z)}{(z - z_0)^{-m+1}} dz\,. \tag{9A1.1.16}$$

This is the Laurent series expansion of a function $f(z_0 + h)$, $z_0 + h$ being a point in the annular space where the function is analytic while z_0 is the point where the function is not analytic. It can be seen that Laurent series expansion is convergent. The series with negative powers of h is called the principal part of the function $f(z)$ at z_0. If this series terminates at a finite value of m then the singularity at $z = z_0$ is called a *pole of order m.* If this series terminates at $m = 1$, the *pole* is called a *simple pole.* If the series in negative powers of m is infinite, then the singularity at $z = z_0$ is called an *essential singularity.*

Residue at a Pole

In the Laurent series expansion if we let $z = z_0 + h$, or $h = z - z_0$, where z_0 is a singular point and $z = z_0 + h$ is a point in the annular space where the function is analytic, then we may write the expansion as

$$f(z) = \sum_{n=0}^{\infty} a_n (z - z_0)^n + \sum_{m=1}^{\infty} b_m (z - z_0)^m\,,$$

where the coefficients a_n and b_m are defined in Eqs. (9A1.15) and (9A1.16). The coefficient

$$b_1 = \frac{1}{2\pi\, i} \oint_{C_1} f(z) dz\,, \tag{9A1.1.17}$$

where C_1 is a closed curve (circle) surrounding the pole at $z = z_0$ is generally a complex number and is called the *residue* at the pole $z = z_0$. It follows that

$$\oint_{C_1} f(z) dz = 2\pi\, i\, b_1 = 2\pi\, i \times \text{Residue at the pole at } z = z_0\,. \tag{9A1.1.18}$$

Cauchy's Residue Theorem

If $f(z)$ is analytic on and inside all points within a closed contour C except at a finite number of points (poles) $z_1,\ z_2, \cdots, z_n$ [Fig. 9A1.5] and if these points are surrounded by small circles $C_1,\ C_2, \cdots, C_n$, respectively, then according to Cauchy's theorem

$$\oint_C f(z) dz = \oint_{C_1} f(z) dz + \oint_{C_2} f(z) dz + \cdots + \oint_{C_n} f(z) dz\,,$$

and, using Eq. (9A1.1.18), we can write

$$\oint_C f(z) dz = 2\pi\, i \times \sum_{r=1}^{n} \text{Residue of } f(z) \text{ at } z = z_r\,. \tag{9A1.1.19}$$

Thus, according to Cauchy's residue theorem, the integral $\oint_C f(z)dz$ can be evaluated if its residues at all poles within the closed contour C are known.

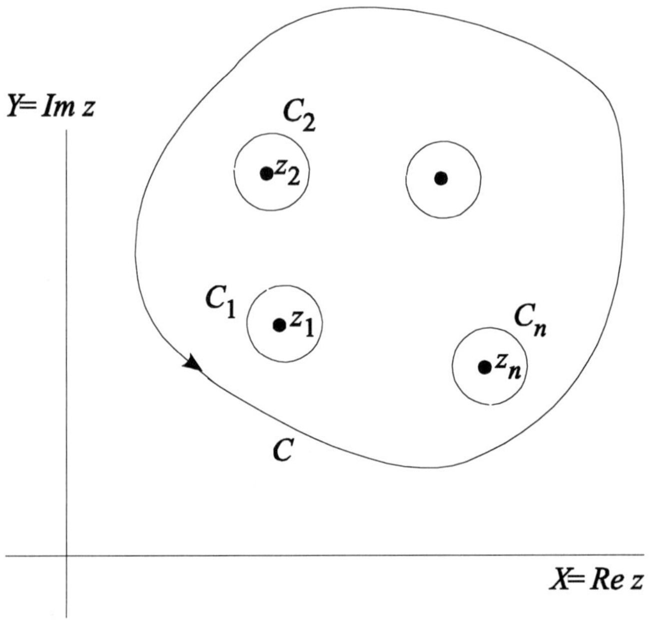

FIGURE 9A1.5
The closed contour C encloses a number of singular points (poles) surrounded by small circles $C_1, C_2, \cdots, C_n$. Note that there may be other singular points outside the region enclosed by C.

Calculation of Residues

If there is a simple pole at $z = z_r$ so that

$$f(z) = \sum_{n=0}^{\infty} a_n (z - z_r)^n + \frac{b_1}{z - z_r},$$

then the residue at $z = z_r$ is

$$b_1 = \text{Residue at } z_r = \lim_{z \to z_r} (z - z_r) f(z). \tag{9A1.1.20}$$

If there is a pole of order m at z_r, then

$$b_1 = \text{Residue at } z_r = \lim_{z \to z_r} \left[\frac{1}{(m-1)!} \frac{d^{m-1}}{dz^{m-1}} \{(z - z_r)^m f(z)\} \right]. \tag{9A1.1.21}$$

Evaluation of Real Integrals Using the Calculus of Residues

To evaluate an integral $\int_{-\infty}^{\infty} f(x)dx$, where the real function $f(x)$ is singular at a number of points $x_1, x_2, \cdots, x_n$, we consider the function $f(z)$ of a complex variable z, which has the same functional form as $f(x)$. If the function $f(z)$ satisfies the following conditions

1. It is analytic in the upper half of the z-plane except at a finite number of poles[10].

2. $z\,f(z) \to 0$ uniformly as $z \to \infty$.

Then, the integral $\oint_C f(z)dz$ may be evaluated by choosing the closed contour C to be a semi-circle of radius R in the upper half of the complex z-plane, bounded by the real axis [Figs. 9.10 and 9.11] and using Cauchy's residue theorem (9A1.1.19) to evaluate the residues at the poles of $f(z)$ in the upper half of the complex z-plane. In the limit $R \to \infty$, the integral $\oint\limits_C f(z)dz \to \int\limits_{-\infty}^{\infty} f(x)dx$. Thus we have a simple prescription for evaluating real integrals when the integrand has poles at a finite number of points in the complex plane. Other integrals may require closed contours different from the one considered here, the idea being that on a part of the contour the complex integral should coincide with the given integral and the contribution from the rest of the contour should be known (preferably vanish).

[10]We assume that there are no poles on the real axis. If there are poles on the real axis, we have an improper integral, which is not uniquely defined. In such cases we may choose to include or exclude them, depending on the physics of the problem under consideration.

10

TIME-DEPENDENT PERTURBATION METHODS

10.1 Introduction

We now consider systems for which the Hamiltonian contains time-dependent interaction terms. In such cases we cannot reduce the time-dependent Schrödinger equation to an eigenvalue equation. However, if the time-dependent terms can be regarded as small perturbations, we can develop a form of perturbation theory that can be used to calculate the effects of time-dependent terms. Consider a system for which the Hamiltonian may be written as the sum of a dominant unperturbed part $\hat{H}_0$ and a small perturbation $\hat{H}'$ as

$$\hat{H} = \hat{H}_0 + \lambda \hat{H}'(t)\,, \tag{10.1.1}$$

where the unperturbed Hamiltonian $\hat{H}_0$ is independent of time, whereas the perturbation $\hat{H}'$ may depend on time. The parameter $0 < \lambda < 1$ has been introduced to keep track of various orders of approximations as in the case of time-independent perturbation theory discussed in Chapter 8. The basic idea is that a small perturbation to the Hamiltonian will produce a small change in the wave function. For a time-dependent perturbation, this change means that a system, initially in a particular eigenstate of $\hat{H}_0$, will find itself in a time-dependent admixture of the eigenstates. Thus the time-dependent perturbation causes the system to undergo transition to different eigenstates of $\hat{H}_0$. So for time-dependent perturbations our interest lies in calculating the probabilities for transition to different states.

To proceed further, we assume that for the unperturbed Hamiltonian, the exact solutions for the time-independent Schrödinger equation

$$\hat{H}_0 \left|\psi_n\right\rangle = E_n \left|\psi_n\right\rangle . \tag{10.1.2}$$

are known. Then the equation of motion for the unperturbed system

$$i\hbar\frac{\partial}{\partial t}\left|\psi(t)\right\rangle = \hat{H}_0 \left|\psi(t)\right\rangle , \tag{10.1.3}$$

has the solution

$$\left|\psi_m(t)\right\rangle = \left|\psi_m\right\rangle e^{-iE_m t/\hbar} . \tag{10.1.4}$$

These solutions form an orthonormal complete set. The wave function $\left|\Psi(t)\right\rangle$ for the complete time-dependent Hamiltonian (10.1.1) obeys the following equation of motion

$$i\hbar\frac{\partial}{\partial t}\left|\Psi(t)\right\rangle = (\hat{H}_0 + \lambda \hat{H}'(t)) \left|\Psi(t)\right\rangle . \tag{10.1.5}$$

It is usually not possible to solve this equation exactly. If the perturbation $\hat{H}'$ is weak, it is sufficient to find the lowest order effects of $\hat{H}'$ on the system behavior. To do this we expand the wave function $\Psi(\boldsymbol{r}, t)$ in terms of the complete set of time-dependent eigenfunctions of the unperturbed Hamiltonian

$$\left|\Psi(t)\right\rangle = \sum_m a_m(t) \left|\psi_m\right\rangle e^{-iE_m t/\hbar} , \tag{10.1.6}$$

where the summation extends over all the eigenstates of the unperturbed system. Since the expansion coefficients $a_m(t)$ depend on time, this is called the method of variation of constants and is due to Dirac (1926). Substituting this expansion into the equation of motion (10.1.5) for the full Hamiltonian we get

$$\sum_m i\hbar\,\dot{a}_m\,|\psi_m\rangle\,e^{-iE_mt/\hbar} = \sum_m a_m(t)\lambda\hat{H}'\,|\psi_m\rangle\,e^{-iE_mt/\hbar}\,, \tag{10.1.7}$$

where Eq. (10.1.2) has been used. Pre-multiplying both sides of Eq. (10.1.7) by $\langle\psi_n|$, we get

$$\sum_m i\,\hbar\dot{a}_m\delta_{mn}e^{-iE_mt/\hbar} = \sum_m a_m(t)\lambda H'_{nm}e^{-iE_mt/\hbar} \tag{10.1.8}$$

where the orthogonality condition for the states $|\psi_n\rangle$ has been used and where

$$H'_{nm} \equiv \langle\psi_n|\,\hat{H}'(t)\,|\psi_m\rangle\,. \tag{10.1.9}$$

Equation (10.1.8) then leads to

$$i\,\hbar\dot{a}_n = \sum_m a_m(t)\lambda H'_{nm}(t)\exp(i\omega_{nm}t) \tag{10.1.10}$$

where $\omega_{nm} = (E_n - E_m)/\hbar$. If we assume that at the initial time $t = 0$, the system is in the unperturbed state $|\psi_{n_o}\rangle$, then Eq. (10.1.10) must be solved subject to the initial condition

$$a_n(0) = \delta_{nn_o}\,. \tag{10.1.11}$$

To solve Eq. (10.1.10) we express the state amplitude $a_n(t)$ as a series in powers (the order of perturbation) of λ[1]

$$a_n(t) = a_n^{(0)}(t) + \lambda a_n^{(1)}(t) + \lambda^2 a_n^{(2)}(t) + \cdots + \lambda^s a_n^{(s)}(t)\cdots \tag{10.1.12}$$

Such an expansion seems reasonable if the perturbation $\lambda\hat{H}'$ is small compared to $\hat{H}_0$. Then we expect successive corrections to state amplitudes to depend on successively higher powers of the perturbation and the series to converge rapidly. The smallness of the perturbation will be made more precise below. Substituting this expansion into Eq. (10.1.10), we get

$$i\hbar(\dot{a}_n^{(0)} + \lambda\dot{a}_n^{(1)} + \lambda^3\dot{a}_n^{(3)} + \cdots) = \sum_m (a_m{}^{(0)} + \lambda a_m^{(1)} + \lambda^3 a_m^{(3)} + \cdots)\lambda H'_{nm}e^{i\omega_{nm}t}\,. \tag{10.1.13}$$

Equating terms independent of λ from both sides we get $\dot{a}_n^{(0)} = 0$, which can be integrated immediately to give

$$a_n^{(0)}(t) = a_n(0) = \delta_{nn_o}\,. \tag{10.1.14}$$

Equating the coefficients of λ^s $(s = 1, 2, 3, \cdots)$ from both sides, we get

$$\dot{a}_n^{(s)} = \frac{1}{i\hbar}\sum_m a_m^{(s-1)}(t)\,e^{i\omega_{nm}t}H'_{nm}\,. \tag{10.1.15}$$

By putting $s = 1$ in this equation we find the equation of motion for the first order correction

$$\dot{a}_n^{(1)}(t) = \frac{1}{i\hbar}\sum_m \delta_{mn_o}\,H'_{nm}\,e^{i\omega_{nmt}} = \frac{1}{i\hbar}H'_{nn_o}(t)e^{i\omega_{nn_o}t}\,, \tag{10.1.16}$$

[1]The actual expansion is in powers of $\hat{H}'$ (strength of perturbation). The parameter λ allows us to keep track of the powers $\hat{H}'$.

which can be integrated formally to give

$$a_n^{(1)}(t) = \frac{1}{i\hbar}\int_0^t H'_{nn_o}(t')\, e^{i\omega_{nn_o}t'}\, dt' \,. \tag{10.1.17}$$

If desired, the second and higher order corrections to $a_m(t)$ may similarly be found. We will confine our attention to the first and second order corrections in this chapter.

With these corrections, we can rewrite the series (10.1.12) for $a_m(t)$ as

$$a_n(t) = \delta_{nn_o} + \frac{1}{i\hbar}\int_0^t H'_{nn_o}(t')e^{i\omega_{nn_o}t'}dt'$$
$$+ \frac{1}{(i\hbar)^2}\sum_m \int_0^t dt_2 e^{i\omega_{nm}t_2} H'_{nm}(t_2) \int_0^{t_2} dt_1 e^{i\omega_{mn_o}t_1} H'_{mn_o}(t_1) + \cdots \tag{10.1.18}$$

Physical Significance of the Coefficient $a_n(t)$

If the perturbation lasts only for a finite time T or vanishes sufficiently rapidly as $t \to \pm\infty$ and the Hamiltonian returns to its unperturbed value $\hat{H}_0$, we can interpret $\lim_{t\to\infty} |a_n(t)|^2$ as the probability that the system will be found in the stationary state $|\psi_n\rangle$, if initially it was in the stationary state $|\psi_{n_o}\rangle$ of the unperturbed Hamiltonian. In other words $|a_n(t)|^2$ after the perturbation has ceased, can be regarded as the probability of transition $|\psi_{n_o}\rangle \to |\psi_n\rangle$. For $n = n_o$, this probability is ≈ 1. For $n \neq n_o$, this probability is $\approx |a_n^{(1)}|^2$ to the first order of perturbation. If $|a_n^{(1)}|^2$ does not vanish, the transition $|\psi_i\rangle \to |\psi_n\rangle$ is said to be allowed in the first order. If it vanishes the transition is referred to as a *first-order forbidden transition.* This does not mean that the transition cannot occur at all; it may still occur in the second (or higher) order of perturbation but with much reduced probability.

So far we have not specified the time dependence of $\hat{H}'$, which can be of many different types. For example, we could imagine a perturbation that is turned on suddenly, a perturbation with periodic (harmonic) time dependence, or one that varies slowly. These time variations can produce very different time evolution of the wave function. We shall consider two examples of these time-dependent perturbations which are of considerable physical interest

(i) Perturbation is *switched on* for a small time and is constant over this time interval, say $0 \le t \le T$.

(ii) Perturbation varies harmonically with time.

Other types of time dependencies will be considered toward the end of this chapter.

10.2 Perturbation Constant over an Interval of Time

In this case the perturbation has the following time dependence

$$H'(t) = \begin{cases} 0 & \text{for } -\infty < t < 0\,, \\ H' & \text{for } 0 < t < T\,, \\ 0 & \text{for } 0 < t < \infty\,. \end{cases} \tag{10.2.1}$$

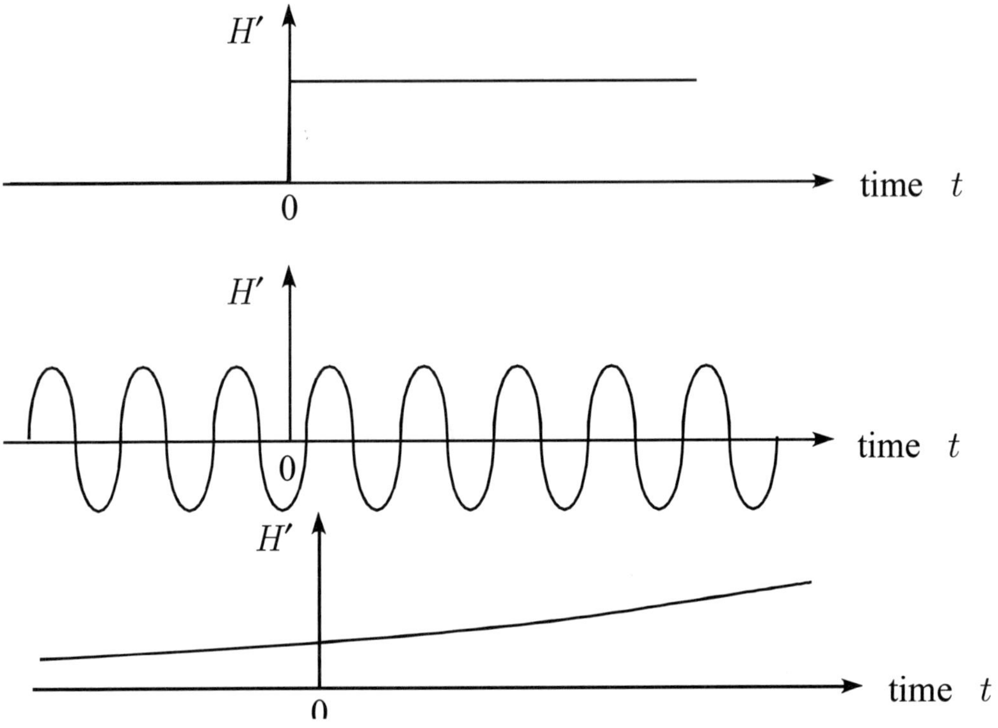

FIGURE 10.1
Examples of different types of time dependence of the perturbation: (a) sudden turn on, (b) periodic variation, (c) slow variation. Each time dependence produces different time evolution of the wave function.

Then the transition amplitude for $n \neq n_o$ (final state different from the initial state) in the first order is given by

$$\begin{aligned} a_n(T) \approx a_n^{(1)}(T) &= \frac{1}{i\hbar}\int_0^T H'_{nn_o} e^{i\omega_{nn_o}t'}\,dt' = -\frac{H'_{nn_o}}{\hbar\omega_{nn_o}}\left[e^{i\omega_{nn_o}T} - 1\right] \\ &= -iH'_{nn_o}\,e^{i\omega_{nn_o}T/2}\left[\frac{\sin[(E_n - E_{n_o})T/2\hbar]}{(E_n - E_{n_o})/2}\right], \quad \omega_{nn_o} = \frac{(E_n - E_{n_o})}{\hbar}. \end{aligned} \tag{10.2.2}$$

The probability of transition to state $|n\rangle$ in time T is then

$$\begin{aligned} |a_n(T)|^2 &= |H'_{nn_o}|^2 \frac{\sin^2[(E_n - E_{n_o})T/2\hbar]}{[(E_n - E_{n_o})/2\hbar]^2} \equiv \frac{|H'_{nn_o}|^2 T^2}{\hbar^2}\frac{\sin^2\phi}{\phi^2}, \\ \phi &= \frac{(E_n - E_{n_o})T}{2\hbar}. \end{aligned} \tag{10.2.3}$$

We note that for very short times, the probability grows as T^2 for transition to all states. For longer times, a plot of the transition probability $|a_n(T)|^2$ in Fig. 10.2 shows that although the transition probability is an oscillatory[2] function of E_n, it is significant only

[2]The oscillatory behavior of transition probability is an artifact of the sharp turn-on of the perturbation. It is possible to consider smoother turn-on of perturbation, which does not exhibit ringing of the probability. The results, however, do not depend on such details.

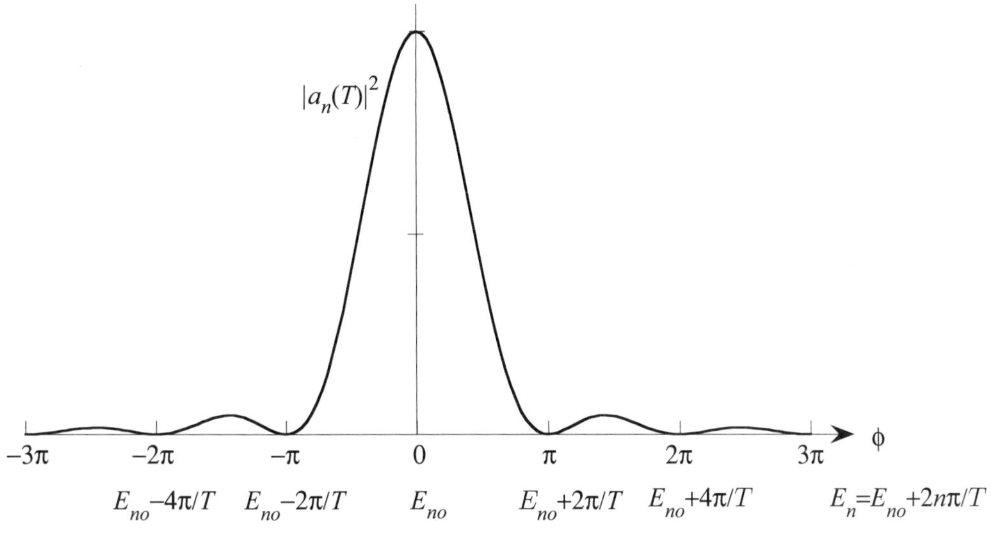

FIGURE 10.2
Probability $|a_n(T)|^2$ of a first-order transition from state $|n_o\rangle$ to state $|n\rangle$ as a function of ϕ (or E_n) assuming H'_{nn_o} to be a slowly varying function of E_n.

for states with energies that fall under the central peak. This means from the initial state n_o the system can make a transition to any one of the final states $|k\rangle$ with energy $|E_n - E_{n_o}| < 2\pi/T$. This is a form of energy-time uncertainty principle and says that when a constant perurbation acts for a duration $\Delta t = T$ the probability is significant for transitions that conserve energy to within $\Delta E = 2\pi\hbar/T$.

The total probability for first-order transition is given by the sum $\sum_n |a_n(T)|^2$. This sum is an oscillatory function of time if we are dealing with transitions to a discrete set of final states. It must be kept in mind that this probability cannot exceed unity. Indeed, it should stay small compared to unity in view of the perturbative nature of the calculation. For long times, higher order effects of perturbation as well as the depletion of initial state population must be taken into account.

A useful formula can be derivedif the final state belongs to a continuum or a group of closely spaced energy levels and we are interested in the probability of transition to a small group ΔE_n of states around $|n\rangle$. The matrix element H'_{nn_o} is not expected to vary significantly over such a group of final states. We can then write the sum of transition probabilities over the final states as an integral over energy E_n

$$\int_{\Delta E_n} |a_n(T)|^2 \rho(E_n) dE_n = \frac{|H'_{nn_o}|^2}{\hbar^2} \int_{\Delta E_n} \frac{\sin^2[(E_n - E_{n_o})T/2\hbar]}{[(E_n - E_{n_o})/2\hbar]^2} \rho(E_n) dE_n \,, \qquad (10.2.4)$$

where $\rho(E_n)$ is the density of final states near energy E_n. Now the width of the central maximum decreases with increasing time as $\Delta E \sim 2\pi/T$. When T is sufficiently large, the central peak becomes very narrow and falls entirely within the interval ΔE_n. The density of state $\rho(E_n)$ can then be evaluated at the center of the peak $E_n = E_{n_o}$ and taken out of the integral. The remaining integral is then simply the area under the curve in Fig. 10.2

$$\int\limits_{\Delta E_n} \frac{\sin^2[(E_n - E_{n_o})T/2\hbar]}{[(E_n - E_{n_o})/2\hbar]^2} dE_n \xrightarrow{\phi=(E_n-E_{n_o})T/2\hbar} 2\hbar T \int\limits_{-\infty}^{\infty} \frac{\sin^2\phi}{\phi^2} d\phi = 2\hbar T\pi \,, \tag{10.2.5}$$

where we have extended the limits of integration from $-\infty$ to ∞ with little error since the function $\frac{\sin^2\phi}{\phi^2}$ lies essentially inside the group ΔE_n of final states. Total transition probability then depends linearly on the duration T of the perturbation according to

$$P_{n_o \to n} = \frac{2\pi T}{\hbar} |H'_{nn_o}|^2 \rho(E_n = E_{n_o}) \,. \tag{10.2.6}$$

We can define a transition rate (probability of transition per unit time) by writing $P_{n_o\to n} \equiv w\, T$ where

$$w = \frac{2\pi}{\hbar} |H'_{nn_o}|^2 \rho(E_n = E_{n_o}) \,. \tag{10.2.7}$$

This formula for the transition rate is called *Fermi's golden rule* and has wide applications in quantum mechanics.

Born Approximation Formula

As a simple application of Fermi's golden rule, consider the elastic scattering of two particles in the center-of-mass frame when the energy of relative motion is so high that the interaction $\hat{H}' = \hat{V}$ may be treated as a perturbation. This problem conforms to the constant perturbation we have just considered for when the particles are far apart their interaction can be ignored; and as they approach one another, their interaction can be considered to be turned on.

Suppose the initial momentum and energy are $\hbar\boldsymbol{k}_0$ and $E_{k_0} = \hbar^2 k_0^2/2\mu$ and the final momentum and energy are $\hbar\boldsymbol{k}$ and $E_k = \hbar^2 k^2/2\mu$, where $|\boldsymbol{k}| = |\boldsymbol{k}_0|$ so that the intial and final energies are the same. μ is the reduced mass of the particles. The momentum transferred in the scattering is $\hbar\boldsymbol{q} = \hbar(\boldsymbol{k}_0 - \boldsymbol{k})$. It is easily seen that $q = |\boldsymbol{q}| = 2k\sin(\theta/2)$, where θ is the angle between $\boldsymbol{k}_0$ and $\boldsymbol{k}$ [Fig. 10.3]. Let us consider the transition from initial momentum state $|E_{k_0}, \boldsymbol{k}_0\rangle$ to any one of the final states $|E_k, \boldsymbol{k}\rangle$, such that the direction of the final momentum lies within the solid angle $d\Omega$ and the magnitude of $\boldsymbol{k}$ is the same as that of $\boldsymbol{k}_0$, within the limits of uncertainty. Note that the states are labeled not only by the energy eigenvalue but also by the momentum vector. To use Fermi's golden rule (10.2.7) we put,

$$\langle \boldsymbol{r}|\, E_{k_0}, \hbar\boldsymbol{k}_0\rangle = u_{\boldsymbol{k}_0}(\boldsymbol{r}) = \frac{1}{\sqrt{L^3}} e^{i\boldsymbol{k}_0\cdot\boldsymbol{r}} \,, \tag{10.2.8a}$$

$$\langle \boldsymbol{r}|\, E_k, \hbar\boldsymbol{k}\rangle = u_{\boldsymbol{k}}(\boldsymbol{r}) = \frac{1}{\sqrt{L^3}} e^{i\boldsymbol{k}\cdot\boldsymbol{r}} \,, \tag{10.2.8b}$$

$$\langle \boldsymbol{r}|\, \hat{H}'\, |\boldsymbol{r}'\rangle = \langle \boldsymbol{r}|\, \hat{V}\, |\boldsymbol{r}'\rangle = V(\boldsymbol{r})\delta^3(\boldsymbol{r} - \boldsymbol{r}') \,, \tag{10.2.8c}$$

where L^3 is the quantization volume and $V(\boldsymbol{r})$ is the interaction potential. Using these we find that the matrix element and first order transition rate are given by

$$\begin{aligned}\langle E_k, \boldsymbol{k}|\, \hat{H}'\, |E_{k_0}, \boldsymbol{k}_0\rangle &= \langle \boldsymbol{k}|\, \hat{V}\, |\boldsymbol{k}_0\rangle = \iint \langle \boldsymbol{k}|\, \boldsymbol{r}\rangle\, d^3r\, \langle \boldsymbol{r}|\, \hat{V}\, |\boldsymbol{r}'\rangle\, d^3r'\, \langle \boldsymbol{r}'|\, \boldsymbol{k}_0\rangle \\ &= \frac{1}{L^3}\int e^{-i\boldsymbol{k}\cdot\boldsymbol{r}} V(\boldsymbol{r}) e^{i\boldsymbol{k}_0\cdot\boldsymbol{r}}\, d^3r = \frac{1}{L^3}\int e^{i\boldsymbol{q}\cdot\boldsymbol{r}} V(\boldsymbol{r}) d^3r \,,\end{aligned} \tag{10.2.9}$$

and

$$w = \frac{2\pi}{\hbar L^6} \left| \int \exp(i\boldsymbol{q}\cdot\boldsymbol{r}) V(\boldsymbol{r}) d^3r \right|^2 \rho(E_k) \,, \tag{10.2.10}$$

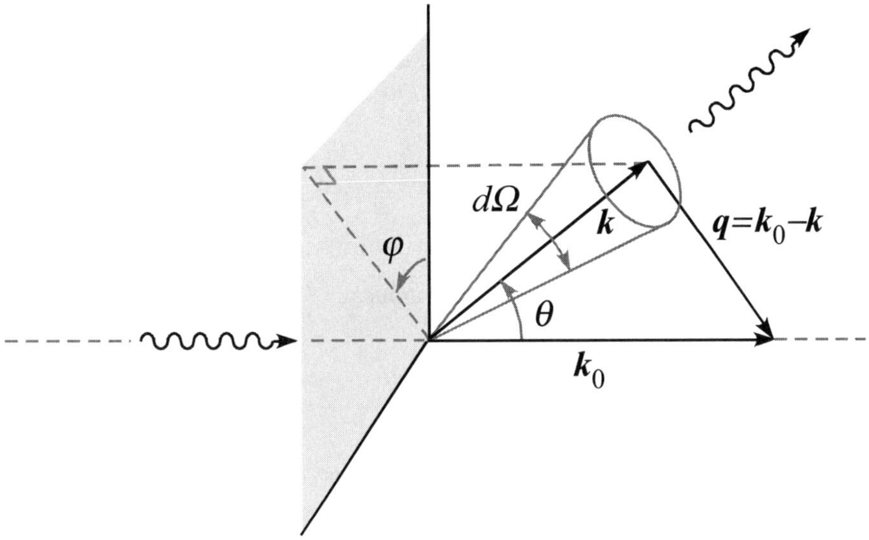

FIGURE 10.3
Scattering in the center-of-mass frame. $\boldsymbol{k}_0$ and $\boldsymbol{k}$ are the relative momenta before and after scattering and $d\Omega = \sin\theta d\theta\, d\varphi$ is the elementary solid angle in the direction of $\boldsymbol{k}$ $(\theta,\,\varphi)$ into which the particle with momentum $\hbar\boldsymbol{k}$ is scattered.

where w is the transition rate (probability per unit time) for scattering from state $|E_{k_0}, \hbar\boldsymbol{k}_0\rangle$ to states $|E_k = E_{k_0}, \hbar\boldsymbol{k}\rangle$ with momentum vectors $\hbar\boldsymbol{k}$ pointing into the solid angle $d\Omega$. The density of such final states is

$$\rho(E_k) = \rho_k \frac{dk}{dE_k} = \frac{k^2 d\Omega}{(2\pi/L)^3} \frac{dk}{dE_k} \xrightarrow{E=\frac{\hbar^2 k^2}{2\mu}} \frac{\mu k L^3 d\Omega}{8\pi^3\hbar^2} \,. \tag{10.2.11}$$

Using this in Eq. (10.2.10), the transition rate can be written as

$$w = \frac{\mu k d\Omega}{4\pi^2\hbar^3 L^3} \left| \int \exp(i\boldsymbol{q}\cdot\boldsymbol{r}) V(\boldsymbol{r}) d^3 r \right|^2 . \tag{10.2.12}$$

This expression has an undesirable dependence on the volume of quantization L^3. A way out of this is to note that w is the rate at which particles scatter into the solid angle $d\Omega$ for an incident particle flux density corresponding to one particle in volume L^3. A single incident particle per L^3 moving with speed $v = \hbar k/\mu$ constitutes a particle flux density of $v/L^3 = \hbar k/\mu L^3$. If we divide the transition rate w by the incident flux we obtain the rate of transition into the solid angle $d\Omega$ per unit incident flux density or the differential scattering cross-section [see Sec. 9.1]

$$\begin{aligned} d\sigma &= \frac{\text{Number of particles scattered into solid angle } d\Omega}{\text{Incident flux}} = \frac{w}{\hbar k/\mu L^3} \\ &= \frac{\mu L^3}{\hbar k} \frac{2}{\hbar L^6} \left| \int e^{i\boldsymbol{q}\cdot\boldsymbol{r}} V(\boldsymbol{r}) d^3 r \right|^2 \frac{k\,\mu\, L^3 d\Omega}{8\pi^3\hbar^2} \end{aligned}$$

or

$$\frac{d\sigma}{d\Omega} = \frac{\mu^2}{4\pi^2\hbar^4} \left| \int e^{i\boldsymbol{q}\cdot\boldsymbol{r}} V(\boldsymbol{r}) d^3 r \right|^2 . \tag{10.2.13}$$

This is independent of L^3. A differential cross-section has units of area per unit solid angle ($\text{cm}^2/\text{steradian}$) and is the quantity measured in scattering experiments. Equation (10.2.13) is the Born Approximation formula and agrees with the results of Sec. 9.10.

10.3 Harmonic Perturbation: Semi-classical Theory of Radiation

Let us consider a perturbation that varies harmonically (time dependence as $e^{\pm i\omega t}$)

$$\hat{H}'(t') = \begin{cases} 0\,, & \text{for } t' < 0, \\ \hat{H}'_o e^{-i\omega t'} + \hat{H}'^{\dagger}_o e^{i\omega t'}\,, & \text{for } 0 \le t' \le t \\ 0\,, & \text{for } t' > t. \end{cases} \tag{10.3.1}$$

Such a perturbation arises when an atomic system interacts with an external electromagnetic radiation field. Electromagnetic field is described in terms of the magnetic and electric field vectors $\boldsymbol{B}$ and $\boldsymbol{E}$ obeying Maxwell's equations [Appendix 12A1]:

$$\boldsymbol{\nabla}\cdot\boldsymbol{E}(\boldsymbol{r},t) = \frac{\rho}{\epsilon_0} \qquad \boldsymbol{\nabla}\cdot\boldsymbol{B}(\boldsymbol{r},t) = 0 \tag{10.3.2}$$

$$\boldsymbol{\nabla}\times\boldsymbol{E}(\boldsymbol{r},t) + \frac{\partial\boldsymbol{B}(\boldsymbol{r},t)}{\partial t} = 0 \qquad \boldsymbol{\nabla}\times\boldsymbol{B}(\boldsymbol{r},t) - \frac{1}{c^2}\frac{\partial\boldsymbol{E}(\boldsymbol{r},t)}{\partial t} = \mu_0\boldsymbol{j}\,. \tag{10.3.3}$$

Alternatively, the electromagnetic field may be specified by the vector and scalar potentials $\boldsymbol{A}(\boldsymbol{r},t)$ and $\phi(\boldsymbol{r},t)$, defined by

$$\boldsymbol{B}(\boldsymbol{r},t) = \boldsymbol{\nabla}\times\boldsymbol{A}(\boldsymbol{r},t)\,, \tag{10.3.4}$$

and
$$\boldsymbol{E}(\boldsymbol{r},t) = -\boldsymbol{\nabla}\phi(\boldsymbol{r},t) - \frac{\partial\boldsymbol{A}(\boldsymbol{r},t)}{\partial t}\,. \tag{10.3.5}$$

These equations do not determine the vector and scalar potentals uniquely for another set of potential (gauge transformation), $\boldsymbol{A}' = \boldsymbol{A} + \boldsymbol{\nabla}\chi$ and $\phi' = \phi - \frac{\partial\chi}{\partial t}$, yields the same fields. Using this freedom, we choose to work in the Coulomb gauge, in which the vector potential satisfies[3]

$$\boldsymbol{\nabla}\cdot\boldsymbol{A} = 0\,. \tag{10.3.6a}$$

Then the transverse parts of the electric and magnetic fields are given by

$$\boldsymbol{B}(\boldsymbol{r},t) = \boldsymbol{\nabla}\times\boldsymbol{A}(\boldsymbol{r},t) \quad \text{and} \quad \boldsymbol{E}(\boldsymbol{r},t) = -\frac{\partial\boldsymbol{A}(\boldsymbol{r},t)}{\partial t}\,, \tag{10.3.6b}$$

where we have retained symbol $\boldsymbol{E}$ to denote the transverse part of the electric field. Note that since the magnetic field always satisfies $\boldsymbol{\nabla}\cdot\boldsymbol{B} = 0$, it is a pure transverse field. The longitudinal part of the electric field gives the Coulomb interaction energy, which is already included in the atomic Hamiltonian. Using Eq. (10.3.6b) in Maxwell's equation we find the vector and scalar potentials satisfy the inhomogeneous wave equation [see Appendix 12A1]

$$\nabla^2\boldsymbol{A}(\boldsymbol{r},t) - \frac{1}{c^2}\frac{\partial^2\boldsymbol{A}(\boldsymbol{r},t)}{\partial t^2} = -\mu_0\boldsymbol{j}_\perp \tag{10.3.7}$$

and
$$\nabla^2\phi(\boldsymbol{r},t) = -\frac{\rho}{\epsilon_0}\,, \tag{10.3.8}$$

[3] As is well known, there can be several choices of vector and scalar potentials connected by gauge transformations, which correspond to the same field vectors $\boldsymbol{E}$ and $\boldsymbol{B}$. Of these the Coulomb gauge (sometimes also referred to as the radiation gauge) is a specific choice. See Appendix 12A1 for details.

where $\boldsymbol{j}_\perp$ is the transverse current density satisfying $\boldsymbol{\nabla}\cdot\boldsymbol{j}_\perp = 0$ in the Coulomb gauge.

When the atomic system interacts with an electromagnetic field we follow the classical procedure so that $\hat{\boldsymbol{p}} \to \hat{\boldsymbol{p}} + e\boldsymbol{A}(\hat{\boldsymbol{r}})$ and $\hat{H} \to \hat{H} + e\phi(\hat{\boldsymbol{r}})$, where we have taken electronic charge to be $-e$ and $\boldsymbol{A}$ and ϕ are now functions of atomic operator $\hat{\boldsymbol{r}}$. Then the Hamiltonian for this system (atom + field) may be written as

$$\left(\hat{H} + e\phi\right) = \frac{1}{2m}(\hat{\boldsymbol{p}} + e\boldsymbol{A})^2$$

or

$$\hat{H} = \left[\left\{\frac{\hat{p}^2}{2m} + V(\hat{\boldsymbol{r}})\right\} + \frac{e}{m}\boldsymbol{A}\cdot\hat{\boldsymbol{p}} + \frac{e^2A^2}{2m}\right], \tag{10.3.9}$$

where the potential $V(\hat{\boldsymbol{r}}) = -e\phi(\hat{\boldsymbol{r}})$ and we have used the result $\hat{\boldsymbol{p}}\cdot\boldsymbol{A}(\hat{\boldsymbol{r}},t) = \boldsymbol{A}(\hat{\boldsymbol{r}},t)\cdot\hat{\boldsymbol{p}}$ in the Coulomb gauge.[4] Assuming that the second order term $\frac{e^2A^2}{2m}$ is small compared to the first order term $\boldsymbol{A}\cdot\hat{\boldsymbol{p}}$, we can drop it and write the Hamiltonian (10.3.9) as

$$\hat{H} = \hat{H}_o + \lambda\hat{H}'(t), \tag{10.3.10}$$

where

$$\hat{H}_o = \frac{\hat{p}^2}{2m} + V(\hat{\boldsymbol{r}}), \tag{10.3.11a}$$

$$\hat{H}'(t) = \frac{e}{m}\boldsymbol{A}(\hat{\boldsymbol{r}},t)\cdot\hat{\boldsymbol{p}}. \tag{10.3.11b}$$

On substituting the general solution of Eq. (10.3.7) corresponding to a traveling wave of single-frequency ω in free space,

$$\boldsymbol{A}(\hat{\boldsymbol{r}},t) = \boldsymbol{A}_o\, e^{i(\boldsymbol{k}\cdot\hat{\boldsymbol{r}} - \omega\, t)} + A_o^*\, e^{-i(\boldsymbol{k}\cdot\hat{\boldsymbol{r}} - \omega\, t)} \tag{10.3.12}$$

in Eq. (10.3.11b), we get time dependent perturbation of the form (10.3.1), where

$$\hat{H}'_o = \frac{e}{m}e^{i\boldsymbol{k}\cdot\hat{\boldsymbol{r}}}\boldsymbol{A}_o\cdot\hat{\boldsymbol{p}}, \tag{10.3.13}$$

and

$$\hat{H}'^{\dagger}_o = \frac{e}{m}e^{-i\boldsymbol{k}\cdot\hat{\boldsymbol{r}}}\boldsymbol{A}^*_o\cdot\hat{\boldsymbol{p}}. \tag{10.3.14}$$

Once again, we recall that we are working in the Coulomb gauge so that $\hat{\boldsymbol{p}}\cdot\boldsymbol{A}(\boldsymbol{r},t) = \boldsymbol{A}(\boldsymbol{r},t)\cdot\hat{\boldsymbol{p}}$, even though $\hat{\boldsymbol{p}}$ and $\hat{\boldsymbol{r}}$ do not commute. This means $\hat{H}'_o$ and $\hat{H}'^{\dagger}_o$ are adjoints of one another. Thus the interaction of the atomic system with radiation field gives rise to harmonic perturbation. This perturbation will induce transitions between different states of the atom.

We have seen in Sec. 10.1, the probability of transition from an initial state $|i\rangle$ to a final state $|f\rangle$ in time t, is given by $|a_f(t)|^2$, where (to first order in perturbation) $a_f(t)$ is given by

$$a_f(t) = \delta_{fi} + \frac{1}{i\hbar}\int_0^t e^{i\omega_{fi}t'}H'_{fi}(t')dt'$$

$$= \delta_{fi} + \frac{1}{i\hbar}\int_0^t e^{i\omega_{fi}t'}\langle f|\,\hat{H}'_o e^{-i\omega\, t'} + \hat{H}'^{\dagger}_o e^{i\omega\, t'}\,|i\rangle\, dt', \tag{10.3.15}$$

[4]To see this consider any state $|\Psi\rangle$ of the system. Then the matrix element

$$\langle \boldsymbol{r}|\,\hat{\boldsymbol{p}}\cdot\boldsymbol{A}(\hat{\boldsymbol{r}},t)\,|\Psi\rangle = -i\hbar\boldsymbol{\nabla}\cdot[\boldsymbol{A}(\boldsymbol{r},t)\Psi(\boldsymbol{r})] = -i\hbar[\boldsymbol{\nabla}\cdot\boldsymbol{A}(\boldsymbol{r},t)]\Psi(\boldsymbol{r},t) - i\hbar\boldsymbol{A}(\boldsymbol{r},t)\cdot\boldsymbol{\nabla}\Psi(\boldsymbol{r},t)$$
$$= -i\hbar\boldsymbol{A}(\boldsymbol{r},t)\cdot\boldsymbol{\nabla}\Psi(\boldsymbol{r},t) = \boldsymbol{A}(\boldsymbol{r},t)\cdot\hat{\boldsymbol{p}}\,\Psi(\boldsymbol{r},t),$$

by virtue of the fact that in Coulomb gauge $\boldsymbol{\nabla}\cdot\boldsymbol{A}(\boldsymbol{r},t) = 0$. Since this holds for any state $|\Psi\rangle$ we can write the Hamiltonian in the form (10.3.9) in the Coulomb gauge.

where $\omega_{fi} = (E_f - E_i)/\hbar$. Then the probability of transition $|i\rangle \to |f\rangle$ $(f \neq i)$ in time t is given by

$$|a_f(t)|^2 \approx \left| \frac{1}{i\hbar} \int_0^t e^{i\omega_{fi}t'} \langle f| \hat{H}'_o e^{-i\omega t'} + \hat{H}'^{\dagger}_o e^{i\omega t'} |i\rangle \right|^2 . \tag{10.3.16}$$

On simplifying this we obtain,

$$|a_f(t)|^2 = \frac{1}{\hbar^2} \left| (H'_o)_{fi} \left(\frac{e^{i(\omega_{fi}-\omega)t} - 1}{\omega_{fi} - \omega} \right) + (H'^{\dagger}_o)_{fi} \left(\frac{e^{i(\omega_{fi}+\omega)t} - 1}{\omega_{fi} + \omega} \right) \right|^2 . \tag{10.3.17}$$

where the matrix elements $(H'_o)_{fi}$ and $(H'^{\dagger}_o)_{fi}$ are given by

$$(H'_o)_{fi} = \frac{e}{m} \langle f| e^{i\boldsymbol{k}\cdot\hat{\boldsymbol{r}}} \boldsymbol{A}_o \cdot \hat{\boldsymbol{p}} |i\rangle \tag{10.3.18}$$

and

$$(H'^{\dagger}_o)_{fi} = \frac{e}{m} \langle f| e^{-i\boldsymbol{k}\cdot\hat{\boldsymbol{r}}} \boldsymbol{A}^*_o \cdot \hat{\boldsymbol{p}} |i\rangle . \tag{10.3.19}$$

An inspection of Eq. (10.3.17) and the discussion following Eq. (10.2.3) induced by constant perturbation show that the probability of transition $|i\rangle \to |f\rangle$ is appreciable when the denominator of one of the terms in Eq. (10.3.17) tends to zero, i.e., when either (i) $\hbar\omega \approx E_f - E_i$ or (ii) $\hbar\omega \approx -(E_f - E_i) = E_i - E_f$. In the first case, $E_f \approx E_i + \hbar\omega$, the first term in Eq. (10.3.17) dominates and the second term makes a negligible contribution. In the second case, $E_f \approx E_i - \hbar\omega$, the second term in Eq. (10.3.17) dominates and the first term contributes negligibly. The first condition is satisfied when the final level E_f lies above E_i and the transition is an upward transition resulting in the absorption of a photon [Fig. 10.4(a)]. On the other hand, the second condition is satisfied when the final level lies below the initial level ($E_f < E_i$) so that the transition is a downward transition, resulting in the emission of a photon [Fig. 10.4(b)]. This is unlike the case of perturbation constant in time, where, during a transition, the energy was required to be conserved within the limits allowed by the uncertainty principle. In the case of harmonic perturbation, the energy needed for upward transition comes from the perturbing field while the energy released in the downward transition goes to the radiation field. The radiation field remains almost unchanged by these processes because, in the semi-classical limit, the radiation field is regarded as infinite *source* and *sink* of quanta.

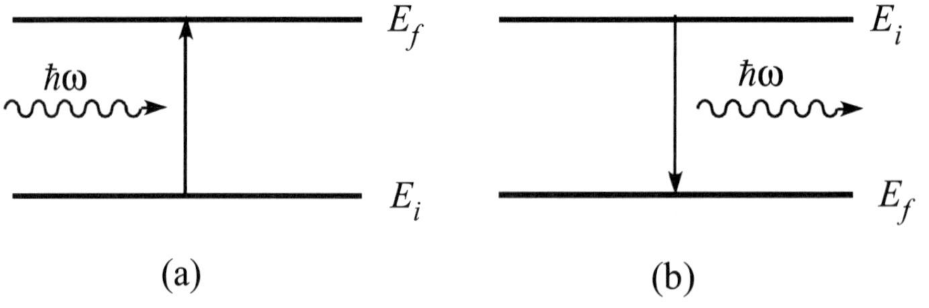

FIGURE 10.4
(a) An upward transition ($E_f > E_I$) is accompanied by the absorption of a quantum of energy $\hbar\omega \approx (E_f - E_i) = \hbar\omega_{fi}$. (b) A downward transition ($E_f < E_i$) is accompanied by the emission of a quantum of energy $\hbar\omega \approx -(E_f - E_i) = -\hbar\omega_{fi}$.

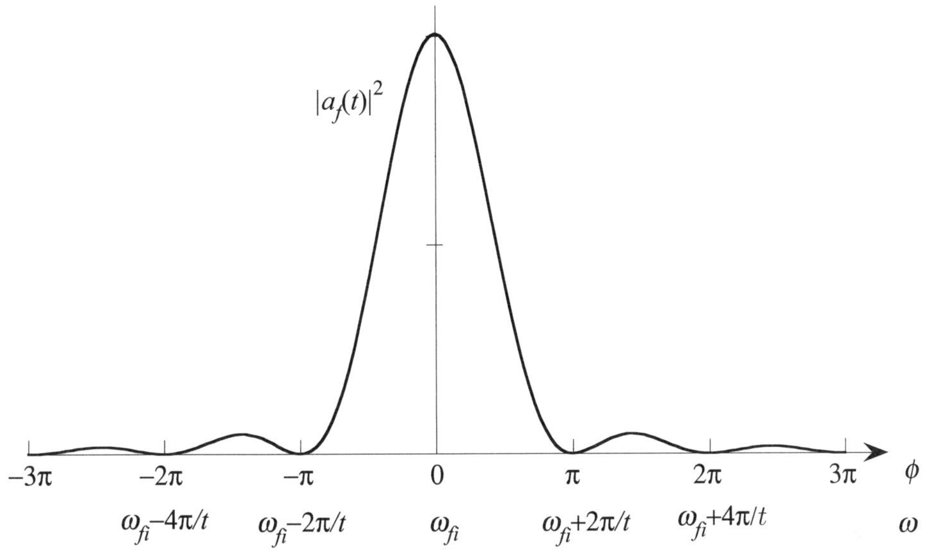

FIGURE 10.5
Probability $|a_f(t)|^2$ of absorptive transition to level $|f\rangle$ for a system initially starting in state $|i\rangle$ as a function of ϕ or ω. Note that for absorption $\omega_{fi} = (E_f - E_i)/\hbar > 0$ $[E_f > E_i]$. The same curve applies to emission if $\omega_{fi} = (E_f - E_i)/\hbar < 0$ $[E_f < E_i]$ in the label is replaced by $-\omega_{fi}$.

It is straightforward to work out the dependence of the transition probability $|a_f(t)|^2$ on the frequency ω of radiation field for both absorption $(E_f > E_i)$ and emission $(E_f < E_i)$. In the case of absorption, the first term in Eq. (10.3.17) dominates over the second. Neglecting the second term we obtain

$$|a_f(t)|^2 = \frac{t^2}{\hbar^2} |(H_o')_{fi}|^2 \frac{\sin^2[(\omega - \omega_{fi})t/2]}{[(\omega - \omega_{fi})t/2]^2} . \tag{10.3.20}$$

Thus the probability of transition in time t is proportional to t^2. This has been observed with monochromatic radiation from a laser incident on a two-level atomic system.

The time dependence of transition probability is modified drastically for a broadband perturbing radiation field or when the final state belongs to a continuum of states. To see this we note from Fig. 10.5 that, for a transition to take place, it is not necessary that ω exactly equal ω_{fi} but can have a spread of the order of $\Delta\omega = 4\pi/t$. Qualitatively, this means when the transition is induced by a broadband radiation field, all frequency components that lie under the central peak (within $\Delta\omega$ of ω_{fi}) contribute to the transition. Since the height of the peak is proportional to t^2 and its width is proportional to $1/t$, the area of the principal peak (or the transition probability) is proportional to time t, leading to a transition rate that is constant in time for broadband excitation.

To calculate the transition rate for broadband excitation, we use Eq. (10.3.18) in the coordinate representation and write the transition probability (10.3.20) for absorption from a monochromatic perturbation of freqeuncy ω as

$$|a_f(t)|^2 = \frac{t^2}{\hbar^2} |A_\omega|^2 \left| \frac{i\hbar e}{m} \int u_f^*(\boldsymbol{r}) e^{i\boldsymbol{k}\cdot\boldsymbol{r}} \boldsymbol{\varepsilon}_o \cdot \boldsymbol{\nabla} u_i(\boldsymbol{r}) d\tau \right|^2 \frac{\sin^2(\omega - \omega_{fi})t/2}{(\omega - \omega_{fi})t/2)^2} . \tag{10.3.21}$$

Here we have written the vector potential amplitude as $\boldsymbol{A}_o = A_\omega \boldsymbol{\varepsilon}_o$ so that A_ω is the vector potential amplitude at frequency ω and $\boldsymbol{\varepsilon}_o$ is the polarization of the incident field. The amplitude A_ω can be expressed in terms of energy density I_ω as[5]

$$|A_\omega|^2 \to \frac{I_\omega}{2\epsilon_0\omega^2}\,. \tag{10.3.22}$$

For a distribution of frequencies, we sum the contributions from all frequency components[6] to obtain

$$|a_f(t)|^2 = \frac{t^2}{\hbar^2}\sum_\omega \frac{I_\omega}{2\epsilon_0\omega^2}\left|\frac{i\hbar e}{m}\int u_f^* e^{i\boldsymbol{k}\cdot\boldsymbol{r}}\boldsymbol{\varepsilon}_o\cdot\boldsymbol{\nabla}u_i d\tau\right|^2 \frac{\sin^2(\omega-\omega_{fi})t/2}{(\omega-\omega_{fi})t/2)^2}\,. \tag{10.3.23}$$

For a continuous distribution of frequencies we replace the frequency sum by an integral $\sum_\omega I_\omega \to \int d\omega I(\omega)$, where $I(\omega)$ is the spectral energy density (energy per unit volume per unit frequency interval) of the radiation field so that $I(\omega)d\omega$ is the energy density of the radiation field within a frequency band $d\omega$ in the neighborhood of ω. For finding the total rate of transition we can assume the spectrum of the perturbing radiation field has a bandwidth $\Delta\omega$ centered at $\omega \approx \omega_{fi}$.

The total transition rate is then given by

$$|a_f(t)|^2 = \frac{t^2e^2}{2\epsilon_0 m^2}\int d\omega\, \frac{I(\omega)}{\omega^2}\left|\int u_f^* e^{i\boldsymbol{k}\cdot\boldsymbol{r}}\boldsymbol{\varepsilon}_o\cdot\boldsymbol{\nabla}u_i d\tau\right|^2 \frac{\sin^2(\omega-\omega_{fi})t/2}{[(\omega-\omega_{fi})t/2]^2}\,. \tag{10.3.24}$$

The last factor in the integrand is sharply peaked at $\omega \approx \omega_{fi}$. We can therefore take the factor $I(\omega_{fi})/\omega_{fi}^2$ and the matrix element out of the integral. The remaining integral can be evaluated by extending the limits of integration from $-\infty$ to ∞ with little error

$$\int_{-\infty}^{\infty} d\omega \frac{\sin^2(\omega-\omega_{fi})t/2}{[(\omega-\omega_{fi})t/2]^2} = \frac{2\pi}{t}\,. \tag{10.3.25}$$

Using this result in Eq. (10.3.24), we find the probability of transition increases linearly with time

$$|a_f(t)|^2 = \frac{\pi t e^2}{\epsilon_0 m^2\omega_{fi}^2} I(\omega_{fi})\left|\int u_f^* e^{i\boldsymbol{k}\cdot\boldsymbol{r}}\boldsymbol{\varepsilon}_o\cdot\boldsymbol{\nabla}u_i d\tau\right|^2\,. \tag{10.3.26}$$

Thus absorption from a broadband field results in a constant transition rate

$$w_{i\to f}^{\text{abs}} = \frac{\pi e^2}{\epsilon_0 m^2\omega_{fi}^2} I(\omega_{fi})\left|\int u_f^* e^{i\boldsymbol{k}\cdot\boldsymbol{r}}\boldsymbol{\varepsilon}_o\cdot\boldsymbol{\nabla}u_i d\tau\right|^2\,. \tag{10.3.27}$$

---

[5]Classically, the power flow across a unit area normal to the propagation vector is given by the Poynting vector $\boldsymbol{S} = \frac{1}{\mu_0}(\boldsymbol{E}\times\boldsymbol{B})$. Expressing the fields in terms of $\boldsymbol{A}$ [Eqs. (10.3.4) and (10.3.5)], with $\boldsymbol{A}$ given by Eq. (10.3.12), we find the Poynting vector averaged over one oscillation is given by

$$\overline{\boldsymbol{S}} = c\epsilon_0 2\omega^2|A_\omega|^2\boldsymbol{e_k}$$

where $\boldsymbol{e_k}$ is a unit vector in the direction of propagation of the incident radiation. If I_ω is the energy density (energy per unit volume) at frequency ω, then $|\overline{\boldsymbol{S}}| \equiv cI_\omega = c\epsilon_0 2\omega^2|A_\omega|^2\boldsymbol{e_k}$ or $|\boldsymbol{A}_\omega|^2 = \frac{I_\omega}{2\epsilon_0\omega^2}$

[6]In calculating the probability from the amplitude $a_f(t)$ we not only generate terms like Eq. (10.3.21) for each frequency component, but also cross terms (interference terms). The interference terms, however, make negligible contribution over time scales large compared with $2\pi/\Delta\omega$.

Similarly, the transition rate for emission ($E_f < E_i$) when the spectrum of a broadband perturbing radiation is centered at $\hbar\omega \approx -(E_f - E_i) \equiv -\hbar\omega_{fi}$ is given by

$$w_{i\to f}^{em} = \frac{\pi e^2}{\epsilon_0 m^2 \omega_{fi}^2} I(\omega_{fi}) \left| \int u_f^* \, e^{-i\boldsymbol{k}\cdot\boldsymbol{r}} \boldsymbol{\varepsilon}_o^* \cdot \boldsymbol{\nabla} u_i d\tau \right|^2 . \tag{10.3.28}$$

If we consider absorption and emission between two fixed levels say, $|n\rangle$ and $|m\rangle$ with $E_m < E_n$ [Fig. 10.4], then for absorption the system must start intially in the lower state $|m\rangle$ and end up in the excited state $|n\rangle$. The rate for this transition is

$$w_{m\to n}^{\text{ab}} = \frac{\pi e^2}{\epsilon_0 m^2 \omega_{nm}^2} I(\omega_{nm}) \left| \int u_n^* e^{i\boldsymbol{k}\cdot\boldsymbol{r}} \boldsymbol{\varepsilon}_o \cdot \boldsymbol{\nabla} u_m d\tau \right|^2 . \tag{10.3.29}$$

For emission, the system must start in the upper state $|n\rangle$ and end up in the lower state $|m\rangle$. The rate for this transition is

$$w_{n\to m}^{\text{em}} = \frac{\pi e^2}{\epsilon_0 m^2 \omega_{nm}^2} I(\omega_{nm}) \left| \int u_m^* \, e^{-i\boldsymbol{k}\cdot\boldsymbol{r}} \boldsymbol{\varepsilon}_o^* \cdot \boldsymbol{\nabla} u_n d\tau \right|^2 . \tag{10.3.30}$$

Since the rates of absorption and emission of radiation are proportional to the spectral energy density of the perturbing electromagnetic field, these processes are referred to as *induced absorption* and *induced emission* (or stimulated emission), respectively.

By writing absorption and emission rates as $w_{m\to n}^{\text{ab}} = I(\omega_{nm}) B_{m\to n}^{\text{ab}}$ and $w_{n\to m}^{\text{em}} = I(\omega_{nm}) B_{n\to m}^{\text{em}}$, we find

$$B_{m\to n}^{\text{ab}} = \frac{\pi e^2}{\epsilon_0 m^2 \omega^2} \left| \int u_n^* e^{i\boldsymbol{k}\cdot\boldsymbol{r}} \boldsymbol{\varepsilon}_o \cdot \boldsymbol{\nabla} u_m d\tau \right|^2 , \tag{10.3.31}$$

$$B_{n\to m}^{\text{em}} = \frac{\pi e^2}{\epsilon_0 m^2 \omega^2} \left| \int u_m^* \, e^{-i\boldsymbol{k}\cdot\boldsymbol{r}} \boldsymbol{\varepsilon}_o^* \cdot \boldsymbol{\nabla} u_n d\tau \right|^2 . \tag{10.3.32}$$

The coefficients $B_{m\to n}^{ab}$ and $B_{n\to m}^{em}$, defined by Eqs. (10.3.31) and (10.3.32), are called the coefficient of induced absorption and coefficient of induced (or stimulated) emission, respectively. An inspection of these equations shows that the integrals in the expressions for $w_{n\to m}^{\text{ab}}$ and $w_{n\to m}^{\text{em}}$ are complex conjugates of one another so that $B_{m\to n}^{ab} = B_{n\to m}^{em}$ and the rate of induced transition $|m\rangle \to |n\rangle$ is the same as the rate of induced transition $|n\rangle \to |m\rangle$ provided the perturbing radiation field has the same intensity at the transition frequency.

10.4 Einstein Coefficients

According to semi-classical theory of radiation, absorption and emission of radiation by an atomic system can take place only in the presence of a perturbing radiation field. However, the transition $n \to m$ ($E_m < E_n$) is known to occur even in the absence of a perturbing radiation field. This happens on account of the interaction of the atomic electron with the vacuum of the electromagnetic field. This emission, which occurs without the perturbing influence of the external radiation field, is called *spontaneous emission*. It cannot be explained on the basis of semi-classical theory of interaction of radiation field with atomic systems, but would result from a theory in which the radiation field is also quantized [Chapter 13].

Presently, without going into the quantization of the radiation field, one may define the *coefficient of spontaneous emission* $A^{\text{em}}_{n\to m}$ as the probability per unit time of spontaneous emission (in the absence of external radiation field) when the system is initially in the excited state $|n\rangle$:

$$w^{\text{sp}}_{n\to m} = A^{\text{em}}_{n\to m}\,. \tag{10.4.1}$$

From a very simple consideration of detailed balancing in an assembly of identical atoms in different excited states, and in statistical equilibrium with the radiation field at a certain temperature, Einstein was able to derive a relationship between the coefficient of *spontaneous emission* A and the coefficient of induced emission B (We have already seen that the coefficients of induced emission $B^{\text{ab}}_{m\to n}$ and induced absorption $B^{\text{em}}_{n\to m}$ are equal). Let p_m and p_n be the probabilities for the atom to be in the states $|m\rangle$ and $|n\rangle$, respectively. Then the rate of absorption of radiation will be

$$R_{m\to n} = p_m w^{\text{ab}}_{m\to n} = p_m I(\omega_{nm}) B^{\text{ab}}_{m\to n}\,, \tag{10.4.2}$$

and the rate of emission (rate of induced emission + rate of spontaneous emission) will be

$$R_{n\to m} = p_n w^{\text{em}}_{m\to n} + p_n A^{\text{em}}_{n\to m} = p_n I(\omega_{nm}) B^{\text{em}}_{n\to m} + p_n A^{\text{em}}_{n\to m}\,. \tag{10.4.3}$$

If the assembly of atomic systems and radiation is in equilibrium at a certain temperature T then rates of upward and downward transition between levels $|n\rangle$ and $|m\rangle$ must be equal. With Eqs. (10.4.2) and (10.4.3) this gives us

$$\frac{p_n}{p_m} = \frac{I(\omega_{nm})B_{m\to n}}{I(\omega_{nm})B_{n\to m} + A_{n\to m}} = \frac{I(\omega_{nm})B_{n\to m}}{I(\omega_{nm})B_{n\to m} + B_{n\to m}}\,, \tag{10.4.4}$$

where we have used $B_{n\to m} = B_{m\to n}$. But according to Maxwell-Boltzmann distribution, in thermal equilibrium the level probabilities p_n and p_m are related by

$$\frac{p_n}{p_m} = \frac{e^{-E_n/k_B T}}{e^{-E_m/k_B T}} = e^{-\hbar\omega_{nm}/k_B T}\,, \tag{10.4.5}$$

where k_B is the Boltzmann constant and $\omega_{nm} = (E_n - E_m)/\hbar$ is the transition frequency. Using this result in Eq. (10.4.4), we find the coefficients of spontaneous and induced emission are related by

$$\frac{A_{n\to m}}{B_{n\to m}} = I(\omega_{nm})\left[e^{\hbar\omega_{nm}/k_B T} - 1\right]. \tag{10.4.6}$$

Now, according to Planck's law [Chapter 1, Eq. (1.1.2)], the spectral energy density for radiation in equilibrium at temperature T is

$$I(\omega)d\omega = \frac{\hbar\omega^3}{\pi^2 c^3}\frac{d\omega}{e^{\hbar\omega/k_B T} - 1}\,. \tag{1.1.2*}$$

Using this result in Eq. (10.4.4), we find the ratio of spontaneous to induced emission is

$$\frac{A_{n\to m}}{B_{n\to m}} = \frac{\hbar\omega^3_{nm}}{\pi^2 c^3}. \tag{10.4.7}$$

Thus, from very simple considerations, Einstein was able to deduce the ratio between the coefficients of spontaneous and induced emission. These two coefficients are known as Einstein A and B coefficients.

10.5 Multipole Transitions

The transition rates involve the matrix element [see Eq. (10.3.27)]

$$\alpha_{fi} \equiv \int u_f^* e^{i\boldsymbol{k}\cdot\boldsymbol{r}} \varepsilon_0 \cdot \boldsymbol{\nabla} u_i \, d\tau \,. \tag{10.5.1}$$

For atomic systems interacting with visible or ultra-violet light, the quantity

$$\boldsymbol{k}\cdot\boldsymbol{r} \approx \frac{2\pi a}{\lambda} \,,$$

where a is of the order of atomic size and λ is the wavelength of radiation. For atomic size $a \approx 0.1$ nm and the wavelegth of light ranging from ultra-violet to infra-red (approximately, 100 nm to 1000 nm) this ratio varies from 10^{-2} to 10^{-3}. Hence in these cases, the factor $e^{i\boldsymbol{k}\cdot\boldsymbol{r}} \approx 1$ to an excellent approximation, and we can write the matrix element as

$$\alpha_{fi} \equiv \int u_f^* \varepsilon_o \cdot \boldsymbol{\nabla} u_i \, d\tau \,. \tag{10.5.2}$$

This is known as the *electric dipole approximation* and the corresponding transition is referred to as electric dipole transition.

In the case of nuclear transitions, the quantity $\boldsymbol{k}\cdot\boldsymbol{r}$ is small but not negligible. In such cases we can consider a few terms in the expansion of $e^{i\boldsymbol{k}\cdot\boldsymbol{r}}$, and express the matrix element in Eq. (10.5.1) as a series. The successive terms of this series are referred to as matrix elements of the electric and magnetic multipoles of order $\ell = 1, 2, 3, \cdots$ Accordingly, the transition is referred to as the electric 2^ℓ-pole [dipole ($E1$), quadrupole ($E2$), octupole ($E3$), in general ($E\ell$)] transition if it is brought about by the corresponding electric multipole operator $Q_{\ell m}$ [parity: $(-1)^\ell$], or the magnetic 2^ℓ-pole (or $M\ell$) transition if it is brought about by the corresponding magnetic multipole operator $M_{\ell m}$ [parity: $(-1)^{\ell+1}$]. The electromagnetic transitions conserve parity and angular momentum. If J_i and J_f are the spins of the initial and final states and π_i and π_f are their parities then conservation of angular momentum and parity requires

$$|J_i - J_f| \le \ell \le J_i + J_f \tag{10.5.3}$$

$$\pi_f = \begin{cases} \pi_i(-1)^\ell & \text{for E}\ell \text{ transition} \\ \pi_i(-1)^{\ell+1} & \text{for M}\ell \text{ transition.} \end{cases} \tag{10.5.4}$$

Accordingly, transitions of even parity (i.e., with no difference between the parities of the initial and final states) can be $M1$ (magnetic dipole), $E2$ (electric quadrupole), $M3$ (magnetic octupole), $E4$ (electric hextapole), $\cdots$ transitions. On the other hand, transitions of odd parity (i.e. those in which the initial and final states of the nucleus have opposite parities) are $E1$ (electric dipole), $M2$ (magnetic quadrupole), $E3$ (electric octupole). $\cdots$ transitions. Hence if the spins (used in the sense of total angular momentum) and parities of the initial and final states of a nucleus are given, then one can identify the nature of electromagnetic transitions that could occur between them. For example, if the spin parity (using the notation J^π, where J denotes the spin and π denotes the parity) of the initial nuclear state is 1^+ ($J_i = 1, \pi = +$) and the final state spin parity is 0^+, then $\ell = 1$ and $\Delta\pi = 0$ so that this transition is a pure $M1$ (magnetic dipole) transition. As another example, if $J_i^{\pi_i} = \frac{3}{2}^+$ and $J_f^{\pi_f} = \frac{5}{2}^+$, then $1 \le \ell \le 4$ and $\Delta\pi = 0$ and it can be a mixed $M1$, $E2$, $M3$, $E4$ transition.[7]

10.6 Electric Dipole Transitions in Atoms and Selection Rules

We have seen that for atomic transitions, we may put $e^{i\boldsymbol{k}\cdot\boldsymbol{r}} \approx 1$ since atomic size is small compared to the wavelength of radiation ($|\boldsymbol{k}\cdot\boldsymbol{r}| \ll 1$). In this (electric dipole) approximation, if the field polarization is denoted by unit vector $\boldsymbol{\varepsilon}_o$, the transition matrix element α_{fi} [Eq. (10.6.6)] can be written as

$$\alpha_{fi} \approx \int u_f^* \boldsymbol{\varepsilon}_o \cdot \boldsymbol{\nabla} u_i d\tau = \frac{i}{\hbar}\int u_f^* \boldsymbol{\varepsilon}_o \cdot \boldsymbol{p} u_i d\tau = \frac{m}{\hbar^2}(E_i - E_f)\boldsymbol{\varepsilon}_o \cdot \boldsymbol{r}_{fi}, \tag{10.6.5}$$

where the matrix element $\boldsymbol{r}_{fi}$ is given by

$$\boldsymbol{r}_{fi} = \langle f|\, \hat{\boldsymbol{r}}\, |i\rangle = \int u_f^* \boldsymbol{r} u_i d\tau\,. \tag{10.6.6}$$

This result is conveniently obtained by evaluating α_{fi} in the Heisenberg picture

$$\alpha_{fi} = \frac{i}{\hbar}\boldsymbol{\varepsilon}_o \cdot \langle f|\, \hat{\boldsymbol{p}}\, |i\rangle = \frac{i}{\hbar}\boldsymbol{\varepsilon}_o \cdot \langle f^H|\, \hat{\boldsymbol{p}}^H\, |i^H\rangle\,.$$

Using the Heisenberg equation of motion for $\hat{\boldsymbol{r}}^H$, we find that the matrix element

$$\begin{aligned}\alpha_{fi} &= \frac{i}{\hbar}\boldsymbol{\varepsilon}_o \cdot \langle f^H|\, \hat{\boldsymbol{p}}^H\, |i^H\rangle = \frac{i}{\hbar}\boldsymbol{\varepsilon}_o \cdot \langle f^H|\, m\frac{d\hat{\boldsymbol{r}}^H}{dt}\, |i^H\rangle = \frac{m}{\hbar^2}\boldsymbol{\varepsilon}_o \cdot \langle f^H|\, [\hat{\boldsymbol{r}}, \hat{H}]\, |i^H\rangle \\ &= \frac{m}{\hbar^2}(E_i - E_f)\boldsymbol{\varepsilon}_o \cdot \langle f^H|\, \hat{\boldsymbol{r}}^H\, |i^H\rangle = \frac{m}{\hbar^2}(E_i - E_f)\boldsymbol{\varepsilon}_o \cdot \boldsymbol{r}_{fi}\,.\end{aligned}$$

It follows that a transition (emissive or absorptive) can occur only if $\boldsymbol{\varepsilon}_o \cdot \boldsymbol{r}_{fi}$ is non-zero. The conditions necessary for this requirement to be met give rise to what are known as the electric dipole ($E1$) selection rules. To explore these conditions, consider first the case of polarization parallel to the x-axis, $\boldsymbol{\varepsilon}_o = \boldsymbol{e}_x$. Then the matrix element $\alpha_{fi} = x_{fi}$. Expressing x in terms of spherical harmonics as $x = r\,\sin\theta\,\cos\varphi = r\sqrt{\frac{2\pi}{3}}\,[Y_{1,1}(\theta,\varphi) + Y_{1,-1}(\theta,\varphi)]$ and using the coordinate representation for the states

$$u_f(\boldsymbol{r}) \equiv \langle \boldsymbol{r}|\, f\rangle = \langle \boldsymbol{r}|\, n_f, \ell_f, m_f\rangle = R_{n_f\ell_f}(r)Y_{\ell_f m_f}(\theta,\varphi) \tag{10.6.7}$$

and

$$u_i(\boldsymbol{r}) \equiv \langle \boldsymbol{r}|\, i\rangle = \langle \boldsymbol{r}|\, n_i, \ell_i, m_i\rangle = R_{n_i\ell_i}(r)Y_{\ell_i m_i}(\theta,\varphi) \tag{10.6.8}$$

where n is the principal quantum number and $\ell(\ell+1)$ and m are the eigenvalues of the orbital angular momentum operator $\hat{L}^2$ and its z-component $\hat{L}_z$. The subscripts i and f

[7]Electric and magnetic multi-pole transitions of the same order cannot interfere with each other because they cannot occur simultaneously between the same two levels for parity considerations. But electric and magnetic multipole transitions of orders differing by one may interfere.

refer to the initial and final states. Then the matrix element x_{fi} can be written as

$$x_{fi} = \langle n_f, \ell_f, m_f| \, x \, |n_i, \ell_i, m_i\rangle$$
$$= \int_0^\infty R_{n_f\ell_f}(r) r R_{n_i\ell_i}(r) r^2 dr \times$$
$$\sqrt{\frac{2\pi}{3}} \int_0^\pi \int_0^{2\pi} Y^*_{\ell_f m_f}(\theta,\varphi)\left[Y_{1,1}(\theta,\varphi) + Y_{1,-1}(\theta,\varphi)\right] Y_{\ell_i m_i}(\theta,\varphi) d\Omega \,. \quad (10.6.9)$$

The angular integral in the last step may be evaluated in terms of the Clebsch-Gordan coefficients defined in Chapter 7 as

$$\int Y^*_{\ell_f m_f}(\theta,\varphi)\left[Y_{1,1}(\theta,\varphi) + Y_{1,-1}(\theta,\varphi)\right] Y_{\ell_i\, m_i}(\theta,\varphi) d\Omega$$
$$= \sqrt{\frac{3(2\ell_i+1)}{4\pi(2\ell_f+1)}} \, (1\, 0\, \ell_i\, 0|\ell 0)\left[(1\, 1\, \ell_i\, m_i|\ell_f m_f) + (1, -1\, \ell_i\, m_i|\ell_f m_f)\right] , \quad (10.6.10)$$

where we have used the identity

$$\int Y^*_{\ell m}(\theta,\varphi) Y_{LM}(\theta,\varphi) Y_{\ell' \, m'}(\theta,\varphi) d\Omega = \sqrt{\frac{(2L+1)(2\ell'+1)}{4\pi(2\ell+1)}} (L 0 \ell' 0|\ell 0)(L M \ell' m'|\ell m) \,, \quad (10.6.11)$$

and $(LM\ell'm'|\ell m)$ are the Clebsch-Gordan coefficients. From the properties of Clebsch-Gordan coefficients we can see that the angular integral and, therefore x_{fi}, is non-zero only if,

(i) ℓ_i, ℓ_f, and 1 satisfy the Δ-condition, i.e., $\ell_i = \ell_f \pm 1$ or $\Delta\ell \equiv |\ell_i - \ell_f| = 1$ (The case $\Delta\ell = 0$ is ruled out because the initial and final states have to have opposite parities for electric dipole transition), and

(ii) $m_f = m_i \pm 1$ or $\Delta m = \pm 1$.

Consider now the case in which the incident radiation polarized in an arbitrary direction $\boldsymbol{\varepsilon}_o$ (or is unpolarized) and find the conditions for the matrix element $\alpha_{fi} = \boldsymbol{\varepsilon}_o \cdot \boldsymbol{r}_{fi}$ to be non-zero. Writing the incident polarization as $\boldsymbol{\varepsilon}_o = \boldsymbol{e}_x \cos\alpha + \boldsymbol{e}_y \sin\beta + \boldsymbol{e}_z \cos\gamma$ in terms of its direction cosines $\cos\alpha$, $\cos\beta$, and $\cos\gamma$, the matrix element can be written as

$$\boldsymbol{\varepsilon}_o \cdot \boldsymbol{r} = x\cos\alpha + y\cos\beta + z\cos\gamma = r(\sin\theta\cos\varphi\cos\alpha + \sin\theta\sin\varphi\cos\beta + \cos\theta\cos\gamma)$$
$$\equiv r(C_1 Y_{11}(\theta,\varphi) + C_2 Y_{1,-1}(\theta,\varphi) + C_3 Y_{10}(\theta,\varphi) \,, \quad (10.6.12)$$

where C's are constants involving the direction cosines. Then the matrix element can be written as

$$\boldsymbol{\varepsilon}_o \cdot \boldsymbol{r}_{fi} = \int_0^\infty R_{n_f\ell_f}(r) r R_{n_i\ell_i}(r) dr \int Y_{\ell_f m_f}(\theta,\varphi)$$
$$\left[C_1 Y_{11}(\theta,\varphi) + C_2 Y_{1,-1}(\theta,\varphi) + C_3 Y_{10}(\theta,\varphi)\right] Y_{\ell_i m_i}(\theta,\varphi) d\Omega \,. \quad (10.6.13)$$

The angular integral can be evaluated in terms of Clebsch-Gordan coefficients as

$$\sqrt{\frac{3(2\ell_i+1)}{4\pi(2\ell_f+1)}} \, (10\ell_i 0|\ell_f 0)\left[C_1(1\, 1\ell_i\, m_i|\ell_f m_f) + C_2(1, -1\, \ell_i\, m_i|\ell_f m_f) + C_3(10\ell_i\, m_i|\ell_f m_f)\right] . \quad (10.6.14)$$

It follows from the properties of the Clebsch-Gordan coefficients, that this is non-zero (and so $\boldsymbol{\varepsilon}_o \cdot \boldsymbol{r}_{fi}$ is non-zero) only if (i) ℓ_f, ℓ_i and 1 satisfy the Δ-condition, viz., $\Delta\ell = \pm 1$ and (ii) $\Delta m = 0, \pm 1$. The case $\Delta\ell = 0$ is ruled out because the initial and final atomic states must have opposite parities to allow an electric dipole transition. These selection rules are consistent with those we stated earlier for electric dipole transitions [Eqs. (10.5.3) and (10.5.4)].

10.7 Photo-electric Effect

Due to the perturbing effect of the external radiation field, the absorption of a light quantum of energy $\hbar\omega$ by an atom may result in the emission of an electron, in which case the atom makes a transition from its initial bound state $|i\rangle$ to an ionized state $|f\rangle$ (represented by $u_{\boldsymbol{k}}(\boldsymbol{r}) = e^{i\boldsymbol{k}\cdot\boldsymbol{r}}/L^{3/2}$ in the coordinate space), belonging to a continuum of states. In this case ($E_f > E_i$) the probability of transition in time t is given by the first term of Eq. (10.3.17):

$$|a_f(t)|^2 = \left| \frac{(H_o')_{fi}}{\hbar} \frac{e^{i(\omega_{fi}-\omega)t} - 1}{\omega_{fi} - \omega} \right|^2 . \tag{10.7.1}$$

The matrix element $(H_o')_{fi}$ is given by

$$(H_0')_{fi} \equiv \frac{i\hbar e}{m} \int u_f{}^*(\boldsymbol{r}) e^{i\boldsymbol{k}\cdot\boldsymbol{r}} \boldsymbol{A}_0 \cdot \boldsymbol{\nabla} u_i d\tau \tag{10.7.2}$$

where e is the magnitude of the electronic charge. In this case we assume that the incident photon has a definite energy $\hbar\omega$. On the other hand, the final state $|f\rangle \equiv |E_k, \hbar\boldsymbol{k}\rangle$ is part of a continuum of states with $|\boldsymbol{k}| = k$ lying between k and $k + dk$ and the direction of $\boldsymbol{k}$ within a solid angle $d\Omega$. For the transition probability per unit time from $|i\rangle$ to one of the $|E_k, \hbar\boldsymbol{k}\rangle$ states we apply Fermi's golden rule [Eq. (10.2.5)]

$$w_{i\to f} = \frac{2\pi}{\hbar} |H_{fi}|^2 \rho(E_f) , \tag{10.2.5*}$$

where the density of final states $\rho(E_f)$ near the energy $E_f = E_k = \hbar^2 k^2/2m$ is given by

$$\rho(E_f) = \frac{d\Omega\, kmL^3}{8\pi^3\hbar^2} . \tag{10.2.8*}$$

In the present context this leads to the transition rate

$$w_{fi} = \frac{2\pi}{\hbar} \left| \frac{ie\hbar}{m} \int u_f^*(\boldsymbol{r}) e^{i\boldsymbol{k}\cdot\boldsymbol{r}} \boldsymbol{A}_0 \cdot \boldsymbol{\nabla} u_i(\boldsymbol{r}) d\tau \right|^2 \frac{d\Omega\, kmL^3}{8\pi^3\hbar^2} . \tag{10.7.3a}$$

The incident photon flux density for the monochromatic field is given by

$$\frac{\text{Intensity}}{\hbar\omega} = \frac{c\epsilon_0 2\omega^2 |A_0|^2}{\hbar\omega} . \tag{10.7.3b}$$

Dividing Eq. (10.7.3a) by Eq. (10.7.3b), we get the transition rate per unit incident flux density (differential cross-section)

$$d\sigma = \frac{w}{\text{flux density}} = \frac{2\pi}{\hbar} \left| -\frac{ie\hbar}{m} \int u_i^*(\boldsymbol{r}) e^{-i\boldsymbol{k}\cdot\boldsymbol{r}} \boldsymbol{A}_0 \cdot \boldsymbol{\nabla} u_f(\boldsymbol{r}) d\tau \right|^2 \frac{d\Omega k m L^3}{8\pi^3\hbar^2} \frac{\hbar\omega}{c 2\epsilon_0 \omega^2 |A_0|^2}$$

or

$$\frac{d\sigma}{d\Omega} = \frac{e^2 k\, |\boldsymbol{\varepsilon}_o \cdot \boldsymbol{k}|^2}{8\pi^2 \epsilon_0 m c\omega} \left| \int u_i^*(\boldsymbol{r}) e^{i(\boldsymbol{k}_0 - \boldsymbol{k})\cdot\boldsymbol{r}} d^3 r \right|^2 , \tag{10.7.4}$$

where we have taken the final state to be a plane wave state and we have used the fact that the integrals are complex conjugates of each other, except for sign. If the bound state function $u_{n_o}(r)$ is known, the last integral can be evaluated. Note that the cross-section is proportional to the absolute value of the Fourier transform squared of the bound state radial function with respect to $\boldsymbol{q} = \boldsymbol{k}_0 - \boldsymbol{k}$.

10.8 Sudden and Adiabatic Approximations

Finally, we consider two extreme cases of the time variation of the perturbation where the Hamiltonian of a system changes from $\hat{H}_o$ at $t = t_0$ to $\hat{H}_o + \hat{H}'$ at time $t = t_1$. The nature of approximations that are applicable in dealing with such changes depends on whether the change is sudden or gradual (or adiabatic). The terms *sudden* and *adiabatic* are defined relative to some characteristic time scale of the system, which may vary from one system to another and from one type of interaction to another for the same system. For example, for a harmonic oscillator of natural frequency ω_o, the characteristic time would be the classical period $2\pi/\omega_o$ of oscillation; for a spin system in a magnetic field, it would be the inverse of the classical cyclotron frequency. If the Hamiltonian changes in time $T = t_1 - t_o$ that is short compared to the classical period $2\pi/\omega_o$ so that $\omega_o T \ll 1$, we say the change is sudden and if it changes so slowly that $\omega_o T \gg 1$, we say the change is slow or adiabatic. Let us consider these cases separately and examine what kind of approximations can be used to deal with such perturbations.

Adiabatic Approximation

Consider a system subjected to a perturbation $\hat{H}'(t)$ which vanishes for $t < t_o$ and varies slowly with time so that the Hamiltonian changes from $\hat{H}_o$ at time t_o to $\hat{H}_o + \hat{H}'(t)$ at time t, the change being small due to slow variation. Assuming that the system is initially in state $|\psi_{n_o}(t_o)\rangle \equiv e^{-iE_{n_o}t_o}\,|\Psi_{n_o}(t_o)\rangle$ — an eigenstate of $\hat{H}_o$ with eigenvalue $E_{n_o}^{(0)} \equiv E_{n_o}(t_o)$ — the state $|\Psi_{n_o}(t)\rangle$ into which it evolves in time t may still be labeled by n_o and expanded as

$$|\Psi_{n_o}(t)\rangle = \sum_n a_n(t) e^{-iE_n^{(0)}t/\hbar}\,|\psi_n(t_o)\rangle\,, \tag{10.8.1}$$

where $E_n^{(0)} = E_n(t_o)$ denote the eigenvalues of the permissible orthogonal eigenstates $|\psi_n(t_o)\rangle$ of $\hat{H}_o$ at time t_o. It may also be pointed out here that, in general, the states $|\Psi_n(t)\rangle \equiv e^{-iE_n(t)t/\hbar}\,|\psi_n(t)\rangle$ are not strictly stationary as $|\Psi_n(t)\rangle$ as well as $E_n(t)$ may have slow dependence on time. Now, to first order in $\hat{H}'$

$$a_n(t) = \frac{1}{i\hbar}\int_{t_o}^{t} dt' H'_{nn_o}(t') e^{i\omega_{nn_o}t'}\,, \quad \omega_{nn_o} = (E_n^{(0)} - E_{n_o}^{(0)})/\hbar \tag{10.8.2a}$$

$$a_{n_o}(t) = 1 - \frac{i}{\hbar}\int_{t_o}^{t} dt' H'_{n_on_o}(t')\,. \tag{10.8.2b}$$

Integrating these by parts we obtain

$$a_n(t) = -\left[\frac{H'_{nn_o}(t')}{\hbar\omega_{nn_o}}\,e^{i\omega_{nn_o}t}\right]_{t_o}^{t} + \frac{1}{\hbar\omega_{nn_o}}\int_{t_o}^{t} dt' e^{i\omega_{nn_o}t'}\frac{\partial H'_{nn_o}}{\partial t'}\,. \tag{10.8.3}$$

The first term vanishes at the lower limit. If $\hat{H}'(t)$ is slowly varying, the second term is small compared to the first, and we have approximately

$$a_n(t) \approx -\frac{H'_{nn_o}(t)}{\hbar\omega_{nn_o}}\, e^{i\omega_{nn_o}t} = -\frac{H'_{nn_o}(t)}{E_n^{(0)} - E_{n_o}^{(0)}}\, e^{i\omega_{nn_o}t}\,. \tag{10.8.4}$$

Similarly for a_{n_o} we have

$$a_{n_o}(t) \approx e^{-iH'_{n_on_o}(t-t_o)/\hbar} \equiv a_{n_o}(t_o)e^{-iH'_{n_on_o}t/\hbar}\,. \tag{10.8.5}$$

Thus to first order in $\hat{H}'$, the wave function $|\Psi_{n_o}(t)\rangle$ becomes

$$\begin{aligned} |\Psi_{n_o}(t)\rangle &= a_{n_o}(t_o)\, |\psi_{n_o}(t_o)\rangle\, e^{-i(E_{n_o}^{(0)}+H'_{n_on_o})t/\hbar} + \sum_{n\neq n_o} a_n(t)\, |\psi_n(t_o)\rangle\, e^{-iE_n^{(0)}t/\hbar} \\ &= e^{-i(E_{n_o}^{(0)}+H_{n_on_o})t/\hbar} a_{n_o}(t_o)\, |\psi_{n_o}(t_o)\rangle + e^{-iE_{n_o}^{(0)}t/\hbar}\left[\sum_{n\neq n_o} \frac{H'_{nn_o}(t)}{E_{n_o}^{(0)} - E_n^{(0)}}\, |\psi_n(t_o)\rangle\right]. \end{aligned} \tag{10.8.6}$$

The second term is proportional to H'. If we multiply the second term, which is of first order in H', by $\exp[-iH'_{n_on_o}t/\hbar]$, we cause an error at most of order H'^2. Hence to first order in the perturbation we have

$$|\Psi_{n_o}(t)\rangle \approx e^{-i(E_{n_o}^{(0)}+H'_{n_on_o})t/\hbar}\left[a_{n_o}(t_o)\, |\psi_{n_o}(t_o)\rangle + \sum_{n\neq n_o} \frac{H'_{nn_o}(t)}{E_{n_o}^{(0)} - E_n^{(0)}}\, |\psi_n(t_o)\rangle\right]. \tag{10.8.7}$$

We recognize that the quantity inside the square brackets is simply an eigenstate of the Hamiltonian $\hat{H}_o + \hat{H}'(t)$ at time t with eigenvalue $E_{n_o} = E_{n_o}^{(0)} + H'_{n_on_o}(t)$ with t as a parameter. This means the system originally in an eigenstate $|\Psi_{n_o}(t_0)\rangle$ of $\hat{H}_o$ with eigenvalue $E_{n_o}^{(0)}$ evolves smoothely into an eigenstate $|\Psi_{n_o}(t)\rangle$ of the new Hamiltonian with eigenvalue $E_{n_o}(t) = E_{n_o}^{(0)} + H'_{n_on_o}(t)$ under adiabatic perturbation. Since the system evolves continuously we say it stays in state n_o-th state. Extending this to other states (assuming them to be discrete), we can say that under adiabatic perturbation the eigenvalues $E_1(t_o)$, $E_2(t_o), E_3(t_o), \cdots$ and the eigenfunctions $|\Psi_1(t_o)\rangle$, $|\Psi_2(t_o)\rangle$, $|\Psi_3(t_o)\rangle \cdots$ of the Hamiltonian $\hat{H}_0$ at time t_o, evolve continuously into the eigenvalues $E_1(t)$, $E_2(t), E_3(t), \cdots$ and the eigenfunctions $|\Psi_1(t)\rangle$, $|\Psi_2(t)\rangle$, $|\Psi_3(t\rangle, \cdots$ of the Hamiltonian $\hat{H}(t) = \hat{H}_o + \hat{H}'(t)$ at time t. This is the physical content of the *adiabatic theorem*.

The conditions for the adiabatic theorem to hold are easily derived from Eq. (10.8.3). If the perturbation changes slowly, we can take $\partial H'/\partial t$ outside the integral in Eq. (10.8.3), so that

$$\left|\int_{t_o}^{t} dt' e^{i\omega_{nn_o}t'} \frac{\partial H'_{nn_o}}{\partial t'}\right| \approx \left|\frac{\partial H'_{nn_o}}{\partial t}\int_{t_o}^{t} dt' e^{i\omega_{nn_o}t'}\right| = \left|\frac{\partial H'_{nn_o}}{\partial t}\frac{2\sin[\omega_{nn_o}(t-t_o)/2]}{\omega_{nn_o}}\right|.$$

Hence the condition for the first term in Eq. (10.8.3) to dominate, and the adiabatic theorem to hold is

$$\left|\frac{1}{\omega_{nn_o}}\frac{\partial H'_{nn_o}}{\partial t}\right| \ll \left|H'_{nn_o}\right| \quad \text{or} \quad \frac{1}{\omega_{nn_o}}\left|\frac{1}{H'_{nn_o}}\frac{\partial H'_{nn_o}}{\partial t}\right| \ll 1\,. \tag{10.8.8}$$

That is, the fractional change in the perturbation $\hat{H}'$ during periods $\sim 2\pi/\omega_{nn_o}$ of transition from state n_o to other states n, should be small for the adiabatic approximation to hold. In this limit the probability of transition to a state with $\omega_{nn_o} \neq 0$ is determined by the second term in Eq. (10.8.3)

$$p_{n_o \to n} = \left| \frac{1}{\hbar\omega_{nn_o}} \int_{t_o}^{t} dt' e^{i\omega_{nn_o}t'} \frac{\partial H'_{nn_o}}{\partial t'} \right|^2 , \tag{10.8.9}$$

which in view of the inequality (10.8.9), would be very small.

We have considered the adiabatic approximation for perturbations that are so weak that perturbation theory can be used. However, the conclusions are valid even when the perturbation changes by large amounts as long as the change takes over a time long enough that the condition (10.8.9) is met. The basic idea of adiabatic approximation is that if $\partial H'/\partial t$ is small enough then the wave function at any instant $t = \tau$ is given by the eigenvalue equation

$$[\hat{H}_o + \hat{H}'(\tau)] \left|\psi_n(\tau)\right\rangle = E_n(\tau) \left|\psi_n(\tau)\right\rangle . \tag{10.8.10}$$

An important application of adiabatic approximation is in the collision of molecules in a gas. Intermolecular forces come into play as the molecules approach one another. Since the average speed of gas molecules is small compared to the electronic speeds in atoms, the molecules do not move much during one electronic period. Hence the change in intermolecular forces in one electronic period is small and we may view intermolecular force as being adiabatically switched on with regard to electronic transitions. On other hand the adiabatic approximation may not be satisfied with respect to rotational or vibrational periods. Consequently one finds that while rotational and vibrational states are altered in the collision, electronic states remain largely unchanged.

As a way of illustrating adiabatic and sudden approximations and their relation to perturbation theory, consider a charged linear oscillator, which is subjected to a perturbing electric field

$$\mathcal{E}(t) = \frac{\mathcal{E}_o}{\sqrt{\pi}} e^{-t^2/\tau^2} , \tag{10.8.11}$$

parallel to its axis. Here $\mathcal{E}_o$ is the peak value of the electric field and and τ characterizes the time scale for the variation of the electric field. Suppose the oscillator is in the ground state initially. The Hamiltonian for the linear oscillator can be written as

$$\hat{H} = \frac{\hat{p}^2}{2m} + \frac{1}{2}m\omega^2\hat{x}^2 - ex\mathcal{E}_o \frac{e^{-t^2/\tau^2}}{\sqrt{\pi}} = \hat{H}_o + H'(t) . \tag{10.8.12}$$

Then the probability of transition (in first order of perturbation) to state n is given by

$$p_n = \frac{1}{\hbar^2} \left| \int_{-\infty}^{\infty} dt \left\langle n\right| \hat{H}' \left|0\right\rangle e^{i\omega_{n0}t} \right|^2 = \frac{e^2\mathcal{E}_0^2 |\left\langle n\right| \hat{x} \left|0\right\rangle|^2}{\hbar^2} \left| \int_{-\infty}^{\infty} dte^{-t^2/\tau^2 + i\omega_{n0}t} \right|^2 . \tag{10.8.13}$$

The matrix element $\langle n| \hat{x} |0\rangle$ according to Eq. (5.4.32) is given by

$$\left\langle n\right| \hat{x} \left|0\right\rangle = \sqrt{\frac{\hbar}{2m\omega}} \delta_{n1} \tag{10.8.14}$$

so that the only nonzero transition probability in first order is to state $n = 1$

$$p_1 = \frac{e^2\mathcal{E}_0^2}{2m\hbar\omega} \left| \int_{-\infty}^{\infty} dte^{-t^2/\tau^2 + i\omega_{10}t} \right|^2 = \frac{e^2\mathcal{E}_0^2\tau^2}{2m\hbar\omega} e^{-(\omega\tau)^2/2} . \tag{10.8.15}$$

Note that the quantity $e\mathcal{E}_0\tau$ is the impulse received by the oscillator. Hence the quantity in front of the exponential is essentially the ratio of energy received by the oscillator to the quantum of energy. Transition to states with $n > 1$ is possible only in higher orders of perturbation.

We can see that if $\tau \gg 1/\omega$, the probability for transition from the ground state is exponentially small. This is the adiabatic limit. On the other hand if $\tau \ll 1/\omega$, the perturbation aproaches a delta function and this probability is approximately constant and given by

$$p_n = \frac{e^2 \mathcal{E}_0^2 \tau^2}{2m\hbar\omega} . \tag{10.8.16}$$

This, of course, has to be small since we have used a perturbative treatment.

Sudden Approximation

In the other extreme case of time variation, where the perturbation is applied suddenly, the derivative $\frac{\partial H'}{\partial t}$ becomes large at the instant of application. Then we can ignore the first term (again working with small perturbation so that the perturbative treatment is applicable) in Eq. (10.8.3). Taking out the comparatively slowly varying phase factor $e^{i\omega_{nn_o}t}$ we can carry out the integration to yield

$$a_n(t) = \frac{H'_{nn_o}}{\hbar\omega_{nn_o}} e^{i\omega_{nn_o}t} \tag{10.8.17}$$

with the transition probability given by

$$p_{n_o \to n} = \left| \frac{H'_{nn_o}}{\hbar\omega_{nn_o}} \right|^2 , \tag{10.8.18}$$

where we have assumed that the interaction is zero for $t < t_o$ and changes to value H' within a very short time interval thereafter. The condition for the sudden approximation to be valid is, from Eq. (10.8.3),

$$\frac{1}{\omega_{nn_o}} \left| \frac{1}{H'_{nn_o}} \frac{\partial H'_{nn_o}}{\partial t} \right| \gg 1 . \tag{10.8.19}$$

This result is consistent with the formula (10.8.18) of Sec. 10.2 for the case of a perturbation which is turned on suddenly at $t = t_0$. The *maximum* value of transition probability (10.2.3) for $(n \neq n_o)$ is indeed given by Eq. (10.8.18) and for short times the transition to an orthogonal state is negligible, being $\sim |H_{nn_o}|^2 T^2$. The probability that the system will make a transition to an orthogonal state, as a result of sudden change in the Hamiltonian, is very small.

If the change in the Hamiltonian is sudden but small, we can speak of the transition between the states of the original Hamiltonian. However, if the change is sudden and large it is more meaningful to talk about the transition to the states of the new Hamiltonian. These probabilities are also easily calculated. Consider a system initially in an eigenstate $\left|\psi_{n_o}^{(0)}\right\rangle$ of the original Hamiltonian $\hat{H}_o$ and let the Hamiltonian change suddenly to $\hat{H}$. If the change occurs in a time short compared to the characteristic periods $2\pi/\omega_{nn_o}$ for transition to other states, the state of the system is unable to vary and remains the same as before the change. It is, however, not an eigenstate (stationary state) of the new Hamiltonian $\hat{H}$. By expanding the initial state $|\psi_{n_o}\rangle$ in terms of the eigenstates $|\psi_n\rangle$ of the new Hamiltonian $\hat{H}$ as

$$\left|\psi_{n_o}^{(0)}\right\rangle = \sum_m a_m \left|\psi_m\right\rangle , \tag{10.8.20}$$

we find the probability of transition from state $\left|\psi_{n_o}^{(0)}\right\rangle$ of the original Hamiltonian to state $|\psi_m\rangle$ of the new Hamiltonian is given by

$$p_{n_o\to m} = |a_m|^2 = \left|\langle\psi_m|\psi_{n_o}^{(0)}\rangle\right|^2 . \tag{10.8.21}$$

10.9 Second Order Effects

Second and higher order perturbation effects can play a significant role if the first order matrix elements vanish. This is particularly important when periodic perturbations, such as those induced by the electric field of a laser are experienced by atoms and molecules. In many cases such effects are purposely enhanced, for example, in coherent light generation in nonlinear optics. In such cases it becomes necessary to calculate the second order transition probabilities. We shall confine ourselves to calculating second order transition probability with periodic perturbations. Consider a periodic perturbation of the form

$$\hat{H}'(t) = \hat{H}'(0)e^{-i\omega t} + \hat{H}'^{\dagger}e^{i\omega t} , \tag{10.9.22}$$

which is turned on at $t = 0$ and stays on for a long time. We assume, for simplicity of writing, that $\hat{H}'(0) \equiv \hat{H}'$ is Hermitian. Then, from Eq. (10.1.18), the second order transition amplitude for transition from an intial state i to some final state f will be

$$\begin{aligned} a_f^{(2)} &= -\frac{1}{\hbar^2}\sum_m H'_{fm}H'_{mi}\int_0^t dt_2 e^{i\omega_{fm}t_1}(e^{-i\omega t_1}+e^{i\omega t_1})\int_0^{t_1} dt_2 e^{i\omega_{mi}t_2}(e^{-i\omega t_2}+e^{i\omega t_2}) \\ &= -\frac{1}{\hbar^2}\sum_m H'_{fm}H'_{mi}\int_0^t dt_2 e^{i\omega_{fm}t_1}(e^{-i\omega t_1}+e^{i\omega t_1}) \\ &\quad\times\left[\frac{e^{i(\omega_{mi}-\omega)t_1}-1}{i(\omega_{mi}-\omega)}+\frac{e^{i(\omega_{mi}+\omega)t_1}-1}{i(\omega_{mi}+\omega)}\right] . \end{aligned} \tag{10.9.23}$$

On carrying out this integration we get eight terms involving the frequencies $\omega_{fi} \pm 2\omega$, $\omega_{mi} \pm \omega$, $\omega_{fm} \pm \omega$ and ω_{fi}. When the perturbation has been on for a time large compared with the period of any these frequencies, a careful consideration of each of these terms, following the discussion of the first order terms shows, that significant contribution to second order amplitude comes from the term whose frequency vanishes. If we consider transition between nondegenerate levels $\omega_{fi} \neq 0$ and $\omega_{mi} \pm \omega$ and $\omega_{nm} \pm \omega$ are nonzero, i.e., the frequency of periodic perturbation does not match any of the transition frequecies, then the only contribution can come from the terms involving $\omega_{fi} \pm 2\omega$:

$$\begin{aligned} a_f^{(2)} \approx -\frac{it}{\hbar^2}\sum_m H'_{fm}H'_{mi} &\left[\frac{e^{i(\omega_{fi}-2\omega)t/2}}{\omega_{mi}-\omega}\left(\frac{\sin[(\omega_{fi}-2\omega)t/2]}{(\omega_{fi}-2\omega)t/2}\right)\right. \\ &\left.+\frac{e^{i(\omega_{fi}+2\omega)t/2}}{\omega_{mi}+\omega}\left(\frac{\sin[(\omega_{fi}+2\omega)t/2]}{(\omega_{fi}+2\omega)t/2}\right)\right] . \end{aligned} \tag{10.9.24}$$

The first term dominates when $\omega_{fi} \approx 2\omega$ $(E_f > E_i)$ and the second term dominates when $-\omega_{fi} \approx 2\omega$ $(E_f < E_i)$. We will see in Chapter 13 that these terms correspond to two-photon absorption and emission processes respectively. Important observation here is that when the condition $\omega_{fi} \approx 2\omega$ is satisfied, the second order transition amplitude has the same time dependence as the first order contribution. Hence the transition probability for absorption in second order is

$$|a_f|^2 = \frac{t^2}{\hbar^2}\left|\sum_m \frac{H'_{fm}H'_{mi}}{\hbar(\omega_{mi}-\omega)}\right|^2 \left(\frac{\sin(\omega_{fi}-2\omega)t/2}{(\omega_{fi}-2\omega)t/2}\right)^2 . \tag{10.9.25}$$

From this point onward we can use arguments similar to those used for first order transition. For example, when the final level lies in the continuum, where the level density is $\rho(E_f)$, the second order absorption rate is given by

$$\begin{aligned} w^{(2)}_{i\to f} &= \frac{1}{t}\int_{\Delta E_f} dE_f\, \rho(E_f)|a^{(2)}_f|^2 , \\ &= \frac{t}{\hbar^2}\int_{\Delta E_f} \rho(E_f)dE_f \left|\sum_m \frac{H'_{fm}H'_{mi}}{\hbar(\omega_{mi}-\omega)}\right|^2 \left(\frac{\sin[(\omega_{fi}-2\omega)t/2]}{(\omega_{fi}-2\omega)t/2}\right)^2 . \end{aligned} \tag{10.9.26}$$

The integrand is sharply peaked at $\omega_{fi} = 2\hbar\omega$ since $\frac{\sin^2(\omega_{fi}-2\omega)t/2}{(\omega_{fi}-2\omega)t/2} \to (2\pi/t)\delta(\omega_{fi}-2\omega)$ as $t \to \infty$. We can then evaluate the integral at its and get the absorption rate in second order as

$$w^{(2)}_{i\to f} = \frac{2\pi}{\hbar}\left|\sum_m \frac{H'_{fm}H'_{mi}}{E_m - E_i - \hbar\omega}\right|^2 \rho(E_f = E_i + \hbar\omega) . \tag{10.9.27}$$

Note that the second order rate in general is much smaller than the first order rate unless, of course, the first order rate vanishes. The second term in Eq. (10.9.23) describes, for example, the decay of an excited state which is forbidden in the first order. We shall encounter more examples of second order processes in Chapter 13.

The range of processes described by the second order transition amplitude is much richer than what we have described. If the interaction Hamiltonian has two frequencies ω_1 and ω_2 present in it, the second order amplitude will have terms that will dominate when ω_{fi} equals second harmonic $[\pm 2\omega_1, \pm 2\omega_2]$ or sum and difference fequencies [$\pm(\omega_1 \pm \omega_2)$. These terms play an important role in nonlinear optics.

Problems

1. A linear harmonic oscillator has time-dependent perturbation so that its total Hamiltonian is given by

$$\hat{H} = \begin{cases} \frac{\hat{p}^2}{2m} + \frac{1}{2}k\hat{x}^2 \equiv \hat{H}_0 & \text{for } t < 0 \\ \hat{H}_0 + \epsilon\hat{x}e^{-t/\tau} & \text{for } t > 0 \end{cases}$$

where ϵ is a small number. Using the first order time-dependent perturbation theory find the probability of transition of the oscillator from the ground state of its first excited state in time $t > 0$.

2. Calculate the probability/time for the spontaneous radiative transition $2p \to 1s$ in a Hydrogen atom.

3. A system with Hamiltonian $\hat{H}_0$ can exist in two states with energies E_1 and E_2 where E_1 is the ground level. It is subjected to a time-dependent perturbation $\hat{H}'(t) = -\hat{\mu} E_o \cos \omega t$ so that

$$\langle E_1 | \hat{H}'(t) | E_1 \rangle = 0 = \langle E_2 | \hat{H}'(t) | E_2 \rangle$$
$$\langle E_1 | \hat{H}'(t) | E_2 \rangle = -\frac{1}{2} \hbar \Omega_o e^{i\omega t} + c.c. = \langle E_2 | \hat{H}'(t) | E_1 \rangle^*$$

where $\hbar \Omega_o = \mu_{12} E_o = \mu_{21} E_o$. Writing the state of the system as $|\Psi\rangle = a_1(t) e^{-iE_1 t/\hbar} |E_1\rangle + a_2(t) e^{-iE_2 t/\hbar} |E_2\rangle$, show that $a_1(t)$ and $a_2(t)$ obey the coupled equations ($k = 1, 2$)

$$i\hbar \, \dot{a}_k(t) = \sum_{n=1}^{2} a_n(t) \, H'_{kn}(t) \, e^{i(E_k - E_n)t/\hbar} .$$

Write the coupled equations explicitly and argue that the term with dominant oscillations at frequency $\omega + \omega_o$ can be ignored compared to the term containing oscillations at frequency $\omega - \omega_o$ where $\omega_o = (E_2 - E_1)/\hbar$. This is the so-called rotating wave approximation (RWA).

Assuming that the system is initially ($t = 0$) in the ground state, find the probabilities $|a_1(t)|^2$ and $|a_2(t)|^2$ that, the states $|E_1\rangle$ and $|E_2\rangle$ are populated, by solving the coupled equations for a_1 and a_2 in the RWA. Define short time. How do these probabilities depend on time for very short times?

4. Show that when an atom is placed in a radiation field with $\boldsymbol{\nabla} \cdot \boldsymbol{A}(\boldsymbol{r}, t) = 0$, its Hamiltonian is modified (in the coordinate representation) from $H_0 = -\frac{\hbar^2}{2m} \nabla^2 + V(\boldsymbol{r})$ to $H = H_0 + \frac{i\hbar e}{m} \boldsymbol{A} \cdot \boldsymbol{\nabla}$. Show also that the perturbation term is Hermitian.

5. A Hydrogen atom in its ground state ($1s$) is subjected to the following time-dependent homogeneous electric field

$$\boldsymbol{E}(t) = \begin{cases} 0 \, , & t < 0 \\ \boldsymbol{E}_0 e^{-t/\tau} \, , & t > 0 \, . \end{cases}$$

Find the probability that the atom is in $2s$ state after a long time.

Find also the corresponding probability that it is in one of the $2p$ states. Hint: Evaluate $|a_n(t)|^2$ [from Eq. (10.1.17)] for the two cases in the limit $t \to \infty$.

6. A Hydrogen atom in its ground state is subjected to the electric field of a UV laser $E(t) = E_0 \sin(\boldsymbol{k}_o \cdot \boldsymbol{r} - \omega t)$ with $\omega > me^4 / 2(4\pi\epsilon_0)^2 \hbar^3$. What is the probability per unit time that the atom will be ionized?

7. A Hydrogen atom in its first excited state ($2p$) is placed in a cavity at temperature T. At what temperature of the cavity are the probabilities of spontaneous and induced emissions $2p \to 1s$ equal?

8. A Tritium atom (H_3) nucleus has two neutrons and one proton. It can decay by β-emission to a Helium nucleus with two protons and one neutron. If a Tritium atom in its ground state decays by β emission, calculate the probability that the He_3^+ atom is produced in an excited state given that the kinetic energy of the emitted electron (β particle) is about 16 keV and that both H_3 and He_3^+ are hydrogenic atoms.

9. A charged linear harmonic oscilator in its ground state is subjected to a electric field which is suddenly turned on at $t = 0$ from 0 to some value $\mathcal{E}_o$ (not necesarily small). Find the probability of excitation of the n-th level of the final Hamiltonian. Hint: The new Hamiltonian can be written as

$$\hat{H} = \frac{\hat{p}^2}{2m} + \frac{1}{2}m\omega^2\hat{x}^2 - e\hat{x}\mathcal{E}_0 = \frac{\hat{p}^2}{2m} + \frac{1}{2}m\omega^2\left(\hat{x} - \frac{e\mathcal{E}_o}{m\omega^2}\right)^2 - \frac{e^2\mathcal{E}_o^2}{2m\omega^2}\,.$$

Answer: $p_n = \frac{\bar{n}^n}{n!}e^{-\bar{n}}$, where $\bar{n} = e^2\mathcal{E}_o^2/2m\hbar\omega^3$. Show that in the weak field limit this agrees with the perturbative result (10.8.16).

10. Consider a perturbation which is constant H' over an interval $0 \leq t \leq T$ and zero outside this interval. Derive an expression for the second order transition rate when the final state is part of a continuum of states. Show that correct to second order, the transition rate is given by

$$w_{i\to f} = \frac{2\pi}{\hbar}\left|H'_{fi} + \sum_m \frac{H'_{fm}H'_{mi}}{E_i - E_m}\right|^2 \rho(E_f)\,. \qquad (10.9.28)$$

References

[1] P. A. M. Dirac, Proc. Roy. Soc. (London) **A114**, 243 (1927).

[2] E. Merzbacher, *Quantum Mechanics*, Second Edition (John Wiley and Sons Inc., New York, 1970).

[3] A. Messiah, *Quantum Mechanics*, Volume II (North Holland, Amsterdam, 1964).

[4] L. I. Schiff, *Quantum Mechanics*, Third Edition (McGraw Hill Book Company, Inc., New York, 1968).

11

THE THREE-BODY PROBLEM

11.1 Introduction

Systems of many particles are common in nature. The complexity of the corresponding quantum mechanical problem increases rapidly as the number of particles increases. We have seen that it is possible to deal with two-body quantum systems effectively by transforming to center-of-mass coordinates, at least within the framework of non-relativistic quantum mechanics. The complexity of the problem increases enormously when we consider systems of three particles. Not only are we dealing with an expanded phase space, but also the task of selecting, out of many possibilities, the *correct variables* to describe the system. Among the important three-body systems that require a quantum mechanical treatment are the triton (bound state of two neutrons and a proton), the Helium atom (bound state of two electrons and a point nucleus) and numerous nuclear and atomic scattering events in which three particles participate. Despite the complexity of the problem, considerable progress has been made. This chapter describes some approaches to the three-body bound state problem in quantum mechanics.

The early methods to tackle the three-body bound state problem were based on the variational approach. In these methods, a trial function ψ of the coordinate of the particles involving several flexible parameters is chosen. (In some cases the choice may be suggested by the assumed form of two-body interaction between pairs of particles.) The Hamiltonian H of the three-body system being known, the energy integral $\int \Psi^* H \Psi d\tau$ is expressed in terms of the variational parameters, which are adjusted so as to minimize the value of the energy integral. The best set of values for the parameters thus found then gives the most appropriate ground state wave function for the system and the value of this integral for these values of the parameters gives the corresponding ground state energy. From the variational ground state wave function, other quantities relevant to the ground state of the system can be calculated.

11.2 Eyges Approach

Among the earliest non-variational approaches to the three-body bound state were those of Eyges (1961), Mitra (1962) and Faddeev (1962). Eyges considered a system of three identical particles with an attractive interaction between the pairs and wrote down the three-body Schrödinger equation for the bound state in the coordinate representation and proposed the following structure for the total three-body bound state wave function

$$\Psi(\boldsymbol{R}) = \psi(\boldsymbol{r}_{12}, \boldsymbol{\rho}_3) + \psi(\boldsymbol{r}_{23}, \boldsymbol{\rho}_1) + \psi(\boldsymbol{r}_{31}, \boldsymbol{\rho}_2) , \tag{11.2.1}$$

where $\boldsymbol{R}$ specifies the spatial configuration of the three particles which can alternatively be specified by the set of vectors $(\boldsymbol{r}_{12}, \boldsymbol{\rho}_3)$ or $(\boldsymbol{r}_{23}, \boldsymbol{\rho}_1)$ or $(\boldsymbol{r}_{31}, \boldsymbol{\rho}_2)$. Here $\boldsymbol{\rho}_k$ is the coordinate of the k-th particle with respect to the center-of-mass of the other two and $\boldsymbol{r}_{ij} \equiv \boldsymbol{r}_k$ is the relative coordinate of ij-th (or k-th) pair. According to Eyges the function $\psi(\boldsymbol{r}, \boldsymbol{\rho})$ can be assumed have the same form for each set of coordinates. Eyges derived an integral equation for $\psi(\boldsymbol{r}, \boldsymbol{\rho})$ and Fourier transformed his results to write his equations in the momentum representation.

We present a derivation of Eyges equation. Our approach differs slightly from Eyges' original derivation as we work in the momentum representation from the very beginning. The Schrödinger equation for the bound state of three particles having the same mass in the center-of-mass frame is

$$\left(\hat{H}_0 + \sum_{i=1}^{3} \hat{V}_i\right) |\Phi\rangle = E\,|\Phi\rangle\ ,$$

$$\left(\hat{H}_0 + \alpha_0^2\right) |\Phi\rangle = -\sum_{i=1}^{3} \hat{V}_i\,|\Phi\rangle\ . \tag{11.2.2}$$

Here the bound state energy $E = -\hbar^2\alpha_0^2/M)$, where M is the common mass of the particles, has been written as $E = -\alpha_0^2$ by choosing units such that $\hbar = 1 = M$. This equation may be written in the momentum representation by choosing the basis set $\langle \boldsymbol{P}|$ to be the set of continuum states of three particles[1]

$$\langle \boldsymbol{P}| \equiv \langle \boldsymbol{q}_1, \boldsymbol{p}_1| \equiv \langle \boldsymbol{q}_2, \boldsymbol{p}_2| \equiv \langle \boldsymbol{q}_3, \boldsymbol{p}_3|$$

where $\boldsymbol{q}_k$ is the momentum of k-th particle in the center-of-mass frame and $\boldsymbol{p}_k$ is the reduced center-of-mass momentum of the ij (or k-th) pair. Any of the three sets of vectors can be expressed in terms of any other set.

In the momentum basis, the three-particle equations assume the form of an integral equation

$$\left(p_k^2 + \frac{3}{4}q_k^2 + \alpha_0^2\right) \Phi\left(\boldsymbol{q}_k, \boldsymbol{p}_k\right)$$

$$= -\iint \langle \boldsymbol{q}_k, \boldsymbol{p}_k|\ \hat{V}_1 + \hat{V}_2 + \hat{V}_3\ |\boldsymbol{q}_k', \boldsymbol{p}_k'\rangle\, d^3q_k' d^3p_k' \Phi\left(\boldsymbol{q}_k', \boldsymbol{p}_k'\right)\ , \tag{11.2.3a}$$

where $\quad \Phi\left(\boldsymbol{q}_k, \boldsymbol{p}_k\right) = \langle \boldsymbol{q}_k, \boldsymbol{p}_k|\Phi\rangle\ .$ (11.2.3b)

[1]Explicitly, $\boldsymbol{q}_k = \boldsymbol{P}_k$ and $\boldsymbol{p}_k \equiv \boldsymbol{p}_{ij} = \frac{m_j\boldsymbol{P}_i - m_i\boldsymbol{P}_j}{m_i + m_j}$, where $\boldsymbol{P}_1, \boldsymbol{P}_2, \boldsymbol{P}_3$ are the momenta of the three particles in the center-of-mass frame. The free Hamiltonian $\hat{H}_0$ is given by

$$H_0 = \frac{P_1^2}{2m_1} + \frac{P_2}{2m_2} + \frac{P_3^2}{2m_3} = \frac{q_k^2}{2n_k} + \frac{p_k^2}{2\mu_k}\ , \qquad (k = 1,\ 2,\ 3)\ ,$$

where

$$n_k = \frac{m_k(m_i + m_j)}{m_i + m_j + m_k} \quad \text{and} \quad \mu_k \equiv \mu_{ij} = \frac{m_i m_j}{m_i + m_j}\ .$$

If the three particles have the same mass then

$$n_k = \frac{2M}{3}\ ,\ \mu_k = \frac{M}{2} \quad \text{and} \quad H_0 = \frac{1}{M}(p_k^2 + \frac{3}{4}q_k^2) = p_k^2 + \frac{3}{4}q_k^2 \ \ \text{in units}\ \ \hbar = 1 = M.$$

Expression of any one set of vectors in terms of the other set is very simple in this case. For example,

$$\boldsymbol{q}_1 = \boldsymbol{p}_3 - \frac{\boldsymbol{q}_3}{2},\ \boldsymbol{p}_1 = -\frac{1}{2}\boldsymbol{p}_3 - \frac{3}{4}\boldsymbol{q}_3 \ \ ;\ \boldsymbol{q}_2 = -\boldsymbol{p}_3 - \frac{1}{2}\boldsymbol{q}_3\ ,\ \boldsymbol{p}_2 = -\frac{\boldsymbol{p}_3}{2} + \frac{3}{4}\boldsymbol{q}_3\ .$$

In these sets of equations we can permute the indices 1,2,3 in cyclic order.

Following Eyges, we assume the following form for the complete three-body wave function in the momentum space

$$\Phi(\boldsymbol{q}_k, \boldsymbol{p}_k) = \varphi(\boldsymbol{q}_1, \boldsymbol{p}_1) + \varphi(\boldsymbol{q}_2, \boldsymbol{p}_2) + \varphi(\boldsymbol{q}_3, \boldsymbol{p}_3) \ . \tag{11.2.4}$$

If we also assume that

$$\left(p_1^2 + \frac{3}{4}q_1^2 + \alpha_0^2\right)\phi(\boldsymbol{q}_1, \boldsymbol{p}_1) = \\ -\iint \langle \boldsymbol{q}_1, \boldsymbol{p}_1 | \hat{V}_1 | \boldsymbol{q}_1', \boldsymbol{p}_1' \rangle \, d^3q_1' d^3p_1' \, \{\phi(\boldsymbol{q}_1', \boldsymbol{p}_1') + \phi(\boldsymbol{q}_2', \boldsymbol{p}_2') + \phi(\boldsymbol{q}_3', \boldsymbol{p}_3')\} \ , \tag{11.2.5}$$

$$\left(p_2^2 + \frac{3}{4}q_2^2 + \alpha_0^2\right)\phi(\boldsymbol{q}_2, \boldsymbol{p}_2) = \\ -\iint \langle \boldsymbol{q}_2, \boldsymbol{p}_2 | \hat{V}_2 | \boldsymbol{q}_2', \boldsymbol{p}_2' \rangle \, d^3q_2' d^3p_2' \, \{\phi(\boldsymbol{q}_1', \boldsymbol{p}_1') + \phi(\boldsymbol{q}_2', \boldsymbol{p}_2') + \phi(\boldsymbol{q}_3', \boldsymbol{p}_3')\} \ , \tag{11.2.6}$$

$$\left(p_3^2 + \frac{3}{4}q_3^2 + \alpha_0^2\right)\phi(\boldsymbol{q}_3, \boldsymbol{p}_3) = \\ -\iint \langle \boldsymbol{q}_3, \boldsymbol{p}_3 | \hat{V}_3 | \boldsymbol{q}_3', \boldsymbol{p}_3' \rangle \, d^3q_3' d^3p_3' \, \{\phi(\boldsymbol{q}_1', \boldsymbol{p}_1') + \phi(\boldsymbol{q}_2', \boldsymbol{p}_2') + \phi(\boldsymbol{q}_3', \boldsymbol{p}_3')\} \ , \tag{11.2.7}$$

then by adding these equations together we get back the three-particle Schrödinger equation (11.2.3a).

The structure of these three equations being the same, one can choose any one of them (say Eq. (11.2.7)) as the typical integral equation for ϕ. Now

$$\langle \boldsymbol{q}_3, \boldsymbol{p}_3 | \hat{V}_3 | \boldsymbol{q}_3', \boldsymbol{p}_3' \rangle = \iiiint \langle \boldsymbol{q}_3, \boldsymbol{p}_3 | \boldsymbol{\rho}_3, \boldsymbol{r}_3 \rangle \, d^3\rho_3 d^3r_3 \, \langle \boldsymbol{\rho}_3, r_3 | \hat{V}_3 | \boldsymbol{\rho}_3', \boldsymbol{r}' \rangle \\ d^3\rho_3' d^3r_3' \langle \boldsymbol{\rho}'_3, \boldsymbol{r}_3' | \boldsymbol{q}_3', \boldsymbol{p}_3' \rangle$$

where the set of coordinate states $|\boldsymbol{\rho}_3, \boldsymbol{r}_3\rangle$ provides a basis[2] for the coordinate representation of state and observables. Here $\boldsymbol{\rho}_3$ is the position of particle 3 in the center-of-mass of the pair $(1, 2)$ while $\boldsymbol{r}_3 \equiv \boldsymbol{r}_{12}$ is the relative coordinate of the pair. Further, since $\hat{V}_3$ is a two-body operator in a three-body space (i.e., the space scanned by three particle states) and only the pair 3 interacts, $\boldsymbol{\rho}_3$ and $\boldsymbol{\rho}_3'$ dependence in the matrix element $\langle \boldsymbol{\rho}_3, \boldsymbol{r}_3 | \hat{V}_3 | \boldsymbol{\rho}_3', \boldsymbol{r}_3' \rangle$ can be factored out as a three-dimensional delta function and we have

$$\langle \boldsymbol{\rho}_3, \boldsymbol{r}_3 | \hat{V}_3 | \boldsymbol{\rho}_3', \boldsymbol{r}_3' \rangle = \delta^3(\boldsymbol{\rho}_3 - \boldsymbol{\rho}_3') \langle \boldsymbol{r}_3 | \hat{v}_3 | \boldsymbol{r}_3' \rangle$$

where $\hat{v}_3$ is two-body operator in a two-body space.[3] Further if the interaction $\hat{V}_3$ is local, we have

$$\langle \boldsymbol{r}_3 | \hat{v}_3 | \boldsymbol{r}_3' \rangle = v_3(\boldsymbol{r}_3)\delta^3(\boldsymbol{r}_3 - \boldsymbol{r}_3')$$

where $v_3(\boldsymbol{r}_3)$ is the conventional potential, i.e., a function of the relative coordinate $\boldsymbol{r}_3$ of the interacting pair. Also since

$$\langle \boldsymbol{\rho}_3, \boldsymbol{r}_3 | \boldsymbol{q}_3, \boldsymbol{p}_3 \rangle = \frac{e^{i\boldsymbol{q}_3 \cdot \boldsymbol{\rho}_3}}{(2\pi)^{3/2}} \frac{e^{i\boldsymbol{p}_3 \cdot \boldsymbol{r}_3}}{(2\pi)^{3/2}}$$

[2] $|\boldsymbol{\rho}_3, \boldsymbol{r}_3\rangle \equiv |\boldsymbol{\rho}_2, \boldsymbol{r}_2\rangle \equiv |\boldsymbol{\rho}_1, \boldsymbol{r}_1\rangle$ since the three sets of vectors are related.

[3] Operators in three-body space will be denoted by capital letters while those in two-body space will be denoted by small letters.

where we have taken $\hbar = 1$, we can write

$$\langle \boldsymbol{q}_3, \boldsymbol{p}_3 | \hat{V}_3 | \boldsymbol{q}_3', \boldsymbol{p}_3' \rangle = \frac{1}{(2\pi)^6} \iint e^{i\,\boldsymbol{\rho}_3 \cdot (\boldsymbol{q}_3' - \boldsymbol{q}_3)} e^{i\boldsymbol{r}_3 \cdot (\boldsymbol{p}_3' - \boldsymbol{p}_3)}\, v_3(\boldsymbol{r}_3)\, d^3\rho_3 d^3 r_3\,,$$
$$= \frac{\delta^3(\boldsymbol{q}_3' - \boldsymbol{q}_3)}{(2\pi)^3} \int v_3(\boldsymbol{r})\, d^3 r\ e^{i\,\boldsymbol{r}.(\boldsymbol{p}_3' - \boldsymbol{p}_3)}\,. \tag{11.2.8}$$

We can drop the suffix 3 on $v_3(\boldsymbol{r})$ since we expect the interaction between all particle pairs to be similar. Using Eq. (11.2.8) we can write Eq. (11.2.7) as

$$\left(p_3^2 + \frac{3}{4}q_3^2 + \alpha_0^2\right)\ \phi(\boldsymbol{q}_3, \boldsymbol{p}_3) = \frac{-1}{(2\pi)^3} \int v(\boldsymbol{r}) d^3 r \iint e^{i\,\boldsymbol{r}.(\boldsymbol{p}_3' - \boldsymbol{p}_3)}$$
$$\times\, \delta^3(\boldsymbol{q}_3' - \boldsymbol{q}_3) d^3 q_3' d^3 p_3' \left[\phi(\boldsymbol{q}_1', \boldsymbol{p}_1') + \phi(\boldsymbol{q}_2', \boldsymbol{p}_2') + \phi(\boldsymbol{q}_3', \boldsymbol{p}_3')\right]$$

and since $d^3 q_3' d^3 p_3' = d^3 q_2' d^3 p_2' = d^3 q_1' d^3 p_1'$, we can rewrite the above equation as

$$\left(p_3^2 + \frac{3}{4}q_3^2 + \alpha_0^2\right) \phi(\boldsymbol{q}_3, \boldsymbol{p}_3) = -\frac{1}{(2\pi)^3} \int v(\boldsymbol{r}) d^3 r$$
$$\times \left[\iint d^3 q_3' d^3 p_3'\, e^{i\,\boldsymbol{r}\cdot(\boldsymbol{p}_3' - \boldsymbol{p}_3)} \delta^3(\boldsymbol{q}_3' - \boldsymbol{q}_3) \phi(\boldsymbol{q}_3', \boldsymbol{p}_3')\right.$$
$$+ \iint d^3 q_1' d^3 p_1' e^{i\,\boldsymbol{r}\cdot(-\frac{1}{2}\boldsymbol{p}_1' + \frac{3}{4}\boldsymbol{q}_1' + \frac{1}{2}\boldsymbol{p}_1 - \frac{3}{4}\boldsymbol{q}_1)} \delta^3\left(\boldsymbol{p}_1 + \frac{1}{2}\boldsymbol{q}_1 - \boldsymbol{p}_1' - \frac{1}{2}\boldsymbol{q}_1'\right) \phi(\boldsymbol{q}_1', \boldsymbol{p}_1')$$
$$\left.+ \iint d^3 q_2' d^3 p_2' e^{i\,\boldsymbol{r}\cdot(-\frac{1}{2}\boldsymbol{p}_2' - \frac{3}{4}\boldsymbol{q}_2' + \frac{1}{2}\boldsymbol{p}_2 + \frac{3}{4}\boldsymbol{q}_2)} \delta^3\left(-\boldsymbol{p}_2 + \frac{1}{2}\boldsymbol{q}_2 + \boldsymbol{p}_2' - \frac{1}{2}\boldsymbol{q}_2'\right) \phi(\boldsymbol{q}_2', \boldsymbol{p}_2')\right]$$

where, in the second and third integrals we have expressed $\boldsymbol{p}_3', \boldsymbol{q}_3'; \boldsymbol{p}_3, \boldsymbol{q}_3$ in terms of $\boldsymbol{p}_1', \boldsymbol{q}_1'; \boldsymbol{p}_1, \boldsymbol{q}_1$ and $\boldsymbol{p}_2', \boldsymbol{q}_2'; \boldsymbol{p}_2, \boldsymbol{q}_2$, respectively. Using the identity $\delta^3(s/2) = 8\,\delta^3(s)$ and integrating the first, second and third terms over $\boldsymbol{q}_3', \boldsymbol{q}_1', \boldsymbol{q}_2'$, respectively, and replacing, in the three integrals, the dummy variables $\boldsymbol{p}_3', \boldsymbol{p}_1', \boldsymbol{p}_2'$, respectively, by $\boldsymbol{p}'$ we get

$$\left(p_3^2 + \frac{3}{4}q_3^2 + \alpha_0^2\right)\ \phi(\boldsymbol{q}_3, \boldsymbol{p}_3) = -\frac{1}{(2\pi)^3} \int v(\boldsymbol{r}) d^3 r \int d^3 p'$$
$$\times\ \left\{e^{i\,\boldsymbol{r}.(\boldsymbol{p}' - \boldsymbol{p}_3)} \phi\left(\boldsymbol{q}_3, \boldsymbol{p}'\right) + 8\, e^{i\,\boldsymbol{r}\cdot(2\boldsymbol{p}_1 - 2\boldsymbol{p}')} \phi\left(2\boldsymbol{p}_1 - 2\boldsymbol{p}' + \boldsymbol{q}_1,\ \boldsymbol{p}'\right)\right.$$
$$\left. + 8\ e^{i\,\boldsymbol{r}.(2\boldsymbol{p}_2 - 2\boldsymbol{p}')}\, \phi(2\boldsymbol{p}' - 2\boldsymbol{p}_2 + \boldsymbol{q}_2\,,\ \boldsymbol{p}')\right\}\,.$$

We may now express the momenta $\boldsymbol{p}_1, \boldsymbol{q}_1$ and $\boldsymbol{p}_2, \boldsymbol{q}_2$ in terms of $\boldsymbol{p}_3, \boldsymbol{q}_3$ and write

$$\left(p_3^2 + \frac{3}{4}q_3^2 + \alpha_0^2\right)\ \phi(\boldsymbol{q}_3, \boldsymbol{p}_3) = -\frac{1}{(2\pi)^3} \int v(\boldsymbol{r}) d^3 r \int d^3 p'$$
$$\times\ \left\{e^{i\,\boldsymbol{r}.(\boldsymbol{p}' - \boldsymbol{p}_3)} \phi\left(\boldsymbol{q}_3, \boldsymbol{p}'\right) + 8 e^{i\,\boldsymbol{r}.(-2\boldsymbol{p}' - \boldsymbol{p}_3 - \frac{3}{2}\boldsymbol{q}_3)} \phi\left(-2\boldsymbol{q}_3 - 2\boldsymbol{p}',\ \boldsymbol{p}'\right)\right.$$
$$\left. + 8\ e^{i\,\boldsymbol{r}.(-2\boldsymbol{p}' - \boldsymbol{p}_3 + \frac{3}{2}\boldsymbol{q}_3)}\ \phi\left(-2\boldsymbol{q}_3 + 2\boldsymbol{p}',\ \boldsymbol{p}'\right)\right\}.$$

Changing the dummy variables in the second and third integrals from $\boldsymbol{p}'$ to $-\frac{\boldsymbol{p}'}{2}$ ($d^3 p' \to \frac{1}{8} d^3 p'$), and replacing the vectors $\boldsymbol{q}_3, \boldsymbol{p}_3$ on both sides by just $\boldsymbol{q}, \boldsymbol{p}$, we get

$$\left(p^2 + \frac{3}{4}q^2 + \alpha_0^2\right)\ \phi(\boldsymbol{q}, \boldsymbol{p}) = -\frac{1}{(2\pi)^3} \int v(\boldsymbol{r})\ d^3 r \int d^3 p'$$
$$\times\ \left\{e^{i\,\boldsymbol{r}.(\boldsymbol{p}' - \boldsymbol{p})} \phi\left(\boldsymbol{q}, \boldsymbol{p}'\right) + e^{i\,\boldsymbol{r}.(\boldsymbol{p}' - \frac{3}{2}\boldsymbol{q} - \boldsymbol{p})} \phi\left(-2\boldsymbol{q} + \boldsymbol{p}',\ -\frac{\boldsymbol{p}'}{2}\right)\right.$$
$$\left. +\ e^{i\,\boldsymbol{r}\cdot(-\boldsymbol{p} + \frac{3}{2}\boldsymbol{q} + \boldsymbol{p}')}\ \phi\left(-2\boldsymbol{q} - \boldsymbol{p}',\ -\frac{\boldsymbol{p}'}{2}\right)\right\}\,. \tag{11.2.9}$$

Equation (11.2.9) was derived by Eyges[4].

By changing the variable of integration in the second integral by letting $\boldsymbol{p}' - 2\boldsymbol{q} \to \boldsymbol{p}'$ and in the third integral by letting $-\boldsymbol{p}' - 2\boldsymbol{q} \to \boldsymbol{p}'$ and also change the order of integration over the coordinate and momentum space, we can put Eyges' equation (11.2.9) in the form

$$\left(p^2 + \frac{3}{4}q^2 + \alpha_0^2\right)\phi(\boldsymbol{q},\boldsymbol{p}) = -\int d^3p' \,\{\phi(\boldsymbol{q},p')\,\langle\mathbf{p}|\,\hat{v}\,|\mathbf{p}'\rangle \\ +\phi(\mathbf{p}',\, -\mathbf{p}'/2 - \boldsymbol{q})\,\langle\mathbf{p}|\,\hat{v}\,|\mathbf{p}' + \boldsymbol{q}/2\rangle \\ +\phi(\mathbf{p}',\, \mathbf{p}'/2 + \boldsymbol{q})\,\langle\mathbf{p}|\,\hat{v}\,|{-\mathbf{p}'} - \boldsymbol{q}/2\rangle\} \qquad (11.2.10)$$

where

$$\langle\mathbf{p}|\hat{v}|\mathbf{p}'\rangle \equiv \frac{1}{(2\pi)^3}\int v(\boldsymbol{r})\,d^3r\; e^{i\,\boldsymbol{r}.(\mathbf{p}'-\mathbf{p})} = v_t(\mathbf{p}-\mathbf{p}') \qquad (11.2.11)$$

and $v_t(\boldsymbol{p}-\boldsymbol{p}')$ can be looked upon as the Fourier transform of the potential function $v(\boldsymbol{r})$. The form (11.2.10) of Eyges' equation can be derived directly from (11.2.7) by using the identity $\langle\boldsymbol{q}_3,\boldsymbol{p}_3|\,\hat{V}_3\,|\boldsymbol{q}_3',\boldsymbol{p}_3'\rangle = \delta^3(\boldsymbol{q}_3 - \boldsymbol{q}_3')\,\langle\boldsymbol{p}_3|\,\hat{v}_3\,|\boldsymbol{p}_3'\rangle$ and changing dummy variables, without making any assumption about the local character of the two-body potential.

11.3 Mitra's Approach

Mitra also solved the time-independent Schrödinger equation for the bound state of three particles of the same mass using, for the pair interactions, non-local separable potentials introduced by Yamaguchi (1951) [see Sec. 3.6, Chapter 3]. He also worked in momentum representation choosing, for the basis states, the continuum of free states $|\boldsymbol{P}_1,\boldsymbol{P}_2,\boldsymbol{P}_3\rangle$ where $\boldsymbol{P}_1,\boldsymbol{P}_2,\boldsymbol{P}_3$ are the momenta of the three particles in the center-of-mass frame so that $\boldsymbol{P}_1+\boldsymbol{P}_2+\boldsymbol{P}_3 = 0$. Since only two of these momenta are independent, we can choose one of the following alternative sets of momentum variables

$$\begin{aligned} & \boldsymbol{P}_{12} = \boldsymbol{P}_1 + \boldsymbol{P}_2 \quad \text{and} \quad \boldsymbol{p}_{12} = \frac{\boldsymbol{P}_2 - \boldsymbol{P}_1}{2}\,, \\ \text{or} \quad & \boldsymbol{P}_{23} = \boldsymbol{P}_2 + \boldsymbol{P}_3 \quad \text{and} \quad \boldsymbol{p}_{23} = \frac{\boldsymbol{P}_3 - \boldsymbol{P}_2}{2}\,, \\ \text{or} \quad & \boldsymbol{P}_{31} = \boldsymbol{P}_3 + \boldsymbol{P}_1 \quad \text{and} \quad \boldsymbol{p}_{31} = \frac{\boldsymbol{P}_1 - \boldsymbol{P}_3}{2}\,. \end{aligned}$$

These are essentially the sets of momenta for three free particles that we had introduced in the context of Eyges' equations (viz., $\boldsymbol{q}_3,\boldsymbol{p}_3$ or $\boldsymbol{q}_1,\boldsymbol{p}_1$ or $\boldsymbol{q}_2,\boldsymbol{p}_2$), except for a change in the signs of both momentum variables.

The three-body Schrödinger equation in the center-of-mass frame is

$$\hat{H}\,|\Psi\rangle \equiv \left(\frac{\hat{P}_1^2 + \hat{P}_2^2 + \hat{P}_3^2}{2M} + \hat{V}_{12} + \hat{V}_{23} + \hat{V}_{31}\right)|\Psi\rangle = E\,|\Psi\rangle\,.$$

[4] The symbols K^2, $\boldsymbol{k}$, and $\boldsymbol{\kappa}$ in Eyges' notation stand for α_0^2, $\boldsymbol{p}$, and $\boldsymbol{q}$ in our notation. Also $\phi(\boldsymbol{k},\boldsymbol{\kappa})$ of Eyges stands for $\phi(\boldsymbol{\kappa},\boldsymbol{k}) \equiv \phi(\boldsymbol{q},\boldsymbol{p})$ in our notation. Discrepancies in signs are due to his using $\boldsymbol{r}_{13}$ instead of $\boldsymbol{r}_{31}$. (The permutation of indices should be cyclic.)

In the momentum basis $|\boldsymbol{P}_s\rangle$, where $\boldsymbol{P}_s$ stands for either one of the sets of moments $\boldsymbol{P}_{12}, \boldsymbol{p}_{12}$ or $\boldsymbol{P}_{23}, \boldsymbol{p}_{23}$ or $\boldsymbol{P}_{31}, \boldsymbol{p}_{31}$, the above equation may be written as[5]

$$\left(\frac{P_1^2+P_2^2+P_3^2}{2M}+\frac{\alpha_0^2}{M}\right)\langle \boldsymbol{P}_s|\Psi\rangle = -\int \langle \boldsymbol{P}_s|\hat{V}_{12}+\hat{V}_{23}+\hat{V}_{31}|\boldsymbol{P}'_s\rangle\, d^3P'_s\, \langle \boldsymbol{P}'_s|\Psi\rangle\,, \quad (11.3.1)$$

where $-\frac{\alpha_0^2}{M} = E$ is the three-body energy in the center-of-mass frame. Also $d^3P'_s = d^3P'_{12}d^3\,p'_{12} = d^3P'_{23}d^3p'_{23} = d^3P'_{31}d^3p'_{31}$. If we assume that $\hat{V}_{np} \neq \hat{V}_{nn}$, let 1 and 2 denote the neutrons and 3 denote the proton in the triton and assume the two-body interaction to be non-local and separable (Yamaguchi form), then

$$\begin{aligned}\langle \boldsymbol{P}|\hat{V}_{12}|\boldsymbol{P}'\rangle &= \langle \boldsymbol{P}_{12}, \boldsymbol{p}_{12}|\hat{V}_{nn}|\boldsymbol{P}'_{12}, \boldsymbol{p}'_{12}\rangle \\ &= \delta^3\left(\boldsymbol{P}'_{12}-\boldsymbol{P}_{12}\right)\langle \boldsymbol{p}_{12}|\,\hat{v}_{nn}\,|\boldsymbol{p}'_{12}\rangle \\ &= \delta^3\left(\boldsymbol{P}'_{12}-\boldsymbol{P}_{12}\right)\left(\frac{-\lambda_1}{M}\right)\, f(\boldsymbol{p}_{12})\, f(\boldsymbol{p}'_{12})\,. \qquad (11.3.2)\end{aligned}$$

Similarly,

$$\begin{aligned}\langle \boldsymbol{P}|\hat{V}_{31}|\boldsymbol{P}'\rangle &= \delta^3\left(\boldsymbol{P}'_{31}-\boldsymbol{P}_{31}\right)\langle \boldsymbol{p}_{31}|\,\hat{v}_{np}\,|\boldsymbol{p}'_{31}\rangle \\ &= \delta^3\left(\boldsymbol{P}'_{31}-\boldsymbol{P}_{31}\right)\left(\frac{-\lambda_0}{M}\right)\, g(\boldsymbol{p}_{31})\, g(\boldsymbol{p}'_{31})\,, \qquad (11.3.3)\end{aligned}$$

and

$$\langle \boldsymbol{P}|\,\hat{V}_{23}\,|\boldsymbol{P}'\rangle = \delta^3\left(\boldsymbol{P}'_{23}-\boldsymbol{P}_{23}\right)\left(\frac{-\lambda_0}{M}\right)\, g(\boldsymbol{p}_{23})\, g(\boldsymbol{p}'_{23})\,. \qquad (11.3.4)$$

The three-dimensional delta function is factored out because one of the three particles in the interaction is a *spectator*, meaning that it does not interact and is not affected. For example in Eq. (11.3.2) the particle 3 with momentum $\boldsymbol{P}_3 = -\boldsymbol{P}_{12}$ is a spectator, hence the term $\delta^3\left(\boldsymbol{P}_3-\boldsymbol{P}'_3\right) = \delta^3\left(\boldsymbol{P}'_{12}-\boldsymbol{P}_{12}\right)$ is factored out. Substituting the forms of interactions in the three-body Schrödinger equation we get

$$\begin{aligned}&\frac{1}{M}\left(P_1^2+P_2^2+\boldsymbol{P}_1\cdot\boldsymbol{P}_2+\alpha_0^2\right)\Psi(\boldsymbol{P}_1,\boldsymbol{P}_2) \\ &= \left(\frac{\lambda_1}{M}\right)\int d^3p'_{12}\, f(\boldsymbol{p}_{12})\, f(\boldsymbol{p}'_{12})\, \left[\Psi\left(\boldsymbol{P}'_1,\boldsymbol{P}'_2\right)\right]_{\boldsymbol{P}'_{12}=\boldsymbol{P}_{12}} \\ &\quad + \left(\frac{\lambda_0}{M}\right)\int d^3p'_{31}\, g(\boldsymbol{p}_{31})\, \left[g(\boldsymbol{p}'_{31})\,\Psi\left(\boldsymbol{P}'_1,\boldsymbol{P}'_2\right)\right]_{\boldsymbol{P}'_{31}=\boldsymbol{P}_{31}} \\ &\quad + \left(\frac{\lambda_0}{M}\right)\int d^3p'_{23}\, g(\boldsymbol{p}_{23})\, \left[g(\boldsymbol{p}'_{23})\,\Psi\left(\boldsymbol{P}'_1,\boldsymbol{P}'_2\right)\right]_{\boldsymbol{P}'_{23}=\boldsymbol{P}_{23}} \qquad (11.3.5)\end{aligned}$$

where, on the left-hand side we have put $\boldsymbol{P}_3 = -(\boldsymbol{P}_1+\boldsymbol{P}_2)$ and the right-hand side has been simplified using the property of delta functions. Now, let us designate $\boldsymbol{P}_{12}$ as $\boldsymbol{P}$, $\boldsymbol{p}_{12}$ as $\boldsymbol{p}$ and $\boldsymbol{p}'_{12}$ as $\boldsymbol{p}'$ in the first integral, then $\boldsymbol{P}'_1 = \frac{\boldsymbol{P}'_{12}}{2} - \boldsymbol{p}'_{12} \rightarrow \frac{\boldsymbol{P}}{2} - \boldsymbol{p}'$ and $\boldsymbol{P}'_2 = \frac{\boldsymbol{P}_{12}}{2} + \boldsymbol{p}'_{12} \rightarrow \frac{\boldsymbol{P}}{2} + \boldsymbol{p}'$, since $\boldsymbol{P}'_{12} = \boldsymbol{P}_{12} = \boldsymbol{P}$. In the second integral $\boldsymbol{P}'_{31} = \boldsymbol{P}_{31}$ implies $\boldsymbol{P}'_2 = \boldsymbol{P}_2$. Also $\boldsymbol{p}'_{31} = \boldsymbol{P}'_1 + \frac{\boldsymbol{P}'_2}{2} = \boldsymbol{P}'_1 + \frac{\boldsymbol{P}_2}{2}$ and $d^3p'_{31} \rightarrow d^3P'_1$. In the third integral $\boldsymbol{P}'_{23} \rightarrow \boldsymbol{P}_{23}$ implies $\boldsymbol{P}'_1 = \boldsymbol{P}_1$, $\boldsymbol{p}'_{23} = -\boldsymbol{P}'_2 - \frac{\boldsymbol{P}'_1}{2} = -\boldsymbol{P}'_2 - \frac{\boldsymbol{P}_1}{2}$ and $d^3p'_{23} \rightarrow d^3P'_2$. Hence

[5] $|\boldsymbol{P}_s\rangle \equiv |\boldsymbol{P}_1, \boldsymbol{P}_2\rangle \equiv |\boldsymbol{P}_{12}, \boldsymbol{p}_{12}\rangle \equiv |\boldsymbol{P}_{23}, \boldsymbol{p}_{23}\rangle \equiv |\boldsymbol{P}_{31}, \boldsymbol{p}_{31}\rangle$.

Eq. (11.3.5) takes the form

$$(P_1^2+P_2^2+\boldsymbol{P}_1\cdot\boldsymbol{P}_2+\alpha_0^2)\,\Psi\,(\boldsymbol{P}_1,\boldsymbol{P}_2)=\lambda_1\int d^3p'\,f(p)\,f(p')\,\Psi\left(\frac{\boldsymbol{P}}{2}-\boldsymbol{p}',\frac{\boldsymbol{P}}{2}+\boldsymbol{p}'\right)$$
$$+\lambda_0\int d^3P_1'\,g\left(\boldsymbol{P}_1+\frac{\boldsymbol{P}_2}{2}\right)\,g\left(\boldsymbol{P}_1'+\frac{\boldsymbol{P}_2}{2}\right)\,\Psi\,(\boldsymbol{P}_1',\boldsymbol{P}_2)$$
$$+\lambda_0\int d^3P_2'\,g\left(-\boldsymbol{P}_2-\frac{\boldsymbol{P}_1}{2}\right)\,g\left(-\boldsymbol{P}_2'-\frac{\boldsymbol{P}_1}{2}\right)\,\Psi\,(\boldsymbol{P}_1,\boldsymbol{P}_2')\,. \tag{11.3.6}$$

If we consider the two-body interaction to be in the S-state, we can assume $g\,(\boldsymbol{p})=g(p)$ and $f\,(\boldsymbol{p})=f(p)$. By observation and intuition, Mitra found that $\Psi\,(\boldsymbol{P}_1,\boldsymbol{P}_2)$ could have the following structure

$$\Psi\,(\boldsymbol{P}_1,\boldsymbol{P}_2)=D^{-1}\,(\boldsymbol{P}_1,\boldsymbol{P}_2)$$
$$\left[g\left(\boldsymbol{P}_1+\frac{1}{2}\boldsymbol{P}_2\right)\,\phi\,(\boldsymbol{P}_2)+f(\boldsymbol{p})\,\chi(\boldsymbol{P})+g\left(\boldsymbol{P}_2+\frac{1}{2}\boldsymbol{P}_1\right)\,\phi(\boldsymbol{P}_1)\right], \tag{11.3.7}$$

where $D\,(\boldsymbol{P}_1,\boldsymbol{P}_2)=\left(P_1^2+P_2^2+\boldsymbol{P}_1\cdot\boldsymbol{P}_2+\alpha_0^2\right)\,.$ (11.3.8)

The functions $\phi\,(\boldsymbol{P}_1)$, $\phi\,(\boldsymbol{P}_2)$ and $\chi\,(\boldsymbol{P})$ are called the *spectator functions*. By substituting the structure for the three-body wave function Ψ into Eq. (11.3.6) we arrive at the following coupled integral equations for the spectator functions ϕ and χ which involve the form factors g and f of the separable interaction:

$$[\lambda_0^{-1}-h_0(\boldsymbol{P}_2)]\,\phi(\boldsymbol{P}_2)=\int d^3\xi\,D^{-1}(\boldsymbol{P}_2,\boldsymbol{\xi})\left[g(\boldsymbol{P}_2+\frac{1}{2}\boldsymbol{\xi})g(\boldsymbol{\xi}+\frac{1}{2}\boldsymbol{P}_2)\,\phi(\boldsymbol{\xi})\right.$$
$$\left.+g(\boldsymbol{\xi}+\frac{1}{2}\boldsymbol{P}_2)f(\boldsymbol{P}_2+\frac{1}{2}\boldsymbol{\xi})\chi(\boldsymbol{\xi})\right], \tag{11.3.9}$$

$$(\lambda_1^{-1}+h_1(\boldsymbol{P}))\chi(\boldsymbol{P})=2\int d^3\xi\,D^{-1}(\boldsymbol{P},\boldsymbol{\xi})f(\boldsymbol{\xi}+\frac{1}{2}\boldsymbol{P})g(\boldsymbol{P}+\frac{1}{2}\boldsymbol{\xi})\,\phi(\boldsymbol{\xi})\,, \tag{11.3.10}$$

where
$$h_0(\boldsymbol{P})=\int d^3p\left(p^2+\frac{3}{4}P^2+\alpha_0^2\right)^{-1}g^2(\boldsymbol{p}) \tag{11.3.11}$$
and
$$h_1(\boldsymbol{P})=\int d^3p\left(p^2+\frac{3}{4}P^2+\alpha_0^2\right)^{-1}f^2(\boldsymbol{p})\,. \tag{11.3.12}$$

We note that in one of these equations (11.3.10) the function $\chi(\boldsymbol{P})$ is expressed as an integral involving the function ϕ only. Hence by substituting for χ into Eq. (11.3.9) one can get a single integral equation for ϕ which is amenable to solution. Using this solution in Eq. (11.3.10) to determine the function χ one can get finally the three-body wave function from Eq. (11.3.7). Detailed calculations for the three-body bound states were made by Mitra and his collaborators, who also considered spin dependence and tensor components in nuclear interaction (of Yamaguchi form) to calculate binding energy of ^3H, Coulomb energy of ^3He and electro-magnetic form factors of ^3H and ^3He.

The structure (11.3.7) of Mitra's three-body bound state wave function Ψ is of considerable interest. It is a linear combination of three terms and each one turns out to be a product of one- and two-particle wave functions. For instance the first term is the product of two functions

$$g(\boldsymbol{P}_1+\frac{1}{2}\boldsymbol{P}_2)\,D^{-1}\,(\boldsymbol{P}_1,\boldsymbol{P}_2)\equiv g\,(\boldsymbol{p}_{31})\left(\boldsymbol{p}_{31}^2+\frac{3}{4}\boldsymbol{P}_2^2+\alpha_0^2\right)^{-1}\quad\text{and}\quad\phi\,(\boldsymbol{P}_2)\,.$$

The first factor may be regarded as the two-body bound state wave function[6] of particles 1 and 3 with $\left(-\frac{3}{4}P_2^2 - \alpha_0^2\right)$ as the energy associated with the particle pair (1,3). The second factor $\phi(\boldsymbol{P}_2)$ may be regarded as the spectator function of particle 2 in the presence of the pair (1,3).

A similar interpretation holds for the function $\chi(\boldsymbol{P})$ which may be looked upon as the spectator wave function of particle 3 in the presence of the pair (1,2) while

$$D^{-1}\left(\boldsymbol{P}_1, \boldsymbol{P}_2\right)\ f(\boldsymbol{p}) \equiv \frac{f\left(\boldsymbol{p}_{12}\right)}{p_{12}^2 + \frac{3}{4}P_3^2 + \alpha_0^2}$$

plays the role of the two-body wave function of particle pair 1,2 with bound state energy equal to $(-\alpha_0^2 - \frac{3}{4}P_3^2)$. Similarly, in the third term, $\phi(\boldsymbol{P}_1)$ is the spectator wave function of particle 1 in the presence of the pair (2,3) while

$$D^{-1}\left(\boldsymbol{P}_1, \boldsymbol{P}_2\right) g(\boldsymbol{P}_2 + \frac{1}{2}\boldsymbol{P}_1) \equiv \left(p_{23}^2 + \frac{3}{4}P_1^2 + {\alpha_0}^2\right)^{-1} g\left(\boldsymbol{p}_{23}\right)$$

may be looked upon as the two-body bound state wave function of particle 2 and 3 with bound state energy $\left(-\alpha_0^2 - \frac{3}{4}P_1^2\right)$. The complete three-body wave function is a linear combination of the products of one- and two-particle wave functions.

If, in Mitra's treatment of the triton problem, the $n-p$ and $n-n$ interactions are taken to be similar, i.e.,

$$\langle \boldsymbol{p}|\hat{v}_{np}|\boldsymbol{p}'\rangle = \langle \boldsymbol{p}|\hat{v}_{nn}|\boldsymbol{p}'\rangle = -\frac{\lambda}{M}\, g(\boldsymbol{p})g(\boldsymbol{p}'),$$

and the neutron and proton are assigned the same mass, then we can take

$$\phi(\boldsymbol{p}) = \chi(\boldsymbol{p})$$

and

$$\Psi\left(\boldsymbol{P}_1, \boldsymbol{P}_2\right) = D^{-1}\left(\boldsymbol{P}_1, \boldsymbol{P}_2\right)\left\{g\left(\boldsymbol{P}_1 + \frac{1}{2}\boldsymbol{P}_2\right)\chi\left(\boldsymbol{P}_2\right) + g\left(\boldsymbol{p}\right)\chi\left(\boldsymbol{P}\right) + g\left(\boldsymbol{P}_2 + \frac{1}{2}\boldsymbol{P}_1\right)\chi\left(\boldsymbol{P}_1\right)\right\}, \qquad (11.3.13)$$

where the spectator function $\chi(\boldsymbol{P})$ satisfies the integral equation[7]

$$(\lambda_0^{-1} - h_0(\boldsymbol{P}))\chi(\boldsymbol{P}) = 2\int d^3\xi\, D^{-1}(\boldsymbol{P}, \boldsymbol{\xi}) g\left(\boldsymbol{\xi} + \frac{1}{2}\boldsymbol{P}\right) g\left(\boldsymbol{P} + \frac{1}{2}\boldsymbol{\xi}\right)\chi(\boldsymbol{\xi})\,. \qquad (11.3.14)$$

The vectors

$$\boldsymbol{P},\ \boldsymbol{p},\ \ \boldsymbol{P}_1 + \frac{1}{2}\boldsymbol{P}_2,\ \boldsymbol{P}_2 + \frac{1}{2}\boldsymbol{P}_1\ ,\ \frac{\boldsymbol{P}}{2} - \boldsymbol{p},\ \frac{\boldsymbol{P}}{2} + \boldsymbol{p}$$

[6]It may be recalled [Sec. 3.6, Chapter 3] that with the use of the non-local separable potential the two-body bound state wave function in the momentum representation is given by

$$\phi(p) = N\ g(p)/(p^2 + B),$$

where $-B$ is the energy associated with the particle pair, p being the relative momentum of the pair and $g(p)$ is the form factor of the separable interaction.

[7]With the substitutions $\lambda_1 = \lambda_0$, $f = g$, $\phi = \chi$ and $h_1(\boldsymbol{P}) = h_0(\boldsymbol{P})$, both the equations (11.3.9) and (11.3.10) reduce to Eq. (11.3.14).

in Mitra's notation are essentially equal to $-\boldsymbol{q}_3$, $-\boldsymbol{p}_3$, $-\boldsymbol{p}_2$, $\boldsymbol{p}_1$, $\boldsymbol{q}_1$, $\boldsymbol{q}_2$, respectively, of our earlier notation. Hence, according to our earlier notation,

$$\Psi(\boldsymbol{P}_1,\boldsymbol{P}_2) = \left[\frac{g(-\boldsymbol{p}_2)}{p_2^2+\frac{3}{4}q_2^2+\alpha_0^2}\chi(\boldsymbol{q}_2) + \frac{g(-\boldsymbol{p}_3)}{p_3^2+\frac{3}{4}q_3^2+\alpha_0^2}\chi(-\boldsymbol{q}_3) + \frac{g(\boldsymbol{p}_1)}{p_1^2+\frac{3}{4}q_1^2+\alpha_0^2}\chi(\boldsymbol{q}_1)\right] \tag{11.3.15}$$

or

$$\Psi(\boldsymbol{P}_1,\boldsymbol{P}_2) = \phi(\boldsymbol{p}_2,\boldsymbol{q}_2) + \phi(\boldsymbol{p}_3,\boldsymbol{q}_3) + \phi(\boldsymbol{p}_1,\boldsymbol{q}_1), \tag{11.3.16}$$

where

$$\phi(\boldsymbol{p},q) = \frac{g(\boldsymbol{p})}{p^2+\frac{3}{4}q^2+\alpha_0^2}\chi(\boldsymbol{q}),$$

and we have assumed that in the S-state, $g(\boldsymbol{p}) = g(|\boldsymbol{p}|) = g(p)$ and $\chi(\boldsymbol{q}) = \chi(|\boldsymbol{q}|) = \chi(q)$. Equation (11.3.16) has precisely the structure of the three-body wave function suggested by Eyges but with the additional provision due to Mitra that each of the three terms may be looked upon as a product of *bound state* of two particles and the *spectator function* of the third particle.

11.4 Faddeev's Approach

It was thought that a straightforward generalization of the two-body Lippmann-Schwinger (LS) equation [see Chapter 9, Sec. 9.7] for a three-body system might provide a basis for the description of the three-body problem. Thus in analogy with the two-body LS equation [Chapter 9, Eq. (9.9.8)], the three-body LS equation may be written as

$$\hat{T}(z) = \hat{V} + \hat{V}\hat{G}_0(z)\hat{T}(z) \tag{11.4.1}$$

where $\hat{T}(z)$ is the three-body transition operator,

$$\hat{V} = \hat{V}_{12} + \hat{V}_{23} + \hat{V}_{31} \equiv \hat{V}_3 + \hat{V}_1 + \hat{V}_2$$

is the sum of all the two-body interactions in three-body space[8] and $\hat{G}_0(z) = (z - \hat{H}_0)^{-1}$ is the free resolvent operator in three-body space, $\hat{H}_0$ being the free Hamiltonian of three particles.

Written in momentum representation[9]Eq. (11.4.1) assumes the form

$$\langle \boldsymbol{q}_k,\boldsymbol{p}_k|\,\hat{T}(z)\,|\boldsymbol{q}_k',\boldsymbol{p}_k'\rangle = \langle \boldsymbol{q}_k,\boldsymbol{p}_k|\hat{V}|\boldsymbol{q}_k',\boldsymbol{p}_k'\rangle$$
$$\iint \langle \boldsymbol{q}_k,\boldsymbol{p}_k|\hat{V}G_0(z)|\boldsymbol{q}_k'',\boldsymbol{p}_k''\rangle\, d^3q_k''\, d^3p_k''\langle \boldsymbol{q}_k'',\boldsymbol{p}_k''|\hat{T}(z)|\boldsymbol{q}_k',\boldsymbol{p}_k'\rangle. \tag{11.4.2}$$

[8]It may be recalled that $\hat{V}_k$'s are two-body operators in the space spanned by three-body states in the sense that only two of the particles (k-th pair) interact and the third (k-th particle) remains free. $\hat{T}_k(z) = \hat{V}_k + V_kG_0(z)T_k(z)$ is also a two-body operator in a three-body space. On the other hand, $\hat{T}(z)$ and $\hat{G}_0(z)$ are three-body operators in three-body space. As mentioned earlier we use capital letters to denote operators in three-body space, while small letters are used to denote operators in two-body space. Thus $\langle \boldsymbol{q}_k,\boldsymbol{p}_k|\hat{V}_k|\boldsymbol{q}_k',\boldsymbol{p}_k'\rangle = \delta^3(\boldsymbol{q}_k - \boldsymbol{q}_k')\langle \boldsymbol{p}_k|\hat{v}_k|\boldsymbol{p}_k'\rangle$ and $\langle \boldsymbol{p}_k,\boldsymbol{q}_k|\hat{T}_k(z)|\boldsymbol{p}_k',\boldsymbol{q}_k'\rangle = \delta^3(\boldsymbol{q}_k - \boldsymbol{q}_k')\langle \boldsymbol{p}_k|\hat{t}_k|\boldsymbol{p}_k'\rangle$, the delta function being factored out because the kth particle is a spectator.

[9]The momenta $\boldsymbol{p}_k$ and $\boldsymbol{q}_k$ for the basis states are as defined in the context of Eyges' equations.

The kernel of this integral equation is $\langle \boldsymbol{q}_k, \boldsymbol{p}_k | \hat{V}\hat{G}_0(z) | \boldsymbol{q}_k'', \boldsymbol{p}_k'' \rangle$. For the solution of an integral equation like (11.4.2) to exist, the Schmidt norm of its kernel,

$$N = \iiiint \left| \langle \boldsymbol{q}_k, \boldsymbol{p}_k | \, \hat{V}\hat{G}_0(z) \, | \boldsymbol{q}_k'', \boldsymbol{p}_k'' \rangle \right|^2 d^3q_k \, d^3p_k \, d^3q_k'' \, d^3p_k'' \,, \tag{11.4.3}$$

must be bounded. The kernel may be expressed as[10]

$$\langle \boldsymbol{q}_k, \boldsymbol{p}_k | \hat{V}_k G_0(z) | \boldsymbol{q}_k'', \boldsymbol{p}_k'' \rangle = \frac{\langle \boldsymbol{p}_k | \hat{v}_k | \boldsymbol{p}_k'' \rangle \delta^3(\boldsymbol{q}_k - \boldsymbol{q}_k'')}{z - \dfrac{p_k''^2}{2\mu_k} - \dfrac{q_k''^2}{2n_k}} \,. \tag{11.4.4}$$

The integrand $|\langle \boldsymbol{q}_k, \boldsymbol{p}_k | \hat{V}_k G_0(z) | \boldsymbol{q}_k'', \boldsymbol{p}_k'' \rangle|^2$ thus involves the square of the delta function $\delta^3 (\boldsymbol{q}_k - \boldsymbol{q}_k'')$. Hence the Schmidt norm would not be bounded and the three-body LS equation would not be solvable.

Another problem with the three-body LS equation is that the solution, if it existed, would be non-unique. To see this we start with the operator identity[11]

$$\hat{G}(z) = \hat{G}_0(z) + \hat{G}_0(z) \, \hat{V} \, \hat{G}(z) \,, \tag{9.9.9*}$$

where $G(z) = (z - \hat{H})^{-1} = (z - \hat{H}_0 - \hat{V}_1 - \hat{V}_2 - \hat{V}_3)^{-1}$ is the full resolvent operator and $z = E_0 + i\epsilon$. Let us introduce the operator

$$\hat{P}_n = \lim_{\epsilon \to 0} +i\epsilon \, \hat{G}(E_n + i\epsilon) \,. \tag{11.4.5}$$

We can interpret the operator $\hat{P}_n$ as the projection operator for the state $|\Psi_n\rangle$ where $|\Psi_n\rangle$ is the (three-body) eigenstate of the full Hamiltonian $\hat{H} = \hat{H}_0 + \hat{V}$ belonging to the eigenvalue E_n such that

$$\hat{H} \, |\Psi_n\rangle = E_n \, |\Psi_n\rangle \,. \tag{11.4.6}$$

In other words, $\hat{P}_n$ can project out the state $|\Psi_n\rangle$ when it operates on the state $|\Phi_n\rangle$ which is a (three-body) eigenstate of the free Hamiltonian $\hat{H}_0$, belonging to the same eigenvalue E_n. To check this we let $\hat{H}$ operate on the projected state $\hat{P}_n \, |\Phi_n\rangle$:

$$\begin{aligned}
\hat{H}\hat{P}_n \, |\Phi_n\rangle &= \hat{H} \lim_{\epsilon \to 0+} i\epsilon \, G(E_n + i\epsilon) \, |\Phi_n\rangle \\
&= \lim_{\epsilon \to 0+} i\epsilon (E_n + i\epsilon - (E_n + i\epsilon - \hat{H})) \, \hat{G}(E_n + i\epsilon) \, |\Phi_n\rangle \\
&= \lim_{\epsilon \to 0+} \left\{ (E_n + i\epsilon) \hat{P}_n |\Phi_n\rangle - i\epsilon \, |\Phi_n\rangle \right\} \\
&= E_n \, \hat{P}_n \, |\Phi_n\rangle \,.
\end{aligned}$$

Hence the projected state $\hat{P}_n \, |\Phi_n\rangle$ may be identified with the state $|\Psi_n\rangle$. Now, if we let both sides of the operator equation (9.9.4*), pre-multiplied by $i\epsilon$, operate on $|\Phi_n\rangle$, we get

---

[10]Since $|\boldsymbol{q}_k'', \boldsymbol{p}_k''\rangle$ is an eigenstate of $\hat{H}_0$,

$$\hat{G}_0(z) \, |\boldsymbol{q}_k'', \boldsymbol{p}_k''\rangle = \left(z - \frac{p_k''^2}{2\mu_k} - \frac{q_k''^2}{2n_k} \right)^{-1} |\boldsymbol{q}_k'', \boldsymbol{p}_k''\rangle \,.$$

[11]This identity may be obtained from the three-body LS equation $\hat{T}(z) = \hat{V} + \hat{G}_0(z)\hat{V}\hat{T}(z)$ in the same way as the corresponding identity for the two-body operators in two-body space, viz., $\hat{g}(z) = \hat{g}_0(z) + \hat{g}_0(z)\hat{v}\hat{g}(z)$ was derived from the two-body LS equation $\hat{t}(z) = \hat{v} + \hat{v}\hat{g}_0(z) \, \hat{t}(z)$ [Chapter 9, Sec. 9.9]. Other identities like $\hat{G}(z)\hat{V} = \hat{G}_0(z) + \hat{T}(z)$ and $\hat{V}\hat{G}(z) = \hat{T}(z) \, \hat{G}_0(z)$, may be similarly derived.

in the limit $\epsilon \to 0$,

$$\lim_{\epsilon\to 0^+} i\epsilon\hat{G}(E_n+i\epsilon)\,|\Phi_n\rangle = \lim_{\epsilon\to 0^+}\Big\{i\epsilon\hat{G}_0(E_n+i\epsilon)\,|\Phi_n\rangle + \hat{G}_0(E_n+i\epsilon)\hat{V}\; i\epsilon\hat{G}(E_n+i\epsilon)\,|\Phi_n\rangle\;\}$$

or

$$|\Psi_n\rangle = |\Phi_n\rangle + \hat{G}_0(E_n+i\epsilon)\hat{V}\,|\Psi_n\rangle\,. \tag{11.4.7}$$

We can look upon Eq. (11.4.7) as three-body Lippmann-Schwinger equation in terms of three-body states or as state version of the LS equation.

Now, let us define $\left|\Phi_n^{(i)}\right\rangle$ to be an eigenstate of $\hat{H}_i = \hat{H}_0 + \hat{V}_i$ belonging to the same eigenvalue E_n. (This is possible because, to the energy of the bound (i-th) pair, we can add the energy of the i-th particle so that the total energy is E_n.) If we let both sides of the operator equation (9.9.9*)[12], pre-multiplied by $i\epsilon$, operate on $\left|\Phi_n^{(i)}\right\rangle$ then we get

$$i\epsilon\, G(E_n+i\epsilon)\left|\Phi_n^{(i)}\right\rangle = i\epsilon\,\hat{G}_0(E_n+i\epsilon)\left|\Phi_n^{(i)}\right\rangle + \;G_0(E_n+i\epsilon)\,\hat{V} i\epsilon\, G(E_n+i\epsilon)\left|\Phi_n^{(i)}\right\rangle\,.$$

The first term on the right-hand side is zero because

$$\left(\hat{H}_0 - E_n\right)\left|\Phi_n^{(i)}\right\rangle \neq 0\,.$$

Using this result and defining a state

$$\left|\Psi_n^{(i)}\right\rangle \equiv \lim_{\epsilon\to 0^+} i\epsilon G(E_n+i\epsilon)\left|\Phi_n^{(i)}\right\rangle,$$

we can write the above equation as

$$\left|\Psi_n^{(i)}\right\rangle = G_0(E_n+i\epsilon)\,\hat{V}\left|\Psi_n^{(i)}\right\rangle. \tag{11.4.8}$$

Hence, in addition to the inhomogeneous integral equation (11.4.7), we also have a homogeneous integral equation (11.4.8). The existence of the homogeneous equation (11.4.8) implies that the inhomogeneous equation (11.4.7) cannot have a unique solution, because to a solution of the inhomogeneous equation the solution of the homogeneous equation can always be added and the result will still be a solution of the inhomogeneous equation. Such arbitrariness in the solution of an integral equation is undesirable because, while boundary conditions have to be imposed on the solution of a differential equation, they are inherent in an integral equation. The arbitrariness in the solution of Eq. (11.4.7) would imply that, among several available solutions, a solution is to be sought through additional requirements. Such subsidiary conditions would destroy the usefulness of an integral equation.

To circumvent the difficulties associated with the solution of the three-body Lippmann-Schwinger equation, Faddeev decomposed the three-body operator $\hat{T}(z)$ as [13]

$$\hat{T}(z) = \sum_i\sum_j \hat{\tau}_{ij}(z), \tag{11.4.9}$$

where

$$\hat{\tau}_{ij}(z) = \delta_{ij}\,\hat{V}_i + \hat{V}_i\hat{G}(z)\hat{V}_j\,. \tag{11.4.10}$$

[12]See Sec. 9.9.

[13]This derivation follows T. A. Osborn, SLAC Report no.79 (December 1967) prepared under AEC Contract AT(04-3)-515 for the USAEC San Francisco Operations Office.

If we sum both sides of Eq. (11.4.10) over i and j the result is the three-body LS equation:

$$\hat{T}(z) = \hat{V} + \hat{V}\hat{G}(z)\hat{V} = \hat{V} + \hat{V}\hat{G}_0(z)\hat{T}(z) = \hat{V} + \hat{T}(z)\hat{G}_0(z)\hat{V}, \tag{11.4.11}$$

where we have used the operator identities

$$\hat{G}(z)\hat{V} = \hat{G}_0(z)\,\hat{T}(z) \tag{11.4.12a}$$

and

$$\hat{V}\,\hat{G}(z) = \hat{T}(z)\,\hat{G}_0(z)\,. \tag{11.4.12b}$$

The operator identity (11.4.12a) also implies

$$\hat{G}_0(z)\sum_i\sum_j \hat{\tau}_{ij}(z) = \hat{G}(z)\sum_j V_j$$

or

$$\hat{G}_0(z)\sum_i \hat{\tau}_{ij}(z) = \hat{G}(z)\,\hat{V}_j\,. \tag{11.4.13}$$

Using Eqs. (11.4.13) one can rewrite Eq. (11.4.10) as

$$\hat{\tau}_{ij}(z) = \delta_{ij}\,\hat{V}_i + \hat{V}_i\hat{G}_0(z)\sum_k \hat{\tau}_{k\,j}(z).$$

If this equation is iterated, its kernel will involve non-compact terms like $\hat{V}_i\hat{G}_0\hat{V}_i\hat{G}_0$. To eliminate such terms Faddeev transferred $\hat{V}_i\hat{G}_0\hat{\tau}_{ij}(z)$ from the right-hand side to the left-hand side so that the equation takes the form:

$$\left(1 - \hat{V}_i\hat{G}_0(z)\right)\,\hat{\tau}_{i\,j}(z) = \delta_{i\,j}V_i + \hat{V}_i\hat{G}_0(z)\sum_{k\neq i}\hat{\tau}_{k\,j}(z)$$

and summing over j we get

$$\left(1 - \hat{V}_i G_0(z)\right)\,T^{(i)}(z) = V_i + V_i G_0(z)\sum_{k\neq i}\hat{T}^k(z)\,, \tag{11.4.14}$$

where

$$\hat{T}^{(i)}(z) \equiv \sum_j \hat{\tau}_{i\,j}\,. \tag{11.4.15}$$

Now, since the solution of two-body LS equation in three-body space[14], viz.,

$$\hat{T}_i(z) = \hat{V}_i + \hat{V}_i\hat{G}_0(z)\,T_i(z),$$

---

[14]It may be noted that, for the case when only one of the pair interaction is operative, one can have the following version of Eqs. (11.4.12a) and(11.4.12b):

$$\hat{G}_i(z)\hat{V}_i = \hat{G}_0(z)\hat{T}_i(z)$$

and $$\hat{V}_i\hat{G}_i(z) = \hat{T}_i(z)\hat{G}_0(z)\,,$$

where $$\hat{G}_i(z) = (z - \hat{H}_i)^{-1} = (z - \hat{H}_0 - \hat{V}_i)^{-1}$$

and $\hat{T}_i(z)$ is the two-body transition operator in three-body space, defined by the LS equation

$$\hat{T}_i(z) = \hat{V}_i + \hat{V}_i\hat{G}_0(z)\hat{T}_i(z) = \hat{V}_i + \hat{T}_i(z)\hat{G}_0(z)\hat{V}_i.$$

The matrix element of $\hat{T}_i(z)$ in the momentum basis is given by

$$\left\langle \boldsymbol{q}_i, \boldsymbol{p}_i \left|\hat{T}_i(z)\right| \boldsymbol{q}'_i, \boldsymbol{p}'_i\right\rangle = \delta^3\left(\boldsymbol{q}_i - \boldsymbol{q}'_i\right)\left\langle \boldsymbol{p}_i \left|\hat{t}_i\left(z - \frac{q_i^2}{2n_i}\right)\right| \boldsymbol{p}'_i\right\rangle.$$

Here, the energy carried by the i–th particle, $\frac{q_i^2}{2n_i}$, is subtracted from z so that

$$\hat{t}_i\left(z - \frac{q_i^2}{2n_i}\right)$$

is the two-body transition operator in two-body space.

exists, i.e.,

$$\hat{T}_i(z) = \left(1 - \hat{V}_i\hat{G}_0(z)\right)^{-1} \hat{V}_i$$

exists, the reciprocal of the operator $\left(\hat{1} - \hat{V}_i\hat{G}_0(z)\right)$ also exists. So we can pre-multiply both sides of Eq. (11.4.14) by $\left(\hat{1} - \hat{V}_i\hat{G}_0(z)\right)^{-1}$ and write

$$T^{(i)}(z) = \left(\hat{1} - \hat{V}_i\hat{G}_0(z)\right)^{-1} \hat{V}_i + \left(\hat{1} - \hat{V}_i\hat{G}_0(z)\right)^{-1} \hat{V}_i\hat{G}_0(z) \sum_{k\neq i} \hat{T}^k(z),$$

or

$$\hat{T}^{(i)}(z) = \hat{T}_i(z) + \hat{T}_i(z)\,\hat{G}_0(z) \sum_{k\neq i} \hat{T}^{(k)}(z), \tag{11.4.16}$$

where

$$\hat{T}(z) = \sum_i \sum_j \hat{\tau}_{ij}(z) = \sum_{i=1}^{3} \hat{T}^{(i)}(z). \tag{11.4.17}$$

Equation (11.4.16) represents a set of coupled equations, called *Faddeev equations*, which can be explicitly written as

$$\hat{T}^{(1)}(z) = \hat{T}_1(z) + \hat{T}_1(z)\,\hat{G}_0(z)\,\left[\hat{T}^{(2)}(z) + \hat{T}^{(3)}(z)\right] \tag{11.4.16a}$$

$$\hat{T}^{(2)}(z) = \hat{T}_2(z) + \hat{T}_2(z)\,\hat{G}_0(z)\,\left[\hat{T}^{(3)}(z) + \hat{T}^{(1)}(z)\right] \tag{11.4.16b}$$

$$\hat{T}^{(3)}(z) = \hat{T}_3(z) + \hat{T}_3(z)\,\hat{G}_0(z)\,\left[\hat{T}^{(1)}(z) + \hat{T}^{(2)}(z)\right] \tag{11.4.16c}$$

where $\hat{T}(z)$ is given by Eq. (11.4.17)

$$\hat{T}(z) = \hat{T}^{(1)}(z) + \hat{T}^{(2)}(z) + \hat{T}^{(3)}(z). \tag{11.4.17}$$

Written in momentum representation these equations will assume the form of three coupled integral equations. The kernel of these equations is compact and they are, in principle, solvable. Instead of writing Faddeev equations in terms of the transition operator $\hat{T}(z) \equiv \sum_i \hat{T}^{(i)}(z)$, it is more useful to formulate them in terms of state vectors. For this purpose we first obtain the resolvent version of Faddeev equations. From Eqs. (11.4.4) and (11.4.12b) we have

$$\begin{aligned}\hat{G}(z) &= \hat{G}_0(z) + \hat{G}_0(z)\,\hat{T}(z)\,\hat{G}_0(z)\\ &= \hat{G}_0(z) + \hat{G}_0(z) \sum_i \hat{T}^{(i)}(z)\,\hat{G}_0(z).\end{aligned}$$

Let $G^{(i)}(z)$ be defined by

$$\hat{G}^{(i)}(z) \equiv -\hat{G}_0(z)\,\hat{T}^{(i)}(z)\,\hat{G}_0(z) \tag{11.4.18}$$

so that

$$\hat{G}(z) = \hat{G}_0(z) - \sum_i \hat{G}^{(i)}(z). \tag{11.4.19}$$

To obtain the resolvent version of Faddeev equation we pre- and post-multiply both sides of Faddeev equation (11.4.16) by $\hat{G}_0(z)$ and use the definition of $G^{(i)}(z)$ [Eq. (11.4.18)] to get

$$\hat{G}_0(z)\,\hat{T}^{(i)}(z)\hat{G}_0(z) = \hat{G}_0(z)\,\hat{T}_i(z)\,\hat{G}_0(z) + \hat{G}_0(z)\,\hat{T}_i(z) \sum_{k\neq i} \hat{G}_0(z)\hat{T}^{(k)}(z)\hat{G}_0(z),$$

or

$$-\hat{G}^{(i)}(z) = \hat{G}_0(z)\,\hat{T}_i(z)\,\hat{G}_0(z) + \hat{G}_0(z)\,\hat{T}_i(z) \sum_{k\neq i} -\hat{G}^{(k)}(z). \tag{11.4.16a*}$$

Now, in analogy with the identity

$$\hat{G}(z) = \hat{G}_0(z) + \hat{G}_0(z)\hat{T}(z)\hat{G}_0(z)\,, \tag{11.4.20}$$

we can derive the identity

$$\hat{G}_i(z) = \hat{G}_0(z) + \hat{G}_0(z)\hat{T}_i(z)\hat{G}_0(z) \tag{11.4.21}$$

for the case in which only one pair interaction (i-th) is operative. Using Eq. (11.4.21) in Eq. (11.4.16a), we get

$$\hat{G}^{(i)}(z) = \hat{G}_0(z) - \hat{G}_i(z) + \hat{G}_0(z)\,\hat{T}_i(z)\sum_{k\neq i}\hat{G}^{(k)}(z)\,. \tag{11.4.22}$$

This is the resolvent version of Faddeev equations where $\hat{G}(z)$ can be expressed in terms of $G^{(i)}(z)$ by Eq. (11.4.19).

From the resolvent version of Faddeev equations (11.4.19) and (11.4.22) we can derive the state version of Faddeev equations. Let $\left|\Phi_n^{(3)}\right\rangle$ represent an eigenstate of $\hat{H}_3 = \hat{H}_0 + \hat{V}_3$ belonging to the eigenvalue E_n. (In this (asymptotic) state the third pair is bound and the third particle is free.) Also let $|\Psi_{3n}\rangle$ be the eigenstate (scattering state) of the total Hamiltonian $\hat{H} = \hat{H}_0 + \hat{V} = \hat{H}_0 + \sum_i V_i$ belonging to the same eigenvalue E_n. From the definition of the projection operators $\hat{P}_n^{(3)} \equiv i\epsilon\,\hat{G}_3\,(E_n + i\epsilon)$ and $\hat{P}_n = i\epsilon\,\hat{G}\,(E_n + i\epsilon)$, we have[15]

$$i\epsilon\,\hat{G}_i\,(E_n + i\epsilon)\left|\Phi_n^{(j)}\right\rangle = \delta_{ij}\left|\Phi_n^{(j)}\right\rangle\,, \tag{11.4.24}$$

and

$$i\epsilon\,\hat{G}\,(E_n + i\epsilon)\left|\Phi_n^{(3)}\right\rangle = |\Psi_{3n}\rangle\,, \tag{11.4.25}$$

where $\lim \epsilon \to 0+$ is implied. In $|\Psi_{3n}\rangle$, the index 3 signifies the total state of the three-body system when particle 3 is incident on bound pair 3 and eventually all the three pairs are interacting. Further, $i\epsilon\,\hat{G}_0\,(E_n + i\epsilon)\,|\Phi_n^{(i)}\rangle = 0$ since $\hat{H}_0|\Phi_n^{(i)}\rangle \neq E_n|\Phi_n^{(i)}\rangle$.

Now from Eqs. (11.4.25) and (11.4.19) we get

$$|\Psi_{3n}\rangle = i\epsilon\left[\hat{G}_0(E_n + i\epsilon) - \sum_{i=1}^{3}\hat{G}^{(i)}(E_n + i\epsilon)\right]|\Phi_n^{(3)}\rangle = \sum_i |\psi_{3n}^{(i)}\rangle \tag{11.4.26}$$

where $|\psi_{3n}^{(i)}\rangle$ is defined by

$$|\psi_{3n}^{(i)}\rangle \equiv -i\epsilon\,G^{(i)}(E_n + i\epsilon)|\Phi_n^{(3)}\rangle\,. \tag{11.4.27}$$

15

$$i\epsilon\left(E_n + i\epsilon - \hat{H}_i\right)^{-1}\left|\Phi_n^{(j)}\right\rangle = \frac{i\epsilon}{E_n + i\epsilon - \hat{H}_i}\left|\Phi_n^{j)}\right\rangle = 0 \text{ if } i \neq j \text{ as } \hat{H}_i\left|\Phi_n^{(j)}\right\rangle \neq E_n\left|\Phi_n^{(j)}\right\rangle$$

and $$i\epsilon\left(E_n + i\epsilon - \hat{H}_i\right)^{-1}\left|\Phi_n^{(i)}\right\rangle = \frac{i\epsilon}{E_n + i\epsilon - \hat{H}_i}\left|\Phi_n^{(i)}\right\rangle = \frac{i\epsilon}{E_n + i\epsilon - E_n}\left|\Phi_n^{(i)}\right\rangle = \left|\Phi_n^{(i)}\right\rangle \tag{11.4.23}$$

If we let $\hat{H}$ operate on the state $\hat{P}_n\left|\Phi_n^{(3)}\right\rangle$, we get

$$\hat{H}\,\hat{P}_n\left|\Phi_n^{(3)}\right\rangle = i\epsilon\left(E_n + i\epsilon - (E_n + i\epsilon - \hat{H}\right)\,\hat{G}(E_n + i\epsilon)\left|\Phi_n^{(3)}\right\rangle = E_n\,\hat{P}\left|\Phi_n^{(3)}\right\rangle\,.$$

Hence $P_n\left|\Phi_n^{(3)}\right\rangle$ may be identified with $|\Psi_{3n}\rangle$.

Pre-multiplying both sides in the resolvent version of Faddeev Eq. (11.4.22) by $-i\epsilon$ and letting both sides operate on $\left|\Phi_n^{(3)}\right\rangle$, we get

$$\begin{aligned}-i\epsilon\,\hat{G}^{(i)}(E_n+i\epsilon)|\Phi_n^{(3)}\rangle &= -i\epsilon\,\hat{G}_0(E_n+i\epsilon)|\Phi_n^{(3)}\rangle + i\epsilon\,\hat{G}_i(E_n+i\epsilon)|\Phi_n^{(3)}\rangle \\ &\quad - i\epsilon\,\hat{G}_0(E_n+i\epsilon)\,\hat{T}_i(E_n+i\epsilon)\sum_{k\neq i}\hat{G}^{(k)}(E_n+i\epsilon)|\Phi_n^{(3)}\rangle\end{aligned}$$

or

$$|\psi_{3n}^{(i)}\rangle = \delta_{i3}|\Phi_n^{(3)}\rangle + \hat{G}_0(E_n+i\epsilon)\,\hat{T}_i(E_n+i\epsilon)\sum_{k\neq i}\psi_{3n}^{(k)}\rangle\,. \tag{11.4.28}$$

This is the state version of Faddeev equations in which the total state of the three-body system is split as in Eq. (11.4.26). Explicitly, we have

$$|\Psi_{3n}\rangle = |\psi_{3n}^{(1)}\rangle + |\psi_{3n}^{(2)}\rangle + |\psi_{3n}^{(3)}\rangle\,, \tag{11.4.29}$$

where

$$\left|\psi_{3n}^{(1)}\right\rangle = \hat{G}_0(E_n+i\epsilon)\,\hat{T}_1(E_n+i\epsilon)\left[\left|\psi_{3n}^{(2)}\right\rangle + \left|\psi_{3n}^{(3)}\right\rangle\right], \tag{11.4.30}$$

$$\left|\psi_{3n}^{(2)}\right\rangle = \hat{G}_0(E_n+i\epsilon)\,\hat{T}_2(E_n+i\epsilon)\left[\left|\psi_{3n}^{(3)}\right\rangle + \left|\psi_{3n}^{(1)}\right\rangle\right], \tag{11.4.31}$$

$$\left|\psi_{3n}^{(3)}\right\rangle = |\Phi_n^{(3)}\rangle + \hat{G}_0(E_n+i\epsilon)\,\hat{T}_3(E_n+i\epsilon)\left[\left|\psi_{3n}^{(1)}\right\rangle + \left|\psi_{3n}^{(2)}\right\rangle\right]. \tag{11.4.32}$$

11.5 Faddeev Equations in Momentum Representation

In the momentum representation, the three-body state $|\Psi_{3n}\rangle$ may be represented by the function

$$\Psi_{3n}\left(\boldsymbol{q}_k,\boldsymbol{p}_k\right) \equiv \Psi_{3n}(\boldsymbol{P}) = \langle \boldsymbol{q}_k,\boldsymbol{p}_k|\Psi_{3n}\rangle\,, \tag{11.5.1}$$

where $k=1,2,3$. Also, in momentum representation, Eq. (11.4.29) can be written as

$$\Psi_{3n}\left(\boldsymbol{q}_k,\boldsymbol{p}_k\right) \equiv \psi_{3n}^{(1)}(\boldsymbol{q}_1,\boldsymbol{p}_1) + \psi_{3n}^{(2)}(\boldsymbol{q}_2,\boldsymbol{p}_2) + \psi_{3n}^{(3)}(\boldsymbol{q}_3,\boldsymbol{p}_3), \tag{11.5.2}$$

where $\psi_{3n}^{(1)}(\boldsymbol{q}_1,\boldsymbol{p}_1)$, $\psi_{3n}^{(2)}(\boldsymbol{q}_2,\boldsymbol{p}_2)$, and $\psi_{3n}^{(3)}(\boldsymbol{q}_3,\boldsymbol{p}_3)$ are defined by

$$\begin{aligned}\psi_{3n}^{(1)}(\boldsymbol{q}_1,\boldsymbol{p}_1) &= \left\langle \boldsymbol{q}_1,\boldsymbol{p}_1\middle|\psi_{3n}^{(1)}\right\rangle = \left\langle \boldsymbol{q}_2,\boldsymbol{p}_2\middle|\psi_{3n}^{(1)}\right\rangle = \left\langle \boldsymbol{q}_3,\boldsymbol{p}_3\middle|\psi_{3n}^{(1)}\right\rangle = \left\langle \boldsymbol{P}\middle|\psi_{3n}^{(1)}\right\rangle, \\ \psi_{3n}^{(2)}(\boldsymbol{q}_2,\boldsymbol{p}_2) &= \left\langle \boldsymbol{q}_2,\boldsymbol{p}_2\middle|\psi_{3n}^{(2)}\right\rangle = \left\langle \boldsymbol{q}_3,\boldsymbol{p}_3\middle|\psi_{3n}^{(2)}\right\rangle = \left\langle \boldsymbol{q}_1,\boldsymbol{p}_1\middle|\psi_{3n}^{(2)}\right\rangle = \left\langle \boldsymbol{P}\middle|\psi_{3n}^{(2)}\right\rangle, \\ \psi_{3n}^{(3)}(\boldsymbol{q}_3,\boldsymbol{p}_3) &= \left\langle \boldsymbol{q}_3,\boldsymbol{p}_3\middle|\psi_{3n}^{(3)}\right\rangle = \left\langle \boldsymbol{q}_1,\boldsymbol{p}_1\middle|\psi_{3n}^{(3)}\right\rangle = \left\langle \boldsymbol{q}_2,\boldsymbol{p}_2\middle|\psi_{3n}^{(3)}\right\rangle = \left\langle \boldsymbol{P}\middle|\psi_{3n}^{(3)}\right\rangle,\end{aligned} \tag{11.5.2a}$$

and the set of Faddeev equations (11.4.28) can be written as

$$\langle P|\psi_{3n}^{(i)}\rangle = \delta_{i3}\langle \boldsymbol{P}|\Phi_{3n}^{(3)}\rangle + \langle \boldsymbol{P}|\hat{G}_0(E_n+i\epsilon)\,\hat{T}_i(E_n+i\epsilon)\left(\left|\psi_{3n}^{(j)}\right\rangle + \left|\psi_{3n}^{(k)}\right\rangle\right) \tag{11.5.3}$$

where $i,j,k = 1,2,3$ or $2,3,1$ or $3,1,2$ $(i\neq j\neq k)$, and

$$\langle \boldsymbol{P}| \equiv \langle \boldsymbol{q}_1,\boldsymbol{p}_1| \equiv \langle \boldsymbol{q}_2,\boldsymbol{p}_2| = \langle \boldsymbol{q}_3,\boldsymbol{p}_3|\,.$$

Let us consider the simple case when $m_i = m_j = m_k = M$. Also, for ease of writing, we take $M = 1 = \hbar$. Writing Eq. (11.5.3) for $i = 3$, we have

$$\begin{aligned}\psi^3_{3n}(\boldsymbol{q}_3, \boldsymbol{p}_3) &= \Phi^{(3)}_n(\boldsymbol{q}_3, \boldsymbol{p}_3) + \frac{1}{E_n + i\epsilon - p_3^2 - \frac{3}{4}q_3^2} \int\int \langle \boldsymbol{q}_3, \boldsymbol{p}_3 | \hat{T}_3(E_n + i\epsilon) \, | \boldsymbol{q}'_3, \boldsymbol{p}'_3 \rangle \\ &\quad d^3q'_3 \, d^3p'_3 \left[\psi^{(1)}_{3n}(\boldsymbol{q}'_1, \boldsymbol{p}'_1) + \psi^{(2)}_{3n}(\boldsymbol{q}'_2, \boldsymbol{p}'_2) \right] \\ &= \Phi^{(3)}_n(\boldsymbol{q}_3, \boldsymbol{p}_3) + (E_n - \frac{3}{4}q_3^2 + i\epsilon - p_3^2)^{-1} \int d^3q'_3 \delta^3(\boldsymbol{q}'_3 - \boldsymbol{q}_3) \int d^3p'_3 \\ &\quad \times \left\langle \boldsymbol{p}_3 \left| \hat{t}_3 \left(E_n + i\epsilon - \frac{3}{4}q'^2_3 \right) \right| \boldsymbol{p}'_3 \right\rangle \left\{ \psi^{(1)}_{3n}(\boldsymbol{q}'_1, \boldsymbol{p}'_1) + \psi^{(2)}_{3n}(\boldsymbol{q}'_2, \boldsymbol{p}'_2) \right\} .\end{aligned}$$

Putting $E_n - \frac{3}{4}q_3^2 + i\epsilon \equiv s + i\epsilon \equiv z$, expressing $\boldsymbol{q}'_1, \boldsymbol{p}'_1$ and $\boldsymbol{q}'_2, \boldsymbol{p}'_2$ in terms of $\boldsymbol{q}'_3, \boldsymbol{p}'_3$ in $\psi^{(1)}_{3n}$ and $\psi^{(2)}_{3n}$, respectively, integrating over $\boldsymbol{q}'_3$ and finally replacing, on both sides of the equation, $\boldsymbol{q}_3, \boldsymbol{p}_3$ by $\boldsymbol{q}, \boldsymbol{p}$ and the dummy variable $\boldsymbol{p}'_3$ by $\boldsymbol{p}'$ we get

$$\begin{aligned}\psi^{(3)}_{3n}(\boldsymbol{q}, \boldsymbol{p}) &= \Phi^{(3)}_n(\boldsymbol{q}, \boldsymbol{p}) + (s + i\epsilon - p^2)^{-1} \int \langle \boldsymbol{p} | \hat{t}_3(s + i\epsilon) | \boldsymbol{p}' \rangle \\ &\quad \times \left\{ \psi^{(1)}_{3n}\left(\boldsymbol{p}' - \frac{1}{2}\boldsymbol{q}, -\frac{1}{2}\boldsymbol{p}' - \frac{3}{4}\boldsymbol{q} \right) + \psi^{(2)}_{3n}\left(-\boldsymbol{p}' - \frac{1}{2}\boldsymbol{q}, -\frac{1}{2}\boldsymbol{p}' + \frac{3}{4}\boldsymbol{q} \right) \right\} d^3p' .\end{aligned}$$

Now, changing in the above equation the dummy variable $\boldsymbol{p}'$ by $\boldsymbol{p}' + \frac{1}{2}\boldsymbol{q}$ in the first integral and by $-\boldsymbol{p}' - \frac{1}{2}\boldsymbol{q}$ in the second integral, we can rewrite this equation as

$$\begin{aligned}\psi^{(3)}_{3n}(\boldsymbol{q}, \boldsymbol{p}) &= \Phi^{(3)}_n(\boldsymbol{q}, \boldsymbol{p}) \\ &\quad + (s + i\epsilon - p^2) \int d^3p' \left\{ \left\langle \boldsymbol{p} \left| \hat{t}_3(s + i\epsilon) \right| \boldsymbol{p}' + \frac{1}{2}\boldsymbol{q} \right\rangle \psi^{(1)}_{3n}\left(\boldsymbol{p}', -\frac{1}{2}\boldsymbol{p}' - \boldsymbol{q} \right) \right. \\ &\quad \left. + \left\langle \boldsymbol{p} \left| \hat{t}_3(s + i\epsilon) \right| -\boldsymbol{p}' - \frac{1}{2}\boldsymbol{q} \right\rangle \psi^{(2)}_{3n}\left(\boldsymbol{p}', \frac{1}{2}\boldsymbol{p}' + \boldsymbol{q} \right) \right\} . \qquad (11.5.4)\end{aligned}$$

Again, putting $i = 1$ in Eq. (11.5.3) we have, similarly,

$$\langle \boldsymbol{P} | \psi^{(1)}_{3n} \rangle = \langle \boldsymbol{P} | G_0(E_n + i\epsilon) T_1(E_n + i\epsilon) | \left\{ \left| \psi^{(2)}_{3n} \right\rangle + \left| \psi^{(3)}_{3n} \right\rangle \right\} .$$

By a similar set of manipulations,

(1) taking $\langle \boldsymbol{P} | \equiv \langle \boldsymbol{q}_1, \boldsymbol{p}_1 |$,

(2) introducing the unit operator $\iint |\boldsymbol{q}'_1, \boldsymbol{p}'_1\rangle d^3q'_1 \, d^3p'_1 \langle \boldsymbol{q}'_1, \boldsymbol{p}'_1|$ after $\hat{T}_1(E_n + i\epsilon)$ and writing $\left\langle \boldsymbol{q}'_1, \boldsymbol{p}'_1 \middle| \psi^{(3)}_{3n} \right\rangle \equiv \psi^{(3)}_{3n}(\boldsymbol{q}'_3, \boldsymbol{p}'_3)$ and $\left\langle \boldsymbol{q}'_1, \boldsymbol{p}'_1 \middle| \psi^{(2)}_{3n} \right\rangle \equiv \psi^{(2)}_{3n}(\boldsymbol{q}'_2, \boldsymbol{p}'_2)$, in accordance with Eq. (11.5.2a),

(3) expressing $\boldsymbol{q}'_2, \boldsymbol{p}'_2$ and $\boldsymbol{q}'_3, \boldsymbol{p}'_3$ in $\psi^{(2)}_{3n}(\boldsymbol{q}'_2, \boldsymbol{p}'_2)$ and $\psi^{(3)}_{3n}(\boldsymbol{q}'_3, \boldsymbol{p}'_3)$ respectively, in terms of $\boldsymbol{q}'_1, \boldsymbol{p}'_1$,

(4) integrating over $\boldsymbol{q}'_1$ and replacing the dummy variable $\boldsymbol{p}'_1$ by $\boldsymbol{p}' + \boldsymbol{q}/2$ in the first integral and by $-\boldsymbol{p}' - \boldsymbol{q}/2$ in the second integral, we get

$$\begin{aligned}\psi^{(1)}_{3n}(\boldsymbol{q}, \boldsymbol{p}) &= (s - p^2 + i\epsilon)^{-1} \int d^3p' \left\{ \left\langle \boldsymbol{p} \left| \hat{t}_1(s + i\epsilon) \right| \boldsymbol{p}' + \frac{1}{2}\boldsymbol{q} \right\rangle \psi^{(2)}_{3n}\left(\boldsymbol{p}', -\frac{1}{2}\boldsymbol{p}' - \boldsymbol{q} \right) \right. \\ &\quad \left. + \left\langle \boldsymbol{p} \left| \hat{t}_1(s + i\epsilon) \right| -\boldsymbol{p}' - \frac{1}{2}\boldsymbol{q} \right\rangle \psi^{(3)}_{3n}\left(\boldsymbol{p}', \frac{1}{2}\boldsymbol{p}' + \boldsymbol{q} \right) \right\} . \qquad (11.5.5)\end{aligned}$$

By following the same procedure for $i = 2$ in Eq. (11.5.3) we get

$$\psi^{(2)}_{3n}(\boldsymbol{q},\boldsymbol{p}) = (s - p^2 + i\epsilon)^{-1} \int d^3p' \left\{ \left\langle p \left| \hat{t}_2(s+i\epsilon) \right| \boldsymbol{p}' + \frac{1}{2}\boldsymbol{q} \right\rangle \psi^{(3)}_{3n}\left(\boldsymbol{p}', -\frac{1}{2}\boldsymbol{p}' - \boldsymbol{q}\right) \right.$$
$$\left. + \left\langle p \left| \hat{t}_2(s+i\epsilon) \right| - \boldsymbol{p}' - \frac{1}{2}\boldsymbol{q} \right\rangle \psi^{(1)}_{3n}\left(\boldsymbol{p}', \frac{1}{2}\boldsymbol{p}' + \boldsymbol{q}\right) \right\} . \qquad (11.5.6)$$

To summarize, if particle 3 is incident on pair 3 and eventually all the three pairs interact then the total three-body wave function is given by

$$\Psi(\boldsymbol{P}) \equiv \Psi_{3n}\left(\boldsymbol{q}_k, \boldsymbol{p}_k\right) = \psi^{(1)}_{3n}\left(\boldsymbol{q}_1, \boldsymbol{p}_1\right) + \psi^{(2)}_{3n}\left(\boldsymbol{q}_2, \boldsymbol{p}_2\right) + \psi^{(3)}_{3n}\left(\boldsymbol{q}_3, \boldsymbol{p}_3\right)$$

and the functions $\psi^{(1)}_{3n}$, $\psi^{(2)}_{3n}$, $\psi^{(3)}_{3n}$ satisfy the set of coupled integral equations (11.5.5) and (11.5.6) and (11.5.4). Here $\Phi^{(3)}_{3n}(\boldsymbol{q},\boldsymbol{p})$ represents the asymptotic state of the system when only pair 3 is interacting and is bound and particle 3 is free and the total energy of the system is E_n.

11.6 Faddeev Equations for a Three-body Bound System

We can adapt Faddeev equation for a three-body bound system as well. We again consider the simple case in which the three particles are identical and spinless and the interaction is central and attractive. This implies that $\hat{V}_1 = \hat{V}_2 = \hat{V}_3 = \hat{V}$ and $\hat{t}_1(z) = \hat{t}_2(z) = \hat{t}_3(z) = \hat{t}(z)$. For the three-body bound state one may also drop the inhomogeneous term $\Phi^{(3)}_n$ in Eq. (11.5.3) and assume the three-body bound state energy E_n to be $-\frac{\alpha_0^2\hbar^2}{M}$ or $-\alpha_0^2$ in units of $\hbar = 1 = M$. Further we can designate the bound state wave function $\Psi_{3n}(\boldsymbol{q},\boldsymbol{p})$ as simply $\Psi_n(\boldsymbol{q},\boldsymbol{p})$ and the function $\psi^{(i)}_{3n}(\boldsymbol{q},\boldsymbol{p})$ as simply $\psi^{(i)}_n(\boldsymbol{q},\boldsymbol{p})$, since the suffix 3 has no relevance now. Hence

$$\Psi_n\left(\boldsymbol{q}_k, \boldsymbol{p}_k\right) = \psi^{(1)}_n\left(\boldsymbol{q}_1, \boldsymbol{p}_1\right) + \psi^{(2)}_n\left(\boldsymbol{q}_2, \boldsymbol{p}_2\right) + \psi^{(3)}_n\left(\boldsymbol{q}_3, \boldsymbol{p}_3\right)$$

where $\psi^{(1)}_n$, $\psi^{(2)}_n$, $\psi^{(3)}_n$ satisfy the set of coupled equations (11.5.4) through (11.5.6), without the inhomogeneous term. Since these three equations then become similar, except for a cyclic permutation of 1, 2, 3, we may drop the suffix i in $\psi^{(i)}_n$ and simply write

$$\Psi_n\left(\boldsymbol{q}_k, \boldsymbol{p}_k\right) = \psi_n\left(\boldsymbol{q}_1, \boldsymbol{p}_1\right) + \psi_n\left(\boldsymbol{q}_2, \boldsymbol{p}_2\right) + \psi_n\left(\boldsymbol{q}_3, \boldsymbol{p}_3\right) , \qquad (11.6.1)$$

where $\psi_n(\boldsymbol{q},\boldsymbol{p})$ satisfies the integral equation

$$\psi_n(\boldsymbol{q},\boldsymbol{p}) = \left(s - p^2\right)^{-1} \int d^3p' \left\{ \left\langle \boldsymbol{p} \left| \hat{t}(s+i\epsilon) \right| - \boldsymbol{p}' - \frac{1}{2}\boldsymbol{q} \right\rangle \psi_n\left(\boldsymbol{p}', \frac{1}{2}\boldsymbol{p}' + \boldsymbol{q}\right) \right.$$
$$\left. + \left\langle \boldsymbol{p} \left| \hat{t}(s+i\epsilon) \right| \boldsymbol{p}' + \frac{1}{2}\boldsymbol{q} \right\rangle \psi_n\left(\boldsymbol{p}', -\frac{1}{2}\boldsymbol{p}' - \boldsymbol{q}\right) \right\} , \qquad (11.6.2)$$

with $s = -\alpha_0^2 - \frac{3}{4}q^2$. Faddeev equations (11.6.1) and (11.6.2) for the bound state of three identical particles are the same as Eyges' equation (11.2.10). The identity of the two equations can be established if we use the two-body Lippmann-Schwinger equation, written in the momentum representation

$$\langle \boldsymbol{p} | \hat{t}(s+i\epsilon) | \boldsymbol{k}' \rangle = \langle \boldsymbol{p} | \hat{v} | \boldsymbol{k}' \rangle + \int \frac{\langle \boldsymbol{p} | \hat{v} | \boldsymbol{k}'' \rangle d^3k'' \langle \boldsymbol{k}'' | \hat{t}(s+i\epsilon) | \boldsymbol{k}' \rangle}{s - k''^2} ,$$

in Eq. (11.6.2). This gives us

$$\psi_n(\boldsymbol{q},\boldsymbol{p}) = \left(s-p^2\right)^{-1}\left[\int d^3p' \left\{\left\langle \boldsymbol{p}\left|\hat{v}\right| -\boldsymbol{p}'-\frac{1}{2}\boldsymbol{q}\right\rangle \psi_n\left(\boldsymbol{p}',\frac{1}{2}\boldsymbol{p}'+\boldsymbol{q}\right)\right.\right.$$
$$+\left.\left\langle \boldsymbol{p}\left|\hat{v}\right|\boldsymbol{p}'+\frac{\boldsymbol{q}}{2}\right\rangle \psi_n\left(\boldsymbol{p}',-\frac{1}{2}\boldsymbol{p}'-\boldsymbol{q}\right)\right\}$$
$$+\int d^3k'' \left\langle \boldsymbol{p}\left|\hat{v}\right|\boldsymbol{k}''\right\rangle \int \frac{d^3p'}{s-k''^2}\left\{\left\langle \boldsymbol{k}''\left|\hat{t}(s+i\epsilon)\right|\boldsymbol{p}'+\frac{1}{2}\boldsymbol{q}\right\rangle \psi_n\left(\boldsymbol{p}',-\frac{\boldsymbol{p}}{2}-\boldsymbol{q}\right)\right.$$
$$+\left.\left.\left\langle \boldsymbol{k}''\left|\hat{t}(s+i\epsilon)\right|-\boldsymbol{p}'-\frac{1}{2}\boldsymbol{q}\right\rangle \psi_n\left(\boldsymbol{p}',\frac{\boldsymbol{p}'}{2}+\boldsymbol{q}\right)\right\}\right].$$

If we make use of Eq. (11.6.2) again, the second term within square brackets reduces to

$$\int d^3k'' \left\langle \boldsymbol{p}\right|\hat{v}\left|\boldsymbol{k}''\right\rangle \psi_n(\boldsymbol{q},\boldsymbol{k}'') \equiv \int d^3p' \left\langle \boldsymbol{p}\right|\hat{v}\left|\boldsymbol{p}'\right\rangle \psi_n(\boldsymbol{q},\boldsymbol{p}').$$

Hence Eq. (11.6.2) is the same as Eyges' equation (11.2.10).

For further reduction of Faddeev equations (11.6.1) and (11.6.2) to a solvable form, it is useful to employ a separable approximation for the t-matrix elements, which are determined by two-body interaction. The t-matrix is separable [see Sec. 9.10] if the two-body potential is of separable form

$$\langle \boldsymbol{p}'|\hat{v}|\boldsymbol{p}\rangle = -\lambda\, g(\boldsymbol{p}')\, g(\boldsymbol{p}), \quad \text{with } (M=1=\hbar),$$

where the function $g(\boldsymbol{p})$ is referred to as a form factor and λ is a constant related to the strength of the interaction. This form of the two-body potential means that the t-matrix can be written as

$$\langle \boldsymbol{p}'|\hat{t}(z)|\boldsymbol{p}\rangle = -\frac{g(\boldsymbol{p}')g(\boldsymbol{p})}{D(z)}, \tag{11.6.3}$$

where

$$D(z) \equiv \frac{1}{\lambda} + \int \frac{g^2(\boldsymbol{p}')d^3k'}{z-p'^2} \quad \text{and} \quad \lambda^{-1} = \int \frac{g^2(\boldsymbol{p}')d^3p'}{B+p'^2} \tag{11.6.4}$$

and $+B$ is the binding energy of the two-body system and we have assumed that in the the only bound state to exist is in the s-channel.

It may be mentioned here that even if the potential is local and is not separable, it is possible to construct a separable approximation to the t-matrix using the two-body bound state wave function in momentum space[16]. This separable approximation, called the *unitary pole approximation*, is fairly effective. The form factor $g(\boldsymbol{p})$, representing the separable potential or the separable approximation to the t-matrix, is related to the two-body bound state wave function $\phi(\boldsymbol{p})$ by

$$\phi(\boldsymbol{p}) = \frac{N\, g(\boldsymbol{p})}{p^2+B}, \tag{11.6.5}$$

where the normalization constant N is given by

$$\frac{1}{N^2} = \int \frac{g^2(\boldsymbol{p}')d^3p'}{(p'^2+B)^2}. \tag{11.6.6}$$

In addition to substituting the separable t-matrix (11.6.3) in Eq. (11.6.2), we can also adopt the structure proposed by Mitra for the function $\psi_n(\boldsymbol{q}_i,\boldsymbol{p}_i)$ in $\Psi_n(\boldsymbol{q},\boldsymbol{p}) \equiv \Psi_n(\boldsymbol{P})$ in

[16] M. G. Fuda, Nuclear Physics **A116**, 83 (1968).

Eq. (11.6.1). Accordingly, each term $\psi_n(\boldsymbol{q}_i, \boldsymbol{p}_i)$ on the right hand side of Eq. (11.6.1) may be regarded as the product of the bound state of the i-th pair

$$\frac{g(\boldsymbol{p}_i)}{p_i^2 + \alpha_0^2 + \frac{3}{4}q_i^2}$$

and the spectator wave function $\chi(\boldsymbol{q}_i)$ of the i-th particle. One may regard $\frac{3}{4}q_i^2$ as the energy taken away by the i-th particle of the three-body system and $-\alpha_0^2 - \frac{3}{4}q_i^2$ to be the energy associated with the i-th pair. Thus each ψ_n has the form

$$\psi_n(\boldsymbol{q}_i, \boldsymbol{p}_i) = \frac{g(\boldsymbol{p}_i)}{\alpha_0^2 + \frac{3}{4}q_i^2 + p_i^2}\,\chi(\boldsymbol{q}_i)\,.$$

Using this in Eq. (11.6.1) we get

$$\Psi_n(\boldsymbol{q}_k, \boldsymbol{p}_k) = \frac{g(\boldsymbol{p}_1)\,\chi(\boldsymbol{q}_1)}{\alpha_0^2 + \frac{3}{4}q_1^2 + p_1^2} + \frac{g(\boldsymbol{p}_2)\,\chi(\boldsymbol{q}_2)}{\alpha_0^2 + \frac{3}{4}q_2^2 + p_2^2} + \frac{g(\boldsymbol{p}_3)\,\chi(\boldsymbol{q}_3)}{\alpha_0^2 + \frac{3}{4}q_3^2 + p_3^2}\,.$$

This is what Mitra termed the *resonating group structure* of the three-body wave function. Using the separable t-matrix and the resonating group structure of the three-body wave function into Faddeev equation (11.6.2) for the three-body bound state, we can reduce the latter into an integral equation for the *spectator wave function* $\chi(\boldsymbol{q})$[17]

$$\chi(\boldsymbol{q}) = 2D^{-1}\left(-\alpha_0^2 - \frac{3}{4}q^2\right)\int \frac{d^3p'\, g\left(|\boldsymbol{p}' + \frac{1}{2}\boldsymbol{q}|\right)\, g\left(|\frac{1}{2}\boldsymbol{p}' + \boldsymbol{q}|\right)}{\alpha_0^2 + p'^2 + q^2 + \boldsymbol{p}'\cdot\boldsymbol{q}}\,\chi(\boldsymbol{p}'), \qquad (11.6.7)$$

where, $D(z) = D(s)$ is defined by Eq. (11.6.4). For the s-state interaction, we may regard $g(\boldsymbol{k})$ to depend only on the magnitude of $\boldsymbol{k}$. Then the equation for the spectator wave function can be written in a simplified form as

$$\chi(\boldsymbol{q}) = 2D^{-1}\left(-\alpha_0^2 - \frac{3}{4}q^2\right)\int_0^\infty \chi(\boldsymbol{p}')\, Z(q, p', \alpha_0)\, p'^2\, dp'\,, \qquad (11.6.8)$$

where

$$Z(q, p', \alpha_0) \equiv 4\pi \int_{-1}^{1} g\left(\sqrt{p'^2 + \frac{q^2}{4} + p'q\zeta}\right) g\left(\sqrt{\frac{p'^2}{4} + q^2 + p'q\zeta}\right) \frac{d\zeta}{(\alpha_0^2 + p'^2 + q^2 + p'q\zeta)}\,.$$

The kernel $2D^{-1}\left(-\alpha_0^2 - \frac{3}{4}q^2\right)\, Z(q, p', \alpha_0)p'^2$ of the homogeneous integral equation (11.6.8) involves the three-body binding energy α_0^2, which must be found by trial so that the above equation holds.

Sophistications in the solution of three-nucleon bound state problems using the *homogeneous* Faddeev equation were made by Sitenko and Kharchenko (1963) [18]who also took account of spin and isospin of nucleons to construct completely anti-symmetric states of the three particles. Since then many sophisticated calculations have been done by various workers for the three-nucleon bound state.

The problem of scattering of one particle by the other two in bound state can also be tackled within the framework of Faddeev equations (11.5.4) through (11.5.6) if $m_1 = m_2 =$

[17] For details see C. Maheshwari, A. V. Lagu, and V. S. Mathur, Proc Ind. Nat. Sci. Acad. **39**, 151 (1973).
[18] A. G. Sitenko and V. F. Kharchenko, Nucl. Phys. **49**, p.1 (1963).

$m_3 = M$, with Eq. (11.5.4) containing the inhomogeneous term as well. However, for solving the three-body scattering problem, for example, neutron-deuteron ($n-d$) scattering, another version of Faddeev equations due to Alt, Grassberger, and Sandhas[19] is more elegant. This formulation is useful not only for $n - d$ elastic scattering[20] but also for calculating the differential cross-section for any rearrangement process of the kind

$$a\,(b \oplus x) + A = B\,(A \oplus x) + b\,. \tag{11.6.9}$$

This describes a scattering event in which a bound state of particles b and x (designated a) interacts with the target particle A resulting in a bound structure of A and x (designated B) and an outgoing particle b. The overall effect is the exchange of one of the incident particles (in the bound state) with the target particle (stripping). In the three-body treatment, the particles A, b, x are treated as *core particles* in the sense that their internal structures remain intact and do not come into play. Thus deuteron stripping (d,p) reactions and α-transfer (^{6}Li,d) reactions, on closed shell (or even-even) nuclei can be dealt with using the AGS framework[21], if the interactions between pairs (A, b), (A, x), and (b, x) are known in a separable form. In the AGS formulation the matrix elements of the AGS operators between the initial and final asymptotic states[22] have a direct interpretation as the physical transition amplitudes and the differential cross-section for the rearrangement process can be easily expressed in terms of these amplitudes.

11.7 Alt, Grassberger and Sandhas (AGS) Equations

Faddeev operator $T^{(j)}(z)$ [Eqs. (11.4.10) and (11.4.15)] may be expressed as

$$\hat{T}^{(j)}(z) = \sum_{i=1}^{3} \hat{\tau}_{ji} = \sum_{i=1}^{3} \left(\delta_{ji}\hat{V}_j + \hat{V}_j\,\hat{G}(z)\,\hat{V}_i\right), \tag{11.7.1}$$

where the operators $\hat{\tau}_{ji}$ satisfy the set of equations

$$\hat{\tau}_{ji} = \delta_{ji}\,\hat{T}_j + \hat{T}_j\,G_0(z) \sum_{k\neq j} \hat{\tau}_{ki}\,. \tag{11.7.2}$$

Summing both sides of Eq. (11.7.2) over i we get the Faddeev equations

$$\hat{T}^{(j)}(z) = \hat{T}_j(z) + \hat{T}_j(z)\,\hat{G}_0(z) \sum_{k\neq j} T^{(k)}(z)\,. \tag{11.7.3}$$

One may, instead, define an operator $\hat{U}_{ji}(z)$, called the AGS operator, by

$$\hat{U}_{ji}(z) = (1 - \delta_{ij})\left(z - \hat{H}_0\right) + \sum_{q\neq j} \sum_{p\neq i} \hat{\tau}_{qp}(z)\,, \tag{11.7.4}$$

[19] E. O. Alt, P. Grassberger, and W. Sandhas, Nucl. Phys. **B1** 167 (1967).

[20] P. Doleschall, Nucl. Phys. **A201**, 264 (1973).

[21] V. S. Mathur and R. Prasad, Phys. Rev. **C24**, 2593 (1981); J. Phys. **G7**, 1955 (1981).

[22] Initial asymptotic state is the product of the bound state of the initial pair and the free state of the incident particle. Final asymptotic state may be similarly defined.

and then show that the AGS operator $\hat{U}_{ji}(z)$ so defined is consistent with Alt, Grassberger and Sandhas definition of $\hat{U}_{ji}(z)$:

$$\hat{G}(z) = \delta_{ij}\,\hat{G}_j(z) + \hat{G}_j(z)\,\hat{U}_{ji}(z)\,\hat{G}_i(z)\,. \tag{11.7.5}$$

To see this we start with the definition (11.7.4),

$$\begin{aligned}
\hat{U}_{ji}(z) &= (1-\delta_{ij})\left(z-\hat{H}_0\right) + \sum_q \sum_p (1-\delta_{qj})\,(1-\delta_{pi})\left(\delta_{qp}\,\hat{V}_q + \hat{V}_q\,\hat{G}\,\hat{V}_p\right) \\
&= (1-\delta_{ij})(z-\hat{H}_0) + (\hat{V}-\hat{V}_j-\hat{V}_i+\delta_{ij}\hat{V}_j+\hat{V}\hat{G}\hat{V}-\hat{V}_j\hat{G}\hat{V}-\hat{V}\hat{G}\hat{V}_i-\hat{V}_j\hat{G}\hat{V}_i) \\
&= (1-\delta_{ij})\left(z-\hat{H}_j\right) + \left[1+\hat{V}\hat{G}(z)-\hat{V}_j\hat{G}(z)\right]\,\hat{\overline{V}}_i
\end{aligned}$$

where $\hat{\overline{V}}_i \equiv \hat{V} - \hat{V}_i = \hat{V}_j + \hat{V}_k$. So,

$$\begin{aligned}
\hat{G}_j(z)\hat{U}_{ji}(z)\hat{G}_i(z) = {} & (1-\delta_{ij})\,\hat{G}_i(z) + \hat{G}_j(z)\,\hat{\overline{V}}_i\hat{G}_i(z) \\
& + \hat{G}_j(z)\,\hat{V}\hat{G}(z)\,\hat{\overline{V}}_i\hat{G}_i(z) - \hat{G}_j(z)\hat{V}_j\hat{G}(z)\,\hat{\overline{V}}_i\hat{G}_i(z)\,.
\end{aligned}$$

Using the identities[23]

$$\begin{aligned}
\hat{G}(z) &= \hat{G}_i(z) + \hat{G}_i(z)\,\hat{\overline{V}}_i\hat{G}(z) \\
\hat{G}(z) &= \hat{G}_i(z) + \hat{G}(z)\,\hat{\overline{V}}_i\hat{G}_i(z)\,,
\end{aligned}$$

the right-hand side of Eq. (11.7.5) can be written as

$$\begin{aligned}
&\delta_{ij}\hat{G}_j(z) + \hat{G}_j(z)\,\hat{U}_{ji}(z)\hat{G}_i(z) \\
&\qquad = \hat{G}_i(z) + \hat{G}_j(z)\,\hat{\overline{V}}_i\hat{G}_i(z) + \hat{G}_j(z)\,\hat{V}(\hat{G}-\hat{G}_i) - \hat{G}_j\hat{V}_j(\hat{G}-\hat{G}_i) \\
&\qquad = \hat{G}_i(z) + \hat{G}_j(z)\,(\hat{\overline{V}}_i - \hat{V} + \hat{V}_j)\,\hat{G}_i(z) + \hat{G}_j\hat{V}\hat{G} - \hat{G}_j\hat{V}_j\hat{G} \\
&\qquad = \hat{G}_i(z) + \hat{G}_j(z)(\hat{V}_j - \hat{V}_i)\hat{G}_i + \hat{G}_j\hat{\overline{V}}_j\hat{G} \\
&\qquad = \hat{G}_i(z) + \hat{G}_j(z)\,(\hat{G}_i^{-1} - \hat{G}_j^{-1})\hat{G}_i + \hat{G} - \hat{G}_j \\
&\qquad = \hat{G}(z)\,.
\end{aligned}$$

Hence the definitions of the AGS operator in term of Eqs. (11.7.4) and (11.7.5) are consistent. We shall now use the definition of AGS operator in terms of Eq. (11.7.4) and Faddeev's version of three-body equations (11.7.2) to derive the AGS version of three-body

[23] $\hat{G}_i(z)$ pertains to the interaction $\hat{V}_i$, while $\hat{G}(z)$ pertains to $\hat{V}_i$ plus an additional interaction $\hat{\overline{V}}_i$, i.e., total interaction $\hat{V}$.

equations[24]. We have from Eqs. (11.7.4) and (11.7.2)

$$\begin{aligned}\hat{U}_{ji}(z) &= (1-\delta_{ji})(z-\hat{H}_0) + \sum_{\ell\neq j}\sum_{p\neq i}\left\{\delta_{\ell p}\hat{T}_\ell(z) + \hat{T}_\ell(z)\,\hat{G}_0(z)\sum_{s\neq \ell}\hat{\tau}_{sp}\right\}\\ &= (1-\delta_{ji})(z-\hat{H}_0) + \sum_{\ell\neq j}\sum_{\cdot p}(1-\delta_{pi})\,\delta_{\ell p}\hat{T}_\ell(z) + \sum_{\ell\neq j}\sum_{p\neq i}\sum_{s\neq\ell}\hat{T}_\ell\,\hat{G}_0(z)\,\hat{\tau}_{sp}(z)\\ &= (1-\delta_{ji})(z-\hat{H}_0) + \sum_{\ell\neq j}\left[(1-\delta_{\ell i})\,\hat{T}_\ell(z)\hat{G}_0(z)\hat{G}_0^{-1} + \sum_{p\neq i}\sum_{s\neq\ell}\hat{T}_\ell\,\hat{G}_0(z)\,\hat{\tau}_{sp}(z)\right]\\ &= (1-\delta_{ji})(z-\hat{H}_0) + \sum_{\ell\neq j}\hat{T}_\ell(z)\hat{G}_0(z)\left\{(1-\delta_{\ell i})\,\hat{G}_0^{-1} + \sum_{p\neq i}\sum_{s\neq\ell}\hat{\tau}_{sp}\right\}\\ &= (1-\delta_{ji})(z-\hat{H}_0) + \sum_{\ell\neq j}\hat{T}_\ell(z)\,\hat{G}_0(z)\,\hat{U}_{\ell_i}(z).\end{aligned} \quad (11.7.6)$$

This is the AGS version of Faddeev equations. This set of equations, for a given j and $i = 1, 2, 3$, reduces to a set of coupled integral equations when written in the momentum representation.

The suitability of AGS version of Faddeev equations for three-body scattering or for reaction dynamics in the three-body framework can be judged from the fact that the matrix element of the AGS operator between the initial and final asymptotic states can be interpreted as the physical transition amplitude and the differential cross-section of the process can be readily expressed in terms of this amplitude.

Let $|\Phi_{j\,n}(t)\rangle = e^{-iE_{j\,n}t/\hbar}\;|\phi_{j\,n}\rangle$ be the initial ($t\to-\infty$) asymptotic state of the system so that $|\phi_{j\,n}\rangle$ is the eigenstate of the Hamiltonian $\hat{H}_j = \hat{H}_0 + \hat{V}_j$ belonging to the energy E_n. Here j refers to the initial partition of the three-body system with j-th pair bound and j-th particle free and n refers to the quantum state of the initial bound pair. Let $|\Psi_{jn}(t)\rangle = e^{-iE_{jn}t/\hbar}\;|\psi^+_{jn}\rangle$ represent the total state of the system at time t, where $|\psi^+_{jn}\rangle$ is the eigenstate of the total Hamiltonian $\hat{H} = \hat{H}_0 + \sum\hat{V}_j$ when all the three pairs interact. The probability amplitude for the event that the system has made a transition from the initial asymptotic state $|\Phi_{j\,n}(t)\rangle$ to a final asymptotic state $|\Phi_{\,i\,m}(t)\rangle = e^{-iE_{\,im}\,t/\hbar}\;|\phi_{im}\rangle$ is given by

$$\lim_{t\to\infty}\;\langle\Phi_{im}(t)|\Psi_{jn}(t)\rangle = S_{im,jn}\,.$$

Thus

$$S_{im,jn} = \lim_{t\to\infty}\;e^{-i(E_{jn}-E_{im})t/\hbar}\;\langle\phi_{im}|\psi^+_{jn}\rangle\,. \quad (11.7.7)$$

This overlap is referred to as the (im, jn)th element of the S-matrix. From the property of the projection operator $\hat{P}_n$ defined by

$$\hat{P}_n \equiv \lim_{\epsilon\to0^+}\;i\epsilon\,G(E_{jn}+i\epsilon),$$

that it projects out the state $|\psi^+_{jn}\rangle$ out of $|\phi_{jn}\rangle$, i.e., $i\epsilon G(E_{jn}+i\epsilon)|\phi_{jn}\rangle = |\psi_{jn}\rangle$, we have

$$S_{i\,m,j\,n} = \lim_{\substack{t\to\infty\\ \epsilon\to0^+}}\;e^{-i(E_{jn}-E_{im})t/\hbar}\;\langle\phi_{i\,m}|\;i\epsilon G(E_{j\,n}+i\epsilon)\;|\phi_{j\,n}\rangle\,.$$

[24]AGS equations may also be derived using definition of the AGS operator $\hat{U}_{ji}(z)$ in terms of Eq. (11.7.5) as done by Alt, Grassberger and Sandhas.

Now, if we use the definition of the AGS operator in terms of the equation (11.7.5), viz., $\hat{G}(z) = \delta_{ij}\hat{G}_j(z) + \hat{G}_i(z)\hat{U}_{ij}(z)\hat{G}_j(z)$, we get

$$S_{im,jn} = \lim_{\substack{t\to\infty \\ \epsilon\to 0}} e^{-i(E_{jn}-E_{im})t/\hbar} \Big[i\epsilon\, \delta_{ij} \,\langle\phi_{im}|\, (E_{jn} + i\epsilon - E_{jn})^{-1} \,|\phi_{jn}\rangle$$
$$+ \, (E_{jn} - E_{im} + i\epsilon)^{-1} \,\langle\phi_{im}|\, \hat{U}_{ij}(E_{jn} + i\epsilon)\, |\phi_{jn}\rangle \Big] ,$$

or

$$S_{im,jn} = \delta_{ij}\, \delta_{im,jn} + \lim_{t\to\infty} \frac{e^{-i(E_{jn}-E_{im})t/\hbar}}{E_{jn} - E_{im}} \,\langle\phi_{im}|\, \hat{U}_{ij}(E_{jn} + i\epsilon)\, |\phi_{jn}\rangle ,$$
$$= \delta_{ij}\, \delta_{im,jn} - i\pi\, \delta(E_{jn} - E_{im})\, \langle\phi_{im}|Uij(E_{jn} + i\epsilon)|\phi_{jn}\rangle . \tag{11.7.8}$$

The delta function[25] takes care of the energy conservation in the process. The final and initial asymptotic states can be expressed as

$$|\phi_{im}\rangle \equiv \left|\boldsymbol{Q}_i \,,\, d_i \,,\, \phi_{J_i}^{n_i}\right\rangle ,$$
$$|\phi_{jn}\rangle \equiv \left|\boldsymbol{Q}_j \,,\, d_j \,,\, \phi_{J_j}^{n_j}\right\rangle ,$$

where $\boldsymbol{Q}_i$ is the momentum of the i-th particle, d_i specifies its spin orientation and $\phi_{J_i}^{n_i}$ the bound state of the i-th pair. If $i \neq j$ and $im \neq jn$ then the probability of the transition $jn \to im$ is given by

$$|\langle\phi_{im}|U_{ij}(E_{jn} + i\epsilon)|\phi_{jn}\rangle|^2 .$$

The cross-section for the reaction can easily be expressed in term of this probability.

Problems

1. Given

$$\hat{G}(z) = (z - \hat{H})^{-1} \equiv (z - \hat{H}_0 - \hat{V}_1 - \hat{V}_2 - \hat{V}_3)^{-1} ,$$

and

$$\hat{G}_i(z) = (z - \hat{H}_i)^{-1} \equiv (z - \hat{H}_0 - \hat{V}_i)^{-1} ,$$

show that

(a) $\hat{G}(z) = \hat{G}_0(z) + \hat{G}_0(z)\,\hat{T}(z)\,\hat{G}_0(z)$

(b) $\hat{G}(z)\,\hat{V} = \hat{G}_0(z) + \hat{T}(z)$

(c) $\hat{V}\,\hat{G}(z) = \hat{T}(z)\,\hat{G}_0(z)$

(d) $\hat{G}_i(z)\,\hat{V}_i = \hat{G}_0(z)\,\hat{T}_i(z)$

[25] To show that

$$\lim_{t\to\infty} \frac{\exp\{-i(E - E')t/\hbar\}}{(E - E')} = -i\pi\delta(E - E') = -i\pi\delta(E' - E) ,$$

we may consider the Fourier representation of the delta function

$$\frac{1}{2\pi} \int_{-\infty}^{\infty} \exp\{\mp i(E - E')s\} ds = \delta(E - E') = \delta(E' - E) ,$$

put $s = t/\hbar$, and check that the two definitions are consistent. For this we may work out the integral on the left-hand side of the latter equation, invoke the first definition and arrive at the result $\delta(E - E')$.

(e) $\hat{V}_i\hat{G}_i(z) = \hat{T}_i(z)\,\hat{G}_0(z)$

where $\hat{T}(z)$ and $\hat{T}_i(z)$ are, respectively, the three- and two-body transition operators in three-body space, defined by

$$\hat{T}(z) = \hat{V} + \hat{V}\,\hat{G}_0(z)\,\hat{T}(z)$$

and
$$\hat{T}_i(z) = \hat{V}_i + \hat{V}_i\,\hat{G}_0(z)\,\hat{T}_i(z)\,.$$

2. If $\hat{\overline{V}}_i = \hat{V} - \hat{V}_i = \hat{V}_j + \hat{V}_k$, then show that

$$\hat{G}(z) = \hat{G}_i(z) + \hat{G}_i(z)\,\hat{\overline{V}}_i\,\hat{G}(z)\,,$$

and
$$\hat{G}(z) = \hat{G}_i(z) + \hat{G}(z)\,\hat{\overline{V}}_i\,\hat{G}_i(z)\,.$$

3. Using Eq. (11.2.7) of this chapter

$$\left(p_3^2 + \frac{3}{4}q^2 + \alpha_0^2\right)\,\phi\,(\boldsymbol{q}_3, \boldsymbol{p}_3) = -\iint \langle \boldsymbol{q}_3\,\boldsymbol{p}_3|\,\hat{V}_3\,|\boldsymbol{q'}_3\,\boldsymbol{p}'_3\rangle\; d^3q'_3\,d^3p'_3 \times \{\phi\,(\boldsymbol{q}'_1, \boldsymbol{p}'_1) + \phi\,(\boldsymbol{q}'_2, \boldsymbol{p'}_2) + \phi\,(\boldsymbol{q}'_3, \boldsymbol{p}'_3)\,\}$$

and assuming that

$$\langle \boldsymbol{q}_3, \boldsymbol{p}_3|\,\hat{V}_3\,|\boldsymbol{q}'_3, \boldsymbol{p}'_3\rangle = \delta^3\,(\boldsymbol{q}_3 - \boldsymbol{q}'_3)\,\langle \boldsymbol{p}_3|\,\hat{v}_3\,|\boldsymbol{p'}_3\rangle\,,$$

where $\hat{V}_3$ is a two-body interaction in a three-body space and $\hat{v}_3$ is a two-body interaction in two-body space, rewrite this equation in the form

$$\left(p^2 + \frac{3}{4}q^2 + \alpha_0^2\right)\phi(\boldsymbol{q}, \boldsymbol{p}) = -\int d^3p'\,\Big\{\,\phi(\boldsymbol{q}, \boldsymbol{p}')\langle \boldsymbol{p}|\hat{v}|\boldsymbol{p}'\rangle + \phi\left(\boldsymbol{p}', \frac{-\boldsymbol{p}'}{2} - \boldsymbol{q}\right)\,\left\langle \boldsymbol{p}\,|\,\hat{v}\,|\,\boldsymbol{p}' + \frac{\boldsymbol{q}}{2}\right\rangle + \phi\left(\boldsymbol{p}', \frac{\boldsymbol{p}'}{2} + \boldsymbol{q}\right)\,\left\langle \boldsymbol{p}\,|\,\hat{v}\,| - \boldsymbol{p}' - \frac{\boldsymbol{q}}{2}\right\rangle\Big\}\,.$$

4. Show that the operators

$$\hat{P}_n^{(i)} \equiv \lim_{\epsilon\to 0} i\epsilon\hat{G}_i(E_n + i\epsilon)$$

and
$$\hat{P}_n \equiv \lim_{\epsilon\to 0} i\epsilon\hat{G}(E_n + i\epsilon)$$

serve, respectively, as the projection operators for the three-body states $|\Phi_n^{(i)}\rangle$ (when only i-th pair interacts and the same is bound) and $|\Psi_{3n}\rangle$ (when all the pairs interact and particle 3 is incident on the $2-3$ bound system). Hint: Show that $\hat{P}_n^{(i)}|\Phi_n^{(j)}\rangle = \delta_{ij}|\Phi_n^{(j)}\rangle$ and $\hat{P}_n|\Phi_n^{(3)}\rangle = |\Psi_{3n}\rangle$ so that $\hat{H}\,\hat{P}_n|\Phi_n^{(3)}\rangle = E_n\hat{P}_n|\Phi_n^{(3)}\rangle$.

5. Show that Faddeev's equations (11.6.1) and (11.6.2) for the three-body bound system

$$\Psi_n(\boldsymbol{q}_k, \boldsymbol{p}_k) = \psi_n(\boldsymbol{q}_1, \boldsymbol{p}_1) + \psi_n(\boldsymbol{q}_2, \boldsymbol{p}_2) + \psi(\boldsymbol{q}_3, \boldsymbol{p}_3)\,,$$

where

$$\psi_n(\boldsymbol{q}, \boldsymbol{p}) = (s - p^2)^{-1}\int d^3p'\{\langle \boldsymbol{p}|\hat{t}(s + i\epsilon)| - \boldsymbol{p}' - \tfrac{1}{2}\boldsymbol{q}\rangle\psi_n(\boldsymbol{p}', \tfrac{1}{2}\boldsymbol{p}' + \boldsymbol{q}) + \langle \boldsymbol{p}|\hat{t}(s + i\epsilon)|\boldsymbol{p}' + \tfrac{1}{2}\boldsymbol{q}\rangle\psi_n(\boldsymbol{p}', -\tfrac{1}{2}\boldsymbol{p}' - \boldsymbol{q})\}$$

are equivalent to Eyges' equations (11.2.4) and (11.2.10).

6. Show that the matrix element of the AGS operator $\hat{U}_{ij}(z)$ between the initial and final asymptotic states,

$$|\phi_{j\,n}\rangle \equiv |\boldsymbol{Q}_j, d_j, \phi_{J_j}^{n_j}\rangle$$

and

$$|\phi_{i\,m}\rangle \equiv |\boldsymbol{Q}_i, d_i, \phi_{J_i}^{n_i}\rangle$$

of a three-body system serves as the physical transition amplitude in a rearrangement collision,

$$\begin{array}{ccc} a(b \oplus x) + A & \rightarrow & B(A \oplus x) + b \\ \left\{\begin{array}{l} j\text{-th pair and} \\ j\text{-th particle} \end{array}\right\} & \rightarrow & \left\{\begin{array}{l} i\text{-th pair and} \\ i\text{-th particle} \end{array}\right\} \end{array}$$

in the sense that the S-matrix can be expressed as

$$S_{i\,m,j\,n} \equiv \lim_{t\to\infty} \langle \Phi_{i\,m}(t)|\Psi_{j\,n}(t)\rangle = \delta_{ij}\delta_{i\,m,j\,n} - i\pi\delta(E_{j\,n} - E_{i\,m})\langle \phi_{i\,m}|U_{ij}(E_n + i\epsilon)|\phi_{j\,n}\rangle\,.$$

Here $|\Psi_{j\,n}(t)\rangle$ represents the time-dependent state into which the initial state $|\Phi_{j\,n}(t)\rangle \equiv \exp(-iE_n t/\hbar)|\phi_{j\,n}\rangle$ evolves with time.

References

[1] L. Eyges, Phys. Rev. **121**, 1744 (1961).

[2] A. N. Mitra, Nucl. Phys. **32**, 529 (1962).

[3] L. D. Faddeev, Sov. Phys. JETP **12**, 1024 (1961).

12

RELATIVISTIC QUANTUM MECHANICS

12.1 Introduction

We have seen in earlier chapters that non-relativistic quantum mechanics does provide a consistent scheme to explain numerous phenomena in atomic, molecular and nuclear physics, but it does not conform to the equivalence of space and time as envisaged in the theory of relativity.

Non-relativistic quantum mechanics is also unable to explain some observations like fine structure of the energy levels of the Hydrogen and Hydrogen-like atoms and the spin and intrinsic magnetic moment of the electron. In general, non-relativistic quantum mechanics cannot explain the behavior of particles moving with velocities comparable to that of light. The first step toward relativistic[1] generalization of Schrödinger equation was taken by Klein and Gordon.

Klein-Gordon Equation

In non-relativistic quantum mechanics, the state of a particle is described by the Schrödinger equation

$$i\hbar\frac{d}{dt}\,|R(t)\rangle \;=\; \hat{H}\,|R(t)\rangle\,. \tag{12.1.1}$$

In the coordinate representation, this equation reads

$$i\hbar\frac{\partial}{\partial t}\psi(\boldsymbol{r},t) = H\psi(\boldsymbol{r},t)\,, \tag{12.1.2}$$

where H is the Hamiltonian operator in the coordinate representation, given by

$$H = -\frac{\hbar^2}{2m}\nabla^2 + V(\boldsymbol{r})\,. \tag{12.1.3}$$

Obviously, the Schrödinger equation does not satisfy the requirement of symmetry between space and time coordinates as required by the theory of relativity since it involves time and space derivatives of different orders. Another shortcoming is that the *spin* and the intrinsic magnetic moment of the electron, which are well established properties of the electron, are not inherent in the equation. The spin and intrinsic magnetic moment have to be artificially grafted into the equation, in that we multiply the wave function by the spin state, which is either $\alpha = \begin{pmatrix}1\\0\end{pmatrix}$ or $\beta = \begin{pmatrix}0\\1\end{pmatrix}$, and add a term $-\mu_{B}\boldsymbol{\sigma}\cdot\boldsymbol{B}$ to the Hamiltonian, if an external magnetic field $\boldsymbol{B}$ is present.

An attempt to make a relativistic generalization of the Schrödinger equation was made by Klein and Gordon. Instead of using the non-relativistic energy-momentum relation

[1]A brief overview of the special relativity and covariant formulation of electromagnetic theory may be found in Appendix 12A1.

$E = p^2/2m$ for a free particle, which implies $H = \frac{(-i\hbar\nabla)^2}{2m} = -\frac{\hbar^2}{2m}\nabla^2$, they chose to use the relativistic energy-momentum relation

$$E^2 = c^2p^2 + m^2c^4\,, \tag{12.1.4}$$

where E is the total energy, including the rest mass energy. This implies $H^2 = c^2(-i\hbar\nabla)^2 + m^2c^4$ so that the relativistic equation of motion may be written as

$$\left(i\hbar\frac{\partial}{\partial t}\right)^2 \psi = H^2\psi = [c^2(-i\hbar\nabla)^2 + m^2c^4]\psi \tag{12.1.5}$$

or

$$\left(\nabla^2 - \frac{1}{c^2}\frac{\partial^2}{\partial t^2}\right)\psi(r,t) = \frac{m^2c^2}{\hbar^2}\psi(r,t)\,. \tag{12.1.6}$$

This is known as the *Klein-Gordon equation*. It satisfies the relativistic requirement of the equation of motion being symmetric in space and time coordinates as both space and time derivatives are of the second order in this equation. Since the d'Alembertian operator $\left(\nabla^2 - \frac{1}{c^2}\frac{\partial^2}{\partial t^2}\right) = \frac{\partial}{\partial x_\mu}\frac{\partial}{\partial x_\mu}$ is an invariant operator under Lorentz transformations [see Appendix 12A1], it suggests that $\psi(\boldsymbol{r},t) = \psi(x_\mu)$ is a scalar (one component) or *invariant* under Lorentz transformations. Since the solution of Klein-Gordon equation is a one-component wave function, it can describe a particle of zero-spin, e.g., a *pion*, and is not suitable for describing an electron which has spin 1/2 (in units of $\hbar$).

There is, however, one disturbing feature of the Klein-Gordon (KG) equation which delayed its acceptance as the basis for a relativistic quantum theory by physicists for some time. In order to interpret the wave function as a probability amplitude we must be able derive an equation of continuity, which expresses the conservation of probability. When this is done[2] we obtain an equation which can be put in the form of the equation of continuity

$$\nabla\cdot\boldsymbol{j} + \frac{\partial\rho}{\partial t} = 0\,, \tag{12.1.7}$$

where the probability current density is given by

$$\boldsymbol{J} = \frac{\hbar}{2im}\left(\psi^*\nabla\psi - \psi\nabla\psi^*\right)\,, \tag{12.1.8}$$

as in the nonrelativistic theory. The expression for the probability density turns out to be

$$\rho(\boldsymbol{r},t) = \frac{i\hbar}{2mc^2}\left(\psi^*\frac{\partial\psi}{\partial t} - \psi\frac{\partial\psi^*}{\partial t}\right)\,, \tag{12.1.9}$$

which need not be positive because the Klein-Gordon is second order in time so that ψ and $\frac{\partial\psi}{\partial t}$ can be specified independently. The probability density $\rho(\boldsymbol{r},t) = \psi^*\psi$ in the non-relativistic equation is guaranteed to be positive. Hence in the Klein-Gordon equation, $\rho(\boldsymbol{r},t)$ cannot be interpreted as the probability density. Due to this feature, the Klein-Gordon equation was ignored for several years after it was proposed in 1927. In 1934, Pauli

[2]The equation of continuity (12.1.7) may be derived by multiplying the KG equation (12.1.6) by ψ^* from the right, and its complex conjugate equation by ψ from the left, and subtracting the results to get

$$(\psi^*\nabla^2\psi - \psi\nabla^2\psi^*) - \frac{1}{c^2}\left(\psi^*\frac{\partial^2\psi}{\partial t^2} - \psi\frac{\partial^2\psi^*}{\partial t^2}\right) = 0$$

or

$$\nabla\cdot(\psi^*\nabla\psi - \psi\nabla\psi^*) - \frac{1}{c^2}\frac{\partial}{\partial t}(\psi^*\frac{\partial\psi}{\partial t} - \psi\frac{\partial\psi^*}{\partial t}) = 0$$

which may be re-written as $\nabla\cdot\boldsymbol{j} + \frac{\partial\rho}{\partial t} = 0$, where $\boldsymbol{j}$ and ρ are defined by Eqs. (12.1.8) and (12.1.9).

and Weisscopf re-established its validity as a *field equation* for spin zero particles in the same sense as Maxwell's equations are the field equations for the photons. According to the theory of quantization of wave fields, particles called *quanta* can be created or destroyed [Chapter 13]. Consequently, the concept of the conservation of probability, which expresses itself through the equation of continuity, has to be interpreted differently.

12.2 Dirac Equation

Dirac approached the problem of finding a relativistic generalization of the Schrödinger equation differently. Instead of writing Eq. (12.1.5), he looked for an equation of motion first order in time by writing

$$i\hbar\frac{\partial\psi}{\partial t} = H\psi\,, \tag{12.2.1}$$

and for H, he wrote terms linear in $\boldsymbol{p} = -i\hbar\boldsymbol{\nabla}$. This was necessary since a relativistic equation must involve space and time derivatives of the same order. Accordingly, Dirac wrote down H as the most general linear combination of four momentum $(\boldsymbol{p}, imc)$

$$H = c\boldsymbol{\alpha}\cdot\boldsymbol{p} + \beta mc^2 = c(\alpha_1 p_1 + \alpha_2 p_2 + \alpha_3 p_3) + \beta mc^2 \tag{12.2.2}$$

where $\alpha_1, \alpha_2, \alpha_3$, and β are dimensionless quantities [independent of $\boldsymbol{p} \equiv (p_1, p_2, p_3)$, $\boldsymbol{r}$ and t] which are to be determined. To identify the nature of α_k and β, and their mathematical representation, we require that a solution of the Dirac equation for a free particle must also be a solution of the Klein-Gordon relativistic equation. This requires that

$$(c\boldsymbol{\alpha}\cdot\boldsymbol{p} + \beta mc^2)^2 = c^2p^2 + m^2c^4\,. \tag{12.2.3}$$

Allowing the possibility that the quantities α_k and β may not commute, we can re-write this condition as

$$c^2\sum_k\sum_\ell \frac{1}{2}(\alpha_k\alpha_\ell + \alpha_\ell\alpha_k)p_k p_\ell + mc^3\sum_k(\alpha_k\beta + \beta\alpha_k)p_k + \beta^2m^2c^4 = c^2\sum_k p_k^2 + m^2c^4\,.$$

Comparing the two sides of this equation we conclude that α_k's and β must satisfy the following conditions

$$\frac{1}{2}(\alpha_k\alpha_l + \alpha_l\alpha_k) = \delta_{kl}\,, \tag{12.2.4}$$

$$\alpha_k\beta + \beta\alpha_k = 0\,, \tag{12.2.5}$$

and

$$\beta^2 = 1\,. \tag{12.2.6}$$

An inspection of these equations (e.g., the non-commutativity) shows α_k's and β cannot be complex numbers. They can, however, be matrices. Obviously, we need four distinct matrices which anti-commute with each other and the square of each matrix is a unit matrix. To find a representation for these matrices we note that they must be Hermitian because the Hamiltonian (12.2.2) must be Hermitian. Now 2×2 Pauli spin matrices σ_1, σ_2, and σ_3 satisfy these conditions but are only three in number and a unit 2×2 matrix cannot be taken as the fourth because it commutes with all of them. Thus a 2×2 matrix representation for α_k and β is ruled out and we must look at their higher order matrix representation.

Our search for higher order matrix representation is guided by two constraints that emerge from Eqs. (12.2.4) through (12.2.6):

(a) the order of these matrices must be even, and

(b) the matrices must be traceless.

To see the justification for the first, we assume that the matrices are of order $(N \times N)$ and rewrite Eq. (12.2.5) as $\alpha_k \beta = -\beta \alpha_k$. Then the determinants of the matrices must satisfy

$$\det(\alpha_k)\ \det(\beta) = \det(-I)\ \det(\beta)\det(\alpha_k)$$

or

$$1 = \det(-I) = (-1)^N\,, \tag{12.2.7}$$

where in the last step we have used the fact that $\det(-I) = (-1)^N$ for $N \times N$ identity matrix I. To prove the second constraint, multiply Eq. (12.2.5) from the left by α_k and take the trace of the resulting matrix equation to get

$$\text{Tr }(\alpha_k^2\beta)+\text{Tr }(\alpha_k\beta\alpha_k) = \text{Tr }(\alpha_k^2\beta)+\text{Tr }(\alpha_k\alpha_k\beta) = 0 \quad \Rightarrow \quad \text{Tr }(\beta)+\text{Tr }(\beta) = 0\,, \tag{12.2.8}$$

where, in the last step, we have used $\alpha_k^2 = I$ and the cyclic property Tr (AB)=Tr (BA) of the trace of a matrix product. Thus matrix β is traceless. Similarly by multiplying Eq. (12.2.5) by β and taking the trace we can show that the matrices α_k must also be traceless.

Now we have ruled out $N = 2$ matrix representation for α_k, β already. So we look for a representation in terms of the next simplest allowed case ($N = 4$) of 4×4 matrices. There are an infinite number of ways of representing these 4×4 matrices, all satisfying the constraints established above. We will use the *standard representation*, which is outlined below. Since α_k, β matrices anti-commute among themselves, only one of them can have a diagonal representation. Choosing matrix β to be diagonal and traceless, we can have the following representation for it

$$\beta = \begin{pmatrix} 1 & 0 & 0 & 0 \\ 0 & 1 & 0 & 0 \\ 0 & 0 & -1 & 0 \\ 0 & 0 & 0 & -1 \end{pmatrix} \equiv \begin{pmatrix} I & 0 \\ 0 & -I \end{pmatrix} \tag{12.2.9}$$

where I represents a $2{\times}2$ unit matrix. Then a representation for α_k's, which is consistent with the commutation relations (12.2.4), is

$$\alpha_k = \begin{pmatrix} 0 & \sigma_k \\ \sigma_k & 0 \end{pmatrix}\,, \qquad k = 1, 2, 3 \tag{12.2.10}$$

where σ_k are 2×2 Pauli matrices. Using the properties of Pauli matrices, it is easily verified that α_k's and β are Hermitian, anti-commute among themselves, and the square of each of the four matrices is a unit matrix. Thus, explicitly, α_k, β have the following matrix representation

$$\alpha_1 = \begin{pmatrix} 0 & 0 & 0 & 1 \\ 0 & 0 & 1 & 0 \\ 0 & 1 & 0 & 0 \\ 1 & 0 & 0 & 0 \end{pmatrix} \qquad \alpha_2 = \begin{pmatrix} 0 & 0 & 0 & -i \\ 0 & 0 & i & 0 \\ 0 & -i & 0 & 0 \\ i & 0 & 0 & 0 \end{pmatrix} \tag{12.2.11a}$$

$$\alpha_3 = \begin{pmatrix} 0 & 0 & 1 & 0 \\ 0 & 0 & 0 & -1 \\ 1 & 0 & 0 & 0 \\ 0 & -1 & 0 & 0 \end{pmatrix} \qquad \beta = \begin{pmatrix} 1 & 0 & 0 & 0 \\ 0 & 1 & 0 & 0 \\ 0 & 0 & -1 & 0 \\ 0 & 0 & 0 & -1 \end{pmatrix}. \tag{12.2.11b}$$

We can also introduce Dirac matrices γ_μ, where the index μ can take the values $\mu = 1, 2, 3, 4$, by

$$\gamma_k = -i\beta\alpha_k\,, \quad k = 1, 2, 3 \tag{12.2.12}$$

$$\gamma_4 = \beta\,. \tag{12.2.13}$$

It may be easily verified that Dirac matrices are all Hermitian and satisfy the following anti-commutation relations

$$\gamma_\mu\gamma_\nu + \gamma_\nu\gamma_\mu = 2\delta_{\mu\nu}I\,. \tag{12.2.14}$$

Equation of Continuity and Probability Density from Dirac Equation

We can rewrite Dirac equation (12.2.1), with H defined by Eq. (12.2.2) and $\boldsymbol{p}$ replaced by $-i\hbar\nabla$, as

$$\frac{1}{c}\frac{\partial\psi}{\partial t} + \sum_{k=1}^{3}\alpha_k\frac{\partial\psi}{\partial x_k} + \frac{imc}{\hbar}\beta\psi = 0\,. \tag{12.2.15}$$

In view of the 4×4 matrix representation for α_k and β, the solution ψ of this equation must be a column vector with four elements. Taking the Hermitian adjoint of both sides of Eq. (12.2.15), we have

$$\frac{1}{c}\frac{\partial\psi^\dagger}{\partial t} + \sum_{k=1}^{3}\frac{\partial\psi^\dagger}{\partial x_k}\alpha_k^\dagger - \frac{imc}{\hbar}\psi^\dagger\beta^\dagger = 0\,, \tag{12.2.16}$$

where $\psi^\dagger$ is the conjugate transpose of the column matrix ψ and is a row matrix. Since the matrices α_k and β are Hermitian, we can drop the dagger sign on them. Multiplying Eq. (12.2.15) from the left by $\psi^\dagger$ and Eq. (12.2.16) from the right by ψ and adding the resulting equations, we get, after some simplification

$$\frac{\partial(\psi^\dagger\psi)}{\partial t} + \sum_{k=1}^{3}\frac{\partial}{\partial x_k}(c\psi^\dagger\alpha_k\psi) = 0\,. \tag{12.2.17}$$

If we identify the probability density ρ and probability current density $\boldsymbol{\mathcal{J}}$ by

$$\rho = \psi^\dagger\psi\,, \tag{12.2.18}$$

$$\mathcal{J}_k = c\psi^\dagger\alpha_k\psi \quad \text{or} \quad \boldsymbol{\mathcal{J}} = c\psi^\dagger\boldsymbol{\alpha}\psi\,, \tag{12.2.19}$$

then Eq. (12.2.17) has the form of the equation of continuity. Moreover, the Dirac equation gives a probability density (12.2.18) which is guaranteed to be positive definite.

Dirac Equation in a Covariant Form

If we multiply Eq. (12.2.15) on the left by $-i\beta$, put $ict = x_4$ and rearrange we get

$$\gamma_4\frac{\partial\psi}{\partial x_4} + \sum_{k=1}^{3}\gamma_k\frac{\partial\psi}{\partial x_k} + \frac{mc}{\hbar}\psi = 0$$

or

$$\gamma_\mu\frac{\partial\psi}{\partial x_\mu} + \frac{mc}{\hbar}\psi = 0\,, \tag{12.2.20}$$

where, according to the convention, summation over repeated index μ in the first term is implied. This is the Dirac equation written in the covariant form.

Similarly, multiplying the adjoint Dirac equation (12.2.16) by $-i\beta^2$ from the right we can rewrite it as

$$\frac{1}{ic}\frac{\partial\psi^\dagger}{\partial t}\beta^2 + \sum_{k=1}^{3} -i\frac{\partial\psi^\dagger}{\partial t}\beta^2\alpha_k - \frac{mc}{\hbar}\psi^\dagger\beta = 0$$

or

$$\frac{\partial\tilde{\psi}}{\partial x_4}\gamma_4 + \sum_{k=1}^{3}\frac{\partial\tilde{\psi}}{\partial x_k}\gamma_k - \frac{mc}{\hbar}\tilde{\psi} = 0$$

or

$$\frac{\partial\tilde{\psi}}{\partial x_\mu}\gamma_\mu - \frac{mc}{\hbar}\tilde{\psi} = 0 \tag{12.2.21}$$

where

$$\tilde{\psi} = \psi^\dagger\gamma_4\,. \tag{12.2.22}$$

Equation (12.2.20) is the covariant form of the Dirac equationand Eq. (12.2.21) is the covariant form of its adjoint. The covariant form of the equation of continuity may be obtained by multiplying Eq. (12.2.20) on the left by $\tilde{\psi}$ and (12.2.21) on the right by ψ and adding the two. This gives us

$$\tilde{\psi}\gamma_\mu\frac{\partial\psi}{\partial x_\mu} + \frac{\partial\tilde{\psi}}{\partial x_\mu}\gamma_\mu\psi = \frac{\partial}{\partial x_\mu}(\tilde{\psi}\gamma_\mu\psi) \equiv \frac{\partial j_\mu}{\partial x_\mu} = 0\,, \tag{12.2.23a}$$

where

$$j_\mu = ic\tilde{\psi}\gamma_\mu\psi \tag{12.2.23b}$$

may be called the *four-probability current density.* In this form the equation of continuity is manifestly covariant.

12.3 Spin of the Electron

We shall now see that, unlike the Schrödinger equation where the spin of the electron had to be introduced in an ad hoc manner, the Dirac equation naturally leads to the spin of the electron. To see this let us evaluate the commutator bracket of the orbital angular momentum operator $\hat{\boldsymbol{L}} = \hat{\boldsymbol{r}} \times \boldsymbol{p}$ with the Dirac Hamiltonian [Eq. (12.2.2)]. Thus for the commutator of $\hat{L}_1$ we have

$$\begin{aligned}[\hat{L}_1, \hat{H}] &= [(\hat{x}_2\hat{p}_3 - \hat{x}_3\hat{p}_2)\,,\ \{c(\alpha_1\hat{p}_1 + \alpha_2\hat{p}_2 + \alpha_3\hat{p}_3) + \beta mc^2\}] \\ &= i\hbar c(\alpha_2\hat{p}_3 - \alpha_3\hat{p}_2)\end{aligned} \tag{12.3.1}$$

where we used the position-momentum commutation relations to evaluate various terms in this equation. In a similar manner, we obtain

$$[\hat{L}_2\,,\ \hat{H}] = i\hbar c(\alpha_3\hat{p}_1 - \alpha_1\hat{p}_3)\,, \tag{12.3.2}$$

$$[\hat{L}_3,\ \hat{H}] = i\hbar c(\alpha_1\hat{p}_2 - \alpha_2\hat{p}_1)\,. \tag{12.3.3}$$

Thus the angular momentum operator $\hat{\boldsymbol{L}}$ does not commute with the Dirac Hamiltonian. But the total angular momentum, being a conserved quantity for a free particle, must commute with the free Hamiltonian. This means the orbital angular momentum cannot be the total angular momentum. The electron must possess an intrinsic angular momentum, which when added to its orbital angular momentum, gives the total angular momentum, which is the conserved quantity.

To identify this intrinsic angular momentum, let us introduce a matrix operator $\boldsymbol{\Sigma} \equiv (\Sigma_1, \Sigma_2, \Sigma_3)$ defined by

$$\Sigma_k = \begin{pmatrix} \sigma_k & 0 \\ 0 & \sigma_k \end{pmatrix}, \tag{12.3.4}$$

where σ_k's are Pauli spin matrices.[3] Then the commutation

$$[\Sigma_1, \hat{H}] = c\hat{p}_2[\Sigma_1, \alpha_2] + c\hat{p}_3[\Sigma_1, \alpha_3],$$

where we have used the condition that Σ_k commutes with α_k and β. Expressing Σ_1, α_1 and α_3 in terms of Pauli matrices σ_k, using the anti-commutation relations for σ_k, and the relation $\sigma_k\sigma_\ell = i\sigma_m$, where k, ℓ, m are a cyclic permutation of (1,2,3) we find

$$\left[(\hbar/2)\Sigma_1, \hat{H}\right] = i\hbar c(\hat{p}_2\alpha_3 - \hat{p}_3\alpha_2)\,. \tag{12.3.5}$$

Adding this to Eq. (12.3.1), we find

$$[\hat{L}_1, \hat{H}] + (\hbar/2)\Sigma_1, \hat{H}] = [\left(\hat{L}_1 + (\hbar/2)\Sigma_1\right), \hat{H}] = 0\,. \tag{12.3.6}$$

Similarly, by considering the commutators of Σ_2 and Σ_3 with $\hat{H}$, we can show that

$$[\left(\hat{L}_2 + (\hbar/2)\Sigma_2\right), \hat{H}] = 0\,, \tag{12.3.7}$$

$$[\left(\hat{L}_3 + (\hbar/2)\Sigma_3\right), \hat{H}] = 0\,. \tag{12.3.8}$$

By adding Eqs. (12.3.6) through (12.3.8) we find that the observable $\hat{\boldsymbol{L}} + \frac{\hbar}{2}\boldsymbol{\Sigma} \equiv \hat{\boldsymbol{J}}$ commutes with the Hamiltonian and, therefore, is a constant of motion. The observable $\hat{\boldsymbol{J}}$ may be called the total angular momentum of the electron. Thus preserving the conservation of angular momentum Dirac equation requires the electron to possess an intrinsic angular momentum. This instrinsic angular momentum is referred to as the *spin* of the electron. The operator $\frac{\hbar}{2}\boldsymbol{\Sigma} = \frac{\hbar}{2}\begin{pmatrix} \boldsymbol{\sigma} & 0 \\ 0 & \boldsymbol{\sigma} \end{pmatrix}$ may be regarded as the spin operator of the electron, where $\boldsymbol{\Sigma} = \begin{pmatrix} \boldsymbol{\sigma} & 0 \\ 0 & \boldsymbol{\sigma} \end{pmatrix}$. Thus spin, an intrinsic property of the electron, follows naturally from the Dirac equation.

12.4 Free Particle (Plane Wave) Solutions of Dirac Equation

The plane wave solution of the Dirac equation for a free particle should have the form

$$\psi(r, t) = u \, \exp\left[\frac{i}{\hbar}(\boldsymbol{p} \cdot \boldsymbol{r} - Et)\right], \tag{12.4.1}$$

[3]Our convention is that the Roman index k runs from 1 to 3 while the Greek index μ runs from 1 to 4.

where $u = \begin{pmatrix} u_1 \\ u_2 \\ u_3 \\ u_4 \end{pmatrix}$ is called the Dirac spinor. Substituting this form of solution in the Dirac equation

$$i\hbar\frac{\partial}{\partial t}\psi(\boldsymbol{r},t) = \left(c\sum_{k=1}^{3}\alpha_k p_k + \beta mc^2\right)\psi(\boldsymbol{r},t)\,, \tag{12.4.2}$$

we find that u must satisfy

$$Eu = \left(c\sum_{k=1}^{3}\alpha_k p_k + \beta mc^2\right)u\,. \tag{12.4.3}$$

Using the matrix form of u and α_k and β, we find explicitly,

$$E\begin{pmatrix} u_1 \\ u_2 \\ u_3 \\ u_4 \end{pmatrix} = cp_1\begin{pmatrix} u_4 \\ u_3 \\ u_2 \\ u_1 \end{pmatrix} - cp_2\begin{pmatrix} -iu_4 \\ +iu_3 \\ -iu_2 \\ +iu_1 \end{pmatrix} - cp_3\begin{pmatrix} +u_3 \\ -u_4 \\ -u_1 \\ -u_2 \end{pmatrix} - mc^2\begin{pmatrix} u_1 \\ u_2 \\ -u_3 \\ -u_4 \end{pmatrix},$$

or

$$\begin{pmatrix} (E-mc^2) & 0 & -cp_3 & -c(p_1-ip_2) \\ 0 & (E-mc^2) & -c(p_1+ip_2) & cp_3 \\ -cp_3 & -c(p_1-ip_2) & (E+mc^2) & 0 \\ -c(p_1+ip_2) & cp_3 & 0 & (E+mc^2) \end{pmatrix}\begin{pmatrix} u_1 \\ u_2 \\ u_3 \\ u_4 \end{pmatrix} = 0\,. \tag{12.4.4}$$

Thus we have a set of four linear equations in u_1, u_2, u_3, and u_4. For this set of equations to yield a non-trivial solution for u, the determinant of the coefficient matrix must vanish

$$\begin{vmatrix} (E-mc^2) & 0 & -cp_3 & -c(p_1-ip_2) \\ 0 & (E-mc^2) & -c(p_1+ip_2) & cp_3 \\ -cp_3 & -c(p_1-ip_2) & (E+mc^2) & 0 \\ -c(p_1+ip_2) & cp_3 & 0 & (E+mc^2) \end{vmatrix} = (E^2-m^2c^4-c^2p^2)^2 = 0\,. \tag{12.4.5}$$

This yields two solutions for the energy of a free particle

$$E \equiv E_+ = (c^2p^2 + m^2c^4)^{1/2} \geq mc^2\,, \tag{12.4.6a}$$

$$E \equiv E_- = -(c^2p^2 + m^2c^4)^{1/2} \leq -mc^2\,. \tag{12.4.6b}$$

Thus the energy of a free Dirac particle can be either $\geq mc^2$ or $\leq -mc^2$. The energy between $+mc^2$ and $-mc^2$ is forbidden for a free particle. The question is how to interpret the negative energy states, i.e., an electron having energy less than its rest mass energy. In classical physics also we have Eq. (12.4.6) because of the relativistic energy-momentum relation. But we can ignore negative energies because in classical physics we cannot have discontinuous jumps in energy. In quantum mechanics, however, we cannot disregard the negative energy solutions (negative energy states) because discontinuous changes in energy can occur in quantum mechanics In other words if the initial energy of a particle is E_-, it can subsequently be E_+, involving a gap of more than $2mc^2$. So if we accepts the Dirac equation, we must accommodate the concept of *negative energy states* in the framework of this theory.

Dirac postulated that all the negative energy states of the electron ($-\infty < E \leq -mc^2$) are normally occupied by electrons so that no electron with positive energy can jump into any negative energy state on account of the Pauli exclusion principle. Dirac further postulated that no observable properties of matter can be ascribed to the electrons filling negative energy states. In other words the electrons filling negative energy states do not produce any external field and do not contribute to the total charge, momentum or energy of the system. Alternatively, we can say that the *zero points* for the total charge, energy and momentum of the system correspond to that electron distribution in which all the negative energy states are occupied but no positive energy state is occupied. This state is called the *vacuum state* or *electron vacuum*. Now, if sufficient energy $\geq 2mc^2$ be given to a system, this may result in a transition of an electron from a negative energy state to a positive energy state, since some positive energy states are always unoccupied. As a result a void or *hole* will be created in the continuum of negative energy states and an electron will appear in the positive energy continuum. According to Dirac, this hole in the negative energy continuum of states can be identified as a real particle with charge $+e$, mass m and energy $\geq mc^2$.

To visualize this, consider what happens when an electron with energy $E_-(\leq -mc^2)$, charge $-e$ and momentum $\boldsymbol{p}$ is removed from the continuum of negative energy states which were all initially occupied by electrons. Initially we had vacuum, but as a result of removal of the electron from the negative states, the whole system has charge $+e$, energy $-E_- = E_+(\geq mc^2)$ and momentum $-\boldsymbol{p}$, all different from zero. So for all practical purposes the *hole* in the negative energy states continuum behaves like a real particle with charge $+e$, mass m (energy $\geq mc^2$), momentum and spin opposite to that of the negative energy electron. On this basis Dirac was able to predict the existence of a particle which has the same mass as an electron but opposite charge. When such a particle, now called a *positron*, was discovered by Anderson, Dirac's theory of the electron stood on solid ground and he was awarded the Nobel Prize. So, if energy greater than $2mc^2$ is supplied to the vacuum, say through a gamma-ray photon, then this may raise an electron from a negative energy state to a positive energy state, resulting in the production of a pair of electron and positron. This process, which has been observed, is called *pair production*.

Now we come back to the problem of the solution of a set of linear equations (12.4.4) involving u_1, u_2, u_3, and u_4. By introducing two-component spinors Φ and ξ by

$$\Phi = \begin{pmatrix} u_1 \\ u_2 \end{pmatrix} \quad \text{and} \quad \xi = \begin{pmatrix} u_3 \\ u_4 \end{pmatrix}, \tag{12.4.7a}$$

we can write Dirac spinor u as

$$u = \begin{pmatrix} \Phi \\ \xi \end{pmatrix}. \tag{12.4.7b}$$

Substituting this in Eq. (12.4.3) and using the representation of α_k and β matrices in terms of Pauli matrices, we get

$$E \begin{pmatrix} \Phi \\ \xi \end{pmatrix} = \left[\sum_{k=1}^{3} c \begin{pmatrix} 0 & \sigma_k \\ \sigma_k & 0 \end{pmatrix} p_k + \begin{pmatrix} I & 0 \\ 0 & -I \end{pmatrix} mc^2 \right] \begin{pmatrix} \Phi \\ \xi \end{pmatrix}. \tag{12.4.8}$$

This gives us two coupled equations.

$$c \left(\sum_{k=1}^{3} \sigma_k p_k \right) \xi = (E - mc^2) \Phi, \tag{12.4.9}$$

and

$$c \left(\sum_{k=0}^{3} \sigma_k p_k \right) \Phi = (E + mc^2) \xi. \tag{12.4.10}$$

For a state of positive energy $E_+ = (c^2p^2 + m^2c^4)^{1/2} \equiv |E|$, two-component spinors ξ and Φ are related by

$$\xi = \frac{c(\boldsymbol{\sigma} \cdot \boldsymbol{p})}{|E| + mc^2} \Phi. \tag{12.4.11}$$

In the non-relativistic limit ($|\boldsymbol{p}|/mc = v/c \ll 1$) $E_+ \to mc^2$, this relation leads to

$$\xi \approx \frac{\boldsymbol{\sigma}\cdot\boldsymbol{p}}{2mc^2}\Phi = \frac{\boldsymbol{\sigma}\cdot\boldsymbol{v}}{2c}\Phi\,, \tag{12.4.12}$$

where $\boldsymbol{v} = \boldsymbol{p}/m$ is the velocity of the particle. This shows that two-component spinor ξ is much smaller (by factor $v/c \ll 1$) compared to Φ. Hence Φ and ξ may be referred to, respectively, as large and small components of Dirac spinor u for positive energy. On the other hand, for the negative energy state ($E_- = -(c^2p^2 + m^2c^4)^{1/2} = -|E|$), we have

$$\Phi = \frac{c(\boldsymbol{\sigma}\cdot\boldsymbol{p})}{-|E| - mc^2}\xi\,. \tag{12.4.13}$$

In the non-relativistic limit $E_- = -|E| \to -mc^2$ yields

$$\Phi \approx -\frac{\boldsymbol{\sigma}\cdot\boldsymbol{v}}{2c}\xi\,. \tag{12.4.14}$$

Hence for negative energies Φ and ξ may be regarded, respectively, as the small and large components of Dirac u. Because of the non-relativistic limits (12.4.11) and (12.4.13), we may say that Φ is the dominant component of the positive energy solutions ($\xi \to 0$), while ξ is the dominant component of the negative energy solutions ($\Phi \to 0$).

Now, for each energy ($E_+ = |E|$ and $E_- = -|E|$) there exist two solutions for u corresponding to the two spin states of Dirac particle. We have thus four (spinor) solutions $u_\uparrow^+, u_\downarrow^+, u_\uparrow^-$ and $u_\downarrow^-$, which may also be denoted as U_1, U_2, U_3, and U_4, respectively. For positive energy, with spin state indicated by suffix σ ($\uparrow,\downarrow$), we have

$$u_\sigma^+ = C\begin{pmatrix} \chi_\sigma^+ \\ \\ \dfrac{(\boldsymbol{\sigma}\cdot\boldsymbol{p})\chi_\sigma^+}{|E| + mc^2}\end{pmatrix}, \tag{12.4.15}$$

where C is a normalization constant and we are using χ_σ^+ now to denote the dominant part Φ of the positive energy spinor u_σ^+ while the smaller component ξ is given by Eq. (12.4.11). Taking χ_σ^+ to be the spin-up state $\chi_\uparrow^+ = \begin{pmatrix}1\\0\end{pmatrix}$, we have for the corresponding Dirac spinor $u_\uparrow^+$

$$u_\uparrow^+ \equiv U_1 = C\begin{pmatrix} 1 \\ 0 \\ \eta p_3/p \\ \eta(p_1 + ip_2)/p \end{pmatrix}, \tag{12.4.16}$$

where $\eta = cp/(|E| + mc^2)$. If we take χ_σ^+ to be the spin-down state $\chi_\downarrow^+ = \begin{pmatrix}0\\1\end{pmatrix}$, we obtain the corresponding Dirac spinor $u_\downarrow^+$

$$u_\downarrow^+ \equiv U_2 = C\begin{pmatrix} 0 \\ 1 \\ \eta(p_1 - ip_2)/p \\ -\eta p_3/p \end{pmatrix}. \tag{12.4.17}$$

The normalization constant C is to be so chosen so as to conform to the normalization condition $u_\sigma^{+\dagger}u_\sigma^+ = 1$. This gives

$$C = \frac{1}{\sqrt{1+\eta^2}}\,. \tag{12.4.18}$$

For the negative energy states ($E = E_- = -|E|$), we have

$$u_\sigma^- = C \begin{pmatrix} \dfrac{c(\boldsymbol{\sigma}\cdot\mathbf{p})\chi_\sigma^-}{-|E| - mc^2} \\ \\ \chi_\sigma^- \end{pmatrix}, \tag{12.4.19}$$

where we are using χ_σ^- to denote the dominant component ξ of the negative energy spinor, while its smaller component Φ is given by Eq. (12.4.13). By taking χ_σ^- to be spin up spinor $\chi_\uparrow^- = \begin{pmatrix}1\\0\end{pmatrix}$, we obtain the corresponding Dirac spinor

$$u_\uparrow^- \equiv U_3 = C \begin{pmatrix} -\eta p_3/p \\ -\eta(p_1 + ip_2)/p \\ 1 \\ 0 \end{pmatrix}. \tag{12.4.20}$$

Similarly, by taking χ_σ^- to be spin down spinor $\chi_\downarrow^- = \begin{pmatrix}0\\1\end{pmatrix}$, we have for the corresponding Dirac spinor

$$u_\downarrow^- \equiv U_4 = C \begin{pmatrix} -\eta(p_1 - ip_2)/p \\ \eta p_3/p \\ 0 \\ 1 \end{pmatrix}. \tag{12.4.21}$$

Choosing the z-axis in the direction of the momentum of the particle so that ($p_3 = p$ and $p_1 = 0 = p_2$), Dirac spinors may be expressed as

$$u_\uparrow^+ = \frac{1}{\sqrt{1+\eta^2}}\begin{pmatrix}1\\0\\\eta\\0\end{pmatrix} \qquad u_\downarrow^+ = \frac{1}{\sqrt{1+\eta^2}}\begin{pmatrix}0\\1\\0\\-\eta\end{pmatrix} \tag{12.4.22a}$$

$$u_\uparrow^- = \frac{1}{\sqrt{1+\eta^2}}\begin{pmatrix}-\eta\\0\\1\\0\end{pmatrix} \qquad u_\downarrow^- = \frac{1}{\sqrt{1+\eta^2}}\begin{pmatrix}0\\\eta\\0\\1\end{pmatrix}. \tag{12.4.22b}$$

12.5 Dirac Equation for a Zero Mass Particle

For a massless ($m = 0$) Dirac particle, such as the neutrino, $\eta = 1$ and the four spinors become

$$u_\uparrow^+ = \frac{1}{\sqrt{2}}\begin{pmatrix}1\\0\\1\\0\end{pmatrix} \qquad u_\downarrow^+ = \frac{1}{\sqrt{2}}\begin{pmatrix}0\\1\\0\\-1\end{pmatrix} \tag{12.5.1a}$$

$$u_\uparrow^- = \frac{1}{\sqrt{2}}\begin{pmatrix}-1\\0\\1\\0\end{pmatrix} \qquad u_\downarrow^- = \frac{1}{\sqrt{2}}\begin{pmatrix}0\\1\\0\\1\end{pmatrix}. \tag{12.5.1b}$$

We can easily see that, apart from being the eigen-spinors of the Hamiltonian, they are also the eigen-spinors of the *helicity operator*

$$\mathfrak{h} = \frac{\boldsymbol{\Sigma}.\boldsymbol{p}}{E/c} . \tag{12.5.2}$$

For the first two spinors (positive energy solutions) in (12.5.1), $\mathfrak{h} = \Sigma_3$ has eigenvalues $+1$ and -1, respectively,

$$\mathfrak{h}\, u_\uparrow^+ = u_\uparrow^+ \text{ and } \mathfrak{h}\, u_\downarrow^+ = -u_\downarrow^+ . \tag{12.5.3}$$

This means that in the first case (eigenvalue $+1$) the spin is parallel to the momentum and in the second case (eigenvalue -1), the spin is anti-parallel to the momentum. For the last two spinors (negative energy solutions), the helicities are is -1 and $+1$, respectively,

$$\mathfrak{h}\, u_\uparrow^- = -u_\uparrow^- \text{ and } \mathfrak{h}\, u_\downarrow^- = u_\downarrow^- \tag{12.5.4}$$

We can also see that the spinors for a zero mass particle also satisfy the *chirality equation*,

$$\gamma_5 u = \pm u \tag{12.5.5a}$$

where

$$\gamma_5 = \gamma_1\gamma_2\gamma_3\gamma_4 = -\begin{pmatrix} 0 & I \\ I & 0 \end{pmatrix} . \tag{12.5.5b}$$

is the *chirality operator*. Equation (12.5.5a) implies

$$\gamma_5 u_\uparrow^+ = -u_\uparrow^+ , \quad \gamma_5 u_\downarrow^+ = u_\downarrow^+ , \quad \gamma_5 u_\uparrow^- = u_\uparrow^- , \text{ and } \gamma_5 u_\downarrow^- = -u_\downarrow^- \tag{12.5.6}$$

A comparison of this equation with Eqs. (12.5.3) and (12.5.4) shows that for all the four spinors, the helicity and chirality are opposite. It follows from Eq. (12.5.6), that the spinors for massless Dirac particle satisfy

$$\frac{1}{2}(1+\gamma_5)u_\downarrow^+ = u_\downarrow^+ , \; \frac{1}{2}(1+\gamma_5)u_\uparrow^- = u_\uparrow^- \text{ and } \frac{1}{2}(1+\gamma_5)u_\uparrow^+ = 0 = \frac{1}{2}(1+\gamma_5)u_\downarrow^- ,$$

$$\text{and } \frac{1}{2}(1-\gamma_5)u_\uparrow^+ = u_\uparrow^+ , \; \frac{1}{2}(1-\gamma_5)u_\downarrow^- = u_\downarrow^- \text{ and } \frac{1}{2}(1-\gamma_5)u_\downarrow^+ = 0 = \frac{1}{2}(1-\gamma_5)u_\uparrow^- .$$

Hence we can look upon the operator $\frac{1}{2}(1+\gamma_5)$ as a projection operator for a negative helicity or positive chirality state, and the operator $\frac{1}{2}(1-\gamma_5)$ as the projection operator for a positive helicity or negative chirality state.

Weyl's Two-Component Theory

For a massless spin-half particle, the Dirac equation assumes the form

$$i\hbar\frac{\partial\psi}{\partial t} = -i\hbar c\alpha_k\frac{\partial\psi}{\partial x_k} . \tag{12.5.7}$$

For a free particle ψ has the form $\psi = u \exp[\frac{i}{\hbar}(\boldsymbol{p}\cdot\boldsymbol{r} - Et)]$, where u is a four-component Dirac spinor. Substitution of this form in Eq. (12.5.7) yields

$$Eu = c\boldsymbol{\alpha}\cdot\boldsymbol{p}\, u , \tag{12.5.8}$$

where $E = \pm\, cp$ for a massless particle. Using the identity $\alpha_k = -\gamma_5\Sigma_k = -\Sigma_k\gamma_5$, we can rewrite Eq. (12.5.8) as

$$Eu = -c(\Sigma_k p_k)\gamma_5 u , \tag{12.5.9}$$

where $\Sigma_k = \begin{pmatrix}\sigma_k & 0\\ 0 & \sigma_k\end{pmatrix}$ and sum over a repeated index is implied. Since $\gamma_5 u = \mp u$ for negative (positive) chirality state we can rewrite Eq. (12.5.9) as

$$Eu = \pm c(\Sigma_k p_k)u\,, \tag{12.5.10}$$

where the upper sign is for negative chirality (or positive helicity) state and the lower sign is for positive chirality (or negative helicity) state.

Now, if we write $u = \begin{pmatrix}\chi_1\\ \chi_2\end{pmatrix}$, where χ_1 and χ_2 are two-component spinors, then Eq. (12.5.7) separates into two identical two-component equations

$$E\chi_1 = \pm c(\sigma_k p_k)\chi_1$$

and

$$E\chi_2 = \pm c(\sigma_k p_k)\chi_2\,.$$

Taking one of these equations and dropping the suffix on χ, we obtain

$$c(\sigma_k p_k)\chi = \pm E\chi \tag{12.5.11}$$

This is called *Weyl two-component equation* for zero mass spin-half particles. The upper sign is for positive helicity state and the lower sign is for negative helicity state.

Since, experimentally, a neutrino (a massless spin-half particle) is assigned a negative helicity, we may write the Weyl equation as

$$c(\boldsymbol{\sigma}\cdot\boldsymbol{p})\chi = -E\chi \tag{12.5.12a}$$

or

$$\frac{\partial\psi}{\partial t} = c\sigma_k\frac{\partial\psi}{\partial x_k}\,, \tag{12.5.12b}$$

where $\psi = \chi\exp\left[\frac{i}{\hbar}(\boldsymbol{p}\cdot\boldsymbol{r} - Et)\right]$ is a two-component wave function. From Eq. (12.5.12) we can readily see that $\mathfrak{h}\chi = \frac{\boldsymbol{\sigma}\cdot\boldsymbol{p}}{E/c}\chi = -\chi$. It appears that for a neutrino, the only four-component solutions of Dirac equation are $u_\downarrow^+$ and $u_\uparrow^-$, both of which correspond to negative helicity [Eqs. (12.5.3) and (12.5.4)]. While the former corresponds to a neutrino with positive energy (negative helicity, of course), the latter corresponds to a neutrino with negative energy (and negative helicity) or to an anti-neutrino with positive energy because a void in the continuum of negative energy neutrino states corresponds to anti-neutrino with signs of $\boldsymbol{p}$, $\boldsymbol{\sigma}$ and E reversed. Since $\mathfrak{h} = \frac{\boldsymbol{\sigma}\cdot\boldsymbol{p}}{E/c}$, a positive energy anti-neutrino is assigned a positive helicity.

Soon after Weyl derived the two-component equations for zero mass and spin-half particles, Pauli rejected them on the ground that they violated the law of parity, i.e., although they were invariant under proper Lorentz transformations, they were not invariant under improper Lorentz transformations or under space inversions. Weyl's equations were reviewed and revived by Salaam, London, Lee and Yang when it was discovered that parity is violated in weak interactions and that a neutrino is left-handed (helicity $= -1$) and anti-neutrino is right-handed (helicity $= +1$). In the two-component theory the Dirac spinor $u_\downarrow^+$ corresponds to $\chi_\downarrow = \begin{pmatrix}0\\1\end{pmatrix}$ and $u_\uparrow^-$ corresponds to $\chi_\uparrow = \begin{pmatrix}1\\0\end{pmatrix}$.

12.6 Zitterbewegung and Negative Energy Solutions

Zitterbewegung refers to high frequency ($\approx mc^2/\hbar$) oscillations of the position of the electron over a distance of the order of Compton wavelength $\hbar/mc$. We shall see shortly that this is solely due to the *interference of the positive and negative energy components of the wave function.*

Consider a free particle obeying the Dirac equation. In the Heisenberg picture[4], the equation of motion for the components of the position operator is

$$\dot{x}_k = \frac{i}{\hbar}[H, x_k] = \frac{i}{\hbar}\left[\left(c\sum_{j=1}^{3}\alpha_j p_j + \beta mc^2\right), x_k\right]$$

or

$$\dot{x}_k = \frac{ic}{\hbar}\sum_{j=1}^{3}\alpha_j[p_j, x_k] = c\alpha_k\,. \tag{12.6.1}$$

Thus we may regard α_k as the velocity operator in units of c. This relation holds even in the presence of an electromagnetic field. Despite the fact that the particle is free, the velocity operator $\dot{x}_k = c\alpha_k$ is not a constant of motion as is easily seen from the equation of motion for α_k[5],

$$\dot{\alpha}_k = \frac{i}{\hbar}[H, \alpha_k] = \frac{i}{\hbar}\left(2H\alpha_k - [H, \alpha_k]_+\right) = \frac{i}{\hbar}(2H\alpha_k - 2cp_k)\,. \tag{12.6.2}$$

Keeping in mind that p_k and H are constants of motion, we can easily verify by direct substitution that the solution of the differential equation for α_k is

$$\alpha_k(t) = cp_k H^{-1} + e^{2iHt/\hbar}\left[\alpha_k(0) - cp_k H^{-1}\right]\,. \tag{12.6.3}$$

The first term on the right-hand side is easily understood; For an eigenstate with energy E and momentum p_k it gives $\langle cp_k H^{-1}\rangle = cp_k/E = \langle v_k\rangle/c$, where $\langle v_k\rangle$ is the average velocity of the particle corresponding to momentum p_k [see Sec. 1.1 on de Broglie relation].

To interpret the second term, we use $\alpha_k(t)$ given by Eq. (12.6.3) in Eq. (12.6.1) and integrate the resulting equation to find the Heisenberg operator $x_k(t)$ as

$$x_k(t) = x_k(0) + c^2 p_k H^{-1} t - \frac{ic\hbar}{2}H^{-1}(e^{2iHt/\hbar} - 1)(\alpha_k(0) - cp_k H^{-1})\,. \tag{12.6.4}$$

The first and second terms simply describe the uniform motion of a free particle. The third term seems to imply that the electron executes a very rapid oscillation, in addition to uniform motion, with a frequency of the order of $2mc^2/\hbar$ and amplitude of the order of $\hbar/mc$. This motion is referred to as *Zitterbewegung*. It can be seen that Zitterbewegung arises on account of the positive and negative energy components in a wave packet. To elaborate, a wave packet may be expanded in terms of the free particle plane waves, with positive as well as negative energies, which form a complete set of states. When the expectation value of the operator $\alpha - cpH^{-1}$ [Eq. (12.6.4)] is taken for the wave packet, then positive and negative energy components of the wave packet (with the same momentum) interfere in the sense that this operator has non-zero matrix elements only between plane wave states of equal momentum but energy having opposite signs[6].The last term in Eq. (12.6.4) thus, contributes on account of the presence of both positive and negative energy plane waves in the wave packet.

---

[4]To simplify writing we drop the hat (ˆ) and the superscript H to denote Heisenberg operators in these equations. Thus instead of $\hat{x}^H$, we shall simply write x for the position operator.

[5]The anti-commutator $[H, \alpha_k]_+ = H\alpha_k + \alpha_k H = 2cp_k$ is not zero. In contrast to this, momentum p_k is a constant of motion for a free particle. The relation (12.6.2) may be regarded as a differential equation for $\alpha_k(t)$.

[6]To see this, we note that the operator $\Gamma_\pm = \frac{1}{2}\left(1 \pm \frac{H}{|E|}\right)$, where $|E| = \sqrt{m^2c^2 + p^2c^2}$, is the projection operator for the positive (negative) energy states, i.e., $H\Gamma_\pm = \pm|E|\Gamma_\pm$.

12.7 Dirac Equation for an Electron in an Electromagnetic Field

In the classical equations of motion for an electron in an electromagnetic field specified by vector and scalar potentials $\boldsymbol{A}(\boldsymbol{r},t)$ and $\phi(\boldsymbol{r},t)$, energy E is replaced by $(E-e\phi)$ while momentum $\boldsymbol{p}$ is replaced by $(\boldsymbol{p}-e\boldsymbol{A})$, where e is the electron charge. Using the same prescription for the Dirac equation

$$i\hbar\frac{\partial}{\partial t}\psi = (c\boldsymbol{\alpha}\cdot\boldsymbol{p}+\beta mc^2)\psi\,, \tag{12.7.1}$$

we may replace the energy operator $i\hbar\frac{\partial}{\partial t}$ by $i\hbar\frac{\partial}{\partial t}-e\phi$ and the momentum operator $(-i\hbar\nabla)$ by $(-i\hbar\nabla-e\boldsymbol{A})$. Making these replacements in Eq. (12.7.1) we have

$$\left(i\hbar\frac{\partial}{\partial t}-e\phi\right)\psi(\boldsymbol{r},t) = \left[c\boldsymbol{\alpha}\cdot(-i\hbar\nabla-e\boldsymbol{A})+\beta mc^2\right]\psi(\boldsymbol{r},t)\,. \tag{12.7.2a}$$

Pre-multiplying both sides by $-\beta/\hbar c$, we get

$$\beta\left(\frac{1}{ic}\frac{\partial}{\partial t}+\frac{e\phi}{\hbar c}\right)\psi = -\beta\sum_{k=1}^{3}-i\alpha_k\left(\frac{\partial}{\partial x_k}-\frac{ieA_k}{\hbar}\right)\psi-\frac{mc}{\hbar}\psi\,. \tag{12.7.2b}$$

Using the four-potential $A_\mu\equiv(A_1,A_2,A_3,A_4)\equiv(A_1,A_2,A_3,i\phi/c)$ and $\gamma_4=\beta$, $\gamma_k=-i\beta\alpha_k$ we can write the Dirac equation for an electron in an electromagnetic field as

$$\left[\gamma_4\left(\frac{\partial}{\partial x_4}-\frac{ie}{\hbar}A_4\right)+\sum_{k=1}^{3}\gamma_k\left(\frac{\partial}{\partial x_k}-\frac{ie}{\hbar}A_k\right)+\frac{mc}{\hbar}\right]\psi=0\,,$$

or

$$\left[\gamma_\mu\left(\frac{\partial}{\partial x_\mu}-\frac{ie}{\hbar}A_\mu\right)+\frac{mc}{\hbar}\right]\psi\equiv\left[\gamma_\mu\Omega_\mu+\frac{mc}{\hbar}\right]\psi=0\,, \tag{12.7.3}$$

where

$$\Omega_\mu\equiv\left(\frac{\partial}{\partial x_\mu}-\frac{ie}{\hbar}A_\mu\right), \tag{12.7.4}$$

and summation convention (sum over repeated indices) is used[7].

Now, to show that the operator $(\alpha-cpH^{-1})$ with $\boldsymbol{\alpha}\equiv\boldsymbol{\alpha}(0)$, has non-zero matrix elements only between the plane waves of the same momentum but energy having opposite signs, it is necessary only to show that

$$\Gamma_\pm(\alpha-c\mathbf{p}H^{-1})\Gamma_\mp\neq 0\,,$$

while

$$\Gamma_\pm(\alpha-c\mathbf{p}H^{-1})\Gamma_\pm=0\,.$$

With the help of the anti-commutator $H\boldsymbol{\alpha}+\boldsymbol{\alpha}H=2c\mathbf{p}$, we find the commutator of $\boldsymbol{\alpha}$ and $\Gamma_\pm$ is given by

$$[\Gamma_\pm,\boldsymbol{\alpha}]=\mp(\boldsymbol{\alpha}-c\mathbf{p}H^{-1})\frac{H}{|E|}\,.$$

Using this commutator, it is easy to check that

$$\Gamma_\pm[\Gamma_\pm\,\boldsymbol{\alpha}]\Gamma_\mp\neq 0\,,\quad\Rightarrow\quad\Gamma_\pm(\alpha-c\mathbf{p}H^{-1})\Gamma_\mp\neq 0\,,$$

and

$$\Gamma_\pm[\Gamma_\pm\,,\alpha]\Gamma_\pm=0\quad\Rightarrow\quad\Gamma_\pm(\alpha-c\mathbf{p}H^{-1})\Gamma_\pm=0\,.$$

[7]Let us recall that we can write a four-vector $V_\mu\equiv(V_1,V_2,V_3,V_4)\equiv(V_k,V_4)\equiv(\boldsymbol{V},V_4)$, where the Greek index μ takes on values $1,2,3,4$. The first three components $(V_1,V_2,V_3)\equiv V_k$ denote the spatial components and the fourth V_4 denotes the time component of the four-vector. We shall also write $V_\mu\equiv(V_k,V_4)$, where the Roman index $k=1,2,3$.

The adjoint of the Dirac equation (12.7.2a) is

$$\left(-i\hbar\frac{\partial}{\partial t} - e\phi\right)\psi^\dagger\beta^2 = c\sum_{k=1}^{3}\left(i\hbar\frac{\partial}{\partial x_k} - eA_k\right)\psi^\dagger\beta^2\alpha_k + mc^2\psi^\dagger\beta\,, \tag{12.7.5}$$

in which we have used $I = \beta^2$ in the first two terms. Using the four-potential A_μ and γ_μ matrices as before and introducing $\tilde{\psi}$ by $\tilde{\psi} = \psi^\dagger\gamma_4$, we can rewrite the adjoint of the Dirac equation in an electromagnetic field as

$$\left(\frac{\partial}{\partial x_\mu} + \frac{ie}{\hbar}A_\mu\right)\tilde{\psi}\gamma_\mu - \frac{mc}{\hbar}\tilde{\psi} = 0\,. \tag{12.7.6}$$

Klein-Gordon Equation for a Charged Particle in Electromagnetic Field

In the presence of an electromagnetic field, the Klein-Gordon equation takes the form

$$\left(i\hbar\frac{\partial}{\partial t} - e\phi\right)^2\psi = \left[c^2(-i\hbar\boldsymbol{\nabla} - e\boldsymbol{A})^2 + m^2c^4\right]\psi \tag{12.7.7}$$

Multiplying throughout by $-1/c^2\hbar^2$ and introducing the four-potential $A_\mu = (A_k, i\phi/c)$, we can rewrite it as

$$(\Omega_\mu\Omega_\mu - \frac{m^2c^2}{\hbar^2})\psi = 0\,, \tag{12.7.8}$$

where Ω_μ is defined in Eq. (12.7.4).

Intrinsic Magnetic Moment of the Dirac Electron

The Dirac equation also incorporates an intrinsic magnetic dipole moment (and also an imaginary electric dipole moment) for the electron. To see this we start with the covariant form of the Dirac equation in the presence of an electromagnetic field [Eq. (12.7.3)] and pre-multiply it by $(\gamma_\nu\Omega_\nu - \frac{mc}{\hbar})$ to get

$$\left(\gamma_\nu\Omega_\nu - \frac{mc}{\hbar}\right)\left(\gamma_\mu\Omega_\mu + \frac{mc}{\hbar}\right)\psi = 0$$

or
$$\gamma_\nu\Omega_\nu\gamma_\mu\Omega_\mu\psi - \frac{mc}{\hbar}\gamma_\mu\Omega_\mu\psi + \frac{mc}{\hbar}\gamma_\nu\Omega_\nu\psi - \frac{m^2c^2}{\hbar^2}\psi = 0\,. \tag{12.7.9}$$

The second and third terms cancel out. Remembering that γ-matrices do not commute while the differential operators do, we have for the operator $\gamma_\nu\Omega_\nu\gamma_\mu\Omega_\mu$

$$\begin{aligned}\gamma_\nu\gamma_\mu\Omega_\nu\Omega_\mu &= \gamma_\nu\gamma_\mu\frac{\partial}{\partial x_\nu}\frac{\partial}{\partial x_\mu} - \frac{ie}{\hbar}\gamma_\nu\gamma_\mu\left(A_\nu\frac{\partial}{\partial x_\mu} + A_\mu\frac{\partial}{\partial x_\nu} + \frac{\partial A_\mu}{\partial x_\nu}\right) - \frac{e^2}{\hbar^2}\gamma_\nu\gamma_\mu A_\nu A_\mu \\ &= \frac{\partial}{\partial x_\mu}\frac{\partial}{\partial x_\mu} - \frac{2ie}{\hbar}A_\mu\frac{\partial}{\partial x_\mu} - \frac{ie}{\hbar}\left(\frac{\partial A_\mu}{\partial x_\mu}\right) - \frac{ie}{2\hbar}\gamma_\mu\gamma_\nu F_{\mu\nu} - \frac{e^2}{\hbar^2}A_\mu A_\mu\,,\end{aligned} \tag{12.7.10}$$

where $F_{\mu\nu} = \frac{\partial A_\nu}{\partial x_\mu} - \frac{\partial A_\mu}{\partial x_\nu}$ is the electromagnetic field tensor [Appendix 12A1] and we have used the anti-commutation relations for the γ_μ matrices [Eq. (12.2.14)]. Hence Eq. (12.7.9) assumes the form

$$\left[\frac{\partial}{\partial x_\mu}\frac{\partial}{\partial x_\mu} - \frac{ie}{\hbar}\frac{\partial A_\mu}{\partial x_\mu} - \frac{2ie}{\hbar}A_\mu\frac{\partial}{\partial x_\mu} - \frac{e^2}{\hbar^2}A_\mu A_\mu - \frac{m^2c^2}{\hbar^2}\right]\psi - \frac{ie}{2\hbar}\gamma_\mu\gamma_\nu F_{\mu\nu}\psi = 0\,,$$

or
$$\left(\Omega_\mu\Omega_\mu - \frac{m^2c^2}{\hbar^2}\right)\psi - \frac{ie}{2\hbar}\gamma_\mu\gamma_\nu F_{\mu\nu}\psi = 0\,. \tag{12.7.11}$$

Comparing this form of the Dirac equation with the Klein-Gordon equation (12.7.8) for a charged particle in an electromagnetic field, we notice an extra term $-\frac{ie}{2\hbar}\gamma_\mu\gamma_\nu F_{\mu\nu}\psi$ in Eq. (12.7.11).

To unravel the meaning of this term, consider the wave function ψ to be of the plane wave form $\psi = u\,\exp[\frac{i}{\hbar}(\boldsymbol{p}\cdot\boldsymbol{r} - Et)]$, where u is a Dirac spinor. (We can always expand ψ in terms of plane wave solutions.) This substitution in the Dirac equation (12.7.11) results in the following replacements: $(-i\hbar\frac{\partial}{\partial x_4}) \to \frac{iE}{c}$, $(-i\hbar\frac{\partial}{\partial x_k}) \to p_k$ and $A_4 \to (i\phi/c)$. Multiplying the resulting equation by $(-i\hbar c)^2$, we have

$$(E - e\phi)^2\psi = \left[\sum_{k=1}^{3} c^2\left(p_k - eA_k\right)^2 + m^2c^4\right]\psi - \frac{ie\hbar c^2}{2}\gamma_\mu\gamma_\nu F_{\mu\nu}\psi\,.$$

In the non-relativistic limit $(E - e\phi)^2 - m^2c^4 \approx 2mc^2(E' - e\phi)$, where $E' = E - mc^2$ is the total energy, excluding rest energy of the particle, this equation leads to

$$(E' - e\phi)\psi = \frac{1}{2m}\sum_{k-1}^{3}(p_k - eA_k)^2\psi + \frac{ie\hbar}{4m}\gamma_\mu\gamma_\nu F_{\mu\nu}\psi\,. \tag{12.7.12}$$

Recalling that $F_{\mu\nu}$ is an anti-symmetric tensor ($F_{\mu\mu} = 0$, and $F_{\mu\nu} = -F_{\nu\mu}$) and the γ-matrices anti-commute ($\gamma_\mu\gamma_\nu = -\gamma_\nu\gamma_\mu$ when $\mu \neq \nu$), we can write the last term as

$$\frac{ie\hbar}{2m}\left[\gamma_1\gamma_2F_{12} + \gamma_2\gamma_3F_{23} + \gamma_3\gamma_1F_{31} + \gamma_1\gamma_4F_{14} + \gamma_2\gamma_4F_{24} + \gamma_3\gamma_4F_{34}\right]\psi\,.$$

Using the relations $\gamma_k\gamma_\ell = i\Sigma_m$, $F_{k\ell} = B_m$ (mth component of the magnetic field) where k, ℓ, m are cyclic permutations of (1,2,3), $\gamma_k\gamma_4 = i\alpha_k$ and $F_{k4} = -iE_k/c$, (kth component of the electric field) we have

$$\frac{ie\hbar}{4m}\gamma_\mu\gamma_\nu F_{\mu\nu} = -\frac{e\hbar}{2m}\left[\boldsymbol{B}\cdot\boldsymbol{\Sigma} + i\boldsymbol{\alpha}\cdot\frac{\boldsymbol{E}}{c}\right]$$

where $\boldsymbol{B}$ and $\boldsymbol{E}$ are, respectively, the magnetic and electric fields. Thus the non-relativistic limit of the Dirac equation (12.7.12) assumes the form

$$E'\psi = \left[e\phi + \frac{1}{2m}\sum_{k=1}^{3}\left(p_k - \frac{e}{c}A_k\right)^2 - \frac{e\hbar}{2m}\boldsymbol{B}\cdot\boldsymbol{\Sigma} - \frac{e\hbar}{2mc}i\boldsymbol{\alpha}\cdot\boldsymbol{E}\right]\psi\,. \tag{12.7.13}$$

While the first two terms on the right-hand side represent the energy of a charged particle in an electromagnetic field, the last term represents the energy of the particle due to an intrinsic magnetic dipole moment $\mu_B = \frac{e\hbar}{2m}$. The relation between electron spin and magnetic moment is

$$\boldsymbol{\mu} = \frac{e}{m}\frac{\hbar}{2}\boldsymbol{\Sigma} = \frac{e}{m}\boldsymbol{S}\,, \tag{12.7.14}$$

where $\boldsymbol{S} = \frac{\hbar}{2}\boldsymbol{\Sigma}$ is the spin operator for the electron. This equation shows that just as the orbital angular momentum of a charge particle gives rise to a magnetic moment, so too does its intrinsic (spin) angular momentum. However the gyromagnetc ratio $|\boldsymbol{\mu}|/|\boldsymbol{S}| = e/m$ for spin-induced magnetic moment is twice as large as it is for the orbital angular momentum in agreement with the experiments. Thus we see that a charged particle described by the Dirac equation has an intrinsic spin as well as an intrinsic magnetic dipole moment, which in the non-relativistic theory were introduced in an ad hoc manner.

Interestingly, Eq. (12.7.13) also predicts an intrinsic (imaginary) electric dipole moment $\frac{ie\hbar}{2mc}$. However, this term in Eq. (12.7.13) is of the order of $(v/c)^2$ and is usually neglected in the non-relativistic limit.

Dirac Equation in the Non-Relativistic Limit

The four-component Dirac wave function may be written as $\begin{pmatrix}\Phi\\ \chi\end{pmatrix}$, where Φ and χ are two-component wave functions introduced in Sec. 12.4. Substituting this into the stationary state Dirac equation for an electron in an electromagnetic field

$$\left[c\sum_{k=1}^{3}\alpha_k(p_k - eA_k) + \beta mc^2\right]\psi = (E - e\phi)\psi\,, \tag{12.7.15}$$

we obtain the coupled equations

$$\Phi = \frac{cP_k\sigma_k}{E - mc^2 - e\varphi}\chi, \tag{12.7.16}$$

and

$$\chi = \frac{cP_k\sigma_k}{(E + mc^2 - e\phi)}\Phi\,, \tag{12.7.17}$$

where $P_k \equiv p_k - eA_k$ and sum over repeated indices is implied. For positive energies ($E > mc^2$), we can put $E - mc^2 = E'$ and write Eq. (12.7.17) as $\chi = \frac{cP_k\sigma_k}{E' - e\phi + 2mc^2}\Phi$, which in the non-relativistic limit $|E' - e\phi| \ll 2mc^2$ gives

$$\chi \approx \frac{P_k\sigma_k}{2mc}\Phi. \tag{12.7.18}$$

Hence in the non-relativistic limit of positive energy solutions, $\chi \ll \Phi$. Similarly, for negative energies ($E < -mc^2$), we can put $E + mc^2 = E'$ and write Eq. (12.7.16) as $\Phi = \frac{cP_k\sigma_k}{E' - e\phi - 2mc^2}\chi$, which in the non-relativistic limit $|E' - e\phi| \ll 2mc^2$ gives

$$\Phi \approx \frac{P_k\sigma_k}{2mc}\chi\,. \tag{12.7.19}$$

This implies that in the non-relativistic limit of negative energy solutions, $\Phi \ll \chi$.

Thus, for the positive energy states and in the non-relativistic limit, Φ represents the large component of Dirac wave function and χ represents the small component. On the other hand, for the negative energy states and in the non-relativistic limit, χ represents the large component and Φ represents the small component. These conclusions are similar to those reached in Sec. 12.4 for a free Dirac particle.

In what follows we shall confine ourselves to non-relativistic limit for positive energies only. From Eqs. (12.7.16) and (12.7.18) we have

$$(E - mc^2)\Phi = \left[\frac{1}{2m}(\boldsymbol{\sigma}\cdot\boldsymbol{P})^2 + e\phi\right]\Phi\,.$$

Comparing this with the stationary state Schrödinger equation, we see that in the non-relativistic case, the time-dependent equation of motion for the electron state can be written as

$$i\hbar\frac{\partial}{\partial t}\Phi = \left[\frac{1}{2m}(\boldsymbol{\sigma}\cdot\boldsymbol{P})^2 + e\phi\right]\Phi\,. \tag{12.7.20}$$

This equation can be cast in a different form by using the identity

$$(\boldsymbol{\sigma}\cdot\boldsymbol{P})(\boldsymbol{\sigma}\cdot\boldsymbol{P}) = P^2 + i\boldsymbol{\sigma}\cdot(\boldsymbol{P}\times\boldsymbol{P})\,, \tag{12.7.21a}$$

which follows from the properties of Pauli spin operators [Problem 2, Chapter 12], and the relation

$$(\boldsymbol{P}\times\boldsymbol{P})_k = ie\hbar(\nabla\times\boldsymbol{A})_k = ie\hbar B_k\,, \tag{12.7.21b}$$

which follows from the commutation relation

$$[p_k, A_\ell] = -i\hbar \frac{\partial A_\ell}{\partial x_k} \, . \tag{12.7.21c}$$

Equation (12.7.21b) can be written as $\boldsymbol{P} \times \boldsymbol{P} = ie\hbar\boldsymbol{B}$. Using this relation in Eq. (12.7.21a) and substituting the result in Eq. (12.7.20), we find that in the non-relativistic approximation, the Dirac equation for positive energies reduces to

$$i\hbar\frac{\partial \Phi}{\partial t} = \left[\frac{1}{2m}(\boldsymbol{p} - e\boldsymbol{A})^2 - \frac{e\hbar}{2m}(\boldsymbol{\sigma}\cdot\boldsymbol{B}) + e\phi\right]\Phi \, . \tag{12.7.22}$$

This is precisely the non-relativistic Schrödinger-Pauli equation for the electron. As already observed, the operator $\boldsymbol{\mu} \equiv \mu_B\boldsymbol{\sigma} = \frac{e\hbar}{2m}\boldsymbol{\sigma}$ may be regarded as the intrinsic magnetic moment operator for the electron, the intrinsic magnetic moment to be assigned to the electron being μ_B.

The term $-\mu_B\boldsymbol{\sigma}\cdot\boldsymbol{B}$ was artificially introduced by Pauli into the Hamiltonian of an electron in an electromagnetic field. Now we find that it is a natural outcome when we use the Dirac equation and investigate its form for positive energies in the non-relativistic limit. Thus, not only does Dirac theory predict the existence of intrinsic magnetic moment of the electron, but also gives the correct value for it. It is remarkable that, according to Dirac theory the *gyromagnetic ratio* [the ratio of the intrinsic magnetic dipole moment of the electron (in units of Bohr-magneton μ_B), to its intrinsic angular momentum (in units of $\hbar$)] comes out to be 2 which is confirmed by experiments. This is one of the major successes of Dirac theory.

Experimentally, a very small deviation is observed in the electron magnetic dipole moment from the prediction of Dirac theory. This deviation is termed the *anomalous magnetic moment* of the electron. Dehmelt and coworkers have found[8] that the magnetic moment of the electron is $\mu_e = 1.001\,159\,652\,188(4)\ \mu_B$. This anomaly has been explained on the basis of *quantum field theory*, which yields $\mu_e = 1.001\,159\,652\,175(8)\mu_B$!

Foldy-Wouthuysen Transformation

We have seen that, in the non-relativistic limit, the Dirac equation may be separated into positive and negative energy components, each involving two-component wave functions. In this limit, the Dirac equation reduces to the Schrödinger -Pauli equation for positive energies. Such a separation can also be brought about for the Dirac equation even in the relativistic case by invoking a transformation known after Foldy and Wouthuysen. Consider a transformation of the free Dirac Hamiltonian $H = c\boldsymbol{\alpha}\cdot\boldsymbol{p} + \beta mc^2$ by a unitary transformation U such that

$$H \;\to\; UHU^\dagger = H' \, . \tag{12.7.23}$$

We require that under this transformation, the wave function transforms according to

$$\psi \equiv \begin{pmatrix}\Phi \\ \chi\end{pmatrix} \;\to\; U\psi = \psi' \equiv \begin{pmatrix}\Phi' \\ \chi'\end{pmatrix} \tag{12.7.24}$$

such that

$$H'\begin{pmatrix}\Phi' \\ 0\end{pmatrix} \equiv H'\psi'_+ = +E_p\psi'_+ \tag{12.7.25}$$

and $$H'\begin{pmatrix}0 \\ \chi'\end{pmatrix} \equiv H'\psi'_- = -E_p\psi'_- \tag{12.7.26}$$

[8]R. S. Van Dyck, Jr., P. B. Schwinberg, and H. G. Dehmelt, Phys. Rev. Lett. **59**, 26 (1987).

where $E_p = (c^2p^2 + m^2c^4)^{1/2}$ and $\psi'_+ \equiv \begin{pmatrix} \Phi' \\ 0 \end{pmatrix}$ and $\psi'_- \equiv \begin{pmatrix} 0 \\ \chi' \end{pmatrix}$ are four-component wave functions while Φ' and χ' are two-component wave functions. According to Foldy and Wouthuysen, the unitary operator U, which can bring about this transformation, has the form

$$U = \sqrt{\frac{2E_p}{mc^2 + E_p}}\frac{1}{2}\left(1 + \frac{\beta H}{E_p}\right) = \sqrt{\frac{mc^2 + E_p}{2E_p}} + \frac{c\beta\boldsymbol{\alpha}\cdot\boldsymbol{p}}{\sqrt{2E_p(mc^2 + E_p)}} . \tag{12.7.27}$$

It may be checked that U is unitary: $UU^\dagger = U^\dagger U = I$.

Under this transformation, the Dirac Hamiltonian

$$H \equiv c\boldsymbol{\alpha}\cdot\boldsymbol{p} + \beta mc^2 \ \rightarrow \ UHU^\dagger = H' \equiv \beta E_p \tag{12.7.28}$$

while the projection operators $\Gamma_\pm = \frac{1}{2}(1 \pm \frac{H}{E_p})$ for the positive and negative energy states transform as

$$\Gamma_+ \rightarrow U\Gamma_+U^\dagger = \frac{1}{2}(1+\beta) \equiv B_+ \tag{12.7.29a}$$

and

$$\Gamma_- \rightarrow U\Gamma_-U^\dagger = \frac{1}{2}(1-\beta) \equiv B_- . \tag{12.7.29b}$$

Thus the Dirac equation $i\hbar\frac{\partial}{\partial t}\psi = H\psi$ with ψ written in terms of two-component wave functions, is transformed into

$$i\hbar\frac{\partial}{\partial t}U\psi = UHU^\dagger U\psi$$

or

$$i\hbar\frac{\partial}{\partial t}\begin{pmatrix} \Phi' \\ \chi' \end{pmatrix} = \beta E_p \begin{pmatrix} \Phi' \\ \chi' \end{pmatrix}$$

resulting in two two-component equations

$$i\hbar\frac{\partial\Phi'}{\partial t} = +E_p\Phi' \tag{12.7.30}$$

and

$$i\hbar\frac{\partial\chi'}{\partial t} = -E_p\chi' . \tag{12.7.31}$$

The transformed projection operators $B_\pm$ thus separate the transformed Dirac four-component wave function ψ' as

$$B_+\psi' = \frac{1}{2}(1+\beta)\begin{pmatrix} \Phi' \\ \chi' \end{pmatrix} = \begin{pmatrix} \Phi' \\ 0 \end{pmatrix} \tag{12.7.32}$$

and

$$B_-\psi' = \frac{1}{2}(1-\beta)\begin{pmatrix} \Phi' \\ \chi' \end{pmatrix} = \begin{pmatrix} 0 \\ \chi' \end{pmatrix} . \tag{12.7.33}$$

12.8 Invariance of Dirac Equation

Any equation in physics must be form invariant under a Lorentz transformation, since physics expressed by the equation must be independent of the reference frame. This requirement is implicit in the principle of relativity and is referred to as Lorentz invariance or covariance. To discuss the invariance of the Dirac equation under Lorentz transformations, it is convenient to write the Dirac equation for a charged particle (charge e and mass m) in an electromagnetic field $A_\nu(x) \equiv (\boldsymbol{A}(x), i\phi(x)/c)$ as [Eq. (12.7.3)]

$$\left[\gamma_\mu \left(\frac{\partial}{\partial x_\mu} - \frac{ie}{\hbar} A_\mu(x)\right) + \frac{mc}{\hbar}\right] \psi(x) = 0 \tag{12.8.1}$$

and its adjoint equation as [Eq. (12.7.6)]

$$\left[\frac{\partial}{\partial x_\mu} + \frac{ie}{\hbar c} A_\mu(x)\right] \tilde{\psi}(x)\gamma_\mu - \frac{mc}{\hbar}\tilde{\psi}(x) = 0 \tag{12.8.2}$$

where $\tilde{\psi}(x) = \psi^\dagger(x)\gamma_4$. Here and in the rest of this section x in the argument of $A_\mu(x)$ or $\psi(x)$ represents the space time coordinates: $x \equiv (x_1, x_2, x_3, x_4) = (x, y, z, ict) \equiv (\boldsymbol{r}, t)$. From the forms of the Dirac equation and its adjoint, it is still not obvious that these equations are form invariant under a Lorentz transformation for, although ψ has four components (a spinor), it does not transform as a four-vector. So we still need to investigate whether the Dirac equation is invariant under Lorentz transformations.

Consider a homogeneous Lorentz transformation (space-time transformation in four-dimensional space) defined by the relation [see Appendix 12A1]

$$x'_\mu = a_{\mu\nu} x_\nu \,, \tag{12.8.3a}$$

where $a_{\mu\nu}$ represents the $\mu\nu$ element of a 4×4 matrix a and a sum over repeated indices is implied. We will restrict to orthogonal transformations ($a^{-1}_{\mu\nu} = a^T_{\mu\nu} = a_{\nu\mu}$) for which the inverse transformation relation is

$$x_\mu = a_{\nu\mu} x'_\nu \,. \tag{12.8.3b}$$

The orthogonality of transformation is then expressed by

$$a^{-1}a = a^T a = I \quad \Rightarrow \quad a_{\mu\rho}a_{\mu\sigma} = \delta_{\rho\sigma} \,. \tag{12.8.4}$$

Since all four-vectors transform according to Eqs. (12.8.3), we have

$$\frac{\partial}{\partial x_\mu} = a_{\nu\mu}\frac{\partial}{\partial x'_\nu} \tag{12.8.5}$$

and

$$A_\mu = a_{\nu\mu} A'_\nu \,. \tag{12.8.6}$$

Now suppose that under this transformation spinor functions $\psi(x)$ and $\tilde{\psi}(x)$ transform according to

$$\psi(x) \to \psi(x') \equiv \psi'(x) = S\psi(x)\,, \tag{12.8.7}$$

$$\tilde{\psi}(x) \to \tilde{\psi}(x') \equiv \tilde{\psi}'(x) = \tilde{\psi}(x)\,\bar{S}\,, \tag{12.8.8}$$

where S and $\bar{S}$ are 4×4 matrix operators independent of x_μ. From the relation $\tilde{\psi}(x) = \psi^\dagger(x)\gamma_4$, it follows that $\bar{S} = \gamma_4 S^\dagger \gamma_4$. We may assume that γ-matrices do not change as a result of this transformation since they are independent of position, momentum and time variables. Then the invariance of the Dirac equation under Lorentz transformation means that the equation satisfied by the transformed spinor $\psi'(x)$ in the presence of the transformed four-potential $A'_\mu(x)$ has the same form as the original equation (12.8.1)

$$\left[\gamma_\mu \left(\frac{\partial}{\partial x'_\mu} - \frac{ie}{\hbar} A'_\mu(x)\right) + \frac{mc}{\hbar}\right] \psi'(x) = 0\,. \tag{12.8.9}$$

In addition, we must be able to prescribe an explicit linear relationship [Eq. (12.8.7)] connecting spinor $\psi'(x)$ to $\psi(x)$. We now find the conditions for the invariance of the Dirac equation under Lorentz transformations and determine the matrix operator S, which satisfies these conditions for various orthogonal transformations represented by Eq. (12.8.3). We shall consider: (a) pure Lorentz transformations (b) pure rotations in ordinary three-dimensional space (c) space inversion and (d) time reversal.

Using Eqs. (12.8.7) and (12.8.8) in Eq. (12.8.9) and multiplying from the left by S^{-1} we get

$$\left[S^{-1}\gamma_\mu S\left(\frac{\partial}{\partial x'_\mu} - \frac{ie}{\hbar}A'_\mu(x)\right) + \frac{mc}{\hbar}S^{-1}S\right]\psi(x) = 0$$

or

$$\left[S^{-1}\gamma_\nu S\left(\frac{\partial}{\partial x'_\nu} - \frac{ie}{\hbar}A'_\nu(x)\right) + \frac{mc}{\hbar}S^{-1}S\right]\psi(x) = 0\,, \tag{12.8.10}$$

where in the second step we have changed the dummy index μ to ν. Going back to Eq. (12.8.1) and using the transformation relations (12.8.5) and (12.8.6) we get

$$\left[\gamma_\mu a_{\nu\mu}\left(\frac{\partial}{\partial x'_\nu} - \frac{ie}{\hbar}A'_\nu(x)\right) + \frac{mc}{\hbar}\right]\psi(x) = 0\,. \tag{12.8.11}$$

Comparing Eqs. (12.8.10) and (12.8.11) we conclude that the operators S and S^{-1} must satisfy the conditions

$$S^{-1}\gamma_\nu S = \gamma_\mu a_{\nu\mu} \tag{12.8.12}$$

and

$$S^{-1}S = I \tag{12.8.13}$$

for the Dirac equation to be invariant under the transformation (12.8.3). We shall now proceed to find the matrix operators S for various orthogonal transformations within the framework of Eq. (12.8.3), which satisfy the conditions in Eqs. (12.8.12) and (12.8.13).

(a) Pure Lorentz Transformations

A pure Lorentz transformation may be viewed as a rotation in Minkowski space by an imaginary angle. The prototypical transformation is the Lorentz-Einstein transformation corresponding to a rotation in the $x_1 - x_4$ plane with the transformation matrix $[a_{\mu\nu}]$ given by [Appendix 12A1]

$$[a_{\mu\nu}] = \begin{pmatrix} \cos\phi & 0 & 0 & \sin\phi \\ 0 & 1 & 0 & 0 \\ 0 & 0 & 1 & 0 \\ -\sin\phi & 0 & 0 & \cos\phi \end{pmatrix} \tag{12.8.14}$$

where ϕ is given by $\tan\phi = i\frac{V}{c}$. For this transformation the primed inertial frame moves relative to the unprimed reference frame with velocity V along the x_1 axis. To identify the operator $S_{_L}$ we use the conditions (12.8.12) and (12.8.13). Consider the case $\nu = 1$ in Eq. (12.8.12), which leads to

$$\begin{aligned} S_{_L}^{-1}\gamma_1 S_{_L} &= a_{1\mu}\gamma_\mu = a_{11}\gamma_1 + a_{14}\gamma_4 = \gamma_1\cos\phi - \gamma_4\sin\phi \\ &= [\cos(\phi/2) + \gamma_1\gamma_4\sin(\phi/2)]\ \gamma_1\ [\cos(\phi/2) - \gamma_1\gamma_4\sin(\phi/2)] \end{aligned} \tag{12.8.15}$$

where we have used the trigonometric identities $\cos\phi = \cos^2(\phi/2) - \sin^2(\phi/2)$, $\sin\phi = 2\sin(\phi/2)\cos(\phi/2)$ and the anti-commutation relation $\gamma_\mu\gamma_\nu + \gamma_\nu\gamma_\mu = 2\delta_{\mu\nu}$ in the last step. So in this case we may identify the operators $S_{_L}$ and $S_{_L}^{-1}$ as

$$S_{_L} = \cos(\phi/2) - \gamma_1\gamma_4\sin(\phi/2) \tag{12.8.16}$$

and

$$S_{_L}^{-1} = \cos(\phi/2) + \gamma_1\gamma_4\sin(\phi/2)\,. \tag{12.8.17}$$

Using the properties of γ-matrices it is easily verified that $S_L^{-1}S_L = S_L S_L^{-1} = I$. It is also easy to check that with this choice of S_L, the condition (12.8.12) is satisfied for $\nu = 2$, 3, and 4 as well. Thus the Dirac equation is invariant under pure Lorentz transformations if S_L and S_L are chosen according to Eqs. (12.8.16) and (12.8.17).

(b)Rotations in Three-dimensional Space

As a prototypical example of pure spatial rotation we consider a rotation about x_1-axis through a real angle θ. The transformation matrix $[a_{\mu\nu}]$ for this rotation is given by

$$[a_{\mu\nu}] = \begin{pmatrix} 1 & 0 & 0 & 0 \\ 0 & \cos\theta & \sin\theta & 0 \\ 0 & -\sin\theta & \cos\theta & 0 \\ 0 & 0 & 0 & 1 \end{pmatrix} . \tag{12.8.18}$$

To identify the operators S_R and S_R in this case we put $\nu = 2$ in Eq. (12.8.12) and get

$$\begin{aligned} S_R^{-1}\gamma_2\, S_R &= a_{2\mu}\gamma_\mu = a_{22}\gamma_2 + a_{23}\gamma_3 = \gamma_2\cos\theta + \gamma_3\sin\theta \\ &= [\cos(\theta/2) - \gamma_2\gamma_3\sin(\theta/2)]\ \gamma_2\ [\cos(\theta/2) + \gamma_2\gamma_3\sin(\theta/2)]\,. \end{aligned} \tag{12.8.19}$$

We may now identify the operators S_R and S_R^{-1} as

$$S_R = \cos(\theta/2) + \gamma_2\gamma_3\sin(\theta/2) = \exp(\gamma_2\gamma_3\theta/2) \tag{12.8.20}$$

and

$$S_R^{-1} = \cos(\theta/2) - \gamma_2\gamma_3\sin(\theta/2) = \exp(-\gamma_2\gamma_3\theta/2)\,. \tag{12.8.21}$$

Obviously $S_R^{-1}S_R = I = S_R S_R^{-1}$ and it can be verified that the condition (12.8.12) is satisfied for $\nu = 1, 3,$ and 4 as well. Thus the Dirac equation is invariant under spatial rotations also if the operators S and S^{-1} are chosen according to Eqs. (12.8.20) and (12.8.21).

Transformations considered thus far can be obtained by successive applications of transformations which differ infinitesimally from the identity transformation for which the invariance of the Dirac equation is obvious. For this reason, the invariance of the Dirac equation under pure Lorentz transformations and rotations has intuitive appeal. This is not the case with the discrete transformations we are about to consider. Nevertheless the Dirac equation and the field equations of electrodynamics are believed to be invariant under these discrete symmetry transformations as well.

(c) Space Inversion or Parity

The transformation matrix corresponding to the operation of space inversion $x \to x' \equiv (x_1', x_2', x_3', x_4') = (-x_1, -x_2, -x_3, x_4)$ is

$$[a_{\mu\nu}] = \begin{pmatrix} -1 & 0 & 0 & 0 \\ 0 & -1 & 0 & 0 \\ 0 & 0 & -1 & 0 \\ 0 & 0 & 0 & +1 \end{pmatrix} . \tag{12.8.22}$$

In this case, (12.8.12) requires $S_P^{-1}\gamma_k \S_P = -\gamma_k$, $k = 1,2,3$ and $S_P^{-1}\gamma_4\S_P = \gamma_4$. Using the properties of γ-matrices $[\gamma_\mu\gamma_\nu = -\gamma_\nu\gamma_\mu\ (\mu \neq \nu)]$ we easily identify that S_P and S_P^{-1} are given by

$$S_P = \gamma_4\,, \tag{12.8.23}$$

and

$$S_P^{-1} = \gamma_4\,. \tag{12.8.24}$$

The last equation follows from $\gamma_4^2 = I$. So the Dirac equation equation is invariant under space inversion (parity) transformation if the operators S_P and S_P^{-1} are chosen according to Eqs. (12.8.23) and (12.8.24).

(d) Time Reversal

For the operation of time reversal $t \to -t$, the transformation matrix is,

$$[a_{\mu\nu}] = \begin{pmatrix} 1 & 0 & 0 & 0 \\ 0 & 1 & 0 & 0 \\ 0 & 0 & 1 & 0 \\ 0 & 0 & 0 & -1 \end{pmatrix}. \tag{12.8.25}$$

In this case Eq. (12.8.12) requires $S_t^{-1}\gamma_k S = \gamma_k$ and $S_t^{-1}\gamma_4 S_t = -\gamma_4$. Once again from the anti-commuting property of γ-matrices, we find that S_t and S_t^{-1} satisfying these requirements are

$$S_t = \gamma_1\gamma_2\gamma_3 \equiv \gamma_5\gamma_4 \tag{12.8.26}$$

and

$$S_t^{-1} = \gamma_3\gamma_2\gamma_1 \equiv \gamma_4\gamma_5\,. \tag{12.8.27}$$

Using the properties of γ-matrices we can verify that $S_t^{-1}S_t = I = S_t S_t^{-1}$.

The transformation matrices S derived so far have been determined solely on the basis of an assumed linear transformation of the Dirac spinor that leaves the Dirac equation (12.8.1) invariant. However, since we are dealing with a Dirac particle in the presence of an electromagnetic field, various transformation matrices must also be compatible with the invariance of the equations of electrodynamics. In particular, the differential equation for the four-potential [see Appendix 12A1, Eq. (12A1.4.15)]

$$\frac{\partial}{\partial x_\mu}\frac{\partial}{\partial x_\mu}A_\nu = -\mu_o j_\nu = -\mu_0\left[iec\,\tilde{\psi}(x)\gamma_\nu\psi(x)\right] \tag{12.8.28}$$

must be form invariant under these transformations. Since the d'Alembertian $\frac{\partial}{\partial x_\mu}\frac{\partial}{\partial x_\mu}$ is invariant under various classes of transformations considered thus far, this requirement means that the corresponding transformation matrix S for the Dirac spinor must be such that the four-current $j_\mu(x) = ice\tilde{\psi}(x)\gamma_\mu\psi(x) = iceJ_\mu(x)$ has the same transformation properties as the four-potential $A_\mu(x)$, viz.,

$$J'_\mu(x) = a_{\mu\nu}J_\nu(x) \quad \text{or} \quad J_\nu(x) = a_{\mu\nu}J'_\mu(x)\,.$$

Recalling that $\bar{S} = \gamma_4 S^\dagger\gamma_4$ we find that $\bar{S} = S^{-1}$ for $S = S_L, S_R, S_P$ (pure Lorentz, pure rotation, and parity). Hence for these transformations

$$J'_\mu(x) = \tilde{\psi}'(x)\gamma_\mu\psi'(x) = \tilde{\psi}(x)\bar{S}\gamma_\mu S\psi(x) = a_{\mu\nu}J_\nu(x)\,,$$

which is the same as the transformation law (12.8.6) for the vector potential. Hence the transformations S_L, S_R, and S_P are compatible with the invariance of both (12.8.1) and (12.8.28). It is easy to check that in all these cases the adjoint of Dirac equation is also invariant.

On the other hand, the transformation matrix $S_t = \gamma_5\gamma_4$ is not compatible with the invariance of Eq. (12.8.28) for in this case $\bar{S}_t = \gamma_4 S_t^\dagger\gamma_4 = \gamma_5\gamma_4$, which leads to

$$\begin{aligned} J'_\mu(x) &= \tilde{\psi}'(x)\gamma_\mu\psi'(x) = \tilde{\psi}(x)\bar{S}_t\gamma_\mu S_t\psi(x) = \tilde{\psi}(x)\gamma_5\gamma_4\gamma_\mu\gamma_5\gamma_4\psi(x) = \tilde{\psi}(x)\gamma_4\gamma_\mu\gamma_4\psi(x) \\ &= -a_{\mu\nu}J_\nu(x)\,, \end{aligned}$$

where we have made use of the matrix $a_{\mu\nu}$ given by Eq. (12.8.25) and the relation $\gamma_5\gamma_\mu = -\gamma_\mu\gamma_5$ ($\mu \neq 5$). Hence Eq. (12.8.28) is not invariant under this transformation.[9] In

[9]The transformation $S_t = \gamma_5\gamma_4$ was introduced by G. Racah, Nuovo Cimento **14**, 329 (1937).

other words, a linear time-reversal transformation S compatible with the invariance of both the Dirac equation and the equation for four-vector potential does not exist. We shall discuss an antilinear time reversal operator S_T which is compatible with the invariance of both (12.8.1) and (12.8.28) in Sec. 12.11, where another discrete symmetry of electrodynamics, viz., charge conjugation is also discussed.

To sum up, we can say that the Dirac equation as well as its adjoint are invariant under all orthogonal transformations (Lorentz, spatial rotation, space inversion) if the forms of the operators S and $\bar{S} = S^{-1}$ are appropriately chosen in each case. For time-reversal operation the Dirac equation and its adjoint for a free spin half particle are form invariant for the same conditions with S and $\bar{S}$ given by Eqs.(12.8.26) and (12.8.27). The discussion of the time-reversed state which conforms to the form-invariance of the Dirac equation as well as the electromagnetic equation (12.8.28) will be discussed in Sec. 12.11.

12.9 Dirac Bilinear Covariants

By inserting the operators I, γ_μ, $i\gamma_\mu\gamma_\nu$, $i\gamma_5\gamma_\mu$, and γ_5 between the spinors $\tilde{\psi}$ and ψ, we can form the following sets of bilinear combinations (of $\tilde{\psi}$ and ψ) containing, respectively, one, four, sixteen, four, and one elements:

	(i)	$\tilde{\psi}\psi$	(one)
	(ii)	$\tilde{\psi}\gamma_\mu\psi$	(four)
	(iii)	$\tilde{\psi}i\gamma_\mu\gamma_\nu\psi$	(sixteen)
	(iv)	$\tilde{\psi}i\gamma_5\gamma_\mu\psi$	(four)
and	*(v)*	$\tilde{\psi}\gamma_5\psi$	(one).

The number of bilinear quantities in a set is determined by the fact that each Greek index takes on values $1, 2, 3, 4$. The quantities in these sets transform, respectively, as the components of a (i) scalar, (ii) vector, (iii) tensor of second rank, (iv) pseudo-vector, and (v) pseudo-scalar.

Let us examine the single element $\tilde{\psi}\psi$ comprising the first set. Under a Lorentz transformation (12.8.3) this transforms according to

$$\tilde{\psi}\psi \to \tilde{\psi}'\psi' = \tilde{\psi}\bar{S}S\psi = \tilde{\psi}\psi\,. \tag{12.9.1}$$

The transformed quantity is the same as the original quantity. Hence $\tilde{\psi}\psi$ is an invariant or *scalar.*

The four quantities $\tilde{\psi}\gamma_\mu\psi$ comprising the second set transform, under a Lorentz transformation (12.8.3), according to

$$\tilde{\psi}\gamma_\mu\psi \to \tilde{\psi}'\,\gamma_\mu\psi' = \tilde{\psi}\,\bar{S}\gamma_\mu S\psi = \tilde{\psi}\,a_{\mu\nu}\gamma_\nu\psi\,,$$

or

$$\tilde{\psi}'\,\gamma_\mu\psi' = a_{\mu\nu}\tilde{\psi}\gamma_\nu\psi\,, \tag{12.9.2}$$

which is the law of transformation for the components of a four-vector.

The 16 quantities $\tilde{\psi}i\gamma_\mu\gamma_\nu\psi$ transform under a Lorentz transformation according to

$$\begin{aligned}\tilde{\psi}i\gamma_\mu\gamma_\nu\psi \to \tilde{\psi}'(i\gamma_\mu\gamma_\nu)\psi' &= \tilde{\psi}\,\bar{S}(i\gamma_\mu\gamma_\nu)S\psi = i\tilde{\psi}\,\bar{S}\gamma_\mu S\bar{S}\gamma_\nu S\psi \\ &= i\tilde{\psi}\,a_{\mu\lambda}\,\gamma_\lambda\,a_{\nu\rho}\gamma_\rho\psi\end{aligned}$$

or

$$\tilde{\psi}'(i\gamma_\mu\gamma_\nu)\psi' = a_{\mu\lambda}a_{\nu\rho}\{\tilde{\psi}(i\gamma_\lambda\gamma_\rho)\psi\}\,, \tag{12.9.3}$$

which is the law of transformation for the components of a second rank tensor. From the anti-commutation properties of the gamma matrices it follows that $\tilde{\psi} i\gamma_\mu \gamma_\nu \psi$ is an anti-symmetric tensor.

We consider now the fourth set of four quantities $\tilde{\psi} i\gamma_5\gamma_\mu\psi$. Under the transformation (12.8.3) they transform according to

$$\tilde{\psi}(i\gamma_5\gamma_\mu)\psi \to \tilde{\psi}'(i\gamma_5\gamma_\mu)\psi' = i\tilde{\psi}\,\bar{S}\gamma_5 S\bar{S}\gamma_\mu S\psi = i\tilde{\psi}(\bar{S}\gamma_5 S)(\bar{S}\gamma_\mu S)\psi \tag{12.9.4}$$

where we have inserted $S\bar{S} \equiv I$ between γ_5 and γ_μ. Recall now that for space inversion (or parity transformation), we have $\bar{S} = \gamma_4$ and $S = \gamma_4$ [Eqs. (12.8.23) and (12.8.24)], and therefore, $\bar{S}\gamma_5 S = -\gamma_5$, whereas for an identity transformation, $S = \bar{S} = I$, and therefore $\bar{S}\gamma_5 S = \gamma_5$. Hence, for space inversion followed by Lorentz transformation (improper Lorentz transformation) or for space inversion followed by rotation in three-dimensional space, we have

$$\tilde{\psi}'(i\gamma_5\gamma_\mu)\psi' = -a_{\mu\nu}\tilde{\psi}(i\gamma_5\gamma_\nu)\psi \tag{12.9.5}$$

and for proper Lorentz transformation, or proper rotation in three-dimensional space, we have

$$\tilde{\psi}\,'(i\gamma_5\gamma_\mu)\psi' = a_{\mu\nu}\tilde{\psi}(i\gamma_5\gamma_\nu)\psi\,. \tag{12.9.6}$$

This is the transformation characteristic of a pseudo-vector.

Finally it is easy to see that the quantity $\tilde{\psi}\gamma_5\psi$ is a pseudo-scalar under the transformation (12.8.3). As a result of this transformation

$$\tilde{\psi}\gamma_5\psi \to \tilde{\psi}'\gamma_5\psi' = \tilde{\psi}\,\bar{S}\gamma_5 S\psi = \begin{cases} -\tilde{\psi}\gamma_5\psi\,, & \text{for improper Lorentz transformation} \\ \\ \tilde{\psi}\gamma_5\psi\,, & \text{for proper Lorentz transformation} \end{cases} \tag{12.9.7}$$

Dirac bilinear covariants are very useful for the discussion of relativistic weak interactions between particles.

12.10 Dirac Electron in a Spherically Symmetric Potential

Consider an electron in an external potential $V(r)$. Then the Hamiltonian

$$H = c\boldsymbol{\alpha}\cdot\boldsymbol{p} + \beta mc^2 + V(r)\,, \tag{12.10.1}$$

commutes with the total angular momentum operator $\hat{\boldsymbol{J}}$ and its square $\hat{\boldsymbol{J}}^2$ so both are constants of motion. We can therefore choose j and m_j, both half integers, and energy eigenvalues as quantum numbers to specify the states. But these are not sufficient as each value of j may correspond to two non-relativistic states with $\ell = j \pm \frac{1}{2}$. So we search for another constant of motion to differentiate between these two. We note that these states correspond to spin $\hat{\boldsymbol{\Sigma}}$ parallel and antiparallel to $\hat{\boldsymbol{J}}$. So we try an operator proportional to $\hat{\boldsymbol{\Sigma}}\cdot\hat{\boldsymbol{J}}$ and after some some trial we find that

$$K = \beta\left(\boldsymbol{\Sigma}\cdot\boldsymbol{J} - \frac{\hbar}{2}I\right) = \beta\,(\boldsymbol{\Sigma}\cdot\boldsymbol{L} + \hbar I) \tag{12.10.2}$$

also commutes with H as well as $\boldsymbol{J}$ and, therefore, is also a constant of motion. To see that the operator K commutes with H we take note of the following commutation relations

(*i*) $$[H,\beta] = c[\boldsymbol{\alpha},\beta]\cdot\boldsymbol{p} = -2c\beta(\boldsymbol{\alpha}\cdot\boldsymbol{p}) \tag{12.10.3a}$$

(*ii*) $$[H,\Sigma_\ell] = 2ic\epsilon_{\ell km}\alpha_k p_m = 2ic[\boldsymbol{\alpha}\times\boldsymbol{p}]_\ell \tag{12.10.3b}$$

where $\epsilon_{k\ell m}$ is the alternating symbol and summation convention is implied. Using these relations we find

$$\beta[H, \Sigma_\ell]\, J_\ell = 2i\, c\beta\epsilon_{\ell km}\alpha_k p_m J_\ell = 2i\, c\beta\,(\boldsymbol{p}\times\boldsymbol{J})\cdot\boldsymbol{\alpha}\,. \tag{12.10.4}$$

(*iii*)
$$(\boldsymbol{\alpha}\cdot\boldsymbol{p})(\boldsymbol{\Sigma}.\boldsymbol{J}) = -\gamma_5\,\boldsymbol{p}.\boldsymbol{J} + i\boldsymbol{\alpha}\cdot(\boldsymbol{p}\times\boldsymbol{J})\,. \tag{12.10.5}$$

With the help of these results we can evaluate the commutator

$$\begin{aligned}
[H, \beta(\boldsymbol{\Sigma}.\boldsymbol{J})] &= [H,\beta](\boldsymbol{\Sigma}\cdot\boldsymbol{J}) + \beta[H, (\boldsymbol{\Sigma}\cdot\boldsymbol{J})]\\
&= -2c\beta(\boldsymbol{\alpha}.\boldsymbol{p})(\boldsymbol{\Sigma}.\boldsymbol{J}) + \beta[H,\boldsymbol{\Sigma}]\cdot\boldsymbol{J} + \beta\,\boldsymbol{\Sigma}\cdot[H,\boldsymbol{J}]\\
&= -2c\beta\{-\gamma_5\boldsymbol{p}.\boldsymbol{J} + i\alpha.(\boldsymbol{p}\times\boldsymbol{J})\} + 2ic\beta\,(\boldsymbol{p}\times\boldsymbol{J})\cdot\boldsymbol{\alpha}\\
&= 2c\beta\gamma_5\boldsymbol{p}.\left(\boldsymbol{L} + \frac{\hbar\boldsymbol{\Sigma}}{2}\right) = -c\hbar\beta\,(\boldsymbol{p}.\boldsymbol{\alpha})\,,
\end{aligned}$$

where we have used $\gamma_5\boldsymbol{\Sigma} = -\boldsymbol{\alpha}$ and $\boldsymbol{p}\cdot\boldsymbol{L} = 0$. With the help of Eq. (12.10.3a), we can replace $-c\hbar\beta(\boldsymbol{p}\cdot\boldsymbol{\alpha})$ in the previous equation by the commutator $\frac{\hbar}{2}[H,\beta]$, giving us

$$[H, \beta(\boldsymbol{\Sigma}\cdot\boldsymbol{J})] = \frac{\hbar}{2}[H,\beta]\,. \tag{12.10.6a}$$

Transposing the right-hand side to the left-hand side and combining the two we get

$$[H, (\beta\,\boldsymbol{\Sigma}\cdot\boldsymbol{J} - \frac{\hbar}{2}\beta)] \equiv [H, K] = 0\,, \tag{12.10.6b}$$

where $K = \beta\,\boldsymbol{\Sigma}\cdot\boldsymbol{J} - \frac{\hbar}{2}\beta = \beta\,(\boldsymbol{\Sigma}\cdot\boldsymbol{L} + \hbar I)$. Now the square of K can be evaluated as

$$K^2 = [\beta(\boldsymbol{\Sigma}\cdot\boldsymbol{L} + \hbar I)]^2 = L^2 + \hbar\boldsymbol{\Sigma}\cdot\boldsymbol{L} + \hbar^2 I$$

or
$$K^2 = J^2 + \frac{\hbar^2}{4}I\,, \tag{12.10.7}$$

where we have used the following relations

$$\begin{aligned}
\Sigma_i\Sigma_j &= i\epsilon_{ijk}\Sigma_k\,,\\
[L_i, L_j] &= i\hbar\epsilon_{ijk}L_k\,,\\
\Sigma_k\Sigma_k &= 3I\,,\\
[\Sigma_k, \beta] &= 0\,.
\end{aligned} \tag{12.10.8}$$

Since $\boldsymbol{J}$ commutes with β and $\boldsymbol{\Sigma}\cdot\boldsymbol{L}$, it also commutes with K. So H, $\boldsymbol{J}$, and K form a set of commuting observables. Equation(12.10.7) then leads to the following relation between the eigenvalues of K^2 and those of J^2

$$\hbar^2\kappa^2 = \hbar^2 j(j+1) + \frac{\hbar^2}{4}$$

which gives
$$\kappa = \pm\left(j + \frac{1}{2}\right)\,. \tag{12.10.9}$$

Hence κ can be a positive or negative (non-zero) integer. As already noted, the sign of κ determines whether the spin is parallel ($\kappa < 0$) or antiparallel ($\kappa > 0$) to the total angular momentum.

We can now construct the simultaneous eigenstates of H, K, $\boldsymbol{J}^2$, and J_3 belonging to eigenvalues $E\,, -\hbar\kappa\,, j(j+1)\hbar^2$, and $\hbar m_j$, respectively. We first write the operator K as

$$K = \beta(\boldsymbol{\Sigma}\cdot\boldsymbol{L} + \hbar I) = \begin{pmatrix} \boldsymbol{\sigma}\cdot\boldsymbol{L}+\hbar I & 0 \\ 0 & -\boldsymbol{\sigma}\cdot\boldsymbol{L}-\hbar I \end{pmatrix}. \tag{12.10.10}$$

Writing the four-component wave function ψ, representing the simultaneous eigenstate of the observables H, K, J^2, J_3 as

$$\psi = \begin{pmatrix} \Phi \\ \chi \end{pmatrix}, \tag{12.10.11}$$

where Φ and χ are two-component wave functions and substituting this form into time-independent Dirac equation $H\psi = E\psi$ with H given by Eq. (12.10.1), we get

$$\left[-i\hbar c \begin{pmatrix} 0 & \sigma_k \\ \sigma_k & 0 \end{pmatrix} \frac{\partial}{\partial x_k} + \begin{pmatrix} 0 & I \\ -I & 0 \end{pmatrix} mc^2 + V(r)\right] \begin{pmatrix} \Phi \\ \chi \end{pmatrix} = E \begin{pmatrix} \Phi \\ \chi \end{pmatrix}. \tag{12.10.12}$$

Thus yields two coupled equations

$$\left[E - mc^2 - V(r)\right]\Phi - c(\boldsymbol{\sigma}\cdot\boldsymbol{p})\chi = 0\,, \tag{12.10.13}$$

and

$$\left[E + mc^2 - V(r)\right]\chi - c(\boldsymbol{\sigma}\cdot\boldsymbol{p})\Phi = 0\,. \tag{12.10.14}$$

Now, the angular momentum operator L^2 does not commute with the Dirac Hamiltonian. Therefore ℓ is not a good quantum number in the eigenstate represented by the four-component wave function ψ. However, by writing L^2 as

$$L^2 = J^2 - \hbar\boldsymbol{\sigma}\cdot\boldsymbol{L} - \frac{3}{4}\hbar^2 I\,, \tag{12.10.15}$$

and operating it on the component states Φ and χ, we see that it is determinate for each of them with eigenvalues, say, $\hbar^2\ell_A(\ell_A+1)$ and $\hbar^2\ell_B(\ell_B+1)$, respectively. Also, from Eq. (12.10.10) we see that $K = \boldsymbol{\sigma}\cdot\boldsymbol{L} + \hbar I$ for the two-component wave function Φ and $K = -(\boldsymbol{\sigma}\cdot\boldsymbol{L} + \hbar I)$ for the two-component wave function χ.

Recalling that j, m_j, and κ are good quantum numbers for the four-component wave function ψ as well as each of the two-component wave functions Φ and χ, the equations $L^2\Phi = \hbar^2\ell_A(\ell_A+1)\Phi$ and $K\Phi \equiv (\boldsymbol{\sigma}\cdot\boldsymbol{L} + \hbar I)\Phi = -\hbar\kappa\Phi$ yield, according to Eq. (12.10.15),

$$j(j+1) + \kappa + \frac{1}{4} = \ell_A(\ell_A+1)\,. \tag{12.10.16}$$

Similarly, $L^2\chi = \hbar^2\ell_B(\ell_B+1)\chi$ and $K\chi \equiv -(\boldsymbol{\sigma}\cdot\boldsymbol{L} + \hbar I)\chi = -\hbar\kappa\chi$ lead to

$$j(j+1) - \kappa + \frac{1}{4} = \ell_B(\ell_B+1)\,. \tag{12.10.17}$$

Thus for each value of j, orbital angular momentum numbers ℓ_A and ℓ_B can take two values $j \pm 1/2$, depending on κ as summarized below.

$$\kappa = j + 1/2\,, \qquad \ell_A = j + 1/2 \quad \text{and} \quad \ell_B = j - 1/2\,, \tag{12.10.18}$$

$$\kappa = -(j + 1/2) \qquad \ell_A = j - 1/2 \quad \text{and} \quad \ell_B = j + 1/2\,. \tag{12.10.19}$$

Alternatively, we can say that for each value of κ the two-component wave functions Φ and χ have orbital quantum numbers ℓ_A and ℓ_B which differ by one and therefore have opposite parity. This conclusion also follows from the fact that the four-component wave function has a definite parity. Using these facts, we can express the two-component wave functions (belonging to the same E, j, κ,) in the form

$$\Phi = g(r)\mathcal{Y}_{j\,\ell_A}^{m_j}(\theta,\varphi) \tag{12.10.20}$$

and
$$\chi = i\,f(r)\mathcal{Y}_{j\,\ell_B}^{m_j}(\theta,\varphi), \tag{12.10.21}$$

where the factor i multiplying $f(r)$ has been inserted to make the functions f and g real for bound state solutions and $\mathcal{Y}_{j\,\ell_A}$ and $\mathcal{Y}_{j\,\ell_B}$ are the normalized spin-angle functions, representing the coupled states $|(\ell_A\frac{1}{2})jm_j\rangle$ and $|(\ell_B\frac{1}{2})jm_j\rangle$, respectively. These may be constructed according to angular momentum coupling rules. Explicitly, the states $\mathcal{Y}_{j\,\ell}^{m_j}(\theta,\varphi)$ in the coordinate representation for $j = \ell \pm 1/2$ are

$$j = \ell + 1/2 : \quad \mathcal{Y}_{j\ell}^{m_j}(\theta,\varphi) = \sqrt{\frac{\ell + m_j + 1/2}{2\ell+1}}\, Y_{\ell,m_j-1/2}(\theta,\varphi)\begin{pmatrix}1\\0\end{pmatrix} + \sqrt{\frac{\ell - m_j + 1/2}{2\ell+1}}\, Y_{\ell,m_j+1/2}(\theta,\varphi)\begin{pmatrix}0\\1\end{pmatrix}, \tag{12.10.22}$$

$$j = \ell - 1/2 : \quad \mathcal{Y}_{j\ell}^{m_j}(\theta,\varphi) = -\sqrt{\frac{\ell - m_j + 1/2}{2\ell+1}}\, Y_{\ell,m_j-1/2}(\theta,\varphi)\begin{pmatrix}1\\0\end{pmatrix} + \sqrt{\frac{\ell + m_j + 1/2}{2\ell+1}}\, Y_{\ell,m_j+1/2}(\theta,\varphi)\begin{pmatrix}0\\1\end{pmatrix}. \tag{12.10.23}$$

The column vectors represent the spin-up and spin-down states of the electron.

In view of Eqs. (12.10.18) and (12.10.19), for negative κ, the first equation (12.10.22) represents $\mathcal{Y}_{j\,\ell_A}^{m_j}(\theta,\varphi)$ ($\ell_A = j - 1/2$) and the second equation (12.10.23) represents $\mathcal{Y}_{j\,\ell_B}^{m_j}(\theta,\varphi)$ ($\ell_B = j + 1/2$). For positive κ, the first equation represents $\mathcal{Y}_{j\,\ell_B}^{m_j}(\theta,\varphi)$ ($\ell_B = j - 1/2$) and the second equation represents $Y_{j\,\ell_A}^{m_j}(\theta,\varphi)$ ($\ell_A = j + 1/2$).

It is clear that the radial functions $f(r)$ and $g(r)$ depend on the sign of κ. Using the identity

$$\left(\boldsymbol{\sigma}\cdot\frac{\boldsymbol{r}}{r}\right)\left(\boldsymbol{\sigma}\cdot\frac{\boldsymbol{r}}{r}\right) = 1, \tag{12.10.24a}$$

we can express $(\boldsymbol{\sigma}\cdot\boldsymbol{p})$ as

$$\begin{aligned}(\boldsymbol{\sigma}\cdot\boldsymbol{p}) &= \frac{(\boldsymbol{\sigma}\cdot\boldsymbol{r})}{r^2}\,[(\boldsymbol{\sigma}\cdot\boldsymbol{r})(\boldsymbol{\sigma}\cdot\boldsymbol{p})] \\ &= \frac{\boldsymbol{\sigma}\cdot\boldsymbol{r}}{r^2}[\boldsymbol{r}\cdot\boldsymbol{p} + i\boldsymbol{\sigma}\cdot(\boldsymbol{r}\times\boldsymbol{p})] = \frac{\boldsymbol{\sigma}\cdot\boldsymbol{r}}{r^2}\{-i\hbar\boldsymbol{r}\cdot\nabla + i\boldsymbol{\sigma}\cdot\boldsymbol{L}\}\end{aligned}$$

or
$$(\boldsymbol{\sigma}\cdot\boldsymbol{p}) = \frac{\boldsymbol{\sigma}\cdot\boldsymbol{r}}{r^2}\left[-i\hbar r\frac{\partial}{\partial r} + i\boldsymbol{\sigma}\cdot\boldsymbol{L}\right]. \tag{12.10.24b}$$

Since $\frac{(\boldsymbol{\sigma}\cdot\boldsymbol{r})}{r}$ is a pseudo-scalar operator, it preserves j and m_j but changes the parity of the function on which it operates. So for a given j we have

$$\left(\frac{\boldsymbol{\sigma}\cdot\boldsymbol{r}}{r}\right)\mathcal{Y}_{j\,l_A}^{m_j}(\theta,\varphi) = -\mathcal{Y}_{j\,l_B}^{m_j}(\theta,\varphi), \tag{12.10.25}$$

and
$$\left(\frac{\boldsymbol{\sigma}\cdot\boldsymbol{r}}{r}\right)\mathcal{Y}_{j\,l_B}^{m_j}(\theta,\varphi) = -\mathcal{Y}_{j\,l_A}^{m_j}(\theta,\varphi). \tag{12.10.26}$$

These relations follow from the fact that ℓ_A and ℓ_B differ by 1 and $\frac{(\boldsymbol{\sigma}\cdot\boldsymbol{r})}{r}$, being a pseudo-scalar operator, preserves j and m_j. Furthermore, the square of $\frac{(\boldsymbol{\sigma}\cdot\boldsymbol{r})}{r}$ is a unit operator

[Eq.(12.10.24a)], its eigenvalues are ± 1. Hence the sign on the right-hand sides of Eqs. (12.10.25) through (12.10.26) must be + or −. The justification for minus sign can be seen by observing that $\left(\frac{\boldsymbol{\sigma}\cdot\boldsymbol{r}}{r}\right)\mathcal{Y}^{m_j}_{j\,\ell_B}(0,0) = -\mathcal{Y}^{m_j}_{j\,\ell_A}(0,0)$ and $\left(\frac{\boldsymbol{\sigma}\cdot\boldsymbol{r}}{r}\right)\mathcal{Y}^{m_j}_{j\,\ell_A}(0,0) = -\mathcal{Y}^{m_j}_{j\,\ell_B}(0,0)$]. With the help of these results we find

$$(\boldsymbol{\sigma}\cdot\boldsymbol{p})\chi = i\frac{\boldsymbol{\sigma}\cdot\boldsymbol{r}}{r^2}\left[-i\hbar r\frac{df}{dr} + i(\kappa-1)\hbar f\right]\mathcal{Y}^{m_j}_{j\,\ell_B}(\theta,\varphi)$$

or

$$(\boldsymbol{\sigma}\cdot\boldsymbol{p})\chi = -\hbar\frac{df}{dr}\mathcal{Y}^{m_j}_{j\,\ell_A}(\theta,\varphi) + \frac{(\kappa-1)\hbar}{r}f\mathcal{Y}^{m_j}_{j\,\ell_A}(\theta,\varphi)\,. \tag{12.10.27}$$

Similarly,

$$(\boldsymbol{\sigma}\cdot\boldsymbol{p})\Phi = i\hbar\frac{dg}{dr}\mathcal{Y}^{m_j}_{j\,\ell_B}(\theta,\varphi) + i\frac{(\kappa+1)\hbar}{r}g(r)\mathcal{Y}^{m_j}_{j\,\ell_B}(\theta,\varphi)\,. \tag{12.10.28}$$

Substituting these relations into the coupled equations (12.10.13) and (12.10.14), and introducing the functions $F(r)$ and $G(r)$ by

$$F(r) = rf(r)\,, \tag{12.10.29}$$

and

$$G(r) = rg(r), \tag{12.10.30}$$

we obtain two coupled radial equations

$$\hbar c\left(\frac{dF}{dr} - \frac{\kappa}{r}F\right) = -(E - V(r) - mc^2)G(r)\,, \tag{12.10.31}$$

and

$$\hbar c\left(\frac{dG}{dr} + \frac{\kappa}{r}G\right) = (E - V(r) + mc^2)F(r)\,. \tag{12.10.32}$$

For Hydrogen (or Hydrogen-like) atoms, the potential has the Coulomb form

$$V(r) = -\frac{Ze^2}{4\pi\epsilon_0 r}\,. \tag{12.10.33}$$

Let us introduce dimensionless energy ϵ, dimensionless radial coordinate ρ, and potential energy parameter $Z\alpha$ by

$$E = \epsilon\, mc^2\,, \quad \rho = \sqrt{1-\epsilon^2}\,\frac{mc}{\hbar}\,r\,, \quad Z\alpha = Z\frac{e^2}{4\pi\epsilon_0\hbar c}\,, \tag{12.10.34}$$

where α is the fine structure constant. Note that we are interested in bound state energies $E < mc^2$ $(\epsilon < 1)$. With these substitutions, the coupled equations (12.10.31) and (12.10.32) reduce to

$$\left(\frac{d}{d\rho} - \frac{\kappa}{\rho}\right)F(\rho) - \left(\sqrt{\frac{1-\epsilon}{1+\epsilon}} - \frac{Z\alpha}{\rho}\right)G(\rho) = 0\,, \tag{12.10.35}$$

and

$$\left(\frac{d}{d\rho} + \frac{\kappa}{\rho}\right)G(\rho) - \left(\sqrt{\frac{1+\epsilon}{1-\epsilon}} + \frac{Z\alpha}{\rho}\right)F(\rho) = 0\,. \tag{12.10.36}$$

By examining the behavior of $F(\rho)$ and $G(\rho)$ for large ρ $(\rho \to \infty)$ we find that the normalizable solutions behave like $e^{-\rho}$. Similarly, by examining the behavior for $\rho \to 0$ by using F and G of the form

$$F = a_0\rho^s\,, \quad G = b_0\rho^s\,, \tag{12.10.37}$$

we arrive at the coupled linear equations

$$(s-\kappa)\,a_o + Z\alpha\,b_0 = 0\,, \tag{12.10.38a}$$

$$-Z\alpha\,a_0 + (s+\kappa)\,b_0 = 0\,. \tag{12.10.38b}$$

For nontrivial soultions of this equation we require

$$\begin{vmatrix} s-\kappa & Z\alpha \\ Z\alpha & (s+\kappa) \end{vmatrix} = s^2 - \kappa^2 + (Z\alpha)^2 = 0\,, \quad \Rightarrow \quad s = \pm\sqrt{\kappa^2 - (Z\alpha)^2}\,. \tag{12.10.39a}$$

For normalizable regular solutions at $\rho = 0$, only the positive square root for s is acceptable. Hence the index s characterizing the small ρ behavior is

$$s = \sqrt{\kappa^2 - (Z\alpha)^2} = \sqrt{(j+1/2)^2 - (Z\alpha)^2}\,. \tag{12.10.39b}$$

Using these insights we seek solutions of the form

$$F(\rho) = e^{-\rho}\rho^s \sum_{m=0}^{\infty} a_m \rho^m \tag{12.10.40}$$

and

$$G(\rho) = e^{-\rho}\rho^s \sum_{m=0}^{\infty} b_m \rho^m \tag{12.10.41}$$

Substituting these in Eqs. (12.10.35) and (12.10.36), we obtain

$$\sum_{m=0}^{\infty} a_m e^{-\rho}\left[(s+m)\rho^{s+m-1} - \rho^{s+m} - \kappa\rho^{s+m-1}\right] - \sum_{m=0}^{\infty} b_m e^{-\rho}\left[\sqrt{\frac{1-\epsilon}{1+\epsilon}}\rho^{s+m} - Z\alpha\rho^{s+m-1}\right] = 0\,,$$

and

$$\sum_{m=0}^{\infty} b_m e^{-\rho}\left[(s+m)\rho^{s+m-1} - \rho^{s+m} + \kappa\rho^{s+m-1}\right] - \sum_{m=0}^{\infty} a_m e^{-\rho}\left[\sqrt{\frac{1+\epsilon}{1-\epsilon}}\rho^{s+m} + Z\alpha\rho^{s+m-1}\right] = 0\,.$$

Equating, in both equations, the coefficients of $e^{-\rho}\rho^{s+q-1}$ from both sides, we have

$$a_q(s+q-\kappa) - a_{q-1} - \left(b_{q-1}\sqrt{\frac{1-\epsilon}{1+\epsilon}} - b_q Z\alpha\right) = 0 \tag{12.10.42}$$

$$b_q(s+q+\kappa) - b_{q-1} - \left(a_{q-1}\sqrt{\frac{1-\epsilon}{1+\epsilon}} + a_q Z\alpha\right) = 0 \tag{12.10.43}$$

The normalization of the four-component wave function also requires that $\int \psi_e^\dagger \psi_e d\tau$ must be finite. This means the integrals $\int e^{-2\rho}\rho^{2s+2}F^2(\rho)d\rho$ and $\int e^{-2\rho}\rho^{2s+2}G^2(\rho)d\rho$ must be finite. Now the functions F and G diverge as $e^{2\rho}$ unless the power series in both functions terminates at some finite power of ρ. Assuming that the two series terminate with the same index n',

$$a_{n'+1} = 0 = b_{n'+1}\,. \tag{12.10.44}$$

Setting $q = n' + 1$ in Eqs. (12.10.42) and (12.10.43), we find that both equations give

$$a_{n'} = -\sqrt{\frac{1-\epsilon}{1+\epsilon}}\, b_{n'}\,. \tag{12.10.45}$$

This justifies our presumption that the two series for the functions F and G terminate with the same power index.

Using this ratio in Eqs. (12.10.42) and (12.10.43), setting $q = n'$ and eliminating $a_{n'-1}$ or a from the resulting equations we get

$$\left[\frac{2\epsilon Z\alpha}{\sqrt{1-\epsilon^2}} - (n'+s)\right] b_{n'} = 0$$

or
$$\frac{2\epsilon}{\sqrt{1-\epsilon^2}} = \frac{n'+s}{Z\alpha} = \frac{n'+\sqrt{\kappa^2-(Z\alpha)^2}}{Z\alpha}\,, \qquad n' = 0,1,2,3\cdots \tag{12.10.46}$$

Solving for $\epsilon_{n'}\, mc^2 \equiv E_{n'}$ we find the relativistic bound state energy is given by

$$E_{n'} = \frac{mc^2}{\sqrt{1+\frac{Z^2\alpha^2}{(s+n')^2}}} = \frac{mc^2}{\sqrt{1+\frac{Z^2\alpha^2}{\left\{n'+\sqrt{(j+1/2)^2-(Z\alpha)^2}\right\}^2}}}\,. \tag{12.10.47}$$

This formula was first derived by Sommerfeld, using the relativistic version of Bohr's old quantum theory. The principal quantum number n of the non-relativistic theory is given by

$$n = j + 1/2 + n' = |\kappa| + n'\,. \tag{12.10.48}$$

Since the minimum value of n' is zero, it follows that $n \geq |\kappa|$. This can be seen explicitly by expanding the expression of energy $E_{n'}$ in powers of $Z^2\alpha^2$

$$\begin{aligned}
E_{n'} &= mc^2\left[1 - \frac{1}{2}\frac{(Z\alpha)^2}{[\,n'+\sqrt{(j+1/2)^2-(Z\alpha)^2}\,]^2} + \frac{3}{8}\frac{(Z\alpha)^4}{[\,n'+\sqrt{(j+1/2)^2-(Z\alpha)^2}\,]^4} - \cdots\right] \\
&= mc^2\left[1 - \frac{1}{2}\frac{(Z\alpha)^2}{(n'+j+1/2)^2}\left(1+\frac{Z\alpha^2}{(j+1/2)(n'+j+\frac{1}{2})}+\cdots\right)\right. \\
&\qquad\qquad \left. + \frac{3}{8}\frac{(Z\alpha)^4}{(n'+j+1/2)^4} - \cdots\right] \\
&= mc^2\left[1 - \frac{(Z\alpha)^2}{2n^2} - \frac{1}{2}\frac{(Z\alpha)^4}{n^3}\left(\frac{1}{j+1/2}-\frac{3}{4n}\right)+\cdots\right].
\end{aligned}$$

The first term mc^2 is the rest energy of the electron. The second term

$$-\frac{mc^2Z^2\alpha^2}{2n^2} = -\frac{mZ^2e^4}{2(4\pi\epsilon_0\hbar)^2n^2} = -\frac{Ze^2}{8\pi\epsilon_0 a_o n^2}\,,$$

where a_o is the Bohr radius, may be identified as the energy of the non-relativistic Hydrogen atom [Chapter 5]. Hence n has the significance of the principal quantum number. We also note that an important correction to Bohr's formula, the *fine structure* splitting discussed in Chapter 8 using perturbative means in the context of non-relativistic qunatum mechanics, is already contained in the relativistic expression for the bound state energy.

Although ℓ is not a good quantum number in Dirac theory, the quantum state of Hydrogen atom, characterized by quantum numbers j, m_j and κ may still be designated by specifying the orbital angular momentum quantum number ℓ_A of the upper two-component wave function Φ, which becomes the wave function of the Schrödinger -Pauli theory in the non-relativistic limit. Thus, for example,

$$\begin{array}{llll}
2p_{3/2}\,:\ n=2;\ j=3/2; & \ell_A=1=j-1/2; & \kappa=-(j+1/2)=-2; & n'=n-|\kappa|=0\\
2s_{1/2}\,:\ n=2;\ j=1/2; & \ell_A=0=j-1/2; & \kappa=-(j+1/2)=-1; & n'=n-|\kappa|=1\\
2p_{1/2}\,:\ n=2;\ j=1/2; & \ell_A=1=j+1/2; & \kappa=+(j+1/2)=+1; & n'=n-|\kappa|=1\\
3p_{1/2}\,:\ n=3;\ j=1/2; & \ell_A=1=j+1/2; & \kappa=+(j+1/2)=+1; & n'=n-|\kappa|=2\\
3p_{3/2}\,:\ n=3;\ j=3/2; & \ell_A=1=j-1/2; & \kappa=-(j+1/2)=-2; & n'=n-|\kappa|=1\\
3s_{1/2}\,:\ n=3;\ j=1/2; & \ell_A=0=j-1/2; & \kappa=-(j+1/2)=-1; & n'=n-|\kappa|=2\\
3d_{3/2}\,:\ n=3;\ j=3/2; & \ell_A=2=j+1/2; & \kappa=+(j+1/2)=+2; & n'=n-|\kappa|=1\\
3d_{5/2}\,:\ n=3;\ j=5/2; & \ell_A=2=j-1/2; & \kappa=-(j+1/2)=-3; & n'=n-|\kappa|=0\,.
\end{array}$$

Note that when $\ell_A = j - 1/2$, quantum number κ is negative and when $\ell_A = j + 1/2$, it is positive.

It may be pointed out here that according to Dirac theory, the states $2s_{1/2}$ and $2p_{1/2}$ are degenerate, as are the pair of states $3s_{1/2}, 3p_{1/2})$ and $(3p_{3/2}, 3d_{3/2}\cdots$. Experimentally, however, this degeneracy is lifted. Lamb and Retherford (1947) first observed the energy shift, now known as the *Lamb-Retherford shift* between the states $2s_{1/2}$ and $2p_{1/2}$ of the Hydrogen atom. The Lamb-Retherford shift, which is 1058 MHz for the $2s_{1/2} - 2p_{1/2}$ levels of Hydrogen, can be explained by considering the interaction of the electron with the vacuum of the quantized electromagnetic field [Chapter 13]. Further corrections to the energy of a Hydrogen atom arise due to the interaction between the magnetic moment of the electron and the nucleus (giving rise to hyperfine splitting) and finite size of the nucleus. A discussion of these and other corrections to the energy of a Hydrogen atom can be found in books on atomic physics.

To find the radial functions, we must solve Eqs. (12.10.42) and (12.10.43). The ground state wave function is easy to compute. In this case

$$n'=0,\quad \kappa=-1,\quad s=\sqrt{1-(Z\alpha)^2}=\epsilon\,,\quad E_0=mc^2\sqrt{1-(Z\alpha)^2}$$

and $$\rho=\sqrt{1-\epsilon^2}\,\frac{mc}{\hbar}\,r=\frac{Zr}{a_o},\quad \frac{a_0}{b_0}=\sqrt{\frac{1-\epsilon}{1+\epsilon}}=-\frac{(1-\sqrt{1-(Z\alpha)^2}\,)}{Z\alpha}\,.$$

Then from Eqs. (12.10.20) through (12.10.23), we find the ground state wave function is

$$\psi = N\left(\frac{Zr}{a_o}\right)^{\epsilon-1} e^{-Zr/a_o}\begin{pmatrix}\mathcal{Y}^{m_j}_{\frac{1}{2},0}\\ -\frac{Z\alpha}{1+\epsilon}\mathcal{Y}^{m_j}_{\frac{1}{2},1}\end{pmatrix}, \tag{12.10.49}$$

where the normalization constant is given by

$$N=\left(\frac{Z}{a_o}\right)^{3/2} 2^{\epsilon}\sqrt{\frac{1+\epsilon}{\Gamma(1+2\epsilon)}}\,. \tag{12.10.50}$$

Note that this wave function displays a mild singularity $r^{\nu-1}\approx r^{-(Z\alpha)^2/2}$. For real nuclei, which have a finite size, this singularity does not occur. For $Z\alpha > 1$, the index ν becomes imaginary and the solutions become oscillatory. However all stable nuclei have $Z\alpha < 1$. Nuclei with $Z\alpha > 1$ can, in principle, be formed in heavy-ion collisions where interesting new physics emerges. This, however, cannot be described in terms of single particle wave functions.

12.11 Charge Conjugation, Parity and Time Reversal Invariance

As noted in Sec. 12.8, the Dirac equation (12.8.1) and the differential equation (12.8.27) for four-vector potential are invariant under Lorentz transformations, pure spatial rotation and spatial inversion. We found that a linear transformation $S_P = \gamma_4$ of Dirac spinor ensures the invariance of the Dirac equation under space reflection. However, it was not possible to find a linear transformation compatible with the invariance under time reversal. In this section we revisit the invariance of Eqs. (12.8.1) and (12.8.28) under discrete transformations of (a) charge conjugation (b) parity (space inversion) and (c) time reversal.

Charge Conjugation

The operation C of *charge conjugation* can, in principle, be defined as one whose application on the electron (positron) state of momentum $\boldsymbol{p}$, energy E and spin expectation value $\frac{\hbar}{2}\langle\Sigma_3\rangle$, results in a positron (electron) state of the same momentum, energy and spin expectation value. In the presence of an electromagnetic field this means that for each electron state in a potential A_μ there corresponds a positron state (charge conjugate state) in the potential $-A_\mu$. Such an operation, however, cannot be defined consistently within the framework of one-electron relativistic theory. In fact a satisfactory formulation of the charge conjugation operation requires a quantized Dirac field [Chapter 14]. However, we can explore the connection between electron wave function ψ and the charge conjugate wave function characterizing the space-time behavior of the negative energy electron state whose absence appears as the positron state within the framework of single-electron Dirac theory.

Consider a spin-half Dirac particle (electric charge e) in an electromagnetic potential $A_\mu(x)$. The Dirac equation in this case is

$$\left(\frac{\partial}{\partial x_\mu} - \frac{ie}{\hbar}A_\mu(x)\right)\gamma_\mu\psi(x) + \frac{mc}{\hbar}\psi(x) = 0\,. \tag{12.11.1}$$

We define charge conjugate wave function via the transformation

$$C\psi(x) \equiv \psi_C(x) = S_C\psi^*(x)\,, \tag{12.11.2a}$$

$$A^c_\mu(x) = -A_\mu(x)\,, \tag{12.11.2b}$$

where $\psi_C(x)$ represents the charge conjugate state ψ_C and A^c_μ represents charge conjugate four-vector potential. The last relation follows since A^c_μ is supposed to be the four-vector potential generated by charges and currents with sign of e changed. The operation of charge conjugation does not affect space-time coordinates ($x' = x$). Hence we will use the unprimed coordinates even for the transformed quantities. Also, unlike Sec. 12.8, we denote the transformed quantities (wave function, four-vector potential, etc.) with a index denoting the type of transformation under consideration instead of a prime.

Invariance of Dirac equation under charge conjugation means that ψ_C also satisfies the Dirac equation (12.8.1) with A_μ replaced by charge conjugate four-vector potential ($A^c_\mu = -A_\mu$)

$$\left(\frac{\partial}{\partial x_\mu} - \frac{ie}{\hbar}A^c_\mu\right)\gamma_\mu\psi_C + \frac{mc}{\hbar}\psi_C = 0\,. \tag{12.11.3}$$

We are after the relation between the positive energy solution of the Dirac equation $\psi(x)$ and the charge conjugate wave function $\psi_C(x)$.

Taking the complex conjugate of Eq. (12.11.1) we get

$$\left(\frac{\partial}{\partial x_k} - \frac{ie}{\hbar}A_k^c(x)\right)\gamma_k^*\psi^*(x) + \left(\frac{\partial}{\partial x_4} - \frac{ie}{\hbar}A_4^c\right)(-\gamma_4^*)\psi^*(x) + \frac{mc}{\hbar}\psi^*(x) = 0\,, \qquad (12.11.4)$$

where we have used $A_k^* = A_k$, $A_4^* = -A_4$, $x_k^* = x_k$, $x_4^* = -x_4$ followed by Eq. (12.11.2b). Using the relation (12.11.2a) in Eq. (12.11.4) and multiplying from the left by S_C we obtain

$$\left[\left(\frac{\partial}{\partial x_k} + \frac{ie}{\hbar}A_k(x)\right)S_C\gamma_k^*S_C^{-1} + \left(\frac{\partial}{\partial x_4} - \frac{ie}{\hbar}A_4^c(x)\right)S_C(-\gamma_4^*)S_C^{-1} + \frac{mc}{\hbar}\right]\psi_C(x) = 0\,. \qquad (12.11.5)$$

A comparison of Eqs. (12.11.5) with (12.11.3) shows that the charge conjugate wave function ψ_C satisfies the Dirac equation provided that

$$S_C\gamma_k^*S_C^{-1} = \gamma_k\,, \quad k = 1, 2, 3, \quad S_C\gamma_4^*S_C^{-1} = -\gamma_4\,. \qquad (12.11.6)$$

Now γ_1 and γ_3 are purely imaginary matrices $[\gamma_{1,3}^* = -\gamma_{1,3}]$ and γ_2 and γ_4 are real matrices $[\gamma_{2,4}^* = \gamma_{2,4}]$. Therefore, Eq. (12.11.6) requires $S_C^{-1}\gamma_1 S_C = -\gamma_1$, $S_C^{-1}\gamma_2 S_C = \gamma_2$, $S_C^{-1}\gamma_3 S_C = -\gamma_3$ and $S_C^{-1}\gamma^4 S_C = -\gamma_4$. Using the anti-commutation property (12.2.24) of γ-matrices we find that (12.11.6) is satisfied with the choice

$$S_C = \gamma_2 \text{ and } S_C^{-1} = \gamma_2. \qquad (12.11.7)$$

Note that the relation between ψ_C and ψ may be written as $\psi_C = S_C K\psi = \gamma_2 K\psi$, where K is the operation of complex conjugation indicating that charge conjugation is not a linear transformation of the wave function. Transformation S_C is an example of an antilinear transformation. [10]

It may be tempting to look upon ψ_C as representing the state of a positron in a potential A_μ. But $\psi_C = \gamma_2\psi^*$ is the solution of the Dirac equation with the sign of A_μ reversed and characterizes the space-time behavior of a negative energy electron whose absence appears as the charge conjugate (positron) state. If we compute the expectation values of various dynamical variables using the wave functions ψ_C and ψ, then we obtain the results that energy, momentum, and spins have opposite signs: $E_C = -E$, $\langle \mathbf{p}\rangle_C = -\langle \mathbf{p}\rangle$, $\langle\Sigma_3\rangle_C = -\langle\Sigma_3\rangle$. On the other hand the total charge

$$Q = e\int \tilde{\psi}\gamma_4\psi\, d\tau = e\int \psi^\dagger\psi\, d\tau$$

does not change sign when ψ is replaced by ψ_C. All this does not conform to the apparent definition of charge conjugation operation, according to which the momentum and spin direction should be unchanged while the particle should be changed to anti-particle. This only means that if ψ represents the wave function of a positive energy electron in a field A_μ, the charge conjugate wave function ψ_C represents the state of negative energy electron in a field $-A_\mu$, whose absence manifests itself as a positron (with sign of charge changed) with the same momentum and spin expectation value as the positive energy electron in a field A_μ.

---

[10] It can be checked that $j_\mu^C = j_\mu$ while $A_\mu^C = -A_\mu$. Thus the invariance of electromagnetic equation (12.8.28) under charge conjugation cannot be guaranteed within the framework of one electron theory and full charge conjugation operation cannot be consistently implemented without quantizing the electron field.

Parity (Space-Inversion) Operation

The space-inversion transformation $[x \to x' \equiv (x'_k, x_4) = (-x_k, x_4)]$ has already been defined in Sec. 12.8. In the notation of this section we have

$$\psi_P(x) = S_P\psi(x) \quad \text{or} \quad \psi_P(\boldsymbol{r}, t) = S_P\psi(\boldsymbol{r}, t)\,, \tag{12.11.8a}$$

and

$$A_k^P = -A_k\,, \quad k = 1, 2, 3\,; \quad A_4^P = A_4\,, \tag{12.11.8b}$$

where ψ_P is the space-inverted wave function and the spinor transformation matrix S_P is given by

$$S_P = \gamma_4\,. \tag{12.11.9}$$

With this identification the space-inverted wave function

$$\psi_P(x) \equiv \psi_P(\boldsymbol{r}, t) = \gamma_4\psi(\boldsymbol{r}, t) \tag{12.11.10}$$

also satisfies the Dirac equation in terms of time-reversed coordinates and potential.

Time Reversal Operation

Under the time reversal transformation $[x' \equiv (x'_k, x'_4) = (x_k, -x_4)]$, we define the wave function transformation according to

$$\psi_T(x) = S_T\psi^*(x')\,, \tag{12.11.11a}$$

$$\overset{T}{A_k}(x) = -A_k(x')\,, \quad \overset{T}{A_4}(x) = A_4(x') \tag{12.11.11b}$$

where $\psi_T(x)$ is the *time-reversed* wave function and 4×4 spinor transformation matrix S_T is still to be determined. The transformation law for the electromagnetic potential follows from the fact that the time reversed (ordinary) vector potential is generated by current densities, the direction of which is reversed with respect the original current densities. The reason for defining time-reversed wave function in terms of complex conjugate wave function is suggested by wave mechanics where the time reversal transformation is achieved by complex conjugation. To find S_T, we use the time-reversed coordinates $x'_k = x_k$, $x'_4 = -x_4$ and the time-reversed four-vector potential $\overset{T}{A_\mu}(x) \equiv A_\mu(x')$ [Eq. (12.11.11b)] in the Dirac equation (12.8.1) and take the complex conjugate of the resulting equation to obtain

$$\left[\gamma_k^*\left(\frac{\partial}{\partial x_k} - \frac{ie}{\hbar}\overset{T}{A_k}(x)\right) + \gamma_4^*\left(\frac{\partial}{\partial x_4} - \frac{ie}{\hbar}\overset{T}{A_4}(x)\right) + \frac{mc}{\hbar}\right]\psi^*(x') = 0\,, \tag{12.11.12}$$

where we have used $\overset{T}{A_k}{}^* = \overset{T}{A_k}\ \overset{T}{A_4}{}^* = -\overset{T}{A_4}$. Substituting $\psi^*(x') = S_T^{-1}\psi_T(x)$ in Eq. (12.11.12) and multiplying from the left by S_T we find

$$\left[S_T\gamma_k^*S_T^{-1}\left(\frac{\partial}{\partial x_k} - \frac{ie}{\hbar}\overset{T}{A_k}(x)\right) + S_T\gamma_4^*S_T^{-1}\left(\frac{\partial}{\partial x_4} - \frac{ie}{\hbar}\overset{T}{A_4}(x)\right) + \frac{mc}{\hbar}\right]\psi_T(x) = 0\,. \tag{12.11.13}$$

The equation satisfied by $\psi_T(x)$ would have exactly the same form as the Dirac equation provided that we can find S_T such that

$$S_T\gamma_\mu^*S_T^{-1} = \gamma_\mu\,. \tag{12.11.14}$$

Recalling that γ_1, γ_3 are imaginary and γ_2, γ_4 are real matrices, this condition implies $S_T\gamma_1S_T^{-1} = -\gamma_1$, $S_T\gamma_2S_T^{-1} = \gamma_2$, $S_T\gamma_3S_T^{-1} = -\gamma_3$, and $S_T\gamma_4S_T^{-1} = \gamma_4$. An inspection of these equations together with the anti-commutation relations for γ-matrices shows that a matrix satisfying the condition (12.11.15) is

$$S_{_T} = \gamma_1\gamma_3\,. \tag{12.11.15}$$

Thus the time-reversed wave function is given by

$$\psi_{_T}(x) = \gamma_1\gamma_3\psi^*(x') \tag{12.11.16}$$

and it also satisfies the Dirac equation. We can check that the transformation $S_{_T}$ defined this way also ensures the invariance of Eq. (12.8.28) for the four-vector potential. To see this, we take the complex conjugate of Eq. (12.8.28) and use the coordinate transformation under time reversal $x_\mu \to x'_\mu \equiv (\boldsymbol{r}, -ict)$ and Eq. (12.11.11b) to get

$$-\frac{\partial}{\partial x'_\mu}\frac{\partial}{\partial x'_\mu}A^{^T}_\nu(x) = -\mu_o j^*_\nu(x) \tag{12.11.17a}$$

and since the d'Alembertian is an invariant operator, we have

$$-\frac{\partial}{\partial x_\mu}\frac{\partial}{\partial x_\mu}A^{^T}_\nu(x) = -\mu_o j^*_\nu(x)\,. \tag{12.11.17b}$$

Now the right-hand side is

$$\begin{aligned} j^*_\nu(x) &= [iec\tilde{\psi}(x)\gamma_\nu\psi(x)]^* = -iec[\psi^\dagger(x)\gamma_4]^*\gamma^*_\mu\psi^*(x') \\ &= -iec[\psi^*(x')]^\dagger\gamma_4\gamma^*_\nu\psi^*(x') \\ &= -iec\,[S^{-1}_{_T}\psi_{_T}(x)]^\dagger\gamma_4\gamma^*_\nu S^{-1}_{_T}\psi_{_T}(x) \\ &= -iec\,\tilde{\psi}_{_T}(x)\gamma_4(\gamma_3\gamma_1)^\dagger\gamma_4\gamma^*_\nu(\gamma_3\gamma_1)\psi_{_T}(x) = -j^{^T}_\nu(x)\,. \end{aligned}$$

Substituting this in Eq. (12.11.21) we find that the differential equation for four-vector potential is invariant also. By writing Eq. (12.11.11a) as $\psi_{_T} \equiv S_{_T}\psi^* = S_{_T}K\psi$, where K is the operation of complex conjugation, we can see that time reversal, like charge conjugation, is not a linear transformation.

In summary, we can say, that the Dirac equation is invariant under the operations of charge conjugation (C), parity (P) and time reversal (T). It must also be mentioned that the conditions like Eqs. (12.11.6) and (12.11.14) determine S within a phase factor and the exact form of S depends on the particular representation chosen for γ-matrices.

Unlike the Dirac equation, Weyl's equation for a massless particle

$$i\hbar\frac{\partial}{\partial t}\psi(\boldsymbol{r},t) = \mp i\hbar c\Sigma_k\frac{\partial}{\partial x_k}\,\psi(\boldsymbol{r},t)\,, \tag{12.11.18}$$

which may also be written as (12.5.10) with the substitution $\psi = u\,\exp[\frac{i}{\hbar}(\boldsymbol{p}\cdot\boldsymbol{r} - Et)]$, or as the chirality equation

$$\gamma_5\psi(\boldsymbol{r},t) = \mp\psi(\boldsymbol{r},t)\,, \tag{12.11.19}$$

where the upper sign is for a negative chirality (positive helicity) state and the lower sign is for positive chirality (negative helicity) state, is neither invariant under charge conjugation $\psi(\boldsymbol{r},t) \to \psi_{_C}(\boldsymbol{r},t) \equiv \gamma_2\psi^*(\boldsymbol{r},t)$ nor under space inversion $\psi(\boldsymbol{r},t) \to \psi_{_P}(-\boldsymbol{r},t) \equiv \gamma_4\psi(\boldsymbol{r},t)$. This can be checked by observing that, according to the chirality equation (12.11.19),

$$\int \psi^\dagger(\boldsymbol{r},t)\gamma_5\psi(\boldsymbol{r},t)d\tau = \mp 1\,. \tag{12.11.20}$$

Now the overlap of $\gamma_5\psi_C$ with $\psi_C^\dagger$ or of $\gamma_5\psi_P$ with $\psi_P^\dagger$, has a sign opposite to the overlap of $\gamma_5\psi$ with $\psi^\dagger$, while the overlap of $\gamma_5\psi_T$ with $\psi_T^\dagger$ has the same sign, i.e.,

$$\int \psi_C^\dagger \gamma_5 \psi_C \, d\tau \equiv \langle\gamma_5\rangle_C = -\int \psi^\dagger \gamma_5 \psi \, d\tau \equiv -\langle\gamma_5\rangle \,,$$

$$\int \psi_P^\dagger \gamma_5 \psi_P \, d\tau \equiv \langle\gamma_5\rangle_P = -\langle\gamma_5\rangle \,,$$

and

$$\int {\psi_T}^\dagger \gamma_5 \psi_T \, d\tau \equiv \langle\gamma_5\rangle_T = +\langle\gamma_5\rangle \,.$$

Hence the Weyl equation (12.11.18) or (12.11.19) is violated for $\psi \to \psi_C$ or $\psi \to \psi_P$ but not for $\psi \to \psi_T$. However, note that the Weyl equation is invariant under a combined CP operation.

Combined CPT Operation

Under a combined CPT operation, the wave function transforms as

$$\begin{aligned}\psi_{CPT}(\boldsymbol{r},t) \equiv CPT\;\psi(\boldsymbol{r},t) &= CP(\gamma_1\gamma_3)\psi^*(\boldsymbol{r},t)\\ &= C\gamma_4\,\gamma_1\gamma_3\psi^*(\boldsymbol{r},t) = \gamma_2\gamma_4\,\gamma_1\gamma_3\psi(\boldsymbol{r},t)\end{aligned}$$

or

$$\psi_{CPT}(\boldsymbol{r},t) = -\gamma_5\psi(\boldsymbol{r},t)\,. \tag{12.11.21}$$

$\psi_{CPT}(\boldsymbol{r},t)$ also satisfies the Dirac equation. A different order of operators, say PCT, simply results in a change of sign, which is of no consequence. The Weyl equation is also invariant under a combined CPT operation.

Problems

1. Show that the equation of continuity in the covariant form [Eq. (12.2.23)] may also be obtained from Eq. (12.2.17) by writing $\psi^\dagger = \tilde{\psi}\gamma_4$.

2. Derive the equation of continuity if $\psi(\boldsymbol{r},t)$ satisfies the Klein-Gordon equation for a particle of charge e and mass m in an electromagnetic field $(\boldsymbol{A},\phi)$,

$$\left(i\hbar\frac{\partial}{\partial t} - e\phi\right)^2 \psi = \left[c^2(-i\hbar\nabla - eA)^2 + m^2c^4\right]\psi\,.$$

Identify the expressions for the probability density and probability current density.

[Ans: $S = \frac{i\hbar}{2mc^2}\left(\psi^*\frac{\partial\psi}{\partial t} - \psi\frac{\partial\psi^*}{\partial t}\right)$ and $\boldsymbol{j} = \frac{\hbar}{2im}\left[\psi^*(\boldsymbol{\nabla}\psi) - (\boldsymbol{\nabla}\psi^*)\psi\right] - \frac{e\boldsymbol{A}}{m}(\psi^*\psi)]$

3. Find the relativistic bound state energies for a spin zero particle of charge $-e$ moving in a Coulomb potential [such as a pion (π^-) in the field of a nucleus]

$$V(r) = -\frac{Ze^2}{4\pi\epsilon_o r}\,.$$

Compare the result with that for the Hydrogen atom. What is the non-relativisitic limit of this expression? Hint: A spin zero particle of charge q in a time-independent electromagnetic field is described by the time-independent Klein-Gordon equation $(E - q\phi)^2 = c^2(\boldsymbol{p} - q\boldsymbol{A})^2 + m^2c^4$.

4. If $\boldsymbol{P}$ and $\boldsymbol{Q}$ are two three-dimensional vectors, show that

$$(\boldsymbol{\alpha} \cdot \boldsymbol{P})(\boldsymbol{\alpha} \cdot \boldsymbol{Q}) = \boldsymbol{P} \cdot \boldsymbol{Q} I - i\boldsymbol{\Sigma} \cdot (\boldsymbol{P} \times \boldsymbol{Q}),$$

where $\boldsymbol{\alpha} = \begin{pmatrix} 0 & \boldsymbol{\sigma} \\ \boldsymbol{\sigma} & 0 \end{pmatrix}$, $\boldsymbol{\Sigma} = \begin{pmatrix} \boldsymbol{\sigma} & 0 \\ 0 & \boldsymbol{\sigma} \end{pmatrix}$, $\boldsymbol{\sigma}$ being the Pauli spin vector and I a unit 4×4 matrix.

5. Calculate the energy values and wave function of a Dirac particle of charge e and mass m moving in a homogeneous magnetic field of infinite extent. Hint: Take the field to be in the z-direction so that $A_1 = -\frac{1}{2}By$, $A_2 = \frac{1}{2}Bx$, $A_3 = 0$ and use the second order Dirac equation giving stationary states of energy $\pm E$

$$\left(\frac{E^2}{\hbar^2 c^2} - \frac{m^2 c^2}{\hbar^2}\right)\psi + \left(\frac{\partial}{\partial x} + \frac{1}{2}\frac{ieB}{\hbar}y\right)^2 \psi$$
$$+ \left(\frac{\partial}{\partial y} - \frac{1}{2}\frac{ieB}{\hbar}x\right)^2 \psi + \frac{\partial^2}{\partial z^2}\psi + \frac{eB}{\hbar}\Sigma_z \psi = 0\,.$$

6. An electron at time $t = 0$ has the wave function

$$\psi(\boldsymbol{r}, 0) = \begin{pmatrix} a \\ b \\ c \\ d \end{pmatrix} e^{ikz}\,, \tag{12.11.22}$$

where a, b, c, d are independent of $\boldsymbol{r}$ and t. Find the probability for measuring (i) $E > 0$, spin up, (ii) $E > 0$, spin down, (iii) $E < 0$, spin up, and (iv) $E < 0$, spin down.

7. A Dirac particle of mass m and charge e is constrained to move in the z direction in the presence of four-potential $(\boldsymbol{A}, \phi)$ with $\boldsymbol{A} = 0$ and

$$\phi = \begin{cases} V_o/e & \text{for } z > 0 \\ 0 & \text{for } z < 0\,. \end{cases}$$

A positive energy particle with energy $E > 0$ is incident from the left. Write down the solutions to the left and right of the potential step at $z = 0$ assuming that particle spin is not affected.

(i) Show that the reflection coefficient is given by

$$R = \left| \frac{1 - \frac{\kappa}{k}\frac{E + mc^2}{E - V_o + mc^2}}{1 + \frac{\kappa}{k}\frac{E + mc^2}{E - V_o + mc^2}} \right|^2,$$

where $k^2 = [E^2 - m^2c^4]/(\hbar c)^2$, $\kappa^2 = [(E - V_o)^2 - m^2c^4]/(\hbar c)^2$. What is the transmission coefficient?

(ii) What do you expect the reflection coefficient to be as barrier height increases? Discuss the behavior of R as barrier height increases by considering the following cases (i)$V_o < E - mc^2$, (ii) $V_o = E - mc^2$, (iii) $E + mc^2 > V_o > E - mc^2$, (iv) $V_o > E + mc^2$ and comment on your result.

8. Solve the Dirac equation for an attractive square well potential of depth V_o/e and radius $r = r_o$. Determine the minimum depth for a given r_o that will bind a particle of mass m.

9. Prove that $(\boldsymbol{\alpha} \cdot \boldsymbol{\varepsilon})\,(\boldsymbol{\alpha} \cdot \boldsymbol{\varepsilon}) = I$, where $\boldsymbol{\varepsilon}$ is a three-dimensional unit vector and I is a 4×4 unit matrix.

10. Show that $(\boldsymbol{\alpha} \cdot \boldsymbol{\varepsilon})\,(\boldsymbol{\alpha} \cdot \boldsymbol{\varepsilon}') + (\boldsymbol{\alpha} \cdot \boldsymbol{\varepsilon}')\,(\boldsymbol{\alpha} \cdot \boldsymbol{\varepsilon}) = 2\,(\boldsymbol{\varepsilon} \cdot \boldsymbol{\varepsilon}')\,I$, where $\boldsymbol{\varepsilon}$ and $\boldsymbol{\varepsilon}'$ are three-dimensional unit vectors and I is a 4×4 unit matrix.

11. Show that if A_μ and B_μ are four-vectors and a is a scalar, then

    (a) $\not{A}\,\gamma_5 = -\gamma_5\,\not{A}$

    (b) $\gamma_\mu\, a\, \gamma_\mu = 4aI$

    (c) $\gamma_\mu\, \not{A}\, \gamma_\mu = -2\not{A}$

    (d) $\gamma_\mu\, \not{A}\,\not{B}\, \gamma_\mu = 4\,A_\lambda B_\lambda I \equiv 4(AB)$

    (e) $\gamma_\mu\, \not{A}\,\not{B}\not{C}\, \gamma_\mu = -\,2\not{C}\,\not{B}\,\not{A}$

    (f) $\not{A}\not{B} = -\not{B}\not{A} + 2(AB)$

    where $\not{A} \equiv \gamma_\rho\, A_\rho = \gamma_1\, A_1 + \gamma_2\, A_2 + \gamma_3\, A_3 + \gamma_4\, A_4$ and $(AB) \equiv A_\lambda B_\lambda I$.

12. Show that Tr $(\boldsymbol{\alpha} \cdot \boldsymbol{k})\,\big(\boldsymbol{\alpha} \cdot \boldsymbol{k}'\big) = \big(\boldsymbol{k} \cdot \boldsymbol{k}'\big)$ Tr $I = 4\,\big(\boldsymbol{k} \cdot \boldsymbol{k}'\big)$ where $\boldsymbol{k}$ and $\boldsymbol{k}'$ are three-dimensional vectors. Also show that Tr $(\gamma_\mu\gamma_\lambda) = 4\delta_{\mu\nu}$.

13. Show that Tr $(\gamma_\mu\gamma_\nu\cdots\gamma_\rho) = 0$ for a matrix product involving an odd number of γ-matrices.

14. Show that

$$\sum_{\mu=1}^{4} u_\mu^\dagger M u_\mu = \text{Tr } M$$

    where M is a 4×4 matrix and u_μ's are Dirac spinors.

15. Show that Tr $(\gamma_\lambda\gamma_\mu\gamma_\rho\gamma_\upsilon) = 4\,(\delta_{\lambda\mu}\delta_{\rho\upsilon} - \delta_{\lambda\rho}\delta_{\mu\upsilon} + \delta_{\lambda\upsilon}\delta_{\mu\rho})$.

16. If p_μ and k_μ are four-vectors defined by $p_\mu \equiv \big(\boldsymbol{p}, \frac{iE}{c}\big)$, $k_\mu \equiv \big(\boldsymbol{k}\,, \frac{i\omega}{c}\big)$ where $\omega = c\,|\boldsymbol{k}|$, and $\gamma_\mu \equiv (\gamma_1, \gamma_2, \gamma_3, \gamma_4)$ so that $\not{p} = p_\mu\gamma_\mu$ and $\not{k} = k_\mu\gamma_\mu$, show that

    (a) $-i\,\big[i(\not{p} + \hbar\not{k}) - mc\big]\;\big[(\not{p} + \hbar\not{k}) - imc\big] = (p + \hbar k)^2 + m^2c^2$

    (b) $(p + \hbar k)^2 + m^2c^2 = -2m\hbar\omega$

    (c) $p_\mu k_\mu = -m\omega$

17. If A, B, C, D are four-vectors and $\not{A} = \gamma_\mu A_\mu$, $\not{B} = \gamma_\mu B_\mu$, etc., show that

    (a) Tr $\not{A} = 0 =$ Tr $\not{A}\not{B}\not{C})$

    (b) Tr $(\not{A}\not{B}) = 4A_\mu B_\mu$

    (c) Tr $(\not{A}\not{B}\not{C}\not{D}) = 4[(A_\mu B_\mu)(C_\nu D_\nu) - (A_\mu C_\mu)(B_\nu D_\nu) + (A_\mu D_\mu)(B_\nu C_\nu)]$.

18. Show that

    (a) $\gamma_5 = \gamma_1\gamma_2\gamma_3\gamma_4 = -\begin{pmatrix} 0 & I \\ I & 0 \end{pmatrix}$ and

    (b) $\gamma_5\Sigma_k = \Sigma_k\gamma_5 = -\alpha_k$

where $\Sigma_k \equiv \begin{pmatrix} \sigma_k & 0 \\ 0 & \sigma_k \end{pmatrix}$ and $\alpha_k \equiv \begin{pmatrix} 0 & \sigma_k \\ \sigma_k & 0 \end{pmatrix}$.

19. Show that $\hat{\Gamma}_{\pm} = \frac{1}{2}\left(1 \pm \frac{\hat{H}}{E_p}\right)$, where $E_p = \sqrt{p^2c^2 + m^2c^4}$ and $\hat{H}$ is the Dirac Hamiltonian, are the projection operators, respectively, for positive (upper sign)/negative (lower sign) energy states.

20. Show that if $\hat{H}$ is the Dirac Hamiltonian for a central potential, then

 (a) $[H, \beta] = -2c\beta(\boldsymbol{\alpha} \cdot \boldsymbol{p})$
 (b) $[H, \Sigma_\ell] = 2ic\epsilon_{k\ell m}p_k\alpha_m$ where $\epsilon_{k\ell m}$ is Levi-cita antisynmetric tensor,
 (c) $\beta\,[H, \Sigma_\ell]\; J_\ell = 2ic\beta(\boldsymbol{p} \times \boldsymbol{J}) \cdot \boldsymbol{\alpha}$ and
 (d) $(\boldsymbol{\alpha} \cdot \boldsymbol{p})(\boldsymbol{\Sigma} \cdot \boldsymbol{J}) = -\gamma_5(\boldsymbol{p} \cdot \boldsymbol{J}) + i\boldsymbol{\alpha} \cdot (\boldsymbol{p} \times \boldsymbol{J})$.

21. Show that the operator K defned by $K = \beta(\boldsymbol{\Sigma} \cdot \boldsymbol{L} + \hbar I)$ commutes with the Dirac Hamiltonian H as well as with the operator J_k. Show also that if the eigenvalues of K^2 are $\hbar^2\kappa^2$ then κ can take positive or negative (non-zero) integer values.

22. Show that for $\kappa = (j + 1/2)$ or $-(j + 1/2)$

$$\left(\frac{\boldsymbol{\sigma} \cdot \boldsymbol{r}}{r}\right) Y^m_{j\,\ell_B}(0,0) = -Y^m_{j\,\ell_A}(0,0)$$

and
$$\left(\frac{\boldsymbol{\sigma} \cdot \boldsymbol{r}}{r}\right) Y^m_{j\,\ell_A}(0,0) = -Y^m_{j\,\ell_B}(0,0)\,.$$

Hint: (a) $Y_{\ell\,m}(0,0) = \sqrt{\frac{2\ell+1}{4\pi}}\delta_{m0}$, (b) $\theta = 0$ and $\varphi = 0$, $\left(\frac{\boldsymbol{\sigma}\cdot\boldsymbol{r}}{r}\right) = \sigma_z$.

23. If ψ represents the state of a positive energy electron, being the solution of Dirac equation $i\hbar\frac{\partial\psi}{\partial t} = H\psi$, where $H = c\alpha_k(p_k - eA_k) + \beta mc^2$ and $\psi_{_C}$ represents its charge conjugate state, defined by $\psi_{_C} = S_{_C}\psi^* = \gamma_2\psi^*$, being the solution of $i\hbar\frac{\partial\psi_C}{\partial t} = H'\psi_{_C}$, where $H' = c\alpha_k(p_k + eA_k) + \beta mc^2$, then show that

 (i) $\langle H'\rangle_{_C} \equiv \int \psi^\dagger_{_C} H'\psi_{_C} d\tau = -\langle H\rangle \equiv -\int \psi^\dagger H\psi d\tau$
 (ii) $\langle \Sigma_3\rangle_{_C} = -\langle \Sigma_3\rangle$,
 (iii) $\langle \boldsymbol{p}\rangle_{_C} = -\langle \boldsymbol{p}\rangle$, and
 (iv) $\langle Q\rangle_{_C} = \langle Q\rangle$.

References

[1] J. D. Bjorken and S. D. Drell, *Relativistic Quantum Mechanics* (McGraw-Hill, New York, 1964).

[2] W. Greiner, *Relativistic Quantum Mechanics* (Springer-Verlag, Berlin, 1990).

[3] J. M. Jauch and F. Rohrlich, *Theory of Photons and Electrons*, Second Edition (Springer-Verlag, New York, 1976).

[4] A. Messiah, *Quantum Mechanics*, Volume II (North Holland Publishing Company, Amsterdam, 1964).

[5] E. Merzbacher, *Quantum Mechanics*, Second Edition (John Wiley and Sons Inc., New York, 1970).

[6] J. Sakurai, *Advanced Quantum Mechanics* (Addison-Wesley, Reading, MA, 1967).

[7] L. I. Schiff *Quantum Mechanics*, Third edition (McGraw Hill Book Company, New York, 1968).

[8] F. Schwabl, *Advanced Quantum Mechanics* (Springer-Verlag, Berlin, 1999).

[9] C. S. Wu and S. A. Moszkowski, *Beta Decay* (Wiley Interscience, New York, 1966).

APPENDIX 12A1: Theory of Special Relativity

The theory of special relativity was proposed by Einstein in 1905. Before that, there had been considerable speculation regarding the existence of a *lumeniferous ether* (light-carrying medium), which permeated all space and existed solely as a vehicle for the propagation of electromagnetic waves. It was believed that there was a preferred reference frame, the one in which the ether was at rest, in which the laws of electromagnetism were valid. This set electromagnetic phenomena apart from other other physical phenomena such as mechanics whose laws were the same in all coordinate systems moving uniformly relative to one another. However, attempts to detect the existence of lumeniferous ether by studying the effects of earth's motion through this medium on various optical and electromagnetic phenomena gave negative results.

12A1.1 Lorentz Transformation

Against this background Einstein formulated his special theory of relativity. It was based on two postulates:

1. Laws of physics are the same in all inertial frames (frames in uniform motion relative to one another) not only with regard mechanical phenomena, but with regard to all physical phenomena, including optical and electro-magnetic. This means all inertial are frames are equally good for describing physical laws.

2. The speed of light in vacuum is the same in all directions and with respect to any uniformly moving observer.

It may be noted that invariance of laws of mechanical phenomena with repect to inertial frames was known even before Einstein. Thus, if S and S' are two inertial frames of reference with S' moving relative to S with velocity v in the common x-direction and the origins of the two frames coinciding at $t = t' = 0$, then the space-time coordinates of an event according to S and S', are related by Galilean transformation

$$x' = x - vt, \quad y' = y, \quad z' = z, \quad t' = t.$$

It naturally followed that observers in S and S' cannot detect any essential difference between their frames of reference by performing mechanical experiments alone; the laws of mechanics will be the same in both frames. But pre-relativistic concepts did not rule out this possibility if they performed optical or electromagnetic experiments. Einstein, through his postulates, ruled that even by performing electromagnetic experiments, observers in their respective frames S and S' cannot detect any essential difference between their frames of reference.

These postulates led to a new connection between the space-time coordinates of an event, as measured by two observers in different inertial frames S and S' with S' moving relative to S with velocity v in the positive x-direction. This connection is summarized by Lorentz transformation[11]

$$x' = \frac{x - vt}{\sqrt{1 - v^2/c^2}}, \quad y' = y, \quad z' = z, \quad t' = \frac{t - vx/c^2}{\sqrt{1 - v^2/c^2}}. \tag{12A1.1.1}$$

The inverse transformations are

$$x = \frac{x' + vt'}{\sqrt{1 - v^2/c^2}}, \quad y = y', \quad z = z', \quad t = \frac{t' + vx'/c^2}{\sqrt{1 - v^2/c^2}}. \tag{12A1.1.2}$$

It is easily verified

$$x^2 + y^2 + z^2 - c^2t^2 = x'^2 + y'^2 + z'^2 - c^2t'^2. \tag{12A1.1.3}$$

The implication of the principle of equivalence of the inertial frames is that, although the measurements of the two observers S and S' are different, both can verify the laws of physics in their respective frames of reference with equal right and equal success, using their measurements. Lorentz-Einstein transformations have far reaching consequences, some of which are summarized below.

Relativity of Simultaneity

If two events appear to be simultaneous to an observer in one inertial frame of reference, they are not so to an observer in another inertial frame. In other words simultaneity is relative.

Time Dilation

If two events are spatially coincident and separated by a time interval $\Delta t'$ in a frame of reference S', then this interval between the two events is the shortest among the intervals measured by observers in different states of uniform motion and is called the *proper time interval* $\Delta\tau$ between the two events. Another observer S, moving with velocity v (or $-v$) relative to S', will measure, according to Lorentz transformation (12A1.1.1), a longer interval Δt between the two events given by

$$\Delta t = \frac{\Delta t'}{\sqrt{1 - v^2/c^2}} = \frac{\Delta\tau}{\sqrt{1 - v^2/c^2}}. \tag{12A1.1.4}$$

Length Contraction

The length of a rod, as measured by an observer at rest with respect to the rod, is called its *proper length* ℓ_o. Another observer, moving relative to the rod, along its length, will measure a length ℓ, which is smaller than ℓ_o:

$$\ell = \ell_0\sqrt{1 - v^2/c^2}. \tag{12A1.1.5}$$

Relativistic Addition of Velocities

If $\boldsymbol{u} \equiv (u_x, u_y, u_z)$ are the components of the velocity of a body as observed by the observer in S and $\boldsymbol{u}'(u'_x, u'_y, u'_z)$ are the components of the velocity of the same body in the frame of reference S' , then one can derive the following relations between the components of the

[11]Lorentz had discovered that Maxwell's electromagnetic equations remained form invariant, not under Galilean transformations, but under new transformations which came to be known after him. The physical basis of these transformations was, however, understood only when Einstein derived them on the basis of his postulates.

two velocities on the basis of Lorentz-Einstein transformation relations [Eqs. (12A1.1.1) and (12A1.1.2)]:

$$u'_x = \frac{u_x - v}{1 - u_x v/c^2}, \quad u'_y = \frac{u_y\sqrt{(1 - v^2/c^2)}}{1 - u_x v/c^2}, \quad u'_z = \frac{u_z\sqrt{1 - v^2/c^2}}{1 - u_x v/c^2}. \tag{12A1.1.6}$$

Conversely, we can express the unprimed components in terms of primed components, either by solving the three equations simultaneously for u_x, u_y, and u_z or by interchanging the primed and unprimed components and changing the sign of v. Thus,

$$u_x = \frac{u'_x + v}{1 + u'_x v/c^2}, \quad u_y = \frac{u'_y\sqrt{(1 - v^2/c^2)}}{1 + u'_x v/c^2}, \quad u_z = \frac{u'_z\sqrt{1 - v^2/c^2}}{1 + u'_x v/c^2}. \tag{12A1.1.7}$$

It also follows that the quantities

$$\frac{u_x}{\sqrt{1 - u^2/c^2}}, \quad \frac{u_y}{\sqrt{(1 - u^2/c^2)}}, \quad \frac{u_z}{\sqrt{(1 - u^2/c^2)}}, \quad \text{and} \quad \frac{1}{\sqrt{(1 - u^2/c^2)}}$$

transform like x, y, z, and t under Lorentz transformations. The first three components are termed *space-like* components and the fourth component *time-like*.

Variation of Mass with Velocity: Relativistic Mass

If the collision of two balls is observed from two different inertial frames of reference S and S' and it is required that the momentum be conserved in the collision for both observers, then an inevitable consequence is that the momentum of the particle for either observer be defined as

$$\boldsymbol{p} = \frac{m\boldsymbol{u}}{\sqrt{1 - u^2/c^2}} = \frac{m}{\sqrt{1 - u^2/c^2}}\boldsymbol{u} \equiv m(u)\boldsymbol{u}, \tag{12A1.1.8}$$

where m is the *rest mass* (or *proper mass*) of the ball as measured by an observer at rest with respect to the ball. The quantity $m(u) = m/\sqrt{1 - u^2/c^2}$ is called the relativistic mass of the ball and can be thought of as velocity-dependent mass.

Mass-Energy Equivalence

The relativistic energy of a body of rest mass m moving with velocity $\boldsymbol{u}$ is given by

$$E = \frac{mc^2}{\sqrt{1 - u^2/c^2}} \equiv m(u)c^2. \tag{12A1.1.9}$$

A body at rest $\boldsymbol{u} = 0$ has energy $E_o = mc^2$, which is referred to as its rest mass energy. Its kinetic energy or the energy acquired by the body by virtue of its motion can then be defined as the difference between its relativistic energy and the rest mass energy:

$$T = \frac{mc^2}{\sqrt{1 - u^2/c^2}} - mc^2 \equiv [m(u) - m]\,c^2, \tag{12A1.1.10}$$

which can be thought of as the energy due to an increase in its relativistic mass $\Delta m(u)c^2$. Because the relativistic mass is just another name for the energy, it has gradually fallen into disuse[12]. Nevertheless it is a useful concept in certain pedagogical contexts.

[12] Lev. B. Okun, Physics Today **42**, 31 (1989).

12A1.2 Minkowski Space-Time Continuum

An examination of Lorentz transformation equations (12A1.1.1) and (12A1.1.2), shows that space and time coordinates are treated on the same footing in the theory of special relativity and they get mixed up in these transformations. This is not so in the non-relativistic (Galilean) transformations where time maintains a distinct identity ($t' = t$) and the rate of flow of time is the same for observers in all inertial frames. If we denote the space-time coordinates of an event x, y, z, ict by x_1 , x_2 , x_3 , and x_4 then Eq. (12A1.1.3) may be rewritten as $x_1^2 + x_2^2 + x_3^2 + x_4^2 = {x'_1}^2 + {x'_2}^2 + {x'_3}^2 + {x'_4}^2$ or simply as

$$x_\mu x_\mu = x'_\mu x'_\mu \,, \tag{12A1.2.1}$$

where, according to the convention, summation over the repeated index μ is implied. This observation led Minkowski to the concept of *space-time continuum* or the *four-dimensional space*. Accordingly, an event that occurs at a space point x, y, z at time t may be denoted by a point in the four-dimensional space whose coordinates are $x_1, x_2, x_3, x_4 (= ict)$ (or x_μ, where $\mu = 1, 2, 3, 4$). In the four-dimensional space we can visualize four mutually perpendicular axes, say OX_1, OX_2, OX_3, OX_4 , and a point in this space, which is representative of an event, has four coordinates. We may now visualize a Lorentz transformation as a rotation of axes OX_1 and OX_4 in the X_1X_4 plane, so that the coordinates of a point, representing an event, change from x_μ to x'_μ such that the square of the length of the four-radius vector remains the same, that is to say Eq. (12A1.2.1) holds.

The transformation equation (12A1.1.1) may be rewritten in the new notation as

$$\begin{aligned} x'_1 &= (x_1 + i\beta x_4)/\sqrt{1-\beta^2}\,, \\ x'_2 &= x_2\,, \\ x'_3 &= x_3\,, \\ x'_4 &= (-i\beta x_1 + x_4)/\sqrt{1-\beta^2}\,, \end{aligned} \tag{12A1.2.2}$$

where $\beta = v/c$. This set of equations may written in the matrix notation as

$$\begin{pmatrix} x'_1 \\ x'_2 \\ x'_3 \\ x'_4 \end{pmatrix} = \begin{pmatrix} \frac{1}{\sqrt{1-\beta^2}} & 0 & 0 & \frac{i\beta}{\sqrt{1-\beta^2}} \\ 0 & 1 & 0 & 0 \\ 0 & 0 & 1 & 0 \\ \frac{-i\beta}{\sqrt{1-\beta^2}} & 0 & 0 & \frac{1}{\sqrt{1-\beta^2}} \end{pmatrix} \begin{pmatrix} x_1 \\ x_2 \\ x_3 \\ x_4 \end{pmatrix} \tag{12A1.2.3a}$$

or

$$x'_\mu = a_{\mu\nu} x_\nu \,, \quad \mu, \nu = 1, 2, 3, 4 \tag{12A1.2.3b}$$

where sum over the repeated index ν is implied (summation convention). The matrix $[a_{\mu\nu}]$, given by

$$[a_{\mu\nu}] \equiv \begin{pmatrix} \frac{1}{\sqrt{1-\beta^2}} & 0 & 0 & \frac{i\beta}{\sqrt{1-\beta^2}} \\ 0 & 1 & 0 & 0 \\ 0 & 0 & 1 & 0 \\ \frac{-i\beta}{\sqrt{1-\beta^2}} & 0 & 0 & \frac{1}{\sqrt{1-\beta^2}} \end{pmatrix}, \tag{12A1.2.4}$$

is called the Lorentz transformation matrix. Geometrically, this transformation can be thought of as resulting from the rotation of the OX_1, OX_4 axes in the X_1, X_4 plane by an *imaginary angle* ϕ given by

$$\tan\phi = i\beta = i\,v/c\,. \tag{12A1.2.5}$$

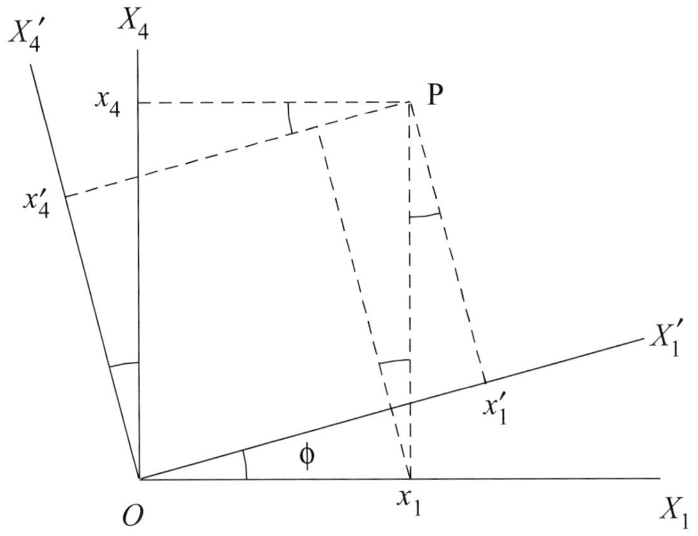

FIGURE 12.1
A physical event may be represented by a point P $(x_1, x_2, x_3, x_4(= ict))$ in the four-dimensional space. Lorentz transformation [Eqs. (12A1.2.2)] may be thought of as resulting from the rotation of OX_1 and OX_4 axes in the (X_1, X_4) plane in four-dimensional space through an imaginary angle ϕ, given by Eq. (12A1.2.5). The other two axes, OX_2 and OX_3 (not shown) are unaffected. Simple geometry gives the relation between the new coordinates x_1', x_4' of the point P in the plane and the old coordinates x_1, x_4 [Eqs. (12A1.2.7)]. The other two coordinates x_2 and x_3 are unchanged.

This transformation changes x_1 and $x_4 \rightarrow x_1'$ and x_4' while x_2 and x_3 are left unchanged [Fig. (12.1)]. In terms of ϕ introduced by Eq. (12A1.2.5), the transformation matrix can be written as

$$[a_{\mu\nu}] = \begin{pmatrix} \cos\phi & 0 & 0 & \sin\phi \\ 0 & 1 & 0 & 0 \\ 0 & 0 & 1 & 0 \\ -\sin\phi & 0 & 0 & \cos\phi \end{pmatrix} \tag{12A1.2.6}$$

and the coordinate transformation relations as

$$\begin{aligned} x_1' &= x_1 \cos\phi + x_4 \sin\phi \\ x_2' &= x_2 \\ x_3' &= x_3 \\ x_4' &= -x_1 \sin\phi + x_4 \cos\phi \,. \end{aligned} \tag{12A1.2.7}$$

The set of four quantities x_ν which transform according to Eqs. (12A1.2.7), under rotation of axes in four-dimensional space form a *four-vector.* A quantity like $x_\nu x_\nu$, which remains unchanged as a result of Lorentz transformation (or under rotation of axes in the four-dimensional space), as in Eq. (12A1.2.1), is called a *four-invariant* or a *four-scalar.* In addition to the square of the length of the four-radius vector (or of any four-vector) other invariant quantities are (i) the proper time interval $\Delta\tau$, (ii) the rest mass m and rest mass energy mc^2, and (iii) proper length ℓ_0. The d'Alembertian operator $\nabla^2 - \frac{1}{c^2}\frac{\partial^2}{\partial t^2} \equiv \frac{\partial}{\partial x_\nu}\frac{\partial}{\partial x_\nu}$

is also an invariant operator in the sense that

$$\frac{\partial}{\partial x'_\mu}\frac{\partial}{\partial x'_\mu} = \frac{\partial}{\partial x_\nu}\frac{\partial}{\partial x_\nu}\,. \tag{12A1.2.8}$$

12A1.3 Four-vectors in Relativistic Mechanics

In addition to the four-radius vector x_ν ($\nu = 1, 2, 3, 4$) there exist other four-vectors. In general if a four-vector is multiplied by an invariant, or differentiated with respect to an invariant variable , e.g., the proper time, then this gives another four-vector. Thus, $q_\nu = \frac{dx_\nu}{d\tau}$ is a four-vector, called four-velocity, $p_\nu = mq_\nu$ is a four-vector, called the four-momentum. We can then introduce a four-vector, called four-force as the proper rate of change of four-momentum

$$K_\nu = \frac{dp_\nu}{d\tau} = m\frac{dq_\nu}{d\tau} = m\frac{d^2x_\nu}{d\tau^2}\,. \tag{12A1.3.1}$$

It may be noted that Eq. (12A1.3.1) will retain its form under Lorentz transformation, wherein the vectors K_ν, p_ν , q_ν, x_ν transform to K'_μ, p'_μ, q'_μ, x'_μ, respectively, so that

$$K'_\mu = \frac{dp'_\mu}{d\tau} = m\frac{dq'_\mu}{d\tau} = m\frac{d^2x'_\mu}{d\tau^2}\,. \tag{12A1.3.2}$$

Such equations are called *form invariant* and such a formulation of equations of physics is called *covariant formulation.* It is easy to work out the components of four-velocity, four-momentum and four-force. In what follows we will take Greek indices to take on values 1, 2, 3, 4 and Roman indices to take values 1, 2, 3. Thus Roman indices will be used to denote the first three components of four-vector. Using this notation we can write the components of four-velocity as

$$q_k = \frac{dx_k}{d\tau} = \frac{dx_k}{dt}\frac{dt}{d\tau} = \frac{u_k}{\sqrt{1-u^2/c^2}}\,, \qquad k = 1, 2, 3\,,$$

$$q_4 = \frac{d(ict)}{d\tau} = \frac{ic}{\sqrt{1-u^2/c^2}}$$

so that

$$q_\nu \equiv \left(\frac{\boldsymbol{u}}{\sqrt{1-u^2/c^2}}\,, \frac{ic}{\sqrt{1-u^2/c^2}}\right)\,, \tag{12A1.3.3}$$

where $\boldsymbol{u}$ is the ordinary velocity (three-vector) of the particle. From the definition of four-velocity it is easily checked that, as expected, $q_\mu q_\mu = -c^2$ is an invariant.

The components of four-momentum are obtained simply by multiplying these equations by the proper mass (rest mass) m as

$$p_k = mq_k = m\frac{u_k}{\sqrt{1-u^2/c^2}}\,, \qquad k = 1, 2, 3$$

$$p_4 = mq_4 = m\frac{ic}{\sqrt{1-u^2/c^2}} = \frac{iE}{c}\,,$$

so that

$$p_\mu \equiv \left(\boldsymbol{p}\,, \frac{iE}{c}\right)\,, \tag{12A1.3.4}$$

where $\boldsymbol{p} = m\boldsymbol{u}/\sqrt{1 - u^2/c^2}$ is the momentum of the particle and E is its total energy. Similarly, for four-force we have

$$K_\nu \equiv \frac{dp_\nu}{d\tau} = \left(\frac{\boldsymbol{F}}{\sqrt{1 - u^2/c^2}}, \frac{i\boldsymbol{F} \cdot \boldsymbol{u}/c}{\sqrt{1 - u^2/c^2}} \right). \tag{12A1.3.5}$$

A four-vector S_ν is called *space-like* if the scalar $A_\nu A_\nu > 0$, *time-like* if $A_\nu A_\nu < 0$, and *light-like* or *null-vector* if $A_\nu A_\nu = 0$.

Relativistic Energy-Momentum Relation

Since the square of the length of any four-vector is invariant, it follows that $p_\nu p_\nu$ is an invariant. Indeed from the definition of four momentum in terms of four-velocity we find

$$p_\mu p_\mu = m^2(q_1^2 + q_2^2 + q_3^2 + q_4^2) = m^2 \frac{u^2 - c^2}{1 - u^2/c^2} = -m^2 c^2, \tag{12A1.3.6}$$

which is an invariant since both the rest mass m and c are invariant. By evaluating this invariant in a frame where particle momentum is $\boldsymbol{p}$ and total energy is E, we have

$$\boldsymbol{p}^2 - \frac{E^2}{c^2} = -m^2 c^2$$

or

$$E^2 = c^2 p^2 + m^2 c^4. \tag{12A1.3.7}$$

This is the relativistic relation between the total energy (kinetic + rest energy) and momentum of a particle in any frame of reference.

Transformation of Electric Current and Charge Densities

The electric charge of a particle is invariant under a Lorentz transformation. Starting from the four-velocity $q_\mu = \left(\boldsymbol{u}/\sqrt{1 - u^2/c^2}, ic/\sqrt{1 - u^2/c^2}\right)$ [Eq. (12A1.3.3)], we can introduce a four-vector electric current density j_μ by multiplying the four-velocity by a Lorentz scalar, ρ_o, the proper charge density (charge density in the frame in which the charge distribution is at rest) by

$$j_\mu = (\rho_o q_k, \rho_o q_4) = (\rho \boldsymbol{u}, ic\rho) \equiv (\boldsymbol{j}, ic\rho), \qquad \rho = \rho_o/\sqrt{1 - u^2/c^2}. \tag{12A1.3.8}$$

It follows that the components of four-current density transform like a four-vector under a Lorentz transformation:

$$\begin{aligned} j_1' &= \frac{j_1 - v\rho}{\sqrt{1 - v^2/c^2}} \\ j_2' &= j_2 \\ j_3' &= j_3 \\ j_4' &= \frac{j_4 - vj_1/c}{\sqrt{1 - v^2/c^2}} \end{aligned} \tag{12A1.3.9}$$

or

$$j_\mu' = a_{\mu\nu} j_\nu. \tag{12A1.3.10}$$

The equation of continuity

$$\boldsymbol{\nabla} \cdot \boldsymbol{j} + \frac{\partial \rho}{\partial t} = 0, \tag{12A1.3.11}$$

which expresses the conservation of charge can be written in a covariant form as

$$\frac{\partial j_\nu}{\partial x_\nu} = 0. \tag{12A1.3.12}$$

12A1.4 Covariant Form of Maxwell's Equations

Maxwell's equations,which relate electric and magnetic fields to charge and current densities are[13]

$$\nabla \cdot \boldsymbol{E} = \rho/\epsilon_0 \,, \tag{12A1.4.1}$$

$$\nabla \cdot \boldsymbol{B} = 0 \,, \tag{12A1.4.2}$$

$$\nabla \times \boldsymbol{B} - \mu_0 \epsilon_0 \frac{\partial \boldsymbol{E}}{\partial t} = \mu_0 \boldsymbol{j} \,, \tag{12A1.4.3}$$

$$\nabla \times \boldsymbol{E} + \frac{\partial \boldsymbol{B}}{\partial t} = 0 \,. \tag{12A1.4.4}$$

Equations (12A1.4.2) and (12A1.4.4) allow us to introduce a vector potential $\boldsymbol{A}$ and a scalar potential ϕ by equations

$$\boldsymbol{B} = \nabla \times \boldsymbol{A} \,, \tag{12A1.4.5}$$

and

$$\boldsymbol{E} = -\nabla \phi - \frac{\partial \boldsymbol{A}}{\partial t} \,. \tag{12A1.4.6}$$

These relations, together with vector identities $\nabla \cdot (\nabla \times \boldsymbol{A}) = 0$ and $\nabla \times (\nabla \phi) = 0$, ensure that Maxwell's equations (12A1.4.2) and (12A1.4.4) are already satisfied. In view of the same vector identities, a *Gauge transformation*,

$$\boldsymbol{A} \to \boldsymbol{A} - \nabla \chi \,, \tag{12A1.4.7}$$

and

$$\phi \to \phi + \frac{\partial \chi}{\partial t} \,, \tag{12A1.4.8}$$

where χ is an arbitrary scalar field, leaves the field quantities $\boldsymbol{E}$ and $\boldsymbol{B}$ unaltered. Of all the possible gauge transformations in relativistic formulation of Maxwell's equations, we use the Lorentz gauge wherein the potentials satisfy the Lorentz condition

$$\nabla \cdot \boldsymbol{A} + \frac{1}{c^2} \frac{\partial \phi}{\partial t} = 0 \,. \tag{12A1.4.9}$$

This is because this gauge condition can be put in a covariant form[14]. Using Eqs. (12A1.4.5) and (12A1.4.6), we can write the first and third of Maxwell's equations in terms of the vector and scalar potentials. Accordingly, Eqs. (12A1.4.1) and (12A1.4.3) yield, respectively,

$$\nabla^2 \phi - \frac{1}{c^2} \frac{\partial^2 \phi}{\partial t^2} = -\rho/\epsilon_0 \tag{12A1.4.10}$$

and

$$\nabla^2 \boldsymbol{A} - \frac{1}{c^2} \frac{\partial^2 \boldsymbol{A}}{\partial t^2} = -\mu_0 \boldsymbol{j} \,, \tag{12A1.4.11}$$

where Lorentz condition (12A1.4.9) has been used. Rewriting these equations as

$$\left[\nabla^2 - \frac{1}{c^2} \frac{\partial^2}{\partial t^2}\right] (i\phi/c) = -\frac{1}{\epsilon_0} (ic\rho)/c^2 = -\mu_0 (ic\rho) \,, \tag{12A1.4.12}$$

and

$$\left[\nabla^2 - \frac{1}{c^2} \frac{\partial^2}{\partial t^2}\right] \boldsymbol{A} = -\mu_0 \boldsymbol{j} \,, \tag{12A1.4.13}$$

[13] In these equations we have used MKS (SI) units to express all quantities.

[14] Another useful choice in non-relativistic radiation problems is the so-called *radiation gauge* also called the Coulomb or transverse gauge, wherein the vector potential is chosen to satisfy $\nabla \cdot \boldsymbol{A} = 0$. This will be used in Chapter 13.

and recalling that the d'Alembertian

$$\left[\nabla^2 - \frac{1}{c^2}\frac{\partial^2}{\partial t^2}\right] \equiv \frac{\partial}{\partial x_\mu}\frac{\partial}{\partial x_\mu}$$

is an invariant operator and the set of four quantities $(\boldsymbol{j},\, ic\rho) \equiv j_\nu$ constitutes the components of a four-vector, we conclude that the set of four quantities

$$(\boldsymbol{A},\ i\phi/c) \equiv A_\nu \qquad (12A1.4.14)$$

must also constitute the components of a four-vector. Then equations (12A1.4.12) and (12A1.4.13) can be written in covariant form as

$$\frac{\partial}{\partial x_\mu}\frac{\partial}{\partial x_\mu} A_\nu = -\mu_0 j_\nu\,. \qquad (12A1.4.15)$$

Further, since A_ν is a four-vector, quantities like $\frac{\partial A_\nu}{\partial x_\mu}$ or $\frac{\partial A_\mu}{\partial x_\nu}$ are four-tensors of rank two, i.e., they transform like the product of two four-vectors. Based on these observations, we may introduce a new tensor $F_{\mu\nu}$ by

$$F_{\mu\nu} \equiv \frac{\partial A_\nu}{\partial x_\mu} - \frac{\partial A_\mu}{\partial x_\nu}\,, \qquad (12A1.4.16)$$

which is an anti-symmetric ($F_{\mu\nu} = -F_{\nu\mu}$) traceless ($F_{\mu\mu} = 0$) tensor of rank two. It has six independent components. The law of transformation for $F_{\mu\nu}$ is characteristic of a tensor of rank two

$$F_{\sigma\rho} \to F'_{\sigma\rho} = a_{\sigma\mu}a_{\rho\nu}F_{\mu\nu}\,, \qquad (12A1.4.17)$$

where $a_{\sigma\nu}$ is the Lorentz transformation matrix defined by Eq. (12A1.2.4).

The six independent components of the electromagnetic field tensor are very simply related to the components of magnetic and electric fields $\boldsymbol{B}$ and $\boldsymbol{E}/c$, for

$$F_{12} = \left(\frac{\partial A_2}{\partial x_1} - \frac{\partial A_1}{\partial x_2}\right) = (\boldsymbol{\nabla} \times \boldsymbol{A})_3 = B_3\,,$$

and similarly, $F_{23} = B_1$ and $F_{31} = B_2$. Further,

$$F_{41} = \frac{\partial A_1}{\partial x_4} - \frac{\partial A_4}{\partial x_1} = \left(\frac{1}{ic}\frac{\partial(A_1)}{\partial t} - \frac{\partial(i\phi/c)}{\partial x_1}\right) = \frac{i}{c}\left[-\frac{\partial \boldsymbol{A}}{\partial t} - \boldsymbol{\nabla}\,\phi\right]_1 = i\frac{E_1}{c}$$

and similarly, $F_{42} = +iE_2/c$ and $F_{43} = +iE_3/c$. Thus the components of the electromagnetic field tensor are as follows

$$[F_{\mu\nu}] = \begin{pmatrix} 0 & B_3 & -B_2 & -iE_1/c \\ -B_3 & 0 & B_1 & -iE_2/c \\ B_2 & -B_1 & 0 & -iE_3/c \\ iE_1/c & iE_2/c & iE_3/c & 0 \end{pmatrix}. \qquad (12A1.4.18)$$

Using the electromagnetic field tensor $F_{\mu\nu}$, it is now possible to write Maxwell's equations in covariant form. Thus Eqs. (12A1.4.2) and (12A1.4.4) can be rewritten as

$$\frac{\partial F_{\mu\nu}}{\partial x_\rho} + \frac{\partial F_{\nu\rho}}{\partial x_\mu} + \frac{\partial F_{\rho\mu}}{\partial x_\nu} = 0\,. \qquad (12A1.4.19)$$

Similarly, Eqs. (12A1.4.3) and (12A1.4.1)) can be rewritten as

$$\frac{\partial F_{\mu\nu}}{\partial x_\nu} = \mu_0 j_\mu \,. \tag{12A1.4.20}$$

This is easily seen by writing the corresponding Maxwell's equations explicitly, in terms of the components of $\boldsymbol{E}$ and $\boldsymbol{B}$ and then expressing these components in terms of components of the field tensor (12A1.4.18).

Thus Maxwell's equations, which predate Einstein's theory of special relativity, can be written in a covariant form without any modification. The form invariance of these equations under Lorentz transformations is an important property of these equations and signifies the physical fact that although the observers in different frames of reference would differ in their measurements of physical quantities (including the components of electric and magnetic fields), they can all formulate the laws of physics in the same way.

13

QUANTIZATION OF RADIATION FIELD

13.1 Introduction

In atomic, nuclear and particle physics we come across several situations where particles or radiation quanta (photons) are created or destroyed. For example in the beta decay of the neutron, a neutron is annihilated and a proton, electron and antineutrino are created. In some situations the interaction of radiation with matter can result in creation or annihilation of a pair of electron and positron. Such processes cannot be understood on the basis of (non-relativistic or relativistic) single particle wave equations because in such theories the probability $\psi^*\psi d\tau$ (of finding the particle in the region of space $d\tau$) when integrated over the whole space, results in unity at all times and so the existence of the particle is always guaranteed. Hence in the framework of single particle wave equation it is not possible to understand the creation or annihilation of particles. Even a simple process in which an atom undergoes a transition from an excited to the ground state by spontaneous emission of a photon in the absence of a perturbing electromagnetic field cannot be explained within this framework.

The first step toward understanding the creation or annihilation of photons was taken by Dirac.[1] He quantized the radiation field which had hitherto been regarded as a classical field. As a result Dirac could explain the *spontaneous* and *induced* emission of radiation in a natural way. This was one of the triumphs of Dirac's quantum theory of radiation and formed the starting point of quantization of wave fields in general. Among other successes of this theory were the explanations of Thomson, Rayleigh, Raman, and Compton scattering processes.

13.2 Radiation Field as a Swarm of Oscillators

To quantize the radiation field we first recall that it is that part of the Hamiltonian, which depends on the transverse components of the fields. The static part of the Hamiltonian expresses the Coulomb interaction between the charges. This separation of the Hamiltonian into radiation and static parts is most conveniently done in the Coulomb gauge [Appendix 13A1]. In the Coulomb gaugethe vector potential satisfies the transversality condition

$$\boldsymbol{\nabla} \cdot \boldsymbol{A} = 0 \,, \tag{13.2.1}$$

[1]P. A. M. Dirac, *The quantum theory of the emission and absorption of radiation*, Proc. Roy. Soc. **A114**, 243 (1927).

and obeys the equation

$$\nabla^2 \boldsymbol{A}(\boldsymbol{r},t) - \frac{1}{c^2}\frac{\partial^2 \boldsymbol{A}(\boldsymbol{r},t)}{\partial t^2} = -\mu_0 \boldsymbol{j}_\perp(\boldsymbol{r},t)\,. \tag{13.2.2}$$

where $\boldsymbol{j}_\perp$ is the transverse part of the electric current [Eq. (13A1.1.16)] due to atomic electrons (or any other charges). In terms of vector potential $\boldsymbol{A}$, the transverse components (radiation fields) of the field are

$$\boldsymbol{B} = \boldsymbol{\nabla} \times \boldsymbol{A}\,, \tag{13.2.3}$$

and

$$\boldsymbol{E}_\perp = -\frac{\partial \boldsymbol{A}}{\partial t}\,. \tag{13.2.4}$$

Note that the $\boldsymbol{B}$ is purely transverse. The Hamiltonian for the radiation field can then be expressesd as

$$H_{rad} = \frac{1}{2}\int \left[\epsilon_0 |\boldsymbol{E}_\perp|^2 + \frac{1}{\mu_0}|\boldsymbol{B}|^2\right] d\tau\,, \tag{13.2.5}$$

or, equivalently,

$$H_{rad} = \frac{1}{2}\int \left[\epsilon_0 \left|\frac{\partial \boldsymbol{A}}{\partial t}\right|^2 + \frac{1}{\mu_0}|\boldsymbol{\nabla} \times \boldsymbol{A}|^2\right] d\tau\,. \tag{13.2.6}$$

From now on we shall drop the subscript '$\perp$' on the field components with the understanding that these are the components we shall be concerned with in quantizing the radiation field.

We first quantize the free field and then take up the interaction of quantized field with matter. Consider the field in a region of space where $\boldsymbol{j}_\perp = 0$ so that the vector potential satisfies

$$\nabla^2 \boldsymbol{A}(\boldsymbol{r},t) - \frac{1}{c^2}\frac{\partial^2 \boldsymbol{A}(\boldsymbol{r},t)}{\partial t^2} = 0\,. \tag{13.2.7}$$

Such a field is said to be free. To quantize the free radiation field we must first express the Hamiltonian of the field in terms of canonical variables. The vector potential is a function defined for all space time points ($\boldsymbol{r}$ and t). If therefore we wish to specify $\boldsymbol{A}(\boldsymbol{r},t)$ in terms of canonical variables, the number of such variables would be infinitely large and non-enumerable. Although it is possible to work with such a decomposition of the vector potential, it is easier to work with a discrete decomposition of $\boldsymbol{A}(\boldsymbol{r},t)$. To do this we imagine the radiation field to be enclosed in a very large box of volume V. The vector potential $\boldsymbol{A}$ can then be expressed in terms of appropriately chosen spatial functions satisfying certain boundary conditions on the surface of V. The imposition of boundary conditions leads to a discrete set of functions, which can be used to express the vector potential. At a suitable stage in the calculation, we allow the volume to tend to infinity. Needless to say that physically meaningful results should not depend on the exact shapes of the boundary surfaces and boundary conditions. The solution of Eq. (13.2.7), separable in space and time variables, can be written in terms of these discrete set of functions as

$$\boldsymbol{A}(\boldsymbol{r},t) = \frac{1}{\sqrt{\epsilon_0}}\sum_\lambda \left(q_\lambda(t)\,\boldsymbol{u}_\lambda(\boldsymbol{r}) + q^*_\lambda(t)\,\boldsymbol{u}^*_\lambda(\boldsymbol{r})\right)\,, \tag{13.2.8}$$

where the functions $\boldsymbol{u}_\lambda(\boldsymbol{r})$ of spatial coordinates, labeled by index λ, form a discrete, though infinite, set as a result of satisfying the boundary conditions. The index λ represents a set of numbers needed to specify the spatial and polarization degrees of freedom of the field. Substituting Eq. (13.2.8) into Eq. (13.2.7) and separating the variables we find $\boldsymbol{u}_\lambda(\boldsymbol{r})$ and

$q_\lambda(t)$ satisfy

$$\nabla^2 \boldsymbol{u}_\lambda(\boldsymbol{r}) + \frac{\omega_\lambda^2}{c^2} \boldsymbol{u}_\lambda(\boldsymbol{r}) = 0\,, \tag{13.2.9}$$

$$\frac{d^2}{dt^2} q_\lambda(t) + \omega_\lambda^2 q_\lambda(t) = 0 \tag{13.2.10}$$

where the the separation constant ω_λ has dimensions of angular frequency. Indeed, the equation satisfied by the time-dependent coefficient $q_\lambda(t)$ may easily be recognized as the equation for a harmonic oscillator of natural (angular) frequency ω_λ.

We solve Eq. (13.2.9) in a cubic region of space of side L ($V = L^3$) with its edges oriented along the coordinate axes, and subject to periodic boundary conditions [see Chapter 5] on $\boldsymbol{u}_\lambda(\boldsymbol{r})$ on the faces of the cube:

$$\begin{aligned} \boldsymbol{u}_\lambda(x+L, y, z) &= \boldsymbol{u}_\lambda(x, y, z)\,, \\ \boldsymbol{u}_\lambda(x, y+L, z) &= \boldsymbol{u}_\lambda(x, y, z)\,, \\ \boldsymbol{u}_\lambda(x, y, z+L) &= \boldsymbol{u}_\lambda(x, y, z)\,. \end{aligned} \tag{13.2.11}$$

Solutions of Eq. (13.2.9) are then given by

$$\boldsymbol{u}_\lambda(\boldsymbol{r}) = \frac{1}{\sqrt{V}} \boldsymbol{\varepsilon}_\lambda\, e^{i\,\boldsymbol{k}_\lambda \cdot \boldsymbol{r}} \tag{13.2.12}$$

where $k_\lambda = \omega_\lambda / c$ and $\boldsymbol{\varepsilon}_\lambda$ is the polarization vector. Periodic boundary conditions lead to a discrete set of functions by giving discrete values to $\boldsymbol{k}_\lambda \equiv (k_{\lambda x}, k_{\lambda y}, k_{\lambda z})$:

$$k_{\lambda x} = \frac{2\pi}{L} n_{\lambda x}\,, \quad k_{\lambda y} = \frac{2\pi}{L} n_{\lambda y}\,, \quad \text{and} \quad k_{\lambda z} = \frac{2\pi}{L} n_{\lambda z} \tag{13.2.13}$$

where $n_{\lambda x}$, $n_{\lambda y}$, and $n_{\lambda z}$ are positive or negative integers. It may be verified that $\boldsymbol{u}_\lambda(\boldsymbol{r})$'s form a set of orthogonal functions satisfying the condition

$$\int \boldsymbol{u}_\lambda(\boldsymbol{r}) \cdot \boldsymbol{u}_\mu^*(\boldsymbol{r}) d\tau = \delta_{\lambda\mu}\,. \tag{13.2.14}$$

Condition $\nabla \cdot \boldsymbol{A}_\lambda = 0$ [Eq. (13.2.1)] now implies

$$\boldsymbol{k}_\lambda \cdot \boldsymbol{\varepsilon}_\lambda = 0, \quad \text{for each } \lambda\,, \tag{13.2.15}$$

which means that the polarization vector $\boldsymbol{\varepsilon}_\lambda$ is perpendicular to the wave-vector $\boldsymbol{k}_\lambda$.

The solution of Eq. (13.2.10) is easily seen to be

$$q_\lambda(t) = q_\lambda(0)\, e^{-i\omega_\lambda t}\,, \tag{13.2.16}$$

where ω_λ is a positive quantity. Since $\boldsymbol{u}_\lambda(\boldsymbol{r})$ are specified orthogonal functions of space coordinates, the field $\boldsymbol{A}(\boldsymbol{r}, t)$ is characterized by the set of variables $q_\lambda(t)$. We can now express the Hamiltonian H_{rad} of the field in terms of $q_\lambda(t)$ and $q_\lambda(t)^*$ by using the expression for it in terms of vector potential

$$H_{rad} = \frac{1}{2} \int \left[\epsilon_0 \left| \left(\frac{\partial \boldsymbol{A}}{\partial t} \right) \right|^2 + \frac{1}{\mu_0} \left| \nabla \times \boldsymbol{A} \right|^2 \right]. \tag{13.2.6*}$$

When we use the expansion (13.2.8), together with Eqs.(13.2.12) and (13.2.16), for the vector potential $\boldsymbol{A}(\boldsymbol{r}, t)$ in this equation, a typical integral encountered in the evaluation of the magnetic energy, which gives nonzero contribution is

$$\frac{1}{2\mu_0\epsilon_0} \int (\nabla \times \mathbf{u}_\lambda(\boldsymbol{r})) \cdot (\nabla \times \boldsymbol{u}_\mu^*(\boldsymbol{r}))\, d\tau = \frac{1}{2} k_\lambda^2 c^2 \delta_{\lambda\mu} = \frac{1}{2} \omega_\lambda^2\, \delta_{\lambda\mu}\,, \tag{13.2.17}$$

where we have used the results

$$\nabla \times \boldsymbol{u}_\lambda = -i\frac{1}{\sqrt{V}}\,(\boldsymbol{\varepsilon}_\lambda \times \boldsymbol{k}_\lambda)\,e^{i\boldsymbol{k}_\lambda\cdot\boldsymbol{r}}\,, \tag{13.2.18}$$

$$\nabla \times u^*_\lambda = i\frac{1}{\sqrt{V}}\,(\boldsymbol{\varepsilon}^*_\lambda \times \boldsymbol{k}_\lambda)\,e^{-i\boldsymbol{k}_\lambda\cdot\boldsymbol{r}}\,, \tag{13.2.19}$$

and

$$[\boldsymbol{\varepsilon}_\lambda \times \boldsymbol{k}_\lambda]\cdot[\boldsymbol{\varepsilon}^*_\lambda \times \boldsymbol{k}_\lambda] = \boldsymbol{\varepsilon}^*_\lambda\cdot[\boldsymbol{k}_\lambda \times (\boldsymbol{\varepsilon}_\lambda \times \boldsymbol{k}_\lambda)] = k^2_\lambda \tag{13.2.20}$$

in evaluating the integral. Terms involving products like $(\nabla \times \boldsymbol{u}_\lambda(\boldsymbol{r})\cdot(\nabla \times \boldsymbol{u}_\mu(\boldsymbol{r})$ give terms rapidly oscillating in space and or time and average out to zero. Similarly, a typical integral in the evaluation of electric energy involves

$$\frac{1}{2}\int\left(\frac{dq_\lambda}{dt}\,\boldsymbol{u}_\lambda\right)\cdot\left(\frac{dq^*_\mu}{dt}\,\boldsymbol{u}^*_\mu\right)\,d\tau = \frac{1}{2}\omega^2_\lambda\,q_\lambda\,q^*_\mu\,\delta_{\lambda\mu}\,. \tag{13.2.21}$$

Using these results in the Hamiltonian (13.2.6) we get, after some simplification,

$$H_{rad} = \sum_\lambda \omega^2_\lambda(q_\lambda q^*_\lambda + q^*_\lambda q_\lambda) \equiv \sum_\lambda H_\lambda \tag{13.2.22}$$

where

$$H_\lambda = \omega^2_\lambda(q_\lambda q^*_\lambda + q^*_\lambda q_\lambda)\,. \tag{13.2.23}$$

The momentum of the field given by

$$\boldsymbol{G} = \frac{1}{c^2\mu_0}\int(\boldsymbol{E}\times\boldsymbol{B})\,d\tau \tag{13.2.24}$$

can also be expressed in terms of the variables q_λ and q^*_λ. Substituting the electric and magnetic fields, $\boldsymbol{E}$ and $\boldsymbol{B}$, in term of the vector potential [Eqs. (13.2.3) and (13.2.6)] and using the expression (13.2.8) for vector potential, we get

$$\boldsymbol{G} = -\epsilon_0\sum_{\lambda,\mu}\int d\tau\,\left[\dot{q}_\lambda q_\mu \boldsymbol{u}_\lambda \times (\nabla \times \boldsymbol{u}_\mu) + \dot{q}_\lambda q^*_\mu \boldsymbol{u}_\lambda \times \left(\nabla \times \boldsymbol{u}^*_\mu\right) + \text{c.c.}\right]\,, \tag{13.2.25}$$

where *c.c.* stands for the complex conjugate of the expression to the left inside the square brackets. Using Eqs. (13.2.18) through (13.2.20) in the integrand for $\boldsymbol{G}$ and carrying out the integration, we obtain

$$\boldsymbol{G} = \sum_\lambda \omega_\lambda \boldsymbol{k}_\lambda(q_\lambda q^*_\lambda + q^*_\lambda q_\lambda) \equiv \sum_\lambda \boldsymbol{G}_\lambda\,, \tag{13.2.26}$$

where

$$\boldsymbol{G}_\lambda = \omega_\lambda \boldsymbol{k}_\lambda(q_\lambda q^*_\lambda + q^*_\lambda q_\lambda)\,. \tag{13.2.27}$$

Note that $|\boldsymbol{G}_\lambda| = H_\lambda/c$, which anticipates the energy-momentum relation for a photon.

If we introduce real variable Q_λ and P_λ by

$$Q_\lambda = q_\lambda(t) + q^*_\lambda(t) \tag{13.2.28}$$

and

$$P_\lambda = -i\omega_\lambda\left[q_\lambda(t) - q^*_\lambda(t)\right]\,, \tag{13.2.29}$$

we can rewrite H_λ and $\boldsymbol{G}_\lambda$ as

$$H_\lambda = \frac{1}{2}\left(P^2_\lambda + \omega^2_\lambda Q^2_\lambda\right)\,, \tag{13.2.30}$$

and

$$\boldsymbol{G}_\lambda = \frac{1}{2}\left(\frac{\boldsymbol{k}_\lambda}{\omega_\lambda}\right)\left(P^2_\lambda + \omega^2_\lambda Q^2_\lambda\right)\,. \tag{13.2.31}$$

We can easily see that Q_λ and P_λ are canonical variables because they satisfy the equations of motion in the canonical form[2]:

$$\dot{P}_\lambda = -\frac{\partial H}{\partial Q_\lambda} \tag{13.2.32}$$

and $$\dot{Q}_\lambda = \frac{\partial H}{\partial P_\lambda}\,. \tag{13.2.33}$$

An inspection of Eqs. (13.2.22) and (13.2.26) together with (13.2.30) and (13.2.31) shows that we can look upon the classical radiation field as equivalent to a swarm of independent oscillators of different frequencies ω_λ. This idea was originally introduced by Planck who also postulated that radiation oscillators can emit or absorb energy in quanta of energy $\hbar\omega_\lambda$.

13.3 Quantization of Radiation Field

To quantize the radiation field we may now consider the canonical variables P_λ and Q_λ of each radiation oscillator as non-commuting operators $\hat{P}_\lambda$ and $\hat{Q}_\lambda$ satisfying the commutation relations

$$\left[\hat{P}_\lambda, \hat{Q}_\mu\right] = -i\hbar\delta_{\lambda\mu} \tag{13.3.1}$$

and $$\left[\hat{P}_\lambda, \hat{P}_\mu\right] = \left[\hat{Q}_\lambda, \hat{Q}_\mu\right] = 0\,. \tag{13.3.2}$$

Obviously the operators $\hat{P}_\lambda$ and $\hat{Q}_\lambda$ are self-adjoint operators. As a result $q_\lambda \to \hat{q}_\lambda$ and $p_\lambda \to \hat{p}_\lambda$. If we now write

$$\hat{P}_\lambda = -i\left(\frac{\hbar\omega_\lambda}{2}\right)^{1/2}\left(\hat{a}_\lambda - \hat{a}_\lambda^\dagger\right) \equiv -i\omega_\lambda(\hat{q}_\lambda - \hat{q}_\lambda^\dagger) \tag{13.3.3}$$

and $$\hat{Q}_\lambda = \left(\frac{\hbar}{2\omega_\lambda}\right)^{1/2}\left(\hat{a}_\lambda + a_\lambda^\dagger\right) \equiv (\hat{q}_\lambda + \hat{q}_\lambda^\dagger) \tag{13.3.4}$$

where $\hat{a}_\lambda$ and $\hat{a}_\lambda^\dagger$ are adjoints of each other and so are $\hat{q}_\lambda$ and $\hat{q}_\lambda^\dagger$, then from the commutation relations (13.3.1) and (13.3.2) we can easily derive the following commutation relations[3]

$$\left[\hat{a}_\lambda, \hat{a}_\mu^\dagger\right] = \delta_{\lambda\mu} \tag{13.3.5}$$

and $$\left[\hat{a}_\lambda, \hat{a}_\mu\right] = \left[\hat{a}_\lambda^\dagger, \hat{a}_\mu^\dagger\right] = 0\,. \tag{13.3.6}$$

An implication of the quantization of the radiation field is that the vector potential $\boldsymbol{A}(\boldsymbol{r}, t)$, which can be expressed in term of the operators $\hat{a}_\lambda$ and $\hat{a}_\lambda^\dagger$, is also to be treated as an operator. Hence the field quantities $\boldsymbol{E}$ and $\boldsymbol{B}$ are also to be treated as operators which do

[2]From the definition of P_λ and Q_λ we find $\dot{P}_\lambda = -i\omega_\lambda\left(\dot{q}_\lambda - \dot{q}_\lambda^*\right) = -\omega_\lambda^2\left(q_\lambda + q_\lambda^*\right) = -\omega_\lambda^2 Q_\lambda = -\frac{\partial H}{\partial Q_\lambda}$ and $\dot{Q}_\lambda = \left(\dot{q}_\lambda^* + \dot{q}_\lambda\right) = -i\omega_\lambda\left(-q_\lambda^* + q_\lambda\right) = P_\lambda = \frac{\partial H}{\partial P_\lambda}$.

[3]Since $\hat{a}_\lambda = \sqrt{\frac{2\omega_\lambda}{\hbar}}\,\frac{1}{2}\left(\hat{Q}_\lambda + \frac{i\hat{P}_\lambda}{\omega_\lambda}\right)$ and $\hat{a}_\mu^\dagger = \left(\frac{2\omega_\mu}{\hbar}\right)^{1/2}\frac{1}{2}\left(\hat{Q}_\mu - \frac{i\hat{P}_\mu}{\omega_\mu}\right)$, the commutators $\left[\hat{a}_\lambda, \hat{a}_\mu^\dagger\right]$, $\left[\hat{a}_\lambda, \hat{a}_\mu\right]$, $\left[a_\lambda^\dagger, a_\mu^\dagger\right]$ are easily worked out using the commutation relation between $\hat{P}_\lambda$ and $\hat{Q}_\lambda$.

not commute. This implies that simultaneous measurement of the field quantities $\hat{\boldsymbol{E}}$ and $\hat{\boldsymbol{B}}$ at a point with arbitrary accuracy is not possible just as it is not possible to measure the position as well as the momentum of a particle with arbitrary accuracy. Only their space average can be known. The Hamiltonian $\hat{H}_{rad}$ of the radiation field as well as momentum $\hat{\boldsymbol{G}}$ of the field are also operators. Now $\hat{H}_\lambda$, defined by

$$\hat{H}_\lambda = \frac{1}{2}\left(\hat{P}_\lambda^2 + \omega_\lambda^2 \hat{Q}_\lambda^2\right) , \tag{13.2.30*}$$

may be expressed in term of the operators $\hat{a}_\lambda$ and $\hat{a}_\lambda^\dagger$ as

$$\hat{H}_\lambda = \frac{\hbar\omega_\lambda}{2}\left(\hat{a}_\lambda\,\hat{a}_\lambda^\dagger + \hat{a}_\lambda^\dagger\,\hat{a}_\lambda\right) = \left(\hat{a}_\lambda^\dagger\,\hat{a}_\lambda + \frac{1}{2}\right)\hbar\omega_\lambda\,. \tag{13.3.7}$$

Similarly, $\hat{\boldsymbol{G}}_\lambda$ defined by

$$\hat{\boldsymbol{G}}_\lambda = \frac{1}{2}\left(\frac{\boldsymbol{k}_\lambda}{\omega_\lambda}\right)\left(\hat{P}_\lambda^2 + \omega_\lambda^2\,\hat{Q}_\lambda^2\right) \tag{13.2.32*}$$

may be expressed as

$$\hat{\boldsymbol{G}}_\lambda = \frac{\hbar\boldsymbol{k}_\lambda}{2}\left(\hat{a}_\lambda\,\hat{a}_\lambda^\dagger + \hat{a}_\lambda^\dagger\,\hat{a}_\lambda\right) = \hbar\boldsymbol{k}_\lambda\left(\hat{a}_\lambda^\dagger\,\hat{a}_\lambda + \frac{1}{2}\right)\,. \tag{13.3.8}$$

Using the commutation relations (13.3.5) and (13.3.6) we can also establish the following commutation relations

$$\left[\hat{a}_\lambda, \hat{H}_\lambda\right] = \hat{a}_\lambda\hbar\omega \qquad \text{or} \qquad [\hat{a}_\lambda, \hat{n}_\lambda] = \hat{a}_\lambda\,, \tag{13.3.9}$$

$$\left[\hat{a}_\lambda^\dagger, \hat{H}_\lambda\right] = -\hat{a}_\lambda^\dagger\hbar\omega \qquad \text{or} \qquad \left[\hat{a}_\lambda^\dagger, \hat{n}_\lambda\right] = -\hat{a}_\lambda^\dagger\,, \tag{13.3.10}$$

where

$$\hat{n}_\lambda \equiv \hat{a}_\lambda^\dagger\,\hat{a}_\lambda \tag{13.3.11}$$

is the number operator. The expression for $\hat{H}_\lambda$ in terms of the operators $\hat{a}_\lambda$ and $\hat{a}_\lambda^\dagger$ and the commutation relations (13.3.9) and(13.3.10) are reminiscent of operators introduced in Chapter 5 (Sec. 5.4) for the linear harmonic oscillator.

From the form of the Hamiltonian of the radiation field $\hat{H}_{rad} = \sum_\lambda \hat{H}_\lambda$ it is clear that we are dealing with an infinite number of oscillators (one oscillator corresponding to each degree of freedom) and that there is no interaction bewteen them since the total Hamiltonian is simply the sum of the Hamiltonians for each oscillator. It folllows that the eigenstate of $\hat{H}_{rad}$ must be the product of eigenstates of $\hat{H}_\lambda$

$$|\Psi\rangle = |\Psi_1\rangle\,|\Psi_2\rangle\,\cdots\quad|\Psi_\lambda\rangle \tag{13.3.12}$$

where $|\Psi_\lambda>$ is the normalized eigenstate of $\hat{H}_\lambda$

$$\hat{H}_\lambda|\Psi_\lambda>= E_\lambda|\Psi_\lambda>\,. \tag{13.3.13}$$

To extract the physical significance of the operators $\hat{a}_\lambda$ and $\hat{a}_\lambda^\dagger$, let us interpret the states $\hat{a}_\lambda^\dagger\,|\Psi_\lambda\rangle$ and $\hat{a}_\lambda\,|\Psi_\lambda\rangle$. Using the commutation relations (13.3.9) and (13.3.10) we find that

$$\hat{H}_\lambda\,\hat{a}_\lambda^\dagger\,|\Psi_\lambda\rangle = (E_\lambda + \hbar\omega_\lambda)\,\hat{a}_\lambda^\dagger\,|\Psi_\lambda\rangle \tag{13.3.14}$$

and

$$\hat{H}_\lambda\,\hat{a}_\lambda\,|\Psi_\lambda\rangle = (E_\lambda - \hbar\omega_\lambda)\,\hat{a}_\lambda\,|\Psi_\lambda\rangle\,. \tag{13.3.15}$$

Thus $\hat{a}_\lambda^\dagger\,|\Psi_\lambda\rangle$ and $\hat{a}_\lambda\,|\Psi_\lambda\rangle$ are eigenstates of $\hat{H}_\lambda$ with eigenvalues $E_\lambda + \hbar\omega_\lambda$ and $E_\lambda - \hbar\omega_\lambda$, respectively. Operators $\hat{a}_\lambda$ and $\hat{a}_\lambda^\dagger$ may then be interpreted as the creation and annihilation operators for a photon of energy $\hbar\omega_\lambda$. It follows that $\hat{a}_\lambda^\dagger$ operating on an n_λ-photon state transforms this state into $(n_\lambda + 1)$-photon state, while $\hat{a}_\lambda$ transforms the same state into $(n_\lambda - 1)$-photon state.

The lowest energy eigenstate $|\Psi_{0\lambda}\rangle$ and the corresponding eigenvalue $E_{0\lambda}$ of $\hat{H}_\lambda$ are easily found by noting that the operation of $\hat{a}_\lambda$ on such state will result in annihilation of the state $\hat{a}_\lambda |\Psi_0\rangle = 0$. It follows from this result that

$$\hat{a}_\lambda^\dagger \hat{a}_\lambda |\Psi_{0\lambda}\rangle = 0$$

and

$$\hat{H}_\lambda |\Psi_{0\lambda}\rangle \equiv \left(\hat{a}_\lambda^\dagger \hat{a}_\lambda + \frac{1}{2}\right) \hbar\omega_\lambda |\Psi_{0\lambda}\rangle = \frac{1}{2}\hbar\omega_\lambda |\Psi_{0\lambda}\rangle$$

or

$$E_{0\lambda} = \frac{1}{2}\hbar\omega_\lambda \,. \tag{13.3.16}$$

Since the eigenvalues of $\hat{H}_\lambda$ can change in steps of $\hbar\omega_\lambda$, we have

$$E_\lambda \equiv E_{n_\lambda} = (n_\lambda + 1/2)\,\hbar\omega_\lambda \tag{13.3.17}$$

where n_λ, called the occupation number for mode λ, takes on positive integer values including zero. The eigenvalue equation for $\hat{H}_\lambda$ can then be written as

$$\hat{H}_\lambda |\Psi_{n_\lambda}\rangle = (n_\lambda + 1/2)\,\hbar\omega_\lambda |\Psi_{n_\lambda}\rangle \,, \tag{13.3.18}$$

and we may represent the n_λ-photon state $|\Psi_{n_\lambda}\rangle$ by just $|n_\lambda\rangle$ so that

$$\hat{H}_\lambda |n_\lambda\rangle = |n_\lambda + 1/2\rangle\, \hbar\omega_\lambda |n_\lambda\rangle$$

and

$$\hat{n}_\lambda |n_\lambda\rangle = n_\lambda |n_\lambda\rangle \,. \tag{13.3.19}$$

The operator $\hat{n}_\lambda$ is called the occupation number operator or simply the number operator. From the interpretation of the operators $\hat{a}_\lambda$ and $\hat{a}_\lambda^\dagger$ as annihilation and creation operators [Eqs. (13.3.14) and (13.3.15)] we may write

$$\hat{a}_\lambda^\dagger |n_\lambda\rangle = C_+ |n_\lambda + 1\rangle \tag{13.3.20}$$

and

$$\hat{a}_\lambda |n_\lambda\rangle = C_- |n_\lambda - 1\rangle \,. \tag{13.3.21}$$

The constants C_+ and C_- are easily seen to be equal to $\sqrt{n_\lambda + 1}$ and $\sqrt{n_\lambda}$, respectively.[4]

The number operators n_λ's (or $\hat{H}_\lambda$) for different modes commute with each other since there is no interaction between the oscillators. Therefore, the eigenstate of $\hat{H}$ can be written as in Eq. (13.3.12) or as

$$|\Psi >= |n_1 > |n_2 > \cdots |n_\lambda >\equiv |n_1, n_2, \cdots, n_\lambda\rangle \prod_\lambda |n_\lambda > \,. \tag{13.3.22}$$

The state of the radiation field may thus be specified in terms of the occupation numbers n_λ for different modes. A representation in which the set of commuting occupation number operators $\hat{n}_1, \hat{n}_2, \cdots, \hat{n}_\lambda$ are diagonal and the basis states are $|n_1, n_2, n_3, \cdots, n_\lambda, \cdots\rangle$ is called the occupation number representation (ONR). In the ONR, the operators $\hat{a}_\lambda, \hat{a}_\lambda^\dagger$, and $\hat{n}_\lambda$ can be represented by matrices of infinite dimensions.

---

[4]Taking the Hermitian conjugate of $\hat{a}_\lambda^\dagger |n_\lambda\rangle = C_+|n_\lambda + 1\rangle$ we find $\langle n_\lambda|\hat{a}_n = C_+^*\langle n_\lambda + 1|$ so that $\langle n_\lambda|\hat{a}_\lambda \hat{a}_\lambda^\dagger|n_\lambda\rangle = |C_+|^2\langle n_\lambda + 1|n_\lambda + 1\rangle = |C_+|^2$. But $\langle n_\lambda|\hat{a}_\lambda a_\lambda^\dagger|n_\lambda\rangle = \langle n_\lambda|\hat{a}_\lambda^\dagger a_\lambda + 1|n_\lambda\rangle = (n_\lambda + 1)$, giving us $|C_+| = \sqrt{n_\lambda + 1}$. Similarly, from $\hat{a}_\lambda|n_\lambda\rangle = C_-|n_\lambda - 1\rangle$ and its Hermitian conjugate we obtain $\langle n_\lambda|\, \hat{a}_\lambda^\dagger \hat{a}_\lambda|n_\lambda\rangle = |C_-|^2\langle n_\lambda - 1|n_\lambda - 1\rangle$. This combined with $\langle n_\lambda|\, \hat{a}_\lambda^\dagger \hat{a}_\lambda|n_\lambda\rangle = n_\lambda$ gives us $|C_-| = \sqrt{n_\lambda}$. We choose C_+ and C_- to be real and positive.

The Electromagnetic Vacuum

The state $|\Psi_o>=|n_1=0, n_2=0, n_3=0, \cdots n_\lambda=0, \cdots>$ in which all modes (corresponding to any wave vector and polarization) have zero occupation number is called the vacuum state. This definition is of utmost importance as any state of the electromagnetic field can be obtained by application of creation operators on it. The application of any destruction operator on it gives zero,

$$\hat{a}_\lambda |\Psi_0\rangle = 0 . \tag{13.3.23}$$

Now the field operators $\hat{\boldsymbol{E}}$ and $\hat{\boldsymbol{B}}$ do not commute with $\hat{H}_\lambda$ or the number operator $\hat{n}_\lambda$ and the number of photons in any mode is precisely known (to be zero) in the vacuum state. It follows that the field amplitudes are indeterminate in the vacuum state. Indeed, while the average values of both $\hat{\boldsymbol{E}}$ and $\hat{\boldsymbol{B}}$ in the vacuum state vanish, average values of fluctuations from the mean $\langle(\Delta\hat{\boldsymbol{E}})^2\rangle_0 \equiv \langle(\hat{\boldsymbol{E}} - \langle\hat{\boldsymbol{E}}\rangle)^2\rangle_0$ and $\langle(\Delta\hat{\boldsymbol{B}})^2\rangle \equiv \langle(\hat{\boldsymbol{B}} - \langle_0\hat{\boldsymbol{B}}\rangle)^2\rangle_0$ do not and, in fact, both diverge. Thus the electromagnetic vacuum is no more a placid region of space characterized by the absence of any quanta but a state of ceaseless activity.

13.4 Interaction of Matter with Quantized Radiation Field

The Hamiltonian of an atomic electron (charge $-e$ and mass m) in an electromagnetic field characterized by vector potential $\hat{\boldsymbol{A}}$ and potential $\hat{V}(\boldsymbol{r})$ (which decsribes all external forces as well as electrostatic Coulomb interaction) is given by

$$\hat{H} = \frac{[\hat{\boldsymbol{p}} + e\hat{\boldsymbol{A}}(\boldsymbol{r}, t)]^2}{2m} + \hat{V}(\boldsymbol{r}) + \hat{H}_{rad} \tag{13.4.1}$$

where we treat the electron non-relativistically. Expanding the first term on the right-hand side and recalling that each operator acts to its right, we obtain

$$\hat{\boldsymbol{p}} \cdot \hat{\boldsymbol{A}} + \hat{\boldsymbol{A}} \cdot \hat{\boldsymbol{p}} = -i\hbar\boldsymbol{\nabla} \cdot \hat{\boldsymbol{A}} + 2\hat{\boldsymbol{A}} \cdot (-i\hbar\boldsymbol{\nabla}) = 2\hat{\boldsymbol{A}} \cdot \hat{\boldsymbol{p}} , \tag{13.4.2}$$

where we have used the transversality condition $\boldsymbol{\nabla} \cdot \hat{\boldsymbol{A}} = 0$ [Eq. (13.2.1)]. Using this result and the Hamiltonian for the quantized radiation field, we find the total Hamiltonian of the system (atom + radiation field) can be written as

$$\hat{H} = \frac{\hat{p}^2}{2m} + \hat{V} + \sum_\lambda \Big(\hat{a}_\lambda^\dagger \hat{a}_\lambda + 1/2\Big)\hbar\omega_\lambda + \frac{e}{m}\hat{\boldsymbol{A}} \cdot \hat{\boldsymbol{p}} + \frac{e^2}{2m}\hat{\boldsymbol{A}}^2 \equiv \hat{H}_0 + \hat{H}' , \tag{13.4.3}$$

where
$$\hat{H}_0 = \frac{\hat{p}^2}{2m} + \hat{V}(r) + \sum_\lambda \Big(\hat{a}_\lambda^\dagger \hat{a}_\lambda + 1/2\Big)\hbar\omega_\lambda \tag{13.4.4}$$

is the unperturbed Hamiltonian of the atomic system and the radiation field and

$$\hat{H}' = \frac{e}{m}\hat{\boldsymbol{A}} \cdot \hat{\boldsymbol{p}} + \frac{e^2}{2m}\hat{\boldsymbol{A}}^2 \tag{13.4.5}$$

may be treated as the perturbation, being the interaction of the atomic system with the electromagnetic field. In what follows we shall drop the $\hat{\boldsymbol{A}}^2$-term compared to the first term since it is of second order in the field.

With the help of Eqs. (13.2.8), (13.2.12), (13.2.32), (13.2.33) (13.3.3), and (13.3.4), the operator $\hat{\boldsymbol{A}}$ representing the vector potential is given by

$$\hat{\boldsymbol{A}}(\boldsymbol{r}) = \sum_{\lambda} \sqrt{\frac{\hbar}{2\epsilon_0 \omega_\lambda V}} \left(\boldsymbol{\varepsilon}_\lambda \hat{a}_\lambda e^{i\boldsymbol{k}_\lambda \cdot \boldsymbol{r}} + \boldsymbol{\varepsilon}^*_\lambda \hat{a}^\dagger_\lambda e^{-i\boldsymbol{k}_\lambda \cdot \boldsymbol{r}} \right). \tag{13.4.6}$$

Using this in the perturbation Hamiltonian and dropping the $\hat{\boldsymbol{A}}^2$ term, we obtain

$$\hat{H}' = \frac{e}{m} \sum_{\lambda} \sqrt{\frac{\hbar}{2\epsilon_0 \omega_\lambda V}} \left(\boldsymbol{\varepsilon}_\lambda \hat{a}_\lambda e^{i\boldsymbol{k}_\lambda \cdot \boldsymbol{r}} + \boldsymbol{\varepsilon}^*_\lambda \hat{a}^\dagger_\lambda e^{-i\boldsymbol{k}_\lambda \cdot \boldsymbol{r}} \right) \cdot \hat{\boldsymbol{p}}. \tag{13.4.7}$$

The eigenstate $|\Psi_i\rangle$ of the unperturbed Hamiltonian $\hat{H}_0$ can be written as the product of the unperturbed state $|u_i\rangle$ of the atomic electron belonging to energy ϵ_i and the unperturbed state of the electromagnetic field $\prod_\lambda |n_\lambda\rangle$ as

$$|\Psi_i\rangle = |u_i\rangle \prod_\lambda |n_\lambda\rangle\,, \tag{13.4.8a}$$

where

$$\hat{H}_0\,|\Psi_i\rangle = E_i\,|\Psi_i\rangle \tag{13.4.8b}$$

and

$$E_i = \varepsilon_i + \sum_\lambda (n_\lambda + 1/2)\,\hbar\omega_\lambda\,. \tag{13.4.8c}$$

Here n_λ is the number of photons in mode λ and ε_i is the energy of the atomic system in the i-th state. To find the probability that the interaction between the atomic electron and the radiation field produces a transition from atomic state $|u_i\rangle$ to some other atomic state $|u_f\rangle$, we must find the matrix element of $\hat{H}'$ between the initial state $|u_i\rangle\,|n_1, n_2, \cdots\rangle \equiv |i;\{n\}\rangle$ and the final state $|u_f\rangle\,|m_1, m_2, \cdots\rangle \equiv |f;\{m\}\rangle$ of the radiation field and the atomic system. This matrix element is given by

$$\begin{aligned}\langle f;\{m\}|\,\hat{H}'\,|i;\{n\}\rangle = \frac{e}{m}\sum_\lambda \sqrt{\frac{\hbar}{2\epsilon_0\omega_\lambda V}} \Big\{ & \int u_f^*(\boldsymbol{r})\, e^{i\boldsymbol{k}_\lambda\cdot\boldsymbol{r}}\, (\boldsymbol{\varepsilon}_\lambda\cdot\hat{\boldsymbol{p}})\; u_i(\boldsymbol{r}) d\tau \times \langle\{m\}|\,\hat{a}_\lambda\,|\{n\}\rangle \\ & + \int u_f^*(\boldsymbol{r})\, e^{-i\boldsymbol{k}_\lambda\cdot\boldsymbol{r}}\, (\boldsymbol{\varepsilon}^*_\lambda\cdot\hat{\boldsymbol{p}})\; u_i(\boldsymbol{r}) d\tau \times \langle\{m\}|\,\hat{a}^\dagger_\lambda\,|\{n\}\rangle \Big\}\,.\end{aligned}$$

The operators $\hat{a}_\lambda$ and $\hat{a}^\dagger_\lambda$ affect only the occupation number of mode λ. The matrix element of $\hat{a}_\lambda$ vanishes unless $m_\lambda = n_\lambda - 1$ while that of $\hat{a}^\dagger_\lambda$ vanishes unless $m_\lambda = n_\lambda + 1$. There are, therefore, only two non-vanishing matrix elements of $\hat{H}'$

$$\langle f; n_\lambda - 1|\,\hat{H}'\,|i; n_\lambda\rangle = \frac{e}{m}\sqrt{\frac{\hbar n_\lambda}{2\epsilon_0\omega_\lambda V}} \int u_f^*\; e^{i\boldsymbol{k}_\lambda\cdot\boldsymbol{r}}\, (\boldsymbol{\varepsilon}_\lambda\cdot\hat{\boldsymbol{p}})\; u_i\, d\tau \tag{13.4.9}$$

and

$$\langle f; n_\lambda + 1|\,\hat{H}'\,|i; n_\lambda\rangle = \frac{e}{m}\sqrt{\frac{\hbar (n_\lambda + 1)}{2\epsilon_0\omega_\lambda V}} \int u_f^*\; e^{-i\boldsymbol{k}_\lambda\cdot\boldsymbol{r}}\, (\boldsymbol{\varepsilon}_\lambda\cdot\hat{\boldsymbol{p}})\; u_i\, d\tau\,. \tag{13.4.10}$$

The first matrix element refers to absorption, as it represents a transition in which the number of quanta in the field is reduced by one. The probability of this process, which is proportional to the square of the matrix element is proportional to n_λ (the number of quanta present) or to the intensity of the radiation field. This is therefore *induced absorption*.

The second matrix element refers to a process in which the number of quanta is increased by one. It represents an emissive process and its probability is proportional to $(n_\lambda + 1)$. Even if the intensity of radiation is reduced to zero ($n_\lambda = 0$), there is a non-zero probability of emission. This part of emission probability, which is independent of n_λ (or which is non-zero even if $n_\lambda = 0$) refers to the process of *spontaneous emission*. The other part of the emission probability which is dependent on n_λ refers to the process of *induced (or stimulated) emission*. The Einstein relationship between the probabilities of spontaneous emission and of induced emission can be verified in the light of this theory.

It must be kept in mind that absorptive and emissive transitions connect the initial state to different final states. In the case of absorption, the final state is higher in energy than the initial state whereas in the case of emission, the final state is lower in energy. To avoid confusion, therefore, we will write the matrix element for absorption and emission as

$$M_{ia}^{(ab)} \equiv \langle a; n_\lambda - 1|\, \hat{H}'\, |i; n_\lambda\rangle = \frac{e}{m}\sqrt{\frac{\hbar n_\lambda}{2\epsilon_0\omega_\lambda V}} \int u_a^*\, e^{i\boldsymbol{k}_\lambda\cdot\boldsymbol{r}}\, (\boldsymbol{\varepsilon}_\lambda \cdot \hat{\boldsymbol{p}})\, u_i\, d\tau \tag{13.4.11}$$

$$M_{ib}^{(em)} \equiv \langle b; n_\lambda + 1|\, \hat{H}'\, |i; n_\lambda\rangle = \frac{e}{m}\sqrt{\frac{\hbar(n_\lambda+1)}{2\epsilon_0\omega_\lambda V}} \int u_b^*\, e^{-i\boldsymbol{k}_\lambda\cdot\boldsymbol{r}}\, (\boldsymbol{\varepsilon}_\lambda \cdot \hat{\boldsymbol{p}})\, u_i\, d\tau \tag{13.4.12}$$

with $E_a > E_i$ and $E_b < E_i$.

It is of interest to compare the expressions (13.4.11) and (13.4.12) with the expressions we obtain from the semi-classical theory in which the vector potential is treated classically [Eqs. (10.3.18) and (10.3.19)]. We note that for absorption and emission processes we may define equivalent classical vector potentials for absorption and emission as

$$A_{ab} = \sqrt{\frac{\hbar n_\lambda}{2\epsilon_0\omega_\lambda V}}\, \boldsymbol{\varepsilon}_\lambda\, e^{i(\boldsymbol{k}_\lambda\cdot\boldsymbol{r} - \omega_\lambda t)} \tag{13.4.13}$$

and

$$A_{em} = \sqrt{\frac{\hbar(n_\lambda+1)}{2\epsilon_0\omega_\lambda V}}\, \boldsymbol{\varepsilon}_\lambda^*\, e^{-i(\boldsymbol{k}_\lambda\cdot\boldsymbol{r} - \omega_\lambda t)}\,. \tag{13.4.14}$$

With the equivalent vector potentials (13.4.13) and (13.4.14), respectively, for the absorption and emission of a light quantum by a charged particle we can obtain from the semi-classical theory matrix elements for one-photon transitions which are identical with those given by a quantized field theory. Caution must be exercised in using them for multi-photon processes as they can lead to incorrect results.

Einstein Coefficients

According to Fermi's golden rule, the probability per unit time for the emissive transition is given by

$$w_{i\to f} = \frac{2\pi}{\hbar}\left|\langle f, n_\lambda + 1|\, \hat{H}'\, |i, n_\lambda\rangle\right|^2 \rho(E_\lambda) \tag{13.4.15}$$

where the matrix element $\langle f, n_\lambda + 1|\, \hat{H}'\, |i, n_\lambda\rangle$ is given by Eq. (13.4.12) with $\langle b| = \langle f|$ and the density of final states $\rho(E_\lambda)$ (number of states of the radiation field per unit energy interval within a volume V of the configuration space) is given by[5]

$$\rho_E(E_\lambda)dE_\lambda = V\rho_\nu(\nu_\lambda)d\nu_\lambda\,. \tag{13.4.16}$$

[5] While atomic states are discrete, the states of the radiation field form a continuum. In our treatment, the radiation field is confined within a volume V and this effectively discretizes the frequencies, the density of state being given by Eq. (13.4.17).

Here $\rho_\nu(\nu_\lambda)d\nu_\lambda$ is the number of frequencies in the range ν_λ and $\nu_\lambda + d\nu_\lambda$ per unit volume and is given by

$$\rho_\nu(\nu_\lambda)d\nu_\lambda = \frac{8\pi\,\nu_\lambda^2\,d\nu_\lambda}{c^3} = \rho_\omega(\omega_\lambda)d\omega_\lambda\,. \tag{13.4.17}$$

Using this in Eq. (13.4.20) we find, with the help of the relation $E_\lambda = h\nu_\lambda = \hbar\omega_\lambda$,

$$\rho_\omega(\omega_\lambda) = \frac{\omega_\lambda^2}{\pi^2 c^3}\,, \tag{13.4.18}$$

$$\rho_E(E_\lambda) = V\frac{\rho_\omega(\omega_\lambda)}{\hbar}\,. \tag{13.4.19}$$

Substituting this density of states in Eq. (13.4.19) we find that the transition rate is given by [6]

$$w_{i\to f} = \frac{\pi\,e^2}{\epsilon_0 m^2\omega_\lambda^2}\,\hbar\omega_\lambda(n_\lambda+1)\rho(\omega_\lambda)\left|\int u_f^*\,e^{-i\,\boldsymbol{k}_\lambda\cdot\boldsymbol{r}}\boldsymbol{\varepsilon}_\lambda^*\cdot\nabla\,u_i\,d\tau\right|^2\,. \tag{13.4.20}$$

Now the energy density $I(\omega_\lambda)d\omega_\lambda$ of the radiation field in the frequency interval $d\omega_\lambda$ centered at ω_λ is proportional to the number of photons[7] n_λ in the mode λ

$$I(\omega_\lambda)d\omega_\lambda = \rho(\omega_\lambda)d\omega_\lambda\,\hbar\omega_\lambda\,n_\lambda \tag{13.4.21}$$

Using Eqs. (13.4.21) and (13.4.18) in Eq. (13.4.20) and dropping the suffix λ for simplicity, we can write the rate of emissive transition as the sum of two terms

$$w_{i\to f} = I(\omega)\,B_{i\to f} + A_{i\to f}\,, \tag{13.4.22}$$

where the first term is proportional to the spectral energy density $I(\omega)$ of radiation at frequency ω while the second term is independent of the spectral energy density. Coefficients $A_{i\to f}$ and $B_{i\to f}$ are precisely the Einstein A and B coefficients given by

$$B_{i\to f} = \frac{\pi\,e^2}{\epsilon_0 m^2\omega^2}\left|\int u_f^*\,e^{-i\,\boldsymbol{k}\cdot\boldsymbol{r}}\boldsymbol{\varepsilon}^*\cdot\boldsymbol{\nabla}u_i\,d\tau\right|^2 \tag{13.4.23}$$

and $$A_{i\to f} = \frac{\pi\,e^2}{\epsilon_0 m^2\omega^2}\,\frac{\hbar\omega^3}{\pi^2c^3}\left|\int u_f^*\,e^{-i\,\boldsymbol{k}\cdot\boldsymbol{r}}\boldsymbol{\varepsilon}^*\cdot\boldsymbol{\nabla}u_i\,d\tau\right|^2\,. \tag{13.4.24}$$

The ratio of Einstein coefficients is given by

$$\frac{A_{i\to f}}{B_{i\to f}} = \frac{\hbar\omega^3}{\pi^2c^3}\,. \tag{13.4.25}$$

[6] Noting that $\hat{\boldsymbol{p}} \to -i\hbar\nabla$ in coordinate representation, we have

$$\left|\int u_f^*\,e^{-i\boldsymbol{k}_\lambda\cdot\boldsymbol{r}}\,(\boldsymbol{\varepsilon}_\lambda^*\cdot\hat{\boldsymbol{p}})\,u_i d\tau\right|^2 = \hbar^2\left|\int u_{n0}^*\,e^{-i\boldsymbol{k}_\lambda\cdot\boldsymbol{r}}\,\boldsymbol{\varepsilon}_\lambda^*\cdot\boldsymbol{\nabla}u_i\,d\tau\right|^2\,.$$

[7] For radiation in thermal equilibrium at temperature T, the occupation number obeys Bose-Einstein distribution law $n_\lambda = 1/[e^{\hbar\omega_\lambda/k_BT} - 1]$. This relation is, however, not used here.

13.5 Applications

The Photo-electric Effect

As a simple application of quantization of radiation field, we consider the photo-emission of electrons when radiation is incident on the surface of a material. In this treatment we shall consider the electromagnetic field to be quantized but treat the electron non-relativistically. For this first order process, in which a photon is absorbed by an atom and an electron is excited from a bound state[8]

$$u_i(\vec{r}) = \sqrt{\frac{Z^3}{\pi a_o^3}} \exp(-Zr/a_o)\,, \tag{13.5.1}$$

where $a_o = \frac{4\pi\epsilon_0\hbar^2}{me^2}$ is the Bohr radius, to a free state

$$u_f(\boldsymbol{r}) = \frac{1}{\sqrt{V}}\, e^{i\boldsymbol{k}\cdot\boldsymbol{r}}\,, \tag{13.5.2}$$

where $\boldsymbol{p} = \hbar\boldsymbol{k}$ is the momentum of the electron in the final state. We have assumed that the final electron energy is large enough that it can be treated like a free particle in the final state unaffected by the long-range Coulomb potential. Since in photo-electron emission a photon is absorbed, the relevant transition matrix element (13.4.11) is given by

$$H'_{fi} = \frac{e}{m}\sqrt{\frac{\hbar n_\lambda}{2\epsilon_0\omega_\lambda V}} \int u_f^*(\vec{r})\, e^{i\boldsymbol{k}_\lambda\cdot\boldsymbol{r}}\, (\boldsymbol{\varepsilon}_\lambda\cdot\hat{\boldsymbol{p}})\, u_i(\boldsymbol{r})\, d\tau\,. \tag{13.5.3}$$

Substituting for u_f and u_i in Eq. (13.5.3), the matrix element can be written as

$$H'_{fi} = \frac{e}{m}\sqrt{\frac{\hbar n_\lambda}{2\epsilon_0\omega_\lambda V}} (\boldsymbol{\varepsilon}_\lambda\cdot\hbar\boldsymbol{k}) \sqrt{\frac{Z^3}{\pi a_o^3 V}}\; 2\pi\, I \tag{13.5.4}$$

where the integral I is given by

$$I \equiv \int_0^\infty e^{-Zr/a_o}\, r^2 dr \int_{-1}^{1} e^{i\beta\, r\,\zeta}\, d\zeta = \frac{4Z/a_o}{[(Z/a_o)^2+\beta^2]^2} \tag{13.5.5}$$

where

$$\boldsymbol{\beta} = \boldsymbol{k}_\lambda - \boldsymbol{k} \quad \text{and} \quad \boldsymbol{\beta}\cdot\boldsymbol{r} = \beta r\zeta\,. \tag{13.5.6}$$

Then the transition probability per unit time $w_{i\to f}$ is

$$w_{i\to f} = \frac{2\pi}{\hbar}|H'_{fi}|^2\rho_E\,, \tag{13.5.7}$$

where the density ρ_E of final electronic states with momentum $\hbar\boldsymbol{k}$ pointing into the solid angle $d\Omega$ is given by [Sec. 10.2]

$$\rho_E = \frac{Vk^2\, d\Omega}{(2\pi)^3}\frac{dk}{dE}\,. \tag{13.5.8}$$

[8]For simplicity we assume the atom to be Hydrogen-like and initially in the ground state with binding energy $W = \frac{Z^2e^2}{8\pi\epsilon_0}\frac{me^2}{4\pi\epsilon_0\hbar^2} = \frac{Z^2e^2}{8\pi\epsilon_0 a_o}$.

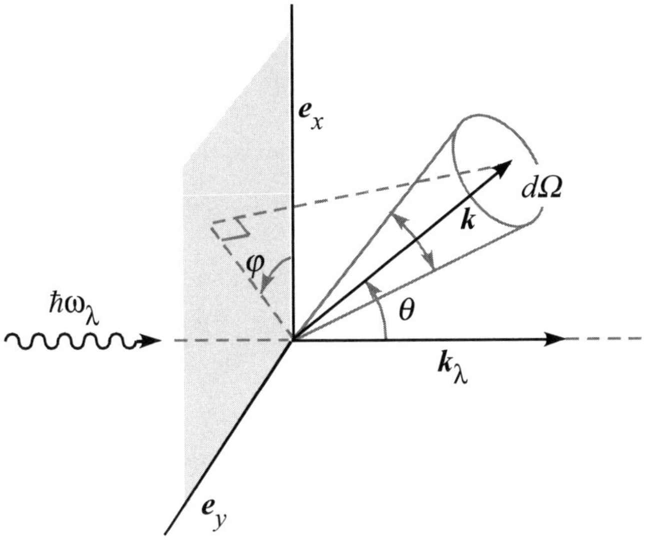

FIGURE 13.1
In the photo-eletric effect an incident photon of energy $\hbar\omega_\lambda$ and polarization ε_λ is absorbed followed by the emission of an electron traveling with a momentum $\hbar\boldsymbol{k}$ and energy $\hbar^2k^2/2m$ given by Einstein's photo-electric equation.

Since $E = \hbar^2k^2/2m$ for the electron, we have $kdk/dE = m/\hbar^2$, giving us

$$\rho_E = \frac{Vkmd\Omega}{(2\pi)^3\hbar^2}\,. \tag{13.5.9}$$

Now n_λ photons in mode volume V correspond to incident photon flux density $n_\lambda\, c/V$. Then introducing the differential cross-section for photo-electron emission by

$$d\sigma = \frac{\text{No. of electrons ejected per/second within solid angle } d\Omega}{\text{Incident photon flux density}} = \frac{Vw_{i\to f}}{n_\lambda c}\,, \tag{13.5.10}$$

and using the expressions for $w_{i\to f}$ and the incident photon flux density, we find

$$\frac{d\sigma}{d\Omega} = \frac{e^2}{4\pi\epsilon_0\hbar c}\,\frac{32\hbar k}{m\omega_\lambda}\,(\varepsilon_\lambda\cdot\boldsymbol{k})^2\,\frac{\left(\frac{Z}{a_0}\right)^5}{\left[\frac{Z^2}{a_0^2}+|\boldsymbol{k}_\lambda-\boldsymbol{k}|^2\right]^4}\,. \tag{13.5.11}$$

This expression contains the angular dependence of photo-electron emisssion relative to the direction of incident photon wave vector $\boldsymbol{k}_\lambda$ and its polarization ε_λ. If photon energy is large compared to the binding energy of the electron but small compared with its rest mass energy ($W \ll \hbar\omega_\lambda \ll mc^2$), we can simplify this expression. According to Einstein's photo-electric equation,

$$\hbar\omega_\lambda = \frac{h^2k^2}{2m} + W = \frac{\hbar^2}{2m}\left[k^2 + \frac{Z^2}{a_o^2}\right] \Rightarrow k^2 + \frac{Z^2}{a_o^2} = \frac{2m\omega_\lambda}{\hbar}\,, \tag{13.5.12a}$$

so that

$$\frac{Z^2}{a_o^2} + |\boldsymbol{k}_\lambda - \boldsymbol{k}|^2 = \frac{2m\omega_\lambda}{\hbar} - 2k_\lambda k\cos\theta + k_\lambda^2 = \frac{2m\omega_\lambda}{\hbar}\left[1 - \frac{v\cos\theta}{c} + \frac{\hbar\omega_\lambda}{2mc^2}\right]$$

$$\approx \frac{2m\omega_\lambda}{\hbar}\left[1 - \frac{v\cos\theta}{c}\right] \qquad (\hbar\omega_\lambda \ll mc^2) \tag{13.5.12b}$$

where the binding energy of the electron is $W = \frac{mZ^2e^4}{2(4\pi\epsilon_0)^2\hbar^2} = \frac{\hbar^2}{2m}\frac{Z^2}{a_o^2}$, $\omega_\lambda = k_\lambda c$, the velocity of the electron is $v = hk/m$ and the polar axis (z-axis) is coincident with the direction of the incident photon $\boldsymbol{k}_\lambda$ so that $\boldsymbol{k}_\lambda \cdot \boldsymbol{k} = k_\lambda k \cos\theta$. The last step in Eq. (13.5.12b) follows since the energy of the photon is much less than the rest energy of the electron. If we take the polarization vector $\boldsymbol{\varepsilon}_\lambda$ to be parallel to the x-axis, then

$$\boldsymbol{\varepsilon}_\lambda \cdot \boldsymbol{k} = k \sin\theta \cos\varphi$$

and the scattering cross-section can be written as

$$\frac{d\sigma}{d\Omega} = \left(\frac{e^2}{4\pi\epsilon_0\hbar c}\right) 64 \left(\frac{ka_o}{Z}\right)^3 \left(\frac{W}{\hbar\omega_\lambda}\right)^5 \frac{(a_o/Z)^2 \sin^2\theta\cos^2\varphi}{\left(1-\frac{v}{c}\cos\theta\right)^4}. \tag{13.5.13}$$

For an unpolarized (equal mixture of polarizations along x and y-directions) photon beam, we replace $|\boldsymbol{\varepsilon}_\lambda \cdot \boldsymbol{k}|^2 = k^2\sin^2\theta\cos^2\varphi$ by its average value $k^2\sin^2\theta\ \frac{1}{2}(\cos^2\varphi + \sin^2\varphi) = \frac{1}{2}k^2\sin^2\theta$. We then obtain

$$\frac{d\sigma}{d\Omega} = \left(\frac{e^2}{4\pi\epsilon_0\hbar c}\right) 32 \left(\frac{ka_o}{Z}\right)^3 \left(\frac{W}{\hbar\omega_\lambda}\right)^5 \frac{(a_o/Z)^2 \sin^2\theta}{\left(1-\frac{v}{c}\cos\theta\right)^4}. \tag{13.5.14}$$

We note that $\frac{d\sigma}{d\Omega}$ vanishes in the forward direction $\theta = 0$ so that no electrons are emitted in the direction of the incident photon. Maximum emission would take place in the direction $\theta = \pi/2, \varphi = 0$, i.e., the direction of polarization of the photon, but for the denominator which tends to shift the maximum slightly in the forward direction. Finally, with the approximation $\hbar^2k^2/2m \approx \hbar\omega_\lambda$, which is valid for $\hbar\omega_\lambda \gg W$, which we are considering here, we obtain

$$\frac{d\sigma}{d\Omega} = \left(\frac{e^2}{4\pi\epsilon_0\hbar c}\right) 64 \left(\frac{W}{\hbar\omega_\lambda}\right)^{7/2} \frac{(a_o/Z)^2 \sin^2\theta\cos^2\varphi}{\left(1-\frac{v}{c}\cos\theta\right)^4}. \tag{13.5.15}$$

Thus the cross-section varies as $Z^5/(\hbar\omega_\lambda)^{7/2}$.

Rayleigh, Thomson and Raman Scattering

We have seen that the emission or absorption of a photon by an atomic system may be investigated by considering the interaction of the atomic system with the time-dependent vector potential $\hat{\boldsymbol{A}}(\boldsymbol{r}_i, t)$, which may be treated as a field operator expressible in terms of the creation and annihilation operators as

$$\hat{\boldsymbol{A}}(\boldsymbol{r}, t) = \sum_\lambda \sqrt{\frac{\hbar}{2\epsilon_0\omega_\lambda V}} \left(\boldsymbol{\varepsilon}_\lambda \hat{a}_\lambda\, e^{i(\boldsymbol{k}_\lambda\cdot\boldsymbol{r}-\omega_\lambda t)} + \boldsymbol{\varepsilon}_\lambda \hat{a}_\lambda^\dagger\, e^{-i(\boldsymbol{k}_\lambda\cdot\boldsymbol{r}-\omega_\lambda t)}\right), \tag{13.5.16}$$

where, for convenience of writing, we choose a real polarization basis. The Hamiltonian for interaction between atomic electrons and the radiation field is given by Eq. (13.4.5)

$$\hat{H}_{\text{int}}(t) = \frac{e}{m}\sum_i \hat{\boldsymbol{A}}(\boldsymbol{r}_i, t)\cdot\hat{\boldsymbol{p}}_i + \frac{e^2}{2m}\sum_i \hat{\boldsymbol{A}}(\boldsymbol{r}_i, t)\cdot\hat{\boldsymbol{A}}(\boldsymbol{r}_i, t) \equiv \hat{H}_{\text{int}}^{(1)} + \hat{H}_{\text{int}}^{(2)}, \tag{13.5.17}$$

where the summation is over atomic electrons, labeled by i, that participate in the interaction. The summation over i and the index i may be dropped if only one electron participates in the interaction.

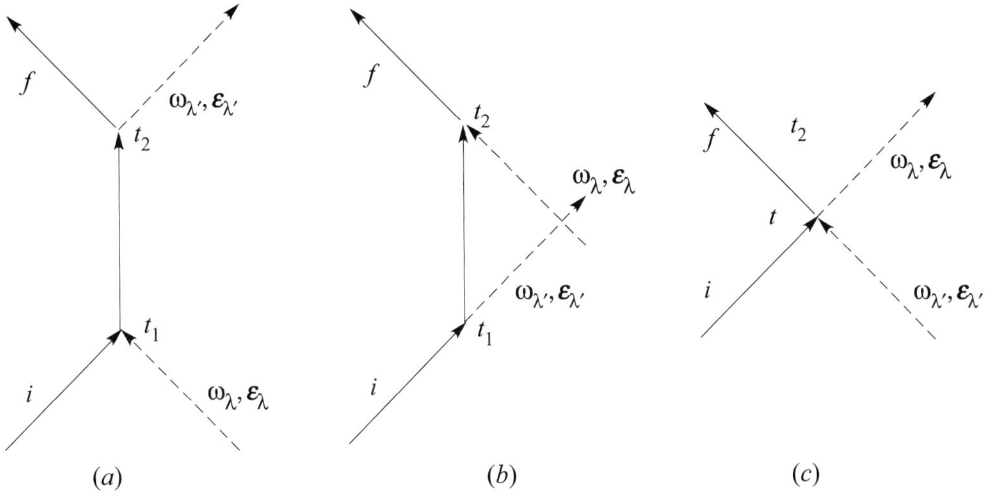

FIGURE 13.2
Figures (a) and (b) represent annihilation and creation (or creation and annihilation) of a photon in succession. Figure (c) indicates this as a one step process.

The processes we are considering here are those in which no change in the number of photons occurs. Since the linear term $\hat{\boldsymbol{A}} \cdot \hat{\boldsymbol{p}}$ changes the number of photons by one, it makes no contribution to the transition matrix element for these processes in the first order of perturbation. Such transition may be brought about either (i) by the linear (in vector potential) term $\hat{\boldsymbol{A}} \cdot \hat{\boldsymbol{p}}$ in second order of perturbation or (ii) by the quadratic (in vector potential) term $\hat{\boldsymbol{A}} \cdot \hat{\boldsymbol{A}}$ in first order of perturbation.

These contributions can be visualized by drawing space-time diagrams (Feynman diagram) [Fig. 13.2]. In such diagrams, which will be discussed in more detail in Chapter 14, a solid line represents the atomic electron and a broken line represents a photon. Time axis runs upward.

The $\hat{\boldsymbol{A}} \cdot \hat{\boldsymbol{p}}$ interaction acting at time t_1 can either annihilate the incoming photon $(\omega_\lambda, \varepsilon_\lambda)$ [or create the outgoing photon $(\omega'_\lambda, \varepsilon'_\lambda)$]. When $\hat{\boldsymbol{A}} \cdot \hat{\boldsymbol{p}}$ interaction acts again at time $t_2 (> t_1)$ it must necessarily create the outgoing photon $(\omega_{\lambda'}, \varepsilon_{\lambda'})$ if the same has not been created [Fig. 13.2(a)] or annihilate the incoming photon $(\omega_\lambda, \varepsilon_\lambda$ if the same has not already been annihilated [Fig. 13.2(b)]; otherwise we get a zero matrix element. The $\hat{\boldsymbol{A}} \cdot \hat{\boldsymbol{A}}$ term involves the operators $\hat{a}\hat{a}^\dagger$, $\hat{a}^\dagger\hat{a}$, $\hat{a}\hat{a}$, and $\hat{a}^\dagger\hat{a}^\dagger$. Obviously, only the first two give a non-vanishing contribution to the matrix element as they can, in succession, annihilate and create (or create and annihilate) photons resulting in no change in the number of photons [Figs. 13.2(a) and 13.2(b)]. The one-step process of annihilation and creation of a photon is indicated by the graph [Fig. 13.2(c)].

To find the probability of transition per unit time, from the atomic state $|i\rangle$ to the state $|f\rangle$, we use the time-dependent perturbation theory by expanding the total wave function of the system as

$$\Psi(\boldsymbol{r}, t) = \sum_j C_j(t)\, u_j(r)\, e^{-iE_jt/\hbar} \tag{13.5.18}$$

and invoking a perturbation expansion for $C_j(t)$

$$C_j(t) = \delta_{ji} + C_j^{(1)}(t) + C_j^{(2)}(t) + \cdots \tag{13.5.19}$$

where the first term corresponds to the atom initially in the state $|i\rangle$ and $C_j^{(1)}(t)$, $C_j^{(2)}(t)$ etc. are, respectively, the first, second and higher order corrections to amplitude of the j-th state. In the present treatment we consider terms only up to the second order. Using the first and second order time-dependent perturbation theory we get the first and second order amplitudes for transition to state $|f\rangle$ [Chapter 10]:

$$C_f^{(1)}(t) = \frac{1}{i\hbar}\int_0^t dt_1 \langle f, \omega_{\lambda'}, \boldsymbol{\varepsilon}_{\lambda'}|\, \hat{H}_{\text{int}}(t_1)\, |i, \omega_\lambda, \boldsymbol{\varepsilon}_\lambda\rangle\; e^{i(E_f - E_i)\, t_1/\hbar}$$

$$C_f^{(2)}(t) = \frac{1}{(i\hbar)^2}\int_0^t dt_2 \sum_I \int_0^{t_2} dt_1 \langle f, \omega_{\lambda'}, \boldsymbol{\varepsilon}_{\lambda'}|\, \hat{H}_{\text{int}}(t_2)\, |I, \omega_{\lambda_I}, \boldsymbol{\varepsilon}_{\lambda_I}\rangle\; e^{i(E_f - E_I)t_2/\hbar} \times \langle I, \omega_{\lambda_I} \boldsymbol{\varepsilon}_{\lambda_I}|\, \hat{H}_{\text{int}}(t_1)\, |i, \omega_\lambda, \boldsymbol{\varepsilon}_\lambda\rangle\; e^{i(E_I - E_i)\, t_1/\hbar}\, .$$

Here the index I labels the intermediate states of the atom and field. Since we are considering only the second order [in vector potential $\hat{\boldsymbol{A}}$] processes, we use $\hat{\boldsymbol{A}} \cdot \hat{\boldsymbol{A}}$ part of the interaction Hamiltonian in the first equation and $\hat{\boldsymbol{A}} \cdot \hat{\boldsymbol{p}}$ part in the second equation. Thus both $C_f^{(1)}(t)$ and $C_f^{(2)}(t)$ are second order terms in vector potential. We then obtain

$$C_f^{(1)}(t) = \frac{1}{i\hbar}\frac{e^2}{2m}\int_0^t dt_1 \langle f, \omega_{\lambda'}, \boldsymbol{\varepsilon}_{\lambda'}|\, \hat{\boldsymbol{A}}(\boldsymbol{r}, t_1) \cdot \hat{\boldsymbol{A}}(\boldsymbol{r}, t_1)\, |i, \omega_\lambda, \boldsymbol{\varepsilon}_\lambda\rangle\; e^{i(E_f - E_i)\, t_1/\hbar}\, ,$$

$$C_f^{(2)}(t) = \frac{1}{(i\hbar)^2}\frac{e^2}{m^2}\int_0^t dt_2 \sum_I \int_0^{t_2} dt_1 \langle f, \omega_{\lambda'}, \boldsymbol{\varepsilon}_{\lambda'}|\, \hat{\boldsymbol{A}}(\boldsymbol{r}, t_2) \cdot \hat{\boldsymbol{p}}\, |I, \omega_{\lambda_I}, \boldsymbol{\varepsilon}_{\lambda_I}\rangle\; e^{i(E_f - E_I)t_2/\hbar} \times \langle I, \omega_{\lambda_I}, \boldsymbol{\varepsilon}_{\lambda_I}|\, \hat{\boldsymbol{A}}(r, t_1) \cdot \hat{\boldsymbol{p}}\, |i, \omega_\lambda, \boldsymbol{\varepsilon}_\lambda\rangle\; e^{i(E_I - E_i)t_1/\hbar}\, .$$

The matrix element for $\hat{\boldsymbol{A}} \cdot \hat{\boldsymbol{A}}$ interaction is given by(considering one-electron atom),

$$\begin{aligned}\langle f, \omega_{\lambda'}, \boldsymbol{\varepsilon}_{\lambda'}|\, &\hat{\boldsymbol{A}}(\boldsymbol{r}, t_1) \cdot \hat{\boldsymbol{A}}(\boldsymbol{r}, t_1)\, |i, \omega_\lambda, \boldsymbol{\varepsilon}_\lambda\rangle \\ &= \frac{\hbar}{2\epsilon_0 V}\langle f, \omega_{\lambda'}, \boldsymbol{\varepsilon}_{\lambda'}| \sum_{\lambda_1, \lambda_2} \frac{\boldsymbol{\varepsilon}_{\lambda_1} \cdot \boldsymbol{\varepsilon}_{\lambda_2}}{(\omega_{\lambda_1}\omega_{\lambda_2})^{1/2}} \Big\{ \hat{a}_{\lambda_1}\hat{a}^\dagger_{\lambda_2}\, e^{i[(\boldsymbol{k}_{\lambda_1} - \boldsymbol{k}_{\lambda_2})\cdot \boldsymbol{r} - (\omega_{\lambda_1} - \omega_{\lambda_2})t_1]} \\ &\qquad + \hat{a}^\dagger_{\lambda_1}\hat{a}_{\lambda_2}\, e^{-i[(\boldsymbol{k}_{\lambda_1} - \boldsymbol{k}_{\lambda_2})\cdot \boldsymbol{r} - (\omega_{\lambda_1} - \omega_{\lambda_2})t_1]}\, |i, \omega_\lambda, \boldsymbol{\varepsilon}_\lambda\rangle \Big\}\end{aligned}$$

where we have dropped terms like $\hat{a}_{\lambda_1}\hat{a}_{\lambda_2}$ and $\hat{a}^\dagger_{\lambda_1}\hat{a}^\dagger_{\lambda_2}$ which do not contribute to the matrix element when the initial and final states have the same number of photons. Inside the summation over λ_1 and λ_2, the first term contributes only when $\lambda_1 = \lambda$ and $\lambda_2 = \lambda'$, while the second contributes only when $\lambda_1 = \lambda'$ and $\lambda_2 = \lambda$. All other terms contribute zero. We can further replace the term $e^{i(\boldsymbol{k}_{\lambda_1} - \boldsymbol{k}_{\lambda_2})\cdot \boldsymbol{r}}$ by 1 in the long wavelength approximation. This means we are assuming that atomic size and, therefore, electron coordinate $\boldsymbol{r}$ are negligible compared to the wavelength. This is an excellent approximation for interaction of optical radiation with atoms. The matrix element for $\hat{\boldsymbol{A}} \cdot \hat{\boldsymbol{A}}$ interaction thus reduces to

$$\begin{aligned}\langle f, \omega_{\lambda'}, \boldsymbol{\varepsilon}_{\lambda'}|\, &\hat{\boldsymbol{A}}(\boldsymbol{r}, t_1) \cdot \hat{\boldsymbol{A}}(\boldsymbol{r}, t_1)\, |i, \omega_\lambda, \boldsymbol{\varepsilon}_\lambda\rangle \\ &= \frac{\hbar}{2\epsilon_0(\omega_\lambda\omega_{\lambda'})^{1/2} V}\, e^{-i(\omega_{\lambda_1} - \omega_{\lambda_2})t_1}\langle f, \omega_{\lambda'}, \boldsymbol{\varepsilon}_{\lambda'}|\, (\boldsymbol{\varepsilon}_\lambda \cdot \boldsymbol{\varepsilon}_{\lambda'})(\hat{a}_\lambda\hat{a}^\dagger_{\lambda'} + \hat{a}^\dagger_{\lambda'}\hat{a}_\lambda)\, |i, \omega_\lambda, \boldsymbol{\varepsilon}_\lambda\rangle\, .\end{aligned}$$

Hence for $\hat{\boldsymbol{A}}\cdot\hat{\boldsymbol{A}}$ interaction, the transition amplitude is

$$C_f^{(1)}(t) = \frac{1}{i\hbar}\frac{e^2}{2m}\frac{\hbar}{2\epsilon_0 V\sqrt{\omega_\lambda\omega_{\lambda'}}}2(\boldsymbol{\varepsilon}_\lambda\cdot\boldsymbol{\varepsilon}_{\lambda'})_{fi}\int_0^t dt_1\, e^{\frac{i}{\hbar}(\hbar\omega_{\lambda'}+E_f-\hbar\omega_\lambda-E_i)\,t_1}$$

where $\omega_\lambda = c|\boldsymbol{k}_\lambda|$ and $\omega_{\lambda'} = c|\boldsymbol{k}_{\lambda'}|$.

The second order transition amplitude due to $\hat{\boldsymbol{A}}\cdot\hat{\boldsymbol{p}}$ interaction can be written as

$$C_f^{(2)}(t) = \frac{1}{(i\hbar)^2}\sum_I\int_0^t dt_2\int_0^{t_2} dt_1\Bigg\{\langle f,\omega_{\lambda'},\boldsymbol{\varepsilon}_{\lambda'}|\,\frac{e}{m}\sum_{\lambda_1}\sqrt{\frac{\hbar}{2\epsilon_0\omega_{\lambda_1}V}}$$
$$\times\left(\hat{a}_{\lambda_1}e^{i(\boldsymbol{k}_{\lambda_1}\cdot\boldsymbol{r}-\omega_{\lambda_1}t_2)}+\hat{a}^\dagger_{\lambda_1}e^{-i(\boldsymbol{k}_{\lambda_1}\cdot\boldsymbol{r}-\omega_{\lambda_1}t_2)}\right)(\boldsymbol{\varepsilon}_{\lambda_1}\cdot\hat{\boldsymbol{p}})\;|I,\omega_{\lambda_I},\boldsymbol{\varepsilon}_{\lambda_I}\rangle\; e^{i(E_f-E_I)t_2/\hbar}$$
$$\times\langle I,\omega_{\lambda_I},\boldsymbol{\varepsilon}_{\lambda_I}|\,\frac{e}{m}\sum_{\lambda_2}\sqrt{\frac{\hbar}{2\epsilon_0\omega_{\lambda_2}V}}\left(\hat{a}_{\lambda_2}e^{i(\boldsymbol{k}_{\lambda_2}\cdot\boldsymbol{r}-\omega_{\lambda_2}t_1)}+\hat{a}^\dagger_{\lambda_2}e^{-i(\boldsymbol{k}_{\lambda_2}\cdot\boldsymbol{r}-\omega_{\lambda_2}t_1)}\right)$$
$$\times(\boldsymbol{\varepsilon}_{\lambda_2}\cdot\hat{\boldsymbol{p}})\;|i,\omega_\lambda,\boldsymbol{\varepsilon}_\lambda\rangle\, e^{i(E_I-E_i)t_1/\hbar}\Bigg\}.$$

An inspection of various terms in this expression shows that nonzero contributions to the matrix element come only from terms like $\langle\cdot|\,\hat{a}^\dagger_{\lambda_1}\,|\cdot\rangle\,\langle\cdot|\,\hat{a}_{\lambda_2}\,|\cdot\rangle$ when $\lambda_1=\lambda'$ and $\lambda_2=\lambda$ and from $\langle\cdot|\,\hat{a}_{\lambda_1}\,|\cdot\rangle\,\langle\cdot|\,\hat{a}^\dagger_{\lambda_2}\,|\cdot\rangle$ when $\lambda_1=\lambda$ and $\lambda_2=\lambda'$. This leads to

$$C_f^{(2)} = \frac{1}{(i\hbar)^2}\left(\frac{e}{m}\right)^2\frac{\hbar}{2\epsilon_0 V(\omega_\lambda\omega_{\lambda'})^{1/2}}\int_0^t dt_2\int_0^{t_2} dt_1$$
$$\Bigg[\sum_I\Big\{\langle f,\omega_{\lambda'},\boldsymbol{\varepsilon}_{\lambda'}|\;\hat{a}^\dagger_{\lambda'}e^{i\omega_{\lambda'}t_2}\;(\boldsymbol{\varepsilon}_{\lambda'}\cdot\hat{\boldsymbol{p}})\;|I\rangle\;e^{i(E_f-E_I)t_2/\hbar}$$
$$\times\langle I|\;\hat{a}_\lambda e^{-i\omega_\lambda t_1}\;(\boldsymbol{\varepsilon}_\lambda\cdot\hat{\boldsymbol{p}})\;|i,\omega_\lambda,\boldsymbol{\varepsilon}_\lambda\rangle\;e^{i(E_I-E_i)t_1/\hbar}\Big\}$$
$$+\sum_I\Big\{\langle f,\omega_{\lambda'},\boldsymbol{\varepsilon}_{\lambda'}|\;\hat{a}_\lambda e^{-i\omega_\lambda t_2}\;(\boldsymbol{\varepsilon}_\lambda\cdot\hat{\boldsymbol{p}})\;|I,\omega_\lambda,\boldsymbol{\varepsilon}_\lambda,\omega_{\lambda'},\boldsymbol{\varepsilon}_{\lambda'}\rangle\;e^{i(E_f-E_I)t_2/\hbar}$$
$$\times\langle I,\omega_\lambda,\boldsymbol{\varepsilon}_\lambda,\omega_{\lambda'},\boldsymbol{\varepsilon}_{\lambda'}|\;\hat{a}^\dagger_{\lambda'}e^{i\omega_{\lambda'}t_1}\;(\boldsymbol{\varepsilon}_{\lambda'}\cdot\hat{\boldsymbol{p}})\;|i,\omega_\lambda,\boldsymbol{\varepsilon}_\lambda\rangle\;e^{i(E_I-E_i)t_1/\hbar}\Big\}\Bigg].$$

In carrying out the time integrals, we recall from Sec. 10.9 [Chapter 10], that only the terms whose frequencies can vanish (resonant or energy conserving terms) make significant contributions in the long time limit. For example, terms with time dependence $e^{i(E_f-E_I+\hbar\omega_{\lambda'})t_2/\hbar}$ and $e^{i(E_f-E_I-\hbar\omega_\lambda)t_2/\hbar}$ give negligible contribution compared to the term $e^{i(E_f-E_i+\hbar\omega_{\lambda'}-\hbar\omega_\lambda)t_2/\hbar}$ since energy conservation $E_j-E_i+\hbar\omega_{\lambda'}-\hbar\omega_\lambda=0$ holds in the overall process. Carrying out the integration with respect to t_1 and dropping the non-resonant terms we obtain

$$c_f^{(2)}(t) = \frac{ie^2}{2m^2\epsilon_0 V(\omega_\lambda\omega_{\lambda'})^{1/2}}\sum_I\left\{\frac{(\boldsymbol{\varepsilon}_{\lambda'}\cdot\boldsymbol{p})_{fI}(\boldsymbol{\varepsilon}_\lambda\cdot\boldsymbol{p})_{Ii}}{(E_I-E_i-\hbar\omega_\lambda)}+\frac{(\boldsymbol{\varepsilon}_\lambda\cdot\boldsymbol{p})_{fI}(\boldsymbol{\varepsilon}_{\lambda'}\cdot\boldsymbol{p})_{Ii}}{(E_I-E_i+\hbar\omega_{\lambda'})}\right\}$$
$$\times\int_0^t dt_2\, e^{i(E_f-E_i+\hbar\omega_{\lambda'}-\hbar\omega_\lambda)t_2/\hbar}.$$

It may also be noted that in the first case [Fig. 13.2(a)] the intermediate state has no photon and in the second case [Fig. 13.2(b)] it has both photons characterized by λ and λ'.

The overall transition amplitude is obtained by adding the second order contribution due to $\hat{\boldsymbol{A}}\cdot\hat{\boldsymbol{p}}$ and the first order contribution due to $\hat{\boldsymbol{A}}\cdot\hat{\boldsymbol{A}}$ since both are of second order in vector potential. Then the transition probability per unit time for $i \to f$ transition is

$$w_{i\to f} = \frac{1}{t}\int_{\Delta E} \left| C_f^{(1)}(t) + C_f^{(2)}(t) \right|^2 \rho_E \, dE\,, \tag{13.5.20}$$

where ΔE represents the energy spread of the set of final states (photons of energy $\hbar\omega_{\lambda'}$ and polarization $\varepsilon_{\lambda'}$. The energy density of ρ_E of final states [scattered photon with energy $\hbar\omega_{\lambda'}$ and momentum $\hbar\omega_{\lambda'}/c$ within a solid angle $d\Omega$] is given by

$$\rho_E = \frac{V\omega_{\lambda'}^2\, d\Omega}{(2\pi)^3\, c^3\hbar}\,. \tag{13.5.21}$$

The time-dependent term in both $C_f^{(1)}(t)$ and $C_f^{(2)}(t)$ is the same [see also Sec. 10.9] and is easily evaluated to be

$$\int_0^t e^{i(\hbar\omega_{\lambda'}+E_f-\hbar\omega_\lambda-E_i)t_2/\hbar}\, dt_2 = 2e^{i(\omega_{fi}+\omega_{\lambda'}-\omega_\lambda)t/2}\frac{\sin(\omega_{fi}+\omega_{\lambda'}-\omega_\lambda)t/2}{\omega_{fi}+\omega_{\lambda'}-\omega_\lambda}\,,$$

where $\omega_{fi} = (E_f - E_i)/\hbar$. Then the transition probability is

$$\left|C_f^{(1)}(t) + C_f^{(2)}(t)\right|^2 = \left| \frac{e^2\,(\varepsilon_\lambda\cdot\varepsilon_{\lambda'})_{fi}}{2iV\epsilon_0 m(\omega_\lambda\omega_{\lambda'})^{1/2}} + \frac{ie^2}{2m^2\epsilon_0(\omega_\lambda\omega_{\lambda'})^{1/2}V} \right.$$
$$\left. \times \sum_I \left\{ \frac{(\varepsilon_{\lambda'}\cdot\boldsymbol{p})_{fI}\,(\varepsilon_\lambda\cdot\boldsymbol{p})_{Ii}}{(E_I - E_i - \hbar\omega_\lambda)} + \frac{(\varepsilon_\lambda\cdot\boldsymbol{p})_{fI}\,(\varepsilon_{\lambda'}\cdot\boldsymbol{p})_{Ii}}{(E_I - E_i + \hbar\omega_{\lambda'})} \right\}\right|^2 \frac{4\sin^2(\omega_{fi}+\omega_{\lambda'}-\omega_\lambda)t/2}{(\omega_{fi}+\omega_{\lambda'}-\omega_\lambda)^2}\,, \tag{13.5.22}$$

and the transition rate is given by

$$w_{i\to f} = \frac{1}{t}\frac{e^4}{4\epsilon_0^2V^2m^2\omega_\lambda\omega_{\lambda'}}\int_{\Delta E}\left|(\varepsilon_\lambda\cdot\varepsilon_{\lambda'})_{fi} - \frac{1}{m}\sum_I\left\{\frac{(\varepsilon_{\lambda'}\cdot\boldsymbol{p})_{fI}(\varepsilon_\lambda\cdot\boldsymbol{p})_{Ii}}{(E_I - E_i - \hbar\omega_\lambda)}\right.\right.$$
$$\left.\left. + \frac{(\boldsymbol{p}\cdot\varepsilon_\lambda)_{fI}(\varepsilon_{\lambda'}\cdot\boldsymbol{p})_{Ii}}{(E_I - E_i + \hbar\omega_{\lambda'})}\right\}\right|^2 \frac{4\sin^2(\omega_{fi}+\omega_{\lambda'}-\omega_\lambda)t/2}{(\omega_{fi}+\omega_{\lambda'}-\omega_\lambda)^2} \times \frac{V\omega_{\lambda'}^2\, d\Omega}{(2\pi)^3c^3\hbar}\hbar d\omega_{\lambda'}\,. \tag{13.5.23}$$

As the function $\frac{4\sin^2(\omega_{fi}+\omega_{\lambda'}-\omega_\lambda)t/2}{(\omega_{fi}+\omega_{\lambda'}-\omega_\lambda)^2}$ has a very sharp peak at $\omega_{fi}+\omega_{\lambda'}-\omega_\lambda = 0$, which gets sharper as $t\to\infty$, [$\lim_{t\to\infty}\frac{4\sin^2(\omega_{fi}+\omega_{\lambda'}-\omega_\lambda)t/2}{(\omega_{fi}+\omega_{\lambda'}-\omega_\lambda)^2} = 2\pi t\delta(\omega_{fi}+\omega_{\lambda'}-\omega_\lambda)$], the value of the integral is simply the integrand evaluated at $\omega_{\lambda'} = E_i + \hbar\omega_\lambda - E_f$, giving us

$$w_{i\to f} = \frac{e^4\omega_{\lambda'}d\Omega}{(4\pi\epsilon_0)^2Vm^2\omega_\lambda c^3}\left|(\varepsilon_\lambda\cdot\varepsilon_{\lambda'})_{fi} - \frac{1}{m}\sum_I\left\{\frac{(\varepsilon_{\lambda'}\cdot\boldsymbol{p})_{fI}(\varepsilon_\lambda\cdot\boldsymbol{p})_{Ii}}{(E_I - E_i - \hbar\omega_\lambda)} + \frac{(\boldsymbol{p}\cdot\varepsilon_\lambda)_{fI}(\varepsilon_{\lambda'}\cdot\boldsymbol{p})_{Ii}}{(E_I - E_i + \hbar\omega_{\lambda'})}\right\}\right|^2$$

The remaining dependence of transition rate on quantization volume can be removed if we work with scattering cross-section. Now one incident photon in quantization volume

V constitutes an incident photon flux density of c/V. Then the differential cross-section $d\sigma = \frac{w_{i\to f}}{(c/V)}$ for the processes under consideration is given by

$$\frac{d\sigma}{d\Omega} = \left(\frac{e^2}{4\pi\epsilon_0 mc^2}\right)^2 \frac{\omega_{\lambda'}}{\omega_\lambda} \left| (\boldsymbol{\varepsilon}_\lambda \cdot \boldsymbol{\varepsilon}_{\lambda'})_{fi} - \frac{1}{m}\sum_I \left\{ \frac{(\boldsymbol{\varepsilon}_{\lambda'} \cdot \boldsymbol{p})_{fI}(\boldsymbol{\varepsilon}_\lambda \cdot \boldsymbol{p})_{Ii}}{(E_I - E_i - \hbar\omega_\lambda)} + \frac{(\boldsymbol{p} \cdot \boldsymbol{\varepsilon}_\lambda)_{fI}(\boldsymbol{\varepsilon}_{\lambda'} \cdot \boldsymbol{p})_{Ii}}{(E_I - E_i + \hbar\omega_{\lambda'})} \right\} \right|^2 . \tag{13.5.24}$$

This formula for scattering of radiation from matter is known as Kramers-Heisenberg formula, and was originally derived by Kramers and Heisenberg in 1925, using the correspondence principle.

As special cases of Kramers-Heisenberg formula we consider the following scattering processes

(a) Rayleigh scattering is an example of elastic scattering of light wherein there is no change in the frequency of light ($\omega_{\lambda'} = \omega_\lambda$), the atom returns to its original state $|i\rangle = |f\rangle$, and the incident photon energy is much less than the energy difference between the atomic levels $\hbar\omega_\lambda \ll E_I - E_i (= \hbar\omega_{Ii})$.

(b) Thomson scattering is also an elastic scattering ($\omega_{\lambda'} = \omega_\lambda$) but the incident photon energy is much larger than the energy difference between atomic levels ($\hbar\omega_\lambda \gg E_I - E_i$).

(c) Raman scattering is an example of inelastic scattering wherein a shift in the energy of the scattered radiation occurs and the atom may be left in a final state different from its initial state in accordance with the conservation of energy $\hbar\omega_\lambda + E_i = \hbar\omega_{\lambda'} + E_f$.

We shall now consider these specific cases in somewhat more detail.

Rayleigh Scattering

Using the commutation relations between $\hat{\boldsymbol{r}}$ and $\hat{\boldsymbol{p}}$ we can derive the following relation

$$(\boldsymbol{\varepsilon}_\lambda \cdot \boldsymbol{\varepsilon}_{\lambda'}) = \frac{1}{i\hbar}\left[(\boldsymbol{\varepsilon}_\lambda \cdot \boldsymbol{r})(\boldsymbol{\varepsilon}_{\lambda'} \cdot \boldsymbol{p}) - (\boldsymbol{\varepsilon}_{\lambda'} \cdot \boldsymbol{p})(\boldsymbol{\varepsilon}_\lambda \cdot \boldsymbol{r})\right] . \tag{13.5.25}$$

Further, using the completeness of intermediate states we have, from the above relation,

$$(\boldsymbol{\varepsilon}_\lambda \cdot \boldsymbol{\varepsilon}_{\lambda'})_{ii} = \frac{1}{i\hbar}\sum_I \left[(\boldsymbol{\varepsilon}_\lambda \cdot \boldsymbol{r})_{iI}(\boldsymbol{\varepsilon}_{\lambda'} \cdot \boldsymbol{p})_{Ii} - (\boldsymbol{\varepsilon}_{\lambda'} \cdot \boldsymbol{p})_{iI}(\boldsymbol{\varepsilon}_\lambda \cdot \boldsymbol{r})_{Ii}\right] \tag{13.5.26}$$

and using the identity

$$(\boldsymbol{r} \cdot \boldsymbol{\varepsilon})_{iI} = \frac{i}{m\omega_{Ii}}(\boldsymbol{\varepsilon}_\lambda \cdot \boldsymbol{p})_{iI} \tag{13.5.27}$$

where $\omega_{Ii} = (E_I - E_i)/\hbar$ we can rewrite the above equation as

$$(\boldsymbol{\varepsilon}_\lambda \cdot \boldsymbol{\varepsilon}_{\lambda'})_{ii} = \frac{1}{m\hbar}\sum_I \frac{1}{\omega_{Ii}}\left[(\boldsymbol{\varepsilon}_\lambda \cdot \boldsymbol{p})_{iI}(\boldsymbol{\varepsilon}_{\lambda'} \cdot \boldsymbol{p})_{Ii} + (\boldsymbol{\varepsilon}_{\lambda'} \cdot \boldsymbol{p})_{iI}(\boldsymbol{\varepsilon}_\lambda \cdot \boldsymbol{p})_{Ii}\right] . \tag{13.5.28}$$

With the help of this equation we can combine the two terms in Eq. (13.5.24) and get

$$\begin{aligned}(\boldsymbol{\varepsilon}_\lambda \cdot \boldsymbol{\varepsilon}_{\lambda'})_{ii} &- \frac{1}{m}\sum_I \left\{ \frac{(\boldsymbol{\varepsilon}_{\lambda'} \cdot \boldsymbol{p})_{iI}(\boldsymbol{\varepsilon}_\lambda \cdot \boldsymbol{p})_{Ii}}{E_I - E_i - \hbar\omega_\lambda} + \frac{(\boldsymbol{\varepsilon}_\lambda \cdot \boldsymbol{p})_{iI}(\boldsymbol{\varepsilon}_{\lambda'} \cdot \boldsymbol{p})_{Ii}}{E_I - E_i + \hbar\omega_{\lambda'}} \right\} \\ &= -\frac{1}{m\hbar}\sum_I \left\{ \frac{\omega_\lambda(\boldsymbol{\varepsilon}_{\lambda'} \cdot \boldsymbol{p})_{iI}(\boldsymbol{\varepsilon}_\lambda \cdot \boldsymbol{p})_{Ii}}{\omega_{Ii}(\omega_{Ii} - \omega_\lambda)} - \frac{\omega_{\lambda'}(\boldsymbol{\varepsilon}_\lambda \cdot \boldsymbol{p})_{iI}(\boldsymbol{\varepsilon}_{\lambda'} \cdot \boldsymbol{p})_{Ii}}{\omega_{Ii}(\omega_{Ii} + \omega_{\lambda'})} \right\} .\end{aligned} \tag{13.5.29}$$

Since, in this case, $\omega_\lambda = \omega_{\lambda'} \ll \omega_{Ii}$, we can use the approximation[9]

$$\frac{1}{(\omega_{Ii} \pm \omega_\lambda)} \approx \frac{1}{\omega_{Ii}}\left(1 \mp \frac{\omega_\lambda}{\omega_{Ii}}\right) \tag{13.5.30}$$

and

$$(\varepsilon_\lambda \cdot \varepsilon_{\lambda'})_{ii} - \frac{1}{m}\sum_I \left\{ \frac{(\varepsilon_{\lambda'} \cdot \boldsymbol{p})_{iI}(\varepsilon_\lambda \cdot \boldsymbol{p})_{Ii}}{E_I - E_i - \hbar\omega_\lambda} + \frac{(\varepsilon_\lambda \cdot \boldsymbol{p})_{iI}(\varepsilon_{\lambda'} \cdot \boldsymbol{p})_{Ii}}{E_I - E_i + \hbar\omega_{\lambda'}} \right\}$$

$$= -\frac{\omega_\lambda^2}{m\hbar}\sum_I \frac{m}{\omega_{Ii}}^2 \{(\varepsilon_{\lambda'} \cdot \boldsymbol{r})_{iI}(\varepsilon_\lambda \cdot \boldsymbol{r})_{Ii} + (\varepsilon_\lambda \cdot \boldsymbol{r})_{iI}(\varepsilon_{\lambda'} \cdot \boldsymbol{r})_{Ii}\}\,. \tag{13.5.31}$$

Using this in Eq. (13.5.24), we finally have the cross-section for Rayleigh scattering

$$\frac{d\sigma}{d\Omega} = \frac{r_o^2 m^2}{\hbar^2}\omega_\lambda^4 \left| \sum_I \frac{1}{\omega_{Ii}} \{(\varepsilon_{\lambda'} \cdot \boldsymbol{r})_{iI}(\varepsilon_\lambda \cdot \boldsymbol{r})_{Ii} + (\varepsilon_\lambda \cdot \boldsymbol{r})_{iI}(\varepsilon_{\lambda'} \cdot \boldsymbol{r})_{Ii}\} \right|^2 \tag{13.5.32}$$

where $r_o = \frac{e^2}{4\pi\epsilon_0 mc^2}$ is the classical radius of the electron. We see that the scattering cross-section for low frequency radiation (ω_λ small compared to atomic transition frequencies) varies as the fourth power of the frequency. So the scattering cross-section increases with frequency. Since the wavelength Λ and (angular) frequency ω_λ of light are related by $\Lambda = 2\pi c/\omega_\lambda$, it follows that shorter wavelengths are scattered more prominently. This result, which coincides with the classical result for scattering of light, explains why the sky is blue and the sunset is red.

Thomson Scattering

At the other extreme, when the incident photon energy $\hbar\omega_\lambda$ is much larger than the energy difference between atomic levels, $\hbar\omega_\lambda \gg \hbar\omega_{Ii}$, we neglect $\hbar\omega_{Ii} = E_I - E_i$ compared to $\hbar\omega_\lambda$ in Kramers-Heisenberg formula (13.5.24). Then the second term, corresponding to graphs in Figs. 13.2(a) and (b) reduces to

$$\frac{1}{m}\frac{1}{\hbar\omega_\lambda}\sum_I \left[(\varepsilon_{\lambda'} \cdot \boldsymbol{p})_{iI}(\varepsilon_\lambda \cdot \boldsymbol{p})_{Ii} - (\varepsilon_\lambda \cdot \boldsymbol{p})_{iI}(\varepsilon_{\lambda'} \cdot \boldsymbol{p})_{Ii}\right],$$

which, of course, is zero. Hence only the matrix element corresponding to the graph in Fig. 13.2(c) contributes to scattering. Further, since $(\varepsilon_\lambda \cdot \varepsilon_{\lambda'})_{ii}$ is insensitive to the nature of binding of atomic electrons, the cross-section coincides with the cross-section of scattering of light by the (unbound) electron obtained by J. J. Thomson

$$\frac{d\sigma}{d\Omega} = r_o^2 \left|(\varepsilon_\lambda \cdot \varepsilon_{\lambda'})_{ii}\right|^2 . \tag{13.5.33}$$

We note that, in contrast to low frequency Rayleigh scattering, high frequency Thomson scattering is insensitive to ω_λ or wavelength of the incident light.

Raman Scattering

The Kramers-Heisenberg formula can also be applied to inelastic scattering of light in which case $\omega_\lambda \neq \omega_{\lambda'}$ and $|i\rangle \neq |f\rangle$. This phenomenon is known as Raman effect after Raman,

[9] For atoms of atmospheric gases, a typical ω_{Ii} is in the ultraviolet region. Hence the approximation $\omega_\lambda \ll \omega_{Ii}$ holds for frequencies ω_λ in the optical region.

who observed a shift in the frequency of light scattered by a liquid. If the initial atomic state $|i\rangle$ is the ground state and the final state is an excited state $|f\rangle$, the energy of the scattered photon $\hbar\omega_{\lambda'}$ has to be less than the initial photon energy $\hbar\omega_\lambda$ (Stokes scattering) to conserve energy. On the other hand, if the initial atomic state $|i\rangle$ is an excited state and the final state is a lower state, then $\hbar\omega_{\lambda'}$ must be larger than $\hbar\omega_\lambda$ (anti-Stokes scattering). Thus in both cases of Raman scattering energy conservation requires

$$\hbar\omega_\lambda + E_i = \hbar\omega_{\lambda'} + E_f \,. \tag{13.5.34}$$

Examples of scattering we have considered correspond to spontaneous scattering. They involve scattering of weak incident light beams such as those produced by thermal sources. In such beams, the occupation number of each mode of the field is small. When strong coherent light beams such as those produced by lasers are present, stimulated scattering processes can become important. The procedure outlined here can be extended to deal with such processes as well.

Resonant Scattering

The treatment given above assumes that the condition for resonance $\hbar\omega_\lambda = E_I - E_i$ is not satisfied for any intermediate state $|I\rangle$. However, if the energy of the incident photon is such that, for some specific state $|R\rangle$ in the summation over I, the condition for resonance is satisfied, the energy denominator in that particular term vanishes. This singularity does not exist in practice because the excited states of an atom are no longer stationary states when the interaction of an atom with quantized radiation field is taken into account. Even if there are no photons in the field, the atom decays spontaneously to a lower state. This decay of an excited can be accounted for if we impart to its energy E_R a negative imaginary part, $E_R - i\hbar\gamma_R$, where γ_R is a real constant. With this change, if the atom is excited to state $I = R$, the probability for it to stay in that state is given by

$$P_R(t) \sim e^{-2\gamma_R t} \,. \tag{13.5.35}$$

Hence $(2\gamma_R)^{-1}$ can be interpreted as the lifetime of the excited state $I = R$. For atomic states, a typical value for $(2\gamma_R)^{-1}$ might be 10 ns corresponding to $\hbar\gamma_R \approx 3 \times 10^{-8}$ eV. This is small compared with typical energy difference $E_R - E_i \approx 1$ eV. Thus if there is an excited state with energy $E_R = E_i + \hbar\omega_\lambda$, then the corresponding term will dominate in the sum in (13.5.24) and all other terms can be neglected for incident photons of energy in this range. Then for elastic scattering $|f\rangle = |i\rangle$ and $\omega_\lambda = \omega_{\lambda'} = \omega$ we have, with the help of Eq. (13.5.27),

$$\left[\frac{d\sigma}{d\Omega}\right]_{resonance} = \left(\frac{e^2}{4\pi\epsilon_0}\right)^2 \frac{\omega^4}{c^4} \frac{|(\boldsymbol{\varepsilon}_{\lambda'} \cdot \boldsymbol{r})_{iR}|^2 \, |(\boldsymbol{\varepsilon}_\lambda \cdot \boldsymbol{r})_{Ri}|^2}{(E_R - E_i - \hbar\omega)^2 + (\hbar\gamma_R)^2} \,. \tag{13.5.36}$$

Thus compared to off-resonance, where the typical energy denominator is $(E_I - E_i)^2 \approx 1 \ \text{eV}^2$, the on-resonance denominator is $\gamma_R^2 \approx 10^{-15}$ eV^2, giving us an on-resonance enhancement of scattering by a factor of 10^{15}!

13.6 Atomic Level Shift due to Self-interaction of the Electron: Lamb-Retherford Shift

Classically the electromagnetic field resulting from the presence of an electron can interact with the electron itself. The energy of this interaction is called the *self-energy* of the electron and, if the electron is taken to be a point particle, this turns out to be infinite. In quantum theory the self-interaction can be viewed as a two-step process in which the electron emits a virtual photon and subsequently absorbs it.

Bethe showed that this self-interaction of the electron gives rise to an energy shift of atomic $2S_{\frac{1}{2}}$ level (in general all S-levels) of the Hydrogen atom while no shift occurs for $2P_{\frac{1}{2}}$ states (in general for all $\ell \neq 0$ states). According to Dirac theory of the electron the $2S_{\frac{1}{2}}$ and $2P_{\frac{1}{2}}$ states of the Hydrogen atom are degenerate. In the non-relativistic theory as well the $2S_{\frac{1}{2}}$ and $2P_{\frac{1}{2}}$ levels are degenerate because energy depends only on the total quantum number n and not on the orbital or total angular quantum numbers ℓ or j. But if we consider the self-energy of the electron then this gives rise to a relative shift between $2S_{\frac{1}{2}}$ and $2P_{\frac{1}{2}}$ levels ($2S_{\frac{1}{2}}$ being higher of the two). This is in conformity with the observation of Lamb and Retherford who found experimentally that the frequency corresponding to $2S_{\frac{1}{2}} - 2P_{\frac{1}{2}}$ separation is about 1000 MHz. Bethe calculated the shift of $2S_{\frac{1}{2}}$ level of the H-atom due to self-interaction of the electron. He treated the electron in the H-atom non-relativistically but considered the electromagnetic field to be quantized. Bethe's work paved way for more accurate calculation by Tomonaga, Schwinger and Feynman for the Lamb shift, based on covariant quantum electrodynamics.

The self-interaction of an atomic electron can be represented by the space-time graphs in Fig. 13.3. According to quantum theory of radiation an atom in state $|i\rangle$ can emit a photon and go to state $|I\rangle$ even if no external field is present. The atom can subsequently absorb the photon and return to the state $|i\rangle$ [Fig. 13.3(a)]. It may also happen that emission is followed by instantaneous absorption [Fig. 13.3(b)]. The second process is not important from the point of view of atomic level shift. So we shall consider only the two-step process shown in Fig. 13.3(a).

The interaction of the electron with the electromagnetic field is given by Eq. (13.4.7):

$$\hat{H}' = \frac{e}{m} \sum_{\lambda} \left(\frac{\hbar}{2\epsilon_0 \omega_\lambda V} \right)^{1/2} (\boldsymbol{\varepsilon}_\lambda \cdot \hat{\boldsymbol{p}}) \left[\hat{a}_\lambda \, e^{i \boldsymbol{k}_\lambda \cdot \boldsymbol{r}} + \hat{a}_\lambda^\dagger \, e^{-i \boldsymbol{k}_\lambda \cdot \boldsymbol{r}} \right] , \tag{13.6.1}$$

where $\hat{a}_\lambda$ and $\hat{a}_\lambda^\dagger$ are, respectively, the annihilation and creation operators for photons of frequency ω_λ and polarization $\boldsymbol{\varepsilon}_\lambda$ and m is the rest mass of the electron. Then the total Hamiltonian of the system may be expressed as

$$\hat{H} = \hat{H}_0 + \hat{H}', \tag{13.6.2}$$

where the unperturbed Hamiltonian $\hat{H}_0 \equiv \hat{H}_a + \hat{H}_{rad}$ is the sum of the atomic Hamiltonian $\hat{H}_a$ and radiation field Hamiltonian $\hat{H}_{rad}$ and $\hat{H}'$ is the interaction Hamiltonian (13.6.1). The stationary states of the unperturbed Hamiltonian are simply the products of the atomic and field states $|j\rangle\, |n_1, n_2, \cdots, n_\lambda, \cdots\rangle = |j\rangle\, |\{n_\lambda\}\rangle \equiv |j; \{n_\lambda\}\rangle$, where $|j\rangle$ is a stationary state of the atom with energy E_j and $|\{n_\lambda\}\rangle$ is the stationary state of the field specified by the set of occupation numbers $\{n_\lambda\}$ with energy $\epsilon_{\{n_\lambda\}} = \sum_\lambda n_\lambda \hbar\omega_\lambda$, where we have dropped the zero point energy terms by choosing the zero of energy appropriately.

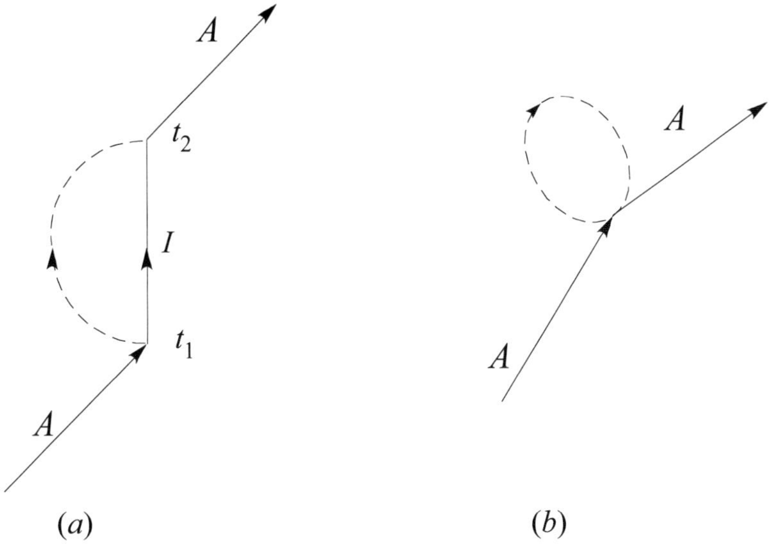

FIGURE 13.3
Space-time graphs for self-interaction of atomic electron.

The time evolution of the overall state of the system in the presence of perturbation is governed by the Schrödinger equation

$$i\hbar\frac{d}{dt}|\Psi(t)\rangle = (\hat{H}_0 + \hat{H}')|\Psi(t)\rangle\,. \qquad (13.6.3)$$

We can expand this state in terms of the stationary states of the unperturbed Hamiltonian $\hat{H}_0$. As a result of self-interaction, the atom initially in state $|A\rangle$ (with the field in vacuum state) may emit a (virtual) photon into a field mode λ, changing the atomic state into an intermediate $|I\rangle$ and field state into $|1_\lambda\rangle$), and subsequently absorb the photon restoring the original atomic and field states. In writing the intermediate single-photon field state we have suppressed the field modes with zero occupation number. Thus the system states relevant to the self-interaction are the initial state $|A\rangle\,|\Phi_0\rangle$, the intermediate states $|I\rangle\,|1_\lambda\rangle$ and the final state $|A\rangle\,|\Phi_0\rangle$, which coincides with the initial state. All these are eigenstates of the unperturbed Hamiltonian $\hat{H}_0$. Then we can expand the system wave function in terms these states as

$$|\Psi(t)\rangle = \left[c_{A0}e^{-iE_At/\hbar}\,|A\rangle\,|\Phi_0\rangle + \sum_{I,\lambda} c_{I\lambda}e^{-i(E_I+\hbar\omega_\lambda)t/\hbar}\,|I\rangle\,|1_\lambda\rangle\right]. \qquad (13.6.4)$$

We note that the matrix elements $\hat{H}'$ between the unperturbed states are

$$\begin{aligned}\langle\Phi_0|\,\langle A|\,\hat{H}'\,|I\rangle\,|1_\lambda\rangle &= \frac{e}{m}\sqrt{\frac{\hbar}{2\epsilon_0\omega_\lambda V}}\,\boldsymbol{\varepsilon}_\lambda\cdot\boldsymbol{p}_{AI}\,e^{i\boldsymbol{k}_\lambda\cdot\boldsymbol{r}} \approx \sqrt{\frac{\hbar}{2\epsilon_0\omega_\lambda V}}\,\boldsymbol{\varepsilon}_\lambda\cdot\boldsymbol{p}_{AI}\\ &= \left(\langle 1_\lambda|\,\langle I|\,\hat{H}'\,|A\rangle\,|\Phi_0\rangle\right)^*, \qquad (13.6.5)\end{aligned}$$

$$\langle 1_{\lambda'}|\,\langle I'|\,\hat{H}'\,|I\rangle\,|1_\lambda\rangle = 0 = \langle\Phi_0|\,\langle A|\,\hat{H}'\,|A\rangle\,|\Phi_0\rangle\,, \qquad (13.6.6)$$

where in the evaluation of the matrix element, we have used the long wavelength (dipole) approximation $e^{i\boldsymbol{k}\cdot\boldsymbol{r}} \approx 1$ since the atomic size is small compared to the wavelengths of photons of interest. Substituting the expansion (13.6.4) into Eq. (13.6.3), pre-multiplying the resulting equation, in succession, by $\langle \Phi_0 A | \exp(iE_A t/\hbar)$ and $\langle 1_{\lambda'}, I' | \exp[i(\hbar\omega_\lambda + E_{I'})t/\hbar]$, respectively, using the orthogonality of unperturbed stationary states and making use of results (13.6.5) and (13.6.6), we get

$$\frac{dC_A}{dt} = -\frac{ie}{m\hbar}\sum_{I,\lambda} c_{I\lambda} e^{-i(\omega_{IA}+\omega_\lambda)t}\sqrt{\frac{\hbar}{2\epsilon_0\omega_\lambda V}}\,\boldsymbol{\varepsilon}_\lambda \cdot \boldsymbol{p}_{AI}\,, \tag{13.6.7}$$

$$\frac{dc_{I\lambda}}{dt} = -\frac{ie}{m\hbar} C_A e^{-i(\omega_{AI}-\omega_\lambda)t}\sqrt{\frac{\hbar}{2\epsilon_0\omega_\lambda V}}\,\boldsymbol{\varepsilon}_\lambda \cdot \boldsymbol{p}_{IA}\,, \tag{13.6.8}$$

where we have eventually replaced I', λ' by I, λ, and written $C_A \equiv c_{A0}$ and $\omega_{IA} = (E_I - E_A)/\hbar$. These equations have to be solved with the initial condition

$$C_A(0) \equiv c_{A0}(0) = 1\,, \quad c_{I\lambda}(0) = 0\,. \tag{13.6.9}$$

To solve these equations we assume $C_A(t)$ to have the form

$$C_A(t) = e^{-i\Delta E_A\, t/\hbar}\,. \tag{13.6.10}$$

This form for $C_A(t)$ implies that the wave function of the atomic state, including self-interaction, is

$$\langle \boldsymbol{r} | A\rangle = \psi_A(\boldsymbol{r}) e^{-i(E_A+\Delta E_A)\,t/\hbar}\,. \tag{13.6.11}$$

Hence ΔE_A may be taken to be the correction to the energy of the initial state on account of the self-interaction. Substituting Eq. (13.6.10) for $C_A(t)$ into Eq. (13.6.8) and integrating over t, we get

$$C_{I\lambda}(t) = \left(\frac{e}{m}\right)\left(\frac{\hbar}{2\epsilon_0\omega_\lambda V}\right)^{1/2} (\boldsymbol{\varepsilon}_\lambda \cdot \hat{\boldsymbol{p}})_{IA}\, \frac{\left[e^{-i(E_A+\Delta E_A - E_I - \hbar\omega_\lambda)t/\hbar} - 1\right]}{(E_A + \Delta E_A - E_I - \hbar\omega_\lambda)}\,. \tag{13.6.12}$$

Substituting this expression for $C_{I\lambda}(t)$ into Eq. (13.6.7) and using the assumed form (13.6.10) for $C_A(t)$ on the left-hand side of the same equation, we getΔE_A, the expression

$$\Delta E_A = \sum_{I,\lambda} \left(\frac{e}{m}\right)^2 \frac{\hbar}{2\epsilon_0\omega_\lambda V}\,(\boldsymbol{\varepsilon}_\lambda \cdot \hat{\boldsymbol{p}})_{AI}\,(\boldsymbol{\varepsilon}_\lambda \cdot \hat{\boldsymbol{p}})_{IA}\left[\frac{1 - e^{i(E_A - E_I - \hbar\omega_\lambda)\,t/\hbar}}{(E_A - E_I - \hbar\omega_\lambda)}\right] \tag{13.6.13}$$

where we have neglected ΔE_A compared to $E_A - E_I$ on the right-hand side of this equation assuming it to be small compared with $E_A - E_I$. Since the time interval for which the perturbation (self-interaction) acts is infinite we may take $t \to \infty$ but, if we do this the exponential term oscillates instead of converging. The result can be made convergent by adding an infinitesimal positive imaginary part $i\epsilon$ (where $\epsilon \to 0^+$) to $E_A - E_I - \hbar\omega_\lambda/\hbar \equiv x$. The last factor on the right-hand side of Eq. (13.6.13) becomes

$$\lim_{\substack{\epsilon\to 0^+\\ t\to\infty}} \frac{1 - e^{i(x+i\epsilon)t}}{x + i\epsilon} = -\lim_{\epsilon\to 0^+} i\int_0^\infty e^{i(x+i\epsilon)t}\,dt = \lim_{\epsilon\to 0^+}\left(\frac{x - i\epsilon}{x^2+\epsilon^2}\right)$$
$$= \frac{1}{x} - i\pi \lim_{\epsilon\to 0^+}\frac{\epsilon/\pi}{x^2+\epsilon^2}\,. \tag{13.6.14}$$

The second term on the right-hand side is obviously zero for $x \neq 0$ and is singular at $x = 0$ such that

$$\int_{-\infty}^{\infty} \frac{\epsilon/\pi}{x^2+\epsilon^2}\,dx = 1\,,$$

and can be identified as delta function. Thus the last factor in Eq. (10.6.13) is defined by

$$\lim_{\substack{\epsilon\to 0^+ \\ t\to\infty}} \frac{1-e^{i(x+i\epsilon)}}{x+i\epsilon} = \frac{1}{x} - i\pi\delta(x). \tag{13.6.15}$$

Writing ΔE_A in terms of its real and imaginary parts

$$\Delta E_A = \mathcal{R}e\,(\Delta E_A) + i\,\mathcal{I}m\,(\Delta E_A)\,, \tag{13.6.16}$$

we find

$$\mathcal{R}e(\Delta E_A) = \sum_{I,\lambda}\left(\frac{e}{m}\right)^2 \frac{\hbar}{2\epsilon_0\omega_\lambda V}\,\frac{(\boldsymbol{\varepsilon}_\lambda\cdot\hat{\boldsymbol{p}})_{AI}\,(\boldsymbol{\varepsilon}_\lambda\cdot\hat{\boldsymbol{p}})_{IA}}{E_A-E_I-\hbar\omega_\lambda}\,, \tag{13.6.17}$$

$$\mathcal{I}m(\Delta E_A) = -\pi\sum_{I,\lambda}\left(\frac{e}{m}\right)^2 \frac{\hbar}{2\epsilon_0\omega_\lambda V}\,(\boldsymbol{\varepsilon}_\lambda\cdot\hat{\boldsymbol{p}})_{AI}\,(\boldsymbol{\varepsilon}_\lambda\cdot\hat{\boldsymbol{p}})_{IA}\,\delta(E_A-E_I-\hbar\omega_\lambda)\,. \tag{13.6.18}$$

The existence of the imaginary part of ΔE_A implies that the state $|A\rangle$ is not stable but decays and the right-hand side of (13.6.18) is proportional to the transition rate for spontaneous emission, with all allowed intermediate states summed over. Note that because of the presence of energy conserving delta function, the sum in the imaginary part (13.6.18) is over real (energy conserving) transitions. Hence the energy shift ΔE_A has an imaginary part only when the atomic state $|A\rangle$ can decay to state $|I\rangle$.

The real part $\mathcal{R}e(\Delta E_A)$ has contributions from energy conserving and non-conserving transitions. It can be looked upon as the energy associated with the emission and re-absorption of the virtual photons and this is what we can call the level shift on account of the self-interaction of the electron. To carry out the sum over modes in the expression for $\mathcal{R}e\,(\Delta E_A)$ we recall that each mode is characterized by a wave vector $\boldsymbol{k}$ and a polarization vector which depends on $\boldsymbol{k}$. In fact for a given wave vector $\boldsymbol{k}$, there are two independent polarization vectors, $\boldsymbol{\varepsilon}_{\boldsymbol{k}1}$ and $\boldsymbol{\varepsilon}_{\boldsymbol{k}2}$, which are orthogonal to $\boldsymbol{k}$ and one another.[10] Then the sum over modes $\sum_\lambda$ implies $\sum_{\boldsymbol{k}}\sum_{\boldsymbol{\varepsilon}_{\boldsymbol{k}s}}$. Using the identity for polarization vectors associated with a given $\boldsymbol{k}$ vector

$$\sum_s(\boldsymbol{\varepsilon}_{\boldsymbol{k}s})_i(\boldsymbol{\varepsilon}_{\boldsymbol{k}s})_j = \delta_{ij}-\kappa_i\kappa_j\,, \tag{13.6.19}$$

the sum over polarizations

$$\sum_s(\boldsymbol{\varepsilon}_{\boldsymbol{k}s}\cdot\boldsymbol{p})\,(\boldsymbol{\varepsilon}_{\boldsymbol{k}s}\cdot\boldsymbol{p}) = \boldsymbol{p}\cdot\boldsymbol{p}-(\boldsymbol{p}\cdot\boldsymbol{\kappa})\,(\boldsymbol{p}\cdot\boldsymbol{\kappa})\,,$$

[10] We can easily write down a set of polarization vectors asociated with a $\boldsymbol{k}$ vector whose polar angles are θ and φ. Then two real orthogonal unit polarization vectors associated with $\boldsymbol{k}$ are

$$\boldsymbol{\varepsilon}_{\boldsymbol{k}1} = \cos\theta\cos\varphi\boldsymbol{e}_x+\cos\theta\sin\varphi\boldsymbol{e}_y-\sin\theta\boldsymbol{e}_z\,,$$
$$\boldsymbol{\varepsilon}_{\boldsymbol{k}2} = -\sin\varphi\boldsymbol{e}_x+\cos\varphi\boldsymbol{e}_y\,.$$

It is easily seen that $\boldsymbol{k}\cdot\boldsymbol{\varepsilon}_{\boldsymbol{k}1}=0=\boldsymbol{k}\cdot\boldsymbol{\varepsilon}_{\boldsymbol{k}2}$, $\boldsymbol{\varepsilon}_{\boldsymbol{k}1}\times\boldsymbol{\varepsilon}_{\boldsymbol{k}2}=\boldsymbol{\kappa}=\frac{\boldsymbol{k}}{k}$, where $\boldsymbol{\kappa}$ is a unit vector in the direction of wave vector $\boldsymbol{k}$. Thus $(\boldsymbol{\varepsilon}_{\boldsymbol{k}1},\boldsymbol{\varepsilon}_{\boldsymbol{k}1},\boldsymbol{\kappa})$ form a right-handed, orthogonal, Cartesian basis. With the help of the expressions for the polarization vector given here, the following identity is also easily established $\sum_s(\boldsymbol{\varepsilon}_{\boldsymbol{k}s})_i(\boldsymbol{\varepsilon}_{\boldsymbol{k}s})_j=\delta_{ij}-\kappa_i\kappa_j$, where i,j denote the Cartesian components of polarization vectors.

The complex basis vectors

$$\boldsymbol{\varepsilon}_{\boldsymbol{k}1} = \frac{1}{\sqrt{2}}\left[(\cos\theta\cos\varphi-i\sin\varphi)\boldsymbol{e}_x+(\cos\theta\sin\varphi+i\cos\varphi)\boldsymbol{e}_y-\sin\theta\boldsymbol{e}_z\right]$$
$$\boldsymbol{\varepsilon}_{\boldsymbol{k}2} = \frac{1}{\sqrt{2}}\left[(i\cos\theta\cos\varphi-\sin\varphi)\boldsymbol{e}_x+(i\cos\theta\sin\varphi+\cos\varphi)\boldsymbol{e}_y-i\sin\theta\boldsymbol{e}_z\right]$$

represent unit right and left circular polarization vectors associated with the wave vector $\boldsymbol{k}$.

where $\boldsymbol{\kappa} = \frac{\boldsymbol{k}}{k}$ is a unit vector in the direction of $\boldsymbol{k}$. Then we have

$$\mathcal{R}e(\Delta E_A) = \sum_{I,\boldsymbol{k}} \left(\frac{e}{m}\right)^2 \frac{\hbar}{2\epsilon_0 \omega_k V} \left[\frac{(\boldsymbol{p}_{IA} \cdot \boldsymbol{p}_{AI})\,(1-\cos^2\theta)}{(E_A - E_I - \hbar\omega_\lambda)}\right],$$

where θ is the angle between $\boldsymbol{p}$ and $\boldsymbol{k}$. Replacing the sum over wave vector $\boldsymbol{k}$ by an integral

$$\frac{1}{V}\sum_{\boldsymbol{k}} \rightarrow \int \frac{d^3k}{(2\pi)^3},$$

we obtain

$$\begin{aligned}\mathcal{R}e(\Delta E_A) &= \sum_I \left(\frac{e}{m}\right)^2 \frac{\hbar}{\epsilon_0} \int \frac{1}{2\omega_k} \frac{(\boldsymbol{p}_{IA} \cdot \boldsymbol{p}_{AI}) \sin^2\theta}{(E_A - E_I - \hbar\omega_\lambda)} \cdot \frac{k^2 dk\, d\Omega}{(2\pi)^3} \\ &= -\frac{2}{3\pi} \frac{e^2}{4\pi\epsilon_0 m^2 \hbar c^3} \sum_I \int_0^\infty \frac{E_\gamma\, dE_\gamma |\boldsymbol{p}_{IA}|^2}{E_\gamma + E_I - E_A} \qquad (13.6.20)\end{aligned}$$

where we have used $E_\gamma = \hbar\omega = \hbar k c$ and $\int d\Omega_k (1-\cos^2\theta) = 8\pi/3$. Unfortunately, the integral over E_γ is divergent. How do we make sense of this? We can try to get around this difficulty by putting an upper limit to the energy of the virtual photons because for sufficiently high energy photons, the non-relativistic approximation for the electron must break down. So we might argue that maximum energy of photon is $E_\gamma \leq mc^2$. When this is done we get a finite result, which depends on the cutoff $E_\gamma^{\max}$, clearly an undesirable feature.

The basic suggestion for the removal of the divergence was made by Kramers who noticed that, even for a free electron of momentum $\boldsymbol{p}$, there is an energy shift ΔE_p which can be obtained from Eq. (13.6.20) by replacing the sum over atomic states by the sum over plane waves for a free particle. Then the matrix element $\langle A|\, \hat{\mathbf{p}}\, |I\rangle$ becomes $\langle \mathbf{p}|\, \hat{\mathbf{p}}\, |\mathbf{p}'\rangle = \mathbf{p}\delta_{\mathbf{p},\mathbf{p}'}$, where we have assumed the system to be enclosed in a certain volume so that the allowed electron momentum values $\boldsymbol{p}$ are discrete. The sum over intermediate states $|I\rangle = |\mathbf{p}'\rangle$ then gives

$$\int_0^{E_\gamma^{max}} \sum_{\mathbf{p}'} \frac{|\langle \mathbf{p}'\,|\hat{\mathbf{p}}\,|\mathbf{p}\rangle\,|^2}{\left(E_\gamma + \frac{p'^2}{2m} - \frac{p^2}{2m}\right)} E_\gamma dE_\gamma = p^2 E_\gamma^{\max}.$$

Thus the energy shift for a free electron due to self interaction is

$$\Delta E_p = -\frac{2}{3\pi}\frac{e^2}{4\pi\epsilon_0 m^2\hbar c^3}\, p^2\, E_\gamma^{\max} = Cp^2, \qquad (13.6.21a)$$

where

$$C = -\frac{2}{3\pi}\frac{e^2}{4\pi\epsilon_0 m^2\hbar c^3}\, E_\gamma^{\max}. \qquad (13.6.21b)$$

Thus the self-interaction correction to the energy for a free electron is proportional to p^2. Since the electromagnetic field of a charge particle is an inalieable part of it, this additional electromagnetic contribution cannot be separated from the usual kinetic energy $p^2/2m$. This means the observed kinetic energy term $p^2/2m$ somehow already accommodates this shift. Kramer introduced the idea of *mass renormalization* by suggesting that electromagnetic self-energy contribution may be regarded as contained in the observed rest mass of the electron. Following this idea, Bethe argued that since this effect exists for a bound as well as free electron, the observed energy shift $(\Delta E_A)_{obs}$ of an atomic level $|A\rangle \equiv |n,\ell,m\rangle$ is the difference between the self-energy $\mathcal{R}e\,(\Delta E_A)$ of the bound electron in state $|A\rangle$ and that of a free electron $\Delta E_p = Cp^2$, provided that in the latter case, the electron is assigned a (momentum)^2 equal to the expectation value of (momentum)^2 of the electron in the atomic state $|A\rangle$:

$$\langle A|\, \hat{\boldsymbol{p}} \cdot \hat{\boldsymbol{p}}\, |A\rangle = \sum_I \boldsymbol{p}_{AI} \cdot \boldsymbol{p}_{IA} = \sum_I |\boldsymbol{p}_{IA}|^2 . \tag{13.6.22}$$

The expression for the observed level shift of an atomic level is then the difference

$$(\Delta E_A)_{obs} = -\left(\frac{e^2}{4\pi\epsilon_0\hbar c}\right)\frac{2}{3\pi m^2 c^2}\sum_I\left[\int_0^{E_\gamma^{\max}} \frac{E_\gamma\, dE_\gamma |\boldsymbol{p}_{IA}|^2}{E_\gamma + E_I - E_A} - \int_0^{E_\gamma^{\max}} |\boldsymbol{p}_{IA}|^2\, dE_\gamma\right]$$

or $$(\Delta E_A)_{obs} = \left(\frac{e^2}{4\pi\epsilon_0\hbar c}\right)\frac{2}{3\pi m^2 c^2}\sum_I |\boldsymbol{p}_{IA}|^2\,(E_I - E_A)\ln\left(\frac{E_I - E_A + E_\gamma^{\max}}{E_I - E_A}\right). \tag{13.6.23}$$

This expression for level shift has much slower divergence (logarithmic compared to the original linear divergence) with cut-off photon energy. In fact we can get a sensible result for level shift if the cutoff energy $E_\gamma^{\max} \gg |(E_I - E_A)|$. Then, since the logarithmic term is slowly varying, we take the average value of $E_I - E_A$ and take it out of the summation sign giving us

$$(\Delta E_A)_{obs} = \left(\frac{e^2}{4\pi\epsilon_0\hbar c}\right)\frac{2}{3\pi m^2 c^2}\ln\left(\frac{E_\gamma^{\max}}{(E_I - E_A)_{avg}}\right)\sum_I |\boldsymbol{p}_{IA}|^2\,(E_I - E_A). \tag{13.6.24}$$

The sum over states can then be evaluated as[11]

$$\sum_I |\boldsymbol{p}_{IA}|^2 (E_I - E_A) = \frac{\hbar^2}{2}\int |\psi_A(r)|^2\, \nabla^2 V\, d^3 r . \tag{13.6.25}$$

In the Hydrogen atom problem, the potential $V(r) = -\frac{e^2}{4\pi\epsilon_0 r}$ gives

$$\nabla^2 V \equiv \nabla^2\left(-\frac{e^2}{4\pi\epsilon_0 r}\right) = +\frac{e^2}{\epsilon_0}\,\delta^3(\boldsymbol{r}). \tag{13.6.26}$$

To see this we recall that, for a point charge q at the origin, the electrostatic field at a point $\boldsymbol{r}$ is given by $\boldsymbol{E} = q\boldsymbol{r}/4\pi\epsilon_0 r^3 = -\boldsymbol{\nabla}(q/4\pi\epsilon_0 r)$, so that $\boldsymbol{\nabla}\cdot\boldsymbol{E} = -q\nabla^2(1/r)$. But, according to Gauss law, $\nabla\cdot\boldsymbol{E} = \frac{1}{\epsilon_0}\rho(\boldsymbol{r}) = \frac{q}{\epsilon_0}\,\delta^3(\boldsymbol{r})$. So $-\nabla^2\left(\frac{1}{r}\right) = 4\pi\delta^3(\boldsymbol{r})$.

Using Eqs. (13.6.25) and (13.6.26) in the expression for level shift (13.6.24), we obtain

$$(\Delta E_A)_{obs} = \left(\frac{e^2}{4\pi\epsilon_0\hbar c}\right)^2\frac{4\hbar^3}{3m^2 c}\ln\left(\frac{E_\gamma^{\max}}{(E_I - E_A)_{avg}}\right)|\psi_A(r=0)|^2 . \tag{13.6.27}$$

Thus the energy shift for level $|A\rangle$ depends on the value of the wave function at the origin. For the Hydrogen atom, the wave function at the origin is

$$|\psi_A(r=0)|^2 = \begin{cases} \dfrac{1}{\pi n^3 a_o^3}, & \text{for } |A\rangle \equiv |n, \ell = 0, m = 0\rangle \\ 0, & \text{for } |A\rangle \equiv |n, \ell \neq 0, m\rangle . \end{cases} \tag{13.6.28}$$

[11] $\left[\hat{\boldsymbol{p}}, \hat{H}\right]_{IA} = \left[\hat{\boldsymbol{p}}, \hat{V}\right]_{IA}\ (E_A - E_I)\ \boldsymbol{p}_{IA} = [\hat{\boldsymbol{p}}, V]_{IA}$. Pre-multiply both sides by $\boldsymbol{p}_{AI}$ and sum over I to get (13.6.25) [see Problem 8, Chapter 2].

Thus for the Hydrogen atom we get

$$(\Delta E_A)_{obs} = \frac{8}{3\pi}\left(\frac{e^2}{4\pi\epsilon_0\hbar c}\right)^3 \frac{e^2}{8\pi\epsilon_0 a_o n^3} \ln\left(\frac{E_\gamma^{\max}}{(E_I - E_A)_{avg}}\right). \tag{13.6.29}$$

This formula was derived by Bethe to account for the Lamb-Retherford observation of $2S_{\frac{1}{2}} - 2P_{\frac{1}{2}}$ splitting in the Hydrogen atom. Note that in view of Eq. (13.6.28) the shift of $2P_{\frac{1}{2}}$ state vanishes. Using Bethe's formula for the splitting of $2S_{\frac{1}{2}} - 2P_{\frac{1}{2}}$ states ($n = 2$) of Hydrogen, we get a transition frequency

$$\begin{aligned}\frac{E_{2S_{\frac{1}{2}}} - E_{2P_{\frac{1}{2}}}}{(2\pi\hbar)} &= \frac{(\Delta E_A)_{obs}}{2\pi\hbar}\,, \\ &= \frac{8}{3\pi}\left(\frac{1}{137}\right)^3 13.53 \times \ln\left(\frac{0.512\times 10^6}{16.6\times 13.53}\right)\frac{1}{8} \times \frac{1.6\times 10^{-19}}{6.63\times 10^{-34}}\, Hz\,, \\ &= 1040 \text{ MHz},\end{aligned} \tag{13.6.30}$$

where $E_\gamma^{\max} \approx 0.512$ MeV = the rest energy of the electron, and $(E_I - E_{n_o})_{Avg} = 16.6$ Rydberg $= 16.6 \times 13.53$ eV. The value of $(E_I - E_A)_{avg}$ is taken from Bethe, Brown, and Stehn[12]. The value 1040 MHz was found to be in remarkable agreement with the Lamb-Retherford observations.

A much more sophisticated calculation for Lamb shift, based on Tomonaga, Schwinger and Feynman's formulation of covariant quantum electrodynamics, yields a value

$$\left(E_{2S_{\frac{1}{2}}} - E_{2P_{\frac{1}{2}}}\right)/h = (1057.70 \pm 0.15)\,\text{MHz}$$

which has even better agreement with the observed value 1057.77 ± 0.10 MHz. This formulation leads to a finite result avoiding even logarithmic divergence. However, the idea of mass renormalization, which was so basic to the success of Bethe's non-relativistic calculation turns out to be just as essential in the relativistic treatment of level shifts.

13.7 Compton Scattering

As another application of the methods discussed in this chapter we consider Compton scattering, in which a high energy photon is scattered by a free electron resulting in the reduction of the photon energy and increase in the energy of the electron. The total energy and the total momentum are finally conserved in this process. Compton scattering may be looked upon as a two-step process involving the absorption of one photon and subsequent emission of another photon. The virtual transitions can occur in two ways: (a) The electron at rest absorbs the incoming photon (energy $\hbar\omega$ and momentum $\hbar\boldsymbol{k}$) and acquires a momentum $\hbar\boldsymbol{k}$ in its virtual state. In the subsequent transition this electron emits the outgoing photon (energy $\hbar\omega'$ and momentum $\hbar\boldsymbol{k}'$), itself being left with momentum $\hbar(\boldsymbol{k} - \boldsymbol{k}')$ [Fig. 13.4(a)]. In each of these virtual transitions the momentum is conserved but not the energy. (b) Alternatively the electron at rest can emit the outgoing photon (energy $\hbar\omega' > 2mc^2$ and momentum $\hbar\boldsymbol{k}'$) and acquire a recoil momentum $-\hbar\boldsymbol{k}'$ and negative energy. In the subsequent transition the electron absorbs the incoming photon (energy $\hbar\omega$ and

[12] H. A. Bethe, L. M. Brown, and J. R. Stehn, Phys. Rev. **77**, 370 (1950).

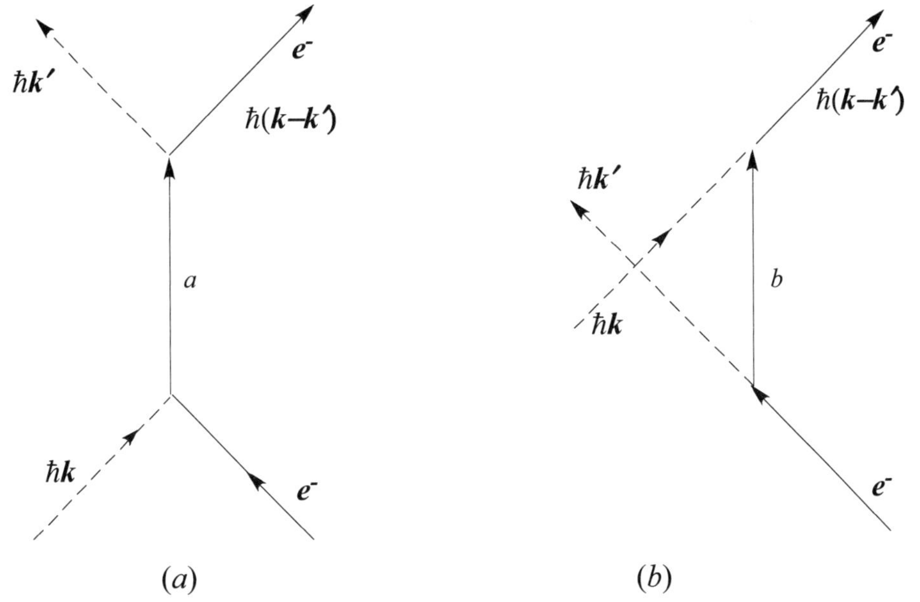

FIGURE 13.4
Space-time diagrams for Compton scattering.

momentum $\hbar\boldsymbol{k}$) and is finally left with momentum $-\hbar\boldsymbol{k}'+\hbar\boldsymbol{k}$ [Fig. 13.4.(b)]. Again, in each one of these virtual transitions, momentum is conserved but not the energy. Violation of energy conservation occurs within the limits of the uncertainty principle. Since high energies are involved in the virtual transitions, the electron has to be treated relativistically. In the present treatment we shall assume the electromagnetic (radiation) field to be quantized while the electron will be assumed to obey the single-particle (Dirac) equation.[13]

The total matrix element M_{fi} for this process consists of the sum of $M_{fi}^{(a)}$ and $M_{fi}^{(b)}$ which are the contributions from both ways (a) and (b) in which this transition can occur. According to second order perturbation theory we have

$$M_{fi}^{(a)} = \sum_a \frac{\langle\psi_f|\,\hat{H}'\,|\psi_a\rangle\,\langle\psi_a|\,\hat{H}'\,|\psi_i\rangle}{(E_i - E_a')} \tag{13.7.1}$$

and

$$M_{fi}^{(b)} = \sum_b \frac{\langle\psi_f|\,\hat{H}'\,|\psi_b\rangle\,\langle\psi_b|\,\hat{H}'\,|\psi_i\rangle}{E_i - E_b'} \tag{13.7.2}$$

where E_a' and E_b' are the energies of the system (electron + photons) in the respective intermediate state $|\psi_a\rangle$ and $|\psi_b\rangle$. The perturbation $\hat{H}'$, which is the interaction of the Dirac electron (charge $-e$) with the electromagnetic field, is given by [14]

$$\hat{H}' = ec\boldsymbol{\alpha}\cdot\hat{\boldsymbol{A}}(\boldsymbol{r})\,, \tag{13.7.3}$$

[13] In the next chapter we shall consider the same phenomenon assuming both the radiation field and Dirac field to be quantized.

[14] As is well known, in the presence of electromagnetic field, $\boldsymbol{p}\to\boldsymbol{p}-q\,\boldsymbol{A}$ and $E\to E-q\phi$. So the single-particle (charge q) Dirac Hamiltonian, in the presence of quantized electromagnetic field, can be written as

$$\hat{H} = c\boldsymbol{\alpha}\cdot(\hat{\boldsymbol{p}}-q\hat{\boldsymbol{A}})+\beta mc^2+q\hat{\phi} = \hat{H}_0+\hat{H}'$$

where $\hat{H}_0\equiv c\boldsymbol{\alpha}\cdot\hat{\boldsymbol{p}}+\beta mc^2$ and $\hat{H}' = -cq\,\boldsymbol{\alpha}\cdot\,\hat{\boldsymbol{A}}$ for the case of pure radiation field, where $\hat{\phi}=0$.

where $\boldsymbol{\alpha} \equiv (\alpha_1, \alpha_2, \alpha_3)$ are the 4×4 matrices of the Dirac equation and $\hat{\boldsymbol{A}}(\boldsymbol{r})$ is the electromagnetic field operator expressible in terms of the creation and annihilation operators. In writing the vector potential we will explicitly use the fact that a mode of the electromagnetic field is characterized by a wave vector $\boldsymbol{k}$ and a polarization vector. There are two orthogonal unit polarization vectors $\boldsymbol{\varepsilon}_{\boldsymbol{k}1}$ and $\boldsymbol{\varepsilon}_{\boldsymbol{k}2}$ associated with each wave vector $\boldsymbol{k}$. Together with unit vector $\boldsymbol{\kappa} = \boldsymbol{k}/k$ in the direction of the wave vector $\boldsymbol{k}$, the vectors $(\boldsymbol{\varepsilon}_{\boldsymbol{k}1}, \boldsymbol{\varepsilon}_{\boldsymbol{k}2}, \boldsymbol{\kappa})$ form a right-handed, orthogonal triad of vectors with $\boldsymbol{\varepsilon}_{\boldsymbol{k}1} \times \boldsymbol{\varepsilon}_{\boldsymbol{k}2} = \boldsymbol{\kappa}$. Thus the sum over mode index λ is replaced by a sum over $\boldsymbol{k}$ and s, where $s(=1,2)$ refers to two orthogonal polarizations associated with each $\boldsymbol{k}$. Then the vector potential is given by

$$\hat{A}(\boldsymbol{r}) = \sum_{\boldsymbol{k},s} \sqrt{\frac{\hbar}{2\epsilon_0 \omega V}} \left(\hat{a}_{\boldsymbol{k}s}\, e^{i\boldsymbol{k}\cdot\boldsymbol{r}} + \hat{a}^{\dagger}_{\boldsymbol{k}s}\, e^{-i\boldsymbol{k}\cdot\boldsymbol{r}} \right) \boldsymbol{\varepsilon}_{\boldsymbol{k}s} \tag{13.7.4}$$

where $\omega = |\boldsymbol{k}|c = kc$ and the matrix elements for the process of Fig. 13.4(a) can be written as

$$\langle \psi_a |\, \hat{H}' \,| \psi_i \rangle = ec \,\langle a |\, \boldsymbol{\alpha} \cdot \sum_{\boldsymbol{k}_1, s_1} \sqrt{\frac{\hbar}{2\epsilon_0 \omega_1 V}} \left(\hat{a}_{\boldsymbol{k}_1 s_1} e^{i\boldsymbol{k}_1\cdot\boldsymbol{r}} + \hat{a}^{\dagger}_{\boldsymbol{k}_1 s_1} e^{-i\boldsymbol{k}_1\cdot\boldsymbol{r}} \right) \boldsymbol{\varepsilon}_{\boldsymbol{k}_1 s_1} \,| i, \boldsymbol{k}, \boldsymbol{\varepsilon}_{\boldsymbol{k}s} \rangle \,,$$

$$\langle \psi_f |\, \hat{H}' \,| \psi_a \rangle = ec \,\langle f, \boldsymbol{k}', \boldsymbol{\varepsilon}_{\boldsymbol{k}'s'} |\, \boldsymbol{\alpha} \cdot \sum_{\boldsymbol{k}_2, s_2} \sqrt{\frac{\hbar}{2\epsilon_0 \omega_2 V}} \left(\hat{a}_{\boldsymbol{k}_2 s_2} e^{i\boldsymbol{k}_2\cdot\boldsymbol{r}} + \hat{a}^{\dagger}_{\boldsymbol{k}_2 s_2} e^{-i\boldsymbol{k}_2\cdot\boldsymbol{r}} \right) \boldsymbol{\varepsilon}_{\boldsymbol{k}_2 s_2} \,| a \rangle \,.$$

Here the initial state $|\psi_i\rangle = |i, \boldsymbol{k}, \boldsymbol{\varepsilon}_{\boldsymbol{k}s}\rangle$ has an electron in state i and a single photon of momentum $\boldsymbol{k}$ (frequency $\omega = kc$) and polarization $\boldsymbol{\varepsilon}_{\boldsymbol{k}s}$ (frequency $\omega = kc$) and the final state $|\psi_f\rangle = |f, \boldsymbol{k}', \boldsymbol{\varepsilon}_{\boldsymbol{k}'s'}\rangle$ has an electron in state f and a photon of wave vector $\boldsymbol{k}'$ (frequency $\omega' = k'c$) and polarization $\boldsymbol{\varepsilon}_{\boldsymbol{k}'s'}$. In the intermediate state $|\psi_a\rangle$ there are no photons. It can be seen that non-zero contributions to these two matrix elements come only from terms corresponding to $(\boldsymbol{k}_1 = \boldsymbol{k}, \boldsymbol{\varepsilon}_{\boldsymbol{k}_1 s_1} = \boldsymbol{\varepsilon}_{\boldsymbol{k}s})$ and $(\boldsymbol{k}_2 = \boldsymbol{k}', \boldsymbol{\varepsilon}_{\boldsymbol{k}_2 s_2} = \boldsymbol{\varepsilon}_{\boldsymbol{k}'s'}0$, respectively. This gives us

$$\langle \psi_a |\, \hat{H}' \,| \psi_i \rangle = ec \sqrt{\frac{\hbar}{2\epsilon_0 \omega V}} \,\langle a |\, e^{i\,\boldsymbol{k}\cdot\boldsymbol{r}} (\boldsymbol{\alpha} \cdot \boldsymbol{\varepsilon}_{\boldsymbol{k}s}) \,| i \rangle \tag{13.7.5}$$

and

$$\langle \psi_f |\, \hat{H}' \,| \psi_a \rangle = ec \sqrt{\frac{\hbar}{2\epsilon_0 \omega' V}} \,\langle f |\, e^{-i\,\boldsymbol{k}'\cdot\boldsymbol{r}} (\boldsymbol{\alpha} \cdot \boldsymbol{\varepsilon}_{\boldsymbol{k}'s'}) \,| a \rangle \,. \tag{13.7.6}$$

Rewriting these matrix elements in the coordinate representation, using

$$\begin{aligned}
\langle \boldsymbol{r} | a \rangle &= u_a\, e^{i\,\boldsymbol{k}_a\cdot\boldsymbol{r}/\hbar} \frac{1}{(2\pi)^{3/2}} \,, \\
\langle \boldsymbol{r} | i \rangle &= u_i\, e^{i\,\boldsymbol{k}_i\cdot\boldsymbol{r}/\hbar} \frac{1}{(2\pi)^{3/2}} \,, \\
\langle \boldsymbol{r} | f \rangle &= u_f\, e^{i\,\boldsymbol{k}_f\cdot\boldsymbol{r}/\hbar} \frac{1}{(2\pi)^{3/2}} \,,
\end{aligned}$$

where $\boldsymbol{k}_a, \boldsymbol{k}_i, \boldsymbol{k}_f$ are the momenta of the electron in the respective states, $\boldsymbol{k}$ and $\boldsymbol{k}'$ denote the wave vectors of the incident and scattered photons, respectively, and u_a, u_i, u_f represent the spinors of the respective states, we get

$$\langle\psi_a|\,\hat{H}'\,|\psi_i\rangle = ec\sqrt{\frac{\hbar}{2\epsilon_0\omega V}}\,\frac{1}{(2\pi)^3}\int e^{i(-\boldsymbol{k}_a+\boldsymbol{k}+\boldsymbol{k}_i)}\,d^3r\,\left[u_a^\dagger(\boldsymbol{\alpha}\cdot\boldsymbol{\varepsilon}_{\boldsymbol{k}s})u_i\right]$$

$$= ec\sqrt{\frac{\hbar}{2\epsilon_0\omega V}}\,\delta^3\left(-\boldsymbol{k}_a+\boldsymbol{k}+\boldsymbol{k}_i\right)\left[u_a^\dagger(\boldsymbol{\alpha}\cdot\boldsymbol{\varepsilon}_{\boldsymbol{k}s})u_i\right]$$

and $$\langle\psi_f|\,\hat{H}'\,|\psi_a\rangle = ec\sqrt{\frac{\hbar}{2\epsilon_0\omega' V}}\,\delta^3\left(-\boldsymbol{k}_f-\boldsymbol{k}'+\boldsymbol{k}_a\right)\left[u_f^\dagger(\boldsymbol{\alpha}\cdot\boldsymbol{\varepsilon}_{\boldsymbol{k}'s'})u_a\right].$$

The three-dimensional δ functions represents conservation of momentum at each step of the process (a). Keeping this in mind we can simply write

$$\langle\psi_a|\,\hat{H}'\,|\psi_i\rangle = ec\sqrt{\frac{\hbar}{2\epsilon_0\omega V}}\,\left[u_a^\dagger(\boldsymbol{\alpha}\cdot\boldsymbol{\varepsilon}_{\boldsymbol{k}s})\,u_i\right]\equiv ec\sqrt{\frac{\hbar}{2\epsilon_0\omega V}}\,(\boldsymbol{\alpha}\cdot\boldsymbol{\varepsilon}_{\boldsymbol{k}s})_{ai}\,,$$

$$\langle\psi_f|\,\hat{H}'\,|\psi_a\rangle = ec\sqrt{\frac{\hbar}{2\epsilon_0\omega' V}}\,\left(u_f^\dagger(\boldsymbol{\alpha}\cdot\boldsymbol{\varepsilon}_{\boldsymbol{k}s'})u_a\right)\equiv ec\sqrt{\frac{\hbar}{2\epsilon_0\omega' V}}\,(\boldsymbol{\alpha}\cdot\boldsymbol{\varepsilon}_{\boldsymbol{k}'s'})_{fa}\,.$$

Hence the second order matrix element is

$$M_{fi}^{(a)} = \frac{e^2c^2\hbar}{2\epsilon_0 V\sqrt{\omega\,\omega'}}\sum_a\frac{(\boldsymbol{\alpha}\cdot\boldsymbol{\varepsilon}_{\boldsymbol{k}'s'})_{fa}\,(\boldsymbol{\alpha}\cdot\boldsymbol{\varepsilon}_{\boldsymbol{k}s})_{ai}}{E_i-E_a}\,, \tag{13.7.7}$$

where E_i = initial energy of the system $= mc^2 + c\hbar k$, E'_a = energy of the system in the intermediate state=energy of the electron in the intermediate state $= E_a$, $E_a^2 = m^2c^4 + c^2(\hbar k)^2$, and $\hbar\boldsymbol{k}$ is the momentum acquired by the electron after absorbing the initial photon. Since E_a can have positive or negative sign we take it to be given by the Dirac equation

$$\left[c\boldsymbol{\alpha}\cdot(\hbar\boldsymbol{k})+\beta mc^2\right]u_a = E_a u_a\,.$$

Similarly, for process (b), the matrix elements $\langle\psi_b|\,\hat{H}'\,|\psi_i\rangle$ and $\langle\psi_f|\,\hat{H}'\,|\psi_b\rangle$ are given by

$$\langle\psi_b|\,\hat{H}'\,|\psi_i\rangle = ec\left\langle b,\boldsymbol{k},\boldsymbol{\varepsilon}_{\boldsymbol{k}s},\boldsymbol{k}',\boldsymbol{\varepsilon}_{\boldsymbol{k}'s'}\right|\boldsymbol{\alpha}\cdot\sum_{\boldsymbol{k}_1,s}\sqrt{\frac{\hbar}{2\epsilon_0\omega_1 V}}$$
$$\times\,\boldsymbol{\varepsilon}_{\boldsymbol{k}_1s_1}\left(\hat{a}_{\boldsymbol{k}_1s_1}e^{i\boldsymbol{k}_1\cdot\boldsymbol{r}}+\hat{a}^\dagger_{\boldsymbol{k}_1s_1}e^{-i\boldsymbol{k}_1\cdot\boldsymbol{r}}\right)\,|i,\boldsymbol{k},\boldsymbol{\varepsilon}_{\boldsymbol{k}s}\rangle$$

and $$\langle\psi_f|\,\hat{H}'\,|\psi_b\rangle = ec\left\langle f,\boldsymbol{k}',\boldsymbol{\varepsilon}_{\boldsymbol{k}'s'}\right|\boldsymbol{\alpha}\cdot\sum_{\boldsymbol{k}_2,s}\sqrt{\frac{\hbar}{2\epsilon_0\omega_2 V}}$$
$$\times\,\boldsymbol{\varepsilon}_{\boldsymbol{k}_2s_2}\left(\hat{a}_{\boldsymbol{k}_2s_2}e^{i\boldsymbol{k}_2\cdot\boldsymbol{r}}+\hat{a}^\dagger_{\boldsymbol{k}_2s_2}e^{-i\boldsymbol{k}_2\cdot\boldsymbol{r}}\right)\left|b,\boldsymbol{k},\boldsymbol{\varepsilon}_{\boldsymbol{k}s},\boldsymbol{k}',\boldsymbol{\varepsilon}_{\boldsymbol{k}'s'}\right\rangle.$$

In this case, the intermediate state contains two photons of wave vector and polarization $\boldsymbol{k},\boldsymbol{\varepsilon}_{\boldsymbol{k}s}$ and $\boldsymbol{k}',\boldsymbol{\varepsilon}_{\boldsymbol{k}'s'}$, respectively, in addition to the electron in state $|b\rangle$. It can be seen that the non-zero contribution to these two matrix elements can come only from $(\boldsymbol{k}_1,\boldsymbol{\varepsilon}_{\boldsymbol{k}_1s_1}) = (\boldsymbol{k}',\boldsymbol{\varepsilon}_{\boldsymbol{k}'s'})$ and $(\boldsymbol{k}_2,\boldsymbol{\varepsilon}_{\boldsymbol{k}_2s_2}) = (\boldsymbol{k},\boldsymbol{\varepsilon}_{\boldsymbol{k}s})$, respectively. Thus

$$\langle\psi_b|\,\hat{H}'\,|\psi_i\rangle = ec\sqrt{\frac{\hbar}{2\epsilon_0\omega' V}}\,\left\langle b,\boldsymbol{k},\boldsymbol{\varepsilon}_{\boldsymbol{k}s},\boldsymbol{k}',\boldsymbol{\varepsilon}_{\boldsymbol{k}'s'}\right|(\boldsymbol{\alpha}\cdot\boldsymbol{\varepsilon}_{\boldsymbol{k}'s'})\,e^{-i\,\boldsymbol{k}'\cdot\boldsymbol{r}}\,\left|i,\boldsymbol{k},\boldsymbol{\varepsilon}_{\boldsymbol{k}s},\boldsymbol{k}',\boldsymbol{\varepsilon}_{\boldsymbol{k}'s'}\right\rangle$$

$$= ec\sqrt{\frac{\hbar}{2\epsilon_0\omega' V}}\,\langle b|\,(\boldsymbol{\alpha}\cdot\boldsymbol{\varepsilon}_{\boldsymbol{k}'s'})\,e^{-i\,\boldsymbol{k}'\cdot\boldsymbol{r}}\,|i\rangle \tag{13.7.8}$$

and $$\langle\psi_f|\,\hat{H}'\,|\psi_b\rangle = ec\sqrt{\frac{\hbar}{2\epsilon_0\omega V}}\,\left\langle f,\boldsymbol{k}',\,\boldsymbol{\varepsilon}_{\boldsymbol{k}'s'}\right|\,(\boldsymbol{\alpha}\cdot\boldsymbol{\varepsilon}_{\boldsymbol{k}s})\,e^{i\,\boldsymbol{k}\cdot\boldsymbol{r}}\,\left|b,\boldsymbol{k}'\,\boldsymbol{\varepsilon}_{\boldsymbol{k}'s'}\right\rangle$$

$$= ec\sqrt{\frac{\hbar}{2\epsilon_0\omega V}}\,\langle f|\,(\boldsymbol{\alpha}\cdot\boldsymbol{\varepsilon}_{\boldsymbol{k}s})\,e^{i\,\boldsymbol{k}\cdot\boldsymbol{r}}\,|b\rangle\,. \tag{13.7.9}$$

Rewriting these matrix elements in the coordinate representation, we have

$$\langle \psi_b | \hat{H}' | \psi_i \rangle = ec\sqrt{\frac{\hbar}{2\epsilon_0 \omega' V}} \left[u_b^\dagger (\boldsymbol{\alpha} \cdot \boldsymbol{\varepsilon}_{\boldsymbol{k}'s'}) \, u_i \right] \frac{1}{(2\pi)^3} \int e^{i(-\boldsymbol{k}_b - \boldsymbol{k}' + \boldsymbol{k}_i)\cdot \boldsymbol{r}} \, d^3 r \,,$$

$$= ec\sqrt{\frac{\hbar}{2\epsilon_0 \omega' V}} \left[u_a^\dagger (\boldsymbol{\alpha} \cdot \boldsymbol{\varepsilon}_{\boldsymbol{k}'s'}) \, u_i \right] \delta^3 \left(-\boldsymbol{k}_b - \boldsymbol{k}' + \boldsymbol{k}_i \right)$$

and

$$\langle f | \hat{H}' | b \rangle = ec\sqrt{\frac{\hbar}{2\epsilon_0 \omega V}} \left[u_f^\dagger (\boldsymbol{\alpha} \cdot \boldsymbol{\varepsilon}_{\boldsymbol{k}s}) \, u_b \right] \delta^3 \left(-\boldsymbol{k}_f + \boldsymbol{k} + \boldsymbol{k}_b \right) .$$

Again keeping in mind that the three-dimensional delta functions merely represent the conservation of momentum at each stage of the process (b), we can write

$$\langle \psi_b | \hat{H}' | \psi_i \rangle = ec\sqrt{\frac{\hbar}{2\epsilon_0 \omega' V}} \, (\boldsymbol{\alpha} \cdot \boldsymbol{\varepsilon}_{\boldsymbol{k}'s'})_{bi} \,, \tag{13.7.10}$$

and

$$\langle \psi_f | \hat{H}' | \psi_b \rangle = ec\sqrt{\frac{\hbar}{2\epsilon_0 \omega V}} \, (\boldsymbol{\alpha} \cdot \boldsymbol{\varepsilon}_{\boldsymbol{k}s})_{fb} \,. \tag{13.7.11}$$

With the help of these, the second order matrix element for process (b) can be written as

$$M_{fi}^{(b)} = \frac{e^2 c^2 \hbar}{2\epsilon_0 V \sqrt{\omega \omega'}} \sum_b \frac{(\boldsymbol{\alpha} \cdot \boldsymbol{\varepsilon}_{\boldsymbol{k}s})_{fb} \, (\boldsymbol{\alpha} \cdot \boldsymbol{\varepsilon}_{\boldsymbol{k}'s'})_{bi}}{E_i - E_b'} \,, \tag{13.7.12}$$

where E_i = initial energy of the system $= mc^2 + c\hbar k$, E_b' = energy of the *system* in the intermediate state $b = E_b + c\hbar k' + c\hbar k$, E_b = the energy of the electron in the intermediate state b, $E_b^2 = m^2c^4 + c^2(-\hbar k)^2$, and $\hbar \boldsymbol{k}$ is the momentum taken away by the emitted photon [Fig. 13.4 (b)]. Since the sign of E_b could be negative, we take E_b to be given by the Dirac equation:

$$\left[c\boldsymbol{\alpha} \cdot (-\hbar \boldsymbol{k}') + \beta mc^2 \right] u_b = E_b u_b \,.$$

In Eq. (13.7.12) we can replace $E_i - E_b'$ by $(mc^2 - c\hbar k') - E_b$.

Now we can write the second order matrix element $M_{fi}^{(a)}$ [Eq. (13.7.7)] for process (a) as

$$M_{fi}^{(a)} = \frac{e^2 c^2 \hbar}{2\epsilon_0 V \sqrt{\omega \omega'}} \sum_a \frac{u_f^\dagger \, (\boldsymbol{\alpha} \cdot \boldsymbol{\varepsilon}_{\boldsymbol{k}'s'}) \, (E_i + E_a) \; u_a u_a^\dagger \, (\boldsymbol{\alpha} \cdot \boldsymbol{\varepsilon}_{\boldsymbol{k}s}) u_i}{E_i^2 - E_a^2}$$

$$= \frac{e^2 c^2 \hbar}{2\epsilon_0 V \sqrt{\omega \omega'}} \, \frac{u_f^\dagger \left[(\boldsymbol{\alpha} \cdot \boldsymbol{\varepsilon}_{\boldsymbol{k}'s'}) \left(c\hbar k + c\hbar (\boldsymbol{\alpha} \cdot \boldsymbol{k}) + (1+\beta) mc^2 \right) (\boldsymbol{\alpha} \cdot \boldsymbol{\varepsilon}_{\boldsymbol{k}s}) \right] u_i}{2mc^2 \hbar k} \,,$$

where we have used the Dirac equation and the completeness condition for the Dirac spinors,

$$E_a u_a = \left[c\hbar (\boldsymbol{\alpha} \cdot \boldsymbol{k}_\lambda) + \beta mc^2 \right] u_a \text{ and } \sum_a u_a{}^\dagger u_a = 1.$$

Further, using the anti-commutation relations of α_i and β matrices we have

$$[(1+\beta) mc^2 (\boldsymbol{\alpha} \cdot \boldsymbol{\varepsilon}_{\boldsymbol{k}s})] u_i = [(\boldsymbol{\alpha} \cdot \boldsymbol{\varepsilon}_{\boldsymbol{k}s}) \, mc^2 (1-\beta)] u_i \,,$$

and since $\boldsymbol{p} = 0$ for the initial electron state u_i, it follows from the Dirac equation that $(1-\beta)mc^2 \, u_i = 0$. Thus the second order matrix element for process (a) can be written as

$$M_{fi}^{(a)} = \frac{e^2 \hbar c}{4\epsilon_0 V m\omega \sqrt{\omega \omega'}} \, u_f{}^\dagger \, (\boldsymbol{\alpha} \cdot \boldsymbol{\varepsilon}_{\boldsymbol{k}'s'}) \, (k + \boldsymbol{\alpha} \cdot \boldsymbol{k}) \, (\boldsymbol{\alpha} \cdot \boldsymbol{\varepsilon}_{\boldsymbol{k}s}) \; u_i \,. \tag{13.7.13}$$

We can deal similarly with the matrix element $M_{fi}^{(b)}$ [Eq. (13.7.12)] for process (b). Since the sign of E_b could be positive or negative, we keep the denominator in the form $(mc^2 - c\hbar k')^2 - E_b^2$ and replace $(mc^2 - c\hbar k' + E_b)\, u_b$ by $[mc^2 - c\hbar k' - c\hbar(\boldsymbol{\alpha}\cdot\boldsymbol{k}') + \beta mc^2]\, u_b$. Then we get, finally,

$$M_{fi}^{(b)} = \frac{e^2\hbar c}{4\epsilon_0 V m\,\omega'\sqrt{\omega'\omega}}\; u_f{}^\dagger\; (\boldsymbol{\alpha}\cdot\boldsymbol{\varepsilon}_{\boldsymbol{k}s})\,(k' + \boldsymbol{\alpha}\cdot\boldsymbol{k}')(\boldsymbol{\alpha}\cdot\boldsymbol{\varepsilon}_{\boldsymbol{k}'s'}))\; u_i\,. \qquad (13.7.14)$$

Adding Eqs. (13.7.13) and (13.7.14), we get the overall second order matrix element

$$\begin{aligned} M_{fi} &= M_{fi}^{(a)} + M_{fi}^{(b)} \\ &= \frac{e^2\hbar c}{4\epsilon_0 V m\sqrt{\omega'\omega}}\, u_f{}^\dagger \left(\frac{1}{\omega}\,[(\boldsymbol{\alpha}\cdot\boldsymbol{\varepsilon}_{\boldsymbol{k}'s'})\,(\boldsymbol{k} + \boldsymbol{\alpha}\cdot\boldsymbol{k})(\boldsymbol{\alpha}\cdot\boldsymbol{\varepsilon}_{\boldsymbol{k}s})]\right. \\ &\qquad \left. + \frac{1}{\omega'}\,[(\boldsymbol{\alpha}\cdot\boldsymbol{\varepsilon}_{\boldsymbol{k}s})\,(\boldsymbol{k}' + \boldsymbol{\alpha}\cdot\boldsymbol{k}')(\boldsymbol{\alpha}\cdot\boldsymbol{\varepsilon}_{\boldsymbol{k}'s'})]\right) u_i\,. \qquad (13.7.15) \end{aligned}$$

With the help of the indentity $(\boldsymbol{\alpha}\cdot\boldsymbol{P})\,(\boldsymbol{\alpha}\cdot\boldsymbol{Q}) = (\boldsymbol{P}\cdot\boldsymbol{Q}) - i\,\boldsymbol{\Sigma}\cdot(\boldsymbol{P}\times\boldsymbol{Q})$ we find that

$$(\boldsymbol{\alpha}\cdot\boldsymbol{\varepsilon}_{\boldsymbol{k}'s'})\,(\boldsymbol{\alpha}\cdot\boldsymbol{\varepsilon}_{\boldsymbol{k}s}) + (\boldsymbol{\alpha}\cdot\boldsymbol{\varepsilon}_{\boldsymbol{k}s})\,(\boldsymbol{\alpha}\cdot\boldsymbol{\varepsilon}_{\boldsymbol{k}'s'}) = 2\,(\boldsymbol{\varepsilon}_{\boldsymbol{k}s}\cdot\boldsymbol{\varepsilon}_{\boldsymbol{k}'s'})\,.$$

Using this result, the second order matrix element and its square can be written as

$$\begin{aligned} M_{fi} &= \left(\frac{e^2\hbar}{4\epsilon_0 V m\sqrt{\omega\omega'}}\right) u_f{}^\dagger \left[2(\boldsymbol{\varepsilon}_{\boldsymbol{k}'s'}\cdot\boldsymbol{\varepsilon}_{\boldsymbol{k}s}) + \frac{1}{k}\,(\boldsymbol{\alpha}\cdot\boldsymbol{\varepsilon}_{\boldsymbol{k}'s'})(\boldsymbol{\alpha}\cdot\boldsymbol{k})(\boldsymbol{\alpha}\cdot\boldsymbol{\varepsilon}_{\boldsymbol{k}s})\right. \\ &\qquad \left. + \frac{1}{k'}\,(\boldsymbol{\alpha}\cdot\boldsymbol{\varepsilon}_{\boldsymbol{k}s})(\boldsymbol{\alpha}\cdot\boldsymbol{k}')(\boldsymbol{\alpha}\cdot\boldsymbol{\varepsilon}_{\boldsymbol{k}'s'})\right] u_i\,, \qquad (13.7.16) \end{aligned}$$

$$\begin{aligned} |M_{fi}|^2 &= M_{fi}^* M_{fi} = \left(\frac{e^2\hbar}{4\epsilon_0 V m\sqrt{\omega'\omega}}\right)^2 \\ &\quad \times \left(u_i{}^\dagger \left\{2(\boldsymbol{\varepsilon}_{\boldsymbol{k}'s'}\cdot\boldsymbol{\varepsilon}_{\boldsymbol{k}s}) + \frac{1}{k}(\boldsymbol{\alpha}\cdot\boldsymbol{\varepsilon}_{\boldsymbol{k}s})(\boldsymbol{\alpha}\cdot\boldsymbol{k})(\boldsymbol{\alpha}\cdot\boldsymbol{\varepsilon}_{\boldsymbol{k}'s'})\right.\right. \\ &\qquad \left.\left. + \frac{1}{k'}(\boldsymbol{\alpha}\cdot\boldsymbol{\varepsilon}_{\boldsymbol{k}s'})(\boldsymbol{\alpha}\cdot\boldsymbol{k}')(\boldsymbol{\alpha}\cdot\boldsymbol{\varepsilon}_{\boldsymbol{k}s})\right\} u_f\right) \\ &\quad \times \left(u_f{}^\dagger \left\{2(\boldsymbol{\varepsilon}_{\boldsymbol{k}'s'}\cdot\boldsymbol{\varepsilon}_{\boldsymbol{k}s}) + \frac{1}{k}(\boldsymbol{\alpha}\cdot\boldsymbol{\varepsilon}_{\boldsymbol{k}'s'})(\boldsymbol{\alpha}\cdot\boldsymbol{k})(\boldsymbol{\alpha}\cdot\boldsymbol{\varepsilon}_{\boldsymbol{k}s})\right.\right. \\ &\qquad \left.\left. + \frac{1}{k'}(\boldsymbol{\alpha}\cdot\boldsymbol{\varepsilon}_{\boldsymbol{k}s})(\boldsymbol{\alpha}\cdot\boldsymbol{k}_{\boldsymbol{k}'s'})(\boldsymbol{\alpha}\cdot\boldsymbol{\varepsilon}_{\boldsymbol{k}'s'})\right\} u_i\right) \\ &= \frac{e^4\hbar^2}{(4\pi\epsilon_0)^2 m^2\,\omega'\omega\,V^2}\,\left(u_i^\dagger\; B\; u_f\right)\left(u_f^\dagger\; C\; u_i\right), \qquad (13.7.17) \end{aligned}$$

where we have used $k = \omega/c$ and B and C are defined by

$$B \equiv 2(\boldsymbol{\varepsilon}_{\boldsymbol{k}'s'}\cdot\boldsymbol{\varepsilon}_{\boldsymbol{k}s}) + \frac{1}{k}(\boldsymbol{\alpha}\cdot\boldsymbol{\varepsilon}_{\boldsymbol{k}s})(\boldsymbol{\alpha}\cdot\boldsymbol{k})(\boldsymbol{\alpha}\cdot\boldsymbol{\varepsilon}_{\boldsymbol{k}'s'}) + \frac{1}{k'}(\boldsymbol{\alpha}\cdot\boldsymbol{\varepsilon}_{\boldsymbol{k}'s'})(\boldsymbol{\alpha}\cdot\boldsymbol{k}')(\boldsymbol{\alpha}\cdot\boldsymbol{\varepsilon}_{\boldsymbol{k}s})\,, \qquad (13.7.18)$$

$$C \equiv 2(\boldsymbol{\varepsilon}_{\boldsymbol{k}'s'}\cdot\boldsymbol{\varepsilon}_{\boldsymbol{k}s}) + \frac{1}{k}(\boldsymbol{\alpha}\cdot\boldsymbol{\varepsilon}_{\boldsymbol{k}'s'})(\boldsymbol{\alpha}\cdot\boldsymbol{k})(\boldsymbol{\alpha}\cdot\boldsymbol{\varepsilon}_{\boldsymbol{k}s}) + \frac{1}{k'}(\boldsymbol{\alpha}\cdot\boldsymbol{\varepsilon}_{\boldsymbol{k}s})(\boldsymbol{\alpha}\cdot\boldsymbol{k}')(\boldsymbol{\alpha}\cdot\boldsymbol{\varepsilon}_{\boldsymbol{k}'s'})\,. \qquad (13.7.19)$$

If the electron spin in the final state is not observed and the electrons are initially unpolarized, we must sum $|M_{fi}|^2$ over the two spin states of the outgoing positive energy electron and average over the two spin states of the initial positive energy electron:

$$\frac{1}{2}\sum_{\sigma_i}\sum_{\sigma_f}|M_{fi}|^2 = \frac{e^4\hbar^2}{(4\pi\epsilon_0)^2 m^2\,\omega'\omega\,V^2}\,\frac{1}{2}\sum_{s_i}\sum_{s_f}\left(u_i^\dagger\, B\, u_f\right)\left(u_f^\dagger\, C\, u_i\right).$$

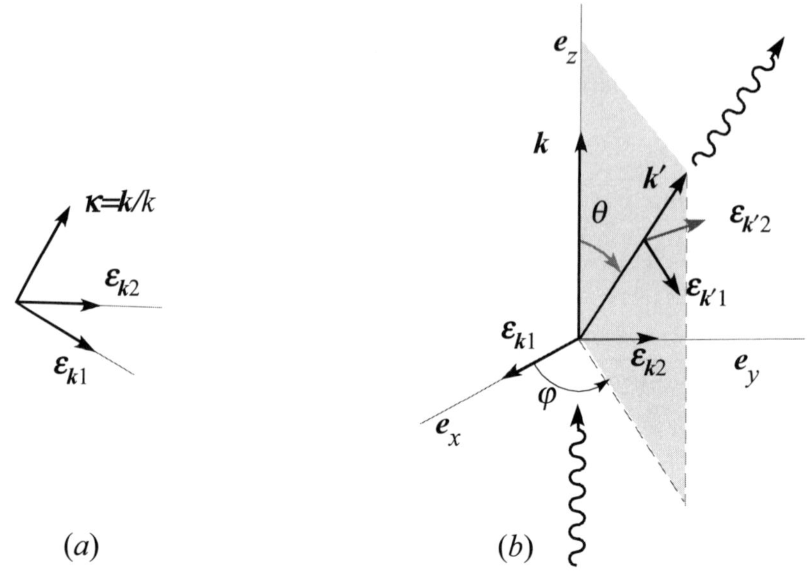

FIGURE 13.5
(a) Two orthogonal linear polarization vectors can be associated with a given $\boldsymbol{k}$ vector. (b) Polarization basis vectors associated with incident and scattered photon in Compton scattering.

To calculate the transition rate or the scattering cross-section we also need to sum over final states since the final state lies in a continuum. The number of states with total energy between E_F and $E_F + dE_F$ and photon momentum within a solid angle $d\Omega_{k'}$ around the direction $\boldsymbol{k}'$ of scattered photon is given by

$$\rho_E dE_f = \frac{V k'^2 dk' d\Omega_{k'}}{(2\pi)^3} .$$

To evaluate ρ_E we need a relation between E_F and k'. Let E_f be the final energy of electron of momentum $\hbar(\boldsymbol{k} - \boldsymbol{k}')$ and E_F be the total energy of the system (electron + photon). Then we have the following relations

$$E_f^2 = m^2c^4 + c^2\hbar^2 \left|\boldsymbol{k} - \boldsymbol{k}'\right|^2 ,$$

and

$$E_F = E_f + c\hbar k' = \sqrt{m^2c^4 + \hbar^2 \left|\boldsymbol{k} - \boldsymbol{k}'\right|^2 c^2} + c\hbar k' ,$$

so that

$$\left(E_F - c\hbar k'\right)^2 = m^2c^4 + c^2\hbar^2 \left(k^2 + k'^2 - 2\boldsymbol{k}\cdot\boldsymbol{k}'\right) .$$

Differentiating the last equation and re-arranging, we find the relation between the spread in final energy dE_F and the spread in final momentum dk',

$$E_f dE_F = c\hbar m c^2 \frac{k}{k'} dk' ,$$

where we have used the Compton relation [Chapter 1, Eq. (1.1.15)]

$$1 - \cos\theta = \frac{mc^2(\omega - \omega')}{2\pi\hbar\omega\omega'} ,$$

where $\cos\theta = \frac{\boldsymbol{k}\cdot\boldsymbol{k}'}{kk'}$. This relation follows from momentum conservation in Compton scattering. This gives the density of states

$$\rho_F = \frac{k'^2}{(2\pi)^3}\frac{\omega'}{\omega}\frac{V}{c\hbar mc^2}E_f\,. \tag{13.7.20}$$

Then the transition probability per unit time (w_{fi}) is given by

$$w_{fi} = \frac{2\pi}{\hbar}\frac{1}{2}\sum_{\sigma_i}\sum_{\sigma_f}|M_{fi}|^2\,\rho_F\,. \tag{13.7.21}$$

Differential cross-section of Compton scattering is given by

$$\begin{aligned}\frac{d\sigma}{d\Omega_{k'}} &= \frac{w_{fi}}{c/V} = \frac{V}{c}\frac{2\pi}{\hbar}\frac{1}{2}\sum_{\sigma_i}\sum_{\sigma_f}|M_{fi}|^2\,\rho_F\\ &= \frac{1}{8}\frac{e^4}{(4\pi\epsilon_0)^2(mc^2)^3}\left(\frac{\omega'}{\omega}\right)^2 E_f\sum_{\sigma_i}\sum_{\sigma_f}\left(u_i^\dagger\,B\,u_f\right)\left(u_f^\dagger\,C\,u_i\right),\end{aligned} \tag{13.7.22}$$

where c/V is the incident photon flux density corresponding to one photon per quantization volume. The sum over spin states of the initial and final positive energy electrons may be replaced by $\sum_i\sum_f$ (the sum over all the four spin and both positive and negative energy states of the initial and final electrons) provided we use the projection operators Λ_+^f and Λ_+^i to project out the positive energy states from the final and initial state spinors u_f and u_i, respectively:

$$\begin{aligned}\sum_{\sigma_i}\sum_{\sigma_f}\left(u_i^\dagger\,B\,u_f\right)\left(u_f^\dagger\,C\,u_i\right) &= \sum_i\sum_f\left(u_i^\dagger\,B\,\Lambda_+^f\,u_f\right)\left(u_f^\dagger\,C\,\Lambda_+^i\,u_i\right)\\ &= \sum_i\left(u_i^\dagger\,B\,\Lambda_+^f\,C\,\Lambda_+^i\,u_i\right)\\ &= \sum_i\left(u_i^\dagger\,D\,\Lambda_+^i\,u_i\right)\,, \quad \text{where} \quad D \equiv B\,\Lambda_+^f\,C\,.\end{aligned}$$

Hence
$$\sum_{\sigma_i}\sum_{\sigma_f}\left(u_i^\dagger\,B\,u_f\right)\left(u_f^\dagger\,C\,u_i\right) = \frac{1}{2}\sum_i u_i^\dagger\,D(I+\beta)\,u_i \tag{13.7.23}$$

since
$$D\Lambda_+^i = \frac{D(E+H)}{2E} = \frac{1}{2}\,D(I+\beta)\,, \tag{13.7.24}$$

and $E = mc^2$ and $H = \beta mc^2$ for the initial state of the electron represented by the spinor u_i. It can be seen [15] that $\sum_i u_i{}^\dagger D\,(I+\beta)\,u_i = \mathrm{Tr}\,\{D(I+\beta)\}$. Now it is shown at the end

[15] Let the column matrix u_j have components u_{j1}, u_{j2}, u_{j3} and u_{j4}, then the adjoint $u_j^\dagger$, a row matrix, has the form

$$u_j^\dagger = (u_{j1}^*\;u_{j2}^*\;u_{j3}^*\;u_{j4}^*)\,.$$

Let a square matrix U be defined by

$$U_{ij} = u_{ji}$$

which means we can write

$$U = (u_1\,u_2\,u_3\,u_4)\,.$$

of this section

$$\sum_{\sigma_i}\sum_{\sigma_f}\left(u_i^\dagger\, B\, u_f\right)\left(u_f^\dagger\, C\, u_i\right) \equiv \frac{1}{2}\ \text{Tr}\ \{D(I+\beta)\}$$

$$= \frac{8mc^2}{4E_f}\left(\frac{\omega}{\omega'} + \frac{\omega'}{\omega} - 2 + 4\cos^2\alpha\right) \qquad (13.7.25)$$

where $\cos\alpha = \boldsymbol{\varepsilon}_{\boldsymbol{k}s}\cdot\boldsymbol{\varepsilon}_{\boldsymbol{k}'s'}$. Using this result, we can write the cross-section for observing photon polarization $\boldsymbol{\varepsilon}_{\boldsymbol{k}'s'}$ in the scattering of incident photons with polarization $\boldsymbol{\varepsilon}_{\boldsymbol{k}s}$

$$\frac{d\sigma(\boldsymbol{\varepsilon}_{\boldsymbol{k}s},\boldsymbol{\varepsilon}_{\boldsymbol{k}'s'})}{d\Omega_{k'}} = \frac{1}{4}r_o^2\left(\frac{\omega'}{\omega}\right)^2\left(\frac{\omega}{\omega'} + \frac{\omega'}{\omega} - 2 + 4\cos^2\alpha\right), \qquad (13.7.26)$$

where $r_o = \frac{e^2}{4\pi\epsilon_0 mc^2}$ = is the classical electron radius. This is the *Klein-Nishina formula.*

If the incident photon beam is unpolarized and the polarization of scattered photons is not observed, then we must average this cross-section (13.7.26) over initial polarization directions and sum over the final polarization states. To do this let us choose the the $z-$axis along the direction of incident $\boldsymbol{k}$ vector. Then the polarization vector of the incident photon may be taken to be

$$\boldsymbol{\varepsilon}_{\boldsymbol{k}}(\phi_o) = \boldsymbol{e}_x\cos\phi_o + \boldsymbol{e}_y\sin\phi_o\,, \qquad (13.7.27)$$

where the angle ϕ_o is uniformly distributed over the range 0 to 2π corresponding to an unpolarized incident beam. If we denote the direction of scattered $\boldsymbol{k}'$ vector by angles θ,φ, then two orthogonal polarization vectors for the scattered photon may be taken to be [see footnote 10]

$$\boldsymbol{\varepsilon}_{\boldsymbol{k}'1} = \cos\theta\cos\varphi\,\boldsymbol{e}_x + \cos\theta\sin\varphi\,\boldsymbol{e}_y - \sin\theta\,\boldsymbol{e}_z\,, \qquad (13.7.28a)$$

$$\boldsymbol{\varepsilon}_{\boldsymbol{k}'2} = -\sin\varphi\boldsymbol{e}_x + \cos\varphi\,\boldsymbol{e}_y\,. \qquad (13.7.28b)$$

Then the sum of $[\boldsymbol{\varepsilon}_{\boldsymbol{k}}(\phi_o)\cdot\boldsymbol{\varepsilon}_{\boldsymbol{k}'s'}]^2$ over final polarizations is

$$\sum_{s'}[\boldsymbol{\varepsilon}_{\boldsymbol{k}}(\phi_o)\cdot\boldsymbol{\varepsilon}_{\boldsymbol{k}'s'}]^2 = (\cos\theta\cos\varphi\cos\phi_0 + \cos\theta\sin\varphi\sin\phi_o)^2 + (-\sin\varphi\cos\phi_o + \cos\varphi\sin\phi_o)^2\,.$$

Expanding and averaging with respect to ϕ_o and using the results

$$\frac{1}{2\pi}\int_0^{2\pi} d\phi_o\cos^2\phi_o = \frac{1}{2} = \frac{1}{2\pi}\int_0^{2\pi} d\phi_o\sin^2\phi_o \quad\text{and}\quad \frac{1}{2\pi}\int_0^{2\pi} d\phi_o\cos\phi_o\sin\phi_o = 0\,,$$

The adjoint $U^\dagger$ of U is then

$$U^\dagger = \begin{pmatrix} u_1^\dagger \\ u_2^\dagger \\ u_3^\dagger \\ u_4^\dagger \end{pmatrix},$$

which means $(U^\dagger)_{ij} = u_{ij}^*$. The orthogonality condition $u_i^\dagger\, u_j = \delta_{ij}$, for the spinors implies $U^\dagger U = I$, i.e., U is a unitary matrix ($UU^\dagger = I$ follows from the completeness condition $\sum_i^4 |u_i\rangle\langle u_i| = \hat{1}$ or $\sum_i u_i\, u_i^\dagger = I$). Now

$$\sum_i u_i^\dagger\, A\, u_i = \sum_i\sum_\ell\sum_k u_{i\ell}^*\, A_{\ell k}\, u_{ik}$$

$$= \sum_i\sum_\ell\sum_k U_{i\ell}^\dagger\, A_{\ell k}\, U_{ki} = \sum_i (U^\dagger A U)_{ii} = \text{Tr}\ (U^\dagger A U) = \text{Tr}\ A\,.$$

we find

$$\frac{1}{2\pi}\int_0^{2\pi} d\phi_o \sum_{s'} [\boldsymbol{\varepsilon}_{\boldsymbol{k}}(\phi_o)\cdot\boldsymbol{\varepsilon}_{\boldsymbol{k}'s'}]^2 = \frac{1}{2}(1+\cos^2\theta)\,. \tag{13.7.29}$$

Summing the cross-section (13.7.26) over final polarization states and averaging over initial polarization states

$$\left[\frac{d\sigma}{d\Omega_{k'}}\right]_{\text{unpolarized}} = \frac{r_o^2}{4}\left(\frac{\omega'}{\omega}\right)^2 \frac{1}{2\pi}\int_0^{2\pi} d\phi_o \sum_{s'}\left[\frac{\omega}{\omega'} + \frac{\omega'}{\omega} - 2 + 4\left(\boldsymbol{\varepsilon}_{\boldsymbol{k}}(\phi_o)\cdot\boldsymbol{\varepsilon}_{\boldsymbol{k}'s'}\right)^2\right]$$

$$= \frac{r_o^2}{4}\left(\frac{\omega'}{\omega}\right)^2\left[2\frac{\omega}{\omega'} + 2\frac{\omega'}{\omega} - 4 + 4\cdot\frac{1}{2}\left(1+\cos^2\theta\right)\right]$$

or

$$\left[\frac{d\sigma}{d\Omega_{k'}}\right]_{\text{unpolarized}} = \frac{r_o^2}{2}\left(\frac{\omega'}{\omega}\right)^2\left[\frac{\omega}{\omega'} + \frac{\omega'}{\omega} - \sin^2\theta\right]. \tag{13.7.30}$$

This equation expresses the unpolarized Compton scattering cross-section as a function of the scattering angle θ. This is in excellent agreement with the experimental results and is regarded as one of the earliest triumphs of Dirac's theory of the electron. A very different formula would have been obtained if we had not allowed the electron to have negative energy in the intermediate state [Fig. 13.4(b)].

Proof of Eq. (13.7.25)

By definition we have

$$D = B\,\Lambda_+^f\,C\,, \tag{13.7.31}$$

where $\Lambda_+^f = \frac{E_f+H_f}{2E_f}$, $E_f = c\hbar\,(k-k') + mc^2$ and $H_f = c\boldsymbol{\alpha}\cdot\left(\hbar\boldsymbol{k} - \hbar\boldsymbol{k}'\right) + \beta mc^2$ and B and C are given by Eqs. (13.7.18) and (13.7.19). With the help of the expressions for E_f and H_f, we can write Λ_+^f as

$$\Lambda_+^f = \frac{c}{2E_f}\,\left[Q + (1+\beta)\,mc\right], \tag{13.7.32a}$$

where

$$Q \equiv \hbar\,(k-k') + \hbar\left(\boldsymbol{\alpha}\cdot\boldsymbol{k} - \boldsymbol{\alpha}\cdot\boldsymbol{k}'\right). \tag{13.7.32b}$$

Using this in the expression (13.7.31) for D, we obtain

$$\text{Tr}\left[\frac{1}{2}D(I+\beta)\right] = \frac{c}{4E_f}\left(\text{Tr}\;BQC + 32\,mc\left[\boldsymbol{\varepsilon}_{\boldsymbol{k}'s'}\cdot\boldsymbol{\varepsilon}_{\boldsymbol{k}s}\right)^2\right], \tag{13.7.33}$$

where we used last three of the following identities:

1. $\text{Tr}\;\left[\alpha_1^\ell\,\alpha_2^m\,\alpha_3^n\,\beta^q\right] = 0$ if $\ell+m+n$ is odd.
2. $(\boldsymbol{\alpha}\cdot\boldsymbol{\varepsilon}_{\boldsymbol{k}s})\,(\boldsymbol{\alpha}\cdot\boldsymbol{\varepsilon}_{\boldsymbol{k}s}) = I$.
3. $(\boldsymbol{\alpha}\cdot\boldsymbol{k})\,(\boldsymbol{\alpha}\cdot\boldsymbol{k}) = k^2$.
4. $(\boldsymbol{\alpha}\cdot\boldsymbol{\varepsilon}_{\boldsymbol{k}s})\,(\boldsymbol{\alpha}\cdot\boldsymbol{\varepsilon}_{\boldsymbol{k}'s'}) + (\boldsymbol{\alpha}\cdot\boldsymbol{\varepsilon}_{\boldsymbol{k}'s'})\,(\boldsymbol{\alpha}\cdot\boldsymbol{\varepsilon}_{\boldsymbol{k}s}) = 2\,(\boldsymbol{\varepsilon}_{\boldsymbol{k}s}\cdot\boldsymbol{\varepsilon}_{\boldsymbol{k}'s'})$.
5. $(\boldsymbol{\alpha}\cdot\boldsymbol{\varepsilon}_{\boldsymbol{k}s})(\boldsymbol{\alpha}\cdot\boldsymbol{k}) + (\boldsymbol{\alpha}\cdot\boldsymbol{k})(\boldsymbol{\alpha}\cdot\boldsymbol{\varepsilon}_{\boldsymbol{k}s}) = 2(\boldsymbol{\varepsilon}_{\boldsymbol{k}s}\cdot\boldsymbol{k}) = 0$.
6. $(\text{Tr}\;X)^* = \text{Tr}\;X^\dagger$.
7. $\text{Tr}\;(\boldsymbol{\alpha}\cdot\boldsymbol{k})\;(\boldsymbol{\alpha}\cdot\boldsymbol{k}') = (\text{Tr}\;I)\;(\boldsymbol{k}\cdot\boldsymbol{k}') = 4\;(\boldsymbol{k}\cdot\boldsymbol{k}')$.

8. Tr $P = 0$.

9. Tr $[BQC\beta] = 0$.

10. $\beta R = -R\beta$.

These identities will be used in the following derivations. By introducing matrix operators P and R by

$$P = B - 2\,(\boldsymbol{\varepsilon}_{\boldsymbol{k}'s'} \cdot \boldsymbol{\varepsilon}_{\boldsymbol{k}s}) \quad \text{and} \quad R = C - 2\,(\boldsymbol{\varepsilon}_{\boldsymbol{k}'s'} \cdot \boldsymbol{\varepsilon}_{\boldsymbol{k}s}),$$

we can write

$$\text{Tr}\,(BQC) = \text{Tr}\,\left[\{2\,(\boldsymbol{\varepsilon}_{\boldsymbol{k}'s'} \cdot \boldsymbol{\varepsilon}_{\boldsymbol{k}s}) + P\}\,\{\hbar\,(k - k') + T\}\,\{2\,(\boldsymbol{\varepsilon}_{\boldsymbol{k}'s'} \cdot \boldsymbol{\varepsilon}_{\boldsymbol{k}s}) + R\}\right],$$

where $T = \hbar\boldsymbol{\alpha} \cdot (\boldsymbol{k} - \boldsymbol{k}')$ and $k \equiv |\boldsymbol{k}|$. On expanding the right-hand side, we obtain

$$\begin{aligned}\text{Tr}\,(BQC) = \text{Tr}\,\Big[\, & \hbar\,(k - k')\,\Big\{4\,(\boldsymbol{\varepsilon}_{\boldsymbol{k}'s'} \cdot \boldsymbol{\varepsilon}_{\boldsymbol{k}s})^2 + PR\Big\} + 2\,(\boldsymbol{\varepsilon}_{\boldsymbol{k}'s'} \cdot \boldsymbol{\varepsilon}_{\boldsymbol{k}s})\,(TR + PT) \\ & + 4\,(\boldsymbol{\varepsilon}_{\boldsymbol{k}'s'} \cdot \boldsymbol{\varepsilon}_{\boldsymbol{k}s})^2\,T + \hbar\,(k - k')\,2\,(\boldsymbol{\varepsilon}_{\boldsymbol{k}'s'} \cdot \boldsymbol{\varepsilon}_{\boldsymbol{k}s})\,P \\ & + 2\,(\boldsymbol{\varepsilon}_{\boldsymbol{k}'s'} \cdot \boldsymbol{\varepsilon}_{\boldsymbol{k}s})\,\hbar\,(k - k')\,R + PTR\Big]\,.\end{aligned}$$

The last four terms vanish because P, T, R, and PTR involve odd number of α-matrices and therefore their traces vanish [identity (1)]. This leads to

$$\begin{aligned}\text{Tr}\,(BQC) = \text{Tr}\,\Big[&\hbar(k - k')\,\Big\{4\,(\boldsymbol{\varepsilon}_{\boldsymbol{k}'s'} \cdot \boldsymbol{\varepsilon}_{\boldsymbol{k}s})^2 + PR\Big\} \\ &+ 2\,(\boldsymbol{\varepsilon}_{\boldsymbol{k}'s'} \cdot \boldsymbol{\varepsilon}_{\boldsymbol{k}s})\,(TR + PT)\Big]\,. \end{aligned} \tag{13.7.34}$$

Now

$$\begin{aligned}\text{Tr}\,(PR) = \text{Tr}\,\Big[&\frac{1}{k^2}\,(\boldsymbol{\alpha} \cdot \boldsymbol{\varepsilon}_{\boldsymbol{k}s})\,(\boldsymbol{\alpha} \cdot \boldsymbol{k})\,(\boldsymbol{\alpha} \cdot \boldsymbol{\varepsilon}_{\boldsymbol{k}'s'})\,(\boldsymbol{\alpha} \cdot \boldsymbol{\varepsilon}_{\boldsymbol{k}'s'})\,(\boldsymbol{\alpha} \cdot \boldsymbol{k})\,(\boldsymbol{\alpha} \cdot \boldsymbol{\varepsilon}_{\boldsymbol{k}s}) \\ &+ \frac{1}{k'^2}\,(\boldsymbol{\alpha} \cdot \boldsymbol{\varepsilon}_{\boldsymbol{k}'s'})\,(\boldsymbol{\alpha} \cdot \boldsymbol{k}')\,(\boldsymbol{\alpha} \cdot \boldsymbol{\varepsilon}_{\boldsymbol{k}s})\,(\boldsymbol{\alpha} \cdot \boldsymbol{\varepsilon}_{\boldsymbol{k}s})\,(\boldsymbol{\alpha} \cdot \boldsymbol{k}')\,(\boldsymbol{\alpha} \cdot \boldsymbol{\varepsilon}_{\boldsymbol{k}'s'}) \\ &+ \frac{1}{kk'}\,(\boldsymbol{\alpha} \cdot \boldsymbol{\varepsilon}_{\boldsymbol{k}s})\,(\boldsymbol{\alpha} \cdot \boldsymbol{k})\,(\boldsymbol{\alpha} \cdot \boldsymbol{\varepsilon}_{\boldsymbol{k}'s'})\,(\boldsymbol{\alpha} \cdot \boldsymbol{\varepsilon}_{\boldsymbol{k}s})\,(\boldsymbol{\alpha} \cdot \boldsymbol{k}')\,(\boldsymbol{\alpha} \cdot \boldsymbol{\varepsilon}_{\boldsymbol{k}'s'}) \\ &+ \frac{1}{kk'}\,(\boldsymbol{\alpha} \cdot \boldsymbol{\varepsilon}_{\boldsymbol{k}'s'})\,(\boldsymbol{\alpha} \cdot \boldsymbol{k}')\,(\boldsymbol{\alpha} \cdot \boldsymbol{\varepsilon}_{\boldsymbol{k}s})\,(\boldsymbol{\alpha} \cdot \boldsymbol{\varepsilon}_{\boldsymbol{k}'s'})\,(\boldsymbol{\alpha} \cdot \boldsymbol{k})\,(\boldsymbol{\alpha} \cdot \boldsymbol{\varepsilon}_{\boldsymbol{k}s})\Big] \\ = \text{Tr}\,\Big[&2I + \frac{2}{kk'}\,(\boldsymbol{\alpha} \cdot \boldsymbol{\varepsilon}_{\boldsymbol{k}s})\,(\boldsymbol{\alpha} \cdot \boldsymbol{k})\,(\boldsymbol{\alpha} \cdot \boldsymbol{\varepsilon}_{\boldsymbol{k}'s'})\,(\boldsymbol{\alpha} \cdot \boldsymbol{\varepsilon}_{\boldsymbol{k}s})\,(\boldsymbol{\alpha} \cdot \boldsymbol{k}')\,(\boldsymbol{\alpha} \cdot \boldsymbol{\varepsilon}_{\boldsymbol{k}'s'})\Big] \\ = 8 + \frac{2}{kk'}\,&\text{Tr}\,\Big[(\boldsymbol{\alpha} \cdot \boldsymbol{\varepsilon}_{\boldsymbol{k}s})\,(\boldsymbol{\alpha} \cdot \boldsymbol{k})\,\Big\{2\,(\boldsymbol{\varepsilon}_{\boldsymbol{k}s} \cdot \boldsymbol{\varepsilon}_{\boldsymbol{k}'s'}) \\ &- (\boldsymbol{\alpha} \cdot \boldsymbol{\varepsilon}_{\boldsymbol{k}s})\,(\boldsymbol{\alpha} \cdot \boldsymbol{\varepsilon}_{\boldsymbol{k}'s'})\Big\}\,(\boldsymbol{\alpha} \cdot \boldsymbol{k}')\,(\boldsymbol{\alpha} \cdot \boldsymbol{\varepsilon}_{\boldsymbol{k}'s'})\Big] \\ = 8 + \frac{2}{kk'}\,&\Big[-2\,(\boldsymbol{\varepsilon}_{\boldsymbol{k}s} \cdot \boldsymbol{\varepsilon}_{\boldsymbol{k}'s'})\,\text{Tr}\,\big\{(\boldsymbol{\alpha} \cdot \boldsymbol{\varepsilon}_{\boldsymbol{k}s})\,(\boldsymbol{\alpha} \cdot \boldsymbol{k})\,(\boldsymbol{\alpha} \cdot \boldsymbol{\varepsilon}_{\boldsymbol{k}'s'})\,(\boldsymbol{\alpha} \cdot \boldsymbol{k}')\big\} \\ &- \text{Tr}\,\{(\boldsymbol{\alpha} \cdot \boldsymbol{k})\,(\boldsymbol{\alpha} \cdot \boldsymbol{k}')\}\Big]\,,\end{aligned}$$

where we have used identities (4) and (5). Using identity (7) in the last step, we get

$$\text{Tr}\,(PR) = 8 + \frac{2}{kk'}\,\Big[-2\,(\boldsymbol{\varepsilon}_{\boldsymbol{k}s} \cdot \boldsymbol{\varepsilon}_{\boldsymbol{k}'s'})\,\text{Tr}\,\big\{(\boldsymbol{\alpha} \cdot \boldsymbol{\varepsilon}_{\boldsymbol{k}s})\,(\boldsymbol{\alpha} \cdot \boldsymbol{k})\,(\boldsymbol{\alpha} \cdot \boldsymbol{\varepsilon}_{\boldsymbol{k}'s'})\,(\boldsymbol{\alpha} \cdot \boldsymbol{k}')\big\} - 4\,(\boldsymbol{k} \cdot \boldsymbol{k}')\Big]\,. \tag{13.7.35}$$

Also

$$
\begin{aligned}
&\mathrm{Tr}\ (PT+TR) \\
&= \mathrm{Tr}\ \left[\left\{\frac{1}{k}(\boldsymbol{\alpha}\cdot\boldsymbol{\varepsilon}_{\boldsymbol{k}s})(\boldsymbol{\alpha}\cdot\boldsymbol{k})(\boldsymbol{\alpha}\cdot\boldsymbol{\varepsilon}_{\boldsymbol{k}'s'}) + \frac{1}{k'}(\boldsymbol{\alpha}\cdot\boldsymbol{\varepsilon}_{\boldsymbol{k}'s'})(\boldsymbol{\alpha}\cdot\boldsymbol{k}')(\boldsymbol{\alpha}\cdot\boldsymbol{\varepsilon}_{\boldsymbol{k}s})\right\}\hbar\boldsymbol{\alpha}\cdot(\boldsymbol{k}-\boldsymbol{k}')\right. \\
&\quad \left. + \hbar\boldsymbol{\alpha}\cdot\ (\boldsymbol{k}-\boldsymbol{k}')\left\{\frac{1}{k}(\boldsymbol{\alpha}\cdot\boldsymbol{\varepsilon}_{\boldsymbol{k}'s'})\ (\boldsymbol{\alpha}\cdot\boldsymbol{k})\ (\boldsymbol{\alpha}\cdot\boldsymbol{\varepsilon}_{\boldsymbol{k}s}) + \frac{1}{k'}\ (\boldsymbol{\alpha}\cdot\boldsymbol{\varepsilon}_{\boldsymbol{k}s})\ (\boldsymbol{\alpha}\cdot\boldsymbol{k}')\ (\boldsymbol{\alpha}\cdot\boldsymbol{\varepsilon}_{\boldsymbol{k}'s'})\right\}\right] \\
&= \mathrm{Tr}\ \left[\frac{\hbar}{k}\ (\boldsymbol{\alpha}\cdot\boldsymbol{\varepsilon}_{\boldsymbol{k}s})\ (\boldsymbol{\alpha}\cdot\boldsymbol{k})\ (\boldsymbol{\alpha}\cdot\boldsymbol{\varepsilon}_{\boldsymbol{k}'s'})\ (\boldsymbol{\alpha}\cdot\boldsymbol{k}) - \frac{\hbar}{k}\ (\boldsymbol{\alpha}\cdot\boldsymbol{\varepsilon}_{\boldsymbol{k}s})\ (\boldsymbol{\alpha}\cdot\boldsymbol{k})\ (\boldsymbol{\alpha}\cdot\boldsymbol{\varepsilon}_{\boldsymbol{k}'s'})\ (\boldsymbol{\alpha}\cdot\boldsymbol{k}')\right. \\
&\quad + \frac{\hbar}{k'}\ (\boldsymbol{\alpha}\cdot\boldsymbol{\varepsilon}_{\boldsymbol{k}'s'})\ (\boldsymbol{\alpha}\cdot\boldsymbol{k}')\ (\boldsymbol{\alpha}\cdot\boldsymbol{\varepsilon}_{\boldsymbol{k}s})\ (\boldsymbol{\alpha}\cdot\boldsymbol{k}) - \frac{\hbar}{k'}\ (\boldsymbol{\alpha}\cdot\boldsymbol{\varepsilon}_{\boldsymbol{k}'s'})\ (\boldsymbol{\alpha}\cdot\boldsymbol{k}')\ (\boldsymbol{\alpha}\cdot\boldsymbol{\varepsilon}_{\boldsymbol{k}s})\ (\boldsymbol{\alpha}\cdot\boldsymbol{k}') \\
&\quad + \frac{\hbar}{k}\ (\boldsymbol{\alpha}\cdot\boldsymbol{k})\ (\boldsymbol{\alpha}\cdot\boldsymbol{\varepsilon}_{\boldsymbol{k}'s'})\ (\boldsymbol{\alpha}\cdot\boldsymbol{k})\ (\boldsymbol{\alpha}\cdot\boldsymbol{\varepsilon}_{\boldsymbol{k}s}) - \frac{\hbar}{k}\ (\boldsymbol{\alpha}\cdot\boldsymbol{k}')\ (\boldsymbol{\alpha}\cdot\boldsymbol{\varepsilon}_{\boldsymbol{k}'s'})\ (\boldsymbol{\alpha}\cdot\boldsymbol{k})(\boldsymbol{\alpha}\cdot\boldsymbol{\varepsilon}_{\boldsymbol{k}s}) \\
&\quad \left. + \frac{\hbar}{k'}\ (\boldsymbol{\alpha}\cdot\boldsymbol{k})\ (\boldsymbol{\alpha}\cdot\boldsymbol{\varepsilon}_{\boldsymbol{k}s})\ (\boldsymbol{\alpha}\cdot\boldsymbol{k}')\ (\boldsymbol{\alpha}\cdot\boldsymbol{\varepsilon}_{\boldsymbol{k}'s'}) - \frac{\hbar}{k'}\ (\boldsymbol{\alpha}\cdot\boldsymbol{k}')\ (\boldsymbol{\alpha}\cdot\boldsymbol{\varepsilon}_{\boldsymbol{k}s})\ (\boldsymbol{\alpha}\cdot\boldsymbol{k}')(\boldsymbol{\alpha}\cdot\boldsymbol{\varepsilon}_{\boldsymbol{k}'s'})\right].
\end{aligned}
$$

Combining 1st and 5th, 4th and 8th, 3rd and 7th, and 2nd and 6th terms on the right-hand side we get:

$$
\begin{aligned}
\mathrm{Tr}\ (PT+TR) = \mathrm{Tr}\ &\left[-2\hbar\,(\boldsymbol{\varepsilon}_{\boldsymbol{k}s}\cdot\boldsymbol{\varepsilon}_{\boldsymbol{k}'s'})\ k\ +\ 2\hbar\,k'\,(\boldsymbol{\varepsilon}_{\boldsymbol{k}s}\cdot\boldsymbol{\varepsilon}_{\boldsymbol{k}'s'})\right. \\
&+ \frac{2\hbar}{k'}\ (\boldsymbol{\alpha}\cdot\boldsymbol{\varepsilon}_{\boldsymbol{k}'s'})\ (\boldsymbol{\alpha}\cdot\boldsymbol{k}')\ (\boldsymbol{\alpha}\cdot\boldsymbol{\varepsilon}_{\boldsymbol{k}s})\ (\alpha\cdot\boldsymbol{k}) \\
&\left. -\frac{2\hbar}{k}\ (\boldsymbol{\alpha}\cdot\boldsymbol{\varepsilon}_{\boldsymbol{k}'s'})\ (\boldsymbol{\alpha}\cdot\boldsymbol{k}')\ (\boldsymbol{\alpha}\cdot\boldsymbol{\varepsilon}_{\boldsymbol{k}s})\ (\boldsymbol{\alpha}\cdot\boldsymbol{k})\right],
\end{aligned}
$$

or

$$
\begin{aligned}
\mathrm{Tr}\ (PT+TR) = \mathrm{Tr}\ &\left[-\hbar\,(k-k')\ 2\,(\boldsymbol{\varepsilon}_{\boldsymbol{k}s}\cdot\boldsymbol{\varepsilon}_{\boldsymbol{k}'s'})\right. \\
&\left. + 2\hbar\left(\frac{1}{k'}-\frac{1}{k}\right)(\boldsymbol{\alpha}\cdot\boldsymbol{\varepsilon}_{\boldsymbol{k}'s'})\ (\boldsymbol{\alpha}\cdot\boldsymbol{k}')\ (\boldsymbol{\alpha}\boldsymbol{\varepsilon}_{\boldsymbol{k}s})\ (\boldsymbol{\alpha}\cdot\boldsymbol{k})\right].
\end{aligned}
\tag{13.7.36}
$$

Using Eqs. (13.7.35) and (13.7.36) in Eq. (13.7.34) we get

$$
\begin{aligned}
\mathrm{Tr}\ (BQC) &= \mathrm{Tr}\ \left[\hbar(k-k')\ \left\{4\,(\boldsymbol{\varepsilon}_{\boldsymbol{k}'s'}\cdot\boldsymbol{\varepsilon}_{\boldsymbol{k}s})^2 + PR\right\} + 2\,(\boldsymbol{\varepsilon}_{\boldsymbol{k}'s'}\cdot\boldsymbol{\varepsilon}_{\boldsymbol{k}s})\,(TR+PT)\right] \\
&= \mathrm{Tr}\ \left[2\hbar(k-k') - 2\hbar\left(\frac{k-k'}{kk'}\right)\ \boldsymbol{k}\cdot\boldsymbol{k}'\right] \\
&= 8\hbar\,(k-k')\,(1-\cos\theta)\ , \\
&= 8mc\,\frac{(\omega-\omega')^2}{\omega\omega'}
\end{aligned}
\tag{13.7.37}
$$

where in the last step we have used $k=\omega/c$ and four-momentum conservation for Compton scattering which leads to $mc^2(\omega-\omega') = \hbar\omega\omega'(1-\cos\theta)$ [cf. Chapter 1, Eq. (1.1.15)]. Using this result in (13.7.33) we get Eq. (13.7.25):

$$
\begin{aligned}
\frac{1}{2}\ \mathrm{Tr}\ [D(I+\beta)] &= \frac{c}{4E_f}\ [\mathrm{Tr}\ BQC\ +\ 32mc(\boldsymbol{\varepsilon}_{\boldsymbol{k}'s'}\cdot\boldsymbol{\varepsilon}_{\boldsymbol{k}s}] \\
&= \frac{8mc^2}{4E_f}\ \left[\frac{\omega}{\omega'} + \frac{\omega'}{\omega} + 4\ (\boldsymbol{\varepsilon}_{\boldsymbol{k}'s'}\cdot\boldsymbol{\varepsilon}_{\boldsymbol{k}s})\ - 2\right].
\end{aligned}
$$

Problems

1. Show that the classical Hamiltonian of pure radiation field

$$H = \frac{1}{2}\int\left(\epsilon_0|\boldsymbol{E}|^2 + \frac{|\boldsymbol{B}|^2}{\mu_0}\right) = \frac{1}{2}\int\left\{\epsilon_0\left|\frac{\partial \boldsymbol{A}}{\partial t}\right|^2 + \frac{|\nabla\times\boldsymbol{A}|^2}{\mu_0}\right\}d\tau$$

may be expressed in terms of the field variables P_λ and Q_λ as

$$H = \sum_\lambda H_\lambda = \sum_\lambda \frac{1}{2}\left(P_\lambda^2 + \omega_\lambda^2 Q_\lambda^2\right)$$

where $Q_\lambda = q_\lambda(t) + q_\lambda^*(t)$, $P_\lambda = -i\omega_\lambda\left(q_\lambda(t) - q_\lambda^*(t)\right)$ and q_λ's occur as coefficients when $\boldsymbol{A}(\boldsymbol{r},t)$ is expanded as

$$\boldsymbol{A}(\boldsymbol{r},t) = \frac{1}{\sqrt{\epsilon_0}}\sum_\lambda [q_\lambda(t)\boldsymbol{u}_\lambda(\boldsymbol{r}) + q_\lambda^*(t)\boldsymbol{u}_\lambda^*(\boldsymbol{r})],$$

and functions $\boldsymbol{u}_\lambda(r)$ form a denumerable, though infinite, set satisfying the equation

$$\nabla^2\boldsymbol{u}_\lambda(r) + \frac{\omega_\lambda^2}{c^2}\boldsymbol{u}_\lambda(r) = 0$$

and the condition of orthonormality

$$\int \boldsymbol{u}_\lambda(r)\cdot\boldsymbol{u}_\mu^*(r)d\tau = \delta_{\lambda\mu}\,.$$

2. Show that P_λ's and Q_λ's, defined in problem 1, satisfy the canonical equations:

$$-\dot{P}_\lambda = \frac{\partial H}{\partial Q_\lambda} \quad\text{and}\quad \dot{Q}_\lambda = \frac{\partial H}{\partial P_\lambda}$$

and may be regarded as a set of canonically conjugate variables.

3. To quantize the radiation field we may regard P_λ and Q_λ as self-adjoint operators $\hat{P}_\lambda$ and $\hat{Q}_\lambda$ and invoke the following commutation relations between them

$$[\hat{P}_\lambda, \hat{Q}_\mu] = -i\hbar\delta_{\lambda\mu}\,,$$

$$[\hat{P}_\lambda, \hat{P}_\mu] = [\hat{Q}_\lambda, \hat{Q}_\mu] = 0\,.$$

Introducing new operators $\hat{a}_\lambda$ and $\hat{a}_\lambda^\dagger$ by

$$\hat{a}_\lambda = \sqrt{\frac{2\omega_\lambda}{\hbar}}\frac{1}{2}\left(\hat{Q}_\lambda + \frac{i\hat{P}_\lambda}{\omega_\lambda}\right) \quad\text{and}\quad \hat{a}_\lambda^\dagger = \sqrt{\frac{2\omega_\lambda}{\hbar}}\frac{1}{2}\left(\hat{Q}_\lambda - \frac{i\hat{P}_\lambda}{\omega_\lambda}\right)$$

show that $\hat{a}_\lambda$ and $\hat{a}_\lambda^\dagger$ satisfy the following commutation relations:

$$[\hat{a}_\lambda, \hat{a}_\mu^\dagger] = \delta_{\lambda\mu}$$

and $$[\hat{a}_\lambda, \hat{a}_\mu] = [\hat{a}_\lambda^\dagger, \hat{a}_\mu^\dagger] = 0.$$

4. Show that, in view of treating the field variables P_λ and Q_λ as operators $\hat{P}_\lambda$ and $\hat{Q}_\lambda$, the vector potential $\boldsymbol{A}(\boldsymbol{r},t)$ and the fields $\boldsymbol{E}$ and $\boldsymbol{B}$ are also operators. Deduce the commutation relations between their components.

5. Express the Hamiltonian operator of the radiation field as

$$\hat{H} = \sum_\lambda H_\lambda = \sum_\lambda \frac{\hbar\omega_\lambda}{2}(\hat{a}_\lambda \hat{a}_\lambda^\dagger + \hat{a}_\lambda^\dagger \hat{a}_\lambda).$$

Interpret the operators $\hat{a}_\lambda^\dagger$ and $\hat{a}_\lambda$ as the creation and annihilation operators for a photon of energy $\hbar\omega_\lambda$.

6. Show that the Heisenberg equations of motion for the electric and magnetic field operators coincide with Maxwell's equations.

7. The eigenstates of the Hamiltonian for a free electromagnetic field are given by $|n_1, n_2, \cdots, n_r, \cdots\rangle \equiv |\{n\}\rangle$. Show that the average values of the field operators $\hat{\boldsymbol{E}}$ and $\hat{\boldsymbol{B}}$ vanish for this state. Evaluate the mean squared deviations for the electric field in such a state and show that they diverge even for the ground state. Interpret your resullt.

8. A single mode electromagnetic field is represented by the vector potential

$$\hat{\boldsymbol{A}}(\boldsymbol{r},t) = \sqrt{\frac{\hbar}{2\epsilon_0 \omega V}}\ \boldsymbol{\varepsilon}\left[\hat{a}e^{i(\boldsymbol{k}\cdot\boldsymbol{r}-\omega t)} + \hat{a}^\dagger e^{-i(\boldsymbol{k}\cdot\boldsymbol{r}-\omega t)}\right]. \tag{13.7.38}$$

Suppose we look for a state which has nonzero expectation value for the electric field. It is easy to see that such a state must satisfy

$$\hat{a}\,|\alpha\rangle = \alpha\,|\alpha\rangle\,, \tag{13.7.39}$$

where α is a complex number. If such a state could be found then

$$\langle\alpha|\,\hat{\boldsymbol{A}}\,|\alpha\rangle = \boldsymbol{A}(\boldsymbol{r},t) = \sqrt{\frac{\hbar}{2\epsilon_0\omega V}}\ \boldsymbol{\varepsilon}\left[\alpha e^{i(\boldsymbol{k}\cdot\boldsymbol{r}-\omega t)} + \alpha^* e^{i(\boldsymbol{k}\cdot\boldsymbol{r}-\omega t)}\right].$$

Determine the state $|\alpha\rangle$. What is the probability of finding n photons in such a state? Such a state is known as a coherent state of the field. [Hint: If such a state exists we can express it in terms of number states as $|\alpha\rangle = \sum_{n=0}^{\infty} c_n\,|n\rangle$.]

9. Consider the interaction of a single-mode quantized radiation field with a classical current distribution $\boldsymbol{j}(\boldsymbol{r},t)$ (You may consider the classical current as being due to a distibution of electrons with a certain momentum distribution.)

   (a) Show that the interaction Hamiltonian (13.5.17) (neglecting the $\boldsymbol{A}^2$ term) in this case can be written as $\hat{H}_{int} = -\int d^3r\boldsymbol{j}(\boldsymbol{r},t)\cdot\hat{\boldsymbol{A}}(\boldsymbol{r},t) \equiv f^*(t)\,\hat{a} + f(t)\,\hat{a}^\dagger$.

   (b) If such an interaction is turned on at time $t = 0$ and if the intial state of the electromagnetic field is the vacuum $|0\rangle$, show that the state of the field evolves into a coherent state with amplitude $\alpha = -\frac{i}{\hbar}\int_0^t f(t')dt'$.

10. Single photon absorption and emission by a charged particle interacting with quantum radiation field is equivalent to that given by the semi-classical theory if the equivalent vector potentials in the two case are taken to be

$$\boldsymbol{A}_{ab} = \sqrt{\frac{\hbar n_\lambda}{2\epsilon_0\omega_\lambda V}}\ \boldsymbol{\varepsilon}_\lambda\, e^{i(\boldsymbol{k}_\lambda\cdot\boldsymbol{r}-\omega_\lambda t)},$$

$$\boldsymbol{A}_{em} = \sqrt{\frac{\hbar(n_\lambda + 1)}{2\epsilon_0\omega_\lambda V}}\, \varepsilon_\lambda\, e^{-i(\boldsymbol{k}_\lambda\cdot\boldsymbol{r}-\omega_\lambda t)}.$$

With this assumption calculate the transition rate for absorption ($w_{n_0\to k}$) and emission ($w_{i\to f}$) and show that the emission can take place even when no photons are present in the radiation field. Work out the ratio of the coefficients of spontaneous and induced emissions (Einstein coefficients).

11. Calculate the cross-section for the photo-electric effect for the ground state of the Hydrogen atom with the assumption that the energy transferred to the photo-electron is large enough for the final state to be treated as a plane wave state and yet its velocity is non-relativistic.

12. By expressing the ground state wave function of a Hydrogen atom in momentum space, show that the calculation of the photo-electric cross-section becomes easier when the matrix element H'_{fi} [Eq. (13.5.4)] is expressed in terms of an integral in momentum space.

13. Show that the transition probability for spontaneous emission is equal to the transition probability for induced emission that would result from an isotopic field of such intensity that there is one quantum per state of the field in the neighbourhood of the transition frequency.

14. Using the commutation relations between components of the vector operators $\hat{\boldsymbol{r}}$ and $\hat{\boldsymbol{p}}$, deduce that

$$\frac{1}{i\hbar}[(\varepsilon\cdot\hat{\boldsymbol{r}}_\lambda)(\varepsilon_{\lambda'}\cdot\hat{\boldsymbol{p}}) - (\varepsilon_{\lambda'}\cdot\hat{\boldsymbol{p}})(\hat{\varepsilon}_\lambda\cdot\hat{\boldsymbol{r}})] = \varepsilon_\lambda\cdot\varepsilon_{\lambda'},$$

where $\hat{\varepsilon}_\lambda$ and $\hat{\varepsilon}_{\lambda'}$ are the polarization vectors for photons of energies $\hbar\omega_\lambda$ and $\hbar\omega_{\lambda'}$. Hence show that

$$\langle i|\varepsilon_\lambda\cdot\varepsilon_{\lambda'}|i\rangle = \frac{1}{i\hbar}\sum_I\{(\varepsilon_\lambda\cdot\hat{\boldsymbol{r}})_{iI}(\varepsilon_{\lambda'}\cdot\hat{\boldsymbol{p}})_{Ii} - (\varepsilon_{\lambda'}\cdot\hat{\boldsymbol{p}})_{iI}(\varepsilon_\lambda\cdot\hat{\boldsymbol{r}})_{Ii}\}$$

where $(\varepsilon_\lambda\cdot\boldsymbol{r})_{iI} \equiv \langle i|\varepsilon_\lambda\cdot\hat{\boldsymbol{r}}|I\rangle$.

15. Prove that

$$\sum_I (E_I - E_i)|\boldsymbol{p}_{Ii}|^2 = \frac{\hbar^2}{2}\int |\psi_i(r)|^2\nabla^2 V(r)d^3r$$

where $\psi_i(\boldsymbol{r}) \equiv \langle\boldsymbol{r}|\,i\rangle$ and $\psi_I(\boldsymbol{r}) \equiv \langle\boldsymbol{r}|\,I\rangle$, $\boldsymbol{p}_{Ii} \equiv \langle I|\,\hat{\boldsymbol{p}}\,|i\rangle$ and $|i\rangle$ and $|I\rangle$ are eigenstates of $\hat{H}$, belonging to the eigenvalues E_i and E_I.

References

[1] P. A. M. Dirac, *Quantum Theory of Emission and Absorption of Radiation*, Proc. Roy. Soc. **A114**, 243 (1927).

[2] J. Sakurai, *Advanced Quantum Mechanics* (Addison-Wesley, Reading, MA, 1967).

[3] W. Heitler, *Quantum Theory of Radiation* (Clarendon Press, Oxford, 1954).

[4] R. Loudon, *The Quantum Theory of Radiation*, Second Edition (Clarendon Press, Oxford, 1983).

Appendix 13A1

Electromagnetic Field in Coulomb Gauge

The Hamiltonian for the electromagnetic field is

$$H_F = \frac{1}{2}\int d^3r \left(\epsilon_0 E^2 + \frac{B^2}{\mu_0}\right) . \tag{13A1.1.1}$$

Using two of Maxwell's equations

$$\boldsymbol{\nabla} \cdot \boldsymbol{B} = 0 , \tag{13A1.1.2}$$

$$\boldsymbol{\nabla} \times \boldsymbol{E} = -\frac{\partial \boldsymbol{B}}{\partial t} , \tag{13A1.1.3}$$

we can introduce the vector potential $\boldsymbol{A}$ and the scalar potential ϕ by

$$\boldsymbol{E} = -\frac{\partial \boldsymbol{A}}{\partial t} - \boldsymbol{\nabla}\Phi , \quad \boldsymbol{B} = -\boldsymbol{\nabla} \times \boldsymbol{A} . \tag{13A1.1.4}$$

The remaining two Maxwell's equations

$$\boldsymbol{\nabla} \cdot \boldsymbol{E} = \frac{\rho}{\epsilon_0} , \tag{13A1.1.5}$$

and

$$\boldsymbol{\nabla} \times \boldsymbol{B} = \mu_0 \boldsymbol{j} + \frac{1}{c^2}\frac{\partial \boldsymbol{E}}{\partial t} , \tag{13A1.1.6}$$

can then be written in terms of the vector and scalar potentials as

$$-\boldsymbol{\nabla}(\boldsymbol{\nabla} \cdot \boldsymbol{A}) + \nabla^2 \boldsymbol{A} - \frac{1}{c^2}\frac{\partial^2 \boldsymbol{A}}{\partial t^2} - \frac{1}{c^2}\frac{\partial}{\partial t}\boldsymbol{\nabla}\phi = -\mu_0 \boldsymbol{j} , \tag{13A1.1.7}$$

and

$$\nabla^2 \phi - \frac{1}{c^2}\frac{\partial^2 \phi}{\partial t^2} + \boldsymbol{\nabla} \cdot \frac{\partial \boldsymbol{A}}{\partial t} = -\frac{\rho}{\epsilon_0} . \tag{13A1.1.8}$$

Equation (13A1.1.4) does not uniquely determine the potentials. For example, another set of potentials $\boldsymbol{A}' = \boldsymbol{A} - \boldsymbol{\nabla}\chi$, and $\phi' = \phi + \frac{\partial \chi}{\partial t}$ also leads to the same fields $\boldsymbol{E}$ and $\boldsymbol{B}$. Using this freedom afforded by Eqs. (13A1.1.4) in the choice of vector and scalar potential, there are two choices made for the potentials. The Lorentz gauge is useful in the relativistic formulation of electromagnetism as discussed in Appendix 12A1. Another gauge, which is particularly useful in the non-relativistic light-matter interaction problems, is the *Coulomb gauge*, also called the *transverse* or *radiation gauge*, where the vector potential satisfies the condition

$$\boldsymbol{\nabla} \cdot \boldsymbol{A} = 0 . \tag{13A1.1.9}$$

Then the equations for the scalar and vector potentials reduce to

$$\nabla^2 \phi(\boldsymbol{r}, t) = -\frac{\rho(\boldsymbol{r}, t)}{\epsilon_0} \tag{13A1.1.10}$$

$$\nabla^2 \boldsymbol{A}(\boldsymbol{r}, t) - \frac{1}{c^2}\frac{\partial^2 \boldsymbol{A}}{\partial t^2} = -\mu_0 \boldsymbol{j}(\boldsymbol{r}, t) + \mu_0 \epsilon_0 \boldsymbol{\nabla}\frac{\partial \phi}{\partial t} . \tag{13A1.1.11}$$

The source terms on the right hand side of Eq. (13A1.1.11) for the vector potential $\boldsymbol{A}(\boldsymbol{r},t)$ can be written in a more elegant form by using the solution to Eq. (13A1.1.10) for ϕ

$$\phi(\boldsymbol{r},t) = \frac{1}{4\pi\epsilon_0}\int \frac{d^3r' \rho(\boldsymbol{r}',t)}{|\boldsymbol{r}-\boldsymbol{r}'|}. \tag{13A1.1.12}$$

With the help of this result we find

$$\begin{aligned}\frac{\partial\phi}{\partial t} &= \frac{1}{4\pi\epsilon_0}\int \frac{d^3r'(\partial\rho/\partial t)}{|\boldsymbol{r}-\boldsymbol{r}'|} = -\frac{1}{4\pi\epsilon_0}\int d^3r' \frac{\boldsymbol{\nabla}'\cdot\boldsymbol{j}(\boldsymbol{r}',t)}{|\boldsymbol{r}-\boldsymbol{r}'|} \\ &= \frac{1}{4\pi\epsilon_0}\int d^3r'\,\boldsymbol{\nabla}'\cdot\left[\frac{\boldsymbol{j}(\boldsymbol{r}',t)}{|\boldsymbol{r}-\boldsymbol{r}'|} - \boldsymbol{j}(\boldsymbol{r}',t)\cdot\boldsymbol{\nabla}'\left(\frac{1}{|\boldsymbol{r}-\boldsymbol{r}'|}\right)\right].\end{aligned}$$

The first term is a surface term which vanishes for a localized source when the surface of integration lies outside the (localized) source. On using the identity

$$\boldsymbol{\nabla}'\left(\frac{1}{|\boldsymbol{r}-\boldsymbol{r}'|}\right) = -\boldsymbol{\nabla}\left(\frac{1}{|\boldsymbol{r}-\boldsymbol{r}'|}\right), \tag{13A1.1.13}$$

we find that

$$\frac{\partial\phi}{\partial t} = -\frac{1}{4\pi\epsilon_0}\boldsymbol{\nabla}\cdot\left(\int\frac{d^3r'\boldsymbol{j}(\boldsymbol{r}',t)}{|\boldsymbol{r}-\boldsymbol{r}'|}\right). \tag{13A1.1.14}$$

The equation for the vector potential becomes

$$\nabla^2\boldsymbol{A}(\boldsymbol{r},t) - \frac{1}{c^2}\frac{\partial^2\boldsymbol{A}}{\partial t^2} = -\mu_0\left[\boldsymbol{j}(\boldsymbol{r},t) + \frac{1}{4\pi}\boldsymbol{\nabla}\boldsymbol{\nabla}\cdot\left(\int\frac{d^3r'\boldsymbol{j}(\boldsymbol{r}',t)}{|\boldsymbol{r}-\boldsymbol{r}'|}\right)\right]. \tag{13A1.1.15}$$

This can be simplified by decomposing vector $\boldsymbol{j}(\boldsymbol{r},t)$ in terms of its transverse and longitudinal parts as $\boldsymbol{j} = \boldsymbol{j}_\perp + \boldsymbol{j}_\parallel$, where

$$\boldsymbol{j}_\perp = \boldsymbol{j} - \boldsymbol{j}_\parallel = \frac{1}{4\pi}\boldsymbol{\nabla}\times\left[\boldsymbol{\nabla}\times\int d^3r\frac{\boldsymbol{j}(\boldsymbol{r}',t)}{|\boldsymbol{r}-\boldsymbol{r}'|}\right] \tag{13A1.1.16}$$

$$\boldsymbol{j}_\parallel = -\frac{1}{4\pi}\boldsymbol{\nabla}\left[\boldsymbol{\nabla}\cdot\int d^3r\frac{\boldsymbol{j}(\boldsymbol{r}',t)}{|\boldsymbol{r}-\boldsymbol{r}'|}\right]. \tag{13A1.1.17}$$

We then find that the vector potential $\boldsymbol{A}$ is determined by the transverse part of $\boldsymbol{j}$ only via the inhomogeneous wave equation

$$\nabla^2\boldsymbol{A}(\boldsymbol{r},t) - \frac{1}{c^2}\frac{\partial^2\boldsymbol{A}}{\partial t^2} = -\mu_0\boldsymbol{j}_\perp. \tag{13A1.1.18}$$

Thus only the transverse part of $\boldsymbol{j}$ drives the potential in the Coulomb gauge. Note that in view of the Coulomb condition the vector potential is purely transverse in Coulomb gauge. Taking advantage of our gauge choice we can write the fields also in terms of their transverse and longitudinal parts as

$$\boldsymbol{B} = \boldsymbol{\nabla}\times\boldsymbol{A}, \quad \boldsymbol{E}_\perp = -\frac{\partial\boldsymbol{A}}{\partial t}, \quad \boldsymbol{E}_\parallel = -\boldsymbol{\nabla}\phi. \tag{13A1.1.19}$$

Thus the magnetic field, which is purely transverse, and the transverse part of the electric field are determined by the vector potential. The Hamiltonian can now be written as

$$H = \frac{1}{2}\int d^3r\left[\epsilon_0({E_\perp}^2 + {E_\parallel}^2 + 2\boldsymbol{E}_\perp\cdot\boldsymbol{E}_\parallel) + \frac{B^2}{\mu_0}\right]. \tag{13A1.1.20}$$

The term involving the dot product of $\boldsymbol{E}_\perp$ and $\boldsymbol{E}_\parallel$ vanishes

$$\int d^3r \boldsymbol{E}_\perp \cdot \boldsymbol{E}_\parallel = -\int d^3r \boldsymbol{E}_\perp \cdot \boldsymbol{\nabla}\phi = -\int d^3r [\boldsymbol{\nabla} \cdot (\phi \boldsymbol{E}_\perp) - \phi \boldsymbol{\nabla} \cdot \boldsymbol{E}_\perp]$$
$$= -\int d^3r \boldsymbol{\nabla} \cdot (\phi \boldsymbol{E}_\perp)$$

where we have used the result $\boldsymbol{\nabla} \cdot \boldsymbol{E}_\perp = 0$. The remaining integral is a surface term and vanishes for localized sources. The term involving $E_\parallel = -\boldsymbol{\nabla}\phi$ can be transformed as follows

$$\epsilon_0 \int d^3r E_\parallel^2 = \epsilon_0 \int d^3r \boldsymbol{\nabla}\phi \cdot \boldsymbol{\nabla}\phi = \epsilon_0 \int d^3r [\boldsymbol{\nabla} \cdot (\phi \boldsymbol{\nabla}\phi) - \phi \nabla^2 \phi]$$
$$= \epsilon_0 \int d^3r \left[\boldsymbol{\nabla} \cdot (\phi \boldsymbol{\nabla}\phi) + \phi \frac{\rho}{\epsilon_0}\right] = \int d^3r \phi(\boldsymbol{r}, t) \rho(\boldsymbol{r}, t)$$
$$= \frac{1}{4\pi\epsilon_0} \iint d^3r' d^3r \frac{\rho(\boldsymbol{r}', t)\rho(\boldsymbol{r}, t)}{|\boldsymbol{r} - \boldsymbol{r}'|} .$$

Thus this term is proportional to the Coulomb interaction energy of the charges. Indeed for a collection of point charges $\rho(\boldsymbol{r}) = \sum_i q_i \delta(\boldsymbol{r} - \boldsymbol{r}_i)$, we find

$$\epsilon_0 \int d^3r {E_\parallel}^2 = \frac{1}{4\pi\epsilon_0} \sum_{i,j}^{N} \frac{q_i q_j}{|\boldsymbol{r}_i - \boldsymbol{r}_j|} . \tag{13A1.1.21}$$

Removing the self energy ($i = j$) terms, which are independent of the locations of the charges, we can write the Hamiltonian as

$$H_{em} = H_{es} + H_{rad} ,$$

where

$$H_{es} = \frac{1}{8\pi\epsilon_0} \sum_{i,j(i \neq j)}^{N} \frac{q_i q_j}{|\boldsymbol{r}_i - \boldsymbol{r}_j|} , \tag{13A1.1.22}$$

and

$$H_{rad} = \frac{1}{2} \int d^3r \left(\epsilon_0 {E_\perp}^2 + \frac{B^2}{\mu_0} \right) . \tag{13A1.1.23}$$

Thus in the Coulomb gauge, the energy associated with the electromagetic field separates into an electrostatic contribution H_{es} arising from the Coulomb interaction between charges and a contribution H_{rad} which refers only to the transverse components of the field (radiation fields), which are determined by the vector potential alone, which in turn is determined by the transverse current $\boldsymbol{j}_\perp$ defined by Eq. (13A1.1.16).

If we have a system of N charges (charge q_i mass m_i) interacting via the electromagnetic field, the total Hamiltonian can be written as

$$H = \sum_{i=1}^{N} \frac{1}{2m_i} [\boldsymbol{p}_i - q_i \boldsymbol{A}(\boldsymbol{r}_i)[^2 + H_{es} + H_F ,$$
$$= \sum_{i=1}^{N} \frac{1}{2m_i} [\boldsymbol{p}_i - q_i \boldsymbol{A}(\boldsymbol{r}_i)]^2 + \frac{1}{8\pi\epsilon_0} \sum_{i,j(i \neq j)}^{N} \frac{q_i q_j}{|\boldsymbol{r}_i - \boldsymbol{r}_j|} + \frac{1}{2} \int d^3r \left(\epsilon_0 {E_\perp}^2 + \frac{B^2}{\mu_0} \right) . \tag{13A1.1.24}$$

14

SECOND QUANTIZATION

14.1 Introduction

As mentioned in previous chapters, processes like electron-positron pair creation or annihilation cannot be understood in the framework of single particle wave equations. Such processes are best understood in terms of a quantum field theory for the electron, similar to the one we have for the photon. The electrons may then be treated as *quanta* of the electron field (or Dirac field) in the same sense that photons are regarded as the quanta of the electromagnetic field. When we treat the Dirac equation (or Klein-Gordon equation) on the same footing as the Maxwell equations and subject the wave field or the wave function to quantization again in order to obtain the quantum field operators from it then the process is called *second quantization*. As a consequence, single-particle probability and current densities formally go over to particle density and current density operators. However, the analogy is purely formal as the former are real functions and the latter operators.[1] Though there is a basic difference between quantizing an electromagnetic field, which is a classical field and quantizing a Dirac field which is not a classical field, second quantization appears to be the only way to understand a large number of phenomenon pertaining to the interaction of radiation and matter. The need for second quantization, in fact, followed from the difficulties faced in the attempts to construct a relativistic single particle wave equation. We may recall, for example, that the relativistic generalization of the Schrödinger equation, the Klein Gordon (KG) equation, yielded an equation of continuity in which the probability density was not a positive definite quantity. For this reason the KG equation was initially discarded; it came to be accepted only after Pauli and Weisskopf suggested that it could well serve as a field equation whose quanta are zero-spin particles of mass m. Once the KG equation was accepted as a field equation, the concepts of probability density and equation of continuity (which represents conservation of total probability) did not have the same significance they did in single-particle theory as particles could be created or destroyed in the framework of field theory. In the Dirac single-particle wave equation we find, of course, that the probability density $\psi^\dagger\psi$ is positive definite. However when we use this equation for finding the energy E of a free particle we find the answer to be[2]

$$E = \pm\sqrt{c^2p^2 + m^2 c^4}\,,$$

so that the existence of states with negative energy ($E < -mc^2$) has to be accommodated in the quantum theory. For this purpose, Dirac devised an ingenious hypothesis that the continuum of negative energy states is normally completely occupied by the electrons and this produces no observable effects. If energy $> 2mc^2$ is made available to a negative

[1] First quantization just implies that one describes the electron by a single-particle wave equation (Dirac equation) or the pion by Klein-Gordon equation. In other words, dynamical variables of classical mechanics become operators.

[2] In this chapter the rest mass of the electron is denoted by m.

energy electron to go to a positive energy state the void or *hole* created in the negative energy continuum manifests itself as a positron and we have an electron positron pair. The subsequent experimental discovery of the positron by Anderson (1932) confirmed Dirac's hypothesis. Dirac's single-particle equation had several other successes as well [Chapter 12]. But the fact remains that the concept of occupied negative energy states was introduced to save single-particle relativistic theory and in this process one had to postulate an infinite number of particles in the negative energy states. Hence it was thought that the Dirac equation, instead of being regarded as a single-particle wave equation should be regarded as a field equation subject to a second quantization.

Before taking up the quantization of wave fields, in general, we first outline a general formulation within the framework of which one can obtain the field equations for various fields by a unique procedure. This formalism for fields, called Lagrangian formalism, was developed by Heisenberg and Pauli (1929).

14.2 Classical Concept of Field

The concept of field was introduced to understand the interaction between two bodies separated by a distance. For instance the interaction between two static charged particles A and B could be viewed as the interaction of the particle B with the electric field created by the particle A or vice versa. The behavior of a field is determined by the interaction of a particle with it. The equations pertaining to a field can be derived using Lagrangian formulation and invoking Hamilton's variational principle on the same lines as the derivation of the equation of motion in particle dynamics. Let us recall that the Lagrangian for a classical system of particles with f degrees of freedom is given by

$$L(q_i\,,\,\dot{q}_i) \;=\; T(q_i\,,\,\dot{q}_i) \;-\; V(q_i)\,, \tag{14.2.1}$$

where $T(q_i, \dot{q}_i)$ and $V(q_i)$ are respectively, the kinetic and potential energy of the system. Hamilton's variational principle implies that, out of all paths available to the system, the actual path followed by the system in moving from a given configuration at time t_1 to another at time t_2 is the one for which the variation in the action function

$$A = \int_{t_1}^{t_2} L(q_i\,,\dot{q}_i\,,t)\,dt$$

as a result of variation in q_i's and $\dot{q}_i$ vanishes. This means

$$\delta A = \int_{t_1}^{t_2} \delta L(q_i\,,\,\dot{q}_i\,,t)\,dt \;=\; 0\,. \tag{14.2.2}$$

The equations of motion that follow from this variational principle are called the Euler-Lagrange equations and are given by [see Appendix 14A1]

$$\frac{d}{dt}\left(\frac{\partial L}{\partial \dot{q}_i}\right) - \frac{\partial L}{\partial q_i} = 0 \qquad i = 1, 2, \cdots, f\,. \tag{14.2.3}$$

As a concrete example of how a field theory arises in this formulation, consider a linear chain of N identical particles, each of mass m, connected by identical springs of force constant k.

Let a be the equilibrium separation between the positions of the successive particles and η_i the displacement of the i-th particle from its equilibrium position along the chain direction. Then the Lagrangian of the whole system may be written as

$$L = T - V = \sum_{i=1}^{N} \left[\frac{1}{2} m\dot{\eta}_i^2 - \frac{1}{2} k \left(-\eta_i + \eta_{i+1} \right)^2 \right] . \tag{14.2.4}$$

From this discrete system of N degrees of freedom, we can pass to a continuous (one-dimensional) system with $N \to \infty$ degrees of freedom by letting $a \to dx$, $m \to dm = \rho\, dx$, where ρ is the linear mass density (mass per unit length), so that

$$\frac{(\eta_{i+1} - \eta_i)}{a} \to \frac{d\eta}{dx} \quad \text{and} \quad \frac{k\eta_i}{\eta_i/a} = ka \equiv Y \text{ (Young's modulus)} .$$

Then the Lagrangian in Eq. (14.2.4) takes the form

$$\begin{aligned} L &\equiv \sum_i a \left[\frac{1}{2} \frac{m}{a} \dot{\eta}_i^2 - \frac{1}{2} ka \left(\frac{-\eta_i + \eta_{i+1}}{a} \right)^2 \right] \\ &\to \int dx \left[\frac{1}{2} \rho \dot{\eta}^2 - \frac{1}{2} Y \left(\frac{d\eta}{dx} \right)^2 \right] \equiv \int \mathcal{L}\, dx , \end{aligned} \tag{14.2.5}$$

where
$$\mathcal{L} \equiv \frac{1}{2} \rho \dot{\eta}^2 - \frac{1}{2} Y \left(\frac{d\eta}{dx} \right)^2 \tag{14.2.6}$$

is called the Lagrangian density. In the continuum limit, particle displacement η has become a function of continuous variables x and t. In other words, $\eta(x,t)$ has become a field, representing an infinite degrees of freedom. In the Lagrangian formulation $\eta(x,t)$ is treated as a generalized coordinate of the field just like q_i in the case of a system of particles. Thus a field variable may be regarded as a generalized coordinate of a system with an infinite number of degrees of freedom.

Generalizing the above considerations to any one-dimensional field, we may take the Lagrangian density to be

$$\mathcal{L} = \mathcal{L} \left(\eta, \dot{\eta}, \frac{d\eta}{dx} \right) , \tag{14.2.7}$$

where the field $\eta(x,t)$ representing an infinite degrees of freedom, may be considered as a *disturbance* pertaining to the field at point x at time t. Application of Hamilton's variational principle [Appendix 14A1]

$$\delta \int dt\, L \equiv \delta \int dt \int dx\, \mathcal{L} \left(\eta, \dot{\eta}, \frac{d\eta}{dx} \right) = 0 , \tag{14.2.2*}$$

leads to Euler-Lagrange equation

$$-\frac{\partial}{\partial x} \left(\frac{\partial \mathcal{L}}{\partial \left(\frac{\partial \eta}{\partial x} \right)} \right) - \frac{d}{dt} \left(\frac{\partial \mathcal{L}}{\partial \dot{\eta}} \right) + \frac{\partial \mathcal{L}}{\partial \eta} = 0 . \tag{14.2.8}$$

This equation may be called the *field equation*. Returning to the case of a linear chain of particles where the Lagrangian density is given by Eq.(14.2.6)

$$\mathcal{L} = \frac{1}{2} \rho \dot{\eta}^2 - \frac{1}{2} Y \left(\frac{d\eta}{dx} \right)^2 ,$$

the corresponding Euler-Lagrange equation (14.2.8) gives the field equation

$$Y\frac{\partial^2\eta}{\partial x^2} - \rho\ddot{\eta} = 0\,. \tag{14.2.9}$$

This is the well-known wave equation for a one-dimensional propagation of a disturbance with velocity $\sqrt{Y/\rho}$. A one-dimensional field is thus specified by its amplitude $\eta(x,t)$ at each point in space at all times.

We can generalize the one-dimensional field to a three-dimensional field $\phi(x,y,z,t)$ which is specified by its values at each space point (x,y,z) at all times. The Lagrangian density in this case will depend on

$$\phi\,,\ \frac{\partial\phi}{\partial x_k}\,,\ \frac{\partial\phi}{\partial t} \equiv \dot{\phi}\,, \qquad k = 1,2,3$$

so that the Lagrangian can be written as

$$L = \int \mathcal{L}\,\left(\phi\,,\dot{\phi}\,,\frac{\partial\phi}{\partial x_k}\right)\,d\tau\,. \tag{14.2.10}$$

Hamilton's principle (14.2.2) then yields Euler-Lagrange equations

$$-\sum_k \frac{\partial}{\partial x_k}\left(\frac{\partial\mathcal{L}}{\partial\phi_k}\right) - \frac{d}{dt}\left(\frac{\partial\mathcal{L}}{\partial\dot{\phi}}\right) + \frac{\partial\mathcal{L}}{\partial\phi} = 0 \tag{14.2.11}$$

where $\phi_k \equiv \frac{\partial\phi}{\partial x_k}$ and $\dot{\phi} = \frac{\partial\phi}{\partial t}$. These field equations lead to explicit equations of motion once the Lagrangian density for the field is known. We saw an example of this for one-dimensional Lagrangian density (14.2.6), which leads to the field equation (14.2.9). Notice that space and time variables appear symmetrically in the Lagrangian formulation of field equations. Indeed, Eq. (14.2.11) can be written readily in the covariant form as

$$\frac{\partial}{\partial x_\mu}\left(\frac{\partial L}{\partial\phi_\mu}\right) - \frac{\partial L}{\partial\phi} = 0 \tag{14.2.12}$$

where $\phi_\mu = \frac{\partial\phi}{\partial x_\mu}$ and $x_\mu \equiv (x_k, x_4) \equiv (\boldsymbol{r}\,, ict)$. Field equations can also be obtained from a Hamiltonian formulation, which has the advantage that the analogy between classical and quantum fields is most direct in this formalism. We will take advantage of both Lagrangian and Hamiltonian formalisms in developing the quantum field theory.

14.3 Analogy of Field and Particle Mechanics

To pursue the analogy of field and particle mechanics further, we write the classical field equations in terms of the Lagrangian L rather than the Lagrangian density $\mathcal{L}$. It may be noted here that while the Lagrangian density $\mathcal{L}\,(\phi\,,\phi_k\,,\dot{\phi})$ is a function of its arguments at specific space-time points, the Lagrangian

$$L \equiv \int \mathcal{L}\,(\phi\,,\phi_k\,,\dot{\phi})\,d\tau$$

depends on the values ϕ and $\dot{\phi}$ at all space points, at a particular time. In other words, L is a *functional* of ϕ and $\dot{\phi}$ [see Appendix 14A2 for more details on functionals and functional derivatives]. To write the classical field equation in terms of the Lagrangian we divide the configuration space into a large number n of tiny cells of volume $\Delta\tau_i$. In the limit $n \to \infty$ and $\Delta\tau_i \to 0$, we can treat the quantities ϕ, $\phi_k \equiv \frac{\partial \phi}{\partial x_k}$, $\dot{\phi}$ approximately constant over each cell and replace the volume integral $L = \int \mathcal{L}\,(\phi\,,\, \phi_k\,,\, \dot{\phi})\, d\tau$ by the sum

$$L = \sum_{i=1}^{n} \mathcal{L}\left(\phi_i\,,\, (\phi_k)_i\,,\, \dot{\phi}_i\right) \Delta\tau_i, \tag{14.3.1}$$

where ϕ_i, $(\phi_k)_i$ and $\dot{\phi}_i$ denote the values of ϕ, ϕ_k, $\dot{\phi}$ in the ith cell. Similarly, the variation

$$\begin{aligned} \delta L &= \int \delta\,\mathcal{L}\,(\phi\,,\, \phi_k\,,\, \dot{\phi})\, d\tau \\ &= \int \left[\left\{\frac{\partial \mathcal{L}}{\partial \phi} - \sum_k \frac{\partial}{\partial x_k}\left(\frac{\partial \mathcal{L}}{\partial \phi_k}\right)\right\} \delta\phi + \frac{\partial \mathcal{L}}{\partial \dot{\phi}}\, \delta\dot{\phi}\right] d\tau\,, \end{aligned} \tag{14.3.2}$$

can be written as the sum

$$\delta L = \sum_{i=1}^{n} \left[\left\{\left(\frac{\partial \mathcal{L}}{\partial \phi}\right)_i - \sum_k \left(\frac{\partial}{\partial x_k}\left[\frac{\partial \mathcal{L}}{\partial \phi_k}\right]\right)_i\right\} \delta\phi_i + \left(\frac{\partial \mathcal{L}}{\partial \dot{\phi}}\right)_i \delta\dot{\phi}_i\right] \Delta\tau_i\,. \tag{14.3.3}$$

If now all variations $\delta\phi_i$ in ϕ_i and $\delta\dot{\phi}_i$ in $\dot{\phi}_i$, except a particular $\delta\dot{\phi}_j$, are taken to be zero then

$$\lim_{\Delta\tau_j \to 0} \frac{\delta L}{\delta\phi_j\, \Delta\tau_j} = \left(\frac{\partial \mathcal{L}}{\partial \phi}\right)_j - \sum_k \left(\frac{\partial}{\partial x_k}\left[\frac{\partial \mathcal{L}}{\partial \phi_k}\right]\right)_j. \tag{14.3.4a}$$

Similar relations hold for each cell j. This means that, if we denote the limit $\lim_{\Delta\tau \to 0} \frac{\delta L}{\delta\phi\, \Delta\tau}$ by $\frac{\eth L}{\eth \phi}$, which is called the *functional derivative* of L with respect to ϕ [see Appendix 14A2], Eq. (14.3.4a) simply gives its value at a point in the jth cell. Thus in the limit $n \to \infty$ or $\Delta\tau \to 0$ we drop the suffix j and define the functional derivative

$$\lim_{\Delta\tau \to 0} \frac{\delta L}{\delta\phi\, \Delta\tau} \equiv \frac{\eth L}{\eth \phi} = \frac{\partial \mathcal{L}}{\partial \phi} - \sum_k \frac{\partial}{\partial x_k}\left(\frac{\partial \mathcal{L}}{\partial \phi_k}\right). \tag{14.3.4b}$$

Similarly if we set all the $\delta\phi_i$'s and all $\delta\dot{\phi}_i$' s except a particular $\delta\dot{\phi}_j$, to be zero then the limit

$$\lim_{\Delta\tau_j \to 0} \left(\frac{\delta L}{\delta\dot{\phi}_j\, \Delta\tau_j}\right) = \left(\frac{\partial \mathcal{L}}{\partial \dot{\phi}}\right)_j$$

gives the value of the functional derivative $\frac{\eth L}{\eth \dot{\phi}}$ at a point in the jth cell. Hence, we may define the functional derivative

$$\lim_{\Delta\tau \to 0} \left(\frac{\delta L}{\delta\dot{\phi}\, \Delta\tau}\right) \equiv \frac{\eth L}{\eth \dot{\phi}} = \frac{\partial \mathcal{L}}{\partial \dot{\phi}}\,. \tag{14.3.5}$$

The difference in Eqs. (14.3.4b) and (14.3.5) arises because $\mathcal{L}$ depends on ϕ_k but not on $\dot{\phi}_k$. The field or Euler-Lagrange equations (14.2.11) can then be rewritten as

$$\frac{\eth L}{\eth \phi} - \frac{d}{dt}\left(\frac{\eth L}{\eth \dot{\phi}}\right) = 0\,. \tag{14.3.6}$$

This form resembles the Lagrange equations of motion for a system of particles with the generalized coordinates q_i replaced by the field quantity ϕ and partial derivatives $\frac{\partial L}{\partial q_i}$ and $\frac{\partial L}{\partial \dot{q}_i}$ replaced by functional derivatives $\frac{\eth L}{\eth \phi}$ and $\frac{\eth L}{\eth \dot{\phi}}$, respectively.

In analogy with particle mechanics (where the generalized momentum is defined as $p_i = \frac{\partial L}{\partial \dot{q}_i}$), we can define the momentum P_j, canonically conjugate to ϕ_j, to be the ratio of the variation δL to the infinitesimal variation $\delta\dot{\phi}_j$ when all other variations $\delta\phi_i$ and $\delta\dot{\phi}_i$ are zero. Thus we have

$$P_j \equiv \frac{\delta L}{\delta \dot{\phi}_j} = \Delta\tau_j \left(\frac{\eth L}{\eth \dot{\phi}}\right)_j = \Delta\tau_j \left(\frac{\partial \mathcal{L}}{\partial \dot{\phi}}\right)_j \equiv \Delta\tau_j \, \pi_j \tag{14.3.7}$$

where

$$\pi_j = \left(\frac{\partial \mathcal{L}}{\partial \dot{\phi}}\right)_j = \left(\frac{\eth L}{\eth \dot{\phi}}\right)_j . \tag{14.3.8}$$

This allows us to define a field variable π, canonically conjugate to ϕ, via the relation

$$\pi = \frac{\partial \mathcal{L}}{\partial \dot{\phi}} = \frac{\eth L}{\eth \dot{\phi}} . \tag{14.3.9}$$

With the help of Eqs. (14.3.6) and (14.3.7) we also have

$$\dot{P}_j = (\Delta\tau_j) \left(\frac{d}{dt}\left(\frac{\eth L}{\eth \dot{\phi}}\right)\right) = \Delta\tau_j \left(\frac{\eth L}{\eth \phi}\right)_j . \tag{14.3.10}$$

Further, in analogy with the definition of Hamiltonian $H = \sum_i p_i \dot{q}_i - L(q_i, \dot{q}_i)$ in particle mechanics, we can define the Hamiltonian for a field by

$$H = \sum_i P_i \, \dot{\phi}_i - L = \sum_i \left(\pi_i \, \dot{\phi}_i - \mathcal{L}_i\right) \Delta\tau_i \equiv \sum_i \mathcal{H}_i \, \Delta\tau_i \tag{14.3.11}$$

where $\mathcal{H}_i = \pi_i \, \dot{\phi}_i - \mathcal{L}_i$ is the Hamiltonian density within the i-th cell. This implies that in the continuum limit, the Hamiltonian density of the field may be expressed as $\mathcal{H} = \pi \, \dot{\phi} - \mathcal{L}$ so that

$$H = \int \left(\pi \, \dot{\phi} - \mathcal{L}\right) d\tau . \tag{14.3.12}$$

Continuing this procedure further, we can also write the field equations in terms of the Hamiltonian. Noting that the field Lagrangian is a functional of ϕ and π, the variation δL, from Eqs. (14.3.2), (14.3.4b) and (14.3.5), can be expressed as

$$\delta L \equiv \int \delta \, \mathcal{L} \; d\tau = \int \left(\frac{\eth L}{\eth \phi} \, \delta\phi + \frac{\eth L}{\eth \dot{\phi}} \, \delta\dot{\phi}\right) d\tau ,$$

which implies that the variation in Lagrangian density is

$$\begin{aligned}
\delta \, \mathcal{L} &= \frac{\eth L}{\eth \phi} \, \delta\phi + \frac{\eth L}{\eth \dot{\phi}} \, \delta\dot{\phi} \\
&= \frac{d}{dt}\left(\frac{\eth L}{\eth \dot{\phi}}\right) \delta\phi + \frac{\eth L}{\eth \dot{\phi}} \delta\dot{\phi} \\
&= \dot{\pi} \, \delta\phi + \pi \delta\dot{\phi} .
\end{aligned}$$

Hence the variation in Hamiltonian density is

$$\delta\,\mathcal{H} = \delta\left(\pi\dot{\phi} - \mathcal{L}\right) = \pi\,\delta\dot{\phi} + \dot{\phi}\delta\pi - \delta\,\mathcal{L} = \dot{\phi}\delta\pi - \dot{\pi}\delta\phi\,, \tag{14.3.13}$$

and the corresponding variation in the Hamiltonian is

$$\delta H = \int \left(\dot{\phi}\,\delta\pi - \dot{\pi}\,\delta\phi\right) d\tau\,. \tag{14.3.14}$$

Since H is a functional of π and ϕ, we can write the variation in H as [see Appendix 14A2]

$$\delta H = \int \left(\frac{\eth H}{\eth \pi}\,\delta\pi + \frac{\eth H}{\eth \phi}\,\delta\phi\right) d\tau\,. \tag{14.3.15}$$

A comparison of Eqs. (14.3.14) and (14.3.15) leads to the field equations in Hamiltonian form

$$\dot{\phi} = \frac{\eth H}{\eth \pi} = \frac{\partial \mathcal{H}}{\partial \pi} - \sum_{k=1}^{3} \frac{\partial}{\partial x_k}\left(\frac{\partial \mathcal{H}}{\partial \pi_k}\right) \tag{14.3.16}$$

and

$$-\dot{\pi} = \frac{\eth H}{\eth \phi} = \frac{\partial \mathcal{H}}{\partial \phi} - \sum_{k=1}^{3} \frac{\partial}{\partial x_k}\left(\frac{\partial \mathcal{H}}{\partial \phi_k}\right). \tag{14.3.17}$$

Equations (14.3.16) and (14.3.17) are referred to as the classical field equations in the canonical form.

14.4 Field Equations from Lagrangian Density

14.4.1 Electromagnetic Field

By a suitable choice for the Lagrangian density for a field, we can derive the field equations using Euler-Lagrange equations (14.2.12). The procedure can be adopted not only for a classical field like electromagnetic but also for fields like Klein-Gordon or Dirac. For example, if we choose the Lagrangian density

$$\begin{aligned}\mathcal{L} &= \frac{1}{\mu_0}\left\{-\frac{1}{4}F_{\mu\nu}\,F_{\mu\nu} - \frac{1}{2}\frac{\partial A_\mu}{\partial x_\nu}\frac{\partial A_\nu}{\partial x_\mu}\right\} + j_\mu A_\mu \\ &= -\frac{1}{2\mu_0}\frac{\partial A_\nu}{\partial x_\mu}\frac{\partial A_\nu}{\partial x_\mu} + j_\mu A_\mu\,,\end{aligned} \tag{14.4.1}$$

where $F_{\mu\nu}$ is the electromagnetic field tensor defined by

$$F_{\mu\nu} = \frac{\partial A_\nu}{\partial x_\mu} - \frac{\partial A_\mu}{\partial x_\nu} \tag{14.4.2}$$

with $A_\nu \equiv (\boldsymbol{A},\, i\phi/c)$ and $j_\nu \equiv (\boldsymbol{j},\, ic\rho)$, the Euler-Lagrange equations[3]

$$\frac{\partial \mathcal{L}}{\partial A_\mu} - \frac{\partial}{\partial x_\mu}\left(\frac{\partial \mathcal{L}}{\partial\left(\frac{\partial A_\mu}{\partial x_\nu}\right)}\right) = 0 \tag{14.4.3}$$

[3]The electromagnetic field is a vector field and we have the four-potential A_μ as the field function.

lead to the field equations, which are in fact Maxwell's equations. This is easily verified since, with the Lagrangian density given by Eq. (14.4.1), we have

$$\begin{aligned}
\frac{\partial}{\partial x_\nu}\left(\frac{\partial \mathcal{L}}{\partial\left(\frac{\partial A_\mu}{\partial x_\nu}\right)}\right) &= \frac{\partial}{\partial x_\nu}\left(\frac{\partial\left\{-\frac{1}{2\mu_0}\left(\frac{\partial A_\sigma}{\partial x_\lambda}\right)\left(\frac{\partial A_\sigma}{\partial x_\lambda}\right)\right\}}{\partial\left(\frac{\partial A_\mu}{\partial x_\nu}\right)}\right) \\
&= -\frac{1}{2\mu_0}\frac{\partial}{\partial x_\nu}\left(2\frac{\partial A_\mu}{\partial x_\nu}\right) \\
&= -\frac{1}{\mu_0}\frac{\partial}{\partial x_\nu}\left(\frac{\partial A_\mu}{\partial x_\nu} - \frac{\partial A_\nu}{\partial x_\mu} + \frac{\partial A_\nu}{\partial x_\mu}\right) \\
&= -\frac{1}{\mu_0}\frac{\partial}{\partial x_\nu}F_{\nu\mu} - \frac{1}{\mu_0}\frac{\partial}{\partial x_\nu}\frac{\partial A_\nu}{\partial x_\mu} \\
&= -\frac{1}{\mu_0}\frac{\partial F_{\nu\mu}}{\partial x_\nu} - \frac{1}{\mu_0}\frac{\partial}{\partial x_\mu}\frac{\partial A_\nu}{\partial x_\nu} \\
&= \frac{1}{\mu_0}\frac{\partial F_{\mu\nu}}{\partial x_\nu},
\end{aligned}$$

where in the last step we have used the Lorentz condition

$$\nabla \cdot \boldsymbol{A} + \frac{1}{c^2}\frac{\partial \phi}{\partial t} \equiv \frac{\partial A_\nu}{\partial x_\nu} = 0\,. \tag{14.4.4}$$

Thus Eq. (14.4.3) gives us

$$\frac{1}{\mu_0}\frac{\partial F_{\mu\nu}}{\partial x_\nu} = j_\mu\,, \tag{14.4.5}$$

which represents two of Maxwell's equations [Gauss's law for the electric field and the Ampere-Maxwell equation, see Appendix 12A1]. The other two equations, which are automatically satisfied with the introduction of scalar and vector potentials, may be written as

$$\frac{\partial F_{\lambda\mu}}{\partial x_\nu} + \frac{\partial F_{\mu\nu}}{\partial x_\lambda} + \frac{\partial F_{\nu\lambda}}{\partial x_\mu} = 0\,. \tag{14.4.6}$$

14.4.2 Klein-Gordon Field (Real and Complex)

If we choose the Lagrangian density to be given by

$$\mathcal{L} = -\frac{1}{2}\left[\frac{\partial \phi}{\partial x_\mu}\frac{\partial \phi}{\partial x_\mu} + \left(\frac{mc}{\hbar}\right)^2 \phi^2\right], \tag{14.4.7}$$

where m is a mass parameter and $\hbar$ is a constant, which in combination with m and speed of light c gives correct dimensions for the Lagragian density. Then Euler-Lagrange equations lead to the field equation

$$\frac{\partial^2 \phi}{\partial x_\mu \partial x_\mu} - \left(\frac{mc}{\hbar}\right)^2 \phi = 0, \tag{14.4.8}$$

where $\frac{\partial^2}{\partial x_\mu \partial x_\mu} \equiv \nabla^2 - \frac{1}{c^2}\frac{\partial^2}{\partial t^2}$. This is the Klein-Gordon equation (or the equation of Klein-Gordon field) for a real scalar (or one-component) field.

For a complex scalar field, we can write the Lagrangian density as

$$\begin{aligned}
\mathcal{L} &= -\left[\frac{\partial \phi^*}{\partial x_\mu}\frac{\partial \phi}{\partial x_\mu} + \left(\frac{mc}{\hbar}\right)^2 \phi^* \phi\right] \\
&= -\left[\sum_k \frac{\partial \phi^*}{\partial x_k}\frac{\partial \phi}{\partial x_k} - \frac{1}{c^2}\frac{\partial \phi^*}{\partial t}\frac{\partial \phi}{\partial t} + \left(\frac{mc}{\hbar}\right)^2 \phi^* \phi\right],
\end{aligned} \tag{14.4.9}$$

where the factor of $\frac{1}{2}$ is omitted since the square of the field quantities no longer appears in the Lagrangian density. Corresponding to ϕ and ϕ^*, we have the canonically conjugate variables

$$\pi = \frac{\partial \mathcal{L}}{\partial \dot{\phi}} = \frac{\dot{\phi}^*}{c^2} \tag{14.4.10}$$

and

$$\pi^* = \frac{\partial \mathcal{L}}{\partial \dot{\phi}^*} = \frac{\dot{\phi}}{c^2}\,. \tag{14.4.11}$$

With the Lagrangian density given by Eq. (14.4.9), Euler-Lagrange equations

$$\frac{\partial \mathcal{L}}{\partial \phi^*} - \sum_k \frac{\partial}{\partial x_k}\left(\frac{\partial \mathcal{L}}{\partial\left(\frac{\partial \phi^*}{\partial x_k}\right)}\right) - \frac{\partial}{\partial t}\left(\frac{\partial \mathcal{L}}{\partial\left(\frac{\partial \phi^*}{\partial t}\right)}\right) = 0 \tag{14.4.12}$$

and

$$\frac{\partial \mathcal{L}}{\partial \phi} - \sum_k \frac{\partial}{\partial x_k}\left(\frac{\partial \mathcal{L}}{\partial\left(\frac{\partial \phi}{\partial x_k}\right)}\right) - \frac{\partial}{\partial t}\left(\frac{\partial \mathcal{L}}{\partial\left(\frac{\partial \phi}{\partial t}\right)}\right) = 0 \tag{14.4.13}$$

lead to the following field equations for the complex KG field:

$$\sum_k \frac{\partial^2 \phi}{\partial x_k^2} - \frac{1}{c^2}\frac{\partial^2 \phi}{\partial t^2} - \left(\frac{mc}{\hbar}\right)^2 \phi \equiv \frac{\partial^2 \phi}{\partial x_\mu \partial x_\mu} - \left(\frac{mc}{\hbar}\right)^2 \phi = 0 \tag{14.4.14}$$

and

$$\sum_k \frac{\partial^2 \phi^*}{\partial x_k^2} - \frac{1}{c^2}\frac{\partial^2 \phi^*}{\partial t^2} - \left(\frac{mc}{\hbar}\right)^2 \phi^* \equiv \frac{\partial^2 \phi^*}{\partial x_\mu \partial x_\mu} - \left(\frac{mc}{\hbar}\right)^2 \phi^* = 0\,. \tag{14.4.15}$$

A real KG field $\phi = \phi^*$ can be used to describe a neutral field whereas a complex KG field can describe a charged field. To see this we multiply Eqs. (14.4.14) and (14.4.15) from the left by ϕ and ϕ^*, respectively and subtract the two; the resulting equation then has the form of the equation of continuity

$$\frac{\partial}{\partial x_\mu}\left[\phi^* \frac{\partial \phi}{\partial x_\mu} - \phi \frac{\partial \phi^*}{\partial x_\mu}\right] = 0\,. \tag{14.4.16}$$

This allows us to introduce a four-current density

$$j_\mu = -\frac{ie}{\hbar}\left[\phi^* \frac{\partial \phi}{\partial x_\mu} - \phi \frac{\partial \phi^*}{\partial x_\mu}\right], \tag{14.4.17}$$

where e may be interpreted as the charge of the field and $\hbar$ is a constant that takes care of the dimensions. Four-current density $j_\mu \equiv (\boldsymbol{j}, ic\rho)$ corresponds to electric current and charge densities given by

$$\boldsymbol{j} = -\frac{ie}{\hbar}\left[\phi^* \boldsymbol{\nabla}\phi - \phi \boldsymbol{\nabla}\phi^*\right], \tag{14.4.18}$$

and

$$\rho = \frac{ie}{\hbar c^2}\left[\phi^* \frac{\partial \phi}{\partial t} - \phi \frac{\partial \phi^*}{\partial t}\right]. \tag{14.4.19}$$

It follows from Eq. (14.4.16) that the four-divergence of j_μ vanishes,

$$\frac{\partial j_\mu}{\partial x_\mu} = 0\,, \tag{14.4.20}$$

conforming to the equation of continuity $\nabla \boldsymbol{j} + \frac{\partial \rho}{\partial t} = 0$, expressing charge conservation. We note that both $\boldsymbol{j}$ and ρ vanish if the field is real. A real field variable thus describes a neutral (uncharged) particle field, say, a neutral pion field. A complex scalar describes a charged particles field.

It may be noted that the charge density ρ is proportional to the probability density in the Klein-Gordon equation and this was found not to be a positive definite [cf. Eqs. (14.4.19) and (12.1.9)] quantity. However, in the present context, with the inclusion of the charge e, ρ represents charge density. Hence both the charge density ρ and the total charge of the field

$$Q = \int \rho \, d\tau \tag{14.4.21}$$

can be positive or negative.

14.4.3 Dirac Field

For the Dirac field, we take the Lagrangian density to be

$$\mathcal{L} = -c\hbar\tilde{\psi}\gamma_\rho \frac{\partial \psi}{\partial x_\rho} - mc^2\tilde{\psi}\psi \tag{14.4.22}$$

where $\tilde{\psi} = \psi^\dagger\gamma_4$, $\gamma_4 = \beta$, and $\gamma_k = -i\beta\alpha_k \;\; (k = 1, 2, 3)$. It is easy to check that Euler-Lagrange equation

$$\frac{\partial \mathcal{L}}{\partial \psi} - \frac{\partial}{\partial x_\mu}\left(\frac{\partial \mathcal{L}}{\partial\left(\frac{\partial \psi}{\partial x_\mu}\right)}\right) = 0$$

yields the equation

$$c\hbar\frac{\partial \tilde{\psi}}{\partial x_\mu}\gamma_\mu - mc^2\tilde{\psi} = 0 \tag{14.4.23}$$

while Euler-Lagrange equation

$$\frac{\partial \mathcal{L}}{\partial \tilde{\psi}} - \frac{\partial}{\partial x_\mu}\left(\frac{\partial \mathcal{L}}{\partial\left(\frac{\partial \tilde{\psi}}{\partial x_\mu}\right)}\right) = 0$$

yields the equation

$$c\hbar\,\gamma_\mu\frac{\partial \psi}{\partial x_\mu} + mc^2\psi = 0\,. \tag{14.4.24}$$

Since the Lagrangian density is real, we have[4]

$$\mathcal{L} = -c\hbar\tilde{\psi}\gamma_\mu\frac{\partial \psi}{\partial x_\mu} - mc^2\tilde{\psi}\psi = c\hbar\frac{\partial \tilde{\psi}}{\partial x_\mu}\gamma_\mu\psi - mc^2\tilde{\psi}\psi\,. \tag{14.4.25}$$

[4]It can be seen that

$$\left(-\tilde{\psi}\gamma_k\frac{\partial \psi}{\partial x_k}\right)^* = \frac{\partial \tilde{\psi}}{\partial x_k}\gamma_k\psi \quad \text{and} \quad \left(-\tilde{\psi}\gamma_4\frac{\partial \psi}{\partial x_4}\right)^* = \frac{\partial \tilde{\psi}}{\partial x_4}\gamma_4\psi\,.$$

Since $\mathcal{L} = -c\hbar\tilde{\psi}\gamma_\mu\frac{\partial \psi}{\partial x_\mu} - mc^2\tilde{\psi}\psi$ is real, $-\tilde{\psi}\gamma_\mu\frac{\partial \psi}{\partial x_\mu}$ is also real and is equal to its complex conjugate $\frac{\partial \tilde{\psi}}{\partial x_\mu}\gamma_\mu\psi$. Hence the result (14.4.25).

We have seen [Chapter 12 Eqs. (12.2.18), (12.2.19), and (12.2.23)] that the Dirac equation yields the equation of continuity in the covariant form, $\frac{\partial j_\mu}{\partial x_\mu} = 0$ where $j_\mu = ic\tilde{\psi}\gamma_\mu\psi$ and $x_\mu = (\boldsymbol{r}, ict)$. This is equivalent to $\boldsymbol{\nabla}\cdot\boldsymbol{j} + \frac{\partial\rho}{\partial t} = 0$ where $\boldsymbol{j} = c(\psi^\dagger\boldsymbol{\alpha}\psi)$ and $\rho = \psi^\dagger\psi$. These equations represent conservation of probability. Unlike the Klein-Gordon equation, in this case the probability density ρ is positive definite. Multiplying the last equation by e we get an equation of continuity, which represents conservation of charge.

14.5 Quantization of a Real Scalar (KG) Field

To quantize a real scalar field we regard the field quantities $\phi(x)$ and $\pi(x)$, $(x = \boldsymbol{r}, ict)$, as also their average values ϕ_i and $\pi_i \left(= \frac{P_i}{\Delta\tau_i}\right)$ in the i-th space cell, to be *operators* and, in analogy between the coordinates and momenta of particles $(\hat{q}_i, \hat{p}_j)$ and the space cell averages for the field and its canonically conjugate momenta $(\hat{\phi}_i, \hat{P}_j)$, impose the quantum conditions

$$\left[\hat{\phi}_i, \hat{\phi}_j\right] = \left[\hat{P}_i, \hat{P}_j\right] = 0 \tag{14.5.1}$$

and

$$\left[\hat{\phi}_i, \hat{P}_j\right] = i\hbar\delta_{ij}\,. \tag{14.5.2}$$

If we take the cell volume $\Delta\tau \to 0$, we can rewrite these commutation relations as

$$\left[\hat{\phi}(\boldsymbol{r},t)\,,\,\hat{\phi}(\boldsymbol{r}',t)\right] = \left[\hat{\pi}(\boldsymbol{r},t)\,,\,\hat{\pi}(\boldsymbol{r}',t)\right] = 0\,, \tag{14.5.3}$$

and

$$\left[\phi(\boldsymbol{r},t)\,,\,\pi(\boldsymbol{r}',t)\right] = i\hbar\,\delta^3(\boldsymbol{r}-\boldsymbol{r}')\,. \tag{14.5.4}$$

Commutation Relations in Terms of Field Operators

We can expand the scalar field $\phi(\boldsymbol{r},t)$ in terms of a complete set of orthonormal functions.[5] We choose this set of functions to be the enumerable set of plane waves satisfying periodic boundary conditions on the surface of a cubical box of side L and volume L^3. At the end of our calculations we can take the limit $L \to \infty$. This parallels our treatment of the electromagnetic field in Chapter 13. Needless to say that any physically meaningful results should be independent of this procedure. Then the field variables $\phi(\boldsymbol{r},t)$ and $\pi(\boldsymbol{r},t)$ can be

[5]We can obtain a discrete (enumerable) set of orthogonal functions by enclosing the system in a cubical box of dimension L and volume $V = L^3$. In this case

$$u_{\boldsymbol{k}}(\boldsymbol{r}) = \frac{1}{\sqrt{V}}e^{i\,\boldsymbol{k}\cdot\boldsymbol{r}}\,;\quad \int d^3r\; u^*_{\boldsymbol{k}'}u_{\boldsymbol{k}} = \delta_{\boldsymbol{k}',\boldsymbol{k}}\,;\quad \sum_{\boldsymbol{k}} u^*_{\boldsymbol{k}}(\boldsymbol{r}')u_{\boldsymbol{k}}(\boldsymbol{r}) = \delta(\boldsymbol{r}-\boldsymbol{r}')\,.$$

In the continuum limit $L \to \infty$, we have

$$\frac{1}{\sqrt{V}}e^{i\,\boldsymbol{k}\cdot\boldsymbol{r}} \to \frac{1}{(2\pi)^{3/2}}e^{i\boldsymbol{k}\cdot\boldsymbol{r}},\quad \delta_{\boldsymbol{k},\boldsymbol{k}'} \to \delta^3(\boldsymbol{k}-\boldsymbol{k}')\,,\quad \text{and}\quad \sum_{\boldsymbol{k}} \to \int d^3k\,.$$

The arrow implies *replaced by* and not equality.

expanded as

$$\phi(\boldsymbol{r},t) = \frac{1}{\sqrt{V}} \sum_{\boldsymbol{k}} q_{\boldsymbol{k}}(t)\, e^{i\,\boldsymbol{k}\cdot\boldsymbol{r}} \tag{14.5.5}$$

and

$$\pi(\boldsymbol{r},t) = \frac{1}{\sqrt{V}} \sum_{\boldsymbol{k}} p_{\boldsymbol{k}}(t)\, e^{-i\,\boldsymbol{k}\cdot\boldsymbol{r}}, \tag{14.5.6}$$

where the expansion coefficients $q_{\boldsymbol{k}}(t)$ and $p_{\boldsymbol{k}}(t)$ are given by

$$q_{\boldsymbol{k}}(t) = \frac{1}{\sqrt{V}} \int d^3r\; e^{-i\,\boldsymbol{k}\cdot\boldsymbol{r}}\, \phi(\boldsymbol{r},t) \tag{14.5.7}$$

and

$$p_{\boldsymbol{k}}(t) = \frac{1}{\sqrt{V}} \int d^3r\; e^{i\,\boldsymbol{k}\cdot\boldsymbol{r}}\, \pi(\boldsymbol{r},t)\,. \tag{14.5.8}$$

Since $\phi(\boldsymbol{r},t)$ and $\pi(\boldsymbol{r},t)$ are real, it follows that $q^*_{\boldsymbol{k}}(t) = q_{-\boldsymbol{k}}(t)$ and $p^*_{\boldsymbol{k}}(t) = p_{-\boldsymbol{k}}(t)$.

To quantize the field, we regard the field variables $\phi(\boldsymbol{r},t)$ and $\pi(r,t)$ as operators with commutation relations given by Eqs. (14.5.3) and (14.5.4). Then it follows from Eqs. (14.5.7) and (14.5.8) that $q_{\boldsymbol{k}}(t)$ and $p_{\boldsymbol{k}}(t)$ must also be regarded as operators $\hat{q}_{\boldsymbol{k}}$ and $\hat{p}_{\boldsymbol{k}}$. With the help of the field commuations (14.5.3) and (14.5.4), we arrive at the following commuation relations for $\hat{q}_{\boldsymbol{k}}$ and $\hat{p}_{\boldsymbol{k}}$:

$$[\hat{q}_{\boldsymbol{k}}(t)\,,\,\hat{p}_{\boldsymbol{k}'}(t)] = i\hbar\,\delta_{\boldsymbol{k},\boldsymbol{k}'}\,, \tag{14.5.9}$$

$$[\hat{q}_{\boldsymbol{k}}(t)\,,\,\hat{q}_{\boldsymbol{k}'}(t)] = [\hat{p}_{\boldsymbol{k}}(t)\,,\,\hat{p}_{\boldsymbol{k}'}(t)] = 0\,. \tag{14.5.10}$$

For a real scalar field, the field operators are Hermitian $\hat{\phi}^\dagger = \hat{\phi}$ and $\hat{\pi}^\dagger = \hat{\pi}$ and we have the operator relations $\hat{q}^\dagger_{\boldsymbol{k}} = \hat{q}_{-\boldsymbol{k}}$ and $\hat{p}^\dagger_{\boldsymbol{k}} = \hat{p}_{-\boldsymbol{k}}$.

With the help of Eqs. (14.4.7), (14.4.10) and (14.4.11), we can express the Hamiltonian density for the quantized field as

$$\hat{\mathcal{H}} = \hat{\pi}\dot{\hat{\phi}} - \hat{\mathcal{L}} = \frac{\dot{\hat{\phi}}^2}{c^2} - \hat{\mathcal{L}} = \frac{1}{2}\left[\boldsymbol{\nabla}\hat{\phi}\cdot\boldsymbol{\nabla}\hat{\phi} + c^2\hat{\pi}^2 + \left(\frac{mc}{\hbar}\right)^2\hat{\phi}^2\right].$$

Using the expansion for the operators $\hat{\phi}$ and $\hat{\pi}$

$$\hat{\phi}(\boldsymbol{r},t) = \frac{1}{\sqrt{V}} \sum_{\boldsymbol{k}} \hat{q}_{\boldsymbol{k}}(t)\, e^{i\,\boldsymbol{k}\cdot\boldsymbol{r}}$$

and

$$\hat{\pi}(\boldsymbol{r},t) = \frac{1}{\sqrt{V}} \sum_{\boldsymbol{k}} \hat{p}_{\boldsymbol{k}}(t)\, e^{-i\,\boldsymbol{k}\cdot\boldsymbol{r}},$$

the Hamiltonian for the quantized field can be expressed as

$$\hat{H} \equiv \int \hat{\mathcal{H}}\,d\tau = \frac{1}{2}\sum_{\boldsymbol{k}}\left[c^2\hat{p}_{\boldsymbol{k}}(t)\,\hat{p}^\dagger_{\boldsymbol{k}}(t) + \frac{\omega_k^2}{c^2}\,\hat{q}_{\boldsymbol{k}}(t)\,\hat{q}^\dagger_{\boldsymbol{k}}(t)\right], \tag{14.5.11}$$

where we have used the orthonormality of plane wave functions [see footnote 5] and put

$$k^2 + \left(\frac{mc}{\hbar}\right)^2 \equiv \frac{\omega_k^2}{c^2},$$

which is easily recognized as the relativistic energy momentum relation

$$E^2_{\boldsymbol{k}} \equiv \hbar^2\omega^2_{\boldsymbol{k}} = c^2\hbar^2k^2 + m^2c^4.$$

If we introduce new operators $\hat{a}_{\boldsymbol{k}}$ and $\hat{a}^\dagger_{\boldsymbol{k}}$ by

$$\hat{q}_{\boldsymbol{k}} = \left(\frac{\hbar c^2}{2\omega_k}\right)^{1/2} \left(\hat{a}_{\boldsymbol{k}} + \hat{a}^\dagger_{-\boldsymbol{k}}\right) \tag{14.5.12}$$

and

$$\hat{p}_{\boldsymbol{k}} = \left(\frac{\hbar \omega_k}{2c^2}\right)^{1/2} i\left(\hat{a}^\dagger_{\boldsymbol{k}} - \hat{a}_{-\boldsymbol{k}}\right), \tag{14.5.13}$$

the Hamiltonian of the field can be written as

$$\hat{H} = \sum_{\boldsymbol{k}} \frac{\hbar\omega_{\boldsymbol{k}}}{2} \left(\hat{a}^\dagger_{\boldsymbol{k}}\hat{a}_{\boldsymbol{k}} + \hat{a}_{\boldsymbol{k}}\hat{a}^\dagger_{\boldsymbol{k}}\right). \tag{14.5.14}$$

In arriving at this form for the Hamiltonian, we have used the identities

$$\sum_{\boldsymbol{k}} \hat{a}^\dagger_{\boldsymbol{k}}\hat{a}^\dagger_{-\boldsymbol{k}} = \sum_{\boldsymbol{k}} \hat{a}^\dagger_{-\boldsymbol{k}}\hat{a}^\dagger_{\boldsymbol{k}}\ , \quad \sum_{\boldsymbol{k}} \hat{a}_{-\boldsymbol{k}}\hat{a}_{\boldsymbol{k}} = \sum_{\boldsymbol{k}} \hat{a}_{\boldsymbol{k}}\hat{a}_{-\boldsymbol{k}}$$

$$\sum_{\boldsymbol{k}} \hat{a}^\dagger_{\boldsymbol{k}}\hat{a}_{\boldsymbol{k}} = \sum_{\boldsymbol{k}} \hat{a}^\dagger_{-\boldsymbol{k}}\hat{a}_{-\boldsymbol{k}}\ , \quad \sum_{\boldsymbol{k}} \hat{a}_{-\boldsymbol{k}}\hat{a}^\dagger_{-\boldsymbol{k}} = \sum_{\boldsymbol{k}} \hat{a}_{\boldsymbol{k}}\hat{a}^\dagger_{\boldsymbol{k}}\,.$$

The Number Representation

The commutation relations (14.5.9) and (14.5.10) for the field operator $\hat{q}_k(t)$ and $\hat{p}_k(t)$ lead to the following commutation relations for the operators $\hat{a}_{\boldsymbol{k}}$ and $\hat{a}^\dagger_{\boldsymbol{k}}$

$$[\hat{a}_{\boldsymbol{k}}, \hat{a}_{\boldsymbol{k}'}] = 0 = \left[\hat{a}^\dagger_{\boldsymbol{k}}, \hat{a}^\dagger_{\boldsymbol{k}'}\right] \ \& \tag{14.5.15}$$

and

$$\left[\hat{a}_{\boldsymbol{k}}, \hat{a}^\dagger_{\boldsymbol{k}'}\right] = \delta_{\boldsymbol{k},\boldsymbol{k}'}\,. \tag{14.5.16}$$

In view of these commutation relations we may write the field Hamiltonian as

$$\hat{H} = \sum_{\boldsymbol{k}} \hbar\omega_k \left(\hat{N}_{\boldsymbol{k}} + \frac{1}{2}\right) \tag{14.5.17}$$

where $\hat{N}_{\boldsymbol{k}} \equiv \hat{a}^\dagger_{\boldsymbol{k}}\,\hat{a}_{\boldsymbol{k}}$ is a Hermitian operator. From the commuation relations for the operators $\hat{a}_{\boldsymbol{k}}$ and $\hat{a}^\dagger_{\boldsymbol{k}}$, the following relations can be established

$$\left[\hat{a}_{\boldsymbol{k}}\,,\ \hat{N}_{\boldsymbol{k}}\right] = \hat{a}_{\boldsymbol{k}}\,, \tag{14.5.18}$$

$$\left[\hat{a}^\dagger_{\boldsymbol{k}}\,,\ \hat{N}_{\boldsymbol{k}}\right] = -\,\hat{a}^\dagger_{\boldsymbol{k}}\,, \tag{14.5.19}$$

and

$$\left[\hat{N}_{\boldsymbol{k}}\,,\ \hat{N}_{\boldsymbol{k}'}\right] = 0\,. \tag{14.5.20}$$

These are precisely the commutation relations obeyed by the operators for a linear oscillator [Chapter 5, Sec. 5.4]. We also encountered these relations in Chapter 13 in the discussion of quantized electromagnetic field. These operators, therefore, have the same interpretation as the corresponding operators for the linear oscillator. Thus $\hat{N}_{\boldsymbol{k}}$ is the number operator for (field) quanta of momentum $\hbar\boldsymbol{k}$ with integer eigenvalues $0, 1, 2, 3\cdots$, and $\hat{a}_{\boldsymbol{k}}$ and $\hat{a}^\dagger_{\boldsymbol{k}}$ are, respectively, the annihilation and creation operators for field quanta of momentum $\hbar\boldsymbol{k}$. Since the set of Hermitian operators $\hat{N}_{\boldsymbol{k}}$ forms a set of commuting observables, we can characterize the field states by the eigenvalues $\{n_{\boldsymbol{k}}\} \equiv n_{\boldsymbol{k}_1}, n_{\boldsymbol{k}_2}, \cdots, n_{\boldsymbol{k}}, \cdots$, of the number operators $\hat{N}_{\boldsymbol{k}_1}, \hat{N}_{\boldsymbol{k}_2}, \cdots, \hat{N}_{\boldsymbol{k}}, \cdots$, as

$$\left|\Psi_{\{n_{\boldsymbol{k}}\}}\right\rangle \equiv |\{n_{\boldsymbol{k}}\}\rangle = |n_{\boldsymbol{k}_1}, n_{\boldsymbol{k}_2}, \cdots, n_{\boldsymbol{k}}, \cdots\rangle . \tag{14.5.21}$$

For this state

$$\hat{N}_{\boldsymbol{k}} |n_{\boldsymbol{k}_1}, n_{\boldsymbol{k}_2}, \cdots, n_{\boldsymbol{k}}, \cdots\rangle = n_{\boldsymbol{k}} |n_{\boldsymbol{k}_1}, n_{\boldsymbol{k}_2}, \cdots, n_{\boldsymbol{k}}, \cdots\rangle , \tag{14.5.22}$$

$$\hat{a}_{\boldsymbol{k}} |n_{\boldsymbol{k}_1}, n_{\boldsymbol{k}_2}, \cdots, n_{\boldsymbol{k}}, \cdots\rangle = \sqrt{n_{\boldsymbol{k}}}\ |n_{\boldsymbol{k}_1}, n_{\boldsymbol{k}_2}, \cdots, (n_{\boldsymbol{k}} - 1), \cdots\rangle , \tag{14.5.23}$$

$$\hat{a}^\dagger_{\boldsymbol{k}} |n_{\boldsymbol{k}_1}, n_{\boldsymbol{k}_2}, \cdots, n_{\boldsymbol{k}}, \cdots\rangle = \sqrt{n_{\boldsymbol{k}} + 1}\ |n_{\boldsymbol{k}_1}, n_{\boldsymbol{k}_2}, \cdots, (n_{\boldsymbol{k}} + 1), \cdots\rangle . \tag{14.5.24}$$

The total energy of the field in this state is given by

$$\begin{aligned}\hat{H} \left|\Psi_{\{n_{\boldsymbol{k}}\}}\right\rangle &= \sum_{\boldsymbol{k}} \hat{H}_{\boldsymbol{k}} \left|\Psi_{\{n_{\boldsymbol{k}}\}}\right\rangle = \sum_{\boldsymbol{k}} \hbar\omega_k \left(\hat{N}_{\boldsymbol{k}} + \frac{1}{2}\right) |n_{\boldsymbol{k}_1}, n_{\boldsymbol{k}_2}, \cdots, n_{\boldsymbol{k}}, \cdots\rangle \\ &= \sum_{\boldsymbol{k}} \hbar\omega_k \left(n_{\boldsymbol{k}} + \frac{1}{2}\right) \left|\Psi_{\{n_{\boldsymbol{k}}\}}\right\rangle. \end{aligned} \tag{14.5.25}$$

The eigenvalue of the total linear momentum for this state is $\sum\limits_{\boldsymbol{k}} \hbar \boldsymbol{k} n_{\boldsymbol{k}}$ We can thus look upon $|\Psi\rangle$ as the eigenstate of the field in which there are $n_{\boldsymbol{k}_1}$ particles (quanta) with energy $\hbar\omega_{k_1}$ and momentum $\hbar\boldsymbol{k}_1$, $n_{\boldsymbol{k}_2}$ particles with energy $\hbar\omega_{k_2}$ and momentum $\hbar\boldsymbol{k}_2$ and so on. The state in which all occupation numbers are zero $\left|\Psi_{\{0\}}\right\rangle \equiv |n_{\boldsymbol{k}_1} = 0,\ n_{\boldsymbol{k}_2} = 0, \cdots, n_{\boldsymbol{k}} = 0 \cdots\rangle$ is called the vacuum state. Any annihilation operator acting on this state results in zero:

$$\hat{a}_{\boldsymbol{k}_i} \left|\Psi_{\{0\}}\right\rangle = 0 . \tag{14.5.26}$$

All other states can be generated from this by operating with appropriate powers of various creation operators. Thus, for example,

$$|0_{\boldsymbol{k}_1}, 0_{\boldsymbol{k}_2}, \cdots, n_{\boldsymbol{k}}, \cdots\rangle = \frac{\hat{a}^{\dagger n_{\boldsymbol{k}}}_{\boldsymbol{k}}}{\sqrt{n_{\boldsymbol{k}}!}} |0_{\boldsymbol{k}_1}, 0_{\boldsymbol{k}_2}, 0_{\boldsymbol{k}}, \cdots\rangle . \tag{14.5.27}$$

For the matrix elements in these basis states $|\{n_{\boldsymbol{k}}\}\rangle$, we have

$$\langle\{n_{\boldsymbol{k}}\}|\ \hat{N}_{\boldsymbol{k}}\ |\{n'_{\boldsymbol{k}}\}\rangle = n_{\boldsymbol{k}}\, \delta_{n'_{\boldsymbol{k}_1} n'_{\boldsymbol{k}_1}} \delta_{n_{\boldsymbol{k}_2} n'_{\boldsymbol{k}_2}}, \cdots, \delta_{n_{\boldsymbol{k}}, n'_{\boldsymbol{k}}} \cdots , \tag{14.5.28}$$

$$\langle\{n_{\boldsymbol{k}}\}|\ \hat{a}^\dagger_{\boldsymbol{k}}\ |\{n'_{\boldsymbol{k}}\}\rangle = \sqrt{n_{\boldsymbol{k}} + 1}\, \delta_{n_{\boldsymbol{k}_1} n'_{\boldsymbol{k}_1}} \delta_{n_{\boldsymbol{k}_2} n'_{\boldsymbol{k}_2}} \cdots \delta_{n_{\boldsymbol{k}} (n'_{\boldsymbol{k}} + 1)} \cdots , \tag{14.5.29}$$

and

$$\langle\{n_{\boldsymbol{k}}\}|\ \hat{a}_{\boldsymbol{k}}\ |\{n'_{\boldsymbol{k}}\}\rangle = \sqrt{n'_{\boldsymbol{k}}}\, \delta_{n_{\boldsymbol{k}_1} n'_{\boldsymbol{k}_1}} \delta_{n_{\boldsymbol{k}_2} n'_{\boldsymbol{k}_2}} \cdots \delta_{n_{\boldsymbol{k}} (n'_{\boldsymbol{k}} - 1)} \cdots . \tag{14.5.30}$$

The occupation number $n_{\boldsymbol{k}}$ and $n'_{\boldsymbol{k}}$ can have any integer value $0, 1, 2, \cdots \infty$. This field corresponds to particles (or field quanta) which obey Bose statistics.

14.6 Quantization of Complex Scalar (KG) Field

With the Lagrangian density (14.4.9) for a complex scalar field ϕ and the field variables π and π^* canonically conjugate to ϕ and ϕ^* given by Eqs. (14.4.10) and (14.4.11), the

Hamiltonian density and the Hamiltonian take the form

$$\mathcal{H} = \pi\dot{\phi} + \pi^*\dot{\phi}^* - \mathcal{L} = c^2\pi^*\pi + \nabla\phi^*\cdot\nabla\phi + \left(\frac{mc}{\hbar}\right)^2 \phi^*\phi, \tag{14.6.1}$$

$$H = \int \mathcal{H}\ d\tau = \int \left[c^2\pi^*\pi + \sum_{k=1}^{3}\frac{\partial\phi^*}{\partial x_k}\frac{\partial\phi}{\partial x_k} + \left(\frac{mc}{\hbar}\right)^2 \phi^*\phi\right] d\tau. \tag{14.6.2}$$

Separating the complex field ϕ into its real and imaginary part by writing

$$\phi = \frac{1}{\sqrt{2}}(\phi_1 - i\phi_2) \tag{14.6.3a}$$

and

$$\phi^* = \frac{1}{\sqrt{2}}(\phi_1 + i\phi_2), \tag{14.6.3b}$$

where ϕ_1 and ϕ_2 are real functions, and using it in the expression for the Lagrangian density (14.4.9) we find that the Euler-Lagrange equations yield the following field equations for ϕ_1 and ϕ_2

$$\left[\frac{\partial}{\partial x_\mu}\frac{\partial}{\partial x_\mu} - \left(\frac{mc}{\hbar}\right)^2\right]\phi_1 = 0 \tag{14.6.4a}$$

and

$$\left[\frac{\partial}{\partial x_\mu}\frac{\partial}{\partial x_\mu} - \left(\frac{mc}{\hbar}\right)^2\right]\phi_2 = 0. \tag{14.6.4b}$$

Thus ϕ_1 and ϕ_2 also satisfy the KG equation. Fields $\phi_1(\boldsymbol{r},t)$ and $\phi_2(\boldsymbol{r},t)$ can be expanded in terms the complete set of orthonormal set of functions $u_{\boldsymbol{k}} = \frac{1}{\sqrt{V}}e^{i\boldsymbol{k}\cdot\boldsymbol{r}}$ [see footnote 5] as

$$\phi_1(\boldsymbol{r},t) = \frac{1}{\sqrt{V}}\sum_k\sqrt{\frac{\hbar c^2}{2\omega_k}}\left[a_{1\boldsymbol{k}}\,e^{i(\boldsymbol{k}\cdot\boldsymbol{r}-\omega_k t)} + a_{1\boldsymbol{k}}^*\,e^{-i(\boldsymbol{k}\cdot\boldsymbol{r}-\omega_k t)}\right] \tag{14.6.5a}$$

and

$$\phi_2(\boldsymbol{r},t) = \frac{1}{\sqrt{V}}\sum_k\sqrt{\frac{\hbar c^2}{2\omega_k}}\left[a_{2\boldsymbol{k}}\,e^{i(\boldsymbol{k}\cdot\boldsymbol{r}-\omega_k t)} + a_{2\boldsymbol{k}}^*\,e^{-i(\boldsymbol{k}\cdot\boldsymbol{r}-\omega_k t)}\right], \tag{14.6.5b}$$

where

$$\frac{\omega_k^2}{c^2} = k^2 + \left(\frac{mc}{\hbar}\right)^2 \quad\text{or}\quad \hbar^2\omega_k^2 \equiv E^2 = c^2p^2 + m^2c^4. \tag{14.6.5c}$$

Here ω_k determined by the last relation ensures that both ϕ_1 and ϕ_2 explicitly satisfy the Klein-Gordon equation. Using these expressions for ϕ_1 and ϕ_2 in Eq. (14.6.3), we find the complex fields ϕ and ϕ^* are given by

$$\phi(\boldsymbol{r},t) = \frac{1}{\sqrt{V}}\sum_{\boldsymbol{k}}\sqrt{\frac{\hbar c^2}{2\omega_k}}\left[a_{\boldsymbol{k}}\,e^{i(\boldsymbol{k}\cdot\boldsymbol{r}-\omega_k t)} + b_{\boldsymbol{k}}^*\,e^{-i(\boldsymbol{k}\cdot\boldsymbol{r}-\omega_k t)}\right] \tag{14.6.6a}$$

and

$$\phi^*(\boldsymbol{r},t) = \frac{1}{\sqrt{V}}\sum_{\boldsymbol{k}}\sqrt{\frac{\hbar c^2}{2\omega_k}}\left[a_{\boldsymbol{k}}^*\,e^{-i(\boldsymbol{k}\cdot\boldsymbol{r}-\omega_k t)} + b_{\boldsymbol{k}}\,e^{i(\boldsymbol{k}\cdot\boldsymbol{r}-\omega_k t)}\right], \tag{14.6.6b}$$

where

$$a_{\boldsymbol{k}} = \frac{1}{\sqrt{2}}(a_{1\boldsymbol{k}} - ia_{2\boldsymbol{k}}) \quad\text{and}\quad b_{\boldsymbol{k}} = \frac{1}{\sqrt{2}}(a_{1\boldsymbol{k}} + ia_{2\boldsymbol{k}}). \tag{14.6.6c}$$

Conjugate field variables π and π^* given by Eqs. (14.4.10) and (14.4.11) can then be written as

$$\pi(\boldsymbol{r},t) = \frac{\dot{\phi}^*}{c^2} = \frac{1}{\sqrt{V}}\frac{i}{c}\sum_{\boldsymbol{k}}\sqrt{\frac{\hbar\omega_k}{2}}\left(a_{\boldsymbol{k}}^*\,e^{-i(\boldsymbol{k}\cdot\boldsymbol{r}-\omega_k t)} - b_{\boldsymbol{k}}\,e^{i(\boldsymbol{k}\cdot\boldsymbol{r}-\omega_k t)}\right), \tag{14.6.7a}$$

$$\pi^*(\boldsymbol{r},t) = \frac{\dot{\phi}}{c^2} = -\frac{1}{\sqrt{V}}\frac{i}{c}\sum_{\boldsymbol{k}}\sqrt{\frac{\hbar\omega_k}{2}}\left(a_{\boldsymbol{k}}\,e^{i(\boldsymbol{k}\cdot\boldsymbol{r}-\omega_k t)} - b_{\boldsymbol{k}}^*\,e^{-i(\boldsymbol{k}\cdot\boldsymbol{r}-\omega_k t)}\right). \tag{14.6.7b}$$

Substituting the field expansions given by Eqs. (14.6.6) and (14.6.7) in Eq. (14.6.2), we find the Hamiltonian is given by

$$H = \frac{1}{V}\int d^3r \sum_{\boldsymbol{k}}\sum_{\boldsymbol{k}'}\left\{\left(\frac{\omega_k\omega_{k'}\hbar^2}{4}\right)^{1/2}\left(a_{\boldsymbol{k}}\,e^{ikx} - b^*_{\boldsymbol{k}}\,e^{-ikx}\right)\left(a^*_{\boldsymbol{k}'}\,e^{-ik'x} - b_{\boldsymbol{k}'}\,e^{ik'x}\right)\right.$$
$$+ \frac{\hbar c^2\boldsymbol{k}\cdot\boldsymbol{k}'}{\sqrt{4\omega_k\omega_{k'}}}\left(a^*_{\boldsymbol{k}}\,e^{-ikx} - b_{\boldsymbol{k}}\,e^{ikx}\right)\left(a_{\boldsymbol{k}'}\,e^{ik'x} - b^*_{\boldsymbol{k}'}\,e^{-ik'x}\right)$$
$$\left.+ \frac{\hbar c^2\left(\frac{mc}{\hbar}\right)^2}{\sqrt{4\omega_k\omega_{k'}}}\left(a^*_{\boldsymbol{k}}\,e^{-ikx} + b_{\boldsymbol{k}}\,e^{ikx}\right)\left(a_{\boldsymbol{k}'}\,e^{ik'x} + b^*_{\boldsymbol{k}'}\,e^{-ik'x}\right)\right\},$$

where $kx \equiv \boldsymbol{k}.\boldsymbol{r} - \omega_k t$, $k \equiv \left(\boldsymbol{k}, \frac{i\omega_k}{c}\right)$ and $x \equiv (\boldsymbol{r}, ict)$. Carrying out the spatial integration with the help of the following results

$$\frac{1}{V}\int e^{i(\boldsymbol{k}\pm\boldsymbol{k}')\cdot\boldsymbol{r}}\,d^3r = \delta_{\boldsymbol{k},\mp\boldsymbol{k}'} = \frac{1}{V}\int e^{-i(\boldsymbol{k}\pm\boldsymbol{k}')\cdot\boldsymbol{r}}\,d^3r$$

and using the fact that $\omega_{k'} = \omega_k$ for $\boldsymbol{k}' = \pm\boldsymbol{k}$, we obtain

$$H = \sum_{\boldsymbol{k}}\sum_{\boldsymbol{k}'}\delta_{\boldsymbol{k},\boldsymbol{k}'}\left[\left(\frac{\hbar(\omega_k\omega_{k'} + c^2\boldsymbol{k}\cdot\boldsymbol{k}')}{\sqrt{4\,\omega_k\,\omega_{k'}}}\right)\left(a_{\boldsymbol{k}}\,a^*_{\boldsymbol{k}'} + b^*_{\boldsymbol{k}}\,b_{\boldsymbol{k}'}\right)\right.$$
$$\left.+ \frac{\left(\frac{mc}{\hbar}\right)^2\hbar c^2}{\sqrt{4\,\omega_k\,\omega_{k'}}}\left(a^*_{\boldsymbol{k}}\,a_{\boldsymbol{k}'} + b_{\boldsymbol{k}}\,b^*_{\boldsymbol{k}'}\right)\right] \tag{14.6.8}$$

where the cross terms, involving the integrals

$$\frac{1}{V}\int d^3r\;e^{\pm i(\boldsymbol{k}\pm\boldsymbol{k}')\cdot\boldsymbol{r}} = \delta_{\boldsymbol{k},-\boldsymbol{k}'}$$

give zero contribution because the factor $\left(\frac{\omega_k^2}{c^2} - k^2 - \left(\frac{mc}{\hbar}\right)^2\right)$ from Eq. (14.6.5c) vanishes. On simplifying Eq. (14.6.8), we get

$$H = \sum_{\boldsymbol{k}}\hbar\omega_k\left[a_{\boldsymbol{k}}\,a^*_{\boldsymbol{k}} + b^*_{\boldsymbol{k}}\,b_{\boldsymbol{k}}\right]. \tag{14.6.9}$$

To quantize the complex scalar field we treat the field quantities $\phi(\boldsymbol{r},t)$ and $\pi(\boldsymbol{r},t)$ as operators and impose the commutation relations

$$\left[\hat{\phi}(\boldsymbol{r},t)\,,\;\hat{\pi}(\boldsymbol{r}',t)\right] = i\hbar\,\delta^3(\boldsymbol{r}-\boldsymbol{r}')\,, \tag{14.6.10a}$$

and

$$\left[\hat{\phi}(\boldsymbol{r},t)\,,\;\hat{\phi}(\boldsymbol{r}',t)\right] = \left[\hat{\pi}(\boldsymbol{r},t)\,,\;\hat{\pi}(\boldsymbol{r}',t)\right] = 0\,. \tag{14.6.10b}$$

The quantized nature of the fields requires that the coefficients $a_{\boldsymbol{k}}$, $b_{\boldsymbol{k}}$, $a^*_{\boldsymbol{k}}$, $b^*_{\boldsymbol{k}}$ also be operators. As usual, we denote the corresponding operators using the same symbols with a caret $\hat{a}_{\boldsymbol{k}}$, $\hat{b}_{\boldsymbol{k}}$, $\hat{a}^\dagger_{\boldsymbol{k}}$, $\hat{b}^\dagger_{\boldsymbol{k}}$. Their commutation relations follow from Eq. (10.6.10).

Using the operator versions of Eqs. (14.6.6) and (14.6.7) and the expansion for the delta function [see footnote 5] in Eq. (14.6.10a), we obtain

$$\sum_{\boldsymbol{k}}\sum_{\boldsymbol{k}'}\frac{i\omega_{k'}/c^2}{\left(4\omega_k\omega_{k'}/\hbar^2c^4\right)^{1/2}}\,\frac{1}{V}\left\{\left[\hat{a}_{\boldsymbol{k}},\,\hat{a}^\dagger_{\boldsymbol{k}'}\right]\,e^{i(\boldsymbol{k}\cdot\boldsymbol{r}-\boldsymbol{k}'\cdot\boldsymbol{r}')}\right.$$
$$+\left[\hat{b}_{\boldsymbol{k}'},\,\hat{b}^\dagger_{\boldsymbol{k}}\right]\,e^{-i(\boldsymbol{k}\cdot\boldsymbol{r}-\boldsymbol{k}'\cdot\boldsymbol{r}')} - \left[\hat{a}_{\boldsymbol{k}},\,\hat{b}_{\boldsymbol{k}'}\right]\,e^{i(\boldsymbol{k}'\cdot\boldsymbol{r}'+\boldsymbol{k}\cdot\boldsymbol{r})}$$
$$\left.+\left[\hat{b}^\dagger_{\boldsymbol{k}},\,\hat{a}^\dagger_{\boldsymbol{k}'}\right]\,e^{-i(\boldsymbol{k}'\cdot\boldsymbol{r}'+\boldsymbol{k}\cdot\boldsymbol{r})}\right\} = i\hbar\frac{1}{V}\sum_{\boldsymbol{k}}e^{i\boldsymbol{k}\cdot(\boldsymbol{r}-\boldsymbol{r}')}. \tag{14.6.11}$$

This equation, together with a similar equation resulting from Eq. (10.6.10b), yields the following commutation relations for operators $\hat{a}_{\boldsymbol{k}}$, $\hat{a}^{\dagger}_{\boldsymbol{k}}$, $\hat{b}_{\boldsymbol{k}}$, and $\hat{b}^{\dagger}_{\boldsymbol{k}}$

$$\left[a_{\boldsymbol{k}} \, , \, \hat{a}^{\dagger}_{\boldsymbol{k}'}\right] \;=\; \left[\hat{b}_{\boldsymbol{k}} \, , \, \hat{b}^{\dagger}_{\boldsymbol{k}'}\right] \;=\; \delta_{\boldsymbol{k},\boldsymbol{k}'} \tag{14.6.12}$$

$$\left[\hat{a}_{\boldsymbol{k}} \, , \, \hat{b}_{\boldsymbol{k}'}\right] \;=\; \left[\hat{a}^{\dagger}_{\boldsymbol{k}'} \, , \, \hat{b}^{\dagger}_{\boldsymbol{k}}\right] \;=\; 0. \tag{14.6.13}$$

As a result of the field quantization, the Hamiltonian operator becomes

$$\hat{H} \;=\; \sum_{\boldsymbol{k}} \hbar\omega_k \left(\hat{a}_{\boldsymbol{k}} \, \hat{a}^{\dagger}_{\boldsymbol{k}} \;+\; \hat{b}^{\dagger}_{\boldsymbol{k}} \, \hat{b}_{\boldsymbol{k}}\right) . \tag{14.6.14}$$

If we introduce Hermitian operators $\hat{N}^{(+)}_{\boldsymbol{k}} = \hat{a}^{\dagger}_{\boldsymbol{k}} \, \hat{a}_{\boldsymbol{k}}$ and $\hat{N}^{(-)}_{\boldsymbol{k}} = \hat{b}^{\dagger}_{\boldsymbol{k}} \, \hat{b}_{\boldsymbol{k}}$, then with the help of Eqs. (14.6.12) and (14.6.13), the following commutation relation can be established

$$\left[\hat{a}_{\boldsymbol{k}} \, , \, \hat{N}^{(+)}_{\boldsymbol{k}}\right] \;=\; \hat{a}_{\boldsymbol{k}} \, , \tag{14.6.15}$$

$$\left[\hat{a}^{\dagger}_{\boldsymbol{k}} \, , \, \hat{N}^{(+)}_{\boldsymbol{k}}\right] \;=\; -\hat{a}^{\dagger}_{\boldsymbol{k}} \, , \tag{14.6.16}$$

$$\left[\hat{b}_{\boldsymbol{k}} \, , \, \hat{N}^{(-)}_{\boldsymbol{k}}\right] \;=\; \hat{b}_{\boldsymbol{k}} \, , \tag{14.6.17}$$

$$\left[\hat{b}^{\dagger}_{\boldsymbol{k}} \, , \, \hat{N}^{(-)}_{\boldsymbol{k}}\right] \;=\; -\hat{b}^{\dagger}_{\boldsymbol{k}} \, , \tag{14.6.18}$$

and

$$\left[b^{\dagger}_{\boldsymbol{k}} \, , \, \hat{N}^{(+)}_{\boldsymbol{k}}\right] \;=\; \left[\hat{b}_{\boldsymbol{k}} \, , \, \hat{N}^{(+)}_{\boldsymbol{k}}\right] \;=\; \left[\hat{a}_{\boldsymbol{k}} \, , \, \hat{N}^{(-)}_{\boldsymbol{k}}\right] \;=\; \left[\hat{a}^{\dagger}_{\boldsymbol{k}} \, , \, \hat{N}^{(-)}_{\boldsymbol{k}}\right] \;=\; 0 \, , \tag{14.6.19}$$

$$\left[N^{(+)}_{\boldsymbol{k}} \, , \, \hat{N}^{(-)}_{\boldsymbol{k}}\right] \;=\; 0 \, . \tag{14.6.20}$$

A comparison of these commutation relations with those for a linear oscillator shows that $\hat{N}^{(+)}_{\boldsymbol{k}}$ is number operator with eigenvalues $0, 1, 2, 3, \cdots$ and $\hat{a}_{\boldsymbol{k}}$ and $\hat{a}^{\dagger}_{\boldsymbol{k}}$ are the corresponding annihilation and creation operators. Similarly, $\hat{N}^{(-)}_{\boldsymbol{k}}$ is also a number operator with eigenvalues $0, 1, 2, \cdots$ and $\hat{b}_{\boldsymbol{k}}$ and $\hat{b}^{\dagger}_{\boldsymbol{k}}$ are the corresponding annihilation and creation operators. With the help of commutation relations (14.6.12), the Hamiltonian can be written in terms of number operators as

$$\hat{H} \;=\; \sum_{\boldsymbol{k}} \hat{H}_{\boldsymbol{k}} \;=\; \sum_{\boldsymbol{k}} \hbar\omega_k \left(\hat{N}^{(+)}_{\boldsymbol{k}} + \hat{N}^{(-)}_{\boldsymbol{k}} + 1\right) . \tag{14.6.21}$$

Charge Operator and Number Operators $\hat{N}^{(+)}_{\boldsymbol{k}}$ and $\hat{N}^{(+)}_{\boldsymbol{k}}$

From Eq. (14.4.19) and the definition of π and π^* from Eqs. (14.4.10) and (14.4.11), the total charge of the complex scalar field is given by

$$Q = \int \rho \, d\tau \;=\; \frac{1}{\hbar} \frac{ie}{c^2} \int \left(\phi^* \, \frac{\partial \phi}{\partial t} \;-\; \frac{\partial \phi^*}{\partial t} \, \phi\right) \; d\tau \;= (-\, ie\,/\hbar) \int \left(\pi\phi - \pi^*\phi^*\right) \, d^3r \, .$$

Using the expansions (14.6.6) and (14.6.7) for ϕ and π we get

$$Q \;=\; e \, \sum_{\boldsymbol{k}} \left(a^*_{\boldsymbol{k}} \, a_{\boldsymbol{k}} \;-\; b^*_{\boldsymbol{k}} \, b_{\boldsymbol{k}}\right) . \tag{14.6.22}$$

For a quantized field, this leads to the total charge operator

$$\hat{Q} \;=\; e \, \sum_{\boldsymbol{k}} \left[\hat{a}^{\dagger}_{\boldsymbol{k}} \, \hat{a}_{\boldsymbol{k}} \;-\; \hat{b}^{\dagger}(\boldsymbol{k}) \, \hat{b}(\boldsymbol{k})\right] \;=\; e \, \sum_{\boldsymbol{k}} \left[N^{(+)}_{\boldsymbol{k}} \;-\; N^{(-)}_{\boldsymbol{k}}\right] . \tag{14.6.23}$$

It follows from this equation that $N_{\boldsymbol{k}}^{(+)}$ may be interpreted as the number operator for field quanta (particles) with charge e and momentum $\hbar\boldsymbol{k}$ and energy $\hbar\omega_k$. Using the commuting set of observables $N_{\boldsymbol{k}}^{(+)}$ and $N_{\boldsymbol{k}}^{(-)}$, we can specify the state of the field by specifying the number of quanta of charge e and $-e$ for each value of momentum $\hbar\boldsymbol{k}$

$$\left|\{n_{\boldsymbol{k}}^{(+)}, n_{\boldsymbol{k}}^{(-)}\}\right\rangle = \left|n_{\boldsymbol{k}_1}^{(+)}, n_{\boldsymbol{k}_1}^{(-)}; n_{\boldsymbol{k}_2}^{(+)} n_{\boldsymbol{k}_2}^{(-)}; n_{\boldsymbol{k}_3}^{(+)}, n_{\boldsymbol{k}_3}^{(-)} \cdots\right\rangle ,$$

where $n_{\boldsymbol{k}}^{(\pm)}$ denotes the number of particles in the field with charge $\pm e$ and momentum $\hbar\boldsymbol{k}$, so that

$$\begin{aligned}\hat{H}\left|\{n_{\boldsymbol{k}}^{(+)}, n_{\boldsymbol{k}}^{(-)}\}\right\rangle &\equiv \sum_{\boldsymbol{k}} \hbar\omega_k \left(\hat{N}_{\boldsymbol{k}}^{(+)} + \hat{N}_{\boldsymbol{k}}^{(-)} + 1\right) \left|\{n_{\boldsymbol{k}}^{(+)}, n_{\boldsymbol{k}}^{(-)}\}\right\rangle \\ &= \sum_{\boldsymbol{k}} \hbar\omega_{\boldsymbol{k}} \left(n_{\boldsymbol{k}}^{(+)} + n_{\boldsymbol{k}}^{(-)} + 1\right) \left|\{n_{\boldsymbol{k}}^{(+)}, n_{\boldsymbol{k}}^{(-)}\}\right\rangle\end{aligned}$$

and

$$\hat{Q}\left|\{n_{\boldsymbol{k}}^{(+)}, n_{\boldsymbol{k}}^{(-)}\}\right\rangle = e \sum_{\boldsymbol{k}} \left(n_k^{(+)} - n_k^{(-)}\right) \left|\{n_{\boldsymbol{k}}^{(+)}, n_{\boldsymbol{k}}^{(-)}\}\right\rangle .$$

With the help of commutation relations (14.6.15) to (14.6.18), one can check the following relations

$$\begin{aligned}\hat{Q}\,\hat{a}_{\boldsymbol{k}}^\dagger \left|\{n_{\boldsymbol{k}}^{(+)}, n_{\boldsymbol{k}}^{(-)}\}\right\rangle &= e \sum_{\boldsymbol{k}} \left(\hat{n}_{\boldsymbol{k}}^{(+)} + 1 - \hat{n}_{\boldsymbol{k}}^{(-)}\right) \hat{a}_{\boldsymbol{k}}^\dagger \left|\{n_{\boldsymbol{k}}^{(+)}, n_{\boldsymbol{k}}^{(-)}\}\right\rangle , \\ \hat{Q}\,\hat{a}_{\boldsymbol{k}} \left|\{n_{\boldsymbol{k}}^{(+)}, n_{\boldsymbol{k}}^{(-)}\}\right\rangle &= e \sum_{\vec{k}} \left(\hat{n}_{\boldsymbol{k}}^{(+)} - 1 - \hat{n}_{\boldsymbol{k}}^{(-)}\right) \hat{a}_{\boldsymbol{k}} \left|\{n_{\boldsymbol{k}}^{(+)}, n_{\boldsymbol{k}}^{(-)}\}\right\rangle , \\ \hat{Q}\,\hat{b}_{\boldsymbol{k}}^\dagger \left|\{n_{\boldsymbol{k}}^{(+)}, n_{\boldsymbol{k}}^{(-)}\}\right\rangle &= e \sum_{k} \left(\hat{n}_{\boldsymbol{k}}^{(+)} - 1 - \hat{n}_{\boldsymbol{k}}^{(-)}\right) \hat{b}_{\boldsymbol{k}}^\dagger \left|\{n_{\boldsymbol{k}}^{(+)}, n_{\boldsymbol{k}}^{(-)}\}\right\rangle , \\ \hat{Q}\,\hat{b}_{\boldsymbol{k}} \left|\{n_{\boldsymbol{k}}^{(+)}, n_{\boldsymbol{k}}^{(-)}\}\right\rangle &= e \sum_{\vec{k}} \left(\hat{n}_{\boldsymbol{k}}^{(+)} + 1 - \hat{n}_{\boldsymbol{k}}^{(-)}\right) \hat{b}_{\boldsymbol{k}} \left|\{n_{\boldsymbol{k}}^{(+)}, n_{\boldsymbol{k}}^{(-)}\}\right\rangle .\end{aligned}$$

Similarly, one can also check the following relations

$$\hat{H}_k\,\hat{a}_{\boldsymbol{k}}^\dagger \left|\{n_{\boldsymbol{k}}^{(+)}, n_{\boldsymbol{k}}^{(-)}\}\right\rangle = \hbar\omega_k \left(n_{\boldsymbol{k}}^{(+)} + n_{\boldsymbol{k}}^{(-)} + 2\right) \hat{a}_{\boldsymbol{k}}^\dagger \left|\{n_{\boldsymbol{k}}^{(+)}, n_{\boldsymbol{k}}^{(-)}\}\right\rangle , \tag{14.6.24}$$

$$\hat{H}_{\boldsymbol{k}}\,\hat{a}_{\boldsymbol{k}} \left|\{n_{\boldsymbol{k}}^{(+)}, n_{\boldsymbol{k}}^{(-)}\}\right\rangle = \hbar\omega_k \left(n_{\boldsymbol{k}}^{(+)} + n_{\boldsymbol{k}}^{(-)}\right) \hat{a}_{\boldsymbol{k}} \left|\Psi\right\rangle , \tag{14.6.25}$$

$$\hat{H}_{\boldsymbol{k}}\,\hat{b}_{\boldsymbol{k}}^\dagger \left|\{n_{\boldsymbol{k}}^{(+)}, n_{\boldsymbol{k}}^{(-)}\}\right\rangle = \hbar\omega_k \left(n_{\vec{k}}^{(+)} + n_{\boldsymbol{k}}^{(-)} + 2\right) \hat{b}_{\boldsymbol{k}} \left|\{n_{\boldsymbol{k}}^{(+)}, n_{\boldsymbol{k}}^{(-)}\}\right\rangle , \tag{14.6.26}$$

and

$$\hat{H}_k\,\hat{b}_{\boldsymbol{k}} \left|\{n_{\boldsymbol{k}}^{(+)}, n_{\boldsymbol{k}}^{(-)}\}\right\rangle = \hbar\omega_k \left(n_{\boldsymbol{k}}^{(+)} + n_{\boldsymbol{k}}^{(-)}\right) \hat{b}_{\boldsymbol{k}} \left|\{n_{\boldsymbol{k}}^{(+)}, n_{\boldsymbol{k}}^{(-)}\}\right\rangle . \tag{14.6.27}$$

These relations explicitly demonstrate that $\hat{a}_{\boldsymbol{k}}^\dagger$ and $\hat{a}_{\boldsymbol{k}}$ may be regarded as creation and annihilation operators, respectively, for particles of charge $+e$, momentum $\boldsymbol{p} = \hbar\boldsymbol{k}$ and energy $\hbar\omega_k$. Similarly, $\hat{b}_{\boldsymbol{k}}^\dagger$ and $\hat{b}_{\boldsymbol{k}}$ may be regarded as creation and annihilation operators, respectively, for particles of charge $-e$, momentum $\hbar\boldsymbol{k}$ and energy $\hbar\omega_k$.

14.7 Dirac Field and Its Quantization

The Lagrangian density (14.4.22) that leads to the correct field equations for the Dirac field is given by

$$\mathcal{L} = -\tilde{\psi}\left(c\hbar\,\gamma_\rho \frac{\partial}{\partial x_\rho} + mc^2\right)\psi = c\hbar \frac{\partial\tilde{\psi}}{\partial x_\rho}\gamma_\rho\psi - mc^2\,\tilde{\psi}\psi\,, \tag{14.4.22*}$$

where sum over repeated indices is implied and each one of the four components of ψ is to be regarded as an independent field variable. It follows from $\pi = \frac{\partial\mathcal{L}}{\partial\dot{\psi}}$, that each component $\pi_s = \frac{\partial\mathcal{L}}{\partial\dot{\psi}_s} = i\hbar\,\psi_s^\dagger$ of π is to be regarded as a field variable canonically conjugate to ψ_s.

From the Lagrangian density (14.4.22), we obtain the Hamiltonian density

$$\begin{aligned}\mathcal{H} &= \frac{\partial\mathcal{L}}{\partial\dot{\psi}}\,\dot{\psi} - \mathcal{L} = \sum_s \pi_s\frac{\partial\psi_s}{\partial t} - \mathcal{L}\\ &= i\hbar\left(\psi^\dagger\frac{\partial\psi}{\partial t} - ic\,\tilde{\psi}\gamma_4\frac{\partial\psi}{\partial(ict)} - ic\tilde{\psi}\gamma_k\frac{\partial\psi}{\partial x_k}\right) + mc^2\,\tilde{\psi}\psi\\ &= \psi^\dagger\,\left(c\boldsymbol{\alpha}\cdot\boldsymbol{p} + \beta mc^2\right)\psi\,.\end{aligned} \tag{14.7.1}$$

This leads to the Hamiltonian

$$H = \int \mathcal{H}\,d^3r = \int d^3r\,[\psi^\dagger\,\left(c\boldsymbol{\alpha}\cdot\boldsymbol{p} + \beta mc^2\right)\psi]\,. \tag{14.7.2}$$

We now make a Fourier expansion of the Dirac field as

$$\psi(\boldsymbol{r},t) = \frac{1}{\sqrt{V}}\sum_{\boldsymbol{p}}\sum_{\nu=1}^{4}\sqrt{\frac{mc^2}{|E_{\boldsymbol{p}}|}}\;b_{\boldsymbol{p}}^{(\nu)}(t)\;u^{(\nu)}(p)\;e^{i\,\boldsymbol{p}\cdot\boldsymbol{r}/\hbar} \tag{14.7.3}$$

and

$$\psi^\dagger(\boldsymbol{r},t) = \frac{1}{\sqrt{V}}\sum_{\boldsymbol{p}}\sum_{\nu=1}^{4}\sqrt{\frac{mc^2}{|E_{\boldsymbol{p}}|}}\;b_{\boldsymbol{p}}^{(\nu)*}(t)\;u^{(\nu)\dagger}(p)\;e^{-i\,\boldsymbol{p}\cdot\boldsymbol{r}/\hbar}\,, \tag{14.7.4}$$

where the time dependence of the field is contained in the coefficients $b_{\boldsymbol{p}}^{(\nu)}(t)$ and $b_{\boldsymbol{p}}^{(\nu)*}(t)$ and $u^{(\nu)}(\boldsymbol{p})$ and $u^{(\nu)\dagger}(\boldsymbol{p})$ represent Dirac spinors with $u^{(1)}(\boldsymbol{p})$ and $u^{(2)}(\boldsymbol{p})$ characterizing the positive energy states and $u^{(3)}$ and $u^{(4)}$ characterizing the negative energy states. Note that $u^\nu(\boldsymbol{p})$ is a column matrix while its adjoint denoted by $u^{(\nu)\dagger}(\boldsymbol{p})$ is a row matrix. Then the Hamiltonian for Dirac field becomes

$$\begin{aligned}H &= \int \psi^\dagger\,(-i\hbar c\,\boldsymbol{\alpha}\cdot\boldsymbol{\nabla} + \gamma_4\,mc^2)\,\psi\,d^3r\\ &= \frac{1}{\sqrt{V}}\int d^3r\sum_{\boldsymbol{p}}\sum_{\boldsymbol{p}'}\sum_{\nu=1}^{4}\sum_{\nu'=1}^{4}\left(\sqrt{\frac{mc^2}{|E_{\boldsymbol{p}}|}}\;b_{\boldsymbol{p}}^{\nu*}\;u^{(\nu)\dagger}(\boldsymbol{p})\;e^{-i\,\boldsymbol{p}\cdot\boldsymbol{r}/\hbar}\right)\\ &\qquad(-i\hbar c\,\boldsymbol{\alpha}\cdot\boldsymbol{\nabla} + \gamma_4\,mc^2)\left(\sqrt{\frac{mc^2}{|E_{p'}|}}\;b_{\boldsymbol{p}'}^{(\nu')}\;u_{\boldsymbol{p}'}^{(\nu')}\;e^{i\,\boldsymbol{p}'\cdot\boldsymbol{r}/\hbar}\right)\end{aligned}$$

or

$$H = \sum_{\boldsymbol{p}}\sum_{\nu} E_{\boldsymbol{p}}^{\nu}\;b_{\boldsymbol{p}}^{(\nu)*}\;b_{\boldsymbol{p}}^{(\nu)}\,. \tag{14.7.5}$$

Here $E_{\boldsymbol{p}}^{\nu} = +|E_{\boldsymbol{p}}|$ for $\nu = 1, 2$ and $= -|E_{\boldsymbol{p}}|$ for $\nu = 3, 4$ and where $E_{\boldsymbol{p}}^2 = c^2p^2 + m^2c^4$. In arriving at Eq. (14.7.5), we have used the identities

$$\frac{1}{V}\int d^3r\, e^{-i(\boldsymbol{p}-\boldsymbol{p}')\cdot\boldsymbol{r}/\hbar} = \delta_{\boldsymbol{p},\boldsymbol{p}'} \tag{14.7.6}$$

and

$$u^{(\nu)\dagger}(\boldsymbol{p})\, u^{(\nu')}(\boldsymbol{p}) = \frac{|E_{\boldsymbol{p}}|}{mc^2}\,\delta_{\nu\nu'}\,. \tag{14.7.7}$$

With $\psi^\dagger$ and ψ given by Eqs. (14.7.3) and (14.7.4), the total charge of the Dirac field is given by

$$\begin{aligned} Q \equiv e\int d^3r\,\psi^\dagger\,\psi &= e\sum_{\boldsymbol{p}}\sum_{\boldsymbol{p}'}\sum_{\nu=1}^{4}\sum_{\nu'=1}^{4}\frac{mc^2}{\sqrt{|E_p||E_{p'}|}}\, b_{\boldsymbol{p}}^{(\nu)*}\, b_{\boldsymbol{p}}^{(\nu)}\, u^{(\nu)\dagger}(\boldsymbol{p})\, u^{(\nu')}(\boldsymbol{p}') \\ &\times \frac{1}{V}\int e^{-i(\boldsymbol{p}-\boldsymbol{p}')\cdot\boldsymbol{r}/\hbar}\, d^3r\,, \\ &= e\sum_{\boldsymbol{p}}\sum_{\nu=1}^{4} b_{\boldsymbol{p}}^{(\nu)*}\, b_{\boldsymbol{p}}^{(\nu)}\,, \end{aligned} \tag{14.7.8}$$

where, once again, we have used the identities (14.7.6) and (14.7.7).

To quantize the Dirac field, we treat the field variables $\psi(\boldsymbol{r}, t)$ and $\tilde{\psi}(\boldsymbol{r}, t)$ as operators which implies that the Fourier expansion coefficients are also operators. Denoting the corresponding operators by the same symbol with a caret, the Hamiltonian operator of the field can be written as

$$\hat{H} = \sum_{\boldsymbol{p}}\sum_{\nu=1}^{4} E_p^\nu\;\; \hat{b}_{\boldsymbol{p}}^{(\nu)\dagger}\,\hat{b}_{\boldsymbol{p}}^{(\nu)}\,, \tag{14.7.9}$$

where

$$E_p^\nu = \begin{cases} +|E_p| & \text{if } \nu = 1, 2 \\ -|E_p| & \text{if } \nu = 3, 4. \end{cases}$$

The total charge operator of the field can be written as

$$Q = e\sum_{\boldsymbol{p}}\sum_{\nu} \hat{b}_{\boldsymbol{p}}^{(\nu)\dagger}\,\hat{b}_{\boldsymbol{p}}^{(\nu)}. \tag{14.7.10}$$

The operator

$$\hat{N}_{\boldsymbol{p}}^{(\nu)} = \hat{b}_{\boldsymbol{p}}^{(\nu)\dagger}\,\hat{b}_{\boldsymbol{p}}^{(\nu)}\,, \tag{14.7.11}$$

can be looked upon as the number operator and $\hat{b}_{\boldsymbol{p}}^{(\nu)\dagger}$ and $\hat{b}_{\boldsymbol{p}}^{(\nu)}$ can be looked upon as creation and annihilation operators for the Dirac field. If we prescribe the same commutation relations for them as we did for the KG field, we cannot satisfy the Pauli exclusion principle according to which, since the quanta of Dirac field are Fermions, the number operator $\hat{N}_{\boldsymbol{p}}^{(\nu)}$ can have only two eigenvalues 0 and 1. The way out of this dilemma is to prescribe anti-commutation relations. Explicitly $\langle n_{\boldsymbol{p}}|\,\hat{N}_{\boldsymbol{p}}^{(\nu)}\,|n'_{\boldsymbol{p}}\rangle = n_{\boldsymbol{p}}\,\delta_{n_{\boldsymbol{p}},n'_{\boldsymbol{p}}}$ where $n_{\boldsymbol{p}} = 0$ or 1. So in the occupation number representation, $\hat{N}_{\boldsymbol{p}}^{(\nu)}$ can be represented by $\begin{pmatrix}0 & 0\\ 0 & 1\end{pmatrix}$. This implies that $\hat{b}_{\boldsymbol{p}}^{(\nu)\dagger}$ and $\hat{b}_{\boldsymbol{p}}^{(\nu)}$ can also be represented by 2×2 matrices $\begin{pmatrix}0\,0\\ 1\,0\end{pmatrix}$ and $\begin{pmatrix}0\,1\\ 0\,0\end{pmatrix}$, respectively. This representation yields anti-commutation relations for $\hat{b}_{\boldsymbol{p}}^{(\nu)\dagger}$ and $\hat{b}_{\boldsymbol{p}}^{(\nu)}$,

$$\left[\hat{b}_{\boldsymbol{p}}^{(\nu)\dagger}\,,\,\hat{b}_{\boldsymbol{p}}^{(\nu)}\right]_+ = \left(\hat{b}_{\boldsymbol{p}}^{(\nu)\dagger}\,\hat{b}_{\boldsymbol{p}}^{(\nu)} + \hat{b}_{\boldsymbol{p}}^{(\nu)}\,\hat{b}_{\boldsymbol{p}}^{(\nu)\dagger}\right) = \hat{1} \tag{14.7.12}$$

$$\left[\hat{b}_{\boldsymbol{p}}^{(\nu)\dagger}\,,\,\hat{b}_{\boldsymbol{p}'}^{(\nu')}\right]_+ = \hat{1}\,\delta_{\boldsymbol{p},\boldsymbol{p}'}\,\delta_{\nu\nu'} \tag{14.7.13}$$

$$\left[\hat{b}_{\boldsymbol{p}}^{(\nu)}\,,\,\hat{b}_{\boldsymbol{p}}^{(\nu')}\right]_+ = \left[\hat{b}_{\boldsymbol{p}}^{(\nu)\dagger}\,,\,\hat{b}_{\boldsymbol{p}}^{(\nu')\dagger}\right]_+ = 0. \tag{14.7.14}$$

These anti-commutation relations lead to the interpretation of the operators $\hat{b}_{\boldsymbol{p}}^{\dagger}$ and $\hat{b}_{\boldsymbol{p}}$ as the creation and annihilation operator, respectively, for the quanta of Dirac field. We can justify this interpretation with the help of identities

$$\hat{N}_{\boldsymbol{p}}^{(\nu)}\,\hat{b}_{\boldsymbol{p}}^{(\nu)\dagger} = \hat{b}_{\boldsymbol{p}}^{(\nu)\dagger}\left(\hat{1} - \hat{N}_{\boldsymbol{p}}^{(\nu)}\right) \tag{14.7.15}$$

and

$$\hat{N}_{\boldsymbol{p}}^{(\nu)}\,\hat{b}_{\boldsymbol{p}}^{(\nu)} = \hat{b}_{\boldsymbol{p}}^{(\nu)}\left(\hat{1} - \hat{N}_{\boldsymbol{p}}^{(\nu)}\right), \tag{14.7.16}$$

which follow from the anti-commutation relations (14.7.12) through (14.7.14). Now consider the field state $\left|n_{\boldsymbol{p}}^{(\nu)} = 1\right\rangle$ in which there is an electron with momentum $\boldsymbol{p}$ and spin and sign of energy indicated by ν. Also consider a field state $\left|n_{\boldsymbol{p}}^{(\nu)} = 0\right\rangle$ in which there is no electron with the above specification. Using the identities we can verify the following equations

$$\hat{N}_{\boldsymbol{p}}^{(\nu)}\,\hat{b}_{\boldsymbol{p}}^{(\nu)\dagger}\left|n_{\boldsymbol{p}}^{(\nu)} = 0\right\rangle = b_{\boldsymbol{p}}^{(\nu)\dagger}\left|n_{\boldsymbol{p}}^{(\nu)} = 0\right\rangle \tag{14.7.17}$$

$$\hat{N}_{\boldsymbol{p}}^{(\nu)}\,\hat{b}_{\boldsymbol{p}}^{(\nu)\dagger}\left|n_{\boldsymbol{p}}^{(\nu)} = 1\right\rangle = 0 \tag{14.7.18}$$

$$\hat{N}_{\boldsymbol{p}}^{(\nu)}\,\hat{b}_{\boldsymbol{p}}^{(\nu)}\left|n_{\boldsymbol{p}}^{(\nu)} = 1\right\rangle = 0 \tag{14.7.19}$$

$$\hat{N}_{\boldsymbol{p}}^{(\nu)}\,\hat{b}_{\boldsymbol{p}}^{(\nu)}\left|n_{\boldsymbol{p}}^{(\nu)} = 0\right\rangle = 0. \tag{14.7.20}$$

We can interpret Eq. (14.7.17) to mean that $\hat{b}_{\boldsymbol{p}}^{(\nu)\dagger}\left|n_{\boldsymbol{p}}^{(\nu)} = 0\right\rangle \equiv \left|n_{\boldsymbol{p}}^{(\nu)} = 1\right\rangle$ and for this state $\hat{N}_{\boldsymbol{p}}^{(\nu)}$ has eigenvalue one. Hence $\hat{b}_{\boldsymbol{p}}^{(\nu)\dagger}$ is the creation operator for the electron with momentum $\boldsymbol{p}$ and spin and sign of energy given by ν. Similarly, Eq. (14.7.18) can be interpreted to mean that $\hat{b}_{\boldsymbol{p}}^{(\nu)\dagger}\left|n_{\boldsymbol{p}} = 1\right\rangle$ is a null state in accordance with the Pauli exclusion principle which forbids another (identical) Fermion in the same state. Equation (14.7.19) means that the state of the field $\hat{b}_{\boldsymbol{p}}^{(\nu)}\left|n_{\boldsymbol{p}}^{(\nu)} = 1\right\rangle \equiv \left|n_{\boldsymbol{p}}^{(\nu)} = 0\right\rangle$. Hence $\hat{N}_{\boldsymbol{p}}^{(\nu)}$ operating on this state gives zero. So $\hat{b}_{\boldsymbol{p}}^{(\nu)}$ may be regarded as annihilation operator for the electron with the specification mentioned above. Finally, Eq. (14.7.20) can be interpreted to mean that $\hat{b}_{\boldsymbol{p}}^{(\nu)}\left|n_{\boldsymbol{p}}^{(\nu)} = 0\right\rangle$ is a null state, since $\hat{b}_{\boldsymbol{p}}^{(\nu)}$ cannot destroy a particle if it did not exist in the original state. For a general field state $\left|\cdots, n_{\boldsymbol{p}}^{(\nu)}, \cdots\right\rangle$ the application of these operators gives the following results

$$\begin{aligned}
\hat{b}_{\boldsymbol{p}}^{(\nu)\dagger}\left|\cdots, n_{\boldsymbol{p}}^{(\nu)} = 1, \cdots\right\rangle &= 0\\
\hat{b}_{\boldsymbol{p}}^{(\nu)\dagger}\left|\cdots, n_{\boldsymbol{p}}^{(\nu)} = 0, \cdots\right\rangle &= \left|\cdots, n_{\boldsymbol{p}}^{(\nu)} = 1, \cdots\right\rangle\\
\hat{b}_{\boldsymbol{p}}^{(\nu)}\left|\cdots, n_{\boldsymbol{p}}^{(\nu)} = 1, \cdots\right\rangle &= \left|\cdots, n_{\boldsymbol{p}}^{(\nu)} = 0, \cdots\right\rangle\\
\hat{b}_{\boldsymbol{p}}^{(\nu)}\left|\cdots, n_{\boldsymbol{p}}^{(\nu)} = 0, \cdots\right\rangle &= 0.
\end{aligned}$$

From these relations we have, for the number operator $\hat{N}_{\boldsymbol{p}}^{(\nu)} \equiv \hat{b}_{\boldsymbol{p}}^{(\nu)\dagger}\,\hat{b}_{\boldsymbol{p}}^{(\nu)}$,

$$\hat{b}_{\boldsymbol{p}}^{(\nu)\dagger}\,\hat{b}_{\boldsymbol{p}}^{(\nu)}\left|n_{\boldsymbol{p}}^{(\nu)}=1\right\rangle = \left|n_{\boldsymbol{p}}^{(\nu)}=1\right\rangle$$

and

$$\hat{b}_{\boldsymbol{p}}^{(\nu)\dagger}\,\hat{b}_{\boldsymbol{p}}^{(\nu)}\left|n_{\boldsymbol{p}}^{(\nu)}=0\right\rangle = 0\,.$$

The anti-commutation relations for the field operators $\hat{b}_{\boldsymbol{p}}^{(\nu)}$ and $\hat{b}_{\boldsymbol{p}}^{(\nu)\dagger}$ lead to anti-commutation relations for Dirac field operators

$$\left[\hat{\psi}_\nu(\boldsymbol{r},t)\,,\,\psi_{\nu'}^\dagger(\boldsymbol{r}',t)\right]_+ = i\hbar\,\delta^3(\boldsymbol{r}-\boldsymbol{r}')\,\delta_{\nu\nu'} \tag{14.7.21}$$

and

$$[\psi_\nu(\boldsymbol{r},t)\,,\,\psi_{\nu'}(\boldsymbol{r}',t)]_+ = \left[\psi_\nu^\dagger(\boldsymbol{r},t)\,,\,\psi_{\nu'}^\dagger(\boldsymbol{r}',t)\right]_+ = 0\,. \tag{14.7.22}$$

The Pauli exclusion principle for the field quanta is thus guaranteed if we prescribe anti-commutation relations, instead of commutation relations, for the field operators.

The time dependence of the operators $\hat{b}_{\boldsymbol{p}}^{(\nu)\dagger}$ and $\hat{b}_{\boldsymbol{p}}^{(\nu)}$ can be inferred from Heisenberg equations of motion

$$\frac{d\hat{b}_{\boldsymbol{p}}^{(\nu)}}{dt} = \frac{i}{\hbar}\left[\hat{H}\,,\,\hat{b}_{\boldsymbol{p}}^{(\nu)}\right] = \begin{cases} -\frac{i\,|E|}{\hbar}\,\hat{b}_{\boldsymbol{p}}^{(\nu)} & \text{for } \nu = 1,2 \\[2ex] \frac{i\,|E|}{\hbar}\,\hat{b}_{\boldsymbol{p}}^{(\nu)} & \text{for } \nu = 3,4 \end{cases}$$

$$\frac{d\hat{b}_{\boldsymbol{p}}^{(\nu)\dagger}}{dt} = \frac{i}{\hbar}\left[\hat{H}\,,\,\hat{b}_{\boldsymbol{p}}^{(\nu)\dagger}\right] = \begin{cases} \frac{i\,|E|}{\hbar}\,\hat{b}_{\boldsymbol{p}}^{(\nu)\dagger} & \text{for } \nu = 1,2 \\[2ex] -\frac{i|E|}{\hbar}\,\hat{b}_{\boldsymbol{p}}^{(\nu)\dagger} & \text{for } \nu = 3,4. \end{cases}$$

These equations imply that

$$\hat{b}_{\boldsymbol{p}}^{(\nu)}(t) = b_{\boldsymbol{p}}^{(\nu)}(t=0)\,e^{\mp i\,|E|\,t/\hbar} \tag{14.7.23}$$

and

$$\hat{b}_{\boldsymbol{p}}^{(\nu)\dagger}(t) = \hat{b}_{\boldsymbol{p}}^{(\nu)\dagger}(t=0)\,e^{\pm i\,|E|\,t/\hbar}\,, \tag{14.7.24}$$

where the upper sign is for $\nu = 1,2$ and lower for $\nu = 3,4$. Then the expansions for the field operators $\hat{\psi}(\boldsymbol{r},t)$ and $\hat{\psi}^\dagger(\boldsymbol{r},t)$ (Eqs. (14.7.3) and (14.7.4)) may be rewritten as

$$\hat{\psi}(\boldsymbol{r},t) = \frac{1}{\sqrt{V}}\sum_{\boldsymbol{p}}\sqrt{\frac{mc^2}{|E|}}\left\{\sum_{\nu}^{1,2}\hat{b}_{\boldsymbol{p}}^{(\nu)}(0)\,u^{(\nu)}(\boldsymbol{p})\,\exp\left[i\,(\boldsymbol{p}\cdot\boldsymbol{r}-|E|t)/\hbar\right]\right.$$
$$\left.+\sum_{\nu}^{3,4}\hat{b}_{\boldsymbol{p}}^{(\nu)}(0)\,u^{(\nu)}(\boldsymbol{p})\,\exp\left[i\,(\boldsymbol{p}\cdot\boldsymbol{r}+|E|t)/\hbar\right]\right\} \tag{14.7.25}$$

and

$$\hat{\psi}^\dagger(\boldsymbol{r},t) = \frac{1}{\sqrt{V}}\sum_{\boldsymbol{p}}\sqrt{\frac{mc^2}{|E|}}\left\{\sum_{\nu}^{1,2}\hat{b}_{\boldsymbol{p}}^{(\nu)\dagger}(0)\,u^{(\nu)\dagger}(\boldsymbol{p})\,\exp\left[-i\,(\boldsymbol{p}\cdot\boldsymbol{r}-|E|t)/\hbar\right]\right.$$
$$\left.+\sum_{\nu}^{3,4}\hat{b}_{\boldsymbol{p}}^{(\nu)\dagger}(0)\,u^{(\nu)\dagger}(\boldsymbol{p})\,\exp\left[-i\,(\boldsymbol{p}\cdot\boldsymbol{r}+|E|t)/\hbar\right]\right\}. \tag{14.7.26}$$

14.8 Positron Operators and Spinors

It is desirable to have a formulation of Dirac field in which the free particle energy is always positive, while the total charge Q can be negative or positive, depending on whether there are excess electrons or positrons in the field. For this we introduce the operators $\hat{b}_{\boldsymbol{p}}^{(s)}$ and $\hat{d}_{\boldsymbol{p}}^{(s)}$ and their adjoints, $\hat{b}_{\boldsymbol{p}}^{(s)\dagger}$ and $\hat{d}_{\boldsymbol{p}}^{(s)\dagger}$, with $s = 1, 2$ in place of the operators $\hat{b}_{\boldsymbol{p}}^{(\nu)}$ and $\hat{b}_{\boldsymbol{p}}^{(\nu)\dagger}$ with $\nu = 1, 2, 3, 4$, such that

$$\hat{b}_{\boldsymbol{p}}^{(s)} = \hat{b}_{\boldsymbol{p}}^{(\nu)} \qquad \text{for } s = \nu = 1, 2 \tag{14.8.1}$$

and
$$\hat{d}_{\boldsymbol{p}}^{(s)\dagger} = \begin{cases} -\hat{b}_{-\boldsymbol{p}}^{(\nu)} & \text{for } s = 1,\ \nu = 4 \\ \hat{b}_{-\boldsymbol{p}}^{(\nu)} & \text{for } s = 2,\ \nu = 3\,. \end{cases} \tag{14.8.2}$$

We also introduce the electron and positron spinors $u^{(s)}(\boldsymbol{p})$ and $v^{(s)}(\boldsymbol{p})$, $s = 1, 2$, such that

$$u^{(s)}(\boldsymbol{p}) = u^{(\nu)}(\boldsymbol{p}) \qquad \text{for } s = \nu = 1, 2 \tag{14.8.3}$$

and
$$v^{(s)}(\boldsymbol{p}) = \begin{cases} -u^{(\nu)}(-\boldsymbol{p}) & \text{for } s = 1,\ \nu = 4 \\ u^{(\nu)}(-\boldsymbol{p}) & \text{for } s = 2,\ \nu = 3\,. \end{cases} \tag{14.8.4}$$

The underlying idea is that annihilation of a negative energy electron with momentum $-\boldsymbol{p}$ and spin down ($\nu = 4$) by the operator $\hat{b}_{-\boldsymbol{p}}^{(4)}$, is equivalent to the creation of a positron with momentum $+\boldsymbol{p}$ and spin up ($s = 1$) by the operator $\hat{d}_{\boldsymbol{p}}^{(1)\dagger}$. Likewise the destruction of a negative energy electron with spin up ($\nu = 3$) and momentum $-\boldsymbol{p}$ by the operator $\hat{b}_{-\boldsymbol{p}}^{(3)}$ can be interpreted as creation of a positron with spin down ($s = 2$), momentum $+\boldsymbol{p}$ and positive energy by the operator $\hat{d}_{\boldsymbol{p}}^{(2)\dagger}$. The spinors $v^{(s)}(\boldsymbol{p})$ represent the state of the positron with spin up (with $s = 1$) or spin down (with $s = 2$). These correspond, respectively, to the states $u^{\nu}(-\boldsymbol{p})$ of the electron with negative energy, momentum $-\boldsymbol{p}$, and spin down ($\nu = 4$) or spin up ($\nu = 3$), whose absence in the continuum of negative energy states can be interpreted as the positron.

The spinors $u^{(s)}(\boldsymbol{p})$ and $v^{(s)}(\boldsymbol{p})$ also conform to the definition of the charge conjugation operator $S_c (\equiv \gamma_2 = -i\beta\alpha_2)$, such that

$$S_c\, u^{(s)*}(\boldsymbol{p}) = v^{(s)}(\boldsymbol{p}) \tag{14.8.5}$$

and
$$S_c\, v^{(s)*}(\boldsymbol{p}) = u^{(s)}(\boldsymbol{p})\,, \tag{14.8.6}$$

as is easily verified[6]. We can also see that the operators $\hat{d}_{\boldsymbol{p}}^{(s)}$ and $\hat{d}_{p}^{(s)\dagger}$ satisfy the same anti-commutation relations as $\hat{b}_{\boldsymbol{p}}^{(\nu)}$ and $\hat{b}_{\boldsymbol{p}}^{(\nu)\dagger}$. Thus

$$\left[\hat{d}_{\boldsymbol{p}}^{(s)}\,,\, \hat{d}_{\boldsymbol{p}'}^{(s')\dagger}\right]_+ = \hat{1}\,\delta_{ss'}\,\delta_{\boldsymbol{p}\boldsymbol{p}'}\,, \tag{14.8.7}$$

$$\left[\hat{d}_{\boldsymbol{p}}^{(s)}\,,\, \hat{d}_{\boldsymbol{p}'}^{(s')}\right]_+ = \left[\hat{d}_{\boldsymbol{p}}^{(s)\dagger}\,,\, \hat{d}_{\boldsymbol{p}'}^{(s')\dagger}\right]_+ = 0\,, \tag{14.8.8}$$

just as
$$\left[\hat{b}_{\boldsymbol{p}}^{(\nu)}\,,\, \hat{b}_{\boldsymbol{p}'}^{(\nu')\dagger}\right]_+ = \hat{1}\,\delta_{\nu\nu'}\,\delta_{\boldsymbol{p}\boldsymbol{p}'}\,, \tag{14.8.9}$$

and
$$\left[\hat{b}_{\boldsymbol{p}}^{(\nu)}\,,\, \hat{b}_{\boldsymbol{p}'}^{(\nu')}\right]_+ = \left[\hat{b}_{\boldsymbol{p}}^{(\nu)\dagger}\,,\, \hat{b}_{\boldsymbol{p}'}^{(\nu')\dagger}\right]_+ = 0\,. \tag{14.8.10}$$

Further,

$$\left[\hat{b}_{\boldsymbol{p}}^{(s)}\,,\, \hat{d}_{\boldsymbol{p}'}^{(s')}\right]_+ = \left[\hat{b}_{\boldsymbol{p}}^{(s)\dagger}\,,\, \hat{d}_{\boldsymbol{p}'}^{(s')\dagger}\right]_+ = \left[\hat{b}_{\boldsymbol{p}}^{(s)\dagger}\,,\, \hat{d}_{\boldsymbol{p}'}^{(s')}\right]_+ = \left[\hat{b}_{\boldsymbol{p}}^{(s)}\,,\, \hat{d}_{\boldsymbol{p}'}^{(s')\dagger}\right] = 0. \tag{14.8.11}$$

---

[6]With $S_c = \gamma_2 = \begin{pmatrix} 0 & 0 & 0 & -1 \\ 0 & 0 & 1 & 0 \\ 0 & 1 & 0 & 0 \\ -1 & 0 & 0 & 0 \end{pmatrix}$, and $u^{(1)} = u_\uparrow^+$, $u^{(2)} = u_\downarrow^+$, $u^{(3)} = u_\uparrow^-$ and $u^{(4)} = u_\downarrow^-$ given by Eqs. (12.4.16), (12.4.17), (12.4.20) and (12.4.21), [see Sec. 12.4] and $v^{(s)}(\boldsymbol{p})$, $u^{(s)}(\boldsymbol{p})$ defined by Eqs. (14.8.3) and (14.8.4), one can check that $S_c\, u^{(s=1)*}(\boldsymbol{p}) = -u^{(\nu=4)}(-\boldsymbol{p}) = v^{(s=1)}(\boldsymbol{p})$; $S_c\, u^{(s=2)*}(\boldsymbol{p}) = u^{(\nu=3)}(-\boldsymbol{p}) = v^{(s=2)}(\boldsymbol{p})$; $S_c\, v^{(s=1)*}(\boldsymbol{p}) = u^{(s=1)}(\boldsymbol{p})$; $S_c\, v^{(s=2)*}(\boldsymbol{p}) = u^{(s=2)}(\boldsymbol{p})$.

14.8.1 Equations Satisfied by Electron and Positron Spinors

Column matrices $u^{(s)}(\boldsymbol{p})$ and $v^{(s)}(\boldsymbol{p})$ may be looked upon as the electron and positron spinors, respectively. With the substitutions $\psi = u^{(s)}(\boldsymbol{p})\, e^{i\, p_\mu x_\mu/\hbar}$ and $\tilde{\psi} = \tilde{u}^{(s)}(\boldsymbol{p})\, e^{-ip_\mu x_\mu/\hbar}$ in the Dirac equation and its adjoint

$$\gamma_\lambda \frac{\partial \psi}{\partial x_\lambda} + \frac{mc}{\hbar}\psi = 0 \qquad \text{and} \qquad \frac{\partial \tilde{\psi}}{\partial x_\lambda}\gamma_\lambda - \frac{mc}{\hbar}\tilde{\psi} = 0\,,$$

we find that electron spinors $u^{(s)}(\boldsymbol{p})$ and $\tilde{u}^{(s)}(\boldsymbol{p}) = u^{(s)\dagger}(\boldsymbol{p})\,\gamma_4$ satisfy the equations

$$(i\gamma_\lambda\, p_\lambda + mc)\; u^{(s)}(\boldsymbol{p}) \;=\; 0 \tag{14.8.12}$$

and

$$\tilde{u}^{(s)}(\boldsymbol{p})\;(ip_\lambda\,\gamma_\lambda + mc) \;=\; 0\,. \tag{14.8.13}$$

The free positron spinor $v^{(s)}(\boldsymbol{p})$ satisfies the equation[7]

$$(-i\gamma_\mu\, p_\mu + mc)\; v^{(s)}(\boldsymbol{p}) \;=\; 0\,. \tag{14.8.14}$$

Using Eq. (14.8.14) it may be verified that the spinor

$$\tilde{v}^{(s)}(\boldsymbol{p}) \;\equiv\; v^{(s)\dagger}(\boldsymbol{p})\,\gamma_4$$

satisfies the equation

$$\tilde{v}^{(s)}(\boldsymbol{p})\;(-i\gamma_\mu\, p_\mu + mc) \;=\; 0\,. \tag{14.8.15}$$

If we choose the normalization[8]

$$\tilde{u}^{(s')}(\boldsymbol{p}')\; u^{(s)}(\boldsymbol{p}) \;=\; \delta_{ss'}\,\delta_{\boldsymbol{p}\boldsymbol{p}'} \tag{14.8.16}$$

then it is easy to show that

$$\tilde{v}^{(s')}(\boldsymbol{p}')\, v^{(s)}(\boldsymbol{p}) \;=\; -\delta_{ss'}\,\delta_{\boldsymbol{p}\boldsymbol{p}'} \tag{14.8.17}$$

and

$$\tilde{v}^{(s')}(\boldsymbol{p}')\; u^{(s)}(\boldsymbol{p}) \;=\; \tilde{u}^{(s')}(\boldsymbol{p}')\, v^{(s)}(\boldsymbol{p}) = 0\,. \tag{14.8.18}$$

This can be seen using the charge conjugation relations

$$u^{(s)}(\boldsymbol{p}) = \gamma_2\, v^{(s)*}(\boldsymbol{p}) \quad \text{and} \quad v^{(s)}(\boldsymbol{p}) = \gamma_2\, u^{(s)*}(\boldsymbol{p})$$

in Eq. (14.8.16), where $\gamma_2 = S_C$ is the charge conjugation operator. From the convention (14.8.16) it follows that

$$u^{(s')\dagger}(\boldsymbol{p}')\, u^{(s)}(\boldsymbol{p}) \;=\; \delta_{\boldsymbol{p}\boldsymbol{p}'}\,\delta_{ss'}\,\frac{E}{mc^2} \tag{14.8.19}$$

and

$$v^{(s')\dagger}(\boldsymbol{p}')\, v^{(s)}(\boldsymbol{p}) \;=\; \delta_{\boldsymbol{p}\boldsymbol{p}'}\,\delta_{ss'}\,\frac{E}{mc^2}\,, \tag{14.8.20}$$

---

[7]We have $u^{(s)}(\boldsymbol{p}) \;=\; \gamma_2\, v^{(s)}(\boldsymbol{p})$ Now $(i\gamma_\mu\, p_\mu + mc)\; u^{(s)}(\boldsymbol{p}) \;=\; 0$ implies $(i\gamma_\mu\, p_\mu + mc)\; \gamma_2\, v^{(s)*}(\boldsymbol{p}) \;=\; 0$. Taking complex conjugates of both sides, and recalling that $\gamma_1^* = -\gamma_1\,,\ \gamma_2^* = \gamma_2\,,\ \gamma_3^* = -\gamma_3\,,\ \gamma_4^* = \gamma_4$ and $p_1^* = p_1\,,\ p_2^* = p_2\,,\ p_3^* = p_3\,,\ p_4^* = -p_4$, and using the anti-commutation relations for the gamma matrices, we get Eq. (14.8.14) [see Sec. 12.11].

[8]This choice is not unique. One can as well choose the condition

$$u^{(s')\dagger}(\boldsymbol{p}')u^{(s)}(\boldsymbol{p}) = \delta_{ss'}\delta_{\boldsymbol{p}\boldsymbol{p}'}$$

in which case $\tilde{u}^{(s')}(\boldsymbol{p}')u^{(s)}(\boldsymbol{p}) = \frac{mc^2}{E}\delta_{ss'}\delta_{\boldsymbol{p}\boldsymbol{p}'}$.

where $E = +\sqrt{p^2 + m^2c^4}$. To see this, multiply Eq. (14.8.12) on the left by $\tilde{u}^{(s')}(\boldsymbol{p}')\gamma_\mu$ and Eq. (14.8.13) on the right by $\gamma_\mu u^{(s)}(\boldsymbol{p})$, add the two equations and simplify to get

$$\tilde{u}^{(s')}(\boldsymbol{p}')\gamma_\mu u^{(s)}(\boldsymbol{p}) = -\frac{ip_\mu}{mc}\tilde{u}^{(s')}(\boldsymbol{p}')u^{(s)}(\boldsymbol{p}) .$$

Substituting $\mu = 4$ in this equation and using Eq. (14.8.16), we obtain Eq. (14.8.19). Similarly, using the relation $u^{(s)}(\boldsymbol{p}) = \gamma_2 v^{(s)*}(\boldsymbol{p})$ in Eq. (14.8.19), we obtain Eq. (14.8.20).

14.8.2 Projection Operators

Projection operators

$$\Lambda^+ = \frac{-i\gamma_\mu p_\mu + mc}{2mc} \tag{14.8.21}$$

and

$$\Lambda^- = \frac{i\gamma_\mu p_\mu + mc}{2mc} \tag{14.8.22}$$

project out, respectively, the electron and positron states. Thus

$$\begin{aligned}
\Lambda^+ u^{(s)}(\boldsymbol{p}) &= u^{(s)}(\boldsymbol{p}) , \\
\Lambda^+ v^{(s)}(\boldsymbol{p}) &= 0 , \\
\Lambda^- v^{(s)}(\boldsymbol{p}) &= v^{(s)}(\boldsymbol{p}) , \\
\Lambda^- u^{(s)}(\boldsymbol{p}) &= 0, \qquad\qquad s = 1, 2 .
\end{aligned}$$

These relations are easily established with the help of Eqs. (14.8.14) and (14.8.15) satisfied by the free electron and positron spinors. We also note that

$$\Lambda^+ + \Lambda^- = \hat{1} . \tag{14.8.23}$$

The expansions of the field operators $\hat{\tilde{\psi}}$ and $\hat{\psi}$ now take the form [cf. Eqs. (14.7.25) and (14.7.26)]

$$\hat{\psi}(\boldsymbol{r}, t) = \hat{\psi}^{(+)}(\boldsymbol{r}, t) + \hat{\psi}^{(-)}(\boldsymbol{r}, t) \tag{14.8.24}$$

and

$$\hat{\tilde{\psi}}(\boldsymbol{r}, t) = \hat{\tilde{\psi}}^{(+)}(\boldsymbol{r}, t) + \hat{\tilde{\psi}}^{(-)}(\boldsymbol{r}, t) , \tag{14.8.25}$$

where

$$\hat{\psi}^{(+)}(\boldsymbol{r}, t) = \frac{1}{\sqrt{V}}\sum_{\boldsymbol{p}}\sum_{s=1,2}\sqrt{\frac{mc^2}{E}}\left\{\hat{b}_{\boldsymbol{p}}^{(s)}\, u^{(s)}(\boldsymbol{p}) \exp\left[i\,(\boldsymbol{p}\cdot\boldsymbol{r} - Et)/\hbar\right]\right\} , \tag{14.8.26}$$

$$\hat{\psi}^{(-)}(\boldsymbol{r}, t) = \frac{1}{\sqrt{V}}\sum_{\boldsymbol{p}}\sum_{s=1,2}\sqrt{\frac{mc^2}{E}}\left\{\hat{d}_{\boldsymbol{p}}^{(s)\dagger}\, v^{(s)}(\boldsymbol{p}) \exp\left[-i\,(\boldsymbol{p}\cdot\boldsymbol{r} - Et)/\hbar\right]\right\} , \tag{14.8.27}$$

and

$$\hat{\tilde{\psi}}^{(+)}(\boldsymbol{r}, t) = \frac{1}{\sqrt{V}}\sum_{\boldsymbol{p}}\sum_{s=1,2}\sqrt{\frac{mc^2}{E}}\left\{\hat{d}_{\boldsymbol{p}}^{(s)}\, \tilde{v}^{(s)}(\boldsymbol{p}) \exp\left[i\,(\boldsymbol{p}\cdot\boldsymbol{r} - Et)/\hbar\right]\right\} , \tag{14.8.28}$$

$$\hat{\tilde{\psi}}^{(-)}(\boldsymbol{r}, t) = \frac{1}{\sqrt{V}}\sum_{\boldsymbol{p}}\sum_{s=1,2}\sqrt{\frac{mc^2}{E}}\left\{\hat{b}_{\boldsymbol{p}}^{(s)\dagger}\, \tilde{u}^{(s)}(\boldsymbol{p}) \exp\left[-i\,(\boldsymbol{p}\cdot\boldsymbol{r} - Et)/\hbar\right]\right\} . \tag{14.8.29}$$

We note that $\hat{\psi}^{(+)}$ is linear in electron annihilation operators. It can therefore annihilate an electron. Likewise $\hat{\psi}^{(-)}$ involves $\hat{d}_{\boldsymbol{p}}^{(s)\dagger}$ and can create a positron. Similarly, $\hat{\bar{\psi}}^{(+)}$ involving $\hat{d}_{\boldsymbol{p}}^{(s)}$ and $\hat{\bar{\psi}}^{(-)}$ involving $\hat{b}_{\boldsymbol{p}}^{(s)\dagger}$ can respectively, annihilate a positron and create an electron. Equal time anti-commutation relations[9] satisfied by the field operators $\hat{\psi}$ and $\hat{\bar{\psi}}$ are

$$\left[\hat{\psi}_\rho(\boldsymbol{r},t)\,,\,\hat{\psi}_\lambda(\boldsymbol{r}',t)\right]_+ = 0 = \left[\hat{\psi}_\rho^\dagger(\boldsymbol{r},t)\,,\,\hat{\psi}_\lambda^\dagger(\boldsymbol{r}',t)\right]_+ ,$$

$$\left[\hat{\bar{\psi}}_\rho(\boldsymbol{r},t)\,,\,\hat{\bar{\psi}}_\lambda(\boldsymbol{r}',t)\right]_+ = 0\,.$$

For $\hat{\psi}(x)$ and $\hat{\psi}^\dagger(x)$ we have equal time anti-commutation relation

$$\left[\hat{\psi}_\rho(\boldsymbol{r},t)\,,\,\hat{\psi}_\lambda^\dagger(\boldsymbol{r}',t)\right]_+ = \delta_{\rho\lambda}\,\delta^3(\boldsymbol{r}-\boldsymbol{r}')\,, \tag{14.8.30}$$

which gives

$$\left[\hat{\psi}_\rho(\boldsymbol{r},t)\,,\,\hat{\bar{\psi}}_\lambda(\boldsymbol{r}',t)\right]_+ = (\gamma_4)_{\rho\lambda}\,\delta^3(\boldsymbol{r}-\boldsymbol{r}')\,. \tag{14.8.31}$$

The Hamiltonian operator $\hat{H}$ [Eq. (14.7.9)] of the electron field can now be written as

$$\hat{H} = \sum_{\boldsymbol{p}}\sum_{s=1,2} |E|\,\left\{\hat{b}_{\boldsymbol{p}}^{(s)\dagger}\,\hat{b}_{\boldsymbol{p}}^{(s)} + \hat{d}_{\boldsymbol{p}}^{(s)\dagger}\,\hat{d}_{\boldsymbol{p}}^{(s)} - 1\right\}, \tag{14.8.32}$$

where anti-commutation relation $\left[\hat{d}_{\boldsymbol{p}}^{(s)}\,,\,\hat{d}_{\boldsymbol{p}'}^{(s')\dagger}\right]_+ = \delta_{ss'}\,\delta_{\boldsymbol{p}\boldsymbol{p}'}$ has been used. Similarly, the charge operator [Eq. (14.7.10)] can be written as

$$\begin{aligned}\hat{Q} &= e\sum_{\boldsymbol{p}}\sum_{\nu=1}^{4}\hat{b}_{\boldsymbol{p}}^{(\nu)\dagger}\,\hat{b}_{\boldsymbol{p}}^{(\nu)}\\ &= e\sum_{\boldsymbol{p}}\left(\sum_{s=1,2}\hat{b}_{\boldsymbol{p}}^{(s)\dagger}\,\hat{b}_{\boldsymbol{p}}^{(s)} + \sum_{s=1,2}\hat{d}_{-\boldsymbol{p}}^{(s)}\,\hat{d}_{-\boldsymbol{p}}^{(s)\dagger}\right)\end{aligned}$$

or

$$\hat{Q} = e\sum_{\boldsymbol{p}}\sum_{s=1,2}\left\{\hat{b}_{\boldsymbol{p}}^{(s)\dagger}\,\hat{b}_{\boldsymbol{p}}^{(s)} - \hat{d}_{\boldsymbol{p}}^{(s)\dagger}\,\hat{d}_{\boldsymbol{p}}^{(s)} + 1\right\}, \tag{14.8.33}$$

where again we have used the anti-commutation relations cited above.

From the expressions for the Hamiltonian and charge operators, we see that we can interpret $\hat{d}_{\boldsymbol{p}}^{(s)\dagger} d_{\boldsymbol{p}}^{s}$ as the occupation number for a positive energy positron ($s = 1$ corresponds to spin up state and $s = 2$ to spin down state). Thus we have the number operators for the electrons and positrons with momentum $\boldsymbol{p}$ and spin defined by s:

$$\hat{b}_{\boldsymbol{p}}^{(s)\dagger}\,\hat{b}_{\boldsymbol{p}}^{(s)} = \hat{N}_{\boldsymbol{p}}^{(e^-,s)}\,, \tag{14.8.34}$$

and

$$\hat{d}_{\boldsymbol{p}}^{(s)\dagger}\,\hat{d}_{\boldsymbol{p}}^{(s)} = \hat{N}_{\boldsymbol{p}}^{(e^+,s)}\,. \tag{14.8.35}$$

The expressions for $\hat{H}$ and $\hat{Q}$ are not satisfactory in the sense that if we apply $\hat{H}$ (or $\hat{Q}$) [Eqs. (14.8.32) and (14.8.33)] to vacuum state we get $-\infty$ (and $+\infty$), respectively. However,

[9] For $t' \neq t$, the commutators or anti-commutators (depending on whether the field quanta are Bosons or Fermions) are functions of the four vector $(x - x')$ where $x \equiv (\boldsymbol{r}, ict)$. It is possible to work out the commutation (or anti-commutation) relations for $t' \neq t$ in a covariant fashion. These are useful in the completely covariant formulation of quantum electrodynamics as developed by Tomonaga, Schwinger and Feynman.

since only the differences in energy and charge are observable, we can subtract the (infinite) vacuum contribution and redefine the operators $\hat{H}$ and $\hat{Q}$ as

$$\hat{H} = \sum_{\boldsymbol{p}} \sum_{s=1,2} |E| \left[\hat{b}_{\boldsymbol{p}}^{(s)\dagger} \, \hat{b}_{\boldsymbol{p}}^{(s)} + \hat{d}_{\boldsymbol{p}}^{(s)\dagger} \, d_{\boldsymbol{p}}^{(s)} \right] \tag{14.8.36}$$

and

$$\hat{Q} = e \sum_{\boldsymbol{p}} \sum_{s=1,2} \left[\hat{b}_{\boldsymbol{p}}^{(s)\dagger} \, \hat{b}_{\boldsymbol{p}}^{(s)} - \hat{d}_{\boldsymbol{p}}^{(s)\dagger} \, d_{\boldsymbol{p}}^{(s)} \right] . \tag{14.8.37}$$

With $\hat{H}$ and $\hat{Q}$ redefined this way, their expectation value for the vacuum state is zero. For any other state, while the eigenvalue of $\hat{H}$ is positive, the eigenvalue of $\hat{Q}$ can be positive or negative. This subtraction procedure obviously amounts to redefining the charge density as

$$\hat{\rho} = e \, \hat{\psi}^{\dagger} \, \hat{\psi} \; - \; e \, \langle \Psi_0 | \, \hat{\psi}^{\dagger} \, \hat{\psi} \, | \Psi_0 \rangle \tag{14.8.38}$$

where the expectation value of the charge density for the vacuum state has been subtracted out.

14.8.3 Electron Vacuum

The vacuum state of the electron field is characterized by the eigenvalues of all number operators $\hat{N}_{\boldsymbol{p}}^{(e^+,s)}$ and $\hat{N}_{\boldsymbol{p}}^{(e^-,s)}$ being equal to zero. It can be checked that the electron field Hamiltonian given by Eq. (14.8.36) or by

$$\hat{H} = \sum_{\boldsymbol{p}} \sum_{s=1}^{2} |E| \left[\hat{N}_{\boldsymbol{p}}^{(e^-,s)} + \hat{N}_{\boldsymbol{p}}^{(e^+,s)} \right] \tag{14.8.39}$$

does not commute with the charge density operator given by Eq. (14.8.38) or with the current density operator given by

$$\hat{j}_k = ec \, \hat{\psi}^{\dagger} \, \alpha_k \hat{\psi} \; - \; ec \, \langle \Psi_0 | \, \hat{\psi}^{\dagger} \, \alpha_k \hat{\psi} \, | \Psi_0 \rangle \, . \tag{14.8.40}$$

Therefore, for the vacuum state the charge density or the charge current density at any point of space at any instant of time is indeterminate.

14.9 Interacting Fields and the Covariant Perturbation Theory

(a) Interaction Picture

Physical phenomena involving fundamental particles can be best described in terms of interacting quantum fields. For example a system of electrons and photons is described by the total Hamiltonian

$$\hat{H} = \hat{H}_0 + \hat{H}_I \tag{14.9.1}$$

where

$$\hat{H}_0 = \hat{H}_{rad} + \hat{H}_{el} \tag{14.9.2}$$

is the Hamiltonian of the free radiation field and the free electron field and

$$\hat{H}_I = \int \mathcal{H}_I d\tau = - \int i \, e \, c \hat{\bar{\psi}} \gamma_\mu \hat{\psi} \, \hat{A}_\mu d\tau \tag{14.9.3}$$

is the interaction between the two basic fields [see Appendix 14A3] and e is the charge of the electron. In the Schrödinger picture the state of the field $|\Psi^s\rangle$, at any instant t, is given by the equation of motion

$$i\hbar\frac{\partial}{\partial t}|\Psi^s(t)\rangle = (\hat{H}_0 + \hat{H}_I)|\Psi^s(t)\rangle. \tag{14.9.4}$$

From the Schrödinger picture we can go to the Heisenberg picture by the transformation $|\Psi^s\rangle \to \hat{U}\,|\Psi^s\rangle = |\Psi^H\rangle$ and $\hat{F} \to \hat{U}\hat{F}\hat{U}^\dagger = \hat{F}^H$ where $\hat{U} = \exp\left[i(\hat{H}_0 + \hat{H}_I)(t-t_0)/\hbar\right]$, t_0 being the initial time. In the Heisenberg picture the state of the field has no time dependence and the equation of motion for the field operator $\hat{F}^H$ is given by [see Sec. 3.8]

$$i\hbar\frac{\partial}{\partial t}\hat{F}^H = [\hat{F}^H, \hat{H}_0 + \hat{H}_I]. \tag{14.9.5}$$

Thus, in the Heisenberg picture, the field operators $\hat{\psi}$ (pertaining to the electron field) and $\hat{A}_\mu$ (pertaining to the radiation field) become time-dependent and satisfy the Heisenberg equations of motion

$$\frac{d\hat{\psi}}{dt} = \frac{i}{\hbar}\left[(\hat{H}_0 + \hat{H}_I), \hat{\psi}\right] \tag{14.9.6}$$

and

$$\frac{d\hat{A}_\mu}{dt} = \frac{i}{\hbar}\left[(\hat{H}_0 + \hat{H}_I), \hat{A}_\mu\right], \tag{14.9.7}$$

which are equivalent to the corresponding field equations

$$\left[\gamma_\mu\left\{\frac{\partial}{\partial x_\mu} - \frac{ie}{\hbar}\hat{A}_\mu\right\} + \frac{mc}{\hbar}\right]\hat{\psi} = 0 \tag{14.9.8}$$

and

$$\frac{\partial^2}{\partial x_\nu \partial x_\nu}\hat{A}_\mu = -\mu_0\hat{j}_\mu = -\mu_0\left(iec\hat{\bar{\psi}}\gamma_\mu\hat{\psi}\right). \tag{14.9.9}$$

An alternative, called the *Interaction Picture* or Dirac picture [see Sec. 3.9], is very useful in the formulation of quantum electrodynamics. In this picture the state $|\Psi^{(s)}(t)\rangle$ of the field, pertaining to the Schrödinger picture, is transformed to the state $|\Psi^{\text{int}}(t)\rangle$ in the interaction picture and an operator $O^{(s)}$ in the Schrödinger picture is transformed to O^{int} by the following unitary transformation:

$$|\Psi^{(s)}(t)\rangle \to \exp\left[i\hat{H}_0(t-t_0)/\hbar\right]|\Psi^{(s)}(t)\rangle = |\Psi^{\text{int}}(t)\rangle, \tag{14.9.10}$$

and

$$\hat{O}^{(s)} \to \exp\left[i\hat{H}_0(t-t_0)/\hbar\right]O^{(s)}\exp\left[-i\hat{H}_0(t-t_0)/\hbar\right] = \hat{O}^{\text{int}}(t), \tag{14.9.11}$$

where t_0 is the initial time. Equation (14.9.11) is equivalent to the operator equation

$$i\hbar\frac{d}{dt}\hat{O}^{\text{int}}(t) = \left[\hat{O}^{\text{int}}, \hat{H}_0\right]. \tag{14.9.11a}$$

Thus, an operator in the interaction picture satisfies interaction-free Heisenberg equation despite the presence of interaction. If the operator $\hat{O}$ in Eq. (14.9.11) represents the interaction $\hat{H}_I^{(s)}$, then in the interaction picture[10]

$$\hat{H}_I^{(s)} \to \exp\left[i\hat{H}_0(t-t_0)/\hbar\right]\hat{H}_I^{(s)}\exp\left[-i\hat{H}_0(t-t_0)/\hbar\right] \equiv \hat{H}_I^{(\text{int})}(t). \tag{14.9.12}$$

[10]In $\hat{H}_I^{\text{int}}$, I as subscript implies interaction part of the Hamiltonian while 'int' as superscript implies *in the interaction picture.*

Obviously, the unperturbed Hamiltonian in the interaction picture is the same as that in the Schrödinger picture. Equation (14.9.10) yields the following equation of motion for the state $|\Psi^{\rm int}(t)\rangle$ in the interaction picture

$$i\hbar\frac{d}{dt}|\Psi^{\rm int}(t)\rangle = \hat{H}_I^{\rm int}|\Psi^{\rm int}(t)\rangle\,, \tag{14.9.13}$$

where $\hat{H}_I^{\rm int}(t)$ is defined by Eq. (14.9.12).

Thus we see that while the state of the field is time-dependent in the Schrödinger picture and time-independent in the Heisenberg picture, in the interaction picture, the time dependence is split into two parts. A part of the time dependence, due to the interaction, is taken up by the state vector $|\Psi^{\rm int}(t)\rangle$, while $\hat{H}_I^{\rm int}(t)$ retains the residual time dependence of the interaction [Eq. (14.9.12)]. To see this more formally, we note that the Heisenberg state

$$|\Psi^H\rangle = \exp[i(\hat{H}_0 + \hat{H}_i)(t - t_0)/\hbar]|\Psi^S(t)\rangle$$

is independent of time. Hence

$$|\Psi^{\rm int}(t)\rangle \; = \exp(i\hat{H}_0(t - t_0)/\hbar)|\Psi^S\rangle \; = \exp\left[-i\hat{H}_I(t - t_0)/\hbar\right]|\Psi^H\rangle$$

has time dependence imparted to it by the interaction. The field operators occurring in $H_I^{\rm int}$ are time-dependent free-field operators given, not by Eqs. (14.9.8) and (14.9.9), but by [see Eq. (14.9.11a)]

$$\frac{d\hat{\psi}}{dt} = \frac{i}{\hbar}\left[\hat{H}_0, \hat{\psi}\right]$$

and

$$\frac{d\hat{A}_\mu}{dt} = \frac{i}{\hbar}[\hat{H}_0, \hat{A}_\mu]\,,$$

which are equivalent to the corresponding free-field equations

$$\left(\gamma_\mu\frac{\partial}{\partial x_\mu} + \frac{mc}{\hbar}\right)\hat{\psi} = 0\,,$$

and

$$\frac{\partial^2}{\partial x_\nu \partial x_\nu}\hat{A}_\mu = 0\,.$$

The Schrödinger-like equation in the interaction picture [Eq. (14.9.13)] is better suited for the description of interacting fields. By transforming away $\hat{H}_0$, the interaction picture focuses on the time dependence induced by the interaction Hamiltonian, which causes transitions between the eigenstates (field states) of the unperturbed field Hamiltonian.

14.9.1 *U* Matrix

Let the solution of the equation of motion in the interaction picture [Eq. (14.9.13)] be formally given by

$$|\Psi^{\rm int}(t)\rangle \; = \hat{U}(t, t_0)|\Psi^{\rm int}(t_0)\rangle \tag{14.9.14}$$

where $|\Psi^{\rm int}(t_0)\rangle$ is the state vector in the interaction picture characterizing the state of the field at some fixed time. Operator $\hat{U}$ must satisfy the intial condition $\hat{U}(t_0, t_0) = \hat{1}$. Equation of motion (14.9.14) in the interaction picture is equivalent to the operator differential equation

$$i\hbar\frac{d}{dt}\hat{U}(t, t_0) = \hat{H}^{\rm int}(t)\hat{U}(t, t_0) \tag{14.9.15}$$

or the integral equation

$$\hat{U}(t,t_0) = \hat{1} - \frac{i}{\hbar}\int_{t_0}^{t} \hat{H}_I^{\text{int}}(t_1,t_0)\hat{U}(t_1,t_0)dt_1 \,.$$

The integral equation may be iterated a number of times to obtain a perturbation expansion

$$\begin{aligned}
\hat{U}(t,t_0) &= \hat{1} - \frac{i}{\hbar}\int_{t_0}^{t} H_I^{\text{int}}(t_1)dt_1 \left\{ 1 - \frac{i}{\hbar}\int_{t_0}^{t_1} \hat{H}_I^{\text{int}}(t_2)\hat{U}(t_2,t_0)dt_2 \right\}, \\
&= \hat{1} + \left(-\frac{i}{\hbar}\right)\int_{t_0}^{t} dt_1 \hat{H}_I^{\text{int}}(t_1) + \left(-\frac{i}{\hbar}\right)^2 \int_{t_0}^{T} dt_1 \hat{H}_I^{\text{int}}(t_1)\int_{t_0}^{t_1} dt_2 \hat{H}_I^{\text{int}}(t_2) + \cdots \\
&\quad + \left(-\frac{i}{\hbar}\right)^n \int_{t_0}^{t} dt_1 \int_{t_0}^{t_1} dt_2 \cdots \int_{t_0}^{t_{n-1}} dt_n \left[\hat{H}_I^{\text{int}}(t_1)\hat{H}_I^{\text{int}}(t_2)\cdots \hat{H}_I^{\text{int}}(t_n)\right] + \cdots \\
&= \hat{U}_0(t,t_0) + \hat{U}_1(t,t_0) + \cdots + \hat{U}_n(t,t_0) + \cdots
\end{aligned} \tag{14.9.16}$$

where

$$\hat{U}_n(t,t_0) = \left(-\frac{i}{\hbar}\right)^n \int_{t_0}^{t} dt_1 \int_{t_0}^{t_1} dt_2. \cdots \int_{t_0}^{t_{n-1}} dt_n \left[\hat{H}_I^{\text{int}}(t_1)\hat{H}_I^{\text{int}}(t_2)\cdots \hat{H}_I^{\text{int}}(t_n)\right] \tag{14.9.17}$$

with $t \geq t_1 \geq t_2 \geq \cdots \geq t_{n-1} \geq t_n \geq t_0$. Operator $\hat{U}_n(t,t_0)$ may also be written as

$$\hat{U}_n(t,t_0) = \left(-\frac{i}{\hbar}\right)^n \frac{1}{n!}\int_{t_0}^{t} dt_1 \int_{t_0}^{t} dt_2 \cdots \int_{t_0}^{t} dt_n \boldsymbol{P}\left[\hat{H}_I^{\text{int}}(t_1)\hat{H}_I^{\text{int}}(t_2)\cdots \hat{H}_I^{\text{int}}(t_n)\right] \tag{14.9.18}$$

where $\boldsymbol{P}$ is Dyson's chronological [also *time ordering*] operator which rearranges the product of operators $\hat{H}_I^{\text{int}}$'s in such a way that operators $\hat{H}_i^{\text{int}}$ involving later times stand to the left of those involving earlier times. The factor $\frac{1}{n!}$ arises because $\hat{U}_n$ is completely symmetric with respect to $t_1, t_2, \cdots, t_n$ and so there will be $n!$ ways in which to make this ordering. As an illustration, consider the second term in the expansion of $\hat{U}(t,t_0)$

$$\hat{U}_2(t,t_0) = \left(\frac{-i}{\hbar}\right)^2 \int_{t_0}^{t} \hat{H}_I^{\text{int}}(t_1)dt_1 \int_{t_0}^{t_1} \hat{H}_I^{\text{int}}(t_2)dt_2 \,,$$

which can also be written as

$$\hat{U}_2(t,t_0) = \left(\frac{-i}{\hbar}\right)^2 \int_{t_0}^{t} \hat{H}_I^{\text{int}}(t_2)dt_2 \int_{t_0}^{t_2} \hat{H}_I^{\text{int}}(t_1)dt_1$$

where t_1 and t_2 are merely interchanged. Hence $\hat{U}_2(t,t_0)$ may also be written as

$$\hat{U}_2(t,t_0) = \left(\frac{-i}{\hbar}\right)^2 \frac{1}{2!}\int_{t_0}^{t} dt_1 \int_{t_0}^{t} dt_2 \boldsymbol{P}\left[\hat{H}_I^{\text{int}}(t_1)\hat{H}_I^{\text{int}}(t_2)\right],$$

where

$$\boldsymbol{P}\{\hat{H}_I^{\text{int}}(t_1)\hat{H}_i^{\text{int}}(t_2)\} = \hat{H}_I^{\text{int}}(t_1)\hat{H}_I^{\text{int}}(t_2), \quad \text{if } t_1 > t_2$$
$$= \hat{H}_I^{\text{int}}(t_2)\hat{H}_I^{\text{int}}(t_1), \quad \text{if } t_2 > t_1.$$

This may be generalized to any value of n and hence to any order of expansion of $\hat{U}(t, t_0)$.

The physical meaning of the operator $\hat{U}(t, t_0)$ is easy to extract. Let $|\Psi_i\rangle = |\Psi^{\text{int}}(t_0)\rangle$ be the state of the system at time t_0 and $|\Psi^{\text{int}}(t)\rangle$ be the state at time t. Then the probability $P_f(t)$ of finding the system in the final state $|\Psi_f\rangle$ at time t is given by

$$\begin{aligned} P_f(t) &= |\langle\Psi_f|\Psi^{\text{int}}(t)\rangle|^2 = |\langle\Psi_f|\hat{U}(t,t_0)|\Psi^{\text{int}}(t_0)\rangle|^2 \\ &= |\langle\Psi_f|\hat{U}(t,t_0)|\Psi_i\rangle|^2 \\ &= |U_{fi}(t,t_0)|^2 \end{aligned} \tag{14.9.19}$$

where $U_{fi}(t, t_0)$ is the matrix element of the operator $\hat{U}(t, t_0)$ between the initial and final unperturbed states of the system. The interaction thus causes a transition between the eigenstates of the unperturbed field Hamiltonian. The field states $|\Psi_i\rangle$ and $|\Psi_f\rangle$ represent the eigenstates of the unperturbed field Hamiltonian in the interaction picture. Even if they represented the states of the unperturbed Hamiltonian in the Schrödinger picture, U_{fi} would contain a phase term and $P_f(t)$ would be unaltered.

14.9.2 S Matrix and Iterative Expansion of S Operator

The S operator is defined by

$$\hat{S} = \hat{U}(t = \infty, t_0 = -\infty)$$

which implies that

$$S_{fi} = U_{fi}(t = \infty, t_0 = -\infty)\,. \tag{14.9.20}$$

Hence $|S_{fi}|^2$ may be regarded as the probability of finding the system finally (at $t = \infty$), in the state $|\Psi_f\rangle$ as a result of the interaction, if it was initially (at $t = -\infty$) in the state $|\Psi_i\rangle$.

From the iterative expansion of the U operator we can write down the iterative expansion of the S operator as

$$\hat{S} = \hat{S}^{(0)} + \hat{S}^{(1)} + \hat{S}^{(2)} + \cdots + \hat{S}^{(n)} \cdots\,, \tag{14.9.21}$$

where

$$\hat{S}^{(n)} = \left(-\frac{i}{\hbar}\right)^n \frac{1}{n!} \int_{-\infty}^{\infty} dt_1 \int_{-\infty}^{\infty} dt_2 \cdots \int_{-\infty}^{\infty} dt_n \boldsymbol{P}\left[\hat{H}_I^{\text{int}}(t_1)\hat{H}_I^{\text{int}}(t_2)\cdots \hat{H}_I^{\text{int}}(t_n)\right]\,, \tag{14.9.22}$$

and $\boldsymbol{P}$ is Dyson time-ordering operator. By expressing the interaction Hamiltonian $\hat{H}_I^{\text{int}}(t)$ as $\int \hat{\mathcal{H}}_I\, d\tau$, where $\hat{\mathcal{H}}_I$ is the interaction Hamiltonian density, the S operator can be written in terms of covariant integrals as

$$\hat{S}^{(n)} = \left(-\frac{1}{\hbar}\right)^n \frac{1}{n!}\frac{1}{c^n} \int d^4x_1 \int d^4x_2 \cdots \int d^4x_n \boldsymbol{P}\left[\hat{\mathcal{H}}_I(x_1)\hat{\mathcal{H}}_I(x_2)\cdots\hat{\mathcal{H}}_I(x_n)\right] \tag{14.9.23}$$

where $x_n = (\boldsymbol{r}_n, ict_n)$ and $d^4x = d^3r\,c\,dt$ and $\boldsymbol{P}$ is Dyson's time ordering operator, which reshuffles $\hat{\mathcal{H}}_I$'s in such a way that $\hat{\mathcal{H}}_I$'s involving later times stand to the left of those involving earlier times.

For an operator product $\left[\hat{\phi}(x_1)\hat{\phi}(x_2)\hat{\phi}(x_3)\cdots\right]$ involving Boson fields $\hat{\phi}(x)$, we have

$$\begin{aligned}\boldsymbol{P}\left[\hat{\phi}(x_1)\hat{\phi}(x_2)\hat{\phi}(x_3)\right] &= \hat{\phi}(x_1)\hat{\phi}(x_2)\hat{\phi}(x_3), \quad \text{if} \quad t_1 > t_2 > t_3 \\ &= \hat{\phi}(x_2)\hat{\phi}(x_1)\hat{\phi}(x_3), \quad \text{if} \quad t_2 > t_1 > t_3 \\ &= \hat{\phi}(x_3)\hat{\phi}(x_2)\hat{\phi}(x_1), \quad \text{if} \quad t_3 > t_2 > t_1 .\end{aligned}$$

For an operator product $\{\hat{\psi}(x_1)\hat{\psi}(x_2)\hat{\psi}(x_3)\}$ involving Fermi fields $\hat{\psi}(x)$, the ordering of the field operators may involve a change of sign if an odd number of exchanges of the field operators are required to go from initial order to final order. In this case we introduce *Wick's time ordering* operator $\boldsymbol{T}$. Thus for an operator product involving three Fermi fields we have

$$\begin{aligned}\boldsymbol{T}\left[\hat{\psi}(x_1)\hat{\psi}(x_2)\hat{\psi}(x_3)\right] &= \hat{\psi}(x_1)\hat{\psi}(x_2)\hat{\psi}(x_3)\,, & \text{if} \quad t_1 > t_2 > t_3 \\ &= -\hat{\psi}(x_2)\hat{\psi}(x_1)\hat{\psi}(x_3)\,, & \text{if} \quad t_2 > t_1 > t_3 \\ &= \hat{\psi}(x_3)\hat{\psi}(x_1)\hat{\psi}(x_2)\,, & \text{if} \quad t_3 > t_1 > t_2 .\end{aligned}$$

For the interaction of the electron-photon fields, the interaction Hamiltonian density, $\hat{\mathcal{H}}_I = -iec\hat{\bar{\psi}}\gamma_\mu\hat{\psi}\,\hat{A}_\mu$ is bilinear in Fermi fields. Hence the $\boldsymbol{P}$ product and $\boldsymbol{T}$ product are identical in this case as only pairs of Fermi fields are involved.

14.9.3 Decomposition of Time-ordered Operator Product in Terms of Normal Constituents

In order to calculate the transition amplitude from the initial state $|i\rangle$ to the final state $|f\rangle$ in, say, the n-th order of perturbation theory we need to determine the matrix element $\langle f|\,\hat{S}_n\,|i\rangle$. This involves the evaluation of the matrix element of a time-ordered product $\boldsymbol{T}\left[\hat{\mathcal{H}}_I(x_1)\hat{\mathcal{H}}_I(x_2)\cdots\hat{\mathcal{H}}_I(x_n)\right]$ of interaction Hamiltonians. Of the many terms in the product, only those whose application on the intial state $|i\rangle$ leads to final state $|f\rangle$ would contribute. For example, if the initial state contains an electron with momentum and spin $(\boldsymbol{p}, s)$ and a photon with momentum vector and polarization $(\boldsymbol{k}, \boldsymbol{\varepsilon_k})$ and the final state contains an electron $(\boldsymbol{p}', s')$ and a photon $(\boldsymbol{k}', \boldsymbol{\varepsilon_{k'}})$, then the nonzero contribution to the transition amplitude comes from those term of $\hat{S}^{(n)}$ that contain those annihilation operators that annihilate the particles in state $|i\rangle$ and those creation operators that subsequently create the particles leading to the final state $|f\rangle$. In addition, a general term may contain creation and annihilation operators responsible for the creation and subsequent annihilation of *virtual particles*, which are not present in the initial and final states. The contribution of such terms can be calculated by using the operator commutation (or anti-commutation) relations to move the annihilation operators to the right so that all annihilation operators are to the left of creation operators. Such an ordering of operators in which the annihilation operators are all on the right and creation operators are on the left is called *normal ordering*. It is clear that using operator commutation relations, we can write any operator product as a sum of normal products. Each term in the sum is referred to as a *normal constituent* (of the given operator product). Instead of going through tedious algebra of using commutation and anti-commutation rules each time we calculate a particular term, we can use simple rules afforded by *Wick's decomposition theorem* to express an arbitrary time-ordered product as a sum of normal constituents. To formulate Wick's theorem we define the contraction $\langle\hat{A}\hat{B}\rangle_0 \equiv \langle\Psi_0|\,\hat{A}\hat{B}\,|\Psi_0\rangle$ of two operators $\hat{A}$ and $\hat{B}$ by

$$\boldsymbol{T}(\hat{A}\hat{B}) \equiv \langle\hat{A}\hat{B}\rangle_0 + :\hat{A}\hat{B}: = \langle\Psi_0|\,\hat{A}\hat{B}\,|\Psi_0\rangle + :\hat{A}\hat{B}:\,, \tag{14.9.24}$$

where $|\Psi_0\rangle$ is the vacuum state and colons enclosing the operator product imply normal ordering of the operator product. A normal ordered product is either denoted by $\hat{N}(\cdots)$ or enclosed between two colons. Note that the vacuum expectation value of a normal ordered product vanishes. The operation of normal ordering of an operator product is distributive

$$: \hat{A}\hat{B} + \hat{C}\hat{D} :=: \hat{A}\hat{B} : + : \hat{C}\hat{D} : . \tag{14.9.25}$$

According to Wick's theorem an operator product $\boldsymbol{T}(\hat{A}\hat{B}\hat{C}\hat{D}\cdots\hat{P}\hat{Q})$ can be written as a sum of normal constituents as follows. We choose an even number (0,2,4,$\cdots$) of operator factors from the given operator product and contract them in pairs. Thus for the product $\hat{A}\hat{B}$ there are only two *factor pairings*, either we choose the pair $\hat{A}\hat{B}$ or none at all. To each factor pairing corresponds a normal product, which contains the contractions of paired operators and unpaired operators arranged in normal order. Then according to Wick's theorem

$$\begin{aligned} \hat{T}(\hat{A}\hat{B}\hat{C}\hat{D}\cdots\hat{P}\hat{Q}) &= : \hat{A}\hat{B}\hat{C}\hat{D}\cdots\hat{P}\hat{Q} : \\ &\quad + \delta_P\langle AB\rangle_0 : \hat{C}\hat{D}\cdots\hat{P}\hat{Q} : + \delta_P\langle BC\rangle_0 : \hat{A}\hat{D}\cdots\hat{P}\hat{Q} : \\ &\quad + \delta_P\langle AC\rangle_0 : \hat{B}\hat{D}\cdots\hat{P}\hat{Q} : + \cdots \\ &\quad + \delta_P\langle\hat{A}\hat{B}\rangle_0\langle\hat{C}\hat{D}\rangle_0 : \hat{E}\hat{F}\cdots\hat{P}\hat{Q} : + \cdots , \end{aligned} \tag{14.9.26}$$

where the sum on the right-hand side includes all possible sets of contractions. The factor δ_P is $+1$ or -1 depending on whether even or odd numbers of exchanges of Fermion operators are needed to rearrange the given product into normal product. As an example, the decomposition of the chronological product of four operators $\boldsymbol{T}(\hat{A}\hat{B}\hat{C}\hat{D})$ into normal constituents is

$$\begin{aligned} \boldsymbol{T}(\hat{A}\hat{B}\hat{C}\hat{D}) &= : \hat{A}\hat{B}\hat{C}\hat{D} : \\ &\quad + \delta_P\langle AB\rangle_0 : \hat{C}\hat{D} : + \delta_P\langle BC\rangle_0 : \hat{A}\hat{D} : + \delta_P\langle AC\rangle_0 : \hat{B}\hat{D} : \\ &\quad + \delta_P\langle CD\rangle_0 : \hat{A}\hat{B} : + \delta_P\langle AD\rangle_0 : \hat{B}\hat{C} : + \langle AB\rangle_0\langle\hat{C}\hat{D}\rangle_0 \\ &\quad + \langle BC\rangle_0\langle AD\rangle_0 + \langle BD\rangle_0\langle AC\rangle_0 . \end{aligned}$$

The operators appearing in the factor pairing have the same relative order as in the original product.

The decomposition theorem may be proved by induction on m, the number of factors in the operator product. The theorem is (obviously) true for $m = 1$ and for $m = 2$ we have, rewriting Eq. (14.9.24),

$$\hat{T}(\hat{A}(x_1)\hat{B}(x_2)) = \langle\Psi_0|\hat{A}(x_1)\hat{B}(x_2)|\Psi_0\rangle + : [\hat{A}(x_1)\hat{B}(x_2)] : \tag{14.9.27}$$

irrespective of whether $\hat{A}$ and $\hat{B}$ are four-potential or Dirac field operators or one four-potential and one Dirac field operator. It can be shown that if the theorem holds for a product of $m-2$ operators, then it holds for a product of m operators as well.[11] It follows that the decomposition theorem holds for all integer values of m.

It may also be recalled that while in Wick's chronological product (WCP), the ordering of the operators is such that the earliest operators are on the right so that they are the first to operate on the state vector of the field, in a normal product the operators are arranged in such a way that the annihilation operators, removing particles in the initial (earliest)

[11] A detailed proof of Wick's decomposition theorem may be found in advanced texts on quantum field theory.

state are all on the right and creation operators (which are latest in time) are all on the left. So time ordering is implicit in a normal product.

For electromagnetic interactions, the normal ordered operator product involves field operators like $\hat{\psi}, \hat{\bar{\psi}}, \hat{A}_\mu$. So each normal constituent : $\hat{C}\hat{D}\cdots\hat{P}\hat{Q}$: may be further split as a sum of many terms by the substitutions

$$\hat{\bar{\psi}} = \hat{\bar{\psi}}^{(+)} + \hat{\bar{\psi}}^{(-)}, \tag{14.9.28}$$

$$\hat{\psi} = \hat{\psi}^{(+)} + \hat{\psi}^{(-)}, \tag{14.9.29}$$

and

$$\hat{A}_\mu = \hat{A}_\mu^{(+)} + \hat{A}_\mu^{(-)}. \tag{14.9.30}$$

The operators on the right-hand side of these equations are either creation or annihilation operators [Sec. 14.8, Eqs. (14.8.24), (14.8.25) and Sec. 13.5, Eq. (13.5.16)].[12] Thus each term of normal constituent contains products of creation and annihilation operators and normal ordering arranges them so that the annihilation operators (which remove particles from the initial state) stand to the right of creation operators (which act subsequently to add new particles to the state). Once this rearrangement is brought about only one term in the normal constituent contributes to the transition amplitude $\langle f|\,\hat{O}\,|i\rangle$ between the specified initial and final states. This term has the right combination and the right order of creation and annihilation operators to remove particles from the initial state $|i\rangle$ and add the required particles to it so that it has a complete overlap with the final state $|f\rangle$.

The decomposition of $S^{(n)}$ into normal constituents and the decomposition of each normal constituent into normal ordered terms thus amounts to cataloguing various pairs of the initial and final states of the field which can have a non-zero matrix element for the interaction. In other words it amounts to listing various processes which can be brought about by the interaction in a given order of perturbation. Of course, the same process may be brought about in different orders of expansion of $\hat{S}$.

14.10 Second Order Processes in Electrodynamics

We shall now restrict to processes in electrodynamics brought about by the second order term $S^{(2)}$ in the expansion of the S operator. The zero order term $\hat{1}$ in the expansion of the $\hat{S}$ operator has a non-zero matrix element only between identical initial and final states. The first order term in the expansion of the $\hat{S}$ operator also cannot give rise to any physical

[12] In Eq. (13.5.16) we replace $\sum_\lambda$ by summation over all possible wave vectors $\boldsymbol{k}$ and polarization states $\boldsymbol{\varepsilon}_{\boldsymbol{k}n}$ $(n = 1, 2)$. In covariant formulation, where $A_\mu \equiv (\boldsymbol{A},\, i\phi/c)$, we write $A_\mu(x) = A_\mu^{(+)}(x) + A_\mu^{(-)}(x)$, where

$$A_\mu^{(+)} \equiv \sum_{\boldsymbol{k},n} \sqrt{\frac{\hbar}{2\epsilon_0\omega_k V}}\, \boldsymbol{\varepsilon}_\mu^{(n)}\, \hat{a}_{\boldsymbol{k}n}\, \exp(ikx)\,,$$

$$A_\mu^{(-)} \equiv \sum_{\boldsymbol{k},n} \sqrt{\frac{\hbar}{2\epsilon_0\omega V}}\, \boldsymbol{\varepsilon}_\mu^{(n)}\, \hat{a}_{\boldsymbol{k}n}^\dagger\, \exp(-ikx),$$

for $\mu = 1, 2, 3$. The four-vectors k_μ and x_μ are defined by $k_\mu \equiv (\boldsymbol{k},\, i\omega/c)$ with $\omega = c|\boldsymbol{k}|$ and $x_\mu \equiv (\boldsymbol{r},\, ict)$. For the radiation field we assume that $\boldsymbol{\nabla}\cdot\boldsymbol{A} = 0$. This can be taken care of by taking the fourth component of the polarization vector to be zero: $\varepsilon_\mu = (\boldsymbol{\varepsilon}_{\boldsymbol{k}n}, 0)$. So these equations hold for $\mu = 1, 2, 3, 4$.

process. This is because

$$\hat{S}^{(1)} = \frac{ie}{\hbar}\int \hat{T}\left(\hat{\bar{\psi}}(x)\gamma_\mu \hat{A}_\mu(x)\,\hat{\psi}(x)\right) d^4x = S_a^{(1)} + S_b^{(1)} + S_c^{(1)} + S_d^{(1)}$$

where

$$\begin{aligned}
\hat{S}_a^{(1)} &= \frac{ie}{\hbar}\int \hat{T}\left(\hat{\bar{\psi}}^{(+)}(x)\gamma_\mu \hat{A}_\mu(x)\hat{\psi}^{(+)}(x)\right) d^4x \\
\hat{S}_b^{(1)} &= \frac{ie}{\hbar}\int \hat{T}\left(\hat{\bar{\psi}}^{(+)}(x)\,\gamma_\mu \hat{A}_\mu(x)\hat{\psi}^{(-)}(x)\right) d^4x \\
\hat{S}_c^{(1)} &= \frac{ie}{\hbar}\int \hat{T}\left(\hat{\bar{\psi}}^{(-)}(x)\gamma_\mu\, \hat{A}_\mu(x)\hat{\psi}^{(+)}(x)\right) d^4x \\
\hat{S}_d^{(1)} &= \frac{ie}{\hbar}\int \hat{T}\left(\hat{\bar{\psi}}^{(-)}(x)\gamma_\mu\, \hat{A}_\mu(x)\hat{\psi}^{(-)}(x)\right) d^4x\,.
\end{aligned}$$

These operators correspond to processes that are unphysical. The operators $\hat{S}_a^{(1)}$ and $\hat{S}_d^{(1)}$ have non-zero matrix elements between initial and final field states which differ in an electron-positron (e^-e^+) pair; they correspond to pair annihilation and pair creation, respectively. The operators $\hat{S}_b^{(1)}$ and $\hat{S}_c^{(1)}$ correspond, respectively, to electron and positron scattering. These processes are not permitted on account of violation of energy and momentum conservation.

The lowest order term in the S matrix expansion, which corresponds to physical processes, is of second order. Higher order terms in the expansion of S matrix may contribute but will not be considered. With $\hat{\mathcal{H}}_I = -iec\hat{\bar{\psi}}\gamma_\mu\hat{\psi}\hat{A}_\mu$ the second order term $\hat{S}^{(2)}$ is given by

$$\hat{S}^{(2)} = \frac{e^2}{\hbar^2}\frac{1}{2!}\int d^4x_2 \int d^4x_1 \boldsymbol{T}\left[\left\{\hat{\bar{\psi}}(x_2)\gamma_\mu \hat{A}_\mu(x_2)\psi(x_2)\right\}\left\{\hat{\bar{\psi}}(x_1)\gamma_\nu \hat{A}_\nu(x_1)\hat{\psi}(x_1)\right\}\right]. \tag{14.10.1}$$

Following Wick's theorem (14.9.26) for decomposing a time-ordered product into its normal constituents we have

$$\hat{G} = \mathbf{T}\left\{\hat{\bar{\psi}}(x_2)\gamma_\mu \hat{A}_\mu(x_2)\hat{\psi}(x_2)\hat{\bar{\psi}}(x_1)\gamma_\nu \hat{A}_\nu(x_1)\hat{\psi}(x_1)\right\} = \sum_{i=1}^{8}\hat{G}_i,$$

so that

$$\hat{S}^{(2)} = \frac{e^2}{2\hbar^2}\iint d^4x_2 d^4x_1 \sum_{i=1}^{8}\hat{G}_i \equiv \sum_{i=1}^{8}\hat{S}_i^{(2)}. \tag{14.10.2}$$

The number of normal constituents of $\hat{S}^{(2)}$ (or $\hat{G}$) is only eight despite the presence of as many as six operators in the time-ordered product. This is because many factor pairs like $\langle\Psi_0|\hat{\psi}(x_2)\hat{\psi}(x_1)|\Psi_0\rangle$ or $\langle\Psi_0|\hat{\bar{\psi}}(x_2)\hat{\bar{\psi}}(x_1)|\Psi_0\rangle$ are zero. The only non-zero $\hat{G}_i$'s will be those in which the factor pairs contain either a $\hat{\psi}$ and $\hat{\bar{\psi}}$ pair of operators or two $\hat{A}_\mu$ operators. Following the rules for the decomposition of a time-ordered product[13]

$$\boldsymbol{T}\left[\left\{\hat{\bar{\psi}}(x_2)\gamma_\mu \hat{A}_\mu(x_2)\hat{\psi}(x_2)\right\}\left\{\hat{\bar{\psi}}(x_1)\gamma_\nu \hat{A}_\nu(x_1)\hat{\psi}(x_1)\right\}\right]$$

or

$$\boldsymbol{T}\left[\left\{\hat{\bar{\psi}}(x_1)\gamma_\nu \hat{A}_\nu(x_1)\hat{\psi}(x_1)\right\}\left\{\hat{\bar{\psi}}(x_2)\gamma_\mu \hat{A}_\mu(x_2)\hat{\psi}(x_2)\right\}\right]$$

[13]Since we have to integrate over x_1 and x_2 eventually, in $\boldsymbol{T}[\mathcal{H}_I(x_2)\mathcal{H}_I(x_1)]$, the order of factors may be interchanged. Time ordering is taken care of by the operator $\boldsymbol{T}$.

into normal constituents we have

$$\hat{G}_1 = \delta_P : \left[(\hat{\bar{\psi}}(x_2)\gamma_\mu \hat{A}_\mu(x_2)\hat{\psi}(x_2))(\hat{\bar{\psi}}(x_1)\gamma_\nu \hat{A}_\nu(x_1)\hat{\psi}(x_1))\right] : \tag{14.10.3}$$

$$\hat{G}_2 = \delta_P \langle \Psi_0 | \hat{\psi}(x_2)\hat{\bar{\psi}}(x_1) | \Psi_0 \rangle : \left[\hat{\bar{\psi}}(x_2)\gamma_\mu \hat{A}_\mu(x_2)\gamma_\nu \hat{A}_\nu(x_1)\hat{\psi}(x_1)\right] : \tag{14.10.4}$$

$$\hat{G}_3 = \delta_P \langle \Psi_0 | \hat{\psi}(x_1)\hat{\bar{\psi}}(x_2) | \Psi_0 \rangle : \left[\hat{\bar{\psi}}(x_1)\gamma_\nu \hat{A}_\nu(x_1)\gamma_\mu \hat{A}_\mu(x_2)\hat{\psi}(x_2)\right] : \tag{14.10.5}$$

$$\hat{G}_4 = \delta_P \langle \Psi_0 | \gamma_\mu \hat{A}_\mu(x_2)\gamma_\nu \hat{A}_\nu(x_1) | \Psi_0 \rangle : \left[\hat{\bar{\psi}}(x_2)\hat{\psi}(x_2)\hat{\bar{\psi}}(x_1)\hat{\psi}(x_1)\right] : \tag{14.10.6}$$

$$\hat{G}_5 = \delta_P \langle \Psi_0 | \hat{\psi}(x_2)\hat{\bar{\psi}}(x_1) | \Psi_0 \rangle \langle \Psi_0 | \gamma_\mu \hat{A}_\mu(x_2)\gamma_\nu \hat{A}_\nu(x_1) | \Psi_0 \rangle : \left[\hat{\bar{\psi}}(x_2)\hat{\psi}(x_1)\right] : \tag{14.10.7}$$

$$\hat{G}_6 = \delta_P \langle \Psi_0 | \hat{\psi}(x_1)\hat{\bar{\psi}}(x_2) | \Psi_0 \rangle \langle \Psi_0 | \gamma_\nu \hat{A}_\nu(x_1)\gamma_\mu \hat{A}_\mu(x_2) | \Psi_0 \rangle : \left[\hat{\bar{\psi}}(x_1)\hat{\psi}(x_2)\right] : \tag{14.10.8}$$

$$\hat{G}_7 = \delta_P \langle \Psi_0 | \hat{\psi}(x_1)\hat{\bar{\psi}}(x_2) | \Psi_0 \rangle \langle \Psi_0 | \hat{\psi}(x_2)\hat{\bar{\psi}}(x_1) | \Psi_0 \rangle : \left[\gamma_\mu \hat{A}_\mu(x_2)\gamma_\nu \hat{A}_\nu(x_1)\right] : \tag{14.10.9}$$

$$\hat{G}_8 = \langle \Psi_0 | \hat{\psi}(x_1)\hat{\bar{\psi}}(x_2) | \Psi_0 \rangle \langle \Psi_0 | \hat{\psi}(x_2)\hat{\bar{\psi}}(x_1) | \Psi_0 \rangle \langle \Psi_0 | \gamma_\mu \hat{A}_\mu(x_2)\gamma_\nu \hat{A}_\nu(x_1) | \Psi_0 \rangle . \tag{14.10.10}$$

The terms $\hat{G}_2$ and $\hat{G}_3$ are equivalent because, with the interchange of x_1 and x_2 they become identical. If we add the two, we can do away with the factor $\frac{1}{2!}$ in $\hat{S}_2^{(2)}$. Similarly the normal constituents involving $\hat{G}_5$ and $\hat{G}_6$ are equivalent.

Consider now the processes initiated by $\hat{G}_2$ (or $\hat{G}_3$) and $\hat{G}_4$. Writing the operators $\hat{\bar{\psi}}$, $\hat{\psi}$ and $\hat{\Phi}_\mu$ in terms of creation and annihilation operators according to Eqs. (14.9.28) through (14.9.30), and substituting them into the expressions for $\hat{G}_2$ and $\hat{G}_4$, we can write each normal constituent in terms of components, each of which can be associated with a specific physical process.

14.10.1 Feynman Diagrams

Feynman introduced a diagramatic representation of the physical processes corresponding to different components of a normal constituent. These diagrams are known as Feynman diagrams. They may be regarded as pictures of actual processes occurring in space-time. Each Feynman diagram can be correlated with a physical process (or with a specific component) of a normal constituent according to the following rules [see Fig. 14.1]:

(a) $\hat{\psi}^{(+)}(x)$, which represents an electron (e^-) *in* (destroyed) at space-time point x (see Sec. 14.8) is denoted by a straight line segment terminating at a vertex x [Fig. 14.1(a)].

(b) $\hat{\psi}^{(-)}(x)$, which represents a positron (e^+) *out* (created) at space-time point x is denoted by a straight line segment terminating at a vertex x [Fig. 14.1(b)].

(c) $\hat{\bar{\psi}}^{(+)}(x)$, which represents a positron (e^+) *in* (or destroyed) at point x is denoted by a straight line segment emanating from a vertex x [Fig. 14.1(c)].

(d) $\hat{\bar{\psi}}^{(-)}(x)$, which represents an electron (e^-) *out* (created) at point x is denoted by a straight line segment emanating at a vertex x [Fig. 14.1(d)].

(e) $\hat{A}_\mu^{(-)}(x)$ represents a photon created at x and $\hat{A}_\mu^{(+)}(x)$ represents a photon destroyed at x. These are denoted by dashed lines emanating from (or terminating at) x [Fig. 14.1(e)].

Further, the convention is adopted that time increases upwards in the diagram. This implies that one can look upon a positron as an *electron moving backward* in time.

(f) A *factor pair* like $\langle\Psi_0|\hat{\psi}(x_2)\hat{\bar{\psi}}(x_1)|\Psi_0\rangle$, also called *virtual electron propagator*, is denoted by a continuous line between space-time points x_1 and x_2 [Fig. 14.1(f)].

(g) A *factor pair*, like $\langle\Psi_0|\gamma_\mu\hat{A}_\mu(x_2)\gamma_\nu\hat{A}_\nu(x_1)|\Psi_0\rangle$, also called a *virtual photon propagator*, is denoted by a dotted line between the space-time points x_1 and x_2 [Fig. 14.1(g)].

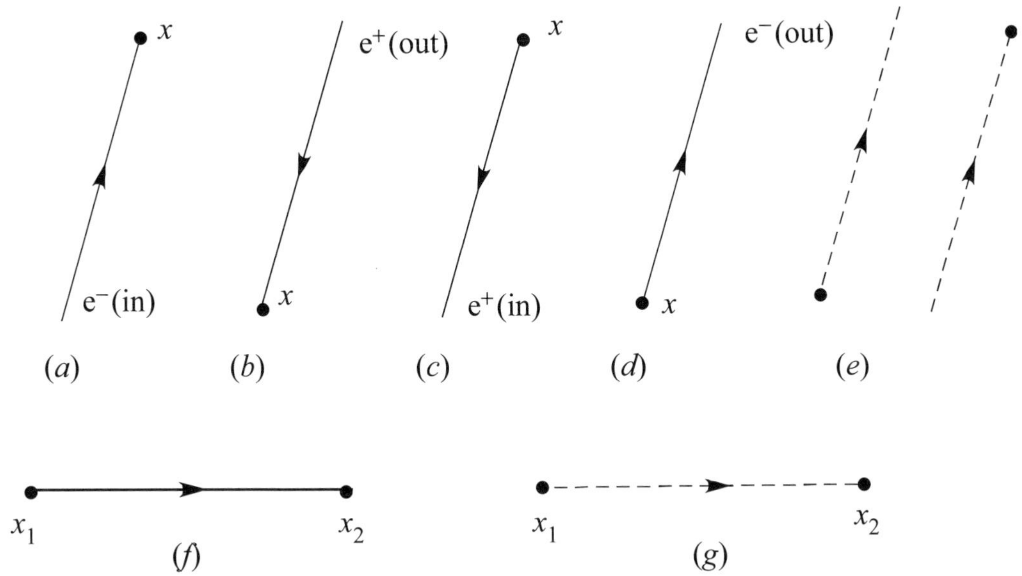

FIGURE 14.1
Rules for Feynman diagram.

With the help of these rules, we can represent each component of a normal constituent of $\hat{G}$ by a Feynman diagram corresponding to a physical process.

Let us consider the components of the normal constituent $\hat{G}_2$. One such component is

$$\hat{G}_2^{(a_1)} = \delta_P\langle\Psi_0|\hat{\psi}(x_2)\hat{\bar{\psi}}(x_1)|\Psi_0\rangle\left[\hat{\bar{\psi}}^{(-)}(x_2)\gamma_\nu\hat{A}_\nu^{(-)}(x_2)\gamma_\mu\hat{A}_\mu^{(+)}(x_1)\hat{\psi}^{(+)}(x_1)\right] . \quad (14.10.11)$$

This is represented by the Feynman diagram shown in Fig. 14.2(a). To get the contribution of this diagram to $S^{(2)}$, we integrate $\hat{G}_2^{(a_1)}$ over x_1 and x_2, and multiply by $\frac{e^2}{2\hbar^2}$. To get the transition amplitude we take the matrix element of the resulting operator between the initial and final field states.

Similarly, the component $G_2^{(a_2)}$ of the normal constituent G_2 given by

$$\hat{G}_2^{(a_2)} = \delta_P\langle\Psi_0|\hat{\psi}(x_2)\hat{\bar{\psi}}(x_1)|\Psi_0\rangle\left[\hat{\bar{\psi}}^{(-)}(x_2)\gamma_\nu\hat{A}_\nu^{(-)}(x_1)\gamma_\mu\hat{A}_\mu^{(+)}(x_2)\hat{\psi}^{(+)}(x_1)\right] , \quad (14.10.12)$$

is represented by the Feynman diagram in Fig. 14.2(b). The diagrams in Fig. 14.2 together represent Compton scattering ($e^- + \gamma \to e^- + \gamma$).

The component $\hat{G}_2^{(b_1)}$ given by

$$\hat{G}_2^{(b_1)} = \delta_P\langle\Psi_0|\hat{\psi}(x_2)\hat{\bar{\psi}}(x_1)|\Psi_0\rangle\left[\hat{\psi}^{(-)}(x_2)\gamma_\nu\hat{A}_\nu^{(-)}(x_2)\gamma_\mu\hat{A}_\mu^{(+)}(x_1)\hat{\bar{\psi}}^{(+)}(x_1)\right] \quad (14.10.13)$$

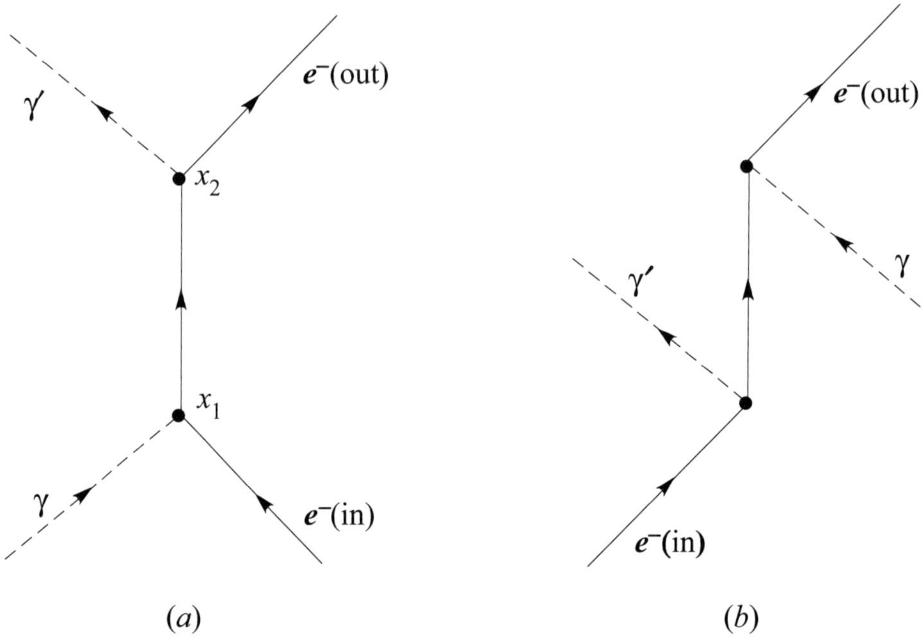

FIGURE 14.2
Feynman diagrams contributing to Compton scattering by electrons. (a) At space time point x_1 an electron and a photon are destroyed and at x_2 an electron and a photon (with different energies) are created. The line between x_1 and x_2 represents a virtual electron propagation and is equivalent to $\langle\Psi_0|\hat{\psi}(x_2)\hat{\bar{\psi}}(x_1)|\Psi_0\rangle$. (b) At x_1 an electron is destroyed and a photon γ' is created. At x_2 an electron is created and a photon γ is destroyed. The line between x_1 and x_2 again represents a virtual electron propagation and contributes a factor $\langle\Psi_0|\psi(x_2)\psi(x_1)|\Psi_0\rangle$.

is represented by the Feynman diagram shown in Fig. 14.3(a) and component $\hat{G}_2^{(b_2)}$ given by

$$\hat{G}_2^{(b_2)} = \delta_P\langle\Psi_0|\hat{\psi}(x_2)\hat{\bar{\psi}}(x_1)|\Psi_0\rangle\left[\hat{\psi}^{(-)}(x_2)\gamma_\nu\hat{A}_\nu^{(-)}(x_1)\gamma_\mu\hat{A}_\mu^{(+)}(x_2)\hat{\bar{\psi}}^{(+)}(x_1)\right] \qquad (14.10.14)$$

is represented by the Feynman diagram shown in Fig. 14.3(b). The two diagrams of Fig. 14.3 together represent Compton scattering by a positron ($e^+ + \gamma \to e^+ + \gamma$).

The component $\hat{G}_2^{(c_1)}$ given by

$$\hat{G}_2^{(c_1)} = \delta_P\langle\Psi_0|\hat{\psi}(x_2)\hat{\bar{\psi}}(x_1)|\Psi_0\rangle\left[\gamma_\mu\hat{A}_\mu^{(-)}(x_1)\gamma_\nu\hat{A}_\nu^{(-)}(x_2)\hat{\bar{\psi}}^{(+)}(x_2)\hat{\psi}^{(+)}(x_1)\right] \qquad (14.10.15)$$

is represented by the Feynman diagram shown in Fig. 14.4(a) and $\hat{G}_2^{(c_2)}$ given by

$$\hat{G}_2^{(c_2)} = \delta_P < \Psi_0|\hat{\psi}(x_2)\hat{\bar{\psi}}(x_1)|\Psi_0\rangle\left[\gamma_\nu A_\nu^{(-)}(x_1)\gamma_\mu\hat{A}_\mu^{(-)}(x_2)\hat{\bar{\psi}}^{(+)}(x_2)\hat{\psi}^{(+)}(x_1)\right] \qquad (14.10.16)$$

is represented by the Feynman diagram in Fig. 14.4(b). The diagrams of Fig. 14.4 together represent electron-positron pair annihilation into a pair of photons.

The component $\hat{G}_2^{(d_1)}$ given by

$$\hat{G}_2^{d_1} = \delta_P\langle\Psi_0|\hat{\psi}(x_2)\hat{\bar{\psi}}(x_1)|\Psi_0\rangle\left[\hat{\bar{\psi}}^{(-)}(x_1)\hat{\psi}^{(-)}(x_2)\gamma_\mu\hat{A}_\mu^{(+)}(x_1)\gamma_\nu\hat{A}_\nu^{(+)}(x_2)\right] \qquad (14.10.17)$$

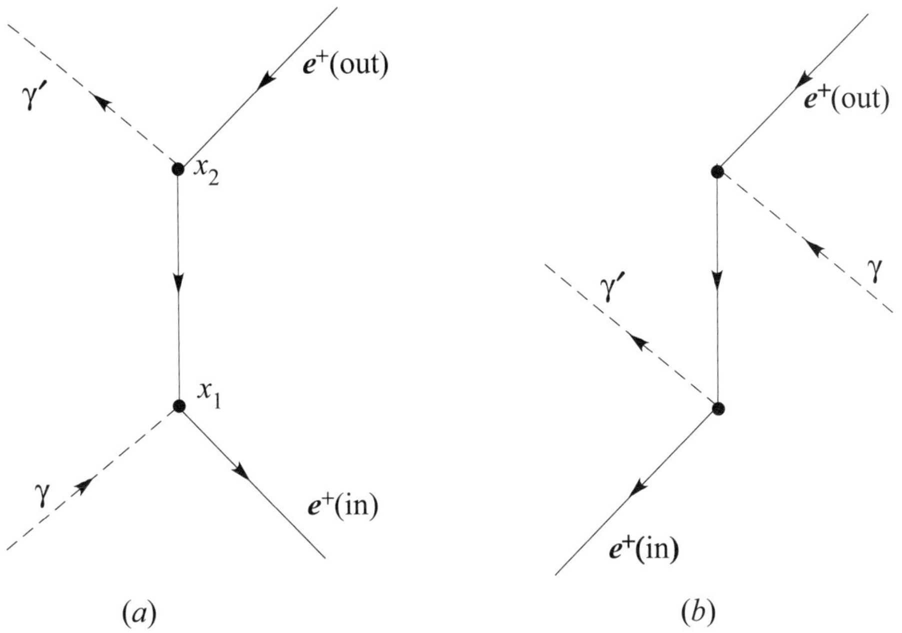

FIGURE 14.3
Feynman diagrams for Compton scattering by a positron. (a) At x_1 a positron and a photon (γ) are destroyed and at x_2 a positron and a photon (γ') are created; (b) At x_1 a positron is destroyed and a photon (γ') is created and at x_2 a positron is created and a photon (γ) is destroyed.

is represented by the Feynman diagram in Fig. 14.5(a). The component $\hat{G}_2^{(d_2)}$ given by

$$\hat{G}_2^{(d_2)} = \delta_P \langle \Psi_0 | \hat{\psi}(x_2) \hat{\bar{\psi}}(x_1) | \Psi_0 \rangle \left[\hat{\psi}^{(-)}(x_1) \hat{\bar{\psi}}^{(-)}(x_2) \gamma_\mu \hat{A}_\mu^{(+)}(x_1) \gamma_\nu \hat{A}_\nu^{(+)}(x_2) \right] \tag{14.10.18}$$

is represented by the Feynman diagram in Fig. 14.5(b). Together, the two diagrams in Fig. 14.5 represent two-photon annihilation and creation of an electron-positron pair (e^-e^+).

We can similarly split the normal constituent $\hat{G}_4$ (or $\hat{S}_4^{(2)}$) into components by the substitutions as in Eqs. (14.9.28) to (14.9.30). Each one of these components corresponds to a specific physical process and to one Feynman diagram. For example, the component $\hat{G}_4^{(a)}$ given by

$$\hat{G}_4^{(a)} = \delta_P \langle \Psi_0 | \gamma_\mu \hat{A}_\mu(x_1) \gamma_\nu \hat{A}_\nu(x_2) | \Psi_0 \rangle \left[\hat{\bar{\psi}}^{(-)}(x_1) \hat{\bar{\psi}}^{(-)}(x_2) \hat{\psi}^{(+)}(x_1) \hat{\psi}^{(+)}(x_2) \right] \tag{14.10.19}$$

is represented by the Feynman diagram shown in Fig. 14.6. This represents Möller (e^-e^-) scattering. The component $\hat{G}_4^{(b)}$ given by

$$G_4^{(b)} = \delta_P \langle \Psi_0 | \gamma_\mu \hat{A}_\mu(x_1) \gamma_\nu \hat{A}_\nu(x_2) | \Psi_0 \rangle \left[\hat{\psi}^{(-)}(x_1) \hat{\psi}^{(-)}(x_2) \hat{\bar{\psi}}^{(+)}(x_1) \hat{\bar{\psi}}^{(+)}(x_2) \right] \tag{14.10.20}$$

is represented by the Feynman diagram shown in Fig. 14.6(b). This represents Möller (e^+e^+) scattering.

The component $\hat{G}_4^{(c)}$ given by

$$\hat{G}_4^{(c)} = \delta_P \langle \Psi_0 | \gamma_\mu \hat{A}_\mu(x_1) \gamma_\nu \hat{A}_\nu(x_2) | \Psi_0 \rangle \left[\hat{\psi}^{(-)}(x_2) \hat{\bar{\psi}}^{(-)}(x_1) \hat{\bar{\psi}}^{(+)}(x_2) \hat{\psi}^{(+)}(x_1) \right] \tag{14.10.21}$$

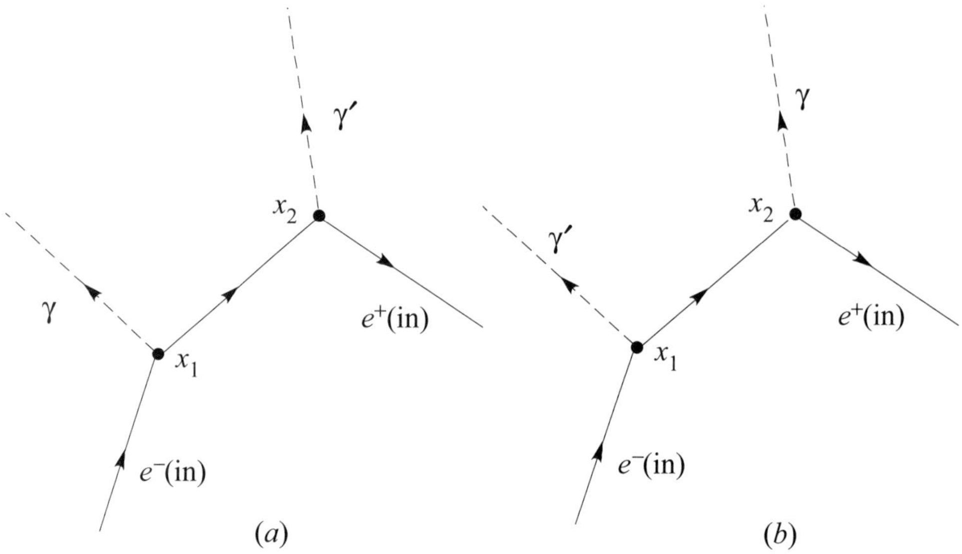

FIGURE 14.4
Feynman diagrams for electron-positron pair annihilation. (a) At x_1 an electron is destroyed and a photon γ is created. At x_2 a positron is destroyed and a photon γ' is created. This represents annihilation of an electron-positron pair($e^-e^+ \to \gamma\gamma'$). (b) At space-time point x_1 an electron is destroyed and a photon γ' is created. At x_2 a positron is destroyed and a photon γ is created. This diagram also represents e^-e^+ annihilation.

is represented by the Feynman diagram in Fig. 14.7(a). Similarly the component $\hat{G}_4^{(d)}$ given by

$$\hat{G}_4^{(d)} = \delta_P \langle \Psi_0 | \gamma_\mu \hat{A}_\mu(x_1) \gamma_\nu \hat{A}_\nu(x_2) | \Psi_0 \rangle \left[\hat{\psi}^{(-)}(x_2) \hat{\bar{\psi}}^{(-)}(x_2) \hat{\bar{\psi}}^{(+)}(x_1) \hat{\psi}^{(+)}(x_1) \right] \quad (14.10.22)$$

is represented by Fig 14.7(b). Both diagrams in Fig. 14.7 represent Bhabha (e^+e^-) scattering.

In all of these diagrams the dotted lines represent the virtual photon propagator

$$\langle \Psi_0 | \gamma_\mu \hat{A}_\mu(x_1) \gamma_o \hat{A}_\nu(x_2) | \Psi_0 \rangle.$$

14.11 Amplitude for Compton Scattering

In the case of Compton scattering ($e^-\gamma \to e^-\gamma$), we have

$$\left(S^{(2)} \right)_{fi} = \langle \Psi_f | \hat{S}^{(2)a_1} + \hat{S}^{(2)a_2} | \Psi_i \rangle = \frac{e^2}{\hbar^2} \langle \Psi_f | \iint d^4x_1 d^4x_2 (\hat{G}_2^{(a_1)} + \hat{G}_2^{(a_2)}) | \Psi_i \rangle \quad (14.11.1)$$

where $|\Psi_i\rangle = \hat{a}^\dagger_{\boldsymbol{k}m} \hat{b}^\dagger_r(\boldsymbol{p}) |\Psi_0\rangle$ is the initial state of the field with one electron with spin state r and momentum $\boldsymbol{p}$ and one photon of energy $\hbar c|\boldsymbol{k}|$ and polarization m. Also$|\Psi_f\rangle =$

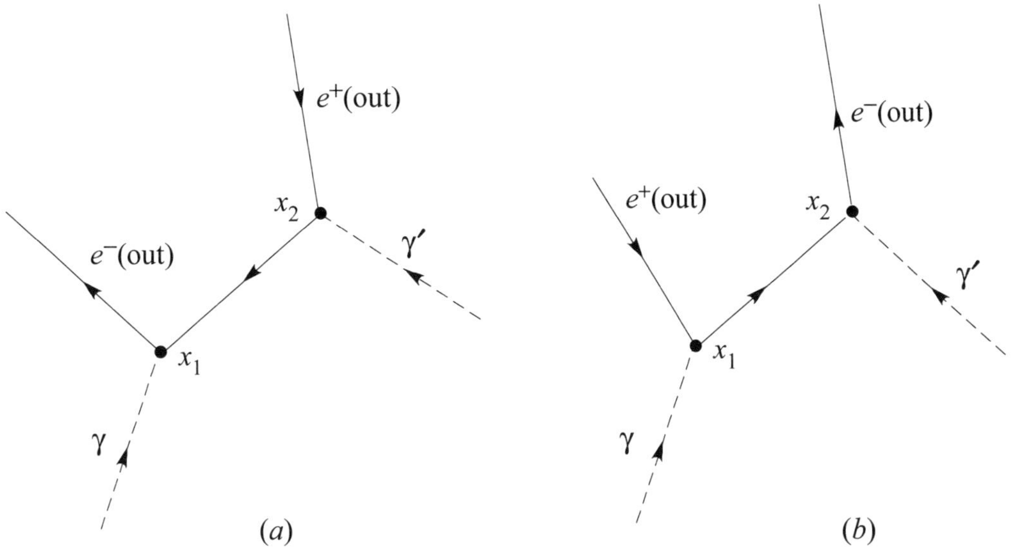

FIGURE 14.5
Electron-positron pair creation. (a) At x_1 and x_2, respectively, photons γ and γ' are annihilated and electron e^- and positron e^+ are created. (b) At x_1 and x_2 the photons γ and γ' are annihilated and positron e^+ and electron e^- are created.

$\hat{a}^\dagger_{\boldsymbol{k}'n}\hat{b}^\dagger_s(\boldsymbol{p}')|\Psi_0\rangle$ is the final state of the field with an electron in spin state s and momentum $\boldsymbol{p}'$ and one photon of energy $\hbar c|\boldsymbol{k}'|$ and polarization n. For $S^{(2)}_{fi}$, the contribution comes only from $\hat{G}^{(a_1)}_2$ and $\hat{G}^{(a_2)}_2$ represented by Feynman diagrams in Figs. 14.2(a) and 14.2(b), respectively, where

$$\hat{G}^{(a_1)}_2 = \langle\Psi_0|\hat{\psi}(x_2)\hat{\bar{\psi}}(x_1)|\Psi_0\rangle\left[\hat{\bar{\psi}}^{(-)}(x_2)\gamma_\nu\hat{A}^{(-)}_\nu(x_2)\gamma_\mu\hat{A}^{(+)}_\mu(x_1)\hat{\psi}^{(+)}(x_1)\right], \tag{14.11.2}$$

$$\hat{G}^{(a_2)}_2 = \langle\Psi_0|\hat{\psi}(x_2)\hat{\bar{\psi}}(x_1)|\Psi_0\rangle\left[\hat{\bar{\psi}}^{(-)}(x_2)\gamma_\mu\hat{A}^{(+)}_\mu(x_2)\gamma_\nu\hat{A}^{(-)}_\nu(x_1)\hat{\psi}^{(+)}(x_1)\right]. \tag{14.11.3}$$

We can now use the explicit expressions for $\hat{\psi}^{(+)}(x)$ and $\hat{\bar{\psi}}^{(-)}$ given by Eqs. (14.8.26) and (14.8.29) and for $\hat{A}^{(+)}_\mu(x)$ and $\hat{A}^{(-)}_\mu(x)$ [Eq. (14.9.30)] the following expressions

$$\hat{A}^{(+)}_\nu(x) = \sum_{\boldsymbol{k}'}\sum_n\sqrt{\frac{\hbar}{2\epsilon_0\omega_{k'}V}}\,\varepsilon^{(n)}_\nu\exp(i\,k'x)\hat{a}_{\boldsymbol{k}'n}\,, \tag{14.11.4}$$

and

$$\hat{A}^{(-)}_\mu(x) = \sum_{\boldsymbol{k}}\sum_m\sqrt{\frac{\hbar}{2\epsilon_0\omega_k V}}\,\varepsilon^{(m)}_\mu\exp(-i\,\kappa x)\hat{a}^\dagger_{\boldsymbol{k}m}\,, \tag{14.11.5}$$

where $x = (\boldsymbol{r}, ict)$, $p = (\boldsymbol{p}, iE/c)$, $k = (\boldsymbol{k}, i\omega_k/c)$. Since the letters x, p, κ denote four vectors, the magnitudes of the three-dimensional vectors $\boldsymbol{r}$, $\boldsymbol{p}$ and $\boldsymbol{k}$ will be denoted by r, $|\boldsymbol{p}| \equiv P$ and $|\boldsymbol{k}| \equiv K$, respectively. The index μ specifies the components of the four-vector ε^m_μ, while the indices m and n refer to the states of polarization, where $\varepsilon^{(m)}_4$ and $\varepsilon^{(n)}_4$ are taken to be zero.

Now the propagator $\langle\Psi_0|\hat{\psi}(x_2)\hat{\bar{\psi}}(x_1)|\Psi_0\rangle$ equals $\langle\,\Psi_0|\,\hat{\psi}^{(+)}(x_2)\hat{\bar{\psi}}^{(-)}(x_1)\,|\Psi_0\rangle$ as the other terms in the expansion of $(\hat{\psi}(x_2)\hat{\bar{\psi}}(x_1))$ do not contribute. Substituting the explicit

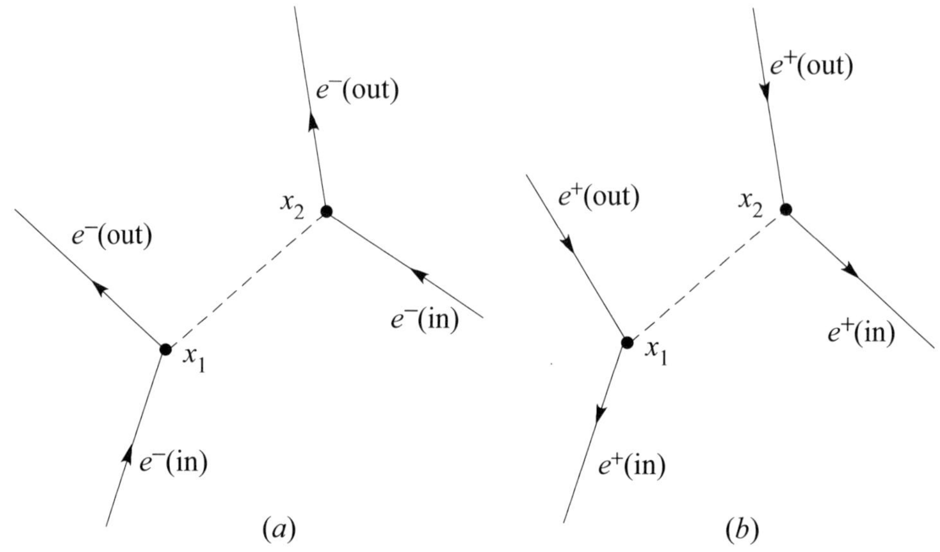

FIGURE 14.6
At x_1 an electron is annihilated and an electron is created. At x_2 again an electron is annihilated and an electron is created. This represents Möller ($e^- e^-$) scattering. (b) At x_1 a positron is annihilated and a positron is created. At x_2 again a positron is annihilated and a positron is created. This represents Möller ($e^+ e^+$) scattering.

expressions for $\hat{\psi}^{(+)}(x_2)$ and $\hat{\tilde{\psi}}^{(-)}(x_1)$ from Eqs. (14.8.26) through (14.8.29), we get, on simplification,

$$\langle\Psi_0|\hat{\psi}(x_2)\hat{\tilde{\psi}}(x_1)\Psi_0\rangle = \frac{1}{V}\sum_p \frac{mc^2}{E}\left\{\frac{-i\gamma p + mc}{2mc}\right\}\exp\left(\frac{ip(x_2 - x_1)}{\hbar}\right), \tag{14.11.6}$$

where we have used the identities

$$\langle\Psi_0|b^{(s)}(\boldsymbol{p})b^{(s')\dagger}(\boldsymbol{p}')|\Psi_0\rangle = \delta_{\boldsymbol{p},\boldsymbol{p}'}\delta_{ss'} \tag{14.11.7}$$

and

$$\sum_s u^{(s)}(\boldsymbol{p})\tilde{u}^{(s)}(\boldsymbol{p}) = \frac{-i\gamma\, p + mc}{2mc}, \tag{14.11.8}$$

with $\tilde{u}^{(s)} = u^{(s)\dagger}\gamma_4$. Replacing the summation over the momentum $\boldsymbol{p}$ by integration over the momentum space

$$\frac{1}{V}\sum_p \rightarrow \frac{1}{(2\pi\,\hbar)^3}\int d^3p,$$

we have, for the propagator, the integral

$$\langle\Psi_0|\hat{\psi}(x_2)\hat{\tilde{\psi}}(x_1)|\Psi_0\rangle = \frac{1}{(2\pi\,\hbar)^3}\int \frac{c}{2E}(-i\gamma\, p + mc)\exp[ip(x_2 - x_1)/\hbar]d^3p\,. \tag{14.11.9}$$

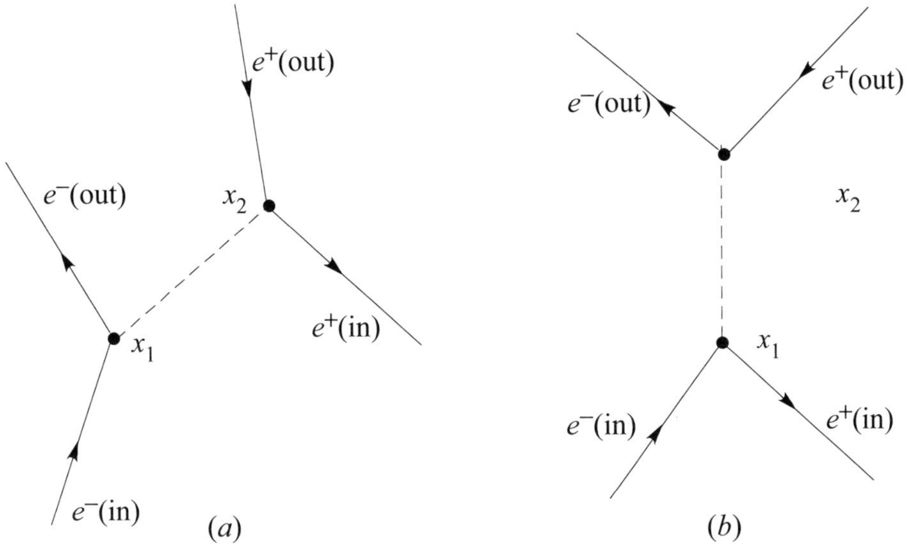

FIGURE 14.7
Feynman diagrams for the Bhabha scattering. (a) At x_1 an electron is annihilated and an electron is created. At x_2 a positron is annihilated and a positron is created. This (direct) diagram represents Bhabha (e^-e^+) scattering. (b) At x_1 the e^-e^+ pair is annihilated and a virtual photon is created. At x_2 the e^-e^+ pair is again created. This (annihilation) diagram also represents Bhabha scattering.

The propagator may also be expressed as a four-dimensional integral[14]

$$\langle\Psi_0|\hat{\psi}(x_2)\hat{\bar{\psi}}(x_1)|\Psi_0\rangle = \frac{1}{(2\pi)^4\hbar^3}\int d^4q\, \frac{(-i\gamma\, q + mc)\exp[iq(x_2 - x_1)/\hbar]}{q^2 + m^2c^2 + i\epsilon}, \tag{14.11.10}$$

where ϵ is an infinitesimal positive quantity $\to 0^+$. Also

$$q \equiv (\boldsymbol{p}, ip_0), \quad q^2 = |\boldsymbol{p}|^2 - p_0^2$$

and $d^4q = d^3p\, dp_0$. Note that since p_0 is a variable of integration, it is not necessarily equal to $\frac{E}{c} = (|\boldsymbol{p}|^2 + m^2c^2)^{1/2}$. When $p_0 \to E/c$, the four-vector q tends to p. Using the identity

$$(-i\gamma\, q + mc)(i\gamma\, q + mc) = q^2 + m^2c^2$$

in Eq. (14.11.10), the propagator can be written as

$$\langle\Psi_0|\hat{\psi}(x_2)\,\hat{\bar{\psi}}(x_1)|\Psi_0\rangle = \frac{1}{(2\pi)^4\hbar^3}\int d^4q\, \frac{\exp[iq(x_2 - x_1)/\hbar]}{\gamma\, q - i\, mc + i\epsilon}. \tag{14.11.11}$$

[14]One can see that

$$\frac{i}{2\pi}\int_{-\infty}^{\infty} dp_0 \frac{(-i\boldsymbol{\gamma}\cdot\boldsymbol{p} - i\gamma_4 ip_0 + mc)\exp[iq(x_2 - x_1)/\hbar]}{p^2 - p_0^2 + m^2c^2 + i\epsilon} = \frac{c}{2E}(-i\gamma\, p + mc)\exp[ip(x_2 - x_1)/\hbar]$$

where $\boldsymbol{\gamma} \equiv (\gamma_1, \gamma_2, \gamma_3)$.

Now

$$S_{fi}^{(2)a_1} = \frac{e^2}{\hbar^2} \langle\Psi_f| \int d^4x_1 \int d^4x_2 \hat{G}_2^{(a_1)}|\Psi_i\rangle$$
$$= \frac{e^2}{\hbar^2} \langle\Psi_0|\hat{b}^{(s)}(\boldsymbol{p}')\hat{a}_{\boldsymbol{k}'n} \int d^4x_1 \int d^4x_2 \, \langle\Psi_0|\hat{\psi}(x_2)\hat{\bar{\psi}}(x_1)|\Psi_0\rangle \times$$
$$\left(\hat{\bar{\psi}}^{(-)}(x_2)\gamma_\nu \hat{A}_\nu^{(-)}(x_2)\gamma_\mu \hat{A}_\mu{}^{(+)}(x_1)\hat{\psi}^{(+)}(x_1)\hat{a}^\dagger_{\boldsymbol{k}m}\hat{b}^{(r)\dagger}(\boldsymbol{p})\right)|\Psi_0\rangle\,. \tag{14.11.12}$$

Substituting the explicit expressions for the field operators $\hat{\psi}^{(+)}(x)$, $\hat{\bar{\psi}}^{(-)}(x)$, $\hat{A}_\mu^{(+)}(x)$ and $\hat{A}_\mu^{(-)}(x)$, simplifying, and using the relations

$$\hat{b}^{(r')}(\boldsymbol{p}')\hat{b}^{(r)\dagger}(\boldsymbol{p})\,|\Psi_0\rangle = \delta_{\boldsymbol{p},\boldsymbol{p}'}\delta_{r,r'}\,|\Psi_0\rangle \tag{14.11.13a}$$

and

$$\hat{a}_{\boldsymbol{k}'m'}\hat{a}^\dagger_{\boldsymbol{k}m}\,|\Psi_0\rangle = \delta_{\boldsymbol{k}',\boldsymbol{k}}\delta_{m',m}\,|\Psi_0\rangle\,, \tag{14.11.13b}$$

we get

$$\langle\Psi_f|\,\hat{S}^{(2)a_1}\,|\Psi_i\rangle = -\frac{e^2}{\epsilon_0(2\pi\,\hbar)^4V^2}\frac{mc^2}{\sqrt{4EE'\omega_k\omega_{k'}}}\int d^4q \int d^4x_1 \exp[i(-q+\hbar k+p)x_1/\hbar]$$
$$\times \int d^4x_2 \exp\left[i(q-\hbar k'-p')x_2/\hbar\right]\left[\tilde{u}_s(\boldsymbol{p}')\gamma\,\varepsilon_{k'}^{(n)}\frac{1}{\gamma\,q-i\,mc}\gamma\varepsilon_k^{(m)}u_r(\boldsymbol{p})\right]$$

or

$$\langle\Psi_f|\hat{S}^{(2)a_1}|\Psi_i\rangle = -\frac{e^2}{\epsilon_0(2\pi\hbar)^4V^2}\frac{mc^2(2\pi\,\hbar)^8}{\sqrt{(4EE'\omega_k\omega_{k'})}}\int d^4q\delta^4(-q+\hbar k+p)\delta^4(q-\hbar k'-p')$$
$$\left[\tilde{u}_s(\boldsymbol{p}')\,\gamma\,\varepsilon_{k'}^{(n)}\frac{1}{\gamma\,q-i\,mc}\gamma\,\varepsilon_k^{(m)}u_r(\boldsymbol{p})\right] \tag{14.11.14a}$$
$$= -\frac{e^2}{\epsilon_0V^2}\frac{mc^2(2\pi\,\hbar)^4}{\sqrt{(4EE'\omega_k\omega_{k'})}}\delta^4(p+\hbar k-\hbar k'-p')$$
$$\left[\tilde{u}_s(\boldsymbol{p}')\not{\varepsilon}_{k'}^{(n)}\left(\frac{1}{\hbar\not{k}+\not{p}-i\,mc}\right)\not{\varepsilon}_k^{(m)}u_r(\boldsymbol{p})\right] \tag{14.11.14b}$$

where we have used the identity

$$\frac{1}{(2\pi\hbar)^4}\int \exp(ipx/\hbar)d^4x = \delta^4(p),$$

and introduced the notation $\not{\varepsilon}^{(n)} = \gamma\,\varepsilon^{(n)} = \gamma_\mu\varepsilon_\mu^{(n)}$, $\not{k} = \gamma\,k = \gamma_\mu k_\mu$ and $\not{p} = \gamma\,p = \gamma_\mu p_\mu$. Eq. (14.11.14a) suggests that in the intermediate state the electron has four-momentum

$$q = p+\hbar k = p'+\hbar k'\,. \tag{14.11.15}$$

The second term $S_{fi}^{(2)a_2} = \langle\Psi_f|\frac{e^2}{\hbar^2}\int d^4x_1\int d^4x_2\hat{G}_2^{(a_2)}|\Psi_i\rangle$, represented by the Feynman diagram in Fig. 14.2(b), may be treated exactly in the same way as the first, the only difference being the order in which the two photons, γ and γ' are absorbed and emitted, respectively. The four-momentum q' of the virtual Fermion in this case is given by $q' = p-\hbar k' = p'-\hbar k$. Hence for $S_{fi}^{(2)a_2}$ we have the same expression with $k \leftrightarrow -k'$ and $n \leftrightarrow m$

$$S_{fi}^{(2)a_2} = -\frac{e^2}{\epsilon_0V^2}(2\pi\,\hbar)^4\frac{mc^2\delta^4(p-\hbar k'-p'+\hbar k)}{\sqrt{4EE'\omega_k\omega_{k'}}}\times$$
$$\left[\tilde{u}_s(\boldsymbol{p}')\not{\varepsilon}_k^{(m)}\frac{1}{\not{p}-\hbar\not{k}'-i\,mc}\not{\varepsilon}_{k'}^{(n)}u_r(\boldsymbol{p})\right]\,. \tag{14.11.16}$$

Hence the total amplitude for Compton scattering is given by

$$\langle\Psi_f|S^{(2)a_1} + S^{(2)a_2}|\Psi_i\rangle = -i(2\pi\hbar)^4 4\pi\delta^4(p - \hbar k' - p' + \hbar k)M_{fi} \tag{14.11.17}$$

where

$$M_{fi} = \frac{e^2}{4\pi\epsilon_0 V^2}\frac{mc^2}{\sqrt{4EE'\omega_k\omega_{k'}}}\tilde{u}_s(\boldsymbol{p}')\left[\not\epsilon_{k'}^{(n)}\frac{-i}{\hbar\not k + \not p - i\,mc}\not\epsilon_k^{(m)} + \not\epsilon_k^{(m)}\frac{-i}{\not p - \hbar\not k' - i\,mc}\not\epsilon_{k'}^{(n)}\right]u_r(\boldsymbol{p})\,. \tag{14.11.18}$$

14.12 Feynman Graphs

We have seen that each Feynman diagram represents a contribution to the $S^{(2)}$ (or G_2) operator, delineating a process involving annihilation or creation of particles at vertices representing space-time points x_1 and x_2. To calculate the corresponding transition amplitudes these have to be integrated over the region spanned by the four vectors x_1 and x_2 and the resulting operators have to be sandwiched between the final and initial states $\langle\Psi_f|$ and $|\Psi_i\rangle$. We have already done this while calculating the transition amplitude for Compton scattering. These matrix elelments can be obtained in a much simpler and direct way if we re-draw them in momentum space and follow certain rules devised by Feynman. This means the vertices are no longer labeled by x_1 and x_2. Also, since the initial and final states are now defined, the momentum and polarizations of particles created or annihilated should be specified. To distinguish the new figures (which look similar but have to be labeled differently) from the previous ones, we shall call them *Feynman graphs*. Feynman graphs for Compton scattering are shown in Figs. 14.8(a) and 14.8(b).

Rules for Writing Transition Amplitude from Feynman Graphs

Feynman devised a set of rules by virtue of which one can write down the expression for the transition matrix element (or amplitude) from the graph itself. These rules can be derived rigorously from quantum field theory. Instead of deriving these rules, summarized in Table 14.1, we shall illustrate their use in calculating the transition amplitude for second order processes and, at least in the case of Compton scattering, check them against our detailed calculations of the previous section.

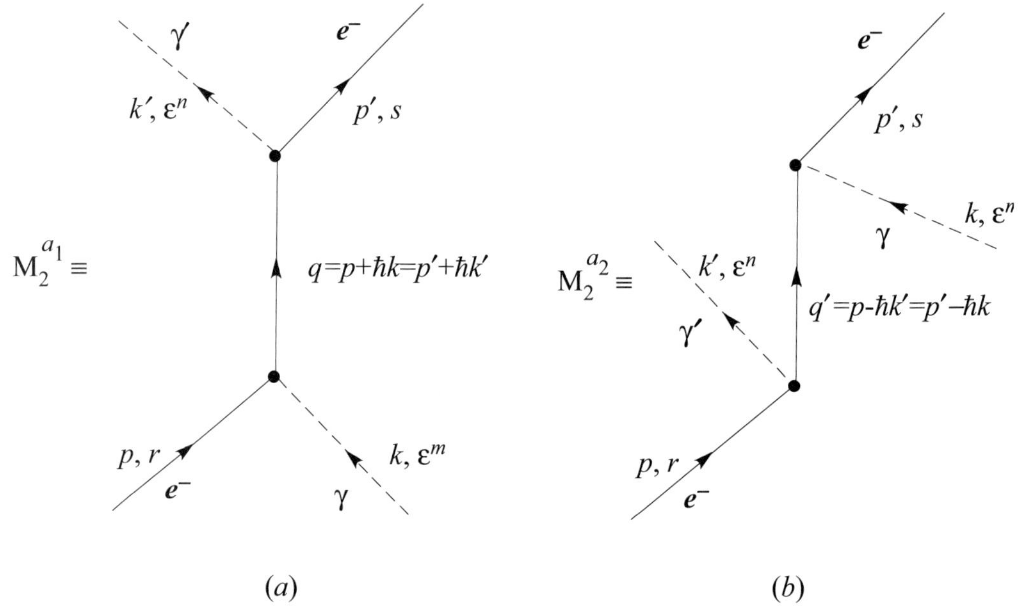

FIGURE 14.8
Feynman graphs for Compton scattering in momentum space. The vertices are no longer labeled by space-time points x_1 and x_2, for we eventually have to integrate the operators $\hat{G}_2^{(a_1)}$ and $\hat{G}_2^{(a_2)}$ over the four-spaces spanned by vectors x_1 and x_2. Further, the initial and final momenta and polarizations or spins of particles are specified. These are indicated on the respective lines.

14.12.1 Compton Scattering Amplitude Using Feynman Rules

Using Feynman rules one can associate, with the Feynman graphs representing Compton scattering [Fig. 14.8], the amplitudes

$$M_{2fi}^{(a_1)} = \sqrt{\frac{mc^2}{E'V}}\,\tilde{u}_s(\boldsymbol{p}')\frac{1}{\sqrt{(2\omega_{\kappa'})V}}i\frac{e}{\sqrt{4\pi\epsilon_0}}\gamma_\nu\varepsilon_\nu^{(n)}$$
$$\frac{i}{\not{q}-imc}i\frac{e}{\sqrt{4\pi\epsilon_0}}\gamma_\mu\varepsilon_\mu^{(m)}\frac{1}{\sqrt{(2\omega_k)V}}\sqrt{\frac{mc^2}{EV}}u_r(\boldsymbol{p})$$

and

$$M_{2fi}^{(a_2)} = \sqrt{\frac{mc^2}{E'V}}\,\tilde{u}_s(\boldsymbol{p}')\frac{1}{\sqrt{(2\omega_k)V}}i\frac{e}{\sqrt{4\pi\epsilon_0}}\gamma_\mu\varepsilon_\mu^{(m)}$$
$$\frac{i}{\not{q}'-imc}i\frac{e}{\sqrt{4\pi\epsilon_0}}\gamma_\nu\varepsilon_\nu^{(n)}\frac{1}{\sqrt{(2\omega_{k'})V}}\sqrt{\frac{mc^2}{EV}}u_r(\boldsymbol{p}),$$

respectively. Then the matrix element $M_{fi} = M_{2fi}^{(a_1)} + M_{2fi}^{(a_2)}$ can be expressed as

$$M_{fi} = \frac{mc^2e^2}{4\pi\epsilon_0V^2\sqrt{4EE'\omega_k\omega_{k'}}}\tilde{u}_s(\boldsymbol{p}')\left[\not{\varepsilon}_{k'}^{(n)}\frac{-i}{\not{p}+\hbar\not{k}-imc}\not{\varepsilon}_k^{(m)}\right.$$
$$\left.+\not{\varepsilon}_k^{(m)}\frac{-i}{\not{p}-\hbar\not{k}'-imc}\not{\varepsilon}_{k'}^{(n)}\right]u_r(\boldsymbol{p})\,. \qquad (14.12.1)$$

This expression for M_{fi} is exactly the same as Eq. (14.11.18), which we derived in the last section. Thus, without going into the intricacies of the detailed calculation for the amplitude M_{fi} of an electromagnetic process, one can write it from the Feynman graph by following Feynman rules.

This method of writing down transition amplitudes holds good not only for Compton scattering, but for other electromagnetic processes as well. In what follows, we will use this method for writing down the matrix elements for the other second order processes.

14.12.2 Electron-positron (e^-e^+) Pair Annihilation

The Feynman graphs for this process are given in Fig. 14.9. Following Feynman rules, we can write down the corresponding amplitudes:

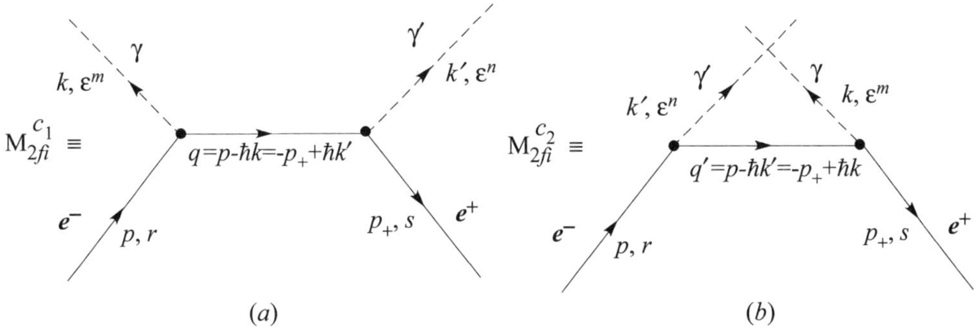

FIGURE 14.9
Feynman graphs for e^-e^+ pair annihilation resulting in the creation of two photons. In the Feynman graphs the vertices are no longer labeled by space-time points while the lines are labeled by the momenta and polarizations of the particles created or annihilated.

$$M_{2fi}^{(c_1)} = \sqrt{\frac{mc^2}{E_+V}}\,\tilde{v}_s(\boldsymbol{p}_+)i\frac{e}{\sqrt{4\pi\epsilon_0}}$$

$$\frac{\gamma_\nu\,\varepsilon_\nu^{(m)}}{\sqrt{V(2\omega_{k'})}}\frac{i}{\not{p}-\hbar\not{k}-imc}i\frac{e}{\sqrt{4\pi\epsilon_0}}\frac{\gamma_\mu\varepsilon_\mu^{(n)}}{\sqrt{V(2\omega_k)}}\sqrt{\frac{mc^2}{EV}}\,u_r(\boldsymbol{p})$$

and

$$M_{2fi}^{(c_2)} = \sqrt{\frac{mc^2}{E_+V}}\tilde{v}_s(\boldsymbol{p}_+)i\frac{e}{\sqrt{4\pi\epsilon_0}}$$

$$\frac{\gamma_\nu\varepsilon_\nu^{(n)}}{\sqrt{V(2\omega_k)}}\frac{i}{\not{p}-\hbar\not{k}'-imc}i\frac{e}{\sqrt{4\pi\epsilon_0}}\frac{\gamma_\mu\varepsilon_\mu^{(m)}}{\sqrt{V(2\omega_{k'})}}\sqrt{\frac{mc^2}{EV}}\,u_r(\boldsymbol{p})\,.$$

TABLE 14.1
Rules for writing transition amplitude from Feynman graphs

With a free Fermion (anti-Fermion) line (p, r) entering a vertex $(\bullet)$, we associate a factor $\sqrt{\frac{mc^2}{EV}}u_r(\boldsymbol{p}) \quad \left(\sqrt{\frac{mc^2}{EV}}\tilde{v}_r(\boldsymbol{p})\right)$	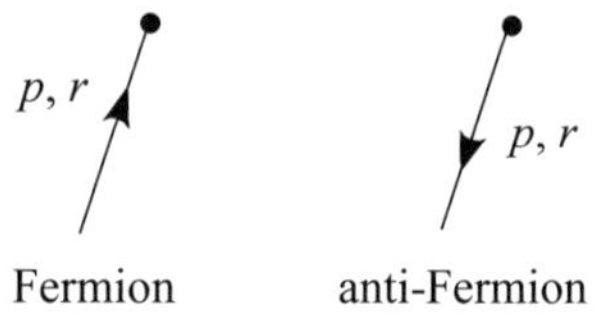
With a free Fermion (anti-Fermion) line (p', s) leaving a vertex $(\bullet)$, we associate a factor $\sqrt{\frac{mc^2}{E'V}}\tilde{u}_s(\boldsymbol{p}') \quad \left(\sqrt{\frac{mc^2}{E'V}}v_s(\boldsymbol{p}')\right)$	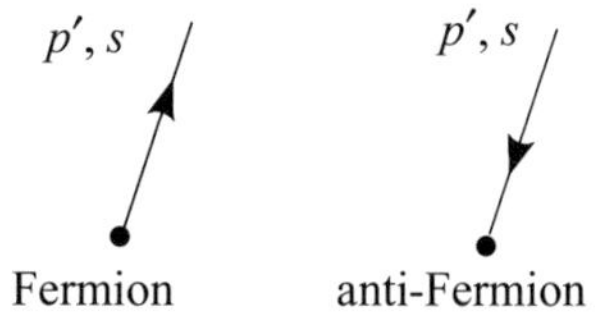
With a photon line entering (leaving) a vertex, we associate a factor $\frac{1}{\sqrt{(2\omega_k)V}}\varepsilon_\mu^{(m)} \quad \left(\frac{1}{\sqrt{(2\omega_{k'})V}}\varepsilon_\nu^{(n)}\right)$	
With each vertex in an electrodynamic process, we associate a term $i\,(e/\sqrt{4\pi\epsilon_0}\,)\,\gamma_\mu$.	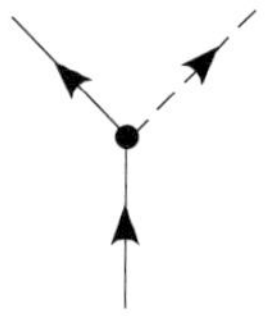
For the Fermion propagator between two vertices we associate a factor $\frac{i}{\not{q}-imc}$, where $\not{q} = \gamma\, q$ and q is the four-momentum of the virtual Fermion state. For the virtual photon propagator between two vertices we associate a term $q^{-2}\delta_{\mu\nu}$.	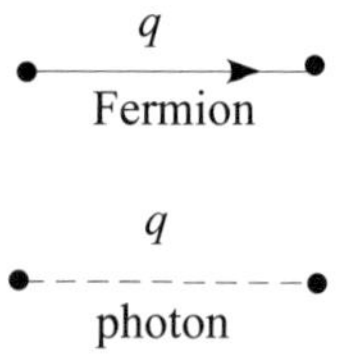

The total amplitude for this process is given by

$$M_{fi} = M^{(c_1)}_{2fi} + M^{(c_2)}_{2fi},$$

or

$$M_{fi} = \frac{e^2 mc^2}{4\pi\epsilon_0 V^2\sqrt{4EE_+\omega_k\omega_{k'}}}\tilde{v}_s\left(\boldsymbol{p}_+\right)\left[\not{\epsilon}^{(m)}\frac{-i}{\not{p}-\hbar\not{k}-imc}\not{\epsilon}^{(n)} + \not{\epsilon}^{(n)}\frac{-i}{\not{p}-\hbar\not{k}'-imc}\not{\epsilon}^{(m)}\right] u_r(\boldsymbol{p}) \qquad (14.12.2)$$

so that

$$S^{(2)}_{fi} = \delta_{fi} - i(2\pi\hbar)^4 4\pi\delta^4(p+p_+-\hbar k-\hbar k')M_{fi}\,.$$

14.12.3 Two-photon Annihilation Leading to $(e^- e^+)$ Pair Creation

This process is represented by the graphs shown in Fig. (14.10). According to Feynman rules the corresponding amplitudes are

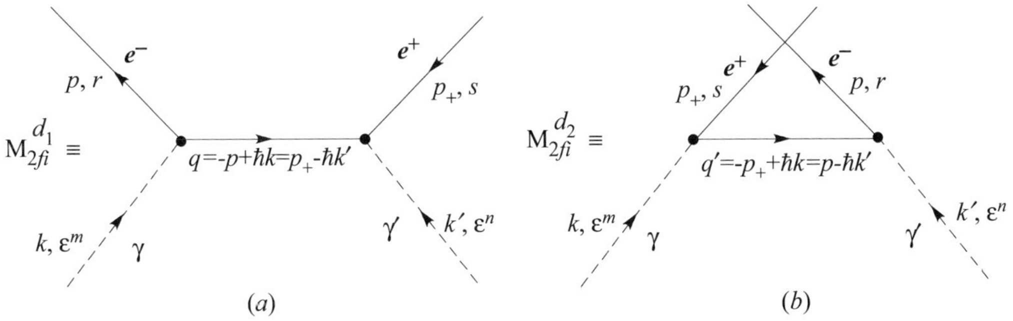

FIGURE 14.10
Feynman graphs for two-photon annihilation resulting in e^-e^+ pair production.

$$M^{(d_1)}_{2fi} = \sqrt{\frac{mc^2}{EV}}\,\tilde{u}_r(\boldsymbol{p})\frac{ie}{\sqrt{4\pi\epsilon_0}}\frac{\gamma_\mu\varepsilon^{(m)}_\mu}{\sqrt{V(2\omega_{k'})}}\frac{i}{\hbar\not{k}-\not{p}-imc}\frac{ie}{\sqrt{4\pi\epsilon_0}}\frac{\gamma_\nu\varepsilon^{(n)}_\nu}{\sqrt{V(2\omega_{k'})}}\sqrt{\frac{mc^2}{E_+V}}\,v_s(\boldsymbol{p}_+)$$

$$M^{(d_2)}_{2fi} = -\sqrt{\frac{mc^2}{EV}}\,\tilde{u}_r(\boldsymbol{p})\frac{ie}{\sqrt{4\pi\epsilon_0}}\frac{\gamma_\nu\varepsilon^{(n)}_\nu}{\sqrt{V(2\omega_{k'})}}\frac{i}{\hbar\not{k}'-\not{p}_+-imc}\frac{ie}{\sqrt{4\pi\epsilon_0}}\frac{\gamma_\mu\varepsilon^{(m)}_\mu}{\sqrt{V(2\omega_k)}}\sqrt{\frac{mc^2}{E_+V}}\,v_s(\boldsymbol{p}_+)$$

so that the overal matrix element $M_{fi} = M^{(d_1)}_{2fi} + M^{(d2)}_{2fi}$ is given by

$$M_{fi} = \frac{e^2 mc^2}{4\pi\epsilon_0 V^2\sqrt{EE_+\omega_k\omega_{k'}}}\tilde{u}_r(\boldsymbol{p})\left[\not{\epsilon}^{(m)}\frac{-i}{\hbar\not{k}-\not{p}-imc}\not{\epsilon}^{(n)} - \not{\epsilon}^{(n)}\frac{-i}{\hbar\not{k}'-\not{p}_+-imc}\not{\epsilon}^{(m)}\right] v_s(\boldsymbol{p}_+)\,. \qquad (14.12.3)$$

The minus sign in $M^{(d_2)}_{2fi}$ arises on account of different signs of δ_P in $\hat{G}^{(d_1)}_2$ and $\hat{G}^{(d_2)}_2$ due to different numbers of exchange of Fermion operators.

14.12.4 Möller (e^-e^-) Scattering

The Feynman graphs that represent Möller (e^-e^-) scattering are shown in Figs. 14.11. According to Feynman rules the corresponding amplitudes are

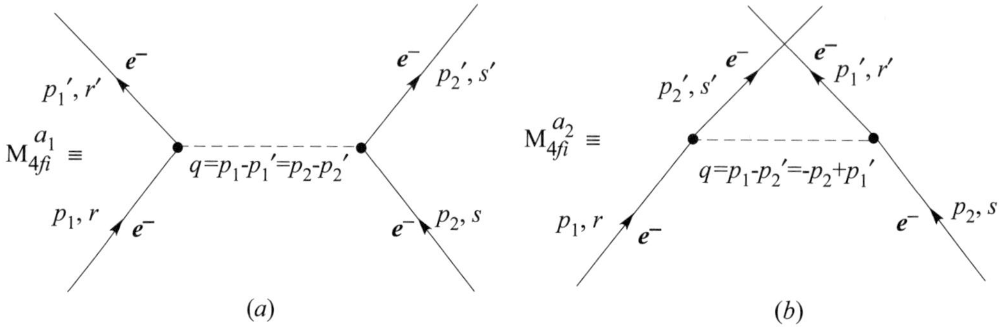

FIGURE 14.11
Feynman graphs for Möller (e^-e^-) scattering.

$$
\begin{aligned}
M_{4fi}^{(a_1)} &= \sqrt{\frac{mc^2}{E_1'V}}\tilde{u}_{r'}(\boldsymbol{p}'_1)i\frac{e}{\sqrt{4\pi\epsilon_0}}\gamma_\mu\frac{\delta_{\mu\nu}}{(p_1-p_1')^2}\sqrt{\frac{mc^2}{E_1V}}\,u_r(\boldsymbol{p}_1) \\
&\times\sqrt{\frac{mc^2}{E_2'V}}\,\tilde{u}_{s'}(\boldsymbol{p}'_2)i\frac{e}{\sqrt{4\pi\epsilon_0}}\gamma_\nu\sqrt{\frac{mc^2}{E_2V}}\,u_r(\boldsymbol{p}_2)\,,
\end{aligned}
$$

$$
\begin{aligned}
M_{4fi}^{(a_2)} &= -\sqrt{\frac{mc^2}{E_2'}}\,\tilde{u}_{s'}(\boldsymbol{p}'_2)i\frac{e}{\sqrt{4\pi\epsilon_0}}\gamma_\mu\frac{\delta_{\mu\nu}}{(p_1-p_2')^2}\sqrt{\frac{mc^2}{E_1V}}\,u_r(\boldsymbol{p}_1) \\
&\times\sqrt{\frac{mc^2}{E_1'V}}\,\tilde{u}_{s'}(\boldsymbol{p}'_1)i\frac{e}{\sqrt{4\pi\epsilon_0}}\gamma_\nu\sqrt{\frac{mc^2}{E_2}}\,u_r(\boldsymbol{p}_1)\,.
\end{aligned}
$$

The complete transition amplitude in this case is given by

$$
\begin{aligned}
M_{fi} &= M_{4fi}^{(a_1)} + M_{4fi}^{(a_2)} \\
&= \frac{-e^2m^2c^4}{4\pi\epsilon_0V^2\sqrt{E_1E_1'E_2E_2'}}\left[\frac{\{\tilde{u}_{r'}(\boldsymbol{p}'_1)\gamma_\mu u_r(\boldsymbol{p}_1)\}\{\tilde{u}_{s'}(\boldsymbol{p}'_2)\gamma_\mu u_s(\boldsymbol{p}_2)\}}{(p_1-p_1')^2}\right. \\
&\qquad\left.-\frac{\{\tilde{u}_{s'}(\boldsymbol{p}'_2)\gamma_\mu u_r(\boldsymbol{p}_1)\}\{\tilde{u}_{r'}(\boldsymbol{p}'_1)\gamma_\mu u_s(\boldsymbol{p}_2)\}}{(p_1-p_2')^2}\right]. \qquad (14.12.4)
\end{aligned}
$$

The minus sign arises because of the different sign of δ_P due to different number of exchanges of Fermion operators in $\hat{G}_4^{(a_1)}$ and $\hat{G}_4^{(a_2)}$.

14.12.5 Bhabha (e^-e^+) Scattering

This scattering process is represented by the Feynman graphs shown in Figs. 14.12. The first Feynman graph (a) represents the direct scattering process while the second Feynman

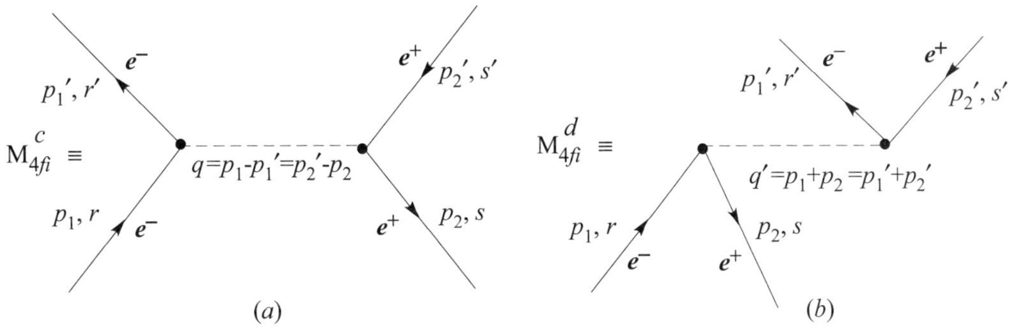

FIGURE 14.12
Feynman graphs for Bhabha (e^-e^+) Scattering.

graph (b) the virtual annihilation and subsequent creation of the e^-e^+ pair. According to Feynman rules, the contributions to the amplitude of the process by the two diagrams are

$$M_{4fi}^{(c)} = \sqrt{\frac{mc^2}{E_1'V}}\,\tilde{u}_{r'}(\boldsymbol{p'}_1)i\frac{e}{\sqrt{4\pi\epsilon_0}}\gamma_\mu\sqrt{\frac{mc^2}{E_1V}}\,u_r(\boldsymbol{p}_1)\frac{\delta_{\mu\nu}}{(p_1-p_1')^2}$$
$$\times\sqrt{\frac{mc^2}{E_2V}}\,\tilde{v}_s(\boldsymbol{p}_2)i\frac{e}{\sqrt{4\pi\epsilon_0}}\gamma_\nu\sqrt{\frac{mc^2}{E_2'V}}\,v_{s'}(\boldsymbol{p'}_2)\,,$$

$$M_{4fi}^{(d)} = -\sqrt{\frac{mc^2}{E_2V}}\,\tilde{v}_s(\boldsymbol{p}_2)i\frac{e}{\sqrt{4\pi\epsilon_0}}\gamma_\mu\sqrt{\frac{mc^2}{E_1V}}\,u_r(\boldsymbol{p}_1)\frac{\delta_{\mu\nu}}{(p_1+p_2)^2}$$
$$\times\sqrt{\frac{mc^2}{E_1'V}}\,\tilde{u}_{r'}(\boldsymbol{p'}_1)i\frac{e}{\sqrt{4\pi\epsilon_0}}\gamma_\nu\sqrt{\frac{mc^2}{E_2'V}}\,v_{s'}(\boldsymbol{p'}_2)\,.$$

The total amplitude is then given by

$$\begin{aligned}M_{fi} &= M_{4fi}^{(c)} + M_{4fi}^{(d)}\\ &= \frac{-e^2}{4\pi\epsilon_0V^2}\frac{m^2c^4}{\sqrt{E_1E_2E_1'E_2'}}\left[\frac{\{\tilde{u}_{r'}(\boldsymbol{p}_1')\gamma_\mu u_r(\boldsymbol{p}_1)\}\{\tilde{v}_s(\boldsymbol{p}_2)\gamma_\mu v_{s'}(\boldsymbol{p}_2')\}}{(p_1-p_1')^2}\right.\\ &\quad\left.-\frac{\{\tilde{v}_s(\boldsymbol{p}_2)\gamma_\mu u_r(\boldsymbol{p}_1)\}\{\tilde{u}_{r'}(\boldsymbol{p}_1')\gamma_\mu v_{s'}(\boldsymbol{p}_2')\}}{(p_1+p_2)^2}\right]\,.\end{aligned} \tag{14.12.5}$$

14.13 Calculation of the Cross-section of Compton Scattering

For any electromagnetic process in second order of perturbation, we can write

$$S_{fi}^{(2)} = \delta_{fi} - i(2\pi\,\hbar)^4 4\pi\,\delta^4(p_i-p_f)M_{fi}\,,$$

where p_i and p_f are the initial and final four-momenta of the system. The expressions for M_{fi}, for different processes, have already been given in Sec. 14.12. From this equation the transition rate w_{fi} (transition probability per unit time) for $i \neq f$ is given by

$$w_{fi} = \frac{1}{t} \left|(2\pi\,\hbar)^4 4\pi\,\delta^4(p_i - p_f)M_{f\,i}\right|^2 .$$

With the help of the identity $[\delta^4(p_i - p_f)]^2 = \delta^4(p_i - p_f)\frac{1}{(2\pi\,\hbar)^4}V(ict)$, the transition rate becomes

$$w_{fi} = (2\pi\hbar)^4(4\pi)^2\delta^4(p_i - p_f)iVc|M_{fi}|^2 . \tag{14.13.1}$$

Now the final state lies in a continuum with the number of final states with the scattered photon momentum between $\hbar\boldsymbol{k}'$ and $\hbar(\boldsymbol{k}'+d\boldsymbol{k}')$ and the recoil electron momentum between $\boldsymbol{p}'$ and $\boldsymbol{p}' + d\boldsymbol{p}'$ given by

$$dN_f = \frac{V\hbar^3 d^3K'}{(2\pi\hbar)^3}\frac{Vd^3P'}{(2\pi\hbar)^3} = \frac{V^2K'^2dK'd\Omega_{K'}d^3P'}{(2\pi)^6\,\hbar^3} , \tag{14.13.2}$$

where $K' = |\boldsymbol{k}'|$ and $P' = |\boldsymbol{p}'|$ (we use lower case letters k' and p' to denote four-vectors and upper case letters to denote the magnitudes of the corresponding three-dimensional vectors). Then the transition rate to this group of states is given by $w_{fi}dN_f = w_{fi}K'^2dK'd\Omega_{K'}d^3P'$. If we consider the scattered photon going into a solid angle $d\Omega_{K'}$ with *any* energy $\hbar c|\boldsymbol{k}'| = \hbar cK' = \hbar\omega_{k'}$ and the recoil electron with *any* momentum $\boldsymbol{p}'$, then we must integrate the transition rate $w_{fi}dN_f$ over K' and $\boldsymbol{P}'$ to obtain the total transition rate for photon scattering into a solid angle $d\Omega_{K'}$. Dividing this rate by the incident photon flux density, we obtain the cross-section for scattering of a photon into a solid angle $d\Omega_{K'}$ around the direction $\boldsymbol{k}'$

$$d\sigma_{fi} = \frac{1}{J_i}d\Omega_{K'}\int\frac{K'dK'}{(2\pi)^6\hbar^3}\int d^3P'w_{fi} = d\Omega_{K'}\,\frac{4iV^4}{\hbar^2}|M_{fi}|^2\int\frac{K'dK'}{(2\pi)^6\hbar^3}\int d^3P'\delta^4(p_i - p_f) \tag{14.13.3}$$

where J_i is the incident photon flux density. In the last step we have used the result that one incident photon in volume V constitutes a photon flux density of

$$J_i = \frac{c}{V} \tag{14.13.4}$$

and assumed that $|M_{fi}|^2$ does not vary strongly over the group of final states involved in the scattering.

If the target electron is unpolarized and polarization of the recoil electron is not observed, we must average the cross-section (14.13.3) over the spin states of the target electron and sum over the spin states of the recoil electron. (We postpone averaging over the polarization states of the incident photon and summing over the polarization states of the scattered photon. This means we consider the incident and scattered photons to be polarized.) Hence the differential cross-section for the Compton scattering of photons in the direction of $\boldsymbol{k}'$ by unpolarized electrons is

$$\frac{d\sigma_{fi}}{d\Omega_{k'}} = \overline{\sum_i}\sum_f V^4 4\hbar^{-2}|M_{fi}|^2 I \tag{14.13.5}$$

where $\overline{\sum_i}$ and $\sum_f$ imply, respectively, averaging over the spin states of the target electron and sum over the spin states of the recoil electron and I stands for the integral

$$\begin{aligned} I &= \int \hbar^3 K'^2 dK' \int d^3P'\delta^4(p_i - p_f) \\ &= \frac{1}{ic^2}\int E_1'^2 dE_1' \int d^3P'\delta^3(p + \hbar k - p' - \hbar k')\delta(E_i - E_f) . \end{aligned} \tag{14.13.6}$$

Here $E_i = c\hbar K + mc^2$ (assuming the target electron to be at rest), $E_f = c\hbar K' + E' = E_1' + E'$ where $E_1' = c\hbar K'$ is the energy of the outgoing photon and E' is the total energy (including rest energy) of the recoil electron. The three-dimensional delta function in Eq. (14.13.6) implies conservation of linear momentum $\boldsymbol{p}' = \boldsymbol{p} + \hbar(\boldsymbol{k} - \boldsymbol{k}') = \hbar(\boldsymbol{k} - \boldsymbol{k}')$, since the target electron is assumed to be at rest ($\boldsymbol{p} = 0$). This results in

$$E'^2 = \hbar^2 c^2 K^2 + E_1'^2 - 2\hbar^2 c^2 (\boldsymbol{k} \cdot \boldsymbol{k}') + m^2 c^4 . \tag{14.13.7}$$

Carrying out the integration over the momentum space, we can write Eq. (14.13.6) as

$$I = \frac{1}{ic^2} \int_0^\infty E_1'^2 dE_1' \delta\{f(E_1')\} \tag{14.13.8}$$

where

$$f(E_1') = E_1' - (E_i - E'), \tag{14.13.9}$$

and $\boldsymbol{p}' = \hbar(\boldsymbol{k} - \boldsymbol{k}')$.

The solution of the equation $f(E_1') = 0$ is obviously the value of E_1' given by Compton relation

$$E_1' = E_0 = \hbar c K' \quad \text{where} \quad \hbar K(1 - \cos\vartheta) + mc = mcK/K' , \tag{14.13.10}$$

which follows from the conservation of linear momentum and energy in this process. To evaluate the integral in Eq. (14.13.8), we use the identity

$$\delta\{f(E_1')\} = \frac{\delta(E_1' - E_0)}{\left|\frac{\partial f(E_1')}{\partial E_1'}\right|_{E_1'=E_0}} , \tag{14.13.11}$$

since $f(E_1')$ involves both E_1' as well as E', which in turn depends on E_1'. Using Eq. (14.13.9) we get

$$I = \frac{1}{ic^2} \int_0^\infty E_1'^2 dE_1' \frac{\delta(E_1' - E_0)}{(1 + \partial E'/\partial E_1')_{E_0}}$$

and, according to Eq. (14.13.7), we have

$$\left(1 + \frac{\partial E'}{\partial E_1'}\right) = \frac{E_1' E_i - (\boldsymbol{k} \cdot \boldsymbol{k}')\hbar^2 c^2}{E_1' E'} ,$$

so that

$$I = \left(\frac{1}{ic^2}\right) \frac{E_0^3 E'}{E_0 E_i - c^2 \hbar^2 (\boldsymbol{k} \cdot \boldsymbol{k}')} .$$

The magnitude of the wave-vector $\boldsymbol{k}'$ is given by $|\boldsymbol{k}'| = K' = E_0/\hbar c$, where E_0 is the energy of the scattered photon given by the Compton relation (14.13.10). Using this we we find

$$I = \frac{1}{ic^2} \left(\frac{\hbar \omega_{k'}^2 E'}{\hbar \omega_k (1 - \cos\theta) + mc^2} \right) = \frac{\hbar^2 \omega_{k'}^3 E'}{imc^4 \omega_k} , \tag{14.13.12}$$

where $\omega_{k'} = cK' = c|\boldsymbol{k}'|$ and $\cos\theta = \frac{\boldsymbol{k} \cdot \boldsymbol{k}'}{|\boldsymbol{k}||\boldsymbol{k}'|}$. Using the result for I and M_{fi} [Eq. (14.12.1)] in Eq. (14.13.5) we get

$$\frac{d\sigma}{d\Omega} = \frac{e^4 \omega_{k'}^2}{(4\pi\epsilon_0)^2 c^2 \omega_k^2} \overline{\sum_i \sum_f} |T_{fi}|^2 \tag{14.13.13}$$

where

$$T_{fi} = \tilde{u}_s(\boldsymbol{p}') \left\{ \not{\varepsilon}^{(n)} \frac{i}{\not{p} + \hbar \not{k} - imc} \not{\varepsilon}^{(m)} + \not{\varepsilon}^{(m)} \frac{i}{\not{p} - \hbar \not{k}' - imc} \not{\varepsilon}^{(n)} \right\} u_r(\boldsymbol{p})$$
$$= \tilde{u}_s(\boldsymbol{p}')\, O\, u_r(\boldsymbol{p}). \tag{14.13.14}$$

The matrix operator O is defined by

$$O = \left[\not{\varepsilon}^{(n)} \frac{i}{\not{p} + \hbar \not{k} - imc} \not{\varepsilon}^{(m)} + \not{\varepsilon}^{(m)} \frac{i}{\not{p}' - \hbar \not{k} - imc} \not{\varepsilon}^{(n)} \right],$$
$$= \frac{1}{2m} \left[\frac{i \not{\varepsilon}^{(n)} \not{\varepsilon}^{(m)} \not{k}}{\omega_k} + \frac{i \not{\varepsilon}^{(m)} \not{\varepsilon}^{(n)} \not{k}'}{\omega_{k'}} \right]. \tag{14.13.15}$$

For putting the operator O in the form (14.13.15) we have used the following identities:

(a) $-i[i\gamma(p + \hbar k) - mc][\gamma(p + \hbar k) - imc] = (p + \hbar k)^2 + m^2c^4$

(b) $(p + \hbar k)^2 + m^2c^2 = -2m\omega_k \hbar$

(c) $[i\gamma(p + \hbar k) - mc]\gamma\, \varepsilon^{(m)} u_r(\boldsymbol{p}) = -i\gamma\, \varepsilon^{(m)} \hbar\gamma\, k\, u_r(\boldsymbol{p})$

(d) $\hbar pk = -\hbar m\omega_k$

Now the matrix element squared can be written as

$$|T_{fi}|^2 = [\tilde{u}_s(\boldsymbol{p}')Ou_r(\boldsymbol{p})]\,[\tilde{u}_s(\boldsymbol{p}')Ou_r(\boldsymbol{p})]^*$$
$$= [\tilde{u}_s(\boldsymbol{p}')Ou_r(\boldsymbol{p})]\,[u_r^\dagger(\boldsymbol{p})\, O^\dagger (\tilde{u}_s(\boldsymbol{p}'))^\dagger]$$

or

$$|T_{fi}|^2 = [\tilde{u}_s(\boldsymbol{p}')Ou_r(\boldsymbol{p})]\left[\tilde{u}_r(\boldsymbol{p})\tilde{O}u_s(\boldsymbol{p}')\right],$$

where $\tilde{O} = \gamma_4 O^\dagger \gamma_4$. Hence

$$\overline{\sum_i} \sum_f |T_{fi}|^2 = \frac{1}{2} \sum_r \sum_s [\tilde{u}_s(\boldsymbol{p}')O\, u_r(\boldsymbol{p})] \left[u_r(\boldsymbol{p})\tilde{O}\, u_s(\boldsymbol{p}')\right]$$
$$= \frac{1}{2} \sum_s \tilde{u}_s(\boldsymbol{p}')O\Lambda^{(+)}(\boldsymbol{p})\tilde{O}\Lambda^{(+)}(\boldsymbol{p}')u_s(\boldsymbol{p}') \tag{14.13.16}$$

where

$$\Lambda^{(+)}(\boldsymbol{p}) = \sum_r u_r(\boldsymbol{p})\tilde{u}_r(\boldsymbol{p}) = \frac{-i\gamma\, p + mc}{2mc}$$

is the projection operator[15] for the electron state $u_s(\boldsymbol{p})$ [Eq. (14.8.21)]. This also implies that $\Lambda^{(+)}(\boldsymbol{p}')u_s(\boldsymbol{p}') = u_s(\boldsymbol{p}')$.

Since $\Lambda^{(+)}(\boldsymbol{p}')v_s(\boldsymbol{p}') = 0$, we can add a term $-\frac{1}{2}\sum_s \tilde{v}_s(\boldsymbol{p}')O\Lambda^{(+)}(\boldsymbol{p})\tilde{O}\Lambda^{(+)}(\boldsymbol{p}')v_s(\boldsymbol{p}')$ to the expression on the right-hand side of Eq. (14.13.16) to get

$$\overline{\sum_i} \sum_f |T_{fi}|^2 = \frac{1}{2} \sum_s [\tilde{u}_s(\boldsymbol{p}')Qu_s(\boldsymbol{p}') - \tilde{v}_s(\boldsymbol{p}')Qv_s(\boldsymbol{p}')],$$

[15] If we let the matrix operator $\sum_r u_r(\boldsymbol{p})\tilde{u}_r(\boldsymbol{p})$ operate on the spinor $u_s(\boldsymbol{p})$ [Eq. (14.8.6)], the result is $u_s(\boldsymbol{p})$. This operator can therefore be identified with $\Lambda^{(+)}(\boldsymbol{p})$.

where $Q = O\Lambda^{(+)}(\boldsymbol{p})\tilde{O}\Lambda^{(+)}(\boldsymbol{p}')$ and $\tilde{O} = \gamma_4 O^\dagger \gamma_4$. Writing out the matrix products on the right-hand side of this equation we get

$$\overline{\sum_i}\sum_f |T_{fi}|^2 = \frac{1}{2}\sum_s\sum_\alpha\sum_\beta [\tilde{u}_{s,\alpha}(\boldsymbol{p}')Q_{\alpha\beta}u_{s,\beta}(\boldsymbol{p}') - \tilde{v}_{s,\alpha}(\boldsymbol{p}')Q_{\alpha\beta}v_{s,\beta}(\boldsymbol{p}')]$$
$$= \frac{1}{2}\ \mathrm{Tr}\ Q\,, \tag{14.13.17}$$

where we have used the identity

$$\sum_s \{\tilde{u}_{s,\alpha}(\boldsymbol{p}')u_{s,\beta}(\boldsymbol{p}') - \tilde{v}_{s,\alpha}(\boldsymbol{p}')v_{s,\beta}(\boldsymbol{p}')\} = \delta_{\alpha,\beta}\,.$$

The suffixes α and β have been used to label the components of the spinors u_s and v_s. Thus

$$\overline{\sum_i}\sum_f |T_{f\,i}|^2 = \frac{1}{2}\ \mathrm{Tr}\ \{O\Lambda^{(+)}(\boldsymbol{p})\tilde{O}\Lambda^{(+)}(\boldsymbol{p}')\},$$

where

$$\tilde{O} = \gamma_4 O^\dagger\gamma_4 = \frac{i}{2m}\left[\frac{\not k\not\varepsilon^{(m)}\not\varepsilon^{(n)}}{\omega_k} + \frac{\not k'\not\varepsilon^{(n)}\not\varepsilon^{(m)}}{\omega_{k'}}\right].$$

Hence

$$\overline{\sum_i}\sum_f |T_{fi}|^2 = \frac{1}{32m^4c^2}\ \mathrm{Tr}\ \left[\left\{\frac{i\not\varepsilon^{(n)}\not\varepsilon^{(m)}\not k}{\omega_k} + \frac{i\not\varepsilon^{(m)}\not\varepsilon^{(n)}\not k'}{\omega_{k'}}\right\}(i\not p - mc)\right.$$
$$\left.\left\{\frac{i\not k\not\varepsilon^{(m)}\not\varepsilon^{(n)}}{\omega_k} + \frac{i\not k'\not\varepsilon^{(n)}\not\varepsilon^{(m)}}{\omega_{k'}}\right\}(i\not p' - mc)\right]. \tag{14.13.18}$$

Let us now introduce a four-vector a defined by

$$a = \frac{k'}{\omega_{k'}} - \frac{k}{\omega_k} \equiv \frac{k'}{\omega'} - \frac{k}{\omega}\,.$$

To simplify writing, we will denote ω_k and $\omega_{k'}$ by ω and ω', respectively, and $\varepsilon^{(n)}$, $\varepsilon^{(m)}$ by ε', ε. Now $p = (0, imc)$, since $|\boldsymbol{p}| = 0$, and $k = (\boldsymbol{k},\ i\omega/c)$, we have

$$a = \left[\left(\frac{\boldsymbol{k}'}{\omega'} - \frac{\boldsymbol{k}}{\omega}\right), 0\right]; \quad pa = 0 \quad \text{and} \quad a^2 = \frac{2m}{\omega\omega'\hbar}(\omega - \omega').$$

We also have

$$\frac{i\not\varepsilon\not\varepsilon'\not k'}{\omega'} + \frac{i\not\varepsilon'\not\varepsilon\not k}{\omega} = i\not\varepsilon\not\varepsilon'\not a + \frac{2i\varepsilon'\varepsilon\not k}{\omega},$$

by virtue of

$$\not\varepsilon'\not\varepsilon = -\not\varepsilon'\not\varepsilon + 2\varepsilon\varepsilon' \quad \text{and} \quad \frac{i\not k\not\varepsilon\not\varepsilon'}{\omega} + \frac{i\not k'\not\varepsilon'\not\varepsilon}{\omega'} = i\not a\not\varepsilon'\not\varepsilon + \frac{2i\not k\varepsilon\varepsilon'}{\omega}.$$

Using these results we find

$$\overline{\sum_i}\sum_f |T_{fi}|^2 = \frac{1}{32m^4c^2}\ \mathrm{Tr}\ \left[\left(i\not\varepsilon\not\varepsilon'\not a + 2i\varepsilon'\varepsilon\frac{\not k}{\omega}\right)(i\not p - mc)\left(i\not a\not\varepsilon'\not\varepsilon + \frac{2i\not k\varepsilon'\varepsilon}{\omega}\right)(i\not p' - mc)\right]$$
$$= \frac{1}{32m^4c^2}\ \mathrm{Tr}\ (P + Q + R + S) \tag{14.13.19}$$

where the matrices P, Q, R, and S are defined by

$$P = i\not\varepsilon\not\varepsilon'\not a(i\not p - mc)i\not a\not\varepsilon'\not\varepsilon(i\not p - mc)\,, \tag{14.13.20}$$

$$Q = 4(\varepsilon\varepsilon')^2\left\{\frac{i\not k}{\omega}(i\not p - mc)\frac{i\not k}{\omega}(i\not p' - mc)\right\}, \tag{14.13.21}$$

$$R = 2(\varepsilon\varepsilon')\left\{i\not\varepsilon\not\varepsilon'\not a(i\not p - mc)\frac{i\not k}{\omega}(i\not p' - mc)\right\}, \tag{14.13.22}$$

$$S = 2(\varepsilon\varepsilon')\left\{\frac{i\not k}{\omega}(i\not p - mc)i\not a\not\varepsilon'\not\varepsilon(i\not p' - mc)\right\}. \tag{14.13.23}$$

Using the identities

$$\not A\not B = -\not B\not A + 2AB$$
$$\text{Tr } \not A = 0 = \text{Tr } \not A\not B\not C$$
$$\text{Tr } \not A\not B = 4AB$$
$$\text{Tr } \not A\not B\not C\not D = 4[(AB)(CD) - (AC)(BD) + (AD)(BC)]$$

where A, B, C, D, are four-vectors, and the relations, $k'^2 = k^2 = ap = \varepsilon\, p = p\varepsilon' = \varepsilon k = 0$ and $p' = p + \hbar(k - k')$, we can show that

$$P = a^2(i\not p + mc)\left[i(\not p + \hbar\not k - \hbar\not k') - mc\right]$$

and

$$Q = \frac{4(\varepsilon\varepsilon')^2}{\omega^2}2(kp)(\not k\not p - \hbar\not k\not k' + imc\not k')\,.$$

Hence

$$\text{Tr } P = -a^2\,\text{Tr }\{\hbar\not p(\not k - \not k')\} = \frac{8m^2(\omega - \omega')^2}{\omega\omega'}, \tag{14.13.24}$$

and

$$\text{Tr } Q = \frac{32(\varepsilon\varepsilon')^2}{\omega^2}(kp)^2\left[1 - \hbar\frac{(kk')}{(kp)}\right]. \tag{14.13.25}$$

Since Tr $R =$ Tr S, we may write

$$\begin{aligned}
\text{Tr }(R+S) &= 4(\varepsilon\varepsilon')\,\text{Tr }[i\not\varepsilon\not\varepsilon'\not a(i\not p - mc)\frac{i\not k}{\omega}(i\not p' - mc)] \\
&= \frac{4(\varepsilon\varepsilon')}{\omega}\,\text{Tr }[(i\not p + mc)\not\varepsilon\not\varepsilon'\not a\not k\{i(\not p + \hbar\not k - \hbar\not k') - mc\}] \\
&= \frac{4(\varepsilon\varepsilon')}{\omega}\,\text{Tr }[(i\not p + mc)\not\varepsilon\not\varepsilon'\not a\not k\{(i\not p - mc) - i\hbar\not k'\}] \\
&= \frac{4(\varepsilon\varepsilon')}{\omega}\hbar\ \text{Tr }(\not p\not\varepsilon\not\varepsilon'\not a\not k\not k')\,.
\end{aligned}$$

The last step follows because Tr $\{(i\not p+mc)\not\varepsilon\not\varepsilon'\not a\not k(i\not p-mc)\} =$ Tr $\{(i\not p-mc)(i\not p+mc)\not\varepsilon\not\varepsilon'\not a\not k\} = 0$, and Tr $(\not\varepsilon\not\varepsilon'\not a\not k\not k') = 0$.

Thus

$$\begin{aligned}
\text{Tr }(R+S) &= \frac{4(\varepsilon\varepsilon')}{\omega}\hbar\,\text{Tr }\left[\not p\not\varepsilon\not\varepsilon'\left(\frac{\not k'}{\omega'} - \frac{\not k}{\omega}\right)\not k\not k'\right] \\
&= \frac{4(\varepsilon\varepsilon')\hbar}{\omega\omega'}\,\text{Tr }(\not p\not\varepsilon\not\varepsilon'\not k'\not k\not k') \\
&= 8\hbar\frac{(kk')(\varepsilon\varepsilon')}{\omega\omega'}\,\text{Tr }(\not p\not\varepsilon\not\varepsilon'\not k') \\
&= 8\hbar\frac{(kk')(\varepsilon\varepsilon')}{\omega\omega'}4[(p\varepsilon)(\varepsilon'k') - (p\varepsilon')(\varepsilon k') + (pk')(\varepsilon\varepsilon')]\,.
\end{aligned}$$

Using the result $(pk)(pk')/m^2 = \omega\omega'$, we can simplify this to yield

$$\text{Tr }(R+S) = 32m^2(\varepsilon\varepsilon')^2\hbar\frac{(kk')}{(pk)}\,. \tag{14.13.26}$$

Thus, finally we have

$$\overline{\sum_i}\sum_f |T_{fi}|^2 = \frac{1}{4m^2c^2}\left[\frac{(\omega-\omega')^2}{\omega\omega'} + 4(\varepsilon\varepsilon')^2\right] \tag{14.13.27}$$

and

$$\frac{d\sigma}{d\Omega_{K'}} = \frac{e^4}{(4\pi\epsilon_0 c)^2}\frac{\omega'^2}{\omega^2}\overline{\sum_i}\sum_f |T_{fi}|^2 = \frac{e^4}{(4\pi\epsilon_0)^2 4m^2c^4}\frac{\omega'^2}{\omega^2}\left[\frac{(\omega-\omega')^2}{\omega\omega'} + 4(\boldsymbol{\varepsilon}\cdot\boldsymbol{\varepsilon}')^2\right]. \tag{14.13.28}$$

The last result agrees with Eq. (13.7.26) of Chapter 13.

The averaging over the initial states of polarization of the incident photon and summing over the states of polarization of the ejected photon can be done exactly as in Chapter 13 [Sec. 13.7] to get the result

$$\left[\frac{d\sigma}{d\Omega}\right]_{\text{unpolarized}} = \frac{r_0^2}{2}\left(\frac{\omega'}{\omega}\right)^2\left\{\frac{\omega}{\omega'} + \frac{\omega'}{\omega} - \sin^2\theta\right\} \tag{14.13.29}$$

where $\cos\theta = \frac{\boldsymbol{k}\cdot\boldsymbol{k}'}{|\boldsymbol{k}||\boldsymbol{k}'|}$ and $r_0 = \frac{e^2}{4\pi\epsilon_0 mc^2}$ = classical radius of the electron.

14.14 Cross-sections for Other Electromagnetic Processes

Trace calculations of a similar nature may be performed to compute the cross-sections of other electromagnetic processes, using the corresponding amplitudes obtained from the respective Feynman graphs.

14.14.1 Electron-Positron Pair Annihilation (Electron at Rest)

Starting from the transition amplitude (14.12.2) corresponding to the Feynman diagrams in Fig. 14.9, one can write the cross-section of the process using Eq. (??). For averaging the cross-section over electron and positron spins and summing over photon polarizations, one may carry out the trace calculations to get the result

$$\frac{d\sigma}{d\Omega} = \left(\frac{r_0^2}{2}\right)\frac{m^2c^4}{c|\boldsymbol{p}_+|E_+ - c^2|\boldsymbol{p}_+|^2\cos\theta}\left[1 - \frac{E_+}{mc^2}\right.$$
$$\left.+\frac{(E_i - c|\boldsymbol{p}_+|\cos\theta)^2}{E_+E_i - E_i c|\boldsymbol{p}_+|\cos\theta} - 2\frac{(E_iE_+ - E_i c|\boldsymbol{p}_+|\cos\theta)}{(E_i - c|\boldsymbol{p}_+|\cos\theta)^2}\right] \tag{14.14.1}$$

where $E_i = E_+ + mc^2$, $\boldsymbol{p}_+$ and E_+ are the initial momentum and energy of the positron and

$$\cos\theta = \frac{\boldsymbol{k}\cdot\boldsymbol{p}_+}{|\boldsymbol{k}||\boldsymbol{p}_+|}\,. \tag{14.14.2}$$

For the total annihilation cross-section for positrons of energy E_+ colliding with electrons at rest, we get the expression

$$\sigma_{total} = \pi r_0^2 \left(\frac{mc^2}{E_i}\right) \left[\left(\frac{E_+^2 + 4mc^2 E_+ + m^2c^4}{c^2|\boldsymbol{p}_+|^2}\right) \ln\left\{\frac{E_+ + c|\boldsymbol{p}_+|}{mc^2}\right\} - \left(\frac{E_+ + 3mc^2}{c|\boldsymbol{p}_+|}\right)\right] . \tag{14.14.3}$$

This result was first obtained by Dirac (1930).

14.14.2 Möller (e^-e^-) and Bhabha (e^-e^+) Scattering

From the amplitudes (14.12.4) and (14.12.5) pertaining to the Feynman graphs of Figs. 14.11 and 14.12, respectively, we can calculate the differential cross-section for Möller and Bhabha scattering. Averaging over the initial spin states and sum over the final spin states can be carried out by performing the trace calculations. One then finds that the cross-section for Möller scattering in the center-of-mass frame is

$$\left[\frac{d\sigma}{d\Omega}\right]_{c.m.} = \frac{r_0^2}{4}\frac{m^2c^4}{E^2c^4p_c^4}\left[\frac{4(E^2 + c^2p_c^2)^2}{\sin^4\theta} - \frac{(8E^4 - 4E^2m^2c^4 - m^4c^8)}{\sin^2\theta} + c^4p_c^4\right] \tag{14.14.4}$$

where p_c is the electron momentum in the center-of-mass frame

$$p_c = |\boldsymbol{p}_1| = |\boldsymbol{p}_2| = |\boldsymbol{p}_1'| = |\boldsymbol{p}_2'| , \tag{14.14.5}$$

$$\cos\theta = \frac{\boldsymbol{p}_1 \cdot \boldsymbol{p}_1'}{p_c^2} , \tag{14.14.6}$$

$$E = E_1 = E_2 = E_1' = E_2' \tag{14.14.7}$$

and vectors $\boldsymbol{p}_1$, $\boldsymbol{p}_2$, $\boldsymbol{p}_1'$, $\boldsymbol{p}_2'$ are as defined in Fig. 14.11.

For Bhabha (e^-e^+) scattering, the expression for the differential cross-section (in the center-of-mass frame) is

$$\left[\frac{d\sigma}{d\Omega}\right]_{cm} = r_0^2\frac{m^2c^4}{64E^2}\left[\frac{A}{16c^4p_c^4\sin^4(\theta/2)} + \frac{B}{E_c^4} + \frac{2C}{4c^2p_c^2E_c^2\sin^2(\theta/2)}\right] , \tag{14.14.8}$$

where

$$A = 32[2m^2c^4(m^2c^4 + p_c^2c^2\cos\theta - E^2) + (c^2p_c^2\cos\theta + E^2)^2 + (c^2p_c^2 + E^2)^2], \tag{14.14.9}$$

$$B = 32c^4[2m^2c^2(m^2c^2 - p_1p_2) + (p_1p_2)^2 + (p_1p_1')^2], \tag{14.14.10}$$

$$C = -32c^4[m^2c^2(m^2c^2 - p_1p_2' + p_1p_1' - p_1p_2) + (p_1p_2')^2]. \tag{14.14.11}$$

Four-vectors p_1, p_2, p'_1, p'_2 have been defined in the expressions (14.12.5) for the amplitude M_{fi} for Bhabha scattering and also specified in the relevant Feynman graphs in Fig.14.12. Also

$$p_c = |\boldsymbol{p}_1| = |\boldsymbol{p}_1'| = |\boldsymbol{p}_2| = |\boldsymbol{p}_2'| , \tag{14.14.12}$$

$$\cos\theta = \frac{\boldsymbol{p}_1 \cdot \boldsymbol{p}_1'}{p_c^2} , \tag{14.14.13}$$

and

$$E = E_1 = E_2 = E_1' = E_2' . \tag{14.14.14}$$

Problems

1. An entity $F(\psi, \pi)$ defined by:

$$F(\psi, \pi) = \int \chi(\psi(\boldsymbol{r}, t), \pi(\boldsymbol{r}, t))\, d\tau$$

is called a functional of ψ and π, if the variation in F, produced by variations $\delta\psi$ and $\delta\pi$ in ψ and π, can be expressed as

$$\delta F = \int \{f_1(\psi, \pi)\,\delta\psi + f_2(\psi, \pi)\,\delta\pi\}\, d\tau.$$

Define functional derivatives $\frac{\eth F}{\eth \psi}$ and $\frac{\eth F}{\eth \pi}$ of F with respect to ψ and π and show that

$$\frac{\eth F}{\eth \psi} = f_1(\psi, \pi)$$

$$\frac{\eth F}{\eth \pi} = f_2(\psi, \pi)\,.$$

2. Using (i) the field equations in the Lagrangian form,

$$\frac{\partial}{\partial x_\mu}\left(\frac{\partial \mathcal{L}}{\partial \phi_\mu}\right) - \frac{\partial \mathcal{L}}{\partial \phi} = 0, \quad \text{or} \quad \frac{d}{dt}\left(\frac{\eth L}{\eth \dot{\phi}}\right) - \frac{\eth L}{\eth \phi} = 0,$$

(ii) the definition of the field variable π canonically conjugate to ϕ, viz., $\pi = \frac{\eth L}{\eth \dot{\psi}}$ such that the momentum P_i canonically conjugate to ϕ_i (mean value of ϕ_i in the i-th cell of volume $\Delta\tau_i$ in the configuration space) is given, in analogy with particle dynamics by

$$P_i = \frac{\delta L}{\delta \dot{\phi}_i} = \left(\frac{\delta \mathcal{L}}{\delta \dot{\phi}}\right)_i \Delta\tau_i = \left(\frac{\eth L}{\eth \dot{\phi}}\right)_i \Delta\tau_i = \pi_i \Delta\tau_i,$$

and (iii) the definition of the Hamiltonian of the field in analogy with particle dynamics,

$$H = \sum_{i=1}^{N} P_i \dot{\phi}_i - L(\phi, \dot{\phi}) = \sum_{i=1}^{N} (\pi_i \dot{\phi}_i - \mathcal{L}_i)\Delta\tau_i$$

or

$$H = \int (\pi\dot{\phi} - \mathcal{L})\, d\tau = \int \mathcal{H}\, d\tau\,,$$

in the limit $\Delta\tau_i \to 0$, $N \to \infty$, show that H is a functional of ϕ and π. Also derive the field equations in the canonical form

$$\dot{\phi} = \frac{\eth H}{\eth \pi} = \frac{\partial \mathcal{H}}{\partial \pi} - \sum_{k=1}^{3} \frac{\partial}{\partial x_k}\left(\frac{\partial \mathcal{H}}{\partial \pi_k}\right),$$

and

$$-\dot{\pi} = \frac{\eth H}{\eth \phi} = \frac{\partial \mathcal{H}}{\partial \phi} - \sum_{k=1}^{3} \frac{\partial}{\partial x_k}\left(\frac{\partial \mathcal{H}}{\partial \phi_k}\right),$$

where $\pi_k \equiv \frac{\partial \pi}{\partial x_k}$ and $\phi_k \equiv \frac{\partial \phi}{\partial x_k}$.

3. Given the Lagrangian density for the electromagnetic field,

$$\mathcal{L}_{em} = \frac{1}{\mu_0}\left(-\frac{1}{4}F_{\mu\nu}F_{\mu\nu} - \frac{1}{2}\frac{\partial A_\mu}{\partial x_\mu}\frac{\partial A_\nu}{\partial x_\nu}\right) + j_\mu A_\mu$$

derive the corresponding field equations.

4. For a real scalar (KG) field the Lagrangian density can be expressed by

$$\mathcal{L} = -\frac{1}{2}\left(\frac{\partial\varphi}{\partial x_\mu}\frac{\partial\varphi}{\partial x_\mu} + \kappa_0^2\varphi^2\right)$$

while for a complex scalar (KG) field it is given by

$$\mathcal{L} = -\left(\frac{\partial\varphi^*}{\partial x_\mu}\frac{\partial\varphi}{\partial x_\mu} + \kappa_0^2\varphi^*\varphi\right)$$

where $\kappa_0 = \frac{mc}{\hbar}$. Derive the corresponding field equations. Show that a real field may be looked upon as a *neutral field* while the complex field can be looked upon as a *charged field.* Derive expressions for the charge density ρ and charge current density $\boldsymbol{j}$ for the field in the latter case.

5. The Lagrangian density for the Dirac field is given as

$$\begin{aligned}\mathcal{L} &= -\tilde{\psi}\left(c\hbar\gamma_\rho\frac{\partial\psi}{\partial x_\rho} + mc^2\psi\right)\\ &= c\hbar\frac{\partial\tilde{\psi}}{\partial x_\rho}\gamma_\rho\psi - mc^2\tilde{\psi}\psi.\end{aligned}$$

Using the Euler-Lagrange equation derive the field equation (Dirac equation). Work out the Hamiltonian density, the total Hamiltonian H and the total charge for the Dirac field.

6. Show that the Hamiltonian density of Dirac field may be expressed as

$$\mathcal{H} = \psi^\dagger(c\,\boldsymbol{\alpha}\cdot\boldsymbol{p} + \beta mc^2)\psi$$

and the total Hamiltonian of the Dirac field as

$$H = \int \mathcal{H}\,d\tau = -c\hbar\int\tilde{\psi}\gamma_4\frac{\partial\psi}{\partial x_4}\,d\tau\,.$$

Using the Fourier expansion of the Dirac field ψ

$$\psi(\boldsymbol{r},t) = \frac{1}{\sqrt{V}}\sum_{\boldsymbol{p}}\sum_{\nu=1}^{4}\sqrt{\frac{mc^2}{|E_p|}}\,b_{\boldsymbol{p}}^\nu\,u^\nu(\boldsymbol{p})\exp(i\boldsymbol{p}\cdot\boldsymbol{r}/\hbar),$$

with similar form for $\psi^\dagger$, express the total Hamiltonian and the total charge of the field in terms of the expansion coefficients as

$$H = \sum_{\boldsymbol{p}}\sum_{\nu=1}^{4}E_p^\nu\,b_{\boldsymbol{p}}^{\nu*}(t)\,b_{\boldsymbol{p}}^\nu(t)\,,$$

$$Q = e\sum_{\boldsymbol{p}}\sum_{\nu=1}^{4}b_{\boldsymbol{p}}^{\nu*}\,b_{\boldsymbol{p}}^\nu\quad.$$

7. To quantize the Dirac field we regard the field functions $\psi(\boldsymbol{r},t)$ and $\tilde{\psi}(\boldsymbol{r},t)$ as operators $\hat{\psi}$ and $\hat{\tilde{\psi}}$. Consequently the Fourier expansion coefficients $b^{\nu}_{\boldsymbol{p}}$ and $b^{\nu *}_{\boldsymbol{p}}$ must also be treated as operators $\hat{b}^{\nu}_{\boldsymbol{p}}$ and $\hat{b}^{\nu\dagger}_{\boldsymbol{p}}$, so that the Hamiltonian and charge operators of the field may be expressed as

$$\hat{H} = \sum_{\boldsymbol{p}} \sum_{\nu=1}^{4} E^{\nu}_{p}\, \hat{b}^{\nu}_{\boldsymbol{p}}\, \hat{b}^{\nu\dagger}_{\boldsymbol{p}}\,,$$

$$\hat{Q} = e \sum_{\boldsymbol{p}} \sum_{\nu=1}^{4} \hat{b}^{\nu}_{\boldsymbol{p}}\, \hat{b}^{\nu\dagger}_{\boldsymbol{p}}.$$

(a) Infer the time dependence of the operators $\hat{b}$ and $\hat{b}^{\dagger}$. (b) What considerations prompt us to assert that the operators $\hat{b}$ and $\hat{b}^{\dagger}$ obey anti-commutation relations

$$[\hat{b}^{\nu\dagger}_{\boldsymbol{p}},\, \hat{b}^{\nu'}_{\boldsymbol{p}'}]_{+} \equiv \hat{b}^{\nu\dagger}_{\boldsymbol{p}}\, \hat{b}^{\nu'}_{\boldsymbol{p}'} + \hat{b}^{\nu'}_{\boldsymbol{p}'}\, \hat{b}^{\nu\dagger}_{\boldsymbol{p}} = \hat{1}\delta_{\boldsymbol{p}\boldsymbol{p}'}\delta_{\nu\nu'}\,,$$

$$[\hat{b}^{\nu}_{\boldsymbol{p}},\, \hat{b}^{\nu'}_{\boldsymbol{p}'}]_{+} = [\hat{b}^{\nu\dagger}_{\boldsymbol{p}},\, \hat{b}^{\nu'\dagger}_{\boldsymbol{p}'}]_{+} = 0,$$

rather than commutation relations.

8. Show that $\hat{b}^{\nu\dagger}_{\boldsymbol{p}}$ and $\hat{b}^{\nu}_{\boldsymbol{p}}$ may be regarded as the creation and annihilation operators for an electron with momentum $\boldsymbol{p}$ and spin and sign of energy specified by ν.

9. In place of Dirac field operators $\hat{b}^{(\nu)}_{\boldsymbol{p}}$ and $\hat{b}^{(\nu)\dagger}_{\boldsymbol{p}}$, (with $\nu = 1(2)$ corresponding to positive energy and spin up (down) and $\nu = 3(4)$ corresponding, negative energy with spin up (down)), one can introduce the electron and positron creation and annihilation operators $\hat{b}^{(s)\dagger}_{\boldsymbol{p}}$, $\hat{b}^{(s)}_{\boldsymbol{p}}$ and $\hat{d}^{(s)\dagger}_{\boldsymbol{p}}$, $\hat{d}^{(s)}_{\boldsymbol{p}}$, with $s = 1, 2$ corresponding to spin up or down, so that

$$\hat{b}^{(s)}_{\boldsymbol{p}} = \hat{b}^{\nu}_{\boldsymbol{p}} \qquad \text{for } s = \nu = 1, 2$$

and

$$\hat{d}^{(s)}_{\boldsymbol{p}} = \begin{cases} -\hat{b}^{(\nu)}_{\boldsymbol{p}} & \text{for } s = 1, \nu = 4 \\ +\hat{b}^{(\nu)}_{\boldsymbol{p}} & \text{for } s = 2, \nu = 3 \end{cases}$$

With this, the energy of a free particle is always positive. One can also introduce electron and positron spinors $u^{s}(\boldsymbol{p})$ and $v^{s}(\boldsymbol{p})$ so that

$$\begin{aligned} u^{s}(\boldsymbol{p}) &= u^{\nu}(\boldsymbol{p}) && \text{with} \quad s = \nu = 1, 2 \\ v^{s}(\boldsymbol{p}) &= \mp\, u^{\nu}(-\boldsymbol{p}) && \text{with} \quad s = 1\,, \quad \nu = 4 \quad \text{(upper sign)} \\ & && \phantom{\text{with}} \quad s = 2\,, \quad \nu = 3 \quad \text{(lower sign)}\,. \end{aligned}$$

Verify that this definition is also consistent with the operation of the charge conjugation operator given by $S_c = \gamma_2 = -i\beta\,\alpha_2$ so that

$$S_c\, u^{(s)*}(\boldsymbol{p}) = v^{s}(\boldsymbol{p})$$

and

$$S_c\, v^{(s)*}(\boldsymbol{p}) = u^{s}(\boldsymbol{p})\,.$$

Dirac spinors $u^{(\nu)}_{\boldsymbol{p}}$ for $\nu = 1, 2, 3, 4$ are given by Eqs. (12.4.16), (12.4.17), (12.4.20) and (12.4.21) of Chapter 12.

10. Show that the anti-commutation bracket has algebraic properties different from those of the commutator bracket (or classical Poisson bracket). Deduce the anti-commutation relations satisfied by the electron and positron creation and annihilation operators $\hat{b}^{(s)}_{\boldsymbol{p}}$, $\hat{b}^{(s)\dagger}_{\boldsymbol{p}}$, $\hat{d}^{(s)}_{\boldsymbol{p}}$ and $\hat{d}^{(s)\dagger}_{\boldsymbol{p}}$.

11. Show that, with the introduction of the electron and positron creation and annihilation operators, the expansions of the field operators $\hat{\psi}$ and $\hat{\tilde{\psi}}$ take the form:

$$\hat{\psi}(\boldsymbol{r},t) = \hat{\psi}^{(+)}(\boldsymbol{r},t) + \hat{\psi}^{(-)}(\boldsymbol{r},t)$$

and
$$\hat{\tilde{\psi}}(\boldsymbol{r},t) = \hat{\tilde{\psi}}^{(+)}(\boldsymbol{r},t) + \hat{\tilde{\psi}}^{(-)}(\boldsymbol{r},t)\,,$$

where
$$\hat{\psi}^{(+)} = \frac{1}{\sqrt{V}}\sum_{\boldsymbol{p}}\sum_{s=1}^{2}\sqrt{\frac{mc^2}{E_p}}\left\{\hat{b}_{\boldsymbol{p}}^{(s)} u^{(s)}(\boldsymbol{p})\exp\left(\frac{i\boldsymbol{p}\cdot\boldsymbol{r}}{\hbar} - \frac{iE_p t}{\hbar}\right)\right\},$$

$$\hat{\psi}^{(-)} = \frac{1}{\sqrt{V}}\sum_{\boldsymbol{p}}\sum_{s=1}^{2}\sqrt{\frac{mc^2}{E_p}}\left\{\hat{d}_{\boldsymbol{p}}^{(s)\dagger} v^{(s)}(\boldsymbol{p})\exp\left(-\frac{i\boldsymbol{p}\cdot\boldsymbol{r}}{\hbar} + \frac{iE_p t}{\hbar}\right)\right\},$$

$$\hat{\tilde{\psi}}^{(+)} = \frac{1}{\sqrt{V}}\sum_{\boldsymbol{p}}\sum_{s=1}^{2}\sqrt{\frac{mc^2}{E_p}}\left\{\hat{d}_{\boldsymbol{p}}^{(s)} \tilde{v}^{(s)}(\boldsymbol{p})\exp\left(\frac{i\boldsymbol{p}.\boldsymbol{r}}{\hbar} - \frac{iE_p t}{\hbar}\right)\right\},$$

$$\hat{\tilde{\psi}}^{(-)} = \frac{1}{\sqrt{V}}\sum_{\boldsymbol{p}}\sum_{s=1}^{2}\sqrt{\frac{mc^2}{E_p}}\left\{\hat{b}_{\boldsymbol{p}}^{(s)\dagger} \tilde{u}^{(s)}(\boldsymbol{p})\exp\left(-\frac{i\boldsymbol{p}.\boldsymbol{r}}{\hbar} + \frac{iE_p}{\hbar}\right)\right\}.$$

12. Show that, in view of the anti-commutation relations satisfied by the electron creation and annihilation operators [Problem 10], the equal time ($t = t'$) anti-commutation relations, satisfied by the field operators, are as follows

$$[\hat{\psi}_\mu(x), \hat{\psi}_\nu(x)]_+ = [\hat{\psi}_\mu^\dagger(x), \hat{\psi}_\nu^\dagger(x')]_+ = [\hat{\tilde{\psi}}_\mu(x), \hat{\tilde{\psi}}_\nu(x')]_+ = 0$$

and
$$[\hat{\psi}_\mu(x), \hat{\psi}_\nu^\dagger(x')]_+ = \delta_{\mu\nu}\delta^3(\boldsymbol{r} - \boldsymbol{r}')$$

which implies
$$[\hat{\psi}_\mu(x), \hat{\tilde{\psi}}_\nu(x')]_+ = (\gamma_4)_{\mu\nu}\delta^3(\boldsymbol{r} - \boldsymbol{r}')\,.$$

Here μ and ν label the four components of the spinors $\psi, \psi^\dagger$ and $\tilde{\psi}$ and $x \equiv (\boldsymbol{r}, ict)$.

13. Show that the interaction Hamiltonian, pertaining to the interaction between the electron field and the radiation field, is given by

$$\hat{H}_I = \int \hat{\mathcal{H}} d\tau = \int -iec\hat{\tilde{\psi}}\gamma_\mu\hat{\psi}\hat{A}_\mu d\tau\,.$$

14. Show that, in the interaction picture, the state of the field $|\Phi^{\text{int}}(t)\rangle$ at time t is given by a Schrödinger-like equation

$$i\hbar\frac{d}{dt}|\Phi^{\text{int}}(t)\rangle = \hat{H}_I^{\text{int}}(t)|\Phi^{\text{int}}(t)\rangle\,,$$

where $\hat{H}_I^{\text{int}}(t)$ is the interaction Hamiltonian in the interaction picture, given by

$$\hat{H}_I^{\text{int}}(t) = \exp(i\hat{H}_0 t/\hbar)\hat{H}_I^{(S)}\exp(-i\hat{H}_0 t/\hbar)\,,$$

$\hat{H}_I^{(S)}$ being the interaction Hamiltonian in the Schrödinger picture. Thus both $|\Phi^{\text{int}}\rangle$ and $\hat{H}_I^{\text{int}}$ have time dependence.

15. Define the U matrix and S matrix. Write down the iterative expansion of the S operator. Show that if the interaction density pertains to the interaction of the electron field and the radiation field,

$$\mathcal{H}_I^{\text{int}}(t) = -iec\hat{\bar{\psi}}\gamma_\mu\hat{\psi}\,\hat{A}_\mu$$

then the first order term in the expansion of the S operator, viz.,

$$\hat{S}^{(1)} = \frac{ie}{\hbar}\int \hat{\bar{\psi}}\gamma_\mu\hat{\psi}\hat{A}_\mu d^4x$$

cannot give rise to any physical process. Draw the relevant Feynman diagrams and explain what prevents these processes from taking place.

16. What physical processes may be brought about by the second order term in the expansion of the S operator. Draw the relevant Feynman diagrams.

17. Explain the difference between Feynman diagrams and Feynman graphs. Enunciate the rules for writing matrix elements (transition amplitudes) for any electrodynamic process from Feynman graphs. Illustrate this with the example of Compton scattering from atomic electrons.

18. If the matrices R and S are defined as

$$R = 2(\varepsilon\varepsilon')\left[i\not{\varepsilon}\not{\varepsilon}'\not{a}(i\not{p} - mc)\frac{i\not{k}}{\omega}(i\not{p}' - mc)\right]$$

$$S = 2(\varepsilon\varepsilon')\left[i\frac{\not{k}}{\omega}(i\not{p} - mc)i\not{a}\not{\varepsilon}'\not{\varepsilon}(i\not{p}' - mc)\right]$$

where $\not{p} \equiv \gamma p = \gamma_\mu p_\mu$; $\not{a} = \gamma_\mu a_\mu$ with $a \equiv \left(\left(\frac{k'}{\omega'} - \frac{k}{\omega}\right), 0\right)$ and $\varepsilon\varepsilon' = \varepsilon_\mu\varepsilon'_\mu$, then show that Tr $R =$ Tr S and calculate Tr $(R + S)$.

19. Given $P = a^2(i\not{p} + mc)\{i(\not{p} + \hbar\not{k} - \hbar\not{k}') - mc\}$ and $Q = 4(\varepsilon\varepsilon')\frac{2kp}{\omega^2}(\not{k}\not{p} - \hbar\not{k}\not{k}' + mc\not{k}')$, calculate Trace P and Trace Q. The four-vector a has been defined in problem 18

References

[1] J. D. Bjorken and S. D. Drell, *Relativistic Quantum Mechanics* (McGraw Hill , New York, 1964).

[2] N. N. Bogoliubov and D. V. Shirkov, *Introduction to the Theory of Quantized Fields*, Third Edition (John Wiley & Sons, New York, 1980).

[3] W. Heitler, *Quantum Theory of Radiation*, Third Edition (Clarendon Press, Oxford, 1954).

[4] H. Muirhead, *Physics of Elementary Particles* (Pergamon Press, Oxford, 1968).

[5] J. Sakurai, *Advanced Quantum Mechanics* (Addison-Wesley, Reading, MA, 1967).

[6] L. I. Schiff, *Quantum Mechanics*, Third Edition (McGraw Hill Book Company, Inc., New York, 1968).

[7] F. Schwabl, *Advanced Quantum Mechanics*, (Springer-Verlag, Berlin, 1999).

[8] S. S. Schweber, *An introduction to Relativistic Quantum Field Theory* (Harper & Row, New York, 1961).

Appendix 14A1: Calculus of Variation and Euler-Lagrange Equations

The calculus of variations deals with the problem of finding a function $y = f(x)$, for which the integral

$$A = \int_{x_1}^{x_2} I(x, y, \frac{dy}{dx})dx \tag{14A1.1}$$

is stationary (maximum or minimum) for an arbitrary small variation in y. The meaning of *variation in* y is illustrated in Fig. 14A1.1. If LQH is the curve $y = f(x)$ joining fixed points L and B and LSH is the varied curve $Y = F(x)$, joining the same end points, then $\delta\, y = Y(x) - y(x) = F(x) - f(x)$ is called the variation in y. Obviously $\delta\, y$ is a function of x which vanishes at the end points x_1 (L) and x_2 (H). Thus we can speak of the *variation* of a function such as $y(x)$ and therefore, also of a constant such as the integral A of (14A1.1), because the latter will change as $y(x)$ is varied.

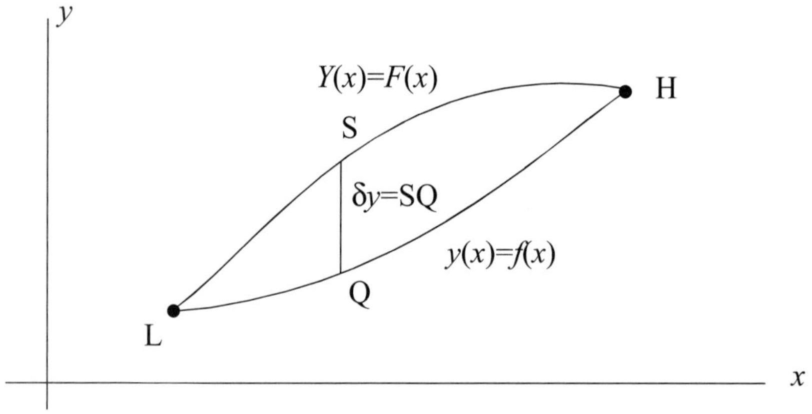

FIGURE 14A1.1
The function $y = f(x)$ is represented by the curve LQH and the varied function $Y = F(x)$ is represented by the curve LSH. $\delta y = SQ$ represents the variation in y at the point x.

Let us now compute the variation in A when $y(x)$ is varied. Let the derivative $\frac{dy}{dx}$ be denoted by y_x. Then the variation in the derivative is given by

$$\delta\, y_x = \delta \left(\frac{dy}{dx}\right) = \frac{dY}{dx} - \frac{dy}{dx} = \frac{d}{dx}(Y - y) = \frac{d}{dx}\delta y\,. \tag{14A1.2}$$

This means that the operations of variation δ and differentiation $\frac{d}{dx}$ may be interchanged. In the same way, the operations of variation and integration may also be interchanged, giving us

$$\delta A \equiv \delta \int I dx = \int \delta\, I\, dx, \tag{14A1.3}$$

which means that the variation in the integral equals the integral of the variation. Now the variation in the function $I(x, y, dy/dx)$ as a result of the variation δy in y is given by

$$\begin{aligned}\delta I &= I\left(x, Y, \frac{dY}{dx}\right) - I\left(x, y, \frac{dy}{dx}\right) \\ &= I(x, y + \delta y, y_x + \delta y_x) - I(x, y, y_x) \\ &= \left\{ I(x, y, y_x) + \frac{\partial I}{\partial y}\delta y + \frac{\partial I}{\partial y_x}\delta y_x + \cdots \right\} - I(x, y, y_x) \\ &= \frac{\partial I}{\partial y}\delta y + \frac{\partial I}{\partial y_x}\delta y_x \,. \end{aligned} \tag{14A1.4}$$

Then the condition that the integral A be stationary with respect to a variation of y means that its variation as a result of arbitrary δy must vanish,

$$\delta A = \delta \int_{x_1}^{x_2} I dx = \int_{x_1}^{x_2} \delta I dx = 0$$

or

$$\int_{x_1}^{x_2} \left(\frac{\partial I}{\partial y}\delta y + \frac{\partial I}{\partial y_x}\delta y_x \right) dx = 0,$$

or

$$\int_{x_1}^{x_2} \frac{\partial I}{\partial y}\delta y dx + \left[\frac{\partial I}{\partial y_x}\delta y \right]_{x_1}^{x_2} - \int_{x_1}^{x_2} \frac{d}{dx}\left(\frac{\partial I}{\partial y_x} \right)\delta y dx = 0 \,,$$

where the last step follows from the preceding one by integration by parts. The integrated term gives zero contribution since the variation at the end points vanishes. The condition for A to be stationary with respect to a variation of $y(x)$ then takes the form

$$\delta A \equiv \int_{x_1}^{x_2} \delta y \left\{ \frac{\partial I}{\partial y} - \frac{d}{dx}\left(\frac{\partial I}{\partial y_x} \right) \right\} dx = 0 \,.$$

Since the variation in y is arbitrary, it follows that the condition for the integral A to be stationary is

$$\frac{\partial I}{\partial y} - \frac{d}{dx}\left(\frac{\partial I}{\partial y_x} \right) = 0 \,. \tag{14A1.5}$$

This equation is called *Euler-Lagrange equation.*

We may as well consider the case in which the integrand I is a function of one independent variable x and several dependent variables $y_1(x)$, $y_2(x)$, $\cdots$, $y_n(x)$ and their derivatives y_{1x}, y_{2x}, $\cdots$, y_{nx}, where $y_{nx} \equiv \frac{dy_n}{dx}$. Then the condition for the integral

$$A = \int I(x; y_1, y_2, \cdots, y_n; y_{1x}, y_{2x}, \cdots, y_{nx}) dx, \tag{14A1.6}$$

to be stationary, for arbitrary variations of y_1, y_2, $\cdots$, y_n, is again

$$\delta A = \int_{x_1}^{x_2} \delta I dx = 0.$$

Now the change in I, when y_1, y_2, $\cdots$, y_n, are varied can be expressed as

$$\begin{aligned}\delta I &= \left(\frac{\partial I}{\partial y_1}\delta y_1 + \frac{\partial I}{\partial y_2}\delta y_2 + \cdots \frac{\partial I}{\partial y_n}\delta y_n \right) + \left(\frac{\partial I}{\partial y_{1x}}\delta y_{1x} + \frac{\partial I}{\partial y_{2x}}\delta y_{2x} + \cdots + \frac{\partial I}{\partial y_{nx}}\delta y_{nx} \right) \\ &= \sum_i^n \left\{ \left(\frac{\partial I}{\partial y_i} \right)\delta y_i + \left(\frac{\partial I}{\partial y_{ix}} \right)\delta y_{ix} \right\} \,. \end{aligned} \tag{14A1.7}$$

The condition $\delta A = 0$ then takes the form

$$\sum_{i=1}^{n} \int_{x_1}^{x_2} \left\{ \frac{\partial I}{\partial y_i} \delta\, y_i + \frac{\partial I}{\partial y_{ix}} \delta\, y_{ix} \right\} dx = 0\,.$$

Integrating the second term by parts once, we obtain

$$\sum_{i=1}^{n} \left[\int_{x_1}^{x_2} \frac{\partial I}{\partial y_i} \delta\, y_i dx + \left\{ \frac{\partial I}{\partial y_{ix}} \delta\, y_i \right\}_{x_2}^{x_1} - \int_{x_1}^{x_2} \frac{d}{dx} \left(\frac{\partial I}{\partial y_{ix}} \right) \delta y_i dx \right] = 0\,.$$

The integrated term vanishes, since the variation in y_i is zero at the end points. Thus the condition for A to be stationary with respect to variations of y_i is

$$\delta A \equiv \sum_{i=1}^{n} \int_{x_1}^{x_2} \left\{ \frac{\partial I}{\partial y_i} - \frac{d}{dx} \left(\frac{\partial I}{\partial y_{ix}} \right) \right\} \delta y_i dx = 0\,.$$

As the variations δy_i are arbitrary, the integrand for each term must vanish separately, leading to the following set of conditions (equations) on y_i for the integral A to be stationary

$$\frac{\partial I}{d y_i} - \frac{d}{dx} \left(\frac{\partial I}{\partial y_{ix}} \right) = 0, \quad \text{where } i = 1, 2, \cdots, n. \tag{14A1.8}$$

These equations, called the *Euler-Lagrange equations*, determine the set of functions $y_i(x)$ which make the integral A stationary.

Hamilton's Principle

Hamilton's principle describes the motion of a system of particles in terms of a *stationarity condition*. It states that the actual motion of a system of particles, with f degrees of freedom, from an initial configuration at time t_1 to a final configuration at time t_2, is determined by the condition that the *action*, defined by

$$A = \int_{t_1}^{t_2} L \left(t, q_1, q_2, \cdots, q_f, \frac{dq_1}{dt}, \frac{dq_2}{dt}, \cdots, \frac{dq_f}{dt} \right) dt$$

be stationary, i.e., the variation in A must vanish. Here the Lagrangian L of the system is defined to be the difference ($L = T - V$) between the kinetic energy T and the potential energy V of the system of particles and can be expressed in terms of the generalized coordinates q_i and generalized velocities $q_{it} \equiv dq_i/dt$ [Appendix 1A1]. Then the stationarity of action implies

$$\delta A = \int_{t_1}^{t_2} \delta L dt = 0,$$

for small arbitrary variations of particle coordinates q_is. In this case, the Euler-Lagrange equations may be written down from Eq. (14A1.8) with the following replacements

$$x \to t,\ y_i \to q_i, \quad y_{ix} \to q_{it} = \frac{dq_i}{dt} = \dot{q}_i,\ I \to L \quad \text{and } n \to f\,.$$

This gives Lagrange equations of motion for a system of particles with f degrees of freedom

$$\frac{\partial L}{\partial q_i} - \frac{d}{dt}\left(\frac{\partial L}{\partial \dot{q}_i}\right) = 0\,. \tag{14A1.9}$$

Hamilton's principle is a very general principle and may also be used to derive equations for fields as well as particles.

Appendix 14A2: Functionals and Functional Derivatives

The integral

$$F(g) = \int_{x_1}^{x_2} I(x, g)dx \tag{14A2.1}$$

where x is the independent variable and $g(x)$ is a dependent variable, depends on the values of the function $g(x)$ at all points in the interval x_1 and x_2. We say that F is a functional of g. A variation of the function $g(x)$ will cause a variation in F.

Now consider a variation of $g(x) \to g(x) + \delta g(x)$ such that $\delta g(x_1) = 0 = \delta g(x_2)$. By dividing the interval $[x_1, x_2]$ in a large number of cells of length Δx_i, where $i = 1, 2, 3, \cdots, n$, the variation in F can be written as

$$\begin{aligned}\delta F \equiv [F(g+\delta g) - F(g)] &= \sum_{i=1}^{n} [I(x_i, g_i + \delta g_i) - I(x_i, g_i)]\Delta x_i \\ &= \sum_{i=1}^{n} \left[\frac{I(x_i, g_i + \delta g_i) - I(x_i, g_i)}{\delta g_i}\right] \delta g_i \Delta x_i\end{aligned}$$

where g_i is the value of g in the ith cell and δg_i is the change in the value of g in the ith cell. If we consider the variation of g such only the change δg_i in the ith cell is nonzero, then the limit $\lim_{\substack{\Delta x_i \to 0 \\ \delta g_i \to 0}} \frac{\delta F}{\Delta x_i \delta g_i}$, if it exists, defines the value of the functional derivative of F with respect to g at point x_i. Since the cell i is arbitrary we can drop the subscript i and define the functional derivative of F with respect to g by

$$\lim_{\substack{\Delta x_i \to 0 \\ \delta g_i \to 0}} \frac{\delta F}{\Delta x_i \delta g_i} = \frac{\eth F}{\eth g}\,. \tag{14A2.2}$$

For our example, it is easy to see that

$$\frac{\eth F}{\eth g} = \frac{\partial I}{\partial g}\,.$$

As another example, consider the case in which the integrand depends on the function $g(x)$ as well as its first derivative $g_x(x) = \frac{dg}{dx}$, as in Eq. (14A1.1) of Appendix 14A1. Then the variation of F due to a variation of g can be written as

$$\delta F = \int_{x_1}^{x_2} [\delta I(x, g, g_x)]\, dx = \int_{x_1}^{x_2} \delta g \left\{\frac{\partial I}{\partial g} - \frac{d}{dx}\left(\frac{\partial I}{\partial g_x}\right)\right\} dx\,. \tag{14A2.3}$$

Hence the functional derivative of F with respect to g, in this case is

$$\frac{\eth F}{\eth g} = \frac{\partial I}{\partial g} - \frac{d}{dx}\left(\frac{\partial I}{\partial g_x}\right)\,. \tag{14A2.4}$$

It is possible to consider functionals which depend on higher (than first) order derivatives but they are seldom useful.

In terms of functional derivative, the variation in F [Eq. (14A2.4)] can be written simply as

$$\delta F = \int_{\mathcal{A}} \left[\frac{\eth F}{\eth g} \delta g \right] dx, \tag{14A2.5}$$

and the condition for F to be stationary with respect to a variation of g can be written as

$$\frac{\eth F}{\eth g} = 0\,. \tag{14A2.6}$$

This is analogous to the condition for an ordinary function to be stationary with respect to a variation of its argument. If F is given by an equation like (14A1.1), then Eq. (14A2.6), together with Eq. (14A2.4) immediately leads to the Euler-Lagrange equation.

Generalization to functionals that depend on a function of several variables and parameters is straightforward. For example, consider the functional

$$F(g,t) = \int_{\mathcal{A}} \chi(x,y,g,g_x,g_y;t) d\tau\,, \tag{14A2.7}$$

which depends on a function $g = g(x,y;t)$ of two independent variables x and y (and a parameter t), as well as its first order derivatives $g_x(x,y;t) = \frac{\partial g}{\partial x}$, $g_y(x,y;t) = \frac{\partial g}{\partial y}$. Here $d\tau = dxdy$ is a volume element in a two-dimensional region $\mathcal{A}$ spanned by x and y.

Then the variation of F due to a variation in g can be written as

$$\delta F = \int_{\mathcal{A}} \left\{ \frac{\partial \chi}{\partial g} \delta g - \frac{\partial \chi}{\partial g_x} \delta g_x - \frac{\partial \chi}{\partial g_y} \delta g_y \right\} \delta\tau\,.$$

Using the results $\delta g_x = \frac{\partial}{\partial x}\delta g$ and $\delta g_x = \frac{\partial}{\partial x}\delta g$, analogous to Eq. (14A1.2), in this equation and integrating the second and third terms by parts once, dropping the boundary terms which vanish since the variation of g at the boundary vanishes, we obtain

$$\delta F = \int_{\mathcal{A}} \delta g \left\{ \frac{\partial \chi}{\partial g} - \frac{\partial}{\partial x}\frac{\partial \chi}{\partial g_x} - \frac{\partial}{\partial y}\frac{\partial \chi}{\partial g_y} \right\} \delta\tau\,. \tag{14A2.8}$$

We can define functional derivative in this two-dimensional case by

$$\lim_{\substack{\Delta\tau\to 0 \\ \Delta g \to 0}} \frac{\delta F}{\delta g \Delta\tau} \equiv \frac{\eth F}{\eth g}\,,$$

which holds for every point inside the region $\mathcal{A}$. For our particular case we find

$$\frac{\eth F}{\eth g} = \frac{\partial \chi}{\partial g} - \frac{\partial}{\partial x}\frac{\partial \chi}{\partial g_x} - \frac{\partial}{\partial y}\frac{\partial \chi}{\partial g_y}\,. \tag{14A2.9}$$

If χ does not depend on g_x and g_y, the last two terms drop out and we have

$$\frac{\eth F}{\eth g} = \frac{\partial \chi}{\partial g}\,. \tag{14A2.10}$$

Finally, if F depends on two functions, $g(x,y,z)$ and $h(x,y,z)$, we may write the variation in F as

$$\delta F = \int \left[\frac{\eth F}{\eth g} \delta g + \frac{\eth F}{\eth h} \delta h \right] d\tau \tag{14A2.11}$$

where $\frac{\eth F}{\eth g}$ and $\frac{\eth F}{\eth h}$ are, respectively, the functional derivatives of F with respect to g and h. It is clear that the procedure can be generalized to F depending on any number of functions.

Appendix 14A3: Interaction of the Electron and Radiation Fields

The total Lagrangian density of the electron field interacting with the radiation field may be written as

$$\mathcal{L} = \mathcal{L}_e + \mathcal{L}_\gamma + \mathcal{L}_I \tag{14A3.1}$$

where

$$\mathcal{L}_e = -\tilde{\psi}\left(c\hbar\gamma_\mu \frac{\partial \psi}{\partial x_\mu} + mc^2\psi\right) = c\hbar \frac{\partial \tilde{\psi}}{\partial x_\mu}\gamma_\mu \psi - mc^2\tilde{\psi}\psi \tag{14.4.26*}$$

represents the Lagrangian density for pure electron field,

$$\mathcal{L}_\gamma = -\frac{1}{2\mu_0}\frac{\partial A_\nu}{\partial x_\mu}\frac{\partial A_\nu}{\partial x_\mu}$$

with

$$A_\nu \equiv (\boldsymbol{A},\, i\phi/c), \tag{14.4.1*}$$

represents the Lagrangian density for pure radiation field, while

$$\mathcal{L}_I = iec\tilde{\psi}\gamma_\mu \psi A_\mu = j_\mu A_\mu \tag{14A3.3}$$

represents the Lagrangian density for the interaction between the two fields. The last expression can be justified because, for L_I also we need a scalar in four-space, and the simplest scalar which can be constructed with the electromagnetic field vector A_μ and the electron current density vector j_μ is $j_\mu A_\mu = iec\tilde{\psi}\gamma_\mu\psi A_\mu$. To further justify the form for $\mathcal{L}_I$ given by Eq. (14A3.3), we may consider variations in $\tilde{\psi}$, ψ and A_μ and use Euler-Lagrange equations,

$$\frac{\partial \mathcal{L}}{\partial \tilde{\psi}} - \frac{\partial}{\partial x_\mu}\left(\frac{\partial \mathcal{L}}{\partial\left(\frac{\partial \tilde{\psi}}{\partial x_\mu}\right)}\right) = 0,$$

$$\frac{\partial \mathcal{L}}{\partial \psi} - \frac{\partial}{\partial x_\mu}\left(\frac{\partial \mathcal{L}}{\partial\left(\frac{\partial \psi}{\partial x_\mu}\right)}\right) = 0,$$

and

$$\frac{\partial \mathcal{L}}{\partial A_\nu} - \frac{\partial}{\partial x_\mu}\left(\frac{\partial \mathcal{L}}{\partial\left(\frac{\partial A_\nu}{\partial x_\mu}\right)}\right) = 0,$$

where $\mathcal{L}$ is given by Eq. (14A3.1), to derive the field equations for $\psi, \tilde{\psi}$ and A_ν. We can easily see that the results are, respectively,

$$\gamma_\mu\left(\frac{\partial}{\partial x_\mu} - \frac{ie}{\hbar}A_\mu\right)\psi + \frac{mc}{\hbar}\psi = 0, \tag{12.7.4*}$$

$$\left(\frac{\partial}{\partial x_\mu} + \frac{ie}{\hbar}A_\mu\right)\tilde{\psi}\gamma_\mu - \frac{mc}{\hbar}\tilde{\psi} = 0, \tag{12.7.5*}$$

and

$$\frac{\partial}{\partial x_\mu}\frac{\partial}{\partial x_\mu}A_\nu = -\mu_0\, j_\nu, \tag{14.4.4*}$$

which are the well known field equations. Hence we accept the form for the Lagrangian density for the interaction, as given by Eq. (14A3.3). Then the the Hamiltonian density for the interaction is

$$\mathcal{H}_I = \frac{\partial \mathcal{L}_I}{\partial \dot{\psi}_\sigma}\dot{\psi}_\sigma - \mathcal{L}_I$$

or

$$\mathcal{H}_I = -\mathcal{L}_I = -iec\tilde{\psi}\gamma_\mu \psi A_\mu \tag{14A3.4}$$

since $\mathcal{L}_I$ does not depend on $\dot{\psi}$.

Appendix 14A4: On the Convergence of Iterative Expansion of the S Operator

The usefulness of the *interaction picture* in the calculation of transition rates depends on the convergence of the series in Eq. (14.9.21). One might expect that, in problems of quantum electrodynamics (QED), the S matrix expansion would converge because the coupling constant (the fine structure constant) $\alpha = \dfrac{e^2}{4\pi\epsilon_0\hbar c} \ll 1$ whereas in the realm of strong interaction, where the coupling constant $g_\pi \approx 1$, the convergence may not be achieved. Unfortunately, even for problems in quantum electrodynamics, this series is not convergent.

If one started with a hypothetical electron of *bare mass* m_0 and *bare charge* e_0 and then made it physical by giving it the property of interaction, it would come to acquire an extra mass, called *self mass* $\delta m = \delta E/c^2$, and an extra charge, called *self charge* δe. The self energy, or self mass δm, of the electron arises due to its interaction with its self field and also its interaction with the fluctuating *electromagnetic vacuum*. The self charge δe arises due to *vacuum polarization*. Now the total mass $m_0 + \delta\, m$ and the total charge $e_0 + \delta e$ could be made to coincide with the observed rest mass m_{obs} and observed charge e_{obs}, respectively. This however, is not the end of the story.

If we calculate the self mass and self charge by going to higher orders in the iterative expansion of the S matrix to include all intermediate processes which involve virtual quanta of unlimited energy, one encounters the most disturbing feature of quantum field theory, viz., the occurrence of infinities. It turns out that, for all quantities of physical interest which should be finite, one gets expressions which tend to infinity, some times quadratically and some times logarithmically. The way out of this difficulty is suggested by the fact that all these infinities can be expressed in terms of two basic infinities, viz., the self mass $\delta\, m$ and self charge δe. So if things are so arranged that wherever the diverging terms δm and δe occur in the theory, they put together with m_0 and e_0, respectively, are re-normalized to the finite observed mass m_{obs} and observed charge e_{obs} of the electron, it is possible to eliminate all the infinities occurring in the theory. This re-normalization procedure must, however, be Lorentz-invariant for, if it were not so, the normalization of infinities would have to be carried out all over again every time one changed the Lorentz frame, and this would render the theory meaningless. For this one needs a completely covariant formulation of electrodynamics from the very start. The invariants — bare mass m_0 and bare charge e_0 — occur as unknown parameters in the covariant QED calculation. Such a covariant formulation was achieved in the pioneering works of Tomonaga (1946), Schwinger (1948) and Feynman (1949). This finally led to an accurate prediction of Lamb-Retherford shift in $^2S_{1/2} - ^2P_{1/2}$ levels in the Hydrogen atom. It also led to the explanation for anomalous magnetic moment of the electron. Tomonaga, Schwinger and Feynman thus demonstrated that, with re-normalization, it was possible to calculate experimentally measurable quantities with unprecedented accuracy and won the Nobel prize for their discovery.

In the present context, if one restricts the calculation of cross-sections for electrodynamic processes only to second order expansion of S matrix, one may use for m and e occurring in the expressions of probability amplitudes, the observed rest mass and charge of the electron.

15

EPILOGUE

15.1 Introduction

Quantum mechanics has been enormously successful in explaining the behavior of microsystems such as molecules, atoms, and nuclei. We have seen that quantum mechanics describes a quantum system by the wave function ψ, which is the solution of the Schrödinger equation. The wave function determines the probability amplitudes (whose modulus squared gives the probability) for all possible outcomes of a measurement corresponding to an observable on the system. In general, the wave function does not uniquely predict the result of a measurement; it only provides a probability distribution for all possible results of the measurement on a quantum system. Only if the system happens to be in an eigenstate corresponding to the observable being measured, is the result of a measurement of this observable completely predictable. On the other hand, the results of measurements corresponding to a conjugate or non-commuting observable, on the same system in the same state, are *indeterminate*. Quantum mechanical description of physical systems raises profound questions: How are the wave function and the probabilities for different outcomes of a measurement to be interpreted? Does the wave function represent a single system or an ensemble of systems? Does the physical system actually have the attribute corresponding to the measurement in question or does the measurement itself create the attribute? These questions led to different interpretations of quantum mechanics, which were vigorously debated during the years of its discovery as also in the following decades. The new idea of indeterminacy was especially difficult to reconcile with the long-held view that a physical theory should aim at predicting accurately, and without ambiguity, the result of any measurement on any system, big or small. These debates not only gave us a deeper understanding of quantum mechanics but also changed profoundly the way we view the micro-world.

The *orthodox* interpretation of quantum mechanics, also known as the *Copenhagen interpretation*, due mainly to Bohr, Heisenberg and Born (as developed in Chapers I - III of this text) is the one to which most physicists subscribe. Although never laid down in detail by any of its proponents, it can be summarized as follows:

The wave function describes a single quantum system and determines the probability amplitudes (whose modulus squared gives the probabilities) for all possible outcomes (eigenvalues) of a measurement of an observable on the system. Prior to a measurement, the system does not actually have an attribute corresponding to any specific eigenvalue of the observable being measured. Its wave function is a coherent superposition of different eigenstates of the observable being measured. The act of measurement on the system forces the system from a coherent superposition to a particular eigenstate limited only by the statistical weight of the eigenstate in the wave function. This means an immediately repeated measurement on the system will yield the same result. This discontinuous evolution (*quantum jump* or *collpase*) of the wave function due to an observation is a fundamental aspect of quantum description of physical systems. The Schrödinger equation of motion prescribes continuous evolution of a system in time, provided the system is not disturbed by an observation.

Quantum mechanical wave function of a quantum system does not allow two non-commuting variables (such as the position and momentum observables) to have definite values simultaneously. The Heisenberg uncertainty principle then reflects the fact that non-commuting variables cannot be measured simultaneously with arbitrary precision. Similarly, the complementarity principle reflects the fact that the quantum mechanical description in terms of a wave function implies that an object has both wave and particle aspects, which are complementary in the sense that their manifestations depend on mutually exclusive measurements and the information gained through these various experiments exhausts all possible objective knowledge of the object.

Quantum systems that have interacted in the past, if left undisturbed, remain *non-separable*, i.e., remain *correlated* even if separated by space-like intervals. Thus the quantum state constitutes an indivisible whole.

Another group of scientists, led primarily by Einstein and Schrödinger, disagreed with the *orthodox* interpretation of quantum mechanics. According to Einstein, the wave function did not describe the behavior of a single quantum system but that of an ensemble of identical systems, all in the same state. Einstein was troubled by the indeterministic aspect of quantum mechanics. According to Einstein[1], when a certain attribute (such as the position or momentum of an electron) of a system can be predicted with certainty, without disturbing the system, then this attribute has a physical reality. This means the system actually possesses this attribute whether or not we choose to measure it and a complete physical theory ought to be able to predict with certainty the outcome of a measurement of this quantity. This must be contrasted with the orthodox interpretation, according to which the measurement creates the *reality* of attributes, in a sense. Since the wave function predicts not a definite outcome for a measurement but only the probabilities for different outcomes, it is an incomplete description of physical reality. This argument was made in a dramatic way by Einstein, Podolsky and Rosen (EPR) in a classic paper[2] by means of a gedanken (thought) experiment involving certain two-particle states, now known as the entangled states, in which, measurements of shared variables on two particles are strongly correlated.

15.2 EPR Gedanken Experiment

Consider two particles in one-dimensional motion that, having interacted in the past, have flown apart such that their state is represented by

$$\psi(x_1, x_2) = \frac{1}{2\pi\hbar}\int_{-\infty}^{\infty} \exp\left[\frac{i}{\hbar}(x_1 - x_2 + x_0)p\right] dp\,, \tag{15.2.1}$$

where x_1 and x_2 are coordinates of the two particles and x_o is some constant. Such a state that cannot be written as the product of states of individual particles (non-factorizable) is called an entangled state.

[1] The EPR definition of reality is best described in their own words: "If, without in any way disturbing a system, we can predict with certainty (i.e., with probability equal to unity) the value of a physical quantity, then there exists an element of physical reality corresponding to this physical quantity."

[2] A. Einstein, B. Podolsky, and N. Rosen, Phys. Rev. **47**, 777 (1935).

Consider now measurements of two non-commuting observables, say, the momentum and position observables $\hat{p}_1$ and $\hat{x}_1$ of particle 1. Now, the entangled wave function $\psi(x_1, x_2)$ may be looked upon as a continuous superposition of momentum states of particle 1

$$u_p(x_1) = \frac{1}{\sqrt{2\pi\hbar}} e^{ipx_1/\hbar} . \tag{15.2.2}$$

Hence, if an observation for $\hat{p}_1$ is made on this system in the state $\psi(x_1, x_2)$ and a result P is obtained, we conclude that the wave function has collapsed from $\psi(x_1, x_2)$ to

$$\frac{e^{iP(x_1-x_2+x_o)/\hbar}}{2\pi\hbar} = \frac{e^{iP(-x_2+x_o)/\hbar}}{\sqrt{2\pi\hbar}} \cdot \frac{e^{iPx_1/\hbar}}{\sqrt{2\pi\hbar}} \equiv \phi_P(x_2) u_p(x_1) , \tag{15.2.3}$$

so that while the particle 1 is in an eigenstate $u_P(x_1)$ of observable $\hat{p}_1$ with eigenvalue P, particle 2 is left in the state

$$\phi_P(x_2) = \frac{e^{i(-P)(x_2-x_0)/\hbar}}{\sqrt{2\pi\hbar}} \tag{15.2.4}$$

which is an eigenstate of momentum $\hat{p}_2$ of particle 2 with eigenvalue $-P$.

Alternatively, if we choose to measure the position observable $\hat{x}_1$, we can regard $\psi(x_1, x_2)$ as a superposition of position states

$$\begin{aligned}\psi(x_1, x_2) &= \frac{1}{2\pi\hbar} \int_{-\infty}^{\infty} \exp\left[\frac{i}{\hbar}(x_1 - x_2 + x_o)p\right] dp \\ &= \delta(x_1 - x_2 + x_o) = \int \delta(x - x_1)\delta(x - x_2 + x_o) dx \\ &\equiv \int_{-\infty}^{\infty} v_x(x_1)\chi_x(x_2) dx \end{aligned} \tag{15.2.5}$$

where

$$\begin{aligned} v_x(x_1) &= \delta(x - x_1) = \delta(x_1 - x) \\ \chi_x(x_2) &= \delta(x - x_2 + x_o) = \delta(x_2 - x - x_o) . \end{aligned} \tag{15.2.6}$$

Hence, if an observation for the observable $\hat{x}_1$ is made on the system in the entangled state and a result X is obtained, we conclude that the wave function ψ has collapsed from ψ to the state

$$v_X(x_1)\chi_X(x_2) = \delta(x_1 - X)\delta(x_2 - X - x_o)$$

in which the particle 1 has position X and particle 2 has position $x_o + X$.

To summarize, if the two-particle system is in the state $\psi(x_1, x_2)$ given by Eq. (15.2.1), and an observation for $\hat{p}_1$ is made with the result P, then particle 2 is left in an eigenstate of momentum observable $\hat{p}_2$ with eigenvalue $-P$. On the other hand if, on the same state, an observation for the position of particle 1 is made and the result is X then the particle 2 is left in the eigenstate of $\hat{x}_2$ with eigenvalue $X + x_o$. Now the decision whether to measure $\hat{p}_1$ or $\hat{x}_1$ on particle 1 can be made when the two particles are so far apart that no influence resulting from the measurement on particle 1 can possibly propagate to particle 2 in the available time. EPR argued that since we can predict with certainly, and without disturbing particle 2, the position (or momentum) of particle 2, according to their criterion of *reality*, both position and momentum of particle 2 have objective reality, i.e., particle 2 must already have had definite values for both position and momentum simultaneously, even though no simultaneous measurement of position and momentum is allowed. Since the wave function contains no elements corresponding to the reality of position and momentum of particle 2, they concluded that the wave function is not a complete description of a physical system.

Note that the EPR argument is based on twin assumptions of particle 2 having its *reality* (separate from or independent of particle 1) even though it is connected to particle 1 and impossibility of any change in the reality of particle 2 as a result of a measurement on particle 1 because of their space-like separation (locality). These assumptions are referred to as *local realism*. It is important to realize that although the *reality* of position and momentum variables of particle 2 is established in specific contexts (measurements of position and momentum, respectively, on particle 1), locality ensures that particle 2 would have the same *reality* even in the absence of the measurements on particle 1. This combined with separability implies that particle 2 has definite values for both position and momentum *simultaneously*, i.e., they are pre-determined (at the time of its interaction with particle 1) attributes of particle 2.

The EPR conclusion regarding the incompleteness of wave function description follows also without any reference to the simultaneous existence of definite values for position and momentum for particle 2; if we re-examine the EPR gedanken experiment, we find that, depending on whether we measure the position or momentum of particle 1, particle 2 is left either in a position eigenstate or in a momentum eigenstate. Now if particle 2 has some physical reality, independent from particle 1, and locality holds, then the measurement on particle 1 cannot disturb the assumed *reality* of particle 2. However, that reality appears to be represented by two different wave functions, depending on which measurement on particle 1 is carried out. A complete description would not permit the same physical state to be represented by wave functions with distinct physical implications. This not being the case, EPR conclude that the wave function provides an incomplete description of physical reality.

Einstein did not doubt that quantum mechanics is *correct*, so far as it goes but considered it to be an incomplete and probabilistic description of physical reality. He believed that a complete physical theory ought to predict or describe those attributes of nature that are independent of measurements or observers. The EPR paper shows that in the case of interacting systems satisfying the twin conditions of local realism, the description of systems provided by wave function is not complete. The EPR argument exposes the dilemma that we have to choose between local realism and quantum mechanical wave function as the basis for the description of an individual physical system. Einstein referred to this dilemma as a paradox. Theories which conform to local realism are referred to as local realistic theories. Such theories require certain parameters to specify the reality of each particle's variables and have come to be called hidden variable theories.

According to the standard interpretation of quantum mechanics the EPR experiment leads to no contradiction since quantum systems that have interacted in the past, if left undisturbed, remain as one indivisible whole even if they have space-like separation. There is single reality describing the whole system and the wave function is its complete description. It should be noted that in the EPR Gedanken experiment, the inference that measurement of momentum of particle 1 leaves particle 2 in a definite momentum state and, alternatively, the observation of position of particle 1 leaves the particle 2 in a definite position state leads to no contradiction in quantum mechanics because $\hat{x}_1$ and $\hat{x}_2$ commute as do $\hat{p}_1$ and $\hat{p}_2$.

For decades the debate whether quantum mechanics or local hidden variable theories are the correct descriptions of physical reality remained a mere epistemological (epistemology refers to how we know things) debate. This changed when in a remarkable paper Bell showed, that the two views of reality have measurable consequences. He derived a set of inequalities which express the limitations that local realism places on the EPR correlations. Corresponding limits placed by quantum mechanics on EPR correlations can also be worked out. These inequalities have been tested in a series of experiments; the results are in agreement with quantum mechanical predictions and in clear violation of the limits placed by local realistic theories.

15.3 Einstein-Podolsky-Rosen-Bohm Gedanken Experiment

The EPR argument was re-cast by Bohm in terms of measurements of spin components on a spin-entangled state. Bohm's version of the EPR experiment, described a plausible experimental scheme where the EPR correlations could be verified. Experimental arrangements involving measurements of spin components or photon polarization for spatially separated systems are referred to as as EPRB experiments.

Bohm considered a system of two spin-half particles, initially in the spin singlet state ($J = 0$), decays by a process that does not change the total angular momentum. As a result of angular momentum conservation, the particles fly apart in opposite directions. Measurements of spin components (which take the place of position and momentum variables in the EPR experiment) of the particle in any direction $\boldsymbol{a}$ can be carried out by a Stern-Gerlach magnet followed by a detector D as shown in Fig. 15.1. Like the position and momentum operators, operators for spin along different directions do not commute. Likewise, if the spin of particle 1 is measured in some direction $\boldsymbol{a}$ and is found to be $+\frac{1}{2}$ (in units of $\hbar$), particle 2 would be found to have spin $-\frac{1}{2}$ relative to the same direction and vice versa.

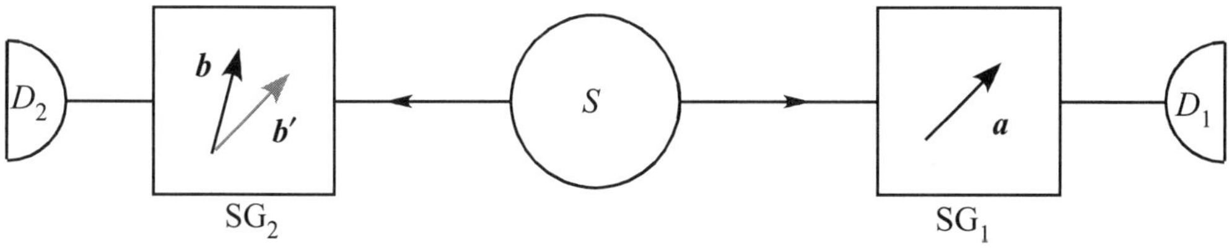

FIGURE 15.1
The spins of the particles which fly apart from source S are analyzed by two Stern-Gerlach magnets SG_1 and SG_2, placed on opposite sides at equal distances from S. The orientations of the magnets may be adjusted so as to measure the spin components in any specified direction. D_1 and D_2 are detectors whose outputs are fed to a coincidence counter.

This is because, given a direction $\boldsymbol{a}$, the entangled (singlet) state of the two particles may be written as

$$\chi_o(1,2) = \frac{1}{\sqrt{2}}\left[\chi^{(1)}_{1/2}(\boldsymbol{a})\,\chi^{(2)}_{-1/2}(\boldsymbol{a}) - \chi^{(1)}_{-1/2}(\boldsymbol{a})\,\chi^{(2)}_{1/2}(\boldsymbol{a})\right], \tag{15.3.1}$$

where $\chi^{(1)}_{\pm 1/2}(\boldsymbol{a})$ represents the state of the particle 1 in which the component of its spin along the direction $\boldsymbol{a}$ is $\pm\frac{1}{2}$ (in units of $\hbar$). The spatial part of the wave function plays no role here and has been suppressed. According to quantum mechanics, if the spin of particle 1 is measured to be $+1/2$ in direction $\boldsymbol{a}$, the spin of the second particle in the same direction must be $-1/2$ and vice versa. This is because when the spin of the particle 1 is observed to be $\frac{1}{2}$, the state of the system collapses from $\chi_o(1,2)$ to $\chi^{(1)}_{1/2}(\boldsymbol{a})\,\chi^{(2)}_{-1/2}(\boldsymbol{a})$ and when it is observed to be $-\frac{1}{2}$, it collapses to $\chi^{(1)}_{-1/2}(\boldsymbol{a})\chi^{(2)}_{1/2}(\boldsymbol{a})$. And no matter which component of the spin of particle 1 is measured, the spin component of the second particle in the same direction would be equal and opposite. This is because the spin singlet state always has the form (15.3.1) for any direction $\boldsymbol{a}$.

Now, the particles can be allowed to move far apart so that at the time of measurement any disturbance resulting from a determination of spin of particle 1 could not affect particle 2. Then the spin measurements exhibit correlations similar to those in the EPR experiment allowing similar arguments and conclusions involving locality, separability, and completeness of quantum mechanical wave function.

Another variant of EPRB experiment involves an atom undergoing a two-photon decay without a net change in its angular momentum. The decay scheme is shown in Fig. 15.2.

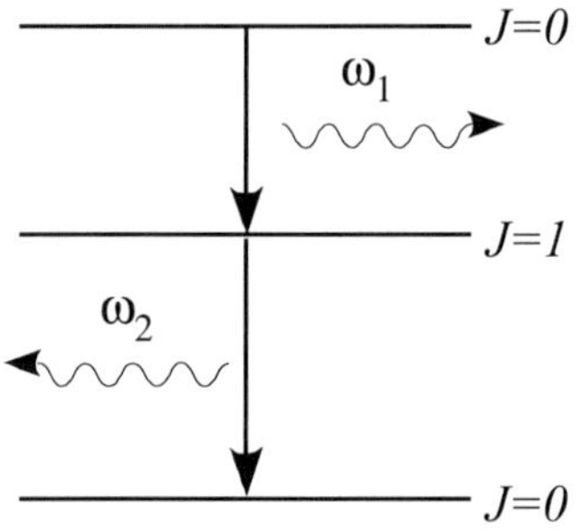

FIGURE 15.2
An atom undergoes a decay in cascades ($J = 0 \to J = 1 \to J = 0$) emitting photons of frequencies ω_1 and ω_2. The correlation between the polarization states of the two photons, propagating in opposite directions may be investigated theoretically as well as experimentally. The simple experimental set-up for this can be as shown in Fig. 15.3.

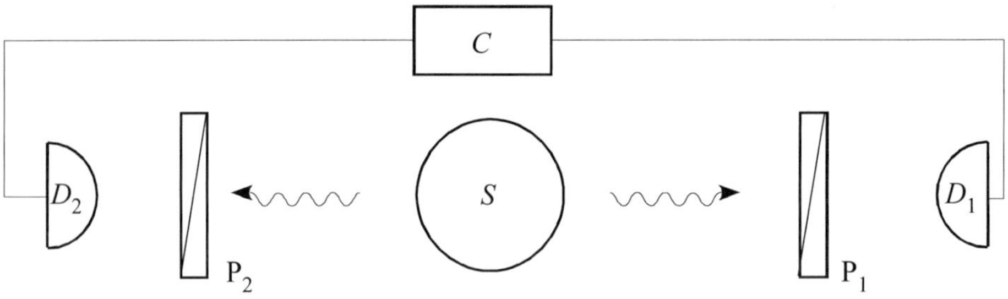

FIGURE 15.3
Photons ω_1 and ω_2 emitted in opposite directions from atomic source A are analyzed by polarizers P_1 and P_2. D_1 and D_2 are the photon detectors and C is a coincident counter.

P_1 and P_2 are two single-channel polarizers which are so set that P_1 passes only a photon linearly polarized at angle φ_1 to the x-axis while P_2 passes a photon polarized at angle φ_2 to the x-axis [see Fig. 15.3] (x, y-axes lie in a plane perpendicular to the direction of propagation of the photons). The coincidence counter C responds positively when there is a simultaneous detection of photons in D_1 and D_2.

15.4 Theory of Hidden Variables and Bell's Inequality

Bell[3] showed that a physical theory based on local realism places limits on certain correlations that can be measured in the EPRB type of experiments. He constructed an explicit theoretical framework within which these correlations can be calculated. This framework satisfied the criteria of locality and realism as envisioned by Einstein and incorporated some type of hidden variables. Such theories are referred to as *local realistic* or *local hidden variable* theories. Bell showed that the correlations based on local hidden variable theories obeyed a certain inequality, now known as Bell's inequality. The term is now used for a family of inequalities based on theories similar to but more general than Bell's original construct. By computing the expectation values for products of the form $(\boldsymbol{\sigma}_1 \cdot \boldsymbol{a})(\boldsymbol{\sigma}_2 \cdot \boldsymbol{b})$ quantum mechanically, where $\boldsymbol{\sigma}_i$ represents the spin (for spin-half particles) of the two particles and $\boldsymbol{a}$ and $\boldsymbol{b}$ are unit vectors, Bell showed that quantum mechanical expectation values violate Bell's inequality. In other words, Bell proved that no local realistic theory can agree with all of the statistical predictions of quantum theory.

We now outline a derivation of Bell's inequality based on a local hidden variable theory appropriate for EPRB experiments summarized in the previous section. Bell considered an ensemble of pairs of particles 1 and 2. Each pair of particles is characterized by a variable λ, which completely determines the properties of the pair at the moment of their interaction. The variable λ is referred to as a hidden variable. The variable λ may be a single parameter or it may represent a set of parameters; this point is not relevant to the establishment of the constraints. It may differ from pair to pair, but the mechanism of interaction establishes a probability distribution $\rho(\lambda)$ with the normalization

$$\int \rho(\lambda) d\lambda = 1 . \tag{15.4.1}$$

Different types (e.g., measurents of spin in different directions) measurements now can be performed on the particles. Let us designate measurements on particle 1 by α, $\alpha', \cdots$ and the corresponding outcomes by $A(\alpha,\lambda)$, $A(\alpha',\lambda), \cdots$ and that on particle 2 by β, $\beta', \cdots$ and the corresponding outcomes by $B(\beta,\lambda)$, $B(\beta,\lambda')$, $\cdots$ in a micro-state characterized by the *hidden variable*. We assume that $A(\alpha,\lambda)$ and $B(\beta,\lambda)$ can take only two possible values $+1$ or -1. For example $A(\alpha,\lambda) = +1$ might represent the emergence of particle 1 with spin projection $+\frac{1}{2}$ from SG_1 in the EPRB set-up of Fig. 15.1 and $A(\alpha,\lambda) = -1$ might represent the emergence of particle 1 with spin projection $-\frac{1}{2}$ along a certain direction perpendicular to the flight path of particle 1. Then the correlation between the two variables $A(\alpha,\lambda)$ and $B(\beta,\lambda)$ can be expressed as

$$C_{LHV}(\alpha,\beta) = \int A(\alpha,\lambda) B(\beta,\lambda)\, \rho(\lambda) d\lambda . \tag{15.4.2}$$

Note that this incorporates both realism (dependence on lambda (pre)-determines the outcome) and locality (outcome A depends only on α and B only on β). Then given two measurements α and α' on particle 1 and β and β' on partice 2 and the fact $|A(\alpha,\lambda)| = 1$,

[3]J. S. Bell, Physics (NY), **1**, 195 (1965); reprinted in J. Bell, *Speakable and Unspeakable in Quantum Mechanics* (Cambridge University Press, 2004).

the following inequalities are easily established

$$|C_{LHV}(\alpha,\beta) - C_{LHV}(\alpha,\beta')| \leq \int |A(\alpha,\lambda)[B(\beta,\lambda) - B(\beta',\lambda)]|\rho(\lambda)\, d\lambda$$

or

$$|C_{LHV}(\alpha,\beta) - C_{LHV}(\alpha,\beta')| \leq \int |B(\beta,\lambda) - B(\beta',\lambda)|\rho(\lambda)\, d\lambda \tag{15.4.3}$$

and

$$|C_{LHV}(\alpha',\beta) + C_{LHV}(\alpha',\beta')| \leq \int |A(\alpha',\lambda)[B(\beta,\lambda) + B(\beta',\lambda)]|\rho(\lambda)\, d\lambda$$

or

$$|C_{LHV}(\alpha',\beta) + C_{LHV}(\alpha',\beta')| \leq \int |B(\beta,\lambda) + B(\beta',\lambda)|\rho(\lambda)\, d\lambda\,. \tag{15.4.4}$$

From the fact that B takes on only the values ± 1, it follows that

$$|B(\beta,\lambda) - B(\beta',\lambda)| + |B(\beta,\lambda) + B(\beta',\lambda)| = 2\,. \tag{15.4.5}$$

Adding Eqs. (15.3.3) and (15.3.4) and using Eqs. (15.3.1) and (15.3.5) we obtain the Bell inequality

$$\begin{aligned}|C_{LHV}(\alpha,\beta) - C_{LHV}(\alpha,\beta')| &+ |C_{LHV}(\alpha',\beta) + C_{LHV}(\alpha',\beta')| \\ &\leq \int \left[|B(\beta,\lambda) - B(\beta',\lambda)| + |B(\beta,\lambda) + B(\beta',\lambda)|\right] \rho(\lambda) d\lambda\end{aligned}$$

or

$$S \equiv |C_{LHV}(\alpha,\beta) - C_{LHV}(\alpha,\beta')| + |C_{LHV}(\alpha',\beta) + C_{LHV}(\alpha',\beta')| \leq 2\,. \tag{15.4.6}$$

This is the constraint that EPR correlations must satisfy according to a local realistic theory.

According to quantum mechanics the measurement on particle 1, designated by α, α', $\cdots$ may be represented by the observables $\hat{A}(\alpha)$, $\hat{A}(\alpha')$, $\cdots$ while those on particle 2 designated by β, β', $\cdots$ may be represented by the observables $\hat{B}(\beta)$, $\hat{B}(\beta')$, $\cdots$. The quantum correlation between the observations on the two particles in an entangled state $|\psi\rangle$ is then given by

$$C_{QM}(\alpha,\beta) = \langle\psi|\, \hat{A}(\alpha)\hat{B}(\beta)\, |\psi\rangle\,. \tag{15.4.7}$$

Let us apply the constraint (15.4.6) to quantum mechanical expression for the EPR correlations in the EPRB set-up in Fig. 15.1. Identifying α with the measurement of spin 1 in direction of unit vectors $\boldsymbol{a}$ and β with measurement of spin 2 in direction of unit vector $\boldsymbol{b}$, we may represent $\hat{A}(\alpha)$ by $\boldsymbol{\sigma}_1 \cdot \boldsymbol{a}$ and $\hat{B}(\beta)$ by $\boldsymbol{\sigma}_2 \cdot \boldsymbol{b}$ and write

$$C_{QM}(\boldsymbol{a},\boldsymbol{b}) = \chi_o^\dagger(\boldsymbol{\sigma}_1 \cdot \boldsymbol{a})(\boldsymbol{\sigma}_2 \cdot \boldsymbol{b})\chi_o\,, \tag{15.4.6a}$$

where χ_o is the spin-singlet state of the two particles and we have used matrix representation of the states and observables.

Using the identity

$$[\boldsymbol{\sigma}_1 \cdot \boldsymbol{a} + \boldsymbol{\sigma}_2 \cdot \boldsymbol{b}] = \frac{1}{2}\left[(\boldsymbol{\sigma}_1 + \boldsymbol{\sigma}_2)\cdot(\boldsymbol{a}+\boldsymbol{b}) + (\boldsymbol{\sigma}_1 - \boldsymbol{\sigma}_2)\cdot(\boldsymbol{a}-\boldsymbol{b})\right]\,,$$

and squaring the two sides of this equation, we obtain

$$(\boldsymbol{\sigma}_1 \cdot \boldsymbol{a})^2 + (\boldsymbol{\sigma}_2 \cdot \boldsymbol{b})^2 + 2(\boldsymbol{\sigma}_1 \cdot \boldsymbol{a})(\boldsymbol{\sigma}_2 \cdot \boldsymbol{b}) = \frac{1}{4}\left[(\boldsymbol{\sigma}_1 + \boldsymbol{\sigma}_2)\cdot(\boldsymbol{a}+\boldsymbol{b}) + (\boldsymbol{\sigma}_1 - \boldsymbol{\sigma}_2)\cdot(\boldsymbol{a}-\boldsymbol{b})\right]^2\,.$$

Taking the expectation value of the two sides of this equation in the singlet state (15.3.1) and using the results $(\boldsymbol{\sigma}_1 \cdot \boldsymbol{a})^2 = 1$ and $(\boldsymbol{\sigma}_2 \cdot \boldsymbol{b})^2 = 1$ and the fact that $\boldsymbol{\sigma}_1 + \boldsymbol{\sigma}_2 = 0$ for the singlet state, we find

$$C_{QM}(\boldsymbol{a},\boldsymbol{b}) = -\boldsymbol{a}\cdot\boldsymbol{b} = -\cos\varphi\,, \tag{15.4.8}$$

where φ is the angle between unit vectors $\boldsymbol{a}$ and $\boldsymbol{b}$. This correlation is -1 for $\boldsymbol{a} = \boldsymbol{b}$ and $+1$ for $\boldsymbol{b} = -\boldsymbol{a}$ showing complete anti-correlations of spins as expected for a singlet state.

Choosing $\boldsymbol{b} = \boldsymbol{a}'$ and $\boldsymbol{a} \cdot \boldsymbol{b} = \boldsymbol{b} \cdot \boldsymbol{b}' = \cos\varphi$ and using the quantum mechanical expectation values, we have

$$\begin{aligned} S_{QM}(\varphi) &\equiv |C_{QM}(\boldsymbol{a}, \boldsymbol{b}) - C_{QM}(\boldsymbol{a}, \boldsymbol{b}')| + |C_{QM}(\boldsymbol{a}', \boldsymbol{b}) + C_{QM}(\boldsymbol{a}', \boldsymbol{b}')| \\ &= |-\cos\varphi + \cos 2\varphi| + |-1 - \cos\varphi| \,. \end{aligned} \tag{15.4.9}$$

A plot of $S_{QM}(\varphi)$ as a function of φ is shown by the continuous curve in Fig. 15.4. Bell's inequality (15.4.6) is violated where the curve exceeds 2.

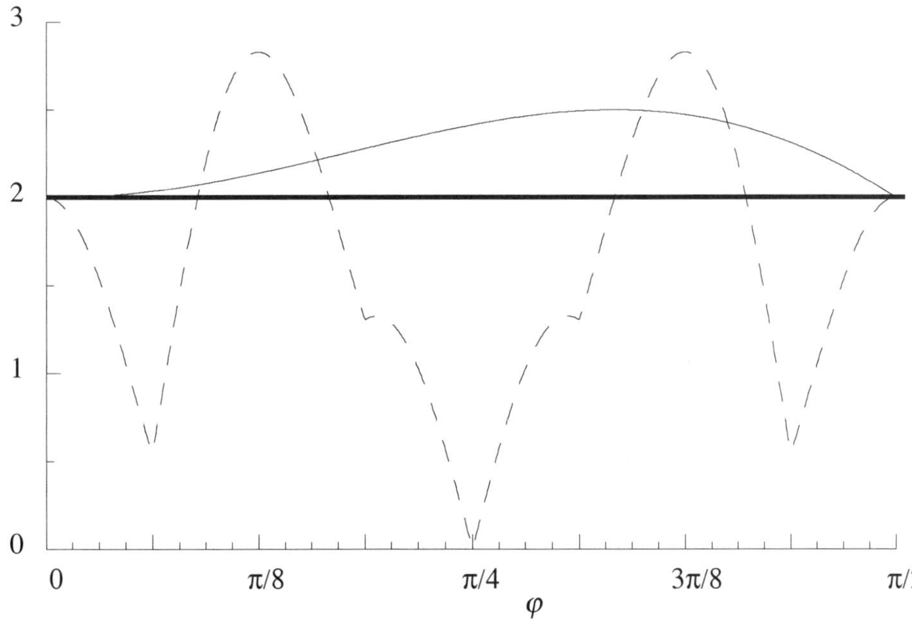

FIGURE 15.4
Variation of Bell parameter according to quantum mechanics. The continuous curve corresponds to Eq. (15.4.9) and the dashed curve corresponds to Eq. (15.4.27). The regions where the curves lie above the thick horizontal line correspond to violations of Bell's inequality.

With the availability of quantitative predictions from both local hidden variable theory and quantum mechanics, the question as to which of the two descriptions conforms to the laws of nature can be settled by the experimental observations. A number of experiments have been carried out to detect EPR correlations. These experiments involve measurements of photon polarization in the EPRB set-up sketched in Fig. 15.3 rather than the spin system described in the preceding paragraphs. The photon pairs are generated in the two-photon cascade decay of an atom with no net change in its angular momentum ($J = 0 \to J = 1 \to J = 0$). Since there is no net change in the angular momentum of the atom in this cascade decay and photons counter propagating along the z-axis are detected,

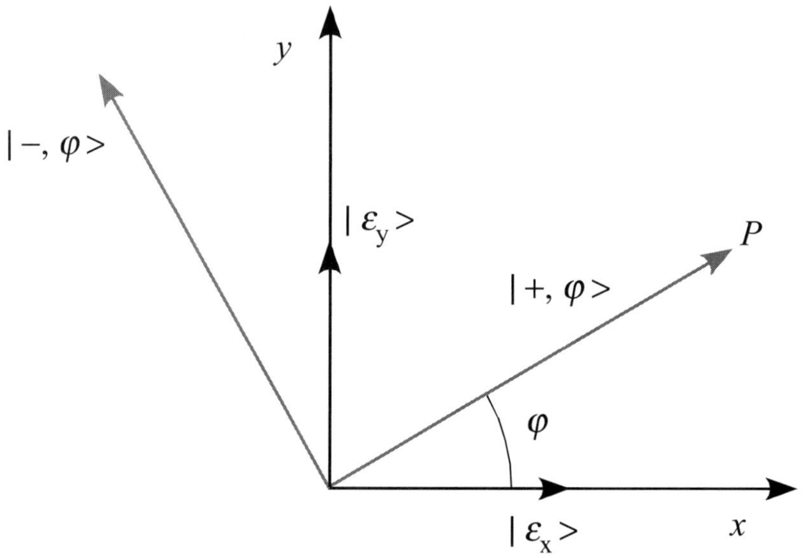

FIGURE 15.5
Rotation of polarization basis vectors through an angle φ.

the polarization-entangled two photon state can be written as (in linear polarization basis)

$$|\psi(1,2)\rangle = \frac{1}{\sqrt{2}}\left[|e_{1x}\rangle\,|e_{2x}\rangle + |e_{1y}\rangle\,|e_{2y}\rangle\right] . \tag{15.4.10}$$

If φ_1 and φ_2 are, respectively, the angles that linear polarizers P_1 and P_2 make with the x-axis, we can express the states $|e_{jx}\rangle$ (photon j polarization along x-axis) and $|e_{jy}\rangle$ (photon j polarization along y-axis) in terms of polarization states $|+,\varphi_j\rangle$ and $|-,\varphi_j\rangle$, where $+$ means photon j polarization parallel to φ_j (the direction of polarizer P_j) and $-$ means polarization orthogonal to φ_j (photon polarization in direction $\varphi_j + \pi/2$). It is clear from Fig. 15.5 that

$$|+,\varphi_1\rangle = \cos\varphi_1\,|e_{1x}\rangle + \sin\varphi_1\,|e_{1y}\rangle , \tag{15.4.11}$$
$$|-,\varphi_1\rangle = -\sin\varphi_1\,|e_{1x}\rangle + \cos\varphi_1\,|e_{1y}\rangle . \tag{15.4.12}$$

The inverse relation is

$$|e_{1x}\rangle = \cos\varphi_1\,|+,\varphi_1\rangle - \sin\varphi_1\,|-,\varphi_1\rangle , \tag{15.4.13}$$
$$|e_{1y}\rangle = \sin\varphi_1\,|+,\varphi_1\rangle + \cos\varphi_1\,|-,\varphi_1\rangle . \tag{15.4.14}$$

Similar relations for photon 2 can be written. Using Eqs. (15.3.13) and (15.3.14) for photon 1 and similar relations for photon 2, we can express the polarization entangled state $|\psi(1,2)\rangle$ as

$$\begin{aligned}|\psi(1,2)\rangle = \frac{1}{\sqrt{2}}\big[&|+,\varphi_1;+,\varphi_2\rangle\cos(\varphi_2-\varphi_1) + |-,\varphi_1;-,\varphi_2\rangle\cos(\varphi_2-\varphi_1)\\ &- |+,\varphi_1;-,\varphi_2\rangle\sin(\varphi_2-\varphi_1) + |-,\varphi_1;+,\varphi_2\rangle\sin(\varphi_2-\varphi_1)\big] .\end{aligned} \tag{15.4.15}$$

From this expression, the joint probability $P(+,\varphi_1,+,\varphi_2)$ that photon 1 emerges from P_1 set at angle φ_1 and photon 2 emerges from P_2 set at φ_2 is given by

$$P(+,\varphi_1;+,\varphi_2) = |\langle +,\varphi_1;+,\varphi_2|\;|\psi(1,2)\rangle|^2 = \frac{1}{2}\cos^2(\varphi_2-\varphi_1) . \tag{15.4.16}$$

The probability $P(-,\varphi_1;-,\varphi_2)$ that no photon emerges either from P_1 or P_2 is given by

$$P(-,\varphi_1;-,\varphi_2) = |\langle -,\varphi_1;-\varphi_2|\psi(1,2)\rangle|^2 = \frac{1}{2}\cos^2(\varphi_2-\varphi_1)\,. \tag{15.4.17}$$

Note that this is equal to $P(+,\varphi_1;+,\varphi_2)$. This is because the probability $P(+,\varphi_1+\pi/2;+,\varphi_+\pi/2)$ that photons will emerge from P_1 set at $\varphi_1+\pi/2$ and P_2 set at $\varphi_2+\pi/2$ can be looked upon as $P(-,\varphi_1;-,\varphi_2)$ that no photons will emerge from P_1 and P_2 set at φ_1 and φ_2, respectively.

Similarly, the probability $P(+,\varphi_1;-,\varphi_2)$, that a photon emerges from P_1 and no photon emerges from P_2, and the probability $P(-,\varphi_1,+,\varphi_2)$ that no photon emerges from P_1 and a photon emerges from P_2 are given by

$$P(+,\varphi_1;-,\varphi_2) = |\langle +,\varphi_1;-,\varphi_2|\psi(1,2)\rangle|^2 = \frac{1}{2}\sin^2(\varphi_2-\varphi_1)\,, \tag{15.4.18}$$

$$P(-,\varphi_1;+,\varphi_2) = |\langle -,\varphi_1;+,\varphi_2|\psi(1,2)\rangle|^2 = \frac{1}{2}\sin^2(\varphi_2-\varphi_1)\,. \tag{15.4.19}$$

From Eqs. (15.3.16) through (15.3.19), we can see that the probability $P_1(\varphi_1)$ of photon 1 emerging from P_1 set at φ_1 when P_2 is set at φ_2 and the probability $P_2(\varphi_2)$ of photon 2 emerging from P_2 set at φ_2 when P_1 is set at φ_1 are

$$P_1(\varphi_1) \equiv P(+,\varphi_1;+,\varphi_2) + P(+,\varphi_1;-,\varphi_2) = \frac{1}{2}\,, \tag{15.4.20}$$

$$P_2(\varphi_2) \equiv P(+,\varphi_1;+,\varphi_2) + P(-,\varphi_1;+,\varphi_2) = \frac{1}{2}\,. \tag{15.4.21}$$

These probabilities are equal as expected since the photons are produced in pairs.

The joint probability $P(+,\varphi_1;+\varphi_2)$ given by Eq. (15.3.16) depends on both φ_1 and φ_2 but cannot be factored into a product of two probabilities, one depending on φ_1 and the other on φ_2, reflecting the correlations between the photons. The conditional probability $P_c(\varphi_2|\varphi_1)$ of detecting photon 2, conditioned on the detection of photon 1, is given by

$$P_c(\varphi_2|\varphi_1) \equiv \frac{P(+,\varphi_1;+,\varphi_2)}{P_1(\varphi_1)} = \cos^2(\varphi_2-\varphi_1)\,. \tag{15.4.22}$$

This is unity for $\varphi_2=\varphi_1$ and zero for $\varphi_2=\varphi_1\pm\pi/2$, showing that photon 2 polarization is definitely parallel to the polarization of photon 1. Equation (15.4.22) shows that the outcome of polarization measurement on photon 2 of appears to be influenced by the measurement of polarization on photon 1 even though the two may have a space-like separation at the time of measurement, reflecting nonlocal character of EPR correlations. However, seemingly non-local character of EPR correlations cannot be used to influence the outcome of a measurement on photon 2 by adjusting the angle φ_1 of polarizer P_1 because setting the polarizer angle φ_1 gurantees photon 1 polarization to be in direction φ_1 only in the special case when the photon emerges from P_1. Indeed the probability $P_2(\varphi_2)$ [Eq. (15.3.21)] that photon 2 emerges from polarizer P_2 set at φ_2 when polarizer P_1 is set at φ_1 is independent of φ_1. So the setting of polarizer P_1 has no influence on the outcome of polarizer 2, indicating that there is no violation of causality due to EPR correlations.

To compare the quantum mechanical results with those of hidden variable theory we identify measurement α and β in Eq. (15.3.2) with the measurements of polarization components in directions φ_1 and φ_2 or with the settings φ_1 and φ_2, respectively, of polarizers P_1 and P_2. Further we take the outcome $A(\varphi_1) = +1$ to represent the emergence of photon from P_1 and $A(\varphi_1) = -1$ to represent its failure to emergence from P_1. Similarly

$B(\varphi_2) = \pm 1$ correspond, respectively, to the emergence and non-emergence of photon from P_2. Then the quantum mechanical correlation between the output of two polarizers will be

$$C_{QM}(\varphi_1, \varphi_2) = \langle \psi | \, A(\varphi_1) B(\varphi_2) \, | \psi \rangle \,, \tag{15.4.23}$$

where $|\psi\rangle$ is the entangled state (15.4.10) of the two photons. This correlation function is the average of all possible outcomes of polarization measurements weighted by the corresponding probabilities. We can write it explicitly as

$$\begin{aligned} C_{QM}(\varphi_1, \varphi_2) &= \sum_{A,B} A(\varphi_1) B(\varphi_2) \times P(A, \varphi_1; B, \varphi_2)\,, \\ &= P(+, \varphi_1; +, \varphi_2) + P(-, \varphi_1; -, \varphi_2) - P(+, \varphi_1; -, \varphi_2) - P(-, \varphi_2; +, \varphi_1)\,. \end{aligned} \tag{15.4.24}$$

Using the expressions (15.4.16) through (15.4.19) for these probabilities, we find

$$C_{QM}(\varphi_1, \varphi_2) = \cos^2(\varphi_2 - \varphi_1) - \sin^2(\varphi_1 - \varphi_2) = \cos 2(\varphi_2 - \varphi_1)\,. \tag{15.4.25}$$

From this we see that there is complete correlation $C_{QM}(\varphi_1, \varphi_2) = 1$ when $\varphi_2 = \varphi_1$ and complete anti-correlation $C_{QM} = -1$ when $\varphi_2 = \varphi_1 \pm \pi/2$.

According to quantum mechanics, the Bell parameter

$$\begin{aligned} S_{QM} &\equiv |C_{QM}(\varphi_1, \varphi_2) - C_{QM}(\varphi_1, \varphi_2')| + |C_{QM}(\varphi_1', \varphi_2) + C_{QM}(\varphi_1', \varphi_2')| \\ &= |\cos 2(\varphi_2 - \varphi_1) - \cos 2(\varphi_2' - \varphi_1)| + |\cos 2(\varphi_2 - \varphi_1') + \cos 2(\varphi_2' - \varphi_1')| \end{aligned} \tag{15.4.26}$$

for $\varphi_1 = 0, \varphi_2 = 3\Phi$, $\varphi_1' = -2\Phi$, $\varphi_2' = \Phi$ takes the form

$$S_{QM}(\Phi) = |\cos(6\Phi) - \cos(2\Phi)| + |\cos(10\Phi) + \cos(6\Phi)|\,. \tag{15.4.27}$$

This parameter is shown by the dashed curve in Fig. 15.4 as a function of Φ. We see that S_{QM} violates Bell inequality by significant amount for a range of values of ϕ.

Experimental tests of Bell inequality in the form (15.4.6) using single-channel polarization measurements just described would be inconclusive because due to finite detector efficiencies, it is impossible to discriminate directly between a photon that fails to pass through the analyzer and one which does pass through the analyzer but is not detected because of the inefficiency of the photo-detectors. For this reason another form of Bell's inequality was derived by Clauser and Horne[4].

15.5 Clauser-Horne Form of Bell's Inequality and Its Violation in Two-photon Correlation Experiments

The Clauser-Horne form of Bell's inequality, which does not depend on detector quantum efficiencies α_1 and α_2 directly, is more suitable for experimental tests of local hidden variable theories. Let $p_j(\varphi_j, \lambda)$ be the probability that the photon in the jth arm reaches the detector D_j when the polarizer P_j is set at angle φ_j and when the hidden variable is λ and let $p_j(\infty, \lambda)$ be the corresponding probability when there is no polarizer in the jth

[4] J. F. Clauser and M. A. Horne, Phys. Rev. **D10**, 526 (1974).

arm. Also, according to *local* hidden variable theory, the joint probability factorizes as $p_{12}(\varphi_1, \varphi_2, \lambda) = p_1(\varphi_1, \lambda)p_2(\varphi_2, \lambda)$. This condition is a statement of *Bell locality*. Note that the assumed forms for p_j and p_{12} satisfy both the locality $[p_{12}(\varphi_1, \varphi_2, \lambda)$ is a product of individual detection probabilities$]$ and (separate) reality $[p_j(\varphi_j, \lambda)$ depends only on $\varphi_j]$ conditions.

Now, the probability of detection without a polarizer is at least as great as when a polarizer is used

$$p_j(\varphi_j, \lambda) \leq p_j(\infty, \lambda) . \tag{15.5.1}$$

This condition is known as *no enhancement assumption*[4]. It follows from Eq. (15.5.1) that the ratios

$$\begin{aligned} x &= p_1(\varphi_1, \lambda)/p_1(\infty, \lambda), \quad & x' &= p_1(\varphi_1', \lambda)/p_1(\infty, \lambda), \\ y &= p_2(\varphi_2, \lambda)/p_2(\infty, \lambda), \quad & y' &= p_2(\varphi_2', \lambda)/p_2(\infty, \lambda), \end{aligned} \tag{15.5.2}$$

all lie in the interval $0 \leq x, x', y, y' \leq 1$ and satisfy the algebraic inequality

$$-1 \leq xy - xy' + x'y + x'y' - x' - y \leq 0 . \tag{15.5.3}$$

The joint probability $P_{12}(\varphi_1, \varphi_2)$ for the detection of photons by both the detectors D_1 and D_2 when the polarizers in the two arms are set at φ_1, and φ_2, respectively, is given by

$$P_{12}(\varphi_1, \varphi_2) \equiv \eta_1\eta_2 \int p_{12}(\varphi_1, \varphi_2, \lambda)\rho(\lambda)d\lambda = \eta_1\eta_2 \int p_1(\varphi_1, \lambda)p_2(\varphi_2, \lambda)\rho(\lambda)d\lambda , \tag{15.5.4}$$

where η_1 and η_2 are the quantum efficiencies of the two detectors. Similarly, if $P_{12}(\varphi_1, \infty)$ and $P_{12}(\infty, \varphi_2)$, respectively, denote the probabilities of joint detection of photons by D_1 and D_2 when one, or the other polarizer is removed, then

$$P_{12}(\varphi_1, \infty) = \eta_1\eta_2 \int p_1(\varphi_1, \lambda)p_2(\infty, \lambda)\rho(\lambda)d\lambda , \tag{15.5.5}$$

$$P_{12}(\infty, \varphi_2) = \eta_1\eta_2 \int p_1(\infty, \lambda)p_2(\varphi_2, \lambda)\rho(\lambda)d\lambda . \tag{15.5.6}$$

By using the definition of x, x', y, y' from Eq. (15.5.2) in the inequality (15.5.3), multiplying throughout by $\eta_1\eta_2 p_1(\infty, \lambda)p_2(\infty, \lambda)\rho(\lambda)d\lambda$ and integrating over λ, we get

$$\begin{aligned} - P_{12}(\infty, \infty) \leq &[P_{12}(\varphi_1, \varphi_2) - P_{12}(\varphi_1, \varphi_2') + P_{12}(\varphi_1', \varphi_2) \\ &+ P_{12}(\varphi,', \varphi_2') - P_{12}(\varphi_1', \infty) - P_{12}(\infty, \varphi_2)] \leq 0 . \end{aligned} \tag{15.5.7}$$

This is the Clauser-Horne version of Bell's inequality. The advantage of this inequality is that it involves only the joint probabilities for coincidence counts by the two detectors, all of which have the same dependence on detector efficiencies. So if we work with the ratios of probabilities we get a relation independent of detector efficiencies. Dividing through by $P_{12}(\infty, \infty)$ we obtain a relation independent of detector efficiencies

$$\begin{aligned} -1 \leq &S \equiv [P_{12}(\varphi_1, \varphi_2) - P_{12}(\varphi_1, \varphi_2') + P_{12}(\varphi_1', \varphi_2) \\ &+ P_{12}(\varphi,', \varphi_2') - P_{12}(\varphi_1', \infty) - P_{12}(\infty, \varphi_2)] / P_{12}(\infty, \infty) \leq 0 . \end{aligned} \tag{15.5.8}$$

Any local realistic theory must satisfy this inequality.

With the help of Eqs. (15.4.16), (15.4.20) and (15.4.21), we find

$$P_{12}(\varphi_1, \varphi_2) = \frac{1}{2}\eta_1\eta_2 \cos^2(\varphi_2 - \varphi_1), \tag{15.5.9a}$$

$$P_{12}(\infty, \infty) = \eta_1\eta_2 , \tag{15.5.9b}$$

$$P_{12}(\varphi_1, \infty) = \frac{1}{2}\eta_1\eta_2 = P_{12}(\infty, \varphi_2) . \tag{15.5.9c}$$

Using the quantum mechanical expressions (15.4.16), (15.4.20) and (15.4.21) for these probabilities, we find the Bell parameter in Eq. (15.5.8) is

$$S_{QM} \equiv \frac{1}{2}\left[\cos^2(\varphi_2-\varphi_1)-\cos^2(\varphi_2'-\varphi_1)+\cos^2(\varphi_2-\varphi_1')+\cos^2(\varphi_2'-\varphi_1')\right]-1\,. \quad (15.5.10)$$

We can easily check that for the choice $\varphi_1 = 0$, $\varphi_2 = 3\pi/8$, $\varphi_1' = -\pi/4$, $\varphi_2' = \pi/8$ the parameter S_{QM} defined by Eq. (15.5.10) is

$$S_{QM} = -\frac{\sqrt{2}+1}{2} \approx -1.207\,, \quad (15.5.11)$$

which is less than -1 and for the choice $\varphi_1 = 0$, $\varphi = \pi/8$, $\varphi' = \pi/4$, $\varphi = 3\pi/8$

$$S_{QM} = \frac{\sqrt{2}-1}{2} \approx 0.207 \quad (15.5.12)$$

which is greater than zero. Both Eqs. (15.5.11) and (15.5.12) violate the inequality (15.5.8).

Bell inequalities like (15.4.6) or (15.5.8) have been tested in increasingly sophisticated experiments with entangled photon pairs derived from two-photon cascade decay of Ca atoms[5] or optical parametric down-conversion[6]. Experimental results are found to be in clear violation of the Bell inequality by several standard deviations and in excellent agreement with the predictions of quantum mechanics.

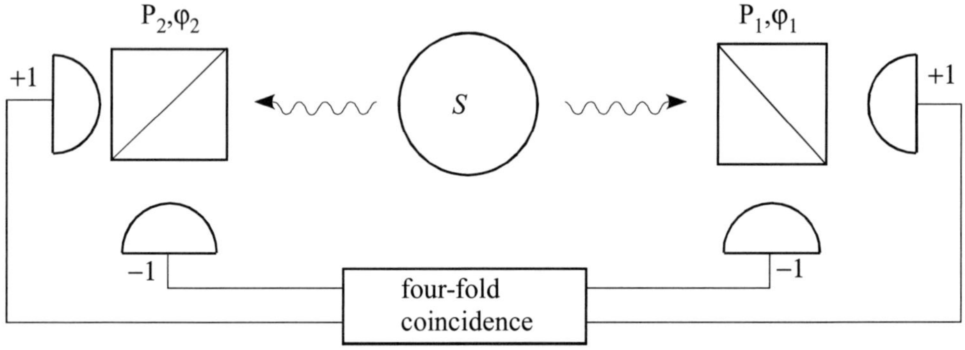

FIGURE 15.6
By replacing the single channel polarizers in Fig. 15.3 with polarizing beams splitters we obtain an optical analog of Stern-Gerlach set-up of Fig. 15.1.

Experiments involving entangled photon pairs derived from two-photon cascade decay

[5] S. J. Freedman and J. F. Clauser, Phys. Rev. Lett. **28**, 938 (1972).
E. S. Fry and R. C. Thomson, Phys. Rev. Lett. **37**, 465 (1976).
A. Aspect, P. Grangier, and G. Roger, Phys. Rev. Lett. **47**, 460 (1981).
A. Aspect, J. Dalibard, and G. Roger, Phys. Rev. Lett. **49**, 1804 (1982).

[6] Z. Y. Ou and L. Mandel, Phys. Rev. Lett. **61**, 50 (1988).
T. E. Kiess, Y. H. Shih, A. V. Sergienko, and C. O. Alley, Phys. Rev. Lett. **71**, 3893 (1993).

of Ca atoms have been carried out using the single-channel polarizer[7] set-up of Fig. 15.3 as well as the two-channel polarizer set up shown in Fig. 15.6. In two-channel polarizer experiments of Aspect and coworkers (1982)[5], photons in both channels of each polarizer are detected, i.e., photons with polarization parallel as well as perpendicular to the optic axis are detected and four-fold coincidence between the detection of photons in the + as well as − channels, of the two polarizers is measured resulting in the estimate of the correlation coefficient, $C(\varphi_1,\varphi_2) = P(+,\varphi_1;+,\varphi_2) + P(-,\varphi_1;-,\varphi_2) - P(+,\varphi_1;+,\varphi_2) - P(+,\varphi_1;+,\varphi_2) = \cos 2(\varphi_2 - \varphi_1)$ [see Problem 2]. The data are collected for several other settings of the polarizers as well. Experiments are compared with the predictions of quantum mechanics using the Bell-Clauser-Horne-Shimnoy-Holt (BCHSH) form of Bell inequality [see Problem 3]

$$-2 \leq S_{BCHSH} \equiv C(\varphi_1,\varphi_2) + C(\varphi_1,\varphi_2') + C(\varphi_1',\varphi_2) - C(\varphi_1',\varphi_2') \leq 2\,. \qquad (15.5.13)$$

According to quantum mechanics this Bell parameter $S_{QM} = 2\sqrt{2}$ for $\varphi_2 - \varphi_1 = -(\varphi_2' - \varphi_1) = -(\varphi_2 - \varphi_1') = \pi/8$ (and $\varphi_2' - \varphi_1' = 3\pi/8$) and $S_{QM} = -2\sqrt{2}$ for $\varphi_2 - \varphi_1 = -(\varphi_2' - \varphi_1) = -(\varphi_2 - \varphi_1') = 3\pi/8$ (and $\varphi_2' - \varphi_1' = \pi/8$), both values giving maximum violation of the inequality (15.5.13). Experimentally too, the violations are maximum for these parameters and have the values predicted by quantum mechanics as seen in Fig. 15.7

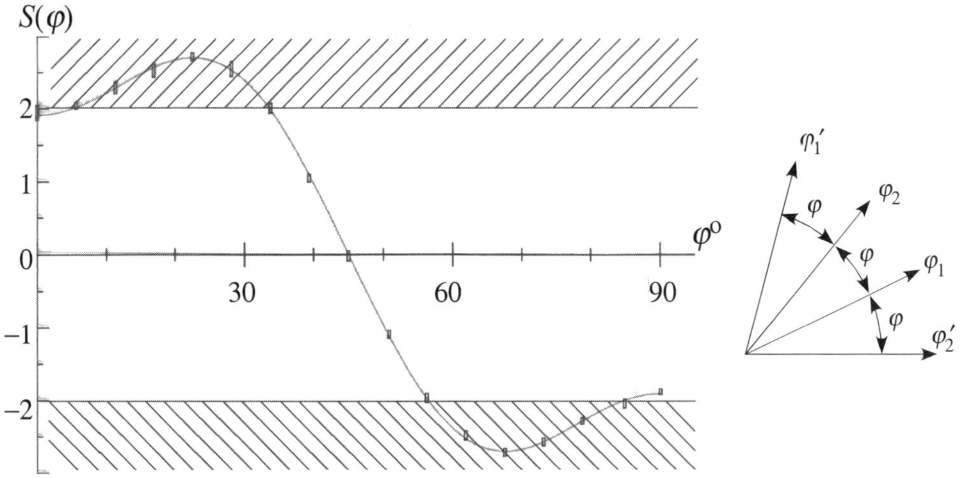

FIGURE 15.7
Violation of Bells inequality $-2 \leq S_{BCHSH} \leq 2$ in experiment with two-channel polarizers. The relative orientation of the detectors is shown in the figure to the right of the plot. The violation (by more than 30 standard deviations) is maximum for $\varphi = \pi/8 \equiv 22.5^o$ and $\varphi = 3\pi/8 \equiv 67.5^o$. For these angles, $S_{QM} = 2.828$ and -2.828, respectively. (Figure courtesy: A. Aspect).

[7]In the single-channel polarizer experiment, only the + (or −) channel of each polarizer is monitored, i.e., only the photons with polarization parallel (or perpendicular) to the optic axis are detected and coincidence between the detection of photons in the + (or −) channels, is measured. Photons in the second channel of each polarizer are either absorbed or not detected.

From a logical standpoint, these experiments still do not rule out a local realistic explanation because of two loopholes. The first loophole, known as the *communication loophole* or *locality loophole*, comes into play if the possibility of communication between the two observers at or less than the speed of light cannot be ruled out because the locality assumption in the derivation of Bell's inequality requires that the measurements on the two photons be space-like events. This means that the duration of an individual measurement has to be shorter than the time it would take any communication to travel, via any channel (known or unknown) at the speed of light, from one observer to the other. The second loophole, known as the *detection loophole*, comes into play because only a small number of all photon pairs created are collected and detected. Hence it is necessary to assume that they represent a fair sample of all photon pairs produced. This assumption, in principle, could be false. Experiments closing one or the other of these two loopholes (but not both) have also been reported[8]. The results are in agreement with quantum mechanics and violate Bell's inequality by up to 30 standard deviations.

Strong violations of Bell's inequalities in these experiments, lead us to abandon local realism as the basis of a microscopic theory to describe physical reality. Although local realism appears to be an intuitive and reasonable assumption about how the world should behave according to classical physics, it is not supported by experiments.

To the contrary, excellent agreement with quantum mechanical predictions strongly supports the quantum mechanical description as a correct and complete description of physical reality. This means that an entangled EPR photon pair is a non-separable object; that is, it is impossible to assign individual local properties (local physical reality) to each photon.

Like the the concept of wave-particle duality of the early quantum revolution, non-separability emerges as the most emblematic feature of quantum mechanics of entangled systems. It means that in an entangled quantum state, while the global state of two (or more) particles is perfectly defined, even if the two particles are far apart, the states of the individual particles remain indeterminate. The entangled state contains complete information about the correlation between the two particles but says nothing and, more importantly, the failure of local realistic theories suggests that nothing can be known about the states of the individual particles.

The experiments have given us a deeper understanding of the implications of quantum mechanical description of nature, and it has taken physicists a long time to understand them. The experiments provide compelling evidence that many basic ideas inherited from classical physics must be abandoned in favor of quantum way of looking at nature. Quantum mechanics is consistent within itself. The wave function is a *complete description* of a physical system in the sense that it contains all that is knowable about a physical system. The indeterminacy associated with the probabilistic interpretation of the wave function in quantum mechanics is not due to a lack of missing information (hidden variable) but is an intrinsic feature of nature.

The final chapter on the interpretations of quantum mechanics remains a work in progress. Although the violations of Bell's inequalities are generally taken to mean that quantum mechanics is nonlocal, one could argue that these violations imply only that locality and realism are not compatible simultaneously. There is no a priori logical way to decide whether one should drop locality or realism or both. These questions continue to be explored and as these explorations are reduced to quantitative predictions that can be tested in experiments,

[8] G. Weihs, T. Jennewein, C. Simon, H. Weinfurter, and A. Zeilinger, Phys. Rev. Lett. **81**, 5037 (1998). M. A. Rowe, D. Kielpinski, V. Meyer, C. A. Sackett, W. M. Itano, C. Monroe, and D. J. Wineland, Nature **409**, 791 (2001).

new insights into the meaning of quantum mechanical descriptions of nature will continue to emerge.

Problems

1. Derive the result (15.4.8). Hint: The rotation operator for a spin-1/2 spinor when coordinate axes are rotated through an angle φ about an axis parallel to unit vector $\boldsymbol{u}$ is $R_{\boldsymbol{u}}(\varphi) = \exp[-i\varphi \boldsymbol{u} \cdot \boldsymbol{S}]$, where $\boldsymbol{S}$ in terms of Pauli spin operators is $\boldsymbol{S} = \frac{1}{2}\boldsymbol{\sigma}$ (in units of $\hbar$).

2. Consider the photon pairs produced in the two-photon cascade decay $J = 0 \rightarrow J+1 \rightarrow J = 0$ of Calcium atoms. Suppose the polarizers in Fig. 15.3 are replaced by polarizing beam splitters [Fig. 15.6] (two-channel polarizer) so that the photons emerge polarized either as the ordinary ray or as the extraordinary ray from each beam splitter and are subsequently detected and joint detection probabilities (coincidence probabilities) $P(+,\varphi_1;+\varphi_2)$, $P(-,\varphi_1;-,\varphi_2)$, $P(+,\varphi_1;-\varphi_2)$, and $P(-,\varphi_1;+,\varphi_2)$ are measured. Here $P(+,\varphi_1;+\varphi_2)$ is the probability that photon 1 is detected with polarization in direction φ_1 and photon 2 is detected with polarization in direction φ_2, $P(+,\varphi_1;-\varphi_2)$ is the probability that photon 1 is detected with polarization φ_1 and photon 2 is detected with polarization orthogonal to φ_2. Define the outcome of measurement on photon 1 by $A(\varphi_1) = 1$ if photon polarization is found to be parallel to φ_1 (channel 1) and $A(\varphi_1) = -1$ if photon polarization is found to be perpendicular to φ_1 (channel 2). Similarly, the outcome of polarization measurement on photon 2 is defined by $B(\varphi_2) = \pm 1$, respectively, for photon polarization parallel and perpendicular to φ_2. With this detection scheme [Fig. 15.6], the two-photon polarization experiment becomes analogous to Stern-Gerlach experiment of Fig. 15.1.

(a) The polarization-entangled state for a pair of photons traveling in opposite directions can be written as [Eq. (15.4.10)] $|\psi(1,2)\rangle = \frac{1}{\sqrt{2}}[|e_{1x}\rangle\,|e_{2x}\rangle + |e_{1y}\rangle\,|e_{2y}\rangle]$. Find the probabilities $P(\pm,\varphi_1;\pm,\varphi_2)$ and show that, according to quantum mechanics, the correlation function

$$C_{QM}(\varphi_1,\varphi_2) = \sum_{A,B} A(\varphi_1)B(\varphi_2)P_{AB}(\varphi_1,\varphi_2) = \cos 2(\varphi_2 - \varphi_1)\,.$$

(b) Compute the Bell parameter S_{BCHSH} of Eq. (15.5.13) according to quantum mechanics and show that for $\varphi_1 = 0$, $\varphi_1' = 2\varphi$, $\varphi_2 = \varphi = -\varphi_2'$, [Fig. 15.7] it takes the form

$$S_{QM} \equiv 3\cos 2\varphi - \cos 6\varphi\,.$$

Plot it as a function of φ and identify regions where BCHSH inequality of (15.5.13) [see Problem 3] is violated.

3. Show that if x, x', y, y are four real numbers in the closed interval $[-1,1]$, the sum $S = xy + xy' + x'y - x'y'$ belongs to the interval

$$-2 \leq xy + xy' + x'y - x'y' \leq 2\,.$$

Using this inequality, establish another form of Bell's inequality for local realistic theories

$$-2 \leq\ S \equiv C(a,b) + C(a,b') + C(a',b) - C(a',b') \leq 2\,,$$

where $C(a,b)$ is the correlation function given by

$$C(a,b) = \int A(a,\lambda)B(b,\lambda)\rho(\lambda)d\lambda\,.$$

This form of Bell's inequality is known as the Bell-Clauser-Horne-Shimony-Holt (BCHSH) inequality.

General References

1. P. A. M. Dirac, *Principles of Quantum Mechanics*, Third Edition, (Clarendon Press, Oxford, 1971).
2. L. I. Schiff, *Quantum Mechanics*, Third Edition (McGraw Hill Book Company, Inc., New York, 1968).
3. L. D. Landau and E. M. Lishitz, *Quantum Mechanics*, (Pergamon Press, New York, 1976).
4. D. Bohm, *Quantum Theory*, (Prentice Hall Inc., New York, 1951), Available also as a Dover reprint.
5. L. Pauling and E. B. Wilson, *Introduction to Quantum Mechanics* (McGraw Hill Book Company, New York, 1935), also available as a Dover reprint.
6. E. M. Corson, *Perturbation Methods in Quantum Mechanics of N-electron Systems* (Hafner Publishing Co., New York).
7. E. C. Kemble, *Fundamental Principles of Quantum Mechanics* (McGraw Hill Book Company, Inc. New York).
8. R. G. Newton, *Scattering Theory of Waves and Particles*, Second Edition (Springer-Verlag, New York, 1982).
9. W. Heitler, *Quantum Theory of Radiation* (Clarendon Press, Oxford, 1954), also avialable as a Dover reprint.
10. Ta-You Wu and Takashi Ohmura, *Quantum Theory of Scattering* (Prentice Hall, Englewood Cliffs, NJ, 1962).
11. J. Sakurai, *Advanced Quantum Mechanics* (Addison-Wesley, Reading, MA, 1967).
12. H. Muirhead, *Physics of Elementary Particles* (Pergamon Press, Oxford, 1968).
13. M. E. Rose, *Elementary Theory of Angular Momentum* (John Wiley, New York, 1957), also available as a Dover reprint.
14. C. S. Wu and S. A. Moszkowski, *Beta Decay* (Wiley Interscience, New York, 1966).
15. H. Eyring, J. Walter, and G. E. Kimball, *Quantum Chemistry* (John Wiley and Sons, New York, NY, 1947).
16. L. R. B. Elton, *Introductory Nuclear Theory* (Sir Isaac Pitman and Sons, Ltd., London, 1965).

Index